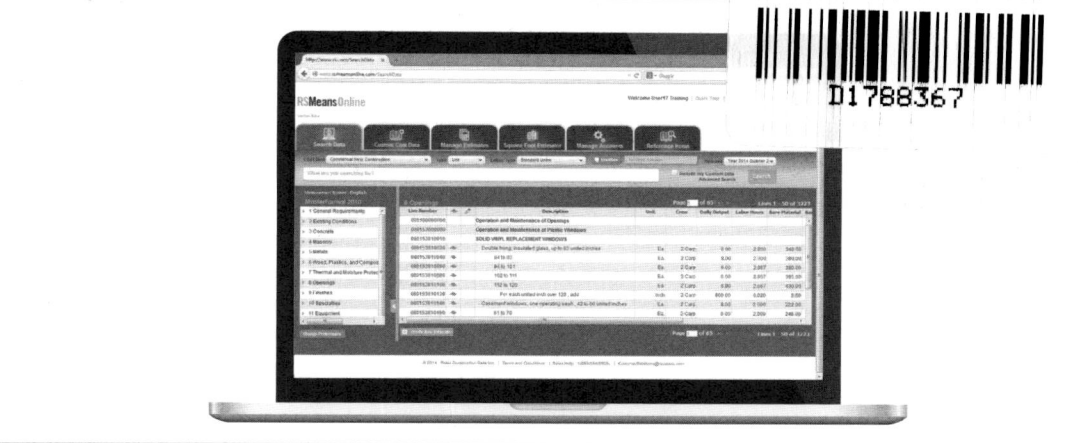

WIN the complete library of RSMeans Online data!

Register your book below to receive your FREE quarterly updates, PLUS you will be entered into a quarterly drawing to win the COMPLETE RSMeans Online library of 2015 data!

Be sure to keep up-to-date in 2015!

Fill out the card below for RSMeans' free quarterly updates,
as well as a chance to win the complete RSMeans Online library.
Please provide your name, address, and email below and return
this card by mail, or to register online:

http://info.thegordiangroup.com/RSMeans.html

Name _____

email _____ Title _____

Company _____

Street _____

City/Town _____ State/Prov. _____ Zip/Postal Code _____

RSMeans

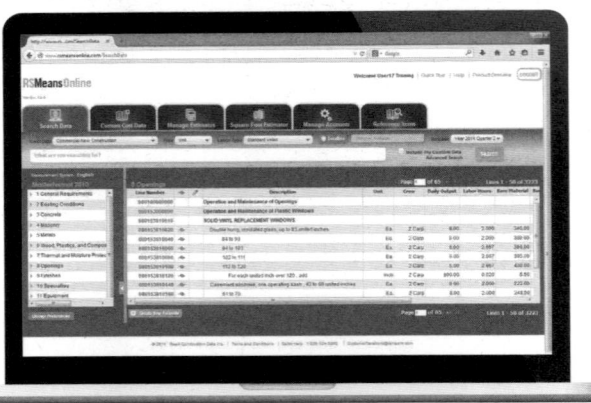

WIN the complete library of RSMeans Online 2015 data!

See other side for details.

BUSINESS REPLY MAIL
FIRST-CLASS MAIL   PERMIT NO. 31   NORWELL, MA

POSTAGE WILL BE PAID BY ADDRESSEE

RSMEANS
700 LONGWATER DRIVE
NORWELL MA 02061-1624

NO POSTAGE
NECESSARY
IF MAILED
IN THE
UNITED STATES

RSMeans Residential Cost Data

2015 — 34th annual edition

Bob Mewis, CCC, Senior Editor

Engineering Director
Bob Mewis, CCC (*1, 14, 41*)

Contributing Editors
Christopher Babbitt
Adrian C. Charest, PE (*26, 27, 28, 48*)
Cheryl Elsmore
Robert Fortier, PE (*2, 31, 32, 33, 34, 35, 44, 46*)
Robert J. Kuchta (*8*)
Thomas Lane (*6, 7*)
Robert C. McNichols

*Numbers in italics are the divisional responsibilities for each editor. Please contact the designated editor directly with any questions.

Genevieve Medeiros
Melville J. Mossman, PE (*21, 22, 23*)
Jeannene D. Murphy
Marilyn Phelan, AIA (*9, 10, 11, 12*)
Stephen C. Plotner (*3, 5*)
Siobhan Soucie (*4, 13*)

Vice President Data & Engineering
Chris Anderson

Cover Design
Sara Rutan

Product Manager
Andrea Sillah

Production Manager
Debbie Panarelli

Production
Jill Goodman
Jonathan Forgit
Sara Rutan
Mary Lou Geary

Technical Support
Gary L. Hoitt
Kathryn S. Rodriguez

RSMeans
Construction Publishers & Consultants
700 Longwater Drive
Norwell, MA 02061
United States of America
1-877-713-2617
www.rsmeans.com

Copyright 2014 by RSMeans
All rights reserved.
Cover photo ©
Hemera Technologoies/
AbleStock.com/Thinkstock

Printed in the United States of America
ISSN 0896-8624
ISBN 978-1-940238-65-4

The authors, editors, and engineers of RSMeans apply diligence and judgment in locating and using reliable sources for the information published. However, RSMeans makes no express or implied warranty or guarantee in connection with the content of the information contained herein, including the accuracy, correctness, value, sufficiency, or completeness of the data, methods, and other information contained herein. RSMeans makes no express or implied warranty of merchantability or fitness for a particular purpose. RSMeans shall have no liability to any customer or third party for any loss, expense, or damage, including consequential, incidental, special, or punitive damage, including lost profits or lost revenue, caused directly or indirectly by any error or omission, or arising out of, or in connection with, the information contained herein. For the purposes of this paragraph, "RSMeans" shall include RS Means Company LLC, and its divisions, subsidiaries, successors, parent companies, and their employees, partners, principals, agents and representatives, and any third-party providers or sources of information or data.

No part of this publication may be reproduced, stored in an information storage or retrieval system, or transmitted in any form or by any means, electronic or mechanical, including photocopying or scanning, without prior written permission of RSMeans. The cost data contained in this publication is valuable and proprietary information of RSMeans and others, and there is no transfer to you of any ownership rights in the cost data or any license granted to you to create derivative works based on or which utilize the cost data.

This book is printed on 50% recycled paper using soy-based ink and is entirely recyclable.

0175 $148.95 per copy (in United States)
Price is subject to change without prior notice.

Related RSMeans Products and Services

This book is aimed primarily at single family and small multi-family housing projects costing up to $1,000,000.

The engineers at RSMeans suggest the following products and services as companion information resources to *RSMeans Residential Cost Data:*

Construction Cost Data Books
Square Foot Costs 2015
Residential Repair and Remodeling Costs 2015

Reference Books
Estimating Building Costs
RSMeans Estimating Handbook
Green Building: Project Planning & Estimating
How to Estimate with RSMeans Data
Plan Reading & Material Takeoff
Project Scheduling & Management for Construction
Universal Design Ideas for Style, Comfort & Safety

Seminars and In-House Training
Unit Price Estimating
RSMeans Online® Training
Practical Project Management for Construction Professionals
Scheduling with MSProject for Construction Professionals
Mechanical & Electrical Estimating

RSMeans Online Bookstore
Visit RSMeans at **www.rsmeans.com** for a one-stop portal for the most reliable and current resources available on the market. Here you can learn about our more than 20 annual cost data books and eBooks, also available on CD and online versions, as well as our libary of reference books and RSMeans seminars.

RSMeans Electronic Data
Get the information found in RSMeans cost books electronically at **RSMeansOnline.com**.

RSMeans Business Solutions
Cost engineers and analysts conduct facility life cycle and benchmark research studies and predictive cost modeling, as well as offer consultation services for conceptual estimating and real property management. Research studies are designed with the application of proprietary cost and project data from RSMeans' extensive North American databases. Analysts offer building product manufacturers qualitative and quantitative market research, as well as time/motion studies for new products, and Web-based dashboards for market opportunity analysis. Clients are from both the public and private sectors, including federal, state, and municipal agencies; corporations; institutions; construction management firms; hospitals and associations.

Construction Costs for Software Applications
More than 25 unit price and assemblies cost databases are available through a number of leading estimating and facilities management software partners. For more information see the "Other RSMeans Products and Services" pages at the back of this publication.

RSMeans data is also available to federal, state, and local government agencies as multi-year, multi-seat licenses.

For information on our current partners, visit:
www.rsmeans.com/partners or call 877.713.2617.

Table of Contents

Foreword	iv
How the Book Is Built: An Overview	v
Absolute Essentials for a Quick Start	vi
Estimating with RSMeans Unit Price Cost Data	vii
How to Use the Book: The Details	ix
Square Foot Cost Section	**1**
Assemblies Cost Section	**97**
Unit Price Section	**283**
Reference Section	**659**
Construction Equipment Rental Costs	661
Crew Listings	673
Location Factors	707
Reference Tables	713
Abbreviations	734
Estimating Forms	738
Index	**740**
Other RSMeans Products and Services	**762**
Labor Trade Rates including Overhead & Profit	**Inside Back Cover**

Foreword

Our Mission

Since 1942, RSMeans has been actively engaged in construction cost publishing and consulting throughout North America.

Today, more than 70 years after RSMeans began, our primary objective remains the same: to provide you, the construction and facilities professional, with the most current and comprehensive construction cost data possible.

Whether you are a contractor, owner, architect, engineer, facilities manager, or anyone else who needs a reliable construction cost estimate, you'll find this publication to be a highly useful and necessary tool.

With the constant flow of new construction methods and materials today, it's difficult to find the time to look at and evaluate all the different construction cost possibilities. In addition, because labor and material costs keep changing, last year's cost information is not a reliable basis for today's estimate or budget.

That's why so many construction professionals turn to RSMeans. We keep track of the costs for you, along with a wide range of other key information, from city cost indexes . . . to productivity rates . . . to crew composition . . . to contractor's overhead and profit rates.

RSMeans performs these functions by collecting data from all facets of the industry and organizing it in a format that is instantly accessible to you. From the preliminary budget to the detailed unit price estimate, you'll find the data in this book useful for all phases of construction cost determination.

The Staff, the Organization, and Our Services

When you purchase one of RSMeans' publications, you are, in effect, hiring the services of a full-time staff of construction and engineering professionals.

Our thoroughly experienced and highly qualified staff works daily at collecting, analyzing, and disseminating comprehensive cost information for your needs. These staff members have years of practical construction experience and engineering training prior to joining the firm. As a result, you can count on them not only for accurate cost figures, but also for additional background reference information that will help you create a realistic estimate.

The RSMeans organization is always prepared to help you solve construction problems through its variety of data solutions, including online, CD, and print book formats, as well as cost estimating expertise available via our business solutions, training, and seminars.

Besides a full array of construction cost estimating books, RSMeans also publishes a number of other reference works for the construction industry. Subjects include construction estimating and project and business management, green building, and a library of facility management references.

In addition, you can access all of our construction cost data electronically in convenient CD format or on the web. Visit **RSMeansOnline.com** for more information on our 24/7 online cost data.

What's more, you can increase your knowledge and improve your construction estimating and management performance with an RSMeans construction seminar or in-house training program. These two-day seminar programs offer unparalleled opportunities for everyone in your organization to become updated on a wide variety of construction-related issues.

RSMeans is also a worldwide provider of construction cost management and analysis services for commercial and government owners.

In short, RSMeans can provide you with the tools and expertise for constructing accurate and dependable construction estimates and budgets in a variety of ways.

Robert Snow Means Established a Tradition of Quality That Continues Today

Robert Snow Means spent years building RSMeans, making certain he always delivered a quality product.

Today, at RSMeans, we do more than talk about the quality of our data and the usefulness of our books. We stand behind all of our data, from historical cost indexes to construction materials and techniques to current costs.

If you have any questions about our products or services, please call us toll-free at 1-877-713-2617. Our customer service representatives will be happy to assist you. You can also visit our website at **www.rsmeans.com**.

How the Book Is Built: An Overview

The Construction Specifications Institute (CSI) and Construction Specifications Canada (CSC) have produced the 2014 edition of MasterFormat®, a system of titles and numbers used extensively to organize construction information.

All unit price data in the RSMeans cost data books is now arranged in the 50-division MasterFormat® 2014 system.

A Powerful Construction Tool

You have in your hands one of the most powerful construction tools available today. A successful project is built on the foundation of an accurate and dependable estimate. This book will enable you to construct just such an estimate.

For the casual user the book is designed to be:

- quickly and easily understood so you can get right to your estimate.
- filled with valuable information so you can understand the necessary factors that go into the cost estimate.

For the regular user, the book is designed to be:

- a handy desk reference that can be quickly referred to for key costs.
- a comprehensive, fully reliable source of current construction costs and productivity rates so you'll be prepared to estimate any project.
- a source book for preliminary project cost, product selections, and alternate materials and methods.

To meet all of these requirements, we have organized the book into the following clearly defined sections.

Quick Start

See our "Quick Start" instructions on the following page to get started right away.

Estimating with RSMeans Unit Price Cost Data

Please refer to these steps for guidance on completing an estimate using RSMeans unit price cost data.

Square Foot Cost Section

This section lists Square Foot costs for typical residential construction projects. The organizational format used divides the projects into basic building classes. These classes are defined at the beginning of the section. The individual projects are further divided into ten common components of construction.

The Table of Contents, an explanation of square foot prices, and an outline of a typical page layout are located at the beginning of this section.

Assemblies Cost Section

This section uses an "Assemblies" (sometimes referred to as "systems") format grouping all the functional elements of a building into nine construction divisions.

At the top of each "Assembly" cost table is an illustration, a brief description, and the design criteria used to develop the cost. Each of the components and its contributing cost to the system is shown.

Material: These cost figures include a standard 10% markup for "handling." They are national average material costs as of January of the current year and include delivery to the job site.

Installation: The installation costs include labor and equipment, plus a markup for the installing contractor's overhead and profit.

For a complete breakdown and explanation of a typical "Assemblies" page, see "How to Use the Assemblies Cost Tables" at the beginning of the Assemblies Section.

Unit Price Section

All cost data has been divided into the 50 divisions according to the MasterFormat system of classification and numbering. For a listing of these divisions and an outline of their subdivisions, see the Unit Price Section Table of Contents.

Estimating tips are included at the beginning of each division.

Reference Section

This section includes information on Equipment Rental Costs, Crew Listings, Location Factors, Reference Tables, and a listing of Abbreviations.

Equipment Rental Costs: This section contains the average costs to rent and operate hundreds of pieces of construction equipment.

Crew Listings: This section lists all of the crews referenced in the book. For the purposes of this book, a crew is composed of more than one trade classification and/or the addition of power equipment to any trade classification. Power equipment is included in the cost of the crew. Costs are shown both with bare labor rates and with the installing contractor's overhead and profit added. For each, the total crew cost per eight-hour day and the composite cost per labor-hour are listed.

Location Factors: You can adjust total project costs to over 900 locations throughout the U.S. and Canada by using the data in this section.

Reference Tables: At the beginning of selected major classifications in the Unit Price section are reference numbers shown in a shaded box. These numbers refer you to related information in the Reference Section. In this section, you'll find reference tables, explanations, estimating information that support how we develop the unit price data, technical data, and estimating procedures.

Abbreviations: A listing of abbreviations used throughout this book, along with the terms they represent, is included in this section.

Index

A comprehensive listing of all terms and subjects in this book will help you quickly find what you need when you are not sure where it occurs in MasterFormat.

The Scope of This Book

This book is designed to be as comprehensive and as easy to use as possible. To that end we have made certain assumptions and limited its scope in two key ways:

1. We have established material prices based on a national average.
2. We have computed labor costs based on a 7 major region average of union wage rates.

Project Size/Type

The material prices in RSMeans cost data books are "contractor's prices." They are the prices that contractors can expect to pay at the lumberyards, suppliers'/distributors warehouses, etc. Small orders of speciality items would be higher than the costs shown, while very large orders, such as truckload lots, would be less. The variation would depend on the size, timing, and negotiating power of the contractor. The labor costs are primarily for new construction or major renovation rather than repairs or minor alterations.

With reasonable exercise of judgment, the figures can be used for any building work.

Absolute Essentials for a Quick Start

If you feel you are ready to use this book and don't think you will need the detailed instructions that begin on the following page, this Absolute Essentials for a Quick Start page is for you. These steps will allow you to get started estimating in a matter of minutes.

1 Scope
Think through the project that you will be estimating, and identify the many individual work tasks that will need to be covered in your estimate.

2 Quantify
Determine the number of units that will be required for each work task that you identified.

3 Pricing
Locate individual Unit Price line items that match the work tasks you identified. The Unit Price Section Table of Contents that begins on page 1 and the Index in the back of the book will help you find these line items.

4 Multiply
Multiply the Total Incl O&P cost for a Unit Price line item in the book by your quantity for that item. The price you calculate will be an estimate for a completed item of work performed by a subcontractor. Keep adding line items in this manner to build your estimate.

5 Project Overhead
Include project overhead items in your estimate. These items are needed to make the job run and are typically, but not always, provided by the General Contractor. They can be found in Division 1. An alternate method of estimating project overhead costs is to apply a percentage of the total project cost.

Include rented tools not included in crews, waste, rubbish handling, and cleanup.

6 Estimate Summary
Include General Contractor's markup on subcontractors, General Contractor's office overhead and profit, and sales tax on materials and equipment.

Adjust your estimate to the project's location by using the City Cost Indexes or Location Factors found in the Reference Section.

Editors' Note: We urge you to spend time reading and understanding the supporting material in the front of this book. An accurate estimate requires experience, knowledge, and careful calculation. The more you know about how we at RSMeans developed the data, the more accurate your estimate will be. In addition, it is important to take into consideration the reference material in the back of the book such as Equipment Listings, Crew Listings, Location Factors, and Reference Tables.

Estimating with RSMeans Unit Price Cost Data

Following these steps will allow you to complete an accurate estimate using RSMeans Unit Price cost data.

1 Scope Out the Project

- Think through the project and identify those CSI Divisions needed in your estimate.
- Identify the individual work tasks that will need to be covered in your estimate.
- The Unit Price data in this book has been divided into 50 Divisions according to the CSI MasterFormat® 2014—their titles are listed on the back cover of your book.
- The Unit Price Section Table of Contents on page 1 may also be helpful when scoping out your project.
- Experienced estimators find it helpful to begin with Division 2 and continue through completion. Division 1 can be estimated after the full project scope is known.

2 Quantify

- Determine the number of units required for each work task that you identified.
- Experienced estimators include an allowance for waste in their quantities. (Waste is not included in RSMeans Unit Price line items unless so stated.)

3 Price the Quantities

- Use the Unit Price Table of Contents, and the Index, to locate individual Unit Price line items for your estimate.
- Reference Numbers indicated within a Unit Price section refer to additional information that you may find useful.
- The crew indicates who is performing the work for that task. Crew codes are expanded in the Crew Listings in the Reference Section to include all trades and equipment that comprise the crew.
- The Daily Output is the amount of work the crew is expected to do in one day.
- The Labor-Hours value is the amount of time it will take for the crew to install one unit of work.
- The abbreviated Unit designation indicates the unit of measure upon which the crew, productivity, and prices are based.
- Bare Costs are shown for materials, labor, and equipment needed to complete the Unit Price line item. Bare costs do not include waste, project overhead, payroll insurance, payroll taxes, main office overhead, or profit.
- The Total Incl O&P cost is the billing rate or invoice amount of the installing contractor or subcontractor who performs the work for the Unit Price line item.

4 Multiply

- Multiply the total number of units needed for your project by the Total Incl O&P cost for each Unit Price line item.
- Be careful that your take off unit of measure matches the unit of measure in the Unit column.
- The price you calculate is an estimate for a completed item of work.
- Keep scoping individual tasks, determining the number of units required for those tasks, matching each task with individual Unit Price line items in the book, and multiply quantities by Total Incl O&P costs.
- An estimate completed in this manner is priced as if a subcontractor, or set of subcontractors, is performing the work. The estimate does not yet include Project Overhead or Estimate Summary components such as General Contractor markups on subcontracted work, General Contractor office overhead and profit, contingency, and location factor.

5 Project Overhead

- Include project overhead items from Division 1 – General Requirements.
- These items are needed to make the job run. They are typically, but not always, provided by the General Contractor. Items include, but are not limited to, field personnel, insurance, performance bond, permits, testing, temporary utilities, field office and storage facilities, temporary scaffolding and platforms, equipment mobilization and demobilization, temporary roads and sidewalks, winter protection, temporary barricades and fencing, temporary security, temporary signs, field engineering and layout, final cleaning and commissioning.
- Each item should be quantified, and matched to individual Unit Price line items in Division 1, and then priced and added to your estimate.
- An alternate method of estimating project overhead costs is to apply a percentage of the total project cost, usually 5% to 15% with an average of 10% (see General Conditions, page x).
- Include other project related expenses in your estimate such as:
  - Rented equipment not itemized in the Crew Listings
  - Rubbish handling throughout the project (see 02 41 19.19)

6 Estimate Summary

- Include sales tax as required by laws of your state or county.
- Include the General Contractor's markup on self-performed work, usually 5% to 15% with an average of 10%.
- Include the General Contractor's markup on subcontracted work, usually 5% to 15% with an average of 10%.
- Include General Contractor's main office overhead and profit:
  - RSMeans gives general guidelines on the General Contractor's main office overhead (see section 01 31 13.60 and Reference Number R013113-50).
  - RSMeans gives no guidance on the General Contractor's profit.
  - Markups will depend on the size of the General Contractor's operations, his projected annual revenue, the level of risk he is taking on, and on the level of competition in the local area and for this project in particular.
- Include a contingency, usually 3% to 5%, if appropriate.
- Adjust your estimate to the project's location by using the Location Factors in the Reference Section:
  - When the proper Factor has been identified for the project's location, convert it to a multiplier by dividing it by 100, and then multiply that multiplier by your estimated total cost. The original estimated total cost will now be adjusted up or down from the national average to a total that is appropriate for your location.

Editors' Notes:
We urge you to spend time reading and understanding the supporting material in the front of this book. An accurate estimate requires experience, knowledge, and careful calculation. The more you know about how we at RSMeans developed the data, the more accurate your estimate will be. In addition, it is important to take into consideration the reference material in the back of the book such as Equipment Listings, Crew Listings, Location Factors, and Reference Tables.

How to Use the Book: The Details

What's Behind the Numbers? The Development of Cost Data

The staff at RSMeans continually monitors developments in the construction industry in order to ensure reliable, thorough, and up-to-date cost information. While overall construction costs may vary relative to general economic conditions, price fluctuations within the industry are dependent upon many factors. Individual price variations may, in fact, be opposite to overall economic trends. Therefore, costs are constantly tracked and complete updates are published yearly. Also, new items are frequently added in response to changes in materials and methods.

Costs—$ (U.S.)

All costs represent U.S. national averages and are given in U.S. dollars. The RSMeans Location Factors can be used to adjust costs to a particular location. The Location Factors for Canada can be used to adjust U.S. national averages to local costs in Canadian dollars. No exchange rate conversion is necessary.

G The processes or products identified by the green symbol in our publications have been determined to be environmentally responsible and/or resource-efficient solely by the RSMeans engineering staff. The inclusion of the green symbol does not represent compliance with any specific industry association or standard.

Material Costs

The RSMeans staff contacts manufacturers, dealers, distributors, and contractors all across the U.S. and Canada to determine national average material costs. If you have access to current material costs for your specific location, you may wish to make adjustments to reflect differences from the national average. Included within material costs are fasteners for a normal installation. RSMeans engineers use manufacturers' recommendations, written specifications, and/or standard construction practice for size and spacing of fasteners. Adjustments to material costs may be required for your specific application or location. The manufacturer's warranty is assumed. Extended warranties are not included in the material costs. Material costs do not include sales tax.

Labor Costs

Labor costs are based on the average of wage rates from 7 major U.S. regions for construction trades for the current year. Rates, along with overhead and profit markups, are listed on the inside back cover of this book.

- If wage rates in your area vary from those used in this book, or if rate increases are expected within a given year, labor costs should be adjusted accordingly.

Labor costs reflect productivity based on actual working conditions. In addition to actual installation, these figures include time spent during a normal weekday on tasks such as, material receiving and handling, mobilization at site, site movement, breaks, and cleanup.

Productivity data is developed over an extended period so as not to be influenced by abnormal variations and reflects a typical average.

Equipment Costs

Equipment costs include not only rental, but also operating costs for equipment under normal use. The operating costs include parts and labor for routine servicing such as repair and replacement of pumps, filters, and worn lines. Normal operating expendables, such as fuel, lubricants, tires, and electricity (where applicable), are also included. Extraordinary operating expendables with highly variable wear patterns, such as diamond bits and blades, are excluded. These costs are included under materials. Equipment rental rates are obtained from industry sources throughout North America—contractors, suppliers, dealers, manufacturers, and distributors.

Rental rates can also be treated as reimbursement costs for contractor-owned equipment. Owned equipment costs include depreciation, loan payments, interest, taxes, insurance, storage and major repairs.

Equipment costs do not include operators' wages; nor do they include the cost to move equipment to a job site (mobilization) or from a job site (demobilization).

Equipment Cost/Day—The cost of power equipment required for each crew is included in the Crew Listings in the Reference Section (small tools that are considered as essential everyday tools are not listed out separately). The Crew Listings itemize specialized tools and heavy equipment along with labor trades. The daily cost of itemized equipment included in a crew is based on dividing the weekly bare rental rate by 5 (number of working days per week) and then adding the hourly operating cost times 8 (the number of hours per day). This Equipment Cost/Day is shown in the last column of the Equipment Rental Cost pages in the Reference Section.

Mobilization/Demobilization—The cost to move construction equipment from an equipment yard or rental company to the job site and back again is not included in equipment costs. Mobilization (to the site) and demobilization (from the site) costs can be found in the Unit Price Section. If a piece of equipment is already at the job site, it is not appropriate to utilize mob./demob. costs again in an estimate.

Factors Affecting Costs

Costs can vary depending upon a number of variables. Here's how we have handled the main factors affecting costs.

Quality——The prices for materials and the workmanship upon which productivity is based represent sound construction work. They are also in line with U.S. government specifications.

Overtime——We have made no allowance for overtime. If you anticipate premium time or work beyond normal working hours, be sure to make an appropriate adjustment to your labor costs.

Productivity——The productivity, daily output, and labor-hour figures for each line item are based on working an eight-hour day in daylight hours in moderate temperatures. For work that extends beyond normal work hours or is performed under adverse conditions, productivity may decrease. (See "How RSMeans Data Works" for more on productivity.)

Size of Project——The size, scope of work, and type of construction project will have a significant impact on cost. Economies of scale can reduce costs for large projects. Unit costs can often run higher for small projects. Costs in this book are intended for the size and type of project as previously described in "How the Book Is Built: An Overview." Costs for projects of a significantly different size or type should be adjusted accordingly.

Location——Material prices in this book are for metropolitan areas. However, in dense urban areas, traffic and site storage limitations may increase costs. Beyond a 20-mile radius of large cities, extra trucking or transportation charges may also increase the material costs slightly. On the other hand, lower wage rates may be in effect. Be sure to consider both of these factors when preparing an estimate, particularly if the job site is located in a central city or remote rural location.

In addition, highly specialized subcontract items may require travel and per-diem expenses for mechanics.

Other Factors——
- season of year
- contractor management
- weather conditions
- local union restrictions
- building code requirements
- availability of:
    - adequate energy
    - skilled labor
    - building materials
- owner's special requirements/restrictions
- safety requirements
- environmental considerations

General Conditions——The "Square Foot" and "Assemblies" sections of this book use costs that include the installing contractor's overhead and profit (O&P). The Unit Price Section presents cost data in two ways: Bare Costs and Total Cost including O&P (Overhead and Profit). General Conditions, when applicable, should also be added to the Total Cost including O&P. The costs for General Conditions are listed in Division 1 of the Unit Price Section and the Reference Section of this book. General Conditions for the Installing Contractor may range from 0% to 10% of the Total Cost including O&P. For the General or Prime Contractor, costs for General Conditions may range from 5% to 15% of the Total Cost including O&P, with a figure of 10% as the most typical allowance.

Overhead & Profit——Total Cost, including O&P, for the Installing Contractor is shown in the last column on both the Unit Price and the Assemblies pages of this book. This figure is the sum of the bare material cost plus 10% for profit, the bare labor cost plus overhead and profit, and the bare equipment cost plus 10% for profit. Details for the calculation of Overhead and Profit on labor are shown on the inside back cover and in the Reference Section of this book. (See "How RSMeans Unit Price Data Works" for an example of this calculation.)

Unpredictable Factors——General business conditions influence "in-place" costs of all items. Substitute materials and construction methods may have to be employed. These may affect the installed cost and/or life cycle costs. Such factors may be difficult to evaluate and cannot necessarily be predicted on the basis of the job's location in a particular section of the country. Thus, where these factors apply, you may find significant, but unavoidable cost variations for which you will have to apply a measure of judgment to your estimate.

Rounding of Costs

In general, all unit prices in excess of $5.00 have been rounded to make them easier to use and still maintain adequate precision of the results. The rounding rules we have chosen are in the following table.

Prices from . . .	Rounded to the nearest . . .
$.01 to $5.00	$.01
$5.01 to $20.00	$.05
$20.01 to $100.00	$.50
$100.01 to $300.00	$1.00
$300.01 to $1,000.00	$5.00
$1,000.01 to $10,000.00	$25.00
$10,000.01 to $50,000.00	$100.00
$50,000.01 and above	$500.00

Final Checklist

Estimating can be a straightforward process provided you remember the basics. Here's a checklist of some of the steps you should remember to complete before finalizing your estimate.

Did you remember to . . .

- factor in the Location Factor for your locale?
- take into consideration which items have been marked up and by how much?
- mark up the entire estimate sufficiently for your purposes?
- read the background information on techniques and technical matters that could impact your project time span and cost?
- include all components of your project in the final estimate?
- double check your figures for accuracy?
- call RSMeans if you have any questions about your estimate or the data you've found in our publications?

Remember, RSMeans stands behind its publications. If you have any questions about your estimate . . . about the costs you've used from our books . . . or even about the technical aspects of the job that may affect your estimate, feel free to call the RSMeans editors at 1-877-713-2617.

Free Quarterly Updates

Stay up-to-date throughout 2015 with RSMeans' free cost data updates four times a year. Sign up online to make sure you have access to the newest data. Every quarter we provide the city cost adjustment factors for hundreds of cities and key materials. Register at:
http://info.thegordiangroup.com/RSMeans.html.

Square Foot Cost Section

Table of Contents

	Page
Introduction	
General	3
How to Use	4
Building Classes	8
Configurations	9
Building Types	10
Garage Types	12
Building Components	13
Exterior Wall Construction	14
Residential Cost Estimate Worksheets	
Instructions	15
Model Residence Example	17
Economy	
Illustrations	23
1 Story	24
1-1/2 Story	26
2 Story	28
Bi-Level	30
Tri-Level	32
Wings & Ells	34
Average	
Illustrations	37
1 Story	38
1-1/2 Story	40
2 Story	42
2-1/2 Story	44
3 Story	46
Bi-Level	48
Tri-Level	50
Solid Wall (log home) 1 Story	52
Solid Wall (log home) 2 Story	54
Wings & Ells	56
Custom	
Illustrations	59
1 Story	60
1-1/2 Story	62
2 Story	64
2-1/2 Story	66
3 Story	68
Bi-Level	70
Tri-Level	72
Wings & Ells	74
Luxury	
Illustrations	77
1 Story	78
1-1/2 Story	80
2 Story	82
2-1/2 Story	84
3 Story	86
Bi-Level	88
Tri-Level	90
Wings & Ells	92
Residential Modifications	
Kitchen Cabinets	93
Kitchen Countertops	93
Vanity Bases	94
Solid Surface Vanity Tops	94
Fireplaces & Chimneys	94
Windows & Skylights	94
Dormers	94
Appliances	95
Breezeway	95
Porches	95
Finished Attic	95
Alarm System	95
Sauna, Prefabricated	95
Garages	96
Swimming Pools	96
Wood & Coal Stoves	96
Sidewalks	96
Fencing	96
Carport	96

Introduction to the Square Foot Cost Section

The Square Foot Cost Section of this manual contains costs per square foot for four classes of construction in seven building types. Costs are listed for various exterior wall materials which are typical of the class and building type. There are cost tables for Wings and Ells with modification tables to adjust the base cost of each class of building. Non-standard items can easily be added to the standard structures.

Cost estimating for a residence is a three-step process:
1. Identification
2. Listing dimensions
3. Calculations

Guidelines and a sample cost estimating procedure are shown on the following pages.

Identification

To properly identify a residential building, the class of construction, type, and exterior wall material must be determined. Page 8 has drawings and guidelines for determining the class of construction. There are also detailed specifications and additional drawings at the beginning of each set of tables to further aid in proper building class and type identification.

Sketches for eight types of residential buildings and their configurations are shown on pages 10 and 11. Definitions of living area are next to each sketch. Sketches and definitions of garage types are on page 12.

Living Area

Base cost tables are prepared as costs per square foot of living area. The living area of a residence is that area which is suitable and normally designed for full time living. It does not include basement recreation rooms or finished attics, although these areas are often considered full time living areas by the owners.

Living area is calculated from the exterior dimensions without the need to adjust for exterior wall thickness. When calculating the living area of a 1-1/2 story, two story, three story or tri-level residence, overhangs and other differences in size and shape between floors must be considered.

Only the floor area with a ceiling height of 7 feet or more in a 1-1/2 story residence is considered living area. In bi-levels and tri-levels, the areas that are below grade are considered living area, even when these areas may not be completely finished.

Base Tables and Modifications

Base cost tables show the base cost per square foot without a basement, with one full bath and one full kitchen for economy and average homes, and an additional half bath for custom and luxury models. Adjustments for finished and unfinished basements are part of the base cost tables. Adjustments for multi-family residences, additional bathrooms, townhouses, alternative roofs, and air conditioning and heating systems are listed in Modifications, Adjustments and Alternatives tables below the base cost tables.

Costs for other modifications, adjustments and alternatives, including garages, breezeways and site improvements, are on pages 93 to 96.

Listing of Dimensions

To use this section of the manual, only the dimensions used to calculate the horizontal area of the building and additions, modifications, adjustments and alternatives are needed. The dimensions, normally the length and width, can come from drawings or field measurements. For ease in calculation, consider measuring in tenths of feet, i.e., 9 ft. 6 in. = 9.5 ft., 9 ft. 4 in. = 9.3 ft.

In all cases, make a sketch of the building. Any protrusions or other variations in shape should be noted on the sketch with dimensions.

Calculations

The calculations portion of the estimate is a two-step activity:
1. The selection of appropriate costs from the tables
2. Computations

Selection of Appropriate Costs

To select the appropriate cost from the base tables, the following information is needed:
1. Class of construction
   - Economy
   - Average
   - Custom
   - Luxury
2. Type of residence
   - 1 story
   - 1-1/2 story
   - 2 story
   - 3 story
   - Bi-level
   - Tri-level
3. Occupancy
   - One family
   - Two family
   - Three family
4. Building configuration
   - Detached
   - Town/Row house
   - Semi-detached
5. Exterior wall construction
   - Wood frame
   - Brick veneer
   - Solid masonry
6. Living areas

Modifications are classified by class, type, and size.

Computations

The computation process should take the following sequence:
1. Multiply the base cost by the area.
2. Add or subtract the modifications, adjustments, and alternatives.
3. Apply the location modifier.

When selecting costs, interpolate or use the cost that most nearly matches the structure under study. This applies to size, exterior wall construction, and class.

How to Use the Residential Square Foot Cost Pages

The following is a detailed explanation of a sample entry in the Residential Square Foot Cost Section. Each bold number below corresponds to the item being described on the facing page with the appropriate component of the sample entry following in parenthesis.

Prices listed are costs that include overhead and profit of the installing contractor. Total model costs include an additional markup for General Contractors' overhead and profit, and fees specific to class of construction.

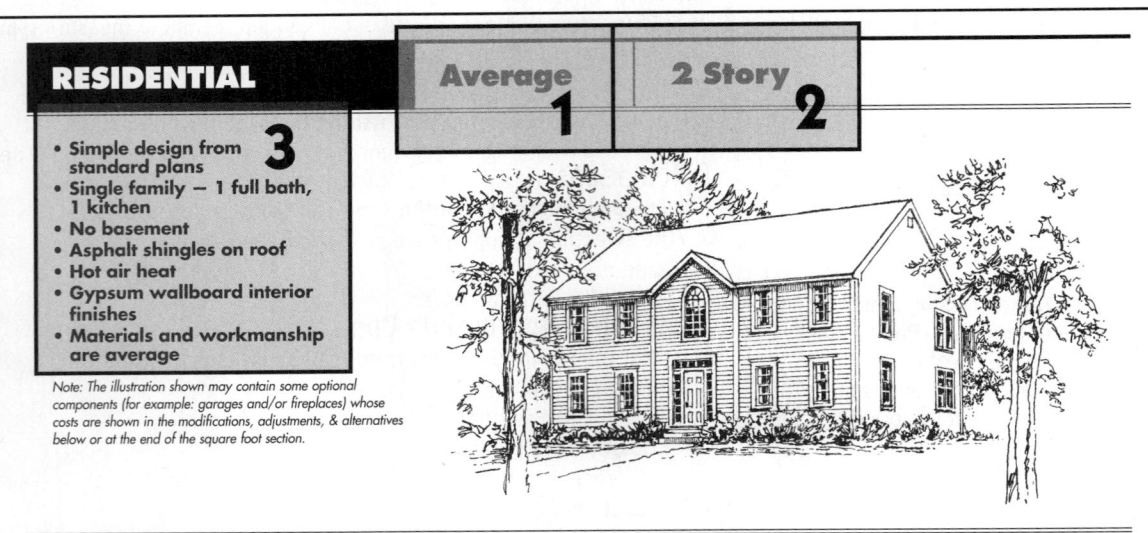

RESIDENTIAL — Average **1** — 2 Story **2**

3
- Simple design from standard plans
- Single family — 1 full bath, 1 kitchen
- No basement
- Asphalt shingles on roof
- Hot air heat
- Gypsum wallboard interior finishes
- Materials and workmanship are average

Note: The illustration shown may contain some optional components (for example: garages and/or fireplaces) whose costs are shown in the modifications, adjustments, & alternatives below or at the end of the square foot section.

Base cost per square foot of living area

Exterior Wall **4**	1000	1200	1400	1600	1800	Living Area **5** 2000	2200	2600	3000	3400	3800
Wood Siding - Wood Frame	136.80	124.05	117.95	113.80	109.55	104.95 **6**	101.95	96.20	90.55	88.10	85.80
Brick Veneer - Wood Frame	144.00	130.75	124.25	119.80	115.20	110.40	107.10	100.90	94.85	92.15	89.70
Stucco on Wood Frame	133.10	120.65	114.80	110.75	106.70	102.20	99.30	93.85	88.35	85.95	83.75
Solid Masonry	160.70	146.20	138.70	133.55	128.15	122.90	118.95	111.65	104.80	101.60	98.65
Finished Basement, Add **7**	22.50	22.15	21.40	20.95	20.40	20.05	19.65 **8**	18.90	18.40	18.10	17.80
Unfinished Basement, Add	8.85	8.25	7.85	7.55	7.15	6.95	6.70	6.25	5.95	5.75	5.55

Modifications

Add to the total cost **9**
Upgrade Kitchen Cabinets	$ +4158
Solid Surface Countertops (Included)	
Full Bath - including plumbing, wall and floor finishes	+7329
Half Bath - including plumbing, wall and floor finishes	+4275
One Car Attached Garage	+13,759
One Car Detached Garage	+17,994
Fireplace & Chimney	+6840

Adjustments

For multi family - add to total cost
Additional Kitchen	$ +8034
Additional Bath	+7329
Additional Entry & Exit	+1658
Separate Heating	+1133
Separate Electric	+2036

For Townhouse/Rowhouse - Multiply cost per square foot by **10**
Inner Unit	.90
End Unit	.95

Alternatives

Add to or deduct from the cost per square foot of living area **11**
Cedar Shake Roof	+ 1.70
Clay Tile Roof	+ 3.70
Slate Roof	+ 3.25
Upgrade Walls to Skim Coat Plaster	+ .52
Upgrade Ceilings to Textured Finish	+ .50
Air Conditioning, in Heating Ductwork	+ 2.94
In Separate Ductwork	+ 5.68
Heating Systems, Hot Water	+ 1.84
Heat Pump	+ 3.16
Electric Heat	− .53
Not Heated	− 2.99

Additional upgrades or components **12**

Kitchen Cabinets & Countertops	Page 93
Bathroom Vanities	94
Fireplaces & Chimneys	94
Windows, Skylights & Dormers	94
Appliances	95
Breezeways & Porches	95
Finished Attic	95
Garages	96
Site Improvements	96
Wings & Ells	56

1. Class of Construction (Average)
The class of construction depends upon the design and specifications of the plan. The four classes are economy, average, custom, and luxury.

2. Type of Residence (2 Story)
The building type describes the number of stories or levels in the model. The seven building types are 1 story, 1-1/2 story, 2 story, 2-1/2 story, 3 story, bi-level, and tri-level.

3. Specification Highlights (Hot Air Heat)
These specifications include information concerning the components of the model, including the number of baths, roofing types, HVAC systems, and materials and workmanship. If the components listed are not appropriate, modifications can be made by consulting the information shown lower on the page or the Assemblies Section.

4. Exterior Wall System (Wood Siding -Wood Frame)
This section includes the types of exterior wall systems and the structural frame used. The exterior wall systems shown are typical of the class of construction and the building type shown.

5. Living Areas (2000 SF)
The living area is that area of the residence which is suitable and normally designed for full time living. It does not include basement recreation rooms or finished attics. Living area is calculated from the exterior dimensions without the need to adjust for exterior wall thickness. When calculating the living area of a 1-1/2 story, 2 story, 3 story or tri-level residence, overhangs and other differences in size and shape between floors must be considered. Only the floor area with a ceiling height of seven feet or more in a 1-1/2 story residence is considered living area. In bi-levels and tri-levels, the areas that are below grade are considered living area, even when these areas may not be completely finished. A range of various living areas for the residential model are shown to aid in selection of values from the matrix.

6. Base Costs per Square Foot of Living Area ($104.95)
Base cost tables show the cost per square foot of living area without a basement, with one full bath and one full kitchen for economy and average homes, and an additional half bath for custom and luxury models. When selecting costs, interpolate or use the cost that most nearly matches the residence under consideration for size, exterior wall system, and class of construction. Prices listed are costs that include overhead and profit of the installing contractor, a general contractor markup, and an allowance for plans that vary by class of construction. For additional information on contractor overhead and architectural fees, see the Reference Section.

7. Basement Types (Finished)
The two types of basements are finished and unfinished. The specifications and components for both are shown on the Building Classes page in the Introduction to this section.

8. Additional Costs for Basements ($20.05 or $6.95)
These values indicate the additional cost per square foot of living area for either a finished or an unfinished basement.

9. Modifications and Adjustments (Upgrade Kitchen Cabinets $4158)
Modifications and Adjustments are costs added to or subtracted from the total cost of the residence. The total cost of the residence is equal to the cost per square foot of living area times the living area. Typical modifications and adjustments include kitchens, baths, garages and fireplaces.

10. Multiplier for Townhouse/ Rowhouse (Inner Unit .90)
The multipliers shown adjust the base costs per square foot of living area for the common wall condition encountered in townhouses or rowhouses.

11. Alternatives (Skim Coat Plaster $0.52)
Alternatives are costs added to or subtracted from the base cost per square foot of living area. Typical alternatives include variations in kitchens, baths, roofing, and air conditioning and heating systems.

12. Additional Upgrades or Components (Wings & Ells page 56)
Costs for additional upgrades or components, including wings or ells, breezeways, porches, finished attics, and site improvements, are shown in other locations in the Residential Square Foot Cost Section.

How to Use the Residential Square Foot Cost Pages (Continued)

1 Average 2 Story
Living Area - 2000 S.F.
Perimeter - 135 L.F.

			Cost Per Square Foot Of Living Area			% of Total
			Mat.	Inst.	Total	(rounded)
1	Site Work	Site preparation for slab; 4' deep trench excavation for foundation wall.		.81	.81	0.8%
2	Foundation	Continuous reinforced concrete footing 8" deep x 18" wide, dampproofed and insulated reinforced concrete foundation wall, 8" thick, 4' deep, 4" concrete slab on 4" crushed stone base and polyethylene vapor barrier, trowel finish.	3.29	4.22	7.51	7.2%
3	Framing	Exterior walls - 2" x 4" wood studs, 16" O.C.; 1/2" plywood sheathing; 2" x 6" rafters 16" O.C. with 1/2" plywood sheathing, 4 in 12 pitch; 2" x 6" ceiling joists 16" O.C.; 2" x 8" floor joists 16" O.C. with 5/8" plywood subfloor; 1/2" plywood subfloor on 1" x 2" wood sleepers 16" O.C.	7.62	9.28	16.90	16.1%
4	Exterior Walls	Beveled wood siding and building paper on insulated wood frame walls; attic insulation; double hung windows; 3 flush solid core wood exterior doors with storms.	13.75	6.00	19.75	18.8%
5	Roofing	25 year asphalt shingles; #15 felt building paper; aluminum gutters, downspouts, drip edge and flashings.	1.08	1.26	2.34	2.2%
6	Interiors	Walls and ceilings, 1/2" taped and finished gypsum wallboard, primed and painted with 2 coats; painted baseboard and trim, finished hardwood floor 40%, carpet with 1/2" underlayment 40%, vinyl tile with 1/2" underlayment 15%, ceramic tile with 1/2" underlayment 5%; hollow core and louvered doors.	14.49	15.02	29.51	28.1%
7	Specialties	Average grade kitchen cabinets - 14 L.F. wall and base with solid surface counter top and kitchen sink; 40 gallon electric water heater.	2.67	.95	3.62	3.4%
8	Mechanical	1 lavatory, white, wall hung; 1 water closet, white; 1 bathtub with shower; enameled steel, white; gas fired warm air heat.	3.14	3.25	6.39	6.1%
9	Electrical	200 Amp. service; Romex wiring; incandescent lighting fixtures, switches, receptacles.	1.13	1.75	2.88	2.7%
10	Overhead	Contractor's overhead and profit and plans.	8.03	7.21	15.24	14.5%
		Total	55.20	49.75	104.95	

1. Specifications
The parameters for an example dwelling from the previous pages are listed here. Included are the square foot dimensions of the proposed building. Living Area takes into account the number of floors and other factors needed to define a building's total square footage. Perimeter dimensions are defined in terms of linear feet.

2. Building Type
This is a sketch of a cross section view through the dwelling. It is shown to help define the living area for the building type. For more information, see the Building Types pages in the Introduction.

3. Components (3 Framing)
This page contains the ten components needed to develop the complete square foot cost of the typical dwelling specified. All components are defined with a description of the materials and/or task involved. Use cost figures from each component to estimate the cost per square foot of that section of the project. The components listed on this page are typical of all sizes of residences from the facing page. Specific quantities of components required would vary with the size of the dwelling and the exterior wall system.

4. Materials (7.62)
This column gives the amount needed to develop the cost of materials. The figures given here are not bare costs. Ten percent has been added to bare material cost for profit.

5. Installation (9.28)
Installation includes labor and equipment costs. The labor rates included here incorporate the total overhead and profit costs for the installing contractor. The average mark-up used to create these figures is 72.0% over and above bare labor costs. The equipment rates include 10% for profit.

6. Total (16.90)
This column lists the sum of two figures. Use this total to determine the sum of material cost plus installation cost. The result is a convenient total cost for each of the ten components.

7. % of Total (rounded) (16.1%)
This column represents the percent of the total cost for this component group.

8. Overhead
The costs in components 1 through 9 include overhead and profit for the installing contractor. Item 10 is overhead and profit for the general contractor. This is typically a percentage mark-up of all other costs. The amount depends on size and type of dwelling, building class, and economic conditions. An allowance for plans or design has been included where appropriate.

9. Bottom Line Total (104.95)
This figure is the complete square foot cost for the construction project and equals the sum of total material and total labor costs. To determine total project cost, multiply the bottom line total by the living area.

Building Classes

Economy Class

An economy class residence is usually built from stock plans. The materials and workmanship are sufficient to satisfy building codes. Low construction cost is more important than distinctive features. The overall shape of the foundation and structure is seldom other than square or rectangular.

An unfinished basement includes 7' high 8" thick foundation wall composed of either concrete block or cast-in-place concrete.

Included in the finished basement cost are inexpensive paneling or drywall as the interior finish on the foundation walls, a low cost sponge backed carpeting adhered to the concrete floor, a drywall ceiling, and overhead lighting.

Custom Class

A custom class residence is usually built from plans and specifications with enough features to give the building a distinction of design. Materials and workmanship are generally above average with obvious attention given to construction details. Construction normally exceeds building code requirements.

An unfinished basement includes a 7'-6" high 10" thick cast-in-place concrete foundation wall or a 7'-6" high 12" thick concrete block foundation wall.

A finished basement includes painted drywall on insulated 2" x4" wood furring as the interior finish to the concrete walls, a suspended ceiling, carpeting adhered to the concrete floor, overhead lighting and heating.

Class

An average class residence is a simple design and built from standard plans. Materials and workmanship are average, but often exceed minimum building codes. There are frequently special features that give the residence some distinctive characteristics.

An unfinished basement includes 7'-6" high 8" thick foundation wall composed of either cast-in-place concrete or concrete block.

Included in the finished basement are plywood paneling or drywall on furring that is fastened to the foundation walls, sponge backed carpeting adhered to the concrete floor, a suspended ceiling, overhead lighting and heating.

Luxury Class

A luxury class residence is built from an architect's plan for a specific owner. It is unique both in design and workmanship. There are many special features, and construction usually exceeds all building codes. It is obvious that primary attention is placed on the owner's comfort and pleasure. Construction is supervised by an architect.

An unfinished basement includes 8' high 12" thick foundation wall that is composed of cast-in-place concrete or concrete block.

A finished basement includes painted drywall on 2" x4" wood furring as the interior finish, suspended ceiling, tackless carpet on wood subfloor with sleepers, overhead lighting and heating.

Configurations

Detached House
This category of residence is a free-standing separate building with or without an attached garage. It has four complete walls.

Semi-Detached House
This category of residence has two living units side-by-side. The common wall is a fireproof wall. Semi-detached residences can be treated as a row house with two end units. Semi-detached residences can be any of the building types.

Town/Row House
This category of residence has a number of attached units made up of inner units and end units. The units are joined by common walls. The inner units have only two exterior walls. The common walls are fireproof. The end units have three walls and a common wall. Town houses/row houses can be any of the building types.

Building Types

One Story
This is an example of a one-story dwelling. The living area of this type of residence is confined to the ground floor. The headroom in the attic is usually too low for use as a living area.

One-and-a-half Story
The living area in the upper level of this type of residence is 50% to 90% of the ground floor. This is made possible by a combination of this design's high-peaked roof and/or dormers. Only the upper level area with a ceiling height of 7 feet or more is considered living area. The living area of this residence is the sum of the ground floor area plus the area on the second level with a ceiling height of 7 feet or more.

One Story with Finished Attic
The main living area in this type of residence is the ground floor. The upper level or attic area has sufficient headroom for use as a living area. This is made possible by a high peaked roof. The living area in the attic is less than 50% of the ground floor. The living area of this type of residence is the ground floor area only. The finished attic is considered an adjustment.

Two Story
This type of residence has a second floor or upper level area which is equal or nearly equal to the ground floor area. The upper level of this type of residence can range from 90% to 110% of the ground floor area, depending on setbacks or overhangs. The living area is the sum of the ground floor area and the upper level floor area.

Two-and-one-half Story

This type of residence has two levels of equal or nearly equal area and a third level which has a living area that is 50% to 90% of the ground floor. This is made possible by a high peaked roof, extended wall heights and/or dormers. Only the upper level area with a ceiling height of 7 feet or more is considered living area. The living area of this residence is the sum of the ground floor area, the second floor area and the area on the third level with a ceiling height of 7 feet or more.

Bi-level

This type of residence has two living areas, one above the other. One area is about 4 feet below grade and the second is about 4 feet above grade. Both areas are equal in size. The lower level in this type of residence is originally designed and built to serve as a living area and not as a basement. Both levels have full ceiling heights. The living area is the sum of the lower level area and the upper level area.

Three Story

This type of residence has three levels which are equal or nearly equal. As in the 2 story residence, the second and third floor areas may vary slightly depending on setbacks or overhangs. The living area is the sum of the ground floor area and the two upper level floor areas.

Tri-level

This type of residence has three levels of living area: one at grade level, one about 4 feet below grade, and one about 4 feet above grade. All levels are originally designed to serve as living areas. All levels have full ceiling heights. The living area is a sum of the areas of each of the three levels.

Garage Types

Attached Garage
Shares a common wall with the dwelling. Access is typically through a door between dwelling and garage.

Basement Garage
Constructed under the roof of the dwelling but below the living area.

Built-In Garage
Constructed under the second floor living space and above basement level of dwelling. Reduces gross square feet of living area.

Detached Garage
Constructed apart from the main dwelling. Shares no common area or wall with the dwelling.

Building Components

1. Excavation
2. Sill Plate
3. Basement Window
4. Floor Joist
5. Shoe Plate
6. Studs
7. Drywall
8. Plate
9. Ceiling Joists
10. Rafters
11. Collar Ties
12. Ridge Rafter
13. Roof Sheathing
14. Roof Felt
15. Roof Shingles
16. Flashing
17. Flue Lining
18. Chimney
19. Roof Shingles
20. Gutter
21. Fascia
22. Downspout
23. Shutter
24. Window
25. Wall Shingles
26. Weather Barrier
27. Wall Sheathing
28. Fire Stop
29. Dampproofing
30. Foundation Wall
31. Backfill
32. Drainage Stone
33. Drainage Tile
34. Wall Footing
35. Gravel
36. Concrete Slab
37. Column Footing
38. Pipe Column
39. Expansion Joint
40. Girder
41. Sub-floor
42. Finish Floor
43. Attic Insulation
44. Soffit
45. Ceiling Strapping
46. Wall Insulation
47. Cross Bridging
48. Bulkhead Stairs

Exterior Wall Construction

Typical Frame Construction

Typical wood frame construction consists of wood studs with insulation between them. A typical exterior surface is made up of sheathing, building paper and exterior siding consisting of wood, vinyl, aluminum or stucco over the wood sheathing.

Brick Veneer

Typical brick veneer construction consists of wood studs with insulation between them. A typical exterior surface is sheathing, building paper and an exterior of brick tied to the sheathing, with metal strips.

Stone

Typical solid masonry construction consists of a stone or block wall covered on the exterior with brick, stone or other masonry.

Residential Cost Estimate Worksheet

Worksheet Instructions

The residential cost estimate worksheet can be used as an outline for developing a residential construction or replacement cost. It is also useful for insurance appraisals. The design of the worksheet helps eliminate errors and omissions. To use the worksheet, follow the example below.

1. Fill out the owner's name, residence address, the estimator or appraiser's name, some type of project identifying number or code, and the date.
2. Determine from the plans, specifications, owner's description, photographs or any other means possible the class of construction. The models in this book use economy, average, custom, and luxury as classes. Fill in the appropriate box.
3. Fill in the appropriate box for the residence type, configuration, occupancy, and exterior wall. If you require clarification, the pages preceding this worksheet describe each of these.
4. Next, the living area of the residence must be established. The heated or air conditioned space of the residence, not including the basement, should be measured. It is easiest to break the structure up into separate components as shown in the example. The main house (A), a one-and-one-half story wing (B), and a one story wing (C). The breezeway (D), garage (E), and open covered porch (F) will be treated differently. Data entry blocks for the living area are included on the worksheet for your use. Keep each level of each component separate, and fill out the blocks as shown.
5. By using the information on the worksheet, find the model, wing or ell in the following square foot cost pages that best matches the class, type, exterior finish and size of the residence being estimated. Use the Modifications, Adjustments, and Alternatives to determine the adjusted cost per square foot of living area for each component.
6. For each component, multiply the cost per square foot by the living area square footage. If the residence is a townhouse/rowhouse, a multiplier should be applied based upon the configuration.
7. The second page of the residential cost estimate worksheet has space for the additional components of a house. The cost for additional bathrooms, finished attic space, breezeways, porches, fireplaces, appliance or cabinet upgrades, and garages should be added on this page. The information for each of these components is found with the model being used, or in the Modifications, Adjustments, and Alternatives pages.
8. Add the total from page one of the estimate worksheet and the items listed on page two. The sum is the adjusted total building cost.
9. Depending on the use of the final estimated cost, one of the remaining two boxes should be filled out. Any additional items or exclusions should be added or subtracted at this time. The data contained in this book is a national average. Construction costs are different throughout the country. To allow for this difference, a location factor based upon the first three digits of the residence's zip code must be applied. The location factor is a multiplier that increases or decreases the adjusted total building cost. Find the appropriate location factor and calculate the local cost. If depreciation is a concern, a dollar figure should be subtracted at this point.
10. No residence will match a model exactly. Many differences will be found. At this level of estimating, a variation of plus or minus 10% should be expected.

Adjustments Instructions

No residence matches a model exactly in shape, material, or specifications. The common differences are:

1. Two or more exterior wall systems:
   - Partial basement
   - Partly finished basement
2. Specifications or features that are between two classes
3. Crawl space instead of a basement

Examples

Below are quick examples. See pages 17–19 for complete examples of cost adjustments for these differences:

1. Residence "A" is an average one-story structure with 1,600 S.F. of living area and no basement. Three walls are wood siding on wood frame, and the fourth wall is brick veneer on wood frame. The brick veneer wall is 35% of the exterior wall area.

   Use page 38 to calculate the Base Cost per S.F. of Living Area. Wood Siding for 1,600 S.F. = $107.95 per S.F. Brick Veneer for 1,600 S.F. = $124.65 per S.F.
   .65 ($107.95) + .35 ($124.65) = $113.80 per S.F. of Living Area.

2a. Residence "B" is the same as Residence "A"; However, it has an unfinished basement under 50% of the building. To adjust the $113.80 per S.F. of living area for this partial basement, use page 38.

   $113.80 + .5($11.25) = $119.43 per S.F. of Living Area.

   b. Residence "C" is the same as Residence "A"; However, it has a full basement under the entire building. 640 S.F. or 40% of the basement area is finished.

   Using Page 38:
   $113.80 + .40 ($32.35) + .60 ($11.25) = $133.49 per S.F. of Living Area.

3. When specifications or features of a building are between classes, estimate the percent deviation, and use two tables to calculate the cost per S.F.

   A two-story residence with wood siding and 1,800 S.F. of living area has features 30% better than Average, but 70% less than Custom.

   From pages 42 and 64:

Custom 1,800 S.F. Base Cost	=	$134.25 per S.F.
Average 1,800 S.F. Base Cost	=	$109.55 per S.F.
DIFFERENCE	=	$24.70 per S.F.
Cost is $109.55 + .30 ($24.70)	=	$116.96 per S.F. of Living Area.

4. To add the cost of a crawl space, use the cost of an unfinished basement as a maximum. For specific costs of components to be added or deducted, such as vapor barrier, underdrain, and floor, see the "Assemblies" section, pages 97 to 281.

Model Residence Example

First Floor Plan

Second Floor Plan

A = Main House
B = 1-1/2 Story Wing
C = 1 Story Wing
D = Breezeway
E = Garage
F = Open Covered Porch

RESIDENTIAL COST ESTIMATE

OWNERS NAME:	Albert Westenberg		APPRAISER:	Nicole Wojtowicz
RESIDENCE ADDRESS:	300 Sygiel Road		PROJECT:	# 55
CITY, STATE, ZIP CODE:	Three Rivers, MA 01080		DATE:	Jan. 1, 2015

CLASS OF CONSTRUCTION
- ☐ ECONOMY
- ☑ AVERAGE
- ☐ CUSTOM
- ☐ LUXURY

RESIDENCE TYPE
- ☐ 1 STORY
- ☐ 1 1/2 STORY
- ☑ 2 STORY
- ☐ 2 1/2 STORY
- ☐ 3 STORY
- ☐ BI-LEVEL
- ☐ TRI-LEVEL

CONFIGURATION
- ☑ DETACHED
- ☐ TOWN/ROW HOUSE
- ☐ SEMI-DETACHED

OCCUPANCY
- ☑ ONE FAMILY
- ☐ TWO FAMILY
- ☐ THREE FAMILY
- ☐ OTHER

EXTERIOR WALL SYSTEM
- ☑ WOOD SIDING - WOOD FRAME
- ☐ BRICK VENEER - WOOD FRAME
- ☐ STUCCO ON WOOD FRAME
- ☐ PAINTED CONCRETE BLOCK
- ☐ SOLID MASONRY (AVERAGE & CUSTOM)
- ☐ STONE VENEER - WOOD FRAME
- ☐ SOLID BRICK (LUXURY)
- ☐ SOLID STONE (LUXURY)

* LIVING AREA (Main Building)
First Level	1288	S.F.
Second level	1288	S.F.
Third Level		S.F.
Total	2576	S.F.

* LIVING AREA (Wing or Ell) (B)
First Level	360	S.F.
Second level	310	S.F.
Third Level		S.F.
Total	670	S.F.

* LIVING AREA (Wing or Ell) (C)
First Level	192	S.F.
Second level		S.F.
Third Level		S.F.
Total	192	S.F.

* Basement Area is not part of living area.

MAIN BUILDING | COSTS PER S.F. LIVING AREA

Cost per Square Foot of Living Area, from Page	42	$	104.95
Basement Addition: ____ % Finished, 100 % Unfinished		+	6.95
Roof Cover Adjustment: Cedar Shake Type, Page 42 (Add or Deduct)		( )	1.70
Central Air Conditioning: ☐ Separate Ducts ☑ Heating Ducts, Page 42		+	2.94
Heating System Adjustment: ____ Type, Page ____ (Add or Deduct)		( )	
Main Building: Adjusted Cost per S.F. of Living Area		$	# 116.54

MAIN BUILDING TOTAL COST	$ 116.54 /S.F.	x	2,576 S.F.	x	____	=	$ 300,207 TOTAL COST
	Cost per S.F. Living Area		Living Area		Town/Row House Multiplier (Use 1 for Detached)		

WING OR ELL (B) 1 - 1/2 STORY | COSTS PER S.F. LIVING AREA

Cost per Square Foot of Living Area, from Page	56	$	93.30
Basement Addition: 100 % Finished, ____ % Unfinished		+	29.25
Roof Cover Adjustment: ____ Type, Page ____ (Add or Deduct)		( )	—
Central Air Conditioning: ☐ Separate Ducts ☑ Heating Ducts, Page 42		+	2.94
Heating System Adjustment: ____ Type, Page ____ (Add or Deduct)		( )	—
Wing or Ell (B): Adjusted Cost per S.F. of Living Area		$	# 125.49

WING OR ELL (B) TOTAL COST	$ 125.49 /S.F.	x	670 S.F.	=	$ 84,078 TOTAL COST

WING OR ELL (C) 1 STORY | COSTS PER S.F. LIVING AREA

Cost per Square Foot of Living Area, from Page	# 56 (WOOD SIDING)	$	# 157.35
Basement Addition: ____ % Finished, ____ % Unfinished		+	—
Roof Cover Adjustment: ____ Type, Page ____ (Add or Deduct)		( )	—
Central Air Conditioning: ☐ Separate Ducts ☐ Heating Ducts, Page ____		+	—
Heating System Adjustment: ____ Type, Page ____ (Add or Deduct)		( )	—
Wing or Ell (C) Adjusted Cost per S.F. of Living Area		$	157.35

WING OR ELL (C) TOTAL COST	$ 157.35 /S.F.	x	192 S.F.	=	$ 30,211 TOTAL COST

TOTAL THIS PAGE	414,496

Page 1 of 2

RESIDENTIAL COST ESTIMATE

Total Page 1							$	414,496
					QUANTITY	UNIT COST		
Additional Bathrooms:	**2** Full, **1** Half	2 @ 7329	1 @ 4275				+	18,933
Finished Attic:	**N/A** Ft. x _____ Ft.					S.F.		
Breezeway: ☑ Open ☐ closed	**12** Ft. x **12** Ft.				144	S.F. 38.49	+	5,543
Covered Porch: ☑ Open ☐ Enclosed	**18** Ft. x **12** Ft.				216	S.F. 36.47	+	7,878
Fireplace: ☑ Interior Chimney ☐ Exterior Chimney								
☑ No. of Flues (**2**) ☑ Additional Fireplaces	**1 - 2nd Story**						+	11,467
Appliances:							+	—
Kitchen Cabinets Adjustments:				(±)				—
☑ Garage ☐ Carport: **2** Car(s) Description	**Wood, Attached**			(±)				23,447
Miscellaneous:							+	
				ADJUSTED TOTAL BUILDING COST			$	481,764

REPLACEMENT COST		
ADJUSTED TOTAL BUILDING COST	$	481,764
Site Improvements		
(A) Paving & Sidewalks	$	
(B) Landscaping	$	
(C) Fences	$	
(D) Swimming Pools	$	
(E) Miscellaneous	$	
TOTAL	$	481,764
Location Factor	x	1.080
Location Replacement Cost	$	520,305
Depreciation	-$	52,030
LOCAL DEPRECIATED COST	$	468,275

INSURANCE COST		
ADJUSTED TOTAL BUILDING COST	$	
Insurance Exclusions		
(A) Footings, sitework, Underground Piping	-$	
(B) Architects Fees	-$	
Total Building Cost Less Exclusion	$	
Location Factor	x	
LOCAL INSURABLE REPLACEMENT COST	$	

SKETCH AND ADDITIONAL CALCULATIONS

Economy Class

RESIDENTIAL | Economy | Illustrations

1 Story

1-1/2 Story

2 Story

Bi-Level

Tri-Level

For customer support on your Residential Cost Data, call 877.713.2617.

RESIDENTIAL | Economy | 1 Story

- Mass produced from stock plans
- Single family — 1 full bath, 1 kitchen
- No basement
- Asphalt shingles on roof
- Hot air heat
- Gypsum wallboard interior finishes
- Materials and workmanship are sufficient to meet codes

Note: The illustration shown may contain some optional components (for example: garages and/or fireplaces) whose costs are shown in the modifications, adjustments, & alternatives below or at the end of the square foot section.

©Home Planners, Inc.

Base cost per square foot of living area

Exterior Wall	600	800	1000	1200	1400	1600	1800	2000	2400	2800	3200
Wood Siding - Wood Frame	127.95	115.85	106.60	99.20	92.65	88.50	86.50	83.75	78.15	73.95	71.20
Brick Veneer - Wood Frame	134.45	121.65	111.85	103.95	97.00	92.55	90.40	87.45	81.45	77.05	74.10
Stucco on Wood Frame	121.25	109.90	101.20	94.30	88.25	84.35	82.45	79.95	74.70	70.75	68.25
Painted Concrete Block	129.05	116.80	107.55	100.05	93.45	89.25	87.15	84.35	78.65	74.50	71.70
Finished Basement, Add	32.35	30.50	29.15	27.95	26.95	26.30	25.95	25.40	24.70	24.15	23.65
Unfinished Basement, Add	14.30	12.80	11.70	10.75	10.00	9.50	9.20	8.75	8.20	7.75	7.40

Modifications

Add to the total cost

Upgrade Kitchen Cabinets	$ + 946
Solid Surface Countertops	+ 678
Full Bath - including plumbing, wall and floor finishes	+ 5863
Half Bath - including plumbing, wall and floor finishes	+ 3420
One Car Attached Garage	+ 12,758
One Car Detached Garage	+ 16,393
Fireplace & Chimney	+ 5874

Adjustments

For multi family - add to total cost

Additional Kitchen	$ + 4417
Additional Bath	+ 5863
Additional Entry & Exit	+ 1658
Separate Heating	+ 1133
Separate Electric	+ 1138

For Townhouse/Rowhouse - Multiply cost per square foot by

Inner Unit	.95
End Unit	.97

Alternatives

Add to or deduct from the cost per square foot of living area

Composition Roll Roofing	– 1.10
Cedar Shake Roof	+ 3.95
Upgrade Walls and Ceilings to Skim Coat Plaster	+ .70
Upgrade Ceilings to Textured Finish	+ .50
Air Conditioning, in Heating Ductwork	+ 4.70
In Separate Ductwork	+ 6.89
Heating Systems, Hot Water	+ 1.93
Heat Pump	+ 2.61
Electric Heat	– 1.23
Not Heated	– 3.84

Additional upgrades or components

Kitchen Cabinets & Countertops	Page 93
Bathroom Vanities	94
Fireplaces & Chimneys	94
Windows, Skylights & Dormers	94
Appliances	95
Breezeways & Porches	95
Finished Attic	95
Garages	96
Site Improvements	96
Wings & Ells	34

Economy 1 Story

Living Area - 1200 S.F.
Perimeter - 146 L.F.

#	Category	Description	Cost Per Square Foot Of Living Area			% of Total
			Mat.	Inst.	Total	(rounded)
1	Site Work	Site preparation for slab; 4' deep trench excavation for foundation wall.		1.31	1.31	1.4%
2	Foundation	Continuous reinforced concrete footing, 8" deep x 18" wide; dampproofed and insulated 8" thick reinforced concrete block foundation wall, 4' deep; 4" concrete slab on 4" crushed stone base and polyethylene vapor barrier, trowel finish.	5.96	7.58	13.54	14.4%
3	Framing	Exterior walls - 2" x 4" wood studs, 16" O.C.; 1/2" insulation board sheathing; wood truss roof frame, 24" O.C. with 1/2" plywood sheathing, 4 in 12 pitch.	5.09	6.42	11.51	12.2%
4	Exterior Walls	Metal lath reinforced stucco exterior on insulated wood frame walls; attic insulation; sliding sash wood windows; 2 flush solid core wood exterior doors with storms.	7.65	6.87	14.52	15.4%
5	Roofing	20 year asphalt shingles; #15 felt building paper; aluminum gutters, downspouts, drip edge and flashings.	2.10	2.46	4.56	4.8%
6	Interiors	Walls and ceilings, 1/2" taped and finished gypsum wallboard, primed and painted with 2 coats; painted baseboard and trim; rubber backed carpeting 80%, asphalt tile 20%; hollow core wood interior doors.	10.18	12.53	22.71	24.1%
7	Specialties	Economy grade kitchen cabinets - 6 L.F. wall and base with plastic laminate counter top and kitchen sink; 30 gallon electric water heater.	2.51	1.01	3.52	3.7%
8	Mechanical	1 lavatory, white, wall hung; 1 water closet, white; 1 bathtub, enameled steel, white; gas fired warm air heat.	4.12	3.80	7.92	8.4%
9	Electrical	100 Amp. service; Romex wiring; incandescent lighting fixtures, switches, receptacles.	.88	1.55	2.43	2.6%
10	Overhead	Contractor's overhead and profit.	5.76	6.52	12.28	13.0%
		Total	44.25	50.05	**94.30**	

RESIDENTIAL — Economy — 1-1/2 Story

- Mass produced from stock plans
- Single family — 1 full bath, 1 kitchen
- No basement
- Asphalt shingles on roof
- Hot air heat
- Gypsum wallboard interior finishes
- Materials and workmanship are sufficient to meet codes

Note: The illustration shown may contain some optional components (for example: garages and/or fireplaces) whose costs are shown in the modifications, adjustments, & alternatives below or at the end of the square foot section.

Base cost per square foot of living area

Exterior Wall	Living Area										
	600	800	1000	1200	1400	1600	1800	2000	2400	2800	3200
Wood Siding - Wood Frame	149.60	124.05	110.65	104.35	99.80	92.95	89.60	86.20	79.00	76.25	73.25
Brick Veneer - Wood Frame	155.70	128.35	115.00	108.55	104.00	96.80	93.30	89.75	82.15	79.30	76.15
Stucco on Wood Frame	137.25	115.00	102.60	96.95	92.95	86.75	83.75	80.70	74.20	71.75	69.20
Painted Concrete Block	148.15	122.90	109.90	103.85	99.50	92.65	89.45	86.05	78.90	76.20	73.30
Finished Basement, Add	24.85	21.15	20.15	19.45	18.95	18.25	17.80	17.45	16.70	16.35	15.90
Unfinished Basement, Add	12.60	9.65	8.85	8.35	7.90	7.35	7.00	6.70	6.05	5.85	5.50

Modifications

Add to the total cost

Upgrade Kitchen Cabinets	$ + 946
Solid Surface Countertops	+ 678
Full Bath - including plumbing, wall and floor finishes	+ 5863
Half Bath - including plumbing, wall and floor finishes	+ 3420
One Car Attached Garage	+ 12,758
One Car Detached Garage	+ 16,393
Fireplace & Chimney	+ 5874

Adjustments

For multi family - add to total cost

Additional Kitchen	$ + 4417
Additional Bath	+ 5863
Additional Entry & Exit	+ 1658
Separate Heating	+ 1133
Separate Electric	+ 1138

For Townhouse/Rowhouse - Multiply cost per square foot by

Inner Unit	.95
End Unit	.97

Alternatives

Add to or deduct from the cost per square foot of living area

Composition Roll Roofing	– .75
Cedar Shake Roof	+ 2.85
Upgrade Walls and Ceilings to Skim Coat Plaster	+ .71
Upgrade Ceilings to Textured Finish	+ .50
Air Conditioning, in Heating Ductwork	+ 3.50
In Separate Ductwork	+ 6.05
Heating Systems, Hot Water	+ 1.84
Heat Pump	+ 2.87
Electric Heat	– .98
Not Heated	– 3.53

Additional upgrades or components

Kitchen Cabinets & Countertops	Page 93
Bathroom Vanities	94
Fireplaces & Chimneys	94
Windows, Skylights & Dormers	94
Appliances	95
Breezeways & Porches	95
Finished Attic	95
Garages	96
Site Improvements	96
Wings & Ells	34

Economy 1-1/2 Story

Living Area - 1600 S.F.
Perimeter - 135 L.F.

			Cost Per Square Foot Of Living Area			% of Total
			Mat.	Inst.	Total	(rounded)
1	Site Work	Site preparation for slab; 4' deep trench excavation for foundation wall.		1.97	1.97	2.1%
2	Foundation	Continuous reinforced concrete footing, 8" deep x 18" wide; dampproofed and insulated 8" thick reinforced concrete block foundation wall, 4' deep; 4" concrete slab on 4" crushed stone base and polyethylene vapor barrier, trowel finish.	3.99	5.14	9.13	9.8%
3	Framing	Exterior walls - 2" x 4" wood studs, 16" O.C.; 1/2" insulation board sheathing; 2" x 6" rafters, 16" O.C. with 1/2" plywood sheathing, 8 in 12 pitch; 2" x 8" floor joists 16" O.C. with bridging and 5/8" plywood subfloor.	5.34	7.26	12.60	13.6%
4	Exterior Walls	Beveled wood siding and building paper on insulated wood frame walls; attic insulation; double hung windows; 2 flush solid core wood exterior doors with storms.	12.69	5.42	18.11	19.5%
5	Roofing	20 year asphalt shingles; #15 felt building paper; aluminum gutters, downspouts, drip edge and flashings.	1.31	1.54	2.85	3.1%
6	Interiors	Walls and ceilings, 1/2" taped and finished gypsum wallboard, primed and painted with 2 coats; painted baseboard and trim; rubber backed carpeting 80%, asphalt tile 20%; hollow core wood interior doors.	11.31	13.29	24.60	26.5%
7	Specialties	Economy grade kitchen cabinets - 6 L.F. wall and base with plastic laminate counter top and kitchen sink; 30 gallon electric water heater.	1.88	.76	2.64	2.8%
8	Mechanical	1 lavatory, white, wall hung; 1 water closet, white; 1 bathtub, enameled steel, white; gas fired warm air heat.	3.33	3.41	6.74	7.3%
9	Electrical	100 Amp. service; Romex wiring; incandescent lighting fixtures, switches, receptacles.	.80	1.39	2.19	2.4%
10	Overhead	Contractor's overhead and profit.	6.10	6.02	12.12	13.0%
		Total	46.75	46.20	**92.95**	

RESIDENTIAL — Economy — 2 Story

- Mass produced from stock plans
- Single family — 1 full bath, 1 kitchen
- No basement
- Asphalt shingles on roof
- Hot air heat
- Gypsum wallboard interior finishes
- Materials and workmanship are sufficient to meet codes

Note: The illustration shown may contain some optional components (for example: garages and/or fireplaces) whose costs are shown in the modifications, adjustments, & alternatives below or at the end of the square foot section.

Base cost per square foot of living area

Exterior Wall	1000	1200	1400	1600	1800	2000	2200	2600	3000	3400	3800
Wood Siding - Wood Frame	115.70	104.65	99.55	95.95	92.40	88.35	85.70	80.65	75.70	73.50	71.50
Brick Veneer - Wood Frame	122.75	111.15	105.65	101.80	97.90	93.55	90.75	85.20	79.85	77.50	75.30
Stucco on Wood Frame	108.50	97.95	93.30	90.00	86.75	82.90	80.60	76.00	71.35	69.40	67.60
Painted Concrete Block	116.95	105.75	100.60	96.95	93.35	89.25	86.55	81.45	76.35	74.10	72.15
Finished Basement, Add	16.95	16.25	15.65	15.30	14.90	14.65	14.35	13.75	13.35	13.10	12.90
Unfinished Basement, Add	7.75	7.15	6.70	6.40	6.05	5.85	5.65	5.20	4.90	4.70	4.50

Modifications

Add to the total cost

Upgrade Kitchen Cabinets	$ + 946
Solid Surface Countertops	+ 678
Full Bath - including plumbing, wall and floor finishes	+ 5863
Half Bath - including plumbing, wall and floor finishes	+ 3420
One Car Attached Garage	+ 12,758
One Car Detached Garage	+ 16,393
Fireplace & Chimney	+ 6491

Adjustments

For multi family - add to total cost

Additional Kitchen	$ + 4417
Additional Bath	+ 5863
Additional Entry & Exit	+ 1658
Separate Heating	+ 1133
Separate Electric	+ 1138

For Townhouse/Rowhouse - Multiply cost per square foot by

Inner Unit	.93
End Unit	.96

Alternatives

Add to or deduct from the cost per square foot of living area

Composition Roll Roofing	– .55
Cedar Shake Roof	+ .2
Upgrade Walls and Ceilings to Skim Coat Plaster	+ .72
Upgrade Ceilings to Textured Finish	+ .50
Air Conditioning, in Heating Ductwork	+ 2.85
In Separate Ductwork	+ 5.55
Heating Systems, Hot Water	+ 1.79
Heat Pump	+ 3.05
Electric Heat	– .85
Not Heated	– 3.34

Additional upgrades or components

Kitchen Cabinets & Countertops	Page 93
Bathroom Vanities	94
Fireplaces & Chimneys	94
Windows, Skylights & Dormers	94
Appliances	95
Breezeways & Porches	95
Finished Attic	95
Garages	96
Site Improvements	96
Wings & Ells	34

Economy 2 Story

Living Area - 2000 S.F.
Perimeter - 135 L.F.

			Cost Per Square Foot Of Living Area			% of Total
			Mat.	Inst.	Total	(rounded)
1	Site Work	Site preparation for slab; 4' deep trench excavation for foundation wall.		.79	.79	0.9%
2	Foundation	Continuous reinforced concrete footing, 8" deep x 18" wide; dampproofed and insulated 8" thick reinforced concrete block foundation wall, 4' deep; 4" concrete slab on 4" crushed stone base and polyethylene vapor barrier, trowel finish.	3.20	4.11	7.31	8.3%
3	Framing	Exterior walls - 2" x 4" wood studs, 16" O.C.; 1/2" insulation board sheathing; wood truss roof frame, 24" O.C. with 1/2" plywood sheathing, 4 in 12 pitch; 2" x 8" floor joists 16" O.C. with bridging and 5/8" plywood subfloor.	5.49	7.64	13.13	14.9%
4	Exterior Walls	Beveled wood siding and building paper on insulated wood frame walls; attic insulation; double hung windows; 2 flush solid core wood exterior doors with storms.	13.00	5.53	18.53	21.0%
5	Roofing	20 year asphalt shingles; #15 felt building paper; aluminum gutters, downspouts, drip edge and flashings.	1.05	1.23	2.28	2.6%
6	Interiors	Walls and ceilings, 1/2" taped and finished gypsum wallboard, primed and painted with 2 coats; painted baseboard and trim; rubber backed carpeting 80%, asphalt tile 20%; hollow core wood interior doors.	11.20	13.32	24.52	27.8%
7	Specialties	Economy grade kitchen cabinets - 6 L.F. wall and base with plastic laminate counter top and kitchen sink; 30 gallon electric water heater.	1.51	.61	2.12	2.4%
8	Mechanical	1 lavatory, white, wall hung; 1 water closet, white; 1 bathtub, enameled steel, white; gas fired warm air heat.	2.87	3.19	6.06	6.9%
9	Electrical	100 Amp. service; Romex wiring; incandescent lighting fixtures; switches, receptacles.	.75	1.30	2.05	2.3%
10	Overhead	Contractor's overhead and profit.	5.88	5.68	11.56	13.1%
		Total	44.95	43.40	**88.35**	

RESIDENTIAL — Economy | Bi-Level

- Mass produced from stock plans
- Single family — 1 full bath, 1 kitchen
- No basement
- Asphalt shingles on roof
- Hot air heat
- Gypsum wallboard interior finishes
- Materials and workmanship are sufficient to meet codes

Note: The illustration shown may contain some optional components (for example: garages and/or fireplaces) whose costs are shown in the modifications, adjustments, & alternatives below or at the end of the square foot section.

Base cost per square foot of living area

Exterior Wall	1000	1200	1400	1600	1800	2000	2200	2600	3000	3400	3800
Wood Siding - Wood Frame	107.25	96.80	92.20	88.95	85.80	82.00	79.70	75.20	70.65	68.70	66.95
Brick Veneer - Wood Frame	112.50	101.75	96.80	93.35	89.95	85.95	83.45	78.60	73.80	71.65	69.75
Stucco on Wood Frame	101.80	91.80	87.55	84.50	81.60	77.95	75.85	71.70	67.40	65.60	64.00
Painted Concrete Block	108.20	97.65	93.00	89.70	86.55	82.60	80.30	75.80	71.15	69.15	67.45
Finished Basement, Add	16.95	16.25	15.65	15.30	14.90	14.65	14.35	13.75	13.35	13.10	12.90
Unfinished Basement, Add	7.75	7.15	6.70	6.40	6.05	5.85	5.65	5.20	4.90	4.70	4.50

Modifications

Add to the total cost

Upgrade Kitchen Cabinets	$ + 946
Solid Surface Countertops	+ 678
Full Bath - including plumbing, wall and floor finishes	+ 5863
Half Bath - including plumbing, wall and floor finishes	+ 3420
One Car Attached Garage	+ 12,758
One Car Detached Garage	+ 16,393
Fireplace & Chimney	+ 5874

Adjustments

For multi family - add to total cost

Additional Kitchen	$ + 4417
Additional Bath	+ 5863
Additional Entry & Exit	+ 1658
Separate Heating	+ 1133
Separate Electric	+ 1138

For Townhouse/Rowhouse - Multiply cost per square foot by

Inner Unit	.94
End Unit	.97

Alternatives

Add to or deduct from the cost per square foot of living area

Composition Roll Roofing	– .55
Cedar Shake Roof	+ 2
Upgrade Walls and Ceilings to Skim Coat Plaster	+ .68
Upgrade Ceilings to Textured Finish	+ .50
Air Conditioning, in Heating Ductwork	+ 2.85
In Separate Ductwork	+ 5.55
Heating Systems, Hot Water	+ 1.79
Heat Pump	+ 3.05
Electric Heat	– .85
Not Heated	– 3.34

Additional upgrades or components

Kitchen Cabinets & Countertops	Page 93
Bathroom Vanities	94
Fireplaces & Chimneys	94
Windows, Skylights & Dormers	94
Appliances	95
Breezeways & Porches	95
Finished Attic	95
Garages	96
Site Improvements	96
Wings & Ells	34

Economy Bi-Level

Living Area - 2000 S.F.
Perimeter - 135 L.F.

			Cost Per Square Foot Of Living Area			% of Total
			Mat.	Inst.	Total	(rounded)
1	Site Work	Excavation for lower level, 4' deep. Site preparation for slab.		.79	.79	1.0%
2	Foundation	Continuous reinforced concrete footing, 8" deep x 18" wide; dampproofed and insulated 8" thick reinforced concrete block foundation wall, 4' deep; 4" concrete slab on 4" crushed stone base and polyethylene vapor barrier, trowel finish.	3.20	4.11	7.31	8.9%
3	Framing	Exterior walls - 2" x 4" wood studs, 16" O.C.; 1/2" insulation board sheathing; wood truss roof frame, 24" O.C. with 1/2" plywood sheathing, 4 in 12 pitch; 2" x 8" floor joists 16" O.C. with bridging and 5/8" plywood subfloor.	5.16	7.19	12.35	15.1%
4	Exterior Walls	Beveled wood siding and building paper on insulated wood frame walls; attic insulation; double hung windows; 2 flush solid core wood exterior doors with storms.	10.17	4.34	14.51	17.7%
5	Roofing	20 year asphalt shingles; #15 felt building paper; aluminum gutters, downspouts, drip edge and flashings.	1.05	1.23	2.28	2.8%
6	Interiors	Walls and ceilings, 1/2" taped and finished gypsum wallboard, primed and painted with 2 coats; painted baseboard and trim, rubber backed carpeting 80%, asphalt tile 20%; hollow core wood interior doors.	10.94	12.90	23.84	29.1%
7	Specialties	Economy grade kitchen cabinets - 6 L.F. wall and base with plastic laminate counter top and kitchen sink; 30 gallon electric water heater.	1.51	.61	2.12	2.6%
8	Mechanical	1 lavatory, white, wall hung; 1 water closet, white; 1 bathtub, enameled steel, white; gas fired warm air heat.	2.87	3.19	6.06	7.4%
9	Electrical	100 Amp. service; Romex wiring; incandescent lighting fixtures; switches, receptacles.	.75	1.30	2.05	2.5%
10	Overhead	Contractor's overhead and profit.	5.35	5.34	10.69	13.0%
		Total	41.00	41.00	**82.00**	

RESIDENTIAL — Economy | Tri-Level

- Mass produced from stock plans
- Single family — 1 full bath, 1 kitchen
- No basement
- Asphalt shingles on roof
- Hot air heat
- Gypsum wallboard interior finishes
- Materials and workmanship are sufficient to meet codes

Note: The illustration shown may contain some optional components (for example: garages and/or fireplaces) whose costs are shown in the modifications, adjustments, & alternatives below or at the end of the square foot section.

©Design Basics, Inc.

Base cost per square foot of living area

Exterior Wall	Living Area										
	1200	1500	1800	2000	2200	2400	2800	3200	3600	4000	4400
Wood Siding - Wood Frame	99.10	90.90	84.90	82.50	78.85	76.05	73.70	70.70	67.20	65.90	63.15
Brick Veneer - Wood Frame	103.90	95.30	88.85	86.25	82.45	79.40	76.95	73.75	70.00	68.65	65.70
Stucco on Wood Frame	94.10	86.40	80.80	78.60	75.20	72.55	70.40	67.60	64.35	63.10	60.50
Solid Masonry	99.90	91.65	85.55	83.15	79.50	76.65	74.30	71.25	67.65	66.35	63.60
Finished Basement, Add*	20.30	19.45	18.65	18.30	17.95	17.55	17.30	16.85	16.50	16.30	16.05
Unfinished Basement, Add*	8.50	7.80	7.15	6.90	6.60	6.30	6.05	5.75	5.45	5.30	5.10

*Basement under middle level only.

Modifications
Add to the total cost

Upgrade Kitchen Cabinets	$ + 946
Solid Surface Countertops	+ 678
Full Bath - including plumbing, wall and floor finishes	+ 5863
Half Bath - including plumbing, wall and floor finishes	+ 3420
One Car Attached Garage	+ 12,758
One Car Detached Garage	+ 16,393
Fireplace & Chimney	+ 5874

Adjustments
For multi family - add to total cost

Additional Kitchen	$ + 4417
Additional Bath	+ 5863
Additional Entry & Exit	+ 1658
Separate Heating	+ 1133
Separate Electric	+ 1138

For Townhouse/Rowhouse -
Multiply cost per square foot by

Inner Unit	.93
End Unit	.96

Alternatives
Add to or deduct from the cost per square foot of living area

Composition Roll Roofing	– .75
Cedar Shake Roof	+ 2.85
Upgrade Walls and Ceilings to Skim Coat Plaster	+ .62
Upgrade Ceilings to Textured Finish	+ .50
Air Conditioning, in Heating Ductwork	+ 2.44
In Separate Ductwork	+ 5.16
Heating Systems, Hot Water	+ 1.72
Heat Pump	+ 3.16
Electric Heat	– .74
Not Heated	– 3.24

Additional upgrades or components

Kitchen Cabinets & Countertops	Page 93
Bathroom Vanities	94
Fireplaces & Chimneys	94
Windows, Skylights & Dormers	94
Appliances	95
Breezeways & Porches	95
Finished Attic	95
Garages	96
Site Improvements	96
Wings & Ells	34

Economy Tri-Level

Living Area - 2400 S.F.
Perimeter - 163 L.F.

			Cost Per Square Foot Of Living Area			% of Total
			Mat.	Inst.	Total	(rounded)
1	Site Work	Site preparation for slab; 4' deep trench excavation for foundation wall, excavation for lower level, 4' deep.		.66	.66	0.9%
2	Foundation	Continuous reinforced concrete footing, 8" deep x 18" wide; dampproofed and insulated 8" thick reinforced concrete block foundation wall, 4' deep; 4" concrete slab on 4" crushed stone base and polyethylene vapor barrier, trowel finish.	3.57	4.48	8.05	10.6%
3	Framing	Exterior walls - 2" x 4" wood studs, 16" O.C.; 1/2" insulation board sheathing; wood truss roof frame, 24" O.C. with 1/2" plywood sheathing, 4 in 12 pitch; 2" x 8" floor joists 16" O.C. with bridging and 5/8" plywood subfloor.	4.82	6.45	11.27	14.8%
4	Exterior Walls	Beveled wood siding and building paper on insulated wood frame walls; attic insulation; double hung windows; 2 flush solid core wood exterior doors with storms.	8.91	3.76	12.67	16.7%
5	Roofing	20 year asphalt shingles; #15 felt building paper; aluminum gutters, downspouts, drip edge and flashings.	1.40	1.64	3.04	4.0%
6	Interiors	Walls and ceilings, 1/2" taped and finished gypsum wallboard, primed and painted with 2 coats; painted baseboard and trim, rubber backed carpeting 80%, asphalt tile 20%; hollow core wood interior doors.	9.67	11.44	21.11	27.8%
7	Specialties	Economy grade kitchen cabinets - 6 L.F. wall and base with plastic laminate counter top and kitchen sink; 30 gallon electric water heater.	1.26	.51	1.77	2.3%
8	Mechanical	1 lavatory, white, wall hung; 1 water closet, white; 1 bathtub, enameled steel, white; gas fired warm air heat.	2.55	3.03	5.58	7.3%
9	Electrical	100 Amp. service; Romex wiring; incandescent lighting fixtures, switches, receptacles.	.72	1.24	1.96	2.6%
10	Overhead	Contractor's overhead and profit.	4.95	4.99	9.94	13.1%
		Total	37.85	38.20	76.05	

RESIDENTIAL Economy | Wings & Ells

1 Story — Base cost per square foot of living area

Exterior Wall	Living Area							
	50	100	200	300	400	500	600	700
Wood Siding - Wood Frame	178.75	136.40	118.05	97.95	91.95	88.30	85.90	86.45
Brick Veneer - Wood Frame	195.15	148.15	127.80	104.45	97.85	93.75	91.10	91.50
Stucco on Wood Frame	161.85	124.40	108.00	91.25	85.90	82.70	80.60	81.30
Painted Concrete Block	181.55	138.45	119.70	99.00	92.95	89.25	86.80	87.30
Finished Basement, Add	49.50	40.40	36.65	30.35	29.05	28.30	27.80	27.45
Unfinished Basement, Add	27.40	20.35	17.40	12.50	11.50	10.95	10.55	10.30

1-1/2 Story — Base cost per square foot of living area

Exterior Wall	Living Area							
	100	200	300	400	500	600	700	800
Wood Siding - Wood Frame	140.85	113.50	96.65	86.20	81.15	78.65	75.55	74.70
Brick Veneer - Wood Frame	155.50	125.20	106.40	93.85	88.20	85.30	81.85	80.90
Stucco on Wood Frame	125.80	101.40	86.60	78.40	73.95	71.85	69.10	68.30
Painted Concrete Block	143.35	115.50	98.35	87.50	82.35	79.85	76.65	75.80
Finished Basement, Add	33.05	29.30	26.75	24.00	23.25	22.75	22.30	22.20
Unfinished Basement, Add	16.90	13.95	12.00	9.80	9.20	8.85	8.50	8.40

2 Story — Base cost per square foot of living area

Exterior Wall	Living Area							
	100	200	400	600	800	1000	1200	1400
Wood Siding - Wood Frame	143.80	107.05	91.05	74.80	69.60	66.40	64.35	65.15
Brick Veneer - Wood Frame	160.20	118.85	100.80	81.35	75.45	71.90	69.55	70.15
Stucco on Wood Frame	126.90	95.05	81.00	68.10	63.60	60.80	59.05	59.95
Painted Concrete Block	146.60	109.05	92.70	75.95	70.65	67.35	65.25	66.00
Finished Basement, Add	24.75	20.25	18.35	15.15	14.60	14.20	13.95	13.75
Unfinished Basement, Add	13.70	10.15	8.70	6.30	5.80	5.50	5.30	5.15

Base costs do not include bathroom or kitchen facilities. Use Modifications/Adjustments/Alternatives on pages 93–96 where appropriate.

Average Class

RESIDENTIAL | Average | Illustrations

1 Story

1-1/2 Story

2 Story

2-1/2 Story

Bi-Level

Tri-Level

RESIDENTIAL — Average — 1 Story

- Simple design from standard plans
- Single family — 1 full bath, 1 kitchen
- No basement
- Asphalt shingles on roof
- Hot air heat
- Gypsum wallboard interior finishes
- Materials and workmanship are average

Note: The illustration shown may contain some optional components (for example: garages and/or fireplaces) whose costs are shown in the modifications, adjustments, & alternatives below or at the end of the square foot section.

©Home Planners, Inc.

Base cost per square foot of living area

Exterior Wall	600	800	1000	1200	1400	1600	1800	2000	2400	2800	3200
Wood Siding - Wood Frame	155.00	140.20	129.15	120.35	112.85	107.95	105.40	102.25	95.80	91.10	87.90
Brick Veneer - Wood Frame	174.20	158.70	147.05	137.75	129.85	124.65	121.90	118.50	111.75	106.80	103.40
Stucco on Wood Frame	164.10	149.65	138.85	130.35	123.05	118.30	115.85	112.80	106.50	101.90	98.90
Solid Masonry	189.65	172.45	159.55	149.05	140.05	134.25	131.15	127.20	119.70	114.15	110.20
Finished Basement, Add	39.25	37.90	36.15	34.50	33.20	32.35	31.85	31.15	30.20	29.45	28.80
Unfinished Basement, Add	16.25	14.75	13.60	12.60	11.80	11.25	10.95	10.55	9.95	9.50	9.10

Modifications

Add to the total cost

Upgrade Kitchen Cabinets	$ + 4158
Solid Surface Countertops (Included)	
Full Bath - including plumbing, wall and floor finishes	+ 7329
Half Bath - including plumbing, wall and floor finishes	+ 4275
One Car Attached Garage	+ 13,759
One Car Detached Garage	+ 17,994
Fireplace & Chimney	+ 6193

Adjustments

For multi family - add to total cost

Additional Kitchen	$ + 8034
Additional Bath	+ 7329
Additional Entry & Exit	+ 1658
Separate Heating	+ 1133
Separate Electric	+ 2036

For Townhouse/Rowhouse - Multiply cost per square foot by

Inner Unit	.92
End Unit	.96

Alternatives

Add to or deduct from the cost per square foot of living area

Cedar Shake Roof	+ 3.35
Clay Tile Roof	+ 7.45
Slate Roof	+ 6.45
Upgrade Walls to Skim Coat Plaster	+ .44
Upgrade Ceilings to Textured Finish	+ .50
Air Conditioning, in Heating Ductwork	+ 4.86
In Separate Ductwork	+ 7.15
Heating Systems, Hot Water	+ 1.97
Heat Pump	+ 2.71
Electric Heat	− .68
Not Heated	− 3.21

Additional upgrades or components

Kitchen Cabinets & Countertops	Page 93
Bathroom Vanities	94
Fireplaces & Chimneys	94
Windows, Skylights & Dormers	94
Appliances	95
Breezeways & Porches	95
Finished Attic	95
Garages	96
Site Improvements	96
Wings & Ells	56

Average 1 Story

Living Area - 1600 S.F.
Perimeter - 163 L.F.

			Cost Per Square Foot Of Living Area			% of Total
			Mat.	Inst.	Total	(rounded)
1	Site Work	Site preparation for slab; 4' deep trench excavation for foundation wall.		1.01	1.01	0.9%
2	Foundation	Continuous reinforced concrete footing 8" deep x 18" wide; dampproofed and insulated reinforced concrete foundation wall, 8" thick, 4' deep; 4" concrete slab on 4" crushed stone base and polyethylene vapor barrier, trowel finish.	5.53	6.89	12.42	11.5%
3	Framing	Exterior walls - 2" x 4" wood studs, 16" O.C.; 1/2" plywood sheathing; 2" x 6" rafters 16" O.C. with 1/2" plywood sheathing, 4 in 12 pitch; 2" x 6" ceiling joists 16" O.C.; 1/2" plywood subfloor on 1" x 2" wood sleepers 16" O.C.	6.76	9.67	16.43	15.2%
4	Exterior Walls	Beveled wood siding and building paper on insulated wood frame walls; attic insulation; double hung windows; 3 flush solid core wood exterior doors with storms.	11.78	5.11	16.89	15.6%
5	Roofing	25 year asphalt shingles; #15 felt building paper; aluminum gutters, downspouts, drip edge and flashings.	2.16	2.53	4.69	4.3%
6	Interiors	Walls and ceilings, 1/2" taped and finished gypsum wallboard, primed and painted with 2 coats; painted baseboard and trim, finished hardwood floor 40%, carpet with 1/2" underlayment 40%, vinyl tile with 1/2" underlayment 15%, ceramic tile with 1/2" underlayment 5%; hollow core and louvered doors.	12.63	13.37	26.00	24.1%
7	Specialties	Average grade kitchen cabinets - 14 L.F. wall and base with solid surface counter top and kitchen sink; 40 gallon electric water heater.	3.36	1.18	4.54	4.2%
8	Mechanical	1 lavatory, white, wall hung; 1 water closet, white; 1 bathtub with shower, enameled steel, white; gas fired warm air heat.	3.67	3.47	7.14	6.6%
9	Electrical	200 Amp. service; Romex wiring; incandescent lighting fixtures, switches, receptacles.	1.25	1.89	3.14	2.9%
10	Overhead	Contractor's overhead and profit and plans.	8.01	7.68	15.69	14.5%
		Total	55.15	52.80	**107.95**	

For customer support on your Residential Cost Data, call 877.713.2617.

RESIDENTIAL | Average | 1-1/2 Story

- Simple design from standard plans
- Single family — 1 full bath, 1 kitchen
- No basement
- Asphalt shingles on roof
- Hot air heat
- Gypsum wallboard interior finishes
- Materials and workmanship are average

Note: The illustration shown may contain some optional components (for example: garages and/or fireplaces) whose costs are shown in the modifications, adjustments, & alternatives below or at the end of the square foot section.

©By Designer

Base cost per square foot of living area

Exterior Wall	Living Area										
	600	800	1000	1200	1400	1600	1800	2000	2400	2800	3200
Wood Siding - Wood Frame	173.40	144.90	129.85	122.65	117.55	109.90	106.15	102.25	94.30	91.20	87.90
Brick Veneer - Wood Frame	182.75	151.75	136.10	128.55	123.15	115.05	110.95	106.85	98.35	95.05	91.45
Stucco on Wood Frame	168.60	141.50	126.65	119.70	114.70	107.30	103.65	99.90	92.25	89.30	86.10
Solid Masonry	204.30	167.40	150.60	142.10	136.05	126.75	122.10	117.40	107.65	103.85	99.60
Finished Basement, Add	31.95	27.80	26.60	25.75	25.10	24.10	23.50	23.05	22.00	21.55	21.00
Unfinished Basement, Add	14.00	11.00	10.25	9.65	9.25	8.60	8.25	8.00	7.35	7.05	6.75

Modifications

Add to the total cost

Upgrade Kitchen Cabinets	$ + 4158
Solid Surface Countertops (Included)	
Full Bath - including plumbing, wall and floor finishes	+ 7329
Half Bath - including plumbing, wall and floor finishes	+ 4275
One Car Attached Garage	+ 13,759
One Car Detached Garage	+ 17,994
Fireplace & Chimney	+ 6193

Adjustments

For multi family - add to total cost

Additional Kitchen	$ + 8034
Additional Bath	+ 7329
Additional Entry & Exit	+ 1658
Separate Heating	+ 1133
Separate Electric	+ 2036

For Townhouse/Rowhouse - Multiply cost per square foot by

Inner Unit	.92
End Unit	.96

Alternatives

Add to or deduct from the cost per square foot of living area

Cedar Shake Roof	+ 2.45
Clay Tile Roof	+ 5.40
Slate Roof	+ 4.65
Upgrade Walls to Skim Coat Plaster	+ .50
Upgrade Ceilings to Textured Finish	+ .50
Air Conditioning, in Heating Ductwork	+ 3.69
In Separate Ductwork	+ 6.26
Heating Systems, Hot Water	+ 1.88
Heat Pump	+ 3
Electric Heat	− .62
Not Heated	− 3.08

Additional upgrades or components

Kitchen Cabinets & Countertops	Page 93
Bathroom Vanities	94
Fireplaces & Chimneys	94
Windows, Skylights & Dormers	94
Appliances	95
Breezeways & Porches	95
Finished Attic	95
Garages	96
Site Improvements	96
Wings & Ells	56

Average 1-1/2 Story

Living Area - 1800 S.F.
Perimeter - 144 L.F.

			Cost Per Square Foot Of Living Area			% of Total
			Mat.	Inst.	Total	(rounded)
1	Site Work	Site preparation for slab; 4' deep trench excavation for foundation wall.		.90	.90	0.8%
2	Foundation	Continuous reinforced concrete footing 8" deep x 18" wide; dampproofed and insulated reinforced concrete foundation wall, 8" thick, 4' deep; 4" concrete slab on 4" crushed stone base and polyethylene vapor barrier, trowel finish.	3.95	5.07	9.02	8.5%
3	Framing	Exterior walls - 2" x 4" wood studs, 16" O.C.; 1/2" plywood sheathing; 2" x 6" rafters 16" O.C. with 1/2" plywood sheathing, 8 in 12 pitch; 2" x 8" floor joists 16" O.C. with 5/8" plywood subfloor; 1/2" plywood subfloor on 1" x 2" wood sleepers 16" O.C.	7.23	9.00	16.23	15.3%
4	Exterior Walls	Beveled wood siding and building paper on insulated wood frame walls; attic insulation; double hung windows; 3 flush solid core wood exterior doors with storms.	12.69	5.54	18.23	17.2%
5	Roofing	25 year asphalt shingles; #15 felt building paper; aluminum gutters, downspouts, drip edge and flashings.	1.35	1.58	2.93	2.8%
6	Interiors	Walls and ceilings, 1/2" taped and finished gypsum wallboard, primed and painted with 2 coats; painted baseboard and trim, finished hardwood floor 40%, carpet with 1/2" underlayment 40%, vinyl tile with 1/2" underlayment 15%, ceramic tile with 1/2" underlayment 5%; hollow core and louvered doors.	14.67	15.02	29.69	28.0%
7	Specialties	Average grade kitchen cabinets - 14 L.F. wall and base with solid surface counter top and kitchen sink; 40 gallon electric water heater.	2.98	1.05	4.03	3.8%
8	Mechanical	1 lavatory, white, wall hung; 1 water closet, white; 1 bathtub with shower, enameled steel, white; gas fired warm air heat.	3.36	3.35	6.71	6.3%
9	Electrical	200 Amp. service; Romex wiring; incandescent lighting fixtures, switches, receptacles.	1.18	1.81	2.99	2.8%
10	Overhead	Contractor's overhead and profit and plans.	8.04	7.38	15.42	14.5%
		Total	55.45	50.70	**106.15**	

RESIDENTIAL | Average | 2 Story

- Simple design from standard plans
- Single family — 1 full bath, 1 kitchen
- No basement
- Asphalt shingles on roof
- Hot air heat
- Gypsum wallboard interior finishes
- Materials and workmanship are average

Note: The illustration shown may contain some optional components (for example: garages and/or fireplaces) whose costs are shown in the modifications, adjustments, & alternatives below or at the end of the square foot section.

Base cost per square foot of living area

Exterior Wall	Living Area										
	1000	1200	1400	1600	1800	2000	2200	2600	3000	3400	3800
Wood Siding - Wood Frame	136.80	124.05	117.95	113.80	109.55	104.95	101.95	96.20	90.55	88.10	85.80
Brick Veneer - Wood Frame	144.00	130.75	124.25	119.80	115.20	110.40	107.10	100.90	94.85	92.15	89.70
Stucco on Wood Frame	133.10	120.65	114.80	110.75	106.70	102.20	99.30	93.85	88.35	85.95	83.75
Solid Masonry	160.70	146.20	138.70	133.55	128.15	122.90	118.95	111.65	104.80	101.60	98.65
Finished Basement, Add	22.50	22.15	21.40	20.95	20.40	20.05	19.65	18.90	18.40	18.10	17.80
Unfinished Basement, Add	8.85	8.25	7.85	7.55	7.15	6.95	6.70	6.25	5.95	5.75	5.55

Modifications

Add to the total cost

Upgrade Kitchen Cabinets	$ + 4158
Solid Surface Countertops (Included)	
Full Bath - including plumbing, wall and floor finishes	+ 7329
Half Bath - including plumbing, wall and floor finishes	+ 4275
One Car Attached Garage	+ 13,759
One Car Detached Garage	+ 17,994
Fireplace & Chimney	+ 6840

Adjustments

For multi family - add to total cost

Additional Kitchen	$ + 8034
Additional Bath	+ 7329
Additional Entry & Exit	+ 1658
Separate Heating	+ 1133
Separate Electric	+ 2036

For Townhouse/Rowhouse - Multiply cost per square foot by

Inner Unit	.90
End Unit	.95

Alternatives

Add to or deduct from the cost per square foot of living area

Cedar Shake Roof	+ 1.70
Clay Tile Roof	+ 3.70
Slate Roof	+ 3.25
Upgrade Walls to Skim Coat Plaster	+ .52
Upgrade Ceilings to Textured Finish	+ .50
Air Conditioning, in Heating Ductwork	+ 2.94
In Separate Ductwork	+ 5.68
Heating Systems, Hot Water	+ 1.84
Heat Pump	+ 3.16
Electric Heat	− .53
Not Heated	− 2.99

Additional upgrades or components

Kitchen Cabinets & Countertops	Page 93
Bathroom Vanities	94
Fireplaces & Chimneys	94
Windows, Skylights & Dormers	94
Appliances	95
Breezeways & Porches	95
Finished Attic	95
Garages	96
Site Improvements	96
Wings & Ells	56

Average 2 Story

Living Area - 2000 S.F.
Perimeter - 135 L.F.

#	Category	Description	Cost Per Square Foot Of Living Area			% of Total
			Mat.	Inst.	Total	(rounded)
1	Site Work	Site preparation for slab; 4' deep trench excavation for foundation wall.		.81	.81	0.8%
2	Foundation	Continuous reinforced concrete footing 8" deep x 18" wide; dampproofed and insulated reinforced concrete foundation wall, 8" thick, 4' deep, 4" concrete slab on 4" crushed stone base and polyethylene vapor barrier, trowel finish.	3.29	4.22	7.51	7.2%
3	Framing	Exterior walls - 2" x 4" wood studs, 16" O.C.; 1/2" plywood sheathing; 2" x 6" rafters 16" O.C. with 1/2" plywood sheathing, 4 in 12 pitch; 2" x 6" ceiling joists 16" O.C.; 2" x 8" floor joists 16" O.C. with 5/8" plywood subfloor; 1/2" plywood subfloor on 1" x 2" wood sleepers 16" O.C.	7.62	9.28	16.90	16.1%
4	Exterior Walls	Beveled wood siding and building paper on insulated wood frame walls; attic insulation; double hung windows; 3 flush solid core wood exterior doors with storms.	13.75	6.00	19.75	18.8%
5	Roofing	25 year asphalt shingles; #15 felt building paper; aluminum gutters, downspouts, drip edge and flashings.	1.08	1.26	2.34	2.2%
6	Interiors	Walls and ceilings, 1/2" taped and finished gypsum wallboard, primed and painted with 2 coats; painted baseboard and trim, finished hardwood floor 40%, carpet with 1/2" underlayment 40%, vinyl tile with 1/2" underlayment 15%, ceramic tile with 1/2" underlayment 5%; hollow core and louvered doors.	14.49	15.02	29.51	28.1%
7	Specialties	Average grade kitchen cabinets - 14 L.F. wall and base with solid surface counter top and kitchen sink; 40 gallon electric water heater.	2.67	.95	3.62	3.4%
8	Mechanical	1 lavatory, white, wall hung; 1 water closet, white; 1 bathtub with shower; enameled steel, white; gas fired warm air heat.	3.14	3.25	6.39	6.1%
9	Electrical	200 Amp. service; Romex wiring; incandescent lighting fixtures, switches, receptacles.	1.13	1.75	2.88	2.7%
10	Overhead	Contractor's overhead and profit and plans.	8.03	7.21	15.24	14.5%
	Total		55.20	49.75	**104.95**	

RESIDENTIAL — Average — 2-1/2 Story

- Simple design from standard plans
- Single family — 1 full bath, 1 kitchen
- No basement
- Asphalt shingles on roof
- Hot air heat
- Gypsum wallboard interior finishes
- Materials and workmanship are average

Note: The illustration shown may contain some optional components (for example: garages and/or fireplaces) whose costs are shown in the modifications, adjustments, & alternatives below or at the end of the square foot section.

Base cost per square foot of living area

Exterior Wall	Living Area										
	1200	1400	1600	1800	2000	2400	2800	3200	3600	4000	4400
Wood Siding - Wood Frame	135.80	127.55	116.45	114.35	110.25	103.75	98.60	93.40	90.75	85.95	84.45
Brick Veneer - Wood Frame	143.50	134.50	122.90	120.80	116.25	109.25	103.75	98.15	95.25	90.10	88.50
Stucco on Wood Frame	131.90	123.95	113.25	111.10	107.15	101.00	95.95	91.00	88.45	83.80	82.40
Solid Masonry	161.25	150.60	137.65	135.60	130.15	121.80	115.75	109.00	105.50	99.65	97.75
Finished Basement, Add	18.95	18.60	17.90	17.75	17.35	16.55	16.25	15.75	15.40	15.05	14.90
Unfinished Basement, Add	7.45	6.85	6.40	6.35	6.05	5.60	5.40	5.05	4.85	4.65	4.55

Modifications

Add to the total cost

Upgrade Kitchen Cabinets	$ + 4158
Solid Surface Countertops (Included)	
Full Bath - including plumbing, wall and floor finishes	+ 7329
Half Bath - including plumbing, wall and floor finishes	+ 4275
One Car Attached Garage	+ 13,759
One Car Detached Garage	+ 17,994
Fireplace & Chimney	+ 7504

Adjustments

For multi family - add to total cost

Additional Kitchen	$ + 8034
Additional Bath	+ 7329
Additional Entry & Exit	+ 1658
Separate Heating	+ 1133
Separate Electric	+ 2036

For Townhouse/Rowhouse - Multiply cost per square foot by

Inner Unit	.90
End Unit	.95

Alternatives

Add to or deduct from the cost per square foot of living area

Cedar Shake Roof	+ 1.45
Clay Tile Roof	+ 3.25
Slate Roof	+ 2.80
Upgrade Walls to Skim Coat Plaster	+ .50
Upgrade Ceilings to Textured Finish	+ .50
Air Conditioning, in Heating Ductwork	+ 2.67
In Separate Ductwork	+ 5.49
Heating Systems, Hot Water	+ 1.66
Heat Pump	+ 3.25
Electric Heat	− .95
Not Heated	− 3.49

Additional upgrades or components

Kitchen Cabinets & Countertops	Page 93
Bathroom Vanities	94
Fireplaces & Chimneys	94
Windows, Skylights & Dormers	94
Appliances	95
Breezeways & Porches	95
Finished Attic	95
Garages	96
Site Improvements	96
Wings & Ells	56

Average 2-1/2 Story

Living Area - 3200 S.F.
Perimeter - 150 L.F.

#	Category	Description	Cost Per Square Foot Of Living Area			% of Total (rounded)
			Mat.	Inst.	Total	
1	Site Work	Site preparation for slab; 4' deep trench excavation for foundation wall.		.50	.50	0.5%
2	Foundation	Continuous reinforced concrete footing 8" deep x 18" wide, dampproofed and insulated reinforced concrete foundation wall, 8" thick, 4' deep; 4" concrete slab on 4" crushed stone base and polyethylene vapor barrier, trowel finish.	2.36	3.00	5.36	5.7%
3	Framing	Exterior walls - 2" x 4" wood studs, 16" O.C.; 1/2" plywood sheathing; 2" x 6" rafters 16" O.C. with 1/2" plywood sheathing, 4 in 12 pitch; 2" x 6" ceiling joists 16" O.C.; 2" x 8" floor joists 16" O.C. with 5/8" plywood subfloor; 1/2" plywood subfloor on 1" x 2" wood sleepers 16" O.C.	7.64	9.10	16.74	17.9%
4	Exterior Walls	Beveled wood siding and building paper on insulated wood frame walls; attic insulation; double hung windows; 3 flush solid core wood exterior doors with storms.	11.62	5.05	16.67	17.8%
5	Roofing	25 year asphalt shingles; #15 felt building paper; aluminum gutters, downspouts, drip edge and flashings.	.83	.98	1.81	1.9%
6	Interiors	Walls and ceilings, 1/2" taped and finished gypsum wallboard, primed and painted with 2 coats; painted baseboard and trim, finished hardwood floor 40%, carpet with 1/2" underlayment 40%, vinyl tile with 1/2" underlayment 15%, ceramic tile with 1/2" underlayment 5%; hollow core and louvered doors.	14.26	14.48	28.74	30.8%
7	Specialties	Average grade kitchen cabinets - 14 L.F. wall and base with solid surface counter top and kitchen sink; 40 gallon electric water heater.	1.68	.60	2.28	2.4%
8	Mechanical	1 lavatory, white, wall hung; 1 water closet, white; 1 bathtub with shower, enameled steel, white; gas fired warm air heat.	2.34	2.91	5.25	5.6%
9	Electrical	200 Amp. service; Romex wiring; incandescent lighting fixtures, switches, receptacles.	.96	1.52	2.48	2.7%
10	Overhead	Contractor's overhead and profit and plans.	7.11	6.46	13.57	14.5%
		Total	48.80	44.60	**93.40**	

For customer support on your Residential Cost Data, call 877.713.2617.

RESIDENTIAL — Average — 3 Story

- Simple design from standard plans
- Single family — 1 full bath, 1 kitchen
- No basement
- Asphalt shingles on roof
- Hot air heat
- Gypsum wallboard interior finishes
- Materials and workmanship are average

Note: The illustration shown may contain some optional components (for example: garages and/or fireplaces) whose costs are shown in the modifications, adjustments, & alternatives below or at the end of the square foot section.

Base cost per square foot of living area

Exterior Wall	\multicolumn{11}{c}{Living Area}										
	1500	1800	2100	2500	3000	3500	4000	4500	5000	5500	6000
Wood Siding - Wood Frame	125.85	114.30	109.20	105.20	97.60	94.35	89.70	84.70	83.15	81.35	79.45
Brick Veneer - Wood Frame	133.05	121.05	115.45	111.15	103.00	99.50	94.40	89.05	87.35	85.35	83.25
Stucco on Wood Frame	122.15	110.90	106.00	102.15	94.80	91.75	87.35	82.50	81.00	79.30	77.55
Solid Masonry	149.75	136.45	129.90	124.95	115.50	111.30	105.15	98.95	96.95	94.60	91.95
Finished Basement, Add	16.25	16.05	15.60	15.30	14.65	14.40	13.95	13.60	13.40	13.25	13.05
Unfinished Basement, Add	6.00	5.65	5.30	5.15	4.75	4.55	4.30	4.10	3.95	3.85	3.75

Modifications

Add to the total cost

Upgrade Kitchen Cabinets	$ + 4158
Solid Surface Countertops (Included)	
Full Bath - including plumbing, wall and floor finishes	+ 7329
Half Bath - including plumbing, wall and floor finishes	+ 4275
One Car Attached Garage	+ 13,759
One Car Detached Garage	+ 17,994
Fireplace & Chimney	+ 7504

Adjustments

For multi family - add to total cost

Additional Kitchen	$ + 8034
Additional Bath	+ 7329
Additional Entry & Exit	+ 1658
Separate Heating	+ 1133
Separate Electric	+ 2036

For Townhouse/Rowhouse - Multiply cost per square foot by

Inner Unit	.88
End Unit	.94

Alternatives

Add to or deduct from the cost per square foot of living area

Cedar Shake Roof	+ 1.10
Clay Tile Roof	+ 2.50
Slate Roof	+ 2.15
Upgrade Walls to Skim Coat Plaster	+ .52
Upgrade Ceilings to Textured Finish	+ .50
Air Conditioning, in Heating Ductwork	+ 2.67
In Separate Ductwork	+ 5.49
Heating Systems, Hot Water	+ 1.66
Heat Pump	+ 3.25
Electric Heat	– .74
Not Heated	– 3.25

Additional upgrades or components

Kitchen Cabinets & Countertops	Page 93
Bathroom Vanities	94
Fireplaces & Chimneys	94
Windows, Skylights & Dormers	94
Appliances	95
Breezeways & Porches	95
Finished Attic	95
Garages	96
Site Improvements	96
Wings & Ells	56

Average 3 Story

Living Area - 3000 S.F.
Perimeter - 135 L.F.

			Cost Per Square Foot Of Living Area			% of Total
			Mat.	Inst.	Total	(rounded)
1	Site Work	Site preparation for slab; 4' deep trench excavation for foundation wall.		.54	.54	0.6%
2	Foundation	Continuous reinforced concrete footing 8" deep x 18" wide, dampproofed and insulated reinforced concrete foundation wall, 8" thick, 4' deep; 4" concrete slab on 4" crushed stone base and polyethylene vapor barrier, trowel finish.	2.19	2.83	5.02	5.1%
3	Framing	Exterior walls - 2" x 4" wood studs, 16" O.C.; 1/2" plywood sheathing; 2" x 6" rafters 16" O.C. with 1/2" plywood sheathing, 4 in 12 pitch; 2" x 6" ceiling joists 16" O.C.; 2" x 8" floor joists 16" O.C. with 5/8" plywood subfloor; 1/2" plywood subfloor on 1" x 2" wood sleepers 16" O.C.	7.90	9.37	17.27	17.7%
4	Exterior Walls	Horizontal beveled wood siding; building paper; insulation; wood double hung windows; 3 flush solid core wood exterior doors; storms and screens.	13.20	5.74	18.94	19.4%
5	Roofing	25 year asphalt shingles; #15 felt building paper; aluminum gutters, downspouts, drip edge and flashings.	.72	.84	1.56	1.6%
6	Interiors	Walls and ceilings, 1/2" taped and finished gypsum wallboard, primed and painted with 2 coats; painted baseboard and trim, finished hardwood floor 40%, carpet with 1/2" underlayment 40%, vinyl tile with 1/2" underlayment 15%, ceramic tile with 1/2" underlayment 5%; hollow core and louvered doors.	14.72	15.04	29.76	30.5%
7	Specialties	Average grade kitchen cabinets - 14 L.F. wall and base with solid surface counter top and kitchen sink; 40 gallon electric water heater.	1.78	.64	2.42	2.5%
8	Mechanical	1 lavatory, white, wall hung; 1 water closet, white; 1 bathtub with shower, enameled steel, white; gas fired warm air heat.	2.43	2.94	5.37	5.5%
9	Electrical	200 Amp. service; Romex wiring; incandescent lighting fixtures, switches, receptacles.	.98	1.54	2.52	2.6%
10	Overhead	Contractor's overhead and profit and plans.	7.48	6.72	14.20	14.5%
		Total	51.40	46.20	**97.60**	

RESIDENTIAL   Average   Bi-Level

- Simple design from standard plans
- Single family — 1 full bath, 1 kitchen
- No basement
- Asphalt shingles on roof
- Hot air heat
- Gypsum wallboard interior finishes
- Materials and workmanship are average

Note: The illustration shown may contain some optional components (for example: garages and/or fireplaces) whose costs are shown in the modifications, adjustments, & alternatives below or at the end of the square foot section.

Base cost per square foot of living area

Exterior Wall	1000	1200	1400	1600	1800	2000	2200	2600	3000	3400	3800
Wood Siding - Wood Frame	127.70	115.55	110.10	106.30	102.45	98.15	95.50	90.30	85.15	82.90	80.85
Brick Veneer - Wood Frame	133.10	120.65	114.80	110.80	106.65	102.20	99.35	93.85	88.35	86.00	83.75
Stucco on Wood Frame	124.95	113.05	107.75	104.00	100.30	96.05	93.50	88.55	83.50	81.40	79.40
Solid Masonry	145.65	132.20	125.65	121.05	116.45	111.60	108.25	101.90	95.85	93.10	90.55
Finished Basement, Add	22.50	22.15	21.40	20.95	20.40	20.05	19.65	18.90	18.40	18.10	17.80
Unfinished Basement, Add	8.85	8.25	7.85	7.55	7.15	6.95	6.70	6.25	5.95	5.75	5.55

Modifications
Add to the total cost

Upgrade Kitchen Cabinets	$ + 4158
Solid Surface Countertops (Included)	
Full Bath - including plumbing, wall and floor finishes	+ 7329
Half Bath - including plumbing, wall and floor finishes	+ 4275
One Car Attached Garage	+ 13,759
One Car Detached Garage	+ 17,994
Fireplace & Chimney	+ 6193

Adjustments
For multi family - add to total cost

Additional Kitchen	$ + 8034
Additional Bath	+ 7329
Additional Entry & Exit	+ 1658
Separate Heating	+ 1133
Separate Electric	+ 2036

For Townhouse/Rowhouse - Multiply cost per square foot by

Inner Unit	.91
End Unit	.96

Alternatives
Add to or deduct from the cost per square foot of living area

Cedar Shake Roof	+ 1.70
Clay Tile Roof	+ 3.70
Slate Roof	+ 3.25
Upgrade Walls to Skim Coat Plaster	+ .48
Upgrade Ceilings to Textured Finish	+ .50
Air Conditioning, in Heating Ductwork	+ 2.94
In Separate Ductwork	+ 5.68
Heating Systems, Hot Water	+ 1.84
Heat Pump	+ 3.16
Electric Heat	- .53
Not Heated	- 2.99

Additional upgrades or components

Kitchen Cabinets & Countertops	Page 93
Bathroom Vanities	94
Fireplaces & Chimneys	94
Windows, Skylights & Dormers	94
Appliances	95
Breezeways & Porches	95
Finished Attic	95
Garages	96
Site Improvements	96
Wings & Ells	56

Average Bi-Level

Living Area - 2000 S.F.
Perimeter - 135 L.F.

			Cost Per Square Foot Of Living Area			% of Total
			Mat.	Inst.	Total	(rounded)
1	Site Work	Excavation for lower level, 4' deep. Site preparation for slab.		.81	.81	0.8%
2	Foundation	Continuous reinforced concrete footing 8" deep x 18" wide, dampproofed and insulated reinforced concrete foundation wall, 8" thick, 4' deep; 4" concrete slab on 4" crushed stone base and polyethylene vapor barrier, trowel finish.	3.29	4.22	7.51	7.7%
3	Framing	Exterior walls - 2" x 4" wood studs, 16" O.C.; 1/2" plywood sheathing; 2" x 6" rafters 16" O.C. with 1/2" plywood sheathing, 4 in 12 pitch; 2" x 6" ceiling joists 16" O.C.; 2" x 8" floor joists 16" O.C. with 5/8" plywood subfloor; 1/2" plywood subfloor on 1" x 2" wood sleepers 16" O.C.	7.28	8.82	16.10	16.4%
4	Exterior Walls	Horizontal beveled wood siding; building paper; insulation; wood double hung windows; 3 flush solid core wood exterior doors; storms and screens.	10.74	4.69	15.43	15.7%
5	Roofing	25 year asphalt shingles; #15 felt building paper; aluminum gutters, downspouts, drip edge and flashings.	1.08	1.26	2.34	2.4%
6	Interiors	Walls and ceilings, 1/2" taped and finished gypsum wallboard, primed and painted with 2 coats; painted baseboard and trim, finished hardwood floor 40%, carpet with 1/2" underlayment 40%, vinyl tile with 1/2" underlayment 15%, ceramic tile with 1/2" underlayment 5%; hollow core and louvered doors.	14.22	14.59	28.81	29.4%
7	Specialties	Average grade kitchen cabinets - 14 L.F. wall and base with solid surface counter top and kitchen sink; 40 gallon electric water heater.	2.67	.95	3.62	3.7%
8	Mechanical	1 lavatory, white, wall hung; 1 water closet, white; 1 bathtub with shower, enameled steel, white; gas fired warm air heat.	3.14	3.25	6.39	6.5%
9	Electrical	200 Amp. service; Romex wiring; incandescent lighting fixtures, switches, receptacles.	1.13	1.75	2.88	2.9%
10	Overhead	Contractor's overhead and profit and plans.	7.40	6.86	14.26	14.5%
		Total	50.95	47.20	98.15	

RESIDENTIAL Average Tri-Level

- Simple design from standard plans
- Single family — 1 full bath, 1 kitchen
- No basement
- Asphalt shingles on roof
- Hot air heat
- Gypsum wallboard interior finishes
- Materials and workmanship are average

Note: The illustration shown may contain some optional components (for example: garages and/or fireplaces) whose costs are shown in the modifications, adjustments, & alternatives below or at the end of the square foot section.

©Design Basics, Inc.

Base cost per square foot of living area

Exterior Wall	Living Area										
	1200	1500	1800	2100	2400	2700	3000	3400	3800	4200	4600
Wood Siding - Wood Frame	121.90	112.45	105.40	99.25	95.30	93.10	90.65	88.30	84.35	81.25	79.65
Brick Veneer - Wood Frame	126.90	116.95	109.45	102.95	98.80	96.40	93.75	91.35	87.20	83.90	82.25
Stucco on Wood Frame	119.35	110.10	103.25	97.30	93.55	91.35	89.00	86.75	82.90	79.90	78.40
Solid Masonry	138.40	127.40	118.80	111.45	106.80	104.15	101.00	98.35	93.75	90.05	88.20
Finished Basement, Add*	25.90	25.45	24.35	23.45	22.85	22.55	22.15	21.85	21.40	20.95	20.70
Unfinished Basement, Add*	9.90	9.15	8.45	7.95	7.60	7.40	7.15	6.95	6.70	6.45	6.30

*Basement under middle level only.

Modifications
Add to the total cost

Upgrade Kitchen Cabinets	$ + 4158
Solid Surface Countertops (Included)	
Full Bath - including plumbing, wall and floor finishes	+ 7329
Half Bath - including plumbing, wall and floor finishes	+ 4275
One Car Attached Garage	+ 13,759
One Car Detached Garage	+ 17,994
Fireplace & Chimney	+ 6193

Adjustments
For multi family - add to total cost

Additional Kitchen	$ + 8034
Additional Bath	+ 7329
Additional Entry & Exit	+ 1658
Separate Heating	+ 1133
Separate Electric	+ 2036

For Townhouse/Rowhouse - Multiply cost per square foot by

Inner Unit	.90
End Unit	.95

Alternatives
Add to or deduct from the cost per square foot of living area

Cedar Shake Roof	+ 2.45
Clay Tile Roof	+ 5.40
Slate Roof	+ 4.65
Upgrade Walls to Skim Coat Plaster	+ .42
Upgrade Ceilings to Textured Finish	+ .50
Air Conditioning, in Heating Ductwork	+ 2.48
In Separate Ductwork	+ 5.35
Heating Systems, Hot Water	+ 1.78
Heat Pump	+ 3.30
Electric Heat	– .45
Not Heated	– 2.88

Additional upgrades or components

Kitchen Cabinets & Countertops	Page 93
Bathroom Vanities	94
Fireplaces & Chimneys	94
Windows, Skylights & Dormers	94
Appliances	95
Breezeways & Porches	95
Finished Attic	95
Garages	96
Site Improvements	96
Wings & Ells	56

Average Tri-Level

Living Area - 2400 S.F.
Perimeter - 163 L.F.

			Cost Per Square Foot Of Living Area			% of Total
			Mat.	Inst.	Total	(rounded)
1	Site Work	Site preparation for slab; 4' deep trench excavation for foundation wall, excavation for lower level, 4' deep.		.68	.68	0.7%
2	Foundation	Continuous reinforced concrete footing 8" deep x 18" wide; dampproofed and insulated reinforced concrete foundation wall, 8" thick, 4' deep; 4" concrete slab on 4" crushed stone base and polyethylene vapor barrier, trowel finish.	3.67	4.60	8.27	8.7%
3	Framing	Exterior walls - 2" x 4" wood studs, 16" O.C.; 1/2" plywood sheathing; 2" x 6" rafters 16" O.C. with 1/2" plywood sheathing, 4 in 12 pitch; 2" x 6" ceiling joists 16" O.C.; 2" x 8" floor joists 16" O.C. with 5/8" plywood subfloor; 1/2" plywood subfloor on 1" x 2" wood sleepers 16" O.C.	6.71	8.31	15.02	15.8%
4	Exterior Walls	Horizontal beveled wood siding: building paper; insulation; wood double hung windows; 3 flush solid core wood exterior doors; storms and screens.	9.42	4.08	13.50	14.2%
5	Roofing	25 year asphalt shingles; #15 felt building paper; aluminum gutters, downspouts, drip edge and flashings.	1.44	1.69	3.13	3.3%
6	Interiors	Walls and ceilings, 1/2" taped and finished gypsum wallboard, primed and painted with 2 coats; painted baseboard and trim, finished hardwood floor 40%, carpet with 1/2" underlayment 40%, vinyl tile with 1/2" underlayment 15%, ceramic tile with 1/2" underlayment 5%; hollow core and louvered doors.	15.11	14.21	29.32	30.8%
7	Specialties	Average grade kitchen cabinets - 14 L.F. wall and base with solid surface counter top and kitchen sink; 40 gallon electric water heater.	2.24	.79	3.03	3.2%
8	Mechanical	1 lavatory, white, wall hung; 1 water closet, white; 1 bathtub with shower, enameled steel, white; gas fired warm air heat.	2.71	3.10	5.81	6.1%
9	Electrical	200 Amp. service; Romex wiring; incandescent lighting fixtures, switches, receptacles.	1.05	1.64	2.69	2.8%
10	Overhead	Contractor's overhead and profit and plans.	7.20	6.65	13.85	14.5%
		Total	49.55	45.75	**95.30**	

For customer support on your Residential Cost Data, call 877.713.2617.

RESIDENTIAL — Solid Wall — 1 Story

- Post and beam frame
- Log exterior walls
- Simple design from standard plans
- Single family — 1 full bath, 1 kitchen
- No basement
- Asphalt shingles on roof
- Hot air heat
- Gypsum wallboard interior finishes
- Materials and workmanship are average

Note: The illustration shown may contain some optional components (for example: garages and/or fireplaces) whose costs are shown in the modifications, adjustments, & alternatives below or at the end of the square foot section.

Base cost per square foot of living area

Exterior Wall	Living Area										
	600	800	1000	1200	1400	1600	1800	2000	2400	2800	3200
6" Log - Solid Wall	173.55	157.60	145.70	136.15	127.95	122.65	119.85	116.35	109.35	104.30	100.75
8" Log - Solid Wall	163.05	148.25	137.20	128.45	120.95	116.10	113.55	110.40	103.95	99.25	96.15
Finished Basement, Add	39.25	37.90	36.15	34.50	33.20	32.35	31.85	31.15	30.20	29.45	28.80
Unfinished Basement, Add	16.25	14.75	13.60	12.60	11.80	11.25	10.95	10.55	9.95	9.50	9.10

Modifications

Add to the total cost

Upgrade Kitchen Cabinets	$ + 4158
Solid Surface Countertops (Included)	
Full Bath - including plumbing, wall and floor finishes	+ 7329
Half Bath - including plumbing, wall and floor finishes	+ 4275
One Car Attached Garage	+ 13,759
One Car Detached Garage	+ 17,994
Fireplace & Chimney	+ 6193

Adjustments

For multi family - add to total cost

Additional Kitchen	$ + 8034
Additional Bath	+ 7329
Additional Entry & Exit	+ 1658
Separate Heating	+ 1133
Separate Electric	+ 2036

For Townhouse/Rowhouse - Multiply cost per square foot by

Inner Unit	.92
End Unit	.96

Alternatives

Add to or deduct from the cost per square foot of living area

Cedar Shake Roof	+ 3.35
Air Conditioning, in Heating Ductwork	+ 4.86
In Separate Ductwork	+ 7.14
Heating Systems, Hot Water	+ 1.97
Heat Pump	+ 2.70
Electric Heat	− .71
Not Heated	− 3.21

Additional upgrades or components

Kitchen Cabinets & Countertops	Page 93
Bathroom Vanities	94
Fireplaces & Chimneys	94
Windows, Skylights & Dormers	94
Appliances	95
Breezeways & Porches	95
Finished Attic	95
Garages	96
Site Improvements	96
Wings & Ells	56

Solid Wall 1 Story

Living Area - 1600 S.F.
Perimeter - 163 L.F.

			Cost Per Square Foot Of Living Area			% of Total
			Mat.	Inst.	Total	(rounded)
1	Site Work	Site preparation for slab; 4' deep trench excavation for foundation wall.		1.01	1.01	0.8%
2	Foundation	Continuous reinforced concrete footing 8" deep x 18" wide; dampproofed and insulated reinforced concrete foundation wall, 8" thick, 4' deep; 4" concrete slab on 4" crushed stone base and polyethylene vapor barrier, trowel finish.	5.53	6.89	12.42	10.1%
3	Framing	Exterior walls - Precut traditional log home. Handcrafted white cedar or pine logs. Delivery included.	24.01	14.88	38.89	31.7%
4	Exterior Walls	Wood double hung windows, solid wood exterior doors.	6.24	2.91	9.15	7.5%
5	Roofing	25 year asphalt shingles; #15 felt building paper; aluminum gutters, downspouts, drip edge and flashings.	2.16	2.53	4.69	3.8%
6	Interiors	Walls and ceilings, 1/2" taped and finished gypsum wallboard, primed and painted with 2 coats; painted baseboard and trim, finished hardwood floor 40%, carpet with 1/2" underlayment 40%, vinyl tile with 1/2" underlayment 15%, ceramic tile with 1/2" underlayment 5%; hollow core and louvered doors.	11.80	12.04	23.84	19.4%
7	Specialties	Average grade kitchen cabinets - 14 L.F. wall and base with solid surface counter top and kitchen sink; 40 gallon electric water heater.	3.36	1.18	4.54	3.7%
8	Mechanical	1 lavatory, white, wall hung; 1 water closet, white; 1 bathtub with shower, enameled steel, white; gas fired warm air heat.	3.67	3.47	7.14	5.8%
9	Electrical	200 Amp. service; Romex wiring; incandescent lighting fixtures, switches, receptacles.	1.25	1.89	3.14	2.6%
10	Overhead	Contractor's overhead and profit and plans.	9.88	7.95	17.83	14.5%
		Total	67.90	54.75	**122.65**	

RESIDENTIAL — Solid Wall — 2 Story

- Post and beam frame
- Log exterior walls
- Simple design from standard plans
- Single family — 1 full bath, 1 kitchen
- No basement
- Asphalt shingles on roof
- Hot air heat
- Gypsum wallboard interior finishes
- Materials and workmanship are average

Note: The illustration shown may contain some optional components (for example: garages and/or fireplaces) whose costs are shown in the modifications, adjustments, & alternatives below or at the end of the square foot section.

Base cost per square foot of living area

Exterior Wall	Living Area										
	1000	1200	1400	1600	1800	2000	2200	2600	3000	3400	3800
6" Log - Solid Wall	154.10	140.45	133.75	129.20	124.40	119.50	116.05	109.55	103.35	100.60	97.95
8" Log - Solid Wall	142.70	129.95	123.95	119.80	115.55	110.95	108.00	102.25	96.65	94.15	91.85
Finished Basement, Add	22.50	22.15	21.40	20.95	20.40	20.05	19.65	18.90	18.40	18.10	17.80
Unfinished Basement, Add	8.85	8.25	7.85	7.55	7.15	6.95	6.70	6.25	5.95	5.75	5.55

Modifications

Add to the total cost

Upgrade Kitchen Cabinets	$ + 4158
Solid Surface Countertops (Included)	
Full Bath - including plumbing, wall and floor finishes	+ 7329
Half Bath - including plumbing, wall and floor finishes	+ 4275
One Car Attached Garage	+ 13,759
One Car Detached Garage	+ 17,994
Fireplace & Chimney	+ 6840

Adjustments

For multi family - add to total cost

Additional Kitchen	$ + 8034
Additional Bath	+ 7329
Additional Entry & Exit	+ 1658
Separate Heating	+ 1133
Separate Electric	+ 2036

For Townhouse/Rowhouse - Multiply cost per square foot by

Inner Unit	.92
End Unit	.96

Alternatives

Add to or deduct from the cost per square foot of living area

Cedar Shake Roof	+ 1.70
Air Conditioning, in Heating Ductwork	+ 2.94
In Separate Ductwork	+ 5.68
Heating Systems, Hot Water	+ 1.84
Heat Pump	+ 3.16
Electric Heat	– .53
Not Heated	– 2.99

Additional upgrades or components

Kitchen Cabinets & Countertops	Page 93
Bathroom Vanities	94
Fireplaces & Chimneys	94
Windows, Skylights & Dormers	94
Appliances	95
Breezeways & Porches	95
Finished Attic	95
Garages	96
Site Improvements	96
Wings & Ells	56

Solid Wall 2 Story

Living Area - 2000 S.F.
Perimeter - 135 L.F.

#	Category	Description	Cost Per Square Foot Of Living Area			% of Total (rounded)
			Mat.	Inst.	Total	
1	Site Work	Site preparation for slab; 4' deep trench excavation for foundation wall.		.81	.81	0.7%
2	Foundation	Continuous reinforced concrete footing 8" deep x 18" wide; dampproofed and insulated reinforced concrete foundation wall, 8" thick, 4' deep, 4" concrete slab on 4" crushed stone base and polyethylene vapor barrier, trowel finish.	3.29	4.22	7.51	6.3%
3	Framing	Exterior walls - Precut traditional log home. Handcrafted white cedar or pine logs. Delivery included.	26.78	15.42	42.20	35.3%
4	Exterior Walls	Wood double hung windows, solid wood exterior doors.	6.54	3.14	9.68	8.1%
5	Roofing	25 year asphalt shingles; #15 felt building paper; aluminum gutters, downspouts, drip edge and flashings.	1.08	1.26	2.34	2.0%
6	Interiors	Walls and ceilings, 1/2" taped and finished gypsum wallboard, primed and painted with 2 coats; painted baseboard and trim, finished hardwood floor 40%, carpet with 1/2" underlayment 40%, vinyl tile with 1/2" underlayment 15%, ceramic tile with 1/2" underlayment 5%; hollow core and louvered doors.	13.41	13.29	26.70	22.3%
7	Specialties	Average grade kitchen cabinets - 14 L.F. wall and base with solid surface counter top and kitchen sink; 40 gallon electric water heater.	2.67	.95	3.62	3.0%
8	Mechanical	1 lavatory, white, wall hung; 1 water closet, white; 1 bathtub with shower; enameled steel, white; gas fired warm air heat.	3.14	3.25	6.39	5.3%
9	Electrical	200 Amp. service; Romex wiring; incandescent lighting fixtures, switches, receptacles.	1.13	1.75	2.88	2.4%
10	Overhead	Contractor's overhead and profit and plans.	9.86	7.51	17.37	14.5%
		Total	67.90	51.60	**119.50**	

RESIDENTIAL | Average | Wings & Ells

1 Story — Base cost per square foot of living area

Exterior Wall	Living Area							
	50	100	200	300	400	500	600	700
Wood Siding - Wood Frame	203.65	157.35	137.20	115.90	109.30	105.35	102.75	103.55
Brick Veneer - Wood Frame	208.00	156.90	134.75	110.10	102.85	98.50	95.60	96.20
Stucco on Wood Frame	194.75	150.90	131.80	112.25	106.00	102.20	99.75	100.70
Solid Masonry	259.40	197.15	170.45	138.05	129.25	123.95	122.05	122.05
Finished Basement, Add	62.90	52.10	46.85	38.15	36.40	35.35	34.65	34.15
Unfinished Basement, Add	29.75	22.45	19.40	14.40	13.40	12.80	12.40	12.10

1-1/2 Story — Base cost per square foot of living area

Exterior Wall	Living Area							
	100	200	300	400	500	600	700	800
Wood Siding - Wood Frame	163.00	131.70	112.90	101.75	95.95	93.30	89.75	88.80
Brick Veneer - Wood Frame	222.45	165.95	137.70	120.65	112.10	107.50	102.55	100.75
Stucco on Wood Frame	199.60	147.70	122.50	108.75	101.15	97.15	92.80	91.10
Solid Masonry	257.10	193.70	160.85	138.70	128.75	123.20	117.50	115.50
Finished Basement, Add	42.40	38.30	34.80	30.95	29.95	29.25	28.60	28.50
Unfinished Basement, Add	18.65	15.65	13.60	11.40	10.80	10.40	10.00	9.95

2 Story — Base cost per square foot of living area

Exterior Wall	Living Area							
	100	200	400	600	800	1000	1200	1400
Wood Siding - Wood Frame	162.80	122.45	104.75	87.45	81.70	78.25	75.90	76.95
Brick Veneer - Wood Frame	226.05	157.65	126.35	101.95	93.50	88.50	85.15	85.45
Stucco on Wood Frame	200.50	139.40	111.10	91.75	84.45	80.00	77.05	77.65
Solid Masonry	264.90	185.40	149.50	117.35	107.45	101.45	97.50	97.35
Finished Basement, Add	33.40	28.05	25.45	21.10	20.20	19.70	19.35	19.10
Unfinished Basement, Add	15.00	11.40	9.90	7.35	6.85	6.55	6.35	6.20

Base costs do not include bathroom or kitchen facilities. Use Modifications/Adjustments/Alternatives on pages 93–96 where appropriate.

Custom Class

RESIDENTIAL | Custom | Illustrations

1 Story

1-1/2 Story

2 Story

2-1/2 Story

Bi-Level

Tri-Level

RESIDENTIAL — Custom — 1 Story

- A distinct residence from designer's plans
- Single family — 1 full bath, 1 half bath, 1 kitchen
- No basement
- Asphalt shingles on roof
- Forced hot air heat/air conditioning
- Gypsum wallboard interior finishes
- Materials and workmanship are above average

Note: The illustration shown may contain some optional components (for example: garages and/or fireplaces) whose costs are shown in the modifications, adjustments, & alternatives below or at the end of the square foot section.

©Design Basics, Inc.

Base cost per square foot of living area

Exterior Wall	800	1000	1200	1400	1600	1800	2000	2400	2800	3200	3600
Wood Siding - Wood Frame	186.65	168.40	154.40	143.00	135.25	130.95	125.95	116.45	109.70	104.95	100.05
Brick Veneer - Wood Frame	208.10	189.05	174.30	162.25	154.20	149.60	144.25	134.35	127.25	122.15	116.90
Stone Veneer - Wood Frame	217.00	197.15	181.60	168.90	160.35	155.60	149.90	139.50	132.00	126.50	121.00
Solid Masonry	220.45	200.35	184.45	171.45	162.85	157.95	152.05	141.50	133.80	128.25	122.65
Finished Basement, Add	60.95	60.95	58.45	56.50	55.15	54.40	53.30	51.90	50.75	49.75	48.85
Unfinished Basement, Add	25.00	23.60	22.35	21.30	20.60	20.25	19.65	18.95	18.30	17.80	17.35

Modifications

Add to the total cost

Upgrade Kitchen Cabinets	$ + 1314
Solid Surface Countertops (Included)	
Full Bath - including plumbing, wall and floor finishes	+ 8795
Half Bath - including plumbing, wall and floor finishes	+ 5130
Two Car Attached Garage	+ 27,346
Two Car Detached Garage	+ 31,142
Fireplace & Chimney	+ 6494

Adjustments

For multi family - add to total cost

Additional Kitchen	$ + 16,920
Additional Full Bath & Half Bath	+ 13,925
Additional Entry & Exit	+ 1658
Separate Heating & Air Conditioning	+ 7328
Separate Electric	+ 2036

For Townhouse/Rowhouse - Multiply cost per square foot by

Inner Unit	.90
End Unit	.95

Alternatives

Add to or deduct from the cost per square foot of living area

Cedar Shake Roof	+ 2.45
Clay Tile Roof	+ 6.55
Slate Roof	+ 5.60
Upgrade Ceilings to Textured Finish	+ .50
Air Conditioning, in Heating Ductwork	Base System
Heating Systems, Hot Water	+ 2.02
Heat Pump	+ 2.69
Electric Heat	− 1.99
Not Heated	− 4.02

Additional upgrades or components

Kitchen Cabinets & Countertops	Page 93
Bathroom Vanities	94
Fireplaces & Chimneys	94
Windows, Skylights & Dormers	94
Appliances	95
Breezeways & Porches	95
Finished Attic	95
Garages	96
Site Improvements	96
Wings & Ells	74

Custom 1 Story

Living Area - 2400 S.F.
Perimeter - 207 L.F.

			Cost Per Square Foot Of Living Area			% of Total
			Mat.	Inst.	Total	(rounded)
1	Site Work	Site preparation for slab; 4' deep trench excavation for foundation wall.		.80	.80	0.7%
2	Foundation	Continuous reinforced concrete footing 8" deep x 18" wide; dampproofed and insulated concrete block foundation wall, 12" thick, 4' deep, with grout and reinforcing; 4" concrete slab on 4" crushed stone base and polyethylene vapor barrier, trowel finish.	6.09	7.58	13.67	11.7%
3	Framing	Exterior walls - 2" x 6" wood studs, 16" O.C.; 1/2" plywood sheathing; 2" x 8" rafters 16" O.C. with 1/2" plywood sheathing, 4 in 12 pitch; 2" x 6" ceiling joists 16" O.C.; 5/8" plywood subfloor on 1" x 3" wood sleepers 16" O.C.	4.59	6.98	11.57	9.9%
4	Exterior Walls	Horizontal beveled wood siding; building paper; insulation; wood double hung windows; 3 solid core wood exterior doors; storms and screens.	10.95	3.90	14.85	12.8%
5	Roofing	30 year asphalt shingles; #15 felt building paper; aluminum gutters, downspouts and drip edge; copper flashings.	4.61	3.67	8.28	7.1%
6	Interiors	Walls and ceilings - 5/8" gypsum wallboard, skim coat plaster, painted with primer and 2 coats; hardwood baseboard and trim, sanded and finished; hardwood floor 70%, ceramic tile with underlayment 20%, vinyl tile with underlayment 10%; wood panel interior doors, primed and painted with 2 coats.	13.76	13.16	26.92	23.1%
7	Specialties	Custom grade kitchen cabinets - 20 L.F. wall and base with solid surface counter top and kitchen sink; 4 L.F. bathroom vanity; 75 gallon electric water heater, medicine cabinet.	6.14	1.34	7.48	6.4%
8	Mechanical	Gas fired warm air heat/air conditioning; one full bath including: bathtub, corner shower, built in lavatory and water closet; one 1/2 bath including: built in lavatory and water closet.	6.85	3.59	10.44	9.0%
9	Electrical	200 Amp. service; Romex wiring; fluorescent and incandescent lighting fixtures, switches, receptacles.	1.18	1.87	3.05	2.6%
10	Overhead	Contractor's overhead and profit and design.	10.83	8.56	19.39	16.7%
		Total	65.00	51.45	**116.45**	

RESIDENTIAL — Custom — 1-1/2 Story

- A distinct residence from designer's plans
- Single family — 1 full bath, 1 half bath, 1 kitchen
- No basement
- Asphalt shingles on roof
- Forced hot air heat/air conditioning
- Gypsum wallboard interior finishes
- Materials and workmanship are above average

Note: The illustration shown may contain some optional components (for example: garages and/or fireplaces) whose costs are shown in the modifications, adjustments, & alternatives below or at the end of the square foot section.

Base cost per square foot of living area

Exterior Wall	Living Area										
	1000	1200	1400	1600	1800	2000	2400	2800	3200	3600	4000
Wood Siding - Wood Frame	167.80	155.80	147.30	137.10	131.20	125.60	114.75	109.80	105.20	101.75	96.85
Brick Veneer - Wood Frame	176.65	164.10	155.20	144.25	138.00	132.05	120.45	115.20	110.20	106.60	101.35
Stone Veneer - Wood Frame	186.00	172.85	163.55	151.85	145.20	138.85	126.45	120.90	115.50	111.65	106.05
Solid Masonry	189.65	176.25	166.80	154.85	148.05	141.55	128.80	123.15	117.55	113.70	107.90
Finished Basement, Add	40.30	40.60	39.65	38.05	37.20	36.40	34.80	34.20	33.30	32.90	32.20
Unfinished Basement, Add	16.75	16.05	15.55	14.75	14.30	13.90	13.05	12.75	12.30	12.10	11.75

Modifications

Add to the total cost

Upgrade Kitchen Cabinets	$ + 1314
Solid Surface Countertops (Included)	
Full Bath - including plumbing, wall and floor finishes	+ 8795
Half Bath - including plumbing, wall and floor finishes	+ 5130
Two Car Attached Garage	+ 27,346
Two Car Detached Garage	+ 31,142
Fireplace & Chimney	+ 6494

Adjustments

For multi family - add to total cost

Additional Kitchen	$ + 16,920
Additional Full Bath & Half Bath	+ 13,925
Additional Entry & Exit	+ 1658
Separate Heating & Air Conditioning	+ 7328
Separate Electric	+ 2036

For Townhouse/Rowhouse - Multiply cost per square foot by

Inner Unit	.90
End Unit	.95

Alternatives

Add to or deduct from the cost per square foot of living area

Cedar Shake Roof	+ 1.80
Clay Tile Roof	+ 4.75
Slate Roof	+ 4
Upgrade Ceilings to Textured Finish	+ .50
Air Conditioning, in Heating Ductwork	Base System
Heating Systems, Hot Water	+ 1.93
Heat Pump	+ 2.81
Electric Heat	- 1.76
Not Heated	- 3.71

Additional upgrades or components

Kitchen Cabinets & Countertops	Page 93
Bathroom Vanities	94
Fireplaces & Chimneys	94
Windows, Skylights & Dormers	94
Appliances	95
Breezeways & Porches	95
Finished Attic	95
Garages	96
Site Improvements	96
Wings & Ells	74

Custom 1-1/2 Story

Living Area - 2800 S.F.
Perimeter - 175 L.F.

#	Category	Description	Cost Per Square Foot Of Living Area			% of Total
			Mat.	Inst.	Total	(rounded)
1	Site Work	Site preparation for slab; 4' deep trench excavation for foundation wall.		.69	.69	0.6%
2	Foundation	Continuous reinforced concrete footing 8" deep x 18" wide; dampproofed and insulated concrete block foundation wall, 12" thick, 4' deep, with grout and reinforcing; 4" concrete slab on 4" crushed stone base and polyethylene vapor barrier, trowel finish.	4.28	5.42	9.70	8.8%
3	Framing	Exterior walls - 2" x 6" wood studs, 16" O.C.; 1/2" plywood sheathing; 2" x 8" rafters 16" O.C. with 1/2" plywood sheathing, 8 in 12 pitch; 2" x 10" floor joists 16" O.C. with 5/8" plywood subfloor; 5/8" plywood subfloor on 1" x 3" wood sleepers 16" O.C.	5.92	7.36	13.28	12.1%
4	Exterior Walls	Horizontal beveled wood siding; building paper; insulation; wood double hung windows; 3 solid core wood exterior doors; storms and screens.	10.65	3.87	14.52	13.2%
5	Roofing	30 year asphalt shingles; #15 felt building paper; aluminum gutters, downspouts and drip edge; copper flashings.	2.89	2.29	5.18	4.7%
6	Interiors	Walls and ceilings - 5/8" gypsum wallboard, skim coat plaster, painted with primer and 2 coats; hardwood baseboard and trim, sanded and finished; hardwood floor 70%, ceramic tile with underlayment 20%, vinyl tile with underlayment 10%; wood panel interior doors, primed and painted with 2 coats.	15.24	14.35	29.59	26.9%
7	Specialties	Custom grade kitchen cabinets - 20 L.F. wall and base with solid surface counter top and kitchen sink; 4 L.F. bathroom vanity; 75 gallon electric water heater, medicine cabinet.	5.25	1.14	6.39	5.8%
8	Mechanical	Gas fired warm air heat/air conditioning; one full bath including: bathtub, corner shower, built in lavatory and water closet; one 1/2 bath including: built in lavatory and water closet.	5.90	3.34	9.24	8.4%
9	Electrical	200 Amp. service; Romex wiring; fluorescent and incandescent lighting fixtures, switches, receptacles.	1.12	1.80	2.92	2.7%
10	Overhead	Contractor's overhead and profit and design.	10.25	8.04	18.29	16.7%
	Total		61.50	48.30	**109.80**	

RESIDENTIAL — Custom — 2 Story

- A distinct residence from designer's plans
- Single family — 1 full bath, 1 half bath, 1 kitchen
- No basement
- Asphalt shingles on roof
- Forced hot air heat/air conditioning
- Gypsum wallboard interior finishes
- Materials and workmanship are above average

Note: The illustration shown may contain some optional components (for example: garages and/or fireplaces) whose costs are shown in the modifications, adjustments, & alternatives below or at the end of the square foot section.

Base cost per square foot of living area

Exterior Wall	Living Area										
	1200	1400	1600	1800	2000	2400	2800	3200	3600	4000	4400
Wood Siding - Wood Frame	156.70	147.20	140.35	134.25	127.90	118.65	110.80	105.65	102.70	99.45	96.65
Brick Veneer - Wood Frame	166.15	156.05	148.80	142.20	135.55	125.55	117.10	111.50	108.40	104.75	101.85
Stone Veneer - Wood Frame	176.15	165.40	157.75	150.60	143.65	132.85	123.70	117.75	114.35	110.40	107.30
Solid Masonry	180.00	169.00	161.20	153.90	146.80	135.70	126.30	120.25	116.75	112.55	109.45
Finished Basement, Add	32.40	32.75	31.90	31.05	30.50	29.25	28.30	27.60	27.25	26.70	26.40
Unfinished Basement, Add	13.45	12.90	12.50	12.05	11.80	11.20	10.65	10.30	10.10	9.85	9.70

Modifications

Add to the total cost

Upgrade Kitchen Cabinets	$ + 1314
Solid Surface Countertops (Included)	
Full Bath - including plumbing, wall and floor finishes	+ 8795
Half Bath - including plumbing, wall and floor finishes	+ 5130
Two Car Attached Garage	+ 27,346
Two Car Detached Garage	+ 31,142
Fireplace & Chimney	+ 7329

Adjustments

For multi family - add to total cost

Additional Kitchen	$ + 16,920
Additional Full Bath & Half Bath	+ 13,925
Additional Entry & Exit	+ 1658
Separate Heating & Air Conditioning	+ 7328
Separate Electric	+ 2036

For Townhouse/Rowhouse - Multiply cost per square foot by

Inner Unit	.87
End Unit	.93

Alternatives

Add to or deduct from the cost per square foot of living area

Cedar Shake Roof	+ 1.25
Clay Tile Roof	+ 3.30
Slate Roof	+ 2.80
Upgrade Ceilings to Textured Finish	+ .50
Air Conditioning, in Heating Ductwork	Base System
Heating Systems, Hot Water	+ 1.88
Heat Pump	+ 3.14
Electric Heat	- 1.76
Not Heated	- 3.50

Additional upgrades or components

Kitchen Cabinets & Countertops	Page 93
Bathroom Vanities	94
Fireplaces & Chimneys	94
Windows, Skylights & Dormers	94
Appliances	95
Breezeways & Porches	95
Finished Attic	95
Garages	96
Site Improvements	96
Wings & Ells	74

Custom 2 Story

Living Area - 2800 S.F.
Perimeter - 156 L.F.

#	Section	Description	Cost Per Square Foot Of Living Area			% of Total (rounded)
			Mat.	Inst.	Total	
1	Site Work	Site preparation for slab; 4' deep trench excavation for foundation wall.		.69	.69	0.6%
2	Foundation	Continuous reinforced concrete footing 8" deep x 18" wide; dampproofed and insulated concrete block foundation wall, 12" thick, 4' deep, with grout and reinforcing; 4" concrete slab on 4" crushed stone base and polyethylene vapor barrier, trowel finish.	3.65	4.64	8.29	7.5%
3	Framing	Exterior walls - 2" x 6" wood studs, 16" O.C.; 1/2" plywood sheathing; 2" x 8" rafters 16" O.C. with 1/2" plywood sheathing, 6 in 12 pitch; 2" x 8" ceiling joists 16" O.C.; 2" x 10" floor joists 16" O.C. with 5/8" plywood subfloor; 5/8" plywood subfloor on 1" x 3" wood sleepers 16" O.C.	6.47	7.60	14.07	12.7%
4	Exterior Walls	Horizontal beveled wood siding; building paper; insulation; wood double hung windows; 3 solid core wood exterior doors; storms and screens.	11.89	4.32	16.21	14.6%
5	Roofing	30 year asphalt shingles; #15 felt building paper; aluminum gutters, downspouts and drip edge; copper flashings.	2.31	1.84	4.15	3.7%
6	Interiors	Walls and ceilings - 5/8" gypsum wallboard, skim coat plaster, painted with primer and 2 coats; hardwood baseboard and trim, sanded and finished; hardwood floor 70%, ceramic tile with underlayment 20%, vinyl tile with underlayment 10%; wood panel interior doors, primed and painted with 2 coats.	15.50	14.67	30.17	27.2%
7	Specialties	Custom grade kitchen cabinets - 20 L.F. wall and base with solid surface counter top and kitchen sink; 4 L.F. bathroom vanity; 75 gallon electric water heater, medicine cabinet.	5.25	1.14	6.39	5.8%
8	Mechanical	Gas fired warm air heat/air conditioning; one full bath including: bathtub, corner shower; built in lavatory and water closet; one 1/2 bath including: built in lavatory and water closet.	6.05	3.39	9.44	8.5%
9	Electrical	200 Amp. service; Romex wiring; fluorescent and incandescent lighting fixtures, switches, receptacles.	1.12	1.80	2.92	2.6%
10	Overhead	Contractor's overhead and profit and design.	10.46	8.01	18.47	16.7%
	Total		62.70	48.10	**110.80**	

RESIDENTIAL Custom 2-1/2 Story

- A distinct residence from designer's plans
- Single family — 1 full bath, 1 half bath, 1 kitchen
- No basement
- Asphalt shingles on roof
- Forced hot air heat/air conditioning
- Gypsum wallboard interior finishes
- Materials and workmanship are above average

Note: The illustration shown may contain some optional components (for example: garages and/or fireplaces) whose costs are shown in the modifications, adjustments, & alternatives below or at the end of the square foot section.

Base cost per square foot of living area

Exterior Wall	Living Area										
	1500	1800	2100	2400	2800	3200	3600	4000	4500	5000	5500
Wood Siding - Wood Frame	153.75	138.35	129.10	123.50	116.35	109.55	105.90	100.05	96.90	94.10	91.30
Brick Veneer - Wood Frame	163.60	147.40	137.25	131.25	123.65	116.25	112.25	105.85	102.45	99.40	96.35
Stone Veneer - Wood Frame	173.95	156.95	145.80	139.35	131.40	123.25	118.85	112.05	108.25	105.00	101.70
Solid Masonry	178.05	160.65	149.15	142.50	134.45	126.00	121.45	114.50	110.60	107.10	103.75
Finished Basement, Add	25.70	25.70	24.45	23.85	23.30	22.40	21.95	21.45	21.00	20.65	20.35
Unfinished Basement, Add	10.80	10.25	9.55	9.25	9.00	8.60	8.35	8.05	7.80	7.65	7.50

Modifications
Add to the total cost

Upgrade Kitchen Cabinets	$ + 1314
Solid Surface Countertops (Included)	
Full Bath - including plumbing, wall and floor finishes	+ 8795
Half Bath - including plumbing, wall and floor finishes	+ 5130
Two Car Attached Garage	+ 27,346
Two Car Detached Garage	+ 31,142
Fireplace & Chimney	+ 8277

Adjustments
For multi family - add to total cost

Additional Kitchen	$ + 16,920
Additional Full Bath & Half Bath	+ 13,925
Additional Entry & Exit	+ 1658
Separate Heating & Air Conditioning	+ 7328
Separate Electric	+ 2036

For Townhouse/Rowhouse - Multiply cost per square foot by

Inner Unit	.87
End Unit	.94

Alternatives
Add to or deduct from the cost per square foot of living area

Cedar Shake Roof	+ 1.10
Clay Tile Roof	+ 2.85
Slate Roof	+ 2.40
Upgrade Ceilings to Textured Finish	+ .50
Air Conditioning, in Heating Ductwork	Base System
Heating Systems, Hot Water	+ 1.69
Heat Pump	+ 3.25
Electric Heat	- 3.09
Not Heated	- 3.50

Additional upgrades or components

Kitchen Cabinets & Countertops	Page 93
Bathroom Vanities	94
Fireplaces & Chimneys	94
Windows, Skylights & Dormers	94
Appliances	95
Breezeways & Porches	95
Finished Attic	95
Garages	96
Site Improvements	96
Wings & Ells	74

Custom 2-1/2 Story

Living Area - 3200 S.F.
Perimeter - 150 L.F.

			Cost Per Square Foot Of Living Area			% of Total
			Mat.	Inst.	Total	(rounded)
1	Site Work	Site preparation for slab; 4' deep trench excavation for foundation wall.		.61	.61	0.6%
2	Foundation	Continuous reinforced concrete footing 8" deep x 18" wide; dampproofed and insulated concrete block foundation wall, 12" thick, 4' deep, with grout and reinforcing; 4" concrete slab on 4" crushed stone base and polyethylene vapor barrier, trowel finish.	2.99	3.86	6.85	6.3%
3	Framing	Exterior walls - 2" x 6" wood studs, 16" O.C.; 1/2" plywood sheathing; 2" x 8" rafters 16" O.C. with 1/2" plywood sheathing, 6 in 12 pitch; 2" x 8" ceiling joists 16" O.C.; 2" x 10" floor joists 16" O.C. with 5/8" plywood subfloor; 5/8" plywood subfloor on 1" x 3" wood sleepers 16" O.C.	7.00	7.91	14.91	13.6%
4	Exterior Walls	Horizontal beveled wood siding; building paper; insulation; wood double hung windows; 3 solid core wood exterior doors; storms and screens.	12.18	4.41	16.59	15.1%
5	Roofing	30 year asphalt shingles; #15 felt building paper; aluminum gutters, downspouts and drip edge; copper flashings.	1.78	1.41	3.19	2.9%
6	Interiors	Walls and ceilings - 5/8" gypsum wallboard, skim coat plaster, painted with primer and 2 coats; hardwood baseboard and trim, sanded and finished; hardwood floor 70%, ceramic tile with underlayment 20%, vinyl tile with underlayment 10%; wood panel interior doors, primed and painted with 2 coats.	16.46	15.55	32.01	29.2%
7	Specialties	Custom grade kitchen cabinets - 20 L.F. wall and base with solid surface counter top and kitchen sink; 4 L.F. bathroom vanity; 75 gallon electric water heater, medicine cabinet.	4.60	1.00	5.60	5.1%
8	Mechanical	Gas fired warm air heat/air conditioning; one full bath including: bathtub, corner shower; built in lavatory and water closet; one 1/2 bath including: built in lavatory and water closet.	5.45	3.26	8.71	8.0%
9	Electrical	200 Amp. service; Romex wiring; fluorescent and incandescent lighting fixtures, switches, receptacles.	1.08	1.75	2.83	2.6%
10	Overhead	Contractor's overhead and profit and design.	10.31	7.94	18.25	16.7%
		Total	61.85	47.70	**109.55**	

RESIDENTIAL | Custom | 3 Story

- A distinct residence from designer's plans
- Single family — 1 full bath, 1 half bath, 1 kitchen
- No basement
- Asphalt shingles on roof
- Forced hot air heat/air conditioning
- Gypsum wallboard interior finishes
- Materials and workmanship are above average

Note: The illustration shown may contain some optional components (for example: garages and/or fireplaces) whose costs are shown in the modifications, adjustments, & alternatives below or at the end of the square foot section.

Base cost per square foot of living area

Exterior Wall	Living Area										
	1500	1800	2100	2500	3000	3500	4000	4500	5000	5500	6000
Wood Siding - Wood Frame	153.05	137.75	130.30	124.05	114.30	109.55	103.70	97.60	95.35	93.05	90.65
Brick Veneer - Wood Frame	163.25	147.25	139.05	132.40	121.95	116.80	110.35	103.70	101.30	98.70	96.00
Stone Veneer - Wood Frame	174.05	157.20	148.40	141.40	130.05	124.40	117.30	110.10	107.45	104.70	101.60
Solid Masonry	178.30	161.10	152.05	144.80	133.20	127.45	120.00	112.60	109.95	107.00	103.80
Finished Basement, Add	22.50	22.55	21.75	21.20	20.30	19.80	19.10	18.60	18.35	18.10	17.75
Unfinished Basement, Add	9.45	8.95	8.55	8.30	7.80	7.60	7.25	7.00	6.85	6.75	6.55

Modifications
Add to the total cost

Upgrade Kitchen Cabinets	$ + 1314
Solid Surface Countertops (Included)	
Full Bath - including plumbing, wall and floor finishes	+ 8795
Half Bath - including plumbing, wall and floor finishes	+ 5130
Two Car Attached Garage	+ 27,346
Two Car Detached Garage	+ 31,142
Fireplace & Chimney	+ 8277

Adjustments
For multi family - add to total cost

Additional Kitchen	$ + 16,920
Additional Full Bath & Half Bath	+ 13,925
Additional Entry & Exit	+ 1658
Separate Heating & Air Conditioning	+ 7328
Separate Electric	+ 2036

For Townhouse/Rowhouse - Multiply cost per square foot by

Inner Unit	.85
End Unit	.93

Alternatives
Add to or deduct from the cost per square foot of living area

Cedar Shake Roof	+ .80
Clay Tile Roof	+ 2.20
Slate Roof	+ 1.85
Upgrade Ceilings to Textured Finish	+ .50
Air Conditioning, in Heating Ductwork	Base System
Heating Systems, Hot Water	+ 1.69
Heat Pump	+ 3.25
Electric Heat	– 3.09
Not Heated	– 3.39

Additional upgrades or components

Kitchen Cabinets & Countertops	Page 93
Bathroom Vanities	94
Fireplaces & Chimneys	94
Windows, Skylights & Dormers	94
Appliances	95
Breezeways & Porches	95
Finished Attic	95
Garages	96
Site Improvements	96
Wings & Ells	74

Custom 3 Story

Living Area - 3000 S.F.
Perimeter - 135 L.F.

#	Category	Description	Cost Per Square Foot Of Living Area			% of Total
			Mat.	Inst.	Total	(rounded)
1	Site Work	Site preparation for slab; 4' deep trench excavation for foundation wall.		.65	.65	0.6%
2	Foundation	Continuous reinforced concrete footing 8" deep x 18" wide; dampproofed and insulated concrete block foundation wall, 12" thick, 4' deep, with grout and reinforcing; 4" concrete slab on 4" crushed stone base and polyethylene vapor barrier, trowel finish.	2.81	3.64	6.45	5.6%
3	Framing	Exterior walls - 2" x 6" wood studs, 16" O.C.; 1/2" plywood sheathing; 2" x 8" rafters 16" O.C. with 1/2" plywood sheathing, 6 in 12 pitch; 2" x 8" ceiling joists 16" O.C.; 2" x 10" floor joists 16" O.C. with 5/8" plywood subfloor; 5/8" plywood subfloor on 1" x 3" wood sleepers 16" O.C.	7.36	8.14	15.50	13.6%
4	Exterior Walls	Horizontal beveled wood siding; building paper; insulation; wood double hung windows; 3 solid core wood exterior doors; storms and screens.	13.81	5.00	18.81	16.5%
5	Roofing	30 year asphalt shingles; #15 felt building paper; aluminum gutters, downspouts and drip edge; copper flashings.	1.54	1.22	2.76	2.4%
6	Interiors	Walls and ceilings - 5/8" gypsum wallboard, skim coat plaster, painted with primer and 2 coats; hardwood baseboard and trim, sanded and finished; hardwood floor 70%, ceramic tile with underlayment 20%, vinyl tile with underlayment 10%; wood panel interior doors, primed and painted with 2 coats.	17.04	16.10	33.14	29.0%
7	Specialties	Custon grade kitchen cabinets - 20 L.F. wall and base with solid surface counter top and kitchen sink; 4 L.F. bathroom vanity; 75 gallon electric water heater, medicine cabinet.	4.92	1.06	5.98	5.2%
8	Mechanical	Gas fired warm air heat/air conditioning; one full bath including: bathtub, corner shower; built in lavatory and water closet; one 1/2 bath including: built in lavatory and water closet.	5.74	3.33	9.07	7.9%
9	Electrical	200 Amp. service; Romex wiring; fluorescent and incandescent lighting fixtures, switches, receptacles.	1.10	1.77	2.87	2.5%
10	Overhead	Contractor's overhead and profit and design.	10.88	8.19	19.07	16.7%
		Total	65.20	49.10	**114.30**	

RESIDENTIAL Custom Bi-Level

- A distinct residence from designer's plans
- Single family — 1 full bath, 1 half bath, 1 kitchen
- No basement
- Asphalt shingles on roof
- Forced hot air heat/air conditioning
- Gypsum wallboard interior finishes
- Materials and workmanship are above average

Note: The illustration shown may contain some optional components (for example: garages and/or fireplaces) whose costs are shown in the modifications, adjustments, & alternatives below or at the end of the square foot section.

Base cost per square foot of living area

Exterior Wall	Living Area										
	1200	1400	1600	1800	2000	2400	2800	3200	3600	4000	4400
Wood Siding - Wood Frame	148.15	139.20	132.70	127.15	121.00	112.40	105.15	100.35	97.60	94.60	92.05
Brick Veneer - Wood Frame	155.25	145.85	139.05	133.10	126.75	117.65	109.85	104.75	101.85	98.60	95.85
Stone Veneer - Wood Frame	162.70	152.80	145.75	139.40	132.85	123.10	114.80	109.40	106.35	102.80	100.00
Solid Masonry	165.65	155.60	148.35	141.85	135.15	125.25	116.75	111.20	108.05	104.45	101.60
Finished Basement, Add	32.40	32.75	31.90	31.05	30.50	29.25	28.30	27.60	27.25	26.70	26.40
Unfinished Basement, Add	13.45	12.90	12.50	12.05	11.80	11.20	10.65	10.30	10.10	9.85	9.70

Modifications

Add to the total cost

Upgrade Kitchen Cabinets	$ + 1314
Solid Surface Countertops (Included)	
Full Bath - including plumbing, wall and floor finishes	+ 8795
Half Bath - including plumbing, wall and floor finishes	+ 5130
Two Car Attached Garage	+ 27,346
Two Car Detached Garage	+ 31,142
Fireplace & Chimney	+ 6494

Adjustments

For multi family - add to total cost

Additional Kitchen	$ + 16,920
Additional Full Bath & Half Bath	+ 13,925
Additional Entry & Exit	+ 1658
Separate Heating & Air Conditioning	+ 7328
Separate Electric	+ 2036

For Townhouse/Rowhouse - Multiply cost per square foot by

Inner Unit	.89
End Unit	.95

Alternatives

Add to or deduct from the cost per square foot of living area

Cedar Shake Roof	+ 1.80
Clay Tile Roof	+ 3.30
Slate Roof	+ 2.80
Upgrade Ceilings to Textured Finish	+ .50
Air Conditioning, in Heating Ductwork	Base System
Heating Systems, Hot Water	+ 1.88
Heat Pump	+ 3.14
Electric Heat	- 1.76
Not Heated	- 3.39

Additional upgrades or components

Kitchen Cabinets & Countertops	Page 93
Bathroom Vanities	94
Fireplaces & Chimneys	94
Windows, Skylights & Dormers	94
Appliances	95
Breezeways & Porches	95
Finished Attic	95
Garages	96
Site Improvements	96
Wings & Ells	74

Important: See the Reference Section for Location Factors (to adjust for your city) and Estimating Forms.

Custom Bi-Level

Living Area - 2800 S.F.
Perimeter - 156 L.F.

			Cost Per Square Foot Of Living Area			% of Total
			Mat.	Inst.	Total	(rounded)
1	Site Work	Excavation for lower level, 4' deep. Site preparation for slab.		.69	.69	0.7%
2	Foundation	Continuous reinforced concrete footing 8" deep x 18" wide; dampproofed and insulated concrete block foundation wall, 12" thick, 4' deep, with grout and reinforcing; 4" concrete slab on 4" crushed stone base and polyethylene vapor barrier, trowel finish.	3.65	4.64	8.29	7.9%
3	Framing	Exterior walls - 2" x 6" wood studs, 16" O.C.; 1/2" plywood sheathing; 2" x 8" rafters 16" O.C. with 1/2" plywood sheathing, 6 in 12 pitch; 2" x 8" ceiling joists 16" O.C.; 2" x 10" floor joists 16" O.C. with 5/8" plywood subfloor; 5/8" plywood subfloor on 1" x 3" wood sleepers 16" O.C.	6.14	7.25	13.39	12.7%
4	Exterior Walls	Horizontal beveled wood siding; building paper; insulation; wood double hung windows; 3 solid core wood exterior doors; storms and screens.	9.41	3.38	12.79	12.2%
5	Roofing	30 year asphalt shingles; #15 felt building paper; aluminum gutters, downspouts and drip edge; copper flashings.	2.31	1.84	4.15	3.9%
6	Interiors	Walls and ceilings - 5/8" gypsum wallboard, skim coat plaster, painted with primer and 2 coats; hardwood baseboard and trim, sanded and finished; hardwood floor 70%, ceramic tile with underlayment 20%, vinyl tile with underlayment 10%; wood panel interior doors, primed and painted with 2 coats.	15.26	14.28	29.54	28.1%
7	Specialties	Custom grade kitchen cabinets - 20 L.F. wall and base with solid surface counter top and kitchen sink; 4 L.F. bathroom vanity; 75 gallon electric water heater, medicine cabinet.	5.25	1.14	6.39	6.1%
8	Mechanical	Gas fired warm air heat/air conditioning; one full bath including: bathtub, corner shower, built in lavatory and water closet; one 1/2 bath including: built in lavatory and water closet.	6.05	3.39	9.44	9.0%
9	Electrical	200 Amp. service; Romex wiring; fluorescent and incandescent lighting fixtures, switches, receptacles.	1.12	1.80	2.92	2.8%
10	Overhead	Contractor's overhead and profit and design.	9.86	7.69	17.55	16.7%
		Total	59.05	46.10	**105.15**	

RESIDENTIAL Custom Tri-Level

- A distinct residence from designer's plans
- Single family — 1 full bath, 1 half bath, 1 kitchen
- No basement
- Asphalt shingles on roof
- Forced hot air heat/air conditioning
- Gypsum wallboard interior finishes
- Materials and workmanship are above average

Note: The illustration shown may contain some optional components (for example: garages and/or fireplaces) whose costs are shown in the modifications, adjustments, & alternatives below or at the end of the square foot section.

Base cost per square foot of living area

Exterior Wall	Living Area										
	1200	1500	1800	2100	2400	2800	3200	3600	4000	4500	5000
Wood Siding - Wood Frame	152.60	138.30	127.70	118.90	112.95	108.60	103.55	98.30	95.95	90.85	88.15
Brick Veneer - Wood Frame	159.65	144.70	133.50	124.20	117.90	113.30	107.95	102.40	99.90	94.55	91.60
Stone Veneer - Wood Frame	167.05	151.45	139.50	129.70	123.00	118.30	112.50	106.65	104.05	98.35	95.20
Solid Masonry	170.00	154.05	141.90	131.80	125.00	120.20	114.30	108.30	105.65	99.80	96.60
Finished Basement, Add*	40.65	40.65	38.95	37.65	36.75	36.15	35.30	34.55	34.25	33.45	32.95
Unfinished Basement, Add*	16.65	15.70	14.90	14.20	13.75	13.45	13.00	12.65	12.45	12.05	11.80

*Basement under middle level only.

Modifications
Add to the total cost

Upgrade Kitchen Cabinets	$ + 1314
Solid Surface Countertops (Included)	
Full Bath - including plumbing, wall and floor finishes	+ 8795
Half Bath - including plumbing, wall and floor finishes	+ 5130
Two Car Attached Garage	+ 27,346
Two Car Detached Garage	+ 31,142
Fireplace & Chimney	+ 6494

Adjustments
For multi family - add to total cost

Additional Kitchen	$ + 16,920
Additional Full Bath & Half Bath	+ 13,925
Additional Entry & Exit	+ 1658
Separate Heating & Air Conditioning	+ 7328
Separate Electric	+ 2036

For Townhouse/Rowhouse - Multiply cost per square foot by

Inner Unit	.87
End Unit	.94

Alternatives
Add to or deduct from the cost per square foot of living area

Cedar Shake Roof	+ 1.25
Clay Tile Roof	+ 4.75
Slate Roof	+ 4
Upgrade Ceilings to Textured Finish	+ .50
Air Conditioning, in Heating Ductwork	Base System
Heating Systems, Hot Water	+ 1.81
Heat Pump	+ 3.28
Electric Heat	- 1.56
Not Heated	- 3.39

Additional upgrades or components

Kitchen Cabinets & Countertops	Page 93
Bathroom Vanities	94
Fireplaces & Chimneys	94
Windows, Skylights & Dormers	94
Appliances	95
Breezeways & Porches	95
Finished Attic	95
Garages	96
Site Improvements	96
Wings & Ells	74

Custom Tri-Level

Living Area - 3200 S.F.
Perimeter - 198 L.F.

			Cost Per Square Foot Of Living Area			% of Total
			Mat.	Inst.	Total	(rounded)
1	Site Work	Site preparation for slab; 4' deep trench excavation for foundation wall, excavation for lower level, 4' deep.		.61	.61	0.6%
2	Foundation	Continuous reinforced concrete footing 8" deep x 18" wide; dampproofed and insulated concrete block foundation wall, 12" thick, 4' deep, with grout and reinforcing; 4" concrete slab on 4" crushed stone base and polyethylene vapor barrier, trowel finish.	4.27	5.35	9.62	9.3%
3	Framing	Exterior walls - 2" x 6" wood studs, 16" O.C.; 1/2" plywood sheathing; 2" x 8" rafters 16" O.C. with 1/2" plywood sheathing, 6 in 12 pitch; 2" x 8" ceiling joists 16" O.C.; 2" x 10" floor joists 16" O.C. with 5/8" plywood subfloor; 5/8" plywood subfloor on 1" x 3" wood sleepers 16" O.C.	5.56	7.09	12.65	12.2%
4	Exterior Walls	Horizontal beveled wood siding; building paper; insulation; wood double hung windows; 3 solid core wood exterior doors; storms and screens.	9.27	3.30	12.57	12.1%
5	Roofing	30 year asphalt shingles; #15 felt building paper; aluminum gutters, downspouts and drip edge; copper flashings.	3.07	2.45	5.52	5.3%
6	Interiors	Walls and ceilings - 5/8" gypsum wallboard, skim coat plaster, painted with primer and 2 coats; hardwood baseboard and trim, sanded and finished; hardwood floor 70%, ceramic tile with underlayment 20%, vinyl tile with underlayment 10%; wood panel interior doors, primed and painted with 2 coats.	14.53	13.65	28.18	27.2%
7	Specialties	Custom grade kitchen cabinets - 20 L.F. wall and base with solid surface counter top and kitchen sink; 4 L.F. bathroom vanity; 75 gallon electric water heater, medicine cabinet.	4.60	1.00	5.60	5.4%
8	Mechanical	Gas fired warm air heat/air conditioning; one full bath including: bathtub, corner shower, built in lavatory and water closet; one 1/2 bath including: built in lavatory and water closet.	5.45	3.26	8.71	8.4%
9	Electrical	200 Amp. service; Romex wiring; fluorescent and incandescent lighting fixtures, switches, receptacles.	1.08	1.75	2.83	2.7%
10	Overhead	Contractor's overhead and profit and design.	9.57	7.69	17.26	16.7%
		Total	57.40	46.15	**103.55**	

RESIDENTIAL Custom Wings & Ells

1 Story — Base cost per square foot of living area

Exterior Wall	Living Area							
	50	100	200	300	400	500	600	700
Wood Siding - Wood Frame	237.35	184.95	162.20	137.75	130.30	125.75	122.80	123.70
Brick Veneer - Wood Frame	261.20	202.00	176.45	147.20	138.80	133.70	130.35	130.95
Stone Veneer - Wood Frame	286.35	219.95	191.35	157.20	147.80	142.10	138.35	138.65
Solid Masonry	296.20	226.95	197.25	161.05	151.25	145.35	141.45	141.70
Finished Basement, Add	95.80	82.35	74.95	62.60	60.15	58.65	57.65	56.95
Unfinished Basement, Add	68.05	47.05	37.45	28.80	26.45	25.10	24.20	23.50

1-1/2 Story — Base cost per square foot of living area

Exterior Wall	Living Area							
	100	200	300	400	500	600	700	800
Wood Siding - Wood Frame	188.65	155.05	134.60	122.30	116.15	113.15	109.40	108.35
Brick Veneer - Wood Frame	209.95	172.05	148.80	133.35	126.35	122.80	118.50	117.40
Stone Veneer - Wood Frame	232.40	190.05	163.75	145.00	137.15	132.95	128.15	126.95
Solid Masonry	241.15	197.10	169.65	149.60	141.40	136.95	131.90	130.65
Finished Basement, Add	64.45	60.00	55.05	49.60	48.10	47.10	46.25	46.10
Unfinished Basement, Add	40.70	31.10	26.50	22.55	21.20	20.30	19.50	19.25

2 Story — Base cost per square foot of living area

Exterior Wall	Living Area							
	100	200	400	600	800	1000	1200	1400
Wood Siding - Wood Frame	187.70	143.65	124.30	105.55	99.25	95.40	92.95	94.15
Brick Veneer - Wood Frame	211.50	160.65	138.50	115.05	107.80	103.35	100.45	101.45
Stone Veneer - Wood Frame	236.65	178.65	153.45	125.00	116.75	111.70	108.45	109.10
Solid Masonry	246.55	185.65	159.30	128.85	120.30	115.00	111.55	112.15
Finished Basement, Add	48.00	41.25	37.50	31.35	30.05	29.35	28.85	28.50
Unfinished Basement, Add	34.05	23.55	18.70	14.40	13.25	12.55	12.10	11.80

Base costs do not include bathroom or kitchen facilities. Use Modifications/Adjustments/Alternatives on pages 93–96 where appropriate.

Luxury Class

RESIDENTIAL | Luxury | Illustrations

1 Story

1-1/2 Story

2 Story

2-1/2 Story

Bi-Level

Tri-Level

RESIDENTIAL | Luxury | 1 Story

- Unique residence built from an architect's plan
- Single family — 1 full bath, 1 half bath, 1 kitchen
- No basement
- Cedar shakes on roof
- Forced hot air heat/air conditioning
- Gypsum wallboard interior finishes
- Many special features
- Extraordinary materials and workmanship

©Home Planners, Inc.

Note: The illustration shown may contain some optional components (for example: garages and/or fireplaces) whose costs are shown in the modifications, adjustments, & alternatives below or at the end of the square foot section.

Base cost per square foot of living area

Exterior Wall	1000	1200	1400	1600	1800	2000	2400	2800	3200	3600	4000
Wood Siding - Wood Frame	215.90	199.20	185.70	176.60	171.30	165.40	154.40	146.50	140.90	135.20	130.50
Brick Veneer - Wood Frame	224.70	207.15	192.90	183.35	177.85	171.50	160.00	151.65	145.70	139.70	134.65
Solid Brick	239.65	220.70	205.20	194.85	188.90	181.95	169.55	160.55	153.85	147.25	141.80
Solid Stone	246.60	226.95	210.90	200.25	194.10	186.75	174.00	164.55	157.65	150.80	145.10
Finished Basement, Add	58.80	63.40	60.95	59.30	58.35	57.05	55.20	53.80	52.50	51.40	50.55
Unfinished Basement, Add	25.60	24.05	22.85	22.05	21.55	20.90	19.95	19.20	18.60	18.00	17.55

Modifications

Add to the total cost

Upgrade Kitchen Cabinets	$ + 1735
Solid Surface Countertops (Included)	
Full Bath - including plumbing, wall and floor finishes	+ 10,554
Half Bath - including plumbing, wall and floor finishes	+ 6157
Two Car Attached Garage	+ 31,466
Two Car Detached Garage	+ 35,593
Fireplace & Chimney	+ 9357

Adjustments

For multi family - add to total cost

Additional Kitchen	$ + 21,730
Additional Full Bath & Half Bath	+ 16,711
Additional Entry & Exit	+ 2306
Separate Heating & Air Conditioning	+ 7328
Separate Electric	+ 2036

For Townhouse/Rowhouse - Multiply cost per square foot by

Inner Unit	.90
End Unit	.95

Alternatives

Add to or deduct from the cost per square foot of living area

Heavyweight Asphalt Shingles	– 2.45
Clay Tile Roof	+ 4.10
Slate Roof	+ 3.10
Upgrade Ceilings to Textured Finish	+ .50
Air Conditioning, in Heating Ductwork	Base System
Heating Systems, Hot Water	+ 2.18
Heat Pump	+ 2.90
Electric Heat	– 1.76
Not Heated	– 4.38

Additional upgrades or components

Kitchen Cabinets & Countertops	Page 93
Bathroom Vanities	94
Fireplaces & Chimneys	94
Windows, Skylights & Dormers	94
Appliances	95
Breezeways & Porches	95
Finished Attic	95
Garages	96
Site Improvements	96
Wings & Ells	92

Luxury 1 Story

Living Area - 2800 S.F.
Perimeter - 219 L.F.

			Cost Per Square Foot Of Living Area			% of Total
			Mat.	Inst.	Total	(rounded)
1	Site Work	Site preparation for slab; 4' deep trench excavation for foundation wall.		.75	.75	0.5%
2	Foundation	Continuous reinforced concrete footing 12" deep x 24" wide; dampproofed and insulated concrete block foundation wall, 12" thick, 4' deep, grouted and reinforced; 4" concrete slab on 4" crushed stone base and polyethylene vapor barrier, trowel finish.	7.12	7.86	14.98	10.2%
3	Framing	Exterior walls - 2" x 6" wood studs, 16" O.C.; 5/8" plywood sheathing; 2" x 10" rafters 16" O.C. with 5/8" plywood sheathing, 6 in 12 pitch; 2" x 8" ceiling joists 16" O.C.; 5/8" plywood subfloor on 1" x 3" wood sleepers 16" O.C.	12.64	14.46	27.10	18.5%
4	Exterior Walls	Horizontal beveled wood siding; building paper; insulation; wood double hung windows; 3 solid core wood exterior doors; storms and screens.	10.52	3.87	14.39	9.8%
5	Roofing	Red cedar shingles; #15 felt building paper; aluminum gutters, downspouts and drip edge; copper flashings.	5.38	4.28	9.66	6.6%
6	Interiors	Walls and ceilings - 5/8" gypsum wallboard, skim coat plaster, painted with primer and 2 coats; hardwood baseboard and trim, sanded and finished; hardwood floor 70%, ceramic tile with underlayment 20%, vinyl tile with underlayment 10%; wood panel interior doors, primed and painted with 2 coats.	14.48	14.49	28.97	19.8%
7	Specialties	Luxury grade kitchen cabinets - 25 L.F. wall and base with solid surface counter top and kitchen sink; 6 L.F. bathroom vanity; 75 gallon electric water heater; medicine cabinet.	6.86	1.45	8.31	5.7%
8	Mechanical	Gas fired warm air heat/air conditioning; one full bath including: bathtub, corner shower; built in lavatory and water closet; one 1/2 bath including: built in lavatory and water closet.	6.80	3.73	10.53	7.2%
9	Electrical	200 Amp. service; Romex wiring; fluorescent and incandescent lighting fixtures; intercom, switches, receptacles.	1.34	2.14	3.48	2.4%
10	Overhead	Contractor's overhead and profit and architect's fees.	15.61	12.72	28.33	19.3%
		Total	80.75	65.75	**146.50**	

RESIDENTIAL — Luxury — 1-1/2 Story

- Unique residence built from an architect's plan
- Single family — 1 full bath, 1 half bath, 1 kitchen
- No basement
- Cedar shakes on roof
- Forced hot air heat/air conditioning
- Gypsum wallboard interior finishes
- Many special features
- Extraordinary materials and workmanship

Note: The illustration shown may contain some optional components (for example: garages and/or fireplaces) whose costs are shown in the modifications, adjustments, & alternatives below or at the end of the square foot section.

Base cost per square foot of living area

Exterior Wall	\multicolumn{11}{c}{Living Area}										
	1000	1200	1400	1600	1800	2000	2400	2800	3200	3600	4000
Wood Siding - Wood Frame	199.70	185.25	174.95	162.95	155.80	149.25	136.50	130.70	125.25	121.20	115.50
Brick Veneer - Wood Frame	209.90	194.75	184.00	171.20	163.65	156.65	143.10	136.90	131.00	126.75	120.65
Solid Brick	227.20	211.00	199.50	185.20	176.95	169.35	154.25	147.45	140.80	136.20	129.50
Solid Stone	235.30	218.55	206.60	191.70	183.15	175.15	159.40	152.40	145.35	140.55	133.55
Finished Basement, Add	41.15	44.70	43.45	41.55	40.55	39.50	37.55	36.75	35.60	35.10	34.30
Unfinished Basement, Add	18.40	17.55	16.95	16.00	15.50	15.00	14.00	13.55	13.05	12.80	12.40

Modifications

Add to the total cost

Upgrade Kitchen Cabinets	$ + 1735
Solid Surface Countertops (Included)	
Full Bath - including plumbing, wall and floor finishes	+ 10,554
Half Bath - including plumbing, wall and floor finishes	+ 6157
Two Car Attached Garage	+ 31,466
Two Car Detached Garage	+ 35,593
Fireplace & Chimney	+ 9357

Adjustments

For multi family - add to total cost

Additional Kitchen	$ + 21,730
Additional Full Bath & Half Bath	+ 16,711
Additional Entry & Exit	+ 2306
Separate Heating & Air Conditioning	+ 7328
Separate Electric	+ 2036

For Townhouse/Rowhouse - Multiply cost per square foot by

Inner Unit	.90
End Unit	.95

Alternatives

Add to or deduct from the cost per square foot of living area

Heavyweight Asphalt Shingles	– 1.80
Clay Tile Roof	+ 2.95
Slate Roof	+ 2.25
Upgrade Ceilings to Textured Finish	+ .50
Air Conditioning, in Heating Ductwork	Base System
Heating Systems, Hot Water	+ 2.09
Heat Pump	+ 3.23
Electric Heat	– 1.76
Not Heated	– 4.04

Additional upgrades or components

Kitchen Cabinets & Countertops	Page 93
Bathroom Vanities	94
Fireplaces & Chimneys	94
Windows, Skylights & Dormers	94
Appliances	95
Breezeways & Porches	95
Finished Attic	95
Garages	96
Site Improvements	96
Wings & Ells	92

Luxury 1-1/2 Story

Living Area - 2800 S.F.
Perimeter - 175 L.F.

#	Category	Description	Mat.	Inst.	Total	% of Total (rounded)
1	Site Work	Site preparation for slab; 4' deep trench excavation for foundation wall.		.75	.75	0.6%
2	Foundation	Continuous reinforced concrete footing 12" deep x 24" wide; dampproofed and insulated concrete block foundation wall, 12" thick, 4' deep, grouted and reinforced; 4" concrete slab on 4" crushed stone base and polyethylene vapor barrier, trowel finish.	5.17	5.95	11.12	8.5%
3	Framing	Exterior walls - 2" x 6" wood studs, 16" O.C.; 5/8" plywood sheathing; 2" x 10" rafters 16" O.C. with 5/8" plywood sheathing, 8 in 12 pitch; 2" x 8" ceiling joists 16" O.C.; 2" x 12" floor joists 16" O.C. with 5/8" plywood subfloor; 5/8" plywood subfloor on 1" x 3" wood sleepers 16" O.C.	7.62	9.36	16.98	13.0%
4	Exterior Walls	Horizontal beveled wood siding; building paper; insulation; wood double hung windows; 3 solid core wood exterior doors; storms and screens.	11.05	4.15	15.20	11.6%
5	Roofing	Red cedar shingles; #15 felt building paper; aluminum gutters, downspouts and drip edge; copper flashings.	3.37	2.68	6.05	4.6%
6	Interiors	Walls and ceilings - 5/8" gypsum wallboard, skim coat plaster, painted with primer and 2 coats; hardwood baseboard and trim, sanded and finished; hardwood floor 70%, ceramic tile with underlayment 20%, vinyl tile with underlayment 10%; wood panel interior doors, primed and painted with 2 coats.	16.55	16.43	32.98	25.2%
7	Specialties	Luxury grade kitchen cabinets - 25 L.F. wall and base with solid surface counter top and kitchen sink; 6 L.F. bathroom vanity; 75 gallon electric water heater; medicine cabinet.	6.86	1.45	8.31	6.4%
8	Mechanical	Gas fired warm air heat/air conditioning; one full bath including: bathtub, corner shower; built in lavatory and water closet; one 1/2 bath including: built in lavatory and water closet.	6.80	3.73	10.53	8.1%
9	Electrical	200 Amp. service; Romex wiring; fluorescent and incandescent lighting fixtures; intercom, switches, receptacles.	1.34	2.14	3.48	2.7%
10	Overhead	Contractor's overhead and profit and architect's fees.	14.09	11.21	25.30	19.4%
	Total		72.85	57.85	**130.70**	

RESIDENTIAL | Luxury | 2 Story

- **Unique residence built from an architect's plan**
- **Single family — 1 full bath, 1 half bath, 1 kitchen**
- **No basement**
- **Cedar shakes on roof**
- **Forced hot air heat/air conditioning**
- **Gypsum wallboard interior finishes**
- **Many special features**
- **Extraordinary materials and workmanship**

Note: The illustration shown may contain some optional components (for example: garages and/or fireplaces) whose costs are shown in the modifications, adjustments, & alternatives below or at the end of the square foot section.

Base cost per square foot of living area

Exterior Wall	Living Area										
	1200	1400	1600	1800	2000	2400	2800	3200	3600	4000	4400
Wood Siding - Wood Frame	184.50	173.10	165.00	157.80	150.30	139.45	130.35	124.35	120.90	116.95	113.95
Brick Veneer - Wood Frame	195.40	183.25	174.70	166.95	159.10	147.40	137.60	131.15	127.40	123.15	119.80
Solid Brick	213.85	200.55	191.20	182.45	174.15	160.90	149.85	142.65	138.45	133.55	129.90
Solid Stone	222.50	208.50	198.85	189.65	181.05	167.15	155.60	148.05	143.60	138.40	134.60
Finished Basement, Add	33.10	36.00	35.10	34.00	33.30	31.80	30.55	29.75	29.30	28.60	28.25
Unfinished Basement, Add	14.75	14.10	13.65	13.15	12.80	12.05	11.40	11.00	10.80	10.45	10.25

Modifications

Add to the total cost

Upgrade Kitchen Cabinets	$ + 1735
Solid Surface Countertops (Included)	
Full Bath - including plumbing, wall and floor finishes	+ 10,554
Half Bath - including plumbing, wall and floor finishes	+ 6157
Two Car Attached Garage	+ 31,466
Two Car Detached Garage	+ 35,593
Fireplace & Chimney	+ 10,256

Adjustments

For multi family - add to total cost

Additional Kitchen	$ + 21,730
Additional Full Bath & Half Bath	+ 16,711
Additional Entry & Exit	+ 2306
Separate Heating & Air Conditioning	+ 7328
Separate Electric	+ 2036

For Townhouse/Rowhouse - Multiply cost per square foot by

Inner Unit	.86
End Unit	.93

Alternatives

Add to or deduct from the cost per square foot of living area

Heavyweight Asphalt Shingles	– 1.25
Clay Tile Roof	+ 2.05
Slate Roof	+ 1.55
Upgrade Ceilings to Textured Finish	+ .50
Air Conditioning, in Heating Ductwork	Base System
Heating Systems, Hot Water	+ 2.02
Heat Pump	+ 3.40
Electric Heat	– 1.59
Not Heated	– 3.82

Additional upgrades or components

Kitchen Cabinets & Countertops	Page 93
Bathroom Vanities	94
Fireplaces & Chimneys	94
Windows, Skylights & Dormers	94
Appliances	95
Breezeways & Porches	95
Finished Attic	95
Garages	96
Site Improvements	96
Wings & Ells	92

Luxury 2 Story

Living Area - 3200 S.F.
Perimeter - 163 L.F.

#	Category	Description	Cost Per Square Foot Of Living Area			% of Total (rounded)
			Mat.	Inst.	Total	
1	Site Work	Site preparation for slab; 4' deep trench excavation for foundation wall.		.66	.66	0.5%
2	Foundation	Continuous reinforced concrete footing 12" deep x 24" wide; dampproofed and insulated concrete block foundation wall, 12" thick, 4' deep, grouted and reinforced; 4" concrete slab on 4" crushed stone base and polyethylene vapor barrier, trowel finish.	4.18	4.83	9.01	7.2%
3	Framing	Exterior walls - 2" x 6" wood studs, 16" O.C.; 5/8" plywood sheathing; 2" x 10" rafters 16" O.C. with 5/8" plywood sheathing, 6 in 12 pitch; 2" x 8" ceiling joists 16" O.C.; 2" x 12" floor joists 16" O.C. with 5/8" plywood subfloor; 5/8" plywood subfloor on 1" x 3" wood sleepers 16" O.C.	7.72	9.27	16.99	13.7%
4	Exterior Walls	Horizontal beveled wood siding, building paper; insulation; wood double hung windows; 3 solid core wood exterior doors; storms and screens.	11.49	4.32	15.81	12.7%
5	Roofing	Red cedar shingles; #15 felt building paper; aluminum gutters, downspouts and drip edge; copper flashings.	2.69	2.15	4.84	3.9%
6	Interiors	Walls and ceilings - 5/8" gypsum wallboard, skim coat plaster, painted with primer and 2 coats; hardwood baseboard and trim, sanded and finished; hardwood floor 70%, ceramic tile with underlayment 20%, vinyl tile with underlayment 10%; wood panel interior doors, primed and painted with 2 coats.	16.31	16.32	32.63	26.2%
7	Specialties	Luxury grade kitchen cabinets - 25 L.F. wall and base with solid surface counter top and kitchen sink; 6 L.F. bathroom vanity; 75 gallon electric water heater; medicine cabinet.	6.00	1.26	7.26	5.8%
8	Mechanical	Gas fired warm air heat/air conditioning; one full bath including: bathtub, corner shower; built in lavatory and water closet; one 1/2 bath including: built in lavatory and water closet.	6.13	3.59	9.72	7.8%
9	Electrical	200 Amp. service; Romex wiring; fluorescent and incandescent lighting fixtures; intercom, switches, receptacles.	1.30	2.09	3.39	2.7%
10	Overhead	Contractor's overhead and profit and architect's fee.	13.38	10.66	24.04	19.3%
	Total		69.20	55.15	**124.35**	

For customer support on your Residential Cost Data, call 877.713.2617.

RESIDENTIAL | Luxury | 2-1/2 Story

- Unique residence built from an architect's plan
- Single family — 1 full bath, 1 half bath, 1 kitchen
- No basement
- Cedar shakes on roof
- Forced hot air heat/air conditioning
- Gypsum wallboard interior finishes
- Many special features
- Extraordinary materials and workmanship

©Larry W. Garnett & Associates, Inc

Note: The illustration shown may contain some optional components (for example: garages and/or fireplaces) whose costs are shown in the modifications, adjustments, & alternatives below or at the end of the square foot section.

Base cost per square foot of living area

Exterior Wall	\multicolumn{11}{c	}{Living Area}
	1500	1800	2100	2500	3000	3500	4000	4500	5000	5500	6000
Wood Siding - Wood Frame	179.05	161.05	150.35	142.40	131.65	123.95	116.55	112.85	109.65	106.45	102.95
Brick Veneer - Wood Frame	190.35	171.50	159.65	151.20	139.60	131.20	123.25	119.20	115.70	112.25	108.45
Solid Brick	209.55	189.20	175.50	166.25	153.20	143.60	134.70	130.05	126.00	122.05	117.90
Solid Stone	218.50	197.40	182.90	173.15	159.55	149.30	140.00	135.05	130.85	126.65	122.25
Finished Basement, Add	26.35	28.40	26.85	26.05	24.85	23.80	23.05	22.60	22.10	21.75	21.40
Unfinished Basement, Add	11.90	11.25	10.45	10.05	9.45	9.00	8.60	8.35	8.15	7.95	7.75

Modifications
Add to the total cost

Upgrade Kitchen Cabinets	$ + 1735
Solid Surface Countertops (Included)	
Full Bath - including plumbing, wall and floor finishes	+ 10,554
Half Bath - including plumbing, wall and floor finishes	+ 6157
Two Car Attached Garage	+ 31,466
Two Car Detached Garage	+ 35,593
Fireplace & Chimney	+ 11,225

Adjustments
For multi family - add to total cost

Additional Kitchen	$ + 21,730
Additional Full Bath & Half Bath	+ 16,711
Additional Entry & Exit	+ 2306
Separate Heating & Air Conditioning	+ 7328
Separate Electric	+ 2036

For Townhouse/Rowhouse - Multiply cost per square foot by

Inner Unit	.86
End Unit	.93

Alternatives
Add to or deduct from the cost per square foot of living area

Heavyweight Asphalt Shingles	– 1.10
Clay Tile Roof	+ 1.80
Slate Roof	+ 1.35
Upgrade Ceilings to Textured Finish	+ .50
Air Conditioning, in Heating Ductwork	Base System
Heating Systems, Hot Water	+ 1.83
Heat Pump	+ 3.50
Electric Heat	– 3.12
Not Heated	– 3.82

Additional upgrades or components

Kitchen Cabinets & Countertops	Page 93
Bathroom Vanities	94
Fireplaces & Chimneys	94
Windows, Skylights & Dormers	94
Appliances	95
Breezeways & Porches	95
Finished Attic	95
Garages	96
Site Improvements	96
Wings & Ells	92

Luxury 2-1/2 Story

Living Area - 3000 S.F.
Perimeter - 148 L.F.

#	Category	Description	Cost Per Square Foot Of Living Area			% of Total
			Mat.	Inst.	Total	(rounded)
1	Site Work	Site preparation for slab; 4' deep trench excavation for foundation wall.		.70	.70	0.5%
2	Foundation	Continuous reinforced concrete footing 12" deep x 24" wide; dampproofed and insulated concrete block foundation wall, 12" thick, 4' deep, grouted and reinforced; 4" concrete slab on 4" crushed stone base and polyethylene vapor barrier, trowel finish.	3.69	4.39	8.08	6.1%
3	Framing	Exterior walls - 2" x 6" wood studs, 16" O.C.; 5/8" plywood sheathing; 2" x 10" rafters 16" O.C. with 5/8" plywood sheathing, 8 in 12 pitch; 2" x 8" ceiling joists 16" O.C.; 2" x 12" floor joists 16" O.C. with 5/8" plywood subfloor; 5/8" plywood subfloor on 1" x 3" wood sleepers 16" O.C.	8.00	9.59	17.59	13.4%
4	Exterior Walls	Horizontal beveled wood siding; building paper; insulation; wood double hung windows; 3 solid core wood exterior doors; storms and screens.	13.17	4.97	18.14	13.8%
5	Roofing	Red cedar shingles; #15 felt building paper; aluminum gutters, downspouts and drip edge; copper flashings.	2.07	1.65	3.72	2.8%
6	Interiors	Walls and ceilings - 5/8" gypsum wallboard, skim coat plaster, painted with primer and 2 coats; hardwood baseboard and trim, sanded and finished; hardwood floor 70%, ceramic tile with underlayment 20%, vinyl tile with underlayment 10%; wood panel interior doors, primed and painted with 2 coats.	18.44	18.16	36.60	27.8%
7	Specialties	Luxury grade kitchen cabinets - 25 L.F. wall and base with solid surface counter top and kitchen sink; 6 L.F. bathroom vanity; 75 gallon electric water heater; medicine cabinet.	6.41	1.35	7.76	5.9%
8	Mechanical	Gas fired warm air heat/air conditioning; one full bath including: bathtub, corner shower; built in lavatory and water closet; one 1/2 bath including: built in lavatory and water closet.	6.45	3.68	10.13	7.7%
9	Electrical	200 Amp. service; Romex wiring; fluorescent and incandescent lighting fixtures; intercom, switches, receptacles.	1.32	2.11	3.43	2.6%
10	Overhead	Contractor's overhead and profit and architect's fee.	14.30	11.20	25.50	19.4%
	Total		73.85	57.80	**131.65**	

RESIDENTIAL | Luxury | 3 Story

- Unique residence built from an architect's plan
- Single family — 1 full bath, 1 half bath, 1 kitchen
- No basement
- Cedar shakes on roof
- Forced hot air heat/air conditioning
- Gypsum wallboard interior finishes
- Many special features
- Extraordinary materials and workmanship

Note: The illustration shown may contain some optional components (for example: garages and/or fireplaces) whose costs are shown in the modifications, adjustments, & alternatives below or at the end of the square foot section.

Base cost per square foot of living area

Exterior Wall	Living Area										
	1500	1800	2100	2500	3000	3500	4000	4500	5000	5500	6000
Wood Siding - Wood Frame	177.70	160.00	151.10	143.75	132.50	126.90	120.15	113.15	110.60	107.85	105.00
Brick Veneer - Wood Frame	189.45	170.90	161.25	153.45	141.30	135.20	127.70	120.20	117.35	114.35	111.15
Solid Brick	209.45	189.40	178.50	169.95	156.30	149.35	140.65	132.05	128.85	125.40	121.60
Solid Stone	218.75	198.00	186.55	177.60	163.25	155.90	146.65	137.60	134.20	130.55	126.40
Finished Basement, Add	23.15	24.95	24.00	23.30	22.15	21.55	20.75	20.10	19.75	19.45	19.00
Unfinished Basement, Add	10.45	9.85	9.40	9.05	8.55	8.25	7.80	7.45	7.35	7.15	6.95

Modifications

Add to the total cost

Upgrade Kitchen Cabinets	$ + 1735
Solid Surface Countertops (Included)	
Full Bath - including plumbing, wall and floor finishes	+ 10,554
Half Bath - including plumbing, wall and floor finishes	+ 6157
Two Car Attached Garage	+ 31,466
Two Car Detached Garage	+ 35,593
Fireplace & Chimney	+ 11,225

Adjustments

For multi family - add to total cost

Additional Kitchen	$ + 21,730
Additional Full Bath & Half Bath	+ 16,711
Additional Entry & Exit	+ 2306
Separate Heating & Air Conditioning	+ 7328
Separate Electric	+ 2036

For Townhouse/Rowhouse - Multiply cost per square foot by

Inner Unit	.84
End Unit	.92

Alternatives

Add to or deduct from the cost per square foot of living area

Heavyweight Asphalt Shingles	– .80
Clay Tile Roof	+ 1.35
Slate Roof	+ 1.05
Upgrade Ceilings to Textured Finish	+ .50
Air Conditioning, in Heating Ductwork	Base System
Heating Systems, Hot Water	+ 1.83
Heat Pump	+ 3.50
Electric Heat	– 3.12
Not Heated	– 3.71

Additional upgrades or components

Kitchen Cabinets & Countertops	Page 93
Bathroom Vanities	94
Fireplaces & Chimneys	94
Windows, Skylights & Dormers	94
Appliances	95
Breezeways & Porches	95
Finished Attic	95
Garages	96
Site Improvements	96
Wings & Ells	92

Luxury 3 Story

Living Area - 3000 S.F.
Perimeter - 135 L.F.

			Cost Per Square Foot Of Living Area			% of Total
			Mat.	Inst.	Total	(rounded)
1	Site Work	Site preparation for slab; 4' deep trench excavation for foundation wall.		.70	.70	0.5%
2	Foundation	Continuous reinforced concrete footing 12" deep x 24" wide; dampproofed and insulated concrete block foundation wall, 12" thick, 4' deep, grouted and reinforced; 4" concrete slab on 4" crushed stone base and polyethylene vapor barrier, trowel finish.	3.32	3.98	7.30	5.5%
3	Framing	Exterior walls - 2" x 6" wood studs, 16" O.C.; 5/8" plywood sheathing; 2" x 10" rafters 16" O.C. with 5/8" plywood sheathing, 6 in 12 pitch; 2" x 8" ceiling joists 16" O.C.; 2" x 12" floor joists 16" O.C. with 5/8" plywood subfloor; 5/8" plywood subfloor on 1" x 3" wood sleepers 16" O.C.	8.14	9.71	17.85	13.5%
4	Exterior Walls	Horizontal beveled wood siding; building paper; insulation; wood double hung windows; 3 solid core wood exterior doors; storms and screens.	14.28	5.39	19.67	14.8%
5	Roofing	Red cedar shingles; #15 felt building paper; aluminum gutters, downspouts and drip edge; copper flashings.	1.80	1.42	3.22	2.4%
6	Interiors	Walls and ceilings - 5/8" gypsum wallboard, skim coat plaster, painted with primer and 2 coats; hardwood baseboard and trim, sanded and finished; hardwood floor 70%, ceramic tile with underlayment 20%, vinyl tile with underlayment 10%; wood panel interior doors, primed and painted with 2 coats.	18.50	18.30	36.80	27.8%
7	Specialties	Luxury grade kitchen cabinets - 25 L.F. wall and base with solid surface counter top and kitchen sink; 6 L.F. bathroom vanity; 75 gallon electric water heater; medicine cabinet.	6.41	1.35	7.76	5.9%
8	Mechanical	Gas fired warm air heat/air conditioning; one full bath including: bathtub, corner shower; built in lavatory and water closet; one 1/2 bath including: built in lavatory and water closet.	6.45	3.68	10.13	7.6%
9	Electrical	200 Amp. service; Romex wiring; fluorescent and incandescent lighting fixtures; intercom, switches, receptacles.	1.32	2.11	3.43	2.6%
10	Overhead	Contractor's overhead and profit and architect's fees.	14.43	11.21	25.64	19.4%
		Total	74.65	57.85	**132.50**	

RESIDENTIAL — Luxury — Bi-Level

- Unique residence built from an architect's plan
- Single family — 1 full bath, 1 half bath, 1 kitchen
- No basement
- Cedar shakes on roof
- Forced hot air heat/air conditioning
- Gypsum wallboard interior finishes
- Many special features
- Extraordinary materials and workmanship

Note: The illustration shown may contain some optional components (for example: garages and/or fireplaces) whose costs are shown in the modifications, adjustments, & alternatives below or at the end of the square foot section.

Base cost per square foot of living area

Exterior Wall	Living Area										
	1200	1400	1600	1800	2000	2400	2800	3200	3600	4000	4400
Wood Siding - Wood Frame	174.95	164.20	156.45	149.75	142.65	132.50	124.05	118.45	115.10	111.60	108.70
Brick Veneer - Wood Frame	183.10	171.80	163.75	156.65	149.20	138.45	129.45	123.55	120.00	116.25	113.15
Solid Brick	197.00	184.85	176.15	168.35	160.45	148.60	138.70	132.20	128.30	124.05	120.75
Solid Stone	203.45	190.80	181.90	173.75	165.60	153.25	143.00	136.20	132.20	127.70	124.30
Finished Basement, Add	33.10	36.00	35.10	34.00	33.30	31.80	30.55	29.75	29.30	28.60	28.25
Unfinished Basement, Add	14.75	14.10	13.65	13.15	12.80	12.05	11.40	11.00	10.80	10.45	10.25

Modifications
Add to the total cost

Upgrade Kitchen Cabinets	$ + 1735
Solid Surface Countertops (Included)	
Full Bath - including plumbing, wall and floor finishes	+ 10,554
Half Bath - including plumbing, wall and floor finishes	+ 6157
Two Car Attached Garage	+ 31,466
Two Car Detached Garage	+ 35,593
Fireplace & Chimney	+ 9357

Adjustments
For multi family - add to total cost

Additional Kitchen	$ + 21,730
Additional Full Bath & Half Bath	+ 16,711
Additional Entry & Exit	+ 2306
Separate Heating & Air Conditioning	+ 7328
Separate Electric	+ 2036

For Townhouse/Rowhouse - Multiply cost per square foot by

Inner Unit	.89
End Unit	.94

Alternatives
Add to or deduct from the cost per square foot of living area

Heavyweight Asphalt Shingles	– 1.25
Clay Tile Roof	+ 2.05
Slate Roof	+ 1.55
Upgrade Ceilings to Textured Finish	+ .50
Air Conditioning, in Heating Ductwork	Base System
Heating Systems, Hot Water	+ 2.02
Heat Pump	+ 3.40
Electric Heat	– 1.59
Not Heated	– 3.82

Additional upgrades or components

Kitchen Cabinets & Countertops	Page 93
Bathroom Vanities	94
Fireplaces & Chimneys	94
Windows, Skylights & Dormers	94
Appliances	95
Breezeways & Porches	95
Finished Attic	95
Garages	96
Site Improvements	96
Wings & Ells	92

Luxury Bi-Level

Living Area - 3200 S.F.
Perimeter - 163 L.F.

			Cost Per Square Foot Of Living Area			% of Total
			Mat.	Inst.	Total	(rounded)
1	Site Work	Excavation for lower level, 4' deep. Site preparation for slab.		.66	.66	0.6%
2	Foundation	Continuous reinforced concrete footing 12" deep x 24" wide; dampproofed and insulated concrete block foundation wall, 12" thick, 4' deep, grouted and reinforced; 4" concrete slab on 4" crushed stone base and polyethylene vapor barrier, trowel finish.	4.18	4.83	9.01	7.6%
3	Framing	Exterior walls - 2" x 6" wood studs, 16" O.C.; 5/8" plywood sheathing; 2" x 10" rafters 16" O.C. with 5/8" plywood sheathing, 6 in 12 pitch; 2" x 8" ceiling joists 16" O.C.; 2" x 12" floor joists 16" O.C. with 5/8" plywood subfloor; 5/8" plywood subfloor on 1" x 3" wood sleepers 16" O.C.	7.36	8.82	16.18	13.7%
4	Exterior Walls	Horizontal beveled wood siding; building paper; insulation; wood double hung windows; 3 solid core wood exterior doors; storms and screens.	9.10	3.37	12.47	10.5%
5	Roofing	Red cedar shingles: #15 felt building paper; aluminum gutters, downspouts and drip edge; copper flashings.	2.69	2.15	4.84	4.1%
6	Interiors	Walls and ceilings - 5/8" gypsum wallboard, skim coat plaster, painted with primer and 2 coats; hardwood baseboard and trim, sanded and finished; hardwood floor 70%, ceramic tile with underlayment 20%; vinyl tile with underlayment 10%; wood panel interior doors, primed and painted with 2 coats.	16.07	15.94	32.01	27.0%
7	Specialties	Luxury grade kitchen cabinets - 25 L.F. wall and base with solid surface counter top and kitchen sink; 6 L.F. bathroom vanity; 75 gallon electric water heater; medicine cabinet.	6.00	1.26	7.26	6.1%
8	Mechanical	Gas fired warm air heat/air conditioning; one full bath including: bathtub, corner shower; built in lavatory and water closet; one 1/2 bath including: built in lavatory and water closet.	6.13	3.59	9.72	8.2%
9	Electrical	200 Amp. service; Romex wiring; fluorescent and incandescent lighting fixtures; intercom, switches, receptacles.	1.30	2.09	3.39	2.9%
10	Overhead	Contractor's overhead and profit and architect's fees.	12.67	10.24	22.91	19.3%
		Total	65.50	52.95	**118.45**	

RESIDENTIAL | Luxury | Tri-Level

- Unique residence built from an architect's plan
- Single family — 1 full bath, 1 half bath, 1 kitchen
- No basement
- Cedar shakes on roof
- Forced hot air heat/air conditioning
- Gypsum wallboard interior finishes
- Many special features
- Extraordinary materials and workmanship

Note: The illustration shown may contain some optional components (for example: garages and/or fireplaces) whose costs are shown in the modifications, adjustments, & alternatives below or at the end of the square foot section.

©Home Planners, Inc.

Base cost per square foot of living area

Exterior Wall	Living Area										
	1500	1800	2100	2400	2800	3200	3600	4000	4500	5000	5500
Wood Siding - Wood Frame	164.45	151.95	141.65	134.65	129.50	123.60	117.50	114.75	108.95	105.80	102.25
Brick Veneer - Wood Frame	171.85	158.60	147.70	140.35	134.90	128.65	122.20	119.25	113.15	109.70	105.95
Solid Brick	184.35	169.85	157.90	149.95	144.15	137.20	130.10	126.95	120.25	116.45	112.30
Solid Stone	190.10	175.00	162.65	154.40	148.45	141.10	133.80	130.50	123.55	119.55	115.20
Finished Basement, Add*	39.20	42.25	40.60	39.50	38.80	37.75	36.80	36.40	35.45	34.85	34.25
Unfinished Basement, Add*	17.10	16.05	15.20	14.65	14.30	13.80	13.35	13.10	12.65	12.30	12.05

*Basement under middle level only.

Modifications

Add to the total cost

Upgrade Kitchen Cabinets	$ + 1735
Solid Surface Countertops (Included)	
Full Bath - including plumbing, wall and floor finishes	+ 10,554
Half Bath - including plumbing, wall and floor finishes	+ 6157
Two Car Attached Garage	+ 31,466
Two Car Detached Garage	+ 35,593
Fireplace & Chimney	+ 9357

Adjustments

For multi family - add to total cost

Additional Kitchen	$ + 21,730
Additional Full Bath & Half Bath	+ 16,711
Additional Entry & Exit	+ 2306
Separate Heating & Air Conditioning	+ 7328
Separate Electric	+ 2036

For Townhouse/Rowhouse - Multiply cost per square foot by

Inner Unit	.86
End Unit	.93

Alternatives

Add to or deduct from the cost per square foot of living area

Heavyweight Asphalt Shingles	– 1.80
Clay Tile Roof	+ 2.95
Slate Roof	+ 2.25
Upgrade Ceilings to Textured Finish	+ .50
Air Conditioning, in Heating Ductwork	Base System
Heating Systems, Hot Water	+ 1.96
Heat Pump	+ 3.53
Electric Heat	– 1.40
Not Heated	– 3.71

Additional upgrades or components

Kitchen Cabinets & Countertops	Page 93
Bathroom Vanities	94
Fireplaces & Chimneys	94
Windows, Skylights & Dormers	94
Appliances	95
Breezeways & Porches	95
Finished Attic	95
Garages	96
Site Improvements	96
Wings & Ells	92

Important: See the Reference Section for Location Factors (to adjust for your city) and Estimating Forms.

Luxury Tri-Level

Living Area - 3600 S.F.
Perimeter - 207 L.F.

			Cost Per Square Foot Of Living Area			% of Total
			Mat.	Inst.	Total	(rounded)
1	Site Work	Site preparation for slab; 4' deep trench excavation for foundation wall, excavation for lower level, 4' deep.		.58	.58	0.5%
2	Foundation	Continuous reinforced concrete footing 12" deep x 24" wide; dampproofed and insulated concrete block foundation wall, 12" thick, 4' deep, grouted and reinforced; 4" concrete slab on 4" crushed stone base and polyethylene vapor barrier, trowel finish.	4.98	5.57	10.55	9.0%
3	Framing	Exterior walls - 2" x 6" wood studs, 16" O.C.; 5/8" plywood sheathing; 2" x 10" rafters 16" O.C. with 5/8" plywood sheathing, 6 in 12 pitch; 2" x 8" ceiling joists 16" O.C.; 2" x 12" floor joists 16" O.C. with 5/8" plywood subfloor; 5/8" plywood subfloor on 1" x 3" wood sleepers 16" O.C.	7.23	8.85	16.08	13.7%
4	Exterior Walls	Horizontal beveled wood siding; building paper; insulation; wood double hung windows; 3 solid core wood exterior doors; storms and screens.	8.96	3.30	12.26	10.4%
5	Roofing	Red cedar shingles; #15 felt building paper; aluminum gutters, downspouts and drip edge; copper flashings.	3.59	2.86	6.45	5.5%
6	Interiors	Walls and ceilings - 5/8" gypsum wallboard, skim coat plaster, painted with primer and 2 coats; hardwood baseboard and trim, sanded and finished; hardwood floor 70%, ceramic tile with underlayment 20%, vinyl tile with underlayment 10%; wood panel interior doors, primed and painted with 2 coats.	15.09	14.91	30.00	25.5%
7	Specialties	Luxury grade kitchen cabinets - 25 L.F. wall and base with solid surface counter top and kitchen sink; 6 L.F. bathroom vanity; 75 gallon electric water heater; medicine cabinet.	5.34	1.11	6.45	5.5%
8	Mechanical	Gas fired warm air heat/air conditioning; one full bath including: bathtub, corner shower; built in lavatory and water closet; one 1/2 bath including: built in lavatory and water closet.	5.60	3.50	9.10	7.7%
9	Electrical	200 Amp. service; Romex wiring; fluorescent and incandescent lighting fixtures; intercom, switches, receptacles.	1.26	2.04	3.30	2.8%
10	Overhead	Contractor's overhead and profit and architect's fees.	12.50	10.23	22.73	19.3%
		Total	64.55	52.95	**117.50**	

RESIDENTIAL | Luxury | Wings & Ells

1 Story — Base cost per square foot of living area

Exterior Wall	Living Area							
	50	100	200	300	400	500	600	700
Wood Siding - Wood Frame	261.15	202.65	177.15	149.80	141.50	136.45	133.20	134.10
Brick Veneer - Wood Frame	288.55	222.20	193.50	160.75	151.25	145.60	141.90	142.45
Solid Brick	335.15	255.50	221.20	179.25	167.90	161.15	156.65	156.75
Solid Stone	356.80	270.95	234.10	187.85	175.70	168.35	163.50	163.35
Finished Basement, Add	103.45	94.30	84.80	69.05	65.95	64.00	62.80	61.85
Unfinished Basement, Add	53.05	41.00	35.95	27.60	25.90	24.90	24.25	23.75

1-1/2 Story — Base cost per square foot of living area

Exterior Wall	Living Area							
	100	200	300	400	500	600	700	800
Wood Siding - Wood Frame	206.75	169.30	146.40	132.70	125.80	122.50	118.30	117.10
Brick Veneer - Wood Frame	231.20	188.85	162.75	145.45	137.50	133.60	128.75	127.50
Solid Brick	272.80	222.15	190.45	167.05	157.50	152.45	146.65	145.10
Solid Stone	292.20	237.60	203.35	177.10	166.75	161.20	154.90	153.35
Finished Basement, Add	68.40	67.80	61.55	54.65	52.75	51.50	50.35	50.20
Unfinished Basement, Add	33.80	28.75	25.40	21.75	20.75	20.05	19.45	19.35

2 Story — Base cost per square foot of living area

Exterior Wall	Living Area							
	100	200	400	600	800	1000	1200	1400
Wood Siding - Wood Frame	202.90	153.70	132.05	111.10	104.05	99.80	97.05	98.35
Brick Veneer - Wood Frame	230.30	173.25	148.35	122.00	113.85	108.95	105.75	106.70
Solid Brick	276.95	206.50	176.10	140.50	130.45	124.50	120.50	121.00
Solid Stone	298.55	221.95	189.00	149.10	138.25	131.75	127.35	127.60
Finished Basement, Add	51.75	47.25	42.50	34.60	33.05	32.10	31.45	31.05
Unfinished Basement, Add	26.55	20.50	18.00	13.75	12.95	12.45	12.10	11.85

Base costs do not include bathroom or kitchen facilities. Use Modifications/Adjustments/Alternatives on pages 93-96 where appropriate.

RESIDENTIAL

Modifications/Adjustments/Alternatives

Kitchen cabinets - Base units, hardwood *(Cost per Unit)*

	Economy	Average	Custom	Luxury
24" deep, 35" high,				
One top drawer,				
One door below				
12" wide	$ 248	$ 330	$ 439	$ 578
15" wide	255	340	452	595
18" wide	278	370	492	648
21" wide	285	380	505	665
24" wide	330	440	585	770
Four drawers				
12" wide	255	340	452	595
15" wide	259	345	459	604
18" wide	289	385	512	674
24" wide	315	420	559	735
Two top drawers,				
Two doors below				
27" wide	353	470	625	823
30" wide	383	510	678	893
33" wide	394	525	698	919
36" wide	409	545	725	954
42" wide	431	575	765	1006
48" wide	461	615	818	1076
Range or sink base				
(Cost per unit)				
Two doors below				
30" wide	319	425	565	744
33" wide	341	455	605	796
36" wide	360	480	638	840
42" wide	375	500	665	875
48" wide	394	525	698	919
Corner Base Cabinet				
(Cost per unit)				
36" wide	555	740	984	1295
Lazy Susan *(Cost per unit)*				
With revolving door	735	980	1303	1715

Kitchen cabinets - Wall cabinets, hardwood *(Cost per Unit)*

	Economy	Average	Custom	Luxury
12" deep, 2 doors				
12" high				
30" wide	$ 223	$ 297	$ 395	$ 520
36" wide	259	345	459	604
15" high				
30" wide	225	300	399	525
33" wide	278	370	492	648
36" wide	270	360	479	630
24" high				
30" wide	296	395	525	691
36" wide	323	430	572	753
42" wide	368	490	652	858
30" high, 1 door				
12" wide	209	278	370	487
15" wide	229	305	406	534
18" wide	251	335	446	586
24" wide	289	385	512	674
30" high, 2 doors				
27" wide	315	420	559	735
30" wide	330	440	585	770
36" wide	371	495	658	866
42" wide	405	540	718	945
48" wide	450	600	798	1050
Corner wall, 30" high				
24" wide	330	440	585	770
30" wide	349	465	618	814
36" wide	394	525	698	919
Broom closet				
84" high, 24" deep				
18" wide	604	805	1071	1409
Oven Cabinet				
84" high, 24" deep				
27" wide	900	1200	1596	2100

Kitchen countertops *(Cost per L.F.)*

	Economy	Average	Custom	Luxury
Solid Surface				
24" wide, no backsplash	$ 110	$ 147	$ 196	$ 257
with backsplash	130	173	230	303
Stock plastic laminate, 24" wide				
with backsplash	25	33	45	59
Custom plastic laminate, no splash				
7/8" thick, alum. molding	36	47	63	83
1-1/4" thick, no splash	41	55	73	96
Marble				
1/2" - 3/4" thick w/splash	54	71	95	125
Maple, laminated				
1-1/2" thick w/splash	86	114	152	200
Stainless steel				
(per S.F.)	140	186	247	326
Cutting blocks, recessed				
16" x 20" x 1" (each)	117	156	207	273

RESIDENTIAL — Modifications/Adjustments/Alternatives

Vanity bases *(Cost per Unit)*

2 door, 30" high, 21" deep	Economy	Average	Custom	Luxury
24" wide	$ 289	$ 385	$ 512	$ 674
30" wide	345	460	612	805
36" wide	345	460	612	805
48" wide	446	595	791	1041

Solid surface vanity tops *(Cost Each)*

Center bowl	Economy	Average	Custom	Luxury
22" x 25"	$ 425	$ 459	$ 496	$ 536
22" x 31"	490	529	571	617
22" x 37"	565	610	659	712
22" x 49"	700	756	816	881

Fireplaces & Chimneys *(Cost per Unit)*

	1-1/2 Story	2 Story	3 Story
Economy (prefab metal)			
Exterior chimney & 1 fireplace	$ 5873	$ 6489	$ 7118
Interior chimney & 1 fireplace	5628	6257	6546
Average (masonry)			
Exterior chimney & 1 fireplace	6191	6841	7504
Interior chimney & 1 fireplace	5933	6596	6901
For more than 1 flue, add	421	717	1202
For more than 1 fireplace, add	4154	4154	4154
Custom (masonry)			
Exterior chimney & 1 fireplace	6493	7329	8276
Interior chimney & 1 fireplace	6088	6891	7427
For more than 1 flue, add	509	881	1475
For more than 1 fireplace, add	4664	4664	4664
Luxury (masonry)			
Exterior chimney & 1 fireplace	9357	10257	11224
Interior chimney & 1 fireplace	8927	9760	10331
For more than 1 flue, add	772	1289	1800
For more than 1 fireplace, add	7375	7375	7375

Windows and Skylights *(Cost Each)*

	Economy	Average	Custom	Luxury
Fixed Picture Windows				
3'-6" x 4'-0"	$ 358	$ 477	$ 635	$ 836
4'-0" x 6'-0"	620	827	1100	1447
5'-0" x 6'-0"	705	939	1250	1645
6'-0" x 6'-0"	705	939	1250	1645
Bay/Bow Windows				
8'-0" x 5'-0"	874	1165	1550	2039
10'-0" x 5'-0"	930	1240	1650	2171
10'-0" x 6'-0"	1466	1954	2600	3421
12'-0" x 6'-0"	1903	2537	3375	4441
Palladian Windows				
3'-2" x 6'-4"		1484	1975	2599
4'-0" x 6'-0"		1729	2300	3026
5'-5" x 6'-10"		2255	3000	3947
8'-0" x 6'-0"		2669	3550	4671
Skylights				
46" x 21-1/2"	595	915	1120	1232
46" x 28"	618	950	1160	1276
57" x 44"	679	1045	1175	1293

Dormers *(Cost/S.F. of plan area)*

	Economy	Average	Custom	Luxury
Framing and Roofing Only				
Gable dormer, 2" x 6" roof frame	$ 30	$ 34	$ 36	$ 60
2" x 8" roof frame	31	35	37	63
Shed dormer, 2" x 6" roof frame	19	22	25	38
2" x 8" roof frame	21	23	26	39
2" x 10" roof frame	22	25	27	40

RESIDENTIAL Modifications/Adjustments/Alternatives

Appliances (Cost per Unit)

	Economy	Average	Custom	Luxury
Range				
30" free standing, 1 oven	$ 550	$ 1475	$ 1938	$ 2400
2 oven	1400	2138	2507	2875
30" built-in, 1 oven	830	1390	1670	1950
2 oven	1675	1927	2053	2178
21" free standing				
1 oven	565	613	637	660
Counter Top Ranges				
4 burner standard	405	1003	1302	1600
As above with griddle	1450	2838	3532	4225
Microwave Oven	252	501	626	750
Compactor				
4 to 1 compaction	800	1038	1157	1275
Deep Freeze				
15 to 23 C.F.	735	880	953	1025
30 C.F.	855	1040	1133	1225
Dehumidifier, portable, auto.				
15 pint	287	330	352	373
30 pint	360	414	441	468
Washing Machine, automatic	700	1438	1807	2175
Water Heater				
Electric, glass lined				
30 gal.	595	730	798	865
80 gal.	1675	2075	2275	2475
Water Heater, Gas, glass lined				
30 gal.	1100	1338	1457	1575
50 gal.	1325	1638	1794	1950
Dishwasher, built-in				
2 cycles	420	608	702	795
4 or more cycles	575	710	1118	1525
Dryer, automatic	680	1578	2027	2475
Garage Door Opener	525	628	679	730
Garbage Disposal	159	232	269	305
Heater, Electric, built-in				
1250 watt ceiling type	238	294	322	350
1250 watt wall type	310	338	352	365
Wall type w/blower				
1500 watt	310	357	380	403
3000 watt	545	627	668	709
Hood For Range, 2 speed				
30" wide	190	620	835	1050
42" wide	297	1349	1875	2400
Humidifier, portable				
7 gal. per day	146	168	179	190
15 gal. per day	232	267	285	302
Ice Maker, automatic				
13 lb. per day	1050	1208	1287	1365
51 lb. per day	2000	2300	2450	2600
Refrigerator, no frost				
10-12 C.F.	530	568	587	605
14-16 C.F.	630	723	769	815
18-20 C.F.	770	1260	1505	1750
21-29 C.F.	1175	2550	3238	3925
Sump Pump, 1/3 H.P.	320	448	512	575

Breezeway (Cost per S.F.)

Class	Type	Area (S.F.)			
		50	100	150	200
Economy	Open	$ 45.40	$ 37.91	$ 33.41	$ 31.17
	Enclosed	165.23	117.54	96.91	83.44
Average	Open	53.61	43.76	38.49	35.86
	Enclosed	169.10	122.41	101.69	91.10
Custom	Open	64.56	53.10	46.86	43.74
	Enclosed	227.04	164.06	135.81	121.44
Luxury	Open	72.93	60.18	53.50	50.16
	Enclosed	268.54	191.24	156.86	139.42

Porches (Cost per S.F.)

Class	Type	Area (S.F.)				
		25	50	100	200	300
Economy	Open	$ 81.31	$ 61.19	$ 46.11	$ 35.43	$ 31.98
	Enclosed	173.76	138.22	90.89	68.90	60.25
Average	Open	85.36	65.72	48.30	36.47	32.63
	Enclosed	187.75	149.79	95.97	70.72	60.99
Custom	Open	137.43	100.35	77.02	57.89	56.56
	Enclosed	230.93	186.65	122.76	92.87	99.67
Luxury	Open	156.39	114.20	86.50	63.27	57.39
	Enclosed	272.13	246.00	158.04	115.21	95.51

Finished attic (Cost per S.F.)

Class	Area (S.F.)				
	400	500	600	800	1000
Economy	$ 20.57	$ 19.88	$ 19.05	$ 18.73	$ 18.04
Average	31.89	31.20	30.43	30.05	29.21
Custom	40.89	39.97	39.06	38.46	37.68
Luxury	52.48	51.20	50.01	48.82	48.01

Alarm system (Cost per System)

	Burglar Alarm	Smoke Detector
Economy	$ 425	$ 71
Average	485	84
Custom	835	199
Luxury	1200	258

Sauna, prefabricated
(Cost per unit, including heater and controls—7' high)

Size	Cost
6' x 4'	$ 5750
6' x 5'	6450
6' x 6'	6900
6' x 9'	8700
8' x 10'	11,200
8' x 12'	12,700
10' x 12'	14,700

RESIDENTIAL — Modifications/Adjustments/Alternatives

Garages *

(Costs include exterior wall systems comparable with the quality of the residence. Included in the cost is an allowance for one personnel door, manual overhead door(s) and electrical fixture.)

Class	Detached			Attached			Built-in		Basement	
	One Car	Two Car	Three Car	One Car	Two Car	Three Car	One Car	Two Car	One Car	Two Car
Economy										
Wood	$16,393	$25,160	$33,926	$12,758	$22,043	$30,810	$-1720	$-3439	$1798	$2508
Masonry	22,375	32,646	42,916	16,501	27,290	37,561	-2454	-4908		
Average										
Wood	17,994	27,163	36,332	13,759	23,447	32,616	-1916	-3832	2034	2980
Masonry	22,416	32,697	42,978	16,527	27,326	37,607	-2459	-3896		
Custom										
Wood	20,221	31,142	42,063	15,749	27,346	38,266	-3109	-2889	2930	4773
Masonry	24,500	36,496	48,493	18,426	31,099	43,095	-3635	-3940		
Luxury										
Wood	22,777	35,593	48,409	17,974	31,466	44,282	-3218	-3106	3909	6280
Masonry	29,636	44,176	58,717	22,266	37,483	52,023	-4060	-4791		

*See the Introduction to this section for definitions of garage types.

Swimming pools (Cost per S.F.)

Residential	
In-ground	$ 34.00 - 83.50
Deck equipment	1.30
Paint pool, preparation & 3 coats (epoxy)	5.31
Rubber base paint	4.56
Pool Cover	1.79
Swimming Pool Heaters	
(not including wiring, external piping, base or pad)	
Gas	
155 MBH	$ 3100.00
190 MBH	3700.00
500 MBH	13,200.
Electric	
15 KW 7200 gallon pool	2825.00
24 KW 9600 gallon pool	3275.00
54 KW 24,000 gallon pool	5100.00

Wood and coal stoves

Wood Only	
Free Standing (minimum)	$ 1975
Fireplace Insert (minimum)	2087
Coal Only	
Free Standing	$ 2359
Fireplace Insert	2582
Wood and Coal	
Free Standing	$ 4851
Fireplace Insert	4972

Sidewalks (Cost per S.F.)

Concrete, 3000 psi with wire mesh	4" thick	$ 3.78
	5" thick	4.58
	6" thick	5.13
Precast concrete patio blocks (natural)	2" thick	6.55
Precast concrete patio blocks (colors)	2" thick	6.95
Flagstone, bluestone	1" thick	16.75
Flagstone, bluestone	1-1/2" thick	22.10
Slate (natural, irregular)	3/4" thick	18.65
Slate (random rectangular)	1/2" thick	29.50
Seeding		
Fine grading & seeding includes lime, fertilizer & seed per S.Y.		2.99
Lawn Sprinkler System	per S.F.	1.10

Fencing (Cost per L.F.)

Chain Link, 4' high, galvanized	$ 16.10
Gate, 4' high (each)	209.00
Cedar Picket, 3' high, 2 rail	14.70
Gate (each)	230.00
3 Rail, 4' high	18.00
Gate (each)	249.00
Cedar Stockade, 3 Rail, 6' high	17.95
Gate (each)	251.00
Board & Battens, 2 sides 6' high, pine	27.50
6' high, cedar	35.50
No. 1 Cedar, basketweave, 6' high	36.50
Gate, 6' high (each)	255.00

Carport (Cost per S.F.)

Economy	$ 9.41
Average	14.09
Custom	20.70
Luxury	24.20

Assemblies Section

Table of Contents

Table No.	Page
Site Work	
1-04 Footing Excavation	102
1-08 Foundation Excavation	104
1-12 Utility Trenching	106
1-16 Sidewalk	108
1-20 Driveway	110
1-24 Septic	112
1-60 Chain Link Fence	114
1-64 Wood Fence	115
Foundations	
2-04 Footing	118
2-08 Block Wall	120
2-12 Concrete Wall	122
2-16 Wood Wall Foundation	124
2-20 Floor Slab	126
Framing	
3-02 Floor (Wood)	130
3-04 Floor (Wood)	132
3-06 Floor (Wood)	134
3-08 Exterior Wall	136
3-12 Gable End Roof	138
3-16 Truss Roof	140
3-20 Hip Roof	142
3-24 Gambrel Roof	144
3-28 Mansard Roof	146
3-32 Shed/Flat Roof	148
3-40 Gable Dormer	150
3-44 Shed Dormer	152
3-48 Partition	154
Exterior Walls	
4-02 Masonry Block Wall	158
4-04 Brick/Stone Veneer	160
4-08 Wood Siding	162
4-12 Shingle Siding	164

Table No.	Page
4-16 Metal & Plastic Siding	166
4-20 Insulation	168
4-28 Double Hung Window	170
4-32 Casement Window	172
4-36 Awning Window	174
4-40 Sliding Window	176
4-44 Bow/Bay Window	178
4-48 Fixed Window	180
4-52 Entrance Door	182
4-53 Sliding Door	184
4-56 Residential Overhead Door	186
4-58 Aluminum Window	188
4-60 Storm Door & Window	190
4-64 Shutters/Blinds	191
5 Roofing	
5-04 Gable End Roofing	194
5-08 Hip Roof Roofing	196
5-12 Gambrel Roofing	198
5-16 Mansard Roofing	200
5-20 Shed Roofing	202
5-24 Gable Dormer Roofing	204
5-28 Shed Dormer Roofing	206
5-32 Skylight/Skywindow	208
5-34 Built-up Roofing	210
6 Interiors	
6-04 Drywall & Thincoat Wall	214
6-08 Drywall & Thincoat Ceiling	216
6-12 Plaster & Stucco Wall	218
6-16 Plaster & Stucco Ceiling	220
6-18 Suspended Ceiling	222
6-20 Interior Door	224
6-24 Closet Door	226
6-60 Carpet	228

Table No.	Page
6-64 Flooring	229
6-90 Stairways	230
7 Specialties	
7-08 Kitchen	234
7-12 Appliances	236
7-16 Bath Accessories	237
7-24 Masonry Fireplace	238
7-30 Prefabricated Fireplace	240
7-32 Greenhouse	242
7-36 Swimming Pool	243
7-40 Wood Deck	244
8 Mechanical	
8-04 Two Fixture Lavatory	248
8-12 Three Fixture Bathroom	250
8-16 Three Fixture Bathroom	252
8-20 Three Fixture Bathroom	254
8-24 Three Fixture Bathroom	256
8-28 Three Fixture Bathroom	258
8-32 Three Fixture Bathroom	260
8-36 Four Fixture Bathroom	262
8-40 Four Fixture Bathroom	264
8-44 Five Fixture Bathroom	266
8-60 Gas Fired Heating/Cooling	268
8-64 Oil Fired Heating/Cooling	270
8-68 Hot Water Heating	272
8-80 Rooftop Heating/Cooling	274
9 Electrical	
9-10 Electric Service	278
9-20 Electric Heating	279
9-30 Wiring Devices	280
9-40 Light Fixtures	281

How to Use the Assemblies Cost Tables

The following is a detailed explanation of a sample Assemblies Cost Table. Included are an illustration and accompanying system descriptions. Additionally, related systems and price sheets may be included. Next to each bold number below is the item being described with the appropriate component of the sample entry following in parenthesis. General contractors should add an additional markup to the figures shown in the Assemblies section. Note: Throughout this section, the words assembly and system are used interchangeably.

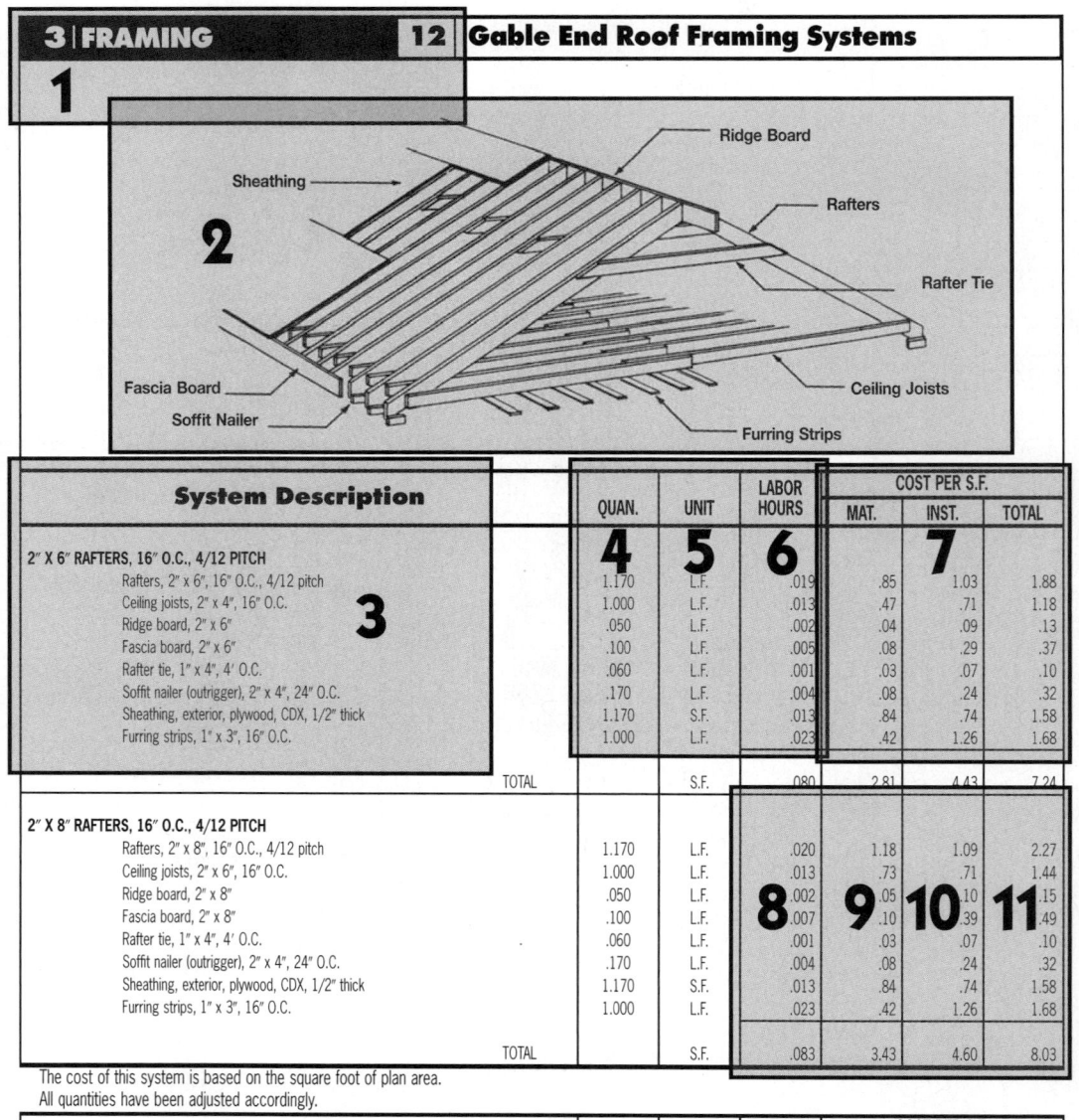

1 System Identification (3 Framing 12)
Each Assemblies section has been assigned a unique identification number, component category, system number and system description.

2 Illustration
At the top of most assembly pages is an illustration with individual components labeled. Elements involved in the total system function are shown.

3 System Description (2" x 6" Rafters, 16" O.C., 4/12 Pitch)
The components of a typical system are listed separately to show what has been included in the development of the total system price. Each page includes a brief outline of any special conditions to be used when pricing a system. Alternative components can be found on the opposite page. Simply insert any chosen new element into the chart to develop a custom system.

4 Quantities for Each Component
Each material in a system is shown with the quantity required for the system unit. For example, there are 1.170 L.F. of rafter per S.F. of plan area.

5 Unit of Measure for Each Component
The abbreviated designation indicates the unit of measure, as defined by industry standards, upon which the individual component has been priced. In this example, items are priced by the linear foot (L.F.) or the square foot (S.F.).

6 Labor Hours
This is the amount of time it takes to install the quantity of the individual component.

7 Unit of Measure (Cost per S.F.)
In the three right-hand columns, each cost figure is adjusted to agree with the unit of measure for the entire system. In this case, cost per S.F. is the common unit of measure.

8 Labor Hours (.083)
The labor hours column shows the amount of time necessary to install the system per the unit of measure. For example, it takes .083 labor hours to install one square foot (plan area) of this roof framing system.

9 Materials (3.43)
This column contains the material cost of each element. These cost figures include 10% for profit.

10 Installation (4.60)
This column contains labor and equipment costs. Labor rates include bare cost and the installing contractor's overhead and profit. On the average, the labor cost will be 69.5% over the bare labor cost. Equipment costs include 10% for profit.

11 Totals (8.03)
The figure in this column is the sum of the material and installation costs.

12 Work Sheet
Using the selective price sheet on the page opposite each system, it is possible to create estimates with alternative items for any number of systems.

Division 1 - Site Work

1 | SITEWORK   04 | Footing Excavation Systems

System Description	QUAN.	UNIT	LABOR HOURS	COST EACH MAT.	COST EACH INST.	COST EACH TOTAL
BUILDING, 24' X 38', 4' DEEP						
Cut & chip light trees to 6" diam.	.190	Acre	9.120		750.50	750.50
Excavator, hydraulic, crawler mtd., 1 C.Y. cap. = 100 C.Y./hr.	174.000	C.Y.	3.480		368.88	368.88
Backfill, dozer, 4" lifts, no compaction	87.000	C.Y.	.580		144.42	144.42
Rough grade, dozer, 30' from building	87.000	C.Y.	.580		144.42	144.42
TOTAL		Ea.	13.760		1,408.22	1,408.22
BUILDING, 26' X 46', 4' DEEP						
Cut & chip light trees to 6" diam.	.210	Acre	10.080		829.50	829.50
Excavator, hydraulic, crawler mtd., 1 C.Y. cap. = 100 C.Y./hr.	201.000	C.Y.	4.020		426.12	426.12
Backfill, dozer, 4" lifts, no compaction	100.000	C.Y.	.667		166	166
Rough grade, dozer, 30' from building	100.000	C.Y.	.667		166	166
TOTAL		Ea.	15.434		1,587.62	1,587.62
BUILDING, 26' X 60', 4' DEEP						
Cut & chip light trees to 6" diam.	.240	Acre	11.520		948	948
Excavator, hydraulic, crawler mtd., 1 C.Y. cap. = 100 C.Y./hr.	240.000	C.Y.	4.800		508.80	508.80
Backfill, dozer, 4" lifts, no compaction	120.000	C.Y.	.800		199.20	199.20
Rough grade, dozer, 30' from building	120.000	C.Y.	.800		199.20	199.20
TOTAL		Ea.	17.920		1,855.20	1,855.20
BUILDING, 30' X 66', 4' DEEP						
Cut & chip light trees to 6" diam.	.260	Acre	12.480		1,027	1,027
Excavator, hydraulic, crawler mtd., 1 C.Y. cap. = 100 C.Y./hr.	268.000	C.Y.	5.360		568.16	568.16
Backfill, dozer, 4" lifts, no compaction	134.000	C.Y.	.894		222.44	222.44
Rough grade, dozer, 30' from building	134.000	C.Y.	.894		222.44	222.44
TOTAL		Ea.	19.628		2,040.04	2,040.04

The costs in this system are on a cost each basis.
Quantities are based on 1'-0" clearance on each side of footing.

Description	QUAN.	UNIT	LABOR HOURS	COST EACH MAT.	COST EACH INST.	COST EACH TOTAL

Footing Excavation Price Sheet	QUAN.	UNIT	LABOR HOURS	COST EACH		
				MAT.	INST.	TOTAL
Clear and grub, medium brush, 30' from building, 24' x 38'	.190	Acre	9.120		750	750
26' x 46'	.210	Acre	10.080		830	830
26' x 60'	.240	Acre	11.520		950	950
30' x 66'	.260	Acre	12.480		1,025	1,025
Light trees, to 6" dia. cut & chip, 24' x 38'	.190	Acre	9.120		750	750
26' x 46'	.210	Acre	10.080		830	830
26' x 60'	.240	Acre	11.520		950	950
30' x 66'	.260	Acre	12.480		1,025	1,025
Medium trees, to 10" dia. cut & chip, 24' x 38'	.190	Acre	13.029		1,075	1,075
26' x 46'	.210	Acre	14.400		1,200	1,200
26' x 60'	.240	Acre	16.457		1,350	1,350
30' x 66'	.260	Acre	17.829		1,475	1,475
Excavation, footing, 24' x 38', 2' deep	68.000	C.Y.	.906		144	144
4' deep	174.000	C.Y.	2.319		370	370
8' deep	384.000	C.Y.	5.119		815	815
26' x 46', 2' deep	79.000	C.Y.	1.053		168	168
4' deep	201.000	C.Y.	2.679		425	425
8' deep	404.000	C.Y.	5.385		855	855
26' x 60', 2' deep	94.000	C.Y.	1.253		199	199
4' deep	240.000	C.Y.	3.199		510	510
8' deep	483.000	C.Y.	6.438		1,025	1,025
30' x 66', 2' deep	105.000	C.Y.	1.400		223	223
4' deep	268.000	C.Y.	3.572		570	570
8' deep	539.000	C.Y.	7.185		1,150	1,150
Backfill, 24' x 38', 2" lifts, no compaction	34.000	C.Y.	.227		56.50	56.50
Compaction, air tamped, add	34.000	C.Y.	2.267		470	470
4" lifts, no compaction	87.000	C.Y.	.580		144	144
Compaction, air tamped, add	87.000	C.Y.	5.800		1,200	1,200
8" lifts, no compaction	192.000	C.Y.	1.281		320	320
Compaction, air tamped, add	192.000	C.Y.	12.801		2,650	2,650
26' x 46', 2" lifts, no compaction	40.000	C.Y.	.267		66.50	66.50
Compaction, air tamped, add	40.000	C.Y.	2.667		550	550
4" lifts, no compaction	100.000	C.Y.	.667		166	166
Compaction, air tamped, add	100.000	C.Y.	6.667		1,375	1,375
8" lifts, no compaction	202.000	C.Y.	1.347		335	335
Compaction, air tamped, add	202.000	C.Y.	13.467		2,800	2,800
26' x 60', 2" lifts, no compaction	47.000	C.Y.	.313		78	78
Compaction, air tamped, add	47.000	C.Y.	3.133		650	650
4" lifts, no compaction	120.000	C.Y.	.800		199	199
Compaction, air tamped, add	120.000	C.Y.	8.000		1,650	1,650
8" lifts, no compaction	242.000	C.Y.	1.614		400	400
Compaction, air tamped, add	242.000	C.Y.	16.134		3,325	3,325
30' x 66', 2" lifts, no compaction	53.000	C.Y.	.354		88	88
Compaction, air tamped, add	53.000	C.Y.	3.534		730	730
4" lifts, no compaction	134.000	C.Y.	.894		223	223
Compaction, air tamped, add	134.000	C.Y.	8.934		1,850	1,850
8" lifts, no compaction	269.000	C.Y.	1.794		445	445
Compaction, air tamped, add	269.000	C.Y.	17.934		3,725	3,725
Rough grade, 30' from building, 24' x 38'	87.000	C.Y.	.580		144	144
26' x 46'	100.000	C.Y.	.667		166	166
26' x 60'	120.000	C.Y.	.800		199	199
30' x 66'	134.000	C.Y.	.894		223	223

1 | SITEWORK
08 | Foundation Excavation Systems

System Description	QUAN.	UNIT	LABOR HOURS	COST EACH		
				MAT.	INST.	TOTAL
BUILDING, 24' X 38', 8' DEEP						
Medium clearing	.190	Acre	2.027		294.50	294.50
Excavate, track loader, 1-1/2 C.Y. bucket	550.000	C.Y.	7.860		1,023	1,023
Backfill, dozer, 8" lifts, no compaction	180.000	C.Y.	1.201		298.80	298.80
Rough grade, dozer, 30' from building	280.000	C.Y.	1.868		464.80	464.80
TOTAL		Ea.	12.956		2,081.10	2,081.10
BUILDING, 26' X 46', 8' DEEP						
Medium clearing	.210	Acre	2.240		325.50	325.50
Excavate, track loader, 1-1/2 C.Y. bucket	672.000	C.Y.	9.603		1,249.92	1,249.92
Backfill, dozer, 8" lifts, no compaction	220.000	C.Y.	1.467		365.20	365.20
Rough grade, dozer, 30' from building	340.000	C.Y.	2.268		564.40	564.40
TOTAL		Ea.	15.578		2,505.02	2,505.02
BUILDING, 26' X 60', 8' DEEP						
Medium clearing	.240	Acre	2.560		372	372
Excavate, track loader, 1-1/2 C.Y. bucket	829.000	C.Y.	11.846		1,541.94	1,541.94
Backfill, dozer, 8" lifts, no compaction	270.000	C.Y.	1.801		448.20	448.20
Rough grade, dozer, 30' from building	420.000	C.Y.	2.801		697.20	697.20
TOTAL		Ea.	19.008		3,059.34	3,059.34
BUILDING, 30' X 66', 8' DEEP						
Medium clearing	.260	Acre	2.773		403	403
Excavate, track loader, 1-1/2 C.Y. bucket	990.000	C.Y.	14.147		1,841.40	1,841.40
Backfill dozer, 8" lifts, no compaction	320.000	C.Y.	2.134		531.20	531.20
Rough grade, dozer, 30' from building	500.000	C.Y.	3.335		830	830
TOTAL		Ea.	22.389		3,605.60	3,605.60

The costs in this system are on a cost each basis.
Quantities are based on 1'-0" clearance beyond footing projection.

Description	QUAN.	UNIT	LABOR HOURS	COST EACH		
				MAT.	INST.	TOTAL

For customer support on your Residential Cost Data, call 877.713.2617.

Foundation Excavation Price Sheet	QUAN.	UNIT	LABOR HOURS	COST EACH MAT.	COST EACH INST.	COST EACH TOTAL
Clear & grub, medium brush, 30' from building, 24' x 38'	.190	Acre	2.027		295	295
26' x 46'	.210	Acre	2.240		325	325
26' x 60'	.240	Acre	2.560		370	370
30' x 66'	.260	Acre	2.773		405	405
Light trees, to 6" dia. cut & chip, 24' x 38'	.190	Acre	9.120		750	750
26' x 46'	.210	Acre	10.080		830	830
26' x 60'	.240	Acre	11.520		950	950
30' x 66'	.260	Acre	12.480		1,025	1,025
Medium trees, to 10" dia. cut & chip, 24' x 38'	.190	Acre	13.029		1,075	1,075
26' x 46'	.210	Acre	14.400		1,200	1,200
26' x 60'	.240	Acre	16.457		1,350	1,350
30' x 66'	.260	Acre	17.829		1,475	1,475
Excavation, basement, 24' x 38', 2' deep	98.000	C.Y.	1.400		183	183
4' deep	220.000	C.Y.	3.144		410	410
8' deep	550.000	C.Y.	7.860		1,025	1,025
26' x 46', 2' deep	123.000	C.Y.	1.758		229	229
4' deep	274.000	C.Y.	3.915		510	510
8' deep	672.000	C.Y.	9.603		1,250	1,250
26' x 60', 2' deep	157.000	C.Y.	2.244		292	292
4' deep	345.000	C.Y.	4.930		640	640
8' deep	829.000	C.Y.	11.846		1,550	1,550
30' x 66', 2' deep	192.000	C.Y.	2.744		355	355
4' deep	419.000	C.Y.	5.988		780	780
8' deep	990.000	C.Y.	14.147		1,850	1,850
Backfill, 24' x 38', 2" lifts, no compaction	32.000	C.Y.	.213		53	53
Compaction, air tamped, add	32.000	C.Y.	2.133		445	445
4" lifts, no compaction	72.000	C.Y.	.480		120	120
Compaction, air tamped, add	72.000	C.Y.	4.800		995	995
8" lifts, no compaction	180.000	C.Y.	1.201		299	299
Compaction, air tamped, add	180.000	C.Y.	12.001		2,475	2,475
26' x 46', 2" lifts, no compaction	40.000	C.Y.	.267		66.50	66.50
Compaction, air tamped, add	40.000	C.Y.	2.667		550	550
4" lifts, no compaction	90.000	C.Y.	.600		149	149
Compaction, air tamped, add	90.000	C.Y.	6.000		1,250	1,250
8" lifts, no compaction	220.000	C.Y.	1.467		365	365
Compacton, air tamped, add	220.000	C.Y.	14.667		3,050	3,050
26' x 60', 2" lifts, no compaction	50.000	C.Y.	.334		83	83
Compaction, air tamped, add	50.000	C.Y.	3.334		695	695
4" lifts, no compaction	110.000	C.Y.	.734		183	183
Compaction, air tamped, add	110.000	C.Y.	7.334		1,500	1,500
8" lifts, no compaction	270.000	C.Y.	1.801		450	450
Compaction, air tamped, add	270.000	C.Y.	18.001		3,725	3,725
30' x 66', 2" lifts, no compaction	60.000	C.Y.	.400		99.50	99.50
Compaction, air tamped, add	60.000	C.Y.	4.000		830	830
4" lifts, no compaction	130.000	C.Y.	.867		216	216
Compaction, air tamped, add	130.000	C.Y.	8.667		1,800	1,800
8" lifts, no compaction	320.000	C.Y.	2.134		530	530
Compaction, air tamped, add	320.000	C.Y.	21.334		4,425	4,425
Rough grade, 30' from building, 24' x 38'	280.000	C.Y.	1.868		465	465
26' x 46'	340.000	C.Y.	2.268		565	565
26' x 60'	420.000	C.Y.	2.801		700	700
30' x 66'	500.000	C.Y.	3.335		830	830

1 | SITEWORK
12 | Utility Trenching Systems

System Description	QUAN.	UNIT	LABOR HOURS	COST PER L.F. MAT.	COST PER L.F. INST.	COST PER L.F. TOTAL
2' DEEP						
Excavation, backhoe	.296	C.Y.	.032		2.34	2.34
Alternate pricing method, 4" deep	.111	C.Y.	.044	3.83	2.11	5.94
Utility, sewer, 6" cast iron	1.000	L.F.	.283	29.67	15.65	45.32
Compaction in 12" layers, hand tamp, add to above	.185	C.Y.	.044		1.79	1.79
TOTAL		L.F.	.403	33.50	21.89	55.39
4' DEEP						
Excavation, backhoe	.889	C.Y.	.095		7.04	7.04
Alternate pricing method, 4" deep	.111	C.Y.	.044	3.83	2.11	5.94
Utility, sewer, 6" cast iron	1.000	L.F.	.283	29.67	15.65	45.32
Compaction in 12" layers, hand tamp, add to above	.778	C.Y.	.183		7.51	7.51
TOTAL		L.F.	.605	33.50	32.31	65.81
6' DEEP						
Excavation, backhoe	1.770	C.Y.	.189		14.02	14.02
Alternate pricing method, 4" deep	.111	C.Y.	.044	3.83	2.11	5.94
Utility, sewer, 6" cast iron	1.000	L.F.	.283	29.67	15.65	45.32
Compaction in 12" layers, hand tamp, add to above	1.660	C.Y.	.391		16.02	16.02
TOTAL		L.F.	.907	33.50	47.80	81.30
8' DEEP						
Excavation, backhoe	2.960	C.Y.	.316		23.44	23.44
Alternate pricing method, 4" deep	.111	C.Y.	.044	3.83	2.11	5.94
Utility, sewer, 6" cast iron	1.000	L.F.	.283	29.67	15.65	45.32
Compaction in 12" layers, hand tamp, add to above	2.850	C.Y.	.671		27.50	27.50
TOTAL		L.F.	1.314	33.50	68.70	102.20

The costs in this system are based on a cost per linear foot of trench, and based on 2' wide at bottom of trench up to 6' deep.

Description	QUAN.	UNIT	LABOR HOURS	COST PER L.F. MAT.	COST PER L.F. INST.	COST PER L.F. TOTAL

Utility Trenching Price Sheet	QUAN.	UNIT	LABOR HOURS	COST PER UNIT		
				MAT.	INST.	TOTAL
Excavation, bottom of trench 2' wide, 2' deep	.296	C.Y.	.032		2.34	2.34
4' deep	.889	C.Y.	.095		7.05	7.05
6' deep	1.770	C.Y.	.142		10.80	10.80
8' deep	2.960	C.Y.	.105		20	20
Bedding, sand, bottom of trench 2' wide, no compaction, pipe, 2" diameter	.070	C.Y.	.028	2.42	1.33	3.75
4" diameter	.084	C.Y.	.034	2.90	1.60	4.50
6" diameter	.105	C.Y.	.042	3.62	2	5.62
8" diameter	.122	C.Y.	.049	4.21	2.33	6.54
Compacted, pipe, 2" diameter	.074	C.Y.	.030	2.55	1.41	3.96
4" diameter	.092	C.Y.	.037	3.17	1.75	4.92
6" diameter	.111	C.Y.	.044	3.83	2.11	5.94
8" diameter	.129	C.Y.	.052	4.45	2.46	6.91
3/4" stone, bottom of trench 2' wide, pipe, 4" diameter	.082	C.Y.	.033	2.83	1.56	4.39
6" diameter	.099	C.Y.	.040	3.42	1.88	5.30
3/8" stone, bottom of trench 2' wide, pipe, 4" diameter	.084	C.Y.	.034	2.90	1.60	4.50
6" diameter	.102	C.Y.	.041	3.52	1.95	5.47
Utilities, drainage & sewerage, corrugated plastic, 6" diameter	1.000	L.F.	.069	3.61	2.89	6.50
8" diameter	1.000	L.F.	.072	7.60	3.02	10.62
Concrete, non-reinforced, 6" diameter	1.000	L.F.	.181	6.60	9.45	16.05
8" diameter	1.000	L.F.	.214	7.25	11.25	18.50
PVC, SDR 35, 4" diameter	1.000	L.F.	.064	1.60	2.70	4.30
6" diameter	1.000	L.F.	.069	3.61	2.89	6.50
8" diameter	1.000	L.F.	.072	7.60	3.02	10.62
Gas & service, polyethylene, 1-1/4" diameter	1.000	L.F.	.059	1.78	2.95	4.73
Steel sched.40, 1" diameter	1.000	L.F.	.107	3.30	6.70	10
2" diameter	1.000	L.F.	.114	5.20	7.20	12.40
Sub-drainage, PVC, perforated, 3" diameter	1.000	L.F.	.064	1.60	2.70	4.30
4" diameter	1.000	L.F.	.064	1.60	2.70	4.30
5" diameter	1.000	L.F.	.069	3.61	2.89	6.50
6" diameter	1.000	L.F.	.069	3.61	2.89	6.50
Water service, copper, type K, 3/4"	1.000	L.F.	.083	6.60	5.30	11.90
1" diameter	1.000	L.F.	.093	8.80	5.95	14.75
PVC, 3/4"	1.000	L.F.	.121	4.70	7.75	12.45
1" diameter	1.000	L.F.	.134	5.40	8.55	13.95
Backfill, bottom of trench 2' wide no compact, 2' deep, pipe, 2" diameter	.226	L.F.	.053		2.18	2.18
4" diameter	.212	L.F.	.050		2.05	2.05
6" diameter	.185	L.F.	.044		1.79	1.79
4' deep, pipe, 2" diameter	.819	C.Y.	.193		7.90	7.90
4" diameter	.805	C.Y.	.189		7.75	7.75
6" diameter	.778	C.Y.	.183		7.50	7.50
6' deep, pipe, 2" diameter	1.700	C.Y.	.400		16.40	16.40
4" diameter	1.690	C.Y.	.398		16.30	16.30
6" diameter	1.660	C.Y.	.391		16	16
8' deep, pipe, 2" diameter	2.890	C.Y.	.680		28	28
4" diameter	2.870	C.Y.	.675		27.50	27.50
6" diameter	2.850	C.Y.	.671		27.50	27.50

1 | SITEWORK   16 | Sidewalk Systems

System Description	QUAN.	UNIT	LABOR HOURS	COST PER S.F. MAT.	COST PER S.F. INST.	COST PER S.F. TOTAL
ASPHALT SIDEWALK SYSTEM, 3' WIDE WALK						
Gravel fill, 4" deep	1.000	S.F.	.001	.36	.05	.41
Compact fill	.012	C.Y.			.02	.02
Handgrade	1.000	S.F.	.004		.17	.17
Walking surface, bituminous paving, 2" thick	1.000	S.F.	.007	.90	.36	1.26
Edging, brick, laid on edge	.670	L.F.	.079	1.80	3.92	5.72
TOTAL		S.F.	.091	3.06	4.52	7.58
CONCRETE SIDEWALK SYSTEM, 3' WIDE WALK						
Gravel fill, 4" deep	1.000	S.F.	.001	.36	.05	.41
Compact fill	.012	C.Y.			.02	.02
Handgrade	1.000	S.F.	.004		.17	.17
Walking surface, concrete, 4" thick	1.000	S.F.	.040	1.86	1.96	3.82
Edging, brick, laid on edge	.670	L.F.	.079	1.80	3.92	5.72
TOTAL		S.F.	.124	4.02	6.12	10.14
PAVERS, BRICK SIDEWALK SYSTEM, 3' WIDE WALK						
Sand base fill, 4" deep	1.000	S.F.	.001	.54	.09	.63
Compact fill	.012	C.Y.			.02	.02
Handgrade	1.000	S.F.	.004		.17	.17
Walking surface, brick pavers	1.000	S.F.	.160	3.50	7.85	11.35
Edging, redwood, untreated, 1" x 4"	.670	L.F.	.032	1.65	1.78	3.43
TOTAL		S.F.	.197	5.69	9.91	15.60

The costs in this system are based on a cost per square foot of sidewalk area. Concrete used is 3000 p.s.i.

Description	QUAN.	UNIT	LABOR HOURS	COST PER S.F. MAT.	COST PER S.F. INST.	COST PER S.F. TOTAL

Sidewalk Price Sheet	QUAN.	UNIT	LABOR HOURS	COST PER S.F. MAT.	COST PER S.F. INST.	COST PER S.F. TOTAL
Base, crushed stone, 3" deep	1.000	S.F.	.001	.42	.10	.52
6" deep	1.000	S.F.	.001	.85	.11	.96
9" deep	1.000	S.F.	.002	1.24	.15	1.39
12" deep	1.000	S.F.	.002	1.44	.18	1.62
Bank run gravel, 6" deep	1.000	S.F.	.001	.53	.08	.61
9" deep	1.000	S.F.	.001	.77	.11	.88
12" deep	1.000	S.F.	.001	1.06	.13	1.19
Compact base, 3" deep	.009	C.Y.	.001		.01	.01
6" deep	.019	C.Y.	.001		.03	.03
9" deep	.028	C.Y.	.001		.05	.05
Handgrade	1.000	S.F.	.004		.17	.17
Surface, brick, pavers dry joints, laid flat, running bond	1.000	S.F.	.160	3.50	7.85	11.35
Basket weave	1.000	S.F.	.168	3.97	8.30	12.27
Herringbone	1.000	S.F.	.174	3.97	8.55	12.52
Laid on edge, running bond	1.000	S.F.	.229	4.25	11.25	15.50
Mortar jts. laid flat, running bond	1.000	S.F.	.192	4.20	9.40	13.60
Basket weave	1.000	S.F.	.202	4.76	9.95	14.71
Herringbone	1.000	S.F.	.209	4.76	10.25	15.01
Laid on edge, running bond	1.000	S.F.	.274	5.10	13.50	18.60
Bituminous paving, 1-1/2" thick	1.000	S.F.	.006	.68	.26	.94
2" thick	1.000	S.F.	.007	.90	.36	1.26
2-1/2" thick	1.000	S.F.	.008	1.15	.39	1.54
Sand finish, 3/4" thick	1.000	S.F.	.001	.35	.12	.47
1" thick	1.000	S.F.	.001	.44	.15	.59
Concrete, reinforced, broom finish, 4" thick	1.000	S.F.	.040	1.82	1.96	3.78
5" thick	1.000	S.F.	.044	2.42	2.16	4.58
6" thick	1.000	S.F.	.047	2.82	2.31	5.13
Crushed stone, white marble, 3" thick	1.000	S.F.	.009	.49	.39	.88
Bluestone, 3" thick	1.000	S.F.	.009	.14	.39	.53
Flagging, bluestone, 1"	1.000	S.F.	.198	7.05	9.70	16.75
1-1/2"	1.000	S.F.	.188	12.85	9.25	22.10
Slate, natural cleft, 3/4"	1.000	S.F.	.174	10.10	8.55	18.65
Random rect., 1/2"	1.000	S.F.	.152	22	7.50	29.50
Granite blocks	1.000	S.F.	.174	13.20	8.55	21.75
Edging, corrugated aluminum, 4", 3' wide walk	.666	L.F.	.008	1.41	.45	1.86
4' wide walk	.500	L.F.	.006	1.06	.34	1.40
6", 3' wide walk	.666	L.F.	.010	1.76	.53	2.29
4' wide walk	.500	L.F.	.007	1.33	.40	1.73
Redwood-cedar-cypress, 1" x 4", 3' wide walk	.666	L.F.	.021	.85	1.17	2.02
4' wide walk	.500	L.F.	.016	.64	.88	1.52
2" x 4", 3' wide walk	.666	L.F.	.032	1.65	1.78	3.43
4' wide walk	.500	L.F.	.024	1.24	1.34	2.58
Brick, dry joints, 3' wide walk	.666	L.F.	.079	1.80	3.92	5.72
4' wide walk	.500	L.F.	.059	1.35	2.93	4.28
Mortar joints, 3' wide walk	.666	L.F.	.095	2.16	4.70	6.86
4' wide walk	.500	L.F.	.071	1.61	3.51	5.12

1 | SITEWORK — 20 | Driveway Systems

System Description	QUAN.	UNIT	LABOR HOURS	COST PER S.F. MAT.	COST PER S.F. INST.	COST PER S.F. TOTAL
ASPHALT DRIVEWAY TO 10' WIDE						
Excavation, driveway to 10' wide, 6" deep	.019	C.Y.			.04	.04
Base, 6" crushed stone	1.000	S.F.	.001	.56	.11	.67
Handgrade base	1.000	S.F.	.004		.17	.17
2" thick base	1.000	S.F.	.002	.91	.19	1.10
1" topping	1.000	S.F.	.001	.44	.15	.59
Edging, brick pavers	.200	L.F.	.024	.54	1.17	1.71
TOTAL		S.F.	.032	2.45	1.83	4.28
CONCRETE DRIVEWAY TO 10' WIDE						
Excavation, driveway to 10' wide, 6" deep	.019	C.Y.			.04	.04
Base, 6" crushed stone	1.000	S.F.	.001	.56	.11	.67
Handgrade base	1.000	S.F.	.004		.17	.17
Surface, concrete, 4" thick	1.000	S.F.	.040	1.86	1.96	3.82
Edging, brick pavers	.200	L.F.	.024	.54	1.17	1.71
TOTAL		S.F.	.069	2.96	3.45	6.41
PAVERS, BRICK DRIVEWAY TO 10' WIDE						
Excavation, driveway to 10' wide, 6" deep	.019	C.Y.			.04	.04
Base, 6" sand	1.000	S.F.	.001	.86	.14	1
Handgrade base	1.000	S.F.	.004		.17	.17
Surface, pavers, brick laid flat, running bond	1.000	S.F.	.160	3.50	7.85	11.35
Edging, redwood, untreated, 2" x 4"	.200	L.F.	.010	.49	.53	1.02
TOTAL		S.F.	.175	4.85	8.73	13.58

Description	QUAN.	UNIT	LABOR HOURS	COST PER S.F. MAT.	COST PER S.F. INST.	COST PER S.F. TOTAL

Driveway Price Sheet

	QUAN.	UNIT	LABOR HOURS	COST PER S.F. MAT.	COST PER S.F. INST.	COST PER S.F. TOTAL
Excavation, by machine, 10' wide, 6" deep	.019	C.Y.	.001		.04	.04
12" deep	.037	C.Y.	.001		.08	.08
18" deep	.055	C.Y.	.001		.12	.12
20' wide, 6" deep	.019	C.Y.	.001		.04	.04
12" deep	.037	C.Y.	.001		.08	.08
18" deep	.055	C.Y.	.001		.12	.12
Base, crushed stone, 10' wide, 3" deep	1.000	S.F.	.001	.43	.06	.49
6" deep	1.000	S.F.	.001	.85	.11	.96
9" deep	1.000	S.F.	.002	1.24	.15	1.39
20' wide, 3" deep	1.000	S.F.	.001	.43	.06	.49
6" deep	1.000	S.F.	.001	.85	.11	.96
9" deep	1.000	S.F.	.002	1.24	.15	1.39
Bank run gravel, 10' wide, 3" deep	1.000	S.F.	.001	.27	.05	.32
6" deep	1.000	S.F.	.001	.53	.08	.61
9" deep	1.000	S.F.	.001	.77	.11	.88
20' wide, 3" deep	1.000	S.F.	.001	.27	.05	.32
6" deep	1.000	S.F.	.001	.53	.08	.61
9" deep	1.000	S.F.	.001	.77	.11	.88
Handgrade, 10' wide	1.000	S.F.	.004		.17	.17
20' wide	1.000	S.F.	.004		.17	.17
Surface, asphalt, 10' wide, 3/4" topping, 1" base	1.000	S.F.	.002	1.05	.25	1.30
2" base	1.000	S.F.	.003	1.26	.31	1.57
1" topping, 1" base	1.000	S.F.	.002	1.14	.28	1.42
2" base	1.000	S.F.	.003	1.35	.34	1.69
20' wide, 3/4" topping, 1" base	1.000	S.F.	.002	1.05	.25	1.30
2" base	1.000	S.F.	.003	1.26	.31	1.57
1" topping, 1" base	1.000	S.F.	.002	1.14	.28	1.42
2" base	1.000	S.F.	.003	1.35	.34	1.69
Concrete, 10' wide, 4" thick	1.000	S.F.	.040	1.82	1.96	3.78
6" thick	1.000	S.F.	.047	2.82	2.31	5.13
20' wide, 4" thick	1.000	S.F.	.040	1.82	1.96	3.78
6" thick	1.000	S.F.	.047	2.82	2.31	5.13
Paver, brick 10' wide dry joints, running bond, laid flat	1.000	S.F.	.160	3.50	7.85	11.35
Laid on edge	1.000	S.F.	.229	4.25	11.25	15.50
Mortar joints, laid flat	1.000	S.F.	.192	4.20	9.40	13.60
Laid on edge	1.000	S.F.	.274	5.10	13.50	18.60
20' wide, running bond, dry jts., laid flat	1.000	S.F.	.160	3.50	7.85	11.35
Laid on edge	1.000	S.F.	.229	4.25	11.25	15.50
Mortar joints, laid flat	1.000	S.F.	.192	4.20	9.40	13.60
Laid on edge	1.000	S.F.	.274	5.10	13.50	18.60
Crushed stone, 10' wide, white marble, 3"	1.000	S.F.	.009	.49	.39	.88
Bluestone, 3"	1.000	S.F.	.009	.14	.39	.53
20' wide, white marble, 3"	1.000	S.F.	.009	.49	.39	.88
Bluestone, 3"	1.000	S.F.	.009	.14	.39	.53
Soil cement, 10' wide	1.000	S.F.	.007	.35	1.04	1.39
20' wide	1.000	S.F.	.007	.35	1.04	1.39
Granite blocks, 10' wide	1.000	S.F.	.174	13.20	8.55	21.75
20' wide	1.000	S.F.	.174	13.20	8.55	21.75
Asphalt block, solid 1-1/4" thick	1.000	S.F.	.119	10.20	5.85	16.05
Solid 3" thick	1.000	S.F.	.123	14.30	6.05	20.35
Edging, brick, 10' wide	.200	L.F.	.024	.54	1.17	1.71
20' wide	.100	L.F.	.012	.27	.59	.86
Redwood, untreated 2" x 4", 10' wide	.200	L.F.	.010	.49	.53	1.02
20' wide	.100	L.F.	.005	.25	.27	.52
Granite, 4 1/2" x 12" straight, 10' wide	.200	L.F.	.032	1.28	2.02	3.30
20' wide	.100	L.F.	.016	.64	1.01	1.65
Finishes, asphalt sealer, 10' wide	1.000	S.F.	.023	.99	.96	1.95
20' wide	1.000	S.F.	.023	.99	.96	1.95
Concrete, exposed aggregate 10' wide	1.000	S.F.	.013	.51	.65	1.16
20' wide	1.000	S.F.	.013	.51	.65	1.16

For customer support on your Residential Cost Data, call 877.713.2617.

1 | SITEWORK — 24 | Septic Systems

System Description	QUAN.	UNIT	LABOR HOURS	COST EACH MAT.	COST EACH INST.	COST EACH TOTAL
SEPTIC SYSTEM WITH 600 S.F. LEACHING FIELD, 1000 GALLON TANK						
Mobilization	2.000	Ea.	16.000		1,340	1,340
Tank, precast, 1000 gallon	1.000	Ea.	3.500	1,100	174.95	1,274.95
Effluent filter	1.000	Ea.	1.000	46	56	102
Distribution box, precast	1.000	Ea.	1.000	88.50	41	129.50
Flow leveler	3.000	Ea.	.480	7.44	19.65	27.09
PVC pipe, 4" diameter	25.000	L.F.	1.600	40	67.50	107.50
Tee	1.000	Ea.	1.000	14.80	56	70.80
Elbow	2.000	Ea.	1.333	21.60	75	96.60
Viewport cap	1.000	Ea.	.333	8.40	18.75	27.15
Filter fabric	67.000	S.Y.	.447	110.55	18.09	128.64
Detectable marking tape	1.600	C.L.F.	.085	11.44	3.50	14.94
PVC perforated pipe, 4" diameter	135.000	L.F.	8.640	216	364.50	580.50
Excavation	160.000	C.Y.	17.654		1,675.20	1,675.20
Backfill	133.000	L.C.Y.	2.660		263.34	263.34
Spoil	55.000	L.C.Y.	3.056		316.25	316.25
Compaction	113.000	E.C.Y.	15.066		690.43	690.43
Stone fill	39.000	C.Y.	6.240	1,560	390.78	1,950.78
TOTAL		Ea.	80.094	3,224.73	5,570.94	8,795.67
SEPTIC SYSTEM WITH 750 S.F. LEACHING FIELD, 1500 GALLON TANK						
Mobilization	2.000	Ea.	16.000		1,340	1,340
Tank, precast, 1500 gallon	1.000	Ea.	4.000	1,475	199.50	1,674.50
Effluent filter	1.000	Ea.	1.000	46	56	102
Distribution box, precast	1.000	Ea.	1.000	88.50	41	129.50
Flow leveler	3.000	Ea.	.480	7.44	19.65	27.09
PVC pipe, 4" diameter	25.000	L.F.	1.600	40	67.50	107.50
Tee	125.000	Ea.	1.000	14.80	56	70.80
Elbow	2.000	Ea.	1.333	21.60	75	96.60
Viewport cap	1.000	Ea.	.333	8.40	18.75	27.15
Filter fabric	84.000	S.Y.	.560	138.60	22.68	161.28
Detectable marking tape	1.900	C.L.F.	.101	13.59	4.16	17.75
PVC perforated pipe, 4" diameter	165.000	L.F.	10.560	264	445.50	709.50
Excavation	199.000	C.Y.	21.958		2,083.53	2,083.53
Backfill	162.000	L.C.Y.	3.240		320.76	320.76
Spoil	73.500	L.C.Y.	4.084		422.63	422.63
Compaction	137.000	E.C.Y.	18.266		837.07	837.07
Stone fill	48.500	C.Y.	7.760	1,940	485.98	2,425.98
TOTAL		Ea.	93.275	4,057.93	6,495.71	10,553.64

The costs in this system include all necessary piping and excavation.

1 | SITEWORK 24 | Septic Systems

System Description	QUAN.	UNIT	LABOR HOURS	COST EACH		
				MAT.	INST.	TOTAL
SEPTIC SYSTEM WITH 1000 S.F. LEACHING FIELD, 1500 GALLON TANK						
Mobilization	2.000	Ea.	16.000		1,340	1,340
Tank, precast, 1500 gallon	1.000	Ea.	4.000	1,475	199.50	1,674.50
Effluent filter	1.000	Ea.	1.000	46	56	102
Distribution box, precast	1.000	Ea.	1.000	88.50	41	129.50
Flow leveler	4.000	Ea.	.640	9.92	26.20	36.12
PVC pipe, 4" diameter	25.000	L.F.	1.600	40	67.50	107.50
Tee	1.000	Ea.	1.000	14.80	56	70.80
Elbow	4.000	Ea.	2.667	43.20	150	193.20
Viewport cap	1.000	Ea.	.333	8.40	18.75	27.15
Filter fabric	111.000	S.Y.	.740	183.15	29.97	213.12
Detectable marking tape	2.400	C.L.F.	.128	17.16	5.26	22.42
PVC perforated pipe, 4" diameter	215.000	L.F.	13.760	344	580.50	924.50
Excavation	229.000	C.Y.	25.268		2,397.63	2,397.63
Backfill	178.000	L.C.Y.	3.560		352.44	352.44
Spoil	91.500	L.C.Y.	5.084		526.13	526.13
Compaction	151.000	E.C.Y.	20.133		922.61	922.61
Stone fill	64.000	C.Y.	10.240	2,560	641.28	3,201.28
TOTAL		Ea.	107.153	4,830.13	7,410.77	12,240.90

The costs in this system include all necessary piping and excavation.

Septic Systems Price Sheet	QUAN.	UNIT	LABOR HOURS	COST EACH		
				MAT.	INST.	TOTAL
Tank, precast concrete, 1000 gallon	1.000	Ea.	3.500	1,100	175	1,275
1500 gallon	1.000	Ea.	4.000	1,475	200	1,675
Distribution box, concrete, 5 outlets	1.000	Ea.	1.000	88.50	41	129.50
12 outlets	1.000	Ea.	2.000	540	82	622
4" pipe, PVC, solid	25.000	L.F.	1.600	40	67.50	107.50
Tank and field excavation, 600 S.F. field	160.000	C.Y.	17.654		1,675	1,675
750 S.F. field	199.000	C.Y.	21.958		2,100	2,100
1000 S.F. field	229.000	C.Y.	25.268		2,400	2,400
Tank excavation only, 1000 gallon tank	20.000	C.Y.	2.206		210	210
1500 gallon tank	26.000	C.Y.	2.869		272	272
Backfill, crushed stone, 600 S.F. field	39.000	C.Y.	12.160	1,550	390	1,940
750 S.F. field	48.500	C.Y.	22.400	1,950	485	2,435
1000 S.F. field	64.000	C.Y.	.240	2,550	640	3,190
Backfill with excavated material, 600 S.F. field	133.000	L.C.Y.	.400		263	263
750 S.F. field	162.000	L.C.Y.	.367		320	320
1000 S.F. field	178.000	L.C.Y.	.280		350	350
Filter fabric, 600 S.F. field	67.000	S.Y.	2.376	111	18.10	129.10
750 S.F. field	84.000	S.Y.	4.860	139	22.50	161.50
1000 S.F. field	111.000	S.Y.	.740	183	30	213
4" pipe, PVC, perforated, 600 S.F. field	135.000	L.F.	9.280	216	365	581
750 S.F. field	165.000	L.F.	16.960	264	445	709
1000 S.F. field	215.000	L.F.	1.939	345	580	925
Pipe fittings, PVC, 600 S.F. field	2.000	Ea.	3.879	21.50	75	96.50
750 S.F. field	2.000	Ea.		21.50	75	96.50
1000 S.F. field	4.000	Ea.		43	150	193
Mobilization	2.000	Ea.	16.000		1,350	1,350
Effluent filter	1.000	Ea.	1.000	46	56	102
Flow leveler, 600 S.F. field	3.000	Ea.	.480	7.45	19.65	27.10
750 S.F. field	3.000	Ea.	.480	7.45	19.65	27.10
1000 S.F. field	4.000	Ea.	.640	9.90	26	35.90
Viewport cap	1.000	Ea.	.333	8.40	18.75	27.15
Detectable marking tape, 600 S.F. field	1.600	C.L.F.	.085	11.45	3.50	14.95
750 S.F. field	1.900	C.L.F.	.101	13.60	4.16	17.76
1000 S.F. field	2.400	C.L.F.	.128	17.15	5.25	22.40

1 | SITEWORK — 60 | Chain Link Fence Price Sheet

System Description	QUAN.	UNIT	LABOR HOURS	COST PER UNIT MAT.	COST PER UNIT INST.	COST PER UNIT TOTAL
Chain link fence						
Galv.9ga. wire, 1-5/8"post 10'O.C., 1-3/8"top rail, 2"corner post, 3'hi	1.000	L.F.	.130	10.05	5.45	15.50
4' high	1.000	L.F.	.141	10.15	5.95	16.10
6' high	1.000	L.F.	.209	12.50	8.80	21.30
Add for gate 3' wide 1-3/8" frame 3' high	1.000	Ea.	2.000	98	84.50	182.50
4' high	1.000	Ea.	2.400	108	101	209
6' high	1.000	Ea.	2.400	129	101	230
Add for gate 4' wide 1-3/8" frame 3' high	1.000	Ea.	2.667	103	112	215
4' high	1.000	Ea.	2.667	112	112	224
6' high	1.000	Ea.	3.000	136	127	263
Alum.9ga. wire, 1-5/8"post, 10'O.C., 1-3/8"top rail, 2"corner post,3'hi	1.000	L.F.	.130	8.20	5.45	13.65
4' high	1.000	L.F.	.141	8.60	5.95	14.55
6' high	1.000	L.F.	.209	11.20	8.80	20
Add for gate 3' wide 1-3/8" frame 3' high	1.000	Ea.	2.000	137	84.50	221.50
4' high	1.000	Ea.	2.400	148	101	249
6' high	1.000	Ea.	2.400	186	101	287
Add for gate 4' wide 1-3/8" frame 3' high	1.000	Ea.	2.400	143	101	244
4' high	1.000	Ea.	2.667	154	112	266
6' high	1.000	Ea.	3.000	196	127	323
Vinyl 9ga. wire, 1-5/8"post 10'O.C., 1-3/8"top rail, 2"corner post,3'hi	1.000	L.F.	.130	8	5.45	13.45
4' high	1.000	L.F.	.141	7.80	5.95	13.75
6' high	1.000	L.F.	.209	10.45	8.80	19.25
Add for gate 3' wide 1-3/8" frame 3' high	1.000	Ea.	2.000	113	84.50	197.50
4' high	1.000	Ea.	2.400	115	101	216
6' high	1.000	Ea.	2.400	142	101	243
Add for gate 4' wide 1-3/8" frame 3' high	1.000	Ea.	2.400	113	101	214
4' high	1.000	Ea.	2.667	118	112	230
6' high	1.000	Ea.	3.000	153	127	280
Tennis court, chain link fence, 10' high						
Galv.11ga.wire, 2"post 10'O.C., 1-3/8"top rail, 2-1/2"corner post	1.000	L.F.	.253	8.35	10.65	19
Add for gate 3' wide 1-3/8" frame	1.000	Ea.	2.400	252	101	353
Alum.11ga.wire, 2"post 10'O.C., 1-3/8"top rail, 2-1/2"corner post	1.000	L.F.	.253	10	10.65	20.65
Add for gate 3' wide 1-3/8" frame	1.000	Ea.	2.400	153	101	254
Vinyl 11ga.wire,2"post 10' O.C.,1-3/8"top rail,2-1/2"corner post	1.000	L.F.	.253	8.65	10.65	19.30
Add for gate 3' wide 1-3/8" frame	1.000	Ea.	2.400	250	101	351
Railings, commercial						
Aluminum balcony rail, 1-1/2" posts with pickets	1.000	L.F.	.164	67.50	11.85	79.35
With expanded metal panels	1.000	L.F.	.164	88	11.85	99.85
With porcelain enamel panel inserts	1.000	L.F.	.164	82.50	11.85	94.35
Mild steel, ornamental rounded top rail	1.000	L.F.	.164	76.50	11.85	88.35
As above, but pitch down stairs	1.000	L.F.	.183	82.50	13.20	95.70
Steel pipe, welded, 1-1/2" round, painted	1.000	L.F.	.160	30	11.55	41.55
Galvanized	1.000	L.F.	.160	42	11.55	53.55
Residential, stock units, mild steel, deluxe	1.000	L.F.	.102	17.25	7.35	24.60
Economy	1.000	L.F.	.102	12.85	7.35	20.20

1 | SITEWORK — 64 | Wood Fence Price Sheet

System Description	QUAN.	UNIT	LABOR HOURS	COST PER UNIT		
				MAT.	INST.	TOTAL
Basketweave, 3/8"x4" boards, 2"x4" stringers on spreaders, 4"x4" posts						
No. 1 cedar, 6' high	1.000	L.F.	.150	28.50	8.15	36.65
Treated pine, 6' high	1.000	L.F.	.160	38	8.65	46.65
Board fence, 1"x4" boards, 2"x4" rails, 4"x4" posts						
Preservative treated, 2 rail, 3' high	1.000	L.F.	.166	10.20	9	19.20
4' high	1.000	L.F.	.178	11.95	9.65	21.60
3 rail, 5' high	1.000	L.F.	.185	13.05	10	23.05
6' high	1.000	L.F.	.192	14.70	10.45	25.15
Western cedar, No. 1, 2 rail, 3' high	1.000	L.F.	.166	11.20	9	20.20
3 rail, 4' high	1.000	L.F.	.178	12.35	9.65	22
5' high	1.000	L.F.	.185	13.90	10	23.90
6' high	1.000	L.F.	.192	14.80	10.45	25.25
No. 1 cedar, 2 rail, 3' high	1.000	L.F.	.166	13.50	9	22.50
4' high	1.000	L.F.	.178	15.35	9.65	25
3 rail, 5' high	1.000	L.F.	.185	18.70	10	28.70
6' high	1.000	L.F.	.192	20.50	10.45	30.95
Shadow box, 1"x6" boards, 2"x4" rails, 4"x4" posts						
Fir, pine or spruce, treated, 3 rail, 6' high	1.000	L.F.	.160	18.90	8.65	27.55
No. 1 cedar, 3 rail, 4' high	1.000	L.F.	.185	19.80	10	29.80
6' high	1.000	L.F.	.192	25	10.45	35.45
Open rail, split rails, No. 1 cedar, 2 rail, 3' high	1.000	L.F.	.150	12.90	8.15	21.05
3 rail, 4' high	1.000	L.F.	.160	13.30	8.65	21.95
No. 2 cedar, 2 rail, 3' high	1.000	L.F.	.150	11.85	8.15	20
3 rail, 4' high	1.000	L.F.	.160	9.40	8.65	18.05
Open rail, rustic rails, No. 1 cedar, 2 rail, 3' high	1.000	L.F.	.150	10.25	8.15	18.40
3 rail, 4' high	1.000	L.F.	.160	9.85	8.65	18.50
No. 2 cedar, 2 rail, 3' high	1.000	L.F.	.150	9.65	8.15	17.80
3 rail, 4' high	1.000	L.F.	.160	8.05	8.65	16.70
Rustic picket, molded pine pickets, 2 rail, 3' high	1.000	L.F.	.171	8.50	9.30	17.80
3 rail, 4' high	1.000	L.F.	.197	9.80	10.70	20.50
No. 1 cedar, 2 rail, 3' high	1.000	L.F.	.171	9.85	9.30	19.15
3 rail, 4' high	1.000	L.F.	.197	11.35	10.70	22.05
Picket fence, fir, pine or spruce, preserved, treated						
2 rail, 3' high	1.000	L.F.	.171	7.90	9.30	17.20
3 rail, 4' high	1.000	L.F.	.185	9	10	19
Western cedar, 2 rail, 3' high	1.000	L.F.	.171	9.05	9.30	18.35
3 rail, 4' high	1.000	L.F.	.185	9.20	10	19.20
No. 1 cedar, 2 rail, 3' high	1.000	L.F.	.171	13.80	9.30	23.10
3 rail, 4' high	1.000	L.F.	.185	18	10	28
Stockade, No. 1 cedar, 3-1/4" rails, 6' high	1.000	L.F.	.150	13.25	8.15	21.40
8' high	1.000	L.F.	.155	18.25	8.40	26.65
No. 2 cedar, treated rails, 6' high	1.000	L.F.	.150	13.60	8.15	21.75
Treated pine, treated rails, 6' high	1.000	L.F.	.150	14.10	8.15	22.25
Gates, No. 2 cedar, picket, 3'-6" wide 4' high	1.000	Ea.	2.667	79.50	145	224.50
No. 2 cedar, rustic round, 3' wide, 3' high	1.000	Ea.	2.667	112	145	257
No. 2 cedar, stockade screen, 3'-6" wide, 6' high	1.000	Ea.	3.000	90	163	253
General, wood, 3'-6" wide, 4' high	1.000	Ea.	2.400	94.50	130	224.50
6' high	1.000	Ea.	3.000	118	163	281

Division 2 - Foundations

2 | FOUNDATIONS — 04 Footing Systems

System Description	QUAN.	UNIT	LABOR HOURS	COST PER L.F. MAT.	COST PER L.F. INST.	COST PER L.F. TOTAL
8" THICK BY 18" WIDE FOOTING						
Concrete, 3000 psi	.040	C.Y.		4.48		4.48
Place concrete, direct chute	.040	C.Y.	.016		.71	.71
Forms, footing, 4 uses	1.330	SFCA	.103	1.09	4.95	6.04
Reinforcing, 1/2" diameter bars, 2 each	1.380	Lb.	.011	.73	.63	1.36
Keyway, 2" x 4", beveled, 4 uses	1.000	L.F.	.015	.25	.83	1.08
Dowels, 1/2" diameter bars, 2' long, 6' O.C.	.166	Ea.	.006	.13	.33	.46
TOTAL		L.F.	.151	6.68	7.45	14.13
12" THICK BY 24" WIDE FOOTING						
Concrete, 3000 psi	.070	C.Y.		7.84		7.84
Place concrete, direct chute	.070	C.Y.	.028		1.25	1.25
Forms, footing, 4 uses	2.000	SFCA	.155	1.64	7.44	9.08
Reinforcing, 1/2" diameter bars, 2 each	1.380	Lb.	.011	.73	.63	1.36
Keyway, 2" x 4", beveled, 4 uses	1.000	L.F.	.015	.25	.83	1.08
Dowels, 1/2" diameter bars, 2' long, 6' O.C.	.166	Ea.	.006	.13	.33	.46
TOTAL		L.F.	.215	10.59	10.48	21.07
12" THICK BY 36" WIDE FOOTING						
Concrete, 3000 psi	.110	C.Y.		12.32		12.32
Place concrete, direct chute	.110	C.Y.	.044		1.96	1.96
Forms, footing, 4 uses	2.000	SFCA	.155	1.64	7.44	9.08
Reinforcing, 1/2" diameter bars, 2 each	1.380	Lb.	.011	.73	.63	1.36
Keyway, 2" x 4", beveled, 4 uses	1.000	L.F.	.015	.25	.83	1.08
Dowels, 1/2" diameter bars, 2' long, 6' O.C.	.166	Ea.	.006	.13	.33	.46
TOTAL		L.F.	.231	15.07	11.19	26.26

The footing costs in this system are on a cost per linear foot basis.

Description	QUAN.	UNIT	LABOR HOURS	COST PER S.F. MAT.	COST PER S.F. INST.	COST PER S.F. TOTAL

Footing Price Sheet	QUAN.	UNIT	LABOR HOURS	COST PER L.F.		
				MAT.	INST.	TOTAL
Concrete, 8" thick by 18" wide footing						
2000 psi concrete	.040	C.Y.		4.28		4.28
2500 psi concrete	.040	C.Y.		4.36		4.36
3000 psi concrete	.040	C.Y.		4.36		4.36
3500 psi concrete	.040	C.Y.		4.60		4.60
4000 psi concrete	.040	C.Y.		4.72		4.72
12" thick by 24" wide footing						
2000 psi concrete	.070	C.Y.		7.50		7.50
2500 psi concrete	.070	C.Y.		7.65		7.65
3000 psi concrete	.070	C.Y.		7.65		7.65
3500 psi concrete	.070	C.Y.		8.05		8.05
4000 psi concrete	.070	C.Y.		8.25		8.25
12" thick by 36" wide footing						
2000 psi concrete	.110	C.Y.		11.75		11.75
2500 psi concrete	.110	C.Y.		12		12
3000 psi concrete	.110	C.Y.		12		12
3500 psi concrete	.110	C.Y.		12.65		12.65
4000 psi concrete	.110	C.Y.		13		13
Place concrete, 8" thick by 18" wide footing, direct chute	.040	C.Y.	.016		.71	.71
Pumped concrete	.040	C.Y.	.017		.99	.99
Crane & bucket	.040	C.Y.	.032		2.05	2.05
12" thick by 24" wide footing, direct chute	.070	C.Y.	.028		1.25	1.25
Pumped concrete	.070	C.Y.	.030		1.74	1.74
Crane & bucket	.070	C.Y.	.056		3.60	3.60
12" thick by 36" wide footing, direct chute	.110	C.Y.	.044		1.96	1.96
Pumped concrete	.110	C.Y.	.047		2.73	2.73
Crane & bucket	.110	C.Y.	.088		5.65	5.65
Forms, 8" thick footing, 1 use	1.330	SFCA	.140	3.34	6.70	10.04
4 uses	1.330	SFCA	.103	1.09	4.95	6.04
12" thick footing, 1 use	2.000	SFCA	.211	5	10.10	15.10
4 uses	2.000	SFCA	.155	1.64	7.45	9.09
Reinforcing, 3/8" diameter bar, 1 each	.400	Lb.	.003	.21	.18	.39
2 each	.800	Lb.	.006	.42	.37	.79
3 each	1.200	Lb.	.009	.64	.55	1.19
1/2" diameter bar, 1 each	.700	Lb.	.005	.37	.32	.69
2 each	1.380	Lb.	.011	.73	.63	1.36
3 each	2.100	Lb.	.016	1.11	.97	2.08
5/8" diameter bar, 1 each	1.040	Lb.	.008	.55	.48	1.03
2 each	2.080	Lb.	.016	1.10	.96	2.06
Keyway, beveled, 2" x 4", 1 use	1.000	L.F.	.030	.50	1.66	2.16
2 uses	1.000	L.F.	.023	.38	1.25	1.63
2" x 6", 1 use	1.000	L.F.	.032	.72	1.76	2.48
2 uses	1.000	L.F.	.024	.54	1.32	1.86
Dowels, 2 feet long, 6' O.C., 3/8" bar	.166	Ea.	.005	.07	.31	.38
1/2" bar	.166	Ea.	.006	.13	.33	.46
5/8" bar	.166	Ea.	.006	.20	.37	.57
3/4" bar	.166	Ea.	.006	.20	.37	.57

For customer support on your Residential Cost Data, call 877.713.2617.

2 | FOUNDATIONS — 08 | Block Wall Systems

System Description	QUAN.	UNIT	LABOR HOURS	COST PER S.F. MAT.	COST PER S.F. INST.	COST PER S.F. TOTAL
8" WALL, GROUTED, FULL HEIGHT						
Concrete block, 8" x 16" x 8"	1.000	S.F.	.094	3.80	4.71	8.51
Masonry reinforcing, every second course	.750	L.F.	.002	.19	.11	.30
Parging, plastering with portland cement plaster, 1 coat	1.000	S.F.	.014	.26	.74	1
Dampproofing, bituminous coating, 1 coat	1.000	S.F.	.012	.25	.61	.86
Insulation, 1" rigid polystyrene	1.000	S.F.	.010	.58	.55	1.13
Grout, solid, pumped	1.000	S.F.	.059	1.25	2.91	4.16
Anchor bolts, 1/2" diameter, 8" long, 4' O.C.	.060	Ea.	.004	.09	.20	.29
Sill plate, 2" x 4", treated	.250	L.F.	.007	.14	.40	.54
TOTAL		S.F.	.202	6.56	10.23	16.79
12" WALL, GROUTED, FULL HEIGHT						
Concrete block, 8" x 16" x 12"	1.000	S.F.	.160	5.15	7.85	13
Masonry reinforcing, every second course	.750	L.F.	.003	.13	.17	.30
Parging, plastering with portland cement plaster, 1 coat	1.000	S.F.	.014	.26	.74	1
Dampproofing, bituminous coating, 1 coat	1.000	S.F.	.012	.25	.61	.86
Insulation, 1" rigid polystyrene	1.000	S.F.	.010	.58	.55	1.13
Grout, solid, pumped	1.000	S.F.	.063	2.05	3.08	5.13
Anchor bolts, 1/2" diameter, 8" long, 4' O.C.	.060	Ea.	.004	.09	.20	.29
Sill plate, 2" x 4", treated	.250	L.F.	.007	.14	.40	.54
TOTAL		S.F.	.273	8.65	13.60	22.25

The costs in this system are based on a square foot of wall. Do not subtract for window or door openings.

Description	QUAN.	UNIT	LABOR HOURS	COST PER S.F. MAT.	COST PER S.F. INST.	COST PER S.F. TOTAL

Block Wall Systems

Block Wall Systems	QUAN.	UNIT	LABOR HOURS	COST PER S.F. MAT.	COST PER S.F. INST.	COST PER S.F. TOTAL
Concrete, block, 8" x 16" x, 6" thick	1.000	S.F.	.089	3.17	4.40	7.57
8" thick	1.000	S.F.	.093	3.80	4.71	8.51
10" thick	1.000	S.F.	.095	3.92	5.70	9.62
12" thick	1.000	S.F.	.122	5.15	7.85	13
Solid block, 8" x 16" x, 6" thick	1.000	S.F.	.091	3.36	4.55	7.91
8" thick	1.000	S.F.	.096	4.57	4.83	9.40
10" thick	1.000	S.F.	.096	4.57	4.83	9.40
12" thick	1.000	S.F.	.126	6.55	6.75	13.30
Masonry reinforcing, wire strips, to 8" wide, every course	1.500	L.F.	.004	.38	.21	.59
Every 2nd course	.750	L.F.	.002	.19	.11	.30
Every 3rd course	.500	L.F.	.001	.13	.07	.20
Every 4th course	.400	L.F.	.001	.10	.06	.16
Wire strips to 12" wide, every course	1.500	L.F.	.006	.26	.33	.59
Every 2nd course	.750	L.F.	.003	.13	.17	.30
Every 3rd course	.500	L.F.	.002	.09	.11	.20
Every 4th course	.400	L.F.	.002	.07	.09	.16
Parging, plastering with portland cement plaster, 1 coat	1.000	S.F.	.014	.26	.74	1
2 coats	1.000	S.F.	.022	.40	1.14	1.54
Dampproofing, bituminous, brushed on, 1 coat	1.000	S.F.	.012	.25	.61	.86
2 coats	1.000	S.F.	.016	.49	.82	1.31
Sprayed on, 1 coat	1.000	S.F.	.010	.25	.49	.74
2 coats	1.000	S.F.	.016	.48	.82	1.30
Troweled on, 1/16" thick	1.000	S.F.	.016	.40	.82	1.22
1/8" thick	1.000	S.F.	.020	.70	1.02	1.72
1/2" thick	1.000	S.F.	.023	2.27	1.17	3.44
Insulation, rigid, fiberglass, 1.5#/C.F., unfaced						
1-1/2" thick R 6.2	1.000	S.F.	.008	.44	.44	.88
2" thick R 8.5	1.000	S.F.	.008	.50	.44	.94
3" thick R 13	1.000	S.F.	.010	.62	.55	1.17
Perlite, 1" thick R 2.77	1.000	S.F.	.010	.46	.55	1.01
2" thick R 5.55	1.000	S.F.	.011	.83	.60	1.43
Polystyrene, extruded, 1" thick R 5.4	1.000	S.F.	.010	.58	.55	1.13
2" thick R 10.8	1.000	S.F.	.011	1.65	.60	2.25
Molded 1" thick R 3.85	1.000	S.F.	.010	.28	.55	.83
2" thick R 7.7	1.000	S.F.	.011	.83	.60	1.43
Grout, concrete block cores, 6" thick	1.000	S.F.	.044	.94	2.19	3.13
8" thick	1.000	S.F.	.059	1.25	2.91	4.16
10" thick	1.000	S.F.	.061	1.65	2.99	4.64
12" thick	1.000	S.F.	.063	2.05	3.08	5.13
Anchor bolts, 2' on center, 1/2" diameter, 8" long	.120	Ea.	.005	.18	.39	.57
12" long	.120	Ea.	.005	.20	.39	.59
3/4" diameter, 8" long	.120	Ea.	.006	.48	.41	.89
12" long	.120	Ea.	.006	.61	.41	1.02
4' on center, 1/2" diameter, 8" long	.060	Ea.	.002	.09	.20	.29
12" long	.060	Ea.	.003	.10	.20	.30
3/4" diameter, 8" long	.060	Ea.	.003	.24	.20	.44
12" long	.060	Ea.	.003	.30	.21	.51
Sill plates, treated, 2" x 4"	.250	L.F.	.007	.14	.40	.54
4" x 4"	.250	L.F.	.007	.36	.38	.74

2 | FOUNDATIONS — 12 | Concrete Wall Systems

System Description	QUAN.	UNIT	LABOR HOURS	COST PER S.F. MAT.	COST PER S.F. INST.	COST PER S.F. TOTAL
8" THICK, POURED CONCRETE WALL						
Concrete, 8" thick, 3000 psi	.025	C.Y.		2.80		2.80
Forms, prefabricated plywood, up to 8' high	2.000	SFCA	.120	2.06	5.84	7.90
Reinforcing, light	.670	Lb.	.004	.36	.21	.57
Placing concrete, direct chute	.025	C.Y.	.013		.60	.60
Dampproofing, brushed on, 2 coats	1.000	S.F.	.016	.49	.82	1.31
Rigid insulation, 1" polystyrene	1.000	S.F.	.010	.58	.55	1.13
Anchor bolts, 1/2" diameter, 12" long, 4' O.C.	.060	Ea.	.004	.10	.20	.30
Sill plates, 2" x 4", treated	.250	L.F.	.007	.14	.40	.54
TOTAL		S.F.	.174	6.53	8.62	15.15
12" THICK, POURED CONCRETE WALL						
Concrete, 12" thick, 3000 psi	.040	C.Y.		4.48		4.48
Forms, prefabricated plywood, up to 8' high	2.000	SFCA	.120	2.06	5.84	7.90
Reinforcing, light	1.000	Lb.	.005	.53	.32	.85
Placing concrete, direct chute	.040	C.Y.	.019		.87	.87
Dampproofing, brushed on, 2 coats	1.000	S.F.	.016	.49	.82	1.31
Rigid insulation, 1" polystyrene	1.000	S.F.	.010	.58	.55	1.13
Anchor bolts, 1/2" diameter, 12" long, 4' O.C.	.060	Ea.	.004	.10	.20	.30
Sill plates, 2" x 4" treated	.250	L.F.	.007	.14	.40	.54
TOTAL		S.F.	.181	8.38	9	17.38

The costs in this system are based on sq. ft. of wall. Do not subtract for window and door openings. The costs assume a 4' high wall.

Description	QUAN.	UNIT	LABOR HOURS	COST PER S.F. MAT.	COST PER S.F. INST.	COST PER S.F. TOTAL

Concrete Wall Price Sheet

	QUAN.	UNIT	LABOR HOURS	COST PER S.F. MAT.	COST PER S.F. INST.	COST PER S.F. TOTAL
Formwork, prefabricated plywood, up to 8' high	2.000	SFCA	.081	2.06	5.85	7.91
Over 8' to 16' high	2.000	SFCA	.076	2.18	7.80	9.98
Job built forms, 1 use per month	2.000	SFCA	.320	6	12.60	18.60
4 uses per month	2.000	SFCA	.221	2.28	9.25	11.53
Reinforcing, 8" wall, light reinforcing	.670	Lb.	.004	.36	.21	.57
Heavy reinforcing	1.500	Lb.	.008	.80	.48	1.28
10" wall, light reinforcing	.850	Lb.	.005	.45	.27	.72
Heavy reinforcing	2.000	Lb.	.011	1.06	.64	1.70
12" wall light reinforcing	1.000	Lb.	.005	.53	.32	.85
Heavy reinforcing	2.250	Lb.	.012	1.19	.72	1.91
Placing concrete, 8" wall, direct chute	.025	C.Y.	.013		.60	.60
Pumped concrete	.025	C.Y.	.016		.93	.93
Crane & bucket	.025	C.Y.	.023		1.45	1.45
10" wall, direct chute	.030	C.Y.	.016		.71	.71
Pumped concrete	.030	C.Y.	.019		1.12	1.12
Crane & bucket	.030	C.Y.	.027		1.73	1.73
12" wall, direct chute	.040	C.Y.	.019		.87	.87
Pumped concrete	.040	C.Y.	.023		1.35	1.35
Crane & bucket	.040	C.Y.	.032		2.05	2.05
Dampproofing, bituminous, brushed on, 1 coat	1.000	S.F.	.012	.25	.61	.86
2 coats	1.000	S.F.	.016	.49	.82	1.31
Sprayed on, 1 coat	1.000	S.F.	.010	.25	.49	.74
2 coats	1.000	S.F.	.016	.48	.82	1.30
Troweled on, 1/16" thick	1.000	S.F.	.016	.40	.82	1.22
1/8" thick	1.000	S.F.	.020	.70	1.02	1.72
1/2" thick	1.000	S.F.	.023	2.27	1.17	3.44
Insulation rigid, fiberglass, 1.5#/C.F., unfaced 1-1/2" thick, R 6.2	1.000	S.F.	.008	.44	.44	.88
2" thick, R 8.3	1.000	S.F.	.008	.50	.44	.94
3" thick, R 12.4	1.000	S.F.	.010	.62	.55	1.17
Perlite, 1" thick R 2.77	1.000	S.F.	.010	.46	.55	1.01
2" thick R 5.55	1.000	S.F.	.011	.83	.60	1.43
Polystyrene, extruded, 1" thick R 5.40	1.000	S.F.	.010	.58	.55	1.13
2" thick R 10.8	1.000	S.F.	.011	1.65	.60	2.25
Molded, 1" thick R 3.85	1.000	S.F.	.010	.28	.55	.83
2" thick R 7.70	1.000	S.F.	.011	.83	.60	1.43
Anchor bolts, 2' on center, 1/2" diameter, 8" long	.120	Ea.	.005	.18	.39	.57
12" long	.120	Ea.	.005	.20	.39	.59
3/4" diameter, 8" long	.120	Ea.	.006	.48	.41	.89
12" long	.120	Ea.	.006	.61	.41	1.02
Sill plates, treated lumber, 2" x 4"	.250	L.F.	.007	.14	.40	.54
4" x 4"	.250	L.F.	.007	.36	.38	.74

2 | FOUNDATIONS — 16 | Wood Wall Foundation Systems

System Description	QUAN.	UNIT	LABOR HOURS	COST PER S.F. MAT.	COST PER S.F. INST.	COST PER S.F. TOTAL
2" X 4" STUDS, 16" O.C., WALL						
Studs, 2" x 4", 16" O.C., treated	1.000	L.F.	.015	.56	.80	1.36
Plates, double top plate, single bottom plate, treated, 2" x 4"	.750	L.F.	.011	.42	.60	1.02
Sheathing, 1/2", exterior grade, CDX, treated	1.000	S.F.	.014	.90	.78	1.68
Asphalt paper, 15# roll	1.100	S.F.	.002	.07	.13	.20
Vapor barrier, 4 mil polyethylene	1.000	S.F.	.002	.04	.12	.16
Fiberglass insulation, 3-1/2" thick	1.000	S.F.	.007	.33	.38	.71
TOTAL		S.F.	.051	2.32	2.81	5.13
2" X 6" STUDS, 16" O.C., WALL						
Studs, 2" x 6", 16" O.C., treated	1.000	L.F.	.016	.86	.88	1.74
Plates, double top plate, single bottom plate, treated, 2" x 6"	.750	L.F.	.012	.65	.66	1.31
Sheathing, 5/8" exterior grade, CDX, treated	1.000	S.F.	.015	1.39	.84	2.23
Asphalt paper, 15# roll	1.100	S.F.	.002	.07	.13	.20
Vapor barrier, 4 mil polyethylene	1.000	S.F.	.002	.04	.12	.16
Fiberglass insulation, 6" thick	1.000	S.F.	.007	.46	.38	.84
TOTAL		S.F.	.054	3.47	3.01	6.48
2" X 8" STUDS, 16" O.C., WALL						
Studs, 2" x 8", 16" O.C. treated	1.000	L.F.	.018	1.16	.98	2.14
Plates, double top plate, single bottom plate, treated, 2" x 8"	.750	L.F.	.013	.87	.74	1.61
Sheathing, 3/4" exterior grade, CDX, treated	1.000	S.F.	.016	1.46	.91	2.37
Asphalt paper, 15# roll	1.100	S.F.	.002	.07	.13	.20
Vapor barrier, 4 mil polyethylene	1.000	S.F.	.002	.04	.12	.16
Fiberglass insulation, 9" thick	1.000	S.F.	.006	.78	.33	1.11
TOTAL		S.F.	.057	4.38	3.21	7.59

The costs in this system are based on a sq. ft. of wall area. Do not subtract for window or door openings. The costs assume a 4' high wall.

Description	QUAN.	UNIT	LABOR HOURS	COST PER S.F. MAT.	COST PER S.F. INST.	COST PER S.F. TOTAL

Wood Wall Foundation Price Sheet	QUAN.	UNIT	LABOR HOURS	COST PER S.F.		
				MAT.	INST.	TOTAL
Studs, treated, 2" x 4", 12" O.C.	1.250	L.F.	.018	.70	1	1.70
16" O.C.	1.000	L.F.	.015	.56	.80	1.36
2" x 6", 12" O.C.	1.250	L.F.	.020	1.08	1.10	2.18
16" O.C.	1.000	L.F.	.016	.86	.88	1.74
2" x 8", 12" O.C.	1.250	L.F.	.022	1.45	1.23	2.68
16" O.C.	1.000	L.F.	.018	1.16	.98	2.14
Plates, treated double top single bottom, 2" x 4"	.750	L.F.	.011	.42	.60	1.02
2" x 6"	.750	L.F.	.012	.65	.66	1.31
2" x 8"	.750	L.F.	.013	.87	.74	1.61
Sheathing, treated exterior grade CDX, 1/2" thick	1.000	S.F.	.014	.90	.78	1.68
5/8" thick	1.000	S.F.	.015	1.39	.84	2.23
3/4" thick	1.000	S.F.	.016	1.46	.91	2.37
Asphalt paper, 15# roll	1.100	S.F.	.002	.07	.13	.20
Vapor barrier, polyethylene, 4 mil	1.000	S.F.	.002	.03	.12	.15
10 mil	1.000	S.F.	.002	.10	.12	.22
Insulation, rigid, fiberglass, 1.5#/C.F., unfaced	1.000	S.F.	.008	.30	.44	.74
1-1/2" thick, R 6.2	1.000	S.F.	.008	.44	.44	.88
2" thick, R 8.3	1.000	S.F.	.008	.50	.44	.94
3" thick, R 12.4	1.000	S.F.	.010	.63	.56	1.19
Perlite 1" thick, R 2.77	1.000	S.F.	.010	.46	.55	1.01
2" thick, R 5.55	1.000	S.F.	.011	.83	.60	1.43
Polystyrene, extruded, 1" thick, R 5.40	1.000	S.F.	.010	.58	.55	1.13
2" thick, R 10.8	1.000	S.F.	.011	1.65	.60	2.25
Molded 1" thick, R 3.85	1.000	S.F.	.010	.28	.55	.83
2" thick, R 7.7	1.000	S.F.	.011	.83	.60	1.43
Non rigid, batts, fiberglass, paper backed, 3-1/2" thick roll, R 11	1.000	S.F.	.005	.33	.38	.71
6", R 19	1.000	S.F.	.006	.46	.38	.84
9", R 30	1.000	S.F.	.006	.78	.33	1.11
12", R 38	1.000	S.F.	.006	1.06	.33	1.39
Mineral fiber, paper backed, 3-1/2", R 13	1.000	S.F.	.005	.66	.28	.94
6", R 19	1.000	S.F.	.005	1.04	.28	1.32
10", R 30	1.000	S.F.	.006	1.37	.33	1.70

2 | FOUNDATIONS   20 | Floor Slab Systems

System Description	QUAN.	UNIT	LABOR HOURS	COST PER S.F. MAT.	COST PER S.F. INST.	COST PER S.F. TOTAL
4" THICK SLAB						
Concrete, 4" thick, 3000 psi concrete	.012	C.Y.		1.34		1.34
Place concrete, direct chute	.012	C.Y.	.005		.24	.24
Bank run gravel, 4" deep	1.000	S.F.	.001	.40	.06	.46
Polyethylene vapor barrier, .006" thick	1.000	S.F.	.002	.04	.12	.16
Edge forms, expansion material	.100	L.F.	.005	.04	.26	.30
Welded wire fabric, 6 x 6, 10/10 (W1.4/W1.4)	1.100	S.F.	.005	.18	.30	.48
Steel trowel finish	1.000	S.F.	.014		.70	.70
TOTAL		S.F.	.032	2	1.68	3.68
6" THICK SLAB						
Concrete, 6" thick, 3000 psi concrete	.019	C.Y.		2.13		2.13
Place concrete, direct chute	.019	C.Y.	.008		.37	.37
Bank run gravel, 4" deep	1.000	S.F.	.001	.40	.06	.46
Polyethylene vapor barrier, .006" thick	1.000	S.F.	.002	.04	.12	.16
Edge forms, expansion material	.100	L.F.	.005	.04	.26	.30
Welded wire fabric, 6 x 6, 10/10 (W1.4/W1.4)	1.100	S.F.	.005	.18	.30	.48
Steel trowel finish	1.000	S.F.	.014		.70	.70
TOTAL		S.F.	.035	2.79	1.81	4.60

The slab costs in this section are based on a cost per square foot of floor area.

Description	QUAN.	UNIT	LABOR HOURS	COST PER S.F. MAT.	COST PER S.F. INST.	COST PER S.F. TOTAL

Floor Slab Price Sheet

	QUAN.	UNIT	LABOR HOURS	COST PER S.F. MAT.	INST.	TOTAL
Concrete, 4" thick slab, 2000 psi concrete	.012	C.Y.		1.28		1.28
2500 psi concrete	.012	C.Y.		1.31		1.31
3000 psi concrete	.012	C.Y.		1.31		1.31
3500 psi concrete	.012	C.Y.		1.38		1.38
4000 psi concrete	.012	C.Y.		1.42		1.42
4500 psi concrete	.012	C.Y.		1.45		1.45
5" thick slab, 2000 psi concrete	.015	C.Y.		1.61		1.61
2500 psi concrete	.015	C.Y.		1.64		1.64
3000 psi concrete	.015	C.Y.		1.64		1.64
3500 psi concrete	.015	C.Y.		1.73		1.73
4000 psi concrete	.015	C.Y.		1.77		1.77
4500 psi concrete	.015	C.Y.		1.82		1.82
6" thick slab, 2000 psi concrete	.019	C.Y.		2.03		2.03
2500 psi concrete	.019	C.Y.		2.07		2.07
3000 psi concrete	.019	C.Y.		2.07		2.07
3500 psi concrete	.019	C.Y.		2.19		2.19
4000 psi concrete	.019	C.Y.		2.24		2.24
4500 psi concrete	.019	C.Y.		2.30		2.30
Place concrete, 4" slab, direct chute	.012	C.Y.	.005		.24	.24
Pumped concrete	.012	C.Y.	.006		.34	.34
Crane & bucket	.012	C.Y.	.008		.51	.51
5" slab, direct chute	.015	C.Y.	.007		.29	.29
Pumped concrete	.015	C.Y.	.007		.43	.43
Crane & bucket	.015	C.Y.	.010		.63	.63
6" slab, direct chute	.019	C.Y.	.008		.37	.37
Pumped concrete	.019	C.Y.	.009		.55	.55
Crane & bucket	.019	C.Y.	.012		.80	.80
Gravel, bank run, 4" deep	1.000	S.F.	.001	.40	.06	.46
6" deep	1.000	S.F.	.001	.53	.08	.61
9" deep	1.000	S.F.	.001	.77	.11	.88
12" deep	1.000	S.F.	.001	1.06	.13	1.19
3/4" crushed stone, 3" deep	1.000	S.F.	.001	.43	.06	.49
6" deep	1.000	S.F.	.001	.85	.11	.96
9" deep	1.000	S.F.	.002	1.24	.15	1.39
12" deep	1.000	S.F.	.002	1.44	.18	1.62
Vapor barrier polyethylene, .004" thick	1.000	S.F.	.002	.03	.12	.15
.006" thick	1.000	S.F.	.002	.04	.12	.16
Edge forms, expansion material, 4" thick slab	.100	L.F.	.004	.03	.17	.20
6" thick slab	.100	L.F.	.005	.04	.26	.30
Welded wire fabric 6 x 6, 10/10 (W1.4/W1.4)	1.100	S.F.	.005	.18	.30	.48
6 x 6, 6/6 (W2.9/W2.9)	1.100	S.F.	.006	.28	.36	.64
4 x 4, 10/10 (W1.4/W1.4)	1.100	S.F.	.006	.24	.34	.58
Finish concrete, screed finish	1.000	S.F.	.009		.29	.29
Float finish	1.000	S.F.	.011		.29	.29
Steel trowel, for resilient floor	1.000	S.F.	.013		.91	.91
For finished floor	1.000	S.F.	.015		.70	.70

Division 3 - Framing

3 | FRAMING — 02 | Floor Framing Systems

System Description	QUAN.	UNIT	LABOR HOURS	COST PER S.F. MAT.	COST PER S.F. INST.	COST PER S.F. TOTAL
2" X 8", 16" O.C.						
Wood joists, 2" x 8", 16" O.C.	1.000	L.F.	.015	1.01	.80	1.81
Bridging, 1" x 3", 6' O.C.	.080	Pr.	.005	.06	.27	.33
Box sills, 2" x 8"	.150	L.F.	.002	.15	.12	.27
Concrete filled steel column, 4" diameter	.125	L.F.	.002	.27	.13	.40
Girder, built up from three 2" x 8"	.125	L.F.	.013	.38	.73	1.11
Sheathing, plywood, subfloor, 5/8" CDX	1.000	S.F.	.012	.86	.65	1.51
Furring, 1" x 3", 16" O.C.	1.000	L.F.	.023	.42	1.26	1.68
Joist hangers	.036	Ea.	.002	.06	.10	.16
TOTAL		S.F.	.074	3.21	4.06	7.27
2" X 10", 16" O.C.						
Wood joists, 2" x 10", 16" OC	1.000	L.F.	.018	1.47	.98	2.45
Bridging, 1" x 3", 6' OC	.080	Pr.	.005	.06	.27	.33
Box sills, 2" x 10"	.150	L.F.	.003	.22	.15	.37
Concrete filled steel column, 4" diameter	.125	L.F.	.002	.27	.13	.40
Girder, built up from three 2" x 10"	.125	L.F.	.014	.55	.78	1.33
Sheathing, plywood, subfloor, 5/8" CDX	1.000	S.F.	.012	.86	.65	1.51
Furring, 1" x 3", 16" OC	1.000	L.F.	.023	.42	1.26	1.68
Joist hangers	.036	Ea.	.002	.06	.10	.16
TOTAL		S.F.	.079	3.91	4.32	8.23
2" X 12", 16" O.C.						
Wood joists, 2" x 12", 16" O.C.	1.000	L.F.	.018	1.90	1.01	2.91
Bridging, 1" x 3", 6' O.C.	.080	Pr.	.005	.06	.27	.33
Box sills, 2" x 12"	.150	L.F.	.003	.29	.15	.44
Concrete filled steel column, 4" diameter	.125	L.F.	.002	.27	.13	.40
Girder, built up from three 2" x 12"	.125	L.F.	.015	.71	.83	1.54
Sheathing, plywood, subfloor, 5/8" CDX	1.000	S.F.	.012	.86	.65	1.51
Furring, 1" x 3", 16" O.C.	1.000	L.F.	.023	.42	1.26	1.68
Joist hangers	.036	Ea.	.002	.06	.10	.16
TOTAL		S.F.	.080	4.57	4.40	8.97

Floor costs on this page are given on a cost per square foot basis.

Description	QUAN.	UNIT	LABOR HOURS	COST PER S.F. MAT.	COST PER S.F. INST.	COST PER S.F. TOTAL

Floor Framing Price Sheet (Wood)

Floor Framing Price Sheet (Wood)	QUAN.	UNIT	LABOR HOURS	COST PER S.F.		
				MAT.	INST.	TOTAL
Joists, #2 or better, pine, 2" x 4", 12" O.C.	1.250	L.F.	.016	.59	.89	1.48
16" O.C.	1.000	L.F.	.013	.47	.71	1.18
2" x 6", 12" O.C.	1.250	L.F.	.016	.91	.89	1.80
16" O.C.	1.000	L.F.	.013	.73	.71	1.44
2" x 8", 12" O.C.	1.250	L.F.	.018	1.26	1	2.26
16" O.C.	1.000	L.F.	.015	1.01	.80	1.81
2" x 10", 12" O.C.	1.250	L.F.	.022	1.84	1.23	3.07
16" O.C.	1.000	L.F.	.018	1.47	.98	2.45
2" x 12", 12" O.C.	1.250	L.F.	.023	2.38	1.26	3.64
16" O.C.	1.000	L.F.	.018	1.90	1.01	2.91
Bridging, wood 1" x 3", joists 12" O.C.	.100	Pr.	.006	.07	.34	.41
16" O.C.	.080	Pr.	.005	.06	.27	.33
Metal, galvanized, joists 12" O.C.	.100	Pr.	.006	.10	.34	.44
16" O.C.	.080	Pr.	.005	.08	.27	.35
Compression type, joists 12" O.C.	.100	Pr.	.004	.10	.22	.32
16" O.C.	.080	Pr.	.003	.08	.18	.26
Box sills, #2 or better pine, 2" x 4"	.150	L.F.	.002	.07	.11	.18
2" x 6"	.150	L.F.	.002	.11	.11	.22
2" x 8"	.150	L.F.	.002	.15	.12	.27
2" x 10"	.150	L.F.	.003	.22	.15	.37
2" x 12"	.150	L.F.	.003	.29	.15	.44
Girders, including lally columns, 3 pieces spiked together, 2" x 8"	.125	L.F.	.015	.65	.86	1.51
2" x 10"	.125	L.F.	.016	.82	.91	1.73
2" x 12"	.125	L.F.	.017	.98	.96	1.94
Solid girders, 3" x 8"	.040	L.F.	.004	.39	.22	.61
3" x 10"	.040	L.F.	.004	.42	.23	.65
3" x 12"	.040	L.F.	.004	.45	.24	.69
4" x 8"	.040	L.F.	.004	.40	.24	.64
4" x 10"	.040	L.F.	.004	.46	.25	.71
4" x 12"	.040	L.F.	.004	.50	.26	.76
Steel girders, bolted & including fabrication, wide flange shapes						
12" deep, 14#/l.f.	.040	L.F.	.003	1.04	.27	1.31
10" deep, 15#/l.f.	.040	L.F.	.003	1.04	.27	1.31
8" deep, 10#/l.f.	.040	L.F.	.003	.70	.27	.97
6" deep, 9#/l.f.	.040	L.F.	.003	.63	.27	.90
5" deep, 16#/l.f.	.040	L.F.	.003	1.04	.27	1.31
Sheathing, plywood exterior grade CDX, 1/2" thick	1.000	S.F.	.011	.72	.63	1.35
5/8" thick	1.000	S.F.	.012	.86	.65	1.51
3/4" thick	1.000	S.F.	.013	1.02	.71	1.73
Boards, 1" x 8" laid regular	1.000	S.F.	.016	2.09	.88	2.97
Laid diagonal	1.000	S.F.	.019	2.09	1.04	3.13
1" x 10" laid regular	1.000	S.F.	.015	2.14	.80	2.94
Laid diagonal	1.000	S.F.	.018	2.14	.98	3.12
Furring, 1" x 3", 12" O.C.	1.250	L.F.	.029	.53	1.58	2.11
16" O.C.	1.000	L.F.	.023	.42	1.26	1.68
24" O.C.	.750	L.F.	.017	.32	.95	1.27

3 | FRAMING — 04 | Floor Framing Systems

Diagram labels: CWJ Rim Joist, Plywood Sheathing, Temporary Strut Lines 1" x 4", 8'-0" O.C., Web Stiffener, Girder, Composite Wood Joists (CWJ)

System Description	QUAN.	UNIT	LABOR HOURS	COST PER S.F. MAT.	COST PER S.F. INST.	COST PER S.F. TOTAL
9-1/2" COMPOSITE WOOD JOISTS, 16" O.C.						
CWJ, 9-1/2", 16" O.C., 15' span	1.000	L.F.	.018	1.98	.98	2.96
Temp. strut line, 1" x 4", 8' O.C.	.160	L.F.	.003	.08	.18	.26
CWJ rim joist, 9-1/2"	.150	L.F.	.003	.30	.15	.45
Concrete filled steel column, 4" diameter	.125	L.F.	.002	.27	.13	.40
Girder, built up from three 2" x 8"	.125	L.F.	.013	.38	.73	1.11
Sheathing, plywood, subfloor, 5/8" CDX	1.000	S.F.	.012	.86	.65	1.51
TOTAL		S.F.	.051	3.87	2.82	6.69
11-1/2" COMPOSITE WOOD JOISTS, 16" O.C.						
CWJ, 11-1/2", 16" O.C., 18' span	1.000	L.F.	.018	2.18	1	3.18
Temp. strut line, 1" x 4", 8' O.C.	.160	L.F.	.003	.08	.18	.26
CWJ rim joist, 11-1/2"	.150	L.F.	.003	.33	.15	.48
Concrete filled steel column, 4" diameter	.125	L.F.	.002	.27	.13	.40
Girder, built up from three 2" x 10"	.125	L.F.	.014	.55	.78	1.33
Sheathing, plywood, subfloor, 5/8" CDX	1.000	S.F.	.012	.86	.65	1.51
TOTAL		S.F.	.052	4.27	2.89	7.16
14" COMPOSITE WOOD JOISTS, 16" O.C.						
CWJ, 14", 16" O.C., 22' span	1.000	L.F.	.020	2.30	1.08	3.38
Temp. strut line, 1" x 4", 8' O.C.	.160	L.F.	.003	.08	.18	.26
CWJ rim joist, 14"	.150	L.F.	.003	.35	.16	.51
Concrete filled steel column, 4" diameter	.600	L.F.	.002	.27	.13	.40
Girder, built up from three 2" x 12"	.600	L.F.	.015	.71	.83	1.54
Sheathing, plywood, subfloor, 5/8" CDX	1.000	S.F.	.012	.86	.65	1.51
TOTAL		S.F.	.055	4.57	3.03	7.60

Floor costs on this page are given on a cost per square foot basis.

Description	QUAN.	UNIT	LABOR HOURS	COST PER S.F. MAT.	COST PER S.F. INST.	COST PER S.F. TOTAL

Floor Framing Price Sheet (Wood)

	QUAN.	UNIT	LABOR HOURS	COST PER S.F. MAT.	COST PER S.F. INST.	COST PER S.F. TOTAL
Composite wood joist 9-1/2" deep, 12" O.C.	1.250	L.F.	.022	2.47	1.23	3.70
16" O.C.	1.000	L.F.	.018	1.98	.98	2.96
11-1/2" deep, 12" O.C.	1.250	L.F.	.023	2.72	1.25	3.97
16" O.C.	1.000	L.F.	.018	2.18	1	3.18
14" deep, 12" O.C.	1.250	L.F.	.024	2.88	1.34	4.22
16" O.C.	1.000	L.F.	.020	2.30	1.08	3.38
16" deep, 12" O.C.	1.250	L.F.	.026	4.78	1.41	6.19
16" O.C.	1.000	L.F.	.021	3.83	1.13	4.96
CWJ rim joist, 9-1/2"	.150	L.F.	.003	.30	.15	.45
11-1/2"	.150	L.F.	.003	.33	.15	.48
14"	.150	L.F.	.003	.35	.16	.51
16"	.150	L.F.	.003	.57	.17	.74
Girders, including lally columns, 3 pieces spiked together, 2" x 8"	.125	L.F.	.015	.65	.86	1.51
2" x 10"	.125	L.F.	.016	.82	.91	1.73
2" x 12"	.125	L.F.	.017	.98	.96	1.94
Solid girders, 3" x 8"	.040	L.F.	.004	.39	.22	.61
3" x 10"	.040	L.F.	.004	.42	.23	.65
3" x 12"	.040	L.F.	.004	.45	.24	.69
4" x 8"	.040	L.F.	.004	.40	.24	.64
4" x 10"	.040	L.F.	.004	.46	.25	.71
4" x 12"	.040	L.F.	.004	.50	.26	.76
Steel girders, bolted & including fabrication, wide flange shapes						
12" deep, 14#/l.f.	.040	L.F.	.061	24	6.10	30.10
10" deep, 15#/l.f.	.040	L.F.	.067	26	6.70	32.70
8" deep, 10#/l.f.	.040	L.F.	.067	17.50	6.70	24.20
6" deep, 9#/l.f.	.040	L.F.	.067	15.75	6.70	22.45
5" deep, 16#/l.f.	.040	L.F.	.064	25	6.45	31.45
Sheathing, plywood exterior grade CDX, 1/2" thick	1.000	S.F.	.011	.72	.63	1.35
5/8" thick	1.000	S.F.	.012	.86	.65	1.51
3/4" thick	1.000	S.F.	.013	1.02	.71	1.73
Boards, 1" x 8" laid regular	1.000	S.F.	.016	2.09	.88	2.97
Laid diagonal	1.000	S.F.	.019	2.09	1.04	3.13
1" x 10" laid regular	1.000	S.F.	.015	2.14	.80	2.94
Laid diagonal	1.000	S.F.	.018	2.14	.98	3.12
Furring, 1" x 3", 12" O.C.	1.250	L.F.	.029	.53	1.58	2.11
16" O.C.	1.000	L.F.	.023	.42	1.26	1.68
24" O.C.	.750	L.F.	.017	.32	.95	1.27

3 | FRAMING 06 | Floor Framing Systems

System Description	QUAN.	UNIT	LABOR HOURS	COST PER S.F. MAT.	COST PER S.F. INST.	COST PER S.F. TOTAL
12" OPEN WEB JOISTS, 16" O.C.						
OWJ 12", 16" O.C., 21' span	1.000	L.F.	.018	3.88	1	4.88
Continuous ribbing, 2" x 4"	.150	L.F.	.002	.07	.11	.18
Concrete filled steel column, 4" diameter	.125	L.F.	.002	.27	.13	.40
Girder, built up from three 2" x 8"	.125	L.F.	.013	.38	.73	1.11
Sheathing, plywood, subfloor, 5/8" CDX	1.000	S.F.	.012	.86	.65	1.51
Furring, 1" x 3", 16" O.C.	1.000	L.F.	.023	.42	1.26	1.68
TOTAL		S.F.	.070	5.88	3.88	9.76
14" OPEN WEB WOOD JOISTS, 16" O.C.						
OWJ 14", 16" O.C., 22' span	1.000	L.F.	.020	4.10	1.08	5.18
Continuous ribbing, 2" x 4"	.150	L.F.	.002	.07	.11	.18
Concrete filled steel column, 4" diameter	.125	L.F.	.002	.27	.13	.40
Girder, built up from three 2" x 10"	.125	L.F.	.014	.55	.78	1.33
Sheathing, plywood, subfloor, 5/8" CDX	1.000	S.F.	.012	.86	.65	1.51
Furring, 1" x 3",16" O.C.	1.000	L.F.	.023	.42	1.26	1.68
TOTAL		S.F.	.073	6.27	4.01	10.28
16" OPEN WEB WOOD JOISTS, 16" O.C.						
OWJ 16", 16" O.C., 24' span	1.000	L.F.	.021	4.20	1.13	5.33
Continuous ribbing, 2" x 4"	.150	L.F.	.002	.07	.11	.18
Concrete filled steel column, 4" diameter	.125	L.F.	.002	.27	.13	.40
Girder, built up from three 2" x 12"	.125	L.F.	.015	.71	.83	1.54
Sheathing, plywood, subfloor, 5/8" CDX	1.000	S.F.	.012	.86	.65	1.51
Furring, 1" x 3", 16" O.C.	1.000	L.F.	.023	.42	1.26	1.68
TOTAL		S.F.	.075	6.53	4.11	10.64

Floor costs on this page are given on a cost per square foot basis.

Description	QUAN.	UNIT	LABOR HOURS	COST PER S.F. MAT.	COST PER S.F. INST.	COST PER S.F. TOTAL

Floor Framing Price Sheet (Wood)

	QUAN.	UNIT	LABOR HOURS	COST PER S.F. MAT.	INST.	TOTAL
Open web joists, 12" deep, 12" O.C.	1.250	L.F.	.023	4.84	1.25	6.09
16" O.C.	1.000	L.F.	.018	3.88	1	4.88
14" deep, 12" O.C.	1.250	L.F.	.024	5.25	1.34	6.59
16" O.C.	1.000	L.F.	.020	4.20	1.08	5.28
16" deep, 12" O.C.	1.250	L.F.	.026	5.25	1.41	6.66
16" O.C.	1.000	L.F.	.021	4.20	1.13	5.33
18" deep, 12" O.C.	1.250	L.F.	.027	5.50	1.50	7
16" O.C.	1.000	L.F.	.022	4.40	1.20	5.60
Continuous ribbing, 2" x 4"	.150	L.F.	.002	.07	.11	.18
2" x 6"	.150	L.F.	.002	.11	.11	.22
2" x 8"	.150	L.F.	.002	.15	.12	.27
2" x 10"	.150	L.F.	.003	.22	.15	.37
2" x 12"	.150	L.F.	.003	.29	.15	.44
Girders, including lally columns, 3 pieces spiked together, 2" x 8"	.125	L.F.	.015	.65	.86	1.51
2" x 10"	.125	L.F.	.016	.82	.91	1.73
2" x 12"	.125	L.F.	.017	.98	.96	1.94
Solid girders, 3" x 8"	.040	L.F.	.004	.39	.22	.61
3" x 10"	.040	L.F.	.004	.42	.23	.65
3" x 12"	.040	L.F.	.004	.45	.24	.69
4" x 8"	.040	L.F.	.004	.40	.24	.64
4" x 10"	.040	L.F.	.004	.46	.25	.71
4" x 12"	.040	L.F.	.004	.50	.26	.76
Steel girders, bolted & including fabrication, wide flange shapes						
12" deep, 14#/l.f.	.040	L.F.	.061	24	6.10	30.10
10" deep, 15#/l.f.	.040	L.F.	.067	26	6.70	32.70
8" deep, 10#/l.f.	.040	L.F.	.067	17.50	6.70	24.20
6" deep, 9#/l.f.	.040	L.F.	.067	15.75	6.70	22.45
5" deep, 16#/l.f.	.040	L.F.	.064	25	6.45	31.45
Sheathing, plywood exterior grade CDX, 1/2" thick	1.000	S.F.	.011	.72	.63	1.35
5/8" thick	1.000	S.F.	.012	.86	.65	1.51
3/4" thick	1.000	S.F.	.013	1.02	.71	1.73
Boards, 1" x 8" laid regular	1.000	S.F.	.016	2.09	.88	2.97
Laid diagonal	1.000	S.F.	.019	2.09	1.04	3.13
1" x 10" laid regular	1.000	S.F.	.015	2.14	.80	2.94
Laid diagonal	1.000	S.F.	.018	2.14	.98	3.12
Furring, 1" x 3", 12" O.C.	1.250	L.F.	.029	.53	1.58	2.11
16" O.C.	1.000	L.F.	.023	.42	1.26	1.68
24" O.C.	.750	L.F.	.017	.32	.95	1.27

3 | FRAMING   08 | Exterior Wall Framing Systems

System Description	QUAN.	UNIT	LABOR HOURS	COST PER S.F. MAT.	COST PER S.F. INST.	COST PER S.F. TOTAL
2" X 4", 16" O.C.						
2" x 4" studs, 16" O.C.	1.000	L.F.	.015	.46	.80	1.26
Plates, 2" x 4", double top, single bottom	.375	L.F.	.005	.17	.30	.47
Corner bracing, let-in, 1" x 6"	.063	L.F.	.003	.05	.19	.24
Sheathing, 1/2" plywood, CDX	1.000	S.F.	.011	.72	.63	1.35
TOTAL		S.F.	.034	1.40	1.92	3.32
2" X 4", 24" O.C.						
2" x 4" studs, 24" O.C.	.750	L.F.	.011	.35	.60	.95
Plates, 2" x 4", double top, single bottom	.375	L.F.	.005	.17	.30	.47
Corner bracing, let-in, 1" x 6"	.063	L.F.	.002	.05	.12	.17
Sheathing, 1/2" plywood, CDX	1.000	S.F.	.011	.72	.63	1.35
TOTAL		S.F.	.029	1.29	1.65	2.94
2" X 6", 16" O.C.						
2" x 6" studs, 16" O.C.	1.000	L.F.	.016	.73	.88	1.61
Plates, 2" x 6", double top, single bottom	.375	L.F.	.006	.27	.33	.60
Corner bracing, let-in, 1" x 6"	.063	L.F.	.003	.05	.19	.24
Sheathing, 1/2" plywood, CDX	1.000	S.F.	.014	.72	.78	1.50
TOTAL		S.F.	.039	1.77	2.18	3.95
2" X 6", 24" O.C.						
2" x 6" studs, 24" O.C.	.750	L.F.	.012	.55	.66	1.21
Plates, 2" x 6", double top, single bottom	.375	L.F.	.006	.27	.33	.60
Corner bracing, let-in, 1" x 6"	.063	L.F.	.002	.05	.12	.17
Sheathing, 1/2" plywood, CDX	1.000	S.F.	.011	.72	.63	1.35
TOTAL		S.F.	.031	1.59	1.74	3.33

The wall costs on this page are given in cost per square foot of wall.
For window and door openings see below.

Description	QUAN.	UNIT	LABOR HOURS	COST PER S.F. MAT.	COST PER S.F. INST.	COST PER S.F. TOTAL

Exterior Wall Framing Price Sheet	QUAN.	UNIT	LABOR HOURS	COST PER S.F. MAT.	COST PER S.F. INST.	COST PER S.F. TOTAL
Studs, #2 or better, 2" x 4", 12" O.C.	1.250	L.F.	.018	.58	1	1.58
16" O.C.	1.000	L.F.	.015	.46	.80	1.26
24" O.C.	.750	L.F.	.011	.35	.60	.95
32" O.C.	.600	L.F.	.009	.28	.48	.76
2" x 6", 12" O.C.	1.250	L.F.	.020	.91	1.10	2.01
16" O.C.	1.000	L.F.	.016	.73	.88	1.61
24" O.C.	.750	L.F.	.012	.55	.66	1.21
32" O.C.	.600	L.F.	.010	.44	.53	.97
2" x 8", 12" O.C.	1.250	L.F.	.025	1.48	1.38	2.86
16" O.C.	1.000	L.F.	.020	1.18	1.10	2.28
24" O.C.	.750	L.F.	.015	.89	.83	1.72
32" O.C.	.600	L.F.	.012	.71	.66	1.37
Plates, #2 or better, double top, single bottom, 2" x 4"	.375	L.F.	.005	.17	.30	.47
2" x 6"	.375	L.F.	.006	.27	.33	.60
2" x 8"	.375	L.F.	.008	.44	.41	.85
Corner bracing, let-in 1" x 6" boards, studs, 12" O.C.	.070	L.F.	.004	.06	.21	.27
16" O.C.	.063	L.F.	.003	.05	.19	.24
24" O.C.	.063	L.F.	.002	.05	.12	.17
32" O.C.	.057	L.F.	.002	.05	.11	.16
Let-in steel ("T" shape), studs, 12" O.C.	.070	L.F.	.001	.06	.05	.11
16" O.C.	.063	L.F.	.001	.06	.05	.11
24" O.C.	.063	L.F.	.001	.06	.05	.11
32" O.C.	.057	L.F.	.001	.05	.04	.09
Sheathing, plywood CDX, 3/8" thick	1.000	S.F.	.010	.66	.58	1.24
1/2" thick	1.000	S.F.	.011	.72	.63	1.35
5/8" thick	1.000	S.F.	.012	.86	.68	1.54
3/4" thick	1.000	S.F.	.013	1.02	.74	1.76
Boards, 1" x 6", laid regular	1.000	S.F.	.025	1.73	1.36	3.09
Laid diagonal	1.000	S.F.	.027	1.73	1.51	3.24
1" x 8", laid regular	1.000	S.F.	.021	2.09	1.15	3.24
Laid diagonal	1.000	S.F.	.025	2.09	1.36	3.45
Wood fiber, regular, no vapor barrier, 1/2" thick	1.000	S.F.	.013	.67	.74	1.41
5/8" thick	1.000	S.F.	.013	.85	.74	1.59
Asphalt impregnated 25/32" thick	1.000	S.F.	.013	.36	.74	1.10
1/2" thick	1.000	S.F.	.013	.28	.74	1.02
Polystyrene, regular, 3/4" thick	1.000	S.F.	.010	.58	.55	1.13
2" thick	1.000	S.F.	.011	1.65	.60	2.25
Fiberglass, foil faced, 1" thick	1.000	S.F.	.008	.99	.44	1.43
2" thick	1.000	S.F.	.009	1.86	.50	2.36

Window & Door Openings	QUAN.	UNIT	LABOR HOURS	COST EACH MAT.	COST EACH INST.	COST EACH TOTAL
The following costs are to be added to the total costs of the wall for each opening. Do not subtract the area of the openings.						
Headers, 2" x 6" double, 2' long	4.000	L.F.	.178	2.92	9.80	12.72
3' long	6.000	L.F.	.267	4.38	14.70	19.08
4' long	8.000	L.F.	.356	5.85	19.60	25.45
5' long	10.000	L.F.	.444	7.30	24.50	31.80
2" x 8" double, 4' long	8.000	L.F.	.376	8.10	21	29.10
5' long	10.000	L.F.	.471	10.10	26	36.10
6' long	12.000	L.F.	.565	12.10	31	43.10
8' long	16.000	L.F.	.753	16.15	41.50	57.65
2" x 10" double, 4' long	8.000	L.F.	.400	11.75	22	33.75
6' long	12.000	L.F.	.600	17.65	33	50.65
8' long	16.000	L.F.	.800	23.50	44	67.50
10' long	20.000	L.F.	1.000	29.50	55	84.50
2" x 12" double, 8' long	16.000	L.F.	.853	30.50	47	77.50
12' long	24.000	L.F.	1.280	45.50	70.50	116

3 | FRAMING

12 | Gable End Roof Framing Systems

System Description	QUAN.	UNIT	LABOR HOURS	COST PER S.F. MAT.	COST PER S.F. INST.	COST PER S.F. TOTAL
2" X 6" RAFTERS, 16" O.C., 4/12 PITCH						
Rafters, 2" x 6", 16" O.C., 4/12 pitch	1.170	L.F.	.019	.85	1.03	1.88
Ceiling joists, 2" x 4", 16" O.C.	1.000	L.F.	.013	.47	.71	1.18
Ridge board, 2" x 6"	.050	L.F.	.002	.04	.09	.13
Fascia board, 2" x 6"	.100	L.F.	.005	.08	.29	.37
Rafter tie, 1" x 4", 4' O.C.	.060	L.F.	.001	.03	.07	.10
Soffit nailer (outrigger), 2" x 4", 24" O.C.	.170	L.F.	.004	.08	.24	.32
Sheathing, exterior, plywood, CDX, 1/2" thick	1.170	S.F.	.013	.84	.74	1.58
Furring strips, 1" x 3", 16" O.C.	1.000	L.F.	.023	.42	1.26	1.68
TOTAL		S.F.	.080	2.81	4.43	7.24
2" X 8" RAFTERS, 16" O.C., 4/12 PITCH						
Rafters, 2" x 8", 16" O.C., 4/12 pitch	1.170	L.F.	.020	1.18	1.09	2.27
Ceiling joists, 2" x 6", 16" O.C.	1.000	L.F.	.013	.73	.71	1.44
Ridge board, 2" x 8"	.050	L.F.	.002	.05	.10	.15
Fascia board, 2" x 8"	.100	L.F.	.007	.10	.39	.49
Rafter tie, 1" x 4", 4' O.C.	.060	L.F.	.001	.03	.07	.10
Soffit nailer (outrigger), 2" x 4", 24" O.C.	.170	L.F.	.004	.08	.24	.32
Sheathing, exterior, plywood, CDX, 1/2" thick	1.170	S.F.	.013	.84	.74	1.58
Furring strips, 1" x 3", 16" O.C.	1.000	L.F.	.023	.42	1.26	1.68
TOTAL		S.F.	.083	3.43	4.60	8.03

The cost of this system is based on the square foot of plan area.
All quantities have been adjusted accordingly.

Description	QUAN.	UNIT	LABOR HOURS	COST PER S.F. MAT.	COST PER S.F. INST.	COST PER S.F. TOTAL

Gable End Roof Framing Price Sheet

Gable End Roof Framing Price Sheet	QUAN.	UNIT	LABOR HOURS	COST PER S.F. MAT.	COST PER S.F. INST.	COST PER S.F. TOTAL
Rafters, #2 or better, 16" O.C., 2" x 6", 4/12 pitch	1.170	L.F.	.019	.85	1.03	1.88
8/12 pitch	1.330	L.F.	.027	.97	1.46	2.43
2" x 8", 4/12 pitch	1.170	L.F.	.020	1.18	1.09	2.27
8/12 pitch	1.330	L.F.	.028	1.34	1.57	2.91
2" x 10", 4/12 pitch	1.170	L.F.	.030	1.72	1.64	3.36
8/12 pitch	1.330	L.F.	.043	1.96	2.37	4.33
24" O.C., 2" x 6", 4/12 pitch	.940	L.F.	.015	.69	.83	1.52
8/12 pitch	1.060	L.F.	.021	.77	1.17	1.94
2" x 8", 4/12 pitch	.940	L.F.	.016	.95	.87	1.82
8/12 pitch	1.060	L.F.	.023	1.07	1.25	2.32
2" x 10", 4/12 pitch	.940	L.F.	.024	1.38	1.32	2.70
8/12 pitch	1.060	L.F.	.034	1.56	1.89	3.45
Ceiling joist, #2 or better, 2" x 4", 16" O.C.	1.000	L.F.	.013	.47	.71	1.18
24" O.C.	.750	L.F.	.010	.35	.53	.88
2" x 6", 16" O.C.	1.000	L.F.	.013	.73	.71	1.44
24" O.C.	.750	L.F.	.010	.55	.53	1.08
2" x 8", 16" O.C.	1.000	L.F.	.015	1.01	.80	1.81
24" O.C.	.750	L.F.	.011	.76	.60	1.36
2" x 10", 16" O.C.	1.000	L.F.	.018	1.47	.98	2.45
24" O.C.	.750	L.F.	.013	1.10	.74	1.84
Ridge board, #2 or better, 1" x 6"	.050	L.F.	.001	.04	.07	.11
1" x 8"	.050	L.F.	.001	.07	.08	.15
1" x 10"	.050	L.F.	.002	.09	.09	.18
2" x 6"	.050	L.F.	.002	.04	.09	.13
2" x 8"	.050	L.F.	.002	.05	.10	.15
2" x 10"	.050	L.F.	.002	.07	.11	.18
Fascia board, #2 or better, 1" x 6"	.100	L.F.	.004	.06	.22	.28
1" x 8"	.100	L.F.	.005	.06	.25	.31
1" x 10"	.100	L.F.	.005	.07	.28	.35
2" x 6"	.100	L.F.	.006	.08	.31	.39
2" x 8"	.100	L.F.	.007	.10	.39	.49
2" x 10"	.100	L.F.	.004	.29	.20	.49
Rafter tie, #2 or better, 4' O.C., 1" x 4"	.060	L.F.	.001	.03	.07	.10
1" x 6"	.060	L.F.	.001	.03	.07	.10
2" x 4"	.060	L.F.	.002	.04	.09	.13
2" x 6"	.060	L.F.	.002	.05	.11	.16
Soffit nailer (outrigger), 2" x 4", 16" O.C.	.220	L.F.	.006	.10	.31	.41
24" O.C.	.170	L.F.	.004	.08	.24	.32
2" x 6", 16" O.C.	.220	L.F.	.006	.12	.36	.48
24" O.C.	.170	L.F.	.005	.09	.28	.37
Sheathing, plywood CDX, 4/12 pitch, 3/8" thick.	1.170	S.F.	.012	.77	.68	1.45
1/2" thick	1.170	S.F.	.013	.84	.74	1.58
5/8" thick	1.170	S.F.	.014	1.01	.80	1.81
8/12 pitch, 3/8"	1.330	S.F.	.014	.88	.77	1.65
1/2" thick	1.330	S.F.	.015	.96	.84	1.80
5/8" thick	1.330	S.F.	.016	1.14	.90	2.04
Boards, 4/12 pitch roof, 1" x 6"	1.170	S.F.	.026	2.02	1.43	3.45
1" x 8"	1.170	S.F.	.021	2.45	1.18	3.63
8/12 pitch roof, 1" x 6"	1.330	S.F.	.029	2.30	1.62	3.92
1" x 8"	1.330	S.F.	.024	2.78	1.34	4.12
Furring, 1" x 3", 12" O.C.	1.200	L.F.	.027	.50	1.51	2.01
16" O.C.	1.000	L.F.	.023	.42	1.26	1.68
24" O.C.	.800	L.F.	.018	.34	1.01	1.35

3 | FRAMING — 16 | Truss Roof Framing Systems

System Description	QUAN.	UNIT	LABOR HOURS	COST PER S.F. MAT.	COST PER S.F. INST.	COST PER S.F. TOTAL
TRUSS, 16" O.C., 4/12 PITCH, 1' OVERHANG, 26' SPAN						
Truss, 40# loading, 16" O.C., 4/12 pitch, 26' span	.030	Ea.	.021	2.64	1.45	4.09
Fascia board, 2" x 6"	.100	L.F.	.005	.08	.29	.37
Sheathing, exterior, plywood, CDX, 1/2" thick	1.170	S.F.	.013	.84	.74	1.58
Furring, 1" x 3", 16" O.C.	1.000	L.F.	.023	.42	1.26	1.68
TOTAL		S.F.	.062	3.98	3.74	7.72
TRUSS, 16" O.C., 8/12 PITCH, 1' OVERHANG, 26' SPAN						
Truss, 40# loading, 16" O.C., 8/12 pitch, 26' span	.030	Ea.	.023	3.30	1.58	4.88
Fascia board, 2" x 6"	.100	L.F.	.005	.08	.29	.37
Sheathing, exterior, plywood, CDX, 1/2" thick	1.330	S.F.	.015	.96	.84	1.80
Furring, 1" x 3", 16" O.C.	1.000	L.F.	.023	.42	1.26	1.68
TOTAL		S.F.	.066	4.76	3.97	8.73
TRUSS, 24" O.C., 4/12 PITCH, 1' OVERHANG, 26' SPAN						
Truss, 40# loading, 24" O.C., 4/12 pitch, 26' span	.020	Ea.	.014	1.76	.96	2.72
Fascia board, 2" x 6"	.100	L.F.	.005	.08	.29	.37
Sheathing, exterior, plywood, CDX, 1/2" thick	1.170	S.F.	.013	.84	.74	1.58
Furring, 1" x 3", 16" O.C.	1.000	L.F.	.023	.42	1.26	1.68
TOTAL		S.F.	.055	3.10	3.25	6.35
TRUSS, 24" O.C., 8/12 PITCH, 1' OVERHANG, 26' SPAN						
Truss, 40# loading, 24" O.C., 8/12 pitch, 26' span	.020	Ea.	.015	2.20	1.05	3.25
Fascia board, 2" x 6"	.100	L.F.	.005	.08	.29	.37
Sheathing, exterior, plywood, CDX, 1/2" thick	1.330	S.F.	.015	.96	.84	1.80
Furring, 1" x 3", 16" O.C.	1.000	L.F.	.023	.42	1.26	1.68
TOTAL		S.F.	.058	3.66	3.44	7.10

The cost of this system is based on the square foot of plan area.
A one foot overhang is included.

Description	QUAN.	UNIT	LABOR HOURS	COST PER S.F. MAT.	COST PER S.F. INST.	COST PER S.F. TOTAL

Truss Roof Framing Price Sheet

	QUAN.	UNIT	LABOR HOURS	COST PER S.F. MAT.	COST PER S.F. INST.	COST PER S.F. TOTAL
Truss, 40# loading, including 1' overhang, 4/12 pitch, 24' span, 16" O.C.	.033	Ea.	.022	2.79	1.51	4.30
24" O.C.	.022	Ea.	.015	1.86	1	2.86
26' span, 16" O.C.	.030	Ea.	.021	2.64	1.45	4.09
24" O.C.	.020	Ea.	.014	1.76	.96	2.72
28' span, 16" O.C.	.027	Ea.	.020	2.78	1.40	4.18
24" O.C.	.019	Ea.	.014	1.96	.98	2.94
32' span, 16" O.C.	.024	Ea.	.019	2.95	1.31	4.26
24" O.C.	.016	Ea.	.013	1.97	.87	2.84
36' span, 16" O.C.	.022	Ea.	.019	3.12	1.30	4.42
24" O.C.	.015	Ea.	.013	2.13	.88	3.01
8/12 pitch, 24' span, 16" O.C.	.033	Ea.	.024	3.47	1.63	5.10
24" O.C.	.022	Ea.	.016	2.31	1.09	3.40
26' span, 16" O.C.	.030	Ea.	.023	3.30	1.58	4.88
24" O.C.	.020	Ea.	.015	2.20	1.05	3.25
28' span, 16" O.C.	.027	Ea.	.022	3.38	1.51	4.89
24" O.C.	.019	Ea.	.016	2.38	1.06	3.44
32' span, 16" O.C.	.024	Ea.	.021	3.43	1.45	4.88
24" O.C.	.016	Ea.	.014	2.29	.97	3.26
36' span, 16" O.C.	.022	Ea.	.021	3.87	1.47	5.34
24" O.C.	.015	Ea.	.015	2.64	1	3.64
Fascia board, #2 or better, 1" x 6"	.100	L.F.	.004	.06	.22	.28
1" x 8"	.100	L.F.	.005	.06	.25	.31
1" x 10"	.100	L.F.	.005	.07	.28	.35
2" x 6"	.100	L.F.	.006	.08	.31	.39
2" x 8"	.100	L.F.	.007	.10	.39	.49
2" x 10"	.100	L.F.	.009	.15	.49	.64
Sheathing, plywood CDX, 4/12 pitch, 3/8" thick	1.170	S.F.	.012	.77	.68	1.45
1/2" thick	1.170	S.F.	.013	.84	.74	1.58
5/8" thick	1.170	S.F.	.014	1.01	.80	1.81
8/12 pitch, 3/8" thick	1.330	S.F.	.014	.88	.77	1.65
1/2" thick	1.330	S.F.	.015	.96	.84	1.80
5/8" thick	1.330	S.F.	.016	1.14	.90	2.04
Boards, 4/12 pitch, 1" x 6"	1.170	S.F.	.026	2.02	1.43	3.45
1" x 8"	1.170	S.F.	.021	2.45	1.18	3.63
8/12 pitch, 1" x 6"	1.330	S.F.	.029	2.30	1.62	3.92
1" x 8"	1.330	S.F.	.024	2.78	1.34	4.12
Furring, 1" x 3", 12" O.C.	1.200	L.F.	.027	.50	1.51	2.01
16" O.C.	1.000	L.F.	.023	.42	1.26	1.68
24" O.C.	.800	L.F.	.018	.34	1.01	1.35

3 | FRAMING — 20 | Hip Roof Framing Systems

System Description	QUAN.	UNIT	LABOR HOURS	COST PER S.F. MAT.	COST PER S.F. INST.	COST PER S.F. TOTAL
2" X 6", 16" O.C., 4/12 PITCH						
Hip rafters, 2" x 8", 4/12 pitch	.160	L.F.	.004	.16	.20	.36
Jack rafters, 2" x 6", 16" O.C., 4/12 pitch	1.430	L.F.	.038	1.04	2.10	3.14
Ceiling joists, 2" x 6", 16" O.C.	1.000	L.F.	.013	.73	.71	1.44
Fascia board, 2" x 8"	.220	L.F.	.016	.22	.86	1.08
Soffit nailer (outrigger), 2" x 4", 24" O.C.	.220	L.F.	.006	.10	.31	.41
Sheathing, 1/2" exterior plywood, CDX	1.570	S.F.	.018	1.13	.99	2.12
Furring strips, 1" x 3", 16" O.C.	1.000	L.F.	.023	.42	1.26	1.68
TOTAL		S.F.	.118	3.80	6.43	10.23
2" X 8", 16" O.C., 4/12 PITCH						
Hip rafters, 2" x 10", 4/12 pitch	.160	L.F.	.004	.24	.25	.49
Jack rafters, 2" x 8", 16" O.C., 4/12 pitch	1.430	L.F.	.047	1.44	2.57	4.01
Ceiling joists, 2" x 6", 16" O.C.	1.000	L.F.	.013	.73	.71	1.44
Fascia board, 2" x 8"	.220	L.F.	.012	.17	.67	.84
Soffit nailer (outrigger), 2" x 4", 24" O.C.	.220	L.F.	.006	.10	.31	.41
Sheathing, 1/2" exterior plywood, CDX	1.570	S.F.	.018	1.13	.99	2.12
Furring strips, 1" x 3", 16" O.C.	1.000	L.F.	.023	.42	1.26	1.68
TOTAL		S.F.	.123	4.23	6.76	10.99

The cost of this system is based on S.F. of plan area. Measurement is area under the hip roof only. See gable roof system for added costs.

Description	QUAN.	UNIT	LABOR HOURS	COST PER S.F. MAT.	COST PER S.F. INST.	COST PER S.F. TOTAL

Hip Roof Framing Price Sheet

Hip Roof Framing Price Sheet	QUAN.	UNIT	LABOR HOURS	COST PER S.F. MAT.	COST PER S.F. INST.	COST PER S.F. TOTAL
Hip rafters, #2 or better, 2" x 6", 4/12 pitch	.160	L.F.	.003	.12	.19	.31
8/12 pitch	.210	L.F.	.006	.15	.32	.47
2" x 8", 4/12 pitch	.160	L.F.	.004	.16	.20	.36
8/12 pitch	.210	L.F.	.006	.21	.34	.55
2" x 10", 4/12 pitch	.160	L.F.	.004	.24	.25	.49
8/12 pitch roof	.210	L.F.	.008	.31	.42	.73
Jack rafters, #2 or better, 16" O.C., 2" x 6", 4/12 pitch	1.430	L.F.	.038	1.04	2.10	3.14
8/12 pitch	1.800	L.F.	.061	1.31	3.35	4.66
2" x 8", 4/12 pitch	1.430	L.F.	.047	1.44	2.57	4.01
8/12 pitch	1.800	L.F.	.075	1.82	4.12	5.94
2" x 10", 4/12 pitch	1.430	L.F.	.051	2.10	2.80	4.90
8/12 pitch	1.800	L.F.	.082	2.65	4.54	7.19
24" O.C., 2" x 6", 4/12 pitch	1.150	L.F.	.031	.84	1.69	2.53
8/12 pitch	1.440	L.F.	.048	1.05	2.68	3.73
2" x 8", 4/12 pitch	1.150	L.F.	.038	1.16	2.07	3.23
8/12 pitch	1.440	L.F.	.060	1.45	3.30	4.75
2" x 10", 4/12 pitch	1.150	L.F.	.041	1.69	2.25	3.94
8/12 pitch	1.440	L.F.	.066	2.12	3.63	5.75
Ceiling joists, #2 or better, 2" x 4", 16" O.C.	1.000	L.F.	.013	.47	.71	1.18
24" O.C.	.750	L.F.	.010	.35	.53	.88
2" x 6", 16" O.C.	1.000	L.F.	.013	.73	.71	1.44
24" O.C.	.750	L.F.	.010	.55	.53	1.08
2" x 8", 16" O.C.	1.000	L.F.	.015	1.01	.80	1.81
24" O.C.	.750	L.F.	.011	.76	.60	1.36
2" x 10", 16" O.C.	1.000	L.F.	.018	1.47	.98	2.45
24" O.C.	.750	L.F.	.013	1.10	.74	1.84
Fascia board, #2 or better, 1" x 6"	.220	L.F.	.009	.12	.47	.59
1" x 8"	.220	L.F.	.010	.14	.55	.69
1" x 10"	.220	L.F.	.011	.16	.62	.78
2" x 6"	.220	L.F.	.013	.18	.69	.87
2" x 8"	.220	L.F.	.016	.22	.86	1.08
2" x 10"	.220	L.F.	.020	.32	1.08	1.40
Soffit nailer (outrigger), 2" x 4", 16" O.C.	.280	L.F.	.007	.13	.40	.53
24" O.C.	.220	L.F.	.006	.10	.31	.41
2" x 8", 16" O.C.	.280	L.F.	.007	.21	.37	.58
24" O.C.	.220	L.F.	.005	.17	.30	.47
Sheathing, plywood CDX, 4/12 pitch, 3/8" thick	1.570	S.F.	.016	1.04	.91	1.95
1/2" thick	1.570	S.F.	.018	1.13	.99	2.12
5/8" thick	1.570	S.F.	.019	1.35	1.07	2.42
8/12 pitch, 3/8" thick	1.900	S.F.	.020	1.25	1.10	2.35
1/2" thick	1.900	S.F.	.022	1.37	1.20	2.57
5/8" thick	1.900	S.F.	.023	1.63	1.29	2.92
Boards, 4/12 pitch, 1" x 6" boards	1.450	S.F.	.032	2.51	1.77	4.28
1" x 8" boards	1.450	S.F.	.027	3.03	1.46	4.49
8/12 pitch, 1" x 6" boards	1.750	S.F.	.039	3.03	2.14	5.17
1" x 8" boards	1.750	S.F.	.032	3.66	1.77	5.43
Furring, 1" x 3", 12" O.C.	1.200	L.F.	.027	.50	1.51	2.01
16" O.C.	1.000	L.F.	.023	.42	1.26	1.68
24" O.C.	.800	L.F.	.018	.34	1.01	1.35

3 | FRAMING — 24 | Gambrel Roof Framing Systems

System Description	QUAN.	UNIT	LABOR HOURS	COST PER S.F. MAT.	COST PER S.F. INST.	COST PER S.F. TOTAL
2" X 6" RAFTERS, 16" O.C.						
Roof rafters, 2" x 6", 16" O.C.	1.430	L.F.	.029	1.04	1.57	2.61
Ceiling joists, 2" x 6", 16" O.C.	.710	L.F.	.009	.52	.50	1.02
Stud wall, 2" x 4", 16" O.C., including plates	.790	L.F.	.012	.37	.69	1.06
Furring strips, 1" x 3", 16" O.C.	.710	L.F.	.016	.30	.89	1.19
Ridge board, 2" x 8"	.050	L.F.	.002	.05	.10	.15
Fascia board, 2" x 6"	.100	L.F.	.006	.08	.31	.39
Sheathing, exterior grade plywood, 1/2" thick	1.450	S.F.	.017	1.04	.91	1.95
TOTAL		S.F.	.091	3.40	4.97	8.37
2" X 8" RAFTERS, 16" O.C.						
Roof rafters, 2" x 8", 16" O.C.	1.430	L.F.	.031	1.44	1.69	3.13
Ceiling joists, 2" x 6", 16" O.C.	.710	L.F.	.009	.52	.50	1.02
Stud wall, 2" x 4", 16" O.C., including plates	.790	L.F.	.012	.37	.69	1.06
Furring strips, 1" x 3", 16" O.C.	.710	L.F.	.016	.30	.89	1.19
Ridge board, 2" x 8"	.050	L.F.	.002	.05	.10	.15
Fascia board, 2" x 8"	.100	L.F.	.007	.10	.39	.49
Sheathing, exterior grade plywood, 1/2" thick	1.450	S.F.	.017	1.04	.91	1.95
TOTAL		S.F.	.094	3.82	5.17	8.99

The cost of this system is based on the square foot of plan area on the first floor.

Description	QUAN.	UNIT	LABOR HOURS	COST PER S.F. MAT.	COST PER S.F. INST.	COST PER S.F. TOTAL

Gambrel Roof Framing Price Sheet	QUAN.	UNIT	LABOR HOURS	COST PER S.F.		
				MAT.	INST.	TOTAL
Roof rafters, #2 or better, 2" x 6", 16" O.C.	1.430	L.F.	.029	1.04	1.57	2.61
24" O.C.	1.140	L.F.	.023	.83	1.25	2.08
2" x 8", 16" O.C.	1.430	L.F.	.031	1.44	1.69	3.13
24" O.C.	1.140	L.F.	.024	1.15	1.35	2.50
2" x 10", 16" O.C.	1.430	L.F.	.046	2.10	2.55	4.65
24" O.C.	1.140	L.F.	.037	1.68	2.03	3.71
Ceiling joist, #2 or better, 2" x 4", 16" O.C.	.710	L.F.	.009	.33	.50	.83
24" O.C.	.570	L.F.	.007	.27	.40	.67
2" x 6", 16" O.C.	.710	L.F.	.009	.52	.50	1.02
24" O.C.	.570	L.F.	.007	.42	.40	.82
2" x 8", 16" O.C.	.710	L.F.	.010	.72	.57	1.29
24" O.C.	.570	L.F.	.008	.58	.46	1.04
Stud wall, #2 or better, 2" x 4", 16" O.C.	.790	L.F.	.012	.37	.69	1.06
24" O.C.	.630	L.F.	.010	.30	.55	.85
2" x 6", 16" O.C.	.790	L.F.	.014	.58	.78	1.36
24" O.C.	.630	L.F.	.011	.46	.62	1.08
Furring, 1" x 3", 16" O.C.	.710	L.F.	.016	.30	.89	1.19
24" O.C.	.590	L.F.	.013	.25	.74	.99
Ridge board, #2 or better, 1" x 6"	.050	L.F.	.001	.04	.07	.11
1" x 8"	.050	L.F.	.001	.07	.08	.15
1" x 10"	.050	L.F.	.002	.09	.09	.18
2" x 6"	.050	L.F.	.002	.04	.09	.13
2" x 8"	.050	L.F.	.002	.05	.10	.15
2" x 10"	.050	L.F.	.002	.07	.11	.18
Fascia board, #2 or better, 1" x 6"	.100	L.F.	.004	.06	.22	.28
1" x 8"	.100	L.F.	.005	.06	.25	.31
1" x 10"	.100	L.F.	.005	.07	.28	.35
2" x 6"	.100	L.F.	.006	.08	.31	.39
2" x 8"	.100	L.F.	.007	.10	.39	.49
2" x 10"	.100	L.F.	.009	.15	.49	.64
Sheathing, plywood, exterior grade CDX, 3/8" thick	1.450	S.F.	.015	.96	.84	1.80
1/2" thick	1.450	S.F.	.017	1.04	.91	1.95
5/8" thick	1.450	S.F.	.018	1.25	.99	2.24
3/4" thick	1.450	S.F.	.019	1.48	1.07	2.55
Boards, 1" x 6", laid regular	1.450	S.F.	.032	2.51	1.77	4.28
Laid diagonal	1.450	S.F.	.036	2.51	1.97	4.48
1" x 8", laid regular	1.450	S.F.	.027	3.03	1.46	4.49
Laid diagonal	1.450	S.F.	.032	3.03	1.77	4.80

For customer support on your Residential Cost Data, call 877.713.2617.

3 | FRAMING — 28 Mansard Roof Framing Systems

System Description	QUAN.	UNIT	LABOR HOURS	MAT.	INST.	TOTAL
2" X 6" RAFTERS, 16" O.C.						
Roof rafters, 2" x 6", 16" O.C.	1.210	L.F.	.033	.88	1.82	2.70
Rafter plates, 2" x 6", double top, single bottom	.364	L.F.	.010	.27	.55	.82
Ceiling joists, 2" x 4", 16" O.C.	.920	L.F.	.012	.43	.65	1.08
Hip rafter, 2" x 6"	.070	L.F.	.002	.05	.12	.17
Jack rafter, 2" x 6", 16" O.C.	1.000	L.F.	.039	.73	2.15	2.88
Ridge board, 2" x 6"	.018	L.F.	.001	.01	.03	.04
Sheathing, exterior grade plywood, 1/2" thick	2.210	S.F.	.025	1.59	1.39	2.98
Furring strips, 1" x 3", 16" O.C.	.920	L.F.	.021	.39	1.16	1.55
TOTAL		S.F.	.143	4.35	7.87	12.22
2" X 8" RAFTERS, 16" O.C.						
Roof rafters, 2" x 8", 16" O.C.	1.210	L.F.	.036	1.22	1.97	3.19
Rafter plates, 2" x 8", double top, single bottom	.364	L.F.	.011	.37	.59	.96
Ceiling joists, 2" x 6", 16" O.C.	.920	L.F.	.012	.67	.65	1.32
Hip rafter, 2" x 8"	.070	L.F.	.002	.07	.13	.20
Jack rafter, 2" x 8", 16" O.C.	1.000	L.F.	.048	1.01	2.63	3.64
Ridge board, 2" x 8"	.018	L.F.	.001	.02	.04	.06
Sheathing, exterior grade plywood, 1/2" thick	2.210	S.F.	.025	1.59	1.39	2.98
Furring strips, 1" x 3", 16" O.C.	.920	L.F.	.021	.39	1.16	1.55
TOTAL		S.F.	.156	5.34	8.56	13.90

The cost of this system is based on the square foot of plan area.

Description	QUAN.	UNIT	LABOR HOURS	MAT.	INST.	TOTAL

Mansard Roof Framing Price Sheet	QUAN.	UNIT	LABOR HOURS	COST PER S.F. MAT.	INST.	TOTAL
Roof rafters, #2 or better, 2" x 6", 16" O.C.	1.210	L.F.	.033	.88	1.82	2.70
24" O.C.	.970	L.F.	.026	.71	1.46	2.17
2" x 8", 16" O.C.	1.210	L.F.	.036	1.22	1.97	3.19
24" O.C.	.970	L.F.	.029	.98	1.58	2.56
2" x 10", 16" O.C.	1.210	L.F.	.046	1.78	2.52	4.30
24" O.C.	.970	L.F.	.037	1.43	2.02	3.45
Rafter plates, #2 or better double top single bottom, 2" x 6"	.364	L.F.	.010	.27	.55	.82
2" x 8"	.364	L.F.	.011	.37	.59	.96
2" x 10"	.364	L.F.	.014	.54	.76	1.30
Ceiling joist, #2 or better, 2" x 4", 16" O.C.	.920	L.F.	.012	.43	.65	1.08
24" O.C.	.740	L.F.	.009	.35	.53	.88
2" x 6", 16" O.C.	.920	L.F.	.012	.67	.65	1.32
24" O.C.	.740	L.F.	.009	.54	.53	1.07
2" x 8", 16" O.C.	.920	L.F.	.013	.93	.74	1.67
24" O.C.	.740	L.F.	.011	.75	.59	1.34
Hip rafter, #2 or better, 2" x 6"	.070	L.F.	.002	.05	.12	.17
2" x 8"	.070	L.F.	.002	.07	.13	.20
2" x 10"	.070	L.F.	.003	.10	.16	.26
Jack rafter, #2 or better, 2" x 6", 16" O.C.	1.000	L.F.	.039	.73	2.15	2.88
24" O.C.	.800	L.F.	.031	.58	1.72	2.30
2" x 8", 16" O.C.	1.000	L.F.	.048	1.01	2.63	3.64
24" O.C.	.800	L.F.	.038	.81	2.10	2.91
Ridge board, #2 or better, 1" x 6"	.018	L.F.	.001	.01	.03	.04
1" x 8"	.018	L.F.	.001	.02	.03	.05
1" x 10"	.018	L.F.	.001	.03	.03	.06
2" x 6"	.018	L.F.	.001	.01	.03	.04
2" x 8"	.018	L.F.	.001	.02	.04	.06
2" x 10"	.018	L.F.	.001	.03	.04	.07
Sheathing, plywood exterior grade CDX, 3/8" thick	2.210	S.F.	.023	1.46	1.28	2.74
1/2" thick	2.210	S.F.	.025	1.59	1.39	2.98
5/8" thick	2.210	S.F.	.027	1.90	1.50	3.40
3/4" thick	2.210	S.F.	.029	2.25	1.64	3.89
Boards, 1" x 6", laid regular	2.210	S.F.	.049	3.82	2.70	6.52
Laid diagonal	2.210	S.F.	.054	3.82	3.01	6.83
1" x 8", laid regular	2.210	S.F.	.040	4.62	2.23	6.85
Laid diagonal	2.210	S.F.	.049	4.62	2.70	7.32
Furring, 1" x 3", 12" O.C.	1.150	L.F.	.026	.48	1.45	1.93
24" O.C.	.740	L.F.	.017	.31	.93	1.24

3 | FRAMING — 32 | Shed/Flat Roof Framing Systems

System Description	QUAN.	UNIT	LABOR HOURS	COST PER S.F. MAT.	COST PER S.F. INST.	COST PER S.F. TOTAL
2" X 6", 16" O.C., 4/12 PITCH						
Rafters, 2" x 6", 16" O.C., 4/12 pitch	1.170	L.F.	.019	.85	1.03	1.88
Fascia, 2" x 6"	.100	L.F.	.006	.08	.31	.39
Bridging, 1" x 3", 6' O.C.	.080	Pr.	.005	.06	.27	.33
Sheathing, exterior grade plywood, 1/2" thick	1.230	S.F.	.014	.89	.77	1.66
TOTAL		S.F.	.044	1.88	2.38	4.26
2" X 6", 24" O.C., 4/12 PITCH						
Rafters, 2" x 6", 24" O.C., 4/12 pitch	.940	L.F.	.015	.69	.83	1.52
Fascia, 2" x 6"	.100	L.F.	.006	.08	.31	.39
Bridging, 1" x 3", 6' O.C.	.060	Pr.	.004	.04	.20	.24
Sheathing, exterior grade plywood, 1/2" thick	1.230	S.F.	.014	.89	.77	1.66
TOTAL		S.F.	.039	1.70	2.11	3.81
2" X 8", 16" O.C., 4/12 PITCH						
Rafters, 2" x 8", 16" O.C., 4/12 pitch	1.170	L.F.	.020	1.18	1.09	2.27
Fascia, 2" x 8"	.100	L.F.	.007	.10	.39	.49
Bridging, 1" x 3", 6' O.C.	.080	Pr.	.005	.06	.27	.33
Sheathing, exterior grade plywood, 1/2" thick	1.230	S.F.	.014	.89	.77	1.66
TOTAL		S.F.	.046	2.23	2.52	4.75
2" X 8", 24" O.C., 4/12 PITCH						
Rafters, 2" x 8", 24" O.C., 4/12 pitch	.940	L.F.	.016	.95	.87	1.82
Fascia, 2" x 8"	.100	L.F.	.007	.10	.39	.49
Bridging, 1" x 3", 6' O.C.	.060	Pr.	.004	.04	.20	.24
Sheathing, exterior grade plywood, 1/2" thick	1.230	S.F.	.014	.89	.77	1.66
TOTAL		S.F.	.041	1.98	2.23	4.21

The cost of this system is based on the square foot of plan area.
A 1' overhang is assumed. No ceiling joists or furring are included.

Description	QUAN.	UNIT	LABOR HOURS	COST PER S.F. MAT.	COST PER S.F. INST.	COST PER S.F. TOTAL

Shed/Flat Roof Framing Price Sheet

Description	QUAN.	UNIT	LABOR HOURS	COST PER S.F. MAT.	COST PER S.F. INST.	COST PER S.F. TOTAL
Rafters, #2 or better, 16" O.C., 2" x 4", 0 - 4/12 pitch	1.170	L.F.	.014	.64	.77	1.41
5/12 - 8/12 pitch	1.330	L.F.	.020	.73	1.10	1.83
2" x 6", 0 - 4/12 pitch	1.170	L.F.	.019	.85	1.03	1.88
5/12 - 8/12 pitch	1.330	L.F.	.027	.97	1.46	2.43
2" x 8", 0 - 4/12 pitch	1.170	L.F.	.020	1.18	1.09	2.27
5/12 - 8/12 pitch	1.330	L.F.	.028	1.34	1.57	2.91
2" x 10", 0 - 4/12 pitch	1.170	L.F.	.030	1.72	1.64	3.36
5/12 - 8/12 pitch	1.330	L.F.	.043	1.96	2.37	4.33
24" O.C., 2" x 4", 0 - 4/12 pitch	.940	L.F.	.011	.52	.62	1.14
5/12 - 8/12 pitch	1.060	L.F.	.021	.77	1.17	1.94
2" x 6", 0 - 4/12 pitch	.940	L.F.	.015	.69	.83	1.52
5/12 - 8/12 pitch	1.060	L.F.	.021	.77	1.17	1.94
2" x 8", 0 - 4/12 pitch	.940	L.F.	.016	.95	.87	1.82
5/12 - 8/12 pitch	1.060	L.F.	.023	1.07	1.25	2.32
2" x 10", 0 - 4/12 pitch	.940	L.F.	.024	1.38	1.32	2.70
5/12 - 8/12 pitch	1.060	L.F.	.034	1.56	1.89	3.45
Fascia, #2 or better, 1" x 4"	.100	L.F.	.003	.04	.16	.20
1" x 6"	.100	L.F.	.004	.06	.22	.28
1" x 8"	.100	L.F.	.005	.06	.25	.31
1" x 10"	.100	L.F.	.005	.07	.28	.35
2" x 4"	.100	L.F.	.005	.07	.27	.34
2" x 6"	.100	L.F.	.006	.08	.31	.39
2" x 8"	.100	L.F.	.007	.10	.39	.49
2" x 10"	.100	L.F.	.009	.15	.49	.64
Bridging, wood 6' O.C., 1" x 3", rafters, 16" O.C.	.080	Pr.	.005	.06	.27	.33
24" O.C.	.060	Pr.	.004	.04	.20	.24
Metal, galvanized, rafters, 16" O.C.	.080	Pr.	.005	.08	.27	.35
24" O.C.	.060	Pr.	.003	.13	.19	.32
Compression type, rafters, 16" O.C.	.080	Pr.	.003	.08	.18	.26
24" O.C.	.060	Pr.	.002	.06	.13	.19
Sheathing, plywood, exterior grade, 3/8" thick, flat 0 - 4/12 pitch	1.230	S.F.	.013	.81	.71	1.52
5/12 - 8/12 pitch	1.330	S.F.	.014	.88	.77	1.65
1/2" thick, flat 0 - 4/12 pitch	1.230	S.F.	.014	.89	.77	1.66
5/12 - 8/12 pitch	1.330	S.F.	.015	.96	.84	1.80
5/8" thick, flat 0 - 4/12 pitch	1.230	S.F.	.015	1.06	.84	1.90
5/12 - 8/12 pitch	1.330	S.F.	.016	1.14	.90	2.04
3/4" thick, flat 0 - 4/12 pitch	1.230	S.F.	.016	1.25	.91	2.16
5/12 - 8/12 pitch	1.330	S.F.	.018	1.36	.98	2.34
Boards, 1" x 6", laid regular, flat 0 - 4/12 pitch	1.230	S.F.	.027	2.13	1.50	3.63
5/12 - 8/12 pitch	1.330	S.F.	.041	2.30	2.26	4.56
Laid diagonal, flat 0 - 4/12 pitch	1.230	S.F.	.030	2.13	1.67	3.80
5/12 - 8/12 pitch	1.330	S.F.	.044	2.30	2.45	4.75
1" x 8", laid regular, flat 0 - 4/12 pitch	1.230	S.F.	.022	2.57	1.24	3.81
5/12 - 8/12 pitch	1.330	S.F.	.034	2.78	1.85	4.63
Laid diagonal, flat 0 - 4/12 pitch	1.230	S.F.	.027	2.57	1.50	4.07
5/12 - 8/12 pitch	1.330	S.F.	.044	2.30	2.45	4.75

3 | FRAMING — 40 | Gable Dormer Framing Systems

System Description	QUAN.	UNIT	LABOR HOURS	COST PER S.F. MAT.	COST PER S.F. INST.	COST PER S.F. TOTAL
2" X 6", 16" O.C.						
Dormer rafter, 2" x 6", 16" O.C.	1.330	L.F.	.036	.97	2	2.97
Ridge board, 2" x 6"	.280	L.F.	.009	.20	.49	.69
Trimmer rafters, 2" x 6"	.880	L.F.	.014	.64	.77	1.41
Wall studs & plates, 2" x 4", 16" O.C.	3.160	L.F.	.056	1.49	3.10	4.59
Fascia, 2" x 6"	.220	L.F.	.012	.17	.67	.84
Valley rafter, 2" x 6", 16" O.C.	.280	L.F.	.009	.20	.48	.68
Cripple rafter, 2" x 6", 16" O.C.	.560	L.F.	.022	.41	1.20	1.61
Headers, 2" x 6", doubled	.670	L.F.	.030	.49	1.64	2.13
Ceiling joist, 2" x 4", 16" O.C.	1.000	L.F.	.013	.47	.71	1.18
Sheathing, exterior grade plywood, 1/2" thick	3.610	S.F.	.041	2.60	2.27	4.87
TOTAL		S.F.	.242	7.64	13.33	20.97
2" X 8", 16" O.C.						
Dormer rafter, 2" x 8", 16" O.C.	1.330	L.F.	.039	1.34	2.17	3.51
Ridge board, 2" x 8"	.280	L.F.	.010	.28	.55	.83
Trimmer rafter, 2" x 8"	.880	L.F.	.015	.89	.82	1.71
Wall studs & plates, 2" x 4", 16" O.C.	3.160	L.F.	.056	1.49	3.10	4.59
Fascia, 2" x 8"	.220	L.F.	.016	.22	.86	1.08
Valley rafter, 2" x 8", 16" O.C.	.280	L.F.	.010	.28	.53	.81
Cripple rafter, 2" x 8", 16" O.C.	.560	L.F.	.027	.57	1.47	2.04
Headers, 2" x 8", doubled	.670	L.F.	.032	.68	1.74	2.42
Ceiling joist, 2" x 4", 16" O.C.	1.000	L.F.	.013	.47	.71	1.18
Sheathing,, exterior grade plywood, 1/2" thick	3.610	S.F.	.041	2.60	2.27	4.87
TOTAL		S.F.	.259	8.82	14.22	23.04

The cost in this system is based on the square foot of plan area.
The measurement being the plan area of the dormer only.

Description	QUAN.	UNIT	LABOR HOURS	COST PER S.F. MAT.	COST PER S.F. INST.	COST PER S.F. TOTAL

Gable Dormer Framing Price Sheet

	QUAN.	UNIT	LABOR HOURS	COST PER S.F. MAT.	COST PER S.F. INST.	COST PER S.F. TOTAL
Dormer rafters, #2 or better, 2" x 4", 16" O.C.	1.330	L.F.	.029	.78	1.60	2.38
24" O.C.	1.060	L.F.	.023	.62	1.28	1.90
2" x 6", 16" O.C.	1.330	L.F.	.036	.97	2	2.97
24" O.C.	1.060	L.F.	.029	.77	1.59	2.36
2" x 8", 16" O.C.	1.330	L.F.	.039	1.34	2.17	3.51
24" O.C.	1.060	L.F.	.031	1.07	1.73	2.80
Ridge board, #2 or better, 1" x 4"	.280	L.F.	.006	.18	.33	.51
1" x 6"	.280	L.F.	.007	.22	.41	.63
1" x 8"	.280	L.F.	.008	.37	.45	.82
2" x 4"	.280	L.F.	.007	.16	.39	.55
2" x 6"	.280	L.F.	.009	.20	.49	.69
2" x 8"	.280	L.F.	.010	.28	.55	.83
Trimmer rafters, #2 or better, 2" x 4"	.880	L.F.	.011	.51	.62	1.13
2" x 6"	.880	L.F.	.014	.64	.77	1.41
2" x 8"	.880	L.F.	.015	.89	.82	1.71
2" x 10"	.880	L.F.	.022	1.29	1.23	2.52
Wall studs & plates, #2 or better, 2" x 4" studs, 16" O.C.	3.160	L.F.	.056	1.49	3.10	4.59
24" O.C.	2.800	L.F.	.050	1.32	2.74	4.06
2" x 6" studs, 16" O.C.	3.160	L.F.	.063	2.31	3.48	5.79
24" O.C.	2.800	L.F.	.056	2.04	3.08	5.12
Fascia, #2 or better, 1" x 4"	.220	L.F.	.006	.09	.35	.44
1" x 6"	.220	L.F.	.008	.11	.43	.54
1" x 8"	.220	L.F.	.009	.13	.50	.63
2" x 4"	.220	L.F.	.011	.15	.59	.74
2" x 6"	.220	L.F.	.014	.19	.74	.93
2" x 8"	.220	L.F.	.016	.22	.86	1.08
Valley rafter, #2 or better, 2" x 4"	.280	L.F.	.007	.16	.39	.55
2" x 6"	.280	L.F.	.009	.20	.48	.68
2" x 8"	.280	L.F.	.010	.28	.53	.81
2" x 10"	.280	L.F.	.012	.41	.65	1.06
Cripple rafter, #2 or better, 2" x 4", 16" O.C.	.560	L.F.	.018	.33	.97	1.30
24" O.C.	.450	L.F.	.014	.26	.77	1.03
2" x 6", 16" O.C.	.560	L.F.	.022	.41	1.20	1.61
24" O.C.	.450	L.F.	.018	.33	.97	1.30
2" x 8", 16" O.C.	.560	L.F.	.027	.57	1.47	2.04
24" O.C.	.450	L.F.	.021	.45	1.18	1.63
Headers, #2 or better double header, 2" x 4"	.670	L.F.	.024	.39	1.32	1.71
2" x 6"	.670	L.F.	.030	.49	1.64	2.13
2" x 8"	.670	L.F.	.032	.68	1.74	2.42
2" x 10"	.670	L.F.	.034	.98	1.85	2.83
Ceiling joist, #2 or better, 2" x 4", 16" O.C.	1.000	L.F.	.013	.47	.71	1.18
24" O.C.	.800	L.F.	.010	.38	.57	.95
2" x 6", 16" O.C.	1.000	L.F.	.013	.73	.71	1.44
24" O.C.	.800	L.F.	.010	.58	.57	1.15
Sheathing, plywood exterior grade, 3/8" thick	3.610	S.F.	.038	2.38	2.09	4.47
1/2" thick	3.610	S.F.	.041	2.60	2.27	4.87
5/8" thick	3.610	S.F.	.044	3.10	2.45	5.55
3/4" thick	3.610	S.F.	.048	3.68	2.67	6.35
Boards, 1" x 6", laid regular	3.610	S.F.	.089	6.25	4.91	11.16
Laid diagonal	3.610	S.F.	.099	6.25	5.45	11.70
1" x 8", laid regular	3.610	S.F.	.076	7.55	4.15	11.70
Laid diagonal	3.610	S.F.	.089	7.55	4.91	12.46

3 | FRAMING — 44 | Shed Dormer Framing Systems

System Description	QUAN.	UNIT	LABOR HOURS	COST PER S.F. MAT.	COST PER S.F. INST.	COST PER S.F. TOTAL
2" X 6" RAFTERS, 16" O.C.						
Dormer rafter, 2" x 6", 16" O.C.	1.080	L.F.	.029	.79	1.62	2.41
Trimmer rafter, 2" x 6"	.400	L.F.	.006	.29	.35	.64
Studs & plates, 2" x 4", 16" O.C.	2.750	L.F.	.049	1.29	2.70	3.99
Fascia, 2" x 6"	.250	L.F.	.014	.19	.74	.93
Ceiling joist, 2" x 4", 16" O.C.	1.000	L.F.	.013	.47	.71	1.18
Sheathing, exterior grade plywood, CDX, 1/2" thick	2.940	S.F.	.034	2.12	1.85	3.97
TOTAL		S.F.	.145	5.15	7.97	13.12
2" X 8" RAFTERS, 16" O.C.						
Dormer rafter, 2" x 8", 16" O.C.	1.080	L.F.	.032	1.09	1.76	2.85
Trimmer rafter, 2" x 8"	.400	L.F.	.007	.40	.37	.77
Studs & plates, 2" x 4", 16" O.C.	2.750	L.F.	.049	1.29	2.70	3.99
Fascia, 2" x 8"	.250	L.F.	.018	.25	.98	1.23
Ceiling joist, 2" x 6", 16" O.C.	1.000	L.F.	.013	.73	.71	1.44
Sheathing, exterior grade plywood, CDX, 1/2" thick	2.940	S.F.	.034	2.12	1.85	3.97
TOTAL		S.F.	.153	5.88	8.37	14.25
2" X 10" RAFTERS, 16" O.C.						
Dormer rafter, 2" x 10", 16" O.C.	1.080	L.F.	.041	1.59	2.25	3.84
Trimmer rafter, 2" x 10"	.400	L.F.	.010	.59	.56	1.15
Studs & plates, 2" x 4", 16" O.C.	2.750	L.F.	.049	1.29	2.70	3.99
Fascia, 2" x 10"	.250	L.F.	.022	.37	1.23	1.60
Ceiling joist, 2" x 6", 16" O.C.	1.000	L.F.	.013	.73	.71	1.44
Sheathing, exterior grade plywood, CDX, 1/2" thick	2.940	S.F.	.034	2.12	1.85	3.97
TOTAL		S.F.	.169	6.69	9.30	15.99

The cost in this system is based on the square foot of plan area.
The measurement is the plan area of the dormer only.

Description	QUAN.	UNIT	LABOR HOURS	COST PER S.F. MAT.	COST PER S.F. INST.	COST PER S.F. TOTAL

Shed Dormer Framing Price Sheet	QUAN.	UNIT	LABOR HOURS	COST PER S.F.		
				MAT.	INST.	TOTAL
Dormer rafters, #2 or better, 2" x 4", 16" O.C.	1.080	L.F.	.023	.63	1.30	1.93
24" O.C.	.860	L.F.	.019	.50	1.03	1.53
2" x 6", 16" O.C.	1.080	L.F.	.029	.79	1.62	2.41
24" O.C.	.860	L.F.	.023	.63	1.29	1.92
2" x 8", 16" O.C.	1.080	L.F.	.032	1.09	1.76	2.85
24" O.C.	.860	L.F.	.025	.87	1.40	2.27
2" x 10", 16" O.C.	1.080	L.F.	.041	1.59	2.25	3.84
24" O.C.	.860	L.F.	.032	1.26	1.79	3.05
Trimmer rafter, #2 or better, 2" x 4"	.400	L.F.	.005	.23	.28	.51
2" x 6"	.400	L.F.	.006	.29	.35	.64
2" x 8"	.400	L.F.	.007	.40	.37	.77
2" x 10"	.400	L.F.	.010	.59	.56	1.15
Studs & plates, #2 or better, 2" x 4", 16" O.C.	2.750	L.F.	.049	1.29	2.70	3.99
24" O.C.	2.200	L.F.	.039	1.03	2.16	3.19
2" x 6", 16" O.C.	2.750	L.F.	.055	2.01	3.03	5.04
24" O.C.	2.200	L.F.	.044	1.61	2.42	4.03
Fascia, #2 or better, 1" x 4"	.250	L.F.	.006	.09	.35	.44
1" x 6"	.250	L.F.	.008	.11	.43	.54
1" x 8"	.250	L.F.	.009	.13	.50	.63
2" x 4"	.250	L.F.	.011	.15	.59	.74
2" x 6"	.250	L.F.	.014	.19	.74	.93
2" x 8"	.250	L.F.	.018	.25	.98	1.23
Ceiling joist, #2 or better, 2" x 4", 16" O.C.	1.000	L.F.	.013	.47	.71	1.18
24" O.C.	.800	L.F.	.010	.38	.57	.95
2" x 6", 16" O.C.	1.000	L.F.	.013	.73	.71	1.44
24" O.C.	.800	L.F.	.010	.58	.57	1.15
2" x 8", 16" O.C.	1.000	L.F.	.015	1.01	.80	1.81
24" O.C.	.800	L.F.	.012	.81	.64	1.45
Sheathing, plywood exterior grade, 3/8" thick	2.940	S.F.	.031	1.94	1.71	3.65
1/2" thick	2.940	S.F.	.034	2.12	1.85	3.97
5/8" thick	2.940	S.F.	.036	2.53	2	4.53
3/4" thick	2.940	S.F.	.039	3	2.18	5.18
Boards, 1" x 6", laid regular	2.940	S.F.	.072	5.10	4	9.10
Laid diagonal	2.940	S.F.	.080	5.10	4.44	9.54
1" x 8", laid regular	2.940	S.F.	.062	6.15	3.38	9.53
Laid diagonal	2.940	S.F.	.072	6.15	4	10.15

Window Openings	QUAN.	UNIT	LABOR HOURS	COST EACH		
				MAT.	INST.	TOTAL
The following are to be added to the total cost of the dormers for window openings. Do not subtract window area from the stud wall quantities.						
Headers, 2" x 6" doubled, 2' long	4.000	L.F.	.178	2.92	9.80	12.72
3' long	6.000	L.F.	.267	4.38	14.70	19.08
4' long	8.000	L.F.	.356	5.85	19.60	25.45
5' long	10.000	L.F.	.444	7.30	24.50	31.80
2" x 8" doubled, 4' long	8.000	L.F.	.376	8.10	21	29.10
5' long	10.000	L.F.	.471	10.10	26	36.10
6' long	12.000	L.F.	.565	12.10	31	43.10
8' long	16.000	L.F.	.753	16.15	41.50	57.65
2" x 10" doubled, 4' long	8.000	L.F.	.400	11.75	22	33.75
6' long	12.000	L.F.	.600	17.65	33	50.65
8' long	16.000	L.F.	.800	23.50	44	67.50
10' long	20.000	L.F.	1.000	29.50	55	84.50

3 | FRAMING — 48 | Partition Framing Systems

System Description	QUAN.	UNIT	LABOR HOURS	COST PER S.F. MAT.	COST PER S.F. INST.	COST PER S.F. TOTAL
2" X 4", 16" O.C.						
2" x 4" studs, #2 or better, 16" O.C.	1.000	L.F.	.015	.46	.80	1.26
Plates, double top, single bottom	.375	L.F.	.005	.17	.30	.47
Cross bracing, let-in, 1" x 6"	.080	L.F.	.004	.06	.24	.30
TOTAL		S.F.	.024	.69	1.34	2.03
2" X 4", 24" O.C.						
2" x 4" studs, #2 or better, 24" O.C.	.800	L.F.	.012	.37	.64	1.01
Plates, double top, single bottom	.375	L.F.	.005	.17	.30	.47
Cross bracing, let-in, 1" x 6"	.080	L.F.	.003	.06	.15	.21
TOTAL		S.F.	.020	.60	1.09	1.69
2" X 6", 16" O.C.						
2" x 6" studs, #2 or better, 16" O.C.	1.000	L.F.	.016	.73	.88	1.61
Plates, double top, single bottom	.375	L.F.	.006	.27	.33	.60
Cross bracing, let-in, 1" x 6"	.080	L.F.	.004	.06	.24	.30
TOTAL		S.F.	.026	1.06	1.45	2.51
2" X 6", 24" O.C.						
2" x 6" studs, #2 or better, 24" O.C.	.800	L.F.	.013	.58	.70	1.28
Plates, double top, single bottom	.375	L.F.	.006	.27	.33	.60
Cross bracing, let-in, 1" x 6"	.080	L.F.	.003	.06	.15	.21
TOTAL		S.F.	.022	.91	1.18	2.09

The costs in this system are based on a square foot of wall area. Do not subtract for door or window openings.

Description	QUAN.	UNIT	LABOR HOURS	COST PER S.F. MAT.	COST PER S.F. INST.	COST PER S.F. TOTAL

Partition Framing Price Sheet	QUAN.	UNIT	LABOR HOURS	COST PER S.F.		
				MAT.	INST.	TOTAL
Wood studs, #2 or better, 2" x 4", 12" O.C.	1.250	L.F.	.018	.58	1	1.58
16" O.C.	1.000	L.F.	.015	.46	.80	1.26
24" O.C.	.800	L.F.	.012	.37	.64	1.01
32" O.C.	.650	L.F.	.009	.30	.52	.82
2" x 6", 12" O.C.	1.250	L.F.	.020	.91	1.10	2.01
16" O.C.	1.000	L.F.	.016	.73	.88	1.61
24" O.C.	.800	L.F.	.013	.58	.70	1.28
32" O.C.	.650	L.F.	.010	.47	.57	1.04
Plates, #2 or better double top single bottom, 2" x 4"	.375	L.F.	.005	.17	.30	.47
2" x 6"	.375	L.F.	.006	.27	.33	.60
2" x 8"	.375	L.F.	.005	.38	.30	.68
Cross bracing, let-in, 1" x 6" boards studs, 12" O.C.	.080	L.F.	.005	.08	.29	.37
16" O.C.	.080	L.F.	.004	.06	.24	.30
24" O.C.	.080	L.F.	.003	.06	.15	.21
32" O.C.	.080	L.F.	.002	.05	.12	.17
Let-in steel (T shaped) studs, 12" O.C.	.080	L.F.	.001	.09	.08	.17
16" O.C.	.080	L.F.	.001	.07	.06	.13
24" O.C.	.080	L.F.	.001	.07	.06	.13
32" O.C.	.080	L.F.	.001	.06	.05	.11
Steel straps studs, 12" O.C.	.080	L.F.	.001	.10	.06	.16
16" O.C.	.080	L.F.	.001	.09	.06	.15
24" O.C.	.080	L.F.	.001	.09	.06	.15
32" O.C.	.080	L.F.	.001	.09	.05	.14
Metal studs, load bearing 24" O.C., 20 ga. galv., 2-1/2" wide	1.000	S.F.	.015	.66	.83	1.49
3-5/8" wide	1.000	S.F.	.015	.78	.84	1.62
4" wide	1.000	S.F.	.016	.82	.86	1.68
6" wide	1.000	S.F.	.016	1.04	.88	1.92
16 ga., 2-1/2" wide	1.000	S.F.	.017	.77	.94	1.71
3-5/8" wide	1.000	S.F.	.017	.91	.96	1.87
4" wide	1.000	S.F.	.018	.96	.98	1.94
6" wide	1.000	S.F.	.018	1.20	1.01	2.21
Non-load bearing 24" O.C., 25 ga. galv., 1-5/8" wide	1.000	S.F.	.011	.20	.58	.78
2-1/2" wide	1.000	S.F.	.011	.25	.59	.84
3-5/8" wide	1.000	S.F.	.011	.30	.60	.90
4" wide	1.000	S.F.	.011	.33	.60	.93
6" wide	1.000	S.F.	.011	.40	.61	1.01
20 ga., 2-1/2" wide	1.000	S.F.	.013	.32	.74	1.06
3-5/8" wide	1.000	S.F.	.014	.37	.75	1.12
4" wide	1.000	S.F.	.014	.44	.75	1.19
6" wide	1.000	S.F.	.014	.51	.76	1.27

Window & Door Openings	QUAN.	UNIT	LABOR HOURS	COST EACH		
				MAT.	INST.	TOTAL
The following costs are to be added to the total costs of the walls. Do not subtract openings from total wall area.						
Headers, 2" x 6" double, 2' long	4.000	L.F.	.178	2.92	9.80	12.72
3' long	6.000	L.F.	.267	4.38	14.70	19.08
4' long	8.000	L.F.	.356	5.85	19.60	25.45
5' long	10.000	L.F.	.444	7.30	24.50	31.80
2" x 8" double, 4' long	8.000	L.F.	.376	8.10	21	29.10
5' long	10.000	L.F.	.471	10.10	26	36.10
6' long	12.000	L.F.	.565	12.10	31	43.10
8' long	16.000	L.F.	.753	16.15	41.50	57.65
2" x 10" double, 4' long	8.000	L.F.	.400	11.75	22	33.75
6' long	12.000	L.F.	.600	17.65	33	50.65
8' long	16.000	L.F.	.800	23.50	44	67.50
10' long	20.000	L.F.	1.000	29.50	55	84.50
2" x 12" double, 8' long	16.000	L.F.	.853	30.50	47	77.50
12' long	24.000	L.F.	1.280	45.50	70.50	116

Division 4 - Exterior Walls

4 | EXTERIOR WALLS 02 | Block Masonry Systems

System Description	QUAN.	UNIT	LABOR HOURS	COST PER S.F. MAT.	COST PER S.F. INST.	COST PER S.F. TOTAL
6" THICK CONCRETE BLOCK WALL						
6" thick concrete block, 6" x 8" x 16"	1.000	S.F.	.100	2.45	5	7.45
Masonry reinforcing, truss strips every other course	.625	L.F.	.002	.16	.09	.25
Furring, 1" x 3", 16" O.C.	1.000	L.F.	.016	.46	.89	1.35
Masonry insulation, poured perlite	1.000	S.F.	.013	1.91	.74	2.65
Stucco, 2 coats	1.000	S.F.	.069	.24	3.49	3.73
Masonry paint, 2 coats	1.000	S.F.	.016	.23	.76	.99
TOTAL		S.F.	.216	5.45	10.97	16.42
8" THICK CONCRETE BLOCK WALL						
8" thick concrete block, 8" x 8" x 16"	1.000	S.F.	.107	3.02	5.35	8.37
Masonry reinforcing, truss strips every other course	.625	L.F.	.002	.16	.09	.25
Furring, 1" x 3", 16" O.C.	1.000	L.F.	.016	.46	.89	1.35
Masonry insulation, poured perlite	1.000	S.F.	.018	2.53	.97	3.50
Stucco, 2 coats	1.000	S.F.	.069	.24	3.49	3.73
Masonry paint, 2 coats	1.000	S.F.	.016	.23	.76	.99
TOTAL		S.F.	.228	6.64	11.55	18.19
12" THICK CONCRETE BLOCK WALL						
12" thick concrete block, 12" x 8" x 16"	1.000	S.F.	.141	4.35	6.95	11.30
Masonry reinforcing, truss strips every other course	.625	L.F.	.003	.11	.14	.25
Furring, 1" x 3", 16" O.C.	1.000	L.F.	.016	.46	.89	1.35
Masonry insulation, poured perlite	1.000	S.F.	.026	3.74	1.44	5.18
Stucco, 2 coats	1.000	S.F.	.069	.24	3.49	3.73
Masonry paint, 2 coats	1.000	S.F.	.016	.23	.76	.99
TOTAL		S.F.	.271	9.13	13.67	22.80

Costs for this system are based on a square foot of wall area. Do not subtract for window openings.

Description	QUAN.	UNIT	LABOR HOURS	COST PER S.F. MAT.	COST PER S.F. INST.	COST PER S.F. TOTAL

Masonry Block Price Sheet

Masonry Block Price Sheet	QUAN.	UNIT	LABOR HOURS	COST PER S.F. MAT.	COST PER S.F. INST.	COST PER S.F. TOTAL
Block concrete, 8" x 16" regular, 4" thick	1.000	S.F.	.093	1.87	4.66	6.53
6" thick	1.000	S.F.	.100	2.45	5	7.45
8" thick	1.000	S.F.	.107	3.02	5.35	8.37
10" thick	1.000	S.F.	.111	3.15	5.55	8.70
12" thick	1.000	S.F.	.141	4.35	6.95	11.30
Solid block, 4" thick	1.000	S.F.	.096	2.15	4.83	6.98
6" thick	1.000	S.F.	.104	2.63	5.20	7.83
8" thick	1.000	S.F.	.111	3.79	5.55	9.34
10" thick	1.000	S.F.	.133	5.20	6.55	11.75
12" thick	1.000	S.F.	.148	5.75	7.25	13
Lightweight, 4" thick	1.000	S.F.	.093	1.87	4.66	6.53
6" thick	1.000	S.F.	.100	2.45	5	7.45
8" thick	1.000	S.F.	.107	3.02	5.35	8.37
10" thick	1.000	S.F.	.111	3.15	5.55	8.70
12" thick	1.000	S.F.	.141	4.35	6.95	11.30
Split rib profile, 4" thick	1.000	S.F.	.116	4.23	5.80	10.03
6" thick	1.000	S.F.	.123	4.81	6.15	10.96
8" thick	1.000	S.F.	.131	5.45	6.70	12.15
10" thick	1.000	S.F.	.157	5.70	7.75	13.45
12" thick	1.000	S.F.	.175	6.35	8.60	14.95
Masonry reinforcing, wire truss strips, every course, 8" block	1.375	L.F.	.004	.34	.19	.53
12" block	1.375	L.F.	.006	.23	.30	.53
Every other course, 8" block	.625	L.F.	.002	.16	.09	.25
12" block	.625	L.F.	.003	.11	.14	.25
Furring, wood, 1" x 3", 12" O.C.	1.250	L.F.	.020	.58	1.11	1.69
16" O.C.	1.000	L.F.	.016	.46	.89	1.35
24" O.C.	.800	L.F.	.013	.37	.71	1.08
32" O.C.	.640	L.F.	.010	.29	.57	.86
Steel, 3/4" channels, 12" O.C.	1.250	L.F.	.034	.40	1.65	2.05
16" O.C.	1.000	L.F.	.030	.36	1.47	1.83
24" O.C.	.800	L.F.	.023	.24	1.11	1.35
32" O.C.	.640	L.F.	.018	.19	.89	1.08
Masonry insulation, vermiculite or perlite poured 4" thick	1.000	S.F.	.009	1.24	.48	1.72
6" thick	1.000	S.F.	.013	1.90	.73	2.63
8" thick	1.000	S.F.	.018	2.53	.97	3.50
10" thick	1.000	S.F.	.021	3.06	1.18	4.24
12" thick	1.000	S.F.	.026	3.74	1.44	5.18
Block inserts polystyrene, 6" thick	1.000	S.F.		1.32		1.32
8" thick	1.000	S.F.		1.49		1.49
10" thick	1.000	S.F.		1.54		1.54
12" thick	1.000	S.F.		1.71		1.71
Stucco, 1 coat	1.000	S.F.	.057	.20	2.88	3.08
2 coats	1.000	S.F.	.069	.24	3.49	3.73
3 coats	1.000	S.F.	.081	.28	4.11	4.39
Painting, 1 coat	1.000	S.F.	.011	.15	.52	.67
2 coats	1.000	S.F.	.016	.23	.76	.99
Primer & 1 coat	1.000	S.F.	.013	.22	.62	.84
2 coats	1.000	S.F.	.018	.30	.86	1.16
Lath, metal lath expanded 2.5 lb/S.Y., painted	1.000	S.F.	.010	.40	.51	.91
Galvanized	1.000	S.F.	.012	.44	.56	1

4 | EXTERIOR WALLS — 04 Brick/Stone Veneer Systems

System Description	QUAN.	UNIT	LABOR HOURS	MAT.	INST.	TOTAL
SELECT COMMON BRICK						
Brick, select common, running bond	1.000	S.F.	.174	4.58	8.70	13.28
Wall ties, 7/8" x 7", 22 gauge	1.000	Ea.	.008	.16	.41	.57
Building paper, spunbonded polypropylene	1.100	S.F.	.002	.17	.12	.29
Trim, pine, painted	.125	L.F.	.004	.07	.22	.29
TOTAL		S.F.	.188	4.98	9.45	14.43
RED FACED COMMON BRICK						
Brick, common, red faced, running bond	1.000	S.F.	.182	4.10	9.10	13.20
Wall ties, 7/8" x 7", 22 gauge	1.000	Ea.	.008	.16	.41	.57
Building paper, spunbonded polypropylene	1.100	S.F.	.002	.17	.12	.29
Trim, pine, painted	.125	L.F.	.004	.07	.22	.29
TOTAL		S.F.	.196	4.50	9.85	14.35
BUFF OR GREY FACE BRICK						
Brick, buff or grey	1.000	S.F.	.182	4.32	9.10	13.42
Wall ties, 7/8" x 7", 22 gauge	1.000	Ea.	.008	.16	.41	.57
Building paper, spunbonded polypropylene	1.100	S.F.	.002	.17	.12	.29
Trim, pine, painted	.125	L.F.	.004	.07	.22	.29
TOTAL		S.F.	.196	4.72	9.85	14.57
STONE WORK, ROUGH STONE, AVERAGE						
Field stone veneer	1.000	S.F.	.223	8.98	11.19	20.17
Wall ties, 7/8" x 7", 22 gauge	1.000	Ea.	.008	.16	.41	.57
Building paper, spunbonded polypropylene	1.000	S.F.	.002	.17	.12	.29
Trim, pine, painted	.125	L.F.	.004	.07	.22	.29
TOTAL		S.F.	.237	9.38	11.94	21.32

The costs in this system are based on a square foot of wall area. Do not subtract area for window & door openings.

Description	QUAN.	UNIT	LABOR HOURS	MAT.	INST.	TOTAL

Brick/Stone Veneer Price Sheet

	QUAN.	UNIT	LABOR HOURS	COST PER S.F. MAT.	COST PER S.F. INST.	COST PER S.F. TOTAL
Brick						
Select common, running bond	1.000	S.F.	.174	4.58	8.70	13.28
Red faced, running bond	1.000	S.F.	.182	4.10	9.10	13.20
Buff or grey faced, running bond	1.000	S.F.	.182	4.32	9.10	13.42
Header every 6th course	1.000	S.F.	.216	4.77	10.85	15.62
English bond	1.000	S.F.	.286	6.10	14.30	20.40
Flemish bond	1.000	S.F.	.195	4.32	9.75	14.07
Common bond	1.000	S.F.	.267	5.45	13.35	18.80
Stack bond	1.000	S.F.	.182	4.32	9.10	13.42
Jumbo, running bond	1.000	S.F.	.092	5.15	4.61	9.76
Norman, running bond	1.000	S.F.	.125	6.50	6.25	12.75
Norwegian, running bond	1.000	S.F.	.107	5.65	5.35	11
Economy, running bond	1.000	S.F.	.129	4.48	6.45	10.93
Engineer, running bond	1.000	S.F.	.154	3.97	7.70	11.67
Roman, running bond	1.000	S.F.	.160	7.60	8	15.60
Utility, running bond	1.000	S.F.	.089	5.10	5.55	10.65
Glazed, running bond	1.000	S.F.	.190	12.95	9.55	22.50
Stone work, rough stone, average	1.000	S.F.	.179	9	11.20	20.20
Maximum	1.000	S.F.	.267	13.40	16.70	30.10
Wall ties, galvanized, corrugated 7/8" x 7", 22 gauge	1.000	Ea.	.008	.16	.41	.57
16 gauge	1.000	Ea.	.008	.31	.41	.72
Cavity wall, every 3rd course 6" long Z type, 1/4" diameter	1.330	L.F.	.010	.58	.53	1.11
3/16" diameter	1.330	L.F.	.010	.42	.53	.95
8" long, Z type, 1/4" diameter	1.330	L.F.	.010	.70	.53	1.23
3/16" diameter	1.330	L.F.	.010	.36	.53	.89
Building paper, aluminum and kraft laminated foil, 1 side	1.000	S.F.	.002	.14	.12	.26
2 sides	1.000	S.F.	.002	.15	.12	.27
#15 asphalt paper	1.100	S.F.	.002	.07	.13	.20
Polyethylene, .002" thick	1.000	S.F.	.002	.02	.12	.14
.004" thick	1.000	S.F.	.002	.03	.12	.15
.006" thick	1.000	S.F.	.002	.04	.12	.16
.010" thick	1.000	S.F.	.002	.10	.12	.22
Trim, 1" x 4", cedar	.125	L.F.	.005	.16	.28	.44
Fir	.125	L.F.	.005	.11	.28	.39
Redwood	.125	L.F.	.005	.16	.28	.44
White pine	.125	L.F.	.005	.11	.28	.39

4 | EXTERIOR WALLS — 08 | Wood Siding Systems

Trim → ... ← *Building Paper* ... ← *Beveled Cedar Siding*

System Description	QUAN.	UNIT	LABOR HOURS	COST PER S.F. MAT.	COST PER S.F. INST.	COST PER S.F. TOTAL
1/2" X 6" BEVELED CEDAR SIDING, "A" GRADE						
1/2" x 6" beveled cedar siding	1.000	S.F.	.027	4.55	1.50	6.05
Building wrap, spundbonded polypropylene	1.100	S.F.	.002	.17	.12	.29
Trim, cedar	.125	L.F.	.005	.16	.28	.44
Paint, primer & 2 coats	1.000	S.F.	.017	.22	.81	1.03
TOTAL		S.F.	.051	5.10	2.71	7.81
1/2" X 8" BEVELED CEDAR SIDING, "A" GRADE						
1/2" x 8" beveled cedar siding	1.000	S.F.	.024	6.50	1.34	7.84
Building wrap, spundbonded polypropylene	1.100	S.F.	.002	.17	.12	.29
Trim, cedar	.125	L.F.	.005	.16	.28	.44
Paint, primer & 2 coats	1.000	S.F.	.017	.22	.81	1.03
TOTAL		S.F.	.048	7.05	2.55	9.60
1" X 4" TONGUE & GROOVE, REDWOOD, VERTICAL GRAIN						
Redwood, clear, vertical grain, 1" x 10"	1.000	S.F.	.020	5.37	1.08	6.45
Building wrap, spundbonded polypropylene	1.100	S.F.	.002	.17	.12	.29
Trim, redwood	.125	L.F.	.005	.16	.28	.44
Sealer, 1 coat, stain, 1 coat	1.000	S.F.	.013	.14	.63	.77
TOTAL		S.F.	.040	5.84	2.11	7.95
1" X 6" TONGUE & GROOVE, REDWOOD, VERTICAL GRAIN						
Redwood, clear, vertical grain, 1" x 10"	1.000	S.F.	.020	5.52	1.11	6.63
Building wrap, spundbonded polypropylene	1.100	S.F.	.002	.17	.12	.29
Trim, redwood	.125	L.F.	.005	.16	.28	.44
Sealer, 1 coat, stain, 1 coat	1.000	S.F.	.013	.14	.63	.77
TOTAL		S.F.	.040	5.99	2.14	8.13

The costs in this system are based on a square foot of wall area.
Do not subtract area for door or window openings.

Description	QUAN.	UNIT	LABOR HOURS	COST PER S.F. MAT.	COST PER S.F. INST.	COST PER S.F. TOTAL

Wood Siding Price Sheet

Item	QUAN.	UNIT	LABOR HOURS	COST PER S.F. MAT.	COST PER S.F. INST.	COST PER S.F. TOTAL
Siding, beveled cedar, "A" grade, 1/2" x 6"	1.000	S.F.	.028	4.55	1.50	6.05
1/2" x 8"	1.000	S.F.	.023	6.50	1.34	7.84
"B" grade, 1/2" x 6"	1.000	S.F.	.032	5.05	1.67	6.72
1/2" x 8"	1.000	S.F.	.029	7.20	1.49	8.69
Clear grade, 1/2" x 6"	1.000	S.F.	.028	5.70	1.88	7.58
1/2" x 8"	1.000	S.F.	.023	8.15	1.68	9.83
Redwood, clear vertical grain, 1/2" x 6"	1.000	S.F.	.036	4.94	1.50	6.44
1/2" x 8"	1.000	S.F.	.032	5.40	1.34	6.74
Clear all heart vertical grain, 1/2" x 6"	1.000	S.F.	.028	5.50	1.67	7.17
1/2" x 8"	1.000	S.F.	.023	6	1.49	7.49
Siding board & batten, cedar, "B" grade, 1" x 10"	1.000	S.F.	.031	4.93	1.05	5.98
1" x 12"	1.000	S.F.	.031	4.93	1.05	5.98
Redwood, clear vertical grain, 1" x 6"	1.000	S.F.	.043	4.94	1.50	6.44
1" x 8"	1.000	S.F.	.018	4.51	1.34	5.85
White pine, #2 & better, 1" x 10"	1.000	S.F.	.029	2.41	1.34	3.75
1" x 12"	1.000	S.F.	.029	2.41	1.34	3.75
Siding vertical, tongue & groove, cedar "B" grade, 1" x 4"	1.000	S.F.	.033	5.25	1.08	6.33
1" x 6"	1.000	S.F.	.024	5.40	1.11	6.51
1" x 8"	1.000	S.F.	.024	5.55	1.15	6.70
1" x 10"	1.000	S.F.	.021	5.70	1.18	6.88
"A" grade, 1" x 4"	1.000	S.F.	.033	4.79	.99	5.78
1" x 6"	1.000	S.F.	.024	4.91	1.02	5.93
1" x 8"	1.000	S.F.	.024	5.05	1.04	6.09
1" x 10"	1.000	S.F.	.021	5.20	1.07	6.27
Clear vertical grain, 1" x 4"	1.000	S.F.	.033	4.42	.91	5.33
1" x 6"	1.000	S.F.	.024	4.52	.94	5.46
1" x 8"	1.000	S.F.	.024	4.63	.96	5.59
1" x 10"	1.000	S.F.	.021	4.75	.98	5.73
Redwood, clear vertical grain, 1" x 4"	1.000	S.F.	.033	5.35	1.08	6.43
1" x 6"	1.000	S.F.	.024	5.50	1.11	6.61
1" x 8"	1.000	S.F.	.024	5.70	1.15	6.85
1" x 10"	1.000	S.F.	.021	5.85	1.18	7.03
Clear all heart vertical grain, 1" x 4"	1.000	S.F.	.033	4.92	.99	5.91
1" x 6"	1.000	S.F.	.024	5.05	1.02	6.07
1" x 8"	1.000	S.F.	.024	5.20	1.04	6.24
1" x 10"	1.000	S.F.	.021	5.30	1.07	6.37
White pine, 1" x 10"	1.000	S.F.	.024	3.50	1.18	4.68
Siding plywood, texture 1-11 cedar, 3/8" thick	1.000	S.F.	.024	1.38	1.31	2.69
5/8" thick	1.000	S.F.	.024	2.83	1.31	4.14
Redwood, 3/8" thick	1.000	S.F.	.024	1.38	1.31	2.69
5/8" thick	1.000	S.F.	.024	2.11	1.31	3.42
Fir, 3/8" thick	1.000	S.F.	.024	.92	1.31	2.23
5/8" thick	1.000	S.F.	.024	1.73	1.31	3.04
Southern yellow pine, 3/8" thick	1.000	S.F.	.024	.92	1.31	2.23
5/8" thick	1.000	S.F.	.024	1.51	1.31	2.82
Paper, #15 asphalt felt	1.100	S.F.	.002	.07	.13	.20
Trim, cedar	.125	L.F.	.005	.16	.28	.44
Fir	.125	L.F.	.005	.11	.28	.39
Redwood	.125	L.F.	.005	.16	.28	.44
White pine	.125	L.F.	.005	.11	.28	.39
Painting, primer, & 1 coat	1.000	S.F.	.013	.14	.63	.77
2 coats	1.000	S.F.	.017	.22	.81	1.03
Stain, sealer, & 1 coat	1.000	S.F.	.017	.16	.82	.98
2 coats	1.000	S.F.	.019	.25	.89	1.14

4 | EXTERIOR WALLS — 12 | Shingle Siding Systems

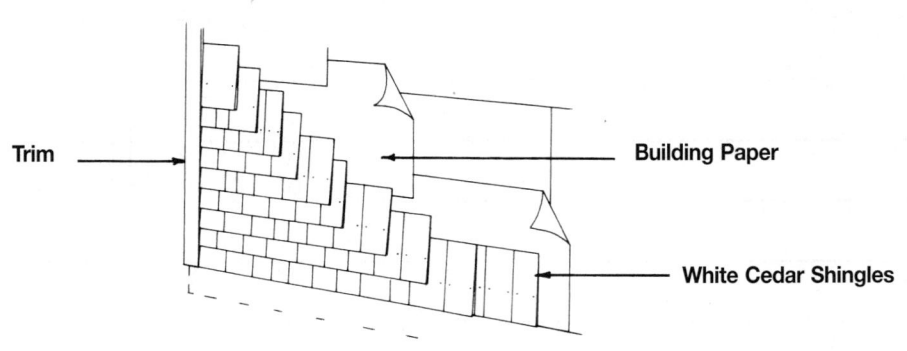

System Description	QUAN.	UNIT	LABOR HOURS	COST PER S.F. MAT.	COST PER S.F. INST.	COST PER S.F. TOTAL
WHITE CEDAR SHINGLES, 5" EXPOSURE						
White cedar shingles, 16" long, grade "A", 5" exposure	1.000	S.F.	.033	2.11	1.84	3.95
Building wrap, spunbonded polypropylene	1.100	S.F.	.002	.17	.12	.29
Trim, cedar	.125	S.F.	.005	.16	.28	.44
Paint, primer & 1 coat	1.000	S.F.	.017	.16	.82	.98
TOTAL		S.F.	.057	2.60	3.06	5.66
RESQUARED & REBUTTED PERFECTIONS, 5-1/2" EXPOSURE						
Resquared & rebutted perfections, 5-1/2" exposure	1.000	S.F.	.027	3.05	1.47	4.52
Building wrap, spunbonded polypropylene	1.100	S.F.	.002	.17	.12	.29
Trim, cedar	.125	S.F.	.005	.16	.28	.44
Stain, sealer & 1 coat	1.000	S.F.	.017	.16	.82	.98
TOTAL		S.F.	.051	3.54	2.69	6.23
HAND-SPLIT SHAKES, 8-1/2" EXPOSURE						
Hand-split red cedar shakes, 18" long, 8-1/2" exposure	1.000	S.F.	.040	1.98	2.21	4.19
Building wrap, spunbonded polypropylene	1.100	S.F.	.002	.17	.12	.29
Trim, cedar	.125	S.F.	.005	.16	.28	.44
Stain, sealer & 1 coat	1.000	S.F.	.017	.16	.82	.98
TOTAL		S.F.	.064	2.47	3.43	5.90

The costs in this system are based on a square foot of wall area.
Do not subtract area for door or window openings.

Description	QUAN.	UNIT	LABOR HOURS	COST PER S.F. MAT.	COST PER S.F. INST.	COST PER S.F. TOTAL

Shingle Siding Price Sheet

Shingle Siding Price Sheet	QUAN.	UNIT	LABOR HOURS	COST PER S.F. MAT.	COST PER S.F. INST.	COST PER S.F. TOTAL
Shingles wood, white cedar 16" long, "A" grade, 5" exposure	1.000	S.F.	.033	2.11	1.84	3.95
7" exposure	1.000	S.F.	.030	1.90	1.66	3.56
8-1/2" exposure	1.000	S.F.	.032	1.21	1.77	2.98
10" exposure	1.000	S.F.	.028	1.06	1.55	2.61
"B" grade, 5" exposure	1.000	S.F.	.040	1.82	2.21	4.03
7" exposure	1.000	S.F.	.028	1.27	1.55	2.82
8-1/2" exposure	1.000	S.F.	.024	1.09	1.33	2.42
10" exposure	1.000	S.F.	.020	.91	1.11	2.02
Fire retardant, "A" grade, 5" exposure	1.000	S.F.	.033	2.73	1.84	4.57
7" exposure	1.000	S.F.	.028	1.70	1.55	3.25
8-1/2" exposure	1.000	S.F.	.032	1.79	1.76	3.55
10" exposure	1.000	S.F.	.025	1.39	1.37	2.76
Fire retardant, 5" exposure	1.000	S.F.	.029	3.33	1.60	4.93
7" exposure	1.000	S.F.	.036	2.61	1.96	4.57
8-1/2" exposure	1.000	S.F.	.032	2.34	1.76	4.10
10" exposure	1.000	S.F.	.025	1.82	1.37	3.19
Resquared & rebutted, 5-1/2" exposure	1.000	S.F.	.027	3.05	1.47	4.52
7" exposure	1.000	S.F.	.024	2.75	1.32	4.07
8-1/2" exposure	1.000	S.F.	.021	2.44	1.18	3.62
10" exposure	1.000	S.F.	.019	2.14	1.03	3.17
Fire retardant, 5" exposure	1.000	S.F.	.027	3.67	1.47	5.14
7" exposure	1.000	S.F.	.024	3.30	1.32	4.62
8-1/2" exposure	1.000	S.F.	.021	2.93	1.18	4.11
10" exposure	1.000	S.F.	.023	2.01	1.26	3.27
Hand-split, red cedar, 24" long, 7" exposure	1.000	S.F.	.045	4.20	2.46	6.66
8-1/2" exposure	1.000	S.F.	.038	3.60	2.11	5.71
10" exposure	1.000	S.F.	.032	3	1.76	4.76
12" exposure	1.000	S.F.	.026	2.40	1.41	3.81
Fire retardant, 7" exposure	1.000	S.F.	.045	5.05	2.46	7.51
8-1/2" exposure	1.000	S.F.	.038	4.34	2.11	6.45
10" exposure	1.000	S.F.	.032	3.62	1.76	5.38
12" exposure	1.000	S.F.	.026	2.89	1.41	4.30
18" long, 5" exposure	1.000	S.F.	.068	3.37	3.76	7.13
7" exposure	1.000	S.F.	.048	2.38	2.65	5.03
8-1/2" exposure	1.000	S.F.	.040	1.98	2.21	4.19
10" exposure	1.000	S.F.	.036	1.78	1.99	3.77
Fire retardant, 5" exposure	1.000	S.F.	.068	4.42	3.76	8.18
7" exposure	1.000	S.F.	.048	3.12	2.65	5.77
8-1/2" exposure	1.000	S.F.	.040	2.60	2.21	4.81
10" exposure	1.000	S.F.	.036	2.33	1.99	4.32
Paper, #15 asphalt felt	1.100	S.F.	.002	.06	.12	.18
Trim, cedar	.125	S.F.	.005	.16	.28	.44
Fir	.125	S.F.	.005	.11	.28	.39
Redwood	.125	S.F.	.005	.16	.28	.44
White pine	.125	S.F.	.005	.11	.28	.39
Painting, primer, & 1 coat	1.000	S.F.	.013	.14	.63	.77
2 coats	1.000	S.F.	.017	.22	.81	1.03
Staining, sealer, & 1 coat	1.000	S.F.	.017	.16	.82	.98
2 coats	1.000	S.F.	.019	.25	.89	1.14

For customer support on your Residential Cost Data, call 877.713.2617.

4 | EXTERIOR WALLS 16 | Metal & Plastic Siding Systems

System Description	QUAN.	UNIT	LABOR HOURS	COST PER S.F.		
				MAT.	INST.	TOTAL
ALUMINUM CLAPBOARD SIDING, 8" WIDE, WHITE						
Aluminum horizontal siding, 8" clapboard	1.000	S.F.	.031	2.81	1.71	4.52
Backer, insulation board	1.000	S.F.	.008	.44	.44	.88
Trim, aluminum	.600	L.F.	.016	1.05	.87	1.92
Building wrap, spunbonded polypropylene	1.100	S.F.	.002	.17	.12	.29
TOTAL		S.F.	.057	4.47	3.14	7.61
ALUMINUM VERTICAL BOARD & BATTEN, WHITE						
Aluminum vertical board & batten	1.000	S.F.	.027	2.40	1.50	3.90
Backer insulation board	1.000	S.F.	.008	.44	.44	.88
Trim, aluminum	.600	L.F.	.016	1.05	.87	1.92
Building wrap, spunbonded polypropylene	1.100	S.F.	.002	.17	.12	.29
TOTAL		S.F.	.053	4.06	2.93	6.99
VINYL CLAPBOARD SIDING, 8" WIDE, WHITE						
Vinyl siding, clabboard profile, smooth texture, .042 thick, single 8	1.000	S.F.	.032	.83	1.78	2.61
Backer, insulation board	1.000	S.F.	.008	.44	.44	.88
Vinyl siding, access., outside corner, woodgrain, 4" face, 3/4" pocket	.600	L.F.	.014	1.27	.76	2.03
Building wrap, spunbonded polypropylene	1.100	S.F.	.002	.17	.12	.29
TOTAL		S.F.	.056	2.71	3.10	5.81
VINYL VERTICAL BOARD & BATTEN, WHITE						
Vinyl siding, vertical pattern, .046 thick, double 5	1.000	S.F.	.029	1.57	1.60	3.17
Backer, insulation board	1.000	S.F.	.008	.44	.44	.88
Vinyl siding, access., outside corner, woodgrain, 4" face, 3/4" pocket	.600	L.F.	.014	1.27	.76	2.03
Building wrap, spunbonded polypropylene	1.100	S.F.	.002	.17	.12	.29
TOTAL		S.F.	.053	3.45	2.92	6.37

The costs in this system are on a square foot of wall basis.
Subtract openings from wall area.

Description	QUAN.	UNIT	LABOR HOURS	COST PER S.F.		
				MAT.	INST.	TOTAL

Metal & Plastic Siding Price Sheet

Metal & Plastic Siding Price Sheet	QUAN.	UNIT	LABOR HOURS	COST PER S.F. MAT.	COST PER S.F. INST.	COST PER S.F. TOTAL
Siding, aluminum, .024" thick, smooth, 8" wide, white	1.000	S.F.	.031	2.81	1.71	4.52
Color	1.000	S.F.	.031	2.96	1.71	4.67
Double 4" pattern, 8" wide, white	1.000	S.F.	.031	2.87	1.71	4.58
Color	1.000	S.F.	.031	3.02	1.71	4.73
Double 5" pattern, 10" wide, white	1.000	S.F.	.029	2.81	1.60	4.41
Color	1.000	S.F.	.029	2.96	1.60	4.56
Embossed, single, 8" wide, white	1.000	S.F.	.031	2.79	1.71	4.50
Color	1.000	S.F.	.031	2.94	1.71	4.65
Double 4" pattern, 8" wide, white	1.000	S.F.	.031	2.87	1.71	4.58
Color	1.000	S.F.	.031	3.02	1.71	4.73
Double 5" pattern, 10" wide, white	1.000	S.F.	.029	2.85	1.60	4.45
Color	1.000	S.F.	.029	3	1.60	4.60
Alum siding with insulation board, smooth, 8" wide, white	1.000	S.F.	.031	2.52	1.71	4.23
Color	1.000	S.F.	.031	2.67	1.71	4.38
Double 4" pattern, 8" wide, white	1.000	S.F.	.031	2.50	1.71	4.21
Color	1.000	S.F.	.031	2.65	1.71	4.36
Double 5" pattern, 10" wide, white	1.000	S.F.	.029	2.50	1.60	4.10
Color	1.000	S.F.	.029	2.65	1.60	4.25
Embossed, single, 8" wide, white	1.000	S.F.	.031	2.90	1.71	4.61
Color	1.000	S.F.	.031	3.05	1.71	4.76
Double 4" pattern, 8" wide, white	1.000	S.F.	.031	2.94	1.71	4.65
Color	1.000	S.F.	.031	3.09	1.71	4.80
Double 5" pattern, 10" wide, white	1.000	S.F.	.029	2.94	1.60	4.54
Color	1.000	S.F.	.029	3.09	1.60	4.69
Aluminum, shake finish, 10" wide, white	1.000	S.F.	.029	3.15	1.60	4.75
Color	1.000	S.F.	.029	3.30	1.60	4.90
Aluminum, vertical, 12" wide, white	1.000	S.F.	.027	2.40	1.50	3.90
Color	1.000	S.F.	.027	2.55	1.50	4.05
Vinyl siding, 8" wide, smooth, white	1.000	S.F.	.032	.83	1.78	2.61
Color	1.000	S.F.	.032	.98	1.78	2.76
10" wide, Dutch lap, smooth, white	1.000	S.F.	.029	1.15	1.60	2.75
Color	1.000	S.F.	.029	1.30	1.60	2.90
Double 4" pattern, 8" wide, white	1.000	S.F.	.032	.83	1.78	2.61
Color	1.000	S.F.	.032	.98	1.78	2.76
Double 5" pattern, 10" wide, white	1.000	S.F.	.029	.83	1.60	2.43
Color	1.000	S.F.	.029	.98	1.60	2.58
Embossed, single, 8" wide, white	1.000	S.F.	.032	1.39	1.78	3.17
Color	1.000	S.F.	.032	1.54	1.78	3.32
10" wide, white	1.000	S.F.	.029	1.67	1.60	3.27
Color	1.000	S.F.	.029	1.82	1.60	3.42
Double 4" pattern, 8" wide, white	1.000	S.F.	.032	1.15	1.78	2.93
Color	1.000	S.F.	.032	1.30	1.78	3.08
Double 5" pattern, 10" wide, white	1.000	S.F.	.029	1.15	1.60	2.75
Color	1.000	S.F.	.029	1.30	1.60	2.90
Vinyl, shake finish, 10" wide, white	1.000	S.F.	.029	3.80	2.21	6.01
Color	1.000	S.F.	.029	3.95	2.21	6.16
Vinyl, vertical, double 5" pattern, 10" wide, white	1.000	S.F.	.029	1.57	1.60	3.17
Color	1.000	S.F.	.029	1.72	1.60	3.32
Backer board, installed in siding panels 8" or 10" wide	1.000	S.F.	.008	.44	.44	.88
4' x 8' sheets, polystyrene, 3/4" thick	1.000	S.F.	.010	.58	.55	1.13
4' x 8' fiberboard, plain	1.000	S.F.	.008	.44	.44	.88
Trim, aluminum, white	.600	L.F.	.016	1.05	.87	1.92
Color	.600	L.F.	.016	1.13	.87	2
Vinyl, white	.600	L.F.	.014	1.27	.76	2.03
Color	.600	L.F.	.014	1.47	.76	2.23
Paper, #15 asphalt felt	1.100	S.F.	.002	.07	.13	.20
Kraft paper, plain	1.100	S.F.	.002	.15	.13	.28
Foil backed	1.100	S.F.	.002	.17	.13	.30

4 | EXTERIOR WALLS — 20 | Insulation Systems

Description	QUAN.	UNIT	LABOR HOURS	COST PER S.F. MAT.	COST PER S.F. INST.	COST PER S.F. TOTAL
Poured insulation, cellulose fiber, R3.8 per inch (1" thick)	1.000	S.F.	.003	.06	.18	.24
Fiberglass, R4.0 per inch (1" thick)	1.000	S.F.	.003	.05	.18	.23
Mineral wool, R3.0 per inch (1" thick)	1.000	S.F.	.003	.04	.18	.22
Polystyrene, R4.0 per inch (1" thick)	1.000	S.F.	.003	.15	.18	.33
Vermiculite, R2.7 per inch (1" thick)	1.000	S.F.	.003	.48	.18	.66
Perlite, R2.7 per inch (1" thick)	1.000	S.F.	.003	.48	.18	.66
Reflective insulation, aluminum foil reinforced with scrim	1.000	S.F.	.004	.16	.23	.39
Reinforced with woven polyolefin	1.000	S.F.	.004	.24	.23	.47
With single bubble air space, R8.8	1.000	S.F.	.005	.31	.30	.61
With double bubble air space, R9.8	1.000	S.F.	.005	.35	.30	.65
Rigid insulation, fiberglass, unfaced,						
1-1/2" thick, R6.2	1.000	S.F.	.008	.44	.44	.88
2" thick, R8.3	1.000	S.F.	.008	.50	.44	.94
2-1/2" thick, R10.3	1.000	S.F.	.010	.62	.55	1.17
3" thick, R12.4	1.000	S.F.	.010	.62	.55	1.17
Foil faced, 1" thick, R4.3	1.000	S.F.	.008	.99	.44	1.43
1-1/2" thick, R6.2	1.000	S.F.	.008	1.49	.44	1.93
2" thick, R8.7	1.000	S.F.	.009	1.86	.50	2.36
2-1/2" thick, R10.9	1.000	S.F.	.010	2.18	.55	2.73
3" thick, R13.0	1.000	S.F.	.010	2.40	.55	2.95
Perlite, 1" thick R2.77	1.000	S.F.	.010	.46	.55	1.01
2" thick R5.55	1.000	S.F.	.011	.83	.60	1.43
Polystyrene, extruded, blue, 2.2#/C.F., 3/4" thick R4	1.000	S.F.	.010	.58	.55	1.13
1-1/2" thick R8.1	1.000	S.F.	.011	1.14	.60	1.74
2" thick R10.8	1.000	S.F.	.011	1.65	.60	2.25
Molded bead board, white, 1" thick R3.85	1.000	S.F.	.010	.28	.55	.83
1-1/2" thick, R5.6	1.000	S.F.	.011	.55	.60	1.15
2" thick, R7.7	1.000	S.F.	.011	.83	.60	1.43
Non-rigid insulation, batts						
Fiberglass, kraft faced, 3-1/2" thick, R13, 11" wide	1.000	S.F.	.005	.33	.38	.71
15" wide	1.000	S.F.	.005	.33	.38	.71
23" wide	1.000	S.F.	.005	.33	.38	.71
6" thick, R19, 11" wide	1.000	S.F.	.006	.46	.38	.84
15" wide	1.000	S.F.	.006	.46	.38	.84
23" wide	1.000	S.F.	.006	.46	.38	.84
9" thick, R30, 15" wide	1.000	S.F.	.006	.78	.33	1.11
23" wide	1.000	S.F.	.006	.78	.33	1.11
12" thick, R38, 15" wide	1.000	S.F.	.006	1.06	.33	1.39
23" wide	1.000	S.F.	.006	1.06	.33	1.39
Fiberglass, foil faced, 3-1/2" thick, R13, 15" wide	1.000	S.F.	.005	.50	.33	.83
23" wide	1.000	S.F.	.005	.50	.33	.83
6" thick, R19, 15" thick	1.000	S.F.	.005	.66	.28	.94
23" wide	1.000	S.F.	.005	.66	.28	.94
9" thick, R30, 15" wide	1.000	S.F.	.006	.99	.33	1.32
23" wide	1.000	S.F.	.006	.99	.33	1.32

Insulation Systems	QUAN.	UNIT	LABOR HOURS	COST PER S.F.		
				MAT.	INST.	TOTAL
Non-rigid insulation batts						
Fiberglass unfaced, 3-1/2" thick, R13, 15" wide	1.000	S.F.	.005	.34	.28	.62
23" wide	1.000	S.F.	.005	.34	.28	.62
6" thick, R19, 15" wide	1.000	S.F.	.006	.40	.33	.73
23" wide	1.000	S.F.	.006	.40	.33	.73
9" thick, R19, 15" wide	1.000	S.F.	.007	.64	.38	1.02
23" wide	1.000	S.F.	.007	.64	.38	1.02
12" thick, R38, 15" wide	1.000	S.F.	.007	.81	.38	1.19
23" wide	1.000	S.F.	.007	.81	.38	1.19
Mineral fiber batts, 3" thick, R11	1.000	S.F.	.005	.66	.28	.94
3-1/2" thick, R13	1.000	S.F.	.005	.66	.28	.94
6" thick, R19	1.000	S.F.	.005	1.04	.28	1.32
6-1/2" thick, R22	1.000	S.F.	.005	1.04	.28	1.32
10" thick, R30	1.000	S.F.	.006	1.37	.33	1.70

4 | EXTERIOR WALLS — 28 | Double Hung Window Systems

System Description	QUAN.	UNIT	LABOR HOURS	COST EACH MAT.	COST EACH INST.	COST EACH TOTAL
BUILDER'S QUALITY WOOD WINDOW 2' X 3', DOUBLE HUNG						
Window, primed, builder's quality, 2' x 3', insulating glass	1.000	Ea.	.800	256	44	300
Trim, interior casing	11.000	L.F.	.352	17.05	19.36	36.41
Paint, interior & exterior, primer & 2 coats	2.000	Face	1.778	2.62	84	86.62
Caulking	10.000	L.F.	.278	2.50	15	17.50
Snap-in grille	1.000	Set	.333	56.50	18.40	74.90
Drip cap, metal	2.000	L.F.	.040	1.22	2.20	3.42
TOTAL		Ea.	3.581	335.89	182.96	518.85
PLASTIC CLAD WOOD WINDOW 3' X 4', DOUBLE HUNG						
Window, plastic clad, premium, 3' x 4', insulating glass	1.000	Ea.	.889	435	49	484
Trim, interior casing	15.000	L.F.	.480	23.25	26.40	49.65
Paint, interior, primer & 2 coats	1.000	Face	.889	1.31	42	43.31
Caulking	14.000	L.F.	.389	3.50	21	24.50
Snap-in grille	1.000	Set	.333	56.50	18.40	74.90
TOTAL		Ea.	2.980	519.56	156.80	676.36
METAL CLAD WOOD WINDOW, 3' X 5', DOUBLE HUNG						
Window, metal clad, deluxe, 3' x 5', insulating glass	1.000	Ea.	1.000	410	55	465
Trim, interior casing	17.000	L.F.	.544	26.35	29.92	56.27
Paint, interior, primer & 2 coats	1.000	Face	.889	1.31	42	43.31
Caulking	16.000	L.F.	.444	4	24	28
Snap-in grille	1.000	Set	.235	144	13	157
Drip cap, metal	3.000	L.F.	.060	1.83	3.30	5.13
TOTAL		Ea.	3.172	587.49	167.22	754.71

The cost of this system is on a cost per each window basis.

Description	QUAN.	UNIT	LABOR HOURS	COST EACH MAT.	COST EACH INST.	COST EACH TOTAL

Double Hung Window Price Sheet

	QUAN.	UNIT	LABOR HOURS	COST EACH MAT.	COST EACH INST.	COST EACH TOTAL
Windows, double-hung, builder's quality, 2' x 3', single glass	1.000	Ea.	.800	231	44	275
Insulating glass	1.000	Ea.	.800	256	44	300
3' x 4', single glass	1.000	Ea.	.889	305	49	354
Insulating glass	1.000	Ea.	.889	320	49	369
4' x 4'-6", single glass	1.000	Ea.	1.000	355	55	410
Insulating glass	1.000	Ea.	1.000	380	55	435
Plastic clad premium insulating glass, 2'-6" x 3'	1.000	Ea.	.800	340	44	384
3' x 3'-6"	1.000	Ea.	.800	370	44	414
3' x 4'	1.000	Ea.	.889	435	49	484
3' x 4'-6"	1.000	Ea.	.889	450	49	499
3' x 5'	1.000	Ea.	1.000	485	55	540
3'-6" x 6'	1.000	Ea.	1.000	540	55	595
Metal clad deluxe insulating glass, 2'-6" x 3'	1.000	Ea.	.800	299	44	343
3' x 3'-6"	1.000	Ea.	.800	345	44	389
3' x 4'	1.000	Ea.	.889	360	49	409
3' x 4'-6"	1.000	Ea.	.889	380	49	429
3' x 5'	1.000	Ea.	1.000	410	55	465
3'-6" x 6'	1.000	Ea.	1.000	495	55	550
Trim, interior casing, window 2' x 3'	11.000	L.F.	.367	17.05	19.35	36.40
2'-6" x 3'	12.000	L.F.	.400	18.60	21	39.60
3' x 3'-6"	14.000	L.F.	.467	21.50	24.50	46
3' x 4'	15.000	L.F.	.500	23.50	26.50	50
3' x 4'-6"	16.000	L.F.	.533	25	28	53
3' x 5'	17.000	L.F.	.567	26.50	30	56.50
3'-6" x 6'	20.000	L.F.	.667	31	35	66
4' x 4'-6"	18.000	L.F.	.600	28	31.50	59.50
Paint or stain, interior or exterior, 2' x 3' window, 1 coat	1.000	Face	.444	.52	21	21.52
2 coats	1.000	Face	.727	1.03	34.50	35.53
Primer & 1 coat	1.000	Face	.727	.87	34.50	35.37
Primer & 2 coats	1.000	Face	.889	1.31	42	43.31
3' x 4' window, 1 coat	1.000	Face	.667	1.15	31.50	32.65
2 coats	1.000	Face	.667	1.80	31.50	33.30
Primer & 1 coat	1.000	Face	.727	1.34	34.50	35.84
Primer & 2 coats	1.000	Face	.889	1.31	42	43.31
4' x 4'-6" window, 1 coat	1.000	Face	.667	1.15	31.50	32.65
2 coats	1.000	Face	.667	1.80	31.50	33.30
Primer & 1 coat	1.000	Face	.727	1.34	34.50	35.84
Primer & 2 coats	1.000	Face	.889	1.31	42	43.31
Caulking, window, 2' x 3'	10.000	L.F.	.323	2.50	15	17.50
2'-6" x 3'	11.000	L.F.	.355	2.75	16.50	19.25
3' x 3'-6"	13.000	L.F.	.419	3.25	19.50	22.75
3' x 4'	14.000	L.F.	.452	3.50	21	24.50
3' x 4'-6"	15.000	L.F.	.484	3.75	22.50	26.25
3' x 5'	16.000	L.F.	.516	4	24	28
3'-6" x 6'	19.000	L.F.	.613	4.75	28.50	33.25
4' x 4'-6"	17.000	L.F.	.548	4.25	25.50	29.75
Grilles, glass size to, 16" x 24" per sash	1.000	Set	.333	56.50	18.40	74.90
32" x 32" per sash	1.000	Set	.235	144	13	157
Drip cap, aluminum, 2' long	2.000	L.F.	.040	1.22	2.20	3.42
3' long	3.000	L.F.	.060	1.83	3.30	5.13
4' long	4.000	L.F.	.080	2.44	4.40	6.84
Wood, 2' long	2.000	L.F.	.067	3.10	3.52	6.62
3' long	3.000	L.F.	.100	4.65	5.30	9.95
4' long	4.000	L.F.	.133	6.20	7.05	13.25

4 | EXTERIOR WALLS — 32 | Casement Window Systems

System Description	QUAN.	UNIT	LABOR HOURS	MAT.	INST.	TOTAL
BUILDER'S QUALITY WINDOW, WOOD, 2' BY 3', CASEMENT						
Window, primed, builder's quality, 2' x 3', insulating glass	1.000	Ea.	.800	285	44	329
Trim, interior casing	11.000	L.F.	.352	17.05	19.36	36.41
Paint, interior & exterior, primer & 2 coats	2.000	Face	1.778	2.62	84	86.62
Caulking	10.000	L.F.	.278	2.50	15	17.50
Snap-in grille	1.000	Ea.	.267	35	14.70	49.70
Drip cap, metal	2.000	L.F.	.040	1.22	2.20	3.42
TOTAL		Ea.	3.515	343.39	179.26	522.65
PLASTIC CLAD WOOD WINDOW, 2' X 4', CASEMENT						
Window, plastic clad, premium, 2' x 4', insulating glass	1.000	Ea.	.889	350	49	399
Trim, interior casing	13.000	L.F.	.416	20.15	22.88	43.03
Paint, interior, primer & 2 coats	1.000	Ea.	.889	1.31	42	43.31
Caulking	12.000	L.F.	.333	3	18	21
Snap-in grille	1.000	Ea.	.267	35	14.70	49.70
TOTAL		Ea.	2.794	409.46	146.58	556.04
METAL CLAD WOOD WINDOW, 2' X 5', CASEMENT						
Window, metal clad, deluxe, 2' x 5', insulating glass	1.000	Ea.	1.000	360	55	415
Trim, interior casing	15.000	L.F.	.480	23.25	26.40	49.65
Paint, interior, primer & 2 coats	1.000	Ea.	.889	1.31	42	43.31
Caulking	14.000	L.F.	.389	3.50	21	24.50
Snap-in grille	1.000	Ea.	.250	47.50	13.80	61.30
Drip cap, metal	12.000	L.F.	.040	1.22	2.20	3.42
TOTAL		Ea.	3.048	436.78	160.40	597.18

The cost of this system is on a cost per each window basis.

Description	QUAN.	UNIT	LABOR HOURS	MAT.	INST.	TOTAL

Casement Window Price Sheet

	QUAN.	UNIT	LABOR HOURS	COST EACH		
				MAT.	INST.	TOTAL
Window, casement, builders quality, 2' x 3', double insulated glass	1.000	Ea.	.800	295	44	339
Insulating glass	1.000	Ea.	.800	285	44	329
2' x 4'-6", double insulated glass	1.000	Ea.	.727	960	40	1,000
Insulating glass	1.000	Ea.	.727	590	40	630
2' x 6', double insulated glass	1.000	Ea.	.889	485	55	540
Insulating glass	1.000	Ea.	.889	525	55	580
Plastic clad premium insulating glass, 2' x 3'	1.000	Ea.	.800	297	44	341
2' x 4'	1.000	Ea.	.889	440	49	489
2' x 5'	1.000	Ea.	1.000	595	55	650
2' x 6'	1.000	Ea.	1.000	420	55	475
Metal clad deluxe insulating glass, 2' x 3'	1.000	Ea.	.800	305	44	349
2' x 4'	1.000	Ea.	.889	305	49	354
2' x 5'	1.000	Ea.	1.000	350	55	405
2' x 6'	1.000	Ea.	1.000	400	55	455
Trim, interior casing, window 2' x 3'	11.000	L.F.	.367	17.05	19.35	36.40
2' x 4'	13.000	L.F.	.433	20	23	43
2' x 4'-6"	14.000	L.F.	.467	21.50	24.50	46
2' x 5'	15.000	L.F.	.500	23.50	26.50	50
2' x 6'	17.000	L.F.	.567	26.50	30	56.50
Paint or stain, interior or exterior, 2' x 3' window, 1 coat	1.000	Face	.444	.52	21	21.52
2 coats	1.000	Face	.727	1.03	34.50	35.53
Primer & 1 coat	1.000	Face	.727	.87	34.50	35.37
Primer & 2 coats	1.000	Face	.889	1.31	42	43.31
2' x 4' window, 1 coat	1.000	Face	.444	.52	21	21.52
2 coats	1.000	Face	.727	1.03	34.50	35.53
Primer & 1 coat	1.000	Face	.727	.87	34.50	35.37
Primer & 2 coats	1.000	Face	.889	1.31	42	43.31
2' x 6' window, 1 coat	1.000	Face	.667	1.15	31.50	32.65
2 coats	1.000	Face	.667	1.80	31.50	33.30
Primer & 1 coat	1.000	Face	.727	1.34	34.50	35.84
Primer & 2 coats	1.000	Face	.889	1.31	42	43.31
Caulking, window, 2' x 3'	10.000	L.F.	.323	2.50	15	17.50
2' x 4'	12.000	L.F.	.387	3	18	21
2' x 4'-6"	13.000	L.F.	.419	3.25	19.50	22.75
2' x 5'	14.000	L.F.	.452	3.50	21	24.50
2' x 6'	16.000	L.F.	.516	4	24	28
Grilles, glass size, to 20" x 36"	1.000	Ea.	.267	35	14.70	49.70
To 20" x 56"	1.000	Ea.	.250	47.50	13.80	61.30
Drip cap, metal, 2' long	2.000	L.F.	.040	1.22	2.20	3.42
Wood, 2' long	2.000	L.F.	.067	3.10	3.52	6.62

4 | EXTERIOR WALLS — 36 | Awning Window Systems

System Description	QUAN.	UNIT	LABOR HOURS	COST EACH MAT.	COST EACH INST.	COST EACH TOTAL
BUILDER'S QUALITY WINDOW, WOOD, 34" X 22", AWNING						
Window, 34" x 22", insulating glass	1.000	Ea.	.800	298	44	342
Trim, interior casing	10.500	L.F.	.336	16.28	18.48	34.76
Paint, interior & exterior, primer & 2 coats	2.000	Face	1.778	2.62	84	86.62
Caulking	9.500	L.F.	.264	2.38	14.25	16.63
Snap-in grille	1.000	Ea.	.267	31	14.70	45.70
Drip cap, metal	3.000	L.F.	.060	1.83	3.30	5.13
TOTAL		Ea.	3.505	352.11	178.73	530.84
PLASTIC CLAD WOOD WINDOW, 40" X 28", AWNING						
Window, plastic clad, premium, 40" x 28", insulating glass	1.000	Ea.	.889	375	49	424
Trim interior casing	13.500	L.F.	.432	20.93	23.76	44.69
Paint, interior, primer & 2 coats	1.000	Face	.889	1.31	42	43.31
Caulking	12.500	L.F.	.347	3.13	18.75	21.88
Snap-in grille	1.000	Ea.	.267	31	14.70	45.70
TOTAL		Ea.	2.824	431.37	148.21	579.58
METAL CLAD WOOD WINDOW, 48" X 36", AWNING						
Window, metal clad, deluxe, 48" x 36", insulating glass	1.000	Ea.	1.000	410	55	465
Trim, interior casing	15.000	L.F.	.480	23.25	26.40	49.65
Paint, interior, primer & 2 coats	1.000	Face	.889	1.31	42	43.31
Caulking	14.000	L.F.	.389	3.50	21	24.50
Snap-in grille	1.000	Ea.	.250	44	13.80	57.80
Drip cap, metal	4.000	L.F.	.080	2.44	4.40	6.84
TOTAL		Ea.	3.088	484.50	162.60	647.10

The cost of this system is on a cost per each window basis.

Description	QUAN.	UNIT	LABOR HOURS	COST EACH MAT.	COST EACH INST.	COST EACH TOTAL

Awning Window Price Sheet

	QUAN.	UNIT	LABOR HOURS	COST EACH MAT.	COST EACH INST.	COST EACH TOTAL
Windows, awning, builder's quality, 34" x 22", insulated glass	1.000	Ea.	.800	297	44	341
Low E glass	1.000	Ea.	.800	298	44	342
40" x 28", insulated glass	1.000	Ea.	.889	345	49	394
Low E glass	1.000	Ea.	.889	375	49	424
48" x 36", insulated glass	1.000	Ea.	1.000	515	55	570
Low E glass	1.000	Ea.	1.000	540	55	595
Plastic clad premium insulating glass, 34" x 22"	1.000	Ea.	.800	290	44	334
40" x 22"	1.000	Ea.	.800	315	44	359
36" x 28"	1.000	Ea.	.889	335	49	384
36" x 36"	1.000	Ea.	.889	375	49	424
48" x 28"	1.000	Ea.	1.000	405	55	460
60" x 36"	1.000	Ea.	1.000	550	55	605
Metal clad deluxe insulating glass, 34" x 22"	1.000	Ea.	.800	271	49	320
40" x 22"	1.000	Ea.	.800	320	49	369
36" x 25"	1.000	Ea.	.889	299	49	348
40" x 30"	1.000	Ea.	.889	375	49	424
48" x 28"	1.000	Ea.	1.000	380	55	435
60" x 36"	1.000	Ea.	1.000	410	55	465
Trim, interior casing window, 34" x 22"	10.500	L.F.	.350	16.30	18.50	34.80
40" x 22"	11.500	L.F.	.383	17.85	20	37.85
36" x 28"	12.500	L.F.	.417	19.40	22	41.40
40" x 28"	13.500	L.F.	.450	21	24	45
48" x 28"	14.500	L.F.	.483	22.50	25.50	48
48" x 36"	15.000	L.F.	.500	23.50	26.50	50
Paint or stain, interior or exterior, 34" x 22", 1 coat	1.000	Face	.444	.52	21	21.52
2 coats	1.000	Face	.727	1.03	34.50	35.53
Primer & 1 coat	1.000	Face	.727	.87	34.50	35.37
Primer & 2 coats	1.000	Face	.889	1.31	42	43.31
36" x 28", 1 coat	1.000	Face	.444	.52	21	21.52
2 coats	1.000	Face	.727	1.03	34.50	35.53
Primer & 1 coat	1.000	Face	.727	.87	34.50	35.37
Primer & 2 coats	1.000	Face	.889	1.31	42	43.31
48" x 36", 1 coat	1.000	Face	.667	1.15	31.50	32.65
2 coats	1.000	Face	.667	1.80	31.50	33.30
Primer & 1 coat	1.000	Face	.727	1.34	34.50	35.84
Primer & 2 coats	1.000	Face	.889	1.31	42	43.31
Caulking, window, 34" x 22"	9.500	L.F.	.306	2.38	14.25	16.63
40" x 22"	10.500	L.F.	.339	2.63	15.75	18.38
36" x 28"	11.500	L.F.	.371	2.88	17.25	20.13
40" x 28"	12.500	L.F.	.403	3.13	18.75	21.88
48" x 28"	13.500	L.F.	.436	3.38	20.50	23.88
48" x 36"	14.000	L.F.	.452	3.50	21	24.50
Grilles, glass size, to 28" by 16"	1.000	Ea.	.267	31	14.70	45.70
To 44" by 24"	1.000	Ea.	.250	44	13.80	57.80
Drip cap, aluminum, 3' long	3.000	L.F.	.060	1.83	3.30	5.13
3'-6" long	3.500	L.F.	.070	2.14	3.85	5.99
4' long	4.000	L.F.	.080	2.44	4.40	6.84
Wood, 3' long	3.000	L.F.	.100	4.65	5.30	9.95
3'-6" long	3.500	L.F.	.117	5.45	6.15	11.60
4' long	4.000	L.F.	.133	6.20	7.05	13.25

4 | EXTERIOR WALLS — 40 Sliding Window Systems

System Description	QUAN.	UNIT	LABOR HOURS	COST EACH MAT.	COST EACH INST.	COST EACH TOTAL
BUILDER'S QUALITY WOOD WINDOW, 3' X 2', SLIDING						
Window, primed, builder's quality, 3' x 3', insul. glass	1.000	Ea.	.800	340	44	384
Trim, interior casing	11.000	L.F.	.352	17.05	19.36	36.41
Paint, interior & exterior, primer & 2 coats	2.000	Face	1.778	2.62	84	86.62
Caulking	10.000	L.F.	.278	2.50	15	17.50
Snap-in grille	1.000	Set	.333	39	18.40	57.40
Drip cap, metal	3.000	L.F.	.060	1.83	3.30	5.13
TOTAL		Ea.	3.601	403	184.06	587.06
PLASTIC CLAD WOOD WINDOW, 4' X 3'-6", SLIDING						
Window, plastic clad, premium, 4' x 3'-6", insulating glass	1.000	Ea.	.889	755	49	804
Trim, interior casing	16.000	L.F.	.512	24.80	28.16	52.96
Paint, interior, primer & 2 coats	1.000	Face	.889	1.31	42	43.31
Caulking	17.000	L.F.	.472	4.25	25.50	29.75
Snap-in grille	1.000	Set	.333	39	18.40	57.40
TOTAL		Ea.	3.095	824.36	163.06	987.42
METAL CLAD WOOD WINDOW, 6' X 5', SLIDING						
Window, metal clad, deluxe, 6' x 5', insulating glass	1.000	Ea.	1.000	800	55	855
Trim, interior casing	23.000	L.F.	.736	35.65	40.48	76.13
Paint, interior, primer & 2 coats	1.000	Face	.889	1.31	42	43.31
Caulking	22.000	L.F.	.611	5.50	33	38.50
Snap-in grille	1.000	Set	.364	47.50	20	67.50
Drip cap, metal	6.000	L.F.	.120	3.66	6.60	10.26
TOTAL		Ea.	3.720	893.62	197.08	1,090.70

The cost of this system is on a cost per each window basis.

Description	QUAN.	UNIT	LABOR HOURS	COST EACH MAT.	COST EACH INST.	COST EACH TOTAL

Sliding Window Price Sheet

	QUAN.	UNIT	LABOR HOURS	COST EACH MAT.	COST EACH INST.	TOTAL
Windows, sliding, builder's quality, 3' x 3', single glass	1.000	Ea.	.800	310	44	354
Insulating glass	1.000	Ea.	.800	340	44	384
4' x 3'-6", single glass	1.000	Ea.	.889	390	49	439
Insulating glass	1.000	Ea.	.889	400	49	449
6' x 5', single glass	1.000	Ea.	1.000	535	55	590
Insulating glass	1.000	Ea.	1.000	580	55	635
Plastic clad premium insulating glass, 3' x 3'	1.000	Ea.	.800	665	44	709
4' x 3'-6"	1.000	Ea.	.889	755	49	804
5' x 4'	1.000	Ea.	.889	990	49	1,039
6' x 5'	1.000	Ea.	1.000	1,250	55	1,305
Metal clad deluxe insulating glass, 3' x 3'	1.000	Ea.	.800	360	44	404
4' x 3'-6"	1.000	Ea.	.889	440	49	489
5' x 4'	1.000	Ea.	.889	525	49	574
6' x 5'	1.000	Ea.	1.000	800	55	855
Trim, interior casing, window 3' x 2'	11.000	L.F.	.367	17.05	19.35	36.40
3' x 3'	13.000	L.F.	.433	20	23	43
4' x 3'-6"	16.000	L.F.	.533	25	28	53
5' x 4'	19.000	L.F.	.633	29.50	33.50	63
6' x 5'	23.000	L.F.	.767	35.50	40.50	76
Paint or stain, interior or exterior, 3' x 2' window, 1 coat	1.000	Face	.444	.52	21	21.52
2 coats	1.000	Face	.727	1.03	34.50	35.53
Primer & 1 coat	1.000	Face	.727	.87	34.50	35.37
Primer & 2 coats	1.000	Face	.889	1.31	42	43.31
4' x 3'-6" window, 1 coat	1.000	Face	.667	1.15	31.50	32.65
2 coats	1.000	Face	.667	1.80	31.50	33.30
Primer & 1 coat	1.000	Face	.727	1.34	34.50	35.84
Primer & 2 coats	1.000	Face	.889	1.31	42	43.31
6' x 5' window, 1 coat	1.000	Face	.889	2.92	42	44.92
2 coats	1.000	Face	1.333	5.35	63	68.35
Primer & 1 coat	1.000	Face	1.333	4.63	63	67.63
Primer & 2 coats	1.000	Face	1.600	7	75.50	82.50
Caulking, window, 3' x 2'	10.000	L.F.	.323	2.50	15	17.50
3' x 3'	12.000	L.F.	.387	3	18	21
4' x 3'-6"	15.000	L.F.	.484	3.75	22.50	26.25
5' x 4'	18.000	L.F.	.581	4.50	27	31.50
6' x 5'	22.000	L.F.	.710	5.50	33	38.50
Grilles, glass size, to 14" x 36"	1.000	Set	.333	39	18.40	57.40
To 36" x 36"	1.000	Set	.364	47.50	20	67.50
Drip cap, aluminum, 3' long	3.000	L.F.	.060	1.83	3.30	5.13
4' long	4.000	L.F.	.080	2.44	4.40	6.84
5' long	5.000	L.F.	.100	3.05	5.50	8.55
6' long	6.000	L.F.	.120	3.66	6.60	10.26
Wood, 3' long	3.000	L.F.	.100	4.65	5.30	9.95
4' long	4.000	L.F.	.133	6.20	7.05	13.25
5' long	5.000	L.F.	.167	7.75	8.80	16.55
6' long	6.000	L.F.	.200	9.30	10.55	19.85

4 | EXTERIOR WALLS 44 | Bow/Bay Window Systems

System Description	QUAN.	UNIT	LABOR HOURS	COST EACH		
				MAT.	INST.	TOTAL
AWNING TYPE BOW WINDOW, BUILDER'S QUALITY, 8' X 5'						
Window, primed, builder's quality, 8' x 5', insulating glass	1.000	Ea.	1.600	1,450	88	1,538
Trim, interior casing	27.000	L.F.	.864	41.85	47.52	89.37
Paint, interior & exterior, primer & 1 coat	2.000	Face	3.200	14	151	165
Caulking	26.000	L.F.	.722	6.50	39	45.50
Snap-in grilles	1.000	Set	1.067	140	58.80	198.80
TOTAL		Ea.	7.453	1,652.35	384.32	2,036.67
CASEMENT TYPE BOW WINDOW, PLASTIC CLAD, 10' X 6'						
Window, plastic clad, premium, 10' x 6', insulating glass	1.000	Ea.	2.286	2,875	126	3,001
Trim, interior casing	33.000	L.F.	1.056	51.15	58.08	109.23
Paint, interior, primer & 1 coat	1.000	Face	1.778	2.62	84	86.62
Caulking	32.000	L.F.	.889	8	48	56
Snap-in grilles	1.000	Set	1.333	175	73.50	248.50
TOTAL		Ea.	7.342	3,111.77	389.58	3,501.35
DOUBLE HUNG TYPE, METAL CLAD, 9' X 5'						
Window, metal clad, deluxe, 9' x 5', insulating glass	1.000	Ea.	2.667	1,575	147	1,722
Trim, interior casing	29.000	L.F.	.928	44.95	51.04	95.99
Paint, interior, primer & 1 coat	1.000	Face	1.778	2.62	84	86.62
Caulking	28.000	L.F.	.778	7	42	49
Snap-in grilles	1.000	Set	1.067	140	58.80	198.80
TOTAL		Ea.	7.218	1,769.57	382.84	2,152.41

The cost of this system is on a cost per each window basis.

Description	QUAN.	UNIT	LABOR HOURS	COST EACH		
				MAT.	INST.	TOTAL

Bow/Bay Window Price Sheet	QUAN.	UNIT	LABOR HOURS	COST EACH		
				MAT.	INST.	TOTAL
Windows, bow awning type, builder's quality, 8' x 5', insulating glass	1.000	Ea.	1.600	1,650	88	1,738
Low E glass	1.000	Ea.	1.600	1,450	88	1,538
12' x 6', insulating glass	1.000	Ea.	2.667	1,500	147	1,647
Low E glass	1.000	Ea.	2.667	1,600	147	1,747
Plastic clad premium insulating glass, 6' x 4'	1.000	Ea.	1.600	1,125	88	1,213
9' x 4'	1.000	Ea.	2.000	1,500	110	1,610
10' x 5'	1.000	Ea.	2.286	2,475	126	2,601
12' x 6'	1.000	Ea.	2.667	3,225	147	3,372
Metal clad deluxe insulating glass, 6' x 4'	1.000	Ea.	1.600	1,325	88	1,413
9' x 4'	1.000	Ea.	2.000	1,700	110	1,810
10' x 5'	1.000	Ea.	2.286	2,325	126	2,451
12' x 6'	1.000	Ea.	2.667	2,950	147	3,097
Bow casement type, builder's quality, 8' x 5', single glass	1.000	Ea.	1.600	2,075	88	2,163
Insulating glass	1.000	Ea.	1.600	2,500	88	2,588
12' x 6', single glass	1.000	Ea.	2.667	2,575	147	2,722
Insulating glass	1.000	Ea.	2.667	2,625	147	2,772
Plastic clad premium insulating glass, 8' x 5'	1.000	Ea.	1.600	1,925	88	2,013
10' x 5'	1.000	Ea.	2.000	2,525	110	2,635
10' x 6'	1.000	Ea.	2.286	2,875	126	3,001
12' x 6'	1.000	Ea.	2.667	3,525	147	3,672
Metal clad deluxe insulating glass, 8' x 5'	1.000	Ea.	1.600	1,825	88	1,913
10' x 5'	1.000	Ea.	2.000	1,975	110	2,085
10' x 6'	1.000	Ea.	2.286	2,325	126	2,451
12' x 6'	1.000	Ea.	2.667	3,225	147	3,372
Bow, double hung type, builder's quality, 8' x 4', single glass	1.000	Ea.	1.600	1,450	88	1,538
Insulating glass	1.000	Ea.	1.600	1,575	88	1,663
9' x 5', single glass	1.000	Ea.	2.667	1,575	147	1,722
Insulating glass	1.000	Ea.	2.667	1,650	147	1,797
Plastic clad premium insulating glass, 7' x 4'	1.000	Ea.	1.600	1,500	88	1,588
8' x 4'	1.000	Ea.	2.000	1,525	110	1,635
8' x 5'	1.000	Ea.	2.286	1,600	126	1,726
9' x 5'	1.000	Ea.	2.667	1,650	147	1,797
Metal clad deluxe insulating glass, 7' x 4'	1.000	Ea.	1.600	1,400	88	1,488
8' x 4'	1.000	Ea.	2.000	1,450	110	1,560
8' x 5'	1.000	Ea.	2.286	1,500	126	1,626
9' x 5'	1.000	Ea.	2.667	1,575	147	1,722
Trim, interior casing, window 7' x 4'	1.000	Ea.	.767	35.50	40.50	76
8' x 5'	1.000	Ea.	.900	42	47.50	89.50
10' x 6'	1.000	Ea.	1.100	51	58	109
12' x 6'	1.000	Ea.	1.233	57.50	65	122.50
Paint or stain, interior, or exterior, 7' x 4' window, 1 coat	1.000	Face	.889	2.92	42	44.92
Primer & 1 coat	1.000	Face	1.333	4.63	63	67.63
8' x 5' window, 1 coat	1.000	Face	.889	2.92	42	44.92
Primer & 1 coat	1.000	Face	1.333	4.63	63	67.63
10' x 6' window, 1 coat	1.000	Face	1.333	2.30	63	65.30
Primer & 1 coat	1.000	Face	1.778	2.62	84	86.62
12' x 6' window, 1 coat	1.000	Face	1.778	5.85	84	89.85
Primer & 1 coat	1.000	Face	2.667	9.25	126	135.25
Drip cap, vinyl moulded window, 7' long	1.000	Ea.	.533	125	24.50	149.50
8' long	1.000	Ea.	.533	143	28	171
Caulking, window, 7' x 4'	1.000	Ea.	.710	5.50	33	38.50
8' x 5'	1.000	Ea.	.839	6.50	39	45.50
10' x 6'	1.000	Ea.	1.032	8	48	56
12' x 6'	1.000	Ea.	1.161	9	54	63
Grilles, window, 7' x 4'	1.000	Set	.800	105	44	149
8' x 5'	1.000	Set	1.067	140	59	199
10' x 6'	1.000	Set	1.333	175	73.50	248.50
12' x 6'	1.000	Set	1.600	210	88	298

4 | EXTERIOR WALLS — 48 Fixed Window Systems

System Description	QUAN.	UNIT	LABOR HOURS	COST EACH MAT.	COST EACH INST.	COST EACH TOTAL
BUILDER'S QUALITY PICTURE WINDOW, 4' X 4'						
Window, primed, builder's quality, 3'-0" x 4', insulating glass	1.000	Ea.	1.333	475	73.50	548.50
Trim, interior casing	17.000	L.F.	.544	26.35	29.92	56.27
Paint, interior & exterior, primer & 2 coats	2.000	Face	1.778	2.62	84	86.62
Caulking	16.000	L.F.	.444	4	24	28
Snap-in grille	1.000	Ea.	.267	132	14.70	146.70
Drip cap, metal	4.000	L.F.	.080	2.44	4.40	6.84
TOTAL		Ea.	4.446	642.41	230.52	872.93
PLASTIC CLAD WOOD WINDOW, 4'-6" X 6'-6"						
Window, plastic clad, prem., 4'-6" x 6'-6", insul. glass	1.000	Ea.	1.455	1,025	80	1,105
Trim, interior casing	23.000	L.F.	.736	35.65	40.48	76.13
Paint, interior, primer & 2 coats	1.000	Face	.889	1.31	42	43.31
Caulking	22.000	L.F.	.611	5.50	33	38.50
Snap-in grille	1.000	Ea.	.267	132	14.70	146.70
TOTAL		Ea.	3.958	1,199.46	210.18	1,409.64
METAL CLAD WOOD WINDOW, 6'-6" X 6'-6"						
Window, metal clad, deluxe, 6'-0" x 6'-0", insulating glass	1.000	Ea.	1.600	765	88	853
Trim interior casing	27.000	L.F.	.864	41.85	47.52	89.37
Paint, interior, primer & 2 coats	1.000	Face	1.600	7	75.50	82.50
Caulking	26.000	L.F.	.722	6.50	39	45.50
Snap-in grille	1.000	Ea.	.267	132	14.70	146.70
Drip cap, metal	6.500	L.F.	.130	3.97	7.15	11.12
TOTAL		Ea.	5.183	956.32	271.87	1,228.19

The cost of this system is on a cost per each window basis.

Description	QUAN.	UNIT	LABOR HOURS	COST EACH MAT.	COST EACH INST.	COST EACH TOTAL

Fixed Window Price Sheet

Fixed Window Price Sheet	QUAN.	UNIT	LABOR HOURS	COST EACH MAT.	COST EACH INST.	COST EACH TOTAL
Window-picture, builder's quality, 4' x 4', single glass	1.000	Ea.	1.333	465	73.50	538.50
Insulating glass	1.000	Ea.	1.333	475	73.50	548.50
4' x 4'-6", single glass	1.000	Ea.	1.455	605	80	685
Insulating glass	1.000	Ea.	1.455	585	80	665
5' x 4', single glass	1.000	Ea.	1.455	640	80	720
Insulating glass	1.000	Ea.	1.455	665	80	745
6' x 4'-6", single glass	1.000	Ea.	1.600	690	88	778
Insulating glass	1.000	Ea.	1.600	700	88	788
Plastic clad premium insulating glass, 4' x 4'	1.000	Ea.	1.333	560	73.50	633.50
4'-6" x 6'-6"	1.000	Ea.	1.455	1,025	80	1,105
5'-6" x 6'-6"	1.000	Ea.	1.600	1,150	88	1,238
6'-6" x 6'-6"	1.000	Ea.	1.600	1,150	88	1,238
Metal clad deluxe insulating glass, 4' x 4'	1.000	Ea.	1.333	410	73.50	483.50
4'-6" x 6'-6"	1.000	Ea.	1.455	600	80	680
5'-6" x 6'-6"	1.000	Ea.	1.600	670	88	758
6'-6" x 6'-6"	1.000	Ea.	1.600	765	88	853
Trim, interior casing, window 4' x 4'	17.000	L.F.	.567	26.50	30	56.50
4'-6" x 4'-6"	19.000	L.F.	.633	29.50	33.50	63
5'-0" x 4'-0"	19.000	L.F.	.633	29.50	33.50	63
4'-6" x 6'-6"	23.000	L.F.	.767	35.50	40.50	76
5'-6" x 6'-6"	25.000	L.F.	.833	39	44	83
6'-6" x 6'-6"	27.000	L.F.	.900	42	47.50	89.50
Paint or stain, interior or exterior, 4' x 4' window, 1 coat	1.000	Face	.667	1.15	31.50	32.65
2 coats	1.000	Face	.667	1.80	31.50	33.30
Primer & 1 coat	1.000	Face	.727	1.34	34.50	35.84
Primer & 2 coats	1.000	Face	.889	1.31	42	43.31
4'-6" x 6'-6" window, 1 coat	1.000	Face	.667	1.15	31.50	32.65
2 coats	1.000	Face	.667	1.80	31.50	33.30
Primer & 1 coat	1.000	Face	.727	1.34	34.50	35.84
Primer & 2 coats	1.000	Face	.889	1.31	42	43.31
6'-6" x 6'-6" window, 1 coat	1.000	Face	.889	2.92	42	44.92
2 coats	1.000	Face	1.333	5.35	63	68.35
Primer & 1 coat	1.000	Face	1.333	4.63	63	67.63
Primer & 2 coats	1.000	Face	1.600	7	75.50	82.50
Caulking, window, 4' x 4'	1.000	Ea.	.516	4	24	28
4'-6" x 4'-6"	1.000	Ea.	.581	4.50	27	31.50
5'-0" x 4'-0"	1.000	Ea.	.581	4.50	27	31.50
4'-6" x 6'-6"	1.000	Ea.	.710	5.50	33	38.50
5'-6" x 6'-6"	1.000	Ea.	.774	6	36	42
6'-6" x 6'-6"	1.000	Ea.	.839	6.50	39	45.50
Grilles, glass size, to 48" x 48"	1.000	Ea.	.267	132	14.70	146.70
To 60" x 68"	1.000	Ea.	.286	202	15.75	217.75
Drip cap, aluminum, 4' long	4.000	L.F.	.080	2.44	4.40	6.84
4'-6" long	4.500	L.F.	.090	2.75	4.95	7.70
5' long	5.000	L.F.	.100	3.05	5.50	8.55
6' long	6.000	L.F.	.120	3.66	6.60	10.26
Wood, 4' long	4.000	L.F.	.133	6.20	7.05	13.25
4'-6" long	4.500	L.F.	.150	7	7.90	14.90
5' long	5.000	L.F.	.167	7.75	8.80	16.55
6' long	6.000	L.F.	.200	9.30	10.55	19.85

4 | EXTERIOR WALLS 52 | Entrance Door Systems

System Description	QUAN.	UNIT	LABOR HOURS	COST EACH MAT.	COST EACH INST.	COST EACH TOTAL
COLONIAL, 6 PANEL, 3' X 6'-8", WOOD						
Door, 3' x 6'-8" x 1-3/4" thick, pine, 6 panel colonial	1.000	Ea.	1.067	490	59	549
Frame, 5-13/16" deep, incl. exterior casing & drip cap	17.000	L.F.	.725	147.05	39.95	187
Interior casing, 2-1/2" wide	18.000	L.F.	.576	27.90	31.68	59.58
Sill, 8/4 x 8" deep	3.000	L.F.	.480	67.50	26.40	93.90
Butt hinges, brass, 4-1/2" x 4-1/2"	1.500	Pr.		32.25		32.25
Average quality	1.000	Ea.	.571	55	31.50	86.50
Weatherstripping, metal, spring type, bronze	1.000	Set	1.053	25.50	58	83.50
Paint, interior & exterior, primer & 2 coats	2.000	Face	1.778	13.90	84	97.90
TOTAL		Ea.	6.250	859.10	330.53	1,189.63
SOLID CORE BIRCH, FLUSH, 3' X 6'-8"						
Door, 3' x 6'-8", 1-3/4" thick, birch, flush solid core	1.000	Ea.	1.067	130	59	189
Frame, 5-13/16" deep, incl. exterior casing & drip cap	17.000	L.F.	.725	147.05	39.95	187
Interior casing, 2-1/2" wide	18.000	L.F.	.576	27.90	31.68	59.58
Sill, 8/4 x 8" deep	3.000	L.F.	.480	67.50	26.40	93.90
Butt hinges, brass, 4-1/2" x 4-1/2"	1.500	Pr.		32.25		32.25
Average quality	1.000	Ea.	.571	55	31.50	86.50
Weatherstripping, metal, spring type, bronze	1.000	Set	1.053	25.50	58	83.50
Paint, Interior & exterior, primer & 2 coats	2.000	Face	1.778	13	84	97
TOTAL		Ea.	6.250	498.20	330.53	828.73

These systems are on a cost per each door basis.

Description	QUAN.	UNIT	LABOR HOURS	COST EACH MAT.	COST EACH INST.	COST EACH TOTAL

Entrance Door Price Sheet	QUAN.	UNIT	LABOR HOURS	COST EACH MAT.	COST EACH INST.	COST EACH TOTAL
Door exterior wood 1-3/4" thick, pine, dutch door, 2'-8" x 6'-8" minimum	1.000	Ea.	1.333	865	73.50	938.50
Maximum	1.000	Ea.	1.600	990	88	1,078
3'-0" x 6'-8", minimum	1.000	Ea.	1.333	865	73.50	938.50
Maximum	1.000	Ea.	1.600	990	88	1,078
Colonial, 6 panel, 2'-8" x 6'-8"	1.000	Ea.	1.000	550	55	605
3'-0" x 6'-8"	1.000	Ea.	1.067	490	59	549
8 panel, 2'-6" x 6'-8"	1.000	Ea.	1.000	635	55	690
3'-0" x 6'-8"	1.000	Ea.	1.067	625	59	684
Flush, birch, solid core, 2'-8" x 6'-8"	1.000	Ea.	1.000	125	55	180
3'-0" x 6'-8"	1.000	Ea.	1.067	130	59	189
Porch door, 2'-8" x 6'-8"	1.000	Ea.	1.000	635	55	690
3'-0" x 6'-8"	1.000	Ea.	1.067	655	59	714
Hand carved mahogany, 2'-8" x 6'-8"	1.000	Ea.	1.067	1,550	59	1,609
3'-0" x 6'-8"	1.000	Ea.	1.067	1,525	59	1,584
Rosewood, 2'-8" x 6'-8"	1.000	Ea.	1.067	825	57.50	882.50
3'-0" x 6-8"	1.000	Ea.	1.067	825	59	884
Door, metal clad wood 1-3/8" thick raised panel, 2'-8" x 6'-8"	1.000	Ea.	1.067	345	52	397
3'-0" x 6'-8"	1.000	Ea.	1.067	297	59	356
Deluxe metal door, 3'-0" x 6'-8"	1.000	Ea.	1.231	297	59	356
3'-0" x 6'-8"	1.000	Ea.	1.231	297	59	356
Frame, pine, including exterior trim & drip cap, 5/4, x 4-9/16" deep	17.000	L.F.	.725	124	40	164
5-13/16" deep	17.000	L.F.	.725	147	40	187
6-9/16" deep	17.000	L.F.	.725	164	40	204
Safety glass lites, add	1.000	Ea.		87		87
Interior casing, 2'-8" x 6'-8" door	18.000	L.F.	.600	28	31.50	59.50
3'-0" x 6'-8" door	19.000	L.F.	.633	29.50	33.50	63
Sill, oak, 8/4 x 8" deep	3.000	L.F.	.480	67.50	26.50	94
8/4 x 10" deep	3.000	L.F.	.533	87	29.50	116.50
Butt hinges, steel plated, 4-1/2" x 4-1/2", plain	1.500	Pr.		32.50		32.50
Ball bearing	1.500	Pr.		58		58
Bronze, 4-1/2" x 4-1/2", plain	1.500	Pr.		45		45
Ball bearing	1.500	Pr.		61		61
Lockset, minimum	1.000	Ea.	.571	55	31.50	86.50
Maximum	1.000	Ea.	1.000	242	55	297
Weatherstripping, metal, interlocking, zinc	1.000	Set	2.667	48.50	147	195.50
Bronze	1.000	Set	2.667	61.50	147	208.50
Spring type, bronze	1.000	Set	1.053	25.50	58	83.50
Rubber, minimum	1.000	Set	1.053	10.05	58	68.05
Maximum	1.000	Set	1.143	12.25	63	75.25
Felt minimum	1.000	Set	.571	4.29	31.50	35.79
Maximum	1.000	Set	.615	4.90	34	38.90
Paint or stain, flush door, interior or exterior, 1 coat	2.000	Face	.941	4.86	44	48.86
2 coats	2.000	Face	1.455	9.70	69	78.70
Primer & 1 coat	2.000	Face	1.455	8.40	69	77.40
Primer & 2 coats	2.000	Face	1.778	13	84	97
Paneled door, interior & exterior, 1 coat	2.000	Face	1.143	5.20	54	59.20
2 coats	2.000	Face	2.000	10.40	94	104.40
Primer & 1 coat	2.000	Face	1.455	8.95	69	77.95
Primer & 2 coats	2.000	Face	1.778	13.90	84	97.90

4 | EXTERIOR WALLS — 53 | Sliding Door Systems

System Description	QUAN.	UNIT	LABOR HOURS	COST EACH MAT.	COST EACH INST.	COST EACH TOTAL
WOOD SLIDING DOOR, 8' WIDE, PREMIUM						
Wood, 5/8" thick tempered insul. glass, 8' wide, premium	1.000	Ea.	5.333	2,000	294	2,294
Interior casing	22.000	L.F.	.704	34.10	38.72	72.82
Exterior casing	22.000	L.F.	.704	34.10	38.72	72.82
Sill, oak, 8/4 x 8" deep	8.000	L.F.	1.280	180	70.40	250.40
Drip cap	8.000	L.F.	.160	4.88	8.80	13.68
Paint, interior & exterior, primer & 2 coats	2.000	Face	2.816	19.36	132.88	152.24
TOTAL		Ea.	10.997	2,272.44	583.52	2,855.96
ALUMINUM SLIDING DOOR, 8' WIDE, PREMIUM						
Aluminum, 5/8" tempered insul. glass, 8' wide, premium	1.000	Ea.	5.333	1,825	294	2,119
Interior casing	22.000	L.F.	.704	34.10	38.72	72.82
Exterior casing	22.000	L.F.	.704	34.10	38.72	72.82
Sill, oak, 8/4 x 8" deep	8.000	L.F.	1.280	180	70.40	250.40
Drip cap	8.000	L.F.	.160	4.88	8.80	13.68
Paint, interior & exterior, primer & 2 coats	2.000	Face	2.816	19.36	132.88	152.24
TOTAL		Ea.	10.997	2,097.44	583.52	2,680.96

The cost of this system is on a cost per each door basis.

Description	QUAN.	UNIT	LABOR HOURS	COST EACH MAT.	COST EACH INST.	COST EACH TOTAL

Sliding Door Price Sheet

	QUAN.	UNIT	LABOR HOURS	COST PER UNIT MAT.	COST PER UNIT INST.	COST PER UNIT TOTAL
Sliding door, wood, 5/8" thick, tempered insul. glass, 6' wide, premium	1.000	Ea.	4.000	1,575	221	1,796
Economy	1.000	Ea.	4.000	1,275	221	1,496
8' wide, wood premium	1.000	Ea.	5.333	2,000	294	2,294
Economy	1.000	Ea.	5.333	1,650	294	1,944
12' wide, wood premium	1.000	Ea.	6.400	3,300	355	3,655
Economy	1.000	Ea.	6.400	2,700	355	3,055
Aluminum, 5/8" thick, tempered insul. glass, 6' wide, premium	1.000	Ea.	4.000	1,725	221	1,946
Economy	1.000	Ea.	4.000	895	221	1,116
8' wide, premium	1.000	Ea.	5.333	1,825	294	2,119
Economy	1.000	Ea.	5.333	1,550	294	1,844
12' wide, premium	1.000	Ea.	6.400	3,275	355	3,630
Economy	1.000	Ea.	6.400	1,700	355	2,055
Interior casing, 6' wide door	20.000	L.F.	.667	31	35	66
8' wide door	22.000	L.F.	.733	34	38.50	72.50
12' wide door	26.000	L.F.	.867	40.50	46	86.50
Exterior casing, 6' wide door	20.000	L.F.	.667	31	35	66
8' wide door	22.000	L.F.	.733	34	38.50	72.50
12' wide door	26.000	L.F.	.867	40.50	46	86.50
Sill, oak, 8/4 x 8" deep, 6' wide door	6.000	L.F.	.960	135	53	188
8' wide door	8.000	L.F.	1.280	180	70.50	250.50
12' wide door	12.000	L.F.	1.920	270	106	376
8/4 x 10" deep, 6' wide door	6.000	L.F.	1.067	174	59	233
8' wide door	8.000	L.F.	1.422	232	78.50	310.50
12' wide door	12.000	L.F.	2.133	350	118	468
Drip cap, 6' wide door	6.000	L.F.	.120	3.66	6.60	10.26
8' wide door	8.000	L.F.	.160	4.88	8.80	13.68
12' wide door	12.000	L.F.	.240	7.30	13.20	20.50
Paint or stain, interior & exterior, 6' wide door, 1 coat	2.000	Face	1.600	6.40	75	81.40
2 coats	2.000	Face	1.600	6.40	75	81.40
Primer & 1 coat	2.000	Face	1.778	12	84	96
Primer & 2 coats	2.000	Face	2.560	17.60	121	138.60
8' wide door, 1 coat	2.000	Face	1.760	7.05	82.50	89.55
2 coats	2.000	Face	1.760	7.05	82.50	89.55
Primer & 1 coat	2.000	Face	1.955	13.20	92.50	105.70
Primer & 2 coats	2.000	Face	2.816	19.35	133	152.35
12' wide door, 1 coat	2.000	Face	2.080	8.30	98	106.30
2 coats	2.000	Face	2.080	8.30	98	106.30
Primer & 1 coat	2.000	Face	2.311	15.60	109	124.60
Primer & 2 coats	2.000	Face	3.328	23	157	180
Aluminum door, trim only, interior & exterior, 6' door, 1 coat	2.000	Face	.800	3.20	37.50	40.70
2 coats	2.000	Face	.800	3.20	37.50	40.70
Primer & 1 coat	2.000	Face	.889	6	42	48
Primer & 2 coats	2.000	Face	1.280	8.80	60.50	69.30
8' wide door, 1 coat	2.000	Face	.880	3.52	41.50	45.02
2 coats	2.000	Face	.880	3.52	41.50	45.02
Primer & 1 coat	2.000	Face	.978	6.60	46	52.60
Primer & 2 coats	2.000	Face	1.408	9.70	66.50	76.20
12' wide door, 1 coat	2.000	Face	1.040	4.16	49	53.16
2 coats	2.000	Face	1.040	4.16	49	53.16
Primer & 1 coat	2.000	Face	1.155	7.80	54.50	62.30
Primer & 2 coats	2.000	Face	1.664	11.45	78.50	89.95

For customer support on your Residential Cost Data, call 877.713.2617.

4 | EXTERIOR WALLS 56 | Residential Garage Door Systems

System Description	QUAN.	UNIT	LABOR HOURS	COST EACH		
				MAT.	INST.	TOTAL
OVERHEAD, SECTIONAL GARAGE DOOR, 9' X 7'						
Wood, overhead sectional door, std., incl. hardware, 9' x 7'	1.000	Ea.	2.000	1,075	110	1,185
Jamb & header blocking, 2" x 6"	25.000	L.F.	.901	18.25	49.75	68
Exterior trim	25.000	L.F.	.800	38.75	44	82.75
Paint, interior & exterior, primer & 2 coats	2.000	Face	3.556	27.80	168	195.80
Weatherstripping, molding type	1.000	Set	.736	35.65	40.48	76.13
Drip cap	9.000	L.F.	.180	5.49	9.90	15.39
TOTAL		Ea.	8.173	1,200.94	422.13	1,623.07
OVERHEAD, SECTIONAL GARAGE DOOR, 16' X 7'						
Wood, overhead sectional, std., incl. hardware, 16' x 7'	1.000	Ea.	2.667	1,800	147	1,947
Jamb & header blocking, 2" x 6"	30.000	L.F.	1.081	21.90	59.70	81.60
Exterior trim	30.000	L.F.	.960	46.50	52.80	99.30
Paint, interior & exterior, primer & 2 coats	2.000	Face	5.333	41.70	252	293.70
Weatherstripping, molding type	1.000	Set	.960	46.50	52.80	99.30
Drip cap	16.000	L.F.	.320	9.76	17.60	27.36
TOTAL		Ea.	11.321	1,966.36	581.90	2,548.26
OVERHEAD, SWING-UP TYPE, GARAGE DOOR, 16' X 7'						
Wood, overhead, swing-up, std., incl. hardware, 16' x 7'	1.000	Ea.	2.667	990	147	1,137
Jamb & header blocking, 2" x 6"	30.000	L.F.	1.081	21.90	59.70	81.60
Exterior trim	30.000	L.F.	.960	46.50	52.80	99.30
Paint, interior & exterior, primer & 2 coats	2.000	Face	5.333	41.70	252	293.70
Weatherstripping, molding type	1.000	Set	.960	46.50	52.80	99.30
Drip cap	16.000	L.F.	.320	9.76	17.60	27.36
TOTAL		Ea.	11.321	1,156.36	581.90	1,738.26

This system is on a cost per each door basis.

Description	QUAN.	UNIT	LABOR HOURS	COST EACH		
				MAT.	INST.	TOTAL

Resi Garage Door Price Sheet	QUAN.	UNIT	LABOR HOURS	COST EACH		
				MAT.	INST.	TOTAL
Overhead, sectional, including hardware, fiberglass, 9' x 7', standard	1.000	Ea.	3.030	1,050	167	1,217
Deluxe	1.000	Ea.	3.030	1,250	167	1,417
16' x 7', standard	1.000	Ea.	2.667	1,725	147	1,872
Deluxe	1.000	Ea.	2.667	2,350	147	2,497
Hardboard, 9' x 7', standard	1.000	Ea.	2.000	690	110	800
Deluxe	1.000	Ea.	2.000	900	110	1,010
16' x 7', standard	1.000	Ea.	2.667	1,350	147	1,497
Deluxe	1.000	Ea.	2.667	1,575	147	1,722
Metal, 9' x 7', standard	1.000	Ea.	3.030	815	167	982
Deluxe	1.000	Ea.	2.000	1,000	110	1,110
16' x 7', standard	1.000	Ea.	5.333	1,025	294	1,319
Deluxe	1.000	Ea.	2.667	1,550	147	1,697
Wood, 9' x 7', standard	1.000	Ea.	2.000	1,075	110	1,185
Deluxe	1.000	Ea.	2.000	2,350	110	2,460
16' x 7', standard	1.000	Ea.	2.667	1,800	147	1,947
Deluxe	1.000	Ea.	2.667	3,325	147	3,472
Overhead swing-up type including hardware, fiberglass, 9' x 7', standard	1.000	Ea.	2.000	1,100	110	1,210
Deluxe	1.000	Ea.	2.000	1,200	110	1,310
16' x 7', standard	1.000	Ea.	2.667	1,375	147	1,522
Deluxe	1.000	Ea.	2.667	1,750	147	1,897
Hardboard, 9' x 7', standard	1.000	Ea.	2.000	605	110	715
Deluxe	1.000	Ea.	2.000	715	110	825
16' x 7', standard	1.000	Ea.	2.667	735	147	882
Deluxe	1.000	Ea.	2.667	935	147	1,082
Metal, 9' x 7', standard	1.000	Ea.	2.000	660	110	770
Deluxe	1.000	Ea.	2.000	1,050	110	1,160
16' x 7', standard	1.000	Ea.	2.667	880	147	1,027
Deluxe	1.000	Ea.	2.667	1,200	147	1,347
Wood, 9' x 7', standard	1.000	Ea.	2.000	770	110	880
Deluxe	1.000	Ea.	2.000	1,225	110	1,335
16' x 7', standard	1.000	Ea.	2.667	990	147	1,137
Deluxe	1.000	Ea.	2.667	2,300	147	2,447
Jamb & header blocking, 2" x 6", 9' x 7' door	25.000	L.F.	.901	18.25	50	68.25
16' x 7' door	30.000	L.F.	1.081	22	59.50	81.50
2" x 8", 9' x 7' door	25.000	L.F.	1.000	25.50	55.50	81
16' x 7' door	30.000	L.F.	1.200	30.50	66.50	97
Exterior trim, 9' x 7' door	25.000	L.F.	.833	39	44	83
16' x 7' door	30.000	L.F.	1.000	46.50	53	99.50
Paint or stain, interior & exterior, 9' x 7' door, 1 coat	1.000	Face	2.286	10.40	108	118.40
2 coats	1.000	Face	4.000	21	188	209
Primer & 1 coat	1.000	Face	2.909	17.90	138	155.90
Primer & 2 coats	1.000	Face	3.556	28	168	196
16' x 7' door, 1 coat	1.000	Face	3.429	15.60	162	177.60
2 coats	1.000	Face	6.000	31	282	313
Primer & 1 coat	1.000	Face	4.364	27	207	234
Primer & 2 coats	1.000	Face	5.333	41.50	252	293.50
Weatherstripping, molding type, 9' x 7' door	1.000	Set	.767	35.50	40.50	76
16' x 7' door	1.000	Set	1.000	46.50	53	99.50
Drip cap, 9' door	9.000	L.F.	.180	5.50	9.90	15.40
16' door	16.000	L.F.	.320	9.75	17.60	27.35
Garage door opener, economy	1.000	Ea.	1.000	470	55	525
Deluxe, including remote control	1.000	Ea.	1.000	675	55	730

4 | EXTERIOR WALLS   58 | Aluminum Window Systems

System Description	QUAN.	UNIT	LABOR HOURS	COST EACH MAT.	COST EACH INST.	COST EACH TOTAL
SINGLE HUNG, 2' X 3' OPENING						
Window, 2' x 3' opening, enameled, insulating glass	1.000	Ea.	1.600	277	106	383
Blocking, 1" x 3" furring strip nailers	10.000	L.F.	.146	4.20	8	12.20
Drywall, 1/2" thick, standard	5.000	S.F.	.040	1.65	2.20	3.85
Corner bead, 1" x 1", galvanized steel	8.000	L.F.	.160	1.28	8.80	10.08
Finish drywall, tape and finish corners inside and outside	16.000	L.F.	.269	1.76	14.88	16.64
Sill, slate	2.000	L.F.	.400	26.80	19.70	46.50
TOTAL		Ea.	2.615	312.69	159.58	472.27
SLIDING, 3' X 2' OPENING						
Window, 3' x 2' opening, enameled, insulating glass	1.000	Ea.	1.600	255	106	361
Blocking, 1" x 3" furring strip nailers	10.000	L.F.	.146	4.20	8	12.20
Drywall, 1/2" thick, standard	5.000	S.F.	.040	1.65	2.20	3.85
Corner bead, 1" x 1", galvanized steel	7.000	L.F.	.140	1.12	7.70	8.82
Finish drywall, tape and finish corners inside and outside	14.000	L.F.	.236	1.54	13.02	14.56
Sill, slate	3.000	L.F.	.600	40.20	29.55	69.75
TOTAL		Ea.	2.762	303.71	166.47	470.18
AWNING, 3'-1" X 3'-2"						
Window, 3'-1" x 3'-2" opening, enameled, insul. glass	1.000	Ea.	1.600	420	106	526
Blocking, 1" x 3" furring strip, nailers	12.500	L.F.	.182	5.25	10	15.25
Drywall, 1/2" thick, standard	4.500	S.F.	.036	1.49	1.98	3.47
Corner bead, 1" x 1", galvanized steel	9.250	L.F.	.185	1.48	10.18	11.66
Finish drywall, tape and finish corners, inside and outside	18.500	L.F.	.312	2.04	17.21	19.25
Sill, slate	3.250	L.F.	.650	43.55	32.01	75.56
TOTAL		Ea.	2.965	473.81	177.38	651.19

Description	QUAN.	UNIT	LABOR HOURS	COST PER S.F. MAT.	COST PER S.F. INST.	COST PER S.F. TOTAL

Aluminum Window Price Sheet

	QUAN.	UNIT	LABOR HOURS	COST EACH MAT.	COST EACH INST.	TOTAL
Window, aluminum, awning, 3'-1" x 3'-2", standard glass	1.000	Ea.	1.600	390	106	496
Insulating glass	1.000	Ea.	1.600	420	106	526
4'-5" x 5'-3", standard glass	1.000	Ea.	2.000	440	133	573
Insulating glass	1.000	Ea.	2.000	515	133	648
Casement, 3'-1" x 3'-2", standard glass	1.000	Ea.	1.600	410	106	516
Insulating glass	1.000	Ea.	1.600	565	106	671
Single hung, 2' x 3', standard glass	1.000	Ea.	1.600	229	106	335
Insulating glass	1.000	Ea.	1.600	277	106	383
2'-8" x 6'-8", standard glass	1.000	Ea.	2.000	400	133	533
Insulating glass	1.000	Ea.	2.000	520	133	653
3'-4" x 5'-0", standard glass	1.000	Ea.	1.778	330	118	448
Insulating glass	1.000	Ea.	1.778	365	118	483
Sliding, 3' x 2', standard glass	1.000	Ea.	1.600	239	106	345
Insulating glass	1.000	Ea.	1.600	255	106	361
5' x 3', standard glass	1.000	Ea.	1.778	365	118	483
Insulating glass	1.000	Ea.	1.778	425	118	543
8' x 4', standard glass	1.000	Ea.	2.667	385	177	562
Insulating glass	1.000	Ea.	2.667	620	177	797
Blocking, 1" x 3" furring, opening 3' x 2'	10.000	L.F.	.146	4.20	8	12.20
3' x 3'	12.500	L.F.	.182	5.25	10	15.25
3' x 5'	16.000	L.F.	.233	6.70	12.80	19.50
4' x 4'	16.000	L.F.	.233	6.70	12.80	19.50
4' x 5'	18.000	L.F.	.262	7.55	14.40	21.95
4' x 6'	20.000	L.F.	.291	8.40	16	24.40
4' x 8'	24.000	L.F.	.349	10.10	19.20	29.30
6'-8" x 2'-8"	19.000	L.F.	.276	8	15.20	23.20
Drywall, 1/2" thick, standard, opening 3' x 2'	5.000	S.F.	.040	1.65	2.20	3.85
3' x 3'	6.000	S.F.	.048	1.98	2.64	4.62
3' x 5'	8.000	S.F.	.064	2.64	3.52	6.16
4' x 4'	8.000	S.F.	.064	2.64	3.52	6.16
4' x 5'	9.000	S.F.	.072	2.97	3.96	6.93
4' x 6'	10.000	S.F.	.080	3.30	4.40	7.70
4' x 8'	12.000	S.F.	.096	3.96	5.30	9.26
6'-8" x 2'	9.500	S.F.	.076	3.14	4.18	7.32
Corner bead, 1" x 1", galvanized steel, opening 3' x 2'	7.000	L.F.	.140	1.12	7.70	8.82
3' x 3'	9.000	L.F.	.180	1.44	9.90	11.34
3' x 5'	11.000	L.F.	.220	1.76	12.10	13.86
4' x 4'	12.000	L.F.	.240	1.92	13.20	15.12
4' x 5'	13.000	L.F.	.260	2.08	14.30	16.38
4' x 6'	14.000	L.F.	.280	2.24	15.40	17.64
4' x 8'	16.000	L.F.	.320	2.56	17.60	20.16
6'-8" x 2'	15.000	L.F.	.300	2.40	16.50	18.90
Tape and finish corners, inside and outside, opening 3' x 2'	14.000	L.F.	.204	1.54	13	14.54
3' x 3'	18.000	L.F.	.262	1.98	16.75	18.73
3' x 5'	22.000	L.F.	.320	2.42	20.50	22.92
4' x 4'	24.000	L.F.	.349	2.64	22.50	25.14
4' x 5'	26.000	L.F.	.378	2.86	24	26.86
4' x 6'	28.000	L.F.	.407	3.08	26	29.08
4' x 8'	32.000	L.F.	.466	3.52	30	33.52
6'-8" x 2'	30.000	L.F.	.437	3.30	28	31.30
Sill, slate, 2' long	2.000	L.F.	.400	27	19.70	46.70
3' long	3.000	L.F.	.600	40	29.50	69.50
4' long	4.000	L.F.	.800	53.50	39.50	93
Wood, 1-5/8" x 6-1/4", 2' long	2.000	L.F.	.128	11.70	7.05	18.75
3' long	3.000	L.F.	.192	17.55	10.60	28.15
4' long	4.000	L.F.	.256	23.50	14.10	37.60

4 | EXTERIOR WALLS — 60 | Storm Door & Window Systems

System Description	QUAN.	UNIT	LABOR HOURS	COST EACH MAT.	COST EACH INST.	TOTAL
Storm door, aluminum, combination, storm & screen, anodized, 2'-6" x 6'-8"	1.000	Ea.	1.067	220	59	279
2'-8" x 6'-8"	1.000	Ea.	1.143	226	63	289
3'-0" x 6'-8"	1.000	Ea.	1.143	195	63	258
Mill finish, 2'-6" x 6'-8"	1.000	Ea.	1.067	259	59	318
2'-8" x 6'-8"	1.000	Ea.	1.143	259	63	322
3'-0" x 6'-8"	1.000	Ea.	1.143	284	63	347
Painted, 2'-6" x 6'-8"	1.000	Ea.	1.067	325	59	384
2'-8" x 6'-8"	1.000	Ea.	1.143	315	63	378
3'-0" x 6'-8"	1.000	Ea.	1.143	340	63	403
Wood, combination, storm & screen, crossbuck, 2'-6" x 6'-9"	1.000	Ea.	1.455	375	80	455
2'-8" x 6'-9"	1.000	Ea.	1.600	330	88	418
3'-0" x 6'-9"	1.000	Ea.	1.778	340	98	438
Full lite, 2'-6" x 6'-9"	1.000	Ea.	1.455	350	80	430
2'-8" x 6'-9"	1.000	Ea.	1.600	350	88	438
3'-0" x 6'-9"	1.000	Ea.	1.778	355	98	453
Windows, aluminum, combination storm & screen, basement, 1'-10" x 1'-0"	1.000	Ea.	.533	38.50	29.50	68
2'-9" x 1'-6"	1.000	Ea.	.533	42	29.50	71.50
3'-4" x 2'-0"	1.000	Ea.	.533	49.50	29.50	79
Double hung, anodized, 2'-0" x 3'-5"	1.000	Ea.	.533	105	29.50	134.50
2'-6" x 5'-0"	1.000	Ea.	.571	127	31.50	158.50
4'-0" x 6'-0"	1.000	Ea.	.640	253	35.50	288.50
Painted, 2'-0" x 3'-5"	1.000	Ea.	.533	121	29.50	150.50
2'-6" x 5'-0"	1.000	Ea.	.571	187	31.50	218.50
4'-0" x 6'-0"	1.000	Ea.	.640	310	35.50	345.50
Fixed window, anodized, 4'-6" x 4'-6"	1.000	Ea.	.640	143	35.50	178.50
5'-8" x 4'-6"	1.000	Ea.	.800	163	44	207
Painted, 4'-6" x 4'-6"	1.000	Ea.	.640	143	35.50	178.50
5'-8" x 4'-6"	1.000	Ea.	.800	165	44	209

4 | EXTERIOR WALLS   64 | Shutters/Blinds Systems

Aluminum Louvered

Wood Louvered

Raised Panel

System Description	QUAN.	UNIT	LABOR HOURS	COST PER PAIR		
				MAT.	INST.	TOTAL
Shutters, exterior blinds, aluminum, louvered, 1'-4" wide, 3"-0" long	1.000	Set	.800	220	44	264
4'-0" long	1.000	Set	.800	264	44	308
5'-4" long	1.000	Set	.800	310	44	354
6'-8" long	1.000	Set	.889	390	49	439
Wood, louvered, 1'-2" wide, 3'-3" long	1.000	Set	.800	238	44	282
4'-7" long	1.000	Set	.800	297	44	341
5'-3" long	1.000	Set	.800	360	44	404
1'-6" wide, 3'-3" long	1.000	Set	.800	268	44	312
4'-7" long	1.000	Set	.800	355	44	399
Polystyrene, louvered, 1'-2" wide, 3'-3" long	1.000	Set	.800	39.50	44	83.50
4'-7" long	1.000	Set	.800	50.50	44	94.50
5'-3" long	1.000	Set	.800	65	44	109
6'-8" long	1.000	Set	.889	75	49	124
Vinyl, louvered, 1'-2" wide, 4'-7" long	1.000	Set	.720	63.50	39.50	103
1'-4" x 6'-8" long	1.000	Set	.889	83.50	49	132.50

Division 5 - Roofing

5 | ROOFING — 04 | Gable End Roofing Systems

System Description	QUAN.	UNIT	LABOR HOURS	COST PER S.F. MAT.	COST PER S.F. INST.	COST PER S.F. TOTAL
ASPHALT, ROOF SHINGLES, CLASS A						
Shingles, inorganic class A, 210-235 lb./sq., 4/12 pitch	1.160	S.F.	.017	1.05	.89	1.94
Drip edge, metal, 5" wide	.150	L.F.	.003	.09	.17	.26
Building paper, #15 felt	1.300	S.F.	.002	.08	.09	.17
Ridge shingles, asphalt	.042	L.F.	.001	.10	.05	.15
Soffit & fascia, white painted aluminum, 1' overhang	.083	L.F.	.012	.40	.66	1.06
Rake trim, 1" x 6"	.040	L.F.	.002	.05	.09	.14
Rake trim, prime and paint	.040	L.F.	.002	.01	.08	.09
Gutter, seamless, aluminum painted	.083	L.F.	.005	.24	.33	.57
Downspouts, aluminum painted	.035	L.F.	.002	.08	.10	.18
TOTAL		S.F.	.046	2.10	2.46	4.56
WOOD, CEDAR SHINGLES NO. 1 PERFECTIONS, 18" LONG						
Shingles, wood, cedar, No. 1 perfections, 4/12 pitch	1.160	S.F.	.035	3.25	1.92	5.17
Drip edge, metal, 5" wide	.150	L.F.	.003	.09	.17	.26
Building paper, #15 felt	1.300	S.F.	.002	.08	.09	.17
Ridge shingles, cedar	.042	L.F.	.001	.21	.07	.28
Soffit & fascia, white painted aluminum, 1' overhang	.083	L.F.	.012	.40	.66	1.06
Rake trim, 1" x 6"	.040	L.F.	.002	.05	.09	.14
Rake trim, prime and paint	.040	L.F.	.002	.01	.08	.09
Gutter, seamless, aluminum, painted	.083	L.F.	.005	.24	.33	.57
Downspouts, aluminum, painted	.035	L.F.	.002	.08	.10	.18
TOTAL		S.F.	.064	4.41	3.51	7.92

The prices in these systems are based on a square foot of plan area.
All quantities have been adjusted accordingly.

Description	QUAN.	UNIT	LABOR HOURS	COST PER S.F. MAT.	COST PER S.F. INST.	COST PER S.F. TOTAL

Gable End Roofing Price Sheet

	QUAN.	UNIT	LABOR HOURS	COST PER S.F. MAT.	COST PER S.F. INST.	COST PER S.F. TOTAL
Shingles, asphalt, inorganic, class A, 210-235 lb./sq., 4/12 pitch	1.160	S.F.	.017	1.05	.89	1.94
8/12 pitch	1.330	S.F.	.019	1.14	.97	2.11
Laminated, multi-layered, 240-260 lb./sq., 4/12 pitch	1.160	S.F.	.021	1.45	1.09	2.54
8/12 pitch	1.330	S.F.	.023	1.57	1.18	2.75
Premium laminated, multi-layered, 260-300 lb./sq., 4/12 pitch	1.160	S.F.	.027	2.02	1.40	3.42
8/12 pitch	1.330	S.F.	.030	2.18	1.52	3.70
Clay tile, Spanish tile, red, 4/12 pitch	1.160	S.F.	.053	7.30	2.68	9.98
8/12 pitch	1.330	S.F.	.058	7.95	2.90	10.85
Mission tile, red, 4/12 pitch	1.160	S.F.	.083	5.40	2.68	8.08
8/12 pitch	1.330	S.F.	.090	5.85	2.90	8.75
French tile, red, 4/12 pitch	1.160	S.F.	.071	15	2.45	17.45
8/12 pitch	1.330	S.F.	.077	16.25	2.65	18.90
Slate, Buckingham, Virginia, black, 4/12 pitch	1.160	S.F.	.055	6.20	2.80	9
8/12 pitch	1.330	S.F.	.059	6.70	3.03	9.73
Vermont, black or grey, 4/12 pitch	1.160	S.F.	.055	6	2.80	8.80
8/12 pitch	1.330	S.F.	.059	6.50	3.03	9.53
Wood, No. 1 red cedar, 5X, 16" long, 5" exposure, 4/12 pitch	1.160	S.F.	.038	3.78	2.11	5.89
8/12 pitch	1.330	S.F.	.042	4.10	2.29	6.39
Fire retardant, 4/12 pitch	1.160	S.F.	.038	4.52	2.11	6.63
8/12 pitch	1.330	S.F.	.042	4.90	2.29	7.19
18" long, No.1 perfections, 5" exposure, 4/12 pitch	1.160	S.F.	.035	3.25	1.92	5.17
8/12 pitch	1.330	S.F.	.038	3.52	2.08	5.60
Fire retardant, 4/12 pitch	1.160	S.F.	.035	3.99	1.92	5.91
8/12 pitch	1.330	S.F.	.038	4.32	2.08	6.40
Resquared & rebutted, 18" long, 6" exposure, 4/12 pitch	1.160	S.F.	.032	3.66	1.76	5.42
8/12 pitch	1.330	S.F.	.035	3.97	1.91	5.88
Fire retardant, 4/12 pitch	1.160	S.F.	.032	4.40	1.76	6.16
8/12 pitch	1.330	S.F.	.035	4.77	1.91	6.68
Wood shakes hand split, 24" long, 10" exposure, 4/12 pitch	1.160	S.F.	.038	3.60	2.11	5.71
8/12 pitch	1.330	S.F.	.042	3.90	2.29	6.19
Fire retardant, 4/12 pitch	1.160	S.F.	.038	4.34	2.11	6.45
8/12 pitch	1.330	S.F.	.042	4.70	2.29	6.99
18" long, 8" exposure, 4/12 pitch	1.160	S.F.	.048	2.38	2.65	5.03
8/12 pitch	1.330	S.F.	.052	2.57	2.87	5.44
Fire retardant, 4/12 pitch	1.160	S.F.	.048	3.12	2.65	5.77
8/12 pitch	1.330	S.F.	.052	3.37	2.87	6.24
Drip edge, metal, 5" wide	.150	L.F.	.003	.09	.17	.26
8" wide	.150	L.F.	.003	.13	.17	.30
Building paper, #15 asphalt felt	1.300	S.F.	.002	.08	.09	.17
Ridge shingles, asphalt	.042	L.F.	.001	.10	.05	.15
Clay	.042	L.F.	.002	.25	.26	.51
Slate	.042	L.F.	.002	.46	.09	.55
Wood, shingles	.042	L.F.	.001	.21	.07	.28
Shakes	.042	L.F.	.001	.21	.07	.28
Soffit & fascia, aluminum, vented, 1' overhang	.083	L.F.	.012	.40	.66	1.06
2' overhang	.083	L.F.	.013	.59	.73	1.32
Vinyl, vented, 1' overhang	.083	L.F.	.011	.39	.61	1
2' overhang	.083	L.F.	.012	.52	.73	1.25
Wood, board fascia, plywood soffit, 1' overhang	.083	L.F.	.004	.02	.17	.19
2' overhang	.083	L.F.	.006	.04	.26	.30
Rake trim, painted, 1" x 6"	.040	L.F.	.004	.06	.17	.23
1" x 8"	.040	L.F.	.004	.24	.15	.39
Gutter, 5" box, aluminum, seamless, painted	.083	L.F.	.006	.24	.33	.57
Vinyl	.083	L.F.	.006	.14	.32	.46
Downspout, 2" x 3", aluminum, one story house	.035	L.F.	.001	.05	.09	.14
Two story house	.060	L.F.	.003	.09	.16	.25
Vinyl, one story house	.035	L.F.	.002	.08	.10	.18
Two story house	.060	L.F.	.003	.09	.16	.25

5 | ROOFING 08 | Hip Roof Roofing Systems

System Description	QUAN.	UNIT	LABOR HOURS	COST PER S.F. MAT.	COST PER S.F. INST.	COST PER S.F. TOTAL
ASPHALT, ROOF SHINGLES, CLASS A						
Shingles, inorganic, class A, 210-235 lb./sq. 4/12 pitch	1.570	S.F.	.023	1.40	1.19	2.59
Drip edge, metal, 5" wide	.122	L.F.	.002	.07	.13	.20
Building paper, #15 asphalt felt	1.800	S.F.	.002	.11	.13	.24
Ridge shingles, asphalt	.075	L.F.	.002	.17	.09	.26
Soffit & fascia, white painted aluminum, 1' overhang	.120	L.F.	.017	.58	.96	1.54
Gutter, seamless, aluminum, painted	.120	L.F.	.008	.35	.47	.82
Downspouts, aluminum, painted	.035	L.F.	.002	.08	.10	.18
TOTAL		S.F.	.056	2.76	3.07	5.83
WOOD, CEDAR SHINGLES, NO. 1 PERFECTIONS, 18" LONG						
Shingles, red cedar, No. 1 perfections, 5" exp., 4/12 pitch	1.570	S.F.	.047	4.34	2.56	6.90
Drip edge, metal, 5" wide	.122	L.F.	.002	.07	.13	.20
Building paper, #15 asphalt felt	1.800	S.F.	.002	.11	.13	.24
Ridge shingles, wood, cedar	.075	L.F.	.002	.37	.12	.49
Soffit & fascia, white painted aluminum, 1' overhang	.120	L.F.	.017	.58	.96	1.54
Gutter, seamless, aluminum, painted	.120	L.F.	.008	.35	.47	.82
Downspouts, aluminum, painted	.035	L.F.	.002	.08	.10	.18
TOTAL		S.F.	.080	5.90	4.47	10.37

The prices in these systems are based on a square foot of plan area.
All quantities have been adjusted accordingly.

Description	QUAN.	UNIT	LABOR HOURS	COST PER S.F. MAT.	COST PER S.F. INST.	COST PER S.F. TOTAL

Hip Roof - Roofing Price Sheet	QUAN.	UNIT	LABOR HOURS	COST PER S.F.		
				MAT.	INST.	TOTAL
Shingles, asphalt, inorganic, class A, 210-235 lb./sq., 4/12 pitch	1.570	S.F.	.023	1.40	1.19	2.59
8/12 pitch	1.850	S.F.	.028	1.66	1.42	3.08
Laminated, multi-layered, 240-260 lb./sq., 4/12 pitch	1.570	S.F.	.028	1.94	1.46	3.40
8/12 pitch	1.850	S.F.	.034	2.30	1.73	4.03
Prem. laminated, multi-layered, 260-300 lb./sq., 4/12 pitch	1.570	S.F.	.037	2.69	1.87	4.56
8/12 pitch	1.850	S.F.	.043	3.19	2.22	5.41
Clay tile, Spanish tile, red, 4/12 pitch	1.570	S.F.	.071	9.75	3.57	13.32
8/12 pitch	1.850	S.F.	.084	11.60	4.24	15.84
Mission tile, red, 4/12 pitch	1.570	S.F.	.111	7.20	3.57	10.77
8/12 pitch	1.850	S.F.	.132	8.55	4.24	12.79
French tile, red, 4/12 pitch	1.570	S.F.	.095	20	3.26	23.26
8/12 pitch	1.850	S.F.	.113	24	3.88	27.88
Slate, Buckingham, Virginia, black, 4/12 pitch	1.570	S.F.	.073	8.25	3.73	11.98
8/12 pitch	1.850	S.F.	.087	9.80	4.43	14.23
Vermont, black or grey, 4/12 pitch	1.570	S.F.	.073	8	3.73	11.73
8/12 pitch	1.850	S.F.	.087	9.50	4.43	13.93
Wood, red cedar, No.1 5X, 16" long, 5" exposure, 4/12 pitch	1.570	S.F.	.051	5.05	2.82	7.87
8/12 pitch	1.850	S.F.	.061	6	3.34	9.34
Fire retardant, 4/12 pitch	1.570	S.F.	.051	6	2.82	8.82
8/12 pitch	1.850	S.F.	.061	7.15	3.34	10.49
18" long, No.1 perfections, 5" exposure, 4/12 pitch	1.570	S.F.	.047	4.34	2.56	6.90
8/12 pitch	1.850	S.F.	.055	5.15	3.04	8.19
Fire retardant, 4/12 pitch	1.570	S.F.	.047	5.30	2.56	7.86
8/12 pitch	1.850	S.F.	.055	6.30	3.04	9.34
Resquared & rebutted, 18" long, 6" exposure, 4/12 pitch	1.570	S.F.	.043	4.88	2.35	7.23
8/12 pitch	1.850	S.F.	.051	5.80	2.79	8.59
Fire retardant, 4/12 pitch	1.570	S.F.	.043	5.85	2.35	8.20
8/12 pitch	1.850	S.F.	.051	6.95	2.79	9.74
Wood shakes hand split, 24" long, 10" exposure, 4/12 pitch	1.570	S.F.	.051	4.80	2.82	7.62
8/12 pitch	1.850	S.F.	.061	5.70	3.34	9.04
Fire retardant, 4/12 pitch	1.570	S.F.	.051	5.80	2.82	8.62
8/12 pitch	1.850	S.F.	.061	6.85	3.34	10.19
18" long, 8" exposure, 4/12 pitch	1.570	S.F.	.064	3.17	3.54	6.71
8/12 pitch	1.850	S.F.	.076	3.76	4.20	7.96
Fire retardant, 4/12 pitch	1.570	S.F.	.064	4.15	3.54	7.69
8/12 pitch	1.850	S.F.	.076	4.93	4.20	9.13
Drip edge, metal, 5" wide	.122	L.F.	.002	.07	.13	.20
8" wide	.122	L.F.	.002	.11	.13	.24
Building paper, #15 asphalt felt	1.800	S.F.	.002	.11	.13	.24
Ridge shingles, asphalt	.075	L.F.	.002	.17	.09	.26
Clay	.075	L.F.	.003	.45	.46	.91
Slate	.075	L.F.	.003	.83	.15	.98
Wood, shingles	.075	L.F.	.002	.37	.12	.49
Shakes	.075	L.F.	.002	.37	.12	.49
Soffit & fascia, aluminum, vented, 1' overhang	.120	L.F.	.017	.58	.96	1.54
2' overhang	.120	L.F.	.019	.85	1.06	1.91
Vinyl, vented, 1' overhang	.120	L.F.	.016	.57	.88	1.45
2' overhang	.120	L.F.	.017	.76	1.06	1.82
Wood, board fascia, plywood soffit, 1' overhang	.120	L.F.	.004	.02	.17	.19
2' overhang	.120	L.F.	.006	.04	.26	.30
Gutter, 5" box, aluminum, seamless, painted	.120	L.F.	.008	.35	.47	.82
Vinyl	.120	L.F.	.009	.21	.46	.67
Downspout, 2" x 3", aluminum, one story house	.035	L.F.	.002	.08	.10	.18
Two story house	.060	L.F.	.003	.09	.16	.25
Vinyl, one story house	.035	L.F.	.001	.05	.09	.14
Two story house	.060	L.F.	.003	.09	.16	.25

5 | ROOFING 12 | Gambrel Roofing Systems

System Description	QUAN.	UNIT	LABOR HOURS	COST PER S.F. MAT.	COST PER S.F. INST.	COST PER S.F. TOTAL
ASPHALT, ROOF SHINGLES, CLASS A						
Shingles, asphalt, inorganic, class A, 210-235 lb./sq.	1.450	S.F.	.022	1.31	1.12	2.43
Drip edge, metal, 5" wide	.146	L.F.	.003	.08	.16	.24
Building paper, #15 asphalt felt	1.500	S.F.	.002	.09	.11	.20
Ridge shingles, asphalt	.042	L.F.	.001	.10	.05	.15
Soffit & fascia, painted aluminum, 1' overhang	.083	L.F.	.012	.40	.66	1.06
Rake trim, 1" x 6"	.063	L.F.	.003	.08	.14	.22
Rake trim, prime and paint	.063	L.F.	.003	.02	.13	.15
Gutter, seamless, alumunum, painted	.083	L.F.	.005	.24	.33	.57
Downspouts, aluminum, painted	.042	L.F.	.002	.09	.12	.21
TOTAL		S.F.	.053	2.41	2.82	5.23
WOOD, CEDAR SHINGLES, NO. 1 PERFECTIONS, 18" LONG						
Shingles, wood, red cedar, No. 1 perfections, 5" exposure	1.450	S.F.	.044	4.07	2.40	6.47
Drip edge, metal, 5" wide	.146	L.F.	.003	.08	.16	.24
Building paper, #15 asphalt felt	1.500	S.F.	.002	.09	.11	.20
Ridge shingles, wood	.042	L.F.	.001	.21	.07	.28
Soffit & fascia, white painted aluminum, 1' overhang	.083	L.F.	.012	.40	.66	1.06
Rake trim, 1" x 6"	.063	L.F.	.003	.08	.14	.22
Rake trim, prime and paint	.063	L.F.	.001	.02	.06	.08
Gutter, seamless, aluminum, painted	.083	L.F.	.005	.24	.33	.57
Downspouts, aluminum, painted	.042	L.F.	.002	.09	.12	.21
TOTAL		S.F.	.073	5.28	4.05	9.33

The prices in this system are based on a square foot of plan area.
All quantities have been adjusted accordingly.

Description	QUAN.	UNIT	LABOR HOURS	COST PER S.F. MAT.	COST PER S.F. INST.	COST PER S.F. TOTAL

Gambrel Roofing Price Sheet	QUAN.	UNIT	LABOR HOURS	COST PER S.F. MAT.	COST PER S.F. INST.	COST PER S.F. TOTAL
Shingles, asphalt, standard, inorganic, class A, 210-235 lb./sq.	1.450	S.F.	.022	1.31	1.12	2.43
Laminated, multi-layered, 240-260 lb./sq.	1.450	S.F.	.027	1.82	1.37	3.19
Premium laminated, multi-layered, 260-300 lb./sq.	1.450	S.F.	.034	2.52	1.76	4.28
Slate, Buckingham, Virginia, black	1.450	S.F.	.069	7.75	3.50	11.25
Vermont, black or grey	1.450	S.F.	.069	7.50	3.50	11
Wood, red cedar, No.1 5X, 16" long, 5" exposure, plain	1.450	S.F.	.048	4.73	2.64	7.37
Fire retardant	1.450	S.F.	.048	5.65	2.64	8.29
18" long, No.1 perfections, 6" exposure, plain	1.450	S.F.	.044	4.07	2.40	6.47
Fire retardant	1.450	S.F.	.044	4.99	2.40	7.39
Resquared & rebutted, 18" long, 6" exposure, plain	1.450	S.F.	.040	4.58	2.21	6.79
Fire retardant	1.450	S.F.	.040	5.55	2.21	7.76
Shakes, hand split, 24" long, 10" exposure, plain	1.450	S.F.	.048	4.50	2.64	7.14
Fire retardant	1.450	S.F.	.048	5.40	2.64	8.04
18" long, 8" exposure, plain	1.450	S.F.	.060	2.97	3.32	6.29
Fire retardant	1.450	S.F.	.060	3.89	3.32	7.21
Drip edge, metal, 5" wide	.146	L.F.	.003	.08	.16	.24
8" wide	.146	L.F.	.003	.13	.16	.29
Building paper, #15 asphalt felt	1.500	S.F.	.002	.09	.11	.20
Ridge shingles, asphalt	.042	L.F.	.001	.10	.05	.15
Slate	.042	L.F.	.002	.46	.09	.55
Wood, shingles	.042	L.F.	.001	.21	.07	.28
Shakes	.042	L.F.	.001	.21	.07	.28
Soffit & fascia, aluminum, vented, 1' overhang	.083	L.F.	.012	.40	.66	1.06
2' overhang	.083	L.F.	.013	.59	.73	1.32
Vinyl vented, 1' overhang	.083	L.F.	.011	.39	.61	1
2' overhang	.083	L.F.	.012	.52	.73	1.25
Wood board fascia, plywood soffit, 1' overhang	.083	L.F.	.004	.02	.17	.19
2' overhang	.083	L.F.	.006	.04	.26	.30
Rake trim, painted, 1" x 6"	.063	L.F.	.006	.10	.27	.37
1" x 8"	.063	L.F.	.007	.12	.35	.47
Gutter, 5" box, aluminum, seamless, painted	.083	L.F.	.006	.24	.33	.57
Vinyl	.083	L.F.	.006	.14	.32	.46
Downspout 2" x 3", aluminum, one story house	.042	L.F.	.002	.06	.11	.17
Two story house	.070	L.F.	.003	.11	.18	.29
Vinyl, one story house	.042	L.F.	.002	.06	.11	.17
Two story house	.070	L.F.	.003	.11	.18	.29

5 | ROOFING 16 | Mansard Roofing Systems

System Description	QUAN.	UNIT	LABOR HOURS	COST PER S.F. MAT.	COST PER S.F. INST.	TOTAL
ASPHALT, ROOF SHINGLES, CLASS A						
Shingles, standard inorganic class A 210-235 lb./sq.	2.210	S.F.	.032	1.93	1.64	3.57
Drip edge, metal, 5" wide	.122	L.F.	.002	.07	.13	.20
Building paper, #15 asphalt felt	2.300	S.F.	.003	.14	.16	.30
Ridge shingles, asphalt	.090	L.F.	.002	.21	.11	.32
Soffit & fascia, white painted aluminum, 1' overhang	.122	L.F.	.018	.59	.98	1.57
Gutter, seamless, aluminum, painted	.122	L.F.	.008	.36	.48	.84
Downspouts, aluminum, painted	.042	L.F.	.002	.09	.12	.21
TOTAL		S.F.	.067	3.39	3.62	7.01
WOOD, CEDAR SHINGLES, NO. 1 PERFECTIONS, 18" LONG						
Shingles, wood, red cedar, No. 1 perfections, 5" exposure	2.210	S.F.	.064	5.96	3.52	9.48
Drip edge, metal, 5" wide	.122	L.F.	.002	.07	.13	.20
Building paper, #15 asphalt felt	2.300	S.F.	.003	.14	.16	.30
Ridge shingles, wood	.090	L.F.	.003	.45	.14	.59
Soffit & fascia, white painted aluminum, 1' overhang	.122	L.F.	.018	.59	.98	1.57
Gutter, seamless, aluminum, painted	.122	L.F.	.008	.36	.48	.84
Downspouts, aluminum, painted	.042	L.F.	.002	.09	.12	.21
TOTAL		S.F.	.100	7.66	5.53	13.19

The prices in these systems are based on a square foot of plan area.
All quantities have been adjusted accordingly.

Description	QUAN.	UNIT	LABOR HOURS	COST PER S.F. MAT.	COST PER S.F. INST.	TOTAL

Mansard Roofing Price Sheet	QUAN.	UNIT	LABOR HOURS	COST PER S.F.		
				MAT.	INST.	TOTAL
Shingles, asphalt, standard, inorganic, class A, 210-235 lb./sq.	2.210	S.F.	.032	1.93	1.64	3.57
Laminated, multi-layered, 240-260 lb./sq.	2.210	S.F.	.039	2.66	2	4.66
Premium laminated, multi-layered, 260-300 lb./sq.	2.210	S.F.	.050	3.70	2.57	6.27
Slate Buckingham, Virginia, black	2.210	S.F.	.101	11.35	5.15	16.50
Vermont, black or grey	2.210	S.F.	.101	11	5.15	16.15
Wood, red cedar, No.1 5X, 16" long, 5" exposure, plain	2.210	S.F.	.070	6.95	3.87	10.82
Fire retardant	2.210	S.F.	.070	8.30	3.87	12.17
18" long, No.1 perfections 6" exposure, plain	2.210	S.F.	.064	5.95	3.52	9.47
Fire retardant	2.210	S.F.	.064	7.30	3.52	10.82
Fire retardant	2.210	S.F.	.059	8.05	3.23	11.28
Shakes, hand split, 24" long 10" exposure, plain	2.210	S.F.	.070	6.60	3.87	10.47
Fire retardant	2.210	S.F.	.070	7.95	3.87	11.82
18" long, 8" exposure, plain	2.210	S.F.	.088	4.36	4.86	9.22
Fire retardant	2.210	S.F.	.088	5.70	4.86	10.56
Drip edge, metal, 5" wide	.122	S.F.	.002	.07	.13	.20
8" wide	.122	S.F.	.002	.11	.13	.24
Building paper, #15 asphalt felt	2.300	S.F.	.003	.14	.16	.30
Ridge shingles, asphalt	.090	L.F.	.002	.21	.11	.32
Slate	.090	L.F.	.004	.99	.18	1.17
Wood, shingles	.090	L.F.	.003	.45	.14	.59
Shakes	.090	L.F.	.003	.45	.14	.59
Soffit & fascia, aluminum vented, 1' overhang	.122	L.F.	.018	.59	.98	1.57
2' overhang	.122	L.F.	.020	.86	1.07	1.93
Vinyl vented, 1' overhang	.122	L.F.	.016	.58	.90	1.48
2' overhang	.122	L.F.	.018	.77	1.07	1.84
Wood board fascia, plywood soffit, 1' overhang	.122	L.F.	.013	.36	.67	1.03
2' overhang	.122	L.F.	.019	.51	1.03	1.54
Gutter, 5" box, aluminum, seamless, painted	.122	L.F.	.008	.36	.48	.84
Vinyl	.122	L.F.	.009	.21	.47	.68
Downspout 2" x 3", aluminum, one story house	.042	L.F.	.002	.06	.11	.17
Two story house	.070	L.F.	.003	.10	.18	.28
Vinyl, one story house	.042	L.F.	.002	.06	.11	.17
Two story house	.070	L.F.	.003	.10	.18	.28

5 | ROOFING — 20 | Shed Roofing Systems

System Description	QUAN.	UNIT	LABOR HOURS	COST PER S.F. MAT.	COST PER S.F. INST.	COST PER S.F. TOTAL
ASPHALT, ROOF SHINGLES, CLASS A						
Shingles, inorganic class A 210-235 lb./sq. 4/12 pitch	1.230	S.F.	.019	1.14	.97	2.11
Drip edge, metal, 5" wide	.100	L.F.	.002	.06	.11	.17
Building paper, #15 asphalt felt	1.300	S.F.	.002	.08	.09	.17
Soffit & fascia, white painted aluminum, 1' overhang	.080	L.F.	.012	.39	.64	1.03
Rake trim, 1" x 6"	.043	L.F.	.002	.05	.10	.15
Rake trim, prime and paint	.043	L.F.	.002	.01	.09	.10
Gutter, seamless, aluminum, painted	.040	L.F.	.003	.12	.16	.28
Downspouts, painted aluminum	.020	L.F.	.001	.05	.05	.10
TOTAL		S.F.	.043	1.90	2.21	4.11
WOOD, CEDAR SHINGLES, NO. 1 PERFECTIONS, 18" LONG						
Shingles, red cedar, No. 1 perfections, 5" exp., 4/12 pitch	1.230	S.F.	.035	3.25	1.92	5.17
Drip edge, metal, 5" wide	.100	L.F.	.002	.06	.11	.17
Building paper, #15 asphalt felt	1.300	S.F.	.002	.08	.09	.17
Soffit & fascia, white painted aluminum, 1' overhang	.080	L.F.	.012	.39	.64	1.03
Rake trim, 1" x 6"	.043	L.F.	.002	.05	.10	.15
Rake trim, prime and paint	.043	L.F.	.001	.01	.04	.05
Gutter, seamless, aluminum, painted	.040	L.F.	.003	.12	.16	.28
Downspouts, painted aluminum	.020	L.F.	.001	.05	.05	.10
TOTAL		S.F.	.058	4.01	3.11	7.12

The prices in these systems are based on a square foot of plan area. All quantities have been adjusted accordingly.

Description	QUAN.	UNIT	LABOR HOURS	COST PER S.F. MAT.	COST PER S.F. INST.	COST PER S.F. TOTAL

Shed Roofing Price Sheet	QUAN.	UNIT	LABOR HOURS	COST PER S.F.		
				MAT.	INST.	TOTAL
Shingles, asphalt, inorganic, class A, 210-235 lb./sq., 4/12 pitch	1.230	S.F.	.017	1.05	.89	1.94
8/12 pitch	1.330	S.F.	.019	1.14	.97	2.11
Laminated, multi-layered, 240-260 lb./sq. 4/12 pitch	1.230	S.F.	.021	1.45	1.09	2.54
8/12 pitch	1.330	S.F.	.023	1.57	1.18	2.75
Premium laminated, multi-layered, 260-300 lb./sq. 4/12 pitch	1.230	S.F.	.027	2.02	1.40	3.42
8/12 pitch	1.330	S.F.	.030	2.18	1.52	3.70
Clay tile, Spanish tile, red, 4/12 pitch	1.230	S.F.	.053	7.30	2.68	9.98
8/12 pitch	1.330	S.F.	.058	7.95	2.90	10.85
Mission tile, red, 4/12 pitch	1.230	S.F.	.083	5.40	2.68	8.08
8/12 pitch	1.330	S.F.	.090	5.85	2.90	8.75
French tile, red, 4/12 pitch	1.230	S.F.	.071	15	2.45	17.45
8/12 pitch	1.330	S.F.	.077	16.25	2.65	18.90
Slate, Buckingham, Virginia, black, 4/12 pitch	1.230	S.F.	.055	6.20	2.80	9
8/12 pitch	1.330	S.F.	.059	6.70	3.03	9.73
Vermont, black or grey, 4/12 pitch	1.230	S.F.	.055	6	2.80	8.80
8/12 pitch	1.330	S.F.	.059	6.50	3.03	9.53
Wood, red cedar, No.1 5X, 16" long, 5" exposure, 4/12 pitch	1.230	S.F.	.038	3.78	2.11	5.89
8/12 pitch	1.330	S.F.	.042	4.10	2.29	6.39
Fire retardant, 4/12 pitch	1.230	S.F.	.038	4.52	2.11	6.63
8/12 pitch	1.330	S.F.	.042	4.90	2.29	7.19
18" long, 6" exposure, 4/12 pitch	1.230	S.F.	.035	3.25	1.92	5.17
8/12 pitch	1.330	S.F.	.038	3.52	2.08	5.60
Fire retardant, 4/12 pitch	1.230	S.F.	.035	3.99	1.92	5.91
8/12 pitch	1.330	S.F.	.038	4.32	2.08	6.40
Resquared & rebutted, 18" long, 6" exposure, 4/12 pitch	1.230	S.F.	.032	3.66	1.76	5.42
8/12 pitch	1.330	S.F.	.035	3.97	1.91	5.88
Fire retardant, 4/12 pitch	1.230	S.F.	.032	4.40	1.76	6.16
8/12 pitch	1.330	S.F.	.035	4.77	1.91	6.68
Wood shakes, hand split, 24" long, 10" exposure, 4/12 pitch	1.230	S.F.	.038	3.60	2.11	5.71
8/12 pitch	1.330	S.F.	.042	3.90	2.29	6.19
Fire retardant, 4/12 pitch	1.230	S.F.	.038	4.34	2.11	6.45
8/12 pitch	1.330	S.F.	.042	4.70	2.29	6.99
18" long, 8" exposure, 4/12 pitch	1.230	S.F.	.048	2.38	2.65	5.03
8/12 pitch	1.330	S.F.	.052	2.57	2.87	5.44
Fire retardant, 4/12 pitch	1.230	S.F.	.048	3.12	2.65	5.77
8/12 pitch	1.330	S.F.	.052	3.37	2.87	6.24
Drip edge, metal, 5" wide	.100	L.F.	.002	.06	.11	.17
8" wide	.100	L.F.	.002	.09	.11	.20
Building paper, #15 asphalt felt	1.300	S.F.	.002	.08	.09	.17
Soffit & fascia, aluminum vented, 1' overhang	.080	L.F.	.012	.39	.64	1.03
2' overhang	.080	L.F.	.013	.56	.70	1.26
Vinyl vented, 1' overhang	.080	L.F.	.011	.38	.59	.97
2' overhang	.080	L.F.	.012	.50	.70	1.20
Wood board fascia, plywood soffit, 1' overhang	.080	L.F.	.010	.24	.50	.74
2' overhang	.080	L.F.	.014	.34	.75	1.09
Rake, trim, painted, 1" x 6"	.043	L.F.	.004	.06	.19	.25
1" x 8"	.043	L.F.	.004	.06	.19	.25
Gutter, 5" box, aluminum, seamless, painted	.040	L.F.	.003	.12	.16	.28
Vinyl	.040	L.F.	.003	.07	.15	.22
Downspout 2" x 3", aluminum, one story house	.020	L.F.	.001	.03	.05	.08
Two story house	.020	L.F.	.001	.05	.09	.14
Vinyl, one story house	.020	L.F.	.001	.03	.05	.08
Two story house	.020	L.F.	.001	.05	.09	.14

5 | ROOFING

24 | Gable Dormer Roofing Systems

System Description	QUAN.	UNIT	LABOR HOURS	COST PER S.F. MAT.	COST PER S.F. INST.	COST PER S.F. TOTAL
ASPHALT, ROOF SHINGLES, CLASS A						
Shingles, standard inorganic class A 210-235 lb./sq	1.400	S.F.	.020	1.23	1.04	2.27
Drip edge, metal, 5" wide	.220	L.F.	.004	.13	.24	.37
Building paper, #15 asphalt felt	1.500	S.F.	.002	.09	.11	.20
Ridge shingles, asphalt	.280	L.F.	.007	.65	.35	1
Soffit & fascia, aluminum, vented	.220	L.F.	.032	1.06	1.76	2.82
Flashing, aluminum, mill finish, .013" thick	1.500	S.F.	.083	1.31	4.23	5.54
TOTAL		S.F.	.148	4.47	7.73	12.20
WOOD, CEDAR, NO. 1 PERFECTIONS						
Shingles, red cedar, No.1 perfections, 18" long, 5" exp.	1.400	S.F.	.041	3.79	2.24	6.03
Drip edge, metal, 5" wide	.220	L.F.	.004	.13	.24	.37
Building paper, #15 asphalt felt	1.500	S.F.	.002	.09	.11	.20
Ridge shingles, wood	.280	L.F.	.008	1.39	.44	1.83
Soffit & fascia, aluminum, vented	.220	L.F.	.032	1.06	1.76	2.82
Flashing, aluminum, mill finish, .013" thick	1.500	S.F.	.083	1.31	4.23	5.54
TOTAL		S.F.	.170	7.77	9.02	16.79
SLATE, BUCKINGHAM, BLACK						
Shingles, Buckingham, Virginia, black	1.400	S.F.	.064	7.21	3.26	10.47
Drip edge, metal, 5" wide	.220	L.F.	.004	.13	.24	.37
Building paper, #15 asphalt felt	1.500	S.F.	.002	.09	.11	.20
Ridge shingles, slate	.280	L.F.	.011	3.08	.57	3.65
Soffit & fascia, aluminum, vented	.220	L.F.	.032	1.06	1.76	2.82
Flashing, copper, 16 oz.	1.500	S.F.	.104	12.68	5.33	18.01
TOTAL		S.F.	.217	24.25	11.27	35.52

The prices in these systems are based on a square foot of plan area under the dormer roof.

Description	QUAN.	UNIT	LABOR HOURS	COST PER S.F. MAT.	COST PER S.F. INST.	COST PER S.F. TOTAL

Gable Dormer Roofing Price Sheet	QUAN.	UNIT	LABOR HOURS	COST PER S.F.		
				MAT.	INST.	TOTAL
Shingles, asphalt, standard, inorganic, class A, 210-235 lb./sq.	1.400	S.F.	.020	1.23	1.04	2.27
Laminated, multi-layered, 240-260 lb./sq.	1.400	S.F.	.025	1.69	1.27	2.96
Premium laminated, multi-layered, 260-300 lb./sq.	1.400	S.F.	.032	2.35	1.64	3.99
Clay tile, Spanish tile, red	1.400	S.F.	.062	8.55	3.12	11.67
Mission tile, red	1.400	S.F.	.097	6.30	3.12	9.42
French tile, red	1.400	S.F.	.083	17.50	2.86	20.36
Slate Buckingham, Virginia, black	1.400	S.F.	.064	7.20	3.26	10.46
Vermont, black or grey	1.400	S.F.	.064	7	3.26	10.26
Wood, red cedar, No.1 5X, 16" long, 5" exposure	1.400	S.F.	.045	4.41	2.46	6.87
Fire retardant	1.400	S.F.	.045	5.25	2.46	7.71
18" long, No.1 perfections, 5" exposure	1.400	S.F.	.041	3.79	2.24	6.03
Fire retardant	1.400	S.F.	.041	4.65	2.24	6.89
Resquared & rebutted, 18" long, 5" exposure	1.400	S.F.	.037	4.27	2.06	6.33
Fire retardant	1.400	S.F.	.037	5.15	2.06	7.21
Shakes hand split, 24" long, 10" exposure	1.400	S.F.	.045	4.20	2.46	6.66
Fire retardant	1.400	S.F.	.045	5.05	2.46	7.51
18" long, 8" exposure	1.400	S.F.	.056	2.77	3.09	5.86
Fire retardant	1.400	S.F.	.056	3.63	3.09	6.72
Drip edge, metal, 5" wide	.220	L.F.	.004	.13	.24	.37
8" wide	.220	L.F.	.004	.19	.24	.43
Building paper, #15 asphalt felt	1.500	S.F.	.002	.09	.11	.20
Ridge shingles, asphalt	.280	L.F.	.007	.65	.35	1
Clay	.280	L.F.	.011	1.68	1.72	3.40
Slate	.280	L.F.	.011	3.08	.57	3.65
Wood	.280	L.F.	.008	1.39	.44	1.83
Soffit & fascia, aluminum, vented	.220	L.F.	.032	1.06	1.76	2.82
Vinyl, vented	.220	L.F.	.029	1.04	1.62	2.66
Wood, board fascia, plywood soffit	.220	L.F.	.026	.68	1.37	2.05
Flashing, aluminum, .013" thick	1.500	S.F.	.083	1.31	4.23	5.54
.032" thick	1.500	S.F.	.083	2.24	4.23	6.47
.040" thick	1.500	S.F.	.083	3.81	4.23	8.04
.050" thick	1.500	S.F.	.083	4.01	4.23	8.24
Copper, 16 oz.	1.500	S.F.	.104	12.70	5.35	18.05
20 oz.	1.500	S.F.	.109	15.55	5.55	21.10
24 oz.	1.500	S.F.	.114	22	5.85	27.85
32 oz.	1.500	S.F.	.120	29.50	6.10	35.60

5 | ROOFING

28 | Shed Dormer Roofing Systems

System Description	QUAN.	UNIT	LABOR HOURS	COST PER S.F. MAT.	COST PER S.F. INST.	COST PER S.F. TOTAL
ASPHALT, ROOF SHINGLES, CLASS A						
Shingles, standard inorganic class A 210-235 lb./sq.	1.100	S.F.	.016	.96	.82	1.78
Drip edge, aluminum, 5" wide	.250	L.F.	.005	.15	.28	.43
Building paper, #15 asphalt felt	1.200	S.F.	.002	.07	.08	.15
Soffit & fascia, aluminum, vented, 1' overhang	.250	L.F.	.036	1.21	2	3.21
Flashing, aluminum, mill finish, 0.013" thick	.800	L.F.	.044	.70	2.26	2.96
TOTAL		S.F.	.103	3.09	5.44	8.53
WOOD, CEDAR, NO. 1 PERFECTIONS, 18" LONG						
Shingles, wood, red cedar, #1 perfections, 5" exposure	1.100	S.F.	.032	2.98	1.76	4.74
Drip edge, aluminum, 5" wide	.250	L.F.	.005	.15	.28	.43
Building paper, #15 asphalt felt	1.200	S.F.	.002	.07	.08	.15
Soffit & fascia, aluminum, vented, 1' overhang	.250	L.F.	.036	1.21	2	3.21
Flashing, aluminum, mill finish, 0.013" thick	.800	L.F.	.044	.70	2.26	2.96
TOTAL		S.F.	.119	5.11	6.38	11.49
SLATE, BUCKINGHAM, BLACK						
Shingles, slate, Buckingham, black	1.100	S.F.	.050	5.67	2.56	8.23
Drip edge, aluminum, 5" wide	.250	L.F.	.005	.15	.28	.43
Building paper, #15 asphalt felt	1.200	S.F.	.002	.07	.08	.15
Soffit & fascia, aluminum, vented, 1' overhang	.250	L.F.	.036	1.21	2	3.21
Flashing, copper, 16 oz.	.800	L.F.	.056	6.76	2.84	9.60
TOTAL		S.F.	.149	13.86	7.76	21.62

The prices in this system are based on a square foot of plan area under the dormer roof.

Description	QUAN.	UNIT	LABOR HOURS	COST PER S.F. MAT.	COST PER S.F. INST.	COST PER S.F. TOTAL

Shed Dormer Roofing Price Sheet

	QUAN.	UNIT	LABOR HOURS	COST PER S.F. MAT.	COST PER S.F. INST.	COST PER S.F. TOTAL
Shingles, asphalt, standard, inorganic, class A, 210-235 lb./sq.	1.100	S.F.	.016	.96	.82	1.78
Laminated, multi-layered, 240-260 lb./sq.	1.100	S.F.	.020	1.33	1	2.33
Premium laminated, multi-layered, 260-300 lb./sq.	1.100	S.F.	.025	1.85	1.29	3.14
Clay tile, Spanish tile, red	1.100	S.F.	.049	6.70	2.45	9.15
Mission tile, red	1.100	S.F.	.077	4.95	2.45	7.40
French tile, red	1.100	S.F.	.065	13.75	2.24	15.99
Slate Buckingham, Virginia, black	1.100	S.F.	.050	5.65	2.56	8.21
Vermont, black or grey	1.100	S.F.	.050	5.50	2.56	8.06
Wood, red cedar, No. 1 5X, 16" long, 5" exposure	1.100	S.F.	.035	3.47	1.94	5.41
Fire retardant	1.100	S.F.	.035	4.15	1.94	6.09
18" long, No.1 perfections, 5" exposure	1.100	S.F.	.032	2.98	1.76	4.74
Fire retardant	1.100	S.F.	.032	3.66	1.76	5.42
Resquared & rebutted, 18" long, 5" exposure	1.100	S.F.	.029	3.36	1.62	4.98
Fire retardant	1.100	S.F.	.029	4.04	1.62	5.66
Shakes hand split, 24" long, 10" exposure	1.100	S.F.	.035	3.30	1.94	5.24
Fire retardant	1.100	S.F.	.035	3.98	1.94	5.92
18" long, 8" exposure	1.100	S.F.	.044	2.18	2.43	4.61
Fire retardant	1.100	S.F.	.044	2.86	2.43	5.29
Drip edge, metal, 5" wide	.250	L.F.	.005	.15	.28	.43
8" wide	.250	L.F.	.005	.22	.28	.50
Building paper, #15 asphalt felt	1.200	S.F.	.002	.07	.08	.15
Soffit & fascia, aluminum, vented	.250	L.F.	.036	1.21	2	3.21
Vinyl, vented	.250	L.F.	.033	1.18	1.84	3.02
Wood, board fascia, plywood soffit	.250	L.F.	.030	.78	1.56	2.34
Flashing, aluminum, .013" thick	.800	L.F.	.044	.70	2.26	2.96
.032" thick	.800	L.F.	.044	1.19	2.26	3.45
.040" thick	.800	L.F.	.044	2.03	2.26	4.29
.050" thick	.800	L.F.	.044	2.14	2.26	4.40
Copper, 16 oz.	.800	L.F.	.056	6.75	2.84	9.59
20 oz.	.800	L.F.	.058	8.30	2.97	11.27
24 oz.	.800	L.F.	.061	11.70	3.11	14.81
32 oz.	.800	L.F.	.064	15.70	3.26	18.96

5 | ROOFING — 32 Skylight/Skywindow Systems

System Description	QUAN.	UNIT	LABOR HOURS	COST EACH MAT.	COST EACH INST.	COST EACH TOTAL
SKYLIGHT, FIXED, 32" X 32"						
Skylight, fixed bubble, insulating, 32" x 32"	1.000	Ea.	1.422	220.44	72.89	293.33
Trimmer rafters, 2" x 6"	28.000	L.F.	.448	20.44	24.64	45.08
Headers, 2" x 6"	6.000	L.F.	.267	4.38	14.70	19.08
Curb, 2" x 4"	12.000	L.F.	.154	5.64	8.52	14.16
Flashing, aluminum, .013" thick	13.500	S.F.	.745	11.75	38.07	49.82
Moldings, casing, ogee, 11/16" x 2-1/2", pine	12.000	L.F.	.384	18.60	21.12	39.72
Trim primer coat, oil base, brushwork	12.000	L.F.	.148	.48	6.96	7.44
Trim paint, 1 coat, brushwork	12.000	L.F.	.148	.72	6.96	7.68
TOTAL		Ea.	3.716	282.45	193.86	476.31
SKYLIGHT, FIXED, 48" X 48"						
Skylight, fixed bubble, insulating, 48" x 48"	1.000	Ea.	1.296	504	66.56	570.56
Trimmer rafters, 2" x 6"	28.000	L.F.	.448	20.44	24.64	45.08
Headers, 2" x 6"	8.000	L.F.	.356	5.84	19.60	25.44
Curb, 2" x 4"	16.000	L.F.	.205	7.52	11.36	18.88
Flashing, aluminum, .013" thick	16.000	S.F.	.883	13.92	45.12	59.04
Moldings, casing, ogee, 11/16" x 2-1/2", pine	16.000	L.F.	.512	24.80	28.16	52.96
Trim primer coat, oil base, brushwork	16.000	L.F.	.197	.64	9.28	9.92
Trim paint, 1 coat, brushwork	16.000	L.F.	.197	.96	9.28	10.24
TOTAL		Ea.	4.094	578.12	214	792.12
SKYWINDOW, OPERATING, 24" X 48"						
Skywindow, operating, thermopane glass, 24" x 48"	1.000	Ea.	3.200	645	164	809
Trimmer rafters, 2" x 6"	28.000	L.F.	.448	20.44	24.64	45.08
Headers, 2" x 6"	8.000	L.F.	.267	4.38	14.70	19.08
Curb, 2" x 4"	14.000	L.F.	.179	6.58	9.94	16.52
Flashing, aluminum, .013" thick	14.000	S.F.	.772	12.18	39.48	51.66
Moldings, casing, ogee, 11/16" x 2-1/2", pine	14.000	L.F.	.448	21.70	24.64	46.34
Trim primer coat, oil base, brushwork	14.000	L.F.	.172	.56	8.12	8.68
Trim paint, 1 coat, brushwork	14.000	L.F.	.172	.84	8.12	8.96
TOTAL		Ea.	5.658	711.68	293.64	1,005.32

The prices in these systems are on a cost each basis.

Description	QUAN.	UNIT	LABOR HOURS	COST EACH MAT.	COST EACH INST.	COST EACH TOTAL

Skylight/Skywindow Price Sheet	QUAN.	UNIT	LABOR HOURS	COST EACH		
				MAT.	INST.	TOTAL
Skylight, fixed bubble insulating, 24" x 24"	1.000	Ea.	.800	124	41	165
32" x 32"	1.000	Ea.	1.422	220	73	293
32" x 48"	1.000	Ea.	.864	335	44.50	379.50
48" x 48"	1.000	Ea.	1.296	505	66.50	571.50
Ventilating bubble insulating, 36" x 36"	1.000	Ea.	2.667	530	137	667
52" x 52"	1.000	Ea.	2.667	735	137	872
28" x 52"	1.000	Ea.	3.200	540	164	704
36" x 52"	1.000	Ea.	3.200	595	164	759
Skywindow, operating, thermopane glass, 24" x 48"	1.000	Ea.	3.200	645	164	809
32" x 48"	1.000	Ea.	3.556	675	183	858
Trimmer rafters, 2" x 6"	28.000	L.F.	.448	20.50	24.50	45
2" x 8"	28.000	L.F.	.472	28.50	26	54.50
2" x 10"	28.000	L.F.	.711	41	39	80
Headers, 24" window, 2" x 6"	4.000	L.F.	.178	2.92	9.80	12.72
2" x 8"	4.000	L.F.	.188	4.04	10.40	14.44
2" x 10"	4.000	L.F.	.200	5.90	11.05	16.95
32" window, 2" x 6"	6.000	L.F.	.267	4.38	14.70	19.08
2" x 8"	6.000	L.F.	.282	6.05	15.60	21.65
2" x 10"	6.000	L.F.	.300	8.80	16.55	25.35
48" window, 2" x 6"	8.000	L.F.	.356	5.85	19.60	25.45
2" x 8"	8.000	L.F.	.376	8.10	21	29.10
2" x 10"	8.000	L.F.	.400	11.75	22	33.75
Curb, 2" x 4", skylight, 24" x 24"	8.000	L.F.	.102	3.76	5.70	9.46
32" x 32"	12.000	L.F.	.154	5.65	8.50	14.15
32" x 48"	14.000	L.F.	.179	6.60	9.95	16.55
48" x 48"	16.000	L.F.	.205	7.50	11.35	18.85
Flashing, aluminum .013" thick, skylight, 24" x 24"	9.000	S.F.	.497	7.85	25.50	33.35
32" x 32"	13.500	S.F.	.745	11.75	38	49.75
32" x 48"	14.000	S.F.	.772	12.20	39.50	51.70
48" x 48"	16.000	S.F.	.883	13.90	45	58.90
Copper 16 oz., skylight, 24" x 24"	9.000	S.F.	.626	76	32	108
32" x 32"	13.500	S.F.	.939	114	48	162
32" x 48"	14.000	S.F.	.974	118	49.50	167.50
48" x 48"	16.000	S.F.	1.113	135	57	192
Trim, interior casing painted, 24" x 24"	8.000	L.F.	.347	13.50	17.85	31.35
32" x 32"	12.000	L.F.	.520	20.50	27	47.50
32" x 48"	14.000	L.F.	.607	23.50	31	54.50
48" x 48"	16.000	L.F.	.693	27	35.50	62.50

5 | ROOFING
34 | Built-up Roofing Systems

System Description	QUAN.	UNIT	LABOR HOURS	COST PER S.F. MAT.	COST PER S.F. INST.	COST PER S.F. TOTAL
ASPHALT, ORGANIC, 4-PLY, INSULATED DECK						
Membrane, asphalt, 4-plies #15 felt, gravel surfacing	1.000	S.F.	.025	1.43	1.51	2.94
Insulation board, 2-layers of 1-1/16" glass fiber	2.000	S.F.	.012	2.22	.62	2.84
Roof deck insulation, fastening alternatives, coated screws, 4" long	1.000	S.F	.003	.11	.13	.24
Wood blocking, 2" x 6"	.040	L.F.	.004	.09	.24	.33
Treated 4" x 4" cant strip	.040	L.F.	.001	.08	.05	.13
Flashing, aluminum, 0.040" thick	.050	S.F.	.003	.13	.14	.27
TOTAL		S.F.	.048	4.06	2.69	6.75
ASPHALT, INORGANIC, 3-PLY, INSULATED DECK						
Membrane, asphalt, 3-plies type IV glass felt, gravel surfacing	1.000	S.F.	.028	1.43	1.66	3.09
Insulation board, 2-layers of 1-1/16" glass fiber	2.000	S.F.	.012	2.22	.62	2.84
Roof deck insulation, fastening alternatives, coated screws, 4" long	1.000	S.F	.003	.11	.13	.24
Wood blocking, 2" x 6"	.040	L.F.	.004	.09	.24	.33
Treated 4" x 4" cant strip	.040	L.F.	.001	.08	.05	.13
Flashing, aluminum, 0.040" thick	.050	S.F.	.003	.13	.14	.27
TOTAL		S.F.	.051	4.06	2.84	6.90
COAL TAR, ORGANIC, 4-PLY, INSULATED DECK						
Membrane, coal tar, 4-plies #15 felt, gravel surfacing	1.000	S.F.	.027	2.57	1.58	4.15
Insulation board, 2-layers of 1-1/16" glass fiber	2.000	S.F.	.012	2.22	.62	2.84
Roof deck insulation, fastening alternatives, coated screws, 4" long	1.000	S.F	.003	.11	.13	.24
Wood blocking, 2" x 6"	.040	L.F.	.004	.09	.24	.33
Treated 4" x 4" cant strip	.040	L.F.	.001	.08	.05	.13
Flashing, aluminum, 0.040" thick	.050	S.F.	.003	.13	.14	.27
TOTAL		S.F.	.050	5.20	2.76	7.96
COAL TAR, INORGANIC, 3-PLY, INSULATED DECK						
Membrane, coal tar, 3-plies type IV glass felt, gravel surfacing	1.000	S.F.	.029	2.12	1.74	3.86
Insulation board, 2-layers of 1-1/16" glass fiber	2.000	S.F.	.012	2.22	.62	2.84
Roof deck insulation, fastening alternatives, coated screws, 4" long	1.000	S.F	.003	.11	.13	.24
Wood blocking, 2" x 6"	.040	L.F.	.004	.09	.24	.33
Treated 4" x 4" cant strip	.040	L.F.	.001	.08	.05	.13
Flashing, aluminum, 0.040" thick	.050	S.F.	.003	.13	.14	.27
TOTAL		S.F.	.052	4.75	2.92	7.67

Built-Up Roofing Price Sheet

	QUAN.	UNIT	LABOR HOURS	COST PER S.F. MAT.	COST PER S.F. INST.	COST PER S.F. TOTAL
Membrane, asphalt, 4-plies #15 organic felt, gravel surfacing	1.000	S.F.	.025	1.43	1.51	2.94
Asphalt base sheet & 3-plies #15 asphalt felt	1.000	S.F.	.025	1.09	1.51	2.60
3-plies type IV glass fiber felt	1.000	S.F.	.028	1.42	1.66	3.08
4-plies type IV glass fiber felt	1.000	S.F.	.028	1.76	1.66	3.42
Coal tar, 4-plies #15 organic felt, gravel surfacing	1.000	S.F.	.027			
4-plies tarred felt	1.000	S.F.	.027	2.56	1.58	4.14
3-plies type IV glass fiber felt	1.000	S.F.	.029	2.11	1.74	3.85
4-plies type IV glass fiber felt	1.000	S.F.	.027	2.90	1.58	4.48
Roll, asphalt, 1-ply #15 organic felt, 2-plies mineral surfaced	1.000	S.F.	.021	.82	1.23	2.05
3-plies type IV glass fiber, 1-ply mineral surfaced	1.000	S.F.	.022	1.38	1.32	2.70
Insulation boards, glass fiber, 1-1/16" thick	1.000	S.F.	.008	1.22	.44	1.66
2-1/16" thick	1.000	S.F.	.010	1.71	.52	2.23
2-7/16" thick	1.000	S.F.	.010	1.96	.52	2.48
Expanded perlite, 1" thick	1.000	S.F.	.010	.66	.52	1.18
1-1/2" thick	1.000	S.F.	.010	.91	.52	1.43
2" thick	1.000	S.F.	.011	1.21	.58	1.79
Fiberboard, 1" thick	1.000	S.F.	.010	.68	.52	1.20
1-1/2" thick	1.000	S.F.	.010	.99	.52	1.51
2" thick	1.000	S.F.	.010	1.28	.52	1.80
Extruded polystyrene, 15 PSI compressive strength, 2" thick R10	1.000	S.F.	.006	.90	.38	1.28
3" thick R15	1.000	S.F.	.008	1.68	.44	2.12
4" thick R20	1.000	S.F.	.008	2.23	.44	2.67
Tapered for drainage	1.000	S.F.	.005	.76	.34	1.10
40 PSI compressive strength, 1" thick R5	1.000	S.F.	.005	.69	.34	1.03
2" thick R10	1.000	S.F.	.006	1.22	.38	1.60
3" thick R15	1.000	S.F.	.008	1.71	.44	2.15
4" thick R20	1.000	S.F.	.008	2.21	.44	2.65
Fiberboard high density, 1/2" thick R1.3	1.000	S.F.	.008	.44	.44	.88
1" thick R2.5	1.000	S.F.	.010	.75	.52	1.27
1 1/2" thick R3.8	1.000	S.F.	.010	1.08	.52	1.60
Polyisocyanurate, 1 1/2" thick	1.000	S.F.	.006	.81	.38	1.19
2" thick	1.000	S.F.	.007	1.02	.42	1.44
3 1/2" thick	1.000	S.F.	.008	2.26	.44	2.70
Tapered for drainage	1.000	S.F.	.006	.81	.35	1.16
Expanded polystyrene, 1" thick	1.000	S.F.	.005	.39	.34	.73
2" thick R10	1.000	S.F.	.006	.66	.38	1.04
3" thick	1.000	S.F.	.006	.94	.38	1.32
Wood blocking, treated, 6" x 2" & 4" x 4" cant	.040	L.F.	.002	.13	.13	.26
6" x 4-1/2" & 4" x 4" cant	.040	L.F.	.005	.21	.29	.50
6" x 5" & 4" x 4" cant	.040	L.F.	.007	.25	.37	.62
Flashing, aluminum, 0.019" thick	.050	S.F.	.003	.06	.14	.20
0.032" thick	.050	S.F.	.003	.07	.14	.21
0.040" thick	.050	S.F.	.003	.13	.14	.27
Copper sheets, 16 oz., under 500 lbs.	.050	S.F.	.003	.42	.18	.60
Over 500 lbs.	.050	S.F.	.003	.42	.13	.55
20 oz., under 500 lbs.	.050	S.F.	.004	.52	.19	.71
Over 500 lbs.	.050	S.F.	.003	.49	.14	.63
Stainless steel, 32 gauge	.050	S.F.	.003	.17	.13	.30
28 gauge	.050	S.F.	.003	.22	.13	.35
26 gauge	.050	S.F.	.003	.23	.13	.36
24 gauge	.050	S.F.	.003	.26	.13	.39

Division 6 - Interiors

6 | INTERIORS — 04 | Drywall & Thincoat Wall Systems

System Description	QUAN.	UNIT	LABOR HOURS	COST PER S.F. MAT.	COST PER S.F. INST.	COST PER S.F. TOTAL
1/2" DRYWALL, TAPED & FINISHED						
Gypsum wallboard, 1/2" thick, standard	1.000	S.F.	.008	.33	.44	.77
Finish, taped & finished joints	1.000	S.F.	.008	.05	.44	.49
Corners, taped & finished, 32 L.F. per 12' x 12' room	.083	L.F.	.002	.01	.08	.09
Painting, primer & 2 coats	1.000	S.F.	.011	.21	.50	.71
Paint trim, to 6" wide, primer + 1 coat enamel	.125	L.F.	.001	.02	.06	.08
Moldings, base, ogee profile, 9/16" x 4-1/2, red oak	.125	L.F.	.005	.49	.25	.74
TOTAL		S.F.	.035	1.11	1.77	2.88
THINCOAT, SKIM-COAT, ON 1/2" BACKER DRYWALL						
Gypsum wallboard, 1/2" thick, thincoat backer	1.000	S.F.	.008	.33	.44	.77
Thincoat plaster	1.000	S.F.	.011	.12	.56	.68
Corners, taped & finished, 32 L.F. per 12' x 12' room	.083	L.F.	.002	.01	.08	.09
Painting, primer & 2 coats	1.000	S.F.	.011	.21	.50	.71
Paint trim, to 6" wide, primer + 1 coat enamel	.125	L.F.	.001	.02	.06	.08
Moldings, base, ogee profile, 9/16" x 4-1/2, red oak	.125	L.F.	.005	.49	.25	.74
TOTAL		S.F.	.038	1.18	1.89	3.07
5/8" DRYWALL, TAPED & FINISHED						
Gypsum wallboard, 5/8" thick, standard	1.000	S.F.	.008	.36	.44	.80
Finish, taped & finished joints	1.000	S.F.	.008	.05	.44	.49
Corners, taped & finished, 32 L.F. per 12' x 12' room	.083	L.F.	.002	.01	.08	.09
Painting, primer & 2 coats	1.000	S.F.	.011	.21	.50	.71
Moldings, base, ogee profile, 9/16" x 4-1/2, red oak	.125	L.F.	.005	.49	.25	.74
Paint trim, to 6" wide, primer + 1 coat enamel	.125	L.F.	.001	.02	.06	.08
TOTAL		S.F.	.035	1.14	1.77	2.91

The costs in this system are based on a square foot of wall.
Do not deduct for openings.

Description	QUAN.	UNIT	LABOR HOURS	COST PER S.F. MAT.	COST PER S.F. INST.	COST PER S.F. TOTAL

Drywall & Thincoat Wall Price Sheet

			LABOR	COST PER S.F.		
	QUAN.	UNIT	HOURS	MAT.	INST.	TOTAL
Gypsum wallboard, 1/2" thick, standard	1.000	S.F.	.008	.33	.44	.77
Fire resistant	1.000	S.F.	.008	.39	.44	.83
Water resistant	1.000	S.F.	.008	.43	.44	.87
5/8" thick, standard	1.000	S.F.	.008	.36	.44	.80
Fire resistant	1.000	S.F.	.008	.37	.44	.81
Water resistant	1.000	S.F.	.008	.46	.44	.90
Gypsum wallboard backer for thincoat system, 1/2" thick	1.000	S.F.	.008	.33	.44	.77
5/8" thick	1.000	S.F.	.008	.36	.44	.80
Gypsum wallboard, taped & finished	1.000	S.F.	.008	.05	.44	.49
Texture spray	1.000	S.F.	.010	.04	.49	.53
Thincoat plaster, including tape	1.000	S.F.	.011	.12	.56	.68
Gypsum wallboard corners, taped & finished, 32 L.F. per 4' x 4' room	.250	L.F.	.004	.03	.23	.26
6' x 6' room	.110	L.F.	.002	.01	.10	.11
10' x 10' room	.100	L.F.	.001	.01	.09	.10
12' x 12' room	.083	L.F.	.001	.01	.08	.09
16' x 16' room	.063	L.F.	.001	.01	.06	.07
Thincoat system, 32 L.F. per 4' x 4' room	.250	L.F.	.003	.03	.14	.17
6' x 6' room	.110	L.F.	.001	.01	.06	.07
10' x 10' room	.100	L.F.	.001	.01	.05	.06
12' x 12' room	.083	L.F.	.001	.01	.04	.05
16' x 16' room	.063	L.F.	.001	.01	.03	.04
Painting, primer, & 1 coat	1.000	S.F.	.008	.14	.39	.53
& 2 coats	1.000	S.F.	.011	.21	.50	.71
Wallpaper, $7/double roll	1.000	S.F.	.013	.67	.60	1.27
$17/double roll	1.000	S.F.	.015	1.22	.72	1.94
$40/double roll	1.000	S.F.	.018	2.42	.88	3.30
Wallcovering, medium weight vinyl		S.F.	.017	.99	.80	1.79
Tile, ceramic adhesive thin set, 4 1/4" x 4 1/4" tiles	1.000	S.F.	.084	2.48	3.64	6.12
6" x 6" tiles	1.000	S.F.	.080	3.61	3.96	7.57
Pregrouted sheets	1.000	S.F.	.067	5.65	2.89	8.54
Trim, painted or stained, baseboard	.125	L.F.	.006	.51	.31	.82
Base shoe	.125	L.F.	.005	.06	.29	.35
Chair rail	.125	L.F.	.005	.24	.26	.50
Cornice molding	.125	L.F.	.004	.15	.26	.41
Cove base, vinyl	.125	L.F.	.003	.14	.15	.29
Paneling, not including furring or trim						
Plywood, prefinished, 1/4" thick, 4' x 8' sheets, vert. grooves						
Birch faced, minimum	1.000	S.F.	.032	1.54	1.76	3.30
Average	1.000	S.F.	.038	1.32	2.10	3.42
Maximum	1.000	S.F.	.046	1.38	2.52	3.90
Mahogany, African	1.000	S.F.	.040	3.03	2.21	5.24
Philippine (lauan)	1.000	S.F.	.032	.72	1.76	2.48
Oak or cherry, minimum	1.000	S.F.	.032	1.54	1.76	3.30
Maximum	1.000	S.F.	.040	2.26	2.21	4.47
Rosewood	1.000	S.F.	.050	3.19	2.76	5.95
Teak	1.000	S.F.	.040	3.19	2.21	5.40
Chestnut	1.000	S.F.	.043	5.30	2.35	7.65
Pecan	1.000	S.F.	.040	2.75	2.21	4.96
Walnut, minimum	1.000	S.F.	.032	2.64	1.76	4.40
Maximum	1.000	S.F.	.040	5.80	2.21	8.01

6 | INTERIORS — 08 | Drywall & Thincoat Ceiling Systems

System Description	QUAN.	UNIT	LABOR HOURS	COST PER S.F. MAT.	COST PER S.F. INST.	COST PER S.F. TOTAL
1/2" GYPSUM WALLBOARD, TAPED & FINISHED						
Gypsum wallboard, 1/2" thick, standard	1.000	S.F.	.008	.33	.44	.77
Finish, taped & finished	1.000	S.F.	.008	.05	.44	.49
Corners, taped & finished, 12' x 12' room	.333	L.F.	.006	.04	.31	.35
Paint, primer & 2 coats	1.000	S.F.	.011	.21	.50	.71
TOTAL		S.F.	.033	.63	1.69	2.32
THINCOAT, SKIM COAT ON 1/2" GYPSUM WALLBOARD						
Gypsum wallboard, 1/2" thick, thincoat backer	1.000	S.F.	.008	.33	.44	.77
Thincoat plaster	1.000	S.F.	.011	.12	.56	.68
Corners, taped & finished, 12' x 12' room	.333	L.F.	.006	.04	.31	.35
Paint, primer & 2 coats	1.000	S.F.	.011	.21	.50	.71
TOTAL		S.F.	.036	.70	1.81	2.51
WATER-RESISTANT GYPSUM WALLBOARD, 1/2" THICK, TAPED & FINISHED						
Gypsum wallboard, 1/2" thick, water-resistant	1.000	S.F.	.008	.43	.44	.87
Finish, taped & finished	1.000	S.F.	.008	.05	.44	.49
Corners, taped & finished, 12' x 12' room	.333	L.F.	.006	.04	.31	.35
Paint, primer & 2 coats	1.000	S.F.	.011	.21	.50	.71
TOTAL		S.F.	.033	.73	1.69	2.42
5/8" GYPSUM WALLBOARD, TAPED & FINISHED						
Gypsum wallboard, 5/8" thick, standard	1.000	S.F.	.008	.36	.44	.80
Finish, taped & finished	1.000	S.F.	.008	.05	.44	.49
Corners, taped & finished, 12' x 12' room	.333	L.F.	.006	.04	.31	.35
Paint, primer & 2 coats	1.000	S.F.	.011	.21	.50	.71
TOTAL		S.F.	.033	.66	1.69	2.35

The costs in this system are based on a square foot of ceiling.

Description	QUAN.	UNIT	LABOR HOURS	COST PER S.F. MAT.	COST PER S.F. INST.	COST PER S.F. TOTAL

For customer support on your Residential Cost Data, call 877.713.2617.

Drywall & Thincoat Ceilings	QUAN.	UNIT	LABOR HOURS	COST PER S.F.		
				MAT.	INST.	TOTAL
Gypsum wallboard ceilings, 1/2" thick, standard	1.000	S.F.	.008	.33	.44	.77
Fire resistant	1.000	S.F.	.008	.39	.44	.83
Water resistant	1.000	S.F.	.008	.43	.44	.87
5/8" thick, standard	1.000	S.F.	.008	.36	.44	.80
Fire resistant	1.000	S.F.	.008	.37	.44	.81
Water resistant	1.000	S.F.	.008	.46	.44	.90
Gypsum wallboard backer for thincoat ceiling system, 1/2" thick	1.000	S.F.	.016	.72	.88	1.60
5/8" thick	1.000	S.F.	.016	.75	.88	1.63
Gypsum wallboard ceilings, taped & finished	1.000	S.F.	.008	.05	.44	.49
Texture spray	1.000	S.F.	.010	.04	.49	.53
Thincoat plaster	1.000	S.F.	.011	.12	.56	.68
Corners taped & finished, 4' x 4' room	1.000	L.F.	.015	.11	.93	1.04
6' x 6' room	.667	L.F.	.010	.07	.62	.69
10' x 10' room	.400	L.F.	.006	.04	.37	.41
12' x 12' room	.333	L.F.	.005	.04	.31	.35
16' x 16' room	.250	L.F.	.003	.02	.17	.19
Thincoat system, 4' x 4' room	1.000	L.F.	.011	.12	.56	.68
6' x 6' room	.667	L.F.	.007	.08	.38	.46
10' x 10' room	.400	L.F.	.004	.05	.23	.28
12' x 12' room	.333	L.F.	.004	.04	.18	.22
16' x 16' room	.250	L.F.	.002	.02	.11	.13
Painting, primer & 1 coat	1.000	S.F.	.008	.14	.39	.53
& 2 coats	1.000	S.F.	.011	.21	.50	.71
Wallpaper, double roll, solid pattern, avg. workmanship	1.000	S.F.	.013	.67	.60	1.27
Basic pattern, avg. workmanship	1.000	S.F.	.015	1.22	.72	1.94
Basic pattern, quality workmanship	1.000	S.F.	.018	2.42	.88	3.30
Tile, ceramic adhesive thin set, 4 1/4" x 4 1/4" tiles	1.000	S.F.	.084	2.48	3.64	6.12
6" x 6" tiles	1.000	S.F.	.080	3.61	3.96	7.57
Pregrouted sheets	1.000	S.F.	.067	5.65	2.89	8.54

6 | INTERIORS — 12 | Plaster & Stucco Wall Systems

System Description	QUAN.	UNIT	LABOR HOURS	COST PER S.F. MAT.	COST PER S.F. INST.	COST PER S.F. TOTAL
PLASTER ON GYPSUM LATH						
Plaster, gypsum or perlite, 2 coats	1.000	S.F.	.053	.44	2.72	3.16
Lath, 3/8" gypsum	1.000	S.F.	.010	.37	.51	.88
Corners, expanded metal, 32 L.F. per 12' x 12' room	.083	L.F.	.002	.01	.09	.10
Painting, primer & 2 coats	1.000	S.F.	.011	.21	.50	.71
Paint trim, to 6" wide, primer + 1 coat enamel	.125	L.F.	.001	.02	.06	.08
Moldings, base, ogee profile, 9/16" x 4-1/2, red oak	.125	L.F.	.005	.49	.25	.74
TOTAL		S.F.	.082	1.54	4.13	5.67
PLASTER ON METAL LATH						
Plaster, gypsum or perlite, 2 coats	1.000	S.F.	.053	.44	2.72	3.16
Lath, 2.5 Lb. diamond, metal	1.000	S.F.	.010	.40	.51	.91
Corners, expanded metal, 32 L.F. per 12' x 12' room	.083	L.F.	.002	.01	.09	.10
Painting, primer & 2 coats	1.000	S.F.	.011	.21	.50	.71
Paint trim, to 6" wide, primer + 1 coat enamel	.125	L.F.	.001	.02	.06	.08
Moldings, base, ogee profile, 9/16" x 4-1/2, red oak	.125	L.F.	.005	.49	.25	.74
TOTAL		S.F.	.082	1.57	4.13	5.70
STUCCO ON METAL LATH						
Stucco, 2 coats	1.000	S.F.	.041	.28	2.08	2.36
Lath, 2.5 Lb. diamond, metal	1.000	S.F.	.010	.40	.51	.91
Corners, expanded metal, 32 L.F. per 12' x 12' room	.083	L.F.	.002	.01	.09	.10
Painting, primer & 2 coats	1.000	S.F.	.011	.21	.50	.71
Paint trim, to 6" wide, primer + 1 coat enamel	.125	L.F.	.001	.02	.06	.08
Moldings, base, ogee profile, 9/16" x 4-1/2, red oak	.125	L.F.	.005	.49	.25	.74
TOTAL		S.F.	.070	1.41	3.49	4.90

The costs in these systems are based on a per square foot of wall area.
Do not deduct for openings.

Description	QUAN.	UNIT	LABOR HOURS	COST PER S.F. MAT.	COST PER S.F. INST.	COST PER S.F. TOTAL

Plaster & Stucco Wall Price Sheet

	QUAN.	UNIT	LABOR HOURS	COST PER S.F. MAT.	INST.	TOTAL
Plaster, gypsum or perlite, 2 coats	1.000	S.F.	.053	.44	2.72	3.16
3 coats	1.000	S.F.	.065	.64	3.28	3.92
Lath, gypsum, standard, 3/8" thick	1.000	S.F.	.010	.37	.51	.88
Fire resistant, 3/8" thick	1.000	S.F.	.013	.30	.62	.92
1/2" thick	1.000	S.F.	.014	.34	.66	1
Metal, diamond, 2.5 Lb.	1.000	S.F.	.010	.40	.51	.91
3.4 Lb.	1.000	S.F.	.012	.47	.58	1.05
Rib, 2.75 Lb.	1.000	S.F.	.012	.37	.58	.95
3.4 Lb.	1.000	S.F.	.013	.45	.62	1.07
Corners, expanded metal, 32 L.F. per 4' x 4' room	.250	L.F.	.005	.04	.28	.32
6' x 6' room	.110	L.F.	.002	.02	.12	.14
10' x 10' room	.100	L.F.	.002	.02	.11	.13
12' x 12' room	.083	L.F.	.002	.01	.09	.10
16' x 16' room	.063	L.F.	.001	.01	.07	.08
Painting, primer & 1 coats	1.000	S.F.	.008	.14	.39	.53
Primer & 2 coats	1.000	S.F.	.011	.21	.50	.71
Wallpaper, low price double roll	1.000	S.F.	.013	.67	.60	1.27
Medium price double roll	1.000	S.F.	.015	1.22	.72	1.94
High price double roll	1.000	S.F.	.018	2.42	.88	3.30
Tile, ceramic thin set, 4-1/4" x 4-1/4" tiles	1.000	S.F.	.084	2.48	3.64	6.12
6" x 6" tiles	1.000	S.F.	.080	3.61	3.96	7.57
Pregrouted sheets	1.000	S.F.	.067	5.65	2.89	8.54
Trim, painted or stained, baseboard	.125	L.F.	.006	.51	.31	.82
Base shoe	.125	L.F.	.005	.06	.29	.35
Chair rail	.125	L.F.	.005	.24	.26	.50
Cornice molding	.125	L.F.	.004	.15	.26	.41
Cove base, vinyl	.125	L.F.	.003	.14	.15	.29
Paneling not including furring or trim						
Plywood, prefinished, 1/4" thick, 4' x 8' sheets, vert. grooves						
Birch faced, minimum	1.000	S.F.	.032	1.54	1.76	3.30
Average	1.000	S.F.	.038	1.32	2.10	3.42
Maximum	1.000	S.F.	.046	1.38	2.52	3.90
Mahogany, African	1.000	S.F.	.040	3.03	2.21	5.24
Philippine (lauan)	1.000	S.F.	.032	.72	1.76	2.48
Oak or cherry, minimum	1.000	S.F.	.032	1.54	1.76	3.30
Maximum	1.000	S.F.	.040	2.26	2.21	4.47
Rosewood	1.000	S.F.	.050	3.19	2.76	5.95
Teak	1.000	S.F.	.040	3.19	2.21	5.40
Chestnut	1.000	S.F.	.043	5.30	2.35	7.65
Pecan	1.000	S.F.	.040	2.75	2.21	4.96
Walnut, minimum	1.000	S.F.	.032	2.64	1.76	4.40
Maximum	1.000	S.F.	.040	5.80	2.21	8.01

6 | INTERIORS — 16 | Plaster & Stucco Ceiling Systems

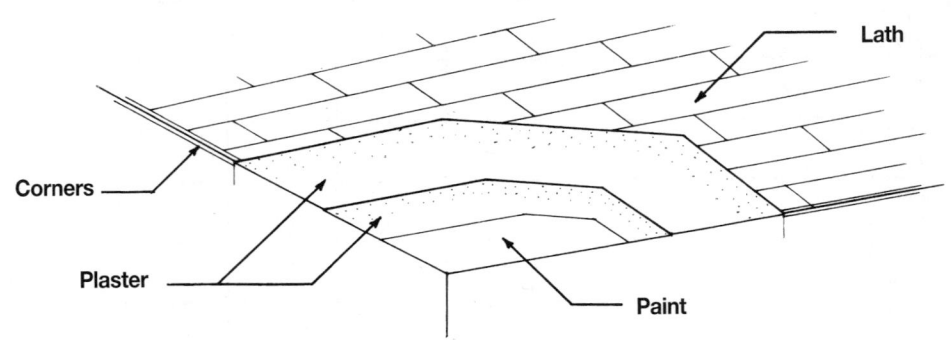

System Description	QUAN.	UNIT	LABOR HOURS	COST PER S.F. MAT.	COST PER S.F. INST.	COST PER S.F. TOTAL
PLASTER ON GYPSUM LATH						
Plaster, gypsum or perlite, 2 coats	1.000	S.F.	.061	.44	3.08	3.52
Gypsum lath, plain or perforated, nailed, 3/8" thick	1.000	S.F.	.010	.37	.51	.88
Gypsum lath, ceiling installation adder	1.000	S.F.	.004		.20	.20
Corners, expanded metal, 12' x 12' room	.330	L.F.	.007	.05	.36	.41
Painting, primer & 2 coats	1.000	S.F.	.011	.21	.50	.71
TOTAL		S.F.	.093	1.07	4.65	5.72
PLASTER ON METAL LATH						
Plaster, gypsum or perlite, 2 coats	1.000	S.F.	.061	.44	3.08	3.52
Lath, 2.5 Lb. diamond, metal	1.000	S.F.	.012	.40	.58	.98
Corners, expanded metal, 12' x 12' room	.330	L.F.	.007	.05	.36	.41
Painting, primer & 2 coats	1.000	S.F.	.011	.21	.50	.71
TOTAL		S.F.	.091	1.10	4.52	5.62
STUCCO ON GYPSUM LATH						
Stucco, 2 coats	1.000	S.F.	.041	.28	2.08	2.36
Gypsum lath, plain or perforated, nailed, 3/8" thick	1.000	S.F.	.010	.37	.51	.88
Gypsum lath, ceiling installation adder	1.000	S.F.	.004		.20	.20
Corners, expanded metal, 12' x 12' room	.330	L.F.	.007	.05	.36	.41
Painting, primer & 2 coats	1.000	S.F.	.011	.21	.50	.71
TOTAL		S.F.	.073	.91	3.65	4.56
STUCCO ON METAL LATH						
Stucco, 2 coats	1.000	S.F.	.041	.28	2.08	2.36
Lath, 2.5 Lb. diamond, metal	1.000	S.F.	.012	.40	.58	.98
Corners, expanded metal, 12' x 12' room	.330	L.F.	.007	.05	.36	.41
Painting, primer & 2 coats	1.000	S.F.	.011	.21	.50	.71
TOTAL		S.F.	.071	.94	3.52	4.46

The costs in these systems are based on a square foot of ceiling area.

Description	QUAN.	UNIT	LABOR HOURS	COST PER S.F. MAT.	COST PER S.F. INST.	COST PER S.F. TOTAL

Plaster & Stucco Ceiling Price Sheet	QUAN.	UNIT	LABOR HOURS	COST PER S.F.		
				MAT.	INST.	TOTAL
Plaster, gypsum or perlite, 2 coats	1.000	S.F.	.061	.44	3.08	3.52
3 coats	1.000	S.F.	.065	.64	3.28	3.92
Lath, gypsum, standard, 3/8" thick	1.000	S.F.	.014	.37	.71	1.08
Fire resistant, 3/8" thick	1.000	S.F.	.017	.30	.82	1.12
1/2" thick	1.000	S.F.	.018	.34	.86	1.20
Metal, diamond, 2.5 Lb.	1.000	S.F.	.012	.40	.58	.98
3.4 Lb.	1.000	S.F.	.015	.47	.72	1.19
Rib, 2.75 Lb.	1.000	S.F.	.012	.37	.58	.95
3.4 Lb.	1.000	S.F.	.013	.45	.62	1.07
Corners expanded metal, 4' x 4' room	1.000	L.F.	.020	.16	1.10	1.26
6' x 6' room	.667	L.F.	.013	.11	.74	.85
10' x 10' room	.400	L.F.	.008	.06	.44	.50
12' x 12' room	.333	L.F.	.007	.05	.36	.41
16' x 16' room	.250	L.F.	.004	.03	.21	.24
Painting, primer & 1 coat	1.000	S.F.	.008	.14	.39	.53
Primer & 2 coats	1.000	S.F.	.011	.21	.50	.71

6 | INTERIORS — 18 | Suspended Ceiling Systems

System Description	QUAN.	UNIT	LABOR HOURS	COST PER S.F. MAT.	COST PER S.F. INST.	COST PER S.F. TOTAL
2' X 2' GRID, FILM FACED FIBERGLASS, 5/8" THICK						
Suspension system, 2' x 2' grid, T bar	1.000	S.F.	.012	1.04	.68	1.72
Ceiling board, film faced fiberglass, 5/8" thick	1.000	S.F.	.013	1.36	.71	2.07
Carrier channels, 1-1/2" x 3/4"	1.000	S.F.	.017	.12	.94	1.06
Hangers, #12 wire	1.000	S.F.	.002		.09	.09
TOTAL		S.F.	.044	2.52	2.42	4.94
2' X 4' GRID, FILM FACED FIBERGLASS, 5/8" THICK						
Suspension system, 2' x 4' grid, T bar	1.000	S.F.	.010	.81	.55	1.36
Ceiling board, film faced fiberglass, 5/8" thick	1.000	S.F.	.013	1.36	.71	2.07
Carrier channels, 1-1/2" x 3/4"	1.000	S.F.	.017	.12	.94	1.06
Hangers, #12 wire	1.000	S.F.	.002		.09	.09
TOTAL		S.F.	.042	2.29	2.29	4.58
2' X 2' GRID, MINERAL FIBER, REVEAL EDGE, 1" THICK						
Suspension system, 2' x 2' grid, T bar	1.000	S.F.	.012	1.04	.68	1.72
Ceiling board, mineral fiber, reveal edge, 1" thick	1.000	S.F.	.013	2.20	.74	2.94
Carrier channels, 1-1/2" x 3/4"	1.000	S.F.	.017	.12	.94	1.06
Hangers, #12 wire	1.000	S.F.	.002		.09	.09
TOTAL		S.F.	.044	3.36	2.45	5.81
2' X 4' GRID, MINERAL FIBER, REVEAL EDGE, 1" THICK						
Suspension system, 2' x 4' grid, T bar	1.000	S.F.	.010	.81	.55	1.36
Ceiling board, mineral fiber, reveal edge, 1" thick	1.000	S.F.	.013	2.20	.74	2.94
Carrier channels, 1-1/2" x 3/4"	1.000	S.F.	.017	.12	.94	1.06
Hangers, #12 wire	1.000	S.F.	.002		.09	.09
TOTAL		S.F.	.042	3.13	2.32	5.45

Description	QUAN.	UNIT	LABOR HOURS	COST PER S.F. MAT.	COST PER S.F. INST.	COST PER S.F. TOTAL

Suspended Ceiling Price Sheet

	QUAN.	UNIT	LABOR HOURS	COST PER S.F. MAT.	INST.	TOTAL
Suspension systems, T bar, 2' x 2' grid	1.000	S.F.	.012	1.04	.68	1.72
2' x 4' grid	1.000	S.F.	.010	.81	.55	1.36
Concealed Z bar, 12" module	1.000	S.F.	.015	.92	.85	1.77
Ceiling boards, fiberglass, film faced, 2' x 2' or 2' x 4', 5/8" thick	1.000	S.F.	.013	1.36	.71	2.07
3/4" thick	1.000	S.F.	.013	3.33	.74	4.07
3" thick thermal R11	1.000	S.F.	.018	3.53	.98	4.51
Glass cloth faced, 3/4" thick	1.000	S.F.	.016	2.92	.88	3.80
1" thick	1.000	S.F.	.016	3.44	.91	4.35
1-1/2" thick, nubby face	1.000	S.F.	.017	2.11	.93	3.04
Mineral fiber boards, 5/8" thick, aluminum face 2' x 2'	1.000	S.F.	.013	3.98	.74	4.72
2' x 4'	1.000	S.F.	.012	3.98	.68	4.66
Standard faced, 2' x 2' or 2' x 4'	1.000	S.F.	.012	.75	.65	1.40
Plastic coated face, 2' x 2' or 2' x 4'	1.000	S.F.	.020	2.56	1.10	3.66
Fire rated, 2 hour rating, 5/8" thick	1.000	S.F.	.012	1.19	.65	1.84
Tegular edge, 2' x 2' or 2' x 4', 5/8" thick, fine textured	1.000	S.F.	.013	1.13	.94	2.07
Rough textured	1.000	S.F.	.015	1.25	.94	2.19
3/4" thick, fine textured	1.000	S.F.	.016	2.35	.98	3.33
Rough textured	1.000	S.F.	.018	1.57	.98	2.55
Luminous panels, prismatic, acrylic	1.000	S.F.	.020	2.22	1.10	3.32
Polystyrene	1.000	S.F.	.020	1.65	1.10	2.75
Flat or ribbed, acrylic	1.000	S.F.	.020	21	1.10	22.10
Polystyrene	1.000	S.F.	.020	2.52	1.10	3.62
Drop pan, white, acrylic	1.000	S.F.	.020	21	1.10	22.10
Polystyrene	1.000	S.F.	.020	1.61	1.10	2.71
Carrier channels, 4'-0" on center, 3/4" x 1-1/2"	1.000	S.F.	.017	.12	.94	1.06
1-1/2" x 3-1/2"	1.000	S.F.	.017	.33	.94	1.27
Hangers, #12 wire	1.000	S.F.	.002		.09	.09

6 | INTERIORS
20 | Interior Door Systems

System Description	QUAN.	UNIT	LABOR HOURS	COST EACH MAT.	COST EACH INST.	COST EACH TOTAL
LAUAN, FLUSH DOOR, HOLLOW CORE						
Door, flush, lauan, hollow core, 2'-8" wide x 6'-8" high	1.000	Ea.	.889	44	49	93
Frame, pine, 4-5/8" jamb	17.000	L.F.	.725	91.80	39.95	131.75
Moldings, casing, ogee, 11/16" x 2-1/2", pine	34.000	L.F.	1.088	52.70	59.84	112.54
Paint trim, to 6" wide, primer + 1 coat enamel	34.000	L.F.	.340	4.76	15.98	20.74
Butt hinges, chrome, 3-1/2" x 3-1/2"	1.500	Pr.		45.75		45.75
Lockset, passage	1.000	Ea.	.500	33	27.50	60.50
Prime door & frame, oil, brushwork	2.000	Face	1.600	7.58	76	83.58
Paint door and frame, oil, 2 coats	2.000	Face	2.667	11	126	137
TOTAL		Ea.	7.809	290.59	394.27	684.86
BIRCH, FLUSH DOOR, HOLLOW CORE						
Door, flush, birch, hollow core, 2'-8" wide x 6'-8" high	1.000	Ea.	.889	55.50	49	104.50
Frame, pine, 4-5/8" jamb	17.000	L.F.	.725	91.80	39.95	131.75
Moldings, casing, ogee, 11/16" x 2-1/2", pine	34.000	L.F.	1.088	52.70	59.84	112.54
Butt hinges, chrome, 3-1/2" x 3-1/2"	1.500	Pr.		45.75		45.75
Lockset, passage	1.000	Ea.	.500	33	27.50	60.50
Prime door & frame, oil, brushwork	2.000	Face	1.600	7.58	76	83.58
Paint door and frame, oil, 2 coats	2.000	Face	2.667	11	126	137
TOTAL		Ea.	7.469	297.33	378.29	675.62
RAISED PANEL, SOLID, PINE DOOR						
Door, pine, raised panel, 2'-8" wide x 6'-8" high	1.000	Ea.	.889	204	49	253
Frame, pine, 4-5/8" jamb	17.000	L.F.	.725	91.80	39.95	131.75
Moldings, casing, ogee, 11/16" x 2-1/2", pine	34.000	L.F.	1.088	52.70	59.84	112.54
Butt hinges, bronze, 3-1/2" x 3-1/2"	1.500	Pr.		53.25		53.25
Lockset, passage	1.000	Ea.	.500	33	27.50	60.50
Prime door & frame, oil, brushwork	2.000	Face	1.600	7.58	76	83.58
Paint door and frame, oil, 2 coats	2.000	Face	2.667	11	126	137
TOTAL		Ea.	7.469	453.33	378.29	831.62

The costs in these systems are based on a cost per each door.

Description	QUAN.	UNIT	LABOR HOURS	COST EACH MAT.	COST EACH INST.	COST EACH TOTAL

Interior Door Price Sheet	QUAN.	UNIT	LABOR HOURS	COST EACH		
				MAT.	INST.	TOTAL
Door, hollow core, lauan 1-3/8" thick, 6'-8" high x 1'-6" wide	1.000	Ea.	.889	38.50	49	87.50
2'-0" wide	1.000	Ea.	.889	38.50	49	87.50
2'-6" wide	1.000	Ea.	.889	42.50	49	91.50
2'-8" wide	1.000	Ea.	.889	44	49	93
3'-0" wide	1.000	Ea.	.941	46.50	52	98.50
Birch 1-3/8" thick, 6'-8" high x 1'-6" wide	1.000	Ea.	.889	45	49	94
2'-0" wide	1.000	Ea.	.889	46.50	49	95.50
2'-6" wide	1.000	Ea.	.889	55	49	104
2'-8" wide	1.000	Ea.	.889	55.50	49	104.50
3'-0" wide	1.000	Ea.	.941	57.50	52	109.50
Louvered pine 1-3/8" thick, 6'-8" high x 1'-6" wide	1.000	Ea.	.842	132	46.50	178.50
2'-0" wide	1.000	Ea.	.889	144	49	193
2'-6" wide	1.000	Ea.	.889	165	49	214
2'-8" wide	1.000	Ea.	.889	177	49	226
3'-0" wide	1.000	Ea.	.941	192	52	244
Paneled pine 1-3/8" thick, 6'-8" high x 1'-6" wide	1.000	Ea.	.842	141	46.50	187.50
2'-0" wide	1.000	Ea.	.889	168	49	217
2'-6" wide	1.000	Ea.	.889	200	49	249
2'-8" wide	1.000	Ea.	.889	204	49	253
3'-0" wide	1.000	Ea.	.941	224	52	276
Frame, pine, 1'-6" thru 2'-0" wide door, 3-5/8" deep	16.000	L.F.	.683	78	37.50	115.50
4-5/8" deep	16.000	L.F.	.683	86.50	37.50	124
5-5/8" deep	16.000	L.F.	.683	82.50	37.50	120
2'-6" thru 3'0" wide door, 3-5/8" deep	17.000	L.F.	.725	83	40	123
4-5/8" deep	17.000	L.F.	.725	92	40	132
5-5/8" deep	17.000	L.F.	.725	87.50	40	127.50
Trim, casing, painted, both sides, 1'-6" thru 2'-6" wide door	32.000	L.F.	1.855	53	93.50	146.50
2'-6" thru 3'-0" wide door	34.000	L.F.	1.971	56	99.50	155.50
Butt hinges 3-1/2" x 3-1/2", steel plated, chrome	1.500	Pr.		46		46
Bronze	1.500	Pr.		53.50		53.50
Locksets, passage, minimum	1.000	Ea.	.500	33	27.50	60.50
Maximum	1.000	Ea.	.575	38	31.50	69.50
Privacy, minimum	1.000	Ea.	.625	41.50	34.50	76
Maximum	1.000	Ea.	.675	44.50	37	81.50
Paint 2 sides, primer & 2 cts., flush door, 1'-6" to 2'-0" wide	2.000	Face	5.547	25.50	262	287.50
2'-6" thru 3'-0" wide	2.000	Face	6.933	31.50	330	361.50
Louvered door, 1'-6" thru 2'-0" wide	2.000	Face	6.400	24	300	324
2'-6" thru 3'-0" wide	2.000	Face	8.000	30	380	410
Paneled door, 1'-6" thru 2'-0" wide	2.000	Face	6.400	24	300	324
2'-6" thru 3'-0" wide	2.000	Face	8.000	30	380	410

6 | INTERIORS — 24 | Closet Door Systems

System Description	QUAN.	UNIT	LABOR HOURS	COST EACH MAT.	COST EACH INST.	COST EACH TOTAL
BI-PASSING, FLUSH, LAUAN, HOLLOW CORE, 4'-0" X 6'-8"						
Door, flush, lauan, hollow core, 4'-0" x 6'-8" opening	1.000	Ea.	1.333	194	73.50	267.50
Frame, pine, 4-5/8" jamb	18.000	L.F.	.768	97.20	42.30	139.50
Moldings, casing, ogee, 11/16" x 2-1/2", pine	36.000	L.F.	1.152	55.80	63.36	119.16
Prime door & frame, oil, brushwork	2.000	Face	1.600	7.58	76	83.58
Paint door and frame, oil, 2 coats	2.000	Face	2.667	11	126	137
TOTAL		Ea.	7.520	365.58	381.16	746.74
BI-PASSING, FLUSH, BIRCH, HOLLOW CORE, 6'-0" X 6'-8"						
Door, flush, birch, hollow core, 6'-0" x 6'-8" opening	1.000	Ea.	1.600	286	88	374
Frame, pine, 4-5/8" jamb	19.000	L.F.	.811	102.60	44.65	147.25
Moldings, casing, ogee, 11/16" x 2-1/2", pine	38.000	L.F.	1.216	58.90	66.88	125.78
Prime door & frame, oil, brushwork	2.000	Face	2.000	9.48	95	104.48
Paint door and frame, oil, 2 coats	2.000	Face	3.333	13.75	157.50	171.25
TOTAL		Ea.	8.960	470.73	452.03	922.76
BI-FOLD, PINE, PANELED, 3'-0" X 6'-8"						
Door, pine, paneled, 3'-0" x 6'-8" opening	1.000	Ea.	1.231	228	68	296
Frame, pine, 4-5/8" jamb	17.000	L.F.	.725	91.80	39.95	131.75
Moldings, casing, ogee, 11/16" x 2-1/2", pine	34.000	L.F.	1.088	52.70	59.84	112.54
Prime door & frame, oil, brushwork	2.000	Face	1.600	7.58	76	83.58
Paint door and frame, oil, 2 coats	2.000	Face	2.667	11	126	137
TOTAL		Ea.	7.311	391.08	369.79	760.87
BI-FOLD, PINE, LOUVERED, 6'-0" X 6'-8"						
Door, pine, louvered, 6'-0" x 6'-8" opening	1.000	Ea.	1.600	320	88	408
Frame, pine, 4-5/8" jamb	19.000	L.F.	.811	102.60	44.65	147.25
Moldings, casing, ogee, 11/16" x 2-1/2", pine	38.000	L.F.	1.216	58.90	66.88	125.78
Prime door & frame, oil, brushwork	2.500	Face	2.000	9.48	95	104.48
Paint door and frame, oil, 2 coats	2.500	Face	3.333	13.75	157.50	171.25
TOTAL		Ea.	8.960	504.73	452.03	956.76

The costs in this system are based on a cost per each door.

Description	QUAN.	UNIT	LABOR HOURS	COST EACH MAT.	COST EACH INST.	COST EACH TOTAL

Closet Door Price Sheet

	QUAN.	UNIT	LABOR HOURS	COST EACH MAT.	COST EACH INST.	TOTAL
Doors, bi-passing, pine, louvered, 4'-0" x 6'-8" opening	1.000	Ea.	1.333	455	73.50	528.50
6'-0" x 6'-8" opening	1.000	Ea.	1.600	585	88	673
Paneled, 4'-0" x 6'-8" opening	1.000	Ea.	1.333	565	73.50	638.50
6'-0" x 6'-8" opening	1.000	Ea.	1.600	670	88	758
Flush, birch, hollow core, 4'-0" x 6'-8" opening	1.000	Ea.	1.333	245	73.50	318.50
6'-0" x 6'-8" opening	1.000	Ea.	1.600	286	88	374
Flush, lauan, hollow core, 4'-0" x 6'-8" opening	1.000	Ea.	1.333	194	73.50	267.50
6'-0" x 6'-8" opening	1.000	Ea.	1.600	230	88	318
Bi-fold, pine, louvered, 3'-0" x 6'-8" opening	1.000	Ea.	1.231	228	68	296
6'-0" x 6'-8" opening	1.000	Ea.	1.600	320	88	408
Paneled, 3'-0" x 6'-8" opening	1.000	Ea.	1.231	228	68	296
6'-0" x 6'-8" opening	1.000	Ea.	1.600	320	88	408
Flush, birch, hollow core, 3'-0" x 6'-8" opening	1.000	Ea.	1.231	78.50	68	146.50
6'-0" x 6'-8" opening	1.000	Ea.	1.600	141	88	229
Flush, lauan, hollow core, 3'-0" x 6'8" opening	1.000	Ea.	1.231	305	68	373
6'-0" x 6'-8" opening	1.000	Ea.	1.600	440	88	528
Frame pine, 3'-0" door, 3-5/8" deep	17.000	L.F.	.725	83	40	123
4-5/8" deep	17.000	L.F.	.725	92	40	132
5-5/8" deep	17.000	L.F.	.725	87.50	40	127.50
4'-0" door, 3-5/8" deep	18.000	L.F.	.768	88	42.50	130.50
4-5/8" deep	18.000	L.F.	.768	97	42.50	139.50
5-5/8" deep	18.000	L.F.	.768	92.50	42.50	135
6'-0" door, 3-5/8" deep	19.000	L.F.	.811	92.50	44.50	137
4-5/8" deep	19.000	L.F.	.811	103	44.50	147.50
5-5/8" deep	19.000	L.F.	.811	98	44.50	142.50
Trim both sides, painted 3'-0" x 6'-8" door	34.000	L.F.	1.971	56	99.50	155.50
4'-0" x 6'-8" door	36.000	L.F.	2.086	59.50	105	164.50
6'-0" x 6'-8" door	38.000	L.F.	2.203	62.50	111	173.50
Paint 2 sides, primer & 2 cts., flush door & frame, 3' x 6'-8" opng	2.000	Face	2.914	13.95	152	165.95
4'-0" x 6'-8" opening	2.000	Face	3.886	18.60	202	220.60
6'-0" x 6'-8" opening	2.000	Face	4.857	23	253	276
Paneled door & frame, 3'-0" x 6'-8" opening	2.000	Face	6.000	22.50	284	306.50
4'-0" x 6'-8" opening	2.000	Face	8.000	30	380	410
6'-0" x 6'-8" opening	2.000	Face	10.000	37.50	475	512.50
Louvered door & frame, 3'-0" x 6'-8" opening	2.000	Face	6.000	22.50	284	306.50
4'-0" x 6'-8" opening	2.000	Face	8.000	30	380	410
6'-0" x 6'-8" opening	2.000	Face	10.000	37.50	475	512.50

6 | INTERIORS — 60 Carpet Systems

System Description	QUAN.	UNIT	LABOR HOURS	COST PER S.F. MAT.	COST PER S.F. INST.	COST PER S.F. TOTAL
Carpet, direct glue-down, nylon, level loop, 26 oz.	1.000	S.F.	.018	3.04	.87	3.91
32 oz.	1.000	S.F.	.018	5.20	.87	6.07
40 oz.	1.000	S.F.	.018	7.30	.87	8.17
Nylon, plush, 20 oz.	1.000	S.F.	.018	2.10	.87	2.97
24 oz.	1.000	S.F.	.018	2.28	.87	3.15
30 oz.	1.000	S.F.	.018	3.61	.87	4.48
42 oz.	1.000	S.F.	.022	5.50	1.05	6.55
48 oz.	1.000	S.F.	.022	6	1.05	7.05
54 oz.	1.000	S.F.	.022	6.70	1.05	7.75
Olefin, 15 oz.	1.000	S.F.	.018	1.84	.87	2.71
22 oz.	1.000	S.F.	.018	1.87	.87	2.74
Tile, foam backed, needle punch	1.000	S.F.	.014	4.75	.68	5.43
Tufted loop or shag	1.000	S.F.	.014	3.41	.68	4.09
Wool, 36 oz., level loop	1.000	S.F.	.018	12.30	.87	13.17
32 oz., patterned	1.000	S.F.	.020	11.15	.97	12.12
48 oz., patterned	1.000	S.F.	.020	12.30	.97	13.27
Padding, sponge rubber cushion, minimum	1.000	S.F.	.006	.55	.29	.84
Maximum	1.000	S.F.	.006	1.07	.29	1.36
Felt, 32 oz. to 56 oz., minimum	1.000	S.F.	.006	.65	.29	.94
Maximum	1.000	S.F.	.006	1.10	.29	1.39
Bonded urethane, 3/8" thick, minimum	1.000	S.F.	.006	.73	.29	1.02
Maximum	1.000	S.F.	.006	.98	.29	1.27
Prime urethane, 1/4" thick, minimum	1.000	S.F.	.006	.39	.29	.68
Maximum	1.000	S.F.	.006	.61	.29	.90
Stairs, for stairs, add to above carpet prices	1.000	Riser	.267		12.95	12.95
Underlayment plywood, 3/8" thick	1.000	S.F.	.011	1.01	.59	1.60
1/2" thick	1.000	S.F.	.011	1.21	.61	1.82
5/8" thick	1.000	S.F.	.011	1.32	.63	1.95
3/4" thick	1.000	S.F.	.012	1.57	.68	2.25
Particle board, 3/8" thick	1.000	S.F.	.011	.44	.59	1.03
1/2" thick	1.000	S.F.	.011	.47	.61	1.08
5/8" thick	1.000	S.F.	.011	.57	.63	1.20
3/4" thick	1.000	S.F.	.012	.69	.68	1.37
Hardboard, 4' x 4', 0.215" thick	1.000	S.F.	.011	.64	.59	1.23

6 | INTERIORS — 64 | Flooring Systems

System Description	QUAN.	UNIT	LABOR HOURS	COST PER S.F. MAT.	COST PER S.F. INST.	COST PER S.F. TOTAL
Resilient flooring, asphalt tile on concrete, 1/8" thick						
Color group B	1.000	S.F.	.020	1.31	.97	2.28
Color group C & D	1.000	S.F.	.020	1.43	.97	2.40
Asphalt tile on wood subfloor, 1/8" thick						
Color group B	1.000	S.F.	.020	1.59	.97	2.56
Color group C & D	1.000	S.F.	.020	1.71	.97	2.68
Vinyl composition tile, 12" x 12", 1/16" thick	1.000	S.F.	.016	1.30	.78	2.08
Embossed	1.000	S.F.	.016	2.43	.78	3.21
Marbleized	1.000	S.F.	.016	2.43	.78	3.21
Plain	1.000	S.F.	.016	3.14	.78	3.92
.080" thick, embossed	1.000	S.F.	.016	1.66	.78	2.44
Marbleized	1.000	S.F.	.016	2.79	.78	3.57
Plain	1.000	S.F.	.016	2.60	.78	3.38
1/8" thick, marbleized	1.000	S.F.	.016	2.64	.78	3.42
Plain	1.000	S.F.	.016	1.68	.78	2.46
Vinyl tile, 12" x 12", .050" thick, minimum	1.000	S.F.	.016	3.95	.78	4.73
Maximum	1.000	S.F.	.016	8	.78	8.78
1/8" thick, minimum	1.000	S.F.	.016	6.20	.78	6.98
Maximum	1.000	S.F.	.016	2.77	.78	3.55
1/8" thick, solid colors	1.000	S.F.	.016	6.50	.78	7.28
Florentine pattern	1.000	S.F.	.016	6.95	.78	7.73
Marbleized or travertine pattern	1.000	S.F.	.016	6.85	.78	7.63
Vinyl sheet goods, backed, .070" thick, minimum	1.000	S.F.	.032	4.50	1.55	6.05
Maximum	1.000	S.F.	.040	5.05	1.94	6.99
.093" thick, minimum	1.000	S.F.	.035	4.51	1.69	6.20
Maximum	1.000	S.F.	.040	6.50	1.94	8.44
.125" thick, minimum	1.000	S.F.	.035	4.68	1.69	6.37
Maximum	1.000	S.F.	.040	8.25	1.94	10.19
Wood, oak, finished in place, 25/32" x 2-1/2" clear	1.000	S.F.	.074	4.50	3.71	8.21
Select	1.000	S.F.	.074	5.25	3.71	8.96
No. 1 common	1.000	S.F.	.074	5.40	3.71	9.11
Prefinished, oak, 2-1/2" wide	1.000	S.F.	.047	5.15	2.60	7.75
3-1/4" wide	1.000	S.F.	.043	5.60	2.38	7.98
Ranch plank, oak, random width	1.000	S.F.	.055	7.85	3.04	10.89
Parquet, 5/16" thick, finished in place, oak, minimum	1.000	S.F.	.077	6.75	3.87	10.62
Maximum	1.000	S.F.	.107	11.55	5.50	17.05
Teak, minimum	1.000	S.F.	.077	7.45	3.87	11.32
Maximum	1.000	S.F.	.107	12.05	5.50	17.55
Sleepers, treated, 16" O.C., 1" x 2"	1.000	S.F.	.007	.30	.38	.68
1" x 3"	1.000	S.F.	.008	.50	.44	.94
2" x 4"	1.000	S.F.	.011	.60	.59	1.19
2" x 6"	1.000	S.F.	.012	.95	.68	1.63
Subfloor, plywood, 1/2" thick	1.000	S.F.	.011	.72	.59	1.31
5/8" thick	1.000	S.F.	.012	.86	.65	1.51
3/4" thick	1.000	S.F.	.013	1.02	.71	1.73
Ceramic tile, color group 2, 1" x 1"	1.000	S.F.	.087	5.70	3.78	9.48
2" x 2" or 2" x 1"	1.000	S.F.	.084	6.70	3.64	10.34
Color group 1, 8" x 8"	1.000	S.F.	.064	4.89	2.31	7.20
12" x 12"	1.000	S.F.	.049	6.90	2.39	9.29
16" x 16"	1.000	S.F.	.029	7.40	2.47	9.87

6 | INTERIORS   90 | Stairways

System Description	QUAN.	UNIT	LABOR HOURS	COST EACH MAT.	COST EACH INST.	COST EACH TOTAL
7 RISERS, OAK TREADS, BOX STAIRS						
Treads, oak, 1-1/4" x 10" wide, 3' long	6.000	Ea.	2.667	636	147	783
Risers, 3/4" thick, beech	7.000	Ea.	.672	196.35	36.96	233.31
30" primed pine balusters	12.000	Ea.	1.000	39.60	55.20	94.80
Newels, 3" wide, plain, paint grade, square	2.000	Ea.	2.286	111	126	237
Handrails, oak laminated	7.000	L.F.	.933	280	51.45	331.45
Stringers, 2" x 10", 3 each	21.000	L.F.	.306	9.66	16.80	26.46
TOTAL		Ea.	7.864	1,272.61	433.41	1,706.02
14 RISERS, OAK TREADS, BOX STAIRS						
Treads, oak, 1-1/4" x 10" wide, 3' long	13.000	Ea.	5.778	1,378	318.50	1,696.50
Risers, 3/4" thick, beech	14.000	Ea.	1.344	392.70	73.92	466.62
30" primed pine balusters	26.000	Ea.	2.167	85.80	119.60	205.40
Newels, 3" wide, plain, paint grade, square	2.000	Ea.	2.286	111	126	237
Handrails, oak, laminated	14.000	L.F.	1.867	560	102.90	662.90
Stair stringers, 2" x 10"	42.000	L.F.	5.169	61.74	285.60	347.34
TOTAL		Ea.	18.611	2,589.24	1,026.52	3,615.76
14 RISERS, PINE TREADS, BOX STAIRS						
Treads, pine, 9-1/2" x 3/4" thick	13.000	Ea.	5.778	220.35	318.50	538.85
Risers, 3/4" thick, pine	14.000	Ea.	1.344	150.36	73.92	224.28
30" primed pine balusters	26.000	Ea.	2.167	85.80	119.60	205.40
Newels, 3" wide, plain, paint grade, square	2.000	Ea.	2.286	111	126	237
Handrails, oak, laminated	14.000	L.F.	1.867	560	102.90	662.90
Stair stringers, 2" x 10"	42.000	L.F.	5.169	61.74	285.60	347.34
TOTAL		Ea.	18.611	1,189.25	1,026.52	2,215.77

Description	QUAN.	UNIT	LABOR HOURS	COST EACH MAT.	COST EACH INST.	COST EACH TOTAL

6 | INTERIORS — 90 Stairways

Stairway Price Sheet

	QUAN.	UNIT	LABOR HOURS	COST EACH MAT.	COST EACH INST.	COST EACH TOTAL
Treads, oak, 1-1/16" x 9-1/2", 3' long, 7 riser stair	6.000	Ea.	2.667	635	147	782
14 riser stair	13.000	Ea.	5.778	1,375	320	1,695
1-1/16" x 11-1/2", 3' long, 7 riser stair	6.000	Ea.	2.667	695	147	842
14 riser stair	13.000	Ea.	5.778	1,500	320	1,820
Pine, 3/4" x 9-1/2", 3' long, 7 riser stair	6.000	Ea.	2.667	102	147	249
14 riser stair	13.000	Ea.	5.778	220	320	540
3/4" x 11-1/4", 3' long, 7 riser stair	6.000	Ea.	2.667	113	147	260
14 riser stair	13.000	Ea.	5.778	245	320	565
Risers, oak, 3/4" x 7-1/2" high, 7 riser stair	7.000	Ea.	2.625	150	145	295
14 riser stair	14.000	Ea.	5.250	300	290	590
Beech, 3/4" x 7-1/2" high, 7 riser stair	7.000	Ea.	2.625	196	145	341
14 riser stair	14.000	Ea.	5.250	395	290	685
Baluster, turned, 30" high, primed pine, 7 riser stair	12.000	Ea.	3.429	39.50	55	94.50
14 riser stair	26.000	Ea.	7.428	86	120	206
30" birch, 7 riser stair	12.000	Ea.	3.429	36	50	86
14 riser stair	26.000	Ea.	7.428	78	109	187
42" pine, 7 riser stair	12.000	Ea.	3.556	58.50	55	113.50
14 riser stair	26.000	Ea.	7.704	126	120	246
42" birch, 7 riser stair	12.000	Ea.	3.556	58.50	55	113.50
14 riser stair	26.000	Ea.	7.704	126	120	246
Newels, 3-1/4" wide, starting, 7 riser stair	2.000	Ea.	2.286	111	126	237
14 riser stair	2.000	Ea.	2.286	111	126	237
Landing, 7 riser stair	2.000	Ea.	3.200	139	176	315
14 riser stair	2.000	Ea.	3.200	139	176	315
Handrails, oak, laminated, 7 riser stair	7.000	L.F.	.933	280	51.50	331.50
14 riser stair	14.000	L.F.	1.867	560	103	663
Stringers, fir, 2" x 10" 7 riser stair	21.000	L.F.	2.585	31	143	174
14 riser stair	42.000	L.F.	5.169	61.50	286	347.50
2" x 12", 7 riser stair	21.000	L.F.	2.585	40	143	183
14 riser stair	42.000	L.F.	5.169	80	286	366

Special Stairways

	QUAN.	UNIT	LABOR HOURS	COST EACH MAT.	COST EACH INST.	COST EACH TOTAL
Basement stairs, open risers	1.000	Flight	4.000	850	220	1,070
Spiral stairs, oak, 4'-6" diameter, prefabricated, 9' high	1.000	Flight	10.667	3,775	590	4,365
Aluminum, 5'-0" diameter stock unit	1.000	Flight	9.956	8,900	720	9,620
Custom unit	1.000	Flight	9.956	16,800	720	17,520
Cast iron, 4'-0" diameter, minimum	1.000	Flight	9.956	8,325	720	9,045
Maximum	1.000	Flight	17.920	11,300	1,300	12,600
Steel, industrial, pre-erected, 3'-6" wide, bar rail	1.000	Flight	7.724	8,475	770	9,245
Picket rail	1.000	Flight	7.724	9,525	770	10,295

Division 7 - Specialties

7 | SPECIALTIES 08 | Kitchen Systems

System Description	QUAN.	UNIT	LABOR HOURS	COST PER L.F. MAT.	COST PER L.F. INST.	COST PER L.F. TOTAL
KITCHEN, ECONOMY GRADE						
Top cabinets, economy grade	1.000	L.F.	.171	61.76	9.44	71.20
Bottom cabinets, economy grade	1.000	L.F.	.256	92.64	14.16	106.80
Square edge, plastic face countertop	1.000	L.F.	.267	36.50	14.70	51.20
Blocking, wood, 2" x 4"	1.000	L.F.	.032	.47	1.76	2.23
Soffit, framing, wood, 2" x 4"	4.000	L.F.	.071	1.88	3.92	5.80
Soffit drywall	2.000	S.F.	.047	.84	2.62	3.46
Drywall painting	2.000	S.F.	.013	.12	.74	.86
TOTAL		L.F.	.857	194.21	47.34	241.55
AVERAGE GRADE						
Top cabinets, average grade	1.000	L.F.	.213	77.20	11.80	89
Bottom cabinets, average grade	1.000	L.F.	.320	115.80	17.70	133.50
Solid surface countertop, solid color	1.000	L.F.	.800	81	44	125
Blocking, wood, 2" x 4"	1.000	L.F.	.032	.47	1.76	2.23
Soffit framing, wood, 2" x 4"	4.000	L.F.	.071	1.88	3.92	5.80
Soffit drywall	2.000	S.F.	.047	.84	2.62	3.46
Drywall painting	2.000	S.F.	.013	.12	.74	.86
TOTAL		L.F.	1.496	277.31	82.54	359.85
CUSTOM GRADE						
Top cabinets, custom grade	1.000	L.F.	.256	174	14.20	188.20
Bottom cabinets, custom grade	1.000	L.F.	.384	261	21.30	282.30
Solid surface countertop, premium patterned color	1.000	L.F.	1.067	140	59	199
Blocking, wood, 2" x 4"	1.000	L.F.	.032	.47	1.76	2.23
Soffit framing, wood, 2" x 4"	4.000	L.F.	.071	1.88	3.92	5.80
Soffit drywall	2.000	S.F.	.047	.84	2.62	3.46
Drywall painting	2.000	S.F.	.013	.12	.74	.86
TOTAL		L.F.	1.870	578.31	103.54	681.85

Description	QUAN.	UNIT	LABOR HOURS	COST PER L.F. MAT.	COST PER L.F. INST.	COST PER L.F. TOTAL

Kitchen Price Sheet

Kitchen Price Sheet	QUAN.	UNIT	LABOR HOURS	COST PER L.F. MAT.	COST PER L.F. INST.	COST PER L.F. TOTAL
Top cabinets, economy grade	1.000	L.F.	.171	62	9.45	71.45
Average grade	1.000	L.F.	.213	77	11.80	88.80
Custom grade	1.000	L.F.	.256	174	14.20	188.20
Bottom cabinets, economy grade	1.000	L.F.	.256	92.50	14.15	106.65
Average grade	1.000	L.F.	.320	116	17.70	133.70
Custom grade	1.000	L.F.	.384	261	21.50	282.50
Counter top, laminated plastic, 7/8" thick, no splash	1.000	L.F.	.267	33	14.70	47.70
With backsplash	1.000	L.F.	.267	32	14.70	46.70
1-1/4" thick, no splash	1.000	L.F.	.286	39	15.75	54.75
With backsplash	1.000	L.F.	.286	46.50	15.75	62.25
Post formed, laminated plastic	1.000	L.F.	.267	11.20	14.70	25.90
Ceramic tile, with backsplash		L.F.	.427	18.50	8.80	27.30
Marble, with backsplash, minimum	1.000	L.F.	.471	46	25.50	71.50
Maximum	1.000	L.F.	.615	116	33	149
Maple, solid laminated, no backsplash	1.000	L.F.	.286	83	15.75	98.75
With backsplash	1.000	L.F.	.286	98.50	15.75	114.25
Solid Surface, with backsplash, minimum		L.F.	.842	89	46.50	135.50
Maximum		L.F.	1.067	140	59	199
Blocking, wood, 2" x 4"	1.000	L.F.	.032	.47	1.76	2.23
2" x 6"	1.000	L.F.	.036	.73	1.99	2.72
2" x 8"	1.000	L.F.	.040	1.01	2.21	3.22
Soffit framing, wood, 2" x 3"	4.000	L.F.	.064	1.72	3.52	5.24
2" x 4"	4.000	L.F.	.071	1.88	3.92	5.80
Soffit, drywall, painted	2.000	S.F.	.060	.96	3.36	4.32
Paneling, standard	2.000	S.F.	.064	3.08	3.52	6.60
Deluxe	2.000	S.F.	.091	2.76	5.05	7.81
Sinks, porcelain on cast iron, single bowl, 21" x 24"	1.000	Ea.	10.334	600	595	1,195
21" x 30"	1.000	Ea.	10.334	965	595	1,560
Double bowl, 20" x 32"	1.000	Ea.	10.810	675	620	1,295
Stainless steel, single bowl, 16" x 20"	1.000	Ea.	10.334	935	595	1,530
22" x 25"	1.000	Ea.	10.334	1,000	595	1,595
Double bowl, 20" x 32"	1.000	Ea.	10.810	830	620	1,450

7 | SPECIALTIES — 12 | Appliance Systems

Kitchen Price Sheet

	QUAN.	UNIT	LABOR HOURS	COST PER L.F. MAT.	COST PER L.F. INST.	COST PER L.F. TOTAL
Range, free standing, minimum	1.000	Ea.	3.600	570	186	756
Maximum	1.000	Ea.	6.000	2,300	284	2,584
Built-in, minimum	1.000	Ea.	3.333	835	200	1,035
Maximum	1.000	Ea.	10.000	1,575	560	2,135
Counter top range, 4-burner, minimum	1.000	Ea.	3.333	410	200	610
Maximum	1.000	Ea.	4.667	1,525	280	1,805
Compactor, built-in, minimum	1.000	Ea.	2.215	730	125	855
Maximum	1.000	Ea.	3.282	1,150	184	1,334
Dishwasher, built-in, minimum	1.000	Ea.	6.735	475	425	900
Maximum	1.000	Ea.	9.235	695	585	1,280
Garbage disposer, minimum	1.000	Ea.	2.810	192	179	371
Maximum	1.000	Ea.	2.810	340	179	519
Microwave oven, minimum	1.000	Ea.	2.615	154	157	311
Maximum	1.000	Ea.	4.615	530	277	807
Range hood, ducted, minimum	1.000	Ea.	4.658	114	267	381
Maximum	1.000	Ea.	5.991	890	340	1,230
Ductless, minimum	1.000	Ea.	2.615	120	149	269
Maximum	1.000	Ea.	3.948	900	224	1,124
Refrigerator, 16 cu.ft., minimum	1.000	Ea.	2.000	580	82	662
Maximum	1.000	Ea.	3.200	930	131	1,061
16 cu.ft. with icemaker, minimum	1.000	Ea.	4.210	795	214	1,009
Maximum	1.000	Ea.	5.410	1,150	263	1,413
19 cu.ft., minimum	1.000	Ea.	2.667	495	109	604
Maximum	1.000	Ea.	4.667	865	191	1,056
19 cu.ft. with icemaker, minimum	1.000	Ea.	5.143	715	252	967
Maximum	1.000	Ea.	7.143	1,075	335	1,410
Sinks, porcelain on cast iron single bowl, 21" x 24"	1.000	Ea.	10.334	600	595	1,195
21" x 30"	1.000	Ea.	10.334	965	595	1,560
Double bowl, 20" x 32"	1.000	Ea.	10.810	675	620	1,295
Stainless steel, single bowl 16" x 20"	1.000	Ea.	10.334	935	595	1,530
22" x 25"	1.000	Ea.	10.334	1,000	595	1,595
Double bowl, 20" x 32"	1.000	Ea.	10.810	830	620	1,450
Water heater, electric, 30 gallon	1.000	Ea.	3.636	525	233	758
40 gallon	1.000	Ea.	4.000	880	256	1,136
Gas, 30 gallon	1.000	Ea.	4.000	985	256	1,241
75 gallon	1.000	Ea.	5.333	1,500	340	1,840
Wall, packaged terminal heater/air conditioner cabinet, wall sleeve, louver, electric heat, thermostat, manual changeover, 208V		Ea.				
6000 BTUH cooling, 8800 BTU heating	1.000	Ea.	2.667	850	156	1,006
9000 BTUH cooling, 13,900 BTU heating	1.000	Ea.	3.200	1,325	187	1,512
12,000 BTUH cooling, 13,900 BTU heating	1.000	Ea.	4.000	1,500	234	1,734
15,000 BTUH cooling, 13,900 BTU heating	1.000	Ea.	5.333	1,600	310	1,910

7 | SPECIALTIES — 16 | Bath Accessories

System Description	QUAN.	UNIT	LABOR HOURS	COST EACH MAT.	COST EACH INST.	COST EACH TOTAL
Curtain rods, stainless, 1" diameter, 3' long	1.000	Ea.	.615	31	34	65
5' long	1.000	Ea.	.615	31	34	65
Grab bar, 1" diameter, 12" long	1.000	Ea.	.283	27	15.65	42.65
36" long	1.000	Ea.	.340	36	18.70	54.70
1-1/4" diameter, 12" long	1.000	Ea.	.333	32	18.40	50.40
36" long	1.000	Ea.	.400	42.50	22	64.50
1-1/2" diameter, 12" long	1.000	Ea.	.383	37	21	58
36" long	1.000	Ea.	.460	49	25.50	74.50
Mirror, 18" x 24"	1.000	Ea.	.400	54	22	76
72" x 24"	1.000	Ea.	1.333	320	73.50	393.50
Medicine chest with mirror, 18" x 24"	1.000	Ea.	.400	213	22	235
36" x 24"	1.000	Ea.	.600	320	33	353
Toilet tissue dispenser, surface mounted, minimum	1.000	Ea.	.267	19.60	14.70	34.30
Maximum	1.000	Ea.	.400	29.50	22	51.50
Flush mounted, minimum	1.000	Ea.	.293	21.50	16.15	37.65
Maximum	1.000	Ea.	.427	31.50	23.50	55
Towel bar, 18" long, minimum	1.000	Ea.	.278	37	15.35	52.35
Maximum	1.000	Ea.	.348	46	19.20	65.20
24" long, minimum	1.000	Ea.	.313	41.50	17.30	58.80
Maximum	1.000	Ea.	.383	50.50	21	71.50
36" long, minimum	1.000	Ea.	.381	123	21	144
Maximum	1.000	Ea.	.419	135	23	158

7 | SPECIALTIES — 24 | Masonry Fireplace Systems

System Description	QUAN.	UNIT	LABOR HOURS	COST EACH MAT.	COST EACH INST.	TOTAL
MASONRY FIREPLACE						
Footing, 8" thick, concrete, 4' x 7'	.700	C.Y.	2.800	128.10	145.77	273.87
Foundation, concrete block, 32" x 60" x 4' deep	1.000	Ea.	5.275	190.20	264	454.20
Fireplace, brick firebox, 30" x 29" opening	1.000	Ea.	40.000	605	1,975	2,580
Damper, cast iron, 30" opening	1.000	Ea.	1.333	137	72	209
Facing brick, standard size brick, 6' x 5'	30.000	S.F.	5.217	137.40	261	398.40
Hearth, standard size brick, 3' x 6'	1.000	Ea.	8.000	225	395	620
Chimney, standard size brick, 8" x 12" flue, one story house	12.000	V.L.F.	12.000	498	588	1,086
Mantle, 4" x 8", wood	6.000	L.F.	1.333	54.60	73.50	128.10
Cleanout, cast iron, 8" x 8"	1.000	Ea.	.667	45	36	81
TOTAL		Ea.	76.625	2,020.30	3,810.27	5,830.57

The costs in this system are on a cost each basis.

Description	QUAN.	UNIT	LABOR HOURS	COST EACH MAT.	COST EACH INST.	TOTAL

7 | SPECIALTIES — 24 | Masonry Fireplace Systems

Masonry Fireplace Price Sheet

	QUAN.	UNIT	LABOR HOURS	COST EACH MAT.	COST EACH INST.	COST EACH TOTAL
Footing 8" thick, 3' x 6'	.440	C.Y.	1.326	79	91.50	170.50
4' x 7'	.700	C.Y.	2.110	126	146	272
5' x 8'	1.000	C.Y.	3.014	180	208	388
1' thick, 3' x 6'	.670	C.Y.	2.020	121	140	261
4' x 7'	1.030	C.Y.	3.105	185	214	399
5' x 8'	1.480	C.Y.	4.461	266	305	571
Foundation-concrete block, 24" x 48", 4' deep	1.000	Ea.	4.267	152	211	363
8' deep	1.000	Ea.	8.533	305	420	725
24" x 60", 4' deep	1.000	Ea.	4.978	178	246	424
8' deep	1.000	Ea.	9.956	355	495	850
32" x 48", 4' deep	1.000	Ea.	4.711	168	233	401
8' deep	1.000	Ea.	9.422	335	465	800
32" x 60", 4' deep	1.000	Ea.	5.333	190	264	454
8' deep	1.000	Ea.	10.845	385	535	920
32" x 72", 4' deep	1.000	Ea.	6.133	219	305	524
8' deep	1.000	Ea.	12.267	435	605	1,040
Fireplace, brick firebox 30" x 29" opening	1.000	Ea.	40.000	605	1,975	2,580
48" x 30" opening	1.000	Ea.	60.000	910	2,975	3,885
Steel fire box with registers, 25" opening	1.000	Ea.	26.667	1,100	1,325	2,425
48" opening	1.000	Ea.	44.000	1,850	2,200	4,050
Damper, cast iron, 30" opening	1.000	Ea.	1.333	137	72	209
36" opening	1.000	Ea.	1.556	160	84	244
Steel, 30" opening	1.000	Ea.	1.333	131	72	203
36" opening	1.000	Ea.	1.556	153	84	237
Facing for fireplace, standard size brick, 6' x 5'	30.000	S.F.	5.217	137	261	398
7' x 5'	35.000	S.F.	6.087	160	305	465
8' x 6'	48.000	S.F.	8.348	220	420	640
Fieldstone, 6' x 5'	30.000	S.F.	5.217	420	261	681
7' x 5'	35.000	S.F.	6.087	490	305	795
8' x 6'	48.000	S.F.	8.348	670	420	1,090
Sheetrock on metal, studs, 6' x 5'	30.000	S.F.	.980	25	54	79
7' x 5'	35.000	S.F.	1.143	29.50	63	92.50
8' x 6'	48.000	S.F.	1.568	40.50	86.50	127
Hearth, standard size brick, 3' x 6'	1.000	Ea.	8.000	225	395	620
3' x 7'	1.000	Ea.	9.280	261	460	721
3' x 8'	1.000	Ea.	10.640	299	525	824
Stone, 3' x 6'	1.000	Ea.	8.000	234	395	629
3' x 7'	1.000	Ea.	9.280	271	460	731
3' x 8'	1.000	Ea.	10.640	310	525	835
Chimney, standard size brick, 8" x 12" flue, one story house	12.000	V.L.F.	12.000	500	590	1,090
Two story house	20.000	V.L.F.	20.000	830	980	1,810
Mantle wood, beams, 4" x 8"	6.000	L.F.	1.333	54.50	73.50	128
4" x 10"	6.000	L.F.	1.371	72	75.50	147.50
Ornate, prefabricated, 6' x 3'-6" opening, minimum	1.000	Ea.	1.600	480	88	568
Maximum	1.000	Ea.	1.600	650	88	738
Cleanout, door and frame, cast iron, 8" x 8"	1.000	Ea.	.667	45	36	81
12" x 12"	1.000	Ea.	.800	119	43	162

7 | SPECIALTIES 30 | Prefabricated Fireplace Systems

System Description	QUAN.	UNIT	LABOR HOURS	COST EACH		
				MAT.	INST.	TOTAL
PREFABRICATED FIREPLACE						
Prefabricated fireplace, metal, painted	1.000	Ea.	6.154	1,525	340	1,865
Framing, 2" x 4" studs, 6' x 5'	35.000	L.F.	.509	16.10	28	44.10
Fire resistant gypsum drywall, unfinished	40.000	S.F.	.320	15.60	17.60	33.20
Drywall finishing adder	40.000	S.F.	.320	2	17.60	19.60
Facing, brick, standard size brick, 6' x 5'	30.000	S.F.	5.217	137.40	261	398.40
Hearth, standard size brick, 3' x 6'	1.000	Ea.	8.000	225	395	620
Chimney, one story house, framing, 2" x 4" studs	80.000	L.F.	1.164	36.80	64	100.80
Sheathing, plywood, 5/8" thick	32.000	S.F.	.758	90.56	41.92	132.48
Flue, 10" metal, insulated pipe	12.000	V.L.F.	4.000	480	222	702
Fittings, ceiling support	1.000	Ea.	.667	153	37	190
Fittings, joist shield	1.000	Ea.	.727	755	40.50	795.50
Fittings, roof flashing	1.000	Ea.	.667	430	37	467
Mantle beam, wood, 4" x 8"	6.000	L.F.	1.333	54.60	73.50	128.10
TOTAL		Ea.	29.836	3,921.06	1,575.12	5,496.18

The costs in this system are on a cost each basis.

Description	QUAN.	UNIT	LABOR HOURS	COST EACH		
				MAT.	INST.	TOTAL

7 | SPECIALTIES — 30 | Prefabricated Fireplace Systems

Prefabricated Fireplace Price Sheet

	QUAN.	UNIT	LABOR HOURS	COST EACH MAT.	COST EACH INST.	COST EACH TOTAL
Prefabricated fireplace, minimum	1.000	Ea.	6.154	1,525	340	1,865
Average	1.000	Ea.	8.000	1,775	440	2,215
Maximum	1.000	Ea.	8.889	3,375	490	3,865
Framing, 2" x 4" studs, fireplace, 6' x 5'	35.000	L.F.	.509	16.10	28	44.10
7' x 5'	40.000	L.F.	.582	18.40	32	50.40
8' x 6'	45.000	L.F.	.655	20.50	36	56.50
Sheetrock, 1/2" thick, fireplace, 6' x 5'	40.000	S.F.	.640	17.60	35	52.60
7' x 5'	45.000	S.F.	.720	19.80	39.50	59.30
8' x 6'	50.000	S.F.	.800	22	44	66
Facing for fireplace, brick, 6' x 5'	30.000	S.F.	5.217	137	261	398
7' x 5'	35.000	S.F.	6.087	160	305	465
8' x 6'	48.000	S.F.	8.348	220	420	640
Fieldstone, 6' x 5'	30.000	S.F.	5.217	400	261	661
7' x 5'	35.000	S.F.	6.087	470	305	775
8' x 6'	48.000	S.F.	8.348	640	420	1,060
Hearth, standard size brick, 3' x 6'	1.000	Ea.	8.000	225	395	620
3' x 7'	1.000	Ea.	9.280	261	460	721
3' x 8'	1.000	Ea.	10.640	299	525	824
Stone, 3' x 6'	1.000	Ea.	8.000	234	395	629
3' x 7'	1.000	Ea.	9.280	271	460	731
3' x 8'	1.000	Ea.	10.640	310	525	835
Chimney, framing, 2" x 4", one story house	80.000	L.F.	1.164	37	64	101
Two story house	120.000	L.F.	1.746	55	96	151
Sheathing, plywood, 5/8" thick	32.000	S.F.	.758	90.50	42	132.50
Stucco on plywood	32.000	S.F.	1.125	54.50	60.50	115
Flue, 10" metal pipe, insulated, one story house	12.000	V.L.F.	4.000	480	222	702
Two story house	20.000	V.L.F.	6.667	800	370	1,170
Fittings, ceiling support	1.000	Ea.	.667	153	37	190
Fittings joist sheild, one story house	1.000	Ea.	.667	755	40.50	795.50
Two story house	2.000	Ea.	1.333	1,500	81	1,581
Fittings roof flashing	1.000	Ea.	.667	430	37	467
Mantle, wood beam, 4" x 8"	6.000	L.F.	1.333	54.50	73.50	128
4" x 10"	6.000	L.F.	1.371	72	75.50	147.50
Ornate prefabricated, 6' x 3'-6" opening, minimum	1.000	Ea.	1.600	480	88	568
Maximum	1.000	Ea.	1.600	650	88	738

7 | SPECIALTIES — 32 | Greenhouse Systems

System Description	QUAN.	UNIT	LABOR HOURS	MAT.	INST.	TOTAL
Economy, lean to, shell only, not including 2' stub wall, fndtn, flrs, heat						
4' x 16'	1.000	Ea.	26.212	2,550	1,450	4,000
4' x 24'	1.000	Ea.	30.259	2,950	1,675	4,625
6' x 10'	1.000	Ea.	16.552	3,300	910	4,210
6' x 16'	1.000	Ea.	23.034	4,600	1,275	5,875
6' x 24'	1.000	Ea.	29.793	5,950	1,650	7,600
8' x 10'	1.000	Ea.	22.069	4,400	1,225	5,625
8' x 16'	1.000	Ea.	38.400	7,650	2,125	9,775
8' x 24'	1.000	Ea.	49.655	9,900	2,725	12,625
Free standing, 8' x 8'	1.000	Ea.	17.356	1,400	955	2,355
8' x 16'	1.000	Ea.	30.211	2,450	1,675	4,125
8' x 24'	1.000	Ea.	39.051	3,175	2,150	5,325
10' x 10'	1.000	Ea.	18.824	4,050	1,050	5,100
10' x 16'	1.000	Ea.	24.095	5,175	1,325	6,500
10' x 24'	1.000	Ea.	31.624	6,800	1,750	8,550
14' x 10'	1.000	Ea.	20.741	5,950	1,150	7,100
14' x 16'	1.000	Ea.	24.889	7,150	1,375	8,525
14' x 24'	1.000	Ea.	33.349	9,575	1,825	11,400
Standard, lean to, shell only, not incl. 2' stub wall, fndtn, flrs, heat 4'x10'	1.000	Ea.	28.235	2,750	1,550	4,300
4' x 16'	1.000	Ea.	39.341	3,850	2,175	6,025
4' x 24'	1.000	Ea.	45.412	4,450	2,500	6,950
6' x 10'	1.000	Ea.	24.827	4,950	1,375	6,325
6' x 16'	1.000	Ea.	34.538	6,875	1,900	8,775
6' x 24'	1.000	Ea.	44.689	8,900	2,450	11,350
8' x 10'	1.000	Ea.	33.103	6,600	1,825	8,425
8' x 16'	1.000	Ea.	57.600	11,500	3,175	14,675
8' x 24'	1.000	Ea.	74.482	14,900	4,100	19,000
Free standing, 8' x 8'	1.000	Ea.	26.034	2,100	1,425	3,525
8' x 16'	1.000	Ea.	45.316	3,675	2,500	6,175
8' x 24'	1.000	Ea.	58.577	4,750	3,225	7,975
10' x 10'	1.000	Ea.	28.236	6,075	1,550	7,625
10' x 16'	1.000	Ea.	36.142	7,775	2,000	9,775
10' x 24'	1.000	Ea.	47.436	10,200	2,625	12,825
14' x 10'	1.000	Ea.	31.112	8,925	1,700	10,625
14' x 16'	1.000	Ea.	37.334	10,700	2,050	12,750
14' x 24'	1.000	Ea.	50.030	14,400	2,750	17,150
Deluxe, lean to, shell only, not incl. 2' stub wall, fndtn, flrs or heat, 4'x10'	1.000	Ea.	20.645	4,650	1,150	5,800
4' x 16'	1.000	Ea.	33.032	7,425	1,825	9,250
4' x 24'	1.000	Ea.	49.548	11,100	2,725	13,825
6' x 10'	1.000	Ea.	30.968	6,950	1,700	8,650
6' x 16'	1.000	Ea.	49.548	11,100	2,725	13,825
6' x 24'	1.000	Ea.	74.323	16,700	4,100	20,800
8' x 10'	1.000	Ea.	41.290	9,275	2,275	11,550
8' x 16'	1.000	Ea.	66.065	14,800	3,650	18,450
8' x 24'	1.000	Ea.	99.097	22,300	5,475	27,775
Freestanding, 8' x 8'	1.000	Ea.	18.618	5,725	1,025	6,750
8' x 16'	1.000	Ea.	37.236	11,500	2,050	13,550
8' x 24'	1.000	Ea.	55.855	17,200	3,075	20,275
10' x 10'	1.000	Ea.	29.091	8,950	1,600	10,550
10' x 16'	1.000	Ea.	46.546	14,300	2,575	16,875
10' x 24'	1.000	Ea.	69.818	21,500	3,850	25,350
14' x 10'	1.000	Ea.	40.727	12,500	2,250	14,750
14' x 16'	1.000	Ea.	65.164	20,000	3,600	23,600
14' x 24'	1.000	Ea.	97.746	30,100	5,400	35,500

7 SPECIALTIES — 36 Swimming Pool Systems

System Description	QUAN.	UNIT	LABOR HOURS	MAT.	INST.	TOTAL
Swimming pools, vinyl lined, metal sides, sand bottom, 12' x 28'	1.000	Ea.	50.177	6,850	2,875	9,725
12' x 32'	1.000	Ea.	55.366	7,575	3,200	10,775
12' x 36'	1.000	Ea.	60.061	8,200	3,475	11,675
16' x 32'	1.000	Ea.	66.798	9,125	3,850	12,975
16' x 36'	1.000	Ea.	71.190	9,725	4,100	13,825
16' x 40'	1.000	Ea.	74.703	10,200	4,325	14,525
20' x 36'	1.000	Ea.	77.860	10,600	4,475	15,075
20' x 40'	1.000	Ea.	82.135	11,200	4,725	15,925
20' x 44'	1.000	Ea.	90.348	12,300	5,200	17,500
24' x 40'	1.000	Ea.	98.562	13,500	5,700	19,200
24' x 44'	1.000	Ea.	108.418	14,800	6,275	21,075
24' x 48'	1.000	Ea.	118.274	16,200	6,825	23,025
Vinyl lined, concrete sides, 12' x 28'	1.000	Ea.	79.447	10,900	4,575	15,475
12' x 32'	1.000	Ea.	88.818	12,100	5,125	17,225
12' x 36'	1.000	Ea.	97.656	13,300	5,650	18,950
16' x 32'	1.000	Ea.	111.393	15,200	6,425	21,625
16' x 36'	1.000	Ea.	121.354	16,600	7,025	23,625
16' x 40'	1.000	Ea.	130.445	17,800	7,525	25,325
28' x 36'	1.000	Ea.	140.585	19,200	8,125	27,325
20' x 40'	1.000	Ea.	149.336	20,400	8,625	29,025
20' x 44'	1.000	Ea.	164.270	22,400	9,500	31,900
24' x 40'	1.000	Ea.	179.203	24,500	10,400	34,900
24' x 44'	1.000	Ea.	197.124	26,900	11,400	38,300
24' x 48'	1.000	Ea.	215.044	29,400	12,400	41,800
Gunite, bottom and sides, 12' x 28'	1.000	Ea.	129.767	16,100	7,500	23,600
12' x 32'	1.000	Ea.	142.164	17,700	8,200	25,900
12' x 36'	1.000	Ea.	153.028	19,000	8,825	27,825
16' x 32'	1.000	Ea.	167.743	20,800	9,675	30,475
16' x 36'	1.000	Ea.	176.421	21,900	10,200	32,100
16' x 40'	1.000	Ea.	182.368	22,700	10,500	33,200
20' x 36'	1.000	Ea.	187.949	23,400	10,900	34,300
20' x 40'	1.000	Ea.	179.200	30,800	10,400	41,200
20' x 44'	1.000	Ea.	197.120	33,900	11,400	45,300
24' x 40'	1.000	Ea.	215.040	37,000	12,400	49,400
24' x 44'	1.000	Ea.	273.244	34,000	15,800	49,800
24' x 48'	1.000	Ea.	298.077	37,000	17,200	54,200

7 | SPECIALTIES   40 | Wood Deck Systems

System Description	QUAN.	UNIT	LABOR HOURS	COST PER S.F. MAT.	COST PER S.F. INST.	COST PER S.F. TOTAL
8' X 12' DECK, PRESSURE TREATED LUMBER, JOISTS 16" O.C.						
4" x 4" Posts	.500	L.F.	.021	.74	1.13	1.87
Ledger, bolted 4' O.C.	.125	L.F.	.005	.21	.29	.50
Joists, 2" x 10", 16" O.C.	1.080	L.F.	.019	1.70	1.06	2.76
5/4" x 6" decking	1.000	S.F.	.025	2.48	1.38	3.86
2" x 12" stair stringers	.375	L.F.	.046	.71	2.55	3.26
5/4" x 6" stair treads	.500	L.F.	.050	.58	2.75	3.33
4' x 4" handrail posts	.210	L.F.	.009	.31	.47	.78
2" x 6" handrail cap	.520	L.F.	.014	.45	.76	1.21
2" x 4" baluster support	1.040	L.F.	.028	.58	1.53	2.11
2" x 2" balusters	3.850	L.F.	.093	1.42	5.16	6.58
Post footings	.042	Ea.	.134	1.09	6.47	7.56
Concrete step	.010	Ea.	.016	.76	.87	1.63
Joist hangers	.210	Ea.	.010	.32	.56	.88
TOTAL		S.F.	.470	11.35	24.98	36.33
12' X 16' DECK, PRESSURE TREATED LUMBER, JOISTS 24" O.C.						
4" x 4" Posts	.312	L.F.	.013	.46	.71	1.17
Ledger, bolted 4' O.C.	.083	L.F.	.003	.11	.19	.30
Joists, 2" x 8", 16" O.C.	.810	L.F.	.014	1.27	.79	2.06
5/4" x 6" decking	1.000	S.F.	.025	2.48	1.38	3.86
2" x 12" stair stringers	.188	L.F.	.023	.35	1.28	1.63
5/4" x 6" stair treads	.250	L.F.	.025	1.22	1.38	2.60
4" x 4" handrail posts	.125	L.F.	.005	.18	.28	.46
2" x 6" handrail cap	.323	L.F.	.009	.28	.47	.75
2" x 4" baluster support	.650	L.F.	.017	.36	.96	1.32
2" x 2" balusters	2.440	L.F.	.059	.90	3.27	4.17
Post footings	.026	Ea.	.083	.68	4	4.68
Concrete step	.005	Ea.	.008	.38	.44	.82
Joist hangers	.135	Ea.	.007	.21	.36	.57
Deck support girder	.083	L.F.	.002	.26	.13	.39
TOTAL		S.F.	.293	9.14	15.64	24.78

The costs in this system are on a square foot basis.

7 | SPECIALTIES — 40 Wood Deck Systems

System Description	QUAN.	UNIT	LABOR HOURS	COST PER S.F. MAT.	COST PER S.F. INST.	COST PER S.F. TOTAL
12' X 24' DECK, REDWOOD OR CEDAR, JOISTS 16" O.C.						
4" x 4" redwood posts	.291	L.F.	.012	2.05	.66	2.71
Redwood joists, 2" x 8", 16" O.C.	.001	M.B.F.	.010	5.47	.56	6.03
5/4" x 6" redwood decking	1.000	S.F.	.025	5.25	1.38	6.63
Stair stringers	.125	L.F.	.015	.24	.85	1.09
Stringer trim	.083	L.F.	.003	.44	.18	.62
Stair treads	.135	S.F.	.003	.71	.19	.90
Handrail supports	.083	L.F.	.003	.59	.19	.78
Handrail cap	.152	L.F.	.004	1.25	.22	1.47
Baluster supports	.416	L.F.	.011	3.43	.61	4.04
Balusters	1.875	L.F.	.050	2.42	2.76	5.18
Post footings	.024	Ea.	.077	.62	3.70	4.32
Concrete step	.003	Ea.	.005	.23	.26	.49
Joist hangers	.132	Ea.	.006	.20	.35	.55
Deck support girder		M.B.F.	.003	1.49	.15	1.64
TOTAL		S.F.	.227	24.39	12.06	36.45

The costs in this system are on a square foot basis.

Wood Deck Price Sheet	QUAN.	UNIT	LABOR HOURS	COST PER S.F. MAT.	COST PER S.F. INST.	COST PER S.F. TOTAL
Decking, treated lumber, 1" x 4"	3.430	L.F.	.031	8.15	5.50	13.65
2" x 4"	3.430	L.F.	.033	6.50	5.05	11.55
2" x 6"	2.200	L.F.	.041	4.09	3.04	7.13
5/4" x 6"	2.200	L.F.	.027	5.45	3.04	8.49
Redwood or cedar, 1" x 4"	3.430	L.F.	.035	6.85	5.50	12.35
2" x 4"	3.430	L.F.	.036	13.65	5.05	18.70
2" x 6"	2.200	L.F.	.028	15.95	3.04	18.99
5/4" x 6"	2.200	L.F.	.027	10.75	3.04	13.79
Joists for deck, treated lumber, 2" x 8", 16" O.C.	1.000	L.F.	.015	1.18	.80	1.98
24" O.C.	.750	L.F.	.012	.89	.60	1.49
2" x 10", 16" O.C.	1.000	L.F.	.018	1.50	.98	2.48
24" O.C.	.750	L.F.	.014	1.13	.74	1.87
Redwood or cedar, 2" x 8", 16" O.C.	1.000	L.F.	.015	5	.80	5.80
24" O.C.	.750	L.F.	.012	3.75	.60	4.35
2" x 10", 16" O.C.	1.000	L.F.	.018	8.30	.98	9.28
24" O.C.	.750	L.F.	.014	6.25	.74	6.99
Girder for joists, treated lumber, 2" x 10", 8' x 12' deck	.250	L.F.	.002	.38	.20	.58
12' x 16' deck	.167	L.F.	.001	.25	.13	.38
12' x 24' deck	.167	L.F.	.001	.25	.13	.38
Redwood or cedar, 2" x 10", 8' x 12' deck	.250	L.F.	.002	2.08	.25	2.33
12' x 16' deck	.167	L.F.	.001	1.39	.16	1.55
12' x 24' deck	.167	L.F.	.001	1.39	.16	1.55
Posts, 4" x 4", including concrete footing, 8' x 12' deck	.500	L.F.	.022	1.79	7.45	9.24
12' x 16' deck	.312	L.F.	.017	1.12	4.71	5.83
12' x 24' deck	.291	L.F.	.017	1.04	4.36	5.40
Stairs 2" x 10" stringers, treated lumber, 8' x 12' deck	1.000	Set	.020	1.53	4.40	5.93
12' x 16' deck	1.000	Set	.012	.77	2.20	2.97
12' x 24' deck	1.000	Set	.008	.54	1.54	2.08
Redwood or cedar, 8' x 12' deck	1.000	Set	.040	3.90	4.40	8.30
12' x 16' deck	1.000	Set	.020	1.95	2.20	4.15
12' x 24' deck	1.000	Set	.012	1.37	1.54	2.91
Railings 2" x 4", treated lumber, 8' x 12' deck	.520	L.F.	.026	2.45	7.45	9.90
12' x 16' deck	.323	L.F.	.017	1.54	4.70	6.24
12' x 24' deck	.210	L.F.	.014	1.11	3.44	4.55
Redwood or cedar, 8' x 12' deck	1.000	L.F.	.009	10.60	7.95	18.55

7 | SPECIALTIES — 40 | Wood Deck Systems

Wood Deck Price Sheet

	QUAN.	UNIT	LABOR HOURS	COST PER S.F. MAT.	COST PER S.F. INST.	COST PER S.F. TOTAL
12' x 16' deck	.670	L.F.	.006	6.60	5	11.60
12' x 24' deck	.540	L.F.	.005	4.70	3.69	8.39
Alternative decking, wood/plastic composite, 5/4" x 6"	2.200	L.F.	.055	7.60	3.04	10.64
1" x 4" square edge fir	3.430	L.F.	.100	8.25	5.50	13.75
1" x 4" tongue and groove fir	3.430	L.F.	.122	5.50	6.70	12.20
1" x 4" mahogany	3.430	L.F.	.100	7.60	5.50	13.10
5/4" x 6" PVC	2.200	L.F.	.064	9	3.52	12.52

Division 8 - Mechanical

8 | MECHANICAL 04 | Two Fixture Lavatory Systems

System Description	QUAN.	UNIT	LABOR HOURS	COST EACH MAT.	COST EACH INST.	COST EACH TOTAL
LAVATORY INSTALLED WITH VANITY, PLUMBING IN 2 WALLS						
Water closet, floor mounted, 2 piece, close coupled, white	1.000	Ea.	3.019	261	174	435
Rough-in, vent, 2" diameter DWV piping	1.000	Ea.	.955	49.60	55	104.60
Waste, 4" diameter DWV piping	1.000	Ea.	.828	63	47.70	110.70
Supply, 1/2" diameter type "L" copper supply piping	1.000	Ea.	.593	24.42	37.80	62.22
Lavatory, 20" x 18", P.E. cast iron white	1.000	Ea.	2.500	370	144	514
Rough-in, vent, 1-1/2" diameter DWV piping	1.000	Ea.	.901	48.60	51.80	100.40
Waste, 2" diameter DWV piping	1.000	Ea.	.955	49.60	55	104.60
Supply, 1/2" diameter type "L" copper supply piping	1.000	Ea.	.988	40.70	63	103.70
Piping, supply, 1/2" diameter type "L" copper supply piping	10.000	L.F.	.988	40.70	63	103.70
Waste, 4" diameter DWV piping	7.000	L.F.	1.931	147	111.30	258.30
Vent, 2" diameter DWV piping	12.000	L.F.	2.866	148.80	165	313.80
Vanity base cabinet, 2 door, 30" wide	1.000	Ea.	1.000	405	55	460
Vanity top, plastic & laminated, square edge	2.670	L.F.	.712	114.81	39.25	154.06
TOTAL		Ea.	18.236	1,763.23	1,061.85	2,825.08
LAVATORY WITH WALL-HUNG LAVATORY, PLUMBING IN 2 WALLS						
Water closet, floor mounted, 2 piece close coupled, white	1.000	Ea.	3.019	261	174	435
Rough-in, vent, 2" diameter DWV piping	1.000	Ea.	.955	49.60	55	104.60
Waste, 4" diameter DWV piping	1.000	Ea.	.828	63	47.70	110.70
Supply, 1/2" diameter type "L" copper supply piping	1.000	Ea.	.593	24.42	37.80	62.22
Lavatory, 20" x 18", P.E. cast iron, wall hung, white	1.000	Ea.	2.000	305	115	420
Rough-in, vent, 1-1/2" diameter DWV piping	1.000	Ea.	.901	48.60	51.80	100.40
Waste, 2" diameter DWV piping	1.000	Ea.	.955	49.60	55	104.60
Supply, 1/2" diameter type "L" copper supply piping	1.000	Ea.	.988	40.70	63	103.70
Piping, supply, 1/2" diameter type "L" copper supply piping	10.000	L.F.	.988	40.70	63	103.70
Waste, 4" diameter DWV piping	7.000	L.F.	1.931	147	111.30	258.30
Vent, 2" diameter DWV piping	12.000	L.F.	2.866	148.80	165	313.80
Carrier, steel for studs, no arms	1.000	Ea.	1.143	105	73	178
TOTAL		Ea.	17.167	1,283.42	1,011.60	2,295.02

Description	QUAN.	UNIT	LABOR HOURS	COST EACH MAT.	COST EACH INST.	COST EACH TOTAL

Two Fixture Lavatory Price Sheet

Description	QUAN.	UNIT	LABOR HOURS	COST EACH MAT.	COST EACH INST.	COST EACH TOTAL
Water closet, close coupled standard 2 piece, white	1.000	Ea.	3.019	261	174	435
Color	1.000	Ea.	3.019	460	174	634
One piece elongated bowl, white	1.000	Ea.	3.019	1,225	174	1,399
Color	1.000	Ea.	3.019	1,575	174	1,749
Low profile, one piece elongated bowl, white	1.000	Ea.	3.019	1,200	174	1,374
Color	1.000	Ea.	3.019	1,575	174	1,749
Rough-in for water closet						
1/2" copper supply, 4" cast iron waste, 2" cast iron vent	1.000	Ea.	2.376	137	141	278
4" PVC waste, 2" PVC vent	1.000	Ea.	2.678	82	158	240
4" copper waste, 2" copper vent	1.000	Ea.	2.520	233	153	386
3" cast iron waste, 1-1/2" cast iron vent	1.000	Ea.	2.244	123	133	256
3" PVC waste, 1-1/2" PVC vent	1.000	Ea.	2.388	74	147	221
3" copper waste, 1-1/2" copper vent	1.000	Ea.	2.524	227	149	376
1/2" PVC supply, 4" PVC waste, 2" PVC vent	1.000	Ea.	2.974	102	177	279
3" PVC waste, 1-1/2" PVC vent	1.000	Ea.	2.684	94	166	260
1/2" steel supply, 4" cast iron waste, 2" cast iron vent	1.000	Ea.	2.545	137	151	288
4" cast iron waste, 2" steel vent	1.000	Ea.	2.590	133	154	287
4" PVC waste, 2" PVC vent	1.000	Ea.	2.847	82	169	251
Lavatory, vanity top mounted, P.E. on cast iron 20" x 18" white	1.000	Ea.	2.500	370	144	514
Color	1.000	Ea.	2.500	445	144	589
Steel, enameled 10" x 17" white	1.000	Ea.	2.759	177	159	336
Color	1.000	Ea.	2.500	210	144	354
Vitreous china 20" x 16", white	1.000	Ea.	2.963	286	171	457
Color	1.000	Ea.	2.963	286	171	457
Wall hung, P.E. on cast iron, 20" x 18", white	1.000	Ea.	2.000	305	115	420
Color	1.000	Ea.	2.000	330	115	445
Vitreous china 19" x 17", white	1.000	Ea.	2.286	228	132	360
Color	1.000	Ea.	2.286	236	132	368
Rough-in supply waste and vent for lavatory						
1/2" copper supply, 2" cast iron waste, 1-1/2" cast iron vent	1.000	Ea.	2.844	139	170	309
2" PVC waste, 1-1/2" PVC vent	1.000	Ea.	2.962	88.50	182	270.50
2" copper waste, 1-1/2" copper vent	1.000	Ea.	2.308	154	147	301
1-1/2" PVC waste, 1-1/4" PVC vent	1.000	Ea.	2.639	87.50	169	256.50
1-1/2" copper waste, 1-1/4" copper vent	1.000	Ea.	2.114	130	135	265
1/2" PVC supply, 2" PVC waste, 1-1/2" PVC vent	1.000	Ea.	3.456	122	214	336
1-1/2" PVC waste, 1-1/4" PVC vent	1.000	Ea.	3.133	121	200	321
1/2" steel supply, 2" cast iron waste, 1-1/2" cast iron vent	1.000	Ea.	3.126	139	188	327
2" cast iron waste, 2" steel vent	1.000	Ea.	3.225	136	194	330
2" PVC waste, 1-1/2" PVC vent	1.000	Ea.	3.244	88.50	200	288.50
1-1/2" PVC waste, 1-1/4" PVC vent	1.000	Ea.	2.921	87.50	187	274.50
Piping, supply, 1/2" copper, type "L"	10.000	L.F.	.988	40.50	63	103.50
1/2" steel	10.000	L.F.	1.270	40.50	81	121.50
1/2" PVC	10.000	L.F.	1.482	74	94.50	168.50
Waste, 4" cast iron	7.000	L.F.	1.931	147	111	258
4" copper	7.000	L.F.	2.800	335	161	496
4" PVC	7.000	L.F.	2.333	77	134	211
Vent, 2" cast iron	12.000	L.F.	2.866	149	165	314
2" copper	12.000	L.F.	2.182	193	140	333
2" PVC	12.000	L.F.	3.254	74.50	187	261.50
2" steel	12.000	Ea.	3.000	136	173	309
Vanity base cabinet, 2 door, 24" x 30"	1.000	Ea.	1.000	405	55	460
24" x 36"	1.000	Ea.	1.200	395	66	461
Vanity top, laminated plastic, square edge 25" x 32"	2.670	L.F.	.712	115	39.50	154.50
25" x 38"	3.170	L.F.	.845	136	46.50	182.50
Post formed, laminated plastic, 25" x 32"	2.670	L.F.	.712	30	39.50	69.50
25" x 38"	3.170	L.F.	.845	35.50	46.50	82
Cultured marble, 25" x 32" with bowl	1.000	Ea.	2.500	236	144	380
25" x 38" with bowl	1.000	Ea.	2.500	270	144	414
Carrier for lavatory, steel for studs	1.000	Ea.	1.143	105	73	178
Wood 2" x 8" blocking	1.330	L.F.	.053	1.34	2.94	4.28

8 | MECHANICAL — 12 Three Fixture Bathroom Systems

System Description	QUAN.	UNIT	LABOR HOURS	MAT.	INST.	TOTAL
BATHROOM INSTALLED WITH VANITY						
Water closet, floor mounted, 2 piece, close coupled, white	1.000	Ea.	3.019	261	174	435
Rough-in, waste, 4" diameter DWV piping	1.000	Ea.	.828	63	47.70	110.70
Vent, 2" diameter DWV piping	1.000	Ea.	.955	49.60	55	104.60
Supply, 1/2" diameter type "L" copper supply piping	1.000	Ea.	.593	24.42	37.80	62.22
Lavatory, 20" x 18", P.E. cast iron with accessories, white	1.000	Ea.	2.500	370	144	514
Rough-in, supply, 1/2" diameter type "L" copper supply piping	1.000	Ea.	.988	40.70	63	103.70
Waste, 1-1/2" diameter DWV piping	1.000	Ea.	1.803	97.20	103.60	200.80
Bathtub, P.E. cast iron, 5' long with accessories, white	1.000	Ea.	3.636	1,250	209	1,459
Rough-in, waste, 4" diameter DWV piping	1.000	Ea.	.828	63	47.70	110.70
Vent, 1-1/2" diameter DWV piping	1.000	Ea.	.593	49.20	37.80	87
Supply, 1/2" diameter type "L" copper supply piping	1.000	Ea.	.988	40.70	63	103.70
Piping, supply, 1/2" diameter type "L" copper supply piping	20.000	L.F.	1.975	81.40	126	207.40
Waste, 4" diameter DWV piping	9.000	L.F.	2.483	189	143.10	332.10
Vent, 2" diameter DWV piping	6.000	L.F.	1.500	68.10	86.40	154.50
Vanity base cabinet, 2 door, 30" wide	1.000	Ea.	1.000	405	55	460
Vanity top, plastic laminated square edge	2.670	L.F.	.712	97.46	39.25	136.71
TOTAL		Ea.	24.401	3,149.78	1,432.35	4,582.13
BATHROOM WITH WALL HUNG LAVATORY						
Water closet, floor mounted, 2 piece, close coupled, white	1.000	Ea.	3.019	261	174	435
Rough-in, vent, 2" diameter DWV piping	1.000	Ea.	.955	49.60	55	104.60
Waste, 4" diameter DWV piping	1.000	Ea.	.828	63	47.70	110.70
Supply, 1/2" diameter type "L" copper supply piping	1.000	Ea.	.593	24.42	37.80	62.22
Lavatory, 20" x 18" P.E. cast iron, wall hung, white	1.000	Ea.	2.000	305	115	420
Rough-in, waste, 1-1/2" diameter DWV piping	1.000	Ea.	1.803	97.20	103.60	200.80
Supply, 1/2" diameter type "L" copper supply piping	1.000	Ea.	.988	40.70	63	103.70
Bathtub, P.E. cast iron, 5' long with accessories, white	1.000	Ea.	3.636	1,250	209	1,459
Rough-in, waste, 4" diameter DWV piping	1.000	Ea.	.828	63	47.70	110.70
Supply, 1/2" diameter type "L" copper supply piping	1.000	Ea.	.988	40.70	63	103.70
Vent, 1-1/2" diameter DWV piping	1.000	Ea.	1.482	123	94.50	217.50
Piping, supply, 1/2" diameter type "L" copper supply piping	20.000	L.F.	1.975	81.40	126	207.40
Waste, 4" diameter DWV piping	9.000	L.F.	2.483	189	143.10	332.10
Vent, 2" diameter DWV piping	6.000	L.F.	1.500	68.10	86.40	154.50
Carrier, steel, for studs, no arms	1.000	Ea.	1.143	105	73	178
TOTAL		Ea.	24.221	2,761.12	1,438.80	4,199.92

The costs in this system are a cost each basis, all necessary piping is included.

Three Fixture Bathroom Price Sheet

Description	QUAN.	UNIT	LABOR HOURS	COST EACH MAT.	COST EACH INST.	TOTAL
Water closet, close coupled standard 2 piece, white	1.000	Ea.	3.019	261	174	435
Color	1.000	Ea.	3.019	460	174	634
One piece, elongated bowl, white	1.000	Ea.	3.019	1,225	174	1,399
Color	1.000	Ea.	3.019	1,575	174	1,749
Low profile, one piece elongated bowl, white	1.000	Ea.	3.019	1,200	174	1,374
Color	1.000	Ea.	3.019	1,575	174	1,749
Rough-in, for water closet						
1/2" copper supply, 4" cast iron waste, 2" cast iron vent	1.000	Ea.	2.376	137	141	278
4" PVC/DWV waste, 2" PVC vent	1.000	Ea.	2.678	82	158	240
4" copper waste, 2" copper vent	1.000	Ea.	2.520	233	153	386
3" cast iron waste, 1-1/2" cast iron vent	1.000	Ea.	2.244	123	133	256
3" PVC waste, 1-1/2" PVC vent	1.000	Ea.	2.388	74	147	221
3" copper waste, 1-1/2" copper vent	1.000	Ea.	2.014	161	123	284
1/2" PVC supply, 4" PVC waste, 2" PVC vent	1.000	Ea.	2.974	102	177	279
3" PVC waste, 1-1/2" PVC supply	1.000	Ea.	2.684	94	166	260
1/2" steel supply, 4" cast iron waste, 2" cast iron vent	1.000	Ea.	2.545	137	151	288
4" cast iron waste, 2" steel vent	1.000	Ea.	2.590	133	154	287
4" PVC waste, 2" PVC vent	1.000	Ea.	2.847	82	169	251
Lavatory, wall hung, P.E. cast iron 20" x 18", white	1.000	Ea.	2.000	305	115	420
Color	1.000	Ea.	2.000	330	115	445
Vitreous china 19" x 17", white	1.000	Ea.	2.286	228	132	360
Color	1.000	Ea.	2.286	236	132	368
Lavatory, for vanity top, P.E. cast iron 20" x 18", white	1.000	Ea.	2.500	370	144	514
Color	1.000	Ea.	2.500	445	144	589
Steel, enameled 20" x 17", white	1.000	Ea.	2.759	177	159	336
Color	1.000	Ea.	2.500	210	144	354
Vitreous china 20" x 16", white	1.000	Ea.	2.963	286	171	457
Color	1.000	Ea.	2.963	286	171	457
Rough-in, for lavatory						
1/2" copper supply, 1-1/2" C.I. waste, 1-1/2" C.I. vent	1.000	Ea.	2.791	138	167	305
1-1/2" PVC waste, 1-1/4" PVC vent	1.000	Ea.	2.639	87.50	169	256.50
1/2" steel supply, 1-1/4" cast iron waste, 1-1/4" steel vent	1.000	Ea.	2.890	120	174	294
1-1/4" PVC waste, 1-1/4" PVC vent	1.000	Ea.	2.794	88	179	267
1/2" PVC supply, 1-1/2" PVC waste, 1-1/2" PVC vent	1.000	Ea.	3.260	120	208	328
Bathtub, P.E. cast iron, 5' long corner with fittings, white	1.000	Ea.	3.636	1,250	209	1,459
Color	1.000	Ea.	3.636	1,500	209	1,709
Rough-in, for bathtub						
1/2" copper supply, 4" cast iron waste, 1-1/2" copper vent	1.000	Ea.	2.409	153	149	302
4" PVC waste, 1-1/2" PVC vent	1.000	Ea.	2.877	97	177	274
1/2" steel supply, 4" cast iron waste, 1-1/2" steel vent	1.000	Ea.	2.898	139	175	314
4" PVC waste, 1-1/2" PVC vent	1.000	Ea.	3.159	97	195	292
1/2" PVC supply, 4" PVC waste, 1-1/2" PVC vent	1.000	Ea.	3.371	130	209	339
Piping, supply 1/2" copper	20.000	L.F.	1.975	81.50	126	207.50
1/2" steel	20.000	L.F.	2.540	81	162	243
1/2" PVC	20.000	L.F.	2.963	148	189	337
Piping, waste, 4" cast iron no hub	9.000	L.F.	2.483	189	143	332
4" PVC/DWV	9.000	L.F.	3.000	99	173	272
4" copper/DWV	9.000	L.F.	3.600	430	207	637
Piping, vent 2" cast iron no hub	6.000	L.F.	1.433	74.50	82.50	157
2" copper/DWV	6.000	L.F.	1.091	96.50	70	166.50
2" PVC/DWV	6.000	L.F.	1.627	37	93.50	130.50
2" steel, galvanized	6.000	L.F.	1.500	68	86.50	154.50
Vanity base cabinet, 2 door, 24" x 30"	1.000	Ea.	1.000	405	55	460
24" x 36"	1.000	Ea.	1.200	395	66	461
Vanity top, laminated plastic square edge 25" x 32"	2.670	L.F.	.712	97.50	39.50	137
25" x 38"	3.160	L.F.	.843	115	46.50	161.50
Cultured marble, 25" x 32", with bowl	1.000	Ea.	2.500	236	144	380
25" x 38", with bowl	1.000	Ea.	2.500	270	144	414
Carrier, for lavatory, steel for studs, no arms	1.000	Ea.	1.143	105	73	178
Wood, 2" x 8" blocking	1.300	L.F.	.052	1.31	2.87	4.18

8 | MECHANICAL
16 Three Fixture Bathroom Systems

System Description	QUAN.	UNIT	LABOR HOURS	COST EACH		
				MAT.	INST.	TOTAL
BATHROOM WITH LAVATORY INSTALLED IN VANITY						
Water closet, floor mounted, 2 piece, close coupled, white	1.000	Ea.	3.019	261	174	435
Rough-in, waste, 4" diameter DWV piping	1.000	Ea.	.828	63	47.70	110.70
Vent, 2" diameter DWV piping	1.000	Ea.	.955	49.60	55	104.60
Supply, 1/2" diameter type "L" copper supply piping	1.000	Ea.	.593	24.42	37.80	62.22
Lavatory, 20" x 18", P.E. cast iron with accessories, white	1.000	Ea.	2.500	370	144	514
Rough-in, waste, 1-1/2" diameter DWV piping	1.000	Ea.	1.803	97.20	103.60	200.80
Supply, 1/2" diameter type "L" copper supply piping	1.000	Ea.	.988	40.70	63	103.70
Bathtub, P.E. cast iron 5' long with accessories, white	1.000	Ea.	3.636	1,250	209	1,459
Rough-in, waste, 4" diameter DWV piping	1.000	Ea.	.828	63	47.70	110.70
Vent, 1-1/2" diameter DWV piping	1.000	Ea.	.593	49.20	37.80	87
Supply, 1/2" diameter type "L" copper supply piping	1.000	Ea.	.988	40.70	63	103.70
Piping, supply, 1/2" diameter type "L" copper supply piping	10.000	L.F.	.988	40.70	63	103.70
Waste, 4" diameter DWV piping	6.000	L.F.	1.655	126	95.40	221.40
Vent, 2" diameter DWV piping	6.000	L.F.	1.500	68.10	86.40	154.50
Vanity base cabinet, 2 door, 30" wide	1.000	Ea.	1.000	405	55	460
Vanity top, plastic laminated square edge	2.670	L.F.	.712	97.46	39.25	136.71
TOTAL		Ea.	22.586	3,046.08	1,321.65	4,367.73
BATHROOM WITH WALL HUNG LAVATORY						
Water closet, floor mounted, 2 piece, close coupled, white	1.000	Ea.	3.019	261	174	435
Rough-in, vent, 2" diameter DWV piping	1.000	Ea.	.955	49.60	55	104.60
Waste, 4" diameter DWV piping	1.000	Ea.	.828	63	47.70	110.70
Supply, 1/2" diameter type "L" copper supply piping	1.000	Ea.	.593	24.42	37.80	62.22
Lavatory, 20" x 18" P.E. cast iron, wall hung, white	1.000	Ea.	2.000	305	115	420
Rough-in, waste, 1-1/2" diameter DWV piping	1.000	Ea.	1.803	97.20	103.60	200.80
Supply, 1/2" diameter type "L" copper supply piping	1.000	Ea.	.988	40.70	63	103.70
Bathtub, P.E. cast iron, 5' long with accessories, white	1.000	Ea.	3.636	1,250	209	1,459
Rough-in, waste, 4" diameter DWV piping	1.000	Ea.	.828	63	47.70	110.70
Supply, 1/2" diameter type "L" copper supply piping	1.000	Ea.	.988	40.70	63	103.70
Vent, 1-1/2" diameter DWV piping	1.000	Ea.	.593	49.20	37.80	87
Piping, supply, 1/2" diameter type "L" copper supply piping	10.000	L.F.	.988	40.70	63	103.70
Waste, 4" diameter DWV piping	6.000	L.F.	1.655	126	95.40	221.40
Vent, 2" diameter DWV piping	6.000	L.F.	1.500	68.10	86.40	154.50
Carrier, steel, for studs, no arms	1.000	Ea.	1.143	105	73	178
TOTAL		Ea.	21.517	2,583.62	1,271.40	3,855.02

The costs in this system are on a cost each basis. All necessary piping is included.

Three Fixture Bathroom Price Sheet	QUAN.	UNIT	LABOR HOURS	COST EACH		
				MAT.	INST.	TOTAL
Water closet, close coupled standard 2 piece, white	1.000	Ea.	3.019	261	174	435
Color	1.000	Ea.	3.019	460	174	634
One piece elongated bowl, white	1.000	Ea.	3.019	1,225	174	1,399
Color	1.000	Ea.	3.019	1,575	174	1,749
Low profile, one piece elongated bowl, white	1.000	Ea.	3.019	1,200	174	1,374
Color	1.000	Ea.	3.019	1,575	174	1,749
Rough-in for water closet						
1/2" copper supply, 4" cast iron waste, 2" cast iron vent	1.000	Ea.	2.376	137	141	278
4" PVC/DWV waste, 2" PVC vent	1.000	Ea.	2.678	82	158	240
4" carrier waste, 2" copper vent	1.000	Ea.	2.520	233	153	386
3" cast iron waste, 1-1/2" cast iron vent	1.000	Ea.	2.244	123	133	256
3" PVC waste, 1-1/2" PVC vent	1.000	Ea.	2.388	74	147	221
3" copper waste, 1-1/2" copper vent	1.000	Ea.	2.014	161	123	284
1/2" PVC supply, 4" PVC waste, 2" PVC vent	1.000	Ea.	2.974	102	177	279
3" PVC waste, 1-1/2" PVC supply	1.000	Ea.	2.684	94	166	260
1/2" steel supply, 4" cast iron waste, 2" cast iron vent	1.000	Ea.	2.545	137	151	288
4" cast iron waste, 2" steel vent	1.000	Ea.	2.590	133	154	287
4" PVC waste, 2" PVC vent	1.000	Ea.	2.847	82	169	251
Lavatory, wall hung, PE cast iron 20" x 18", white	1.000	Ea.	2.000	305	115	420
Color	1.000	Ea.	2.000	330	115	445
Vitreous china 19" x 17", white	1.000	Ea.	2.286	228	132	360
Color	1.000	Ea.	2.286	236	132	368
Lavatory, for vanity top, PE cast iron 20" x 18", white	1.000	Ea.	2.500	370	144	514
Color	1.000	Ea.	2.500	445	144	589
Steel enameled 20" x 17", white	1.000	Ea.	2.759	177	159	336
Color	1.000	Ea.	2.500	210	144	354
Vitreous china 20" x 16", white	1.000	Ea.	2.963	286	171	457
Color	1.000	Ea.	2.963	286	171	457
Rough-in for lavatory						
1/2" copper supply, 1-1/2" cast iron waste, 1-1/2" cast iron vent	1.000	Ea.	2.791	138	167	305
1-1/2" PVC waste, 1-1/4" PVC vent	1.000	Ea.	2.639	87.50	169	256.50
1/2" steel supply, 1-1/4" cast iron waste, 1-1/4" steel vent	1.000	Ea.	2.890	120	174	294
1-1/4" PVC waste, 1-1/4" PVC vent	1.000	Ea.	2.794	88	179	267
1/2" PVC supply, 1-1/2" PVC waste, 1-1/2" PVC vent	1.000	Ea.	3.260	120	208	328
Bathtub, PE cast iron, 5' long corner with fittings, white	1.000	Ea.	3.636	1,250	209	1,459
Color	1.000	Ea.	3.636	1,500	209	1,709
Rough-in for bathtub						
1/2" copper supply, 4" cast iron waste, 1-1/2" copper vent	1.000	Ea.	2.409	153	149	302
4" PVC waste, 1/2" PVC vent	1.000	Ea.	2.877	97	177	274
1/2" steel supply, 4" cast iron waste, 1-1/2" steel vent	1.000	Ea.	2.898	139	175	314
4" PVC waste, 1-1/2" PVC vent	1.000	Ea.	3.159	97	195	292
1/2" PVC supply, 4" PVC waste, 1-1/2" PVC vent	1.000	Ea.	3.371	130	209	339
Piping supply, 1/2" copper	10.000	L.F.	.988	40.50	63	103.50
1/2" steel	10.000	L.F.	1.270	40.50	81	121.50
1/2" PVC	10.000	L.F.	1.482	74	94.50	168.50
Piping waste, 4" cast iron no hub	6.000	L.F.	1.655	126	95.50	221.50
4" PVC/DWV	6.000	L.F.	2.000	66	115	181
4" copper/DWV	6.000	L.F.	2.400	288	138	426
Piping vent 2" cast iron no hub	6.000	L.F.	1.433	74.50	82.50	157
2" copper/DWV	6.000	L.F.	1.091	96.50	70	166.50
2" PVC/DWV	6.000	L.F.	1.627	37	93.50	130.50
2" steel, galvanized	6.000	L.F.	1.500	68	86.50	154.50
Vanity base cabinet, 2 door, 24" x 30"	1.000	Ea.	1.000	405	55	460
24" x 36"	1.000	Ea.	1.200	395	66	461
Vanity top, laminated plastic square edge 25" x 32"	2.670	L.F.	.712	97.50	39.50	137
25" x 38"	3.160	L.F.	.843	115	46.50	161.50
Cultured marble, 25" x 32", with bowl	1.000	Ea.	2.500	236	144	380
25" x 38", with bowl	1.000	Ea.	2.500	270	144	414
Carrier, for lavatory, steel for studs, no arms	1.000	Ea.	1.143	105	73	178
Wood, 2" x 8" blocking	1.300	L.F.	.052	1.31	2.87	4.18

8 | MECHANICAL
20 | Three Fixture Bathroom Systems

System Description	QUAN.	UNIT	LABOR HOURS	COST EACH		
				MAT.	INST.	TOTAL
BATHROOM WITH LAVATORY INSTALLED IN VANITY						
Water closet, floor mounted, 2 piece, close coupled, white	1.000	Ea.	3.019	261	174	435
Rough-in, vent, 2" diameter DWV piping	1.000	Ea.	.955	49.60	55	104.60
Waste, 4" diameter DWV piping	1.000	Ea.	.828	63	47.70	110.70
Supply, 1/2" diameter type "L" copper supply piping	1.000	Ea.	.593	24.42	37.80	62.22
Lavatory, 20" x 18", PE cast iron with accessories, white	1.000	Ea.	2.500	370	144	514
Rough-in, vent, 1-1/2" diameter DWV piping	1.000	Ea.	1.803	97.20	103.60	200.80
Supply, 1/2" diameter type "L" copper supply piping	1.000	Ea.	.988	40.70	63	103.70
Bathtub, P.E. cast iron, 5' long with accessories, white	1.000	Ea.	3.636	1,250	209	1,459
Rough-in, waste, 4" diameter DWV piping	1.000	Ea.	.828	63	47.70	110.70
Supply, 1/2" diameter type "L" copper supply piping	1.000	Ea.	.988	40.70	63	103.70
Vent, 1-1/2" diameter DWV piping	1.000	Ea.	.593	49.20	37.80	87
Piping, supply, 1/2" diameter type "L" copper supply piping	32.000	L.F.	3.161	130.24	201.60	331.84
Waste, 4" diameter DWV piping	12.000	L.F.	3.310	252	190.80	442.80
Vent, 2" diameter DWV piping	6.000	L.F.	1.500	68.10	86.40	154.50
Vanity base cabinet, 2 door, 30" wide	1.000	Ea.	1.000	405	55	460
Vanity top, plastic laminated square edge	2.670	L.F.	.712	97.46	39.25	136.71
TOTAL		Ea.	26.414	3,261.62	1,555.65	4,817.27
BATHROOM WITH WALL HUNG LAVATORY						
Water closet, floor mounted, 2 piece, close coupled, white	1.000	Ea.	3.019	261	174	435
Rough-in, vent, 2" diameter DWV piping	1.000	Ea.	.955	49.60	55	104.60
Waste, 4" diameter DWV piping	1.000	Ea.	.828	63	47.70	110.70
Supply, 1/2" diameter type "L" copper supply piping	1.000	Ea.	.593	24.42	37.80	62.22
Lavatory, 20" x 18" P.E. cast iron, wall hung, white	1.000	Ea.	2.000	305	115	420
Rough-in, waste, 1-1/2" diameter DWV piping	1.000	Ea.	1.803	97.20	103.60	200.80
Supply, 1/2" diameter type "L" copper supply piping	1.000	Ea.	.988	40.70	63	103.70
Bathtub, P.E. cast iron, 5' long with accessories, white	1.000	Ea.	3.636	1,250	209	1,459
Rough-in, waste, 4" diameter DWV piping	1.000	Ea.	.828	63	47.70	110.70
Supply, 1/2" diameter type "L" copper supply piping	1.000	Ea.	.988	40.70	63	103.70
Vent, 1-1/2" diameter DWV piping	1.000	Ea.	.593	49.20	37.80	87
Piping, supply, 1/2" diameter type "L" copper supply piping	32.000	L.F.	3.161	130.24	201.60	331.84
Waste, 4" diameter DWV piping	12.000	L.F.	3.310	252	190.80	442.80
Vent, 2" diameter DWV piping	6.000	L.F.	1.500	68.10	86.40	154.50
Carrier steel, for studs, no arms	1.000	Ea.	1.143	105	73	178
TOTAL		Ea.	25.345	2,799.16	1,505.40	4,304.56

The costs in this system are on a cost each basis. All necessary piping is included.

Three Fixture Bathroom Price Sheet

Description	QUAN.	UNIT	LABOR HOURS	MAT.	INST.	TOTAL
Water closet, close coupled, standard 2 piece, white	1.000	Ea.	3.019	261	174	435
Color	1.000	Ea.	3.019	460	174	634
One piece, elongated bowl, white	1.000	Ea.	3.019	1,225	174	1,399
Color	1.000	Ea.	3.019	1,575	174	1,749
Low profile, one piece, elongated bowl, white	1.000	Ea.	3.019	1,200	174	1,374
Color	1.000	Ea.	3.019	1,575	174	1,749
Rough-in, for water closet						
1/2" copper supply, 4" cast iron waste, 2" cast iron vent	1.000	Ea.	2.376	137	141	278
4" PVC/DWV waste, 2" PVC vent	1.000	Ea.	2.678	82	158	240
4" copper waste, 2" copper vent	1.000	Ea.	2.520	233	153	386
3" cast iron waste, 1-1/2" cast iron vent	1.000	Ea.	2.244	123	133	256
3" PVC waste, 1-1/2" PVC vent	1.000	Ea.	2.388	74	147	221
3" copper waste, 1-1/2" copper vent	1.000	Ea.	2.014	161	123	284
1/2" PVC supply, 4" PVC waste, 2" PVC vent	1.000	Ea.	2.974	102	177	279
3" PVC waste, 1-1/2" PVC supply	1.000	Ea.	2.684	94	166	260
1/2" steel supply, 4" cast iron waste, 2" cast iron vent	1.000	Ea.	2.545	137	151	288
4" cast iron waste, 2" steel vent	1.000	Ea.	2.590	133	154	287
4" PVC waste, 2" PVC vent	1.000	Ea.	2.847	82	169	251
Lavatory wall hung, P.E. cast iron, 20" x 18", white	1.000	Ea.	2.000	305	115	420
Color	1.000	Ea.	2.000	330	115	445
Vitreous china, 19" x 17", white	1.000	Ea.	2.286	228	132	360
Color	1.000	Ea.	2.286	236	132	368
Lavatory, for vanity top, P.E., cast iron, 20" x 18", white	1.000	Ea.	2.500	370	144	514
Color	1.000	Ea.	2.500	445	144	589
Steel, enameled, 20" x 17", white	1.000	Ea.	2.759	177	159	336
Color	1.000	Ea.	2.500	210	144	354
Vitreous china, 20" x 16", white	1.000	Ea.	2.963	286	171	457
Color	1.000	Ea.	2.963	286	171	457
Rough-in, for lavatory						
1/2" copper supply, 1-1/2" C.I. waste, 1-1/2" C.I. vent	1.000	Ea.	2.791	138	167	305
1-1/2" PVC waste, 1-1/4" PVC vent	1.000	Ea.	2.639	87.50	169	256.50
1/2" steel supply, 1-1/4" cast iron waste, 1-1/4" steel vent	1.000	Ea.	2.890	120	174	294
1-1/4" PVC waste, 1-1/4" PVC vent	1.000	Ea.	2.794	88	179	267
1/2" PVC supply, 1-1/2" PVC waste, 1-1/2" PVC vent	1.000	Ea.	3.260	120	208	328
Bathtub, P.E. cast iron, 5' long corner with fittings, white	1.000	Ea.	3.636	1,250	209	1,459
Color	1.000	Ea.	3.636	1,500	209	1,709
Rough-in, for bathtub						
1/2" copper supply, 4" cast iron waste, 1-1/2" copper vent	1.000	Ea.	2.409	153	149	302
4" PVC waste, 1/2" PVC vent	1.000	Ea.	2.877	97	177	274
1/2" steel supply, 4" cast iron waste, 1-1/2" steel vent	1.000	Ea.	2.898	139	175	314
4" PVC waste, 1-1/2" PVC vent	1.000	Ea.	3.159	97	195	292
1/2" PVC supply, 4" PVC waste, 1-1/2" PVC vent	1.000	Ea.	3.371	130	209	339
Piping, supply, 1/2" copper	32.000	L.F.	3.161	130	202	332
1/2" steel	32.000	L.F.	4.063	130	259	389
1/2" PVC	32.000	L.F.	4.741	237	300	537
Piping, waste, 4" cast iron no hub	12.000	L.F.	3.310	252	191	443
4" PVC/DWV	12.000	L.F.	4.000	132	230	362
4" copper/DWV	12.000	L.F.	4.800	575	276	851
Piping, vent, 2" cast iron no hub	6.000	L.F.	1.433	74.50	82.50	157
2" copper/DWV	6.000	L.F.	1.091	96.50	70	166.50
2" PVC/DWV	6.000	L.F.	1.627	37	93.50	130.50
2" steel, galvanized	6.000	L.F.	1.500	68	86.50	154.50
Vanity base cabinet, 2 door, 24" x 30"	1.000	Ea.	1.000	405	55	460
24" x 36"	1.000	Ea.	1.200	395	66	461
Vanity top, laminated plastic square edge, 25" x 32"	2.670	L.F.	.712	97.50	39.50	137
25" x 38"	3.160	L.F.	.843	115	46.50	161.50
Cultured marble, 25" x 32", with bowl	1.000	Ea.	2.500	236	144	380
25" x 38", with bowl	1.000	Ea.	2.500	270	144	414
Carrier, for lavatory, steel for studs, no arms	1.000	Ea.	1.143	105	73	178
Wood, 2" x 8" blocking	1.300	L.F.	.052	1.31	2.87	4.18

8 | MECHANICAL — 24 | Three Fixture Bathroom Systems

System Description	QUAN.	UNIT	LABOR HOURS	COST EACH MAT.	COST EACH INST.	COST EACH TOTAL
BATHROOM WITH LAVATORY INSTALLED IN VANITY						
Water closet, floor mounted, 2 piece, close coupled, white	1.000	Ea.	3.019	261	174	435
Rough-in, vent, 2" diameter DWV piping	1.000	Ea.	.955	49.60	55	104.60
Waste, 4" diameter DWV piping	1.000	Ea.	.828	63	47.70	110.70
Supply, 1/2" diameter type "L" copper supply piping	1.000	Ea.	.593	24.42	37.80	62.22
Lavatory, 20" x 18", P.E. cast iron with fittings, white	1.000	Ea.	2.500	370	144	514
Rough-in, waste, 1-1/2" diameter DWV piping	1.000	Ea.	1.803	97.20	103.60	200.80
Supply, 1/2" diameter type "L" copper supply piping	1.000	Ea.	.988	40.70	63	103.70
Bathtub, P.E. cast iron, corner with fittings, white	1.000	Ea.	3.636	2,900	209	3,109
Rough-in, waste, 4" diameter DWV piping	1.000	Ea.	.828	63	47.70	110.70
Supply, 1/2" diameter type "L" copper supply piping	1.000	Ea.	.988	40.70	63	103.70
Vent, 1-1/2" diameter DWV piping	1.000	Ea.	.593	49.20	37.80	87
Piping, supply, 1/2" diameter type "L" copper supply piping	32.000	L.F.	3.161	130.24	201.60	331.84
Waste, 4" diameter DWV piping	12.000	L.F.	3.310	252	190.80	442.80
Vent, 2" diameter DWV piping	6.000	L.F.	1.500	68.10	86.40	154.50
Vanity base cabinet, 2 door, 30" wide	1.000	Ea.	1.000	405	55	460
Vanity top, plastic laminated, square edge	2.670	L.F.	.712	114.81	39.25	154.06
TOTAL		Ea.	26.414	4,928.97	1,555.65	6,484.62
BATHROOM WITH WALL HUNG LAVATORY						
Water closet, floor mounted, 2 piece, close coupled, white	1.000	Ea.	3.019	261	174	435
Rough-in, vent, 2" diameter DWV piping	1.000	Ea.	.955	49.60	55	104.60
Waste, 4" diameter DWV piping	1.000	Ea.	.828	63	47.70	110.70
Supply, 1/2" diameter type "L" copper supply piping	1.000	Ea.	.593	24.42	37.80	62.22
Lavatory, 20" x 18", P.E. cast iron, with fittings, white	1.000	Ea.	2.000	305	115	420
Rough-in, waste, 1-1/2" diameter DWV piping	1.000	Ea.	1.803	97.20	103.60	200.80
Supply, 1/2" diameter type "L" copper supply piping	1.000	Ea.	.988	40.70	63	103.70
Bathtub, P.E. cast iron, corner, with fittings, white	1.000	Ea.	3.636	2,900	209	3,109
Rough-in, waste, 4" diameter DWV piping	1.000	Ea.	.828	63	47.70	110.70
Supply, 1/2" diameter type "L" copper supply piping	1.000	Ea.	.988	40.70	63	103.70
Vent, 1-1/2" diameter DWV piping	1.000	Ea.	.593	49.20	37.80	87
Piping, supply, 1/2" diameter type "L" copper supply piping	32.000	L.F.	3.161	130.24	201.60	331.84
Waste, 4" diameter DWV piping	12.000	L.F.	3.310	252	190.80	442.80
Vent, 2" diameter DWV piping	6.000	L.F.	1.500	68.10	86.40	154.50
Carrier, steel, for studs, no arms	1.000	Ea.	1.143	105	73	178
TOTAL		Ea.	25.345	4,449.16	1,505.40	5,954.56

The costs in this system are on a cost each basis. All necessary piping is included.

Three Fixture Bathroom Price Sheet	QUAN.	UNIT	LABOR HOURS	COST EACH MAT.	COST EACH INST.	TOTAL
Water closet, close coupled, standard 2 piece, white	1.000	Ea.	3.019	261	174	435
Color	1.000	Ea.	3.019	460	174	634
One piece elongated bowl, white	1.000	Ea.	3.019	1,225	174	1,399
Color	1.000	Ea.	3.019	1,575	174	1,749
Low profile, one piece elongated bowl, white	1.000	Ea.	3.019	1,200	174	1,374
Color	1.000	Ea.	3.019	1,575	174	1,749
Rough-in, for water closet						
1/2" copper supply, 4" cast iron waste, 2" cast iron vent	1.000	Ea.	2.376	137	141	278
4" PVC/DWV waste, 2" PVC vent	1.000	Ea.	2.678	82	158	240
4" copper waste, 2" copper vent	1.000	Ea.	2.520	233	153	386
3" cast iron waste, 1-1/2" cast iron vent	1.000	Ea.	2.244	123	133	256
3" PVC waste, 1-1/2" PVC vent	1.000	Ea.	2.388	74	147	221
3" copper waste, 1-1/2" copper vent	1.000	Ea.	2.014	161	123	284
1/2" PVC supply, 4" PVC waste, 2" PVC vent	1.000	Ea.	2.974	102	177	279
3" PVC waste, 1-1/2" PVC supply	1.000	Ea.	2.684	94	166	260
1/2" steel supply, 4" cast iron waste, 2" cast iron vent	1.000	Ea.	2.545	137	151	288
4" cast iron waste, 2" steel vent	1.000	Ea.	2.590	133	154	287
4" PVC waste, 2" PVC vent	1.000	Ea.	2.847	82	169	251
Lavatory, wall hung P.E. cast iron 20" x 18", white	1.000	Ea.	2.000	305	115	420
Color	1.000	Ea.	2.000	330	115	445
Vitreous china 19" x 17", white	1.000	Ea.	2.286	228	132	360
Color	1.000	Ea.	2.286	236	132	368
Lavatory, for vanity top, P.E., cast iron, 20" x 18", white	1.000	Ea.	2.500	370	144	514
Color	1.000	Ea.	2.500	445	144	589
Steel enameled 20" x 17", white	1.000	Ea.	2.759	177	159	336
Color	1.000	Ea.	2.500	210	144	354
Vitreous china 20" x 16", white	1.000	Ea.	2.963	286	171	457
Color	1.000	Ea.	2.963	286	171	457
Rough-in, for lavatory						
1/2" copper supply, 1-1/2" cast iron waste, 1-1/2" cast iron vent	1.000	Ea.	2.791	138	167	305
1-1/2" PVC waste, 1-1/4" PVC vent	1.000	Ea.	2.639	87.50	169	256.50
1/2" steel supply, 1-1/4" cast iron waste, 1-1/4" steel vent	1.000	Ea.	2.890	120	174	294
1-1/4" PVC waste, 1-1/4" PVC vent	1.000	Ea.	2.794	88	179	267
1/2" PVC supply, 1-1/2" PVC waste, 1-1/2" PVC vent	1.000	Ea.	3.260	120	208	328
Bathtub, P.E. cast iron, corner with fittings, white	1.000	Ea.	3.636	2,900	209	3,109
Color	1.000	Ea.	4.000	3,300	230	3,530
Rough-in, for bathtub						
1/2" copper supply, 4" cast iron waste, 1-1/2" copper vent	1.000	Ea.	2.409	153	149	302
4" PVC waste, 1-1/2" PVC vent	1.000	Ea.	2.877	97	177	274
1/2" steel supply, 4" cast iron waste, 1-1/2" steel vent	1.000	Ea.	2.898	139	175	314
4" PVC waste, 1-1/2" PVC vent	1.000	Ea.	3.159	97	195	292
1/2" PVC supply, 4" PVC waste, 1-1/2" PVC vent	1.000	Ea.	3.371	130	209	339
Piping, supply, 1/2" copper	32.000	L.F.	3.161	130	202	332
1/2" steel	32.000	L.F.	4.063	130	259	389
1/2" PVC	32.000	L.F.	4.741	237	300	537
Piping, waste, 4" cast iron, no hub	12.000	L.F.	3.310	252	191	443
4" PVC/DWV	12.000	L.F.	4.000	132	230	362
4" copper/DWV	12.000	L.F.	4.800	575	276	851
Piping, vent 2" cast iron, no hub	6.000	L.F.	1.433	74.50	82.50	157
2" copper/DWV	6.000	L.F.	1.091	96.50	70	166.50
2" PVC/DWV	6.000	L.F.	1.627	37	93.50	130.50
2" steel, galvanized	6.000	L.F.	1.500	68	86.50	154.50
Vanity base cabinet, 2 door, 24" x 30"	1.000	Ea.	1.000	405	55	460
24" x 36"	1.000	Ea.	1.200	395	66	461
Vanity top, laminated plastic square edge 25" x 32"	2.670	L.F.	.712	115	39.50	154.50
25" x 38"	3.160	L.F.	.843	136	46.50	182.50
Cultured marble, 25" x 32", with bowl	1.000	Ea.	2.500	236	144	380
25" x 38", with bowl	1.000	Ea.	2.500	270	144	414
Carrier, for lavatory, steel for studs, no arms	1.000	Ea.	1.143	105	73	178
Wood, 2" x 8" blocking	1.300	L.F.	.053	1.34	2.94	4.28

8 | MECHANICAL — 28 | Three Fixture Bathroom Systems

System Description	QUAN.	UNIT	LABOR HOURS	COST EACH MAT.	COST EACH INST.	COST EACH TOTAL
BATHROOM WITH SHOWER, LAVATORY INSTALLED IN VANITY						
Water closet, floor mounted, 2 piece, close coupled, white	1.000	Ea.	3.019	261	174	435
Rough-in, vent, 2" diameter DWV piping	1.000	Ea.	.955	49.60	55	104.60
Waste, 4" diameter DWV piping	1.000	Ea.	.828	63	47.70	110.70
Supply, 1/2" diameter type "L" copper supply piping	1.000	Ea.	.593	24.42	37.80	62.22
Lavatory, 20" x 18" P.E. cast iron with fittings, white	1.000	Ea.	2.500	370	144	514
Rough-in, waste, 1-1/2" diameter DWV piping	1.000	Ea.	1.803	97.20	103.60	200.80
Supply, 1/2" diameter type "L" copper supply piping	1.000	Ea.	.988	40.70	63	103.70
Shower, steel enameled, stone base, corner, white	1.000	Ea.	3.200	1,275	184	1,459
Shower mixing valve	1.000	Ea.	1.333	169	85.50	254.50
Shower door	1.000	Ea.	1.000	420	61.50	481.50
Rough-in, vent, 1-1/2" diameter DWV piping	1.000	Ea.	.225	12.15	12.95	25.10
Waste, 2" diameter DWV piping	1.000	Ea.	1.433	74.40	82.50	156.90
Supply, 1/2" diameter type "L" copper supply piping	1.000	Ea.	1.580	65.12	100.80	165.92
Piping, supply, 1/2" diameter type "L" copper supply piping	36.000	L.F.	4.148	170.94	264.60	435.54
Waste, 4" diameter DWV piping	7.000	L.F.	2.759	210	159	369
Vent, 2" diameter DWV piping	6.000	L.F.	2.250	102.15	129.60	231.75
Vanity base 2 door, 30" wide	1.000	Ea.	1.000	405	55	460
Vanity top, plastic laminated, square edge	2.170	L.F.	.712	85.44	39.25	124.69
TOTAL		Ea.	30.326	3,895.12	1,799.80	5,694.92
BATHROOM WITH SHOWER, WALL HUNG LAVATORY						
Water closet, floor mounted, close coupled	1.000	Ea.	3.019	261	174	435
Rough-in, vent, 2" diameter DWV piping	1.000	Ea.	.955	49.60	55	104.60
Waste, 4" diameter DWV piping	1.000	Ea.	.828	63	47.70	110.70
Supply, 1/2" diameter type "L" copper supply piping	1.000	Ea.	.593	24.42	37.80	62.22
Lavatory, 20" x 18" P.E. cast iron with fittings, white	1.000	Ea.	2.000	305	115	420
Rough-in, waste, 1-1/2" diameter DWV piping	1.000	Ea.	1.803	97.20	103.60	200.80
Supply, 1/2" diameter type "L" copper supply piping	1.000	Ea.	.988	40.70	63	103.70
Shower, steel enameled, stone base, white	1.000	Ea.	3.200	1,275	184	1,459
Mixing valve	1.000	Ea.	1.333	169	85.50	254.50
Shower door	1.000	Ea.	1.000	420	61.50	481.50
Rough-in, vent, 1-1/2" diameter DWV piping	1.000	Ea.	.225	12.15	12.95	25.10
Waste, 2" diameter DWV piping	1.000	Ea.	1.433	74.40	82.50	156.90
Supply, 1/2" diameter type "L" copper supply piping	1.000	Ea.	1.580	65.12	100.80	165.92
Piping, supply, 1/2" diameter type "L" copper supply piping	36.000	L.F.	4.148	170.94	264.60	435.54
Waste, 4" diameter DWV piping	7.000	L.F.	2.759	210	159	369
Vent, 2" diameter DWV piping	6.000	L.F.	2.250	102.15	129.60	231.75
Carrier, steel, for studs, no arms	1.000	Ea.	1.143	105	73	178
TOTAL		Ea.	29.257	3,444.68	1,749.55	5,194.23

The costs in this system are on a cost each basis. All necessary piping is included.

Three Fixture Bathroom Price Sheet

	QUAN.	UNIT	LABOR HOURS	COST EACH MAT.	COST EACH INST.	COST EACH TOTAL
Water closet, close coupled, standard 2 piece, white	1.000	Ea.	3.019	261	174	435
Color	1.000	Ea.	3.019	460	174	634
One piece elongated bowl, white	1.000	Ea.	3.019	1,225	174	1,399
Color	1.000	Ea.	3.019	1,575	174	1,749
Low profile, one piece elongated bowl, white	1.000	Ea.	3.019	1,200	174	1,374
Color	1.000	Ea.	3.019	1,575	174	1,749
Rough-in, for water closet						
1/2" copper supply, 4" cast iron waste, 2" cast iron vent	1.000	Ea.	2.376	137	141	278
4" PVC/DWV waste, 2" PVC vent	1.000	Ea.	2.678	82	158	240
4" copper waste, 2" copper vent	1.000	Ea.	2.520	233	153	386
3" cast iron waste, 1-1/2" cast iron vent	1.000	Ea.	2.244	123	133	256
3" PVC waste, 1-1/2" PVC vent	1.000	Ea.	2.388	74	147	221
3" copper waste, 1-1/2" copper vent	1.000	Ea.	2.014	161	123	284
1/2" PVC supply, 4" PVC waste, 2" PVC vent	1.000	Ea.	2.974	102	177	279
3" PVC waste, 1-1/2" PVC supply	1.000	Ea.	2.684	94	166	260
1/2" steel supply, 4" cast iron waste, 2" cast iron vent	1.000	Ea.	2.545	137	151	288
4" cast iron waste, 2" steel vent	1.000	Ea.	2.590	133	154	287
4" PVC waste, 2" PVC vent	1.000	Ea.	2.847	82	169	251
Lavatory, wall hung, P.E. cast iron 20" x 18", white	1.000	Ea.	2.000	305	115	420
Color	1.000	Ea.	2.000	330	115	445
Vitreous china 19" x 17", white	1.000	Ea.	2.286	228	132	360
Color	1.000	Ea.	2.286	236	132	368
Lavatory, for vanity top, P.E. cast iron 20" x 18", white	1.000	Ea.	2.500	370	144	514
Color	1.000	Ea.	2.500	445	144	589
Steel enameled 20" x 17", white	1.000	Ea.	2.759	177	159	336
Color	1.000	Ea.	2.500	210	144	354
Vitreous china 20" x 16", white	1.000	Ea.	2.963	286	171	457
Color	1.000	Ea.	2.963	286	171	457
Rough-in, for lavatory						
1/2" copper supply, 1-1/2" cast iron waste, 1-1/2" cast iron vent	1.000	Ea.	2.791	138	167	305
1-1/2" PVC waste, 1-1/2" PVC vent	1.000	Ea.	2.639	87.50	169	256.50
1/2" steel supply, 1-1/4" cast iron waste, 1-1/4" steel vent	1.000	Ea.	2.890	120	174	294
1-1/4" PVC waste, 1-1/4" PVC vent	1.000	Ea.	2.921	87.50	187	274.50
1/2" PVC supply, 1-1/2" PVC waste, 1-1/2" PVC vent	1.000	Ea.	3.260	120	208	328
Shower, steel enameled stone base, 32" x 32", white	1.000	Ea.	8.000	1,275	184	1,459
Color	1.000	Ea.	7.822	1,950	169	2,119
36" x 36" white	1.000	Ea.	8.889	1,625	192	1,817
Color	1.000	Ea.	8.889	2,225	192	2,417
Rough-in, for shower						
1/2" copper supply, 4" cast iron waste, 1-1/2" copper vent	1.000	Ea.	3.238	152	196	348
4" PVC waste, 1-1/2" PVC vent	1.000	Ea.	3.429	108	209	317
1/2" steel supply, 4" cast iron waste, 1-1/2" steel vent	1.000	Ea.	3.665	148	224	372
4" PVC waste, 1-1/2" PVC vent	1.000	Ea.	3.881	108	237	345
1/2" PVC supply, 4" PVC waste, 1-1/2" PVC vent	1.000	Ea.	4.219	161	259	420
Piping, supply, 1/2" copper	36.000	L.F.	4.148	171	265	436
1/2" steel	36.000	L.F.	5.333	171	340	511
1/2" PVC	36.000	L.F.	6.222	310	395	705
Piping, waste, 4" cast iron no hub	7.000	L.F.	2.759	210	159	369
4" PVC/DWV	7.000	L.F.	3.333	110	192	302
4" copper/DWV	7.000	L.F.	4.000	480	230	710
Piping, vent, 2" cast iron no hub	6.000	L.F.	2.149	112	124	236
2" copper/DWV	6.000	L.F.	1.636	145	105	250
2" PVC/DWV	6.000	L.F.	2.441	56	140	196
2" steel, galvanized	6.000	L.F.	2.250	102	130	232
Vanity base cabinet, 2 door, 24" x 30"	1.000	Ea.	1.000	405	55	460
24" x 36"	1.000	Ea.	1.200	395	66	461
Vanity top, laminated plastic square edge, 25" x 32"	2.170	L.F.	.712	85.50	39.50	125
25" x 38"	2.670	L.F.	.845	101	46.50	147.50
Carrier, for lavatory, steel for studs, no arms	1.000	Ea.	1.143	105	73	178
Wood, 2" x 8" blocking	1.300	L.F.	.052	1.31	2.87	4.18

8 | MECHANICAL   32 | Three Fixture Bathroom Systems

System Description	QUAN.	UNIT	LABOR HOURS	MAT.	INST.	TOTAL
BATHROOM WITH LAVATORY INSTALLED IN VANITY						
Water closet, floor mounted, 2 piece, close coupled, white	1.000	Ea.	3.019	261	174	435
Rough-in, vent, 2" diameter DWV piping	1.000	Ea.	.955	49.60	55	104.60
Waste, 4" diameter DWV piping	1.000	Ea.	.828	63	47.70	110.70
Supply, 1/2" diameter type "L" copper supply piping	1.000	Ea.	.593	24.42	37.80	62.22
Lavatory, 20" x 18", P.E. cast iron with fittings, white	1.000	Ea.	2.500	370	144	514
Rough-in, waste, 1-1/2" diameter DWV piping	1.000	Ea.	1.803	97.20	103.60	200.80
Supply, 1/2" diameter type "L" copper supply piping	1.000	Ea.	.988	40.70	63	103.70
Shower, steel enameled, stone base, corner, white	1.000	Ea.	3.200	1,275	184	1,459
Mixing valve	1.000	Ea.	1.333	169	85.50	254.50
Shower door	1.000	Ea.	1.000	420	61.50	481.50
Rough-in, vent, 1-1/2" diameter DWV piping	1.000	Ea.	.225	12.15	12.95	25.10
Waste, 2" diameter DWV piping	1.000	Ea.	1.433	74.40	82.50	156.90
Supply, 1/2" diameter type "L" copper supply piping	1.000	Ea.	1.580	65.12	100.80	165.92
Piping, supply, 1/2" diameter type "L" copper supply piping	36.000	L.F.	3.556	146.52	226.80	373.32
Waste, 4" diameter DWV piping	7.000	L.F.	1.931	147	111.30	258.30
Vent, 2" diameter DWV piping	6.000	L.F.	1.500	68.10	86.40	154.50
Vanity base, 2 door, 30" wide	1.000	Ea.	1.000	405	55	460
Vanity top, plastic laminated, square edge	2.670	L.F.	.712	97.46	39.25	136.71
TOTAL		Ea.	28.156	3,785.67	1,671.10	5,456.77
BATHROOM, WITH WALL HUNG LAVATORY						
Water closet, floor mounted, 2 piece, close coupled, white	1.000	Ea.	3.019	261	174	435
Rough-in, vent, 2" diameter DWV piping	1.000	Ea.	.955	49.60	55	104.60
Waste, 4" diameter DWV piping	1.000	Ea.	.828	63	47.70	110.70
Supply, 1/2" diameter type "L" copper supply piping	1.000	Ea.	.593	24.42	37.80	62.22
Lavatory, wall hung, 20" x 18" P.E. cast iron with fittings, white	1.000	Ea.	2.000	305	115	420
Rough-in, waste, 1-1/2" diameter DWV piping	1.000	Ea.	1.803	97.20	103.60	200.80
Supply, 1/2" diameter type "L" copper supply piping	1.000	Ea.	.988	40.70	63	103.70
Shower, steel enameled, stone base, corner, white	1.000	Ea.	3.200	1,275	184	1,459
Mixing valve	1.000	Ea.	1.333	169	85.50	254.50
Shower door	1.000	Ea.	1.000	420	61.50	481.50
Rough-in, waste, 1-1/2" diameter DWV piping	1.000	Ea.	.225	12.15	12.95	25.10
Waste, 2" diameter DWV piping	1.000	Ea.	1.433	74.40	82.50	156.90
Supply, 1/2" diameter type "L" copper supply piping	1.000	Ea.	1.580	65.12	100.80	165.92
Piping, supply, 1/2" diameter type "L" copper supply piping	36.000	L.F.	3.556	146.52	226.80	373.32
Waste, 4" diameter DWV piping	7.000	L.F.	1.931	147	111.30	258.30
Vent, 2" diameter DWV piping	6.000	L.F.	1.500	68.10	86.40	154.50
Carrier, steel, for studs, no arms	1.000	Ea.	1.143	105	73	178
TOTAL		Ea.	27.087	3,323.21	1,620.85	4,944.06

The costs in this system are on a cost each basis. All necessary piping is included.

Three Fixture Bathroom Price Sheet

	QUAN.	UNIT	LABOR HOURS	COST EACH MAT.	COST EACH INST.	COST EACH TOTAL
Water closet, close coupled, standard 2 piece, white	1.000	Ea.	3.019	261	174	435
Color	1.000	Ea.	3.019	460	174	634
One piece elongated bowl, white	1.000	Ea.	3.019	1,225	174	1,399
Color	1.000	Ea.	3.019	1,575	174	1,749
Low profile one piece elongated bowl, white	1.000	Ea.	3.019	1,200	174	1,374
Color	1.000	Ea.	3.623	1,575	174	1,749
Rough-in, for water closet						
1/2" copper supply, 4" cast iron waste, 2" cast iron vent	1.000	Ea.	2.376	137	141	278
4" P.V.C./DWV waste, 2" PVC vent	1.000	Ea.	2.678	82	158	240
4" copper waste, 2" copper vent	1.000	Ea.	2.520	233	153	386
3" cast iron waste, 1-1/2" cast iron vent	1.000	Ea.	2.244	123	133	256
3" PVC waste, 1-1/2" PVC vent	1.000	Ea.	2.388	74	147	221
3" copper waste, 1-1/2" copper vent	1.000	Ea.	2.014	161	123	284
1/2" P.V.C. supply, 4" P.V.C. waste, 2" P.V.C. vent	1.000	Ea.	2.974	102	177	279
3" P.V.C. waste, 1-1/2" P.V.C. vent	1.000	Ea.	2.684	94	166	260
1/2" steel supply, 4" cast iron waste, 2" cast iron vent	1.000	Ea.	2.545	137	151	288
4" cast iron waste, 2" steel vent	1.000	Ea.	2.590	133	154	287
4" P.V.C. waste, 2" P.V.C. vent	1.000	Ea.	2.847	82	169	251
Lavatory, wall hung P.E. cast iron 20" x 18", white	1.000	Ea.	2.000	305	115	420
Color	1.000	Ea.	2.000	330	115	445
Vitreous china 19" x 17", white	1.000	Ea.	2.286	228	132	360
Color	1.000	Ea.	2.286	236	132	368
Lavatory, for vanity top P.E. cast iron 20" x 18", white	1.000	Ea.	2.500	370	144	514
Color	1.000	Ea.	2.500	445	144	589
Steel enameled 20" x 17", white	1.000	Ea.	2.759	177	159	336
Color	1.000	Ea.	2.500	210	144	354
Vitreous china 20" x 16", white	1.000	Ea.	2.963	286	171	457
Color	1.000	Ea.	2.963	286	171	457
Rough-in, for lavatory						
1/2" copper supply, 1-1/2" cast iron waste, 1-1/2" cast iron vent	1.000	Ea.	2.791	138	167	305
1-1/2" P.V.C. waste, 1-1/2" P.V.C. vent	1.000	Ea.	2.639	87.50	169	256.50
1/2" steel supply, 1-1/2" cast iron waste, 1-1/4" steel vent	1.000	Ea.	2.890	120	174	294
1-1/2" P.V.C. waste, 1-1/4" P.V.C. vent	1.000	Ea.	2.921	87.50	187	274.50
1/2" P.V.C. supply, 1-1/2" P.V.C. waste, 1-1/2" P.V.C. vent	1.000	Ea.	3.260	120	208	328
Shower, steel enameled stone base, 32" x 32", white	1.000	Ea.	8.000	1,275	184	1,459
Color	1.000	Ea.	7.822	1,950	169	2,119
36" x 36", white	1.000	Ea.	8.889	1,625	192	1,817
Color	1.000	Ea.	8.889	2,225	192	2,417
Rough-in, for shower						
1/2" copper supply, 2" cast iron waste, 1-1/2" copper vent	1.000	Ea.	3.161	152	193	345
2" P.V.C. waste, 1-1/2" P.V.C. vent	1.000	Ea.	3.429	108	209	317
1/2" steel supply, 2" cast iron waste, 1-1/2" steel vent	1.000	Ea.	3.887	200	237	437
2" P.V.C. waste, 1-1/2" P.V.C. vent	1.000	Ea.	3.881	108	237	345
1/2" P.V.C. supply, 2" P.V.C. waste, 1-1/2" P.V.C. vent	1.000	Ea.	4.219	161	259	420
Piping, supply, 1/2" copper	36.000	L.F.	3.556	147	227	374
1/2" steel	36.000	L.F.	4.571	146	292	438
1/2" P.V.C.	36.000	L.F.	5.333	266	340	606
Waste, 4" cast iron, no hub	7.000	L.F.	1.931	147	111	258
4" P.V.C./DWV	7.000	L.F.	2.333	77	134	211
4" copper/DWV	7.000	L.F.	2.800	335	161	496
Vent, 2" cast iron, no hub	6.000	L.F.	1.091	96.50	70	166.50
2" copper/DWV	6.000	L.F.	1.091	96.50	70	166.50
2" P.V.C./DWV	6.000	L.F.	1.627	37	93.50	130.50
2" steel, galvanized	6.000	L.F.	1.500	68	86.50	154.50
Vanity base cabinet, 2 door, 24" x 30"	1.000	Ea.	1.000	405	55	460
24" x 36"	1.000	Ea.	1.200	395	66	461
Vanity top, laminated plastic square edge, 25" x 32"	2.670	L.F.	.712	97.50	39.50	137
25" x 38"	3.170	L.F.	.845	116	46.50	162.50
Carrier, for lavatory, steel, for studs, no arms	1.000	Ea.	1.143	105	73	178
Wood, 2" x 8" blocking	1.300	L.F.	.052	1.31	2.87	4.18

8 | MECHANICAL — 36 | Four Fixture Bathroom Systems

System Description	QUAN.	UNIT	LABOR HOURS	COST EACH MAT.	COST EACH INST.	COST EACH TOTAL
BATHROOM WITH LAVATORY INSTALLED IN VANITY						
Water closet, floor mounted, 2 piece, close coupled, white	1.000	Ea.	3.019	261	174	435
Rough-in, vent, 2" diameter DWV piping	1.000	Ea.	.955	49.60	55	104.60
Waste, 4" diameter DWV piping	1.000	Ea.	.828	63	47.70	110.70
Supply, 1/2" diameter type "L" copper supply piping	1.000	Ea.	.593	24.42	37.80	62.22
Lavatory, 20" x 18" P.E. cast iron with fittings, white	1.000	Ea.	2.500	370	144	514
Shower, steel, enameled, stone base, corner, white	1.000	Ea.	3.333	1,900	192	2,092
Mixing valve	1.000	Ea.	1.333	169	85.50	254.50
Shower door	1.000	Ea.	1.000	420	61.50	481.50
Rough-in, waste, 1-1/2" diameter DWV piping	2.000	Ea.	4.507	243	259	502
Supply, 1/2" diameter type "L" copper supply piping	2.000	Ea.	3.161	130.24	201.60	331.84
Bathtub, P.E. cast iron, 5' long with fittings, white	1.000	Ea.	3.636	1,250	209	1,459
Rough-in, waste, 4" diameter DWV piping	1.000	Ea.	.828	63	47.70	110.70
Supply, 1/2" diameter type "L" copper supply piping	1.000	Ea.	.988	40.70	63	103.70
Vent, 1-1/2" diameter DWV piping	1.000	Ea.	.593	49.20	37.80	87
Piping, supply, 1/2" diameter type "L" copper supply piping	42.000	L.F.	4.148	170.94	264.60	435.54
Waste, 4" diameter DWV piping	10.000	L.F.	2.759	210	159	369
Vent, 2" diameter DWV piping	13.000	L.F.	3.250	147.55	187.20	334.75
Vanity base, 2 doors, 30" wide	1.000	Ea.	1.000	405	55	460
Vanity top, plastic laminated, square edge	2.670	L.F.	.712	97.46	39.25	136.71
TOTAL		Ea.	39.143	6,064.11	2,320.65	8,384.76
BATHROOM WITH WALL HUNG LAVATORY						
Water closet, floor mounted, 2 piece, close coupled, white	1.000	Ea.	3.019	261	174	435
Rough-in, vent, 2" diameter DWV piping	1.000	Ea.	.955	49.60	55	104.60
Waste, 4" diameter DWV piping	1.000	Ea.	.828	63	47.70	110.70
Supply, 1/2" diameter type "L" copper supply piping	1.000	Ea.	.593	24.42	37.80	62.22
Lavatory, 20" x 18" P.E. cast iron with fittings, white	1.000	Ea.	2.000	305	115	420
Shower, steel enameled, stone base, corner, white	1.000	Ea.	3.333	1,900	192	2,092
Mixing valve	1.000	Ea.	1.333	169	85.50	254.50
Shower door	1.000	Ea.	1.000	420	61.50	481.50
Rough-in, waste, 1-1/2" diameter DWV piping	2.000	Ea.	4.507	243	259	502
Supply, 1/2" diameter type "L" copper supply piping	2.000	Ea.	3.161	130.24	201.60	331.84
Bathtub, P.E. cast iron, 5' long with fittings, white	1.000	Ea.	3.636	1,250	209	1,459
Rough-in, waste, 4" diameter DWV piping	1.000	Ea.	.828	63	47.70	110.70
Supply, 1/2" diameter type "L" copper supply piping	1.000	Ea.	.988	40.70	63	103.70
Vent, 1-1/2" diameter copper DWV piping	1.000	Ea.	.593	49.20	37.80	87
Piping, supply, 1/2" diameter type "L" copper supply piping	42.000	L.F.	4.148	170.94	264.60	435.54
Waste, 4" diameter DWV piping	10.000	L.F.	2.759	210	159	369
Vent, 2" diameter DWV piping	13.000	L.F.	3.250	147.55	187.20	334.75
Carrier, steel, for studs, no arms	1.000	Ea.	1.143	105	73	178
TOTAL		Ea.	38.074	5,601.65	2,270.40	7,872.05

The costs in this system are on a cost each basis. All necessary piping is included.

Four Fixture Bathroom Price Sheet	QUAN.	UNIT	LABOR HOURS	COST EACH MAT.	COST EACH INST.	COST EACH TOTAL
Water closet, close coupled, standard 2 piece, white	1.000	Ea.	3.019	261	174	435
Color	1.000	Ea.	3.019	460	174	634
One piece elongated bowl, white	1.000	Ea.	3.019	1,225	174	1,399
Color	1.000	Ea.	3.019	1,575	174	1,749
Low profile, one piece elongated bowl, white	1.000	Ea.	3.019	1,200	174	1,374
Color	1.000	Ea.	3.019	1,575	174	1,749
1/2" copper supply, 4" cast iron waste, 2" cast iron vent	1.000	Ea.	2.376	137	141	278
4" PVC/DWV waste, 2" PVC vent	1.000	Ea.	2.678	82	158	240
4" copper waste, 2" copper vent	1.000	Ea.	2.520	233	153	386
3" cast iron waste, 1-1/2" cast iron vent	1.000	Ea.	2.244	123	133	256
3" P.V.C. waste, 1-1/2" P.V.C. vent	1.000	Ea.	2.388	74	147	221
3" copper waste, 1-1/2" copper vent	1.000	Ea.	2.014	161	123	284
1/2" P.V.C. supply, 4" P.V.C. waste, 2" P.V.C. vent	1.000	Ea.	2.974	102	177	279
3" P.V.C. waste, 1-1/2" P.V.C. vent	1.000	Ea.	2.684	94	166	260
1/2" steel supply, 4" cast iron waste, 2" cast iron vent	1.000	Ea.	2.545	137	151	288
4" cast iron waste, 2" steel vent	1.000	Ea.	2.590	133	154	287
4" P.V.C. waste, 2" P.V.C. vent	1.000	Ea.	2.847	82	169	251
Lavatory, wall hung P.E. cast iron 20" x 18", white	1.000	Ea.	2.000	305	115	420
Color	1.000	Ea.	2.000	330	115	445
Vitreous china 19" x 17", white	1.000	Ea.	2.286	228	132	360
Color	1.000	Ea.	2.286	236	132	368
Lavatory for vanity top, P.E. cast iron 20" x 18", white	1.000	Ea.	2.500	370	144	514
Color	1.000	Ea.	2.500	445	144	589
Steel enameled, 20" x 17", white	1.000	Ea.	2.759	177	159	336
Color	1.000	Ea.	2.500	210	144	354
Vitreous china 20" x 16", white	1.000	Ea.	2.963	286	171	457
Color	1.000	Ea.	2.963	286	171	457
Shower, steel enameled stone base, 36" square, white	1.000	Ea.	8.889	1,900	192	2,092
Color	1.000	Ea.	8.889	1,950	192	2,142
Rough-in, for lavatory or shower						
1/2" copper supply, 1-1/2" cast iron waste, 1-1/2" cast iron vent	1.000	Ea.	3.834	187	230	417
1-1/2" P.V.C. waste, 1-1/4" P.V.C. vent	1.000	Ea.	3.675	124	235	359
1/2" steel supply, 1-1/4" cast iron waste, 1-1/4" steel vent	1.000	Ea.	4.103	169	249	418
1-1/4" P.V.C. waste, 1-1/4" P.V.C. vent	1.000	Ea.	3.937	124	252	376
1/2" P.V.C. supply, 1-1/2" P.V.C. waste, 1-1/2" P.V.C. vent	1.000	Ea.	4.592	176	293	469
Bathtub, P.E. cast iron, 5' long with fittings, white	1.000	Ea.	3.636	1,250	209	1,459
Color	1.000	Ea.	3.636	1,500	209	1,709
Steel, enameled 5' long with fittings, white	1.000	Ea.	2.909	570	167	737
Color	1.000	Ea.	2.909	570	167	737
Rough-in, for bathtub						
1/2" copper supply, 4" cast iron waste, 1-1/2" copper vent	1.000	Ea.	2.409	153	149	302
4" P.V.C. waste, 1-1/2" P.V.C. vent	1.000	Ea.	2.877	97	177	274
1/2" steel supply, 4" cast iron waste, 1-1/2" steel vent	1.000	Ea.	2.898	139	175	314
4" P.V.C. waste, 1-1/2" P.V.C. vent	1.000	Ea.	3.159	97	195	292
1/2" P.V.C. supply, 4" P.V.C. waste, 1-1/2" P.V.C. vent	1.000	Ea.	3.371	130	209	339
Piping, supply, 1/2" copper	42.000	L.F.	4.148	171	265	436
1/2" steel	42.000	L.F.	5.333	171	340	511
1/2" P.V.C.	42.000	L.F.	6.222	310	395	705
Waste, 4" cast iron, no hub	10.000	L.F.	2.759	210	159	369
4" P.V.C./DWV	10.000	L.F.	3.333	110	192	302
4" copper/DWV	10.000	Ea.	4.000	480	230	710
Vent 2" cast iron, no hub	13.000	L.F.	3.105	161	179	340
2" copper/DWV	13.000	L.F.	2.364	209	151	360
2" P.V.C./DWV	13.000	L.F.	3.525	80.50	203	283.50
2" steel, galvanized	13.000	L.F.	3.250	148	187	335
Vanity base cabinet, 2 doors, 30" wide	1.000	Ea.	1.000	405	55	460
Vanity top, plastic laminated, square edge	2.670	L.F.	.712	97.50	39.50	137
Carrier, steel for studs, no arms	1.000	Ea.	1.143	105	73	178
Wood, 2" x 8" blocking	1.300	L.F.	.052	1.31	2.87	4.18

8 | MECHANICAL — 40 | Four Fixture Bathroom Systems

System Description	QUAN.	UNIT	LABOR HOURS	MAT.	INST.	TOTAL
BATHROOM WITH LAVATORY INSTALLED IN VANITY						
Water closet, floor mounted, 2 piece, close coupled, white	1.000	Ea.	3.019	261	174	435
Rough-in, vent, 2" diameter DWV piping	1.000	Ea.	.955	49.60	55	104.60
Waste, 4" diameter DWV piping	1.000	Ea.	.828	63	47.70	110.70
Supply, 1/2" diameter type "L" copper supply piping	1.000	Ea.	.593	24.42	37.80	62.22
Lavatory, 20" x 18" P.E. cast iron with fittings, white	1.000	Ea.	2.500	370	144	514
Shower, steel, enameled, stone base, corner, white	1.000	Ea.	3.333	1,900	192	2,092
Mixing valve	1.000	Ea.	1.333	169	85.50	254.50
Shower door	1.000	Ea.	1.000	420	61.50	481.50
Rough-in, waste, 1-1/2" diameter DWV piping	2.000	Ea.	4.507	243	259	502
Supply, 1/2" diameter type "L" copper supply piping	2.000	Ea.	3.161	130.24	201.60	331.84
Bathtub, P.E. cast iron, 5' long with fittings, white	1.000	Ea.	3.636	1,250	209	1,459
Rough-in, waste, 4" diameter DWV piping	1.000	Ea.	.828	63	47.70	110.70
Supply, 1/2" diameter type "L" copper supply piping	1.000	Ea.	.988	40.70	63	103.70
Vent, 1-1/2" diameter DWV piping	1.000	Ea.	.593	49.20	37.80	87
Piping, supply, 1/2" diameter type "L" copper supply piping	42.000	L.F.	4.939	203.50	315	518.50
Waste, 4" diameter DWV piping	10.000	L.F.	4.138	315	238.50	553.50
Vent, 2" diameter DWV piping	13.000	L.F.	4.500	204.30	259.20	463.50
Vanity base, 2 doors, 30" wide	1.000	Ea.	1.000	405	55	460
Vanity top, plastic laminated, square edge	2.670	L.F.	.712	85.44	39.25	124.69
TOTAL		Ea.	42.563	6,246.40	2,522.55	8,768.95
BATHROOM WITH WALL HUNG LAVATORY						
Water closet, floor mounted, 2 piece, close coupled, white	1.000	Ea.	3.019	261	174	435
Rough-in, vent, 2" diameter DWV piping	1.000	Ea.	.955	49.60	55	104.60
Waste, 4" diameter DWV piping	1.000	Ea.	.828	63	47.70	110.70
Supply, 1/2" diameter type "L" copper supply piping	1.000	Ea.	.593	24.42	37.80	62.22
Lavatory, 20" x 18" P.E. cast iron with fittings, white	1.000	Ea.	2.000	305	115	420
Shower, steel enameled, stone base, corner, white	1.000	Ea.	3.333	1,900	192	2,092
Mixing valve	1.000	Ea.	1.333	169	85.50	254.50
Shower door	1.000	Ea.	1.000	420	61.50	481.50
Rough-in, waste, 1-1/2" diameter DWV piping	2.000	Ea.	4.507	243	259	502
Supply, 1/2" diameter type "L" copper supply piping	2.000	Ea.	3.161	130.24	201.60	331.84
Bathtub, P.E. cast iron, 5" long with fittings, white	1.000	Ea.	3.636	1,250	209	1,459
Rough-in, waste, 4" diameter DWV piping	1.000	Ea.	.828	63	47.70	110.70
Supply, 1/2" diameter type "L" copper supply piping	1.000	Ea.	.988	40.70	63	103.70
Vent, 1-1/2" diameter DWV piping	1.000	Ea.	.593	49.20	37.80	87
Piping, supply, 1/2" diameter type "L" copper supply piping	42.000	L.F.	4.939	203.50	315	518.50
Waste, 4" diameter DWV piping	10.000	L.F.	4.138	315	238.50	553.50
Vent, 2" diameter DWV piping	13.000	L.F.	4.500	204.30	259.20	463.50
Carrier, steel for studs, no arms	1.000	Ea.	1.143	105	73	178
TOTAL		Ea.	41.494	5,795.96	2,472.30	8,268.26

The costs in this system are on a cost each basis. All necessary piping is included.

Four Fixture Bathroom Price Sheet	QUAN.	UNIT	LABOR HOURS	COST EACH		
				MAT.	INST.	TOTAL
Water closet, close coupled, standard 2 piece, white	1.000	Ea.	3.019	261	174	435
Color	1.000	Ea.	3.019	460	174	634
One piece, elongated bowl, white	1.000	Ea.	3.019	1,225	174	1,399
Color	1.000	Ea.	3.019	1,575	174	1,749
Low profile, one piece elongated bowl, white	1.000	Ea.	3.019	1,200	174	1,374
Color	1.000	Ea.	3.019	1,575	174	1,749
Rough-in, for water closet						
1/2" copper supply, 4" cast iron waste, 2" cast iron vent	1.000	Ea.	2.376	137	141	278
4" PVC/DWV waste, 2" PVC vent	1.000	Ea.	2.678	82	158	240
4" copper waste, 2" copper vent	1.000	Ea.	2.520	233	153	386
3" cast iron waste, 1-1/2" cast iron vent	1.000	Ea.	2.244	123	133	256
3" PVC waste, 1-1/2" PVC vent	1.000	Ea.	2.388	74	147	221
3" PVC waste, 1-1/2" PVC vent	1.000	Ea.	2.014	161	123	284
1/2" PVC supply, 4" PVC waste, 2" PVC vent	1.000	Ea.	2.974	102	177	279
3" PVC waste, 1-1/2" PVC vent	1.000	Ea.	2.684	94	166	260
1/2" steel supply, 4" cast iron waste, 2" cast iron vent	1.000	Ea.	2.545	137	151	288
4" cast iron waste, 2" steel vent	1.000	Ea.	2.590	133	154	287
4" PVC waste, 2" PVC vent	1.000	Ea.	2.847	82	169	251
Lavatory wall hung, P.E. cast iron 20" x 18", white	1.000	Ea.	2.000	305	115	420
Color	1.000	Ea.	2.000	330	115	445
Vitreous china 19" x 17", white	1.000	Ea.	2.286	228	132	360
Color	1.000	Ea.	2.286	236	132	368
Lavatory for vanity top, P.E. cast iron, 20" x 18", white	1.000	Ea.	2.500	370	144	514
Color	1.000	Ea.	2.500	445	144	589
Steel, enameled 20" x 17", white	1.000	Ea.	2.759	177	159	336
Color	1.000	Ea.	2.500	210	144	354
Vitreous china 20" x 16", white	1.000	Ea.	2.963	286	171	457
Color	1.000	Ea.	2.963	286	171	457
Shower, steel enameled, stone base 36" square, white	1.000	Ea.	8.889	1,900	192	2,092
Color	1.000	Ea.	8.889	1,950	192	2,142
Rough-in, for lavatory and shower						
1/2" copper supply, 1-1/2" cast iron waste, 1-1/2" cast iron vent	1.000	Ea.	7.668	375	460	835
1-1/2" PVC waste, 1-1/4" PVC vent	1.000	Ea.	7.352	247	470	717
1/2" steel supply, 1-1/4" cast iron waste, 1-1/4" steel vent	1.000	Ea.	8.205	335	495	830
1-1/4" PVC waste, 1-1/4" PVC vent	1.000	Ea.	7.873	248	505	753
1/2" PVC supply, 1-1/2" PVC waste, 1-1/2" PVC vent	1.000	Ea.	9.185	355	585	940
Bathtub, P.E. cast iron, 5' long with fittings, white	1.000	Ea.	3.636	1,250	209	1,459
Color	1.000	Ea.	3.636	1,500	209	1,709
Steel enameled, 5' long with fittings, white	1.000	Ea.	2.909	570	167	737
Color	1.000	Ea.	2.909	570	167	737
Rough-in, for bathtub						
1/2" copper supply, 4" cast iron waste, 1-1/2" copper vent	1.000	Ea.	2.409	153	149	302
4" PVC waste, 1-1/2" PVC vent	1.000	Ea.	2.877	97	177	274
1/2" steel supply, 4" cast iron waste, 1-1/2" steel vent	1.000	Ea.	2.898	139	175	314
4" PVC waste, 1-1/2" PVC vent	1.000	Ea.	3.159	97	195	292
1/2" PVC supply, 4" PVC waste, 1-1/2" PVC vent	1.000	Ea.	3.371	130	209	339
Piping supply, 1/2" copper	42.000	L.F.	4.148	171	265	436
1/2" steel	42.000	L.F.	5.333	171	340	511
1/2" PVC	42.000	L.F.	6.222	310	395	705
Piping, waste, 4" cast iron, no hub	10.000	L.F.	3.586	273	207	480
4" PVC/DWV	10.000	L.F.	4.333	143	250	393
4" copper/DWV	10.000	L.F.	5.200	625	299	924
Piping, vent, 2" cast iron, no hub	13.000	L.F.	3.105	161	179	340
2" copper/DWV	13.000	L.F.	2.364	209	151	360
2" PVC/DWV	13.000	L.F.	3.525	80.50	203	283.50
2" steel, galvanized	13.000	L.F.	3.250	148	187	335
Vanity base cabinet, 2 doors, 30" wide	1.000	Ea.	1.000	405	55	460
Vanity top, plastic laminated, square edge	3.160	L.F.	.843	115	46.50	161.50
Carrier, steel, for studs, no arms	1.000	Ea.	1.143	105	73	178
Wood, 2" x 8" blocking	1.300	L.F.	.052	1.31	2.87	4.18

8 | MECHANICAL   44 | Five Fixture Bathroom Systems

System Description	QUAN.	UNIT	LABOR HOURS	COST EACH MAT.	COST EACH INST.	COST EACH TOTAL
BATHROOM WITH SHOWER, BATHTUB, LAVATORIES IN VANITY						
Water closet, floor mounted, 1 piece, white	1.000	Ea.	3.019	1,200	174	1,374
Rough-in, vent, 2" diameter DWV piping	1.000	Ea.	.955	49.60	55	104.60
Waste, 4" diameter DWV piping	1.000	Ea.	.828	63	47.70	110.70
Supply, 1/2" diameter type "L" copper supply piping	1.000	Ea.	.593	24.42	37.80	62.22
Lavatory, 20" x 16", vitreous china oval, with fittings, white	2.000	Ea.	5.926	572	342	914
Shower, steel enameled, stone base, corner, white	1.000	Ea.	3.333	1,900	192	2,092
Mixing valve	1.000	Ea.	1.333	169	85.50	254.50
Shower door	1.000	Ea.	1.000	420	61.50	481.50
Rough-in, waste, 1-1/2" diameter DWV piping	3.000	Ea.	5.408	291.60	310.80	602.40
Supply, 1/2" diameter type "L" copper supply piping	3.000	Ea.	2.963	122.10	189	311.10
Bathtub, P.E. cast iron, 5' long with fittings, white	1.000	Ea.	3.636	1,250	209	1,459
Rough-in, waste, 4" diameter DWV piping	1.000	Ea.	1.103	84	63.60	147.60
Supply, 1/2" diameter type "L" copper supply piping	1.000	Ea.	.988	40.70	63	103.70
Vent, 1-1/2" diameter copper DWV piping	1.000	Ea.	.593	49.20	37.80	87
Piping, supply, 1/2" diameter type "L" copper supply piping	42.000	L.F.	4.148	170.94	264.60	435.54
Waste, 4" diameter DWV piping	10.000	L.F.	2.759	210	159	369
Vent, 2" diameter DWV piping	13.000	L.F.	3.250	147.55	187.20	334.75
Vanity base, 2 door, 24" x 48"	1.000	Ea.	1.400	520	77	597
Vanity top, plastic laminated, square edge	4.170	L.F.	1.112	152.21	61.30	213.51
TOTAL		Ea.	44.347	7,436.32	2,617.80	10,054.12

The costs in this system are on a cost each basis. All necessary piping is included.

Description	QUAN.	UNIT	LABOR HOURS	COST EACH MAT.	COST EACH INST.	COST EACH TOTAL

Five Fixture Bathroom Price Sheet	QUAN.	UNIT	LABOR HOURS	COST EACH MAT.	COST EACH INST.	COST EACH TOTAL
Water closet, close coupled, standard 2 piece, white	1.000	Ea.	3.019	261	174	435
Color	1.000	Ea.	3.019	460	174	634
One piece elongated bowl, white	1.000	Ea.	3.019	1,225	174	1,399
Color	1.000	Ea.	3.019	1,575	174	1,749
Low profile, one piece elongated bowl, white	1.000	Ea.	3.019	1,200	174	1,374
Color	1.000	Ea.	3.019	1,575	174	1,749
Rough-in, supply, waste and vent for water closet						
1/2" copper supply, 4" cast iron waste, 2" cast iron vent	1.000	Ea.	2.376	137	141	278
4" P.V.C./DWV waste, 2" P.V.C. vent	1.000	Ea.	2.678	82	158	240
4" copper waste, 2" copper vent	1.000	Ea.	2.520	233	153	386
3" cast iron waste, 1-1/2" cast iron vent	1.000	Ea.	2.244	123	133	256
3" P.V.C. waste, 1-1/2" P.V.C. vent	1.000	Ea.	2.388	74	147	221
3" copper waste, 1-1/2" copper vent	1.000	Ea.	2.014	161	123	284
1/2" P.V.C. supply, 4" P.V.C. waste, 2" P.V.C. vent	1.000	Ea.	2.974	102	177	279
3" P.V.C. waste, 1-1/2" P.V.C. supply	1.000	Ea.	2.684	94	166	260
1/2" steel supply, 4" cast iron waste, 2" cast iron vent	1.000	Ea.	2.545	137	151	288
4" cast iron waste, 2" steel vent	1.000	Ea.	2.590	133	154	287
4" P.V.C. waste, 2" P.V.C. vent	1.000	Ea.	2.847	82	169	251
Lavatory, wall hung, P.E. cast iron 20" x 18", white	2.000	Ea.	4.000	610	230	840
Color	2.000	Ea.	4.000	660	230	890
Vitreous china, 19" x 17", white	2.000	Ea.	4.571	455	264	719
Color	2.000	Ea.	4.571	470	264	734
Lavatory, for vanity top, P.E. cast iron, 20" x 18", white	2.000	Ea.	5.000	740	288	1,028
Color	2.000	Ea.	5.000	890	288	1,178
Steel enameled 20" x 17", white	2.000	Ea.	5.517	355	320	675
Color	2.000	Ea.	5.000	420	288	708
Vitreous china 20" x 16", white	2.000	Ea.	5.926	570	340	910
Color	2.000	Ea.	5.926	570	340	910
Shower, steel enameled, stone base 36" square, white	1.000	Ea.	8.889	1,900	192	2,092
Color	1.000	Ea.	8.889	1,950	192	2,142
Rough-in, for lavatory or shower						
1/2" copper supply, 1-1/2" cast iron waste, 1-1/2" cast iron vent	3.000	Ea.	8.371	415	500	915
1-1/2" P.V.C. waste, 1-1/4" P.V.C. vent	3.000	Ea.	7.916	263	505	768
1/2" steel supply, 1-1/4" cast iron waste, 1-1/4" steel vent	3.000	Ea.	8.670	360	525	885
1-1/4" P.V.C. waste, 1-1/4" P.V.C. vent	3.000	Ea.	8.381	263	535	798
1/2" P.V.C. supply, 1-1/2" P.V.C. waste, 1-1/2" P.V.C. vent	3.000	Ea.	9.778	360	625	985
Bathtub, P.E. cast iron 5' long with fittings, white	1.000	Ea.	3.636	1,250	209	1,459
Color	1.000	Ea.	3.636	1,500	209	1,709
Steel, enameled 5' long with fittings, white	1.000	Ea.	2.909	570	167	737
Color	1.000	Ea.	2.909	570	167	737
Rough-in, for bathtub						
1/2" copper supply, 4" cast iron waste, 1-1/2" copper vent	1.000	Ea.	2.684	174	164	338
4" P.V.C. waste, 1-1/2" P.V.C. vent	1.000	Ea.	3.210	108	197	305
1/2" steel supply, 4" cast iron waste, 1-1/2" steel vent	1.000	Ea.	3.173	160	191	351
4" P.V.C. waste, 1-1/2" P.V.C. vent	1.000	Ea.	3.492	108	215	323
1/2" P.V.C. supply, 4" P.V.C. waste, 1-1/2" P.V.C. vent	1.000	Ea.	3.704	141	228	369
Piping, supply, 1/2" copper	42.000	L.F.	4.148	171	265	436
1/2" steel	42.000	L.F.	5.333	171	340	511
1/2" P.V.C.	42.000	L.F.	6.222	310	395	705
Piping, waste, 4" cast iron, no hub	10.000	L.F.	2.759	210	159	369
4" P.V.C./DWV	10.000	L.F.	3.333	110	192	302
4" copper/DWV	10.000	L.F.	4.000	480	230	710
Piping, vent, 2" cast iron, no hub	13.000	L.F.	3.105	161	179	340
2" copper/DWV	13.000	L.F.	2.364	209	151	360
2" P.V.C./DWV	13.000	L.F.	3.525	80.50	203	283.50
2" steel, galvanized	13.000	L.F.	3.250	148	187	335
Vanity base cabinet, 2 doors, 24" x 48"	1.000	Ea.	1.400	520	77	597
Vanity top, plastic laminated, square edge	4.170	L.F.	1.112	152	61.50	213.50
Carrier, steel, for studs, no arms	1.000	Ea.	1.143	105	73	178
Wood, 2" x 8" blocking	1.300	L.F.	.052	1.31	2.87	4.18

8 | MECHANICAL — 60 | Gas Heating/Cooling Systems

System Description	QUAN.	UNIT	LABOR HOURS	COST PER SYSTEM MAT.	COST PER SYSTEM INST.	COST PER SYSTEM TOTAL
HEATING ONLY, GAS FIRED HOT AIR, ONE ZONE, 1200 S.F. BUILDING						
Furnace, gas, up flow	1.000	Ea.	5.000	690	277	967
Intermittent pilot	1.000	Ea.		166		166
Supply duct, rigid fiberglass	176.000	S.F.	12.068	153.12	693.44	846.56
Return duct, sheet metal, galvanized	158.000	Lb.	16.137	113.76	932.20	1,045.96
Lateral ducts, 6" flexible fiberglass	144.000	L.F.	8.862	465.12	491.04	956.16
Register, elbows	12.000	Ea.	6.400	252	354	606
Floor registers, enameled steel	12.000	Ea.	3.000	118.80	184.80	303.60
Floor grille, return air	2.000	Ea.	.727	58	45	103
Thermostat	1.000	Ea.	1.000	56	65	121
Plenum	1.000	Ea.	1.000	98.50	55.50	154
TOTAL		System	54.194	2,171.30	3,097.98	5,269.28
HEATING/COOLING, GAS FIRED FORCED AIR, ONE ZONE, 1200 S.F. BUILDING						
Furnace, including plenum, compressor, coil	1.000	Ea.	14.720	5,865	818.80	6,683.80
Intermittent pilot	1.000	Ea.		166		166
Supply duct, rigid fiberglass	176.000	S.F.	12.068	153.12	693.44	846.56
Return duct, sheet metal, galvanized	158.000	Lb.	16.137	113.76	932.20	1,045.96
Lateral duct, 6" flexible fiberglass	144.000	L.F.	8.862	465.12	491.04	956.16
Register elbows	12.000	Ea.	6.400	252	354	606
Floor registers, enameled steel	12.000	Ea.	3.000	118.80	184.80	303.60
Floor grille return air	2.000	Ea.	.727	58	45	103
Thermostat	1.000	Ea.	1.000	56	65	121
Refrigeration piping, 25 ft. (pre-charged)	1.000	Ea.		300		300
TOTAL		System	62.914	7,547.80	3,584.28	11,132.08

The costs in these systems are based on complete system basis. For larger buildings use the price sheet on the opposite page.

Description	QUAN.	UNIT	LABOR HOURS	COST PER SYSTEM MAT.	COST PER SYSTEM INST.	COST PER SYSTEM TOTAL

Gas Heating/Cooling Price Sheet	QUAN.	UNIT	LABOR HOURS	COST EACH MAT.	COST EACH INST.	COST EACH TOTAL
Furnace, heating only, 100 MBH, area to 1200 S.F.	1.000	Ea.	5.000	690	277	967
120 MBH, area to 1500 S.F.	1.000	Ea.	5.000	690	277	967
160 MBH, area to 2000 S.F.	1.000	Ea.	5.714	735	315	1,050
200 MBH, area to 2400 S.F.	1.000	Ea.	6.154	3,075	340	3,415
Heating/cooling, 100 MBH heat, 36 MBH cool, to 1200 S.F.	1.000	Ea.	16.000	6,375	890	7,265
120 MBH heat, 42 MBH cool, to 1500 S.F.	1.000	Ea.	18.462	6,800	1,050	7,850
144 MBH heat, 47 MBH cool, to 2000 S.F.	1.000	Ea.	20.000	7,850	1,150	9,000
200 MBH heat, 60 MBH cool, to 2400 S.F.	1.000	Ea.	34.286	8,275	1,975	10,250
Intermittent pilot, 100 MBH furnace	1.000	Ea.		166		166
200 MBH furnace	1.000	Ea.		166		166
Supply duct, rectangular, area to 1200 S.F., rigid fiberglass	176.000	S.F.	12.068	153	695	848
Sheet metal insulated	228.000	Lb.	31.331	1,050	1,850	2,900
Area to 1500 S.F., rigid fiberglass	176.000	S.F.	12.068	153	695	848
Sheet metal insulated	228.000	Lb.	31.331	1,050	1,850	2,900
Area to 2400 S.F., rigid fiberglass	205.000	S.F.	14.057	178	810	988
Sheet metal insulated	271.000	Lb.	37.048	1,225	2,175	3,400
Round flexible, insulated 6" diameter, to 1200 S.F.	156.000	L.F.	9.600	505	530	1,035
To 1500 S.F.	184.000	L.F.	11.323	595	625	1,220
8" diameter, to 2000 S.F.	269.000	L.F.	23.911	1,025	1,325	2,350
To 2400 S.F.	248.000	L.F.	22.045	950	1,225	2,175
Return duct, sheet metal galvanized, to 1500 S.F.	158.000	Lb.	16.137	114	930	1,044
To 2400 S.F.	191.000	Lb.	19.507	138	1,125	1,263
Lateral ducts, flexible round 6" insulated, to 1200 S.F.	144.000	L.F.	8.862	465	490	955
To 1500 S.F.	172.000	L.F.	10.585	555	585	1,140
To 2000 S.F.	261.000	L.F.	16.062	845	890	1,735
To 2400 S.F.	300.000	L.F.	18.462	970	1,025	1,995
Spiral steel insulated, to 1200 S.F.	144.000	L.F.	20.067	1,575	1,175	2,750
To 1500 S.F.	172.000	L.F.	23.952	1,875	1,425	3,300
To 2000 S.F.	261.000	L.F.	36.352	2,850	2,150	5,000
To 2400 S.F.	300.000	L.F.	41.825	3,300	2,475	5,775
Rectangular sheet metal galvanized insulated, to 1200 S.F.	228.000	Lb.	39.056	1,875	2,325	4,200
To 1500 S.F.	344.000	Lb.	53.966	2,300	3,175	5,475
To 2000 S.F.	522.000	Lb.	81.926	3,500	4,850	8,350
To 2400 S.F.	600.000	Lb.	94.189	4,025	5,575	9,600
Register elbows, to 1500 S.F.	12.000	Ea.	6.400	252	355	607
To 2400 S.F.	14.000	Ea.	7.470	294	415	709
Floor registers, enameled steel w/damper, to 1500 S.F.	12.000	Ea.	3.000	119	185	304
To 2400 S.F.	14.000	Ea.	4.308	162	265	427
Return air grille, area to 1500 S.F. 12" x 12"	2.000	Ea.	.727	58	45	103
Area to 2400 S.F. 8" x 16"	2.000	Ea.	.444	39.50	27.50	67
Area to 2400 S.F. 8" x 16"	2.000	Ea.	.727	58	45	103
16" x 16"	1.000	Ea.	.364	37.50	22.50	60
Thermostat, manual, 1 set back	1.000	Ea.	1.000	56	65	121
Electric, timed, 1 set back	1.000	Ea.	1.000	41	65	106
2 set back	1.000	Ea.	1.000	205	65	270
Plenum, heating only, 100 M.B.H.	1.000	Ea.	1.000	98.50	55.50	154
120 MBH	1.000	Ea.	1.000	98.50	55.50	154
160 MBH	1.000	Ea.	1.000	98.50	55.50	154
200 MBH	1.000	Ea.	1.000	98.50	55.50	154
Refrigeration piping, 3/8"	25.000	L.F.		33		33
3/4"	25.000	L.F.		73.50		73.50
7/8"	25.000	L.F.		85.50		85.50
Refrigerant piping, 25 ft. (precharged)	1.000	Ea.		300		300
Diffusers, ceiling, 6" diameter, to 1500 S.F.	10.000	Ea.	4.444	99	275	374
To 2400 S.F.	12.000	Ea.	6.000	126	370	496
Floor, aluminum, adjustable, 2-1/4" x 12" to 1500 S.F.	12.000	Ea.	3.000	86.50	185	271.50
To 2400 S.F.	14.000	Ea.	3.500	101	216	317
Side wall, aluminum, adjustable, 8" x 4", to 1500 S.F.	12.000	Ea.	3.000	214	185	399
5" x 10" to 2400 S.F.	12.000	Ea.	3.692	282	227	509

8 | MECHANICAL 64 | Oil Fired Heating/Cooling Systems

System Description	QUAN.	UNIT	LABOR HOURS	COST PER SYSTEM		
				MAT.	INST.	TOTAL
HEATING ONLY, OIL FIRED HOT AIR, ONE ZONE, 1200 S.F. BUILDING						
Furnace, oil fired, atomizing gun type burner	1.000	Ea.	4.571	2,050	254	2,304
3/8" diameter copper supply pipe	1.000	Ea.	2.759	100.80	177	277.80
Shut off valve	1.000	Ea.	.333	15.60	21.50	37.10
Oil tank, 275 gallon, on legs	1.000	Ea.	3.200	540	187	727
Supply duct, rigid fiberglass	176.000	S.F.	12.068	153.12	693.44	846.56
Return duct, sheet metal, galvanized	158.000	Lb.	16.137	113.76	932.20	1,045.96
Lateral ducts, 6" flexible fiberglass	144.000	L.F.	8.862	465.12	491.04	956.16
Register elbows	12.000	Ea.	6.400	252	354	606
Floor register, enameled steel	12.000	Ea.	3.000	118.80	184.80	303.60
Floor grille, return air	2.000	Ea.	.727	58	45	103
Thermostat	1.000	Ea.	1.000	56	65	121
TOTAL		System	59.057	3,923.20	3,404.98	7,328.18
HEATING/COOLING, OIL FIRED, FORCED AIR, ONE ZONE, 1200 S.F. BUILDING						
Furnace, including plenum, compressor, coil	1.000	Ea.	16.000	6,800	890	7,690
3/8" diameter copper supply pipe	1.000	Ea.	2.759	100.80	177	277.80
Shut off valve	1.000	Ea.	.333	15.60	21.50	37.10
Oil tank, 275 gallon on legs	1.000	Ea.	3.200	540	187	727
Supply duct, rigid fiberglass	176.000	S.F.	12.068	153.12	693.44	846.56
Return duct, sheet metal, galvanized	158.000	Lb.	16.137	113.76	932.20	1,045.96
Lateral ducts, 6" flexible fiberglass	144.000	L.F.	8.862	465.12	491.04	956.16
Register elbows	12.000	Ea.	6.400	252	354	606
Floor registers, enameled steel	12.000	Ea.	3.000	118.80	184.80	303.60
Floor grille, return air	2.000	Ea.	.727	58	45	103
Refrigeration piping (precharged)	25.000	L.F.		300		300
TOTAL		System	69.486	8,917.20	3,975.98	12,893.18

Description	QUAN.	UNIT	LABOR HOURS	COST EACH		
				MAT.	INST.	TOTAL

Oil Fired Heating/Cooling	QUAN.	UNIT	LABOR HOURS	COST EACH		
				MAT.	INST.	TOTAL
Furnace, heating, 95.2 MBH, area to 1200 S.F.	1.000	Ea.	4.706	2,075	261	2,336
123.2 MBH, area to 1500 S.F.	1.000	Ea.	5.000	2,400	277	2,677
151.2 MBH, area to 2000 S.F.	1.000	Ea.	5.333	2,475	296	2,771
200 MBH, area to 2400 S.F.	1.000	Ea.	6.154	2,700	340	3,040
Heating/cooling, 95.2 MBH heat, 36 MBH cool, to 1200 S.F.	1.000	Ea.	16.000	6,800	890	7,690
112 MBH heat, 42 MBH cool, to 1500 S.F.	1.000	Ea.	24.000	10,200	1,325	11,525
151 MBH heat, 47 MBH cool, to 2000 S.F.	1.000	Ea.	20.800	8,850	1,150	10,000
184.8 MBH heat, 60 MBH cool, to 2400 S.F.	1.000	Ea.	24.000	4,000	1,375	5,375
Oil piping to furnace, 3/8" dia., copper	1.000	Ea.	3.412	252	216	468
Oil tank, on legs above ground, 275 gallons	1.000	Ea.	3.200	540	187	727
550 gallons	1.000	Ea.	5.926	4,125	345	4,470
Below ground, 275 gallons	1.000	Ea.	3.200	540	187	727
550 gallons	1.000	Ea.	5.926	4,125	345	4,470
1000 gallons	1.000	Ea.	6.400	4,425	375	4,800
Supply duct, rectangular, area to 1200 S.F., rigid fiberglass	176.000	S.F.	12.068	153	695	848
Sheet metal, insulated	228.000	Lb.	31.331	1,050	1,850	2,900
Area to 1500 S.F., rigid fiberglass	176.000	S.F.	12.068	153	695	848
Sheet metal, insulated	228.000	Lb.	31.331	1,050	1,850	2,900
Area to 2400 S.F., rigid fiberglass	205.000	S.F.	14.057	178	810	988
Sheet metal, insulated	271.000	Lb.	37.048	1,225	2,175	3,400
Round flexible, insulated, 6" diameter to 1200 S.F.	156.000	L.F.	9.600	505	530	1,035
To 1500 S.F.	184.000	L.F.	11.323	595	625	1,220
8" diameter to 2000 S.F.	269.000	L.F.	23.911	1,025	1,325	2,350
To 2400 S.F.	269.000	L.F.	22.045	950	1,225	2,175
Return duct, sheet metal galvanized, to 1500 S.F.	158.000	Lb.	16.137	114	930	1,044
To 2400 S.F.	191.000	Lb.	19.507	138	1,125	1,263
Lateral ducts, flexible round, 6", insulated to 1200 S.F.	144.000	L.F.	8.862	465	490	955
To 1500 S.F.	172.000	L.F.	10.585	555	585	1,140
To 2000 S.F.	261.000	L.F.	16.062	845	890	1,735
To 2400 S.F.	300.000	L.F.	18.462	970	1,025	1,995
Spiral steel, insulated to 1200 S.F.	144.000	L.F.	20.067	1,575	1,175	2,750
To 1500 S.F.	172.000	L.F.	23.952	1,875	1,425	3,300
To 2000 S.F.	261.000	L.F.	36.352	2,850	2,150	5,000
To 2400 S.F.	300.000	L.F.	41.825	3,300	2,475	5,775
Rectangular sheet metal galvanized insulated, to 1200 S.F.	288.000	Lb.	45.183	1,925	2,675	4,600
To 1500 S.F.	344.000	Lb.	53.966	2,300	3,175	5,475
To 2000 S.F.	522.000	Lb.	81.926	3,500	4,850	8,350
To 2400 S.F.	600.000	Lb.	94.189	4,025	5,575	9,600
Register elbows, to 1500 S.F.	12.000	Ea.	6.400	252	355	607
To 2400 S.F.	14.000	Ea.	7.470	294	415	709
Floor registers, enameled steel w/damper, to 1500 S.F.	12.000	Ea.	3.000	119	185	304
To 2400 S.F.	14.000	Ea.	4.308	162	265	427
Return air grille, area to 1500 S.F., 12" x 12"	2.000	Ea.	.727	58	45	103
12" x 24"	1.000	Ea.	.444	39.50	27.50	67
Area to 2400 S.F., 8" x 16"	2.000	Ea.	.727	58	45	103
16" x 16"	1.000	Ea.	.364	37.50	22.50	60
Thermostat, manual, 1 set back	1.000	Ea.	1.000	56	65	121
Electric, timed, 1 set back	1.000	Ea.	1.000	41	65	106
2 set back	1.000	Ea.	1.000	205	65	270
Refrigeration piping, 3/8"	25.000	L.F.		33		33
3/4"	25.000	L.F.		73.50		73.50
Diffusers, ceiling, 6" diameter, to 1500 S.F.	10.000	Ea.	4.444	99	275	374
To 2400 S.F.	12.000	Ea.	6.000	126	370	496
Floor, aluminum, adjustable, 2-1/4" x 12" to 1500 S.F.	12.000	Ea.	3.000	86.50	185	271.50
To 2400 S.F.	14.000	Ea.	3.500	101	216	317
Side wall, aluminum, adjustable, 8" x 4", to 1500 S.F.	12.000	Ea.	3.000	214	185	399
5" x 10" to 2400 S.F.	12.000	Ea.	3.692	282	227	509

8 | MECHANICAL — 68 | Hot Water Heating Systems

System Description	QUAN.	UNIT	LABOR HOURS	COST EACH		
				MAT.	INST.	TOTAL
OIL FIRED HOT WATER HEATING SYSTEM, AREA TO 1200 S.F.						
Boiler package, oil fired, 97 MBH, area to 1200 S.F. building	1.000	Ea.	15.000	1,950	845	2,795
3/8" diameter copper supply pipe	1.000	Ea.	2.759	100.80	177	277.80
Shut off valve	1.000	Ea.	.333	15.60	21.50	37.10
Oil tank, 275 gallon, with black iron filler pipe	1.000	Ea.	3.200	540	187	727
Supply piping, 3/4" copper tubing	176.000	L.F.	18.526	1,003.20	1,188	2,191.20
Supply fittings, copper 3/4"	36.000	Ea.	15.158	97.56	972	1,069.56
Supply valves, 3/4"	2.000	Ea.	.800	218	51	269
Baseboard radiation, 3/4"	106.000	L.F.	35.333	731.40	2,072.30	2,803.70
Zone valve	1.000	Ea.	.400	184	26	210
TOTAL		Ea.	91.509	4,840.56	5,539.80	10,380.36
OIL FIRED HOT WATER HEATING SYSTEM, AREA TO 2400 S.F.						
Boiler package, oil fired, 225 MBH, area to 2400 S.F. building	1.000	Ea.	25.105	7,225	1,425	8,650
3/8" diameter copper supply pipe	1.000	Ea.	2.759	100.80	177	277.80
Shut off valve	1.000	Ea.	.333	15.60	21.50	37.10
Oil tank, 550 gallon, with black iron pipe filler pipe	1.000	Ea.	5.926	4,125	345	4,470
Supply piping, 3/4" copper tubing	228.000	L.F.	23.999	1,299.60	1,539	2,838.60
Supply fittings, copper	46.000	Ea.	19.368	124.66	1,242	1,366.66
Supply valves	2.000	Ea.	.800	218	51	269
Baseboard radiation	212.000	L.F.	70.666	1,462.80	4,144.60	5,607.40
Zone valve	1.000	Ea.	.400	184	26	210
TOTAL		Ea.	149.356	14,755.46	8,971.10	23,726.56

The costs in this system are on a cost each basis. The costs represent total cost for the system based on a gross square foot of plan area.

Description	QUAN.	UNIT	LABOR HOURS	COST EACH		
				MAT.	INST.	TOTAL

Hot Water Heating Price Sheet

	QUAN.	UNIT	LABOR HOURS	COST EACH MAT.	INST.	TOTAL
Boiler, oil fired, 97 MBH, area to 1200 S.F.	1.000	Ea.	15.000	2,000	845	2,845
118 MBH, area to 1500 S.F.	1.000	Ea.	16.506	2,075	930	3,005
161 MBH, area to 2000 S.F.	1.000	Ea.	18.405	2,325	1,050	3,375
215 MBH, area to 2400 S.F.	1.000	Ea.	19.704	4,100	2,200	6,300
Oil piping, (valve & filter), 3/8" copper	1.000	Ea.	3.289	116	199	315
1/4" copper	1.000	Ea.	3.242	136	207	343
Oil tank, filler pipe and cap on legs, 275 gallon	1.000	Ea.	3.200	540	187	727
550 gallon	1.000	Ea.	5.926	4,125	345	4,470
Buried underground, 275 gallon	1.000	Ea.	3.200	540	187	727
550 gallon	1.000	Ea.	5.926	4,125	345	4,470
1000 gallon	1.000	Ea.	6.400	4,425	375	4,800
Supply piping copper, area to 1200 S.F., 1/2" tubing	176.000	L.F.	17.384	715	1,100	1,815
3/4" tubing	176.000	L.F.	18.526	1,000	1,200	2,200
Area to 1500 S.F., 1/2" tubing	186.000	L.F.	18.371	755	1,175	1,930
3/4" tubing	186.000	L.F.	19.578	1,050	1,250	2,300
Area to 2000 S.F., 1/2" tubing	204.000	L.F.	20.149	830	1,275	2,105
3/4" tubing	204.000	L.F.	21.473	1,175	1,375	2,550
Area to 2400 S.F., 1/2" tubing	228.000	L.F.	22.520	930	1,425	2,355
3/4" tubing	228.000	L.F.	23.999	1,300	1,550	2,850
Supply pipe fittings copper, area to 1200 S.F., 1/2"	36.000	Ea.	14.400	43.50	920	963.50
3/4"	36.000	Ea.	15.158	97.50	970	1,067.50
Area to 1500 S.F., 1/2"	40.000	Ea.	16.000	48.50	1,025	1,073.50
3/4"	40.000	Ea.	16.842	108	1,075	1,183
Area to 2000 S.F., 1/2"	44.000	Ea.	17.600	53	1,125	1,178
3/4"	44.000	Ea.	18.526	119	1,200	1,319
Area to 2400, S.F., 1/2"	46.000	Ea.	18.400	55.50	1,175	1,230.50
3/4"	46.000	Ea.	19.368	125	1,250	1,375
Supply valves, 1/2" pipe size	2.000	Ea.	.667	210	43	253
3/4"	2.000	Ea.	.800	218	51	269
Baseboard radiation, area to 1200 S.F., 1/2" tubing	106.000	L.F.	28.267	1,300	1,650	2,950
3/4" tubing	106.000	L.F.	35.333	730	2,075	2,805
Area to 1500 S.F., 1/2" tubing	134.000	L.F.	35.734	1,650	2,100	3,750
3/4" tubing	134.000	L.F.	44.666	925	2,625	3,550
Area to 2000 S.F., 1/2" tubing	178.000	L.F.	47.467	2,200	2,775	4,975
3/4" tubing	178.000	L.F.	59.333	1,225	3,475	4,700
Area to 2400 S.F., 1/2" tubing	212.000	L.F.	56.534	2,600	3,300	5,900
3/4" tubing	212.000	L.F.	70.666	1,475	4,150	5,625
Zone valves, 1/2" tubing	1.000	Ea.	.400	184	26	210
3/4" tubing	1.000	Ea.	.400	185	26	211

8 | MECHANICAL — 80 | Rooftop Systems

System Description	QUAN.	UNIT	LABOR HOURS	COST EACH MAT.	COST EACH INST.	COST EACH TOTAL
ROOFTOP HEATING/COOLING UNIT, AREA TO 2000 S.F.						
Rooftop unit, single zone, electric cool, gas heat, to 2000 S.F.	1.000	Ea.	28.521	5,625	1,675	7,300
Gas piping	34.500	L.F.	5.207	191.48	332.93	524.41
Duct, supply and return, galvanized steel	38.000	Lb.	3.881	27.36	224.20	251.56
Insulation, ductwork	33.000	S.F.	1.760	164.67	92.73	257.40
Lateral duct, flexible duct 12" diameter, insulated	72.000	L.F.	11.520	388.80	640.80	1,029.60
Diffusers	4.000	Ea.	4.571	1,184	282	1,466
Return registers	1.000	Ea.	.727	71.50	45	116.50
TOTAL		Ea.	56.187	7,652.81	3,292.66	10,945.47
ROOFTOP HEATING/COOLING UNIT, AREA TO 5000 S.F.						
Rooftop unit, single zone, electric cool, gas heat, to 5000 S.F.	1.000	Ea.	42.032	14,900	2,375	17,275
Gas piping	86.250	L.F.	13.019	478.69	832.31	1,311
Duct supply and return, galvanized steel	95.000	Lb.	9.702	68.40	560.50	628.90
Insulation, ductwork	82.000	S.F.	4.373	409.18	230.42	639.60
Lateral duct, flexible duct, 12" diameter, insulated	180.000	L.F.	28.800	972	1,602	2,574
Diffusers	10.000	Ea.	11.429	2,960	705	3,665
Return registers	3.000	Ea.	2.182	214.50	135	349.50
TOTAL		Ea.	111.537	20,002.77	6,440.23	26,443

Description	QUAN.	UNIT	LABOR HOURS	COST EACH MAT.	COST EACH INST.	COST EACH TOTAL

Rooftop Price Sheet	QUAN.	UNIT	LABOR HOURS	COST EACH		
				MAT.	INST.	TOTAL
Rooftop unit, single zone, electric cool, gas heat to 2000 S.F.	1.000	Ea.	28.521	5,625	1,675	7,300
Area to 3000 S.F.	1.000	Ea.	35.982	10,400	2,025	12,425
Area to 5000 S.F.	1.000	Ea.	42.032	14,900	2,375	17,275
Area to 10000 S.F.	1.000	Ea.	68.376	33,900	4,000	37,900
Gas piping, area 2000 through 4000 S.F.	34.500	L.F.	5.207	191	335	526
Area 5000 to 10000 S.F.	86.250	L.F.	13.019	480	830	1,310
Duct, supply and return, galvanized steel, to 2000 S.F.	38.000	Lb.	3.881	27.50	224	251.50
Area to 3000 S.F.	57.000	Lb.	5.821	41	335	376
Area to 5000 S.F.	95.000	Lb.	9.702	68.50	560	628.50
Area to 10000 S.F.	190.000	Lb.	19.405	137	1,125	1,262
Rigid fiberglass, area to 2000 S.F.	33.000	S.F.	2.263	28.50	130	158.50
Area to 3000 S.F.	49.000	S.F.	3.360	42.50	193	235.50
Area to 5000 S.F.	82.000	S.F.	5.623	71.50	325	396.50
Area to 10000 S.F.	164.000	S.F.	11.245	143	645	788
Insulation, supply and return, blanket type, area to 2000 S.F.	33.000	S.F.	1.508	165	92.50	257.50
Area to 3000 S.F.	49.000	S.F.	2.240	245	138	383
Area to 5000 S.F.	82.000	S.F.	3.748	410	230	640
Area to 10000 S.F.	164.000	S.F.	7.496	820	460	1,280
Lateral ducts, flexible round, 12" insulated, to 2000 S.F.	72.000	L.F.	11.520	390	640	1,030
Area to 3000 S.F.	108.000	L.F.	17.280	585	960	1,545
Area to 5000 S.F.	180.000	L.F.	28.800	970	1,600	2,570
Area to 10000 S.F.	360.000	L.F.	57.600	1,950	3,200	5,150
Rectangular, galvanized steel, to 2000 S.F.	239.000	Lb.	24.409	172	1,400	1,572
Area to 3000 S.F.	360.000	Lb.	36.767	259	2,125	2,384
Area to 5000 S.F.	599.000	Lb.	61.176	430	3,525	3,955
Area to 10000 S.F.	998.000	Lb.	101.926	720	5,900	6,620
Diffusers, ceiling, 1 to 4 way blow, 24" x 24", to 2000 S.F.	4.000	Ea.	4.571	1,175	282	1,457
Area to 3000 S.F.	6.000	Ea.	6.857	1,775	425	2,200
Area to 5000 S.F.	10.000	Ea.	11.429	2,950	705	3,655
Area to 10000 S.F.	20.000	Ea.	22.857	5,925	1,400	7,325
Return grilles, 24" x 24", to 2000 S.F.	1.000	Ea.	.727	71.50	45	116.50
Area to 3000 S.F.	2.000	Ea.	1.455	143	90	233
Area to 5000 S.F.	3.000	Ea.	2.182	215	135	350
Area to 10000 S.F.	5.000	Ea.	3.636	360	225	585

Division 9 - Electrical

9 | ELECTRICAL
10 | Electric Service Systems

System Description	QUAN.	UNIT	LABOR HOURS	COST EACH MAT.	COST EACH INST.	TOTAL
100 AMP SERVICE						
Weather cap	1.000	Ea.	.667	9.65	40	49.65
Service entrance cable	10.000	L.F.	.762	40.90	45.80	86.70
Meter socket	1.000	Ea.	2.500	47.50	150	197.50
Ground rod with clamp	1.000	Ea.	1.455	22	87.50	109.50
Ground cable	5.000	L.F.	.250	7.05	15	22.05
Panel board, 12 circuit	1.000	Ea.	6.667	273	400	673
TOTAL		Ea.	12.301	400.10	738.30	1,138.40
200 AMP SERVICE						
Weather cap	1.000	Ea.	1.000	27	60	87
Service entrance cable	10.000	L.F.	1.143	83	68.50	151.50
Meter socket	1.000	Ea.	4.211	101	253	354
Ground rod with clamp	1.000	Ea.	1.818	42.50	109	151.50
Ground cable	10.000	L.F.	.500	14.10	30	44.10
3/4" EMT	5.000	L.F.	.308	5.20	18.50	23.70
Panel board, 24 circuit	1.000	Ea.	12.308	600	625	1,225
TOTAL		Ea.	21.288	872.80	1,164	2,036.80
400 AMP SERVICE						
Weather cap	1.000	Ea.	2.963	281	178	459
Service entrance cable	180.000	L.F.	5.760	779.40	345.60	1,125
Meter socket	1.000	Ea.	4.211	101	253	354
Ground rod with clamp	1.000	Ea.	2.000	59.50	120	179.50
Ground cable	20.000	L.F.	.485	55	29.20	84.20
3/4" Greenfield	20.000	L.F.	1.000	14.80	60	74.80
Current transformer cabinet	1.000	Ea.	6.154	181	370	551
Panel board, 42 circuit	1.000	Ea.	33.333	3,575	2,000	5,575
TOTAL		Ea.	55.906	5,046.70	3,355.80	8,402.50

9 | ELECTRICAL 20 | Electric Perimeter Heating Systems

System Description	QUAN.	UNIT	LABOR HOURS	COST EACH MAT.	COST EACH INST.	TOTAL
4' BASEBOARD HEATER						
Electric baseboard heater, 4' long	1.000	Ea.	1.194	43	71.50	114.50
Thermostat, integral	1.000	Ea.	.500	33	30	63
Romex, 12-3 with ground	40.000	L.F.	1.600	22.80	96	118.80
Panel board breaker, 20 Amp	1.000	Ea.	.300	11.85	18	29.85
TOTAL		Ea.	3.594	110.65	215.50	326.15
6' BASEBOARD HEATER						
Electric baseboard heater, 6' long	1.000	Ea.	1.600	59.50	96	155.50
Thermostat, integral	1.000	Ea.	.500	33	30	63
Romex, 12-3 with ground	40.000	L.F.	1.600	22.80	96	118.80
Panel board breaker, 20 Amp	1.000	Ea.	.400	15.80	24	39.80
TOTAL		Ea.	4.100	131.10	246	377.10
8' BASEBOARD HEATER						
Electric baseboard heater, 8' long	1.000	Ea.	2.000	73	120	193
Thermostat, integral	1.000	Ea.	.500	33	30	63
Romex, 12-3 with ground	40.000	L.F.	1.600	22.80	96	118.80
Panel board breaker, 20 Amp	1.000	Ea.	.500	19.75	30	49.75
TOTAL		Ea.	4.600	148.55	276	424.55
10' BASEBOARD HEATER						
Electric baseboard heater, 10' long	1.000	Ea.	2.424	209	146	355
Thermostat, integral	1.000	Ea.	.500	33	30	63
Romex, 12-3 with ground	40.000	L.F.	1.600	22.80	96	118.80
Panel board breaker, 20 Amp	1.000	Ea.	.750	29.63	45	74.63
TOTAL		Ea.	5.274	294.43	317	611.43

The costs in this system are on a cost each basis and include all necessary conduit fittings.

Description	QUAN.	UNIT	LABOR HOURS	COST EACH MAT.	COST EACH INST.	TOTAL

9 | ELECTRICAL 30 | Wiring Device Systems

System Description	QUAN.	UNIT	LABOR HOURS	COST EACH MAT.	COST EACH INST.	TOTAL
Air conditioning receptacles						
Using non-metallic sheathed cable	1.000	Ea.	.800	22	48	70
Using BX cable	1.000	Ea.	.964	29	58	87
Using EMT conduit	1.000	Ea.	1.194	43	71.50	114.50
Disposal wiring						
Using non-metallic sheathed cable	1.000	Ea.	.889	19.70	53.50	73.20
Using BX cable	1.000	Ea.	1.067	26	64	90
Using EMT conduit	1.000	Ea.	1.333	42	80	122
Dryer circuit						
Using non-metallic sheathed cable	1.000	Ea.	1.455	39.50	87.50	127
Using BX cable	1.000	Ea.	1.739	42.50	104	146.50
Using EMT conduit	1.000	Ea.	2.162	55	130	185
Duplex receptacles						
Using non-metallic sheathed cable	1.000	Ea.	.615	22	37	59
Using BX cable	1.000	Ea.	.741	29	44.50	73.50
Using EMT conduit	1.000	Ea.	.920	43	55	98
Exhaust fan wiring						
Using non-metallic sheathed cable	1.000	Ea.	.800	24	48	72
Using BX cable	1.000	Ea.	.964	31.50	58	89.50
Using EMT conduit	1.000	Ea.	1.194	45	71.50	116.50
Furnace circuit & switch						
Using non-metallic sheathed cable	1.000	Ea.	1.333	30	80	110
Using BX cable	1.000	Ea.	1.600	40	96	136
Using EMT conduit	1.000	Ea.	2.000	50	120	170
Ground fault						
Using non-metallic sheathed cable	1.000	Ea.	1.000	61	60	121
Using BX cable	1.000	Ea.	1.212	68	73	141
Using EMT conduit	1.000	Ea.	1.481	96	89	185
Heater circuits						
Using non-metallic sheathed cable	1.000	Ea.	1.000	24	60	84
Using BX cable	1.000	Ea.	1.212	28.50	73	101.50
Using EMT conduit	1.000	Ea.	1.481	41.50	89	130.50
Lighting wiring						
Using non-metallic sheathed cable	1.000	Ea.	.500	24.50	30	54.50
Using BX cable	1.000	Ea.	.602	29	36	65
Using EMT conduit	1.000	Ea.	.748	40	45	85
Range circuits						
Using non-metallic sheathed cable	1.000	Ea.	2.000	85	120	205
Using BX cable	1.000	Ea.	2.424	112	146	258
Using EMT conduit	1.000	Ea.	2.963	98.50	178	276.50
Switches, single pole						
Using non-metallic sheathed cable	1.000	Ea.	.500	24	30	54
Using BX cable	1.000	Ea.	.602	31.50	36	67.50
Using EMT conduit	1.000	Ea.	.748	45	45	90
Switches, 3-way						
Using non-metallic sheathed cable	1.000	Ea.	.667	27	40	67
Using BX cable	1.000	Ea.	.800	32	48	80
Using EMT conduit	1.000	Ea.	1.333	53	80	133
Water heater						
Using non-metallic sheathed cable	1.000	Ea.	1.600	32	96	128
Using BX cable	1.000	Ea.	1.905	43	114	157
Using EMT conduit	1.000	Ea.	2.353	48.50	141	189.50
Weatherproof receptacle						
Using non-metallic sheathed cable	1.000	Ea.	1.333	154	80	234
Using BX cable	1.000	Ea.	1.600	159	96	255
Using EMT conduit	1.000	Ea.	2.000	173	120	293

9 | ELECTRICAL 40 | Light Fixture Systems

System Description	QUAN.	UNIT	LABOR HOURS	COST EACH MAT.	COST EACH INST.	COST EACH TOTAL
Fluorescent strip, 4' long, 1 light, average	1.000	Ea.	.941	30	56.50	86.50
Deluxe	1.000	Ea.	1.129	36	68	104
2 lights, average	1.000	Ea.	1.000	42	60	102
Deluxe	1.000	Ea.	1.200	50.50	72	122.50
8' long, 1 light, average	1.000	Ea.	1.194	53	71.50	124.50
Deluxe	1.000	Ea.	1.433	63.50	86	149.50
2 lights, average	1.000	Ea.	1.290	63.50	77.50	141
Deluxe	1.000	Ea.	1.548	76	93	169
Surface mounted, 4' x 1', economy	1.000	Ea.	.914	56.50	55	111.50
Average	1.000	Ea.	1.143	70.50	68.50	139
Deluxe	1.000	Ea.	1.371	84.50	82	166.50
4' x 2', economy	1.000	Ea.	1.208	71	72.50	143.50
Average	1.000	Ea.	1.509	89	90.50	179.50
Deluxe	1.000	Ea.	1.811	107	109	216
Recessed, 4' x 1', 2 lamps, economy	1.000	Ea.	1.123	41.50	67.50	109
Average	1.000	Ea.	1.404	52	84.50	136.50
Deluxe	1.000	Ea.	1.684	62.50	101	163.50
4' x 2', 4' lamps, economy	1.000	Ea.	1.362	51	81.50	132.50
Average	1.000	Ea.	1.702	63.50	102	165.50
Deluxe	1.000	Ea.	2.043	76	122	198
Incandescent, exterior, 150W, single spot	1.000	Ea.	.500	35	30	65
Double spot	1.000	Ea.	1.167	97	70	167
Recessed, 100W, economy	1.000	Ea.	.800	63	48	111
Average	1.000	Ea.	1.000	78.50	60	138.50
Deluxe	1.000	Ea.	1.200	94	72	166
150W, economy	1.000	Ea.	.800	91	48	139
Average	1.000	Ea.	1.000	114	60	174
Deluxe	1.000	Ea.	1.200	137	72	209
Surface mounted, 60W, economy	1.000	Ea.	.800	49	48	97
Average	1.000	Ea.	1.000	55	60	115
Deluxe	1.000	Ea.	1.194	77.50	71.50	149
Metal halide, recessed 2' x 2' 250W	1.000	Ea.	2.500	350	150	500
2' x 2', 400W	1.000	Ea.	2.759	400	166	566
Surface mounted, 2' x 2', 250W	1.000	Ea.	2.963	380	178	558
2' x 2', 400W	1.000	Ea.	3.333	445	200	645
High bay, single, unit, 400W	1.000	Ea.	3.478	450	209	659
Twin unit, 400W	1.000	Ea.	5.000	905	300	1,205
Low bay, 250W	1.000	Ea.	2.500	400	150	550

Unit Price Section

Table of Contents

Sect. No.		Page
	General Requirements	
01 11	Summary of Work	292
01 21	Allowances	292
01 31	Project Management and Coordination	293
01 41	Regulatory Requirements	293
01 54	Construction Aids	293
01 56	Temporary Barriers and Enclosures	296
01 71	Examination and Preparation	297
01 74	Cleaning and Waste Management	297
01 76	Protecting Installed Construction	297
	Existing Conditions	
02 21	Surveys	300
02 32	Geotechnical Investigations	300
02 41	Demolition	300
02 83	Lead Remediation	304
	Concrete	
03 01	Maintenance of Concrete	308
03 05	Common Work Results for Concrete	308
03 11	Concrete Forming	308
03 15	Concrete Accessories	310
03 21	Reinforcement Bars	312
03 22	Fabric and Grid Reinforcing	312
03 23	Stressed Tendon Reinforcing	313
03 24	Fibrous Reinforcing	313
03 30	Cast-In-Place Concrete	314
03 31	Structural Concrete	315
03 35	Concrete Finishing	317
03 39	Concrete Curing	318
03 41	Precast Structural Concrete	318
03 48	Precast Concrete Specialties	318
03 54	Cast Underlayment	319
03 63	Epoxy Grouting	319
03 82	Concrete Boring	319
	Masonry	
04 01	Maintenance of Masonry	322
04 05	Common Work Results for Masonry	322
04 21	Clay Unit Masonry	325
04 22	Concrete Unit Masonry	327
04 23	Glass Unit Masonry	329
04 24	Adobe Unit Masonry	329
04 27	Multiple-Wythe Unit Masonry	330
04 41	Dry-Placed Stone	330
04 43	Stone Masonry	330
04 51	Flue Liner Masonry	332
04 57	Masonry Fireplaces	332
04 72	Cast Stone Masonry	333
	Metals	
05 05	Common Work Results for Metals	336
05 12	Structural Steel Framing	339
05 31	Steel Decking	340
05 41	Structural Metal Stud Framing	341
05 42	Cold-Formed Metal Joist Framing	344
05 44	Cold-Formed Metal Trusses	350
05 51	Metal Stairs	351
05 52	Metal Railings	351
05 58	Formed Metal Fabrications	352
05 71	Decorative Metal Stairs	353
	Wood, Plastics & Composites	
06 05	Common Work Results for Wood, Plastics, and Composites	356
06 11	Wood Framing	370
06 12	Structural Panels	381
06 13	Heavy Timber Construction	382
06 15	Wood Decking	384
06 16	Sheathing	384
06 17	Shop-Fabricated Structural Wood	387
06 18	Glued-Laminated Construction	388
06 22	Millwork	391
06 25	Prefinished Paneling	402
06 26	Board Paneling	404
06 43	Wood Stairs and Railings	404
06 44	Ornamental Woodwork	406
06 48	Wood Frames	408
06 49	Wood Screens and Exterior Wood Shutters	409
06 51	Structural Plastic Shapes and Plates	410
06 63	Plastic Railings	411
06 65	Plastic Trim	411
06 80	Composite Fabrications	412
06 81	Composite Railings	412
	Thermal & Moisture Protection	
07 01	Operation and Maint. of Thermal and Moisture Protection	416
07 05	Common Work Results for Thermal and Moisture Protection	416
07 11	Dampproofing	417
07 19	Water Repellents	418
07 21	Thermal Insulation	418
07 22	Roof and Deck Insulation	421
07 24	Exterior Insulation and Finish Systems	423
07 25	Weather Barriers	423
07 26	Vapor Retarders	423
07 27	Air Barriers	424
07 31	Shingles and Shakes	424
07 32	Roof Tiles	425
07 33	Natural Roof Coverings	426
07 41	Roof Panels	427
07 42	Wall Panels	427
07 46	Siding	428
07 51	Built-Up Bituminous Roofing	431
07 52	Modified Bituminous Membrane Roofing	432
07 57	Coated Foamed Roofing	432
07 58	Roll Roofing	433
07 61	Sheet Metal Roofing	433
07 62	Sheet Metal Flashing and Trim	434
07 65	Flexible Flashing	434
07 71	Roof Specialties	435
07 72	Roof Accessories	439
07 76	Roof Pavers	440
07 91	Preformed Joint Seals	440
07 92	Joint Sealants	441
	Openings	
08 01	Operation and Maintenance of Openings	444
08 05	Common Work Results for Openings	444
08 11	Metal Doors and Frames	446
08 12	Metal Frames	446
08 13	Metal Doors	447
08 14	Wood Doors	448
08 16	Composite Doors	454
08 17	Integrated Door Opening Assemblies	455
08 31	Access Doors and Panels	456
08 32	Sliding Glass Doors	456
08 36	Panel Doors	457
08 51	Metal Windows	458
08 52	Wood Windows	459
08 53	Plastic Windows	468
08 54	Composite Windows	470
08 61	Roof Windows	471
08 62	Unit Skylights	471
08 71	Door Hardware	472
08 75	Window Hardware	477
08 79	Hardware Accessories	477
08 81	Glass Glazing	478
08 83	Mirrors	479
08 87	Glazing Surface Films	479
08 91	Louvers	480
08 95	Vents	480
	Finishes	
09 01	Maintenance of Finishes	482
09 05	Common Work Results for Finishes	482
09 22	Supports for Plaster and Gypsum Board	483
09 23	Gypsum Plastering	486
09 24	Cement Plastering	487
09 25	Other Plastering	487
09 26	Veneer Plastering	487
09 28	Backing Boards and Underlayments	488
09 29	Gypsum Board	488
09 30	Tiling	492
09 34	Waterproofing-Membrane Tiling	494
09 51	Acoustical Ceilings	494
09 53	Acoustical Ceiling Suspension Assemblies	495
09 61	Flooring Treatment	496
09 62	Specialty Flooring	496
09 63	Masonry Flooring	497
09 64	Wood Flooring	497
09 65	Resilient Flooring	498
09 66	Terrazzo Flooring	500
09 68	Carpeting	500
09 72	Wall Coverings	502
09 91	Painting	503
09 93	Staining and Transparent Finishing	516
09 96	High-Performance Coatings	517
	Specialties	
10 28	Toilet, Bath, and Laundry Accessories	520
10 31	Manufactured Fireplaces	521
10 32	Fireplace Specialties	522
10 35	Stoves	523
10 44	Fire Protection Specialties	523
10 55	Postal Specialties	523
10 56	Storage Assemblies	523
10 57	Wardrobe and Closet Specialties	524
10 73	Protective Covers	524
10 74	Manufactured Exterior Specialties	525
10 75	Flagpoles	526
	Equipment	
11 30	Residential Equipment	528
11 32	Unit Kitchens	530
11 41	Foodservice Storage Equipment	530
11 81	Facility Maintenance Equipment	531
	Furnishings	
12 21	Window Blinds	534
12 22	Curtains and Drapes	534
12 23	Interior Shutters	535
12 24	Window Shades	535
12 32	Manufactured Wood Casework	535
12 34	Manufactured Plastic Casework	540
12 36	Countertops	540
	Special Construction	
13 11	Swimming Pools	544
13 12	Fountains	544
13 17	Tubs and Pools	545
13 24	Special Activity Rooms	545
13 34	Fabricated Engineered Structures	546
	Conveying Equipment	
14 21	Electric Traction Elevators	548
14 42	Wheelchair Lifts	548
	Fire Suppression	
21 05	Common Work Results for Fire Suppression	550
21 11	Facility Fire-Suppression Water-Service Piping	550
21 12	Fire-Suppression Standpipes	553
21 13	Fire-Suppression Sprinkler Systems	554

Table of Contents (cont.)

Sect. No.		Page
	Plumbing	
22 05	Common Work Results for Plumbing	558
22 07	Plumbing Insulation	560
22 11	Facility Water Distribution	561
22 13	Facility Sanitary Sewerage	569
22 14	Facility Storm Drainage	572
22 31	Domestic Water Softeners	572
22 33	Electric Domestic Water Heaters	572
22 34	Fuel-Fired Domestic Water Heaters	572
22 41	Residential Plumbing Fixtures	573
22 42	Commercial Plumbing Fixtures	576
22 51	Swimming Pool Plumbing Systems	577
	Heating Ventilation Air Conditioning	
23 05	Common Work Results for HVAC	580
23 07	HVAC Insulation	580
23 09	Instrumentation and Control for HVAC	581
23 13	Facility Fuel-Storage Tanks	581
23 21	Hydronic Piping and Pumps	582
23 23	Refrigerant Piping	582
23 31	HVAC Ducts and Casings	583
23 33	Air Duct Accessories	584
23 34	HVAC Fans	584
23 37	Air Outlets and Inlets	585
23 41	Particulate Air Filtration	586
23 42	Gas-Phase Air Filtration	587
23 43	Electronic Air Cleaners	587
23 51	Breechings, Chimneys, and Stacks	587
23 52	Heating Boilers	588
23 54	Furnaces	589
23 62	Packaged Compressor and Condenser Units	591
23 74	Packaged Outdoor HVAC Equipment	592
23 81	Decentralized Unitary HVAC Equipment	592

Sect. No.		Page
23 82	Convection Heating and Cooling Units	593
23 83	Radiant Heating Units	594
	Electrical	
26 05	Common Work Results for Electrical	598
26 24	Switchboards and Panelboards	607
26 27	Low-Voltage Distribution Equipment	608
26 28	Low-Voltage Circuit Protective Devices	610
26 32	Packaged Generator Assemblies	610
26 33	Battery Equipment	611
26 36	Transfer Switches	611
26 41	Facility Lightning Protection	611
26 51	Interior Lighting	611
26 55	Special Purpose Lighting	613
26 56	Exterior Lighting	613
26 61	Lighting Systems and Accessories	614
	Communications	
27 41	Audio-Video Systems	616
	Electronic Safety & Security	
28 16	Intrusion Detection	618
28 31	Fire Detection and Alarm	618
	Earthwork	
31 05	Common Work Results for Earthwork	620
31 11	Clearing and Grubbing	620
31 13	Selective Tree and Shrub Removal and Trimming	620
31 14	Earth Stripping and Stockpiling	620
31 22	Grading	621
31 23	Excavation and Fill	621
31 25	Erosion and Sedimentation Controls	633
31 31	Soil Treatment	633

Sect. No.		Page
	Exterior Improvements	
32 01	Operation and Maintenance of Exterior Improvements	636
32 06	Schedules for Exterior Improvements	636
32 11	Base Courses	636
32 12	Flexible Paving	637
32 13	Rigid Paving	638
32 14	Unit Paving	638
32 16	Curbs, Gutters, Sidewalks, and Driveways	639
32 31	Fences and Gates	640
32 32	Retaining Walls	642
32 33	Site Furnishings	644
32 84	Planting Irrigation	644
32 91	Planting Preparation	644
32 92	Turf and Grasses	645
32 93	Plants	646
32 94	Planting Accessories	648
32 96	Transplanting	648
	Utilities	
33 05	Common Work Results for Utilities	650
33 11	Water Utility Distribution Piping	650
33 12	Water Utility Distribution Equipment	651
33 21	Water Supply Wells	651
33 31	Sanitary Utility Sewerage Piping	652
33 36	Utility Septic Tanks	652
33 41	Storm Utility Drainage Piping	653
33 44	Storm Utility Water Drains	654
33 46	Subdrainage	654
33 49	Storm Drainage Structures	654
33 51	Natural-Gas Distribution	654
	Electrical Power Generation	
48 15	Wind Energy Electrical Power Generation Equipment	658

How RSMeans Unit Price Data Works

All RSMeans unit price data is organized in the same way.

It is important to understand the structure, so that you can find information easily and use it correctly.

RSMeans **Line Numbers** consist of 12 characters, which identify a unique location in the database for each task. The first 6 or 8 digits conform to the Construction Specifications Institute MasterFormat® 2014. The remainder of the digits are a further breakdown by RSMeans in order to arrange items in understandable groups of similar tasks. Line numbers are consistent across all RSMeans publications, so a line number in any RSMeans product will always refer to the same unit of work.

RSMeans engineers have created **reference** information to assist you in your estimate. If there is information that applies to a section, it will be indicated at the start of the section. The Reference Section is located in the back of the book on the pages with a black edge.

RSMeans **Descriptions** are shown in a hierarchical structure to make them readable. In order to read a complete description, read up through the indents to the top of the section. Include everything that is above and to the left that is not contradicted by information below. For instance, the complete description for line 03 30 53.40 3550 is "Concrete in place, including forms (4 uses), Grade 60 rebar, concrete (Portland cement Type 1), placement and finishing unless otherwise indicated; Equipment pad (3000 psi), 4' x 4' x 6" thick."

03 30 Cast-In-Place Concrete
03 30 53 – Miscellaneous Cast-In-Place Concrete

03 30 53.40 Concrete In Place

0010	**CONCRETE IN PLACE**
0020	Including forms (4 uses), Grade 60 rebar, concrete (Portland cement
0050	Type I), placement and finishing unless otherwise indicated
0500	Chimney foundations (5000 psi), over 5 C.Y.
0510	(3500 psi), under 5 C.Y.
3540	Equipment pad (3000 psi), 3' x 3' x 6" thick
3550	4' x 4' x 6" thick
3560	5' x 5' x 8" thick
3570	6' x 6' x 8" thick
3580	8' x 8' x 10" thick
3590	10' x 10' x 12" thick
3800	Footings (3000 psi), spread under 1 C.Y.
3825	1 C.Y. to 5 C.Y.
3850	Over 5 C.Y.
3900	Footings, strip (3000 psi), 18" x 9", unreinforced

When using **RSMeans data**, it is important to read through an entire section to ensure that you use the data that most closely matches your work. Note that sometimes there is additional information shown in the section that may improve your price. There are frequently lines that further describe, add to, or adjust data for specific situations.

The data published in RSMeans print books represents a "national average" cost. This data should be modified to the project location using the **Location Factors** tables found in the Reference Section (see pages 707–712).

Use the location factors to adjust estimate totals.

Crews include labor or labor and equipment necessary to accomplish each task. In this case, Crew C-14C is used. RSMeans selects a crew to represent the workers and equipment that are typically used for that task. In this case, Crew C-14C consists of one carpenter foreman (outside), two carpenters, one rodman, one laborer, one cement finisher, and one gas engine vibrator. Details of all crews can be found in the reference section.

Crews

Crew No.	Bare Costs		Incl. Subs O&P		Cost Per Labor-Hour	
Crew C-14C	Hr.	Daily	Hr.	Daily	Bare Costs	Incl. O&P
1 Carpenter Foreman (outside)	$34.85	$278.80	$58.50	$468.00	$30.83	$51.75
6 Carpenters	32.85	1576.80	55.15	2647.20		
2 Rodmen (reinf.)	35.20	563.20	59.90	958.40		
4 Laborers	24.45	782.40	41.05	1313.60		
1 Cement Finisher	31.50	252.00	51.05	408.40		
1 Gas Engine Vibrator		31.20		34.32	0.28	0.31
112 L.H., Daily Totals		$3484.40		$5829.92	$31.11	$52.05

The **Daily Output** is the amount of work that the crew can do in a normal 8-hour workday, including mobilization, layout, movement of materials, and cleanup. In this case, crew C-14C can install thirty 4' x 4' x 6" thick concrete pads in a day. Daily output is variable, based on many factors, including the size of the job, location, and environmental conditions. RSMeans data represents work done in daylight (or adequate lighting) and temperate conditions.

Bare Costs are the costs of materials, labor, and equipment that the installing contractor pays. They represent the cost, in U.S. dollars, for one unit of work. They do not include any markups for profit or labor burden.

Crew	Daily Output	Labor-Hours	Unit	Material	2015 Bare Costs Labor	Equipment	Total	Total Incl O&P
C-14C	32.22	3.476	C.Y.	150	107	.97	257.97	345
"	23.71	4.724	"	177	146	1.32	324.32	440
C-14H	45	1.067	Ea.	47	34	.69	81.69	109
	30	1.600		69.50	51	1.04	121.54	163
	18	2.667		122	85	1.73	208.73	279
	14	3.429		164	110	2.23	276.23	365
	8	6		350	192	3.90	545.90	710
	5	9.600		595	305	6.25	906.25	1,175
C-14C	28	4	C.Y.	166	123	1.12	290.12	390
	43	2.605		201	80.50	.73	282.23	355
	75	1.493		185	46	.42	231.42	282
C-14L	40	2.400		125	72	.79	197.79	260

The **Total Incl O&P** column is the total cost, including overhead and profit, that the installing contractor will charge the customer. This represents the cost of materials plus 10% profit, the cost of labor plus labor burden and 10% profit, and the cost of equipment plus 10% profit. It does not include the general contractor's overhead and profit. Note: See the inside back cover for details of how RSMeans calculates labor burden.

The **Total column** represents the total bare cost for the installing contractor, in U.S. dollars. In this case, the sum of $69.50 for material + $51.00 for labor + $1.04 for equipment is $121.54.

The figure in the **Labor Hours** column is the amount of labor required to perform one unit of work—in this case the amount of labor required to construct one 4' x 4' equipment pad. This figure is calculated by dividing the number of hours of labor in the crew by the daily output (48 labor hours divided by 30 pads = 1.6 hours of labor per pad). Multiply 1.600 times 60 to see the value in minutes: 60 x 1.6 = 96 minutes. Note: the labor hour figure is not dependent on the crew size. A change in crew size will result in a corresponding change in daily output, but the labor hours per unit of work will not change.

All RSMeans unit cost data includes the typical **Unit of Measure** used for estimating that item. For concrete-in-place the typical unit is cubic yards (C.Y.) or each (Ea.). For installing broadloom carpet it is square yard, and for gypsum board it is square foot. The estimator needs to take special care that the unit in the data matches the unit in the take-off. Unit conversions may be found in the Reference Section.

How RSMeans Unit Price Data Works (Continued)

Sample Estimate

This sample demonstrates the elements of an estimate, including a tally of the RSMeans data lines, and a summary of the markups on a contractor's work to arrive at a total cost to the owner. The RSMeans Location Factor is added at the bottom of the estimate to adjust the cost of the work to a specific location.

Work Performed: The body of the estimate shows the RSMeans data selected, including line number, a brief description of each item, its take-off unit and quantity, and the bare costs of materials, labor, and equipment. This estimate also includes a column titled "SubContract." This data is taken from the RSMeans column "Total Incl O&P," and represents the total that a subcontractor would charge a general contractor for the work, including the sub's markup for overhead and profit.

Division 1, General Requirements: This is the first division numerically, but the last division estimated. Division 1 includes project-wide needs provided by the general contractor. These requirements vary by project, but may include temporary facilities and utilities, security, testing, project cleanup, etc. For small projects a percentage can be used, typically between 5% and 15% of project cost. For large projects the costs may be itemized and priced individually.

Bonds: Bond costs should be added to the estimate. The figures here represent a typical performance bond, ensuring the owner that if the general contractor does not complete the obligations in the construction contract the bonding company will pay the cost for completion of the work.

Location Adjustment: RSMeans published data is based on national average costs. If necessary, adjust the total cost of the project using a location factor from the "Location Factor" table. Use location factors if the work is general, covering multiple trades.

This estimate is based on an interactive spreadsheet. A copy of this spreadsheet is located on the RSMeans website at http://www.reedconstructiondata.com/rsmeans/extras/546011. You are free to download it and adjust it to your methodology.

Project Name: Pre-Engineered Steel Building			Architect: As Shown	
Location:	Anywhere, USA			
Line Number	**Description**	**Qty**	**Unit**	**Materi**
03 30 53.40 3940	Strip footing, 12" x 24", reinforced	34	C.Y.	$4,726.0
03 30 53.40 3950	Strip footing, 12" x 36", reinforced	15	C.Y.	$1,995.0
03 11 13.65 3000	Concrete slab edge forms	500	L.F.	$175.0
03 22 11.10 0200	Welded wire fabric reinforcing	150	C.S.F.	$2,580.0
03 31 13.35 0300	Ready mix concrete, 4000 psi for slab on grade	278	C.Y.	$29,746.0
03 31 13.70 4300	Place, strike off & consolidate concrete slab	278	C.Y.	$0.0
03 35 13.30 0250	Machine float & trowel concrete slab	15,000	S.F.	$0.0
03 15 16.20 0140	Cut control joints in concrete slab	950	L.F.	$57.0
03 39 23.13 0300	Sprayed concrete curing membrane	150	C.S.F.	$1,657.5
Division 03	**Subtotal**			**$40,936.5**
08 36 13.10 2650	Manual 10' x 10' steel sectional overhead door	8	Ea.	$8,800.0
08 36 13.10 2860	Insulation and steel back panel for OH door	800	S.F.	$3,800.0
Division 08	**Subtotal**			**$12,600.0**
13 34 19.50 1100	Pre-Engineered Steel Building, 100' x 150' x 24'	15,000	SF Flr.	$0.0
13 34 19.50 6050	Framing for PESB door opening, 3' x 7'	4	Opng.	$0.0
13 34 19.50 6100	Framing for PESB door opening, 10' x 10'	8	Opng.	$0.0
13 34 19.50 6200	Framing for PESB window opening, 4' x 3'	6	Opng.	$0.0
13 34 19.50 5750	PESB door, 3' x 7', single leaf	4	Opng.	$2,340.0
13 34 19.50 7750	PESB sliding window, 4' x 3' with screen	6	Opng.	$2,280.0
13 34 19.50 6550	PESB gutter, eave type, 26 ga., painted	300	L.F.	$2,085.0
13 34 19.50 8650	PESB roof vent, 12" wide x 10' long	15	Ea.	$540.0
13 34 19.50 6900	PESB insulation, vinyl faced, 4" thick	27,400	S.F.	$11,508.0
Division 13	**Subtotal**			**$18,753.0**
	Subtotal			$72,289.5
Division 01	General Requirements @ 7%			5,060.2
	Estimate Subtotal			$77,349.7
	Sales Tax @ 5%			3,867.4
	Subtotal A			81,217.2
	GC O & P			8,121.7
	Subtotal B			89,338.9
	Contingency @ 5%			
	Subtotal C			
	Bond @ $12/1000 +10% O&P			
	Subtotal D			
	Location Adjustment Factor			
	Grand Total			

288

This example shows the cost to construct a pre-engineered steel building. The foundation, doors, windows, and insulation will be installed by the general contractor. A subcontractor will install the structural steel, roofing, and siding.

Labor	Equipment	01/01/15 SubContract	RESI Estimate Total	
$2,448.00	$22.10	$0.00		
$862.50	$7.80	$0.00		
$765.00	$0.00	$0.00		
$2,722.50	$0.00	$0.00		
$0.00	$0.00	$0.00		
$3,155.30	$158.46	$0.00		
$6,150.00	$450.00	$0.00		
$266.00	$85.50	$0.00		
$618.00	$0.00	$0.00		
$16,987.30	$723.86	$0.00	$58,647.66	Division 03
$2,336.00	$0.00	$0.00		
$0.00	$0.00	$0.00		
$2,336.00	$0.00	$0.00	$14,936.00	Division 08
$0.00	$0.00	$295,500.00		
$0.00	$0.00	$1,880.00		
$0.00	$0.00	$8,200.00		
$0.00	$0.00	$2,850.00		
$452.00	$0.00	$0.00		
$261.00	$67.20	$0.00		
$531.00	$0.00	$0.00		
$2,115.00	$0.00	$0.00		
$6,302.00	$0.00	$0.00		
$9,661.00	$67.20	$308,430.00	$336,911.20	Division 13
$28,984.30	$791.06	$308,430.00	$410,494.86	Subtotal
2,028.90	55.37	21,590.10		Gen. Requirements
$31,013.20	$846.43	$330,020.10	$410,494.86	Estimate Subtotal
	42.32	8,250.50		Sales tax
31,013.20	888.76	338,270.60		Subtotal
21,244.04	88.88	33,827.06		GC O&P
52,257.24	977.63	372,097.66	$514,671.52	Subtotal
			25,733.58	Contingency
			$540,405.09	Subtotal
			7,133.35	Bond
			$547,538.44	Subtotal
102.30			12,593.38	Location Adjustment
			$560,131.82	Grand Total

Sales Tax: If the work is subject to state or local sales taxes, the amount must be added to the estimate. Sales tax may be added to material costs, equipment costs, and subcontracted work. In this case, sales tax was added in all three categories. It was assumed that approximately half the subcontracted work would be material cost, so the tax was applied to 50% of the subcontract total.

GC O&P: This entry represents the general contractor's markup on material, labor, equipment, and subcontractor costs. RSMeans' standard markup on materials, equipment, and subcontracted work is 10%. In this estimate, the markup on the labor performed by the GC's workers uses "Skilled Workers Average" shown in Column F on the table "Installing Contractor's Overhead & Profit," which can be found on the inside-back cover of the book.

Contingency: A factor for contingency may be added to any estimate to represent the cost of unknowns that may occur between the time that the estimate is performed and the time the project is constructed. The amount of the allowance will depend on the stage of design at which the estimate is done, and the contractor's assessment of the risk involved. Refer to section 01 21 16.50 for contigency allowances.

Estimating Tips
01 20 00 Price and Payment Procedures
- Allowances that should be added to estimates to cover contingencies and job conditions that are not included in the national average material and labor costs are shown in section 01 21.
- When estimating historic preservation projects (depending on the condition of the existing structure and the owner's requirements), a 15%–20% contingency or allowance is recommended, regardless of the stage of the drawings.

01 30 00 Administrative Requirements
- Before determining a final cost estimate, it is a good practice to review all the items listed in Subdivisions 01 31 and 01 32 to make final adjustments for items that may need customizing to specific job conditions.
- Requirements for initial and periodic submittals can represent a significant cost to the General Requirements of a job. Thoroughly check the submittal specifications when estimating a project to determine any costs that should be included.

01 40 00 Quality Requirements
- All projects will require some degree of Quality Control. This cost is not included in the unit cost of construction listed in each division. Depending upon the terms of the contract, the various costs of inspection and testing can be the responsibility of either the owner or the contractor. Be sure to include the required costs in your estimate.

01 50 00 Temporary Facilities and Controls
- Barricades, access roads, safety nets, scaffolding, security, and many more requirements for the execution of a safe project are elements of direct cost. These costs can easily be overlooked when preparing an estimate. When looking through the major classifications of this subdivision, determine which items apply to each division in your estimate.
- Construction Equipment Rental Costs can be found in the Reference Section in section 01 54 33. Operators' wages are not included in equipment rental costs.
- Equipment mobilization and demobilization costs are not included in equipment rental costs and must be considered separately in section 01 54 36.50.

- The cost of small tools provided by the installing contractor for his workers is covered in the "Overhead" column on the "Installing Contractor's Overhead and Profit" table that lists labor trades, base rates and markups and, therefore, is included in the "Total Incl. O&P" cost of any Unit Price line item.

01 70 00 Execution and Closeout Requirements
- When preparing an estimate, thoroughly read the specifications to determine the requirements for Contract Closeout. Final cleaning, record documentation, operation and maintenance data, warranties and bonds, and spare parts and maintenance materials can all be elements of cost for the completion of a contract. Do not overlook these in your estimate.

Reference Numbers
Reference numbers are shown in shaded boxes at the beginning of some major classifications. These numbers refer to related items in the Reference Section. The reference information may be an estimating procedure, an alternate pricing method, or technical information.

Note: Not all subdivisions listed here necessarily appear in this publication. ■

No part of this publication may be reproduced, stored in a retrieval system, or transmitted in any form or by any means without prior written permission of RSMeans.

01 11 Summary of Work

01 11 31 – Professional Consultants

01 11 31.10 Architectural Fees

		Crew	Daily Output	Labor-Hours	Unit	Material	2015 Bare Costs Labor	Equipment	Total	Total Incl O&P
0010	**ARCHITECTURAL FEES** R011110-10									
0020	For new construction				Project					
0060	Minimum								4.90%	4.90%
0090	Maximum								16%	16%
0100	For alteration work, to $500,000, add to new construction fee								50%	50%
0150	Over $500,000, add to new construction fee								25%	25%

01 11 31.20 Construction Management Fees

0010	**CONSTRUCTION MANAGEMENT FEES**									
0060	For work to $100,000				Project				10%	10%
0070	To $250,000								9%	9%
0090	To $1,000,000								6%	6%

01 11 31.75 Renderings

0010	**RENDERINGS** Color, matted, 20" x 30", eye level,									
0050	Average				Ea.	2,875			2,875	3,175

01 21 Allowances

01 21 16 – Contingency Allowances

01 21 16.50 Contingencies

0010	**CONTINGENCIES**, Add to estimate									
0020	Conceptual stage				Project				20%	20%
0150	Final working drawing stage				"				3%	3%

01 21 55 – Job Conditions Allowance

01 21 55.50 Job Conditions

		Crew	Daily Output	Labor-Hours	Unit	Material	Labor	Equipment	Total	Total Incl O&P
0010	**JOB CONDITIONS** Modifications to applicable									
8000	Remove and reset contents of small room	1 Clab	6.50	1.231	Room		30		30	50.50
8010	Average room		4.70	1.702			41.50		41.50	70
8020	Large room		3.50	2.286			56		56	94
8030	Extra large room		2.40	3.333			81.50		81.50	137

01 21 63 – Taxes

01 21 63.10 Taxes

0010	**TAXES** R012909-80									
0020	Sales tax, State, average				%	5.04%				
0050	Maximum R012909-85					7.50%				
0200	Social Security, on first $117,000 of wages						7.65%			
0300	Unemployment, combined Federal and State, minimum						.60%			
0350	Average						7.80%			
0400	Maximum						12.87%			

01 31 Project Management and Coordination

01 31 13 − Project Coordination

01 31 13.30 Insurance		Crew	Daily Output	Labor-Hours	Unit	Material	2015 Bare Costs Labor	Equipment	Total	Total Incl O&P
0010	**INSURANCE** R013113-40									
0020	Builders risk, standard, minimum				Job				.24%	.24%
0050	Maximum R013113-60								.64%	.64%
0200	All-risk type, minimum								.25%	.25%
0250	Maximum								.62%	.62%
0400	Contractor's equipment floater, minimum				Value				.50%	.50%
0450	Maximum				"				1.50%	1.50%
0600	Public liability, average				Job				2.02%	2.02%
0800	Workers' compensation & employer's liability, average									
0850	by trade, carpentry, general				Payroll		14.93%			
0900	Clerical						.50%			
0950	Concrete						12.93%			
1000	Electrical						5.76%			
1050	Excavation						9.70%			
1100	Glazing						13.29%			
1150	Insulation						11.66%			
1200	Lathing						8.38%			
1250	Masonry						13.68%			
1300	Painting & decorating						11.72%			
1350	Pile driving						14.31%			
1400	Plastering						10.90%			
1450	Plumbing						6.98%			
1500	Roofing						31.69%			
1550	Sheet metal work (HVAC)						8.82%			
1600	Steel erection, structural						31.68%			
1650	Tile work, interior ceramic						8.86%			
1700	Waterproofing, brush or hand caulking						6.56%			
1800	Wrecking						26.14%			
2000	Range of 35 trades in 50 states, excl. wrecking, min.						1.31%			
2100	Average						13.50%			
2200	Maximum						132%			

01 41 Regulatory Requirements

01 41 26 − Permit Requirements

01 41 26.50 Permits

		Crew	Daily Output	Labor-Hours	Unit	Material	2015 Bare Costs Labor	Equipment	Total	Total Incl O&P
0010	**PERMITS**									
0020	Rule of thumb, most cities, minimum				Job				.50%	.50%
0100	Maximum				"				2%	2%

01 54 Construction Aids

01 54 16 − Temporary Hoists

01 54 16.50 Weekly Forklift Crew

		Crew	Daily Output	Labor-Hours	Unit	Material	2015 Bare Costs Labor	Equipment	Total	Total Incl O&P
0010	**WEEKLY FORKLIFT CREW**									
0100	All-terrain forklift, 45' lift, 35' reach, 9000 lb. capacity	A-3P	.20	40	Week		1,350	2,625	3,975	5,125

01 54 Construction Aids

01 54 19 – Temporary Cranes

01 54 19.50 Daily Crane Crews

		Crew	Daily Output	Labor-Hours	Unit	Material	2015 Bare Costs Labor	2015 Bare Costs Equipment	Total	Total Incl O&P
0010	**DAILY CRANE CREWS** for small jobs, portal to portal									
0100	12-ton truck-mounted hydraulic crane	A-3H	1	8	Day		285	860	1,145	1,425
0900	If crane is needed on a Saturday, Sunday or Holiday									
0910	At time-and-a-half, add				Day		50%			
0920	At double time, add				"		100%			

01 54 23 – Temporary Scaffolding and Platforms

01 54 23.60 Pump Staging

			Crew	Daily Output	Labor-Hours	Unit	Material	Labor	Equipment	Total	Total Incl O&P
0010	**PUMP STAGING**, Aluminum										
1300	System in place, 50' working height, per use based on 50 uses		2 Carp	84.80	.189	C.S.F.	6.95	6.20		13.15	18
1400	100 uses	R015423-20		84.80	.189		3.46	6.20		9.66	14.20
1500	150 uses			84.80	.189		2.32	6.20		8.52	12.95

01 54 23.70 Scaffolding

			Crew	Daily Output	Labor-Hours	Unit	Material	Labor	Equipment	Total	Total Incl O&P
0010	**SCAFFOLDING**	R015423-10									
0015	Steel tube, regular, no plank, labor only to erect & dismantle										
0091	Building exterior, wall face, 1 to 5 stories, 6'-4" x 5' frames		3 Clab	8	3	C.S.F.		73.50		73.50	123
0201	6 to 12 stories		4 Clab	8	4			98		98	164
0310	13 to 20 stories		5 Carp	8	5			164		164	276
0461	Building interior, walls face area, up to 16' high		3 Clab	12	2			49		49	82
0561	16' to 40' high			10	2.400			58.50		58.50	98.50
0801	Building interior floor area, up to 30' high			150	.160	C.C.F.		3.91		3.91	6.55
0901	Over 30' high		4 Clab	160	.200	"		4.89		4.89	8.20
0906	Complete system for face of walls, no plank, material only rent/mo					C.S.F.	35.50			35.50	39.50
0908	Interior spaces, no plank, material only rent/mo					C.C.F.	4.33			4.33	4.76
0910	Steel tubular, heavy duty shoring, buy										
0920	Frames 5' high 2' wide					Ea.	81.50			81.50	90
0925	5' high 4' wide						93			93	102
0930	6' high 2' wide						93.50			93.50	103
0935	6' high 4' wide						109			109	120
0940	Accessories										
0945	Cross braces					Ea.	15.50			15.50	17.05
0950	U-head, 8" x 8"						19.10			19.10	21
0955	J-head, 4" x 8"						13.90			13.90	15.30
0960	Base plate, 8" x 8"						15.50			15.50	17.05
0965	Leveling jack						33.50			33.50	36.50
1000	Steel tubular, regular, buy										
1100	Frames 3' high 5' wide					Ea.	75			75	82.50
1150	5' high 5' wide						86.50			86.50	95
1200	6'-4" high 5' wide						118			118	129
1350	7'-6" high 6' wide						158			158	173
1500	Accessories cross braces						15.50			15.50	17.05
1550	Guardrail post						16.40			16.40	18.05
1600	Guardrail 7' section						8.70			8.70	9.60
1650	Screw jacks & plates						21.50			21.50	24
1700	Sidearm brackets						30.50			30.50	33.50
1750	8" casters						31			31	34
1800	Plank 2" x 10" x 16'-0"						57.50			57.50	63.50
1900	Stairway section						286			286	315
1910	Stairway starter bar						32			32	35
1920	Stairway inside handrail						58.50			58.50	64
1930	Stairway outside handrail						86.50			86.50	95
1940	Walk-thru frame guardrail						41.50			41.50	46

01 54 Construction Aids

01 54 23 – Temporary Scaffolding and Platforms

01 54 23.70 Scaffolding

		Crew	Daily Output	Labor-Hours	Unit	Material	2015 Bare Costs Labor	Equipment	Total	Total Incl O&P
2000	Steel tubular, regular, rent/mo.									
2100	Frames 3' high 5' wide				Ea.	5			5	5.50
2150	5' high 5' wide					5			5	5.50
2200	6'-4" high 5' wide					5.50			5.50	6.05
2250	7'-6" high 6' wide					10			10	11
2500	Accessories, cross braces					1			1	1.10
2550	Guardrail post					1			1	1.10
2600	Guardrail 7' section					1			1	1.10
2650	Screw jacks & plates					2			2	2.20
2700	Sidearm brackets					2			2	2.20
2750	8" casters					8			8	8.80
2800	Outrigger for rolling tower					3			3	3.30
2850	Plank 2" x 10" x 16'-0"					10			10	11
2900	Stairway section					35			35	38.50
2940	Walk-thru frame guardrail					2.50			2.50	2.75
3000	Steel tubular, heavy duty shoring, rent/mo.									
3250	5' high 2' & 4' wide				Ea.	8.50			8.50	9.35
3300	6' high 2' & 4' wide					8.50			8.50	9.35
3500	Accessories, cross braces					1			1	1.10
3600	U - head, 8" x 8"					2.50			2.50	2.75
3650	J - head, 4" x 8"					2.50			2.50	2.75
3700	Base plate, 8" x 8"					1			1	1.10
3750	Leveling jack					2.50			2.50	2.75
5700	Planks, 2" x 10" x 16'-0", labor only to erect & remove to 50' H	3 Carp	72	.333			10.95		10.95	18.40
5800	Over 50' high	4 Carp	80	.400			13.15		13.15	22

01 54 23.80 Staging Aids

		Crew	Daily Output	Labor-Hours	Unit	Material	Labor	Equipment	Total	Total Incl O&P
0010	**STAGING AIDS** and fall protection equipment									
0100	Sidewall staging bracket, tubular, buy				Ea.	55			55	60.50
0110	Cost each per day, based on 250 days use				Day	.22			.22	.24
0200	Guard post, buy				Ea.	50.50			50.50	55.50
0210	Cost each per day, based on 250 days use				Day	.20			.20	.22
0300	End guard chains, buy per pair				Pair	36			36	39.50
0310	Cost per set per day, based on 250 days use				Day	.22			.22	.24
1010	Cost each per day, based on 250 days use				"	.04			.04	.04
1100	Wood bracket, buy				Ea.	16.80			16.80	18.50
1110	Cost each per day, based on 250 days use				Day	.07			.07	.07
2010	Cost per pair per day, based on 250 days use				"	.47			.47	.52
2100	Steel siderail jack, buy per pair				Pair	97.50			97.50	107
2110	Cost per pair per day, based on 250 days use				Day	.39			.39	.43
3010	Cost each per day, based on 250 days use				"	.23			.23	.25
3100	Aluminum scaffolding plank, 20" wide x 24' long, buy				Ea.	755			755	830
3110	Cost each per day, based on 250 days use				Day	3.01			3.01	3.31
4010	Cost each per day, based on 250 days use				"	.69			.69	.76
4100	Rope for safety line, 5/8" x 100' nylon, buy				Ea.	56			56	61.50
4110	Cost each per day, based on 250 days use				Day	.22			.22	.25
4200	Permanent U-Bolt roof anchor, buy				Ea.	35.50			35.50	39.50
4300	Temporary (one use) roof ridge anchor, buy				"	32			32	35.50
5000	Installation (setup and removal) of staging aids									
5010	Sidewall staging bracket	2 Carp	64	.250	Ea.		8.20		8.20	13.80
5020	Guard post with 2 wood rails	"	64	.250			8.20		8.20	13.80
5030	End guard chains, set	1 Carp	64	.125			4.11		4.11	6.90
5100	Roof shingling bracket		96	.083			2.74		2.74	4.60

01 54 Construction Aids

01 54 23 – Temporary Scaffolding and Platforms

01 54 23.80 Staging Aids

		Crew	Daily Output	Labor-Hours	Unit	Material	2015 Bare Costs Labor	Equipment	Total	Total Incl O&P
5200	Ladder jack	1 Carp	64	.125	Ea.		4.11		4.11	6.90
5300	Wood plank, 2" x 10" x 16'	2 Carp	80	.200			6.55		6.55	11.05
5310	Aluminum scaffold plank, 20" x 24'	"	40	.400			13.15		13.15	22
5410	Safety rope	1 Carp	40	.200			6.55		6.55	11.05
5420	Permanent U-Bolt roof anchor (install only)	2 Carp	40	.400			13.15		13.15	22
5430	Temporary roof ridge anchor (install only)	1 Carp	64	.125			4.11		4.11	6.90

01 54 36 – Equipment Mobilization

01 54 36.50 Mobilization

0010	**MOBILIZATION** (Use line item again for demobilization)	R015436-50								
0015	Up to 25 mi. haul dist. (50 mi. RT for mob/demob crew)									
1200	Small equipment, placed in rear of, or towed by pickup truck	A-3A	4	2	Ea.		67	39	106	154
1300	Equipment hauled on 3-ton capacity towed trailer	A-3Q	2.67	3			101	67	168	241
1400	20-ton capacity	B-34U	2	8			248	237	485	670
1500	40-ton capacity	B-34N	2	8			253	375	628	835
1700	Crane, truck-mounted, up to 75 ton (driver only)	1 Eqhv	4	2			71.50		71.50	118
2500	For each additional 5 miles haul distance, add						10%	10%		
3000	For large pieces of equipment, allow for assembly/knockdown									
3100	For mob/demob of micro-tunneling equip, see Section 33 05 23.19									

01 56 Temporary Barriers and Enclosures

01 56 13 – Temporary Air Barriers

01 56 13.60 Tarpaulins

0010	**TARPAULINS**									
0020	Cotton duck, 10 oz. to 13.13 oz. per S.Y., 6'x8'				S.F.	.83			.83	.91
0050	30'x30'					.59			.59	.65
0200	Reinforced polyethylene 3 mils thick, white					.04			.04	.04
0300	4 mils thick, white, clear or black					.09			.09	.10
0730	Polyester reinforced w/integral fastening system 11 mils thick					.21			.21	.23

01 56 16 – Temporary Dust Barriers

01 56 16.10 Dust Barriers, Temporary

0010	**DUST BARRIERS, TEMPORARY**									
0020	Spring loaded telescoping pole & head, to 12', erect and dismantle	1 Clab	240	.033	Ea.		.81		.81	1.37
0025	Cost per day (based upon 250 days)				Day	.26			.26	.29
0030	To 21', erect and dismantle	1 Clab	240	.033	Ea.		.81		.81	1.37
0035	Cost per day (based upon 250 days)				Day	.44			.44	.48
0040	Accessories, caution tape reel, erect and dismantle	1 Clab	480	.017	Ea.		.41		.41	.68
0045	Cost per day (based upon 250 days)				Day	.36			.36	.40
0060	Foam rail and connector, erect and dismantle	1 Clab	240	.033	Ea.		.81		.81	1.37
0065	Cost per day (based upon 250 days)				Day	.10			.10	.11
0070	Caution tape	1 Clab	384	.021	C.L.F.	2.70	.51		3.21	3.83
0080	Zipper, standard duty		60	.133	Ea.	8	3.26		11.26	14.25
0090	Heavy duty		48	.167	"	9.50	4.08		13.58	17.30
0100	Polyethylene sheet, 4 mil		37	.216	Sq.	2.90	5.30		8.20	12.10
0110	6 mil		37	.216	"	4.02	5.30		9.32	13.30
1000	Dust partition, 6 mil polyethylene, 1" x 3" frame	2 Carp	2000	.008	S.F.	.30	.26		.56	.76
1080	2" x 4" frame	"	2000	.008	"	.35	.26		.61	.83

01 71 Examination and Preparation

01 71 23 – Field Engineering

01 71 23.13 Construction Layout		Crew	Daily Output	Labor-Hours	Unit	Material	2015 Bare Costs Labor	Equipment	Total	Total Incl O&P
0010	**CONSTRUCTION LAYOUT**									
1100	Crew for layout of building, trenching or pipe laying, 2 person crew	A-6	1	16	Day		525	55	580	935
1200	3 person crew	A-7	1	24	"		845	54.50	899.50	1,450

01 74 Cleaning and Waste Management

01 74 13 – Progress Cleaning

01 74 13.20 Cleaning Up

		Crew	Daily Output	Labor-Hours	Unit	Material	Labor	Equipment	Total	Total Incl O&P
0010	**CLEANING UP**									
0020	After job completion, allow, minimum				Job				.30%	.30%
0040	Maximum				"				1%	1%

01 76 Protecting Installed Construction

01 76 13 – Temporary Protection of Installed Construction

01 76 13.20 Temporary Protection

		Crew	Daily Output	Labor-Hours	Unit	Material	Labor	Equipment	Total	Total Incl O&P
0010	**TEMPORARY PROTECTION**									
0020	Flooring, 1/8" tempered hardboard, taped seams	2 Carp	1500	.011	S.F.	.41	.35		.76	1.04
0030	Peel away carpet protection	1 Clab	3200	.003	"	.11	.06		.17	.22

Division Notes

	CREW	DAILY OUTPUT	LABOR-HOURS	UNIT	BARE COSTS				TOTAL INCL O&P
					MAT.	LABOR	EQUIP.	TOTAL	

Estimating Tips

02 30 00 Subsurface Investigation

In preparing estimates on structures involving earthwork or foundations, all information concerning soil characteristics should be obtained. Look particularly for hazardous waste, evidence of prior dumping of debris, and previous stream beds.

02 40 00 Demolition and Structure Moving

The costs shown for selective demolition do not include rubbish handling or disposal. These items should be estimated separately using RSMeans data or other sources.

- Historic preservation often requires that the contractor remove materials from the existing structure, rehab them, and replace them. The estimator must be aware of any related measures and precautions that must be taken when doing selective demolition and cutting and patching. Requirements may include special handling and storage, as well as security.
- In addition to Subdivision 02 41 00, you can find selective demolition items in each division. Example: Roofing demolition is in Division 7.

02 40 00 Building Deconstruction

This section provides costs for the careful dismantling and recycling of most of low-rise building materials.

02 50 00 Containment of Hazardous Waste

This section addresses on-site hazardous waste disposal costs.

02 80 00 Hazardous Material Disposal/ Remediation

This subdivision includes information on hazardous waste handling, asbestos remediation, lead remediation, and mold remediation. See reference R028213-20 and R028319-60 for further guidance in using these unit price lines.

02 90 00 Monitoring Chemical Sampling, Testing Analysis

This section provides costs for on-site sampling and testing hazardous waste.

Reference Numbers

Reference numbers are shown in shaded boxes at the beginning of some major classifications. These numbers refer to related items in the Reference Section. The reference information may be an estimating procedure, an alternate pricing method, or technical information.

Note: Not all subdivisions listed here necessarily appear in this publication. ■

Did you know?
RSMeans Online gives you the same access to RSMeans' data with 24/7 access:
- Quickly locate costs in the searchable database.
- Build cost lists, estimates, and reports in minutes.
- Adjust costs to any location in the U.S. and Canada with the click of a button.

Start your free trial today at **www.rsmeansonline.com**

RSMeansOnline

No part of this publication may be reproduced, stored in a retrieval system, or transmitted in any form or by any means without prior written permission of RSMeans.

02 21 Surveys

02 21 13 – Site Surveys

02 21 13.09 Topographical Surveys		Crew	Daily Output	Labor-Hours	Unit	Material	2015 Bare Costs Labor	2015 Bare Costs Equipment	Total	Total Incl O&P
0010	**TOPOGRAPHICAL SURVEYS**									
0020	Topographical surveying, conventional, minimum	A-7	3.30	7.273	Acre	20	256	16.60	292.60	470
0100	Maximum	A-8	.60	53.333	"	60	1,825	91	1,976	3,250

02 21 13.13 Boundary and Survey Markers										
0010	**BOUNDARY AND SURVEY MARKERS**									
0300	Lot location and lines, large quantities, minimum	A-7	2	12	Acre	35	420	27.50	482.50	775
0320	Average	"	1.25	19.200		55	675	44	774	1,225
0400	Small quantities, maximum	A-8	1	32		75	1,100	54.50	1,229.50	2,000
0600	Monuments, 3' long	A-7	10	2.400	Ea.	40	84.50	5.45	129.95	191
0800	Property lines, perimeter, cleared land	"	1000	.024	L.F.	.05	.84	.05	.94	1.53
0900	Wooded land	A-8	875	.037	"	.07	1.26	.06	1.39	2.25

02 32 Geotechnical Investigations

02 32 13 – Subsurface Drilling and Sampling

02 32 13.10 Boring and Exploratory Drilling		Crew	Daily Output	Labor-Hours	Unit	Material	Labor	Equipment	Total	Total Incl O&P
0010	**BORING AND EXPLORATORY DRILLING**									
0020	Borings, initial field stake out & determination of elevations	A-6	1	16	Day		525	55	580	935
0100	Drawings showing boring details				Total		335		335	425
0200	Report and recommendations from P.E.						775		775	970
0300	Mobilization and demobilization	B-55	4	4			104	271	375	470
0350	For over 100 miles, per added mile		450	.036	Mile		.93	2.41	3.34	4.20
0600	Auger holes in earth, no samples, 2-1/2" diameter		78.60	.204	L.F.		5.30	13.75	19.05	24
0800	Cased borings in earth, with samples, 2-1/2" diameter		55.50	.288	"	14	7.50	19.50	41	49.50
1400	Borings, earth, drill rig and crew with truck mounted auger		1	16	Day		420	1,075	1,495	1,900
1500	For inner city borings add, minimum								10%	10%
1510	Maximum								20%	20%

02 41 Demolition

02 41 13 – Selective Site Demolition

02 41 13.17 Demolish, Remove Pavement and Curb		Crew	Daily Output	Labor-Hours	Unit	Material	Labor	Equipment	Total	Total Incl O&P
0010	**DEMOLISH, REMOVE PAVEMENT AND CURB** R024119-10									
5010	Pavement removal, bituminous roads, up to 3" thick	B-38	690	.035	S.Y.		.96	.79	1.75	2.47
5050	4" to 6" thick		420	.057			1.57	1.30	2.87	4.05
5100	Bituminous driveways		640	.038			1.03	.85	1.88	2.66
5200	Concrete to 6" thick, hydraulic hammer, mesh reinforced		255	.094			2.59	2.14	4.73	6.65
5300	Rod reinforced		200	.120			3.30	2.72	6.02	8.50
5600	With hand held air equipment, bituminous, to 6" thick	B-39	1900	.025	S.F.		.63	.12	.75	1.18
5700	Concrete to 6" thick, no reinforcing		1600	.030			.74	.15	.89	1.41
5800	Mesh reinforced		1400	.034			.85	.17	1.02	1.61
5900	Rod reinforced		765	.063			1.55	.30	1.85	2.94
6000	Curbs, concrete, plain	B-6	360	.067	L.F.		1.83	1.01	2.84	4.17
6100	Reinforced		275	.087			2.40	1.32	3.72	5.45
6200	Granite		360	.067			1.83	1.01	2.84	4.17
6300	Bituminous		528	.045			1.25	.69	1.94	2.85

02 41 13.23 Utility Line Removal										
0010	**UTILITY LINE REMOVAL**									
0015	No hauling, abandon catch basin or manhole	B-6	7	3.429	Ea.		94	52	146	214
0020	Remove existing catch basin or manhole, masonry		4	6			165	91	256	375
0030	Catch basin or manhole frames and covers, stored		13	1.846			50.50	28	78.50	116

02 41 Demolition

02 41 13 – Selective Site Demolition

02 41 13.23 Utility Line Removal

		Crew	Daily Output	Labor-Hours	Unit	Material	2015 Bare Costs Labor	2015 Bare Costs Equipment	Total	Total Incl O&P
0040	Remove and reset	B-6	7	3.429	Ea.		94	52	146	214
2900	Pipe removal, sewer/water, no excavation, 12" diameter	↓	175	.137	L.F.		3.77	2.08	5.85	8.60
2930	15"-18" diameter	B-12Z	150	.160			4.51	10.70	15.21	19.30
2960	21"-24" diameter	"	120	.200	↓		5.65	13.40	19.05	24

02 41 13.30 Minor Site Demolition

		Crew	Daily Output	Labor-Hours	Unit	Material	Labor	Equipment	Total	Total Incl O&P
0010	**MINOR SITE DEMOLITION** R024119-10									
1000	Masonry walls, block, solid	B-5	1800	.022	C.F.		.60	.79	1.39	1.87
1200	Brick, solid		900	.044			1.20	1.58	2.78	3.74
1400	Stone, with mortar		900	.044			1.20	1.58	2.78	3.74
1500	Dry set	↓	1500	.027	↓		.72	.95	1.67	2.24
4000	Sidewalk removal, bituminous, 2" thick	B-6	350	.069	S.Y.		1.88	1.04	2.92	4.29
4010	2-1/2" thick		325	.074			2.03	1.12	3.15	4.62
4050	Brick, set in mortar		185	.130			3.57	1.97	5.54	8.10
4100	Concrete, plain, 4"		160	.150			4.12	2.28	6.40	9.40
4110	Plain, 5"		140	.171			4.71	2.60	7.31	10.70
4120	Plain, 6"		120	.200			5.50	3.04	8.54	12.55
4200	Mesh reinforced, concrete, 4"		150	.160			4.40	2.43	6.83	10
4210	5" thick		131	.183			5.05	2.78	7.83	11.45
4220	6" thick	↓	112	.214	↓		5.90	3.25	9.15	13.45
4300	Slab on grade removal, plain	B-5	45	.889	C.Y.		24	31.50	55.50	74.50
4310	Mesh reinforced		33	1.212			32.50	43	75.50	102
4320	Rod reinforced	↓	25	1.600			43	57	100	135
4400	For congested sites or small quantities, add up to								200%	200%
4450	For disposal on site, add	B-11A	232	.069			2.05	6	8.05	10
4500	To 5 miles, add	B-34D	76	.105	↓		2.99	9.75	12.74	15.65

02 41 13.60 Selective Demolition Fencing

		Crew	Daily Output	Labor-Hours	Unit	Material	Labor	Equipment	Total	Total Incl O&P
0010	**SELECTIVE DEMOLITION FENCING** R024119-10									
1600	Fencing, barbed wire, 3 strand	2 Clab	430	.037	L.F.		.91		.91	1.53
1650	5 strand	"	280	.057			1.40		1.40	2.35
1700	Chain link, posts & fabric, 8' to 10' high, remove only	B-6	445	.054			1.48	.82	2.30	3.38
1750	Remove and reset	"	70	.343	↓		9.40	5.20	14.60	21.50

02 41 16 – Structure Demolition

02 41 16.13 Building Demolition

		Crew	Daily Output	Labor-Hours	Unit	Material	Labor	Equipment	Total	Total Incl O&P
0010	**BUILDING DEMOLITION** Large urban projects, incl. 20 mi. haul R024119-10									
0500	Small bldgs, or single bldgs, no salvage included, steel	B-3	14800	.003	C.F.		.09	.17	.26	.34
0600	Concrete	"	11300	.004	"		.12	.23	.35	.45
0605	Concrete, plain	B-5	33	1.212	C.Y.		32.50	43	75.50	102
0610	Reinforced		25	1.600			43	57	100	135
0615	Concrete walls		34	1.176			31.50	42	73.50	99
0620	Elevated slabs	↓	26	1.538	↓		41.50	54.50	96	130
0650	Masonry	B-3	14800	.003	C.F.		.09	.17	.26	.34
0700	Wood		14800	.003	"		.09	.17	.26	.34
1000	Demoliton single family house, one story, wood 1600 S.F.		1	48	Ea.		1,325	2,575	3,900	5,050
1020	3200 S.F.		.50	96			2,675	5,150	7,825	10,100
1200	Demoliton two family house, two story, wood 2400 S.F.		.67	71.964			2,000	3,850	5,850	7,600
1220	4200 S.F.		.38	128			3,575	6,850	10,425	13,500
1300	Demoliton three family house, three story, wood 3200 S.F.		.50	96			2,675	5,150	7,825	10,100
1320	5400 S.F.	↓	.30	160	↓		4,450	8,575	13,025	16,900

02 41 Demolition

02 41 16 – Structure Demolition

02 41 16.17 Building Demolition Footings and Foundations

		Crew	Daily Output	Labor-Hours	Unit	Material	2015 Bare Costs Labor	2015 Bare Costs Equipment	Total	Total Incl O&P
0010	**BUILDING DEMOLITION FOOTINGS AND FOUNDATIONS** R024119-10									
0200	Floors, concrete slab on grade,									
0240	4" thick, plain concrete	B-13L	5000	.003	S.F.		.11	.41	.52	.64
0280	Reinforced, wire mesh		4000	.004			.14	.51	.65	.81
0300	Rods		4500	.004			.13	.46	.59	.71
0400	6" thick, plain concrete		4000	.004			.14	.51	.65	.81
0420	Reinforced, wire mesh		3200	.005			.18	.64	.82	1.01
0440	Rods		3600	.004			.16	.57	.73	.89
1000	Footings, concrete, 1' thick, 2' wide	B-5	300	.133	L.F.		3.59	4.74	8.33	11.20
1080	1'-6" thick, 2' wide		250	.160			4.31	5.70	10.01	13.45
1120	3' wide		200	.200			5.40	7.10	12.50	16.80
1200	Average reinforcing, add								10%	10%
2000	Walls, block, 4" thick	B-13L	8000	.002	S.F.		.07	.26	.33	.40
2040	6" thick		6000	.003			.10	.34	.44	.54
2080	8" thick		4000	.004			.14	.51	.65	.81
2100	12" thick		3000	.005			.19	.68	.87	1.06
2400	Concrete, plain concrete, 6" thick		4000	.004			.14	.51	.65	.81
2420	8" thick		3500	.005			.16	.59	.75	.92
2440	10" thick		3000	.005			.19	.68	.87	1.06
2500	12" thick		2500	.006			.23	.82	1.05	1.28
2600	For average reinforcing, add								10%	10%
4000	For congested sites or small quantities, add up to								200%	200%
4200	Add for disposal, on site	B-11A	232	.069	C.Y.		2.05	6	8.05	10
4250	To five miles	B-30	220	.109	"		3.34	10.95	14.29	17.55

02 41 19 – Selective Demolition

02 41 19.13 Selective Building Demolition

0010	**SELECTIVE BUILDING DEMOLITION**									
0020	Costs related to selective demolition of specific building components									
0025	are included under Common Work Results (XX 05)									
0030	in the component's appropriate division.									

02 41 19.16 Selective Demolition, Cutout

		Crew	Daily Output	Labor-Hours	Unit	Material	Labor	Equipment	Total	Total Incl O&P
0010	**SELECTIVE DEMOLITION, CUTOUT** R024119-10									
0020	Concrete, elev. slab, light reinforcement, under 6 C.F.	B-9	65	.615	C.F.		15.30	3.58	18.88	29.50
0050	Light reinforcing, over 6 C.F.		75	.533	"		13.25	3.10	16.35	26
0200	Slab on grade to 6" thick, not reinforced, under 8 S.F.		85	.471	S.F.		11.70	2.74	14.44	22.50
0250	8 – 16 S.F.		175	.229	"		5.70	1.33	7.03	11
0255	For over 16 S.F. see Line 02 41 16.17 0400									
0600	Walls, not reinforced, under 6 C.F.	B-9	60	.667	C.F.		16.55	3.88	20.43	32.50
0650	6 – 12 C.F.	"	80	.500	"		12.45	2.91	15.36	24
0655	For over 12 C.F. see Line 02 41 16.17 2500									
1000	Concrete, elevated slab, bar reinforced, under 6 C.F.	B-9	45	.889	C.F.		22	5.15	27.15	42.50
1050	Bar reinforced, over 6 C.F.		50	.800	"		19.90	4.66	24.56	38.50
1200	Slab on grade to 6" thick, bar reinforced, under 8 S.F.		75	.533	S.F.		13.25	3.10	16.35	26
1250	8 – 16 S.F.		150	.267	"		6.65	1.55	8.20	12.85
1255	For over 16 S.F. see Line 02 41 16.17 0440									
1400	Walls, bar reinforced, under 6 C.F.	B-9	50	.800	C.F.		19.90	4.66	24.56	38.50
1450	6 – 12 C.F.	"	70	.571	"		14.20	3.33	17.53	27.50
1455	For over 12 C.F. see Lines 02 41 16.17 2500 and 2600									
2000	Brick, to 4 S.F. opening, not including toothing									
2040	4" thick	B-9	30	1.333	Ea.		33	7.75	40.75	64
2060	8" thick		18	2.222			55	12.95	67.95	107

02 41 Demolition

02 41 19 – Selective Demolition

02 41 19.16 Selective Demolition, Cutout

		Crew	Daily Output	Labor-Hours	Unit	Material	2015 Bare Costs Labor	2015 Bare Costs Equipment	Total	Total Incl O&P
2080	12" thick	B-9	10	4	Ea.		99.50	23.50	123	193
2400	Concrete block, to 4 S.F. opening, 2" thick		35	1.143			28.50	6.65	35.15	55
2420	4" thick		30	1.333			33	7.75	40.75	64
2440	8" thick		27	1.481			37	8.60	45.60	71.50
2460	12" thick		24	1.667			41.50	9.70	51.20	80
2600	Gypsum block, to 4 S.F. opening, 2" thick		80	.500			12.45	2.91	15.36	24
2620	4" thick		70	.571			14.20	3.33	17.53	27.50
2640	8" thick		55	.727			18.05	4.23	22.28	35
2800	Terra cotta, to 4 S.F. opening, 4" thick		70	.571			14.20	3.33	17.53	27.50
2840	8" thick		65	.615			15.30	3.58	18.88	29.50
2880	12" thick		50	.800			19.90	4.66	24.56	38.50
3000	Toothing masonry cutouts, brick, soft old mortar	1 Brhe	40	.200	V.L.F.		5.35		5.35	8.90
3100	Hard mortar		30	.267			7.10		7.10	11.85
3200	Block, soft old mortar		70	.114			3.05		3.05	5.10
3400	Hard mortar		50	.160			4.26		4.26	7.10
6000	Walls, interior, not including re-framing,									
6010	openings to 5 S.F.									
6100	Drywall to 5/8" thick	1 Clab	24	.333	Ea.		8.15		8.15	13.70
6200	Paneling to 3/4" thick		20	.400			9.80		9.80	16.40
6300	Plaster, on gypsum lath		20	.400			9.80		9.80	16.40
6340	On wire lath		14	.571			13.95		13.95	23.50
7000	Wood frame, not including re-framing, openings to 5 S.F.									
7200	Floors, sheathing and flooring to 2" thick	1 Clab	5	1.600	Ea.		39		39	65.50
7310	Roofs, sheathing to 1" thick, not including roofing		6	1.333			32.50		32.50	54.50
7410	Walls, sheathing to 1" thick, not including siding		7	1.143			28		28	47

02 41 19.19 Selective Demolition

		Crew	Daily Output	Labor-Hours	Unit	Material	Labor	Equipment	Total	Total Incl O&P
0010	**SELECTIVE DEMOLITION,** Rubbish Handling R024119-10									
0020	The following are to be added to the demolition prices									
0600	Dumpster, weekly rental, 1 dump/week, 6 C.Y. capacity (2 Tons)				Week	415			415	455
0700	10 C.Y. capacity (3 Tons)					480			480	530
0725	20 C.Y. capacity (5 Tons) R024119-20					565			565	625
0800	30 C.Y. capacity (7 Tons)					730			730	800
0840	40 C.Y. capacity (10 Tons)					775			775	850
2000	Load, haul, dump and return, 0 – 50' haul, hand carried	2 Clab	24	.667	C.Y.		16.30		16.30	27.50
2005	Wheeled		37	.432			10.55		10.55	17.75
2040	0 – 100' haul, hand carried		16.50	.970			23.50		23.50	40
2045	Wheeled		25	.640			15.65		15.65	26.50
2080	Haul and return, add per each extra 100' haul, hand carried		35.50	.451			11		11	18.50
2085	Wheeled		54	.296			7.25		7.25	12.15
2120	For travel in elevators, up to 10 floors, add		140	.114			2.79		2.79	4.69
2130	0 – 50' haul, incl. up to 5 riser stair, hand carried		23	.696			17		17	28.50
2135	Wheeled		35	.457			11.20		11.20	18.75
2140	6 – 10 riser stairs, hand carried		22	.727			17.80		17.80	30
2145	Wheeled		34	.471			11.50		11.50	19.30
2150	11 – 20 riser stairs, hand carried		20	.800			19.55		19.55	33
2155	Wheeled		31	.516			12.60		12.60	21
2160	21 – 40 riser stairs, hand carried		16	1			24.50		24.50	41
2165	Wheeled		24	.667			16.30		16.30	27.50
2170	0 – 100' haul, incl. 5 riser stair, hand carried		15	1.067			26		26	44
2175	Wheeled		23	.696			17		17	28.50
2180	6 – 10 riser stair, hand carried		14	1.143			28		28	47
2185	Wheeled		21	.762			18.65		18.65	31.50

02 41 Demolition

02 41 19 – Selective Demolition

02 41 19.19 Selective Demolition

		Crew	Daily Output	Labor-Hours	Unit	Material	2015 Bare Costs Labor	2015 Bare Costs Equipment	Total	Total Incl O&P
2190	11 – 20 riser stair, hand carried	2 Clab	12	1.333	C.Y.		32.50		32.50	54.50
2195	Wheeled		18	.889			21.50		21.50	36.50
2200	21 – 40 riser stair, hand carried		8	2			49		49	82
2205	Wheeled		12	1.333			32.50		32.50	54.50
2210	Haul and return, add per each extra 100' haul, hand carried		35.50	.451			11		11	18.50
2215	Wheeled		54	.296			7.25		7.25	12.15
2220	For each additional flight of stairs, up to 5 risers, add		550	.029	Flight		.71		.71	1.19
2225	6 – 10 risers, add		275	.058			1.42		1.42	2.39
2230	11 – 20 risers, add		138	.116			2.83		2.83	4.76
2235	21 – 40 risers, add		69	.232			5.65		5.65	9.50
3000	Loading & trucking, including 2 mile haul, chute loaded	B-16	45	.711	C.Y.		18.45	15.35	33.80	48
3040	Hand loading truck, 50' haul	"	48	.667			17.30	14.40	31.70	45
3080	Machine loading truck	B-17	120	.267			7.40	6.45	13.85	19.40
5000	Haul, per mile, up to 8 C.Y. truck	B-34B	1165	.007			.20	.59	.79	.97
5100	Over 8 C.Y. truck	"	1550	.005			.15	.45	.60	.73

02 41 19.20 Selective Demolition, Dump Charges

		Crew	Daily Output	Labor-Hours	Unit	Material	Labor	Equipment	Total	Total Incl O&P
0010	**SELECTIVE DEMOLITION, DUMP CHARGES** R024119-10									
0020	Dump charges, typical urban city, tipping fees only									
0100	Building construction materials				Ton	74			74	81
0200	Trees, brush, lumber					63			63	69.50
0300	Rubbish only					63			63	69.50
0500	Reclamation station, usual charge					74			74	81

02 41 19.21 Selective Demolition, Gutting

		Crew	Daily Output	Labor-Hours	Unit	Material	Labor	Equipment	Total	Total Incl O&P
0010	**SELECTIVE DEMOLITION, GUTTING** R024119-10									
0020	Building interior, including disposal, dumpster fees not included									
0500	Residential building									
0560	Minimum	B-16	400	.080	SF Flr.		2.08	1.73	3.81	5.35
0580	Maximum	"	360	.089	"		2.31	1.92	4.23	5.95
0900	Commercial building									
1000	Minimum	B-16	350	.091	SF Flr.		2.37	1.97	4.34	6.15
1020	Maximum	"	250	.128	"		3.32	2.76	6.08	8.60

02 83 Lead Remediation

02 83 19 – Lead-Based Paint Remediation

02 83 19.22 Preparation of Lead Containment Area

		Crew	Daily Output	Labor-Hours	Unit	Material	Labor	Equipment	Total	Total Incl O&P
0010	**PREPARATION OF LEAD CONTAINMENT AREA**									
0020	Lead abatement work area, test kit, per swab	1 Skwk	16	.500	Ea.	4.21	16.70		20.91	32.50
0025	For dust barriers see Section 01 56 16.10									
0050	Caution sign	1 Skwk	48	.167	Ea.	8.35	5.55		13.90	18.50
0100	Pre-cleaning, HEPA vacuum and wet wipe, floor and wall surfaces	3 Skwk	5000	.005	S.F.	.01	.16		.17	.29
0105	Ceiling, 6' - 11' high		4100	.006		.08	.20		.28	.41
0108	12' - 15' high		3550	.007		.07	.23		.30	.46
0115	Over 15' high		3000	.008		.09	.27		.36	.55
0500	Cover surfaces with polyethylene sheeting									
0550	Floors, each layer, 6 mil	3 Skwk	8000	.003	S.F.	.04	.10		.14	.22
0560	Walls, each layer, 6 mil	"	6000	.004	"	.04	.13		.17	.27
0570	For heights above 12', add						20%			
2400	Post abatement cleaning of protective sheeting, HEPA vacuum & wet wipe	3 Skwk	5000	.005	S.F.	.01	.16		.17	.29
2450	Doff, bag and seal protective sheeting		12000	.002		.01	.07		.08	.12
2500	Post abatement cleaning, HEPA vacuum & wet wipe		5000	.005		.01	.16		.17	.29

02 83 Lead Remediation

02 83 19 – Lead-Based Paint Remediation

02 83 19.23 Encapsulation of Lead-Based Paint	Crew	Daily Output	Labor-Hours	Unit	Material	2015 Bare Costs Labor	Equipment	Total	Total Incl O&P
0010 **ENCAPSULATION OF LEAD-BASED PAINT**									
0020 Interior, brushwork, trim, under 6"	1 Pord	240	.033	L.F.	2.30	.95		3.25	4.10
0030 6" to 12" wide		180	.044		3.06	1.27		4.33	5.45
0040 Balustrades		300	.027		1.84	.76		2.60	3.28
0050 Pipe to 4" diameter		500	.016		1.12	.46		1.58	1.99
0060 To 8" diameter		375	.021		1.48	.61		2.09	2.64
0070 To 12" diameter		250	.032		2.19	.92		3.11	3.92
0080 To 16" diameter		170	.047		3.26	1.35		4.61	5.80
0090 Cabinets, ornate design		200	.040	S.F.	2.81	1.15		3.96	4.98
0100 Simple design		250	.032	"	2.24	.92		3.16	3.97
0110 Doors, 3' x 7', both sides, incl. frame & trim									
0120 Flush	1 Pord	6	1.333	Ea.	28	38		66	94
0130 French, 10 – 15 lite		3	2.667		5.65	76.50		82.15	132
0140 Panel		4	2		34	57.50		91.50	132
0150 Louvered		2.75	2.909		31	83.50		114.50	172
0160 Windows, per interior side, per 15 S.F.									
0170 1 to 6 lite	1 Pord	14	.571	Ea.	19.40	16.35		35.75	48.50
0180 7 to 10 lite		7.50	1.067		21.50	30.50		52	74.50
0190 12 lite		5.75	1.391		29	40		69	97.50
0200 Radiators		8	1		69	28.50		97.50	123
0210 Grilles, vents		275	.029	S.F.	2.04	.83		2.87	3.61
0220 Walls, roller, drywall or plaster		1000	.008		.56	.23		.79	1
0230 With spunbonded reinforcing fabric		720	.011		.63	.32		.95	1.21
0240 Wood		800	.010		.69	.29		.98	1.23
0250 Ceilings, roller, drywall or plaster		900	.009		.63	.25		.88	1.11
0260 Wood		700	.011		.78	.33		1.11	1.40
0270 Exterior, brushwork, gutters and downspouts		300	.027	L.F.	1.84	.76		2.60	3.28
0280 Columns		400	.020	S.F.	1.38	.57		1.95	2.46
0290 Spray, siding		600	.013	"	.93	.38		1.31	1.65
0300 Miscellaneous									
0310 Electrical conduit, brushwork, to 2" diameter	1 Pord	500	.016	L.F.	1.12	.46		1.58	1.99
0320 Brick, block or concrete, spray		500	.016	S.F.	1.12	.46		1.58	1.99
0330 Steel, flat surfaces and tanks to 12"		500	.016		1.12	.46		1.58	1.99
0340 Beams, brushwork		400	.020		1.38	.57		1.95	2.46
0350 Trusses		400	.020		1.38	.57		1.95	2.46

02 83 19.26 Removal of Lead-Based Paint

	Crew	Daily Output	Labor-Hours	Unit	Material	Labor	Equipment	Total	Total Incl O&P
0010 **REMOVAL OF LEAD-BASED PAINT**									
0011 By chemicals, per application									
0050 Baseboard, to 6" wide	1 Pord	64	.125	L.F.	.72	3.58		4.30	6.70
0070 To 12" wide		32	.250	"	1.40	7.15		8.55	13.35
0200 Balustrades, one side		28	.286	S.F.	1.45	8.20		9.65	15.10
1400 Cabinets, simple design		32	.250		1.31	7.15		8.46	13.25
1420 Ornate design		25	.320		1.58	9.15		10.73	16.85
1600 Cornice, simple design		60	.133		1.50	3.82		5.32	7.95
1620 Ornate design		20	.400		5.35	11.45		16.80	25
2800 Doors, one side, flush		84	.095		1.88	2.73		4.61	6.55
2820 Two panel		80	.100		1.31	2.87		4.18	6.15
2840 Four panel		45	.178		1.40	5.10		6.50	9.95
2880 For trim, one side, add		64	.125	L.F.	.71	3.58		4.29	6.70
3000 Fence, picket, one side		30	.267	S.F.	1.31	7.65		8.96	14.05
3200 Grilles, one side, simple design		30	.267		1.33	7.65		8.98	14.05
3220 Ornate design		25	.320		1.43	9.15		10.58	16.65

02 83 Lead Remediation

02 83 19 – Lead-Based Paint Remediation

02 83 19.26 Removal of Lead-Based Paint		Crew	Daily Output	Labor-Hours	Unit	Material	2015 Bare Costs Labor	Equipment	Total	Total Incl O&P
3240	Handrails	1 Pord	90	.089	L.F.	1.33	2.55		3.88	5.65
4400	Pipes, to 4" diameter		90	.089		1.88	2.55		4.43	6.25
4420	To 8" diameter		50	.160		3.75	4.58		8.33	11.70
4440	To 12" diameter		36	.222		5.65	6.35		12	16.70
4460	To 16" diameter		20	.400		7.50	11.45		18.95	27
4500	For hangers, add		40	.200	Ea.	2.49	5.75		8.24	12.20
4800	Siding		90	.089	S.F.	1.24	2.55		3.79	5.55
5000	Trusses, open		55	.145	SF Face	1.96	4.17		6.13	9
6200	Windows, one side only, double hung, 1/1 light, 24" x 48" high		4	2	Ea.	23	57.50		80.50	120
6220	30" x 60" high		3	2.667		31	76.50		107.50	160
6240	36" x 72" high		2.50	3.200		37	91.50		128.50	192
6280	40" x 80" high		2	4		46.50	115		161.50	240
6400	Colonial window, 6/6 light, 24" x 48" high		2	4		46.50	115		161.50	240
6420	30" x 60" high		1.50	5.333		62	153		215	320
6440	36" x 72" high		1	8		93	229		322	480
6480	40" x 80" high		1	8		93	229		322	480
6600	8/8 light, 24" x 48" high		2	4		46.50	115		161.50	240
6620	40" x 80" high		1	8		93	229		322	480
6800	12/12 light, 24" x 48" high		1	8		93	229		322	480
6820	40" x 80" high		.75	10.667		124	305		429	640
6840	Window frame & trim items, included in pricing above									

Division 3 – Concrete

Estimating Tips
General
- Carefully check all the plans and specifications. Concrete often appears on drawings other than structural drawings, including mechanical and electrical drawings for equipment pads. The cost of cutting and patching is often difficult to estimate. See Subdivision 03 81 for Concrete Cutting, Subdivision 02 41 19.16 for Cutout Demolition, Subdivision 03 05 05.10 for Concrete Demolition, and Subdivision 02 41 19.19 for Rubbish Handling (handling, loading and hauling of debris).
- Always obtain concrete prices from suppliers near the job site. A volume discount can often be negotiated, depending upon competition in the area. Remember to add for waste, particularly for slabs and footings on grade.

03 10 00 Concrete Forming and Accessories
- A primary cost for concrete construction is forming. Most jobs today are constructed with prefabricated forms. The selection of the forms best suited for the job and the total square feet of forms required for efficient concrete forming and placing are key elements in estimating concrete construction. Enough forms must be available for erection to make efficient use of the concrete placing equipment and crew.
- Concrete accessories for forming and placing depend upon the systems used. Study the plans and specifications to ensure that all special accessory requirements have been included in the cost estimate, such as anchor bolts, inserts, and hangers.
- Included within costs for forms-in-place are all necessary bracing and shoring.

03 20 00 Concrete Reinforcing
- Ascertain that the reinforcing steel supplier has included all accessories, cutting, bending, and an allowance for lapping, splicing, and waste. A good rule of thumb is 10% for lapping, splicing, and waste. Also, 10% waste should be allowed for welded wire fabric.
- The unit price items in the subdivisions for Reinforcing In Place, Glass Fiber Reinforcing, and Welded Wire Fabric include the labor to install accessories such as beam and slab bolsters, high chairs, and bar ties and tie wire. The material cost for these accessories is not included; they may be obtained from the Accessories Division.

03 30 00 Cast-In-Place Concrete
- When estimating structural concrete, pay particular attention to requirements for concrete additives, curing methods, and surface treatments. Special consideration for climate, hot or cold, must be included in your estimate. Be sure to include requirements for concrete placing equipment, and concrete finishing.
- For accurate concrete estimating, the estimator must consider each of the following major components individually: forms, reinforcing steel, ready-mix concrete, placement of the concrete, and finishing of the top surface. For faster estimating, Subdivision 03 30 53.40 for Concrete-In-Place can be used; here, various items of concrete work are presented that include the costs of all five major components (unless specifically stated otherwise).

03 40 00 Precast Concrete
03 50 00 Cast Decks and Underlayment
- The cost of hauling precast concrete structural members is often an important factor. For this reason, it is important to get a quote from the nearest supplier. It may become economically feasible to set up precasting beds on the site if the hauling costs are prohibitive.

Reference Numbers
Reference numbers are shown in shaded boxes at the beginning of some major classifications. These numbers refer to related items in the Reference Section. The reference information may be an estimating procedure, an alternate pricing method, or technical information.

Note: Not all subdivisions listed here necessarily appear in this publication. ■

No part of this publication may be reproduced, stored in a retrieval system, or transmitted in any form or by any means without prior written permission of RSMeans.

03 01 Maintenance of Concrete

03 01 30 – Maintenance of Cast-In-Place Concrete

03 01 30.64 Floor Patching

		Crew	Daily Output	Labor-Hours	Unit	Material	2015 Bare Costs Labor	2015 Bare Costs Equipment	Total	Total Incl O&P
0010	**FLOOR PATCHING**									
0012	Floor patching, 1/4" thick, small areas, regular	1 Cefi	170	.047	S.F.	2.88	1.48		4.36	5.55
0100	Epoxy	"	100	.080	"	7.45	2.52		9.97	12.30

03 05 Common Work Results for Concrete

03 05 13 – Basic Concrete Materials

03 05 13.85 Winter Protection

		Crew	Daily Output	Labor-Hours	Unit	Material	Labor	Equipment	Total	Total Incl O&P
0010	**WINTER PROTECTION**									
0012	For heated ready mix, add				C.Y.	4.05			4.05	4.46
0100	Temporary heat to protect concrete, 24 hours	2 Clab	50	.320	M.S.F.	535	7.80		542.80	600
0200	Temporary shelter for slab on grade, wood frame/polyethylene sheeting									
0201	Build or remove, light framing for short spans	2 Carp	10	1.600	M.S.F.	310	52.50		362.50	435
0210	Large framing for long spans	"	3	5.333	"	415	175		590	750
0710	Electrically, heated pads, 15 watts/S.F., 20 uses				S.F.	.52			.52	.57

03 11 Concrete Forming

03 11 13 – Structural Cast-In-Place Concrete Forming

03 11 13.25 Forms In Place, Columns

			Crew	Daily Output	Labor-Hours	Unit	Material	Labor	Equipment	Total	Total Incl O&P
0010	**FORMS IN PLACE, COLUMNS**										
1500	Round fiber tube, recycled paper, 1 use, 8" diameter	G	C-1	155	.206	L.F.	1.77	5.90		7.67	11.90
1550	10" diameter	G		155	.206		2.40	5.90		8.30	12.60
1600	12" diameter	G		150	.213		2.83	6.10		8.93	13.35
1650	14" diameter	G		145	.221		3.91	6.30		10.21	14.95
1700	16" diameter	G		140	.229		4.88	6.55		11.43	16.35
1720	18" diameter	G		140	.229		5.70	6.55		12.25	17.30
5000	Job-built plywood, 8" x 8" columns, 1 use			165	.194	SFCA	2.73	5.55		8.28	12.35
5500	12" x 12" columns, 1 use			180	.178	"	2.62	5.10		7.72	11.45
7400	Steel framed plywood, based on 50 uses of purchased										
7420	forms, and 4 uses of bracing lumber										
7500	8" x 8" column		C-1	340	.094	SFCA	2.14	2.70		4.84	6.90
7550	10" x 10"			350	.091		1.87	2.62		4.49	6.45
7600	12" x 12"			370	.086		1.59	2.48		4.07	5.90

03 11 13.45 Forms In Place, Footings

		Crew	Daily Output	Labor-Hours	Unit	Material	Labor	Equipment	Total	Total Incl O&P
0010	**FORMS IN PLACE, FOOTINGS**										
0020	Continuous wall, plywood, 1 use	C-1	375	.085	SFCA	6.80	2.44		9.24	11.60
0150	4 use	"	485	.066	"	2.22	1.89		4.11	5.60
1500	Keyway, 4 use, tapered wood, 2" x 4"	1 Carp	530	.015	L.F.	.22	.50		.72	1.08
1550	2" x 6"	"	500	.016	"	.33	.53		.86	1.24
5000	Spread footings, job-built lumber, 1 use	C-1	305	.105	SFCA	2.28	3		5.28	7.55
5150	4 use	"	414	.077	"	.74	2.21		2.95	4.54

03 11 13.50 Forms In Place, Grade Beam

		Crew	Daily Output	Labor-Hours	Unit	Material	Labor	Equipment	Total	Total Incl O&P
0010	**FORMS IN PLACE, GRADE BEAM**										
0020	Job-built plywood, 1 use	C-2	530	.091	SFCA	3.14	2.62		5.76	7.85
0150	4 use	"	605	.079	"	1.02	2.30		3.32	4.99

03 11 13.65 Forms In Place, Slab On Grade

			Crew	Daily Output	Labor-Hours	Unit	Material	Labor	Equipment	Total	Total Incl O&P
0010	**FORMS IN PLACE, SLAB ON GRADE**											
1000	Bulkhead forms w/keyway, wood, 6" high, 1 use		C-1	510	.063	L.F.	1.04	1.80		2.84	4.17
1400	Bulkhead form for slab, 4-1/2" high, exp metal, incl keyway & stakes	G		1200	.027		.92	.76		1.68	2.29

03 11 Concrete Forming

03 11 13 – Structural Cast-In-Place Concrete Forming

03 11 13.65 Forms In Place, Slab On Grade

			Crew	Daily Output	Labor-Hours	Unit	Material	2015 Bare Costs Labor	Equipment	Total	Total Incl O&P
1410	5-1/2" high	G	C-1	1100	.029	L.F.	1.12	.83		1.95	2.63
1420	7-1/2" high	G		960	.033		1.31	.95		2.26	3.04
1430	9-1/2" high	G		840	.038		1.43	1.09		2.52	3.40
2000	Curb forms, wood, 6" to 12" high, on grade, 1 use			215	.149	SFCA	2.44	4.26		6.70	9.85
2150	4 use			275	.116	"	.79	3.33		4.12	6.45
3000	Edge forms, wood, 4 use, on grade, to 6" high			600	.053	L.F.	.35	1.53		1.88	2.95
3050	7" to 12" high			435	.074	SFCA	.79	2.11		2.90	4.40
4000	For slab blockouts, to 12" high, 1 use			200	.160	L.F.	.80	4.58		5.38	8.55
4100	Plastic (extruded), to 6" high, multiple use, on grade			800	.040	"	8.20	1.15		9.35	10.95
8760	Void form, corrugated fiberboard, 4" x 12", 4' long	G		3000	.011	S.F.	2.87	.31		3.18	3.66
8770	6" x 12", 4' long			3000	.011		3.42	.31		3.73	4.27
8780	1/4" thick hardboard protective cover for void form		2 Carp	1500	.011		.58	.35		.93	1.23

03 11 13.85 Forms In Place, Walls

			Crew	Daily Output	Labor-Hours	Unit	Material	Labor	Equipment	Total	Total Incl O&P
0010	**FORMS IN PLACE, WALLS**										
0100	Box out for wall openings, to 16" thick, to 10 S.F.		C-2	24	2	Ea.	27	58		85	127
0150	Over 10 S.F. (use perimeter)		"	280	.171	L.F.	2.31	4.97		7.28	10.90
0250	Brick shelf, 4" w, add to wall forms, use wall area above shelf										
0260	1 use		C-2	240	.200	SFCA	2.50	5.80		8.30	12.50
0350	4 use			300	.160	"	1	4.64		5.64	8.90
0500	Bulkhead, wood with keyway, 1 use, 2 piece			265	.181	L.F.	2.12	5.25		7.37	11.20
0600	Bulkhead forms with keyway, 1 piece expanded metal, 8" wall	G	C-1	1000	.032		1.31	.92		2.23	2.98
0610	10" wall	G	"	800	.040		1.43	1.15		2.58	3.50
2000	Wall, job-built plywood, to 8' high, 1 use		C-2	370	.130	SFCA	2.74	3.76		6.50	9.30
2050	2 use			435	.110		1.75	3.20		4.95	7.30
2100	3 use			495	.097		1.27	2.81		4.08	6.15
2150	4 use			505	.095		1.03	2.75		3.78	5.75
2400	Over 8' to 16' high, 1 use			280	.171		3.03	4.97		8	11.70
2450	2 use			345	.139		1.34	4.03		5.37	8.25
2500	3 use			375	.128		.95	3.71		4.66	7.30
2550	4 use			395	.122		.78	3.52		4.30	6.75
7800	Modular prefabricated plywood, based on 20 uses of purchased										
7820	forms, and 4 uses of bracing lumber										
7860	To 8' high		C-2	800	.060	SFCA	.94	1.74		2.68	3.95
8060	Over 8' to 16' high		"	600	.080	"	.99	2.32		3.31	4.99

03 11 19 – Insulating Concrete Forming

03 11 19.10 Insulating Forms, Left In Place

			Crew	Daily Output	Labor-Hours	Unit	Material	Labor	Equipment	Total	Total Incl O&P
0010	**INSULATING FORMS, LEFT IN PLACE**										
0020	S.F. is for exterior face, but includes forms for both faces (total R22)										
2000	4" wall, straight block, 16" x 48" (5.33 S.F.)	G	2 Carp	90	.178	Ea.	19.60	5.85		25.45	31.50
2010	90 corner block, exterior 16" x 38" x 22" (6.67 S.F.)	G		75	.213		23	7		30	37.50
2020	45 corner block, exterior 16" x 34" x 18" (5.78 S.F.)	G		75	.213		23	7		30	37.50
2100	6" wall, straight block, 16" x 48" (5.33 S.F.)	G		90	.178		20	5.85		25.85	32
2110	90 corner block, exterior 16" x 32" x 24" (6.22 S.F.)	G		75	.213		25	7		32	39.50
2120	45 corner block, exterior 16" x 26" x 18" (4.89 S.F.)	G		75	.213		22.50	7		29.50	36.50
2130	Brick ledge block, 16" x 48" (5.33 S.F.)	G		80	.200		25	6.55		31.55	38.50
2140	Taper top block, 16" x 48" (5.33 S.F.)	G		80	.200		23.50	6.55		30.05	37
2200	8" wall, straight block, 16" x 48" (5.33 S.F.)	G		90	.178		21.50	5.85		27.35	33.50
2210	90 corner block, exterior 16" x 34" x 26" (6.67 S.F.)	G		75	.213		27.50	7		34.50	42.50
2220	45 corner block, exterior 16" x 28" x 20" (5.33 S.F.)	G		75	.213		23.50	7		30.50	38
2230	Brick ledge block, 16" x 48" (5.33 S.F.)	G		80	.200		26.50	6.55		33.05	40
2240	Taper top block, 16" x 48" (5.33 S.F.)	G		80	.200		24.50	6.55		31.05	38

03 11 Concrete Forming

03 11 23 – Permanent Stair Forming

03 11 23.75 Forms In Place, Stairs		Crew	Daily Output	Labor-Hours	Unit	Material	2015 Bare Costs Labor	Equipment	Total	Total Incl O&P
0010	**FORMS IN PLACE, STAIRS**									
0015	(Slant length x width), 1 use	C-2	165	.291	S.F.	5.90	8.45		14.35	20.50
0150	4 use		190	.253		2.08	7.30		9.38	14.60
2000	Stairs, cast on sloping ground (length x width), 1 use		220	.218		2.36	6.30		8.66	13.25
2025	2 use		232	.207		1.30	6		7.30	11.50
2050	3 use		244	.197		.94	5.70		6.64	10.65
2100	4 use		256	.188		.77	5.45		6.22	10

03 15 Concrete Accessories

03 15 05 – Concrete Forming Accessories

03 15 05.12 Chamfer Strips

		Crew	Daily Output	Labor-Hours	Unit	Material	Labor	Equipment	Total	Total Incl O&P
0010	**CHAMFER STRIPS**									
5000	Wood, 1/2" wide	1 Carp	535	.015	L.F.	.15	.49		.64	.99
5200	3/4" wide		525	.015		.16	.50		.66	1.02
5400	1" wide		515	.016		.33	.51		.84	1.22

03 15 05.15 Column Form Accessories

0010	**COLUMN FORM ACCESSORIES**									
1000	Column clamps, adjustable to 24" x 24", buy	G			Set	145			145	160
1100	Rent per month	G			"	14.55			14.55	16

03 15 05.75 Sleeves and Chases

0010	**SLEEVES AND CHASES**									
0100	Plastic, 1 use, 12" long, 2" diameter	1 Carp	100	.080	Ea.	2.15	2.63		4.78	6.80
0150	4" diameter		90	.089		6.05	2.92		8.97	11.55
0200	6" diameter		75	.107		10.65	3.50		14.15	17.60

03 15 05.80 Snap Ties

0010	**SNAP TIES**, 8-1/4" L&W (Lumber and wedge)									
0100	2250 lb., w/flat washer, 8" wall	G			C	90			90	99
0150	10" wall	G				131			131	144
0200	12" wall	G				136			136	150
0500	With plastic cone, 8" wall	G				80			80	88
0550	10" wall	G				83			83	91.50
0600	12" wall	G				89			89	98

03 15 05.95 Wall and Foundation Form Accessories

0010	**WALL AND FOUNDATION FORM ACCESSORIES**									
4000	Nail stakes, 3/4" diameter, 18" long	G			Ea.	3.30			3.30	3.63
4050	24" long	G				4.26			4.26	4.69
4200	30" long	G				5.45			5.45	6
4250	36" long	G				6.60			6.60	7.25

03 15 13 – Waterstops

03 15 13.50 Waterstops

0010	**WATERSTOPS**, PVC and Rubber									
0020	PVC, ribbed 3/16" thick, 4" wide	1 Carp	155	.052	L.F.	1.35	1.70		3.05	4.34
0050	6" wide		145	.055		2.27	1.81		4.08	5.55
0500	With center bulb, 6" wide, 3/16" thick		135	.059		1.99	1.95		3.94	5.45
0550	3/8" thick		130	.062		3.67	2.02		5.69	7.45
0600	9" wide x 3/8" thick		125	.064		6.05	2.10		8.15	10.20

03 15 Concrete Accessories

03 15 16 – Concrete Construction Joints

03 15 16.20 Control Joints, Saw Cut

		Crew	Daily Output	Labor-Hours	Unit	Material	2015 Bare Costs Labor	2015 Bare Costs Equipment	Total	Total Incl O&P
0010	**CONTROL JOINTS, SAW CUT**									
0100	Sawcut control joints in green concrete									
0120	1" depth	C-27	2000	.008	L.F.	.04	.25	.08	.37	.54
0140	1-1/2" depth		1800	.009		.06	.28	.09	.43	.61
0160	2" depth		1600	.010		.08	.32	.10	.50	.72
0180	Sawcut joint reservoir in cured concrete									
0182	3/8" wide x 3/4" deep, with single saw blade	C-27	1000	.016	L.F.	.06	.50	.17	.73	1.06
0184	1/2" wide x 1" deep, with double saw blades		900	.018		.11	.56	.19	.86	1.24
0186	3/4" wide x 1-1/2" deep, with double saw blades		800	.020		.23	.63	.21	1.07	1.51
0190	Water blast joint to wash away laitance, 2 passes	C-29	2500	.003			.08	.03	.11	.16
0200	Air blast joint to blow out debris and air dry, 2 passes	C-28	2000	.004			.13	.01	.14	.21
0300	For backer rod, see Section 07 91 23.10									
0340	For joint sealant, see Sections 03 15 16.30 or 07 92 13.20									

03 15 16.30 Expansion Joints

			Crew	Daily Output	Labor-Hours	Unit	Material	Labor	Equipment	Total	Total Incl O&P
0010	**EXPANSION JOINTS**										
0020	Keyed, cold, 24 ga., incl. stakes, 3-1/2" high	G	1 Carp	200	.040	L.F.	.81	1.31		2.12	3.10
0050	4-1/2" high	G		200	.040		.92	1.31		2.23	3.22
0100	5-1/2" high	G		195	.041		1.12	1.35		2.47	3.49
2000	Premolded, bituminous fiber, 1/2" x 6"			375	.021		.44	.70		1.14	1.66
2050	1" x 12"			300	.027		1.99	.88		2.87	3.66
2140	Concrete expansion joint, recycled paper and fiber, 1/2" x 6"	G		390	.021		.42	.67		1.09	1.59
2150	1/2" x 12"	G		360	.022		.83	.73		1.56	2.15
2500	Neoprene sponge, closed cell, 1/2" x 6"			375	.021		2.37	.70		3.07	3.79
2550	1" x 12"			300	.027		8.25	.88		9.13	10.55
5000	For installation in walls, add							75%			
5250	For installation in boxouts, add							25%			

03 15 19 – Cast-In Concrete Anchors

03 15 19.05 Anchor Bolt Accessories

		Crew	Daily Output	Labor-Hours	Unit	Material	Labor	Equipment	Total	Total Incl O&P
0010	**ANCHOR BOLT ACCESSORIES**									
0015	For anchor bolts set in fresh concrete, see Section 03 15 19.10									
8150	Anchor bolt sleeve, plastic, 1" diam. bolts	1 Carp	60	.133	Ea.	8.20	4.38		12.58	16.40
8500	1-1/2" diameter		28	.286		14.60	9.40		24	32
8600	2" diameter		24	.333		18.60	10.95		29.55	39
8650	3" diameter		20	.400		34	13.15		47.15	59.50

03 15 19.10 Anchor Bolts

			Crew	Daily Output	Labor-Hours	Unit	Material	Labor	Equipment	Total	Total Incl O&P
0010	**ANCHOR BOLTS**										
0015	Made from recycled materials										
0025	Single bolts installed in fresh concrete, no templates										
0030	Hooked w/nut and washer, 1/2" diameter, 8" long	G	1 Carp	132	.061	Ea.	1.33	1.99		3.32	4.81
0040	12" long	G		131	.061		1.48	2.01		3.49	5
0070	3/4" diameter, 8" long	G		127	.063		3.67	2.07		5.74	7.50
0080	12" long	G		125	.064		4.59	2.10		6.69	8.60

03 15 19.30 Inserts

			Crew	Daily Output	Labor-Hours	Unit	Material	Labor	Equipment	Total	Total Incl O&P
0010	**INSERTS**										
1000	Inserts, slotted nut type for 3/4" bolts, 4" long	G	1 Carp	84	.095	Ea.	19.65	3.13		22.78	27
2100	6" long	G		84	.095		22	3.13		25.13	30
2150	8" long	G		84	.095		29.50	3.13		32.63	38
2200	Slotted, strap type, 4" long	G		84	.095		21	3.13		24.13	29
2300	8" long	G		84	.095		31.50	3.13		34.63	40.50

03 21 Reinforcement Bars

03 21 11 – Plain Steel Reinforcement Bars

03 21 11.60 Reinforcing In Place

			Crew	Daily Output	Labor-Hours	Unit	Material	2015 Bare Costs Labor	Equipment	Total	Total Incl O&P
0010	**REINFORCING IN PLACE**, 50-60 ton lots, A615 Grade 60										
0020	Includes labor, but not material cost, to install accessories										
0030	Made from recycled materials										
0502	Footings, #4 to #7	G	4 Rodm	4200	.008	Lb.	.48	.27		.75	.99
0550	#8 to #18	G		3.60	8.889	Ton	970	315		1,285	1,600
0702	Walls, #3 to #7	G		6000	.005	Lb.	.48	.19		.67	.85
0750	#8 to #18	G	↓	4	8	Ton	970	282		1,252	1,550
0900	For other than 50 – 60 ton lots										
1000	Under 10 ton job, #3 to #7, add						25%	10%			
1010	#8 to #18, add						20%	10%			
1050	10 – 50 ton job, #3 to #7, add						10%				
1060	#8 to #18, add						5%				
1100	60 – 100 ton job, #3 to #7, deduct						5%				
1110	#8 to #18, deduct						10%				
1150	Over 100 ton job, #3 to #7, deduct						10%				
1160	#8 to #18, deduct						15%				
2400	Dowels, 2 feet long, deformed, #3	G	2 Rodm	520	.031	Ea.	.40	1.08		1.48	2.28
2410	#4	G		480	.033		.71	1.17		1.88	2.78
2420	#5	G		435	.037		1.11	1.29		2.40	3.42
2430	#6	G	↓	360	.044	↓	1.60	1.56		3.16	4.42
2600	Dowel sleeves for CIP concrete, 2-part system										
2610	Sleeve base, plastic, for 5/8" smooth dowel sleeve, fasten to edge form		1 Rodm	200	.040	Ea.	.53	1.41		1.94	2.98
2615	Sleeve, plastic, 12" long, for 5/8" smooth dowel, snap onto base			400	.020		1.18	.70		1.88	2.50
2620	Sleeve base, for 3/4" smooth dowel sleeve			175	.046		.53	1.61		2.14	3.32
2625	Sleeve, 12" long, for 3/4" smooth dowel			350	.023		1.33	.80		2.13	2.83
2630	Sleeve base, for 1" smooth dowel sleeve			150	.053		.67	1.88		2.55	3.93
2635	Sleeve, 12" long, for 1" smooth dowel		↓	300	.027		1.40	.94		2.34	3.14
2700	Dowel caps, visual warning only, plastic, #3 to #8		2 Rodm	800	.020		.26	.70		.96	1.49
2720	#8 to #18			750	.021		.63	.75		1.38	1.97
2750	Impalement protective, plastic, #4 to #9		↓	800	.020	↓	1.22	.70		1.92	2.54

03 22 Fabric and Grid Reinforcing

03 22 11 – Plain Welded Wire Fabric Reinforcing

03 22 11.10 Welded Wire Fabric

			Crew	Daily Output	Labor-Hours	Unit	Material	Labor	Equipment	Total	Total Incl O&P
0011	**WELDED WIRE FABRIC**, 6 x 6 - W1.4 x W1.4 (10 x 10)	G	2 Rodm	3500	.005	S.F.	.15	.16		.31	.43
0301	6 x 6 - W2.9 x W2.9 (6 x 6) 42 lb. per C.S.F.	G		2900	.006		.22	.19		.41	.58
0501	4 x 4 - W1.4 x W1.4 (10 x 10) 31 lb. per C.S.F.	G	↓	3100	.005	↓	.20	.18		.38	.53
0750	Rolls										
0901	2 x 2 - #12 galv. for gunite reinforcing	G	2 Rodm	650	.025	S.F.	.63	.87		1.50	2.16

03 22 13 – Galvanized Welded Wire Fabric Reinforcing

03 22 13.10 Galvanized Welded Wire Fabric

0010	**GALVANIZED WELDED WIRE FABRIC**										
0100	Add to plain welded wire pricing for galvanized welded wire					Lb.	.23			.23	.25

03 22 16 – Epoxy-Coated Welded Wire Fabric Reinforcing

03 22 16.10 Epoxy-Coated Welded Wire Fabric

0010	**EPOXY-COATED WELDED WIRE FABRIC**										
0100	Add to plain welded wire pricing for epoxy-coated welded wire					Lb.	.21			.21	.23

03 23 Stressed Tendon Reinforcing

03 23 05 – Prestressing Tendons

03 23 05.50 Prestressing Steel		Crew	Daily Output	Labor-Hours	Unit	Material	2015 Bare Costs Labor	Equipment	Total	Total Incl O&P
0010	**PRESTRESSING STEEL**									
3000	Slabs on grade, 0.5-inch diam. non-bonded strands, HDPE sheathed,									
3050	attached dead-end anchors, loose stressing-end anchors									
3100	25' x 30' slab, strands @ 36" O.C., placing	2 Rodm	2940	.005	S.F.	.60	.19		.79	.99
3105	Stressing	C-4A	3750	.004			.15	.01	.16	.27
3110	42" O.C., placing	2 Rodm	3200	.005		.53	.18		.71	.89
3115	Stressing	C-4A	4040	.004			.14	.01	.15	.25
3120	48" O.C., placing	2 Rodm	3510	.005		.47	.16		.63	.78
3125	Stressing	C-4A	4390	.004			.13	.01	.14	.23
3150	25' x 40' slab, strands @ 36" O.C., placing	2 Rodm	3370	.005		.58	.17		.75	.92
3155	Stressing	C-4A	4360	.004			.13	.01	.14	.23
3160	42" O.C., placing	2 Rodm	3760	.004		.50	.15		.65	.81
3165	Stressing	C-4A	4820	.003			.12	.01	.13	.21
3170	48" O.C., placing	2 Rodm	4090	.004		.45	.14		.59	.72
3175	Stressing	C-4A	5190	.003			.11	.01	.12	.19
3200	30' x 30' slab, strands @ 36" O.C., placing	2 Rodm	3260	.005		.58	.17		.75	.93
3205	Stressing	C-4A	4190	.004			.13	.01	.14	.24
3210	42" O.C., placing	2 Rodm	3530	.005		.52	.16		.68	.85
3215	Stressing	C-4A	4500	.004			.13	.01	.14	.22
3220	48" O.C., placing	2 Rodm	3840	.004		.47	.15		.62	.76
3225	Stressing	C-4A	4850	.003			.12	.01	.13	.21
3230	30' x 40' slab, strands @ 36" O.C., placing	2 Rodm	3780	.004		.56	.15		.71	.87
3235	Stressing	C-4A	4920	.003			.11	.01	.12	.20
3240	42" O.C., placing	2 Rodm	4190	.004		.49	.13		.62	.77
3245	Stressing	C-4A	5410	.003			.10	.01	.11	.19
3250	48" O.C., placing	2 Rodm	4520	.004		.45	.12		.57	.70
3255	Stressing	C-4A	5790	.003			.10	.01	.11	.18
3260	30' x 50' slab, strands @ 36" O.C., placing	2 Rodm	4300	.004		.53	.13		.66	.81
3265	Stressing	C-4A	5650	.003			.10	.01	.11	.18
3270	42" O.C., placing	2 Rodm	4720	.003		.47	.12		.59	.72
3275	Stressing	C-4A	6150	.003			.09	.01	.10	.17
3280	48" O.C., placing	2 Rodm	5240	.003		.42	.11		.53	.64
3285	Stressing	C-4A	6760	.002			.08	.01	.09	.15

03 24 Fibrous Reinforcing

03 24 05 – Reinforcing Fibers

03 24 05.30 Synthetic Fibers

					Unit	Material			Total	Total Incl O&P
0010	**SYNTHETIC FIBERS**									
0100	Synthetic fibers, add to concrete				Lb.	4.40			4.40	4.84
0110	1-1/2 lb. per C.Y.				C.Y.	6.80			6.80	7.50

03 24 05.70 Steel Fibers

						Unit	Material			Total	Total Incl O&P
0010	**STEEL FIBERS**										
0140	ASTM A850, Type V, continuously deformed, 1-1/2" long x 0.045" diam.										
0150	Add to price of ready mix concrete	G				Lb.	1.13			1.13	1.24
0205	Alternate pricing, dosing at 5 lb. per C.Y., add to price of RMC	G				C.Y.	5.65			5.65	6.20
0210	10 lb. per C.Y.	G					11.30			11.30	12.45
0215	15 lb. per C.Y.	G					16.95			16.95	18.65
0220	20 lb. per C.Y.	G					22.50			22.50	25
0225	25 lb. per C.Y.	G					28.50			28.50	31
0230	30 lb. per C.Y.	G					34			34	37.50
0235	35 lb. per C.Y.	G					39.50			39.50	43.50

03 24 Fibrous Reinforcing

03 24 05 – Reinforcing Fibers

03 24 05.70 Steel Fibers

		Crew	Daily Output	Labor-Hours	Unit	Material	2015 Bare Costs Labor	Equipment	Total	Total Incl O&P
0240	40 lb. per C.Y.	G			C.Y.	45			45	49.50
0250	50 lb. per C.Y.	G				56.50			56.50	62
0275	75 lb. per C.Y.	G				85			85	93
0300	100 lb. per C.Y.	G				113			113	124

03 30 Cast-In-Place Concrete

03 30 53 – Miscellaneous Cast-In-Place Concrete

03 30 53.40 Concrete In Place

		Crew	Daily Output	Labor-Hours	Unit	Material	Labor	Equipment	Total	Total Incl O&P
0010	**CONCRETE IN PLACE**									
0020	Including forms (4 uses), Grade 60 rebar, concrete (Portland cement									
0050	Type I), placement and finishing unless otherwise indicated									
0500	Chimney foundations (5000 psi), over 5 C.Y.	C-14C	32.22	3.476	C.Y.	150	107	.97	257.97	345
0510	(3500 psi), under 5 C.Y.	"	23.71	4.724	"	177	146	1.32	324.32	440
3540	Equipment pad (3000 psi), 3' x 3' x 6" thick	C-14H	45	1.067	Ea.	47	34	.69	81.69	109
3550	4' x 4' x 6" thick		30	1.600		69.50	51	1.04	121.54	163
3560	5' x 5' x 8" thick		18	2.667		122	85	1.73	208.73	279
3570	6' x 6' x 8" thick		14	3.429		164	110	2.23	276.23	365
3580	8' x 8' x 10" thick		8	6		350	192	3.90	545.90	710
3590	10' x 10' x 12" thick		5	9.600		595	305	6.25	906.25	1,175
3800	Footings (3000 psi), spread under 1 C.Y.	C-14C	28	4	C.Y.	166	123	1.12	290.12	390
3825	1 C.Y. to 5 C.Y.		43	2.605		201	80.50	.73	282.23	355
3850	Over 5 C.Y.		75	1.493		185	46	.42	231.42	282
3900	Footings, strip (3000 psi), 18" x 9", unreinforced	C-14L	40	2.400		125	72	.79	197.79	260
3920	18" x 9", reinforced	C-14C	35	3.200		148	98.50	.90	247.40	330
3925	20" x 10", unreinforced	C-14L	45	2.133		122	64	.70	186.70	243
3930	20" x 10", reinforced	C-14C	40	2.800		140	86.50	.78	227.28	300
3935	24" x 12", unreinforced	C-14L	55	1.745		120	52.50	.58	173.08	221
3940	24" x 12", reinforced	C-14C	48	2.333		139	72	.65	211.65	275
3945	36" x 12", unreinforced	C-14L	70	1.371		116	41.50	.45	157.95	196
3950	36" x 12", reinforced	C-14C	60	1.867		133	57.50	.52	191.02	243
4000	Foundation mat (3000 psi), under 10 C.Y.		38.67	2.896		204	89.50	.81	294.31	375
4050	Over 20 C.Y.		56.40	1.986		178	61	.56	239.56	300
4520	Handicap access ramp (4000 psi), railing both sides, 3' wide	C-14H	14.58	3.292	L.F.	320	105	2.14	427.14	530
4525	5' wide		12.22	3.928		330	126	2.55	458.55	580
4530	With 6" curb and rails both sides, 3' wide		8.55	5.614		330	179	3.65	512.65	665
4535	5' wide		7.31	6.566		335	210	4.27	549.27	725
4650	Slab on grade (3500 psi), not including finish, 4" thick	C-14E	60.75	1.449	C.Y.	124	45.50	.51	170.01	214
4700	6" thick	"	92	.957	"	119	30	.33	149.33	182
4701	Thickened slab edge (3500 psi), for slab on grade poured									
4702	monolithically with slab; depth is in addition to slab thickness;									
4703	formed vertical outside edge, earthen bottom and inside slope									
4705	8" deep x 8" wide bottom, unreinforced	C-14L	2190	.044	L.F.	3.47	1.32	.01	4.80	6.05
4710	8" x 8", reinforced	C-14C	1670	.067		5.75	2.07	.02	7.84	9.80
4715	12" deep x 12" wide bottom, unreinforced	C-14L	1800	.053		7.05	1.61	.02	8.68	10.45
4720	12" x 12", reinforced	C-14C	1310	.086		11.20	2.64	.02	13.86	16.75
4725	16" deep x 16" wide bottom, unreinforced	C-14L	1440	.067		11.85	2.01	.02	13.88	16.40
4730	16" x 16", reinforced	C-14C	1120	.100		16.80	3.08	.03	19.91	23.50
4735	20" deep x 20" wide bottom, unreinforced	C-14L	1150	.083		17.90	2.51	.03	20.44	24
4740	20" x 20", reinforced	C-14C	920	.122		24	3.75	.03	27.78	33
4745	24" deep x 24" wide bottom, unreinforced	C-14L	930	.103		25	3.11	.03	28.14	33
4750	24" x 24", reinforced	C-14C	740	.151		33.50	4.67	.04	38.21	44.50

03 30 Cast-In-Place Concrete

03 30 53 – Miscellaneous Cast-In-Place Concrete

03 30 53.40 Concrete In Place

		Crew	Daily Output	Labor-Hours	Unit	Material	2015 Bare Costs Labor	Equipment	Total	Total Incl O&P
4751	Slab on grade (3500 psi), incl. troweled finish, not incl. forms									
4760	or reinforcing, over 10,000 S.F., 4" thick	C-14F	3425	.021	S.F.	1.35	.62	.01	1.98	2.51
4820	6" thick	"	3350	.021	"	1.98	.63	.01	2.62	3.22
5000	Slab on grade (3000 psi), incl. broom finish, not incl. forms									
5001	or reinforcing, 4" thick	C-14G	2873	.019	S.F.	1.32	.56	.01	1.89	2.38
5010	6" thick		2590	.022		2.07	.62	.01	2.70	3.30
5020	8" thick		2320	.024		2.70	.69	.01	3.40	4.11
6800	Stairs (3500 psi), not including safety treads, free standing, 3'-6" wide	C-14H	83	.578	LF Nose	5.60	18.50	.38	24.48	37.50
6850	Cast on ground		125	.384	"	4.63	12.25	.25	17.13	26
7000	Stair landings, free standing		200	.240	S.F.	4.52	7.65	.16	12.33	18
7050	Cast on ground		475	.101	"	3.52	3.23	.07	6.82	9.35

03 31 Structural Concrete

03 31 13 – Heavyweight Structural Concrete

03 31 13.25 Concrete, Hand Mix

		Crew	Daily Output	Labor-Hours	Unit	Material	2015 Bare Costs Labor	Equipment	Total	Total Incl O&P
0010	**CONCRETE, HAND MIX** for small quantities or remote areas									
0050	Includes bulk local aggregate, bulk sand, bagged Portland									
0060	cement (Type I) and water, using gas powered cement mixer									
0125	2500 psi	C-30	135	.059	C.F.	3.29	1.45	1.28	6.02	7.45
0130	3000 psi		135	.059		3.52	1.45	1.28	6.25	7.70
0135	3500 psi		135	.059		3.66	1.45	1.28	6.39	7.85
0140	4000 psi		135	.059		3.82	1.45	1.28	6.55	8.05
0145	4500 psi		135	.059		4	1.45	1.28	6.73	8.25
0150	5000 psi		135	.059		4.26	1.45	1.28	6.99	8.50
0300	Using pre-bagged dry mix and wheelbarrow (80-lb. bag = 0.6 C.F.)									
0340	4000 psi	1 Clab	48	.167	C.F.	6.25	4.08		10.33	13.75

03 31 13.30 Concrete, Volumetric Site-Mixed

		Crew	Daily Output	Labor-Hours	Unit	Material	2015 Bare Costs Labor	Equipment	Total	Total Incl O&P
0010	**CONCRETE, VOLUMETRIC SITE-MIXED**									
0015	Mixed on-site in volumetric truck									
0020	Includes local aggregate, sand, Portland cement (Type I) and water									
0025	Excludes all additives and treatments									
0100	3000 psi, 1 C.Y. mixed and discharged				C.Y.	177			177	194
0110	2 C.Y.					138			138	151
0120	3 C.Y.					124			124	136
0130	4 C.Y.					113			113	124
0140	5 C.Y.					106			106	117
0200	For truck holding/waiting time past first 2 on-site hours, add				Hr.	78			78	86
0210	For trip charge beyond first 20 miles, each way, add				Mile	3.50			3.50	3.85
0220	For each additional increase of 500 psi, add				Ea.	4.17			4.17	4.59

03 31 13.35 Heavyweight Concrete, Ready Mix

		Crew	Daily Output	Labor-Hours	Unit	Material	2015 Bare Costs Labor	Equipment	Total	Total Incl O&P
0010	**HEAVYWEIGHT CONCRETE, READY MIX**, delivered									
0012	Includes local aggregate, sand, Portland cement (Type I) and water									
0015	Excludes all additives and treatments									
0020	2000 psi				C.Y.	97			97	107
0100	2500 psi					99.50			99.50	109
0150	3000 psi					102			102	112
0200	3500 psi					104			104	115
0300	4000 psi					107			107	118
0350	4500 psi					110			110	121
0400	5000 psi					113			113	125

03 31 Structural Concrete

03 31 13 – Heavyweight Structural Concrete

03 31 13.35 Heavyweight Concrete, Ready Mix		Crew	Daily Output	Labor-Hours	Unit	Material	2015 Bare Costs Labor	Equipment	Total	Total Incl O&P
0411	6000 psi				C.Y.	116			116	128
0412	8000 psi					123			123	135
0413	10,000 psi					129			129	142
0414	12,000 psi					135			135	149
1000	For high early strength (Portland cement Type III), add					10%				
1300	For winter concrete (hot water), add					4.05			4.05	4.46
1410	For mid-range water reducer, add					3.31			3.31	3.64
1420	For high-range water reducer/superplasticizer, add					5.65			5.65	6.20
1430	For retarder, add					3.23			3.23	3.55
1440	For non-Chloride accelerator, add					5.50			5.50	6.05
1450	For Chloride accelerator, per 1%, add					2.90			2.90	3.19
1460	For fiber reinforcing, synthetic (1 lb./C.Y.), add					6.65			6.65	7.30
1500	For Saturday delivery, add					10.80			10.80	11.90
1510	For truck holding/waiting time past 1st hour per load, add				Hr.	94			94	103
1520	For short load (less than 4 C.Y.), add per load				Ea.	60.50			60.50	66.50
2000	For all lightweight aggregate, add				C.Y.	45%				
03 31 13.70 Placing Concrete										
0010	**PLACING CONCRETE**									
0020	Includes labor and equipment to place, level (strike off) and consolidate									
1900	Footings, continuous, shallow, direct chute	C-6	120	.400	C.Y.		10.40	.52	10.92	17.85
1950	Pumped	C-20	150	.427			11.45	5.20	16.65	25
2000	With crane and bucket	C-7	90	.800			22	13.50	35.50	51.50
2400	Footings, spread, under 1 C.Y., direct chute	C-6	55	.873			22.50	1.13	23.63	39.50
2600	Over 5 C.Y., direct chute		120	.400			10.40	.52	10.92	17.85
2900	Foundation mats, over 20 C.Y., direct chute		350	.137			3.56	.18	3.74	6.15
4300	Slab on grade, up to 6" thick, direct chute		110	.436			11.35	.57	11.92	19.50
4350	Pumped	C-20	130	.492			13.25	6	19.25	28.50
4400	With crane and bucket	C-7	110	.655			17.90	11.05	28.95	42
4900	Walls, 8" thick, direct chute	C-6	90	.533			13.85	.69	14.54	24
4950	Pumped	C-20	100	.640			17.20	7.85	25.05	37
5000	With crane and bucket	C-7	80	.900			24.50	15.15	39.65	57.50
5050	12" thick, direct chute	C-6	100	.480			12.45	.62	13.07	21.50
5100	Pumped	C-20	110	.582			15.65	7.10	22.75	34
5200	With crane and bucket	C-7	90	.800			22	13.50	35.50	51.50
5600	Wheeled concrete dumping, add to placing costs above									
5610	Walking cart, 50' haul, add	C-18	32	.281	C.Y.		6.95	1.88	8.83	13.70
5620	150' haul, add		24	.375			9.25	2.50	11.75	18.30
5700	250' haul, add		18	.500			12.35	3.34	15.69	24
5800	Riding cart, 50' haul, add	C-19	80	.113			2.78	1.25	4.03	6.05
5810	150' haul, add		60	.150			3.70	1.66	5.36	8.05
5900	250' haul, add		45	.200			4.93	2.22	7.15	10.75

03 35 Concrete Finishing

03 35 13 – High-Tolerance Concrete Floor Finishing

03 35 13.30 Finishing Floors, High Tolerance

		Crew	Daily Output	Labor-Hours	Unit	Material	2015 Bare Costs Labor	2015 Bare Costs Equipment	Total	Total Incl O&P
0010	**FINISHING FLOORS, HIGH TOLERANCE**									
0012	Finishing of fresh concrete flatwork requires that concrete									
0013	first be placed, struck off & consolidated									
0015	Basic finishing for various unspecified flatwork									
0100	Bull float only	C-10	4000	.006	S.F.		.17		.17	.29
0125	Bull float & manual float		2000	.012			.35		.35	.57
0150	Bull float, manual float, & broom finish, w/edging & joints		1850	.013			.38		.38	.62
0200	Bull float, manual float & manual steel trowel		1265	.019			.55		.55	.91
0210	For specified Random Access Floors in ACI Classes 1, 2, 3 and 4 to achieve									
0215	Composite Overall Floor Flatness and Levelness values up to FF35/FL25									
0250	Bull float, machine float & machine trowel (walk-behind)	C-10C	1715	.014	S.F.		.41	.03	.44	.70
0300	Power screed, bull float, machine float & trowel (walk-behind)	C-10D	2400	.010			.29	.05	.34	.53
0350	Power screed, bull float, machine float & trowel (ride-on)	C-10E	4000	.006			.17	.06	.23	.36

03 35 23 – Exposed Aggregate Concrete Finishing

03 35 23.30 Finishing Floors, Exposed Aggregate

		Crew	Daily Output	Labor-Hours	Unit	Material	Labor	Equipment	Total	Total Incl O&P
0010	**FINISHING FLOORS, EXPOSED AGGREGATE**									
1600	Exposed local aggregate finish, seeded on fresh concrete, 3 lb. per S.F.	1 Cefi	625	.013	S.F.	.47	.40		.87	1.16
1650	4 lb. per S.F.	"	465	.017	"	.87	.54		1.41	1.84

03 35 29 – Tooled Concrete Finishing

03 35 29.30 Finishing Floors, Tooled

		Crew	Daily Output	Labor-Hours	Unit	Material	Labor	Equipment	Total	Total Incl O&P
0010	**FINISHING FLOORS, TOOLED**									
4400	Stair finish, fresh concrete, float finish	1 Cefi	275	.029	S.F.		.92		.92	1.49
4500	Steel trowel finish	"	200	.040	"		1.26		1.26	2.04

03 35 29.60 Finishing Walls

		Crew	Daily Output	Labor-Hours	Unit	Material	Labor	Equipment	Total	Total Incl O&P
0010	**FINISHING WALLS**									
0020	Break ties and patch voids	1 Cefi	540	.015	S.F.	.04	.47		.51	.80
0050	Burlap rub with grout	"	450	.018		.04	.56		.60	.95
0300	Bush hammer, green concrete	B-39	1000	.048			1.19	.23	1.42	2.26
0350	Cured concrete	"	650	.074			1.83	.36	2.19	3.46
0500	Acid etch	1 Cefi	575	.014		.14	.44		.58	.86
0850	Grind form fins flush	1 Clab	700	.011	L.F.		.28		.28	.47

03 35 33 – Stamped Concrete Finishing

03 35 33.50 Slab Texture Stamping

		Crew	Daily Output	Labor-Hours	Unit	Material	Labor	Equipment	Total	Total Incl O&P
0010	**SLAB TEXTURE STAMPING**									
0050	Stamping requires that concrete first be placed, struck off, consolidated,									
0060	bull floated and free of bleed water. Decorative stamping tasks include:									
0100	Step 1 - first application of dry shake colored hardener	1 Cefi	6400	.001	S.F.	.43	.04		.47	.53
0110	Step 2 - bull float		6400	.001			.04		.04	.06
0130	Step 3 - second application of dry shake colored hardener		6400	.001		.21	.04		.25	.29
0140	Step 4 - bull float, manual float & steel trowel	3 Cefi	1280	.019			.59		.59	.96
0150	Step 5 - application of dry shake colored release agent	1 Cefi	6400	.001		.10	.04		.14	.17
0160	Step 6 - place, tamp & remove mats	3 Cefi	2400	.010		1.43	.32		1.75	2.09
0170	Step 7 - touch up edges, mat joints & simulated grout lines	1 Cefi	1280	.006			.20		.20	.32
0300	Alternate stamping estimating method includes all tasks above	4 Cefi	800	.040		2.17	1.26		3.43	4.43
0400	Step 8 - pressure wash @ 3000 psi after 24 hours	1 Cefi	1600	.005			.16		.16	.26
0500	Step 9 - roll 2 coats cure/seal compound when dry	"	800	.010		.62	.32		.94	1.19

03 39 Concrete Curing

03 39 13 – Water Concrete Curing

03 39 13.50 Water Curing

		Crew	Daily Output	Labor-Hours	Unit	Material	2015 Bare Costs Labor	2015 Bare Costs Equipment	Total	Total Incl O&P
0011	**WATER CURING**									
0020	With burlap, 4 uses assumed, 7.5 oz.	2 Clab	5500	.003	S.F.	.13	.07		.20	.26
0101	10 oz.	"	5500	.003	"	.23	.07		.30	.38

03 39 23 – Membrane Concrete Curing

03 39 23.13 Chemical Compound Membrane Concrete Curing

		Crew	Daily Output	Labor-Hours	Unit	Material	Labor	Equipment	Total	Total Incl O&P
0010	**CHEMICAL COMPOUND MEMBRANE CONCRETE CURING**									
0301	Sprayed membrane curing compound	2 Clab	9500	.002	S.F.	.11	.04		.15	.19

03 39 23.23 Sheet Membrane Concrete Curing

		Crew	Daily Output	Labor-Hours	Unit	Material	Labor	Equipment	Total	Total Incl O&P
0010	**SHEET MEMBRANE CONCRETE CURING**									
0201	Curing blanket, burlap/poly, 2-ply	2 Clab	7000	.002	S.F.	.23	.06		.29	.34

03 41 Precast Structural Concrete

03 41 23 – Precast Concrete Stairs

03 41 23.50 Precast Stairs

		Crew	Daily Output	Labor-Hours	Unit	Material	Labor	Equipment	Total	Total Incl O&P
0010	**PRECAST STAIRS**									
0020	Precast concrete treads on steel stringers, 3' wide	C-12	75	.640	Riser	135	20.50	8.70	164.20	192
0300	Front entrance, 5' wide with 48" platform, 2 risers		16	3	Flight	485	97	41	623	735
0350	5 risers		12	4		765	129	54.50	948.50	1,125
0500	6' wide, 2 risers		15	3.200		535	103	43.50	681.50	810
0550	5 risers		11	4.364		845	141	59.50	1,045.50	1,225
0700	7' wide, 2 risers		14	3.429		685	111	46.50	842.50	985
1200	Basement entrance stairwell, 6 steps, incl. steel bulkhead door	B-51	22	2.182		1,450	55.50	11.15	1,516.65	1,675
1250	14 steps	"	11	4.364		2,400	111	22.50	2,533.50	2,825

03 48 Precast Concrete Specialties

03 48 43 – Precast Concrete Trim

03 48 43.40 Precast Lintels

		Crew	Daily Output	Labor-Hours	Unit	Material	Labor	Equipment	Total	Total Incl O&P
0010	**PRECAST LINTELS**, smooth gray, prestressed, stock units only									
0800	4" wide, 8" high, x 4' long	D-10	28	1.143	Ea.	25	37	16.90	78.90	108
0850	8' long		24	1.333		60.50	43	19.70	123.20	160
1000	6" wide, 8" high, x 4' long		26	1.231		35.50	39.50	18.20	93.20	125
1050	10' long		22	1.455		91	47	21.50	159.50	202

03 48 43.90 Precast Window Sills

		Crew	Daily Output	Labor-Hours	Unit	Material	Labor	Equipment	Total	Total Incl O&P
0010	**PRECAST WINDOW SILLS**									
0600	Precast concrete, 4" tapers to 3", 9" wide	D-1	70	.229	L.F.	11.90	6.75		18.65	24.50
0650	11" wide		60	.267		15.75	7.85		23.60	30.50
0700	13" wide, 3 1/2" tapers to 2 1/2", 12" wall		50	.320		15.50	9.45		24.95	33

03 54 Cast Underlayment

03 54 13 – Gypsum Cement Underlayment

03 54 13.50 Poured Gypsum Underlayment

		Crew	Daily Output	Labor-Hours	Unit	Material	2015 Bare Costs Labor	2015 Bare Costs Equipment	Total	Total Incl O&P
0010	**POURED GYPSUM UNDERLAYMENT**									
0400	Underlayment, gypsum based, self-leveling 2500 psi, pumped, 1/2" thick	C-8	24000	.002	S.F.	.36	.07	.03	.46	.54
0500	3/4" thick		20000	.003		.54	.08	.04	.66	.77
0600	1" thick		16000	.004		.72	.10	.05	.87	1.01
1400	Hand placed, 1/2" thick	C-18	450	.020		.36	.49	.13	.98	1.38
1500	3/4" thick	"	300	.030		.54	.74	.20	1.48	2.06

03 63 Epoxy Grouting

03 63 05 – Grouting of Dowels and Fasteners

03 63 05.10 Epoxy Only

		Crew	Daily Output	Labor-Hours	Unit	Material	2015 Bare Costs Labor	2015 Bare Costs Equipment	Total	Total Incl O&P
0010	**EPOXY ONLY**									
1500	Chemical anchoring, epoxy cartridge, excludes layout, drilling, fastener									
1530	For fastener 3/4" diam. x 6" embedment	2 Skwk	72	.222	Ea.	5.65	7.40		13.05	18.70
1535	1" diam. x 8" embedment		66	.242		8.45	8.10		16.55	23
1540	1-1/4" diam. x 10" embedment		60	.267		16.90	8.90		25.80	33.50
1545	1-3/4" diam. x 12" embedment		54	.296		28	9.90		37.90	47.50
1550	14" embedment		48	.333		34	11.10		45.10	56
1555	2" diam. x 12" embedment		42	.381		45	12.70		57.70	71
1560	18" embedment		32	.500		56.50	16.70		73.20	90

03 82 Concrete Boring

03 82 16 – Concrete Drilling

03 82 16.10 Concrete Impact Drilling

		Crew	Daily Output	Labor-Hours	Unit	Material	2015 Bare Costs Labor	2015 Bare Costs Equipment	Total	Total Incl O&P
0010	**CONCRETE IMPACT DRILLING**									
0020	Includes bit cost, layout and set-up time, no anchors									
0050	Up to 4" deep in concrete/brick floors/walls									
0100	Holes, 1/4" diameter	1 Carp	75	.107	Ea.	.07	3.50		3.57	5.95
0150	For each additional inch of depth in same hole, add		430	.019		.02	.61		.63	1.05
0200	3/8" diameter		63	.127		.06	4.17		4.23	7.05
0250	For each additional inch of depth in same hole, add		340	.024		.01	.77		.78	1.32
0300	1/2" diameter		50	.160		.06	5.25		5.31	8.85
0350	For each additional inch of depth in same hole, add		250	.032		.01	1.05		1.06	1.78
0400	5/8" diameter		48	.167		.09	5.50		5.59	9.30
0450	For each additional inch of depth in same hole, add		240	.033		.02	1.09		1.11	1.86
0500	3/4" diameter		45	.178		.12	5.85		5.97	9.95
0550	For each additional inch of depth in same hole, add		220	.036		.03	1.19		1.22	2.04
0600	7/8" diameter		43	.186		.16	6.10		6.26	10.40
0650	For each additional inch of depth in same hole, add		210	.038		.04	1.25		1.29	2.14
0700	1" diameter		40	.200		.17	6.55		6.72	11.25
0750	For each additional inch of depth in same hole, add		190	.042		.04	1.38		1.42	2.37
0800	1-1/4" diameter		38	.211		.26	6.90		7.16	11.90
0850	For each additional inch of depth in same hole, add		180	.044		.06	1.46		1.52	2.52
0900	1-1/2" diameter		35	.229		.39	7.50		7.89	13.05
0950	For each additional inch of depth in same hole, add		165	.048		.10	1.59		1.69	2.78
1000	For ceiling installations, add						40%			

Division Notes

		CREW	DAILY OUTPUT	LABOR-HOURS	UNIT	BARE COSTS				TOTAL INCL O&P
						MAT.	LABOR	EQUIP.	TOTAL	

Estimating Tips
04 05 00 Common Work Results for Masonry

- The terms mortar and grout are often used interchangeably, and incorrectly. Mortar is used to bed masonry units, seal the entry of air and moisture, provide architectural appearance, and allow for size variations in the units. Grout is used primarily in reinforced masonry construction and is used to bond the masonry to the reinforcing steel. Common mortar types are M(2500 psi), S(1800 psi), N(750 psi), and O(350 psi), and conform to ASTM C270. Grout is either fine or coarse and conforms to ASTM C476, and in-place strengths generally exceed 2500 psi. Mortar and grout are different components of masonry construction and are placed by entirely different methods. An estimator should be aware of their unique uses and costs.

- Mortar is included in all assembled masonry line items. The mortar cost, part of the assembled masonry material cost, includes all ingredients, all labor, and all equipment required. Please see reference number R040513-10.

- Waste, specifically the loss/droppings of mortar and the breakage of brick and block, is included in all masonry assemblies in this division. A factor of 25% is added for mortar and 3% for brick and concrete masonry units.

- Scaffolding or staging is not included in any of the Division 4 costs. Refer to Subdivision 01 54 23 for scaffolding and staging costs.

04 20 00 Unit Masonry

- The most common types of unit masonry are brick and concrete masonry. The major classifications of brick are building brick (ASTM C62), facing brick (ASTM C216), glazed brick, fire brick, and pavers. Many varieties of texture and appearance can exist within these classifications, and the estimator would be wise to check local custom and availability within the project area. For repair and remodeling jobs, matching the existing brick may be the most important criteria.

- Brick and concrete block are priced by the piece and then converted into a price per square foot of wall. Openings less than two square feet are generally ignored by the estimator because any savings in units used is offset by the cutting and trimming required.

- It is often difficult and expensive to find and purchase small lots of historic brick. Costs can vary widely. Many design issues affect costs, selection of mortar mix, and repairs or replacement of masonry materials. Cleaning techniques must be reflected in the estimate.

- All masonry walls, whether interior or exterior, require bracing. The cost of bracing walls during construction should be included by the estimator, and this bracing must remain in place until permanent bracing is complete. Permanent bracing of masonry walls is accomplished by masonry itself, in the form of pilasters or abutting wall corners, or by anchoring the walls to the structural frame. Accessories in the form of anchors, anchor slots, and ties are used, but their supply and installation can be by different trades. For instance, anchor slots on spandrel beams and columns are supplied and welded in place by the steel fabricator, but the ties from the slots into the masonry are installed by the bricklayer. Regardless of the installation method, the estimator must be certain that these accessories are accounted for in pricing.

Reference Numbers

Reference numbers are shown in shaded boxes at the beginning of some major classifications. These numbers refer to related items in the Reference Section. The reference information may be an estimating procedure, an alternate pricing method, or technical information.

Note: Not all subdivisions listed here necessarily appear in this publication. ■

04 01 Maintenance of Masonry

04 01 20 – Maintenance of Unit Masonry

04 01 20.20 Pointing Masonry

		Crew	Daily Output	Labor-Hours	Unit	Material	2015 Bare Costs Labor	Equipment	Total	Total Incl O&P
0010	POINTING MASONRY									
0300	Cut and repoint brick, hard mortar, running bond	1 Bric	80	.100	S.F.	.56	3.23		3.79	6
0320	Common bond		77	.104		.56	3.36		3.92	6.20
0360	Flemish bond		70	.114		.59	3.69		4.28	6.80
0400	English bond		65	.123		.59	3.98		4.57	7.30
0600	Soft old mortar, running bond		100	.080		.56	2.58		3.14	4.92
0620	Common bond		96	.083		.56	2.69		3.25	5.10
0640	Flemish bond		90	.089		.59	2.87		3.46	5.45
0680	English bond		82	.098		.59	3.15		3.74	5.90
0700	Stonework, hard mortar		140	.057	L.F.	.74	1.85		2.59	3.90
0720	Soft old mortar		160	.050	"	.74	1.62		2.36	3.51
1000	Repoint, mask and grout method, running bond		95	.084	S.F.	.74	2.72		3.46	5.35
1020	Common bond		90	.089		.74	2.87		3.61	5.60
1040	Flemish bond		86	.093		.78	3		3.78	5.85
1060	English bond		77	.104		.78	3.36		4.14	6.45
2000	Scrub coat, sand grout on walls, thin mix, brushed		120	.067		2.96	2.15		5.11	6.85
2020	Troweled		98	.082		4.12	2.64		6.76	8.95

04 01 20.30 Pointing CMU

		Crew	Daily Output	Labor-Hours	Unit	Material	Labor	Equipment	Total	Total Incl O&P
0010	POINTING CMU									
0300	Cut and repoint block, hard mortar, running bond	1 Bric	190	.042	S.F.	.23	1.36		1.59	2.52
0310	Stacked bond		200	.040		.23	1.29		1.52	2.40
0600	Soft old mortar, running bond		230	.035		.23	1.12		1.35	2.12
0610	Stacked bond		245	.033		.23	1.05		1.28	2.01

04 01 30 – Unit Masonry Cleaning

04 01 30.60 Brick Washing

			Crew	Daily Output	Labor-Hours	Unit	Material	Labor	Equipment	Total	Total Incl O&P
0010	BRICK WASHING	R040130-10									
0012	Acid cleanser, smooth brick surface		1 Bric	560	.014	S.F.	.04	.46		.50	.82
0050	Rough brick			400	.020		.06	.65		.71	1.14
0060	Stone, acid wash			600	.013		.07	.43		.50	.80
1000	Muriatic acid, price per gallon in 5 gallon lots					Gal.	8.70			8.70	9.55

04 05 Common Work Results for Masonry

04 05 05 – Selective Demolition for Masonry

04 05 05.10 Selective Demolition

			Crew	Daily Output	Labor-Hours	Unit	Material	Labor	Equipment	Total	Total Incl O&P
0010	SELECTIVE DEMOLITION	R024119-10									
0200	Bond beams, 8" block with #4 bar		2 Clab	32	.500	L.F.		12.25		12.25	20.50
0300	Concrete block walls, unreinforced, 2" thick			1200	.013	S.F.		.33		.33	.55
0310	4" thick			1150	.014			.34		.34	.57
0320	6" thick			1100	.015			.36		.36	.60
0330	8" thick			1050	.015			.37		.37	.63
0340	10" thick			1000	.016			.39		.39	.66
0360	12" thick			950	.017			.41		.41	.69
0380	Reinforced alternate courses, 2" thick			1130	.014			.35		.35	.58
0390	4" thick			1080	.015			.36		.36	.61
0400	6" thick			1035	.015			.38		.38	.63
0410	8" thick			990	.016			.40		.40	.66
0420	10" thick			940	.017			.42		.42	.70
0430	12" thick			890	.018			.44		.44	.74
0440	Reinforced alternate courses & vertically 48" OC, 4" thick			900	.018			.43		.43	.73
0450	6" thick			850	.019			.46		.46	.77

04 05 Common Work Results for Masonry

04 05 05 – Selective Demolition for Masonry

04 05 05.10 Selective Demolition		Crew	Daily Output	Labor-Hours	Unit	Material	2015 Bare Costs Labor	2015 Bare Costs Equipment	Total	Total Incl O&P
0460	8" thick	2 Clab	800	.020	S.F.		.49		.49	.82
0480	10" thick		750	.021			.52		.52	.88
0490	12" thick	↓	700	.023	↓		.56		.56	.94
1000	Chimney, 16" x 16", soft old mortar	1 Clab	55	.145	C.F.		3.56		3.56	5.95
1020	Hard mortar		40	.200			4.89		4.89	8.20
1030	16" x 20", soft old mortar		55	.145			3.56		3.56	5.95
1040	Hard mortar		40	.200			4.89		4.89	8.20
1050	16" x 24", soft old mortar		55	.145			3.56		3.56	5.95
1060	Hard mortar		40	.200			4.89		4.89	8.20
1080	20" x 20", soft old mortar		55	.145			3.56		3.56	5.95
1100	Hard mortar		40	.200			4.89		4.89	8.20
1110	20" x 24", soft old mortar		55	.145			3.56		3.56	5.95
1120	Hard mortar		40	.200			4.89		4.89	8.20
1140	20" x 32", soft old mortar		55	.145			3.56		3.56	5.95
1160	Hard mortar		40	.200			4.89		4.89	8.20
1200	48" x 48", soft old mortar		55	.145			3.56		3.56	5.95
1220	Hard mortar	↓	40	.200	↓		4.89		4.89	8.20
1250	Metal, high temp steel jacket, 24" diameter	E-2	130	.369	V.L.F.		13.20	11.60	24.80	37.50
1260	60" diameter	"	60	.800			28.50	25	53.50	80
1280	Flue lining, up to 12" x 12"	1 Clab	200	.040			.98		.98	1.64
1282	Up to 24" x 24"		150	.053			1.30		1.30	2.19
2000	Columns, 8" x 8", soft old mortar		48	.167			4.08		4.08	6.85
2020	Hard mortar		40	.200			4.89		4.89	8.20
2060	16" x 16", soft old mortar		16	.500			12.25		12.25	20.50
2100	Hard mortar		14	.571			13.95		13.95	23.50
2140	24" x 24", soft old mortar		8	1			24.50		24.50	41
2160	Hard mortar		6	1.333			32.50		32.50	54.50
2200	36" x 36", soft old mortar		4	2			49		49	82
2220	Hard mortar		3	2.667	↓		65		65	109
2230	Alternate pricing method, soft old mortar		30	.267	C.F.		6.50		6.50	10.95
2240	Hard mortar	↓	23	.348	"		8.50		8.50	14.30
3000	Copings, precast or masonry, to 8" wide									
3020	Soft old mortar	1 Clab	180	.044	L.F.		1.09		1.09	1.82
3040	Hard mortar	"	160	.050	"		1.22		1.22	2.05
3100	To 12" wide									
3120	Soft old mortar	1 Clab	160	.050	L.F.		1.22		1.22	2.05
3140	Hard mortar	"	140	.057	"		1.40		1.40	2.35
4000	Fireplace, brick, 30" x 24" opening									
4020	Soft old mortar	1 Clab	2	4	Ea.		98		98	164
4040	Hard mortar		1.25	6.400			156		156	263
4100	Stone, soft old mortar		1.50	5.333			130		130	219
4120	Hard mortar		1	8	↓		196		196	330
5000	Veneers, brick, soft old mortar		140	.057	S.F.		1.40		1.40	2.35
5020	Hard mortar		125	.064			1.56		1.56	2.63
5050	Glass block, up to 4" thick		500	.016			.39		.39	.66
5100	Granite and marble, 2" thick		180	.044			1.09		1.09	1.82
5120	4" thick		170	.047			1.15		1.15	1.93
5140	Stone, 4" thick		180	.044			1.09		1.09	1.82
5160	8" thick		175	.046	↓		1.12		1.12	1.88
5400	Alternate pricing method, stone, 4" thick		60	.133	C.F.		3.26		3.26	5.45
5420	8" thick	↓	85	.094	"		2.30		2.30	3.86

04 05 Common Work Results for Masonry

04 05 13 – Masonry Mortaring

04 05 13.10 Cement

		Crew	Daily Output	Labor-Hours	Unit	Material	2015 Bare Costs Labor	Equipment	Total	Total Incl O&P
0010	**CEMENT** R040513-10									
0100	Masonry, 70 lb. bag, T.L. lots				Bag	12.15			12.15	13.35
0150	L.T.L. lots					12.85			12.85	14.15
0200	White, 70 lb. bag, T.L. lots					15.85			15.85	17.45
0250	L.T.L. lots					16.70			16.70	18.40

04 05 16 – Masonry Grouting

04 05 16.30 Grouting

		Crew	Daily Output	Labor-Hours	Unit	Material	Labor	Equipment	Total	Total Incl O&P
0010	**GROUTING**									
0011	Bond beams & lintels, 8" deep, 6" thick, 0.15 C.F. per L.F.	D-4	1480	.027	L.F.	.66	.74	.09	1.49	2.06
0020	8" thick, 0.2 C.F. per L.F.		1400	.029		1.06	.78	.10	1.94	2.56
0050	10" thick, 0.25 C.F. per L.F.		1200	.033		1.11	.91	.11	2.13	2.86
0060	12" thick, 0.3 C.F. per L.F.		1040	.038		1.33	1.05	.13	2.51	3.36
0200	Concrete block cores, solid, 4" thk., by hand, 0.067 C.F./S.F. of wall	D-8	1100	.036	S.F.	.30	1.09		1.39	2.15
0210	6" thick, pumped, 0.175 C.F. per S.F.	D-4	720	.056		.77	1.52	.19	2.48	3.59
0250	8" thick, pumped, 0.258 C.F. per S.F.		680	.059		1.14	1.61	.20	2.95	4.16
0300	10" thick, pumped, 0.340 C.F. per S.F.		660	.061		1.50	1.66	.20	3.36	4.64
0350	12" thick, pumped, 0.422 C.F. per S.F.		640	.063		1.87	1.71	.21	3.79	5.15

04 05 19 – Masonry Anchorage and Reinforcing

04 05 19.05 Anchor Bolts

		Crew	Daily Output	Labor-Hours	Unit	Material	Labor	Equipment	Total	Total Incl O&P
0010	**ANCHOR BOLTS**									
0015	Installed in fresh grout in CMU bond beams or filled cores, no templates									
0020	Hooked, with nut and washer, 1/2" diam., 8" long	1 Bric	132	.061	Ea.	1.33	1.96		3.29	4.73
0030	12" long		131	.061		1.48	1.97		3.45	4.92
0040	5/8" diameter, 8" long		129	.062		2.98	2		4.98	6.60
0050	12" long		127	.063		3.67	2.03		5.70	7.45
0060	3/4" diameter, 8" long		127	.063		3.67	2.03		5.70	7.45
0070	12" long		125	.064		4.59	2.07		6.66	8.50

04 05 19.16 Masonry Anchors

		Crew	Daily Output	Labor-Hours	Unit	Material	Labor	Equipment	Total	Total Incl O&P
0010	**MASONRY ANCHORS**									
0020	For brick veneer, galv., corrugated, 7/8" x 7", 22 Ga.	1 Bric	10.50	.762	C	14.35	24.50		38.85	57
0100	24 Ga.		10.50	.762		9.85	24.50		34.35	52
0150	16 Ga.		10.50	.762		28	24.50		52.50	71.50
0200	Buck anchors, galv., corrugated, 16 ga., 2" bend, 8" x 2"		10.50	.762		56.50	24.50		81	103
0250	8" x 3"		10.50	.762		58.50	24.50		83	106
0660	Cavity wall, Z-type, galvanized, 6" long, 1/8" diam.		10.50	.762		25	24.50		49.50	68.50
0670	3/16" diameter		10.50	.762		29	24.50		53.50	73
0680	1/4" diameter		10.50	.762		40.50	24.50		65	85.50
0850	8" long, 3/16" diameter		10.50	.762		25	24.50		49.50	68.50
0855	1/4" diameter		10.50	.762		49	24.50		73.50	95
1000	Rectangular type, galvanized, 1/4" diameter, 2" x 6"		10.50	.762		72.50	24.50		97	121
1050	4" x 6"		10.50	.762		87.50	24.50		112	138
1100	3/16" diameter, 2" x 6"		10.50	.762		45	24.50		69.50	90.50
1150	4" x 6"		10.50	.762		51.50	24.50		76	97.50
1500	Rigid partition anchors, plain, 8" long, 1" x 1/8"		10.50	.762		235	24.50		259.50	300
1550	1" x 1/4"		10.50	.762		277	24.50		301.50	345
1580	1-1/2" x 1/8"		10.50	.762		259	24.50		283.50	325
1600	1-1/2" x 1/4"		10.50	.762		325	24.50		349.50	395
1650	2" x 1/8"		10.50	.762		305	24.50		329.50	375
1700	2" x 1/4"		10.50	.762		405	24.50		429.50	485

04 05 Common Work Results for Masonry

04 05 19 – Masonry Anchorage and Reinforcing

04 05 19.26 Masonry Reinforcing Bars

		Crew	Daily Output	Labor-Hours	Unit	Material	2015 Bare Costs Labor	Equipment	Total	Total Incl O&P
0010	**MASONRY REINFORCING BARS** R040519-50									
0015	Steel bars A615, placed horiz., #3 & #4 bars	1 Bric	450	.018	Lb.	.48	.57		1.05	1.49
0050	Placed vertical, #3 & #4 bars		350	.023		.48	.74		1.22	1.76
0060	#5 & #6 bars		650	.012		.48	.40		.88	1.19
0200	Joint reinforcing, regular truss, to 6" wide, mill std galvanized		30	.267	C.L.F.	22.50	8.60		31.10	39
0250	12" wide		20	.400		26	12.90		38.90	50
0400	Cavity truss with drip section, to 6" wide		30	.267		22.50	8.60		31.10	39
0450	12" wide		20	.400		26	12.90		38.90	50

04 21 Clay Unit Masonry

04 21 13 – Brick Masonry

04 21 13.13 Brick Veneer Masonry

		Crew	Daily Output	Labor-Hours	Unit	Material	Labor	Equipment	Total	Total Incl O&P
0010	**BRICK VENEER MASONRY**, T.L. lots, excl. scaff., grout & reinforcing									
0015	Material costs incl. 3% brick and 25% mortar waste	R042110-10								
2000	Standard, sel. common, 4" x 2-2/3" x 8", (6.75/S.F.)	R042110-20	D-8	230	.174	S.F.	4.16	5.20		9.36	13.30
2020	Red, 4" x 2-2/3" x 8", running bond			220	.182		3.72	5.45		9.17	13.20
2050	Full header every 6th course (7.88/S.F.)	R042110-50		185	.216		4.34	6.50		10.84	15.60
2100	English, full header every 2nd course (10.13/S.F.)			140	.286		5.55	8.60		14.15	20.50
2150	Flemish, alternate header every course (9.00/S.F.)			150	.267		4.94	8		12.94	18.80
2200	Flemish, alt. header every 6th course (7.13/S.F.)			205	.195		3.93	5.85		9.78	14.05
2250	Full headers throughout (13.50/S.F.)			105	.381		7.40	11.45		18.85	27
2300	Rowlock course (13.50/S.F.)			100	.400		7.40	12		19.40	28
2350	Rowlock stretcher (4.50/S.F.)			310	.129		2.51	3.88		6.39	9.20
2400	Soldier course (6.75/S.F.)			200	.200		3.72	6		9.72	14.10
2450	Sailor course (4.50/S.F.)			290	.138		2.51	4.14		6.65	9.65
2600	Buff or gray face, running bond, (6.75/S.F.)			220	.182		3.93	5.45		9.38	13.40
2700	Glazed face brick, running bond			210	.190		11.75	5.70		17.45	22.50
2750	Full header every 6th course (7.88/S.F.)			170	.235		13.70	7.05		20.75	27
3000	Jumbo, 6" x 4" x 12" running bond (3.00/S.F.)			435	.092		4.67	2.76		7.43	9.75
3050	Norman, 4" x 2-2/3" x 12" running bond, (4.5/S.F.)			320	.125		5.95	3.76		9.71	12.75
3100	Norwegian, 4" x 3-1/5" x 12" (3.75/S.F.)			375	.107		5.15	3.20		8.35	11
3150	Economy, 4" x 4" x 8" (4.50/S.F.)			310	.129		4.08	3.88		7.96	10.95
3200	Engineer, 4" x 3-1/5" x 8" (5.63/S.F.)			260	.154		3.61	4.62		8.23	11.65
3250	Roman, 4" x 2" x 12" (6.00/S.F.)			250	.160		6.95	4.81		11.76	15.60
3300	SCR, 6" x 2-2/3" x 12" (4.50/S.F.)			310	.129		6	3.88		9.88	13.05
3350	Utility, 4" x 4" x 12" (3.00/S.F.)			360	.111		4.65	3.34		7.99	10.65
3360	For less than truck load lots, add					M	15%				
3400	For cavity wall construction, add							15%			
3450	For stacked bond, add							10%			
3500	For interior veneer construction, add							15%			
3550	For curved walls, add							30%			

04 21 13.14 Thin Brick Veneer

		Crew	Daily Output	Labor-Hours	Unit	Material	Labor	Equipment	Total	Total Incl O&P
0010	**THIN BRICK VENEER**									
0015	Material costs incl. 3% brick and 25% mortar waste									
0020	On & incl. metal panel support sys, modular, 2-2/3" x 5/8" x 8", red	D-7	92	.174	S.F.	9	4.65		13.65	17.45
0100	Closure, 4" x 5/8" x 8"		110	.145		8.80	3.89		12.69	15.95
0110	Norman, 2-2/3" x 5/8" x 12"		110	.145		8.85	3.89		12.74	16
0120	Utility, 4" x 5/8" x 12"		125	.128		8.55	3.42		11.97	14.95
0130	Emperor, 4" x 3/4" x 16"		175	.091		9.65	2.44		12.09	14.60
0140	Super emperor, 8" x 3/4" x 16"		195	.082		9.95	2.19		12.14	14.50

04 21 Clay Unit Masonry

04 21 13 – Brick Masonry

		Crew	Daily Output	Labor-Hours	Unit	Material	2015 Bare Costs Labor	Equipment	Total	Total Incl O&P
04 21 13.14	**Thin Brick Veneer**									
0150	For L shaped corners with 4" return, add				L.F.	9.25			9.25	10.20
0200	On masonry/plaster back-up, modular, 2-2/3" x 5/8" x 8", red	D-7	137	.117	S.F.	4.21	3.12		7.33	9.70
0210	Closure, 4" x 5/8" x 8"		165	.097		3.98	2.59		6.57	8.60
0220	Norman, 2-2/3" x 5/8" x 12"		165	.097		4.02	2.59		6.61	8.60
0230	Utility, 4" x 5/8" x 12"		185	.086		3.75	2.31		6.06	7.85
0240	Emperor, 4" x 3/4" x 16"		260	.062		4.85	1.65		6.50	8
0250	Super emperor, 8" x 3/4" x 16"		285	.056		5.15	1.50		6.65	8.10
0260	For L shaped corners with 4" return, add				L.F.	9.25			9.25	10.20
0270	For embedment into pre-cast concrete panels, add				S.F.	14.40			14.40	15.85
04 21 13.15	**Chimney**									
0010	CHIMNEY, excludes foundation, scaffolding, grout and reinforcing									
0100	Brick, 16" x 16", 8" flue	D-1	18.20	.879	V.L.F.	24	26		50	69.50
0150	16" x 20" with one 8" x 12" flue		16	1		37.50	29.50		67	90.50
0200	16" x 24" with two 8" x 8" flues		14	1.143		55	33.50		88.50	117
0250	20" x 20" with one 12" x 12" flue		13.70	1.168		45.50	34.50		80	108
0300	20" x 24" with two 8" x 12" flues		12	1.333		62.50	39.50		102	134
0350	20" x 32" with two 12" x 12" flues		10	1.600		80	47		127	167
04 21 13.18	**Columns**									
0010	COLUMNS, solid, excludes scaffolding, grout and reinforcing									
0050	Brick, 8" x 8", 9 brick per V.L.F.	D-1	56	.286	V.L.F.	4.76	8.40		13.16	19.30
0100	12" x 8", 13.5 brick per V.L.F.		37	.432		7.15	12.75		19.90	29.50
0200	12" x 12", 20 brick per V.L.F.		25	.640		10.60	18.85		29.45	43
0300	16" x 12", 27 brick per V.L.F.		19	.842		14.30	25		39.30	57
0400	16" x 16", 36 brick per V.L.F.		14	1.143		19.05	33.50		52.55	77
0500	20" x 16", 45 brick per V.L.F.		11	1.455		24	43		67	97.50
0600	20" x 20", 56 brick per V.L.F.		9	1.778		29.50	52.50		82	120
0700	24" x 20", 68 brick per V.L.F.		7	2.286		36	67.50		103.50	152
0800	24" x 24", 81 brick per V.L.F.		6	2.667		43	78.50		121.50	178
1000	36" x 36", 182 brick per V.L.F.		3	5.333		96.50	157		253.50	370
04 21 13.30	**Oversized Brick**									
0010	OVERSIZED BRICK, excludes scaffolding, grout and reinforcing									
0100	Veneer, 4" x 2.25" x 16"	D-8	387	.103	S.F.	5.10	3.10		8.20	10.85
0102	8" x 2.25" x 16", multicell		265	.151		16.10	4.53		20.63	25.50
0105	4" x 2.75" x 16"		412	.097		5.30	2.92		8.22	10.70
0107	8" x 2.75" x 16", multicell		295	.136		16.10	4.07		20.17	24.50
0110	4" x 4" x 16"		460	.087		3.48	2.61		6.09	8.20
0120	4" x 8" x 16"		533	.075		4.08	2.25		6.33	8.25
0122	4" x 8" x 16" multi cell		327	.122		15.10	3.67		18.77	23
0125	Loadbearing, 6" x 4" x 16", grouted and reinforced		387	.103		10.30	3.10		13.40	16.55
0130	8" x 4" x 16", grouted and reinforced		327	.122		11.30	3.67		14.97	18.55
0132	10" x 4" x 16", grouted and reinforced		327	.122		23	3.67		26.67	31
0135	6" x 8" x 16", grouted and reinforced		440	.091		13.40	2.73		16.13	19.25
0140	8" x 8" x 16", grouted and reinforced		400	.100		14.30	3		17.30	20.50
0145	Curtainwall/reinforced veneer, 6" x 4" x 16"		387	.103		14.65	3.10		17.75	21.50
0150	8" x 4" x 16"		327	.122		17.80	3.67		21.47	26
0152	10" x 4" x 16"		327	.122		25	3.67		28.67	33.50
0155	6" x 8" x 16"		440	.091		18.20	2.73		20.93	24.50
0160	8" x 8" x 16"		400	.100		25.50	3		28.50	33
0200	For 1 to 3 slots in face, add					15%				
0210	For 4 to 7 slots in face, add					25%				
0220	For bond beams, add					20%				
0230	For bullnose shapes, add					20%				

04 21 Clay Unit Masonry

04 21 13 – Brick Masonry

04 21 13.30 Oversized Brick

		Crew	Daily Output	Labor-Hours	Unit	Material	2015 Bare Costs Labor	Equipment	Total	Total Incl O&P
0240	For open end knockout, add				S.F.	10%				
0250	For white or gray color group, add					10%				
0260	For 135 degree corner, add					250%				

04 21 13.35 Common Building Brick

0010	**COMMON BUILDING BRICK**, C62, TL lots, material only R042110-20									
0020	Standard				M	490			490	540
0050	Select				"	500			500	550

04 21 13.45 Face Brick

0010	**FACE BRICK** Material Only, C216, TL lots R042110-20									
0300	Standard modular, 4" x 2-2/3" x 8"				M	435			435	480
2170	For less than truck load lots, add					15			15	16.50
2180	For buff or gray brick, add					16			16	17.60

04 22 Concrete Unit Masonry

04 22 10 – Concrete Masonry Units

04 22 10.11 Autoclave Aerated Concrete Block

		Crew	Daily Output	Labor-Hours	Unit	Material	Labor	Equipment	Total	Total Incl O&P
0010	**AUTOCLAVE AERATED CONCRETE BLOCK**, excl. scaffolding, grout & reinforcing									
0050	Solid, 4" x 8" x 24", incl. mortar G	D-8	600	.067	S.F.	1.48	2		3.48	4.97
0060	6" x 8" x 24" G		600	.067		2.24	2		4.24	5.80
0070	8" x 8" x 24" G		575	.070		2.98	2.09		5.07	6.75
0080	10" x 8" x 24" G		575	.070		3.64	2.09		5.73	7.50
0090	12" x 8" x 24" G		550	.073		4.47	2.18		6.65	8.55

04 22 10.14 Concrete Block, Back-Up

0010	**CONCRETE BLOCK, BACK-UP**, C90, 2000 psi R042210-20									
0020	Normal weight, 8" x 16" units, tooled joint 1 side									
0050	Not-reinforced, 2000 psi, 2" thick	D-8	475	.084	S.F.	1.50	2.53		4.03	5.85
0200	4" thick		460	.087		1.79	2.61		4.40	6.35
0300	6" thick		440	.091		2.31	2.73		5.04	7.10
0350	8" thick		400	.100		2.84	3		5.84	8.10
0400	10" thick		330	.121		2.96	3.64		6.60	9.30
0450	12" thick	D-9	310	.155		4.05	4.56		8.61	12.05
1000	Reinforced, alternate courses, 4" thick	D-8	450	.089		1.95	2.67		4.62	6.60
1100	6" thick		430	.093		2.48	2.79		5.27	7.40
1150	8" thick		395	.101		3.03	3.04		6.07	8.40
1200	10" thick		320	.125		3.12	3.76		6.88	9.70
1250	12" thick	D-9	300	.160		4.21	4.72		8.93	12.50

04 22 10.16 Concrete Block, Bond Beam

0010	**CONCRETE BLOCK, BOND BEAM**, C90, 2000 psi									
0020	Not including grout or reinforcing									
0125	Regular block, 6" thick	D-8	584	.068	L.F.	2.65	2.06		4.71	6.35
0130	8" high, 8" thick	"	565	.071		2.71	2.13		4.84	6.55
0150	12" thick	D-9	510	.094		4.02	2.77		6.79	9.05
0525	Lightweight, 6" thick	D-8	592	.068		2.73	2.03		4.76	6.40

04 22 10.19 Concrete Block, Insulation Inserts

0010	**CONCRETE BLOCK, INSULATION INSERTS**									
0100	Styrofoam, plant installed, add to block prices									
0200	8" x 16" units, 6" thick				S.F.	1.20			1.20	1.32
0250	8" thick					1.35			1.35	1.49
0300	10" thick					1.40			1.40	1.54
0350	12" thick					1.55			1.55	1.71

04 22 Concrete Unit Masonry

04 22 10 – Concrete Masonry Units

04 22 10.19 Concrete Block, Insulation Inserts

		Crew	Daily Output	Labor-Hours	Unit	Material	Labor	Equipment	Total	Total Incl O&P
0500	8" x 8" units, 8" thick				S.F.	1.20			1.20	1.32
0550	12" thick					1.40			1.40	1.54

04 22 10.23 Concrete Block, Decorative

		Crew	Daily Output	Labor-Hours	Unit	Material	Labor	Equipment	Total	Total Incl O&P
0010	**CONCRETE BLOCK, DECORATIVE**, C90, 2000 psi									
5000	Split rib profile units, 1" deep ribs, 8 ribs									
5100	8" x 16" x 4" thick	D-8	345	.116	S.F.	3.84	3.48		7.32	10.05
5150	6" thick		325	.123		4.37	3.70		8.07	10.95
5200	8" thick		300	.133		4.94	4.01		8.95	12.15
5250	12" thick	D-9	275	.175		5.80	5.15		10.95	14.95
5400	For special deeper colors, 4" thick, add					1.24			1.24	1.37
5450	12" thick, add					1.28			1.28	1.40
5600	For white, 4" thick, add					1.24			1.24	1.37
5650	6" thick, add					1.27			1.27	1.39
5700	8" thick, add					1.29			1.29	1.42
5750	12" thick, add					1.33			1.33	1.47

04 22 10.24 Concrete Block, Exterior

		Crew	Daily Output	Labor-Hours	Unit	Material	Labor	Equipment	Total	Total Incl O&P
0010	**CONCRETE BLOCK, EXTERIOR**, C90, 2000 psi									
0020	Reinforced alt courses, tooled joints 2 sides									
0100	Normal weight, 8" x 16" x 6" thick	D-8	395	.101	S.F.	2.32	3.04		5.36	7.60
0200	8" thick		360	.111		4	3.34		7.34	9.95
0250	10" thick		290	.138		4.24	4.14		8.38	11.55
0300	12" thick	D-9	250	.192		4.84	5.65		10.49	14.75

04 22 10.26 Concrete Block Foundation Wall

		Crew	Daily Output	Labor-Hours	Unit	Material	Labor	Equipment	Total	Total Incl O&P
0010	**CONCRETE BLOCK FOUNDATION WALL**, C90/C145									
0050	Normal-weight, cut joints, horiz joint reinf, no vert reinf.									
0200	Hollow, 8" x 16" x 6" thick	D-8	455	.088	S.F.	2.89	2.64		5.53	7.55
0250	8" thick		425	.094		3.45	2.83		6.28	8.50
0300	10" thick		350	.114		3.57	3.43		7	9.60
0350	12" thick	D-9	300	.160		4.68	4.72		9.40	13
0500	Solid, 8" x 16" block, 6" thick	D-8	440	.091		3.05	2.73		5.78	7.90
0550	8" thick	"	415	.096		4.15	2.90		7.05	9.40
0600	12" thick	D-9	350	.137		5.95	4.04		9.99	13.30

04 22 10.32 Concrete Block, Lintels

		Crew	Daily Output	Labor-Hours	Unit	Material	Labor	Equipment	Total	Total Incl O&P
0010	**CONCRETE BLOCK, LINTELS**, C90, normal weight									
0100	Including grout and horizontal reinforcing									
0200	8" x 8" x 8", 1 #4 bar	D-4	300	.133	L.F.	3.74	3.65	.45	7.84	10.70
0250	2 #4 bars		295	.136		3.96	3.71	.45	8.12	11.05
0400	8" x 16" x 8", 1 #4 bar		275	.145		3.83	3.98	.49	8.30	11.40
0450	2 #4 bars		270	.148		4.05	4.05	.49	8.59	11.75
1000	12" x 8" x 8", 1 #4 bar		275	.145		5.20	3.98	.49	9.67	12.90
1100	2 #4 bars		270	.148		5.40	4.05	.49	9.94	13.25
1150	2 #5 bars		270	.148		5.65	4.05	.49	10.19	13.50
1200	2 #6 bars		265	.151		5.95	4.13	.50	10.58	14
1500	12" x 16" x 8", 1 #4 bar		250	.160		5.80	4.37	.53	10.70	14.30
1600	2 #3 bars		245	.163		5.80	4.46	.55	10.81	14.45
1650	2 #4 bars		245	.163		6	4.46	.55	11.01	14.65
1700	2 #5 bars		240	.167		6.25	4.56	.56	11.37	15.10

04 22 10.33 Lintel Block

		Crew	Daily Output	Labor-Hours	Unit	Material	Labor	Equipment	Total	Total Incl O&P
0010	**LINTEL BLOCK**									
3481	Lintel block 6" x 8" x 8"	D-1	300	.053	Ea.	1.28	1.57		2.85	4.03
3501	6" x 16" x 8"		275	.058		2	1.72		3.72	5.05

04 22 Concrete Unit Masonry

04 22 10 - Concrete Masonry Units

04 22 10.33 Lintel Block		Crew	Daily Output	Labor-Hours	Unit	Material	2015 Bare Costs Labor	Equipment	Total	Total Incl O&P
3521	8" x 8" x 8"	D-1	275	.058	Ea.	1.15	1.72		2.87	4.13
3561	8" x 16" x 8"		250	.064		2.02	1.89		3.91	5.35

04 22 10.34 Concrete Block, Partitions		Crew	Daily Output	Labor-Hours	Unit	Material	Labor	Equipment	Total	Total Incl O&P
0010	**CONCRETE BLOCK, PARTITIONS**, excludes scaffolding									
1000	Lightweight block, tooled joints, 2 sides, hollow									
1100	Not reinforced, 8" x 16" x 4" thick	D-8	440	.091	S.F.	1.80	2.73		4.53	6.55
1150	6" thick		410	.098		2.56	2.93		5.49	7.70
1200	8" thick		385	.104		3.13	3.12		6.25	8.65
1250	10" thick		370	.108		3.79	3.25		7.04	9.55
1300	12" thick	D-9	350	.137		4.01	4.04		8.05	11.15
4000	Regular block, tooled joints, 2 sides, hollow									
4100	Not reinforced, 8" x 16" x 4" thick	D-8	430	.093	S.F.	1.70	2.79		4.49	6.55
4150	6" thick		400	.100		2.22	3		5.22	7.45
4200	8" thick		375	.107		2.74	3.20		5.94	8.35
4250	10" thick		360	.111		2.86	3.34		6.20	8.70
4300	12" thick	D-9	340	.141		3.95	4.16		8.11	11.30

04 23 Glass Unit Masonry

04 23 13 - Vertical Glass Unit Masonry

04 23 13.10 Glass Block		Crew	Daily Output	Labor-Hours	Unit	Material	Labor	Equipment	Total	Total Incl O&P
0010	**GLASS BLOCK**									
0100	Plain, 4" thick, under 1,000 S.F., 6" x 6"	D-8	115	.348	S.F.	26.50	10.45		36.95	46.50
0150	8" x 8"		160	.250		15.35	7.50		22.85	29.50
0160	end block		160	.250		59	7.50		66.50	77.50
0170	90 deg corner		160	.250		53.50	7.50		61	71.50
0180	45 deg corner		160	.250		44	7.50		51.50	61
0200	12" x 12"		175	.229		22	6.85		28.85	36
0210	4" x 8"		160	.250		28	7.50		35.50	43.50
0220	6" x 8"		160	.250		19.90	7.50		27.40	34.50
0700	For solar reflective blocks, add					100%				
1000	Thinline, plain, 3-1/8" thick, under 1,000 S.F., 6" x 6"	D-8	115	.348	S.F.	21	10.45		31.45	40.50
1050	8" x 8"		160	.250		11.60	7.50		19.10	25.50
1400	For cleaning block after installation (both sides), add		1000	.040		.16	1.20		1.36	2.18

04 24 Adobe Unit Masonry

04 24 16 - Manufactured Adobe Unit Masonry

04 24 16.06 Adobe Brick			Crew	Daily Output	Labor-Hours	Unit	Material	Labor	Equipment	Total	Total Incl O&P
0010	**ADOBE BRICK**, Semi-stabilized, with cement mortar										
0060	Brick, 10" x 4" x 14", 2.6/S.F.	G	D-8	560	.071	S.F.	4.50	2.15		6.65	8.50
0080	12" x 4" x 16", 2.3/S.F.	G		580	.069		6.85	2.07		8.92	10.95
0100	10" x 4" x 16", 2.3/S.F.	G		590	.068		6.40	2.04		8.44	10.45
0120	8" x 4" x 16", 2.3/S.F.	G		560	.071		4.89	2.15		7.04	9
0140	4" x 4" x 16", 2.3/S.F.	G		540	.074		4.79	2.23		7.02	8.95
0160	6" x 4" x 16", 2.3/S.F.	G		540	.074		4.44	2.23		6.67	8.60
0180	4" x 4" x 12", 3.0/S.F.	G		520	.077		5.15	2.31		7.46	9.55
0200	8" x 4" x 12", 3.0/S.F.	G		520	.077		4.16	2.31		6.47	8.45

04 27 Multiple-Wythe Unit Masonry

04 27 10 – Multiple-Wythe Masonry

04 27 10.10 Cornices

		Crew	Daily Output	Labor-Hours	Unit	Material	2015 Bare Costs Labor	Equipment	Total	Total Incl O&P
0010	**CORNICES**									
0110	Face bricks, 12 brick/S.F.	D-1	30	.533	SF Face	5.65	15.70		21.35	32
0150	15 brick/S.F.	"	23	.696	"	6.75	20.50		27.25	41.50

04 27 10.30 Brick Walls

		Crew	Daily Output	Labor-Hours	Unit	Material	Labor	Equipment	Total	Total Incl O&P
0010	**BRICK WALLS**, including mortar, excludes scaffolding									
0800	Face brick, 4" thick wall, 6.75 brick/S.F.	D-8	215	.186	S.F.	3.65	5.60		9.25	13.30
0850	Common brick, 4" thick wall, 6.75 brick/S.F.		240	.167		4.03	5		9.03	12.80
0900	8" thick, 13.50 bricks per S.F.		135	.296		8.30	8.90		17.20	24
1000	12" thick, 20.25 bricks per S.F.		95	.421		12.55	12.65		25.20	35
1050	16" thick, 27.00 bricks per S.F.		75	.533		17	16		33	45
1200	Reinforced, face brick, 4" thick wall, 6.75 brick/S.F.		210	.190		3.82	5.70		9.52	13.75
1220	Common brick, 4" thick wall, 6.75 brick/S.F.		235	.170		4.20	5.10		9.30	13.15
1250	8" thick, 13.50 bricks per S.F.		130	.308		8.65	9.25		17.90	25
1300	12" thick, 20.25 bricks per S.F.		90	.444		13.05	13.35		26.40	37
1350	16" thick, 27.00 bricks per S.F.		70	.571		17.65	17.15		34.80	48

04 27 10.40 Steps

		Crew	Daily Output	Labor-Hours	Unit	Material	Labor	Equipment	Total	Total Incl O&P
0010	**STEPS**									
0012	Entry steps, select common brick	D-1	.30	53.333	M	500	1,575		2,075	3,175

04 41 Dry-Placed Stone

04 41 10 – Dry Placed Stone

04 41 10.10 Rough Stone Wall

			Crew	Daily Output	Labor-Hours	Unit	Material	Labor	Equipment	Total	Total Incl O&P
0011	**ROUGH STONE WALL**, Dry										
0012	Dry laid (no mortar), under 18" thick	G	D-1	60	.267	C.F.	12.90	7.85		20.75	27.50
0100	Random fieldstone, under 18" thick	G	D-12	60	.533		12.90	15.70		28.60	40
0150	Over 18" thick	G	"	63	.508		15.50	14.95		30.45	42
0500	Field stone veneer	G	D-8	120	.333	S.F.	12.20	10		22.20	30
0510	Valley stone veneer	G		120	.333		12.20	10		22.20	30
0520	River stone veneer	G		120	.333		12.20	10		22.20	30
0600	Rubble stone walls, in mortar bed, up to 18" thick	G	D-11	75	.320	C.F.	15.55	9.75		25.30	33.50

04 43 Stone Masonry

04 43 10 – Masonry with Natural and Processed Stone

04 43 10.45 Granite

		Crew	Daily Output	Labor-Hours	Unit	Material	Labor	Equipment	Total	Total Incl O&P
0010	**GRANITE**, cut to size									
2500	Steps, copings, etc., finished on more than one surface									
2550	Low price, gray, light gray, etc.	D-10	50	.640	C.F.	91	20.50	9.45	120.95	145
2575	Medium price, pink, brown, etc.		50	.640		118	20.50	9.45	147.95	175
2600	High price, red, black, etc.		50	.640		146	20.50	9.45	175.95	205
2800	Pavers, 4" x 4" x 4" blocks, split face and joints									
2850	Low price, gray, light gray, etc.	D-11	80	.300	S.F.	13.10	9.15		22.25	29.50
2875	Medium price, pinks, browns, etc.		80	.300		21	9.15		30.15	38
2900	High price, red, black, etc.		80	.300		29	9.15		38.15	47

04 43 10.55 Limestone

		Crew	Daily Output	Labor-Hours	Unit	Material	Labor	Equipment	Total	Total Incl O&P
0010	**LIMESTONE**, cut to size									
0020	Veneer facing panels									
0500	Texture finish, light stick, 4-1/2" thick, 5' x 12'	D-4	300	.133	S.F.	19.50	3.65	.45	23.60	28
0750	5" thick, 5' x 14' panels	D-10	275	.116		20.50	3.75	1.72	25.97	30.50

04 43 Stone Masonry

04 43 10 – Masonry with Natural and Processed Stone

04 43 10.55 Limestone		Crew	Daily Output	Labor-Hours	Unit	Material	2015 Bare Costs Labor	Equipment	Total	Total Incl O&P
1000	Sugarcube finish, 2" Thick, 3' x 5' panels	D-10	275	.116	S.F.	28.50	3.75	1.72	33.97	39.50
1050	3" Thick, 4' x 9' panels		275	.116		22.50	3.75	1.72	27.97	33
1200	4" Thick, 5' x 11' panels		275	.116		30	3.75	1.72	35.47	41
1400	Sugarcube, textured finish, 4-1/2" thick, 5' x 12'		275	.116		31	3.75	1.72	36.47	42
1450	5" thick, 5' x 14' panels		275	.116		32	3.75	1.72	37.47	43.50
2000	Coping, sugarcube finish, top & 2 sides		30	1.067	C.F.	67.50	34.50	15.80	117.80	148
2100	Sills, lintels, jambs, trim, stops, sugarcube finish, simple		20	1.600		67.50	51.50	23.50	142.50	186
2150	Detailed		20	1.600		67.50	51.50	23.50	142.50	186
2300	Steps, extra hard, 14" wide, 6" rise		50	.640	L.F.	25	20.50	9.45	54.95	72.50
3000	Quoins, plain finish, 6" x 12" x 12"	D-12	25	1.280	Ea.	37.50	37.50		75	105
3050	6" x 16" x 24"	"	25	1.280	"	50	37.50		87.50	118
04 43 10.60 Marble										
0011	**MARBLE**, ashlar, split face, 4" + or – thick, random									
0040	Lengths 1' to 4' & heights 2" to 7-1/2", average	D-8	175	.229	S.F.	17.45	6.85		24.30	30.50
0100	Base, polished, 3/4" or 7/8" thick, polished, 6" high	D-10	65	.492	L.F.	12	15.85	7.30	35.15	47.50
1000	Facing, polished finish, cut to size, 3/4" to 7/8" thick									
1050	Carrara or equal	D-10	130	.246	S.F.	19	7.95	3.64	30.59	38
1100	Arabescato or equal	"	130	.246	"	39	7.95	3.64	50.59	60
2200	Window sills, 6" x 3/4" thick	D-1	85	.188	L.F.	10.90	5.55		16.45	21.50
2500	Flooring, polished tiles, 12" x 12" x 3/8" thick									
2510	Thin set, Giallo Solare or equal	D-11	90	.267	S.F.	14.75	8.10		22.85	30
2600	Sky Blue or equal		90	.267		16	8.10		24.10	31
2700	Mortar bed, Giallo Solare or equal		65	.369		14.75	11.25		26	35
2740	Sky Blue or equal		65	.369		16	11.25		27.25	36.50
2780	Travertine, 3/8" thick, Sierra or equal	D-10	130	.246		9.25	7.95	3.64	20.84	27.50
2790	Silver or equal	"	130	.246		25.50	7.95	3.64	37.09	45
3500	Thresholds, 3' long, 7/8" thick, 4" to 5" wide, plain	D-12	24	1.333	Ea.	35.50	39.50		75	105
3550	Beveled		24	1.333	"	70.50	39.50		110	143
3700	Window stools, polished, 7/8" thick, 5" wide		85	.376	L.F.	21.50	11.10		32.60	42
04 43 10.75 Sandstone or Brownstone										
0011	**SANDSTONE OR BROWNSTONE**									
0100	Sawed face veneer, 2-1/2" thick, to 2' x 4' panels	D-10	130	.246	S.F.	19	7.95	3.64	30.59	38
0150	4" thick, to 3'-6" x 8' panels		100	.320		19	10.30	4.73	34.03	43.50
0300	Split face, random sizes		100	.320		13.65	10.30	4.73	28.68	37.50
0350	Cut stone trim (limestone)									
0360	Ribbon stone, 4" thick, 5' pieces	D-8	120	.333	Ea.	155	10		165	187
0370	Cove stone, 4" thick, 5' pieces		105	.381		156	11.45		167.45	190
0380	Cornice stone, 10" to 12" wide		90	.444		192	13.35		205.35	234
0390	Band stone, 4" thick, 5' pieces		145	.276		99.50	8.30		107.80	123
0410	Window and door trim, 3" to 4" wide		160	.250		84.50	7.50		92	106
0420	Key stone, 18" long		60	.667		88.50	20		108.50	131
04 43 10.80 Slate										
0010	**SLATE**									
3100	Stair landings, 1" thick, black, clear	D-1	65	.246	S.F.	21	7.25		28.25	35.50
3200	Ribbon	"	65	.246	"	23.50	7.25		30.75	37.50
3500	Stair treads, sand finish, 1" thick x 12" wide									
3600	3 L.F. to 6 L.F.	D-10	120	.267	L.F.	25	8.60	3.94	37.54	46
3700	Ribbon, sand finish, 1" thick x 12" wide									
3750	To 6 L.F.	D-10	120	.267	L.F.	21	8.60	3.94	33.54	42

04 43 Stone Masonry

04 43 10 – Masonry with Natural and Processed Stone

04 43 10.85 Window Sill		Crew	Daily Output	Labor-Hours	Unit	Material	2015 Bare Costs Labor	Equipment	Total	Total Incl O&P
0010	**WINDOW SILL**									
0020	Bluestone, thermal top, 10" wide, 1-1/2" thick	D-1	85	.188	S.F.	10	5.55		15.55	20.50
0050	2" thick		75	.213	"	10	6.30		16.30	21.50
0100	Cut stone, 5" x 8" plain		48	.333	L.F.	12.10	9.85		21.95	30
0200	Face brick on edge, brick, 8" wide		80	.200		5.05	5.90		10.95	15.40
0400	Marble, 9" wide, 1" thick		85	.188		8.65	5.55		14.20	18.75
0900	Slate, colored, unfading, honed, 12" wide, 1" thick		85	.188		8.50	5.55		14.05	18.60
0950	2" thick		70	.229		8.50	6.75		15.25	20.50

04 51 Flue Liner Masonry

04 51 10 – Clay Flue Lining

04 51 10.10 Flue Lining

		Crew	Daily Output	Labor-Hours	Unit	Material	Labor	Equipment	Total	Total Incl O&P
0010	**FLUE LINING**, including mortar									
0020	Clay, 8" x 8"	D-1	125	.128	V.L.F.	5.50	3.77		9.27	12.35
0100	8" x 12"		103	.155		8.15	4.58		12.73	16.60
0200	12" x 12"		93	.172		10.75	5.05		15.80	20.50
0300	12" x 18"		84	.190		21.50	5.60		27.10	33
0400	18" x 18"		75	.213		28	6.30		34.30	41
0500	20" x 20"		66	.242		40	7.15		47.15	56
0600	24" x 24"		56	.286		51.50	8.40		59.90	71
1000	Round, 18" diameter		66	.242		37	7.15		44.15	53
1100	24" diameter		47	.340		72.50	10.05		82.55	96.50

04 57 Masonry Fireplaces

04 57 10 – Brick or Stone Fireplaces

04 57 10.10 Fireplace

		Crew	Daily Output	Labor-Hours	Unit	Material	Labor	Equipment	Total	Total Incl O&P
0010	**FIREPLACE**									
0100	Brick fireplace, not incl. foundations or chimneys									
0110	30" x 29" opening, incl. chamber, plain brickwork	D-1	.40	40	Ea.	550	1,175		1,725	2,575
0200	Fireplace box only (110 brick)	"	2	8	"	157	236		393	565
0300	For elaborate brickwork and details, add					35%	35%			
0400	For hearth, brick & stone, add	D-1	2	8	Ea.	205	236		441	620
0410	For steel, damper, cleanouts, add		4	4		19.70	118		137.70	219
0600	Plain brickwork, incl. metal circulator		.50	32		995	945		1,940	2,675
0800	Face brick only, standard size, 8" x 2-2/3" x 4"		.30	53.333	M	550	1,575		2,125	3,225
0900	Stone fireplace, fieldstone, add				SF Face	8.50			8.50	9.35
1000	Cut stone, add				"	8			8	8.80

04 72 Cast Stone Masonry

04 72 10 – Cast Stone Masonry Features

04 72 10.10 Coping

		Crew	Daily Output	Labor-Hours	Unit	Material	2015 Bare Costs Labor	Equipment	Total	Total Incl O&P
0010	**COPING**, stock units									
0050	Precast concrete, 10" wide, 4" tapers to 3-1/2", 8" wall	D-1	75	.213	L.F.	17.10	6.30		23.40	29.50
0100	12" wide, 3-1/2" tapers to 3", 10" wall		70	.229		18.45	6.75		25.20	32
0110	14" wide, 4" tapers to 3-1/2", 12" wall		65	.246		20.50	7.25		27.75	34.50
0150	16" wide, 4" tapers to 3-1/2", 14" wall		60	.267		22	7.85		29.85	37.50
0300	Limestone for 12" wall, 4" thick		90	.178		14.70	5.25		19.95	25
0350	6" thick		80	.200		22	5.90		27.90	34.50
0500	Marble, to 4" thick, no wash, 9" wide		90	.178		12.05	5.25		17.30	22
0550	12" wide		80	.200		16.15	5.90		22.05	27.50
0700	Terra cotta, 9" wide		90	.178		6.35	5.25		11.60	15.70
0750	12" wide		80	.200		8.65	5.90		14.55	19.35
0800	Aluminum, for 12" wall		80	.200		8.90	5.90		14.80	19.65

04 72 20 – Cultured Stone Veneer

04 72 20.10 Cultured Stone Veneer Components

		Crew	Daily Output	Labor-Hours	Unit	Material	Labor	Equipment	Total	Total Incl O&P
0010	**CULTURED STONE VENEER COMPONENTS**									
0110	On wood frame and sheathing substrate, random sized cobbles, corner stones	D-8	70	.571	V.L.F.	10.15	17.15		27.30	39.50
0120	Field stones		140	.286	S.F.	7.40	8.60		16	22.50
0130	Random sized flats, corner stones		70	.571	V.L.F.	10.05	17.15		27.20	39.50
0140	Field stones		140	.286	S.F.	8.70	8.60		17.30	24
0150	Horizontal lined ledgestones, corner stones		75	.533	V.L.F.	10.15	16		26.15	37.50
0160	Field stones		150	.267	S.F.	7.40	8		15.40	21.50
0170	Random shaped flats, corner stones		65	.615	V.L.F.	10.15	18.50		28.65	42
0180	Field stones		150	.267	S.F.	7.40	8		15.40	21.50
0190	Random shaped/textured face, corner stones		65	.615	V.L.F.	10.15	18.50		28.65	42
0200	Field stones		130	.308	S.F.	7.40	9.25		16.65	23.50
0210	Random shaped river rock, corner stones		65	.615	V.L.F.	10.15	18.50		28.65	42
0220	Field stones		130	.308	S.F.	7.40	9.25		16.65	23.50
0240	On concrete or CMU substrate, random sized cobbles, corner stones		70	.571	V.L.F.	9.55	17.15		26.70	39
0250	Field stones		140	.286	S.F.	7.10	8.60		15.70	22
0260	Random sized flats, corner stones		70	.571	V.L.F.	9.45	17.15		26.60	39
0270	Field stones		140	.286	S.F.	8.40	8.60		17	23.50
0280	Horizontal lined ledgestones, corner stones		75	.533	V.L.F.	9.55	16		25.55	37
0290	Field stones		150	.267	S.F.	7.10	8		15.10	21
0300	Random shaped flats, corner stones		70	.571	V.L.F.	9.55	17.15		26.70	39
0310	Field stones		140	.286	S.F.	7.10	8.60		15.70	22
0320	Random shaped/textured face, corner stones		65	.615	V.L.F.	9.55	18.50		28.05	41.50
0330	Field stones		130	.308	S.F.	7.10	9.25		16.35	23
0340	Random shaped river rock, corner stones		65	.615	V.L.F.	9.55	18.50		28.05	41.50
0350	Field stones		130	.308	S.F.	7.10	9.25		16.35	23
0360	Cultured stone veneer, #15 felt weather resistant barrier	1 Clab	3700	.002	Sq.	5.40	.05		5.45	6.05
0390	Water table or window sill, 18" long	1 Bric	80	.100	Ea.	10.30	3.23		13.53	16.75

Division Notes

	CREW	DAILY OUTPUT	LABOR-HOURS	UNIT	BARE COSTS				TOTAL INCL O&P
					MAT.	LABOR	EQUIP.	TOTAL	

Division 5 – Metals

Estimating Tips
05 05 00 Common Work Results for Metals

- Nuts, bolts, washers, connection angles, and plates can add a significant amount to both the tonnage of a structural steel job and the estimated cost. As a rule of thumb, add 10% to the total weight to account for these accessories.
- Type 2 steel construction, commonly referred to as "simple construction," consists generally of field-bolted connections with lateral bracing supplied by other elements of the building, such as masonry walls or x-bracing. The estimator should be aware, however, that shop connections may be accomplished by welding or bolting. The method may be particular to the fabrication shop and may have an impact on the estimated cost.

05 10 00 Structural Steel

- Steel items can be obtained from two sources: a fabrication shop or a metals service center. Fabrication shops can fabricate items under more controlled conditions than can crews in the field. They are also more efficient and can produce items more economically. Metal service centers serve as a source of long mill shapes to both fabrication shops and contractors.
- Most line items in this structural steel subdivision, and most items in 05 50 00 Metal Fabrications, are indicated as being shop fabricated. The bare material cost for these shop fabricated items is the "Invoice Cost" from the shop and includes the mill base price of steel plus mill extras, transportation to the shop, shop drawings and detailing where warranted, shop fabrication and handling, sandblasting and a shop coat of primer paint, all necessary structural bolts, and delivery to the job site. The bare labor cost and bare equipment cost for these shop fabricated items is for field installation or erection.
- Line items in Subdivision 05 12 23.40 Lightweight Framing, and other items scattered in Division 5, are indicated as being field fabricated. The bare material cost for these field fabricated items is the "Invoice Cost" from the metals service center and includes the mill base price of steel plus mill extras, transportation to the metals service center, material handling, and delivery of long lengths of mill shapes to the job site. Material costs for structural bolts and welding rods should be added to the estimate. The bare labor cost and bare equipment cost for these items is for both field fabrication and field installation or erection, and include time for cutting, welding and drilling in the fabricated metal items. Drilling into concrete and fasteners to fasten field fabricated items to other work are not included and should be added to the estimate.

05 20 00 Steel Joist Framing

- In any given project the total weight of open web steel joists is determined by the loads to be supported and the design. However, economies can be realized in minimizing the amount of labor used to place the joists. This is done by maximizing the joist spacing, and therefore minimizing the number of joists required to be installed on the job. Certain spacings and locations may be required by the design, but in other cases maximizing the spacing and keeping it as uniform as possible will keep the costs down.

05 30 00 Steel Decking

- The takeoff and estimating of metal deck involves more than simply the area of the floor or roof and the type of deck specified or shown on the drawings. Many different sizes and types of openings may exist. Small openings for individual pipes or conduits may be drilled after the floor/roof is installed, but larger openings may require special deck lengths as well as reinforcing or structural support. The estimator should determine who will be supplying this reinforcing. Additionally, some deck terminations are part of the deck package, such as screed angles and pour stops, and others will be part of the steel contract, such as angles attached to structural members and cast-in-place angles and plates. The estimator must ensure that all pieces are accounted for in the complete estimate.

05 50 00 Metal Fabrications

- The most economical steel stairs are those that use common materials, standard details, and most importantly, a uniform and relatively simple method of field assembly. Commonly available A36 channels and plates are very good choices for the main stringers of the stairs, as are angles and tees for the carrier members. Risers and treads are usually made by specialty shops, and it is most economical to use a typical detail in as many places as possible. The stairs should be pre-assembled and shipped directly to the site. The field connections should be simple and straightforward to be accomplished efficiently, and with minimum equipment and labor.

Reference Numbers
Reference numbers are shown in shaded boxes at the beginning of some major classifications. These numbers refer to related items in the Reference Section. The reference information may be an estimating procedure, an alternate pricing method, or technical information.

Note: Not all subdivisions listed here necessarily appear in this publication. ■

05 05 Common Work Results for Metals

05 05 19 – Post-Installed Concrete Anchors

05 05 19.10 Chemical Anchors

		Crew	Daily Output	Labor-Hours	Unit	Material	2015 Bare Costs Labor	Equipment	Total	Total Incl O&P
0010	**CHEMICAL ANCHORS**									
0020	Includes layout & drilling									
1430	Chemical anchor, w/rod & epoxy cartridge, 3/4" diam. x 9-1/2" long	B-89A	27	.593	Ea.	9.80	17.15	4.28	31.23	44.50
1435	1" diameter x 11-3/4" long		24	.667		17.55	19.25	4.82	41.62	57
1440	1-1/4" diameter x 14" long		21	.762		36.50	22	5.50	64	83
1445	1-3/4" diameter x 15" long		20	.800		65	23	5.80	93.80	117
1450	18" long		17	.941		78	27	6.80	111.80	140
1455	2" diameter x 18" long		16	1		103	29	7.25	139.25	169
1460	24" long		15	1.067		134	31	7.70	172.70	208

05 05 19.20 Expansion Anchors

			Crew	Daily Output	Labor-Hours	Unit	Material	Labor	Equipment	Total	Total Incl O&P
0010	**EXPANSION ANCHORS**										
0100	Anchors for concrete, brick or stone, no layout and drilling										
0200	Expansion shields, zinc, 1/4" diameter, 1-5/16" long, single	G	1 Carp	90	.089	Ea.	.43	2.92		3.35	5.35
0300	1-3/8" long, double	G		85	.094		.54	3.09		3.63	5.80
0400	3/8" diameter, 1-1/2" long, single	G		85	.094		.64	3.09		3.73	5.90
0500	2" long, double	G		80	.100		1.18	3.29		4.47	6.80
0600	1/2" diameter, 2-1/16" long, single	G		80	.100		1.18	3.29		4.47	6.80
0700	2-1/2" long, double	G		75	.107		1.91	3.50		5.41	8
0800	5/8" diameter, 2-5/8" long, single	G		75	.107		2.05	3.50		5.55	8.15
0900	2-3/4" long, double	G		70	.114		2.70	3.75		6.45	9.25
1000	3/4" diameter, 2-3/4" long, single	G		70	.114		3.09	3.75		6.84	9.70
1100	3-15/16" long, double	G		65	.123		5	4.04		9.04	12.30
2100	Hollow wall anchors for gypsum wall board, plaster or tile										
2500	3/16" diameter, short	G	1 Carp	150	.053	Ea.	.40	1.75		2.15	3.38
3000	Toggle bolts, bright steel, 1/8" diameter, 2" long	G		85	.094		.19	3.09		3.28	5.40
3100	4" long	G		80	.100		.25	3.29		3.54	5.80
3200	3/16" diameter, 3" long	G		80	.100		.26	3.29		3.55	5.80
3300	6" long	G		75	.107		.36	3.50		3.86	6.30
3400	1/4" diameter, 3" long	G		75	.107		.34	3.50		3.84	6.25
3500	6" long	G		70	.114		.51	3.75		4.26	6.85
3600	3/8" diameter, 3" long	G		70	.114		.84	3.75		4.59	7.20
3700	6" long	G		60	.133		1.36	4.38		5.74	8.85
3800	1/2" diameter, 4" long	G		60	.133		1.88	4.38		6.26	9.40
3900	6" long	G		50	.160		2.45	5.25		7.70	11.50
4000	Nailing anchors										
4100	Nylon nailing anchor, 1/4" diameter, 1" long		1 Carp	3.20	2.500	C	21	82		103	161
4200	1-1/2" long			2.80	2.857		26	94		120	187
4300	2" long			2.40	3.333		30.50	110		140.50	218
4400	Metal nailing anchor, 1/4" diameter, 1" long	G		3.20	2.500		17.85	82		99.85	158
4500	1-1/2" long	G		2.80	2.857		23.50	94		117.50	184
4600	2" long	G		2.40	3.333		27.50	110		137.50	215
5000	Screw anchors for concrete, masonry,										
5100	stone & tile, no layout or drilling included										
5700	Lag screw shields, 1/4" diameter, short	G	1 Carp	90	.089	Ea.	.34	2.92		3.26	5.25
5800	Long	G		85	.094		.38	3.09		3.47	5.60
5900	3/8" diameter, short	G		85	.094		.67	3.09		3.76	5.95
6000	Long	G		80	.100		.87	3.29		4.16	6.45
6100	1/2" diameter, short	G		80	.100		.95	3.29		4.24	6.55
6200	Long	G		75	.107		1.22	3.50		4.72	7.25
6300	5/8" diameter, short	G		70	.114		1.45	3.75		5.20	7.90
6400	Long	G		65	.123		1.86	4.04		5.90	8.85
6600	Lead, #6 & #8, 3/4" long	G		260	.031		.18	1.01		1.19	1.90

05 05 Common Work Results for Metals

05 05 19 – Post-Installed Concrete Anchors

05 05 19.20 Expansion Anchors			Crew	Daily Output	Labor-Hours	Unit	Material	2015 Bare Costs Labor	Equipment	Total	Total Incl O&P
6700	#10 - #14, 1-1/2" long	G	1 Carp	200	.040	Ea.	.37	1.31		1.68	2.62
6800	#16 & #18, 1-1/2" long	G		160	.050		.41	1.64		2.05	3.21
6900	Plastic, #6 & #8, 3/4" long			260	.031		.04	1.01		1.05	1.74
7000	#8 & #10, 7/8" long			240	.033		.04	1.09		1.13	1.88
7100	#10 & #12, 1" long			220	.036		.05	1.19		1.24	2.07
7200	#14 & #16, 1-1/2" long			160	.050		.07	1.64		1.71	2.84
8950	Self-drilling concrete screw, hex washer head, 3/16" diam. x 1-3/4" long	G		300	.027		.20	.88		1.08	1.69
8960	2-1/4" long	G		250	.032		.24	1.05		1.29	2.02
8970	Phillips flat head, 3/16" diam. x 1-3/4" long	G		300	.027		.21	.88		1.09	1.70
8980	2-1/4" long	G		250	.032		.23	1.05		1.28	2.01

05 05 21 – Fastening Methods for Metal

05 05 21.15 Drilling Steel

		Crew	Daily Output	Labor-Hours	Unit	Material	Labor	Equipment	Total	Total Incl O&P
0010	**DRILLING STEEL**									
1910	Drilling & layout for steel, up to 1/4" deep, no anchor									
1920	Holes, 1/4" diameter	1 Sswk	112	.071	Ea.	.09	2.52		2.61	4.83
1925	For each additional 1/4" depth, add		336	.024		.09	.84		.93	1.68
1930	3/8" diameter		104	.077		.09	2.72		2.81	5.20
1935	For each additional 1/4" depth, add		312	.026		.09	.91		1	1.80
1940	1/2" diameter		96	.083		.10	2.94		3.04	5.60
1945	For each additional 1/4" depth, add		288	.028		.10	.98		1.08	1.95
1950	5/8" diameter		88	.091		.14	3.21		3.35	6.15
1955	For each additional 1/4" depth, add		264	.030		.14	1.07		1.21	2.17
1960	3/4" diameter		80	.100		.18	3.53		3.71	6.85
1965	For each additional 1/4" depth, add		240	.033		.18	1.18		1.36	2.41
1970	7/8" diameter		72	.111		.23	3.92		4.15	7.60
1975	For each additional 1/4" depth, add		216	.037		.23	1.31		1.54	2.71
1980	1" diameter		64	.125		.24	4.41		4.65	8.55
1985	For each additional 1/4" depth, add		192	.042		.24	1.47		1.71	3.02
1990	For drilling up, add						40%			

05 05 23 – Metal Fastenings

05 05 23.10 Bolts and Hex Nuts

			Crew	Daily Output	Labor-Hours	Unit	Material	Labor	Equipment	Total	Total Incl O&P
0010	**BOLTS & HEX NUTS**, Steel, A307										
0100	1/4" diameter, 1/2" long	G	1 Sswk	140	.057	Ea.	.06	2.02		2.08	3.86
0200	1" long	G		140	.057		.07	2.02		2.09	3.87
0300	2" long	G		130	.062		.10	2.17		2.27	4.19
0400	3" long	G		130	.062		.15	2.17		2.32	4.25
0500	4" long	G		120	.067		.17	2.35		2.52	4.60
0600	3/8" diameter, 1" long	G		130	.062		.14	2.17		2.31	4.24
0700	2" long	G		130	.062		.18	2.17		2.35	4.28
0800	3" long	G		120	.067		.24	2.35		2.59	4.68
0900	4" long	G		120	.067		.30	2.35		2.65	4.75
1000	5" long	G		115	.070		.38	2.46		2.84	5
1100	1/2" diameter, 1-1/2" long	G		120	.067		.40	2.35		2.75	4.86
1200	2" long	G		120	.067		.46	2.35		2.81	4.93
1300	4" long	G		115	.070		.75	2.46		3.21	5.45
1400	6" long	G		110	.073		1.05	2.57		3.62	5.95
1500	8" long	G		105	.076		1.38	2.69		4.07	6.55
1600	5/8" diameter, 1-1/2" long	G		120	.067		.98	2.35		3.33	5.50
1700	2" long	G		120	.067		1.09	2.35		3.44	5.60
1800	4" long	G		115	.070		1.59	2.46		4.05	6.35
1900	6" long	G		110	.073		2.05	2.57		4.62	7.10
2000	8" long	G		105	.076		3.06	2.69		5.75	8.40

05 05 Common Work Results for Metals

05 05 23 – Metal Fastenings

	05 05 23.10 Bolts and Hex Nuts		Crew	Daily Output	Labor-Hours	Unit	Material	2015 Bare Costs Labor	Equipment	Total	Total Incl O&P
2100	10" long	G	1 Sswk	100	.080	Ea.	3.87	2.82		6.69	9.55
2200	3/4" diameter, 2" long	G		120	.067		1.15	2.35		3.50	5.70
2300	4" long	G		110	.073		1.65	2.57		4.22	6.65
2400	6" long	G		105	.076		2.12	2.69		4.81	7.40
2500	8" long	G		95	.084		3.20	2.97		6.17	9.10
2600	10" long	G		85	.094		4.20	3.32		7.52	10.85
2700	12" long	G		80	.100		4.92	3.53		8.45	12.05
2800	1" diameter, 3" long	G		105	.076		2.69	2.69		5.38	8
2900	6" long	G		90	.089		3.94	3.14		7.08	10.25
3000	12" long	G		75	.107		7.10	3.77		10.87	14.85
3100	For galvanized, add						75%				
3200	For stainless, add						350%				

05 05 23.30 Lag Screws

0010	**LAG SCREWS**										
0020	Steel, 1/4" diameter, 2" long	G	1 Carp	200	.040	Ea.	.09	1.31		1.40	2.31
0100	3/8" diameter, 3" long	G		150	.053		.30	1.75		2.05	3.27
0200	1/2" diameter, 3" long	G		130	.062		.64	2.02		2.66	4.09
0300	5/8" diameter, 3" long	G		120	.067		1.15	2.19		3.34	4.95

05 05 23.50 Powder Actuated Tools and Fasteners

0010	**POWDER ACTUATED TOOLS & FASTENERS**										
0020	Stud driver, .22 caliber, single shot					Ea.	148			148	163
0100	.27 caliber, semi automatic, strip					"	440			440	485
0300	Powder load, single shot, .22 cal, power level 2, brown					C	5.35			5.35	5.90
0400	Strip, .27 cal, power level 4, red						7.70			7.70	8.50
0600	Drive pin, .300 x 3/4" long	G	1 Carp	4.80	1.667		4.11	55		59.11	96.50
0700	.300 x 3" long with washer	G	"	4	2		12.65	65.50		78.15	124

05 05 23.55 Rivets

0010	**RIVETS**										
0100	Aluminum rivet & mandrel, 1/2" grip length x 1/8" diameter	G	1 Carp	4.80	1.667	C	7.65	55		62.65	100
0200	3/16" diameter	G		4	2		11.40	65.50		76.90	123
0300	Aluminum rivet, steel mandrel, 1/8" diameter	G		4.80	1.667		10.25	55		65.25	103
0400	3/16" diameter	G		4	2		16.45	65.50		81.95	128
0500	Copper rivet, steel mandrel, 1/8" diameter	G		4.80	1.667		9.50	55		64.50	102
0800	Stainless rivet & mandrel, 1/8" diameter	G		4.80	1.667		24.50	55		79.50	119
0900	3/16" diameter	G		4	2		39	65.50		104.50	153
1000	Stainless rivet, steel mandrel, 1/8" diameter	G		4.80	1.667		15.70	55		70.70	109
1100	3/16" diameter	G		4	2		25.50	65.50		91	138
1200	Steel rivet and mandrel, 1/8" diameter	G		4.80	1.667		7.40	55		62.40	100
1300	3/16" diameter	G		4	2		12	65.50		77.50	123
1400	Hand riveting tool, standard					Ea.	70.50			70.50	77.50
1500	Deluxe						365			365	400
1600	Power riveting tool, standard						500			500	555
1700	Deluxe						1,650			1,650	1,825

05 12 Structural Steel Framing

05 12 23 – Structural Steel for Buildings

05 12 23.10 Ceiling Supports		Crew	Daily Output	Labor-Hours	Unit	Material	2015 Bare Costs Labor	Equipment	Total	Total Incl O&P
0010	**CEILING SUPPORTS**									
1000	Entrance door/folding partition supports, shop fabricated	G E-4	60	.533	L.F.	26.50	19.10	2.43	48.03	67.50
1100	Linear accelerator door supports	G	14	2.286		121	82	10.40	213.40	298
1200	Lintels or shelf angles, hung, exterior hot dipped galv.	G	267	.120		18.10	4.29	.55	22.94	28.50
1250	Two coats primer paint instead of galv.	G	267	.120		15.65	4.29	.55	20.49	26
1400	Monitor support, ceiling hung, expansion bolted	G	4	8	Ea.	420	286	36.50	742.50	1,050
1450	Hung from pre-set inserts	G	6	5.333		450	191	24.50	665.50	880
1600	Motor supports for overhead doors	G	4	8		214	286	36.50	536.50	815
1700	Partition support for heavy folding partitions, without pocket	G	24	1.333	L.F.	60.50	47.50	6.10	114.10	163
1750	Supports at pocket only	G	12	2.667		121	95.50	12.15	228.65	325
2000	Rolling grilles & fire door supports	G	34	.941		51.50	33.50	4.29	89.29	125
2100	Spider-leg light supports, expansion bolted to ceiling slab	G	8	4	Ea.	172	143	18.25	333.25	480
2150	Hung from pre-set inserts	G	12	2.667	"	186	95.50	12.15	293.65	395
2400	Toilet partition support	G	36	.889	L.F.	60.50	32	4.05	96.55	130
2500	X-ray travel gantry support	G	12	2.667	"	207	95.50	12.15	314.65	420

05 12 23.15 Columns, Lightweight		Crew	Daily Output	Labor-Hours	Unit	Material	Labor	Equipment	Total	Total Incl O&P
0010	**COLUMNS, LIGHTWEIGHT**									
8000	Lally columns, to 8', 3-1/2" diameter	2 Carp	24	.667	Ea.	66	22		88	110
8080	4" diameter	"	20	.800	"	81	26.50		107.50	133

05 12 23.17 Columns, Structural		Crew	Daily Output	Labor-Hours	Unit	Material	Labor	Equipment	Total	Total Incl O&P
0010	**COLUMNS, STRUCTURAL**									
0015	Made from recycled materials									
0020	Shop fab'd for 100-ton, 1-2 story project, bolted connections									
0800	Steel, concrete filled, extra strong pipe, 3-1/2" diameter	E-2	660	.073	L.F.	43.50	2.60	2.29	48.39	55.50
0830	4" diameter		780	.062		48.50	2.20	1.94	52.64	59.50
0890	5" diameter		1020	.047		58	1.68	1.48	61.16	68.50
0930	6" diameter		1200	.040		77	1.43	1.26	79.69	88.50
0940	8" diameter		1100	.044		77	1.56	1.37	79.93	89
1100	For galvanizing, add				Lb.	.25			.25	.28
1300	For web ties, angles, etc., add per added lb.	1 Sswk	945	.008		1.33	.30		1.63	2.02
1500	Steel pipe, extra strong, no concrete, 3" to 5" diameter	G E-2	16000	.003		1.33	.11	.09	1.53	1.76
1600	6" to 12" diameter	G	14000	.003		1.33	.12	.11	1.56	1.81
5100	Structural tubing, rect., 5" to 6" wide, light section	G	8000	.006		1.33	.21	.19	1.73	2.06
5200	Heavy section	G	12000	.004		1.33	.14	.13	1.60	1.86
8090	For projects 75 to 99 tons, add				All	10%				
8092	50 to 74 tons, add					20%				
8094	25 to 49 tons, add					30%	10%			
8096	10 to 24 tons, add					50%	25%			
8098	2 to 9 tons, add					75%	50%			
8099	Less than 2 tons, add					100%	100%			

05 12 23.45 Lintels		Crew	Daily Output	Labor-Hours	Unit	Material	Labor	Equipment	Total	Total Incl O&P
0010	**LINTELS**									
0015	Made from recycled materials									
0020	Plain steel angles, shop fabricated, under 500 lb.	G 1 Bric	550	.015	Lb.	1.02	.47		1.49	1.90
0100	500 to 1000 lb.	G	640	.013	"	.99	.40		1.39	1.76
2000	Steel angles, 3-1/2" x 3", 1/4" thick, 2'-6" long	G	47	.170	Ea.	14.30	5.50		19.80	25
2100	4'-6" long	G	26	.308		26	9.95		35.95	45
2600	4" x 3-1/2", 1/4" thick, 5'-0" long	G	21	.381		33	12.30		45.30	56.50
2700	9'-0" long	G	12	.667		59	21.50		80.50	101

05 12 Structural Steel Framing

05 12 23 – Structural Steel for Buildings

05 12 23.65 Plates

			Crew	Daily Output	Labor-Hours	Unit	Material	Labor	Equipment	Total	Total Incl O&P
0010	**PLATES**										
0015	Made from recycled materials										
0020	For connections & stiffener plates, shop fabricated										
0050	1/8" thick (5.1 lb./S.F.)	G				S.F.	6.75			6.75	7.45
0100	1/4" thick (10.2 lb./S.F.)	G					13.50			13.50	14.85
0300	3/8" thick (15.3 lb./S.F.)	G					20.50			20.50	22.50
0400	1/2" thick (20.4 lb./S.F.)	G					27			27	29.50
0450	3/4" thick (30.6 lb./S.F.)	G					40.50			40.50	44.50
0500	1" thick (40.8 lb./S.F.)	G					54			54	59.50
2000	Steel plate, warehouse prices, no shop fabrication										
2100	1/4" thick (10.2 lb./S.F.)	G				S.F.	7.30			7.30	8

05 12 23.79 Structural Steel

			Crew	Daily Output	Labor-Hours	Unit	Material	Labor	Equipment	Total	Total Incl O&P
0010	**STRUCTURAL STEEL**										
0020	Shop fab'd for 100-ton, 1-2 story project, bolted conn's.										
0050	Beams, W 6 x 9	G	E-2	720	.067	L.F.	14.30	2.38	2.10	18.78	22.50
0100	W 8 x 10	G		720	.067		15.90	2.38	2.10	20.38	24
0200	Columns, W 6 x 15	G		540	.089		26	3.17	2.80	31.97	37.50
0250	W 8 x 31	G		540	.089		53.50	3.17	2.80	59.47	67.50
7990	For projects 75 to 99 tons, add					All	10%				
7992	50 to 75 tons, add						20%				
7994	25 to 49 tons, add						30%	10%			
7996	10 to 24 tons, add						50%	25%			
7998	2 to 9 tons, add						75%	50%			
7999	Less than 2 tons, add						100%	100%			

05 31 Steel Decking

05 31 23 – Steel Roof Decking

05 31 23.50 Roof Decking

			Crew	Daily Output	Labor-Hours	Unit	Material	Labor	Equipment	Total	Total Incl O&P
0010	**ROOF DECKING**										
0015	Made from recycled materials										
2100	Open type, 1-1/2" deep, Type B, wide rib, galv., 22 ga., under 50 sq.	G	E-4	4500	.007	S.F.	2.05	.25	.03	2.33	2.77
2600	20 ga., under 50 squares	G		3865	.008		2.39	.30	.04	2.73	3.23
2900	18 ga., under 50 squares	G		3800	.008		3.09	.30	.04	3.43	4.01
3050	16 ga., under 50 squares	G		3700	.009		4.17	.31	.04	4.52	5.20

05 31 33 – Steel Form Decking

05 31 33.50 Form Decking

			Crew	Daily Output	Labor-Hours	Unit	Material	Labor	Equipment	Total	Total Incl O&P
0010	**FORM DECKING**										
0015	Made from recycled materials										
6100	Slab form, steel, 28 ga., 9/16" deep, Type UFS, uncoated	G	E-4	4000	.008	S.F.	1.55	.29	.04	1.88	2.28
6200	Galvanized	G		4000	.008		1.37	.29	.04	1.70	2.09
6220	24 ga., 1" deep, Type UF1X, uncoated	G		3900	.008		1.49	.29	.04	1.82	2.23
6240	Galvanized	G		3900	.008		1.75	.29	.04	2.08	2.52
6300	24 ga., 1-5/16" deep, Type UFX, uncoated	G		3800	.008		1.58	.30	.04	1.92	2.35
6400	Galvanized	G		3800	.008		1.86	.30	.04	2.20	2.66
6500	22 ga., 1-5/16" deep, uncoated	G		3700	.009		2	.31	.04	2.35	2.82
6600	Galvanized	G		3700	.009		2.04	.31	.04	2.39	2.86
6700	22 ga., 2" deep, uncoated	G		3600	.009		2.60	.32	.04	2.96	3.50
6800	Galvanized	G		3600	.009		2.55	.32	.04	2.91	3.45

05 41 Structural Metal Stud Framing

05 41 13 – Load-Bearing Metal Stud Framing

05 41 13.05 Bracing

			Crew	Daily Output	Labor-Hours	Unit	Material	2015 Bare Costs Labor	Equipment	Total	Total Incl O&P
0010	**BRACING**, shear wall X-bracing, per 10' x 10' bay, one face										
0015	Made of recycled materials										
0120	Metal strap, 20 ga. x 4" wide	G	2 Carp	18	.889	Ea.	17.55	29		46.55	68.50
0130	6" wide	G		18	.889		29.50	29		58.50	81
0160	18 ga. x 4" wide	G		16	1		30.50	33		63.50	88.50
0170	6" wide	G	↓	16	1	↓	45	33		78	105
0410	Continuous strap bracing, per horizontal row on both faces										
0420	Metal strap, 20 ga. x 2" wide, studs 12" O.C.	G	1 Carp	7	1.143	C.L.F.	53	37.50		90.50	121
0430	16" O.C.	G		8	1		53	33		86	113
0440	24" O.C.	G		10	.800		53	26.50		79.50	102
0450	18 ga. x 2" wide, studs 12" O.C.	G		6	1.333		75	44		119	156
0460	16" O.C.	G		7	1.143		75	37.50		112.50	146
0470	24" O.C.	G	↓	8	1		75	33		108	138

05 41 13.10 Bridging

			Crew	Daily Output	Labor-Hours	Unit	Material	Labor	Equipment	Total	Total Incl O&P
0010	**BRIDGING**, solid between studs w/1-1/4" leg track, per stud bay										
0015	Made from recycled materials										
0200	Studs 12" O.C., 18 ga. x 2-1/2" wide	G	1 Carp	125	.064	Ea.	.89	2.10		2.99	4.51
0210	3-5/8" wide	G		120	.067		1.07	2.19		3.26	4.86
0220	4" wide	G		120	.067		1.13	2.19		3.32	4.93
0230	6" wide	G		115	.070		1.48	2.29		3.77	5.45
0240	8" wide	G		110	.073		1.84	2.39		4.23	6.05
0300	16 ga. x 2-1/2" wide	G		115	.070		1.13	2.29		3.42	5.10
0310	3-5/8" wide	G		110	.073		1.38	2.39		3.77	5.50
0320	4" wide	G		110	.073		1.47	2.39		3.86	5.60
0330	6" wide	G		105	.076		1.87	2.50		4.37	6.25
0340	8" wide	G		100	.080		2.34	2.63		4.97	7
1200	Studs 16" O.C., 18 ga. x 2-1/2" wide	G		125	.064		1.14	2.10		3.24	4.78
1210	3-5/8" wide	G		120	.067		1.37	2.19		3.56	5.20
1220	4" wide	G		120	.067		1.45	2.19		3.64	5.30
1230	6" wide	G		115	.070		1.90	2.29		4.19	5.95
1240	8" wide	G		110	.073		2.36	2.39		4.75	6.60
1300	16 ga. x 2-1/2" wide	G		115	.070		1.45	2.29		3.74	5.45
1310	3-5/8" wide	G		110	.073		1.77	2.39		4.16	5.95
1320	4" wide	G		110	.073		1.88	2.39		4.27	6.10
1330	6" wide	G		105	.076		2.39	2.50		4.89	6.85
1340	8" wide	G		100	.080		3	2.63		5.63	7.70
2200	Studs 24" O.C., 18 ga. x 2-1/2" wide	G		125	.064		1.65	2.10		3.75	5.35
2210	3-5/8" wide	G		120	.067		1.98	2.19		4.17	5.85
2220	4" wide	G		120	.067		2.10	2.19		4.29	6
2230	6" wide	G		115	.070		2.75	2.29		5.04	6.85
2240	8" wide	G		110	.073		3.41	2.39		5.80	7.75
2300	16 ga. x 2-1/2" wide	G		115	.070		2.10	2.29		4.39	6.15
2310	3-5/8" wide	G		110	.073		2.55	2.39		4.94	6.80
2320	4" wide	G		110	.073		2.72	2.39		5.11	7
2330	6" wide	G		105	.076		3.46	2.50		5.96	8
2340	8" wide	G	↓	100	.080	↓	4.34	2.63		6.97	9.20
3000	Continuous bridging, per row										
3100	16 ga. x 1-1/2" channel thru studs 12" O.C.	G	1 Carp	6	1.333	C.L.F.	48	44		92	127
3110	16" O.C.	G		7	1.143		48	37.50		85.50	116
3120	24" O.C.	G		8.80	.909		48	30		78	103
4100	2" x 2" angle x 18 ga., studs 12" O.C.	G		7	1.143		75	37.50		112.50	146
4110	16" O.C.	G	↓	9	.889		75	29		104	132

05 41 Structural Metal Stud Framing

05 41 13 – Load-Bearing Metal Stud Framing

05 41 13.10 Bridging

		Crew	Daily Output	Labor-Hours	Unit	Material	2015 Bare Costs Labor	Equipment	Total	Total Incl O&P
4120	24" O.C.	1 Carp	12	.667	C.L.F.	75	22		97	120
4200	16 ga., studs 12" O.C.		5	1.600		94.50	52.50		147	192
4210	16" O.C.		7	1.143		94.50	37.50		132	167
4220	24" O.C.		10	.800		94.50	26.50		121	148

05 41 13.25 Framing, Boxed Headers/Beams

		Crew	Daily Output	Labor-Hours	Unit	Material	Labor	Equipment	Total	Total Incl O&P
0010	**FRAMING, BOXED HEADERS/BEAMS**									
0015	Made from recycled materials									
0200	Double, 18 ga. x 6" deep	2 Carp	220	.073	L.F.	5.10	2.39		7.49	9.60
0210	8" deep		210	.076		5.65	2.50		8.15	10.40
0220	10" deep		200	.080		6.90	2.63		9.53	12
0230	12" deep		190	.084		7.55	2.77		10.32	12.95
0300	16 ga. x 8" deep		180	.089		6.50	2.92		9.42	12.05
0310	10" deep		170	.094		7.90	3.09		10.99	13.85
0320	12" deep		160	.100		8.60	3.29		11.89	14.95
0400	14 ga. x 10" deep		140	.114		9.10	3.75		12.85	16.30
0410	12" deep		130	.123		10	4.04		14.04	17.75
1210	Triple, 18 ga. x 8" deep		170	.094		8.20	3.09		11.29	14.20
1220	10" deep		165	.097		9.85	3.19		13.04	16.20
1230	12" deep		160	.100		10.85	3.29		14.14	17.40
1300	16 ga. x 8" deep		145	.110		9.50	3.62		13.12	16.50
1310	10" deep		140	.114		11.35	3.75		15.10	18.75
1320	12" deep		135	.119		12.40	3.89		16.29	20
1400	14 ga. x 10" deep		115	.139		12.40	4.57		16.97	21.50
1410	12" deep		110	.145		13.70	4.78		18.48	23

05 41 13.30 Framing, Stud Walls

		Crew	Daily Output	Labor-Hours	Unit	Material	Labor	Equipment	Total	Total Incl O&P
0010	**FRAMING, STUD WALLS** w/top & bottom track, no openings,									
0020	Headers, beams, bridging or bracing									
0025	Made from recycled materials									
4100	8' high walls, 18 ga. x 2-1/2" wide, studs 12" O.C.	2 Carp	54	.296	L.F.	8.50	9.75		18.25	25.50
4110	16" O.C.		77	.208		6.80	6.85		13.65	18.90
4120	24" O.C.		107	.150		5.10	4.91		10.01	13.85
4130	3-5/8" wide, studs 12" O.C.		53	.302		10.05	9.90		19.95	27.50
4140	16" O.C.		76	.211		8.05	6.90		14.95	20.50
4150	24" O.C.		105	.152		6.05	5		11.05	15.05
4160	4" wide, studs 12" O.C.		52	.308		10.55	10.10		20.65	28.50
4170	16" O.C.		74	.216		8.45	7.10		15.55	21
4180	24" O.C.		103	.155		6.35	5.10		11.45	15.55
4190	6" wide, studs 12" O.C.		51	.314		13.40	10.30		23.70	32
4200	16" O.C.		73	.219		10.75	7.20		17.95	24
4210	24" O.C.		101	.158		8.10	5.20		13.30	17.65
4220	8" wide, studs 12" O.C.		50	.320		16.30	10.50		26.80	35.50
4230	16" O.C.		72	.222		13.10	7.30		20.40	26.50
4240	24" O.C.		100	.160		9.90	5.25		15.15	19.70
4300	16 ga. x 2-1/2" wide, studs 12" O.C.		47	.340		10.10	11.20		21.30	30
4310	16" O.C.		68	.235		8	7.75		15.75	22
4320	24" O.C.		94	.170		5.90	5.60		11.50	15.90
4330	3-5/8" wide, studs 12" O.C.		46	.348		12.05	11.45		23.50	32.50
4340	16" O.C.		66	.242		9.55	7.95		17.50	24
4350	24" O.C.		92	.174		7.05	5.70		12.75	17.35
4360	4" wide, studs 12" O.C.		45	.356		12.65	11.70		24.35	33.50
4370	16" O.C.		65	.246		10	8.10		18.10	24.50
4380	24" O.C.		90	.178		7.40	5.85		13.25	17.95

05 41 Structural Metal Stud Framing

05 41 13 – Load-Bearing Metal Stud Framing

05 41 13.30 Framing, Stud Walls		Crew	Daily Output	Labor-Hours	Unit	Material	2015 Bare Costs Labor	Equipment	Total	Total Incl O&P
4390	6" wide, studs 12" O.C.	G 2 Carp	44	.364	L.F.	15.80	11.95		27.75	37.50
4400	16" O.C.	G	64	.250		12.55	8.20		20.75	27.50
4410	24" O.C.	G	88	.182		9.30	5.95		15.25	20.50
4420	8" wide, studs 12" O.C.	G	43	.372		19.50	12.20		31.70	42
4430	16" O.C.	G	63	.254		15.50	8.35		23.85	31
4440	24" O.C.	G	86	.186		11.50	6.10		17.60	23
5100	10' high walls, 18 ga. x 2-1/2" wide, studs 12" O.C.	G	54	.296		10.20	9.75		19.95	27.50
5110	16" O.C.	G	77	.208		8.05	6.85		14.90	20.50
5120	24" O.C.	G	107	.150		5.95	4.91		10.86	14.80
5130	3-5/8" wide, studs 12" O.C.	G	53	.302		12.05	9.90		21.95	30
5140	16" O.C.	G	76	.211		9.55	6.90		16.45	22
5150	24" O.C.	G	105	.152		7.05	5		12.05	16.15
5160	4" wide, studs 12" O.C.	G	52	.308		12.65	10.10		22.75	31
5170	16" O.C.	G	74	.216		10.05	7.10		17.15	23
5180	24" O.C.	G	103	.155		7.40	5.10		12.50	16.70
5190	6" wide, studs 12" O.C.	G	51	.314		16	10.30		26.30	35
5200	16" O.C.	G	73	.219		12.70	7.20		19.90	26
5210	24" O.C.	G	101	.158		9.40	5.20		14.60	19.10
5220	8" wide, studs 12" O.C.	G	50	.320		19.50	10.50		30	39
5230	16" O.C.	G	72	.222		15.50	7.30		22.80	29.50
5240	24" O.C.	G	100	.160		11.50	5.25		16.75	21.50
5300	16 ga. x 2-1/2" wide, studs 12" O.C.	G	47	.340		12.20	11.20		23.40	32
5310	16" O.C.	G	68	.235		9.55	7.75		17.30	23.50
5320	24" O.C.	G	94	.170		6.95	5.60		12.55	17.05
5330	3-5/8" wide, studs 12" O.C.	G	46	.348		14.55	11.45		26	35
5340	16" O.C.	G	66	.242		11.40	7.95		19.35	26
5350	24" O.C.	G	92	.174		8.30	5.70		14	18.70
5360	4" wide, studs 12" O.C.	G	45	.356		15.25	11.70		26.95	36.50
5370	16" O.C.	G	65	.246		12	8.10		20.10	27
5380	24" O.C.	G	90	.178		8.70	5.85		14.55	19.40
5390	6" wide, studs 12" O.C.	G	44	.364		19	11.95		30.95	41
5400	16" O.C.	G	64	.250		14.95	8.20		23.15	30.50
5410	24" O.C.	G	88	.182		10.90	5.95		16.85	22
5420	8" wide, studs 12" O.C.	G	43	.372		23.50	12.20		35.70	46.50
5430	16" O.C.	G	63	.254		18.50	8.35		26.85	34.50
5440	24" O.C.	G	86	.186		13.50	6.10		19.60	25
6190	12' high walls, 18 ga. x 6" wide, studs 12" O.C.	G	41	.390		18.65	12.80		31.45	42
6200	16" O.C.	G	58	.276		14.70	9.05		23.75	31.50
6210	24" O.C.	G	81	.198		10.75	6.50		17.25	22.50
6220	8" wide, studs 12" O.C.	G	40	.400		22.50	13.15		35.65	47
6230	16" O.C.	G	57	.281		17.90	9.20		27.10	35
6240	24" O.C.	G	80	.200		13.10	6.55		19.65	25.50
6390	16 ga. x 6" wide, studs 12" O.C.	G	35	.457		22.50	15		37.50	49.50
6400	16" O.C.	G	51	.314		17.40	10.30		27.70	36.50
6410	24" O.C.	G	70	.229		12.55	7.50		20.05	26.50
6420	8" wide, studs 12" O.C.	G	34	.471		27.50	15.45		42.95	56.50
6430	16" O.C.	G	50	.320		21.50	10.50		32	41
6440	24" O.C.	G	69	.232		15.50	7.60		23.10	30
6530	14 ga. x 3-5/8" wide, studs 12" O.C.	G	34	.471		21	15.45		36.45	49.50
6540	16" O.C.	G	48	.333		16.55	10.95		27.50	36.50
6550	24" O.C.	G	65	.246		11.90	8.10		20	26.50
6560	4" wide, studs 12" O.C.	G	33	.485		22.50	15.95		38.45	51
6570	16" O.C.	G	47	.340		17.55	11.20		28.75	38

For customer support on your Residential Cost Data, call 877.713.2617.

05 41 Structural Metal Stud Framing

05 41 13 – Load-Bearing Metal Stud Framing

05 41 13.30 Framing, Stud Walls

		Crew	Daily Output	Labor-Hours	Unit	Material	2015 Bare Costs Labor	2015 Bare Costs Equipment	Total	Total Incl O&P
6580	24" O.C.	G 2 Carp	64	.250	L.F.	12.65	8.20		20.85	27.50
6730	12 ga. x 3-5/8" wide, studs 12" O.C.	G	31	.516		29.50	16.95		46.45	61
6740	16" O.C.	G	43	.372		23	12.20		35.20	45.50
6750	24" O.C.	G	59	.271		16.05	8.90		24.95	32.50
6760	4" wide, studs 12" O.C.	G	30	.533		31.50	17.50		49	64
6770	16" O.C.	G	42	.381		24.50	12.50		37	47.50
6780	24" O.C.	G	58	.276		17.15	9.05		26.20	34
7390	16' high walls, 16 ga. x 6" wide, studs 12" O.C.	G	33	.485		28.50	15.95		44.45	58
7400	16" O.C.	G	48	.333		22.50	10.95		33.45	43
7410	24" O.C.	G	67	.239		15.80	7.85		23.65	30.50
7420	8" wide, studs 12" O.C.	G	32	.500		35.50	16.45		51.95	66.50
7430	16" O.C.	G	47	.340		27.50	11.20		38.70	49.50
7440	24" O.C.	G	66	.242		19.50	7.95		27.45	35
7560	14 ga. x 4" wide, studs 12" O.C.	G	31	.516		29	16.95		45.95	60.50
7570	16" O.C.	G	45	.356		22.50	11.70		34.20	44
7580	24" O.C.	G	61	.262		15.90	8.60		24.50	32
7590	6" wide, studs 12" O.C.	G	30	.533		36.50	17.50		54	69.50
7600	16" O.C.	G	44	.364		28.50	11.95		40.45	51
7610	24" O.C.	G	60	.267		20	8.75		28.75	36.50
7760	12 ga. x 4" wide, studs 12" O.C.	G	29	.552		41	18.10		59.10	75.50
7770	16" O.C.	G	40	.400		31.50	13.15		44.65	56.50
7780	24" O.C.	G	55	.291		22	9.55		31.55	40
7790	6" wide, studs 12" O.C.	G	28	.571		51.50	18.75		70.25	88.50
7800	16" O.C.	G	39	.410		39.50	13.50		53	66
7810	24" O.C.	G	54	.296		27.50	9.75		37.25	47
8590	20' high walls, 14 ga. x 6" wide, studs 12" O.C.	G	29	.552		45	18.10		63.10	79.50
8600	16" O.C.	G	42	.381		34.50	12.50		47	59
8610	24" O.C.	G	57	.281		24	9.20		33.20	42
8620	8" wide, studs 12" O.C.	G	28	.571		48.50	18.75		67.25	85
8630	16" O.C.	G	41	.390		37.50	12.80		50.30	62.50
8640	24" O.C.	G	56	.286		26.50	9.40		35.90	45
8790	12 ga. x 6" wide, studs 12" O.C.	G	27	.593		64	19.45		83.45	103
8800	16" O.C.	G	37	.432		48.50	14.20		62.70	77.50
8810	24" O.C.	G	51	.314		33.50	10.30		43.80	54.50
8820	8" wide, studs 12" O.C.	G	26	.615		77.50	20		97.50	119
8830	16" O.C.	G	36	.444		59	14.60		73.60	89.50
8840	24" O.C.	G	50	.320		41	10.50		51.50	62.50

05 42 Cold-Formed Metal Joist Framing

05 42 13 – Cold-Formed Metal Floor Joist Framing

05 42 13.05 Bracing

		Crew	Daily Output	Labor-Hours	Unit	Material	2015 Bare Costs Labor	2015 Bare Costs Equipment	Total	Total Incl O&P
0010	**BRACING**, continuous, per row, top & bottom									
0015	Made from recycled materials									
0120	Flat strap, 20 ga. x 2" wide, joists at 12" O.C.	G 1 Carp	4.67	1.713	C.L.F.	55	56.50		111.50	155
0130	16" O.C.	G	5.33	1.501		53.50	49.50		103	142
0140	24" O.C.	G	6.66	1.201		51.50	39.50		91	123
0150	18 ga. x 2" wide, joists at 12" O.C.	G	4	2		74	65.50		139.50	192
0160	16" O.C.	G	4.67	1.713		73	56.50		129.50	175
0170	24" O.C.	G	5.33	1.501		71.50	49.50		121	162

05 42 Cold-Formed Metal Joist Framing

05 42 13 – Cold-Formed Metal Floor Joist Framing

05 42 13.10 Bridging

			Crew	Daily Output	Labor-Hours	Unit	Material	2015 Bare Costs Labor	Equipment	Total	Total Incl O&P
0010	**BRIDGING**, solid between joists w/1-1/4" leg track, per joist bay										
0015	Made from recycled materials										
0230	Joists 12" O.C., 18 ga. track x 6" wide	G	1 Carp	80	.100	Ea.	1.48	3.29		4.77	7.15
0240	8" wide	G		75	.107		1.84	3.50		5.34	7.90
0250	10" wide	G		70	.114		2.29	3.75		6.04	8.80
0260	12" wide	G		65	.123		2.60	4.04		6.64	9.65
0330	16 ga. track x 6" wide	G		70	.114		1.87	3.75		5.62	8.35
0340	8" wide	G		65	.123		2.34	4.04		6.38	9.40
0350	10" wide	G		60	.133		2.91	4.38		7.29	10.55
0360	12" wide	G		55	.145		3.35	4.78		8.13	11.70
0440	14 ga. track x 8" wide	G		60	.133		2.93	4.38		7.31	10.60
0450	10" wide	G		55	.145		3.64	4.78		8.42	12
0460	12" wide	G		50	.160		4.20	5.25		9.45	13.40
0550	12 ga. track x 10" wide	G		45	.178		5.35	5.85		11.20	15.65
0560	12" wide	G		40	.200		5.45	6.55		12	17.05
1230	16" O.C., 18 ga. track x 6" wide	G		80	.100		1.90	3.29		5.19	7.60
1240	8" wide	G		75	.107		2.36	3.50		5.86	8.50
1250	10" wide	G		70	.114		2.94	3.75		6.69	9.55
1260	12" wide	G		65	.123		3.33	4.04		7.37	10.45
1330	16 ga. track x 6" wide	G		70	.114		2.39	3.75		6.14	8.95
1340	8" wide	G		65	.123		3	4.04		7.04	10.10
1350	10" wide	G		60	.133		3.73	4.38		8.11	11.45
1360	12" wide	G		55	.145		4.29	4.78		9.07	12.70
1440	14 ga. track x 8" wide	G		60	.133		3.76	4.38		8.14	11.50
1450	10" wide	G		55	.145		4.67	4.78		9.45	13.15
1460	12" wide	G		50	.160		5.40	5.25		10.65	14.70
1550	12 ga. track x 10" wide	G		45	.178		6.85	5.85		12.70	17.30
1560	12" wide	G		40	.200		7	6.55		13.55	18.75
2230	24" O.C., 18 ga. track x 6" wide	G		80	.100		2.75	3.29		6.04	8.50
2240	8" wide	G		75	.107		3.41	3.50		6.91	9.65
2250	10" wide	G		70	.114		4.25	3.75		8	10.95
2260	12" wide	G		65	.123		4.82	4.04		8.86	12.10
2330	16 ga. track x 6" wide	G		70	.114		3.46	3.75		7.21	10.10
2340	8" wide	G		65	.123		4.34	4.04		8.38	11.60
2350	10" wide	G		60	.133		5.40	4.38		9.78	13.30
2360	12" wide	G		55	.145		6.20	4.78		10.98	14.85
2440	14 ga. track x 8" wide	G		60	.133		5.45	4.38		9.83	13.35
2450	10" wide	G		55	.145		6.75	4.78		11.53	15.45
2460	12" wide	G		50	.160		7.80	5.25		13.05	17.35
2550	12 ga. track x 10" wide	G		45	.178		9.90	5.85		15.75	20.50
2560	12" wide	G		40	.200		10.10	6.55		16.65	22

05 42 13.25 Framing, Band Joist

			Crew	Daily Output	Labor-Hours	Unit	Material	Labor	Equipment	Total	Total Incl O&P
0010	**FRAMING, BAND JOIST** (track) fastened to bearing wall										
0015	Made from recycled materials										
0220	18 ga. track x 6" deep	G	2 Carp	1000	.016	L.F.	1.21	.53		1.74	2.21
0230	8" deep	G		920	.017		1.50	.57		2.07	2.61
0240	10" deep	G		860	.019		1.87	.61		2.48	3.09
0320	16 ga. track x 6" deep	G		900	.018		1.52	.58		2.10	2.65
0330	8" deep	G		840	.019		1.91	.63		2.54	3.15
0340	10" deep	G		780	.021		2.37	.67		3.04	3.74
0350	12" deep	G		740	.022		2.73	.71		3.44	4.19
0430	14 ga. track x 8" deep	G		750	.021		2.39	.70		3.09	3.81

05 42 Cold-Formed Metal Joist Framing

05 42 13 – Cold-Formed Metal Floor Joist Framing

05 42 13.25 Framing, Band Joist

			Crew	Daily Output	Labor-Hours	Unit	Material	2015 Bare Costs Labor	Equipment	Total	Total Incl O&P
0440	10" deep	G	2 Carp	720	.022	L.F.	2.97	.73		3.70	4.50
0450	12" deep	G		700	.023		3.42	.75		4.17	5.05
0540	12 ga. track x 10" deep	G		670	.024		4.35	.78		5.13	6.10
0550	12" deep	G		650	.025		4.45	.81		5.26	6.25

05 42 13.30 Framing, Boxed Headers/Beams

			Crew	Daily Output	Labor-Hours	Unit	Material	Labor	Equipment	Total	Total Incl O&P
0010	**FRAMING, BOXED HEADERS/BEAMS**										
0015	Made from recycled materials										
0200	Double, 18 ga. x 6" deep	G	2 Carp	220	.073	L.F.	5.10	2.39		7.49	9.60
0210	8" deep	G		210	.076		5.65	2.50		8.15	10.40
0220	10" deep	G		200	.080		6.90	2.63		9.53	12
0230	12" deep	G		190	.084		7.55	2.77		10.32	12.95
0300	16 ga. x 8" deep	G		180	.089		6.50	2.92		9.42	12.05
0310	10" deep	G		170	.094		7.90	3.09		10.99	13.85
0320	12" deep	G		160	.100		8.60	3.29		11.89	14.95
0400	14 ga. x 10" deep	G		140	.114		9.10	3.75		12.85	16.30
0410	12" deep	G		130	.123		10	4.04		14.04	17.75
0500	12 ga. x 10" deep	G		110	.145		12	4.78		16.78	21
0510	12" deep	G		100	.160		13.25	5.25		18.50	23.50
1210	Triple, 18 ga. x 8" deep	G		170	.094		8.20	3.09		11.29	14.20
1220	10" deep	G		165	.097		9.85	3.19		13.04	16.20
1230	12" deep	G		160	.100		10.85	3.29		14.14	17.40
1300	16 ga. x 8" deep	G		145	.110		9.50	3.62		13.12	16.50
1310	10" deep	G		140	.114		11.35	3.75		15.10	18.75
1320	12" deep	G		135	.119		12.40	3.89		16.29	20
1400	14 ga. x 10" deep	G		115	.139		13.15	4.57		17.72	22
1410	12" deep	G		110	.145		14.50	4.78		19.28	24
1500	12 ga. x 10" deep	G		90	.178		17.50	5.85		23.35	29
1510	12" deep	G		85	.188		19.40	6.20		25.60	32

05 42 13.40 Framing, Joists

			Crew	Daily Output	Labor-Hours	Unit	Material	Labor	Equipment	Total	Total Incl O&P
0010	**FRAMING, JOISTS**, no band joists (track), web stiffeners, headers,										
0020	Beams, bridging or bracing										
0025	Made from recycled materials										
0030	Joists (2" flange) and fasteners, materials only										
0220	18 ga. x 6" deep	G				L.F.	1.58			1.58	1.73
0230	8" deep	G					1.86			1.86	2.04
0240	10" deep	G					2.18			2.18	2.40
0320	16 ga. x 6" deep	G					1.93			1.93	2.13
0330	8" deep	G					2.31			2.31	2.54
0340	10" deep	G					2.70			2.70	2.97
0350	12" deep	G					3.07			3.07	3.37
0430	14 ga. x 8" deep	G					2.90			2.90	3.19
0440	10" deep	G					3.34			3.34	3.67
0450	12" deep	G					3.80			3.80	4.18
0540	12 ga. x 10" deep	G					4.86			4.86	5.35
0550	12" deep	G					5.50			5.50	6.10
1010	Installation of joists to band joists, beams & headers, labor only										
1220	18 ga. x 6" deep		2 Carp	110	.145	Ea.		4.78		4.78	8
1230	8" deep			90	.178			5.85		5.85	9.80
1240	10" deep			80	.200			6.55		6.55	11.05
1320	16 ga. x 6" deep			95	.168			5.55		5.55	9.30
1330	8" deep			70	.229			7.50		7.50	12.60
1340	10" deep			60	.267			8.75		8.75	14.70

05 42 Cold-Formed Metal Joist Framing

05 42 13 – Cold-Formed Metal Floor Joist Framing

05 42 13.40 Framing, Joists		Crew	Daily Output	Labor-Hours	Unit	Material	2015 Bare Costs Labor	Equipment	Total	Total Incl O&P
1350	12" deep	2 Carp	55	.291	Ea.	9.55			9.55	16.05
1430	14 ga. x 8" deep		65	.246		8.10			8.10	13.60
1440	10" deep		45	.356		11.70			11.70	19.60
1450	12" deep		35	.457		15			15	25
1540	12 ga. x 10" deep		40	.400		13.15			13.15	22
1550	12" deep		30	.533		17.50			17.50	29.50

05 42 13.45 Framing, Web Stiffeners

0010	**FRAMING, WEB STIFFENERS** at joist bearing, fabricated from									
0020	Stud piece (1-5/8" flange) to stiffen joist (2" flange)									
0025	Made from recycled materials									
2120	For 6" deep joist, with 18 ga. x 2-1/2" stud	G	1 Carp	120	.067	Ea.	.85	2.19		3.04	4.62
2130	3-5/8" stud	G		110	.073		1	2.39		3.39	5.10
2140	4" stud	G		105	.076		1.05	2.50		3.55	5.35
2150	6" stud	G		100	.080		1.32	2.63		3.95	5.85
2160	8" stud	G		95	.084		1.60	2.77		4.37	6.40
2220	8" deep joist, with 2-1/2" stud	G		120	.067		1.14	2.19		3.33	4.93
2230	3-5/8" stud	G		110	.073		1.34	2.39		3.73	5.50
2240	4" stud	G		105	.076		1.41	2.50		3.91	5.75
2250	6" stud	G		100	.080		1.77	2.63		4.40	6.35
2260	8" stud	G		95	.084		2.14	2.77		4.91	7
2320	10" deep joist, with 2-1/2" stud	G		110	.073		1.41	2.39		3.80	5.55
2330	3-5/8" stud	G		100	.080		1.66	2.63		4.29	6.25
2340	4" stud	G		95	.084		1.74	2.77		4.51	6.55
2350	6" stud	G		90	.089		2.19	2.92		5.11	7.30
2360	8" stud	G		85	.094		2.66	3.09		5.75	8.10
2420	12" deep joist, with 2-1/2" stud	G		110	.073		1.70	2.39		4.09	5.90
2430	3-5/8" stud	G		100	.080		2	2.63		4.63	6.60
2440	4" stud	G		95	.084		2.10	2.77		4.87	6.95
2450	6" stud	G		90	.089		2.64	2.92		5.56	7.80
2460	8" stud	G		85	.094		3.20	3.09		6.29	8.70
3130	For 6" deep joist, with 16 ga. x 3-5/8" stud	G		100	.080		1.25	2.63		3.88	5.80
3140	4" stud	G		95	.084		1.31	2.77		4.08	6.10
3150	6" stud	G		90	.089		1.62	2.92		4.54	6.70
3160	8" stud	G		85	.094		2	3.09		5.09	7.40
3230	8" deep joist, with 3-5/8" stud	G		100	.080		1.68	2.63		4.31	6.25
3240	4" stud	G		95	.084		1.76	2.77		4.53	6.55
3250	6" stud	G		90	.089		2.17	2.92		5.09	7.30
3260	8" stud	G		85	.094		2.68	3.09		5.77	8.15
3330	10" deep joist, with 3-5/8" stud	G		85	.094		2.08	3.09		5.17	7.50
3340	4" stud	G		80	.100		2.17	3.29		5.46	7.90
3350	6" stud	G		75	.107		2.69	3.50		6.19	8.85
3360	8" stud	G		70	.114		3.32	3.75		7.07	9.95
3430	12" deep joist, with 3-5/8" stud	G		85	.094		2.50	3.09		5.59	7.95
3440	4" stud	G		80	.100		2.62	3.29		5.91	8.40
3450	6" stud	G		75	.107		3.24	3.50		6.74	9.45
3460	8" stud	G		70	.114		4	3.75		7.75	10.70
4230	For 8" deep joist, with 14 ga. x 3-5/8" stud	G		90	.089		2.08	2.92		5	7.20
4240	4" stud	G		85	.094		2.20	3.09		5.29	7.60
4250	6" stud	G		80	.100		2.76	3.29		6.05	8.55
4260	8" stud	G		75	.107		2.95	3.50		6.45	9.15
4330	10" deep joist, with 3-5/8" stud	G		75	.107		2.57	3.50		6.07	8.75
4340	4" stud	G		70	.114		2.72	3.75		6.47	9.30

For customer support on your Residential Cost Data, call 877.713.2617.

05 42 Cold-Formed Metal Joist Framing

05 42 13 – Cold-Formed Metal Floor Joist Framing

05 42 13.45 Framing, Web Stiffeners

			Crew	Daily Output	Labor-Hours	Unit	Material	2015 Bare Costs Labor	Equipment	Total	Total Incl O&P
4350	6" stud	G	1 Carp	65	.123	Ea.	3.42	4.04		7.46	10.55
4360	8" stud	G		60	.133		3.65	4.38		8.03	11.35
4430	12" deep joist, with 3-5/8" stud	G		75	.107		3.10	3.50		6.60	9.30
4440	4" stud	G		70	.114		3.28	3.75		7.03	9.90
4450	6" stud	G		65	.123		4.12	4.04		8.16	11.35
4460	8" stud	G		60	.133		4.40	4.38		8.78	12.20
5330	For 10" deep joist, with 12 ga. x 3-5/8" stud	G		65	.123		3.72	4.04		7.76	10.90
5340	4" stud	G		60	.133		3.97	4.38		8.35	11.70
5350	6" stud	G		55	.145		5	4.78		9.78	13.50
5360	8" stud	G		50	.160		6.05	5.25		11.30	15.45
5430	12" deep joist, with 3-5/8" stud	G		65	.123		4.48	4.04		8.52	11.75
5440	4" stud	G		60	.133		4.78	4.38		9.16	12.60
5450	6" stud	G		55	.145		6	4.78		10.78	14.60
5460	8" stud	G		50	.160		7.30	5.25		12.55	16.80

05 42 23 – Cold-Formed Metal Roof Joist Framing

05 42 23.05 Framing, Bracing

			Crew	Daily Output	Labor-Hours	Unit	Material	Labor	Equipment	Total	Total Incl O&P
0010	**FRAMING, BRACING**										
0015	Made from recycled materials										
0020	Continuous bracing, per row										
0100	16 ga. x 1-1/2" channel thru rafters/trusses @ 16" O.C.	G	1 Carp	4.50	1.778	C.L.F.	48	58.50		106.50	151
0120	24" O.C.	G		6	1.333		48	44		92	127
0300	2" x 2" angle x 18 ga., rafters/trusses @ 16" O.C.	G		6	1.333		75	44		119	156
0320	24" O.C.	G		8	1		75	33		108	138
0400	16 ga., rafters/trusses @ 16" O.C.	G		4.50	1.778		94.50	58.50		153	202
0420	24" O.C.	G		6.50	1.231		94.50	40.50		135	172

05 42 23.10 Framing, Bridging

			Crew	Daily Output	Labor-Hours	Unit	Material	Labor	Equipment	Total	Total Incl O&P
0010	**FRAMING, BRIDGING**										
0015	Made from recycled materials										
0020	Solid, between rafters w/1-1/4" leg track, per rafter bay										
1200	Rafters 16" O.C., 18 ga. x 4" deep	G	1 Carp	60	.133	Ea.	1.45	4.38		5.83	8.95
1210	6" deep	G		57	.140		1.90	4.61		6.51	9.85
1220	8" deep	G		55	.145		2.36	4.78		7.14	10.60
1230	10" deep	G		52	.154		2.94	5.05		7.99	11.75
1240	12" deep	G		50	.160		3.33	5.25		8.58	12.45
2200	24" O.C., 18 ga. x 4" deep	G		60	.133		2.10	4.38		6.48	9.65
2210	6" deep	G		57	.140		2.75	4.61		7.36	10.75
2220	8" deep	G		55	.145		3.41	4.78		8.19	11.75
2230	10" deep	G		52	.154		4.25	5.05		9.30	13.15
2240	12" deep	G		50	.160		4.82	5.25		10.07	14.10

05 42 23.50 Framing, Parapets

			Crew	Daily Output	Labor-Hours	Unit	Material	Labor	Equipment	Total	Total Incl O&P
0010	**FRAMING, PARAPETS**										
0015	Made from recycled materials										
0100	3' high installed on 1st story, 18 ga. x 4" wide studs, 12" O.C.	G	2 Carp	100	.160	L.F.	5.30	5.25		10.55	14.65
0110	16" O.C.	G		150	.107		4.52	3.50		8.02	10.85
0120	24" O.C.	G		200	.080		3.73	2.63		6.36	8.50
0200	6" wide studs, 12" O.C.	G		100	.160		6.80	5.25		12.05	16.25
0210	16" O.C.	G		150	.107		5.80	3.50		9.30	12.25
0220	24" O.C.	G		200	.080		4.80	2.63		7.43	9.70
1100	Installed on 2nd story, 18 ga. x 4" wide studs, 12" O.C.	G		95	.168		5.30	5.55		10.85	15.15
1110	16" O.C.	G		145	.110		4.52	3.62		8.14	11.05
1120	24" O.C.	G		190	.084		3.73	2.77		6.50	8.75
1200	6" wide studs, 12" O.C.	G		95	.168		6.80	5.55		12.35	16.75

05 42 Cold-Formed Metal Joist Framing

05 42 23 – Cold-Formed Metal Roof Joist Framing

05 42 23.50 Framing, Parapets

			Crew	Daily Output	Labor-Hours	Unit	Material	2015 Bare Costs Labor	Equipment	Total	Total Incl O&P
1210	16" O.C.	G	2 Carp	145	.110	L.F.	5.80	3.62		9.42	12.45
1220	24" O.C.	G		190	.084		4.80	2.77		7.57	9.95
2100	Installed on gable, 18 ga. x 4" wide studs, 12" O.C.	G		85	.188		5.30	6.20		11.50	16.25
2110	16" O.C.	G		130	.123		4.52	4.04		8.56	11.75
2120	24" O.C.	G		170	.094		3.73	3.09		6.82	9.30
2200	6" wide studs, 12" O.C.	G		85	.188		6.80	6.20		13	17.85
2210	16" O.C.	G		130	.123		5.80	4.04		9.84	13.15
2220	24" O.C.	G		170	.094		4.80	3.09		7.89	10.50

05 42 23.60 Framing, Roof Rafters

			Crew	Daily Output	Labor-Hours	Unit	Material	Labor	Equipment	Total	Total Incl O&P
0010	**FRAMING, ROOF RAFTERS**										
0015	Made from recycled materials										
0100	Boxed ridge beam, double, 18 ga. x 6" deep	G	2 Carp	160	.100	L.F.	5.10	3.29		8.39	11.10
0110	8" deep	G		150	.107		5.65	3.50		9.15	12.10
0120	10" deep	G		140	.114		6.90	3.75		10.65	13.90
0130	12" deep	G		130	.123		7.55	4.04		11.59	15.10
0200	16 ga. x 6" deep	G		150	.107		5.80	3.50		9.30	12.25
0210	8" deep	G		140	.114		6.50	3.75		10.25	13.45
0220	10" deep	G		130	.123		7.90	4.04		11.94	15.45
0230	12" deep	G		120	.133		8.60	4.38		12.98	16.80
1100	Rafters, 2" flange, material only, 18 ga. x 6" deep	G					1.58			1.58	1.73
1110	8" deep	G					1.86			1.86	2.04
1120	10" deep	G					2.18			2.18	2.40
1130	12" deep	G					2.52			2.52	2.77
1200	16 ga. x 6" deep	G					1.93			1.93	2.13
1210	8" deep	G					2.31			2.31	2.54
1220	10" deep	G					2.70			2.70	2.97
1230	12" deep	G					3.07			3.07	3.37
2100	Installation only, ordinary rafter to 4:12 pitch, 18 ga. x 6" deep		2 Carp	35	.457	Ea.		15		15	25
2110	8" deep			30	.533			17.50		17.50	29.50
2120	10" deep			25	.640			21		21	35.50
2130	12" deep			20	.800			26.50		26.50	44
2200	16 ga. x 6" deep			30	.533			17.50		17.50	29.50
2210	8" deep			25	.640			21		21	35.50
2220	10" deep			20	.800			26.50		26.50	44
2230	12" deep			15	1.067			35		35	59
8100	Add to labor, ordinary rafters on steep roofs							25%			
8110	Dormers & complex roofs							50%			
8200	Hip & valley rafters to 4:12 pitch							25%			
8210	Steep roofs							50%			
8220	Dormers & complex roofs							75%			
8300	Hip & valley jack rafters to 4:12 pitch							50%			
8310	Steep roofs							75%			
8320	Dormers & complex roofs							100%			

05 42 23.70 Framing, Soffits and Canopies

			Crew	Daily Output	Labor-Hours	Unit	Material	Labor	Equipment	Total	Total Incl O&P
0010	**FRAMING, SOFFITS & CANOPIES**										
0015	Made from recycled materials										
0130	Continuous ledger track @ wall, studs @ 16" O.C., 18 ga. x 4" wide	G	2 Carp	535	.030	L.F.	.97	.98		1.95	2.71
0140	6" wide	G		500	.032		1.27	1.05		2.32	3.15
0150	8" wide	G		465	.034		1.57	1.13		2.70	3.63
0160	10" wide	G		430	.037		1.96	1.22		3.18	4.20
0230	Studs @ 24" O.C., 18 ga. x 4" wide	G		800	.020		.92	.66		1.58	2.12
0240	6" wide	G		750	.021		1.21	.70		1.91	2.51

05 42 Cold-Formed Metal Joist Framing

05 42 23 – Cold-Formed Metal Roof Joist Framing

05 42 23.70 Framing, Soffits and Canopies

			Crew	Daily Output	Labor-Hours	Unit	Material	2015 Bare Costs Labor	Equipment	Total	Total Incl O&P
0250	8" wide	G	2 Carp	700	.023	L.F.	1.50	.75		2.25	2.91
0260	10" wide	G	↓	650	.025	↓	1.87	.81		2.68	3.42
1000	Horizontal soffit and canopy members, material only										
1030	1-5/8" flange studs, 18 ga. x 4" deep	G				L.F.	1.26			1.26	1.39
1040	6" deep	G					1.58			1.58	1.74
1050	8" deep	G					1.92			1.92	2.11
1140	2" flange joists, 18 ga. x 6" deep	G					1.80			1.80	1.98
1150	8" deep	G					2.12			2.12	2.34
1160	10" deep	G				↓	2.50			2.50	2.75
4030	Installation only, 18 ga., 1-5/8" flange x 4" deep		2 Carp	130	.123	Ea.		4.04		4.04	6.80
4040	6" deep			110	.145			4.78		4.78	8
4050	8" deep			90	.178			5.85		5.85	9.80
4140	2" flange, 18 ga. x 6" deep			110	.145			4.78		4.78	8
4150	8" deep			90	.178			5.85		5.85	9.80
4160	10" deep			80	.200			6.55		6.55	11.05
6010	Clips to attach facia to rafter tails, 2" x 2" x 18 ga. angle	G	1 Carp	120	.067		.88	2.19		3.07	4.65
6020	16 ga. angle	G	"	100	.080	↓	1.12	2.63		3.75	5.65

05 44 Cold-Formed Metal Trusses

05 44 13 – Cold-Formed Metal Roof Trusses

05 44 13.60 Framing, Roof Trusses

			Crew	Daily Output	Labor-Hours	Unit	Material	Labor	Equipment	Total	Total Incl O&P
0010	**FRAMING, ROOF TRUSSES**										
0015	Made from recycled materials										
0020	Fabrication of trusses on ground, Fink (W) or King Post, to 4:12 pitch										
0120	18 ga. x 4" chords, 16' span	G	2 Carp	12	1.333	Ea.	59	44		103	138
0130	20' span	G		11	1.455		73.50	48		121.50	161
0140	24' span	G		11	1.455		88	48		136	177
0150	28' span	G		10	1.600		103	52.50		155.50	201
0160	32' span	G		10	1.600		118	52.50		170.50	217
0250	6" chords, 28' span	G		9	1.778		129	58.50		187.50	240
0260	32' span	G		9	1.778		148	58.50		206.50	261
0270	36' span	G		8	2		166	65.50		231.50	293
0280	40' span	G		8	2		185	65.50		250.50	315
1120	5:12 to 8:12 pitch, 18 ga. x 4" chords, 16' span	G		10	1.600		67	52.50		119.50	162
1130	20' span	G		9	1.778		84	58.50		142.50	191
1140	24' span	G		9	1.778		101	58.50		159.50	209
1150	28' span	G		8	2		118	65.50		183.50	239
1160	32' span	G		8	2		134	65.50		199.50	258
1250	6" chords, 28' span	G		7	2.286		148	75		223	289
1260	32' span	G		7	2.286		169	75		244	310
1270	36' span	G		6	2.667		190	87.50		277.50	355
1280	40' span	G		6	2.667		211	87.50		298.50	380
2120	9:12 to 12:12 pitch, 18 ga. x 4" chords, 16' span	G		8	2		84	65.50		149.50	203
2130	20' span	G		7	2.286		105	75		180	242
2140	24' span	G		7	2.286		126	75		201	265
2150	28' span	G		6	2.667		147	87.50		234.50	310
2160	32' span	G		6	2.667		168	87.50		255.50	330
2250	6" chords, 28' span	G		5	3.200		185	105		290	380
2260	32' span	G		5	3.200		211	105		316	410
2270	36' span	G		4	4		238	131		369	480
2280	40' span	G	↓	4	4		264	131		395	510

05 44 Cold-Formed Metal Trusses

05 44 13 – Cold-Formed Metal Roof Trusses

05 44 13.60 Framing, Roof Trusses

		Crew	Daily Output	Labor-Hours	Unit	Material	2015 Bare Costs Labor	2015 Bare Costs Equipment	Total	Total Incl O&P
5120	Erection only of roof trusses, to 4:12 pitch, 16' span	F-6	48	.833	Ea.		25	13.60	38.60	57
5130	20' span		46	.870			26	14.20	40.20	59
5140	24' span		44	.909			27.50	14.85	42.35	62
5150	28' span		42	.952			28.50	15.55	44.05	65
5160	32' span		40	1			30	16.35	46.35	68.50
5170	36' span		38	1.053			31.50	17.20	48.70	72
5180	40' span		36	1.111			33.50	18.15	51.65	76
5220	5:12 to 8:12 pitch, 16' span		42	.952			28.50	15.55	44.05	65
5230	20' span		40	1			30	16.35	46.35	68.50
5240	24' span		38	1.053			31.50	17.20	48.70	72
5250	28' span		36	1.111			33.50	18.15	51.65	76
5260	32' span		34	1.176			35.50	19.25	54.75	80
5270	36' span		32	1.250			37.50	20.50	58	85.50
5280	40' span		30	1.333			40	22	62	91
5320	9:12 to 12:12 pitch, 16' span		36	1.111			33.50	18.15	51.65	76
5330	20' span		34	1.176			35.50	19.25	54.75	80
5340	24' span		32	1.250			37.50	20.50	58	85.50
5350	28' span		30	1.333			40	22	62	91
5360	32' span		28	1.429			43	23.50	66.50	97.50
5370	36' span		26	1.538			46	25	71	105
5380	40' span		24	1.667			50	27.50	77.50	114

05 51 Metal Stairs

05 51 13 – Metal Pan Stairs

05 51 13.50 Pan Stairs

			Crew	Daily Output	Labor-Hours	Unit	Material	Labor	Equipment	Total	Total Incl O&P
0010	**PAN STAIRS**, shop fabricated, steel stringers										
0015	Made from recycled materials										
1700	Pre-erected, steel pan tread, 3'-6" wide, 2 line pipe rail	G	E-2	87	.552	Riser	550	19.70	17.35	587.05	660
1800	With flat bar picket rail	G	"	87	.552	"	615	19.70	17.35	652.05	735

05 51 23 – Metal Fire Escapes

05 51 23.50 Fire Escape Stairs

		Crew	Daily Output	Labor-Hours	Unit	Material	Labor	Equipment	Total	Total Incl O&P
0010	**FIRE ESCAPE STAIRS**, portable									
0100	Portable ladder					Ea.	112			112	123

05 52 Metal Railings

05 52 13 – Pipe and Tube Railings

05 52 13.50 Railings, Pipe

			Crew	Daily Output	Labor-Hours	Unit	Material	Labor	Equipment	Total	Total Incl O&P
0010	**RAILINGS, PIPE**, shop fab'd, 3'-6" high, posts @ 5' O.C.										
0015	Made from recycled materials										
0020	Aluminum, 2 rail, satin finish, 1-1/4" diameter	G	E-4	160	.200	L.F.	36	7.15	.91	44.06	54
0030	Clear anodized	G		160	.200		44	7.15	.91	52.06	63
0040	Dark anodized	G		160	.200		49	7.15	.91	57.06	68.50
0080	1-1/2" diameter, satin finish	G		160	.200		42.50	7.15	.91	50.56	61
0090	Clear anodized	G		160	.200		47.50	7.15	.91	55.56	66.50
0100	Dark anodized	G		160	.200		52.50	7.15	.91	60.56	72
0140	Aluminum, 3 rail, 1-1/4" diam., satin finish	G		137	.234		54.50	8.35	1.07	63.92	76.50
0150	Clear anodized	G		137	.234		67.50	8.35	1.07	76.92	91.50
0160	Dark anodized	G		137	.234		75	8.35	1.07	84.42	99.50
0200	1-1/2" diameter, satin finish	G		137	.234		64.50	8.35	1.07	73.92	88

For customer support on your Residential Cost Data, call 877.713.2617

05 52 Metal Railings

05 52 13 – Pipe and Tube Railings

05 52 13.50 Railings, Pipe

		Crew	Daily Output	Labor-Hours	Unit	Material	2015 Bare Costs Labor	Equipment	Total	Total Incl O&P
0210	Clear anodized [G]	E-4	137	.234	L.F.	74	8.35	1.07	83.42	98
0220	Dark anodized [G]		137	.234		80.50	8.35	1.07	89.92	105
0500	Steel, 2 rail, on stairs, primed, 1-1/4" diameter [G]		160	.200		25	7.15	.91	33.06	42
0520	1-1/2" diameter [G]		160	.200		27	7.15	.91	35.06	44.50
0540	Galvanized, 1-1/4" diameter [G]		160	.200		34	7.15	.91	42.06	52
0560	1-1/2" diameter [G]		160	.200		38.50	7.15	.91	46.56	56.50
0580	Steel, 3 rail, primed, 1-1/4" diameter [G]		137	.234		37.50	8.35	1.07	46.92	58
0600	1-1/2" diameter [G]		137	.234		39	8.35	1.07	48.42	60
0620	Galvanized, 1-1/4" diameter [G]		137	.234		52.50	8.35	1.07	61.92	74.50
0640	1-1/2" diameter [G]		137	.234		60.50	8.35	1.07	69.92	83.50
0700	Stainless steel, 2 rail, 1-1/4" diam. #4 finish [G]		137	.234		110	8.35	1.07	119.42	138
0720	High polish [G]		137	.234		178	8.35	1.07	187.42	213
0740	Mirror polish [G]		137	.234		223	8.35	1.07	232.42	262
0760	Stainless steel, 3 rail, 1-1/2" diam., #4 finish [G]		120	.267		166	9.55	1.22	176.77	202
0770	High polish [G]		120	.267		275	9.55	1.22	285.77	325
0780	Mirror finish [G]		120	.267		335	9.55	1.22	345.77	390
0900	Wall rail, alum. pipe, 1-1/4" diam., satin finish [G]		213	.150		20	5.40	.69	26.09	33
0905	Clear anodized [G]		213	.150		25	5.40	.69	31.09	38.50
0910	Dark anodized [G]		213	.150		29.50	5.40	.69	35.59	43.50
0915	1-1/2" diameter, satin finish [G]		213	.150		22.50	5.40	.69	28.59	36
0920	Clear anodized [G]		213	.150		28	5.40	.69	34.09	42
0925	Dark anodized [G]		213	.150		35	5.40	.69	41.09	49.50
0930	Steel pipe, 1-1/4" diameter, primed [G]		213	.150		15	5.40	.69	21.09	27.50
0935	Galvanized [G]		213	.150		22	5.40	.69	28.09	35
0940	1-1/2" diameter [G]		176	.182		15.50	6.50	.83	22.83	30
0945	Galvanized [G]		213	.150		22	5.40	.69	28.09	35
0955	Stainless steel pipe, 1-1/2" diam., #4 finish [G]		107	.299		88	10.70	1.36	100.06	119
0960	High polish [G]		107	.299		179	10.70	1.36	191.06	219
0965	Mirror polish [G]		107	.299		212	10.70	1.36	224.06	255
2000	2-line pipe rail (1-1/2" T&B) with 1/2" pickets @ 4-1/2" O.C.,									
2005	attached handrail on brackets									
2010	42" high aluminum, satin finish, straight & level [G]	E-4	120	.267	L.F.	198	9.55	1.22	208.77	237
2050	42" high steel, primed, straight & level [G]	"	120	.267		127	9.55	1.22	137.77	159
4000	For curved and level rails, add					10%	10%			
4100	For sloped rails for stairs, add					30%	30%			

05 58 Formed Metal Fabrications

05 58 25 – Formed Lamp Posts

05 58 25.40 Lamp Posts

		Crew	Daily Output	Labor-Hours	Unit	Material	Labor	Equipment	Total	Total Incl O&P
0010	**LAMP POSTS**									
0020	Aluminum, 7' high, stock units, post only [G]	1 Carp	16	.500	Ea.	82	16.45		98.45	118
0100	Mild steel, plain [G]	"	16	.500	"	73	16.45		89.45	108

05 71 Decorative Metal Stairs

05 71 13 – Fabricated Metal Spiral Stairs

05 71 13.50 Spiral Stairs		Crew	Daily Output	Labor-Hours	Unit	Material	2015 Bare Costs Labor	Equipment	Total	Total Incl O&P
0010	**SPIRAL STAIRS**									
1805	Shop fabricated, custom ordered									
1810	Aluminum, 5'-0" diameter, plain units	G	E-4	45	.711	Riser	575	25.50	3.24	603.74	685
1820	Fancy units	G		45	.711		1,100	25.50	3.24	1,128.74	1,250
1900	Cast iron, 4'-0" diameter, plain units	G		45	.711		540	25.50	3.24	568.74	645
1920	Fancy Units	G	↓	25	1.280	↓	735	46	5.85	786.85	900
3100	Spiral stair kits, 12 stacking risers to fit exact floor height										

Division Notes

	CREW	DAILY OUTPUT	LABOR-HOURS	UNIT	BARE COSTS				TOTAL INCL O&P
					MAT.	LABOR	EQUIP.	TOTAL	

Estimating Tips
06 05 00 Common Work Results for Wood, Plastics, and Composites

- Common to any wood-framed structure are the accessory connector items such as screws, nails, adhesives, hangers, connector plates, straps, angles, and hold-downs. For typical wood-framed buildings, such as residential projects, the aggregate total for these items can be significant, especially in areas where seismic loading is a concern. For floor and wall framing, the material cost is based on 10 to 25 lbs. per MBF. Hold-downs, hangers, and other connectors should be taken off by the piece.

  Included with material costs are fasteners for a normal installation. RSMeans engineers use manufacturer's recommendations, written specifications, and/or standard construction practice for size and spacing of fasteners. Prices for various fasteners are shown for informational purposes only. Adjustments should be made if unusual fastening conditions exist.

06 10 00 Carpentry

- Lumber is a traded commodity and therefore sensitive to supply and demand in the marketplace. Even in "budgetary" estimating of wood-framed projects, it is advisable to call local suppliers for the latest market pricing.

- Common quantity units for wood-framed projects are "thousand board feet" (MBF). A board foot is a volume of wood, 1" x 1' x 1', or 144 cubic inches. Board-foot quantities are generally calculated using nominal material dimensions— dressed sizes are ignored. Board foot per lineal foot of any stick of lumber can be calculated by dividing the nominal cross-sectional area by 12. As an example, 2,000 lineal feet of 2 x 12 equates to 4 MBF by dividing the nominal area, 2 x 12, by 12, which equals 2, and multiplying by 2,000 to give 4,000 board feet. This simple rule applies to all nominal dimensioned lumber.

- Waste is an issue of concern at the quantity takeoff for any area of construction. Framing lumber is sold in even foot lengths, i.e., 10', 12', 14', 16' and, depending on spans, wall heights, and the grade of lumber, waste is inevitable. A rule of thumb for lumber waste is 5%–10% depending on material quality and the complexity of the framing.

- Wood in various forms and shapes is used in many projects, even where the main structural framing is steel, concrete, or masonry. Plywood as a back-up partition material and 2x boards used as blocking and cant strips around roof edges are two common examples. The estimator should ensure that the costs of all wood materials are included in the final estimate.

06 20 00 Finish Carpentry

- It is necessary to consider the grade of workmanship when estimating labor costs for erecting millwork and interior finish. In practice, there are three grades: premium, custom, and economy. The RSMeans daily output for base and case moldings is in the range of 200 to 250 L.F. per carpenter per day. This is appropriate for most average custom-grade projects. For premium projects, an adjustment to productivity of 25%–50% should be made, depending on the complexity of the job.

Reference Numbers
Reference numbers are shown in shaded boxes at the beginning of some major classifications. These numbers refer to related items in the Reference Section. The reference information may be an estimating procedure, an alternate pricing method, or technical information.

Note: Not all subdivisions listed here necessarily appear in this publication. ■

No part of this publication may be reproduced, stored in a retrieval system, or transmitted in any form or by any means without prior written permission of RSMeans.

06 05 Common Work Results for Wood, Plastics, and Composites

06 05 05 – Selective Demolition for Wood, Plastics, and Composites

06 05 05.10 Selective Demolition Wood Framing		Crew	Daily Output	Labor-Hours	Unit	Material	2015 Bare Costs Labor	2015 Bare Costs Equipment	Total	Total Incl O&P
0010	**SELECTIVE DEMOLITION WOOD FRAMING** R024119-10									
0100	Timber connector, nailed, small	1 Clab	96	.083	Ea.		2.04		2.04	3.42
0110	Medium		60	.133			3.26		3.26	5.45
0120	Large		48	.167			4.08		4.08	6.85
0130	Bolted, small		48	.167			4.08		4.08	6.85
0140	Medium		32	.250			6.10		6.10	10.25
0150	Large		24	.333			8.15		8.15	13.70
2958	Beams, 2" x 6"	2 Clab	1100	.015	L.F.		.36		.36	.60
2960	2" x 8"		825	.019			.47		.47	.80
2965	2" x 10"		665	.024			.59		.59	.99
2970	2" x 12"		550	.029			.71		.71	1.19
2972	2" x 14"		470	.034			.83		.83	1.40
2975	4" x 8"	B-1	413	.058			1.46		1.46	2.45
2980	4" x 10"		330	.073			1.83		1.83	3.07
2985	4" x 12"		275	.087			2.19		2.19	3.68
3000	6" x 8"		275	.087			2.19		2.19	3.68
3040	6" x 10"		220	.109			2.74		2.74	4.60
3080	6" x 12"		185	.130			3.26		3.26	5.45
3120	8" x 12"		140	.171			4.31		4.31	7.25
3160	10" x 12"		110	.218			5.50		5.50	9.20
3162	Alternate pricing method		1.10	21.818	M.B.F.		550		550	920
3170	Blocking, in 16" OC wall framing, 2" x 4"	1 Clab	600	.013	L.F.		.33		.33	.55
3172	2" x 6"		400	.020			.49		.49	.82
3174	In 24" OC wall framing, 2" x 4"		600	.013			.33		.33	.55
3176	2" x 6"		400	.020			.49		.49	.82
3178	Alt method, wood blocking removal from wood framing		.40	20	M.B.F.		490		490	820
3179	Wood blocking removal from steel framing		.36	22.222	"		545		545	910
3180	Bracing, let in, 1" x 3", studs 16" OC		1050	.008	L.F.		.19		.19	.31
3181	Studs 24" OC		1080	.007			.18		.18	.30
3182	1" x 4", studs 16" OC		1050	.008			.19		.19	.31
3183	Studs 24" OC		1080	.007			.18		.18	.30
3184	1" x 6", studs 16" OC		1050	.008			.19		.19	.31
3185	Studs 24" OC		1080	.007			.18		.18	.30
3186	2" x 3", studs 16" OC		800	.010			.24		.24	.41
3187	Studs 24" OC		830	.010			.24		.24	.40
3188	2" x 4", studs 16" OC		800	.010			.24		.24	.41
3189	Studs 24" OC		830	.010			.24		.24	.40
3190	2" x 6", studs 16" OC		800	.010			.24		.24	.41
3191	Studs 24" OC		830	.010			.24		.24	.40
3192	2" x 8", studs 16" OC		800	.010			.24		.24	.41
3193	Studs 24" OC		830	.010			.24		.24	.40
3194	"T" shaped metal bracing, studs at 16" OC		1060	.008			.18		.18	.31
3195	Studs at 24" OC		1200	.007			.16		.16	.27
3196	Metal straps, studs at 16" OC		1200	.007			.16		.16	.27
3197	Studs at 24" OC		1240	.006			.16		.16	.26
3200	Columns, round, 8' to 14' tall		40	.200	Ea.		4.89		4.89	8.20
3202	Dimensional lumber sizes	2 Clab	1.10	14.545	M.B.F.		355		355	595
3250	Blocking, between joists	1 Clab	320	.025	Ea.		.61		.61	1.03
3252	Bridging, metal strap, between joists		320	.025	Pr.		.61		.61	1.03
3254	Wood, between joists		320	.025	"		.61		.61	1.03
3260	Door buck, studs, header & access., 8' high 2" x 4" wall, 3' wide		32	.250	Ea.		6.10		6.10	10.25
3261	4' wide		32	.250			6.10		6.10	10.25
3262	5' wide		32	.250			6.10		6.10	10.25

06 05 Common Work Results for Wood, Plastics, and Composites

06 05 05 – Selective Demolition for Wood, Plastics, and Composites

06 05 05.10 Selective Demolition Wood Framing		Crew	Daily Output	Labor-Hours	Unit	Material	2015 Bare Costs Labor	Equipment	Total	Total Incl O&P
3263	6' wide	1 Clab	32	.250	Ea.		6.10		6.10	10.25
3264	8' wide		30	.267			6.50		6.50	10.95
3265	10' wide		30	.267			6.50		6.50	10.95
3266	12' wide		30	.267			6.50		6.50	10.95
3267	2" x 6" wall, 3' wide		32	.250			6.10		6.10	10.25
3268	4' wide		32	.250			6.10		6.10	10.25
3269	5' wide		32	.250			6.10		6.10	10.25
3270	6' wide		32	.250			6.10		6.10	10.25
3271	8' wide		30	.267			6.50		6.50	10.95
3272	10' wide		30	.267			6.50		6.50	10.95
3273	12' wide		30	.267			6.50		6.50	10.95
3274	Window buck, studs, header & access, 8' high 2" x 4" wall, 2' wide		24	.333			8.15		8.15	13.70
3275	3' wide		24	.333			8.15		8.15	13.70
3276	4' wide		24	.333			8.15		8.15	13.70
3277	5' wide		24	.333			8.15		8.15	13.70
3278	6' wide		24	.333			8.15		8.15	13.70
3279	7' wide		24	.333			8.15		8.15	13.70
3280	8' wide		22	.364			8.90		8.90	14.95
3281	10' wide		22	.364			8.90		8.90	14.95
3282	12' wide		22	.364			8.90		8.90	14.95
3283	2" x 6" wall, 2' wide		24	.333			8.15		8.15	13.70
3284	3' wide		24	.333			8.15		8.15	13.70
3285	4' wide		24	.333			8.15		8.15	13.70
3286	5' wide		24	.333			8.15		8.15	13.70
3287	6' wide		24	.333			8.15		8.15	13.70
3288	7' wide		24	.333			8.15		8.15	13.70
3289	8' wide		22	.364			8.90		8.90	14.95
3290	10' wide		22	.364			8.90		8.90	14.95
3291	12' wide		22	.364			8.90		8.90	14.95
3360	Deck or porch decking		825	.010	L.F.		.24		.24	.40
3400	Fascia boards, 1" x 6"		500	.016			.39		.39	.66
3440	1" x 8"		450	.018			.43		.43	.73
3480	1" x 10"		400	.020			.49		.49	.82
3490	2" x 6"		450	.018			.43		.43	.73
3500	2" x 8"		400	.020			.49		.49	.82
3510	2" x 10"		350	.023			.56		.56	.94
3610	Furring, on wood walls or ceiling		4000	.002	S.F.		.05		.05	.08
3620	On masonry or concrete walls or ceiling		1200	.007	"		.16		.16	.27
3800	Headers over openings, 2 @ 2" x 6"		110	.073	L.F.		1.78		1.78	2.99
3840	2 @ 2" x 8"		100	.080			1.96		1.96	3.28
3880	2 @ 2" x 10"		90	.089			2.17		2.17	3.65
3885	Alternate pricing method		.26	30.651	M.B.F.		750		750	1,250
3920	Joists, 1" x 4"		1250	.006	L.F.		.16		.16	.26
3930	1" x 6"		1135	.007			.17		.17	.29
3940	1" x 8"		1000	.008			.20		.20	.33
3950	1" x 10"		895	.009			.22		.22	.37
3960	1" x 12"		765	.010			.26		.26	.43
4200	2" x 4"	2 Clab	1000	.016			.39		.39	.66
4230	2" x 6"		970	.016			.40		.40	.68
4240	2" x 8"		940	.017			.42		.42	.70
4250	2" x 10"		910	.018			.43		.43	.72
4280	2" x 12"		880	.018			.44		.44	.75
4281	2" x 14"		850	.019			.46		.46	.77

For customer support on your Residential Cost Data, call 877.713.2617.

06 05 Common Work Results for Wood, Plastics, and Composites

06 05 05 – Selective Demolition for Wood, Plastics, and Composites

06 05 05.10 Selective Demolition Wood Framing		Crew	Daily Output	Labor-Hours	Unit	Material	2015 Bare Costs Labor	2015 Bare Costs Equipment	Total	Total Incl O&P
4282	Composite joists, 9-1/2"	2 Clab	960	.017	L.F.		.41		.41	.68
4283	11-7/8"		930	.017			.42		.42	.71
4284	14"		897	.018			.44		.44	.73
4285	16"		865	.019			.45		.45	.76
4290	Wood joists, alternate pricing method		1.50	10.667	M.B.F.		261		261	440
4500	Open web joist, 12" deep		500	.032	L.F.		.78		.78	1.31
4505	14" deep		475	.034			.82		.82	1.38
4510	16" deep		450	.036			.87		.87	1.46
4520	18" deep		425	.038			.92		.92	1.55
4530	24" deep		400	.040			.98		.98	1.64
4550	Ledger strips, 1" x 2"	1 Clab	1200	.007			.16		.16	.27
4560	1" x 3"		1200	.007			.16		.16	.27
4570	1" x 4"		1200	.007			.16		.16	.27
4580	2" x 2"		1100	.007			.18		.18	.30
4590	2" x 4"		1000	.008			.20		.20	.33
4600	2" x 6"		1000	.008			.20		.20	.33
4601	2" x 8 or 2" x 10"		800	.010			.24		.24	.41
4602	4" x 6"		600	.013			.33		.33	.55
4604	4" x 8"		450	.018			.43		.43	.73
5400	Posts, 4" x 4"	2 Clab	800	.020			.49		.49	.82
5405	4" x 6"		550	.029			.71		.71	1.19
5410	4" x 8"		440	.036			.89		.89	1.49
5425	4" x 10"		390	.041			1		1	1.68
5430	4" x 12"		350	.046			1.12		1.12	1.88
5440	6" x 6"		400	.040			.98		.98	1.64
5445	6" x 8"		350	.046			1.12		1.12	1.88
5450	6" x 10"		320	.050			1.22		1.22	2.05
5455	6" x 12"		290	.055			1.35		1.35	2.26
5480	8" x 8"		300	.053			1.30		1.30	2.19
5500	10" x 10"		240	.067			1.63		1.63	2.74
5660	Tongue and groove floor planks		2	8	M.B.F.		196		196	330
5682	Rafters, ordinary, 16" OC, 2" x 4"		880	.018	S.F.		.44		.44	.75
5683	2" x 6"		840	.019			.47		.47	.78
5684	2" x 8"		820	.020			.48		.48	.80
5685	2" x 10"		820	.020			.48		.48	.80
5686	2" x 12"		810	.020			.48		.48	.81
5687	24" OC, 2" x 4"		1170	.014			.33		.33	.56
5688	2" x 6"		1117	.014			.35		.35	.59
5689	2" x 8"		1091	.015			.36		.36	.60
5690	2" x 10"		1091	.015			.36		.36	.60
5691	2" x 12"		1077	.015			.36		.36	.61
5795	Rafters, ordinary, 2" x 4" (alternate method)		862	.019	L.F.		.45		.45	.76
5800	2" x 6" (alternate method)		850	.019			.46		.46	.77
5840	2" x 8" (alternate method)		837	.019			.47		.47	.78
5855	2" x 10" (alternate method)		825	.019			.47		.47	.80
5865	2" x 12" (alternate method)		812	.020			.48		.48	.81
5870	Sill plate, 2" x 4"	1 Clab	1170	.007			.17		.17	.28
5871	2" x 6"		780	.010			.25		.25	.42
5872	2" x 8"		586	.014			.33		.33	.56
5873	Alternate pricing method		.78	10.256	M.B.F.		251		251	420
5885	Ridge board, 1" x 4"	2 Clab	900	.018	L.F.		.43		.43	.73
5886	1" x 6"		875	.018			.45		.45	.75
5887	1" x 8"		850	.019			.46		.46	.77

For customer support on your Residential Cost Data, call 877.713.2617.

06 05 Common Work Results for Wood, Plastics, and Composites

06 05 05 – Selective Demolition for Wood, Plastics, and Composites

06 05 05.10 Selective Demolition Wood Framing		Crew	Daily Output	Labor-Hours	Unit	Material	2015 Bare Costs Labor	2015 Bare Costs Equipment	Total	Total Incl O&P
5888	1" x 10"	2 Clab	825	.019	L.F.		.47		.47	.80
5889	1" x 12"		800	.020			.49		.49	.82
5890	2" x 4"		900	.018			.43		.43	.73
5892	2" x 6"		875	.018			.45		.45	.75
5894	2" x 8"		850	.019			.46		.46	.77
5896	2" x 10"		825	.019			.47		.47	.80
5898	2" x 12"		800	.020			.49		.49	.82
6050	Rafter tie, 1" x 4"		1250	.013			.31		.31	.53
6052	1" x 6"		1135	.014			.34		.34	.58
6054	2" x 4"		1000	.016			.39		.39	.66
6056	2" x 6"		970	.016			.40		.40	.68
6070	Sleepers, on concrete, 1" x 2"	1 Clab	4700	.002			.04		.04	.07
6075	1" x 3"		4000	.002			.05		.05	.08
6080	2" x 4"		3000	.003			.07		.07	.11
6085	2" x 6"		2600	.003			.08		.08	.13
6086	Sheathing from roof, 5/16"	2 Clab	1600	.010	S.F.		.24		.24	.41
6088	3/8"		1525	.010			.26		.26	.43
6090	1/2"		1400	.011			.28		.28	.47
6092	5/8"		1300	.012			.30		.30	.51
6094	3/4"		1200	.013			.33		.33	.55
6096	Board sheathing from roof		1400	.011			.28		.28	.47
6100	Sheathing, from walls, 1/4"		1200	.013			.33		.33	.55
6110	5/16"		1175	.014			.33		.33	.56
6120	3/8"		1150	.014			.34		.34	.57
6130	1/2"		1125	.014			.35		.35	.58
6140	5/8"		1100	.015			.36		.36	.60
6150	3/4"		1075	.015			.36		.36	.61
6152	Board sheathing from walls		1500	.011			.26		.26	.44
6158	Subfloor/roof deck, with boards		2200	.007			.18		.18	.30
6159	Subfloor/roof deck, with tongue & groove boards		2000	.008			.20		.20	.33
6160	Plywood, 1/2" thick		768	.021			.51		.51	.86
6162	5/8" thick		760	.021			.51		.51	.86
6164	3/4" thick		750	.021			.52		.52	.88
6165	1-1/8" thick		720	.022			.54		.54	.91
6166	Underlayment, particle board, 3/8" thick	1 Clab	780	.010			.25		.25	.42
6168	1/2" thick		768	.010			.25		.25	.43
6170	5/8" thick		760	.011			.26		.26	.43
6172	3/4" thick		750	.011			.26		.26	.44
6200	Stairs and stringers, straight run	2 Clab	40	.400	Riser		9.80		9.80	16.40
6240	With platforms, winders or curves	"	26	.615	"		15.05		15.05	25.50
6300	Components, tread	1 Clab	110	.073	Ea.		1.78		1.78	2.99
6320	Riser		80	.100	"		2.45		2.45	4.11
6390	Stringer, 2" x 10"		260	.031	L.F.		.75		.75	1.26
6400	2" x 12"		260	.031			.75		.75	1.26
6410	3" x 10"		250	.032			.78		.78	1.31
6420	3" x 12"		250	.032			.78		.78	1.31
6590	Wood studs, 2" x 3"	2 Clab	3076	.005			.13		.13	.21
6600	2" x 4"		2000	.008			.20		.20	.33
6640	2" x 6"		1600	.010			.24		.24	.41
6720	Wall framing, including studs, plates and blocking, 2" x 4"	1 Clab	600	.013	S.F.		.33		.33	.55
6740	2" x 6"		480	.017	"		.41		.41	.68
6750	Headers, 2" x 4"		1125	.007	L.F.		.17		.17	.29
6755	2" x 6"		1125	.007			.17		.17	.29

06 05 Common Work Results for Wood, Plastics, and Composites

06 05 05 – Selective Demolition for Wood, Plastics, and Composites

06 05 05.10 Selective Demolition Wood Framing

		Crew	Daily Output	Labor-Hours	Unit	Material	2015 Bare Costs Labor	2015 Bare Costs Equipment	Total	Total Incl O&P
6760	2" x 8"	1 Clab	1050	.008	L.F.		.19		.19	.31
6765	2" x 10"		1050	.008			.19		.19	.31
6770	2" x 12"		1000	.008			.20		.20	.33
6780	4" x 10"		525	.015			.37		.37	.63
6785	4" x 12"		500	.016			.39		.39	.66
6790	6" x 8"		560	.014			.35		.35	.59
6795	6" x 10"		525	.015			.37		.37	.63
6797	6" x 12"		500	.016			.39		.39	.66
7000	Trusses									
7050	12' span	2 Clab	74	.216	Ea.		5.30		5.30	8.90
7150	24' span	F-3	66	.606			18.20	9.90	28.10	41.50
7200	26' span		64	.625			18.75	10.20	28.95	43
7250	28' span		62	.645			19.35	10.55	29.90	44
7300	30' span		58	.690			20.50	11.30	31.80	47
7350	32' span		56	.714			21.50	11.70	33.20	49
7400	34' span		54	.741			22	12.10	34.10	51
7450	36' span		52	.769			23	12.60	35.60	53
8000	Soffit, T & G wood	1 Clab	520	.015	S.F.		.38		.38	.63
8010	Hardboard, vinyl or aluminum	"	640	.013			.31		.31	.51
8030	Plywood	2 Carp	315	.051			1.67		1.67	2.80
9500	See Section 02 41 19.19 for rubbish handling									

06 05 05.20 Selective Demolition Millwork and Trim

		Crew	Daily Output	Labor-Hours	Unit	Material	2015 Bare Costs Labor	2015 Bare Costs Equipment	Total	Total Incl O&P
0010	**SELECTIVE DEMOLITION MILLWORK AND TRIM** R024119-10									
1000	Cabinets, wood, base cabinets, per L.F.	2 Clab	80	.200	L.F.		4.89		4.89	8.20
1020	Wall cabinets, per L.F.	"	80	.200	"		4.89		4.89	8.20
1060	Remove and reset, base cabinets	2 Carp	18	.889	Ea.		29		29	49
1070	Wall cabinets		20	.800			26.50		26.50	44
1072	Oven cabinet, 7' high		11	1.455			48		48	80
1074	Cabinet door, up to 2' high	1 Clab	66	.121			2.96		2.96	4.98
1076	2' - 4' high	"	46	.174			4.25		4.25	7.15
1100	Steel, painted, base cabinets	2 Clab	60	.267	L.F.		6.50		6.50	10.95
1120	Wall cabinets		60	.267	"		6.50		6.50	10.95
1200	Casework, large area		320	.050	S.F.		1.22		1.22	2.05
1220	Selective		200	.080	"		1.96		1.96	3.28
1500	Counter top, straight runs		200	.080	L.F.		1.96		1.96	3.28
1510	L, U or C shapes		120	.133	"		3.26		3.26	5.45
2000	Paneling, 4' x 8' sheets		2000	.008	S.F.		.20		.20	.33
2100	Boards, 1" x 4"		700	.023			.56		.56	.94
2120	1" x 6"		750	.021			.52		.52	.88
2140	1" x 8"		800	.020			.49		.49	.82
3000	Trim, baseboard, to 6" wide		1200	.013	L.F.		.33		.33	.55
3040	Greater than 6" and up to 12" wide		1000	.016			.39		.39	.66
3080	Remove and reset, minimum	2 Carp	400	.040			1.31		1.31	2.21
3090	Maximum	"	300	.053			1.75		1.75	2.94
3100	Ceiling trim	2 Clab	1000	.016			.39		.39	.66
3120	Chair rail		1200	.013			.33		.33	.55
3140	Railings with balusters		240	.067			1.63		1.63	2.74
3160	Wainscoting		700	.023	S.F.		.56		.56	.94
4000	Curtain rod	1 Clab	80	.100	L.F.		2.45		2.45	4.11

06 05 Common Work Results for Wood, Plastics, and Composites

06 05 23 – Wood, Plastic, and Composite Fastenings

06 05 23.10 Nails		Crew	Daily Output	Labor-Hours	Unit	Material	2015 Bare Costs Labor	Equipment	Total	Total Incl O&P
0010	**NAILS**, material only, based upon 50# box purchase									
0020	Copper nails, plain				Lb.	9.85			9.85	10.80
0400	Stainless steel, plain					9.20			9.20	10.15
0500	Box, 3d to 20d, bright					1.37			1.37	1.51
0520	Galvanized					1.68			1.68	1.85
0600	Common, 3d to 60d, plain					1.30			1.30	1.43
0700	Galvanized					1.97			1.97	2.17
0800	Aluminum					10.80			10.80	11.90
1000	Annular or spiral thread, 4d to 60d, plain					2.63			2.63	2.89
1200	Galvanized					3.15			3.15	3.47
1400	Drywall nails, plain					1.50			1.50	1.65
1600	Galvanized					2.20			2.20	2.42
1800	Finish nails, 4d to 10d, plain					1.25			1.25	1.38
2000	Galvanized					1.86			1.86	2.05
2100	Aluminum					5.85			5.85	6.45
2300	Flooring nails, hardened steel, 2d to 10d, plain					3.38			3.38	3.72
2400	Galvanized					4.04			4.04	4.44
2500	Gypsum lath nails, 1-1/8", 13 ga. flathead, blued					2			2	2.20
2600	Masonry nails, hardened steel, 3/4" to 3" long, plain					1.71			1.71	1.88
2700	Galvanized					3.27			3.27	3.60
2900	Roofing nails, threaded, galvanized					1.76			1.76	1.94
3100	Aluminum					4.85			4.85	5.35
3300	Compressed lead head, threaded, galvanized					2.49			2.49	2.74
3600	Siding nails, plain shank, galvanized					2.02			2.02	2.22
3800	Aluminum					5.85			5.85	6.45
5000	Add to prices above for cement coating					.11			.11	.12
5200	Zinc or tin plating					.14			.14	.15
5500	Vinyl coated sinkers, 8d to 16d					2.50			2.50	2.75

06 05 23.50 Wood Screws		Crew	Daily Output	Labor-Hours	Unit	Material	Labor	Equipment	Total	Total Incl O&P
0010	**WOOD SCREWS**									
0020	#8 x 1" long, steel				C	2.70			2.70	2.97
0100	Brass					12.20			12.20	13.40
0200	#8, 2" long, steel					4.16			4.16	4.58
0300	Brass					22			22	24
0400	#10, 1" long, steel					3.30			3.30	3.63
0500	Brass					15.60			15.60	17.15
0600	#10, 2" long, steel					5.30			5.30	5.85
0700	Brass					27			27	29.50
0800	#10, 3" long, steel					8.85			8.85	9.75
1000	#12, 2" long, steel					7.30			7.30	8.05
1100	Brass					34.50			34.50	38
1500	#12, 3" long, steel					11.15			11.15	12.30
2000	#12, 4" long, steel					18.75			18.75	20.50

06 05 23.60 Timber Connectors		Crew	Daily Output	Labor-Hours	Unit	Material	Labor	Equipment	Total	Total Incl O&P
0010	**TIMBER CONNECTORS**									
0020	Add up cost of each part for total cost of connection									
0100	Connector plates, steel, with bolts, straight	2 Carp	75	.213	Ea.	30.50	7		37.50	45.50
0110	Tee, 7 ga.		50	.320		35	10.50		45.50	56
0120	T-Strap, 14 ga., 12" x 8" x 2"		50	.320		35	10.50		45.50	56
0150	Anchor plates, 7 ga., 9" x 7"		75	.213		30.50	7		37.50	45.50
0200	Bolts, machine, sq. hd. with nut & washer, 1/2" diameter, 4" long	1 Carp	140	.057		.75	1.88		2.63	3.98
0300	7-1/2" long		130	.062		1.37	2.02		3.39	4.90

06 05 Common Work Results for Wood, Plastics, and Composites

06 05 23 – Wood, Plastic, and Composite Fastenings

06 05 23.60 Timber Connectors

		Crew	Daily Output	Labor-Hours	Unit	Material	2015 Bare Costs Labor	Equipment	Total	Total Incl O&P
0500	3/4" diameter, 7-1/2" long	1 Carp	130	.062	Ea.	3.20	2.02		5.22	6.90
0610	Machine bolts, w/nut, washer, 3/4" diam., 15" L, HD's & beam hangers		95	.084	↓	5.95	2.77		8.72	11.20
0720	Machine bolts, sq. hd. w/nut & wash		150	.053	Lb.	3.45	1.75		5.20	6.75
0800	Drilling bolt holes in timber, 1/2" diameter		450	.018	Inch		.58		.58	.98
0900	1" diameter		350	.023	"		.75		.75	1.26
1100	Framing anchor, angle, 3" x 3" x 1-1/2", 12 ga		175	.046	Ea.	2.61	1.50		4.11	5.40
1150	Framing anchors, 18 ga., 4-1/2" x 2-3/4"		175	.046		2.61	1.50		4.11	5.40
1160	Framing anchors, 18 ga., 4-1/2" x 3"		175	.046		2.61	1.50		4.11	5.40
1170	Clip anchors plates, 18 ga., 12" x 1-1/8"		175	.046		2.61	1.50		4.11	5.40
1250	Holdowns, 3 ga. base, 10 ga. body		8	1		25	33		58	82.50
1260	Holdowns, 7 ga. 11-1/16" x 3-1/4"		8	1		25	33		58	82.50
1270	Holdowns, 7 ga. 14-3/8" x 3-1/8"		8	1		25	33		58	82.50
1275	Holdowns, 12 ga. 8" x 2-1/2"		8	1		25	33		58	82.50
1300	Joist and beam hangers, 18 ga. galv., for 2" x 4" joist		175	.046		.73	1.50		2.23	3.32
1400	2" x 6" to 2" x 10" joist		165	.048		1.39	1.59		2.98	4.20
1600	16 ga. galv., 3" x 6" to 3" x 10" joist		160	.050		2.96	1.64		4.60	6
1700	3" x 10" to 3" x 14" joist		160	.050		4.92	1.64		6.56	8.15
1800	4" x 6" to 4" x 10" joist		155	.052		3.07	1.70		4.77	6.25
1900	4" x 10" to 4" x 14" joist		155	.052		5	1.70		6.70	8.35
2000	Two-2" x 6" to two-2" x 10" joists		150	.053		4.19	1.75		5.94	7.55
2100	Two-2" x 10" to two-2" x 14" joists		150	.053		4.68	1.75		6.43	8.10
2300	3/16" thick, 6" x 8" joist		145	.055		65	1.81		66.81	74.50
2400	6" x 10" joist		140	.057		67.50	1.88		69.38	77.50
2500	6" x 12" joist		135	.059		70.50	1.95		72.45	81
2700	1/4" thick, 6" x 14" joist		130	.062		73	2.02		75.02	84
2900	Plywood clips, extruded aluminum H clip, for 3/4" panels					.24			.24	.26
3000	Galvanized 18 ga. back-up clip					.18			.18	.20
3200	Post framing, 16 ga. galv. for 4" x 4" base, 2 piece	1 Carp	130	.062		16.40	2.02		18.42	21.50
3300	Cap		130	.062		23	2.02		25.02	28.50
3500	Rafter anchors, 18 ga. galv., 1-1/2" wide, 5-1/4" long		145	.055		.49	1.81		2.30	3.58
3600	10-3/4" long		145	.055		1.46	1.81		3.27	4.65
3800	Shear plates, 2-5/8" diameter		120	.067		2.38	2.19		4.57	6.30
3900	4" diameter		115	.070		5.65	2.29		7.94	10.05
4000	Sill anchors, embedded in concrete or block, 25-1/2" long		115	.070		12.70	2.29		14.99	17.80
4100	Spike grids, 3" x 6"		120	.067		.96	2.19		3.15	4.74
4400	Split rings, 2-1/2" diameter		120	.067		1.96	2.19		4.15	5.85
4500	4" diameter		110	.073		2.97	2.39		5.36	7.30
4550	Tie plate, 20 ga., 7" x 3 1/8"		110	.073		2.97	2.39		5.36	7.30
4560	Tie plate, 20 ga., 5" x 4 1/8"		110	.073		2.97	2.39		5.36	7.30
4575	Twist straps, 18 ga., 12" x 1 1/4"		110	.073		2.97	2.39		5.36	7.30
4580	Twist straps, 18 ga., 16" x 1 1/4"		110	.073		2.97	2.39		5.36	7.30
4600	Strap ties, 20 ga., 2-1/16" wide, 12 13/16" long		180	.044		.94	1.46		2.40	3.48
4700	Strap ties, 16 ga., 1-3/8" wide, 12" long		180	.044		.94	1.46		2.40	3.48
4800	21-5/8" x 1-1/4"		160	.050		2.96	1.64		4.60	6
5000	Toothed rings, 2-5/8" or 4" diameter		90	.089	↓	1.78	2.92		4.70	6.85
5200	Truss plates, nailed, 20 ga., up to 32' span		17	.471	Truss	12.90	15.45		28.35	40
5400	Washers, 2" x 2" x 1/8"				Ea.	.40			.40	.44
5500	3" x 3" x 3/16"				"	1.06			1.06	1.17
6000	Angles and gussets, painted									
6012	7 ga., 3-1/4" x 3-1/4" x 2-1/2" long	1 Carp	1.90	4.211	C	1,100	138		1,238	1,450
6014	3-1/4" x 3-1/4" x 5" long		1.90	4.211		2,150	138		2,288	2,600
6016	3-1/4" x 3-1/4" x 7-1/2" long		1.85	4.324		4,075	142		4,217	4,725
6018	5-3/4" x 5-3/4" x 2-1/2" long		1.85	4.324	↓	2,625	142		2,767	3,150

06 05 Common Work Results for Wood, Plastics, and Composites

06 05 23 – Wood, Plastic, and Composite Fastenings

06 05 23.60	Timber Connectors	Crew	Daily Output	Labor-Hours	Unit	Material	2015 Bare Costs Labor	Equipment	Total	Total Incl O&P
6020	5-3/4" x 5-3/4" x 5" long	1 Carp	1.85	4.324	C	4,175	142		4,317	4,850
6022	5-3/4" x 5-3/4" x 7-1/2" long		1.80	4.444		6,175	146		6,321	7,050
6024	3 ga., 4-1/4" x 4-1/4" x 3" long		1.85	4.324		2,800	142		2,942	3,325
6026	4-1/4" x 4-1/4" x 6" long		1.85	4.324		6,000	142		6,142	6,850
6028	4-1/4" x 4-1/4" x 9" long		1.80	4.444		6,750	146		6,896	7,675
6030	7-1/4" x 7-1/4" x 3" long		1.80	4.444		4,825	146		4,971	5,575
6032	7-1/4" x 7-1/4" x 6" long		1.80	4.444		6,525	146		6,671	7,425
6034	7-1/4" x 7-1/4" x 9" long		1.75	4.571		14,600	150		14,750	16,400
6036	Gussets									
6038	7 ga., 8-1/8" x 8-1/8" x 2-3/4" long	1 Carp	1.80	4.444	C	4,625	146		4,771	5,325
6040	3 ga., 9-3/4" x 9-3/4" x 3-1/4" long	"	1.80	4.444	"	6,400	146		6,546	7,300
6101	Beam hangers, polymer painted									
6102	Bolted, 3 ga., (W x H x L)									
6104	3-1/4" x 9" x 12" top flange	1 Carp	1	8	C	19,600	263		19,863	22,000
6106	5-1/4" x 9" x 12" top flange		1	8		20,500	263		20,763	22,900
6108	5-1/4" x 11" x 11-3/4" top flange		1	8		23,200	263		23,463	26,000
6110	6-7/8" x 9" x 12" top flange		1	8		21,200	263		21,463	23,700
6112	6-7/8" x 11" x 13-1/2" top flange		1	8		24,400	263		24,663	27,300
6114	8-7/8" x 11" x 15-1/2" top flange		1	8		26,100	263		26,363	29,100
6116	Nailed, 3 ga., (W x H x L)									
6118	3-1/4" x 10-1/2" x 10" top flange	1 Carp	1.80	4.444	C	20,500	146		20,646	22,700
6120	3-1/4" x 10-1/2" x 12" top flange		1.80	4.444		20,500	146		20,646	22,700
6122	5-1/4" x 9-1/2" x 10" top flange		1.80	4.444		20,900	146		21,046	23,200
6124	5-1/4" x 9-1/2" x 12" top flange		1.80	4.444		20,900	146		21,046	23,200
6128	6-7/8" x 8-1/2" x 12" top flange		1.80	4.444		21,400	146		21,546	23,700
6134	Saddle hangers, glu-lam (W x H x L)									
6136	3-1/4" x 10-1/2" x 5-1/4" x 6" saddle	1 Carp	.50	16	C	15,300	525		15,825	17,700
6138	3-1/4" x 10-1/2" x 6-7/8" x 6" saddle		.50	16		16,100	525		16,625	18,600
6140	3-1/4" x 10-1/2" x 8-7/8" x 6" saddle		.50	16		16,900	525		17,425	19,500
6142	3-1/4" x 19-1/2" x 5-1/4" x 10-1/8" saddle		.40	20		15,300	655		15,955	17,900
6144	3-1/4" x 19-1/2" x 6-7/8" x 10-1/8" saddle		.40	20		16,100	655		16,755	18,800
6146	3-1/4" x 19-1/2" x 8-7/8" x 10-1/8" saddle		.40	20		16,900	655		17,555	19,700
6148	5-1/4" x 9-1/2" x 5-1/4" x 12" saddle		.50	16		18,300	525		18,825	21,000
6150	5-1/4" x 9-1/2" x 6-7/8" x 9" saddle		.50	16		20,000	525		20,525	22,900
6152	5-1/4" x 10-1/2" x spec x 12" saddle		.50	16		21,800	525		22,325	24,800
6154	5-1/4" x 18" x 5-1/4" x 12-1/8" saddle		.40	20		18,300	655		18,955	21,200
6156	5-1/4" x 18" x 6-7/8" x 12-1/8" saddle		.40	20		20,000	655		20,655	23,100
6158	5-1/4" x 18" x spec x 12-1/8" saddle		.40	20		21,800	655		22,455	25,000
6160	6-7/8" x 8-1/2" x 6-7/8" x 12" saddle		.50	16		21,800	525		22,325	24,900
6162	6-7/8" x 8-1/2" x 8-7/8" x 12" saddle		.50	16		22,600	525		23,125	25,700
6164	6-7/8" x 10-1/2" x spec x 12" saddle		.50	16		21,800	525		22,325	24,900
6166	6-7/8" x 18" x 6-7/8" x 13-3/4" saddle		.40	20		21,800	655		22,455	25,100
6168	6-7/8" x 18" x 8-7/8" x 13-3/4" saddle		.40	20		22,600	655		23,255	25,900
6170	6-7/8" x 18" x spec x 13-3/4" saddle		.40	20		24,400	655		25,055	27,900
6172	8-7/8" x 18" x spec x 15-3/4" saddle		.40	20		38,000	655		38,655	42,900
6201	Beam and purlin hangers, galvanized, 12 ga.									
6202	Purlin or joist size, 3" x 8"	1 Carp	1.70	4.706	C	2,025	155		2,180	2,475
6204	3" x 10"		1.70	4.706		2,175	155		2,330	2,650
6206	3" x 12"		1.65	4.848		2,500	159		2,659	3,025
6208	3" x 14"		1.65	4.848		2,675	159		2,834	3,200
6210	3" x 16"		1.65	4.848		2,825	159		2,984	3,375
6212	4" x 8"		1.65	4.848		2,025	159		2,184	2,525
6214	4" x 10"		1.65	4.848		2,200	159		2,359	2,700

06 05 Common Work Results for Wood, Plastics, and Composites

06 05 23 – Wood, Plastic, and Composite Fastenings

06 05 23.60 Timber Connectors		Crew	Daily Output	Labor-Hours	Unit	Material	2015 Bare Costs Labor	Equipment	Total	Total Incl O&P
6216	4" x 12"	1 Carp	1.60	5	C	2,600	164		2,764	3,125
6218	4" x 14"		1.60	5		2,750	164		2,914	3,300
6220	4" x 16"		1.60	5		2,925	164		3,089	3,500
6222	6" x 8"		1.60	5		2,625	164		2,789	3,150
6224	6" x 10"		1.55	5.161		2,675	170		2,845	3,225
6226	6" x 12"		1.55	5.161		4,575	170		4,745	5,325
6228	6" x 14"		1.50	5.333		4,850	175		5,025	5,650
6230	6" x 16"		1.50	5.333		5,125	175		5,300	5,950
6250	Beam seats									
6252	Beam size, 5-1/4" wide									
6254	5" x 7" x 1/4"	1 Carp	1.80	4.444	C	7,725	146		7,871	8,725
6256	6" x 7" x 3/8"		1.80	4.444		8,650	146		8,796	9,775
6258	7" x 7" x 3/8"		1.80	4.444		9,250	146		9,396	10,400
6260	8" x 7" x 3/8"		1.80	4.444		10,900	146		11,046	12,200
6262	Beam size, 6-7/8" wide									
6264	5" x 9" x 1/4"	1 Carp	1.80	4.444	C	9,225	146		9,371	10,300
6266	6" x 9" x 3/8"		1.80	4.444		12,000	146		12,146	13,400
6268	7" x 9" x 3/8"		1.80	4.444		12,200	146		12,346	13,600
6270	8" x 9" x 3/8"		1.80	4.444		14,400	146		14,546	16,100
6272	Special beams, over 6-7/8" wide									
6274	5" x 10" x 3/8"	1 Carp	1.80	4.444	C	12,500	146		12,646	13,900
6276	6" x 10" x 3/8"		1.80	4.444		14,600	146		14,746	16,200
6278	7" x 10" x 3/8"		1.80	4.444		15,200	146		15,346	16,900
6280	8" x 10" x 3/8"		1.75	4.571		16,300	150		16,450	18,200
6282	5-1/4" x 12" x 5/16"		1.75	4.571		12,600	150		12,750	14,200
6284	6-1/2" x 12" x 3/8"		1.75	4.571		20,900	150		21,050	23,300
6286	5-1/4" x 16" x 5/16"		1.70	4.706		18,600	155		18,755	20,700
6288	6-1/2" x 16" x 3/8"		1.70	4.706		24,200	155		24,355	27,000
6290	5-1/4" x 20" x 5/16"		1.70	4.706		21,800	155		21,955	24,300
6292	6-1/2" x 20" x 3/8"		1.65	4.848		28,500	159		28,659	31,700
6300	Column bases									
6302	4 x 4, 16 ga.	1 Carp	1.80	4.444	C	780	146		926	1,100
6306	7 ga.		1.80	4.444		2,850	146		2,996	3,400
6308	4 x 6, 16 ga.		1.80	4.444		1,825	146		1,971	2,250
6312	7 ga.		1.80	4.444		2,975	146		3,121	3,525
6314	6 x 6, 16 ga.		1.75	4.571		2,050	150		2,200	2,525
6318	7 ga.		1.75	4.571		4,075	150		4,225	4,725
6320	6 x 8, 7 ga.		1.70	4.706		3,175	155		3,330	3,725
6322	6 x 10, 7 ga.		1.70	4.706		3,400	155		3,555	3,975
6324	6 x 12, 7 ga.		1.70	4.706		3,675	155		3,830	4,300
6326	8 x 8, 7 ga.		1.65	4.848		6,200	159		6,359	7,100
6330	8 x 10, 7 ga.		1.65	4.848		7,425	159		7,584	8,450
6332	8 x 12, 7 ga.		1.60	5		8,075	164		8,239	9,150
6334	10 x 10, 3 ga.		1.60	5		8,225	164		8,389	9,325
6336	10 x 12, 3 ga.		1.60	5		9,475	164		9,639	10,700
6338	12 x 12, 3 ga.		1.55	5.161		10,300	170		10,470	11,600
6350	Column caps, painted, 3 ga.									
6352	3-1/4" x 3-5/8"	1 Carp	1.80	4.444	C	10,400	146		10,546	11,700
6354	3-1/4" x 5-1/2"		1.80	4.444		10,400	146		10,546	11,700
6356	3-5/8" x 3-5/8"		1.80	4.444		8,550	146		8,696	9,650
6358	3-5/8" x 5-1/2"		1.80	4.444		8,550	146		8,696	9,650
6360	5-1/4" x 5-1/2"		1.75	4.571		11,200	150		11,350	12,600
6362	5-1/4" x 7-1/2"		1.75	4.571		11,200	150		11,350	12,600

06 05 Common Work Results for Wood, Plastics, and Composites
06 05 23 – Wood, Plastic, and Composite Fastenings

06 05 23.60 Timber Connectors		Crew	Daily Output	Labor-Hours	Unit	Material	2015 Bare Costs Labor	Equipment	Total	Total Incl O&P
6364	5-1/2" x 3-5/8"	1 Carp	1.75	4.571	C	12,100	150		12,250	13,600
6366	5-1/2" x 5-1/2"		1.75	4.571		12,100	150		12,250	13,600
6368	5-1/2" x 7-1/2"		1.70	4.706		12,100	155		12,255	13,600
6370	6-7/8" x 5-1/2"		1.70	4.706		12,600	155		12,755	14,100
6372	6-7/8" x 6-7/8"		1.70	4.706		12,600	155		12,755	14,100
6374	6-7/8" x 7-1/2"		1.70	4.706		12,600	155		12,755	14,100
6376	7-1/2" x 5-1/2"		1.65	4.848		13,100	159		13,259	14,800
6378	7-1/2" x 7-1/2"		1.65	4.848		13,100	159		13,259	14,800
6380	8-7/8" x 5-1/2"		1.60	5		13,900	164		14,064	15,600
6382	8-7/8" x 7-1/2"		1.60	5		13,900	164		14,064	15,600
6384	9-1/2" x 5-1/2"		1.60	5		18,800	164		18,964	20,900
6400	Floor tie anchors, polymer paint									
6402	10 ga., 3" x 37-1/2"	1 Carp	1.80	4.444	C	5,000	146		5,146	5,750
6404	3-1/2" x 45-1/2"		1.75	4.571		5,250	150		5,400	6,025
6406	3 ga., 3-1/2" x 56"		1.70	4.706		9,250	155		9,405	10,500
6410	Girder hangers									
6412	6" wall thickness, 4" x 6"	1 Carp	1.80	4.444	C	2,800	146		2,946	3,350
6414	4" x 8"		1.80	4.444		3,150	146		3,296	3,700
6416	8" wall thickness, 4" x 6"		1.80	4.444		3,250	146		3,396	3,825
6418	4" x 8"		1.80	4.444		3,250	146		3,396	3,825
6420	Hinge connections, polymer painted									
6422	3/4" thick top plate									
6424	5-1/4" x 12" w/5" x 5" top	1 Carp	1	8	C	35,900	263		36,163	39,900
6426	5-1/4" x 15" w/6" x 6" top		.80	10		38,100	330		38,430	42,500
6428	5-1/4" x 18" w/7" x 7" top		.70	11.429		40,100	375		40,475	44,700
6430	5-1/4" x 26" w/9" x 9" top		.60	13.333		42,700	440		43,140	47,600
6432	1" thick top plate									
6434	6-7/8" x 14" w/5" x 5" top	1 Carp	.80	10	C	43,700	330		44,030	48,700
6436	6-7/8" x 17" w/6" x 6" top		.80	10		48,600	330		48,930	54,000
6438	6-7/8" x 21" w/7" x 7" top		.70	11.429		53,000	375		53,375	59,000
6440	6-7/8" x 31" w/9" x 9" top		.60	13.333		58,000	440		58,440	64,500
6442	1-1/4" thick top plate									
6444	8-7/8" x 16" w/5" x 5" top	1 Carp	.60	13.333	C	54,500	440		54,940	60,500
6446	8-7/8" x 21" w/6" x 6" top		.50	16		60,000	525		60,525	67,000
6448	8-7/8" x 26" w/7" x 7" top		.40	20		68,000	655		68,655	75,500
6450	8-7/8" x 39" w/9" x 9" top		.30	26.667		84,500	875		85,375	94,500
6460	Holddowns									
6462	Embedded along edge									
6464	26" long, 12 ga.	1 Carp	.90	8.889	C	1,375	292		1,667	2,000
6466	35" long, 12 ga.		.85	9.412		1,800	310		2,110	2,525
6468	35" long, 10 ga.		.85	9.412		1,450	310		1,760	2,100
6470	Embedded away from edge									
6472	Medium duty, 12 ga.									
6474	18-1/2" long	1 Carp	.95	8.421	C	825	277		1,102	1,375
6476	23-3/4" long		.90	8.889		965	292		1,257	1,550
6478	28" long		.85	9.412		985	310		1,295	1,600
6480	35" long		.85	9.412		1,350	310		1,660	2,000
6482	Heavy duty, 10 ga.									
6484	28" long	1 Carp	.85	9.412	C	1,750	310		2,060	2,475
6486	35" long	"	.85	9.412	"	1,925	310		2,235	2,625
6490	Surface mounted (W x H)									
6492	2-1/2" x 5-3/4", 7 ga.	1 Carp	1	8	C	2,000	263		2,263	2,650
6494	2-1/2" x 8", 12 ga.		1	8		1,225	263		1,488	1,800

06 05 Common Work Results for Wood, Plastics, and Composites

06 05 23 – Wood, Plastic, and Composite Fastenings

06 05 23.60 Timber Connectors		Crew	Daily Output	Labor-Hours	Unit	Material	2015 Bare Costs Labor	Equipment	Total	Total Incl O&P
6496	2-7/8" x 6-3/8", 7 ga.	1 Carp	1	8	C	4,425	263		4,688	5,325
6498	2-7/8" x 12-1/2", 3 ga.		1	8		4,550	263		4,813	5,450
6500	3-3/16" x 9-3/8", 10 ga.		1	8		3,025	263		3,288	3,800
6502	3-1/2" x 11-5/8", 3 ga.		1	8		5,525	263		5,788	6,550
6504	3-1/2" x 14-3/4", 3 ga.		1	8		7,000	263		7,263	8,150
6506	3-1/2" x 16-1/2", 3 ga.		1	8		8,450	263		8,713	9,750
6508	3-1/2" x 20-1/2", 3 ga.		.90	8.889		8,650	292		8,942	10,000
6510	3-1/2" x 24-1/2", 3 ga.		.90	8.889		10,900	292		11,192	12,500
6512	4-1/4" x 20-3/4", 3 ga.		.90	8.889		7,450	292		7,742	8,700
6520	Joist hangers									
6522	Sloped, field adjustable, 18 ga.									
6524	2" x 6"	1 Carp	1.65	4.848	C	545	159		704	865
6526	2" x 8"		1.65	4.848		980	159		1,139	1,350
6528	2" x 10" and up		1.65	4.848		1,625	159		1,784	2,075
6530	3" x 10" and up		1.60	5		1,225	164		1,389	1,625
6532	4" x 10" and up		1.55	5.161		1,475	170		1,645	1,900
6536	Skewed 45°, 16 ga.									
6538	2" x 4"	1 Carp	1.75	4.571	C	870	150		1,020	1,200
6540	2" x 6" or 2" x 8"		1.65	4.848		885	159		1,044	1,250
6542	2" x 10" or 2" x 12"		1.65	4.848		1,025	159		1,184	1,400
6544	2" x 14" or 2" x 16"		1.60	5		1,825	164		1,989	2,275
6546	(2) 2" x 6" or (2) 2" x 8"		1.60	5		1,650	164		1,814	2,075
6548	(2) 2" x 10" or (2) 2" x 12"		1.55	5.161		1,750	170		1,920	2,225
6550	(2) 2" x 14" or (2) 2" x 16"		1.50	5.333		2,775	175		2,950	3,350
6552	4" x 6" or 4" x 8"		1.60	5		1,375	164		1,539	1,775
6554	4" x 10" or 4" x 12"		1.55	5.161		1,600	170		1,770	2,050
6556	4" x 14" or 4" x 16"		1.55	5.161		2,475	170		2,645	3,000
6560	Skewed 45°, 14 ga.									
6562	(2) 2" x 6" or (2) 2" x 8"	1 Carp	1.60	5	C	1,900	164		2,064	2,350
6564	(2) 2" x 10" or (2) 2" x 12"		1.55	5.161		2,625	170		2,795	3,150
6566	(2) 2" x 14" or (2) 2" x 16"		1.50	5.333		3,850	175		4,025	4,525
6568	4" x 6" or 4" x 8"		1.60	5		2,250	164		2,414	2,750
6570	4" x 10" or 4" x 12"		1.55	5.161		2,375	170		2,545	2,875
6572	4" x 14" or 4" x 16"		1.55	5.161		3,175	170		3,345	3,775
6590	Joist hangers, heavy duty 12 ga., galvanized									
6592	2" x 4"	1 Carp	1.75	4.571	C	1,275	150		1,425	1,650
6594	2" x 6"		1.65	4.848		1,400	159		1,559	1,800
6595	2" x 6", 16 ga.		1.65	4.848		1,325	159		1,484	1,725
6596	2" x 8"		1.65	4.848		2,100	159		2,259	2,600
6597	2" x 8", 16 ga.		1.65	4.848		2,000	159		2,159	2,475
6598	2" x 10"		1.65	4.848		2,175	159		2,334	2,650
6600	2" x 12"		1.65	4.848		2,675	159		2,834	3,200
6602	2" x 14"		1.65	4.848		2,775	159		2,934	3,325
6604	2" x 16"		1.65	4.848		2,925	159		3,084	3,500
6606	3" x 4"		1.65	4.848		1,775	159		1,934	2,225
6608	3" x 6"		1.65	4.848		2,375	159		2,534	2,875
6610	3" x 8"		1.65	4.848		2,400	159		2,559	2,925
6612	3" x 10"		1.60	5		2,775	164		2,939	3,350
6614	3" x 12"		1.60	5		3,350	164		3,514	3,950
6616	3" x 14"		1.60	5		3,900	164		4,064	4,575
6618	3" x 16"		1.60	5		4,300	164		4,464	5,025
6620	(2) 2" x 4"		1.75	4.571		2,075	150		2,225	2,525
6622	(2) 2" x 6"		1.60	5		2,500	164		2,664	3,025

06 05 Common Work Results for Wood, Plastics, and Composites

06 05 23 – Wood, Plastic, and Composite Fastenings

06 05 23.60 Timber Connectors

		Crew	Daily Output	Labor-Hours	Unit	Material	2015 Bare Costs Labor	Equipment	Total	Total Incl O&P
6624	(2) 2" x 8"	1 Carp	1.60	5	C	2,550	164		2,714	3,075
6626	(2) 2" x 10"		1.55	5.161		2,750	170		2,920	3,300
6628	(2) 2" x 12"		1.55	5.161		3,525	170		3,695	4,150
6630	(2) 2" x 14"		1.50	5.333		3,550	175		3,725	4,225
6632	(2) 2" x 16"		1.50	5.333		3,600	175		3,775	4,250
6634	4" x 4"		1.65	4.848		1,500	159		1,659	1,925
6636	4" x 6"		1.60	5		1,650	164		1,814	2,100
6638	4" x 8"		1.60	5		1,900	164		2,064	2,375
6640	4" x 10"		1.55	5.161		2,350	170		2,520	2,850
6642	4" x 12"		1.55	5.161		2,500	170		2,670	3,025
6644	4" x 14"		1.55	5.161		2,975	170		3,145	3,550
6646	4" x 16"		1.55	5.161		3,275	170		3,445	3,875
6648	(3) 2" x 10"		1.50	5.333		3,625	175		3,800	4,275
6650	(3) 2" x 12"		1.50	5.333		4,025	175		4,200	4,750
6652	(3) 2" x 14"		1.45	5.517		4,325	181		4,506	5,050
6654	(3) 2" x 16"		1.45	5.517		4,400	181		4,581	5,150
6656	6" x 6"		1.60	5		1,950	164		2,114	2,425
6658	6" x 8"		1.60	5		2,000	164		2,164	2,475
6660	6" x 10"		1.55	5.161		2,400	170		2,570	2,900
6662	6" x 12"		1.55	5.161		2,725	170		2,895	3,275
6664	6" x 14"		1.50	5.333		3,425	175		3,600	4,050
6666	6" x 16"		1.50	5.333		4,025	175		4,200	4,725
6690	Knee braces, galvanized, 12 ga.									
6692	Beam depth, 10" x 15" x 5' long	1 Carp	1.80	4.444	C	5,225	146		5,371	6,000
6694	15" x 22-1/2" x 7' long		1.70	4.706		6,000	155		6,155	6,850
6696	22-1/2" x 28-1/2" x 8' long		1.60	5		6,450	164		6,614	7,375
6698	28-1/2" x 36" x 10' long		1.55	5.161		6,725	170		6,895	7,675
6700	36" x 42" x 12' long		1.50	5.333		7,425	175		7,600	8,450
6710	Mudsill anchors									
6714	2" x 4" or 3" x 4"	1 Carp	115	.070	C	1,350	2.29		1,352.29	1,475
6716	2" x 6" or 3" x 6"		115	.070		1,350	2.29		1,352.29	1,475
6718	Block wall, 13-1/4" long		115	.070		85	2.29		87.29	97.50
6720	21-1/4" long		115	.070		126	2.29		128.29	143
6730	Post bases, 12 ga. galvanized									
6732	Adjustable, 3-9/16" x 3-9/16"	1 Carp	1.30	6.154	C	1,075	202		1,277	1,525
6734	3-9/16" x 5-1/2"		1.30	6.154		2,025	202		2,227	2,575
6736	4" x 4"		1.30	6.154		930	202		1,132	1,375
6738	4" x 6"		1.30	6.154		2,700	202		2,902	3,325
6740	5-1/2" x 5-1/2"		1.30	6.154		3,325	202		3,527	4,000
6742	6" x 6"		1.30	6.154		3,325	202		3,527	4,000
6744	Elevated, 3-9/16" x 3-1/4"		1.30	6.154		1,150	202		1,352	1,625
6746	5-1/2" x 3-5/16"		1.30	6.154		1,650	202		1,852	2,150
6748	5-1/2" x 5"		1.30	6.154		2,475	202		2,677	3,050
6750	Regular, 3-9/16" x 3-3/8"		1.30	6.154		885	202		1,087	1,325
6752	4" x 3-3/8"		1.30	6.154		1,250	202		1,452	1,725
6754	18 ga., 5-1/4" x 3-1/8"		1.30	6.154		1,300	202		1,502	1,775
6755	5-1/2" x 3-3/8"		1.30	6.154		1,300	202		1,502	1,775
6756	5-1/2" x 5-3/8"		1.30	6.154		1,875	202		2,077	2,400
6758	6" x 3-3/8"		1.30	6.154		2,250	202		2,452	2,825
6760	6" x 5-3/8"		1.30	6.154		2,500	202		2,702	3,125
6762	Post combination cap/bases									
6764	3-9/16" x 3-9/16"	1 Carp	1.20	6.667	C	460	219		679	880
6766	3-9/16" x 5-1/2"		1.20	6.667		1,050	219		1,269	1,525

For customer support on your Residential Cost Data, call 877.713.2617.

06 05 Common Work Results for Wood, Plastics, and Composites

06 05 23 – Wood, Plastic, and Composite Fastenings

06 05 23.60 Timber Connectors		Crew	Daily Output	Labor-Hours	Unit	Material	2015 Bare Costs Labor	Equipment	Total	Total Incl O&P
6768	4" x 4"	1 Carp	1.20	6.667	C	2,075	219		2,294	2,675
6770	5-1/2" x 5-1/2"		1.20	6.667		1,150	219		1,369	1,650
6772	6" x 6"		1.20	6.667		4,175	219		4,394	4,975
6774	7-1/2" x 7-1/2"		1.20	6.667		4,575	219		4,794	5,400
6776	8" x 8"		1.20	6.667		4,725	219		4,944	5,575
6790	Post-beam connection caps									
6792	Beam size 3-9/16"									
6794	12 ga. post, 4" x 4"	1 Carp	1	8	C	2,900	263		3,163	3,625
6796	4" x 6"		1	8		3,850	263		4,113	4,700
6798	4" x 8"		1	8		5,700	263		5,963	6,725
6800	16 ga. post, 4" x 4"		1	8		1,200	263		1,463	1,750
6802	4" x 6"		1	8		2,000	263		2,263	2,650
6804	4" x 8"		1	8		3,350	263		3,613	4,150
6805	18 ga. post, 2-7/8" x 3"		1	8		3,350	263		3,613	4,150
6806	Beam size 5-1/2"									
6808	12 ga. post, 6" x 4"	1 Carp	1	8	C	3,450	263		3,713	4,250
6810	6" x 6"		1	8		5,450	263		5,713	6,450
6812	6" x 8"		1	8		3,750	263		4,013	4,575
6816	16 ga. post, 6" x 4"		1	8		1,875	263		2,138	2,525
6818	6" x 6"		1	8		2,000	263		2,263	2,650
6820	Beam size 7-1/2"									
6822	12 ga. post, 8" x 4"	1 Carp	1	8	C	4,775	263		5,038	5,725
6824	8" x 6"		1	8		5,025	263		5,288	5,975
6826	8" x 8"		1	8		7,550	263		7,813	8,750
6840	Purlin anchors, embedded									
6842	Heavy duty, 10 ga.									
6844	Straight, 28" long	1 Carp	1.60	5	C	1,450	164		1,614	1,850
6846	35" long		1.50	5.333		1,775	175		1,950	2,250
6848	Twisted, 28" long		1.60	5		1,450	164		1,614	1,850
6850	35" long		1.50	5.333		1,775	175		1,950	2,250
6852	Regular duty, 12 ga.									
6854	Straight, 18-1/2" long	1 Carp	1.80	4.444	C	825	146		971	1,150
6856	23-3/4" long		1.70	4.706		1,025	155		1,180	1,400
6858	29" long		1.60	5		1,050	164		1,214	1,450
6860	35" long		1.50	5.333		1,450	175		1,625	1,900
6862	Twisted, 18" long		1.80	4.444		825	146		971	1,150
6866	28" long		1.60	5		980	164		1,144	1,350
6868	35" long		1.50	5.333		1,450	175		1,625	1,900
6870	Straight, plastic coated									
6872	23-1/2" long	1 Carp	1.60	5	C	2,025	164		2,189	2,500
6874	26-7/8" long		1.60	5		2,375	164		2,539	2,875
6876	32-1/2" long		1.50	5.333		2,525	175		2,700	3,075
6878	35-7/8" long		1.50	5.333		2,625	175		2,800	3,200
6890	Purlin hangers, painted									
6892	12 ga., 2" x 6"	1 Carp	1.80	4.444	C	1,850	146		1,996	2,300
6894	2" x 8"		1.80	4.444		2,025	146		2,171	2,475
6896	2" x 10"		1.80	4.444		2,175	146		2,321	2,650
6898	2" x 12"		1.75	4.571		2,350	150		2,500	2,825
6900	2" x 14"		1.75	4.571		2,500	150		2,650	3,000
6902	2" x 16"		1.75	4.571		2,650	150		2,800	3,175
6904	3" x 6"		1.70	4.706		1,875	155		2,030	2,300
6906	3" x 8"		1.70	4.706		2,025	155		2,180	2,475
6908	3" x 10"		1.70	4.706		2,175	155		2,330	2,650

06 05 Common Work Results for Wood, Plastics, and Composites

06 05 23 – Wood, Plastic, and Composite Fastenings

06 05 23.60	Timber Connectors	Crew	Daily Output	Labor-Hours	Unit	Material	2015 Bare Costs Labor	Equipment	Total	Total Incl O&P
6910	3" x 12"	1 Carp	1.65	4.848	C	2,500	159		2,659	3,025
6912	3" x 14"		1.65	4.848		2,675	159		2,834	3,200
6914	3" x 16"		1.65	4.848		2,825	159		2,984	3,375
6916	4" x 6"		1.65	4.848		1,875	159		2,034	2,350
6918	4" x 8"		1.65	4.848		2,025	159		2,184	2,525
6920	4" x 10"		1.65	4.848		2,200	159		2,359	2,700
6922	4" x 12"		1.60	5		2,600	164		2,764	3,125
6924	4" x 14"		1.60	5		2,750	164		2,914	3,300
6926	4" x 16"		1.60	5		2,925	164		3,089	3,500
6928	6" x 6"		1.60	5		2,475	164		2,639	3,000
6930	6" x 8"		1.60	5		2,625	164		2,789	3,150
6932	6" x 10"		1.55	5.161		2,675	170		2,845	3,225
6934	double 2" x 6"		1.70	4.706		2,025	155		2,180	2,475
6936	double 2" x 8"		1.70	4.706		2,175	155		2,330	2,650
6938	double 2" x 10"		1.70	4.706		2,350	155		2,505	2,825
6940	double 2" x 12"		1.65	4.848		2,500	159		2,659	3,025
6942	double 2" x 14"		1.65	4.848		2,675	159		2,834	3,200
6944	double 2" x 16"		1.65	4.848		2,825	159		2,984	3,375
6960	11 ga., 4" x 6"		1.65	4.848		3,700	159		3,859	4,350
6962	4" x 8"		1.65	4.848		3,975	159		4,134	4,650
6964	4" x 10"		1.65	4.848		4,250	159		4,409	4,950
6966	6" x 6"		1.60	5		3,750	164		3,914	4,400
6968	6" x 8"		1.60	5		4,025	164		4,189	4,700
6970	6" x 10"		1.55	5.161		4,300	170		4,470	5,000
6972	6" x 12"		1.55	5.161		4,575	170		4,745	5,325
6974	6" x 14"		1.55	5.161		4,850	170		5,020	5,625
6976	6" x 16"		1.50	5.333		5,125	175		5,300	5,950
6978	7 ga., 8" x 6"		1.60	5		4,075	164		4,239	4,750
6980	8" x 8"		1.60	5		4,350	164		4,514	5,050
6982	8" x 10"		1.55	5.161		4,625	170		4,795	5,375
6984	8" x 12"		1.55	5.161		4,900	170		5,070	5,675
6986	8" x 14"		1.50	5.333		5,175	175		5,350	6,000
6988	8" x 16"		1.50	5.333		5,450	175		5,625	6,300
7000	Strap connectors, galvanized									
7002	12 ga., 2-1/16" x 36"	1 Carp	1.55	5.161	C	1,175	170		1,345	1,575
7004	2-1/16" x 47"		1.50	5.333		1,650	175		1,825	2,100
7005	10 ga., 2-1/16" x 72"		1.50	5.333		1,725	175		1,900	2,200
7006	7 ga., 2-1/16" x 34"		1.55	5.161		2,925	170		3,095	3,500
7008	2-1/16" x 45"		1.50	5.333		3,825	175		4,000	4,500
7010	3 ga., 3" x 32"		1.55	5.161		4,925	170		5,095	5,700
7012	3" x 41"		1.55	5.161		5,125	170		5,295	5,925
7014	3" x 50"		1.50	5.333		7,825	175		8,000	8,900
7016	3" x 59"		1.50	5.333		9,525	175		9,700	10,800
7018	3-1/2" x 68"		1.45	5.517		9,675	181		9,856	10,900
7030	Tension ties									
7032	19-1/8" long, 16 ga., 3/4" anchor bolt	1 Carp	1.80	4.444	C	1,350	146		1,496	1,750
7034	20" long, 12 ga., 1/2" anchor bolt		1.80	4.444		1,750	146		1,896	2,175
7036	20" long, 12 ga., 3/4" anchor bolt		1.80	4.444		1,750	146		1,896	2,175
7038	27-3/4" long, 12 ga., 3/4" anchor bolt		1.75	4.571		3,100	150		3,250	3,675
7050	Truss connectors, galvanized									
7052	Adjustable hanger									
7054	18 ga., 2" x 6"	1 Carp	1.65	4.848	C	530	159		689	850
7056	4" x 6"		1.65	4.848		700	159		859	1,025

For customer support on your Residential Cost Data, call 877.713.2617.

06 05 Common Work Results for Wood, Plastics, and Composites

06 05 23 – Wood, Plastic, and Composite Fastenings

06 05 23.60 Timber Connectors

		Crew	Daily Output	Labor-Hours	Unit	Material	2015 Bare Costs Labor	Equipment	Total	Total Incl O&P
7058	16 ga., 4" x 10"	1 Carp	1.60	5	C	1,025	164		1,189	1,400
7060	(2) 2" x 10"		1.60	5		1,025	164		1,189	1,400
7062	Connectors to plate									
7064	16 ga., 2" x 4" plate	1 Carp	1.80	4.444	C	525	146		671	820
7066	2" x 6" plate	"	1.80	4.444	"	675	146		821	990
7068	Hip jack connector									
7070	14 ga.	1 Carp	1.50	5.333	C	2,750	175		2,925	3,325

06 05 23.80 Metal Bracing

		Crew	Daily Output	Labor-Hours	Unit	Material	Labor	Equipment	Total	Total Incl O&P
0010	**METAL BRACING**									
0302	Let-in, "T" shaped, 22 ga. galv. steel, studs at 16" O.C.	1 Carp	580	.014	L.F.	.81	.45		1.26	1.65
0402	Studs at 24" O.C.		600	.013		.81	.44		1.25	1.63
0502	Steel straps, 16 ga. galv. steel, studs at 16" O.C.		600	.013		1.05	.44		1.49	1.90
0602	Studs at 24" O.C.		620	.013		1.05	.42		1.47	1.87

06 11 Wood Framing

06 11 10 – Framing with Dimensional, Engineered or Composite Lumber

06 11 10.01 Forest Stewardship Council Certification

		Crew	Daily Output	Labor-Hours	Unit	Material	Labor	Equipment	Total	Total Incl O&P
0010	**FOREST STEWARDSHIP COUNCIL CERTIFICATION**									
0020	For Forest Stewardship Council (FSC) cert dimension lumber, add G					65%				

06 11 10.02 Blocking

		Crew	Daily Output	Labor-Hours	Unit	Material	Labor	Equipment	Total	Total Incl O&P
0010	**BLOCKING**									
1790	Bolted to concrete									
1798	Ledger board, 2" x 4"	1 Carp	180	.044	L.F.	4.45	1.46		5.91	7.35
1800	2" x 6"		160	.050		4.48	1.64		6.12	7.70
1810	4" x 6"		140	.057		7.85	1.88		9.73	11.75
1820	4" x 8"		120	.067		7.85	2.19		10.04	12.35
1950	Miscellaneous, to wood construction									
2000	2" x 4"	1 Carp	250	.032	L.F.	.42	1.05		1.47	2.23
2005	Pneumatic nailed		305	.026		.43	.86		1.29	1.92
2050	2" x 6"		222	.036		.67	1.18		1.85	2.72
2055	Pneumatic nailed		271	.030		.67	.97		1.64	2.37
2100	2" x 8"		200	.040		.92	1.31		2.23	3.22
2105	Pneumatic nailed		244	.033		.93	1.08		2.01	2.83
2150	2" x 10"		178	.045		1.34	1.48		2.82	3.95
2155	Pneumatic nailed		217	.037		1.35	1.21		2.56	3.51
2200	2" x 12"		151	.053		1.72	1.74		3.46	4.82
2205	Pneumatic nailed		185	.043		1.73	1.42		3.15	4.29
2300	To steel construction									
2320	2" x 4"	1 Carp	208	.038	L.F.	.42	1.26		1.68	2.59
2340	2" x 6"		180	.044		.67	1.46		2.13	3.18
2360	2" x 8"		158	.051		.92	1.66		2.58	3.81
2380	2" x 10"		136	.059		1.34	1.93		3.27	4.71
2400	2" x 12"		109	.073		1.73	2.41		4.14	5.95

06 11 10.04 Wood Bracing

		Crew	Daily Output	Labor-Hours	Unit	Material	Labor	Equipment	Total	Total Incl O&P
0010	**WOOD BRACING**									
0012	Let-in, with 1" x 6" boards, studs @ 16" O.C.	1 Carp	150	.053	L.F.	.73	1.75		2.48	3.74
0202	Studs @ 24" O.C.	"	230	.035	"	.73	1.14		1.87	2.72

06 11 Wood Framing

06 11 10 – Framing with Dimensional, Engineered or Composite Lumber

06 11 10.06 Bridging

		Crew	Daily Output	Labor-Hours	Unit	Material	2015 Bare Costs Labor	2015 Bare Costs Equipment	Total	Total Incl O&P
0010	**BRIDGING**									
0012	Wood, for joists 16" O.C., 1" x 3"	1 Carp	130	.062	Pr.	.63	2.02		2.65	4.08
0017	Pneumatic nailed		170	.047		.68	1.55		2.23	3.35
0102	2" x 3" bridging		130	.062		.69	2.02		2.71	4.14
0107	Pneumatic nailed		170	.047		.69	1.55		2.24	3.36
0302	Steel, galvanized, 18 ga., for 2" x 10" joists at 12" O.C.		130	.062		.92	2.02		2.94	4.40
0352	16" O.C.		135	.059		.93	1.95		2.88	4.29
0402	24" O.C.		140	.057		1.90	1.88		3.78	5.25
0602	For 2" x 14" joists at 16" O.C.		130	.062		1.40	2.02		3.42	4.93
0902	Compression type, 16" O.C., 2" x 8" joists		200	.040		.93	1.31		2.24	3.23
1002	2" x 12" joists		200	.040		.93	1.31		2.24	3.23

06 11 10.10 Beam and Girder Framing

		Crew	Daily Output	Labor-Hours	Unit	Material	Labor	Equipment	Total	Total Incl O&P
0010	**BEAM AND GIRDER FRAMING** R061110-30									
1000	Single, 2" x 6"	2 Carp	700	.023	L.F.	.67	.75		1.42	1.99
1005	Pneumatic nailed		812	.020		.67	.65		1.32	1.83
1020	2" x 8"		650	.025		.92	.81		1.73	2.37
1025	Pneumatic nailed		754	.021		.93	.70		1.63	2.19
1040	2" x 10"		600	.027		1.34	.88		2.22	2.94
1045	Pneumatic nailed		696	.023		1.35	.76		2.11	2.75
1060	2" x 12"		550	.029		1.72	.96		2.68	3.50
1065	Pneumatic nailed		638	.025		1.73	.82		2.55	3.29
1080	2" x 14"		500	.032		2.01	1.05		3.06	3.97
1085	Pneumatic nailed		580	.028		2.02	.91		2.93	3.75
1100	3" x 8"		550	.029		2.55	.96		3.51	4.41
1120	3" x 10"		500	.032		3.22	1.05		4.27	5.30
1140	3" x 12"		450	.036		3.86	1.17		5.03	6.20
1160	3" x 14"		400	.040		4.51	1.31		5.82	7.15
1170	4" x 6"	F-3	1100	.036		2.84	1.09	.59	4.52	5.60
1180	4" x 8"		1000	.040		2.89	1.20	.65	4.74	5.90
1200	4" x 10"		950	.042		4.28	1.26	.69	6.23	7.60
1220	4" x 12"		900	.044		5.15	1.33	.73	7.21	8.70
1240	4" x 14"		850	.047		6	1.41	.77	8.18	9.80
1250	6" x 8"		525	.076		7.40	2.29	1.25	10.94	13.35
1260	6" x 10"		500	.080		6.45	2.40	1.31	10.16	12.55
1290	8" x 12"		300	.133		15.20	4	2.18	21.38	26
2000	Double, 2" x 6"	2 Carp	625	.026		1.33	.84		2.17	2.88
2005	Pneumatic nailed		725	.022		1.34	.73		2.07	2.70
2020	2" x 8"		575	.028		1.84	.91		2.75	3.56
2025	Pneumatic nailed		667	.024		1.86	.79		2.65	3.36
2040	2" x 10"		550	.029		2.68	.96		3.64	4.55
2045	Pneumatic nailed		638	.025		2.70	.82		3.52	4.35
2060	2" x 12"		525	.030		3.45	1		4.45	5.45
2065	Pneumatic nailed		610	.026		3.47	.86		4.33	5.25
2080	2" x 14"		475	.034		4.02	1.11		5.13	6.30
2085	Pneumatic nailed		551	.029		4.05	.95		5	6.05
3000	Triple, 2" x 6"		550	.029		2	.96		2.96	3.80
3005	Pneumatic nailed		638	.025		2.01	.82		2.83	3.59
3020	2" x 8"		525	.030		2.77	1		3.77	4.72
3025	Pneumatic nailed		609	.026		2.78	.86		3.64	4.51
3040	2" x 10"		500	.032		4.02	1.05		5.07	6.20
3045	Pneumatic nailed		580	.028		4.04	.91		4.95	5.95
3060	2" x 12"		475	.034		5.15	1.11		6.26	7.55

For customer support on your Residential Cost Data, call 877.713.2617.

06 11 Wood Framing

06 11 10 – Framing with Dimensional, Engineered or Composite Lumber

		Crew	Daily Output	Labor-Hours	Unit	Material	2015 Bare Costs Labor	Equipment	Total	Total Incl O&P
06 11 10.10 Beam and Girder Framing										
3065	Pneumatic nailed	2 Carp	551	.029	L.F.	5.20	.95		6.15	7.30
3080	2" x 14"		450	.036		6.70	1.17		7.87	9.30
3085	Pneumatic nailed		522	.031		6.05	1.01		7.06	8.40
06 11 10.12 Ceiling Framing										
0010	CEILING FRAMING									
6000	Suspended, 2" x 3"	2 Carp	1000	.016	L.F.	.39	.53		.92	1.31
6050	2" x 4"		900	.018		.42	.58		1	1.45
6100	2" x 6"		800	.020		.67	.66		1.33	1.83
6150	2" x 8"		650	.025		.92	.81		1.73	2.37
06 11 10.14 Posts and Columns										
0010	POSTS AND COLUMNS									
0100	4" x 4"	2 Carp	390	.041	L.F.	1.73	1.35		3.08	4.16
0150	4" x 6"		275	.058		2.84	1.91		4.75	6.35
0200	4" x 8"		220	.073		2.89	2.39		5.28	7.20
0250	6" x 6"		215	.074		5.05	2.44		7.49	9.65
0300	6" x 8"		175	.091		7.40	3		10.40	13.20
0350	6" x 10"		150	.107		6.45	3.50		9.95	13
06 11 10.18 Joist Framing										
0010	JOIST FRAMING									
2000	Joists, 2" x 4"	2 Carp	1250	.013	L.F.	.42	.42		.84	1.18
2005	Pneumatic nailed		1438	.011		.43	.37		.80	1.08
2100	2" x 6"		1250	.013		.67	.42		1.09	1.44
2105	Pneumatic nailed		1438	.011		.67	.37		1.04	1.35
2150	2" x 8"		1100	.015		.92	.48		1.40	1.81
2155	Pneumatic nailed		1265	.013		.93	.42		1.35	1.72
2200	2" x 10"		900	.018		1.34	.58		1.92	2.45
2205	Pneumatic nailed		1035	.015		1.35	.51		1.86	2.33
2250	2" x 12"		875	.018		1.72	.60		2.32	2.91
2255	Pneumatic nailed		1006	.016		1.73	.52		2.25	2.79
2300	2" x 14"		770	.021		2.01	.68		2.69	3.36
2305	Pneumatic nailed		886	.018		2.02	.59		2.61	3.23
2350	3" x 6"		925	.017		1.90	.57		2.47	3.03
2400	3" x 10"		780	.021		3.22	.67		3.89	4.67
2450	3" x 12"		600	.027		3.86	.88		4.74	5.70
2500	4" x 6"		800	.020		2.84	.66		3.50	4.22
2550	4" x 10"		600	.027		4.28	.88		5.16	6.20
2600	4" x 12"		450	.036		5.15	1.17		6.32	7.60
2605	Sister joist, 2" x 6"		800	.020		.67	.66		1.33	1.83
2606	Pneumatic nailed		960	.017		.67	.55		1.22	1.66
3000	Composite wood joist 9-1/2" deep		.90	17.778	M.L.F.	1,800	585		2,385	2,950
3010	11-1/2" deep		.88	18.182		1,975	595		2,570	3,175
3020	14" deep		.82	19.512		2,100	640		2,740	3,375
3030	16" deep		.78	20.513		3,475	675		4,150	4,950
4000	Open web joist 12" deep		.88	18.182		3,525	595		4,120	4,875
4002	Per linear foot		880	.018	L.F.	3.52	.60		4.12	4.87
4004	Treated, per linear foot		880	.018	"	4.39	.60		4.99	5.85
4010	14" deep		.82	19.512	M.L.F.	3,725	640		4,365	5,175
4012	Per linear foot		820	.020	L.F.	3.72	.64		4.36	5.20
4014	Treated, per linear foot		820	.020	"	4.74	.64		5.38	6.30
4020	16" deep		.78	20.513	M.L.F.	3,825	675		4,500	5,325
4022	Per linear foot		780	.021	L.F.	3.83	.67		4.50	5.35
4024	Treated, per linear foot		780	.021	"	4.99	.67		5.66	6.65

06 11 Wood Framing

06 11 10 – Framing with Dimensional, Engineered or Composite Lumber

06 11 10.18 Joist Framing

		Crew	Daily Output	Labor-Hours	Unit	Material	2015 Bare Costs Labor	Equipment	Total	Total Incl O&P
4030	18" deep	2 Carp	.74	21.622	M.L.F.	4,000	710		4,710	5,600
4032	Per linear foot		740	.022	L.F.	4	.71		4.71	5.60
4034	Treated, per linear foot		740	.022	"	5.30	.71		6.01	7.05
6000	Composite rim joist, 1-1/4" x 9-1/2"		.90	17.778	M.L.F.	1,775	585		2,360	2,925
6010	1-1/4" x 11-1/2"		.88	18.182		2,050	595		2,645	3,250
6020	1-1/4" x 14-1/2"		.82	19.512		2,750	640		3,390	4,100
6030	1-1/4" x 16-1/2"		.78	20.513		3,375	675		4,050	4,850

06 11 10.24 Miscellaneous Framing

		Crew	Daily Output	Labor-Hours	Unit	Material	Labor	Equipment	Total	Total Incl O&P
0010	**MISCELLANEOUS FRAMING**									
2000	Firestops, 2" x 4"	2 Carp	780	.021	L.F.	.42	.67		1.09	1.60
2005	Pneumatic nailed		952	.017		.43	.55		.98	1.40
2100	2" x 6"		600	.027		.67	.88		1.55	2.20
2105	Pneumatic nailed		732	.022		.67	.72		1.39	1.95
5000	Nailers, treated, wood construction, 2" x 4"		800	.020		.51	.66		1.17	1.66
5005	Pneumatic nailed		960	.017		.51	.55		1.06	1.48
5100	2" x 6"		750	.021		.78	.70		1.48	2.04
5105	Pneumatic nailed		900	.018		.78	.58		1.36	1.84
5120	2" x 8"		700	.023		1.05	.75		1.80	2.42
5125	Pneumatic nailed		840	.019		1.06	.63		1.69	2.21
5200	Steel construction, 2" x 4"		750	.021		.51	.70		1.21	1.74
5220	2" x 6"		700	.023		.78	.75		1.53	2.12
5240	2" x 8"		650	.025		1.05	.81		1.86	2.52
7000	Rough bucks, treated, for doors or windows, 2" x 6"		400	.040		.78	1.31		2.09	3.07
7005	Pneumatic nailed		480	.033		.78	1.09		1.87	2.70
7100	2" x 8"		380	.042		1.05	1.38		2.43	3.48
7105	Pneumatic nailed		456	.035		1.06	1.15		2.21	3.10
8000	Stair stringers, 2" x 10"		130	.123		1.34	4.04		5.38	8.25
8100	2" x 12"		130	.123		1.72	4.04		5.76	8.70
8150	3" x 10"		125	.128		3.22	4.20		7.42	10.60
8200	3" x 12"		125	.128		3.86	4.20		8.06	11.30
8870	Laminated structural lumber, 1-1/4" x 11-1/2"		130	.123		2.05	4.04		6.09	9.05
8880	1-1/4" x 14-1/2"		130	.123		2.72	4.04		6.76	9.80

06 11 10.26 Partitions

		Crew	Daily Output	Labor-Hours	Unit	Material	Labor	Equipment	Total	Total Incl O&P
0010	**PARTITIONS**									
0020	Single bottom and double top plate, no waste, std. & better lumber									
0180	2" x 4" studs, 8' high, studs 12" O.C.	2 Carp	80	.200	L.F.	4.65	6.55		11.20	16.15
0185	12" O.C., pneumatic nailed		96	.167		4.69	5.50		10.19	14.35
0200	16" O.C.		100	.160		3.81	5.25		9.06	13
0205	16" O.C., pneumatic nailed		120	.133		3.84	4.38		8.22	11.55
0300	24" O.C.		125	.128		2.96	4.20		7.16	10.30
0305	24" O.C., pneumatic nailed		150	.107		2.98	3.50		6.48	9.20
0380	10' high, studs 12" O.C.		80	.200		5.50	6.55		12.05	17.10
0385	12" O.C., pneumatic nailed		96	.167		5.55	5.50		11.05	15.30
0400	16" O.C.		100	.160		4.44	5.25		9.69	13.70
0405	16" O.C., pneumatic nailed		120	.133		4.47	4.38		8.85	12.25
0500	24" O.C.		125	.128		3.38	4.20		7.58	10.75
0505	24" O.C., pneumatic nailed		150	.107		3.41	3.50		6.91	9.65
0580	12' high, studs 12" O.C.		65	.246		6.35	8.10		14.45	20.50
0585	12" O.C., pneumatic nailed		78	.205		6.40	6.75		13.15	18.35
0600	16" O.C.		80	.200		5.10	6.55		11.65	16.65
0605	16" O.C., pneumatic nailed		96	.167		5.10	5.50		10.60	14.80
0700	24" O.C.		100	.160		3.81	5.25		9.06	13

For customer support on your Residential Cost Data, call 877.713.2617.

06 11 Wood Framing

06 11 10 – Framing with Dimensional, Engineered or Composite Lumber

06 11 10.26 Partitions

		Crew	Daily Output	Labor-Hours	Unit	Material	2015 Bare Costs Labor	Equipment	Total	Total Incl O&P
0705	24" O.C., pneumatic nailed	2 Carp	120	.133	L.F.	3.84	4.38		8.22	11.55
0780	2" x 6" studs, 8' high, studs 12" O.C.		70	.229		7.35	7.50		14.85	20.50
0785	12" O.C., pneumatic nailed		84	.190		7.40	6.25		13.65	18.60
0800	16" O.C.		90	.178		6	5.85		11.85	16.40
0805	16" O.C., pneumatic nailed		108	.148		6.05	4.87		10.92	14.80
0900	24" O.C.		115	.139		4.66	4.57		9.23	12.80
0905	24" O.C., pneumatic nailed		138	.116		4.70	3.81		8.51	11.55
0980	10' high, studs 12" O.C.		70	.229		8.65	7.50		16.15	22
0985	12" O.C., pneumatic nailed		84	.190		8.75	6.25		15	20
1000	16" O.C.		90	.178		7	5.85		12.85	17.50
1005	16" O.C., pneumatic nailed		108	.148		7.05	4.87		11.92	15.90
1100	24" O.C.		115	.139		5.35	4.57		9.92	13.50
1105	24" O.C., pneumatic nailed		138	.116		5.35	3.81		9.16	12.30
1180	12' high, studs 12" O.C.		55	.291		10	9.55		19.55	27
1185	12" O.C., pneumatic nailed		66	.242		10.05	7.95		18	24.50
1200	16" O.C.		70	.229		8	7.50		15.50	21.50
1205	16" O.C., pneumatic nailed		84	.190		8.05	6.25		14.30	19.35
1300	24" O.C.		90	.178		6	5.85		11.85	16.40
1305	24" O.C., pneumatic nailed		108	.148		6.05	4.87		10.92	14.80
1400	For horizontal blocking, 2" x 4", add		600	.027		.42	.88		1.30	1.94
1500	2" x 6", add		600	.027		.67	.88		1.55	2.20
1600	For openings, add	▼	250	.064	▼		2.10		2.10	3.53
1702	Headers for above openings, material only, add				B.F.	.76			.76	.84

06 11 10.28 Porch or Deck Framing

		Crew	Daily Output	Labor-Hours	Unit	Material	Labor	Equipment	Total	Total Incl O&P
0010	**PORCH OR DECK FRAMING**									
0100	Treated lumber, posts or columns, 4" x 4"	2 Carp	390	.041	L.F.	1.34	1.35		2.69	3.73
0110	4" x 6"		275	.058		1.91	1.91		3.82	5.30
0120	4" x 8"		220	.073		3.81	2.39		6.20	8.20
0130	Girder, single, 4" x 4"		675	.024		1.34	.78		2.12	2.78
0140	4" x 6"		600	.027		1.91	.88		2.79	3.57
0150	4" x 8"		525	.030		3.81	1		4.81	5.85
0160	Double, 2" x 4"		625	.026		1.05	.84		1.89	2.56
0170	2" x 6"		600	.027		1.60	.88		2.48	3.23
0180	2" x 8"		575	.028		2.15	.91		3.06	3.90
0190	2" x 10"		550	.029		2.86	.96		3.82	4.75
0200	2" x 12"		525	.030		3.95	1		4.95	6.05
0210	Triple, 2" x 4"		575	.028		1.57	.91		2.48	3.25
0220	2" x 6"		550	.029		2.40	.96		3.36	4.24
0230	2" x 8"		525	.030		3.23	1		4.23	5.25
0240	2" x 10"		500	.032		4.29	1.05		5.34	6.50
0250	2" x 12"		475	.034		5.95	1.11		7.06	8.35
0260	Ledger, bolted 4' O.C., 2" x 4"		400	.040		.66	1.31		1.97	2.94
0270	2" x 6"		395	.041		.93	1.33		2.26	3.25
0280	2" x 8"		390	.041		1.20	1.35		2.55	3.58
0290	2" x 10"		385	.042		1.54	1.37		2.91	3.99
0300	2" x 12"		380	.042		2.08	1.38		3.46	4.61
0310	Joists, 2" x 4"		1250	.013		.52	.42		.94	1.28
0320	2" x 6"		1250	.013		.79	.42		1.21	1.58
0330	2" x 8"		1100	.015		1.07	.48		1.55	1.98
0340	2" x 10"		900	.018		1.42	.58		2	2.55
0350	2" x 12"	▼	875	.018	▼	1.71	.60		2.31	2.89
0360	Railings and trim, 1" x 4"	1 Carp	300	.027		.47	.88		1.35	1.99

06 11 Wood Framing

06 11 10 – Framing with Dimensional, Engineered or Composite Lumber

06 11 10.28 Porch or Deck Framing		Crew	Daily Output	Labor-Hours	Unit	Material	2015 Bare Costs Labor	Equipment	Total	Total Incl O&P
0370	2" x 2"	1 Carp	300	.027	L.F.	.33	.88		1.21	1.84
0380	2" x 4"		300	.027		.51	.88		1.39	2.03
0390	2" x 6"		300	.027		.79	.88		1.67	2.33
0400	Decking, 1" x 4"		275	.029	S.F.	2.16	.96		3.12	3.98
0410	2" x 4"		300	.027		1.73	.88		2.61	3.37
0420	2" x 6"		320	.025		1.69	.82		2.51	3.24
0430	5/4" x 6"		320	.025		2.25	.82		3.07	3.86
0440	Balusters, square, 2" x 2"	2 Carp	660	.024	L.F.	.34	.80		1.14	1.71
0450	Turned, 2" x 2"		420	.038		.45	1.25		1.70	2.59
0460	Stair stringer, 2" x 10"		130	.123		1.42	4.04		5.46	8.35
0470	2" x 12"		130	.123		1.71	4.04		5.75	8.70
0480	Stair treads, 1" x 4"		140	.114		2.17	3.75		5.92	8.70
0490	2" x 4"		140	.114		.52	3.75		4.27	6.85
0500	2" x 6"		160	.100		.78	3.29		4.07	6.35
0510	5/4" x 6"		160	.100		1.05	3.29		4.34	6.65
0520	Turned handrail post, 4" x 4"		64	.250	Ea.	31	8.20		39.20	48
0530	Lattice panel, 4' x 8', 1/2"		1600	.010	S.F.	.81	.33		1.14	1.44
0535	3/4"		1600	.010	"	1.21	.33		1.54	1.88
0540	Cedar, posts or columns, 4" x 4"		390	.041	L.F.	3.93	1.35		5.28	6.60
0550	4" x 6"		275	.058		7.35	1.91		9.26	11.30
0560	4" x 8"		220	.073		10	2.39		12.39	15
0800	Decking, 1" x 4"		550	.029		1.81	.96		2.77	3.59
0810	2" x 4"		600	.027		3.62	.88		4.50	5.45
0820	2" x 6"		640	.025		6.60	.82		7.42	8.65
0830	5/4" x 6"		640	.025		4.45	.82		5.27	6.25
0840	Railings and trim, 1" x 4"		600	.027		1.81	.88		2.69	3.46
0860	2" x 4"		600	.027		3.62	.88		4.50	5.45
0870	2" x 6"		600	.027		6.60	.88		7.48	8.70
0920	Stair treads, 1" x 4"		140	.114		1.81	3.75		5.56	8.30
0930	2" x 4"		140	.114		3.62	3.75		7.37	10.30
0940	2" x 6"		160	.100		6.60	3.29		9.89	12.75
0950	5/4" x 6"		160	.100		4.45	3.29		7.74	10.40
0980	Redwood, posts or columns, 4" x 4"		390	.041		6.45	1.35		7.80	9.30
0990	4" x 6"		275	.058		12.60	1.91		14.51	17.05
1000	4" x 8"		220	.073		23.50	2.39		25.89	30
1240	Decking, 1" x 4"	1 Carp	275	.029	S.F.	3.97	.96		4.93	5.95
1260	2" x 6"		340	.024		7.50	.77		8.27	9.55
1270	5/4" x 6"		320	.025		4.77	.82		5.59	6.65
1280	Railings and trim, 1" x 4"	2 Carp	600	.027	L.F.	1.17	.88		2.05	2.76
1310	2" x 6"		600	.027		7.50	.88		8.38	9.70
1420	Alternative decking, wood/plastic composite, 5/4" x 6" G		640	.025		3.14	.82		3.96	4.83
1440	1" x 4" square edge fir		550	.029		2.18	.96		3.14	4
1450	1" x 4" tongue and groove fir		450	.036		1.46	1.17		2.63	3.56
1460	1" x 4" mahogany		550	.029		2.01	.96		2.97	3.81
1462	5/4" x 6" PVC		550	.029		3.73	.96		4.69	5.70
1465	Framing, porch or deck, alt deck fastening, screws, add	1 Carp	240	.033	S.F.		1.09		1.09	1.84
1470	Accessories, joist hangers, 2" x 4"		160	.050	Ea.	.73	1.64		2.37	3.56
1480	2" x 6" through 2" x 12"		150	.053		1.39	1.75		3.14	4.47
1530	Post footing, incl excav, backfill, tube form & concrete, 4' deep, 8" dia	F-7	12	2.667		12.15	76.50		88.65	141
1540	10" diameter		11	2.909		17.75	83.50		101.25	160
1550	12" diameter		10	3.200		23.50	91.50		115	180

06 11 Wood Framing

06 11 10 – Framing with Dimensional, Engineered or Composite Lumber

06 11 10.30 Roof Framing

		Crew	Daily Output	Labor-Hours	Unit	Material	2015 Bare Costs Labor	Equipment	Total	Total Incl O&P
0010	**ROOF FRAMING**									
1900	Rough fascia, 2" x 6"	2 Carp	250	.064	L.F.	.67	2.10		2.77	4.26
2000	2" x 8"		225	.071		.92	2.34		3.26	4.93
2100	2" x 10"		180	.089		1.34	2.92		4.26	6.35
5002	Rafters, to 4 in 12 pitch, 2" x 6", ordinary		1000	.016		.67	.53		1.20	1.61
5021	On steep roofs		800	.020		.67	.66		1.33	1.83
5041	On dormers or complex roofs		590	.027		.67	.89		1.56	2.23
5062	2" x 8", ordinary		950	.017		.92	.55		1.47	1.94
5081	On steep roofs		750	.021		.92	.70		1.62	2.19
5101	On dormers or complex roofs		540	.030		.92	.97		1.89	2.64
5122	2" x 10", ordinary		630	.025		1.34	.83		2.17	2.87
5141	On steep roofs		495	.032		1.34	1.06		2.40	3.25
5161	On dormers or complex roofs		425	.038		1.34	1.24		2.58	3.55
5182	2" x 12", ordinary		575	.028		1.72	.91		2.63	3.43
5201	On steep roofs		455	.035		1.72	1.16		2.88	3.84
5221	On dormers or complex roofs		395	.041		1.72	1.33		3.05	4.13
5250	Composite rafter, 9-1/2" deep		575	.028		1.79	.91		2.70	3.50
5260	11-1/2" deep		575	.028		1.97	.91		2.88	3.70
5301	Hip and valley rafters, 2" x 6", ordinary		760	.021		.67	.69		1.36	1.89
5321	On steep roofs		585	.027		.67	.90		1.57	2.24
5341	On dormers or complex roofs		510	.031		.67	1.03		1.70	2.46
5361	2" x 8", ordinary		720	.022		.92	.73		1.65	2.24
5381	On steep roofs		545	.029		.92	.96		1.88	2.63
5401	On dormers or complex roofs		470	.034		.92	1.12		2.04	2.89
5421	2" x 10", ordinary		570	.028		1.34	.92		2.26	3.02
5441	On steep roofs		440	.036		1.34	1.19		2.53	3.48
5461	On dormers or complex roofs		380	.042		1.34	1.38		2.72	3.79
5481	2" x 12", ordinary		525	.030		1.72	1		2.72	3.58
5501	On steep roofs		410	.039		1.72	1.28		3	4.05
5521	On dormers or complex roofs		355	.045		1.72	1.48		3.20	4.39
5541	Hip and valley jacks, 2" x 6", ordinary		600	.027		.67	.88		1.55	2.20
5561	On steep roofs		475	.034		.67	1.11		1.78	2.59
5581	On dormers or complex roofs		410	.039		.67	1.28		1.95	2.88
5601	2" x 8", ordinary		490	.033		.92	1.07		1.99	2.81
5621	On steep roofs		385	.042		.92	1.37		2.29	3.30
5641	On dormers or complex roofs		335	.048		.92	1.57		2.49	3.64
5661	2" x 10", ordinary		450	.036		1.34	1.17		2.51	3.43
5681	On steep roofs		350	.046		1.34	1.50		2.84	3.99
5701	On dormers or complex roofs		305	.052		1.34	1.72		3.06	4.36
5721	2" x 12", ordinary		375	.043		1.72	1.40		3.12	4.25
5741	On steep roofs		295	.054		1.72	1.78		3.50	4.89
5762	On dormers or complex roofs		255	.063		1.72	2.06		3.78	5.35
5781	Rafter tie, 1" x 4", #3		800	.020		.47	.66		1.13	1.62
5791	2" x 4", #3		800	.020		.42	.66		1.08	1.57
5801	Ridge board, #2 or better, 1" x 6"		600	.027		.73	.88		1.61	2.27
5821	1" x 8"		550	.029		1.20	.96		2.16	2.92
5841	1" x 10"		500	.032		1.56	1.05		2.61	3.47
5861	2" x 6"		500	.032		.67	1.05		1.72	2.49
5881	2" x 8"		450	.036		.92	1.17		2.09	2.97
5901	2" x 10"		400	.040		1.34	1.31		2.65	3.68
5921	Roof cants, split, 4" x 4"		650	.025		1.73	.81		2.54	3.26
5941	6" x 6"		600	.027		5.05	.88		5.93	7

06 11 Wood Framing

06 11 10 – Framing with Dimensional, Engineered or Composite Lumber

06 11 10.30 Roof Framing

		Crew	Daily Output	Labor-Hours	Unit	Material	2015 Bare Costs Labor	2015 Bare Costs Equipment	Total	Total Incl O&P
5961	Roof curbs, untreated, 2" x 6"	2 Carp	520	.031	L.F.	.67	1.01		1.68	2.43
5981	2" x 12"		400	.040		1.72	1.31		3.03	4.11
6001	Sister rafters, 2" x 6"		800	.020		.67	.66		1.33	1.83
6021	2" x 8"		640	.025		.91	.82		1.73	2.38
6041	2" x 10"		535	.030		1.34	.98		2.32	3.12
6061	2" x 12"		455	.035		1.72	1.16		2.88	3.84

06 11 10.32 Sill and Ledger Framing

		Crew	Daily Output	Labor-Hours	Unit	Material	Labor	Equipment	Total	Total Incl O&P
0010	**SILL AND LEDGER FRAMING**									
2002	Ledgers, nailed, 2" x 4"	2 Carp	755	.021	L.F.	.42	.70		1.12	1.64
2052	2" x 6"		600	.027		.67	.88		1.55	2.20
2102	Bolted, not including bolts, 3" x 6"		325	.049		1.88	1.62		3.50	4.79
2152	3" x 12"		233	.069		3.83	2.26		6.09	8
2602	Mud sills, redwood, construction grade, 2" x 4"		895	.018		2.25	.59		2.84	3.46
2622	2" x 6"		780	.021		3.40	.67		4.07	4.87
4002	Sills, 2" x 4"		600	.027		.42	.88		1.30	1.93
4052	2" x 6"		550	.029		.66	.96		1.62	2.32
4082	2" x 8"		500	.032		.91	1.05		1.96	2.76
4101	2" x 10"		450	.036		1.32	1.17		2.49	3.41
4121	2" x 12"		400	.040		1.70	1.31		3.01	4.08
4202	Treated, 2" x 4"		550	.029		.50	.96		1.46	2.15
4222	2" x 6"		500	.032		.77	1.05		1.82	2.61
4242	2" x 8"		450	.036		1.04	1.17		2.21	3.10
4261	2" x 10"		400	.040		1.38	1.31		2.69	3.73
4281	2" x 12"		350	.046		1.92	1.50		3.42	4.63
4402	4" x 4"		450	.036		1.30	1.17		2.47	3.39
4422	4" x 6"		350	.046		1.85	1.50		3.35	4.56
4462	4" x 8"		300	.053		3.73	1.75		5.48	7.05
4480	4" x 10"		260	.062		4.67	2.02		6.69	8.55

06 11 10.34 Sleepers

		Crew	Daily Output	Labor-Hours	Unit	Material	Labor	Equipment	Total	Total Incl O&P
0010	**SLEEPERS**									
0100	On concrete, treated, 1" x 2"	2 Carp	2350	.007	L.F.	.27	.22		.49	.68
0150	1" x 3"		2000	.008		.45	.26		.71	.94
0200	2" x 4"		1500	.011		.55	.35		.90	1.19
0250	2" x 6"		1300	.012		.86	.40		1.26	1.63

06 11 10.36 Soffit and Canopy Framing

		Crew	Daily Output	Labor-Hours	Unit	Material	Labor	Equipment	Total	Total Incl O&P
0010	**SOFFIT AND CANOPY FRAMING**									
1002	Canopy or soffit framing, 1" x 4"	2 Carp	900	.018	L.F.	.47	.58		1.05	1.50
1021	1" x 6"		850	.019		.73	.62		1.35	1.84
1042	1" x 8"		750	.021		1.20	.70		1.90	2.50
1102	2" x 4"		620	.026		.42	.85		1.27	1.89
1121	2" x 6"		560	.029		.67	.94		1.61	2.31
1142	2" x 8"		500	.032		.92	1.05		1.97	2.77
1202	3" x 4"		500	.032		1.07	1.05		2.12	2.94
1221	3" x 6"		400	.040		1.90	1.31		3.21	4.29
1242	3" x 10"		300	.053		3.22	1.75		4.97	6.50

06 11 10.38 Treated Lumber Framing Material

		Crew	Daily Output	Labor-Hours	Unit	Material	Labor	Equipment	Total	Total Incl O&P
0010	**TREATED LUMBER FRAMING MATERIAL**									
0100	2" x 4"				M.B.F.	755			755	830
0110	2" x 6"					770			770	845
0120	2" x 8"					780			780	855
0130	2" x 10"					830			830	910
0140	2" x 12"					960			960	1,050

For customer support on your Residential Cost Data, call 877.713.2617.

06 11 Wood Framing

06 11 10 – Framing with Dimensional, Engineered or Composite Lumber

06 11 10.38 Treated Lumber Framing Material

		Crew	Daily Output	Labor-Hours	Unit	Material	2015 Bare Costs Labor	Equipment	Total	Total Incl O&P
0200	4" x 4"				M.B.F.	975			975	1,075
0210	4" x 6"					925			925	1,025
0220	4" x 8"					1,400			1,400	1,550

06 11 10.40 Wall Framing

		Crew	Daily Output	Labor-Hours	Unit	Material	Labor	Equipment	Total	Total Incl O&P
0010	**WALL FRAMING** R061110-30									
0100	Door buck, studs, header, access, 8' H, 2" x 4" wall, 3' W	1 Carp	32	.250	Ea.	17.85	8.20		26.05	33.50
0110	4' wide		32	.250		19.15	8.20		27.35	35
0120	5' wide		32	.250		23	8.20		31.20	39.50
0130	6' wide		32	.250		25	8.20		33.20	41.50
0140	8' wide		30	.267		35.50	8.75		44.25	53.50
0150	10' wide		30	.267		48.50	8.75		57.25	68
0160	12' wide		30	.267		62.50	8.75		71.25	83
0170	2" x 6" wall, 3' wide		32	.250		25.50	8.20		33.70	42
0180	4' wide		32	.250		27	8.20		35.20	43.50
0190	5' wide		32	.250		31	8.20		39.20	48
0200	6' wide		32	.250		33	8.20		41.20	50
0210	8' wide		30	.267		43.50	8.75		52.25	62
0220	10' wide		30	.267		56.50	8.75		65.25	76.50
0230	12' wide		30	.267		70.50	8.75		79.25	92
0240	Window buck, studs, header & access, 8' high 2" x 4" wall, 2' wide		24	.333		18.80	10.95		29.75	39
0250	3' wide		24	.333		22	10.95		32.95	43
0260	4' wide		24	.333		24	10.95		34.95	45
0270	5' wide		24	.333		28	10.95		38.95	49.50
0280	6' wide		24	.333		31.50	10.95		42.45	53
0290	7' wide		24	.333		40	10.95		50.95	62.50
0300	8' wide		22	.364		44.50	11.95		56.45	68.50
0310	10' wide		22	.364		59	11.95		70.95	84.50
0320	12' wide		22	.364		75	11.95		86.95	103
0330	2" x 6" wall, 2' wide		24	.333		28.50	10.95		39.45	50
0340	3' wide		24	.333		32	10.95		42.95	53.50
0350	4' wide		24	.333		34.50	10.95		45.45	56.50
0360	5' wide		24	.333		39	10.95		49.95	61.50
0370	6' wide		24	.333		43	10.95		53.95	65.50
0380	7' wide		24	.333		52.50	10.95		63.45	76.50
0390	8' wide		22	.364		57.50	11.95		69.45	83
0400	10' wide		22	.364		72.50	11.95		84.45	99.50
0410	12' wide		22	.364		90.50	11.95		102.45	120
2002	Headers over openings, 2" x 6"	2 Carp	360	.044	L.F.	.67	1.46		2.13	3.18
2007	2" x 6", pneumatic nailed		432	.037		.67	1.22		1.89	2.78
2052	2" x 8"		340	.047		.92	1.55		2.47	3.61
2057	2" x 8", pneumatic nailed		408	.039		.93	1.29		2.22	3.18
2101	2" x 10"		320	.050		1.34	1.64		2.98	4.23
2106	2" x 10", pneumatic nailed		384	.042		1.35	1.37		2.72	3.78
2152	2" x 12"		300	.053		1.72	1.75		3.47	4.84
2157	2" x 12", pneumatic nailed		360	.044		1.73	1.46		3.19	4.36
2180	4" x 8"		260	.062		2.89	2.02		4.91	6.55
2185	4" x 8", pneumatic nailed		312	.051		2.90	1.68		4.58	6
2191	4" x 10"		240	.067		4.28	2.19		6.47	8.40
2196	4" x 10", pneumatic nailed		288	.056		4.30	1.83		6.13	7.80
2202	4" x 12"		190	.084		5.15	2.77		7.92	10.30
2207	4" x 12", pneumatic nailed		228	.070		5.15	2.31		7.46	9.55
2241	6" x 10"		165	.097		6.45	3.19		9.64	12.45

06 11 Wood Framing

06 11 10 – Framing with Dimensional, Engineered or Composite Lumber

06 11 10.40 Wall Framing		Crew	Daily Output	Labor-Hours	Unit	Material	2015 Bare Costs Labor	Equipment	Total	Total Incl O&P
2246	6" x 10", pneumatic nailed	2 Carp	198	.081	L.F.	6.50	2.65		9.15	11.60
2251	6" x 12"		140	.114		8.20	3.75		11.95	15.35
2256	6" x 12", pneumatic nailed		168	.095		8.25	3.13		11.38	14.30
5002	Plates, untreated, 2" x 3"		850	.019		.39	.62		1.01	1.47
5007	2" x 3", pneumatic nailed		1020	.016		.40	.52		.92	1.30
5022	2" x 4"		800	.020		.42	.66		1.08	1.57
5027	2" x 4", pneumatic nailed		960	.017		.43	.55		.98	1.39
5041	2" x 6"		750	.021		.67	.70		1.37	1.91
5045	2" x 6", pneumatic nailed		900	.018		.67	.58		1.25	1.72
5061	Treated, 2" x 3"		850	.019		.46	.62		1.08	1.54
5066	2" x 3", treated, pneumatic nailed		1020	.016		.46	.52		.98	1.37
5081	2" x 4"		800	.020		.51	.66		1.17	1.66
5086	2" x 4", treated, pneumatic nailed		960	.017		.51	.55		1.06	1.48
5101	2" x 6"		750	.021		.78	.70		1.48	2.04
5106	2" x 6", treated, pneumatic nailed		900	.018		.78	.58		1.36	1.84
5122	Studs, 8' high wall, 2" x 3"		1200	.013		.39	.44		.83	1.17
5127	2" x 3", pneumatic nailed		1440	.011		.40	.37		.77	1.04
5142	2" x 4"		1100	.015		.42	.48		.90	1.26
5147	2" x 4", pneumatic nailed		1320	.012		.43	.40		.83	1.14
5162	2" x 6"		1000	.016		.67	.53		1.20	1.61
5167	2" x 6", pneumatic nailed		1200	.013		.67	.44		1.11	1.48
5182	3" x 4"		800	.020		1.07	.66		1.73	2.28
5187	3" x 4", pneumatic nailed		960	.017		1.08	.55		1.63	2.10
5201	Installed on second story, 2" x 3"		1170	.014		.39	.45		.84	1.18
5206	2" x 3", pneumatic nailed		1200	.013		.40	.44		.84	1.17
5221	2" x 4"		1015	.016		.42	.52		.94	1.34
5226	2" x 4", pneumatic nailed		1080	.015		.43	.49		.92	1.29
5241	2" x 6"		890	.018		.67	.59		1.26	1.72
5246	2" x 6", pneumatic nailed		1020	.016		.67	.52		1.19	1.61
5261	3" x 4"		800	.020		1.07	.66		1.73	2.28
5266	3" x 4", pneumatic nailed		960	.017		1.08	.55		1.63	2.10
5281	Installed on dormer or gable, 2" x 3"		1045	.015		.39	.50		.89	1.27
5286	2" x 3", pneumatic nailed		1254	.013		.40	.42		.82	1.13
5301	2" x 4"		905	.018		.42	.58		1	1.45
5306	2" x 4", pneumatic nailed		1086	.015		.43	.48		.91	1.28
5321	2" x 6"		800	.020		.67	.66		1.33	1.83
5326	2" x 6", pneumatic nailed		960	.017		.67	.55		1.22	1.66
5341	3" x 4"		700	.023		1.07	.75		1.82	2.44
5346	3" x 4", pneumatic nailed		840	.019		1.08	.63		1.71	2.23
5361	6' high wall, 2" x 3"		970	.016		.39	.54		.93	1.34
5366	2" x 3", pneumatic nailed		1164	.014		.40	.45		.85	1.19
5381	2" x 4"		850	.019		.42	.62		1.04	1.51
5386	2" x 4", pneumatic nailed		1020	.016		.43	.52		.95	1.34
5401	2" x 6"		740	.022		.67	.71		1.38	1.92
5406	2" x 6", pneumatic nailed		888	.018		.67	.59		1.26	1.73
5421	3" x 4"		600	.027		1.07	.88		1.95	2.65
5426	3" x 4", pneumatic nailed		720	.022		1.08	.73		1.81	2.41
5441	Installed on second story, 2" x 3"		950	.017		.40	.55		.95	1.37
5446	2" x 3", pneumatic nailed		1140	.014		.40	.46		.86	1.20
5461	2" x 4"		810	.020		.42	.65		1.07	1.56
5466	2" x 4", pneumatic nailed		972	.016		.43	.54		.97	1.38
5481	2" x 6"		700	.023		.67	.75		1.42	1.99
5486	2" x 6", pneumatic nailed		840	.019		.67	.63		1.30	1.79

06 11 Wood Framing

06 11 10 – Framing with Dimensional, Engineered or Composite Lumber

06 11 10.40 Wall Framing

		Crew	Daily Output	Labor-Hours	Unit	Material	2015 Bare Costs Labor	Equipment	Total	Total Incl O&P
5501	3" x 4"	2 Carp	550	.029	L.F.	1.07	.96		2.03	2.78
5506	3" x 4", pneumatic nailed		660	.024		1.08	.80		1.88	2.52
5521	Installed on dormer or gable, 2" x 3"		850	.019		.39	.62		1.01	1.47
5526	2" x 3", pneumatic nailed		1020	.016		.40	.52		.92	1.30
5541	2" x 4"		720	.022		.42	.73		1.15	1.70
5546	2" x 4", pneumatic nailed		864	.019		.43	.61		1.04	1.49
5561	2" x 6"		620	.026		.67	.85		1.52	2.15
5566	2" x 6", pneumatic nailed		744	.022		.67	.71		1.38	1.93
5581	3" x 4"		480	.033		1.07	1.09		2.16	3.02
5586	3" x 4", pneumatic nailed		576	.028		1.08	.91		1.99	2.71
5601	3' high wall, 2" x 3"		740	.022		.39	.71		1.10	1.62
5606	2" x 3", pneumatic nailed		888	.018		.40	.59		.99	1.42
5621	2" x 4"		640	.025		.42	.82		1.24	1.85
5626	2" x 4", pneumatic nailed		768	.021		.43	.68		1.11	1.62
5641	2" x 6"		550	.029		.67	.96		1.63	2.33
5646	2" x 6", pneumatic nailed		660	.024		.67	.80		1.47	2.08
5661	3" x 4"		440	.036		1.07	1.19		2.26	3.19
5666	3" x 4", pneumatic nailed		528	.030		1.08	1		2.08	2.85
5681	Installed on second story, 2" x 3"		700	.023		.39	.75		1.14	1.69
5686	2" x 3", pneumatic nailed		840	.019		.40	.63		1.03	1.48
5701	2" x 4"		610	.026		.42	.86		1.28	1.92
5706	2" x 4", pneumatic nailed		732	.022		.43	.72		1.15	1.68
5721	2" x 6"		520	.031		.67	1.01		1.68	2.43
5726	2" x 6", pneumatic nailed		624	.026		.67	.84		1.51	2.15
5741	3" x 4"		430	.037		1.07	1.22		2.29	3.23
5746	3" x 4", pneumatic nailed		516	.031		1.08	1.02		2.10	2.89
5761	Installed on dormer or gable, 2" x 3"		625	.026		.39	.84		1.23	1.84
5766	2" x 3", pneumatic nailed		750	.021		.40	.70		1.10	1.61
5781	2" x 4"		545	.029		.42	.96		1.38	2.09
5786	2" x 4", pneumatic nailed		654	.024		.43	.80		1.23	1.82
5801	2" x 6"		465	.034		.67	1.13		1.80	2.63
5806	2" x 6", pneumatic nailed		558	.029		.67	.94		1.61	2.32
5821	3" x 4"		380	.042		1.07	1.38		2.45	3.50
5826	3" x 4", pneumatic nailed		456	.035		1.08	1.15		2.23	3.12
8250	For second story & above, add						5%			
8300	For dormer & gable, add						15%			

06 11 10.42 Furring

		Crew	Daily Output	Labor-Hours	Unit	Material	Labor	Equipment	Total	Total Incl O&P
0010	**FURRING**									
0012	Wood strips, 1" x 2", on walls, on wood	1 Carp	550	.015	L.F.	.24	.48		.72	1.06
0015	On wood, pneumatic nailed		710	.011		.24	.37		.61	.88
0300	On masonry		495	.016		.26	.53		.79	1.18
0400	On concrete		260	.031		.26	1.01		1.27	1.99
0600	1" x 3", on walls, on wood		550	.015		.39	.48		.87	1.22
0605	On wood, pneumatic nailed		710	.011		.39	.37		.76	1.04
0700	On masonry		495	.016		.42	.53		.95	1.35
0800	On concrete		260	.031		.42	1.01		1.43	2.16
0850	On ceilings, on wood		350	.023		.39	.75		1.14	1.68
0855	On wood, pneumatic nailed		450	.018		.39	.58		.97	1.40
0900	On masonry		320	.025		.42	.82		1.24	1.84
0950	On concrete		210	.038		.42	1.25		1.67	2.56

06 11 Wood Framing

06 11 10 – Framing with Dimensional, Engineered or Composite Lumber

06 11 10.44 Grounds

		Crew	Daily Output	Labor-Hours	Unit	Material	2015 Bare Costs Labor	Equipment	Total	Total Incl O&P
0010	**GROUNDS**									
0020	For casework, 1" x 2" wood strips, on wood	1 Carp	330	.024	L.F.	.24	.80		1.04	1.60
0100	On masonry		285	.028		.26	.92		1.18	1.84
0200	On concrete		250	.032		.26	1.05		1.31	2.05
0400	For plaster, 3/4" deep, on wood		450	.018		.24	.58		.82	1.24
0500	On masonry		225	.036		.26	1.17		1.43	2.25
0600	On concrete		175	.046		.26	1.50		1.76	2.81
0700	On metal lath		200	.040		.26	1.31		1.57	2.50

06 12 Structural Panels

06 12 10 – Structural Insulated Panels

06 12 10.10 OSB Faced Panels

			Crew	Daily Output	Labor-Hours	Unit	Material	Labor	Equipment	Total	Total Incl O&P
0010	**OSB FACED PANELS**										
0100	Structural insul. panels, 7/16" OSB both faces, EPS insul, 3-5/8" T	G	F-3	2075	.019	S.F.	3.60	.58	.32	4.50	5.30
0110	5-5/8" thick	G		1725	.023		4.05	.70	.38	5.13	6.05
0120	7-3/8" thick	G		1425	.028		4.40	.84	.46	5.70	6.75
0130	9-3/8" thick	G		1125	.036		4.70	1.07	.58	6.35	7.60
0140	7/16" OSB one face, EPS insul, 3-5/8" thick	G		2175	.018		3.70	.55	.30	4.55	5.35
0150	5-5/8" thick	G		1825	.022		4.30	.66	.36	5.32	6.20
0160	7-3/8" thick	G		1525	.026		4.80	.79	.43	6.02	7.10
0170	9-3/8" thick	G		1225	.033		5.30	.98	.53	6.81	8.10
0190	7/16" OSB - 1/2" GWB faces, EPS insul, 3-5/8" T	G		2075	.019		3.29	.58	.32	4.19	4.94
0200	5-5/8" thick	G		1725	.023		3.90	.70	.38	4.98	5.90
0210	7-3/8" thick	G		1425	.028		4.42	.84	.46	5.72	6.75
0220	9-3/8" thick	G		1125	.036		5	1.07	.58	6.65	7.95
0240	7/16" OSB - 1/2" MRGWB faces, EPS insul, 3-5/8" T	G		2075	.019		3.39	.58	.32	4.29	5.05
0250	5-5/8" thick	G		1725	.023		4	.70	.38	5.08	6
0260	7-3/8" thick	G		1425	.028		4.52	.84	.46	5.82	6.90
0270	9-3/8" thick	G		1125	.036		5.10	1.07	.58	6.75	8.10
0300	For 1/2" GWB added to OSB skin, add	G					1.28			1.28	1.41
0310	For 1/2" MRGWB added to OSB skin, add	G					1.28			1.28	1.41
0320	For one T1-11 skin, add to OSB-OSB	G					1.90			1.90	2.09
0330	For one 19/32" CDX skin, add to OSB-OSB	G					1.46			1.46	1.61
0500	Structural insulated panel, 7/16" OSB both sides, straw core										
0510	4-3/8" T, walls (w/sill, splines, plates)	G	F-6	2400	.017	S.F.	7.40	.50	.27	8.17	9.25
0520	Floors (w/splines)	G		2400	.017		7.40	.50	.27	8.17	9.25
0530	Roof (w/splines)	G		2400	.017		7.40	.50	.27	8.17	9.25
0550	7-7/8" T, walls (w/sill, splines, plates)	G		2400	.017		11.20	.50	.27	11.97	13.45
0560	Floors (w/splines)	G		2400	.017		11.20	.50	.27	11.97	13.45
0570	Roof (w/splines)	G		2400	.017		11.20	.50	.27	11.97	13.45

06 12 19 – Composite Shearwall Panels

06 12 19.10 Steel and Wood Composite Shearwall Panels

		Crew	Daily Output	Labor-Hours	Unit	Material	Labor	Equipment	Total	Total Incl O&P
0010	**STEEL & WOOD COMPOSITE SHEARWALL PANELS**									
0020	Anchor bolts, 36" long (must be placed in wet concrete)	1 Carp	150	.053	Ea.	32	1.75		33.75	38.50
0030	On concrete, 2 x 4 & 2 x 6 walls, 7' - 10' high, 360 lb. shear, 12" wide	2 Carp	8	2		420	65.50		485.50	570
0040	715 lb. shear, 15" wide		8	2		500	65.50		565.50	660
0050	1860 lb. shear, 18" wide		8	2		520	65.50		585.50	680
0060	2780 lb. shear, 21" wide		8	2		620	65.50		685.50	790
0070	3790 lb. shear, 24" wide		8	2		710	65.50		775.50	890
0080	2 x 6 walls, 11' to 13' high, 1180 lb. shear, 18" wide		6	2.667		635	87.50		722.50	845

06 12 Structural Panels

06 12 19 – Composite Shearwall Panels

	06 12 19.10 Steel and Wood Composite Shearwall Panels	Crew	Daily Output	Labor-Hours	Unit	Material	2015 Bare Costs Labor	Equipment	Total	Total Incl O&P
0090	1555 lb. shear, 21" wide	2 Carp	6	2.667	Ea.	800	87.50		887.50	1,025
0100	2280 lb. shear, 24" wide		6	2.667		895	87.50		982.50	1,125
0110	For installing above on wood floor frame, add									
0120	Coupler nuts, threaded rods, bolts, shear transfer plate kit	1 Carp	16	.500	Ea.	63	16.45		79.45	97
0130	Framing anchors, angle (2 required)	"	96	.083	"	2.61	2.74		5.35	7.45
0140	For blocking see Section 06 11 10.02									
0150	For installing above, first floor to second floor, wood floor frame, add									
0160	Add stack option to first floor wall panel				Ea.	69.50			69.50	76
0170	Threaded rods, bolts, shear transfer plate kit	1 Carp	16	.500		76	16.45		92.45	111
0180	Framing anchors, angle (2 required)	"	96	.083		2.61	2.74		5.35	7.45
0190	For blocking see section 06 11 10.02									
0200	For installing stacked panels, balloon framing									
0210	Add stack option to first floor wall panel				Ea.	69.50			69.50	76
0220	Threaded rods, bolts kit	1 Carp	16	.500	"	44	16.45		60.45	76

06 13 Heavy Timber Construction

06 13 13 – Log Construction

06 13 13.10 Log Structures

		Crew	Daily Output	Labor-Hours	Unit	Material	Labor	Equipment	Total	Total Incl O&P
0010	**LOG STRUCTURES**									
0020	Exterior walls, pine, D logs, with double T & G, 6" x 6"	2 Carp	500	.032	L.F.	4.38	1.05		5.43	6.55
0030	6" x 8"		375	.043		4.98	1.40		6.38	7.80
0040	8" x 6"		375	.043		4.38	1.40		5.78	7.15
0050	8" x 7"		322	.050		4.38	1.63		6.01	7.55
0060	Square/rectangular logs, with double T & G, 6" x 6"		500	.032		4.38	1.05		5.43	6.55
0070	6" x 8"		375	.043		4.38	1.40		5.78	7.15
0080	8" x 6"		375	.043		4.38	1.40		5.78	7.15
0090	8" x 7"		322	.050		4.38	1.63		6.01	7.55
0100	Round logs, with double T & G, 6" x 8"		375	.043		4.38	1.40		5.78	7.15
0110	8" x 7"		322	.050		4.38	1.63		6.01	7.55
0120	Log siding, ship lapped, 2" x 6"		225	.071	S.F.	3.16	2.34		5.50	7.40
0130	2" x 8"		200	.080		2.64	2.63		5.27	7.30
0140	2" x 12"		180	.089		2.97	2.92		5.89	8.15
0150	Foam sealant strip, 3/8" x 3/8"	1 Carp	1920	.004	L.F.	.20	.14		.34	.45
0152	Chinking, 2" - 3" wide joint, 1/4" to 3/8" deep		600	.013		1.28	.44		1.72	2.15
0154	Caulking, 1/4" to 1/2" joint		900	.009		.49	.29		.78	1.03
0156	Backer rod, 1/4"		900	.009		.19	.29		.48	.70
0157	Penetrating wood preservative	1 Pord	2000	.004	S.F.	.13	.11		.24	.34
0158	Insect treatment	"	4000	.002	"	.33	.06		.39	.46
0160	Upper floor framing, pine, posts/columns, 4" x 6"	2 Carp	750	.021	L.F.	2.82	.70		3.52	4.28
0180	4" x 8"		562	.028		2.86	.94		3.80	4.72
0190	6" x 6"		500	.032		5	1.05		6.05	7.25
0200	6" x 8"		375	.043		7.35	1.40		8.75	10.45
0210	8" x 8"		281	.057		8.80	1.87		10.67	12.85
0220	8" x 10"		225	.071		11	2.34		13.34	16
0230	Beams, 4" x 8"		562	.028		2.86	.94		3.80	4.72
0240	4" x 10"		449	.036		4.25	1.17		5.42	6.65
0250	4" x 12"		375	.043		5.10	1.40		6.50	7.95
0260	6" x 8"		375	.043		7.35	1.40		8.75	10.45
0270	6" x 10"		300	.053		6.40	1.75		8.15	10
0280	6" x 12"		250	.064		8.15	2.10		10.25	12.50
0290	8" x 10"		225	.071		11	2.34		13.34	16

06 13 Heavy Timber Construction

06 13 13 – Log Construction

06 13 13.10 Log Structures

		Crew	Daily Output	Labor-Hours	Unit	Material	2015 Bare Costs Labor	Equipment	Total	Total Incl O&P
0300	8" x 12"	2 Carp	188	.085	L.F.	15.10	2.80		17.90	21.50
0310	Joists, 4" x 8"		562	.028		2.86	.94		3.80	4.72
0320	4" x 10"		449	.036		4.25	1.17		5.42	6.65
0330	4" x 12"		375	.043		5.10	1.40		6.50	7.95
0340	6" x 8"		375	.043		7.35	1.40		8.75	10.45
0350	6" x 10"		300	.053		6.40	1.75		8.15	10
0360	6" x 12"		250	.064		8.15	2.10		10.25	12.50
0370	8" x 10"		225	.071		11	2.34		13.34	16
0380	8" x 12"		188	.085		13.20	2.80		16	19.20
0390	Decking, 1" x 6" T & G		964	.017	S.F.	1.64	.55		2.19	2.72
0400	1" x 8" T & G		700	.023		1.47	.75		2.22	2.87
0410	2" x 6" T & G		482	.033		3.47	1.09		4.56	5.65
0420	Gable end roof framing, rafters, 4" x 8"		562	.028	L.F.	2.86	.94		3.80	4.72
0430	4" x 10"		449	.036		4.25	1.17		5.42	6.65
0450	4" x 12"		375	.043		5.10	1.40		6.50	7.95
0460	6" x 8"		375	.043		7.35	1.40		8.75	10.45
0470	6" x 10"		300	.053		6.40	1.75		8.15	10
0480	6" x 12"		250	.064		8.15	2.10		10.25	12.50
0490	8" x 10"		225	.071		11	2.34		13.34	16
0500	8" x 12"		188	.085		13.20	2.80		16	19.20
0510	Purlins, 4" x 8"		562	.028		2.86	.94		3.80	4.72
0520	6" x 8"		375	.043		7.35	1.40		8.75	10.45
0530	Roof decking, 1" x 6" T & G		640	.025	S.F.	1.64	.82		2.46	3.18
0540	1" x 8" T & G		430	.037		1.47	1.22		2.69	3.66
0550	2" x 6" T & G		320	.050		3.47	1.64		5.11	6.55

06 13 23 – Heavy Timber Framing

06 13 23.10 Heavy Framing

		Crew	Daily Output	Labor-Hours	Unit	Material	Labor	Equipment	Total	Total Incl O&P
0010	**HEAVY FRAMING**									
0020	Beams, single 6" x 10"	2 Carp	1.10	14.545	M.B.F.	1,600	480		2,080	2,550
0100	Single 8" x 16"		1.20	13.333	"	2,000	440		2,440	2,925
0202	Built from 2" lumber, multiple 2" x 14"		900	.018	B.F.	.85	.58		1.43	1.92
0212	Built from 3" lumber, multiple 3" x 6"		700	.023		1.25	.75		2	2.64
0222	Multiple 3" x 8"		800	.020		1.27	.66		1.93	2.49
0232	Multiple 3" x 10"		900	.018		1.28	.58		1.86	2.38
0242	Multiple 3" x 12"		1000	.016		1.28	.53		1.81	2.28
0252	Built from 4" lumber, multiple 4" x 6"		800	.020		1.41	.66		2.07	2.65
0262	Multiple 4" x 8"		900	.018		1.07	.58		1.65	2.16
0272	Multiple 4" x 10"		1000	.016		1.28	.53		1.81	2.28
0282	Multiple 4" x 12"		1100	.015		1.28	.48		1.76	2.20
0292	Columns, structural grade, 1500f, 4" x 4"		450	.036	L.F.	1.70	1.17		2.87	3.83
0302	6" x 6"		225	.071		4.21	2.34		6.55	8.55
0402	8" x 8"		240	.067		7.20	2.19		9.39	11.60
0502	10" x 10"		90	.178		12.10	5.85		17.95	23
0602	12" x 12"		70	.229		18.35	7.50		25.85	32.50
0802	Floor planks, 2" thick, T & G, 2" x 6"		1050	.015	B.F.	1.59	.50		2.09	2.59
0902	2" x 10"		1100	.015		1.59	.48		2.07	2.55
1102	3" thick, 3" x 6"		1050	.015		1.59	.50		2.09	2.59
1202	3" x 10"		1100	.015		1.59	.48		2.07	2.55
1402	Girders, structural grade, 12" x 12"		800	.020		1.53	.66		2.19	2.78
1502	10" x 16"		1000	.016		2.40	.53		2.93	3.52
2302	Roof purlins, 4" thick, structural grade		1050	.015		1.07	.50		1.57	2.02

06 15 Wood Decking

06 15 16 – Wood Roof Decking

06 15 16.10 Solid Wood Roof Decking

		Crew	Daily Output	Labor-Hours	Unit	Material	2015 Bare Costs Labor	Equipment	Total	Total Incl O&P
0010	**SOLID WOOD ROOF DECKING**									
0350	Cedar planks, 2" thick	2 Carp	350	.046	S.F.	6.25	1.50		7.75	9.35
0400	3" thick		320	.050		9.35	1.64		10.99	13.05
0500	4" thick		250	.064		12.45	2.10		14.55	17.25
0550	6" thick		200	.080		18.70	2.63		21.33	25
0650	Douglas fir, 2" thick		350	.046		2.69	1.50		4.19	5.50
0700	3" thick		320	.050		4.04	1.64		5.68	7.20
0800	4" thick		250	.064		5.40	2.10		7.50	9.45
0850	6" thick		200	.080		8.05	2.63		10.68	13.30
0950	Hemlock, 2" thick		350	.046		2.74	1.50		4.24	5.55
1000	3" thick		320	.050		4.11	1.64		5.75	7.30
1100	4" thick		250	.064		5.50	2.10		7.60	9.55
1150	6" thick		200	.080		8.20	2.63		10.83	13.45
1250	Western white spruce, 2" thick		350	.046		1.75	1.50		3.25	4.44
1300	3" thick		320	.050		2.62	1.64		4.26	5.65
1400	4" thick		250	.064		3.50	2.10		5.60	7.40
1450	6" thick		200	.080		5.25	2.63		7.88	10.15

06 15 23 – Laminated Wood Decking

06 15 23.10 Laminated Roof Deck

		Crew	Daily Output	Labor-Hours	Unit	Material	Labor	Equipment	Total	Total Incl O&P
0010	**LAMINATED ROOF DECK**									
0020	Pine or hemlock, 3" thick	2 Carp	425	.038	S.F.	6.20	1.24		7.44	8.90
0100	4" thick		325	.049		8.10	1.62		9.72	11.60
0300	Cedar, 3" thick		425	.038		7	1.24		8.24	9.80
0400	4" thick		325	.049		9.40	1.62		11.02	13.05
0600	Fir, 3" thick		425	.038		5.40	1.24		6.64	8.05
0700	4" thick		325	.049		7.40	1.62		9.02	10.85

06 16 Sheathing

06 16 13 – Insulating Sheathing

06 16 13.10 Insulating Sheathing

			Crew	Daily Output	Labor-Hours	Unit	Material	Labor	Equipment	Total	Total Incl O&P
0010	**INSULATING SHEATHING**										
0020	Expanded polystyrene, 1#/C.F. density, 3/4" thick R2.89	G	2 Carp	1400	.011	S.F.	.33	.38		.71	.99
0030	1" thick R3.85	G		1300	.012		.39	.40		.79	1.11
0040	2" thick R7.69	G		1200	.013		.64	.44		1.08	1.45
0050	Extruded polystyrene, 15 PSI compressive strength, 1" thick, R5	G		1300	.012		.69	.40		1.09	1.44
0060	2" thick, R10	G		1200	.013		.86	.44		1.30	1.68
0070	Polyisocyanurate, 2#/C.F. density, 3/4" thick	G		1400	.011		.60	.38		.98	1.29
0080	1" thick	G		1300	.012		.62	.40		1.02	1.37
0090	1-1/2" thick	G		1250	.013		.78	.42		1.20	1.57
0100	2" thick	G		1200	.013		.97	.44		1.41	1.81

06 16 23 – Subflooring

06 16 23.10 Subfloor

		Crew	Daily Output	Labor-Hours	Unit	Material	Labor	Equipment	Total	Total Incl O&P
0010	**SUBFLOOR**									
0011	Plywood, CDX, 1/2" thick	2 Carp	1500	.011	SF Flr.	.66	.35		1.01	1.31
0015	Pneumatic nailed		1860	.009		.66	.28		.94	1.19
0017	Pneumatic nailed		1860	.009		.66	.28		.94	1.19
0102	5/8" thick		1350	.012		.78	.39		1.17	1.51
0107	Pneumatic nailed		1674	.010		.78	.31		1.09	1.39
0202	3/4" thick		1250	.013		.93	.42		1.35	1.73
0207	Pneumatic nailed		1550	.010		.93	.34		1.27	1.59

06 16 Sheathing

06 16 23 – Subflooring

06 16 23.10 Subfloor

		Crew	Daily Output	Labor-Hours	Unit	Material	2015 Bare Costs Labor	Equipment	Total	Total Incl O&P
0302	1-1/8" thick, 2-4-1 including underlayment	2 Carp	1050	.015	SF Flr.	2.03	.50		2.53	3.07
0440	With boards, 1" x 6", S4S, laid regular		900	.018		1.57	.58		2.15	2.71
0452	1" x 8", laid regular		1000	.016		1.90	.53		2.43	2.97
0462	Laid diagonal		850	.019		1.90	.62		2.52	3.13
0502	1" x 10", laid regular		1100	.015		1.95	.48		2.43	2.94
0602	Laid diagonal		900	.018		1.95	.58		2.53	3.12
1500	OSB, 5/8" thick		1330	.012	S.F.	.52	.40		.92	1.23
1600	3/4" thick		1230	.013	"	.63	.43		1.06	1.41
8990	Subfloor adhesive, 3/8" bead	1 Carp	2300	.003	L.F.	.12	.11		.23	.32

06 16 26 – Underlayment

06 16 26.10 Wood Product Underlayment

		Crew	Daily Output	Labor-Hours	Unit	Material	Labor	Equipment	Total	Total Incl O&P
0010	**WOOD PRODUCT UNDERLAYMENT**									
0015	Plywood, underlayment grade, 1/4" thick	2 Carp	1500	.011	S.F.	.83	.35		1.18	1.50
0018	Pneumatic nailed		1860	.009		.83	.28		1.11	1.38
0030	3/8" thick		1500	.011		.92	.35		1.27	1.60
0070	Pneumatic nailed		1860	.009		.92	.28		1.20	1.48
0102	1/2" thick		1450	.011		1.10	.36		1.46	1.82
0107	Pneumatic nailed		1798	.009		1.10	.29		1.39	1.70
0202	5/8" thick		1400	.011		1.20	.38		1.58	1.95
0207	Pneumatic nailed		1736	.009		1.20	.30		1.50	1.83
0302	3/4" thick		1300	.012		1.43	.40		1.83	2.25
0306	Pneumatic nailed		1612	.010		1.43	.33		1.76	2.12
0502	Particle board, 3/8" thick		1500	.011		.40	.35		.75	1.03
0507	Pneumatic nailed		1860	.009		.40	.28		.68	.91
0602	1/2" thick		1450	.011		.43	.36		.79	1.08
0607	Pneumatic nailed		1798	.009		.43	.29		.72	.96
0802	5/8" thick		1400	.011		.52	.38		.90	1.20
0807	Pneumatic nailed		1736	.009		.52	.30		.82	1.08
0902	3/4" thick		1300	.012		.63	.40		1.03	1.37
0907	Pneumatic nailed		1612	.010		.63	.33		.96	1.24
1100	Hardboard, underlayment grade, 4' x 4', .215" thick [G]		1500	.011		.58	.35		.93	1.23

06 16 33 – Wood Board Sheathing

06 16 33.10 Board Sheathing

		Crew	Daily Output	Labor-Hours	Unit	Material	Labor	Equipment	Total	Total Incl O&P
0009	**BOARD SHEATHING**									
0010	Roof, 1" x 6" boards, laid horizontal	2 Carp	725	.022	S.F.	1.57	.73		2.30	2.95
0020	On steep roof		520	.031		1.57	1.01		2.58	3.43
0040	On dormers, hips, & valleys		480	.033		1.57	1.09		2.66	3.57
0050	Laid diagonal		650	.025		1.57	.81		2.38	3.09
0070	1" x 8" boards, laid horizontal		875	.018		1.90	.60		2.50	3.10
0080	On steep roof		635	.025		1.90	.83		2.73	3.48
0090	On dormers, hips, & valleys		580	.028		1.95	.91		2.86	3.66
0100	Laid diagonal		725	.022		1.90	.73		2.63	3.31
0110	Skip sheathing, 1" x 4", 7" OC	1 Carp	1200	.007		.58	.22		.80	1.01
0120	1" x 6", 9" OC		1450	.006		.72	.18		.90	1.09
0180	Tongue and groove sheathing/decking, 1" x 6"		1000	.008		1.80	.26		2.06	2.42
0190	2" x 6"		1000	.008		3.81	.26		4.07	4.63
0200	Walls, 1" x 6" boards, laid regular	2 Carp	650	.025		1.57	.81		2.38	3.09
0210	Laid diagonal		585	.027		1.57	.90		2.47	3.24
0220	1" x 8" boards, laid regular		765	.021		1.90	.69		2.59	3.24
0230	Laid diagonal		650	.025		1.90	.81		2.71	3.45

06 16 Sheathing

06 16 36 – Wood Panel Product Sheathing

06 16 36.10 Sheathing		Crew	Daily Output	Labor-Hours	Unit	Material	2015 Bare Costs Labor	Equipment	Total	Total Incl O&P
0010	**SHEATHING** R061636-20									
0012	Plywood on roofs, CDX									
0032	5/16" thick	2 Carp	1600	.010	S.F.	.56	.33		.89	1.16
0037	Pneumatic nailed		1952	.008		.56	.27		.83	1.06
0052	3/8" thick		1525	.010		.60	.34		.94	1.24
0057	Pneumatic nailed		1860	.009		.60	.28		.88	1.13
0102	1/2" thick		1400	.011		.66	.38		1.04	1.35
0103	Pneumatic nailed		1708	.009		.66	.31		.97	1.24
0202	5/8" thick		1300	.012		.78	.40		1.18	1.54
0207	Pneumatic nailed		1586	.010		.78	.33		1.11	1.42
0302	3/4" thick		1200	.013		.93	.44		1.37	1.76
0307	Pneumatic nailed		1464	.011		.93	.36		1.29	1.62
0502	Plywood on walls, with exterior CDX, 3/8" thick		1200	.013		.60	.44		1.04	1.40
0507	Pneumatic nailed		1488	.011		.60	.35		.95	1.25
0603	1/2" thick		1125	.014		.66	.47		1.13	1.50
0608	Pneumatic nailed		1395	.011		.66	.38		1.04	1.35
0702	5/8" thick		1050	.015		.78	.50		1.28	1.70
0707	Pneumatic nailed		1302	.012		.78	.40		1.18	1.54
0803	3/4" thick		975	.016		.93	.54		1.47	1.93
0808	Pneumatic nailed		1209	.013		.93	.43		1.36	1.75
1000	For shear wall construction, add						20%			
1200	For structural 1 exterior plywood, add				S.F.	10%				
3000	Wood fiber, regular, no vapor barrier, 1/2" thick	2 Carp	1200	.013		.61	.44		1.05	1.41
3100	5/8" thick		1200	.013		.77	.44		1.21	1.59
3300	No vapor barrier, in colors, 1/2" thick		1200	.013		.74	.44		1.18	1.55
3400	5/8" thick		1200	.013		.78	.44		1.22	1.60
3600	With vapor barrier one side, white, 1/2" thick		1200	.013		.60	.44		1.04	1.40
3700	Vapor barrier 2 sides, 1/2" thick		1200	.013		.83	.44		1.27	1.65
3800	Asphalt impregnated, 25/32" thick		1200	.013		.33	.44		.77	1.10
3850	Intermediate, 1/2" thick		1200	.013		.25	.44		.69	1.02
4500	Oriented strand board, on roof, 7/16" thick G		1460	.011		.50	.36		.86	1.15
4505	Pneumatic nailed G		1780	.009		.50	.30		.80	1.05
4550	1/2" thick G		1400	.011		.50	.38		.88	1.18
4555	Pneumatic nailed G		1736	.009		.50	.30		.80	1.06
4600	5/8" thick G		1300	.012		.69	.40		1.09	1.44
4605	Pneumatic nailed G		1586	.010		.69	.33		1.02	1.32
4610	On walls, 7/16" thick		1200	.013		.50	.44		.94	1.29
4615	Pneumatic nailed		1488	.011		.50	.35		.85	1.14
4620	1/2" thick		1195	.013		.50	.44		.94	1.29
4625	Pneumatic nailed		1325	.012		.50	.40		.90	1.22
4630	5/8" thick		1050	.015		.69	.50		1.19	1.60
4635	Pneumatic nailed		1302	.012		.69	.40		1.09	1.44
4700	Oriented strand board, factory laminated W.R. barrier, on roof, 1/2" thick G		1400	.011		.78	.38		1.16	1.49
4705	Pneumatic nailed G		1736	.009		.78	.30		1.08	1.37
4720	5/8" thick G		1300	.012		.93	.40		1.33	1.70
4725	Pneumatic nailed G		1586	.010		.93	.33		1.26	1.58
4730	5/8" thick, T&G G		1150	.014		.93	.46		1.39	1.79
4735	Pneumatic nailed, T&G G		1400	.011		.93	.38		1.31	1.65
4740	On walls, 7/16" thick G		1200	.013		.66	.44		1.10	1.47
4745	Pneumatic nailed G		1488	.011		.66	.35		1.01	1.32
4750	1/2" thick G		1195	.013		.78	.44		1.22	1.60
4755	Pneumatic nailed G		1325	.012		.78	.40		1.18	1.53

For customer support on your Residential Cost Data, call 877.713.2617.

06 16 Sheathing

06 16 36 – Wood Panel Product Sheathing

06 16 36.10 Sheathing		Crew	Daily Output	Labor-Hours	Unit	Material	2015 Bare Costs Labor	Equipment	Total	Total Incl O&P
4800	Joint sealant tape, 3-1/2"	2 Carp	7600	.002	L.F.	.31	.07		.38	.46
4810	Joint sealant tape, 6"	↓	7600	.002	"	.44	.07		.51	.60

06 16 43 – Gypsum Sheathing

06 16 43.10 Gypsum Sheathing

0010	**GYPSUM SHEATHING**									
0020	Gypsum, weatherproof, 1/2" thick	2 Carp	1125	.014	S.F.	.45	.47		.92	1.28
0040	With embedded glass mats	"	1100	.015	"	.72	.48		1.20	1.59

06 17 Shop-Fabricated Structural Wood

06 17 33 – Wood I-Joists

06 17 33.10 Wood and Composite I-Joists

0010	**WOOD AND COMPOSITE I-JOISTS**									
0100	Plywood webs, incl. bridging & blocking, panels 24" O.C.									
1200	15' to 24' span, 50 psf live load	F-5	2400	.013	SF Flr.	2.02	.38		2.40	2.87
1300	55 psf live load		2250	.014		2.23	.41		2.64	3.14
1400	24' to 30' span, 45 psf live load		2600	.012		2.36	.35		2.71	3.18
1500	55 psf live load	↓	2400	.013	↓	3.93	.38		4.31	4.97

06 17 53 – Shop-Fabricated Wood Trusses

06 17 53.10 Roof Trusses

0010	**ROOF TRUSSES**									
5000	Common wood, 2" x 4" metal plate connected, 24" O.C., 4/12 slope									
5010	1' overhang, 12' span	F-5	55	.582	Ea.	41.50	16.65		58.15	73.50
5050	20' span	F-6	62	.645		68	19.40	10.55	97.95	119
5100	24' span		60	.667		77	20	10.90	107.90	130
5150	26' span		57	.702		80	21	11.45	112.45	136
5200	28' span		53	.755		94	22.50	12.35	128.85	155
5240	30' span		51	.784		100	23.50	12.80	136.30	164
5250	32' span		50	.800		112	24	13.10	149.10	177
5280	34' span		48	.833		114	25	13.60	152.60	182
5350	8/12 pitch, 1' overhang, 20' span		57	.702		75	21	11.45	107.45	131
5400	24' span		55	.727		95	22	11.90	128.90	155
5450	26' span		52	.769		100	23	12.60	135.60	162
5500	28' span		49	.816		114	24.50	13.35	151.85	181
5550	32' span		45	.889		130	26.50	14.55	171.05	204
5600	36' span		41	.976		160	29.50	15.95	205.45	243
5650	38' span		40	1		180	30	16.35	226.35	267
5700	40' span	↓	40	1	↓	184	30	16.35	230.35	271

06 18 Glued-Laminated Construction

06 18 13 – Glued-Laminated Beams

06 18 13.10 Laminated Beams

		Crew	Daily Output	Labor-Hours	Unit	Material	2015 Bare Costs Labor	Equipment	Total	Total Incl O&P
0010	LAMINATED BEAMS									
0050	3-1/2" x 18"	F-3	480	.083	L.F.	31	2.50	1.36	34.86	39.50
0100	5-1/4" x 11-7/8"		450	.089		30.50	2.67	1.45	34.62	39.50
0150	5-1/4" x 16"		360	.111		41	3.34	1.82	46.16	52.50
0200	5-1/4" x 18"		290	.138		46	4.14	2.26	52.40	60.50
0250	5-1/4" x 24"		220	.182		61.50	5.45	2.97	69.92	80
0300	7" x 11-7/8"		320	.125		43.50	3.75	2.04	49.29	56
0350	7" x 16"		260	.154		59	4.62	2.52	66.14	75.50
0400	7" x 18"		210	.190		68	5.70	3.11	76.81	88
0500	For premium appearance, add to S.F. prices					5%				
0550	For industrial type, deduct					15%				
0600	For stain and varnish, add					5%				
0650	For 3/4" laminations, add					25%				

06 18 13.20 Laminated Framing

		Crew	Daily Output	Labor-Hours	Unit	Material	Labor	Equipment	Total	Total Incl O&P
0010	LAMINATED FRAMING									
0020	30 lb., short term live load, 15 lb. dead load									
0200	Straight roof beams, 20' clear span, beams 8' O.C.	F-3	2560	.016	SF Flr.	2.04	.47	.26	2.77	3.31
0300	Beams 16' O.C.		3200	.013		1.48	.38	.20	2.06	2.48
0500	40' clear span, beams 8' O.C.		3200	.013		3.90	.38	.20	4.48	5.15
0600	Beams 16' O.C.		3840	.010		3.20	.31	.17	3.68	4.23
0800	60' clear span, beams 8' O.C.	F-4	2880	.014		6.70	.42	.39	7.51	8.55
0900	Beams 16' O.C.	"	3840	.010		5	.31	.29	5.60	6.35
1100	Tudor arches, 30' to 40' clear span, frames 8' O.C.	F-3	1680	.024		8.75	.72	.39	9.86	11.30
1200	Frames 16' O.C.	"	2240	.018		6.85	.54	.29	7.68	8.75
1400	50' to 60' clear span, frames 8' O.C.	F-4	2200	.018		9.45	.55	.51	10.51	11.90
1500	Frames 16' O.C.		2640	.015		8.05	.46	.43	8.94	10.10
1700	Radial arches, 60' clear span, frames 8' O.C.		1920	.021		8.85	.63	.59	10.07	11.40
1800	Frames 16' O.C.		2880	.014		6.80	.42	.39	7.61	8.60
2000	100' clear span, frames 8' O.C.		1600	.025		9.15	.75	.71	10.61	12.10
2100	Frames 16' O.C.		2400	.017		8.05	.50	.47	9.02	10.20
2300	120' clear span, frames 8' O.C.		1440	.028		12.15	.83	.78	13.76	15.60
2400	Frames 16' O.C.		1920	.021		11.10	.63	.59	12.32	13.90
2600	Bowstring trusses, 20' O.C., 40' clear span	F-3	2400	.017		5.50	.50	.27	6.27	7.20
2700	60' clear span	F-4	3600	.011		4.92	.33	.31	5.56	6.30
2800	100' clear span		4000	.010		6.95	.30	.28	7.53	8.45
2900	120' clear span		3600	.011		7.45	.33	.31	8.09	9.10
3100	For premium appearance, add to S.F. prices					5%				
3300	For industrial type, deduct					15%				
3500	For stain and varnish, add					5%				
3900	For 3/4" laminations, add to straight					25%				
4100	Add to curved					15%				
4300	Alternate pricing method: (use nominal footage of									
4310	components). Straight beams, camber less than 6"	F-3	3.50	11.429	M.B.F.	2,925	345	187	3,457	3,975
4400	Columns, including hardware		2	20		3,125	600	325	4,050	4,800
4600	Curved members, radius over 32'		2.50	16		3,200	480	262	3,942	4,625
4700	Radius 10' to 32'		3	13.333		3,175	400	218	3,793	4,400
4900	For complicated shapes, add maximum					100%				
5100	For pressure treating, add to straight					35%				
5200	Add to curved					45%				
6000	Laminated veneer members, southern pine or western species									
6050	1-3/4" wide x 5-1/2" deep	2 Carp	480	.033	L.F.	3.44	1.09		4.53	5.60
6100	9-1/2" deep		480	.033		4.74	1.09		5.83	7.05

06 18 Glued-Laminated Construction

06 18 13 – Glued-Laminated Beams

06 18 13.20 Laminated Framing		Crew	Daily Output	Labor-Hours	Unit	Material	2015 Bare Costs Labor	Equipment	Total	Total Incl O&P
6150	14" deep	2 Carp	450	.036	L.F.	7.85	1.17		9.02	10.60
6200	18" deep	↓	450	.036	↓	10.65	1.17		11.82	13.70
6300	Parallel strand members, southern pine or western species									
6350	1-3/4" wide x 9-1/4" deep	2 Carp	480	.033	L.F.	5.10	1.09		6.19	7.45
6400	11-1/4" deep		450	.036		5.75	1.17		6.92	8.30
6450	14" deep		400	.040		7.85	1.31		9.16	10.85
6500	3-1/2" wide x 9-1/4" deep		480	.033		16.30	1.09		17.39	19.80
6550	11-1/4" deep		450	.036		20.50	1.17		21.67	24.50
6600	14" deep		400	.040		24	1.31		25.31	28.50
6650	7" wide x 9-1/4" deep		450	.036		34.50	1.17		35.67	40
6700	11-1/4" deep		420	.038		43.50	1.25		44.75	49.50
6750	14" deep	↓	400	.040	↓	52	1.31		53.31	59
8000	Straight beams									
8102	20' span									
8104	3-1/8" x 9"	F-3	30	1.333	Ea.	137	40	22	199	242
8106	X 10-1/2"		30	1.333		160	40	22	222	267
8108	X 12"		30	1.333		183	40	22	245	292
8110	X 13-1/2"		30	1.333		205	40	22	267	315
8112	X 15"		29	1.379		228	41.50	22.50	292	345
8114	5-1/8" x 10-1/2"		30	1.333		262	40	22	324	380
8116	X 12"		30	1.333		299	40	22	361	420
8118	X 13-1/2"		30	1.333		335	40	22	397	460
8120	X 15"		29	1.379		375	41.50	22.50	439	505
8122	X 16-1/2"		29	1.379		410	41.50	22.50	474	550
8124	X 18"		29	1.379		450	41.50	22.50	514	590
8126	X 19-1/2"		29	1.379		485	41.50	22.50	549	630
8128	X 21"		28	1.429		525	43	23.50	591.50	675
8130	X 22-1/2"		28	1.429		560	43	23.50	626.50	715
8132	X 24"		28	1.429		600	43	23.50	666.50	760
8134	6-3/4" x 12"		29	1.379		395	41.50	22.50	459	530
8136	X 13-1/2"		29	1.379		445	41.50	22.50	509	585
8138	X 15"		29	1.379		495	41.50	22.50	559	635
8140	X 16-1/2"		28	1.429		540	43	23.50	606.50	695
8142	X 18"		28	1.429		590	43	23.50	656.50	750
8144	X 19-1/2"		28	1.429		640	43	23.50	706.50	805
8146	X 21"		27	1.481		690	44.50	24	758.50	860
8148	X 22-1/2"		27	1.481		740	44.50	24	808.50	915
8150	X 24"		27	1.481		790	44.50	24	858.50	965
8152	X 25-1/2"		27	1.481		840	44.50	24	908.50	1,025
8154	X 27"		26	1.538		885	46	25	956	1,075
8156	X 28-1/2"		26	1.538		935	46	25	1,006	1,125
8158	X 30"	↓	26	1.538	↓	985	46	25	1,056	1,175
8200	30' span									
8250	3-1/8" x 9"	F-3	30	1.333	Ea.	205	40	22	267	315
8252	X 10-1/2"		30	1.333		240	40	22	302	355
8254	X 12"		30	1.333		274	40	22	336	390
8256	X 13-1/2"		30	1.333		310	40	22	372	430
8258	X 15"		29	1.379		340	41.50	22.50	404	470
8260	5-1/8" x 10-1/2"		30	1.333		395	40	22	457	520
8262	X 12"		30	1.333		450	40	22	512	585
8264	X 13-1/2"		30	1.333		505	40	22	567	645
8266	X 15"		29	1.379		560	41.50	22.50	624	710
8268	X 16-1/2"		29	1.379		615	41.50	22.50	679	775

For customer support on your Residential Cost Data, call 877.713.2617.

06 18 Glued-Laminated Construction

06 18 13 – Glued-Laminated Beams

06 18 13.20 Laminated Framing		Crew	Daily Output	Labor-Hours	Unit	Material	2015 Bare Costs Labor	Equipment	Total	Total Incl O&P
8270	X 18"	F-3	29	1.379	Ea.	675	41.50	22.50	739	835
8272	X 19-1/2"		29	1.379		730	41.50	22.50	794	900
8274	X 21"		28	1.429		785	43	23.50	851.50	965
8276	X 22-1/2"		28	1.429		840	43	23.50	906.50	1,025
8278	X 24"		28	1.429		900	43	23.50	966.50	1,100
8280	6-3/4" x 12"		29	1.379		590	41.50	22.50	654	745
8282	X 13-1/2"		29	1.379		665	41.50	22.50	729	825
8284	X 15"		29	1.379		740	41.50	22.50	804	910
8286	X 16-1/2"		28	1.429		815	43	23.50	881.50	995
8288	X 18"		28	1.429		885	43	23.50	951.50	1,075
8290	X 19-1/2"		28	1.429		960	43	23.50	1,026.50	1,150
8292	X 21"		27	1.481		1,025	44.50	24	1,093.50	1,250
8294	X 22-1/2"		27	1.481		1,100	44.50	24	1,168.50	1,325
8296	X 24"		27	1.481		1,175	44.50	24	1,243.50	1,400
8298	X 25-1/2"		27	1.481		1,250	44.50	24	1,318.50	1,475
8300	X 27"		26	1.538		1,325	46	25	1,396	1,575
8302	X 28-1/2"		26	1.538		1,400	46	25	1,471	1,650
8304	X 30"	▼	26	1.538	▼	1,475	46	25	1,546	1,725
8400	40' span									
8402	3-1/8" x 9"	F-3	30	1.333	Ea.	274	40	22	336	390
8404	X 10-1/2"		30	1.333		320	40	22	382	440
8406	X 12"		30	1.333		365	40	22	427	490
8408	X 13-1/2"		30	1.333		410	40	22	472	540
8410	X 15"		29	1.379		455	41.50	22.50	519	595
8412	5-1/8" x 10-1/2"		30	1.333		525	40	22	587	665
8414	X 12"		30	1.333		600	40	22	662	750
8416	X 13-1/2"		30	1.333		675	40	22	737	830
8418	X 15"		29	1.379		750	41.50	22.50	814	920
8420	X 16-1/2"		29	1.379		825	41.50	22.50	889	1,000
8422	X 18"		29	1.379		900	41.50	22.50	964	1,075
8424	X 19-1/2"		29	1.379		975	41.50	22.50	1,039	1,175
8426	X 21"		28	1.429		1,050	43	23.50	1,116.50	1,250
8428	X 22-1/2"		28	1.429		1,125	43	23.50	1,191.50	1,325
8430	X 24"		28	1.429		1,200	43	23.50	1,266.50	1,425
8432	6-3/4" x 12"		29	1.379		790	41.50	22.50	854	960
8434	X 13-1/2"		29	1.379		885	41.50	22.50	949	1,075
8436	X 15"		29	1.379		985	41.50	22.50	1,049	1,175
8438	X 16-1/2"		28	1.429		1,075	43	23.50	1,141.50	1,300
8440	X 18"		28	1.429		1,175	43	23.50	1,241.50	1,400
8442	X 19-1/2"		28	1.429		1,275	43	23.50	1,341.50	1,500
8444	X 21"		27	1.481		1,375	44.50	24	1,443.50	1,625
8446	X 22-1/2"		27	1.481		1,475	44.50	24	1,543.50	1,725
8448	X 24"		27	1.481		1,575	44.50	24	1,643.50	1,825
8450	X 25-1/2"		27	1.481		1,675	44.50	24	1,743.50	1,950
8452	X 27"		26	1.538		1,775	46	25	1,846	2,050
8454	X 28-1/2"		26	1.538		1,875	46	25	1,946	2,150
8456	X 30"	▼	26	1.538		1,975	46	25	2,046	2,275

06 22 Millwork

06 22 13 – Standard Pattern Wood Trim

06 22 13.10 Millwork

		Crew	Daily Output	Labor-Hours	Unit	Material	2015 Bare Costs Labor	2015 Bare Costs Equipment	Total	Total Incl O&P
0010	**MILLWORK**									
0020	Rule of thumb, milled material equals rough lumber cost x 3									
1020	1" x 12", custom birch				L.F.	5.05			5.05	5.55
1040	Cedar					7.20			7.20	7.95
1060	Oak					4.72			4.72	5.20
1080	Redwood					4.71			4.71	5.20
1100	Southern yellow pine					3.98			3.98	4.38
1120	Sugar pine					3.80			3.80	4.18
1140	Teak					33.50			33.50	37
1160	Walnut					7.95			7.95	8.75
1180	White pine					5.25			5.25	5.75

06 22 13.15 Moldings, Base

		Crew	Daily Output	Labor-Hours	Unit	Material	Labor	Equipment	Total	Total Incl O&P
0010	**MOLDINGS, BASE**									
5100	Classic profile, 5/8" x 5-1/2", finger jointed and primed	1 Carp	250	.032	L.F.	1.46	1.05		2.51	3.36
5105	Poplar		240	.033		1.54	1.09		2.63	3.53
5110	Red oak		220	.036		2.36	1.19		3.55	4.60
5115	Maple		220	.036		3.57	1.19		4.76	5.95
5120	Cherry		220	.036		4.24	1.19		5.43	6.65
5125	3/4 x 7-1/2", finger jointed and primed		250	.032		1.85	1.05		2.90	3.79
5130	Poplar		240	.033		2.46	1.09		3.55	4.54
5135	Red oak		220	.036		3.43	1.19		4.62	5.80
5140	Maple		220	.036		4.75	1.19		5.94	7.20
5145	Cherry		220	.036		5.65	1.19		6.84	8.20
5150	Modern profile, 5/8" x 3-1/2", finger jointed and primed		250	.032		.89	1.05		1.94	2.74
5155	Poplar		240	.033		.98	1.09		2.07	2.92
5160	Red oak		220	.036		1.62	1.19		2.81	3.79
5165	Maple		220	.036		2.50	1.19		3.69	4.76
5170	Cherry		220	.036		2.70	1.19		3.89	4.98
5175	Ogee profile, 7/16" x 3", finger jointed and primed		250	.032		.59	1.05		1.64	2.41
5180	Poplar		240	.033		.64	1.09		1.73	2.54
5185	Red oak		220	.036		.90	1.19		2.09	3
5200	9/16" x 3-1/2", finger jointed and primed		250	.032		.66	1.05		1.71	2.48
5205	Pine		240	.033		1.13	1.09		2.22	3.08
5210	Red oak		220	.036		2.38	1.19		3.57	4.63
5215	9/16" x 4-1/2", red oak		220	.036		3.56	1.19		4.75	5.95
5220	5/8" x 3-1/2", finger jointed and primed		250	.032		.89	1.05		1.94	2.74
5225	Poplar		240	.033		.98	1.09		2.07	2.92
5230	Red oak		220	.036		1.62	1.19		2.81	3.79
5235	Maple		220	.036		2.50	1.19		3.69	4.76
5240	Cherry		220	.036		2.70	1.19		3.89	4.98
5245	5/8" x 4", finger jointed and primed		250	.032		1.15	1.05		2.20	3.02
5250	Poplar		240	.033		1.26	1.09		2.35	3.22
5255	Red oak		220	.036		1.81	1.19		3	4
5260	Maple		220	.036		2.79	1.19		3.98	5.10
5265	Cherry		220	.036		2.93	1.19		4.12	5.25
5270	Rectangular profile, oak, 3/8" x 1-1/4"		260	.031		1.25	1.01		2.26	3.08
5275	1/2" x 2-1/2"		255	.031		1.94	1.03		2.97	3.86
5280	1/2" x 3-1/2"		250	.032		2.36	1.05		3.41	4.35
5285	1" x 6"		240	.033		4.09	1.09		5.18	6.35
5290	1" x 8"		240	.033		5.70	1.09		6.79	8.10
5295	Pine, 3/8" x 1-3/4"		260	.031		.46	1.01		1.47	2.20
5300	7/16" x 2-1/2"		255	.031		.74	1.03		1.77	2.54

06 22 Millwork

06 22 13 – Standard Pattern Wood Trim

06 22 13.15 Moldings, Base

		Crew	Daily Output	Labor-Hours	Unit	Material	2015 Bare Costs Labor	Equipment	Total	Total Incl O&P
5305	1" x 6"	1 Carp	240	.033	L.F.	.83	1.09		1.92	2.75
5310	1" x 8"		240	.033		1.06	1.09		2.15	3
5315	Shoe, 1/2" x 3/4", primed		260	.031		.48	1.01		1.49	2.23
5320	Pine		240	.033		.30	1.09		1.39	2.17
5325	Poplar		240	.033		.38	1.09		1.47	2.26
5330	Red oak		220	.036		.52	1.19		1.71	2.58
5335	Maple		220	.036		.69	1.19		1.88	2.77
5340	Cherry		220	.036		.78	1.19		1.97	2.87
5345	11/16" x 1-1/2", pine		240	.033		.70	1.09		1.79	2.61
5350	Caps, 11/16" x 1-3/8", pine		240	.033		.55	1.09		1.64	2.44
5355	3/4" x 1-3/4", finger jointed and primed		260	.031		.70	1.01		1.71	2.47
5360	Poplar		240	.033		.99	1.09		2.08	2.93
5365	Red oak		220	.036		1.11	1.19		2.30	3.23
5370	Maple		220	.036		1.55	1.19		2.74	3.71
5375	Cherry		220	.036		3.21	1.19		4.40	5.55
5380	Combination base & shoe, 9/16" x 3-1/2" & 1/2" x 3/4", pine		125	.064		1.43	2.10		3.53	5.10
5385	Three piece oak, 6" high		80	.100		5.70	3.29		8.99	11.80
5390	Including 3/4" x 1" base shoe		70	.114		5.90	3.75		9.65	12.80
5395	Flooring cant strip, 3/4" x 3/4", pre-finished pine		260	.031		.51	1.01		1.52	2.26
5400	For pre-finished, stain and clear coat, add					.51			.51	.56
5405	Clear coat only, add					.42			.42	.46

06 22 13.30 Moldings, Casings

		Crew	Daily Output	Labor-Hours	Unit	Material	2015 Bare Costs Labor	Equipment	Total	Total Incl O&P
0010	**MOLDINGS, CASINGS**									
0085	Apron, 9/16" x 2-1/2", pine	1 Carp	250	.032	L.F.	1.63	1.05		2.68	3.55
0090	5/8" x 2-1/2", pine		250	.032		1.63	1.05		2.68	3.55
0110	5/8" x 3-1/2", pine		220	.036		1.86	1.19		3.05	4.05
0300	Band, 11/16" x 1-1/8", pine		270	.030		.75	.97		1.72	2.45
0310	11/16" x 1-1/2", finger jointed and primed		270	.030		.76	.97		1.73	2.46
0320	Pine		270	.030		.96	.97		1.93	2.68
0330	11/16" x 1-3/4", finger jointed and primed		270	.030		.95	.97		1.92	2.67
0350	Pine		270	.030		1.16	.97		2.13	2.90
0355	Beaded, 3/4" x 3-1/2", finger jointed and primed		220	.036		1.09	1.19		2.28	3.21
0360	Poplar		220	.036		1.09	1.19		2.28	3.21
0365	Red oak		220	.036		1.62	1.19		2.81	3.79
0370	Maple		220	.036		2.50	1.19		3.69	4.76
0375	Cherry		220	.036		3.02	1.19		4.21	5.35
0380	3/4" x 4", finger jointed and primed		220	.036		1.15	1.19		2.34	3.27
0385	Poplar		220	.036		1.41	1.19		2.60	3.56
0390	Red oak		220	.036		1.98	1.19		3.17	4.19
0395	Maple		220	.036		2.60	1.19		3.79	4.87
0400	Cherry		220	.036		3.29	1.19		4.48	5.65
0405	3/4" x 5-1/2", finger jointed and primed		200	.040		1.15	1.31		2.46	3.47
0410	Poplar		200	.040		1.89	1.31		3.20	4.29
0415	Red oak		200	.040		2.78	1.31		4.09	5.25
0420	Maple		200	.040		3.66	1.31		4.97	6.25
0425	Cherry		200	.040		4.22	1.31		5.53	6.85
0430	Classic profile, 3/4" x 2-3/4", finger jointed and primed		250	.032		.77	1.05		1.82	2.61
0435	Poplar		250	.032		.98	1.05		2.03	2.84
0440	Red oak		250	.032		1.43	1.05		2.48	3.33
0445	Maple		250	.032		2.05	1.05		3.10	4.01
0450	Cherry		250	.032		2.32	1.05		3.37	4.31
0455	Fluted, 3/4" x 3-1/2", poplar		220	.036		1.09	1.19		2.28	3.21

06 22 Millwork

06 22 13 – Standard Pattern Wood Trim

06 22 13.30 Moldings, Casings		Crew	Daily Output	Labor-Hours	Unit	Material	2015 Bare Costs Labor	Equipment	Total	Total Incl O&P
0460	Red oak	1 Carp	220	.036	L.F.	1.62	1.19		2.81	3.79
0465	Maple		220	.036		2.50	1.19		3.69	4.76
0470	Cherry		220	.036		3.02	1.19		4.21	5.35
0475	3/4" x 4", poplar		220	.036		1.41	1.19		2.60	3.56
0480	Red oak		220	.036		1.98	1.19		3.17	4.19
0485	Maple		220	.036		2.60	1.19		3.79	4.87
0490	Cherry		220	.036		3.29	1.19		4.48	5.65
0495	3/4" x 5-1/2", poplar		200	.040		1.89	1.31		3.20	4.29
0500	Red oak		200	.040		2.78	1.31		4.09	5.25
0505	Maple		200	.040		3.66	1.31		4.97	6.25
0510	Cherry		200	.040		4.22	1.31		5.53	6.85
0515	3/4" x 7-1/2", poplar		190	.042		1.27	1.38		2.65	3.72
0520	Red oak		190	.042		3.85	1.38		5.23	6.55
0525	Maple		190	.042		5.55	1.38		6.93	8.40
0530	Cherry		190	.042		6.65	1.38		8.03	9.60
0535	3/4" x 9-1/2", poplar		180	.044		4.04	1.46		5.50	6.90
0540	Red oak		180	.044		6.45	1.46		7.91	9.50
0545	Maple		180	.044		8.80	1.46		10.26	12.15
0550	Cherry		180	.044		9.60	1.46		11.06	13.05
0555	Modern profile, 9/16" x 2-1/4", poplar		250	.032		.56	1.05		1.61	2.37
0560	Red oak		250	.032		.76	1.05		1.81	2.59
0565	11/16" x 2-1/2", finger jointed & primed		250	.032		.84	1.05		1.89	2.68
0570	Pine		250	.032		1.28	1.05		2.33	3.17
0575	3/4" x 2-1/2", poplar		250	.032		.86	1.05		1.91	2.70
0580	Red oak		250	.032		1.17	1.05		2.22	3.05
0585	Maple		250	.032		1.76	1.05		2.81	3.69
0590	Cherry		250	.032		2.23	1.05		3.28	4.21
0595	Mullion, 5/16" x 2", pine		270	.030		.86	.97		1.83	2.57
0600	9/16" x 2-1/2", fingerjointed and primed		250	.032		.98	1.05		2.03	2.84
0605	Pine		250	.032		1.28	1.05		2.33	3.17
0610	Red oak		250	.032		2.86	1.05		3.91	4.90
0615	1-1/16" x 3-3/4", red oak		220	.036		6.95	1.19		8.14	9.60
0620	Ogee, 7/16" x 2-1/2", poplar		250	.032		.56	1.05		1.61	2.37
0625	Red oak		250	.032		.77	1.05		1.82	2.61
0630	9/16" x 2-1/4", finger jointed and primed		250	.032		.49	1.05		1.54	2.30
0635	Poplar		250	.032		.56	1.05		1.61	2.37
0640	Red oak		250	.032		.76	1.05		1.81	2.59
0645	11/16" x 2-1/2", finger jointed and primed		250	.032		.72	1.05		1.77	2.55
0700	Pine		250	.032		1.41	1.05		2.46	3.31
0701	Red oak		250	.032		2.70	1.05		3.75	4.73
0730	11/16" x 3-1/2", finger jointed and primed		220	.036		1.11	1.19		2.30	3.23
0750	Pine		220	.036		1.72	1.19		2.91	3.90
0755	3/4" x 2-1/2", finger jointed and primed		250	.032		.89	1.05		1.94	2.74
0760	Poplar		250	.032		.86	1.05		1.91	2.70
0765	Red oak		250	.032		1.17	1.05		2.22	3.05
0770	Maple		250	.032		1.76	1.05		2.81	3.69
0775	Cherry		250	.032		2.23	1.05		3.28	4.21
0780	3/4" x 3-1/2", finger jointed and primed		220	.036		.89	1.19		2.08	2.99
0785	Poplar		220	.036		1.09	1.19		2.28	3.21
0790	Red oak		220	.036		1.62	1.19		2.81	3.79
0795	Maple		220	.036		2.50	1.19		3.69	4.76
0800	Cherry		220	.036		3.02	1.19		4.21	5.35
4700	Square profile, 1" x 1", teak		215	.037		2.26	1.22		3.48	4.53

06 22 Millwork

06 22 13 – Standard Pattern Wood Trim

06 22 13.30 Moldings, Casings

		Crew	Daily Output	Labor-Hours	Unit	Material	2015 Bare Costs Labor	Equipment	Total	Total Incl O&P
4800	Rectangular profile, 1" x 3", teak	1 Carp	200	.040	L.F.	6.20	1.31		7.51	9.05

06 22 13.35 Moldings, Ceilings

		Crew	Daily Output	Labor-Hours	Unit	Material	Labor	Equipment	Total	Total Incl O&P
0010	**MOLDINGS, CEILINGS**									
0600	Bed, 9/16" x 1-3/4", pine	1 Carp	270	.030	L.F.	1.06	.97		2.03	2.79
0650	9/16" x 2", pine		270	.030		1.14	.97		2.11	2.88
0710	9/16" x 1-3/4", oak		270	.030		1.86	.97		2.83	3.67
1200	Cornice, 9/16" x 1-3/4", pine		270	.030		.96	.97		1.93	2.68
1300	9/16" x 2-1/4", pine		265	.030		1.27	.99		2.26	3.06
1350	Cove, 1/2" x 2-1/4", poplar		265	.030		1	.99		1.99	2.77
1360	Red oak		265	.030		1.33	.99		2.32	3.13
1370	Hard maple		265	.030		1.83	.99		2.82	3.68
1380	Cherry		265	.030		2.24	.99		3.23	4.13
2400	9/16" x 1-3/4", pine		270	.030		.97	.97		1.94	2.70
2401	Oak		270	.030		1.26	.97		2.23	3.01
2500	11/16" x 2-3/4", pine		265	.030		1.80	.99		2.79	3.65
2510	Crown, 5/8" x 5/8", poplar		300	.027		.43	.88		1.31	1.94
2520	Red oak		300	.027		.51	.88		1.39	2.03
2530	Hard maple		300	.027		.69	.88		1.57	2.23
2540	Cherry		300	.027		.73	.88		1.61	2.27
2600	9/16" x 3-5/8", pine		250	.032		2.02	1.05		3.07	3.98
2700	11/16" x 4-1/4", pine		250	.032		2.90	1.05		3.95	4.95
2705	Oak		250	.032		6.40	1.05		7.45	8.80
2710	3/4" x 1-3/4", poplar		270	.030		.70	.97		1.67	2.40
2720	Red oak		270	.030		1.03	.97		2	2.76
2730	Hard maple		270	.030		1.36	.97		2.33	3.12
2740	Cherry		270	.030		1.68	.97		2.65	3.48
2750	3/4" x 2", poplar		270	.030		.92	.97		1.89	2.64
2760	Red oak		270	.030		1.24	.97		2.21	2.99
2770	Hard maple		270	.030		1.65	.97		2.62	3.44
2780	Cherry		270	.030		1.84	.97		2.81	3.65
2790	3/4" x 2-3/4", poplar		265	.030		1	.99		1.99	2.77
2800	Red oak		265	.030		1.54	.99		2.53	3.36
2810	Hard maple		265	.030		2.51	.99		3.50	4.43
2820	Cherry		265	.030		2.35	.99		3.34	4.25
2830	3/4" x 3-1/2", poplar		250	.032		1.27	1.05		2.32	3.15
2840	Red oak		250	.032		1.92	1.05		2.97	3.87
2850	Hard maple		250	.032		2.51	1.05		3.56	4.52
2860	Cherry		250	.032		2.94	1.05		3.99	4.99
2870	FJP poplar		250	.032		.90	1.05		1.95	2.75
2880	3/4" x 5", poplar		245	.033		1.89	1.07		2.96	3.88
2890	Red oak		245	.033		2.79	1.07		3.86	4.87
2900	Hard maple		245	.033		3.67	1.07		4.74	5.85
2910	Cherry		245	.033		4.24	1.07		5.31	6.45
2920	FJP poplar		245	.033		1.29	1.07		2.36	3.22
2930	3/4" x 6-1/4", poplar		240	.033		2.26	1.09		3.35	4.32
2940	Red oak		240	.033		3.40	1.09		4.49	5.55
2950	Hard maple		240	.033		4.44	1.09		5.53	6.70
2960	Cherry		240	.033		5.25	1.09		6.34	7.65
2970	7/8" x 8-3/4", poplar		220	.036		3.51	1.19		4.70	5.85
2980	Red oak		220	.036		5.35	1.19		6.54	7.85
2990	Hard maple		220	.036		7.35	1.19		8.54	10.10
3000	Cherry		220	.036		8.40	1.19		9.59	11.25

06 22 Millwork

06 22 13 – Standard Pattern Wood Trim

06 22 13.35 Moldings, Ceilings

		Crew	Daily Output	Labor-Hours	Unit	Material	2015 Bare Costs Labor	Equipment	Total	Total Incl O&P
3010	1" x 7-1/4", poplar	1 Carp	220	.036	L.F.	3.71	1.19		4.90	6.10
3020	Red oak		220	.036		5.75	1.19		6.94	8.35
3030	Hard maple		220	.036		8	1.19		9.19	10.75
3040	Cherry		220	.036		9.30	1.19		10.49	12.25
3050	1-1/16" x 4-1/4", poplar		250	.032		2.25	1.05		3.30	4.23
3060	Red oak		250	.032		2.64	1.05		3.69	4.66
3070	Hard maple		250	.032		3.49	1.05		4.54	5.60
3080	Cherry		250	.032		4.98	1.05		6.03	7.20
3090	Dentil crown, 3/4" x 5", poplar		250	.032		1.89	1.05		2.94	3.84
3100	Red oak		250	.032		2.79	1.05		3.84	4.83
3110	Hard maple		250	.032		3.67	1.05		4.72	5.80
3120	Cherry		250	.032		4.24	1.05		5.29	6.40
3130	Dentil piece for above, 1/2" x 1/2", poplar		300	.027		2.68	.88		3.56	4.41
3140	Red oak		300	.027		3.08	.88		3.96	4.85
3150	Hard maple		300	.027		3.50	.88		4.38	5.30
3160	Cherry		300	.027		3.81	.88		4.69	5.65

06 22 13.40 Moldings, Exterior

		Crew	Daily Output	Labor-Hours	Unit	Material	2015 Bare Costs Labor	Equipment	Total	Total Incl O&P
0010	**MOLDINGS, EXTERIOR**									
0100	Band board, cedar, rough sawn, 1" x 2"	1 Carp	300	.027	L.F.	.54	.88		1.42	2.07
0110	1" x 3"		300	.027		.81	.88		1.69	2.36
0120	1" x 4"		250	.032		1.07	1.05		2.12	2.94
0130	1" x 6"		250	.032		1.61	1.05		2.66	3.53
0140	1" x 8"		225	.036		2.15	1.17		3.32	4.33
0150	1" x 10"		225	.036		2.67	1.17		3.84	4.90
0160	1" x 12"		200	.040		3.21	1.31		4.52	5.75
0240	STK, 1" x 2"		300	.027		.51	.88		1.39	2.03
0250	1" x 3"		300	.027		.56	.88		1.44	2.09
0260	1" x 4"		250	.032		.65	1.05		1.70	2.47
0270	1" x 6"		250	.032		1.03	1.05		2.08	2.89
0280	1" x 8"		225	.036		1.60	1.17		2.77	3.72
0290	1" x 10"		225	.036		2.14	1.17		3.31	4.31
0300	1" x 12"		200	.040		3.54	1.31		4.85	6.10
0310	Pine, #2, 1" x 2"		300	.027		.25	.88		1.13	1.75
0320	1" x 3"		300	.027		.40	.88		1.28	1.91
0330	1" x 4"		250	.032		.49	1.05		1.54	2.30
0340	1" x 6"		250	.032		.75	1.05		1.80	2.58
0350	1" x 8"		225	.036		1.22	1.17		2.39	3.30
0360	1" x 10"		225	.036		1.59	1.17		2.76	3.71
0370	1" x 12"		200	.040		2.05	1.31		3.36	4.46
0380	D & better, 1" x 2"		300	.027		.41	.88		1.29	1.92
0390	1" x 3"		300	.027		.61	.88		1.49	2.14
0400	1" x 4"		250	.032		.82	1.05		1.87	2.66
0410	1" x 6"		250	.032		1.12	1.05		2.17	2.99
0420	1" x 8"		225	.036		1.62	1.17		2.79	3.74
0430	1" x 10"		225	.036		2.03	1.17		3.20	4.19
0440	1" x 12"		200	.040		2.63	1.31		3.94	5.10
0450	Redwood, clear all heart, 1" x 2"		300	.027		.63	.88		1.51	2.17
0460	1" x 3"		300	.027		.94	.88		1.82	2.51
0470	1" x 4"		250	.032		1.19	1.05		2.24	3.07
0480	1" x 6"		252	.032		1.78	1.04		2.82	3.70
0490	1" x 8"		225	.036		2.36	1.17		3.53	4.55
0500	1" x 10"		225	.036		3.96	1.17		5.13	6.30

06 22 Millwork

06 22 13 – Standard Pattern Wood Trim

	06 22 13.40 Moldings, Exterior	Crew	Daily Output	Labor-Hours	Unit	Material	2015 Bare Costs Labor	Equipment	Total	Total Incl O&P
0510	1" x 12"	1 Carp	200	.040	L.F.	4.75	1.31		6.06	7.45
0530	Corner board, cedar, rough sawn, 1" x 2"		225	.036		.54	1.17		1.71	2.56
0540	1" x 3"		225	.036		.81	1.17		1.98	2.85
0550	1" x 4"		200	.040		1.07	1.31		2.38	3.39
0560	1" x 6"		200	.040		1.61	1.31		2.92	3.98
0570	1" x 8"		200	.040		2.15	1.31		3.46	4.58
0580	1" x 10"		175	.046		2.67	1.50		4.17	5.45
0590	1" x 12"		175	.046		3.21	1.50		4.71	6.05
0670	STK, 1" x 2"		225	.036		.51	1.17		1.68	2.52
0680	1" x 3"		225	.036		.56	1.17		1.73	2.58
0690	1" x 4"		200	.040		.63	1.31		1.94	2.91
0700	1" x 6"		200	.040		1.03	1.31		2.34	3.34
0710	1" x 8"		200	.040		1.60	1.31		2.91	3.97
0720	1" x 10"		175	.046		2.14	1.50		3.64	4.87
0730	1" x 12"		175	.046		3.54	1.50		5.04	6.40
0740	Pine, #2, 1" x 2"		225	.036		.25	1.17		1.42	2.24
0750	1" x 3"		225	.036		.40	1.17		1.57	2.40
0760	1" x 4"		200	.040		.49	1.31		1.80	2.75
0770	1" x 6"		200	.040		.75	1.31		2.06	3.03
0780	1" x 8"		200	.040		1.22	1.31		2.53	3.55
0790	1" x 10"		175	.046		1.59	1.50		3.09	4.27
0800	1" x 12"		175	.046		2.05	1.50		3.55	4.77
0810	D & better, 1" x 2"		225	.036		.41	1.17		1.58	2.41
0820	1" x 3"		225	.036		.61	1.17		1.78	2.63
0830	1" x 4"		200	.040		.82	1.31		2.13	3.11
0840	1" x 6"		200	.040		1.12	1.31		2.43	3.44
0850	1" x 8"		200	.040		1.62	1.31		2.93	3.99
0860	1" x 10"		175	.046		2.03	1.50		3.53	4.75
0870	1" x 12"		175	.046		2.63	1.50		4.13	5.40
0880	Redwood, clear all heart, 1" x 2"		225	.036		.63	1.17		1.80	2.66
0890	1" x 3"		225	.036		.94	1.17		2.11	3
0900	1" x 4"		200	.040		1.19	1.31		2.50	3.52
0910	1" x 6"		200	.040		1.78	1.31		3.09	4.16
0920	1" x 8"		200	.040		2.36	1.31		3.67	4.80
0930	1" x 10"		175	.046		3.96	1.50		5.46	6.90
0940	1" x 12"		175	.046		4.75	1.50		6.25	7.75
0950	Cornice board, cedar, rough sawn, 1" x 2"		330	.024		.54	.80		1.34	1.94
0960	1" x 3"		290	.028		.81	.91		1.72	2.41
0970	1" x 4"		250	.032		1.07	1.05		2.12	2.94
0980	1" x 6"		250	.032		1.61	1.05		2.66	3.53
0990	1" x 8"		200	.040		2.15	1.31		3.46	4.58
1000	1" x 10"		180	.044		2.67	1.46		4.13	5.40
1010	1" x 12"		180	.044		3.21	1.46		4.67	6
1020	STK, 1" x 2"		330	.024		.51	.80		1.31	1.90
1030	1" x 3"		290	.028		.56	.91		1.47	2.14
1040	1" x 4"		250	.032		.65	1.05		1.70	2.47
1050	1" x 6"		250	.032		1.03	1.05		2.08	2.89
1060	1" x 8"		200	.040		1.60	1.31		2.91	3.97
1070	1" x 10"		180	.044		2.14	1.46		3.60	4.80
1080	1" x 12"		180	.044		3.54	1.46		5	6.35
1500	Pine, #2, 1" x 2"		330	.024		.25	.80		1.05	1.62
1510	1" x 3"		290	.028		.27	.91		1.18	1.82
1600	1" x 4"		250	.032		.49	1.05		1.54	2.30

06 22 Millwork

06 22 13 – Standard Pattern Wood Trim

06 22 13.40 Moldings, Exterior		Crew	Daily Output	Labor-Hours	Unit	Material	2015 Bare Costs Labor	Equipment	Total	Total Incl O&P
1700	1" x 6"	1 Carp	250	.032	L.F.	.75	1.05		1.80	2.58
1800	1" x 8"		200	.040		1.22	1.31		2.53	3.55
1900	1" x 10"		180	.044		1.59	1.46		3.05	4.20
2000	1" x 12"		180	.044		2.05	1.46		3.51	4.70
2020	D & better, 1" x 2"		330	.024		.41	.80		1.21	1.79
2030	1" x 3"		290	.028		.61	.91		1.52	2.19
2040	1" x 4"		250	.032		.82	1.05		1.87	2.66
2050	1" x 6"		250	.032		1.12	1.05		2.17	2.99
2060	1" x 8"		200	.040		1.62	1.31		2.93	3.99
2070	1" x 10"		180	.044		2.03	1.46		3.49	4.68
2080	1" x 12"		180	.044		2.63	1.46		4.09	5.35
2090	Redwood, clear all heart, 1" x 2"		330	.024		.63	.80		1.43	2.04
2100	1" x 3"		290	.028		.94	.91		1.85	2.56
2110	1" x 4"		250	.032		1.19	1.05		2.24	3.07
2120	1" x 6"		250	.032		1.78	1.05		2.83	3.71
2130	1" x 8"		200	.040		2.36	1.31		3.67	4.80
2140	1" x 10"		180	.044		3.96	1.46		5.42	6.80
2150	1" x 12"		180	.044		4.75	1.46		6.21	7.70
2160	3 piece, 1" x 2", 1" x 4", 1" x 6", rough sawn cedar		80	.100		3.24	3.29		6.53	9.05
2180	STK cedar		80	.100		2.19	3.29		5.48	7.90
2200	#2 pine		80	.100		1.49	3.29		4.78	7.15
2210	D & better pine		80	.100		2.35	3.29		5.64	8.10
2220	Clear all heart redwood		80	.100		3.60	3.29		6.89	9.45
2230	1" x 8", 1" x 10", 1" x 12", rough sawn cedar		65	.123		8	4.04		12.04	15.60
2240	STK cedar		65	.123		7.25	4.04		11.29	14.80
2300	#2 pine		65	.123		4.82	4.04		8.86	12.10
2320	D & better pine		65	.123		6.25	4.04		10.29	13.70
2330	Clear all heart redwood		65	.123		11.05	4.04		15.09	18.95
2340	Door/window casing, cedar, rough sawn, 1" x 2"		275	.029		.54	.96		1.50	2.20
2350	1" x 3"		275	.029		.81	.96		1.77	2.49
2360	1" x 4"		250	.032		1.07	1.05		2.12	2.94
2370	1" x 6"		250	.032		1.61	1.05		2.66	3.53
2380	1" x 8"		230	.035		2.15	1.14		3.29	4.29
2390	1" x 10"		230	.035		2.67	1.14		3.81	4.86
2395	1" x 12"		210	.038		3.21	1.25		4.46	5.65
2410	STK, 1" x 2"		275	.029		.51	.96		1.47	2.16
2420	1" x 3"		275	.029		.56	.96		1.52	2.22
2430	1" x 4"		250	.032		.65	1.05		1.70	2.47
2440	1" x 6"		250	.032		1.03	1.05		2.08	2.89
2450	1" x 8"		230	.035		1.60	1.14		2.74	3.68
2460	1" x 10"		230	.035		2.14	1.14		3.28	4.27
2470	1" x 12"		210	.038		3.54	1.25		4.79	6
2550	Pine, #2, 1" x 2"		275	.029		.25	.96		1.21	1.88
2560	1" x 3"		275	.029		.40	.96		1.36	2.04
2570	1" x 4"		250	.032		.49	1.05		1.54	2.30
2580	1" x 6"		250	.032		.75	1.05		1.80	2.58
2590	1" x 8"		230	.035		1.22	1.14		2.36	3.26
2600	1" x 10"		230	.035		1.59	1.14		2.73	3.67
2610	1" x 12"		210	.038		2.05	1.25		3.30	4.35
2620	Pine, D & better, 1" x 2"		275	.029		.41	.96		1.37	2.05
2630	1" x 3"		275	.029		.61	.96		1.57	2.27
2640	1" x 4"		250	.032		.82	1.05		1.87	2.66
2650	1" x 6"		250	.032		1.12	1.05		2.17	2.99

For customer support on your Residential Cost Data, call 877.713.2617.

06 22 Millwork

06 22 13 – Standard Pattern Wood Trim

06 22 13.40 Moldings, Exterior

		Crew	Daily Output	Labor-Hours	Unit	Material	2015 Bare Costs Labor	Equipment	Total	Total Incl O&P
2660	1" x 8"	1 Carp	230	.035	L.F.	1.62	1.14		2.76	3.70
2670	1" x 10"		230	.035		2.03	1.14		3.17	4.15
2680	1" x 12"		210	.038		2.63	1.25		3.88	4.99
2690	Redwood, clear all heart, 1" x 2"		275	.029		.63	.96		1.59	2.30
2695	1" x 3"		275	.029		.94	.96		1.90	2.64
2710	1" x 4"		250	.032		1.19	1.05		2.24	3.07
2715	1" x 6"		250	.032		1.78	1.05		2.83	3.71
2730	1" x 8"		230	.035		2.36	1.14		3.50	4.51
2740	1" x 10"		230	.035		3.96	1.14		5.10	6.30
2750	1" x 12"		210	.038		4.75	1.25		6	7.35
3500	Bellyband, pine, 11/16" x 4-1/4"		250	.032		3.23	1.05		4.28	5.30
3610	Brickmold, pine, 1-1/4" x 2"		200	.040		2.43	1.31		3.74	4.89
3620	FJP, 1-1/4" x 2"		200	.040		1.09	1.31		2.40	3.41
5100	Fascia, cedar, rough sawn, 1" x 2"		275	.029		.54	.96		1.50	2.20
5110	1" x 3"		275	.029		.81	.96		1.77	2.49
5120	1" x 4"		250	.032		1.07	1.05		2.12	2.94
5200	1" x 6"		250	.032		1.61	1.05		2.66	3.53
5300	1" x 8"		230	.035		2.15	1.14		3.29	4.29
5310	1" x 10"		230	.035		2.67	1.14		3.81	4.86
5320	1" x 12"		210	.038		3.21	1.25		4.46	5.65
5400	2" x 4"		220	.036		1.05	1.19		2.24	3.17
5500	2" x 6"		220	.036		1.58	1.19		2.77	3.75
5600	2" x 8"		200	.040		2.11	1.31		3.42	4.53
5700	2" x 10"		180	.044		2.62	1.46		4.08	5.35
5800	2" x 12"		170	.047		6.25	1.55		7.80	9.50
6120	STK, 1" x 2"		275	.029		.51	.96		1.47	2.16
6130	1" x 3"		275	.029		.56	.96		1.52	2.22
6140	1" x 4"		250	.032		.65	1.05		1.70	2.47
6150	1" x 6"		250	.032		1.03	1.05		2.08	2.89
6160	1" x 8"		230	.035		1.60	1.14		2.74	3.68
6170	1" x 10"		230	.035		2.14	1.14		3.28	4.27
6180	1" x 12"		210	.038		3.54	1.25		4.79	6
6185	2" x 2"		260	.031		.63	1.01		1.64	2.40
6190	Pine, #2, 1" x 2"		275	.029		.25	.96		1.21	1.88
6200	1" x 3"		275	.029		.40	.96		1.36	2.04
6210	1" x 4"		250	.032		.49	1.05		1.54	2.30
6220	1" x 6"		250	.032		.75	1.05		1.80	2.58
6230	1" x 8"		230	.035		1.22	1.14		2.36	3.26
6240	1" x 10"		230	.035		1.59	1.14		2.73	3.67
6250	1" x 12"		210	.038		2.05	1.25		3.30	4.35
6260	D & better, 1" x 2"		275	.029		.41	.96		1.37	2.05
6270	1" x 3"		275	.029		.61	.96		1.57	2.27
6280	1" x 4"		250	.032		.82	1.05		1.87	2.66
6290	1" x 6"		250	.032		1.12	1.05		2.17	2.99
6300	1" x 8"		230	.035		1.62	1.14		2.76	3.70
6310	1" x 10"		230	.035		2.03	1.14		3.17	4.15
6312	1" x 12"		210	.038		2.63	1.25		3.88	4.99
6330	Southern yellow, 1-1/4" x 5"		240	.033		2.27	1.09		3.36	4.33
6340	1-1/4" x 6"		240	.033		2.52	1.09		3.61	4.61
6350	1-1/4" x 8"		215	.037		3.43	1.22		4.65	5.80
6360	1-1/4" x 12"		190	.042		5.05	1.38		6.43	7.85
6370	Redwood, clear all heart, 1" x 2"		275	.029		.63	.96		1.59	2.30
6380	1" x 3"		275	.029		1.19	.96		2.15	2.91

06 22 Millwork

06 22 13 – Standard Pattern Wood Trim

	06 22 13.40 Moldings, Exterior	Crew	Daily Output	Labor-Hours	Unit	Material	2015 Bare Costs Labor	Equipment	Total	Total Incl O&P
6390	1" x 4"	1 Carp	250	.032	L.F.	1.19	1.05		2.24	3.07
6400	1" x 6"		250	.032		1.78	1.05		2.83	3.71
6410	1" x 8"		230	.035		2.36	1.14		3.50	4.51
6420	1" x 10"		230	.035		3.96	1.14		5.10	6.30
6430	1" x 12"		210	.038		4.75	1.25		6	7.35
6440	1-1/4" x 5"		240	.033		1.86	1.09		2.95	3.88
6450	1-1/4" x 6"		240	.033		2.22	1.09		3.31	4.28
6460	1-1/4" x 8"		215	.037		3.53	1.22		4.75	5.95
6470	1-1/4" x 12"		190	.042		7.10	1.38		8.48	10.15
6580	Frieze, cedar, rough sawn, 1" x 2"		275	.029		.54	.96		1.50	2.20
6590	1" x 3"		275	.029		.81	.96		1.77	2.49
6600	1" x 4"		250	.032		1.07	1.05		2.12	2.94
6610	1" x 6"		250	.032		1.61	1.05		2.66	3.53
6620	1" x 8"		250	.032		2.15	1.05		3.20	4.13
6630	1" x 10"		225	.036		2.67	1.17		3.84	4.90
6640	1" x 12"		200	.040		3.18	1.31		4.49	5.70
6650	STK, 1" x 2"		275	.029		.51	.96		1.47	2.16
6660	1" x 3"		275	.029		.56	.96		1.52	2.22
6670	1" x 4"		250	.032		.65	1.05		1.70	2.47
6680	1" x 6"		250	.032		1.03	1.05		2.08	2.89
6690	1" x 8"		250	.032		1.60	1.05		2.65	3.52
6700	1" x 10"		225	.036		2.14	1.17		3.31	4.31
6710	1" x 12"		200	.040		3.54	1.31		4.85	6.10
6790	Pine, #2, 1" x 2"		275	.029		.25	.96		1.21	1.88
6800	1" x 3"		275	.029		.40	.96		1.36	2.04
6810	1" x 4"		250	.032		.49	1.05		1.54	2.30
6820	1" x 6"		250	.032		.75	1.05		1.80	2.58
6830	1" x 8"		250	.032		1.22	1.05		2.27	3.10
6840	1" x 10"		225	.036		1.59	1.17		2.76	3.71
6850	1" x 12"		200	.040		2.05	1.31		3.36	4.46
6860	D & better, 1" x 2"		275	.029		.41	.96		1.37	2.05
6870	1" x 3"		275	.029		.61	.96		1.57	2.27
6880	1" x 4"		250	.032		.82	1.05		1.87	2.66
6890	1" x 6"		250	.032		1.12	1.05		2.17	2.99
6900	1" x 8"		250	.032		1.62	1.05		2.67	3.54
6910	1" x 10"		225	.036		2.03	1.17		3.20	4.19
6920	1" x 12"		200	.040		2.63	1.31		3.94	5.10
6930	Redwood, clear all heart, 1" x 2"		275	.029		.63	.96		1.59	2.30
6940	1" x 3"		275	.029		.94	.96		1.90	2.64
6950	1" x 4"		250	.032		1.19	1.05		2.24	3.07
6960	1" x 6"		250	.032		1.78	1.05		2.83	3.71
6970	1" x 8"		250	.032		2.36	1.05		3.41	4.35
6980	1" x 10"		225	.036		3.96	1.17		5.13	6.30
6990	1" x 12"		200	.040		4.75	1.31		6.06	7.45
7000	Grounds, 1" x 1", cedar, rough sawn		300	.027		.28	.88		1.16	1.77
7010	STK		300	.027		.31	.88		1.19	1.81
7020	Pine, #2		300	.027		.16	.88		1.04	1.64
7030	D & better		300	.027		.25	.88		1.13	1.75
7050	Redwood		300	.027		.39	.88		1.27	1.89
7060	Rake/verge board, cedar, rough sawn, 1" x 2"		225	.036		.54	1.17		1.71	2.56
7070	1" x 3"		225	.036		.81	1.17		1.98	2.85
7080	1" x 4"		200	.040		1.07	1.31		2.38	3.39
7090	1" x 6"		200	.040		1.61	1.31		2.92	3.98

For customer support on your Residential Cost Data, call 877.713.2617.

06 22 Millwork

06 22 13 – Standard Pattern Wood Trim

06 22 13.40 Moldings, Exterior		Crew	Daily Output	Labor-Hours	Unit	Material	2015 Bare Costs Labor	Equipment	Total	Total Incl O&P
7100	1" x 8"	1 Carp	190	.042	L.F.	2.15	1.38		3.53	4.69
7110	1" x 10"		190	.042		2.67	1.38		4.05	5.25
7120	1" x 12"		180	.044		3.21	1.46		4.67	6
7130	STK, 1" x 2"		225	.036		.51	1.17		1.68	2.52
7140	1" x 3"		225	.036		.56	1.17		1.73	2.58
7150	1" x 4"		200	.040		.65	1.31		1.96	2.92
7160	1" x 6"		200	.040		1.03	1.31		2.34	3.34
7170	1" x 8"		190	.042		1.60	1.38		2.98	4.08
7180	1" x 10"		190	.042		2.14	1.38		3.52	4.67
7190	1" x 12"		180	.044		3.54	1.46		5	6.35
7200	Pine, #2, 1" x 2"		225	.036		.25	1.17		1.42	2.24
7210	1" x 3"		225	.036		.40	1.17		1.57	2.40
7220	1" x 4"		200	.040		.49	1.31		1.80	2.75
7230	1" x 6"		200	.040		.75	1.31		2.06	3.03
7240	1" x 8"		190	.042		1.22	1.38		2.60	3.66
7250	1" x 10"		190	.042		1.59	1.38		2.97	4.07
7260	1" x 12"		180	.044		2.05	1.46		3.51	4.70
7340	D & better, 1" x 2"		225	.036		.41	1.17		1.58	2.41
7350	1" x 3"		225	.036		.61	1.17		1.78	2.63
7360	1" x 4"		200	.040		.82	1.31		2.13	3.11
7370	1" x 6"		200	.040		1.12	1.31		2.43	3.44
7380	1" x 8"		190	.042		1.62	1.38		3	4.10
7390	1" x 10"		190	.042		2.03	1.38		3.41	4.55
7400	1" x 12"		180	.044		2.63	1.46		4.09	5.35
7410	Redwood, clear all heart, 1" x 2"		225	.036		.63	1.17		1.80	2.66
7420	1" x 3"		225	.036		.94	1.17		2.11	3
7430	1" x 4"		200	.040		1.19	1.31		2.50	3.52
7440	1" x 6"		200	.040		1.78	1.31		3.09	4.16
7450	1" x 8"		190	.042		2.36	1.38		3.74	4.91
7460	1" x 10"		190	.042		3.96	1.38		5.34	6.70
7470	1" x 12"		180	.044		4.75	1.46		6.21	7.70
7480	2" x 4"		200	.040		2.28	1.31		3.59	4.71
7490	2" x 6"		182	.044		3.43	1.44		4.87	6.20
7500	2" x 8"	↓	165	.048		4.56	1.59		6.15	7.65
7630	Soffit, cedar, rough sawn, 1" x 2"	2 Carp	440	.036		.54	1.19		1.73	2.61
7640	1" x 3"		440	.036		.81	1.19		2	2.90
7650	1" x 4"		420	.038		1.07	1.25		2.32	3.28
7660	1" x 6"		420	.038		1.61	1.25		2.86	3.87
7670	1" x 8"		420	.038		2.15	1.25		3.40	4.47
7680	1" x 10"		400	.040		2.67	1.31		3.98	5.15
7690	1" x 12"		400	.040		3.21	1.31		4.52	5.75
7700	STK, 1" x 2"		440	.036		.51	1.19		1.70	2.57
7710	1" x 3"		440	.036		.56	1.19		1.75	2.63
7720	1" x 4"		420	.038		.65	1.25		1.90	2.81
7730	1" x 6"		420	.038		1.03	1.25		2.28	3.23
7740	1" x 8"		420	.038		1.60	1.25		2.85	3.86
7750	1" x 10"		400	.040		2.14	1.31		3.45	4.56
7760	1" x 12"		400	.040		3.54	1.31		4.85	6.10
7770	Pine, #2, 1" x 2"		440	.036		.25	1.19		1.44	2.29
7780	1" x 3"		440	.036		.40	1.19		1.59	2.45
7790	1" x 4"		420	.038		.49	1.25		1.74	2.64
7800	1" x 6"		420	.038		.75	1.25		2	2.92
7810	1" x 8"	↓	420	.038		1.22	1.25		2.47	3.44

06 22 Millwork

06 22 13 – Standard Pattern Wood Trim

06 22 13.40 Moldings, Exterior

		Crew	Daily Output	Labor-Hours	Unit	Material	2015 Bare Costs Labor	Equipment	Total	Total Incl O&P
7820	1" x 10"	2 Carp	400	.040	L.F.	1.59	1.31		2.90	3.96
7830	1" x 12"		400	.040		2.05	1.31		3.36	4.46
7840	D & better, 1" x 2"		440	.036		.41	1.19		1.60	2.46
7850	1" x 3"		440	.036		.61	1.19		1.80	2.68
7860	1" x 4"		420	.038		.82	1.25		2.07	3
7870	1" x 6"		420	.038		1.12	1.25		2.37	3.33
7880	1" x 8"		420	.038		1.62	1.25		2.87	3.88
7890	1" x 10"		400	.040		2.03	1.31		3.34	4.44
7900	1" x 12"		400	.040		2.63	1.31		3.94	5.10
7910	Redwood, clear all heart, 1" x 2"		440	.036		.63	1.19		1.82	2.71
7920	1" x 3"		440	.036		.94	1.19		2.13	3.05
7930	1" x 4"		420	.038		1.19	1.25		2.44	3.41
7940	1" x 6"		420	.038		1.78	1.25		3.03	4.05
7950	1" x 8"		420	.038		2.36	1.25		3.61	4.69
7960	1" x 10"		400	.040		3.96	1.31		5.27	6.55
7970	1" x 12"		400	.040		4.75	1.31		6.06	7.45
8050	Trim, crown molding, pine, 11/16" x 4-1/4"	1 Carp	250	.032		4.54	1.05		5.59	6.75
8060	Back band, 11/16" x 1-1/16"		250	.032		.99	1.05		2.04	2.85
8070	Insect screen frame stock, 1-1/16" x 1-3/4"		395	.020		2.39	.67		3.06	3.75
8080	Dentils, 2-1/2" x 2-1/2" x 4", 6" O.C.		30	.267		1.22	8.75		9.97	16.05
8100	Fluted, 5-1/2"		165	.048		5.05	1.59		6.64	8.20
8110	Stucco bead, 1-3/8" x 1-5/8"		250	.032		2.50	1.05		3.55	4.51

06 22 13.45 Moldings, Trim

		Crew	Daily Output	Labor-Hours	Unit	Material	Labor	Equipment	Total	Total Incl O&P
0010	**MOLDINGS, TRIM**									
0200	Astragal, stock pine, 11/16" x 1-3/4"	1 Carp	255	.031	L.F.	1.42	1.03		2.45	3.29
0250	1-5/16" x 2-3/16"		240	.033		2.10	1.09		3.19	4.15
0800	Chair rail, stock pine, 5/8" x 2-1/2"		270	.030		1.58	.97		2.55	3.37
0900	5/8" x 3-1/2"		240	.033		2.40	1.09		3.49	4.48
1000	Closet pole, stock pine, 1-1/8" diameter		200	.040		1.16	1.31		2.47	3.49
1100	Fir, 1-5/8" diameter		200	.040		2.20	1.31		3.51	4.63
1150	Corner, inside, 5/16" x 1"		225	.036		.33	1.17		1.50	2.32
1160	Outside, 1-1/16" x 1-1/16"		240	.033		1.31	1.09		2.40	3.28
1161	1-5/16" x 1-5/16"		240	.033		1.55	1.09		2.64	3.55
3300	Half round, stock pine, 1/4" x 1/2"		270	.030		.24	.97		1.21	1.89
3350	1/2" x 1"		255	.031		.73	1.03		1.76	2.53
3400	Handrail, fir, single piece, stock, hardware not included									
3450	1-1/2" x 1-3/4"	1 Carp	80	.100	L.F.	2.56	3.29		5.85	8.30
3470	Pine, 1-1/2" x 1-3/4"		80	.100		2.40	3.29		5.69	8.15
3500	1-1/2" x 2-1/2"		76	.105		2.46	3.46		5.92	8.50
3600	Lattice, stock pine, 1/4" x 1-1/8"		270	.030		.35	.97		1.32	2.02
3700	1/4" x 1-3/4"		250	.032		.92	1.05		1.97	2.77
3800	Miscellaneous, custom, pine, 1" x 1"		270	.030		.44	.97		1.41	2.11
3850	1" x 2"		265	.030		.87	.99		1.86	2.63
3900	1" x 3"		240	.033		1.31	1.09		2.40	3.28
4100	Birch or oak, nominal 1" x 1"		240	.033		.42	1.09		1.51	2.30
4200	Nominal 1" x 3"		215	.037		1.26	1.22		2.48	3.43
4400	Walnut, nominal 1" x 1"		215	.037		.66	1.22		1.88	2.78
4500	Nominal 1" x 3"		200	.040		1.99	1.31		3.30	4.40
4700	Teak, nominal 1" x 1"		215	.037		2.80	1.22		4.02	5.15
4800	Nominal 1" x 3"		200	.040		8.40	1.31		9.71	11.45
4900	Quarter round, stock pine, 1/4" x 1/4"		275	.029		.24	.96		1.20	1.86
4950	3/4" x 3/4"		255	.031		.50	1.03		1.53	2.28

06 22 Millwork

06 22 13 – Standard Pattern Wood Trim

06 22 13.45 Moldings, Trim

		Crew	Daily Output	Labor-Hours	Unit	Material	2015 Bare Costs Labor	2015 Bare Costs Equipment	Total	Total Incl O&P
5600	Wainscot moldings, 1-1/8" x 9/16", 2' high, minimum	1 Carp	76	.105	S.F.	11.65	3.46		15.11	18.60
5700	Maximum		65	.123	"	15.75	4.04		19.79	24

06 22 13.50 Moldings, Window and Door

		Crew	Daily Output	Labor-Hours	Unit	Material	Labor	Equipment	Total	Total Incl O&P
0010	**MOLDINGS, WINDOW AND DOOR**									
2800	Door moldings, stock, decorative, 1-1/8" wide, plain	1 Carp	17	.471	Set	47.50	15.45		62.95	78.50
2900	Detailed		17	.471	"	91.50	15.45		106.95	127
2960	Clear pine door jamb, no stops, 11/16" x 4-9/16"		240	.033	L.F.	5.20	1.09		6.29	7.55
3150	Door trim set, 1 head and 2 sides, pine, 2-1/2 wide		12	.667	Opng.	24	22		46	63
3170	3-1/2" wide		11	.727	"	29	24		53	72
3250	Glass beads, stock pine, 3/8" x 1/2"		275	.029	L.F.	.33	.96		1.29	1.96
3270	3/8" x 7/8"		270	.030		.42	.97		1.39	2.09
4850	Parting bead, stock pine, 3/8" x 3/4"		275	.029		.44	.96		1.40	2.08
4870	1/2" x 3/4"		255	.031		.42	1.03		1.45	2.19
5000	Stool caps, stock pine, 11/16" x 3-1/2"		200	.040		2.07	1.31		3.38	4.49
5100	1-1/16" x 3-1/4"		150	.053		3.24	1.75		4.99	6.50
5300	Threshold, oak, 3' long, inside, 5/8" x 3-5/8"		32	.250	Ea.	6.50	8.20		14.70	21
5400	Outside, 1-1/2" x 7-5/8"		16	.500	"	45	16.45		61.45	77
5900	Window trim sets, including casings, header, stops,									
5910	stool and apron, 2-1/2" wide, FJP	1 Carp	13	.615	Opng.	31.50	20		51.50	68.50
5950	Pine		10	.800		37	26.50		63.50	85
6000	Oak		6	1.333		64.50	44		108.50	145

06 22 13.60 Moldings, Soffits

		Crew	Daily Output	Labor-Hours	Unit	Material	Labor	Equipment	Total	Total Incl O&P
0010	**MOLDINGS, SOFFITS**									
0200	Soffits, pine, 1" x 4"	2 Carp	420	.038	L.F.	.47	1.25		1.72	2.61
0210	1" x 6"		420	.038		.72	1.25		1.97	2.89
0220	1" x 8"		420	.038		1.19	1.25		2.44	3.41
0230	1" x 10"		400	.040		1.55	1.31		2.86	3.91
0240	1" x 12"		400	.040		2.01	1.31		3.32	4.42
0250	STK cedar, 1" x 4"		420	.038		.62	1.25		1.87	2.78
0260	1" x 6"		420	.038		1	1.25		2.25	3.20
0270	1" x 8"		420	.038		1.57	1.25		2.82	3.83
0280	1" x 10"		400	.040		2.10	1.31		3.41	4.52
0290	1" x 12"		400	.040		3.50	1.31		4.81	6.05
1000	Exterior AC plywood, 1/4" thick		400	.040	S.F.	.88	1.31		2.19	3.18
1050	3/8" thick		400	.040		.92	1.31		2.23	3.22
1100	1/2" thick		400	.040		1.10	1.31		2.41	3.42
1150	Polyvinyl chloride, white, solid	1 Carp	230	.035		2.10	1.14		3.24	4.23
1160	Perforated	"	230	.035		2.10	1.14		3.24	4.23
1170	Accessories, "J" channel 5/8"	2 Carp	700	.023	L.F.	.45	.75		1.20	1.76

06 25 Prefinished Paneling

06 25 13 – Prefinished Hardboard Paneling

06 25 13.10 Paneling, Hardboard

			Crew	Daily Output	Labor-Hours	Unit	Material	Labor	Equipment	Total	Total Incl O&P
0010	**PANELING, HARDBOARD**										
0050	Not incl. furring or trim, hardboard, tempered, 1/8" thick	G	2 Carp	500	.032	S.F.	.41	1.05		1.46	2.21
0100	1/4" thick	G		500	.032		.64	1.05		1.69	2.46
0300	Tempered pegboard, 1/8" thick	G		500	.032		.40	1.05		1.45	2.20
0400	1/4" thick	G		500	.032		.68	1.05		1.73	2.51
0600	Untempered hardboard, natural finish, 1/8" thick	G		500	.032		.42	1.05		1.47	2.22
0700	1/4" thick	G		500	.032		.51	1.05		1.56	2.32

06 25 Prefinished Paneling

06 25 13 – Prefinished Hardboard Paneling

06 25 13.10 Paneling, Hardboard

		Crew	Daily Output	Labor-Hours	Unit	Material	2015 Bare Costs Labor	Equipment	Total	Total Incl O&P
0900	Untempered pegboard, 1/8" thick G	2 Carp	500	.032	S.F.	.44	1.05		1.49	2.24
1000	1/4" thick G		500	.032		.48	1.05		1.53	2.29
1200	Plastic faced hardboard, 1/8" thick G		500	.032		.66	1.05		1.71	2.49
1300	1/4" thick G		500	.032		.89	1.05		1.94	2.74
1500	Plastic faced pegboard, 1/8" thick G		500	.032		.67	1.05		1.72	2.50
1600	1/4" thick G		500	.032		.85	1.05		1.90	2.70
1800	Wood grained, plain or grooved, 1/8" thick G		500	.032		.69	1.05		1.74	2.52
1900	1/4" thick G		425	.038		1.40	1.24		2.64	3.62
2100	Moldings, wood grained MDF		500	.032	L.F.	.41	1.05		1.46	2.21
2200	Pine		425	.038	"	1.40	1.24		2.64	3.62

06 25 16 – Prefinished Plywood Paneling

06 25 16.10 Paneling, Plywood

		Crew	Daily Output	Labor-Hours	Unit	Material	Labor	Equipment	Total	Total Incl O&P
0010	**PANELING, PLYWOOD**									
2400	Plywood, prefinished, 1/4" thick, 4' x 8' sheets									
2410	with vertical grooves. Birch faced, economy	2 Carp	500	.032	S.F.	1.40	1.05		2.45	3.30
2420	Average		420	.038		1.20	1.25		2.45	3.42
2430	Custom		350	.046		1.25	1.50		2.75	3.90
2600	Mahogany, African		400	.040		2.75	1.31		4.06	5.25
2700	Philippine (Lauan)		500	.032		.65	1.05		1.70	2.48
2900	Oak		500	.032		1.40	1.05		2.45	3.30
3000	Cherry		400	.040		2.05	1.31		3.36	4.47
3200	Rosewood		320	.050		2.90	1.64		4.54	5.95
3400	Teak		400	.040		2.90	1.31		4.21	5.40
3600	Chestnut		375	.043		4.80	1.40		6.20	7.65
3800	Pecan		400	.040		2.50	1.31		3.81	4.96
3900	Walnut, average		500	.032		2.40	1.05		3.45	4.40
3950	Custom		400	.040		5.25	1.31		6.56	8
4000	Plywood, prefinished, 3/4" thick, stock grades, economy		320	.050		1.56	1.64		3.20	4.48
4100	Average		224	.071		4.68	2.35		7.03	9.10
4300	Architectural grade, custom		224	.071		5.20	2.35		7.55	9.65
4400	Luxury		160	.100		5.20	3.29		8.49	11.20
4600	Plywood, "A" face, birch, VC, 1/2" thick, natural		450	.036		2.05	1.17		3.22	4.22
4700	Select		450	.036		2.15	1.17		3.32	4.33
4900	Veneer core, 3/4" thick, natural		320	.050		2.24	1.64		3.88	5.20
5000	Select		320	.050		2.44	1.64		4.08	5.45
5200	Lumber core, 3/4" thick, natural		320	.050		3.05	1.64		4.69	6.10
5500	Plywood, knotty pine, 1/4" thick, A2 grade		450	.036		1.70	1.17		2.87	3.83
5600	A3 grade		450	.036		2.10	1.17		3.27	4.27
5800	3/4" thick, veneer core, A2 grade		320	.050		2.15	1.64		3.79	5.15
5900	A3 grade		320	.050		2.42	1.64		4.06	5.40
6100	Aromatic cedar, 1/4" thick, plywood		400	.040		2.20	1.31		3.51	4.63
6200	1/4" thick, particle board		400	.040		1.05	1.31		2.36	3.37

06 25 26 – Panel System

06 25 26.10 Panel Systems

		Crew	Daily Output	Labor-Hours	Unit	Material	Labor	Equipment	Total	Total Incl O&P
0010	**PANEL SYSTEMS**									
0100	Raised panel, eng. wood core w/wood veneer, std., paint grade	2 Carp	300	.053	S.F.	11.20	1.75		12.95	15.25
0110	Oak veneer		300	.053		25.50	1.75		27.25	31
0120	Maple veneer		300	.053		32.50	1.75		34.25	39
0130	Cherry veneer		300	.053		37	1.75		38.75	44
0300	Class I fire rated, paint grade		300	.053		13	1.75		14.75	17.25
0310	Oak veneer		300	.053		30	1.75		31.75	36
0320	Maple veneer		300	.053		40.50	1.75		42.25	47.50

06 25 Prefinished Paneling

06 25 26 – Panel System

06 25 26.10 Panel Systems

		Crew	Daily Output	Labor-Hours	Unit	Material	2015 Bare Costs Labor	Equipment	Total	Total Incl O&P
0330	Cherry veneer	2 Carp	300	.053	S.F.	49	1.75		50.75	56.50
0510	Beadboard, 5/8" MDF, standard, primed		300	.053		8.45	1.75		10.20	12.25
0520	Oak veneer, unfinished		300	.053		13.35	1.75		15.10	17.65
0530	Maple veneer, unfinished		300	.053		14.65	1.75		16.40	19.05
0610	Rustic paneling, 5/8" MDF, standard, maple veneer, unfinished	↓	300	.053	↓	18.15	1.75		19.90	23

06 26 Board Paneling

06 26 13 – Profile Board Paneling

06 26 13.10 Paneling, Boards

		Crew	Daily Output	Labor-Hours	Unit	Material	Labor	Equipment	Total	Total Incl O&P
0010	**PANELING, BOARDS**									
6400	Wood board paneling, 3/4" thick, knotty pine	2 Carp	300	.053	S.F.	1.91	1.75		3.66	5.05
6500	Rough sawn cedar		300	.053		3.20	1.75		4.95	6.45
6700	Redwood, clear, 1" x 4" boards		300	.053		4.96	1.75		6.71	8.40
6900	Aromatic cedar, closet lining, boards	↓	275	.058	↓	2.32	1.91		4.23	5.75

06 43 Wood Stairs and Railings

06 43 13 – Wood Stairs

06 43 13.20 Prefabricated Wood Stairs

		Crew	Daily Output	Labor-Hours	Unit	Material	Labor	Equipment	Total	Total Incl O&P
0010	**PREFABRICATED WOOD STAIRS**									
0100	Box stairs, prefabricated, 3'-0" wide									
0110	Oak treads, up to 14 risers	2 Carp	39	.410	Riser	92.50	13.50		106	125
0600	With pine treads for carpet, up to 14 risers	"	39	.410	"	59.50	13.50		73	88
1100	For 4' wide stairs, add				Flight	25%				
1550	Stairs, prefabricated stair handrail with balusters	1 Carp	30	.267	L.F.	78.50	8.75		87.25	101
1700	Basement stairs, prefabricated, pine treads									
1710	Pine risers, 3' wide, up to 14 risers	2 Carp	52	.308	Riser	59.50	10.10		69.60	82.50
4000	Residential, wood, oak treads, prefabricated		1.50	10.667	Flight	1,200	350		1,550	1,925
4200	Built in place	↓	.44	36.364	"	2,175	1,200		3,375	4,375
4400	Spiral, oak, 4'-6" diameter, unfinished, prefabricated,									
4500	incl. railing, 9' high	2 Carp	1.50	10.667	Flight	3,425	350		3,775	4,375

06 43 13.40 Wood Stair Parts

		Crew	Daily Output	Labor-Hours	Unit	Material	Labor	Equipment	Total	Total Incl O&P
0010	**WOOD STAIR PARTS**									
0020	Pin top balusters, 1-1/4", oak, 34"	1 Carp	96	.083	Ea.	5.10	2.74		7.84	10.20
0030	38"		96	.083		5.85	2.74		8.59	11.05
0040	42"		96	.083		6.35	2.74		9.09	11.60
0050	Poplar, 34"		96	.083		3.05	2.74		5.79	7.95
0060	38"		96	.083		3.88	2.74		6.62	8.85
0070	42"		96	.083		8.85	2.74		11.59	14.30
0080	Maple, 34"		96	.083		4.90	2.74		7.64	10
0090	38"		96	.083		5.60	2.74		8.34	10.80
0100	42"		96	.083		6.50	2.74		9.24	11.75
0130	Primed, 34"		96	.083		3	2.74		5.74	7.90
0140	38"		96	.083		3.62	2.74		6.36	8.60
0150	42"		96	.083		4.42	2.74		7.16	9.45
0180	Box top balusters, 1-1/4", oak, 34"		60	.133		8.95	4.38		13.33	17.20
0190	38"		60	.133		9.95	4.38		14.33	18.30
0200	42"		60	.133		10.95	4.38		15.33	19.40
0210	Poplar, 34"		60	.133		6.25	4.38		10.63	14.25
0220	38"	↓	60	.133		6.95	4.38		11.33	15

06 43 Wood Stairs and Railings

06 43 13 – Wood Stairs

06 43 13.40 Wood Stair Parts		Crew	Daily Output	Labor-Hours	Unit	Material	2015 Bare Costs Labor	Equipment	Total	Total Incl O&P
0230	42"	1 Carp	60	.133	Ea.	7.50	4.38		11.88	15.60
0240	Maple, 34"		60	.133		8.25	4.38		12.63	16.45
0250	38"		60	.133		9	4.38		13.38	17.25
0260	42"		60	.133		10	4.38		14.38	18.35
0290	Primed, 34"		60	.133		7	4.38		11.38	15.05
0300	38"		60	.133		8	4.38		12.38	16.15
0310	42"		60	.133		8.35	4.38		12.73	16.55
0340	Square balusters, cut from lineal stock, pine, 1-1/16" x 1-1/16"		180	.044	L.F.	1.50	1.46		2.96	4.10
0350	1-5/16" x 1-5/16"		180	.044		2.10	1.46		3.56	4.76
0360	1-5/8" x 1-5/8"		180	.044		3.30	1.46		4.76	6.10
0370	Turned newel, oak, 3-1/2" square, 48" high		8	1	Ea.	92	33		125	156
0380	62" high		8	1		92	33		125	156
0390	Poplar, 3-1/2" square, 48" high		8	1		54	33		87	115
0400	62" high		8	1		68	33		101	130
0410	Maple, 3-1/2" square, 48" high		8	1		72	33		105	134
0420	62" high		8	1		92	33		125	156
0430	Square newel, oak, 3-1/2" square, 48" high		8	1		54	33		87	115
0440	58" high		8	1		68	33		101	130
0450	Poplar, 3-1/2" square, 48" high		8	1		35	33		68	93.50
0460	58" high		8	1		42	33		75	101
0470	Maple, 3" square, 48" high		8	1		52	33		85	112
0480	58" high		8	1		64	33		97	126
0490	Railings, oak, economy		96	.083	L.F.	8.65	2.74		11.39	14.10
0500	Average		96	.083		13	2.74		15.74	18.90
0510	Custom		96	.083		16.50	2.74		19.24	23
0520	Maple, economy		96	.083		11	2.74		13.74	16.70
0530	Average		96	.083		13.50	2.74		16.24	19.45
0540	Custom		96	.083		16.95	2.74		19.69	23.50
0550	Oak, for bending rail, economy		48	.167		23.50	5.50		29	35
0560	Average		48	.167		26	5.50		31.50	37.50
0570	Custom		48	.167		29.50	5.50		35	41
0580	Maple, for bending rail, economy		48	.167		32	5.50		37.50	44
0590	Average		48	.167		32	5.50		37.50	44
0600	Custom		48	.167		32	5.50		37.50	44
0610	Risers, oak, 3/4" x 8", 36" long		80	.100	Ea.	13	3.29		16.29	19.80
0620	42" long		70	.114		15.15	3.75		18.90	23
0630	48" long		63	.127		17.30	4.17		21.47	26
0640	54" long		56	.143		19.50	4.69		24.19	29.50
0650	60" long		50	.160		21.50	5.25		26.75	33
0660	72" long		42	.190		26	6.25		32.25	39
0670	Poplar, 3/4" x 8", 36" long		80	.100		12.50	3.29		15.79	19.25
0680	42" long		71	.113		14.55	3.70		18.25	22
0690	48" long		63	.127		16.65	4.17		20.82	25.50
0700	54" long		56	.143		18.70	4.69		23.39	28.50
0710	60" long		50	.160		21	5.25		26.25	32
0720	72" long		42	.190		25	6.25		31.25	38
0730	Pine, 1" x 8", 36" long		80	.100		3.57	3.29		6.86	9.45
0740	42" long		70	.114		4.17	3.75		7.92	10.90
0750	48" long		63	.127		4.76	4.17		8.93	12.25
0760	54" long		56	.143		5.35	4.69		10.04	13.80
0770	60" long		50	.160		5.95	5.25		11.20	15.35
0780	72" long		42	.190		7.15	6.25		13.40	18.35
0790	Treads, oak, no returns, 1-1/32" x 11-1/2" x 36" long		32	.250		27	8.20		35.20	43.50

06 43 Wood Stairs and Railings

06 43 13 – Wood Stairs

06 43 13.40 Wood Stair Parts

		Crew	Daily Output	Labor-Hours	Unit	Material	2015 Bare Costs Labor	2015 Bare Costs Equipment	Total	Total Incl O&P
0800	42" long	1 Carp	32	.250	Ea.	31.50	8.20		39.70	48.50
0810	48" long		32	.250		36	8.20		44.20	53.50
0820	54" long		32	.250		40.50	8.20		48.70	58.50
0830	60" long		32	.250		45	8.20		53.20	63.50
0840	72" long		32	.250		54	8.20		62.20	73.50
0850	Mitred return one end, 1-1/32" x 11-1/2" x 36" long		24	.333		36	10.95		46.95	58
0860	42" long		24	.333		42	10.95		52.95	64.50
0870	48" long		24	.333		48	10.95		58.95	71.50
0880	54" long		24	.333		54	10.95		64.95	78
0890	60" long		24	.333		60	10.95		70.95	84.50
0900	72" long		24	.333		72	10.95		82.95	97.50
0910	Mitred return two ends, 1-1/32" x 11-1/2" x 36" long		12	.667		46	22		68	87.50
0920	42" long		12	.667		53.50	22		75.50	96
0930	48" long		12	.667		61.50	22		83.50	105
0940	54" long		12	.667		69	22		91	113
0950	60" long		12	.667		76.50	22		98.50	122
0960	72" long		12	.667		92	22		114	138
0970	Starting step, oak, 48", bullnose		8	1		172	33		205	244
0980	Double end bullnose		8	1		254	33		287	335
1030	Skirt board, pine, 1" x 10"		55	.145	L.F.	1.55	4.78		6.33	9.70
1040	1" x 12"		52	.154	"	2.01	5.05		7.06	10.70
1050	Oak landing tread, 1-1/16" thick		54	.148	S.F.	9	4.87		13.87	18.05
1060	Oak cove molding		96	.083	L.F.	1	2.74		3.74	5.70
1070	Oak stringer molding		96	.083	"	4	2.74		6.74	9
1090	Rail bolt, 5/16" x 3-1/2"		48	.167	Ea.	2.75	5.50		8.25	12.25
1100	5/16" x 4-1/2"		48	.167		2.75	5.50		8.25	12.25
1120	Newel post anchor		16	.500		13	16.45		29.45	42
1130	Tapered plug, 1/2"		240	.033		1	1.09		2.09	2.94
1140	1"		240	.033		.99	1.09		2.08	2.93

06 43 16 – Wood Railings

06 43 16.10 Wood Handrails and Railings

		Crew	Daily Output	Labor-Hours	Unit	Material	Labor	Equipment	Total	Total Incl O&P
0010	**WOOD HANDRAILS AND RAILINGS**									
0020	Custom design, architectural grade, hardwood, plain	1 Carp	38	.211	L.F.	12.05	6.90		18.95	25
0100	Shaped		30	.267		62.50	8.75		71.25	83.50
0300	Stock interior railing with spindles 4" O.C., 4' long		40	.200		38.50	6.55		45.05	53.50
0400	8' long		48	.167		38.50	5.50		44	51.50

06 44 Ornamental Woodwork

06 44 19 – Wood Grilles

06 44 19.10 Grilles

		Crew	Daily Output	Labor-Hours	Unit	Material	Labor	Equipment	Total	Total Incl O&P
0010	**GRILLES** and panels, hardwood, sanded									
0020	2' x 4' to 4' x 8', custom designs, unfinished, economy	1 Carp	38	.211	S.F.	62	6.90		68.90	80
0050	Average		30	.267		69	8.75		77.75	90.50
0100	Custom		19	.421		72	13.85		85.85	102

06 44 33 – Wood Mantels

06 44 33.10 Fireplace Mantels

		Crew	Daily Output	Labor-Hours	Unit	Material	Labor	Equipment	Total	Total Incl O&P
0010	**FIREPLACE MANTELS**									
0015	6" molding, 6' x 3'-6" opening, plain, paint grade	1 Carp	5	1.600	Opng.	440	52.50		492.50	570
0100	Ornate, oak		5	1.600		590	52.50		642.50	740
0300	Prefabricated pine, colonial type, stock, deluxe		2	4		1,500	131		1,631	1,875

06 44 Ornamental Woodwork

06 44 33 – Wood Mantels

06 44 33.10 Fireplace Mantels

		Crew	Daily Output	Labor-Hours	Unit	Material	2015 Bare Costs Labor	Equipment	Total	Total Incl O&P
0400	Economy	1 Carp	3	2.667	Opng.	690	87.50		777.50	905

06 44 33.20 Fireplace Mantel Beam

		Crew	Daily Output	Labor-Hours	Unit	Material	Labor	Equipment	Total	Total Incl O&P
0010	**FIREPLACE MANTEL BEAM**									
0020	Rough texture wood, 4" x 8"	1 Carp	36	.222	L.F.	8.30	7.30		15.60	21.50
0100	4" x 10"		35	.229	"	10.90	7.50		18.40	24.50
0300	Laminated hardwood, 2-1/4" x 10-1/2" wide, 6' long		5	1.600	Ea.	110	52.50		162.50	209
0400	8' long		5	1.600	"	150	52.50		202.50	253
0600	Brackets for above, rough sawn		12	.667	Pr.	10	22		32	48
0700	Laminated		12	.667	"	15	22		37	53.50

06 44 39 – Wood Posts and Columns

06 44 39.10 Decorative Beams

		Crew	Daily Output	Labor-Hours	Unit	Material	Labor	Equipment	Total	Total Incl O&P
0010	**DECORATIVE BEAMS**									
0020	Rough sawn cedar, non-load bearing, 4" x 4"	2 Carp	180	.089	L.F.	1.19	2.92		4.11	6.20
0100	4" x 6"		170	.094		1.79	3.09		4.88	7.15
0200	4" x 8"		160	.100		2.39	3.29		5.68	8.15
0300	4" x 10"		150	.107		3.63	3.50		7.13	9.90
0400	4" x 12"		140	.114		4.61	3.75		8.36	11.35
0500	8" x 8"		130	.123		4.78	4.04		8.82	12.05
0600	Plastic beam, "hewn finish", 6" x 2"		240	.067		3.20	2.19		5.39	7.20
0601	6" x 4"		220	.073		3.60	2.39		5.99	7.95

06 44 39.20 Columns

		Crew	Daily Output	Labor-Hours	Unit	Material	Labor	Equipment	Total	Total Incl O&P
0010	**COLUMNS**									
0050	Aluminum, round colonial, 6" diameter	2 Carp	80	.200	V.L.F.	19	6.55		25.55	32
0100	8" diameter		62.25	.257		22	8.45		30.45	38
0200	10" diameter		55	.291		22.50	9.55		32.05	41
0250	Fir, stock units, hollow round, 6" diameter		80	.200		29.50	6.55		36.05	43.50
0300	8" diameter		80	.200		35.50	6.55		42.05	50
0350	10" diameter		70	.229		44.50	7.50		52	61.50
0360	12" diameter		65	.246		54.50	8.10		62.60	73.50
0400	Solid turned, to 8' high, 3-1/2" diameter		80	.200		9.60	6.55		16.15	21.50
0500	4-1/2" diameter		75	.213		11.90	7		18.90	25
0600	5-1/2" diameter		70	.229		16	7.50		23.50	30
0800	Square columns, built-up, 5" x 5"		65	.246		14.20	8.10		22.30	29
0900	Solid, 3-1/2" x 3-1/2"		130	.123		9.60	4.04		13.64	17.35
1600	Hemlock, tapered, T & G, 12" diam., 10' high		100	.160		39.50	5.25		44.75	52.50
1700	16' high		65	.246		73	8.10		81.10	93.50
1900	14" diameter, 10' high		100	.160		113	5.25		118.25	134
2000	18' high		65	.246		103	8.10		111.10	127
2200	18" diameter, 12' high		65	.246		165	8.10		173.10	196
2300	20' high		50	.320		118	10.50		128.50	148
2500	20" diameter, 14' high		40	.400		180	13.15		193.15	220
2600	20' high		35	.457		170	15		185	212
2800	For flat pilasters, deduct					33%				
3000	For splitting into halves, add				Ea.	106			106	117
4000	Rough sawn cedar posts, 4" x 4"	2 Carp	250	.064	V.L.F.	3.90	2.10		6	7.80
4100	4" x 6"		235	.068		6.80	2.24		9.04	11.25
4200	6" x 6"		220	.073		9.90	2.39		12.29	14.90
4300	8" x 8"		200	.080		19.20	2.63		21.83	25.50

06 48 Wood Frames

06 48 13 – Exterior Wood Door Frames

06 48 13.10 Exterior Wood Door Frames and Accessories		Crew	Daily Output	Labor-Hours	Unit	Material	2015 Bare Costs Labor	Equipment	Total	Total Incl O&P
0010	**EXTERIOR WOOD DOOR FRAMES AND ACCESSORIES**									
0400	Exterior frame, incl. ext. trim, pine, 5/4 x 4-9/16" deep	2 Carp	375	.043	L.F.	6.60	1.40		8	9.65
0420	5-3/16" deep		375	.043		7.90	1.40		9.30	11
0440	6-9/16" deep		375	.043		8.80	1.40		10.20	12
0600	Oak, 5/4 x 4-9/16" deep		350	.046		19.75	1.50		21.25	24
0620	5-3/16" deep		350	.046		21.50	1.50		23	26.50
0640	6-9/16" deep		350	.046		19.50	1.50		21	24
1000	Sills, 8/4 x 8" deep, oak, no horns		100	.160		6.65	5.25		11.90	16.10
1020	2" horns		100	.160		20.50	5.25		25.75	31.50
1040	3" horns		100	.160		20.50	5.25		25.75	31.50
1100	8/4 x 10" deep, oak, no horns		90	.178		6.40	5.85		12.25	16.85
1120	2" horns		90	.178		26.50	5.85		32.35	39
1140	3" horns		90	.178		26.50	5.85		32.35	39
2000	Wood frame & trim, ext, colonial, 3' opng, fluted pilasters, flat head		22	.727	Ea.	505	24		529	595
2010	Dentil head		21	.762		580	25		605	675
2020	Ram's head		20	.800		695	26.50		721.50	805
2100	5'-4" opening, in-swing, fluted pilasters, flat head		17	.941		440	31		471	535
2120	Ram's head		15	1.067		1,400	35		1,435	1,575
2140	Out swing, fluted pilasters, flat head		17	.941		520	31		551	620
2160	Ram's head		15	1.067		1,475	35		1,510	1,675
2400	6'-0" opening, in-swing, fluted pilasters, flat head		16	1		520	33		553	625
2420	Ram's head		10	1.600		1,475	52.50		1,527.50	1,725
2460	Out-swing, fluted pilasters, flat head		16	1		520	33		553	625
2480	Ram's head		10	1.600		1,475	52.50		1,527.50	1,725
2600	For two sidelights, flat head, add		30	.533	Opng.	226	17.50		243.50	279
2620	Ram's head, add		20	.800	"	840	26.50		866.50	970
2700	Custom birch frame, 3'-0" opening		16	1	Ea.	240	33		273	320
2750	6'-0" opening		16	1		360	33		393	450
2900	Exterior, modern, plain trim, 3' opng., in-swing, FJP		26	.615		46.50	20		66.50	85
2920	Fir		24	.667		55	22		77	97
2940	Oak		22	.727		63	24		87	110

06 48 16 – Interior Wood Door Frames

06 48 16.10 Interior Wood Door Jamb and Frames		Crew	Daily Output	Labor-Hours	Unit	Material	2015 Bare Costs Labor	Equipment	Total	Total Incl O&P
0010	**INTERIOR WOOD DOOR JAMB AND FRAMES**									
3000	Interior frame, pine, 11/16" x 3-5/8" deep	2 Carp	375	.043	L.F.	4.44	1.40		5.84	7.25
3020	4-9/16" deep		375	.043		4.92	1.40		6.32	7.75
3040	5-3/16" deep		375	.043		4.66	1.40		6.06	7.50
3200	Oak, 11/16" x 3-5/8" deep		350	.046		9.85	1.50		11.35	13.30
3220	4-9/16" deep		350	.046		9.95	1.50		11.45	13.45
3240	5-3/16" deep		350	.046		13.90	1.50		15.40	17.75
3400	Walnut, 11/16" x 3-5/8" deep		350	.046		9	1.50		10.50	12.40
3420	4-9/16" deep		350	.046		9.45	1.50		10.95	12.90
3440	5-3/16" deep		350	.046		9.55	1.50		11.05	13
3600	Pocket door frame		16	1	Ea.	81	33		114	144
3800	Threshold, oak, 5/8" x 3-5/8" deep		200	.080	L.F.	3.56	2.63		6.19	8.35
3820	4-5/8" deep		190	.084		4.13	2.77		6.90	9.20
3840	5-5/8" deep		180	.089		6.50	2.92		9.42	12.05

06 49 Wood Screens and Exterior Wood Shutters

06 49 19 – Exterior Wood Shutters

06 49 19.10 Shutters, Exterior		Crew	Daily Output	Labor-Hours	Unit	Material	2015 Bare Costs Labor	Equipment	Total	Total Incl O&P
0010	**SHUTTERS, EXTERIOR**									
0012	Aluminum, louvered, 1'-4" wide, 3'-0" long	1 Carp	10	.800	Pr.	200	26.50		226.50	264
0200	4'-0" long		10	.800		240	26.50		266.50	310
0300	5'-4" long		10	.800		281	26.50		307.50	355
0400	6'-8" long		9	.889		355	29		384	440
1000	Pine, louvered, primed, each 1'-2" wide, 3'-3" long		10	.800		216	26.50		242.50	282
1100	4'-7" long		10	.800		270	26.50		296.50	340
1250	Each 1'-4" wide, 3'-0" long		10	.800		223	26.50		249.50	289
1350	5'-3" long		10	.800		330	26.50		356.50	405
1500	Each 1'-6" wide, 3'-3" long		10	.800		244	26.50		270.50	310
1600	4'-7" long		10	.800		325	26.50		351.50	400
1620	Cedar, louvered, 1'-2" wide, 5'-7" long		10	.800		315	26.50		341.50	390
1630	Each 1'-4" wide, 2'-2" long		10	.800		175	26.50		201.50	237
1640	3'-0" long		10	.800		217	26.50		243.50	283
1650	3'-3" long		10	.800		228	26.50		254.50	295
1660	3'-11" long		10	.800		265	26.50		291.50	335
1670	4'-3" long		10	.800		275	26.50		301.50	350
1680	5'-3" long		10	.800		290	26.50		316.50	365
1690	5'-11" long		10	.800		350	26.50		376.50	430
1700	Door blinds, 6'-9" long, each 1'-3" wide		9	.889		380	29		409	465
1710	1'-6" wide		9	.889		435	29		464	530
1720	Cedar, solid raised panel, each 1'-4" wide, 3'-3" long		10	.800		315	26.50		341.50	390
1730	3'-11" long		10	.800		360	26.50		386.50	440
1740	4'-3" long		10	.800		365	26.50		391.50	445
1750	4'-7" long		10	.800		390	26.50		416.50	475
1760	4'-11" long		10	.800		415	26.50		441.50	505
1770	5'-11" long		10	.800		520	26.50		546.50	615
1800	Door blinds, 6'-9" long, each 1'-3" wide		9	.889		535	29		564	640
1900	1'-6" wide		9	.889		630	29		659	745
2500	Polystyrene, solid raised panel, each 1'-4" wide, 3'-3" long		10	.800		71.50	26.50		98	123
2600	3'-11" long		10	.800		93	26.50		119.50	146
2700	4'-7" long		10	.800		105	26.50		131.50	160
2800	5'-3" long		10	.800		120	26.50		146.50	176
2900	6'-8" long		9	.889		152	29		181	216
4500	Polystyrene, louvered, each 1'-2" wide, 3'-3" long		10	.800		36	26.50		62.50	83.50
4600	4'-7" long		10	.800		46	26.50		72.50	94.50
4750	5'-3" long		10	.800		59	26.50		85.50	109
4850	6'-8" long		9	.889		68	29		97	124
6000	Vinyl, louvered, each 1'-2" x 4'-7" long		10	.800		64	26.50		90.50	115
6200	Each 1'-4" x 6'-8" long		9	.889		76	29		105	133
8000	PVC exterior rolling shutters									
8100	including crank control	1 Carp	8	1	Ea.	520	33		553	630
8500	Insulative - 6' x 6'8" stock unit	"	8	1	"	745	33		778	875

06 51 Structural Plastic Shapes and Plates

06 51 13 - Plastic Lumber

06 51 13.10 Recycled Plastic Lumber		Crew	Daily Output	Labor-Hours	Unit	Material	2015 Bare Costs Labor	Equipment	Total	Total Incl O&P
0010	**RECYCLED PLASTIC LUMBER**									
4000	Sheeting, recycled plastic, black or white, 4' x 8' x 1/8"	G 2 Carp	1100	.015	S.F.	1.26	.48		1.74	2.19
4010	4' x 8' x 3/16"	G	1100	.015		1.90	.48		2.38	2.89
4020	4' x 8' x 1/4"	G	950	.017		2.25	.55		2.80	3.41
4030	4' x 8' x 3/8"	G	950	.017		3.80	.55		4.35	5.10
4040	4' x 8' x 1/2"	G	900	.018		5	.58		5.58	6.50
4050	4' x 8' x 5/8"	G	900	.018		7.50	.58		8.08	9.25
4060	4' x 8' x 3/4"	G	850	.019		9.40	.62		10.02	11.40
4070	Add for colors	G			Ea.	5%				
8500	100% recycled plastic, var colors, NLB, 2" x 2"	G			L.F.	1.76			1.76	1.94
8510	2" x 4"	G				3.65			3.65	4.02
8520	2" x 6"	G				5.75			5.75	6.35
8530	2" x 8"	G				7.90			7.90	8.70
8540	2" x 10"	G				11.50			11.50	12.65
8550	5/4" x 4"	G				4.45			4.45	4.90
8560	5/4" x 6"	G				5			5	5.50
8570	1" x 6"	G				2.87			2.87	3.16
8580	1/2" x 8"	G				3.05			3.05	3.36
8590	2" x 10" T & G	G				11.50			11.50	12.65
8600	3" x 10" T & G	G				15.30			15.30	16.85
8610	Add for premium colors	G				20%				

06 51 13.12 Structural Plastic Lumber		Crew	Daily Output	Labor-Hours	Unit	Material	Labor	Equipment	Total	Total Incl O&P
0010	**STRUCTURAL PLASTIC LUMBER**									
1320	Plastic lumber, posts or columns, 4" x 4"	2 Carp	390	.041	L.F.	9	1.35		10.35	12.15
1325	4" x 6"		275	.058		13.15	1.91		15.06	17.70
1330	4" x 8"		220	.073		19.20	2.39		21.59	25
1340	Girder, single, 4" x 4"		675	.024		9	.78		9.78	11.20
1345	4" x 6"		600	.027		13.15	.88		14.03	15.95
1350	4" x 8"		525	.030		19.20	1		20.20	22.50
1352	Double, 2" x 4"		625	.026		8.25	.84		9.09	10.45
1354	2" x 6"		600	.027		12.65	.88		13.53	15.40
1356	2" x 8"		575	.028		16.10	.91		17.01	19.25
1358	2" x 10"		550	.029		20	.96		20.96	23.50
1360	2" x 12"		525	.030		25	1		26	29
1362	Triple, 2" x 4"		575	.028		12.35	.91		13.26	15.15
1364	2" x 6"		550	.029		19	.96		19.96	22.50
1366	2" x 8"		525	.030		24	1		25	28
1368	2" x 10"		500	.032		30	1.05		31.05	35
1370	2" x 12"		475	.034		37	1.11		38.11	43
1372	Ledger, bolted 4' O.C., 2" x 4"		400	.040		4.26	1.31		5.57	6.90
1374	2" x 6"		550	.029		6.40	.96		7.36	8.65
1376	2" x 8"		390	.041		8.15	1.35		9.50	11.25
1378	2" x 10"		385	.042		10.10	1.37		11.47	13.40
1380	2" x 12"		380	.042		12.50	1.38		13.88	16.05
1382	Joists, 2" x 4"		1250	.013		4.12	.42		4.54	5.25
1384	2" x 6"		1250	.013		6.35	.42		6.77	7.65
1386	2" x 8"		1100	.015		8.05	.48		8.53	9.65
1388	2" x 10"		500	.032		10.10	1.05		11.15	12.85
1390	2" x 12"		875	.018		12.40	.60		13	14.65
1392	Railings and trim, 5/4" x 4"	1 Carp	300	.027		4.25	.88		5.13	6.15
1394	2" x 2"		300	.027		2.35	.88		3.23	4.06
1396	2" x 4"		300	.027		4.10	.88		4.98	6

06 51 Structural Plastic Shapes and Plates

06 51 13 – Plastic Lumber

06 51 13.12 Structural Plastic Lumber		Crew	Daily Output	Labor-Hours	Unit	Material	2015 Bare Costs Labor	Equipment	Total	Total Incl O&P
1398	2" x 6"	1 Carp	300	.027	L.F.	6.30	.88		7.18	8.40

06 63 Plastic Railings

06 63 10 – Plastic (PVC) Railings

06 63 10.10 Plastic Railings

		Crew	Daily Output	Labor-Hours	Unit	Material	Labor	Equipment	Total	Total Incl O&P
0010	**PLASTIC RAILINGS**									
0100	Horizontal PVC handrail with balusters, 3-1/2" wide, 36" high	1 Carp	96	.083	L.F.	24.50	2.74		27.24	31.50
0150	42" high		96	.083		28	2.74		30.74	35.50
0200	Angled PVC handrail with balusters, 3-1/2" wide, 36" high		72	.111		28.50	3.65		32.15	37
0250	42" high		72	.111		31.50	3.65		35.15	40.50
0300	Post sleeve for 4 x 4 post		96	.083		12.80	2.74		15.54	18.65
0400	Post cap for 4 x 4 post, flat profile		48	.167	Ea.	12.80	5.50		18.30	23.50
0450	Newel post style profile		48	.167		23.50	5.50		29	35
0500	Raised corbeled profile		48	.167		35.50	5.50		41	48
0550	Post base trim for 4 x 4 post		96	.083		18.25	2.74		20.99	24.50

06 65 Plastic Trim

06 65 10 – PVC Trim

06 65 10.10 PVC Trim, Exterior

		Crew	Daily Output	Labor-Hours	Unit	Material	Labor	Equipment	Total	Total Incl O&P
0010	**PVC TRIM, EXTERIOR**									
0100	Cornerboards, 5/4" x 6" x 6"	1 Carp	240	.033	L.F.	8.45	1.09		9.54	11.15
0110	Door/window casing, 1" x 4"		200	.040		1.39	1.31		2.70	3.74
0120	1" x 6"		200	.040		2.11	1.31		3.42	4.53
0130	1" x 8"		195	.041		2.78	1.35		4.13	5.30
0140	1" x 10"		195	.041		3.50	1.35		4.85	6.10
0150	1" x 12"		190	.042		4.33	1.38		5.71	7.10
0160	5/4" x 4"		195	.041		1.69	1.35		3.04	4.12
0170	5/4" x 6"		195	.041		2.72	1.35		4.07	5.25
0180	5/4" x 8"		190	.042		3.56	1.38		4.94	6.25
0190	5/4" x 10"		190	.042		4.56	1.38		5.94	7.30
0200	5/4" x 12"		185	.043		5.25	1.42		6.67	8.20
0210	Fascia, 1" x 4"		250	.032		1.39	1.05		2.44	3.29
0220	1" x 6"		250	.032		2.11	1.05		3.16	4.08
0230	1" x 8"		225	.036		2.78	1.17		3.95	5
0240	1" x 10"		225	.036		3.50	1.17		4.67	5.80
0250	1" x 12"		200	.040		4.33	1.31		5.64	6.95
0260	5/4" x 4"		240	.033		1.69	1.09		2.78	3.70
0270	5/4" x 6"		240	.033		2.72	1.09		3.81	4.83
0280	5/4" x 8"		215	.037		3.56	1.22		4.78	5.95
0290	5/4" x 10"		215	.037		4.56	1.22		5.78	7.05
0300	5/4" x 12"		190	.042		5.25	1.38		6.63	8.10
0310	Frieze, 1" x 4"		250	.032		1.39	1.05		2.44	3.29
0320	1" x 6"		250	.032		2.11	1.05		3.16	4.08
0330	1" x 8"		225	.036		2.78	1.17		3.95	5
0340	1" x 10"		225	.036		3.50	1.17		4.67	5.80
0350	1" x 12"		200	.040		4.33	1.31		5.64	6.95
0360	5/4" x 4"		240	.033		1.69	1.09		2.78	3.70
0370	5/4" x 6"		240	.033		2.72	1.09		3.81	4.83
0380	5/4" x 8"		215	.037		3.56	1.22		4.78	5.95

For customer support on your Residential Cost Data, call 877.713.2617.

06 65 Plastic Trim

06 65 10 – PVC Trim

	06 65 10.10 PVC Trim, Exterior	Crew	Daily Output	Labor-Hours	Unit	Material	2015 Bare Costs Labor	Equipment	Total	Total Incl O&P
0390	5/4" x 10"	1 Carp	215	.037	L.F.	4.56	1.22		5.78	7.05
0400	5/4" x 12"		190	.042		5.25	1.38		6.63	8.10
0410	Rake, 1" x 4"		200	.040		1.39	1.31		2.70	3.74
0420	1" x 6"		200	.040		2.11	1.31		3.42	4.53
0430	1" x 8"		190	.042		2.78	1.38		4.16	5.40
0440	1" x 10"		190	.042		3.50	1.38		4.88	6.15
0450	1" x 12"		180	.044		4.33	1.46		5.79	7.20
0460	5/4" x 4"		195	.041		1.69	1.35		3.04	4.12
0470	5/4" x 6"		195	.041		2.72	1.35		4.07	5.25
0480	5/4" x 8"		185	.043		3.56	1.42		4.98	6.30
0490	5/4" x 10"		185	.043		4.56	1.42		5.98	7.40
0500	5/4" x 12"		175	.046		5.25	1.50		6.75	8.30
0510	Rake trim, 1" x 4"		225	.036		1.39	1.17		2.56	3.49
0520	1" x 6"		225	.036		2.11	1.17		3.28	4.28
0560	5/4" x 4"		220	.036		1.69	1.19		2.88	3.87
0570	5/4" x 6"	↓	220	.036		2.72	1.19		3.91	5
0610	Soffit, 1" x 4"	2 Carp	420	.038		1.39	1.25		2.64	3.63
0620	1" x 6"		420	.038		2.11	1.25		3.36	4.42
0630	1" x 8"		420	.038		2.78	1.25		4.03	5.15
0640	1" x 10"		400	.040		3.50	1.31		4.81	6.05
0650	1" x 12"		400	.040		4.33	1.31		5.64	6.95
0660	5/4" x 4"		410	.039		1.69	1.28		2.97	4.01
0670	5/4" x 6"		410	.039		2.72	1.28		4	5.15
0680	5/4" x 8"		410	.039		3.56	1.28		4.84	6.05
0690	5/4" x 10"		390	.041		4.56	1.35		5.91	7.25
0700	5/4" x 12"	↓	390	.041	↓	5.25	1.35		6.60	8.05

06 80 Composite Fabrications

06 80 10 – Composite Decking

06 80 10.10 Woodgrained Composite Decking

		Crew	Daily Output	Labor-Hours	Unit	Material	Labor	Equipment	Total	Total Incl O&P
0010	**WOODGRAINED COMPOSITE DECKING**									
0100	Woodgrained composite decking, 1" x 6"	2 Carp	640	.025	L.F.	3.54	.82		4.36	5.25
0110	Grooved edge		660	.024		3.68	.80		4.48	5.40
0120	2" x 6"		640	.025		3.67	.82		4.49	5.40
0130	Encased, 1" x 6"		640	.025		3.67	.82		4.49	5.40
0140	Grooved edge		660	.024		3.81	.80		4.61	5.55
0150	2" x 6"	↓	640	.025	↓	4.98	.82		5.80	6.90

06 81 Composite Railings

06 81 10 – Encased Railings

06 81 10.10 Encased Composite Railings

		Crew	Daily Output	Labor-Hours	Unit	Material	Labor	Equipment	Total	Total Incl O&P
0010	**ENCASED COMPOSITE RAILINGS**									
0100	Encased composite railing, 6' long, 36" high, incl. balusters	1 Carp	16	.500	Ea.	156	16.45		172.45	199
0110	42" high, incl. balusters		16	.500		172	16.45		188.45	218
0120	8' long, 36" high, incl. balusters		12	.667		156	22		178	209
0130	42" high, incl. balusters		12	.667		172	22		194	227
0140	Accessories, post sleeve, 4" x 4", 39" long		32	.250		23	8.20		31.20	39.50
0150	96" long		24	.333		86.50	10.95		97.45	113
0160	6" x 6", 39" long		32	.250		47	8.20		55.20	65.50

06 81 Composite Railings

06 81 10 – Encased Railings

06 81 10.10 Encased Composite Railings		Crew	Daily Output	Labor-Hours	Unit	Material	2015 Bare Costs Labor	Equipment	Total	Total Incl O&P
0170	96" long	1 Carp	24	.333	Ea.	137	10.95		147.95	168
0180	Accessories, post skirt, 4" x 4"		96	.083		4	2.74		6.74	9
0190	6" x 6"		96	.083		4.85	2.74		7.59	9.95
0200	Post cap, 4" x 4", flat		48	.167		6	5.50		11.50	15.80
0210	Pyramid		48	.167		6	5.50		11.50	15.80
0220	Post cap, 6" x 6", flat		48	.167		9.60	5.50		15.10	19.75
0230	Pyramid		48	.167		9.60	5.50		15.10	19.75

Division Notes

	CREW	DAILY OUTPUT	LABOR-HOURS	UNIT	BARE COSTS				TOTAL INCL O&P
					MAT.	LABOR	EQUIP.	TOTAL	

Estimating Tips

07 10 00 Damproofing and Waterproofing

- Be sure of the job specifications before pricing this subdivision. The difference in cost between waterproofing and dampproofing can be great. Waterproofing will hold back standing water. Dampproofing prevents the transmission of water vapor. Also included in this section are vapor retarding membranes.

07 20 00 Thermal Protection

- Insulation and fireproofing products are measured by area, thickness, volume or R-value. Specifications may give only what the specific R-value should be in a certain situation. The estimator may need to choose the type of insulation to meet that R-value.

07 30 00 Steep Slope Roofing
07 40 00 Roofing and Siding Panels

- Many roofing and siding products are bought and sold by the square. One square is equal to an area that measures 100 square feet.

  This simple change in unit of measure could create a large error if the estimator is not observant. Accessories necessary for a complete installation must be figured into any calculations for both material and labor.

07 50 00 Membrane Roofing
07 60 00 Flashing and Sheet Metal
07 70 00 Roofing and Wall Specialties and Accessories

- The items in these subdivisions compose a roofing system. No one component completes the installation, and all must be estimated. Built-up or single-ply membrane roofing systems are made up of many products and installation trades. Wood blocking at roof perimeters or penetrations, parapet coverings, reglets, roof drains, gutters, downspouts, sheet metal flashing, skylights, smoke vents, and roof hatches all need to be considered along with the roofing material. Several different installation trades will need to work together on the roofing system. Inherent difficulties in the scheduling and coordination of various trades must be accounted for when estimating labor costs.

07 90 00 Joint Protection

- To complete the weather-tight shell, the sealants and caulkings must be estimated. Where different materials meet—at expansion joints, at flashing penetrations, and at hundreds of other locations throughout a construction project—they provide another line of defense against water penetration. Often, an entire system is based on the proper location and placement of caulking or sealants. The detailed drawings that are included as part of a set of architectural plans show typical locations for these materials. When caulking or sealants are shown at typical locations, this means the estimator must include them for all the locations where this detail is applicable. Be careful to keep different types of sealants separate, and remember to consider backer rods and primers if necessary.

Reference Numbers

Reference numbers are shown in shaded boxes at the beginning of some major classifications. These numbers refer to related items in the Reference Section. The reference information may be an estimating procedure, an alternate pricing method, or technical information.

Note: Not all subdivisions listed here necessarily appear in this publication. ∎

Estimate with precision with RSMeans Online...

Quick, intuitive and easy-to-use, RSMeans Online gives you instant access to hundreds of thousands of material, labor, and equipment costs from RSMeans' comprehensive database, delivering the information you need to build competitive estimates every time.

Start your free trial today at **www.rsmeansonline.com**

RSMeansOnline

07 01 Operation and Maint. of Thermal and Moisture Protection

07 01 50 – Maintenance of Membrane Roofing

07 01 50.10 Roof Coatings

		Crew	Daily Output	Labor-Hours	Unit	Material	2015 Bare Costs Labor	Equipment	Total	Total Incl O&P
0010	**ROOF COATINGS**									
0012	Asphalt, brush grade, material only				Gal.	8.95			8.95	9.80
0800	Glass fibered roof & patching cement, 5 gallon					8.25			8.25	9.10
1100	Roof patch & flashing cement, 5 gallon					8.70			8.70	9.55

07 05 Common Work Results for Thermal and Moisture Protection

07 05 05 – Selective Demolition for Thermal and Moisture Protection

07 05 05.10 Selective Demo., Thermal and Moist. Protection

			Crew	Daily Output	Labor-Hours	Unit	Material	Labor	Equipment	Total	Total Incl O&P
0010	**SELECTIVE DEMO., THERMAL AND MOISTURE PROTECTION**										
0020	Caulking/sealant, to 1" x 1" joint	R024119-10	1 Clab	600	.013	L.F.		.33		.33	.55
0120	Downspouts, including hangers			350	.023	"		.56		.56	.94
0220	Flashing, sheet metal			290	.028	S.F.		.67		.67	1.13
0420	Gutters, aluminum or wood, edge hung			240	.033	L.F.		.81		.81	1.37
0520	Built-in			100	.080	"		1.96		1.96	3.28
0620	Insulation, air/vapor barrier			3500	.002	S.F.		.06		.06	.09
0670	Batts or blankets			1400	.006	C.F.		.14		.14	.23
0720	Foamed or sprayed in place		2 Clab	1000	.016	B.F.		.39		.39	.66
0770	Loose fitting		1 Clab	3000	.003	C.F.		.07		.07	.11
0870	Rigid board			3450	.002	B.F.		.06		.06	.10
1120	Roll roofing, cold adhesive			12	.667	Sq.		16.30		16.30	27.50
1170	Roof accessories, adjustable metal chimney flashing			9	.889	Ea.		21.50		21.50	36.50
1325	Plumbing vent flashing			32	.250	"		6.10		6.10	10.25
1375	Ridge vent strip, aluminum			310	.026	L.F.		.63		.63	1.06
1620	Skylight to 10 S.F.			8	1	Ea.		24.50		24.50	41
2120	Roof edge, aluminum soffit and fascia			570	.014	L.F.		.34		.34	.58
2170	Concrete coping, up to 12" wide		2 Clab	160	.100			2.45		2.45	4.11
2220	Drip edge		1 Clab	1000	.008			.20		.20	.33
2270	Gravel stop			950	.008			.21		.21	.35
2370	Sheet metal coping, up to 12" wide			240	.033			.81		.81	1.37
2470	Roof insulation board, over 2" thick		B-2	7800	.005	B.F.		.13		.13	.21
2520	Up to 2" thick		"	3900	.010	S.F.		.26		.26	.43
2620	Roof ventilation, louvered gable vent		1 Clab	16	.500	Ea.		12.25		12.25	20.50
2670	Remove, roof hatch		G-3	15	2.133			65.50		65.50	110
2675	Rafter vents		1 Clab	960	.008			.20		.20	.34
2720	Soffit vent and/or fascia vent			575	.014	L.F.		.34		.34	.57
2775	Soffit vent strip, aluminum, 3" to 4" wide			160	.050			1.22		1.22	2.05
2820	Roofing accessories, shingle moulding, to 1" x 4"			1600	.005			.12		.12	.21
2870	Cant strip		B-2	2000	.020			.50		.50	.83
2920	Concrete block walkway		1 Clab	230	.035			.85		.85	1.43
3070	Roofing, felt paper, 15#			70	.114	Sq.		2.79		2.79	4.69
3125	#30 felt			30	.267	"		6.50		6.50	10.95
3170	Asphalt shingles, 1 layer		B-2	3500	.011	S.F.		.28		.28	.48
3180	2 layers			1750	.023	"		.57		.57	.95
3370	Modified bitumen			26	1.538	Sq.		38		38	64
3420	Built-up, no gravel, 3 ply			25	1.600			40		40	67
3470	4 ply			21	1.905			47.50		47.50	79.50
3620	5 ply			1600	.025	S.F.		.62		.62	1.04
3720	5 ply, with gravel			890	.045			1.12		1.12	1.87
3725	Loose gravel removal			5000	.008			.20		.20	.33
3730	Embedded gravel removal			2000	.020			.50		.50	.83
3870	Fiberglass sheet			1200	.033			.83		.83	1.39

07 05 Common Work Results for Thermal and Moisture Protection

07 05 05 – Selective Demolition for Thermal and Moisture Protection

	07 05 05.10 Selective Demo., Thermal and Moist. Protection	Crew	Daily Output	Labor-Hours	Unit	Material	2015 Bare Costs Labor	Equipment	Total	Total Incl O&P
4120	Slate shingles	B-2	1900	.021	S.F.		.52		.52	.88
4170	Ridge shingles, clay or slate		2000	.020	L.F.		.50		.50	.83
4320	Single ply membrane, attached at seams		52	.769	Sq.		19.10		19.10	32
4370	Ballasted		75	.533			13.25		13.25	22.50
4420	Fully adhered		39	1.026			25.50		25.50	43
4550	Roof hatch, 2'-6" x 3'-0"	1 Clab	10	.800	Ea.		19.55		19.55	33
4670	Wood shingles	B-2	2200	.018	S.F.		.45		.45	.76
4820	Sheet metal roofing	"	2150	.019			.46		.46	.78
4970	Siding, horizontal wood clapboards	1 Clab	380	.021			.51		.51	.86
5025	Exterior insulation finish system	"	120	.067			1.63		1.63	2.74
5070	Tempered hardboard, remove and reset	1 Carp	380	.021			.69		.69	1.16
5120	Tempered hardboard sheet siding	"	375	.021			.70		.70	1.18
5170	Metal, corner strips	1 Clab	850	.009	L.F.		.23		.23	.39
5225	Horizontal strips		444	.018	S.F.		.44		.44	.74
5320	Vertical strips		400	.020			.49		.49	.82
5520	Wood shingles		350	.023			.56		.56	.94
5620	Stucco siding		360	.022			.54		.54	.91
5670	Textured plywood		725	.011			.27		.27	.45
5720	Vinyl siding		510	.016			.38		.38	.64
5770	Corner strips		900	.009	L.F.		.22		.22	.36
5870	Wood, boards, vertical		400	.020	S.F.		.49		.49	.82
5920	Waterproofing, protection/drain board	2 Clab	3900	.004	B.F.		.10		.10	.17
5970	Over 1/2" thick		1750	.009	S.F.		.22		.22	.38
6020	To 1/2" thick		2000	.008	"		.20		.20	.33

07 11 Dampproofing

07 11 13 – Bituminous Dampproofing

07 11 13.10 Bituminous Asphalt Coating

		Crew	Daily Output	Labor-Hours	Unit	Material	2015 Bare Costs Labor	Equipment	Total	Total Incl O&P
0010	**BITUMINOUS ASPHALT COATING**									
0030	Brushed on, below grade, 1 coat	1 Rofc	665	.012	S.F.	.22	.33		.55	.86
0100	2 coat		500	.016		.45	.44		.89	1.31
0300	Sprayed on, below grade, 1 coat		830	.010		.22	.27		.49	.74
0400	2 coat		500	.016		.44	.44		.88	1.30
0500	Asphalt coating, with fibers				Gal.	8.25			8.25	9.10
0600	Troweled on, asphalt with fibers, 1/16" thick	1 Rofc	500	.016	S.F.	.36	.44		.80	1.22
0700	1/8" thick		400	.020		.64	.55		1.19	1.72
1000	1/2" thick		350	.023		2.07	.63		2.70	3.44

07 11 16 – Cementitious Dampproofing

07 11 16.20 Cementitious Parging

		Crew	Daily Output	Labor-Hours	Unit	Material	2015 Bare Costs Labor	Equipment	Total	Total Incl O&P
0010	**CEMENTITIOUS PARGING**									
0020	Portland cement, 2 coats, 1/2" thick	D-1	250	.064	S.F.	.33	1.89		2.22	3.51
0100	Waterproofed Portland cement, 1/2" thick, 2 coats	"	250	.064	"	3.93	1.89		5.82	7.50

07 19 Water Repellents

07 19 19 – Silicone Water Repellents

07 19 19.10 Silicone Based Water Repellents

		Crew	Daily Output	Labor-Hours	Unit	Material	2015 Bare Costs Labor	Equipment	Total	Total Incl O&P
0010	**SILICONE BASED WATER REPELLENTS**									
0020	Water base liquid, roller applied	2 Rofc	7000	.002	S.F.	.53	.06		.59	.70
0200	Silicone or stearate, sprayed on CMU, 1 coat	1 Rofc	4000	.002		.34	.06		.40	.47
0300	2 coats	"	3000	.003		.68	.07		.75	.89

07 21 Thermal Insulation

07 21 13 – Board Insulation

07 21 13.10 Rigid Insulation

			Crew	Daily Output	Labor-Hours	Unit	Material	Labor	Equipment	Total	Total Incl O&P
0010	**RIGID INSULATION**, for walls										
0040	Fiberglass, 1.5#/C.F., unfaced, 1" thick, R4.1	G	1 Carp	1000	.008	S.F.	.27	.26		.53	.74
0060	1-1/2" thick, R6.2	G		1000	.008		.40	.26		.66	.88
0080	2" thick, R8.3	G		1000	.008		.45	.26		.71	.94
0120	3" thick, R12.4	G		800	.010		.56	.33		.89	1.17
0370	3#/C.F., unfaced, 1" thick, R4.3	G		1000	.008		.52	.26		.78	1.01
0390	1-1/2" thick, R6.5	G		1000	.008		.78	.26		1.04	1.30
0400	2" thick, R8.7	G		890	.009		1.05	.30		1.35	1.66
0420	2-1/2" thick, R10.9	G		800	.010		1.10	.33		1.43	1.76
0440	3" thick, R13	G		800	.010		1.59	.33		1.92	2.30
0520	Foil faced, 1" thick, R4.3	G		1000	.008		.90	.26		1.16	1.43
0540	1-1/2" thick, R6.5	G		1000	.008		1.35	.26		1.61	1.93
0560	2" thick, R8.7	G		890	.009		1.69	.30		1.99	2.36
0580	2-1/2" thick, R10.9	G		800	.010		1.98	.33		2.31	2.73
0600	3" thick, R13	G		800	.010		2.18	.33		2.51	2.95
1600	Isocyanurate, 4' x 8' sheet, foil faced, both sides										
1610	1/2" thick	G	1 Carp	800	.010	S.F.	.31	.33		.64	.89
1620	5/8" thick	G		800	.010		.33	.33		.66	.91
1630	3/4" thick	G		800	.010		.36	.33		.69	.95
1640	1" thick	G		800	.010		.52	.33		.85	1.12
1650	1-1/2" thick	G		730	.011		.63	.36		.99	1.29
1660	2" thick	G		730	.011		.80	.36		1.16	1.48
1670	3" thick	G		730	.011		1.80	.36		2.16	2.58
1680	4" thick	G		730	.011		2.05	.36		2.41	2.86
1700	Perlite, 1" thick, R2.77	G		800	.010		.42	.33		.75	1.01
1750	2" thick, R5.55	G		730	.011		.75	.36		1.11	1.43
1900	Extruded polystyrene, 25 PSI compressive strength, 1" thick, R5	G		800	.010		.53	.33		.86	1.13
1940	2" thick R10	G		730	.011		1.04	.36		1.40	1.74
1960	3" thick, R15	G		730	.011		1.50	.36		1.86	2.25
2100	Expanded polystyrene, 1" thick, R3.85	G		800	.010		.25	.33		.58	.83
2120	2" thick, R7.69	G		730	.011		.50	.36		.86	1.15
2140	3" thick, R11.49	G		730	.011		.75	.36		1.11	1.43

07 21 13.13 Foam Board Insulation

			Crew	Daily Output	Labor-Hours	Unit	Material	Labor	Equipment	Total	Total Incl O&P
0010	**FOAM BOARD INSULATION**										
0600	Polystyrene, expanded, 1" thick, R4	G	1 Carp	680	.012	S.F.	.25	.39		.64	.93
0700	2" thick, R8	G	"	675	.012	"	.50	.39		.89	1.20

07 21 16 – Blanket Insulation

07 21 16.10 Blanket Insulation for Floors/Ceilings

			Crew	Daily Output	Labor-Hours	Unit	Material	Labor	Equipment	Total	Total Incl O&P
0010	**BLANKET INSULATION FOR FLOORS/CEILINGS**										
0020	Including spring type wire fasteners										
2000	Fiberglass, blankets or batts, paper or foil backing										
2100	3-1/2" thick, R13	G	1 Carp	700	.011	S.F.	.37	.38		.75	1.04

07 21 Thermal Insulation

07 21 16 – Blanket Insulation

07 21 16.10 Blanket Insulation for Floors/Ceilings

			Crew	Daily Output	Labor-Hours	Unit	Material	2015 Bare Costs Labor	Equipment	Total	Total Incl O&P
2150	6-1/4" thick, R19	G	1 Carp	600	.013	S.F.	.49	.44		.93	1.28
2210	9-1/2" thick, R30	G		500	.016		.71	.53		1.24	1.66
2220	12" thick, R38	G		475	.017		.96	.55		1.51	1.99
3000	Unfaced, 3-1/2" thick, R13	G		600	.013		.31	.44		.75	1.08
3010	6-1/4" thick, R19	G		500	.016		.36	.53		.89	1.28
3020	9-1/2" thick, R30	G		450	.018		.58	.58		1.16	1.62
3030	12" thick, R38	G		425	.019		.74	.62		1.36	1.85

07 21 16.20 Blanket Insulation for Walls

			Crew	Daily Output	Labor-Hours	Unit	Material	Labor	Equipment	Total	Total Incl O&P
0010	**BLANKET INSULATION FOR WALLS**										
0020	Kraft faced fiberglass, 3-1/2" thick, R11, 15" wide	G	1 Carp	1350	.006	S.F.	.27	.19		.46	.63
0030	23" wide	G		1600	.005		.27	.16		.43	.58
0060	R13, 11" wide	G		1150	.007		.30	.23		.53	.71
0080	15" wide	G		1350	.006		.30	.19		.49	.66
0100	23" wide	G		1600	.005		.30	.16		.46	.61
0110	R15, 11" wide	G		1150	.007		.45	.23		.68	.88
0120	15" wide	G		1350	.006		.45	.19		.64	.83
0140	6" thick, R19, 11" wide	G		1150	.007		.42	.23		.65	.84
0160	15" wide	G		1350	.006		.42	.19		.61	.79
0180	23" wide	G		1600	.005		.42	.16		.58	.74
0201	9" thick, R30, 15" wide	G		1350	.006		.71	.19		.90	1.11
0241	12" thick, R38, 15" wide	G		1350	.006		.96	.19		1.15	1.39
0410	Foil faced fiberglass, 3-1/2" thick, R13, 11" wide	G		1150	.007		.45	.23		.68	.88
0420	15" wide	G		1350	.006		.45	.19		.64	.83
0442	R15, 11" wide	G		1150	.007		.47	.23		.70	.90
0444	15" wide	G		1350	.006		.47	.19		.66	.85
0461	6" thick, R19, 15" wide	G		1600	.005		.60	.16		.76	.94
0482	R21, 11" wide	G		1150	.007		.62	.23		.85	1.06
0501	9" thick, R30, 15" wide	G		1350	.006		.90	.19		1.09	1.32
0620	Unfaced fiberglass, 3-1/2" thick, R13, 11" wide	G		1150	.007		.31	.23		.54	.72
0821	3-1/2" thick, R13, 15" wide	G		1600	.005		.31	.16		.47	.62
0832	R15, 11" wide	G		1150	.007		.42	.23		.65	.84
0834	15" wide	G		1350	.006		.42	.19		.61	.79
0861	6" thick, R19, 15" wide	G		1350	.006		.36	.19		.55	.73
0901	9" thick, R30, 15" wide	G		1150	.007		.58	.23		.81	1.02
0941	12" thick, R38, 15" wide	G		1150	.007		.74	.23		.97	1.19
1300	Wall or ceiling insulation, mineral wool batts										
1320	3-1/2" thick, R15	G	1 Carp	1600	.005	S.F.	.60	.16		.76	.94
1340	5-1/2" thick, R23	G		1600	.005		.94	.16		1.10	1.32
1380	7-1/4" thick, R30	G		1350	.006		1.24	.19		1.43	1.70
1700	Non-rigid insul, recycled blue cotton fiber, unfaced batts, R13, 16" wide	G		1600	.005		1.01	.16		1.17	1.39
1710	R19, 16" wide	G		1600	.005		1.38	.16		1.54	1.80
1850	Friction fit wire insulation supports, 16" O.C.			960	.008	Ea.	.07	.27		.34	.54

07 21 19 – Foamed In Place Insulation

07 21 19.10 Masonry Foamed In Place Insulation

			Crew	Daily Output	Labor-Hours	Unit	Material	Labor	Equipment	Total	Total Incl O&P
0010	**MASONRY FOAMED IN PLACE INSULATION**										
0100	Amino-plast foam, injected into block core, 6" block	G	G-2A	6000	.004	Ea.	.15	.10	.12	.37	.47
0110	8" block	G		5000	.005		.19	.12	.14	.45	.57
0120	10" block	G		4000	.006		.23	.15	.18	.56	.72
0130	12" block	G		3000	.008		.31	.19	.24	.74	.95
0140	Injected into cavity wall	G		13000	.002	B.F.	.05	.04	.06	.15	.20
0150	Preparation, drill holes into mortar joint every 4 VLF, 5/8" dia		1 Clab	960	.008	Ea.		.20		.20	.34
0160	7/8" dia			680	.012			.29		.29	.48

07 21 Thermal Insulation

07 21 19 – Foamed In Place Insulation

07 21 19.10 Masonry Foamed In Place Insulation		Crew	Daily Output	Labor-Hours	Unit	Material	2015 Bare Costs Labor	Equipment	Total	Total Incl O&P
0170	Patch drilled holes, 5/8" diameter	1 Clab	1800	.004	Ea.	.03	.11		.14	.22
0180	7/8" diameter		1200	.007		.05	.16		.21	.32

07 21 23 – Loose-Fill Insulation

07 21 23.10 Poured Loose-Fill Insulation

			Crew	Daily Output	Labor-Hours	Unit	Material	Labor	Equipment	Total	Total Incl O&P
0010	**POURED LOOSE-FILL INSULATION**										
0020	Cellulose fiber, R3.8 per inch	G	1 Carp	200	.040	C.F.	.69	1.31		2	2.97
0021	4" thick	G		1000	.008	S.F.	.17	.26		.43	.62
0022	6" thick	G		800	.010	"	.28	.33		.61	.86
0080	Fiberglass wool, R4 per inch	G		200	.040	C.F.	.55	1.31		1.86	2.81
0081	4" thick	G		600	.013	S.F.	.19	.44		.63	.95
0082	6" thick	G		400	.020	"	.26	.66		.92	1.39
0100	Mineral wool, R3 per inch	G		200	.040	C.F.	.41	1.31		1.72	2.66
0101	4" thick	G		600	.013	S.F.	.14	.44		.58	.89
0102	6" thick	G		400	.020	"	.21	.66		.87	1.33
0300	Polystyrene, R4 per inch	G		200	.040	C.F.	1.60	1.31		2.91	3.97
0301	4" thick	G		600	.013	S.F.	.53	.44		.97	1.32
0302	6" thick	G		400	.020	"	.80	.66		1.46	1.98
0400	Perlite, R2.78 per inch	G		200	.040	C.F.	5.20	1.31		6.51	7.95
0401	4" thick	G		1000	.008	S.F.	1.73	.26		1.99	2.35
0402	6" thick	G		800	.010	"	2.61	.33		2.94	3.42

07 21 23.20 Masonry Loose-Fill Insulation

			Crew	Daily Output	Labor-Hours	Unit	Material	Labor	Equipment	Total	Total Incl O&P
0010	**MASONRY LOOSE-FILL INSULATION**, vermiculite or perlite										
0100	In cores of concrete block, 4" thick wall, .115 C.F./S.F.	G	D-1	4800	.003	S.F.	.60	.10		.70	.82
0700	Foamed in place, urethane in 2-5/8" cavity	G	G-2A	1035	.023		1.38	.56	.69	2.63	3.29
0800	For each 1" added thickness, add	G	"	2372	.010		.53	.25	.30	1.08	1.35

07 21 26 – Blown Insulation

07 21 26.10 Blown Insulation

			Crew	Daily Output	Labor-Hours	Unit	Material	Labor	Equipment	Total	Total Incl O&P
0010	**BLOWN INSULATION** Ceilings, with open access										
0020	Cellulose, 3-1/2" thick, R13	G	G-4	5000	.005	S.F.	.24	.12	.08	.44	.55
0030	5-3/16" thick, R19	G		3800	.006		.35	.16	.11	.62	.78
0050	6-1/2" thick, R22	G		3000	.008		.45	.20	.13	.78	.99
0100	8-11/16" thick, R30	G		2600	.009		.61	.23	.15	.99	1.23
0120	10-7/8" thick, R38	G		1800	.013		.78	.33	.22	1.33	1.66
1000	Fiberglass, 5.5" thick, R11	G		3800	.006		.19	.16	.11	.46	.59
1050	6" thick, R12	G		3000	.008		.26	.20	.13	.59	.78
1100	8.8" thick, R19	G		2200	.011		.33	.27	.18	.78	1.02
1200	10" thick, R22	G		1800	.013		.38	.33	.22	.93	1.23
1300	11.5" thick, R26	G		1500	.016		.46	.40	.27	1.13	1.46
1350	13" thick, R30	G		1400	.017		.53	.43	.29	1.25	1.62
1450	16" thick, R38	G		1145	.021		.68	.53	.35	1.56	2.01
1500	20" thick, R49	G		920	.026		.89	.66	.44	1.99	2.56

07 21 27 – Reflective Insulation

07 21 27.10 Reflective Insulation Options

			Crew	Daily Output	Labor-Hours	Unit	Material	Labor	Equipment	Total	Total Incl O&P
0010	**REFLECTIVE INSULATION OPTIONS**										
0020	Aluminum foil on reinforced scrim	G	1 Carp	19	.421	C.S.F.	14.20	13.85		28.05	38.50
0100	Reinforced with woven polyolefin	G		19	.421		22	13.85		35.85	47
0500	With single bubble air space, R8.8	G		15	.533		28	17.50		45.50	60.50
0600	With double bubble air space, R9.8	G		15	.533		32	17.50		49.50	64.50

07 21 Thermal Insulation

07 21 29 – Sprayed Insulation

07 21 29.10 Sprayed-On Insulation

			Crew	Daily Output	Labor-Hours	Unit	Material	2015 Bare Costs Labor	Equipment	Total	Total Incl O&P
0010	**SPRAYED-ON INSULATION**										
0300	Closed cell, spray polyurethane foam, 2 pounds per cubic foot density										
0310	1" thick	G	G-2A	6000	.004	S.F.	.53	.10	.12	.75	.88
0320	2" thick	G		3000	.008		1.05	.19	.24	1.48	1.77
0330	3" thick	G		2000	.012		1.58	.29	.36	2.23	2.64
0335	3-1/2" thick	G		1715	.014		1.84	.34	.42	2.60	3.09
0340	4" thick	G		1500	.016		2.10	.39	.48	2.97	3.54
0350	5" thick	G		1200	.020		2.63	.49	.60	3.72	4.42
0355	5-1/2" thick	G		1090	.022		2.89	.53	.66	4.08	4.86
0360	6" thick	G		1000	.024		3.15	.58	.72	4.45	5.30

07 22 Roof and Deck Insulation

07 22 16 – Roof Board Insulation

07 22 16.10 Roof Deck Insulation

			Crew	Daily Output	Labor-Hours	Unit	Material	Labor	Equipment	Total	Total Incl O&P
0010	**ROOF DECK INSULATION**, fastening excluded										
0016	Asphaltic cover board, fiberglass lined, 1/8" thick		1 Rofc	1400	.006	S.F.	.47	.16		.63	.81
0018	1/4" thick			1400	.006		.94	.16		1.10	1.32
0020	Fiberboard low density, 1/2" thick R1.39	G		1300	.006		.30	.17		.47	.64
0030	1" thick R2.78	G		1040	.008		.52	.21		.73	.96
0080	1-1/2" thick R4.17	G		1040	.008		.80	.21		1.01	1.27
0100	2" thick R5.56	G		1040	.008		1.06	.21		1.27	1.56
0110	Fiberboard high density, 1/2" thick R1.3	G		1300	.006		.30	.17		.47	.64
0120	1" thick R2.5	G		1040	.008		.58	.21		.79	1.03
0130	1-1/2" thick R3.8	G		1040	.008		.88	.21		1.09	1.36
0200	Fiberglass, 3/4" thick R2.78	G		1300	.006		.61	.17		.78	.98
0400	15/16" thick R3.70	G		1300	.006		.81	.17		.98	1.20
0460	1-1/16" thick R4.17	G		1300	.006		1.01	.17		1.18	1.42
0600	1-5/16" thick R5.26	G		1300	.006		1.37	.17		1.54	1.82
0650	2-1/16" thick R8.33	G		1040	.008		1.45	.21		1.66	1.99
0700	2-7/16" thick R10	G		1040	.008		1.68	.21		1.89	2.24
0800	Gypsum cover board, fiberglass mat facer, 1/4" thick			1400	.006		.47	.16		.63	.81
0810	1/2" thick			1300	.006		.57	.17		.74	.94
0820	5/8" thick			1200	.007		.61	.18		.79	1.01
0830	Primed fiberglass mat facer, 1/4" thick			1400	.006		.51	.16		.67	.85
0840	1/2" thick			1300	.006		.62	.17		.79	.99
0850	5/8" thick			1200	.007		.65	.18		.83	1.06
1650	Perlite, 1/2" thick R1.32	G		1365	.006		.28	.16		.44	.61
1655	3/4" thick R2.08	G		1040	.008		.34	.21		.55	.76
1660	1" thick R2.78	G		1040	.008		.50	.21		.71	.94
1670	1-1/2" thick R4.17	G		1040	.008		.73	.21		.94	1.19
1680	2" thick R5.56	G		910	.009		1	.24		1.24	1.55
1685	2-1/2" thick R6.67	G		910	.009		1.30	.24		1.54	1.88
1690	Tapered for drainage	G		1040	.008	B.F.	1.01	.21		1.22	1.50
1700	Polyisocyanurate, 2#/C.F. density, 3/4" thick	G		1950	.004	S.F.	.46	.11		.57	.72
1705	1" thick	G		1820	.004		.48	.12		.60	.75
1715	1-1/2" thick	G		1625	.005		.64	.14		.78	.95
1725	2" thick	G		1430	.006		.83	.15		.98	1.20
1735	2-1/2" thick	G		1365	.006		1.06	.16		1.22	1.47
1745	3" thick	G		1300	.006		1.27	.17		1.44	1.71
1755	3-1/2" thick	G		1300	.006		1.95	.17		2.12	2.46

07 22 Roof and Deck Insulation

07 22 16 – Roof Board Insulation

	07 22 16.10 Roof Deck Insulation		Crew	Daily Output	Labor-Hours	Unit	Material	2015 Bare Costs Labor	Equipment	Total	Total Incl O&P
1765	Tapered for drainage	G	1 Rofc	1820	.004	B.F.	.64	.12		.76	.92
1900	Extruded Polystyrene										
1910	15 PSI compressive strength, 1" thick, R5	G	1 Rofc	1950	.004	S.F.	.55	.11		.66	.82
1920	2" thick, R10	G		1625	.005		.72	.14		.86	1.04
1930	3" thick, R15	G		1300	.006		1.43	.17		1.60	1.88
1932	4" thick, R20	G		1300	.006		1.93	.17		2.10	2.43
1934	Tapered for drainage	G		1950	.004	B.F.	.59	.11		.70	.86
1940	25 PSI compressive strength, 1" thick, R5	G		1950	.004	S.F.	.69	.11		.80	.97
1942	2" thick, R10	G		1625	.005		1.31	.14		1.45	1.69
1944	3" thick, R15	G		1300	.006		2	.17		2.17	2.51
1946	4" thick, R20	G		1300	.006		2.76	.17		2.93	3.35
1948	Tapered for drainage	G		1950	.004	B.F.	.61	.11		.72	.88
1950	40 psi compressive strength, 1" thick, R5	G		1950	.004	S.F.	.53	.11		.64	.79
1952	2" thick, R10	G		1625	.005		1.01	.14		1.15	1.36
1954	3" thick, R15	G		1300	.006		1.46	.17		1.63	1.91
1956	4" thick, R20	G		1300	.006		1.91	.17		2.08	2.41
1958	Tapered for drainage	G		1820	.004	B.F.	.76	.12		.88	1.06
1960	60 PSI compressive strength, 1" thick, R5	G		1885	.004	S.F.	.74	.12		.86	1.03
1962	2" thick, R10	G		1560	.005		1.41	.14		1.55	1.81
1964	3" thick, R15	G		1270	.006		2.29	.17		2.46	2.84
1966	4" thick, R20	G		1235	.006		2.85	.18		3.03	3.46
1968	Tapered for drainage	G		1820	.004	B.F.	.96	.12		1.08	1.28
2010	Expanded polystyrene, 1#/C.F. density, 3/4" thick, R2.89	G		1950	.004	S.F.	.19	.11		.30	.42
2020	1" thick, R3.85	G		1950	.004		.25	.11		.36	.49
2100	2" thick, R7.69	G		1625	.005		.50	.14		.64	.80
2110	3" thick, R11.49	G		1625	.005		.75	.14		.89	1.08
2120	4" thick, R15.38	G		1625	.005		1	.14		1.14	1.35
2130	5" thick, R19.23	G		1495	.005		1.25	.15		1.40	1.65
2140	6" thick, R23.26	G		1495	.005		1.50	.15		1.65	1.92
2150	Tapered for drainage	G		1950	.004	B.F.	.53	.11		.64	.79
2400	Composites with 2" EPS										
2410	1" fiberboard	G	1 Rofc	1325	.006	S.F.	1.40	.17		1.57	1.85
2420	7/16" oriented strand board	G		1040	.008		1.13	.21		1.34	1.63
2430	1/2" plywood	G		1040	.008		1.39	.21		1.60	1.92
2440	1" perlite	G		1040	.008		1.14	.21		1.35	1.64
2450	Composites with 1-1/2" polyisocyanurate										
2460	1" fiberboard	G	1 Rofc	1040	.008	S.F.	1.19	.21		1.40	1.70
2470	1" perlite	G		1105	.007		1.06	.20		1.26	1.54
2480	7/16" oriented strand board	G		1040	.008		.93	.21		1.14	1.41
3000	Fastening alternatives, coated screws, 2" long			3744	.002	Ea.	.05	.06		.11	.17
3010	4" long			3120	.003		.10	.07		.17	.24
3020	6" long			2675	.003		.17	.08		.25	.34
3030	8" long			2340	.003		.25	.09		.34	.45
3040	10" long			1872	.004		.43	.12		.55	.69
3050	Pre-drill and drive wedge spike, 2-1/2"			1248	.006		.37	.18		.55	.74
3060	3-1/2"			1101	.007		.48	.20		.68	.90
3070	4-1/2"			936	.009		.60	.24		.84	1.10
3075	3" galvanized deck plates			7488	.001		.07	.03		.10	.13
3080	Spot mop asphalt		G-1	295	.190	Sq.	5.55	4.93	1.93	12.41	17.30
3090	Full mop asphalt		"	192	.292		11.10	7.55	2.96	21.61	29.50
3110	Low-rise polyurethane adhesive, from 5 gallon kit, 12" OC beads		1 Rofc	45	.178		33	4.92		37.92	45
3120	6" OC beads			32	.250		65.50	6.90		72.40	85
3130	4" OC beads			30	.267		98.50	7.35		105.85	122

07 24 Exterior Insulation and Finish Systems

07 24 13 – Polymer-Based Exterior Insulation and Finish System

07 24 13.10 Exterior Insulation and Finish Systems

		Crew	Daily Output	Labor-Hours	Unit	Material	2015 Bare Costs Labor	Equipment	Total	Total Incl O&P
0010	**EXTERIOR INSULATION AND FINISH SYSTEMS**									
0095	Field applied, 1" EPS insulation	G	J-1	390	.103	S.F.	1.91	2.94	.36	5.21	7.35
0100	With 1/2" cement board sheathing	G		268	.149		2.65	4.28	.52	7.45	10.50
0105	2" EPS insulation	G		390	.103		2.16	2.94	.36	5.46	7.60
0110	With 1/2" cement board sheathing	G		268	.149		2.90	4.28	.52	7.70	10.75
0115	3" EPS insulation	G		390	.103		2.41	2.94	.36	5.71	7.90
0120	With 1/2" cement board sheathing	G		268	.149		3.15	4.28	.52	7.95	11.05
0125	4" EPS insulation	G		390	.103		2.66	2.94	.36	5.96	8.15
0130	With 1/2" cement board sheathing	G		268	.149		4.14	4.28	.52	8.94	12.15
0140	Premium finish add			1265	.032		.33	.91	.11	1.35	1.97
0150	Heavy duty reinforcement add			914	.044		.83	1.26	.15	2.24	3.14
0160	2.5#/S.Y. metal lath substrate add		1 Lath	75	.107	S.Y.	2.72	3.21		5.93	8.20
0170	3.4#/S.Y. metal lath substrate add		"	75	.107	"	4.13	3.21		7.34	9.75
0180	Color or texture change,		J-1	1265	.032	S.F.	.77	.91	.11	1.79	2.46
0190	With substrate leveling base coat		1 Plas	530	.015		.77	.45		1.22	1.59
0210	With substrate sealing base coat		1 Pord	1224	.007		.10	.19		.29	.42
0370	V groove shape in panel face					L.F.	.62			.62	.68
0380	U groove shape in panel face					"	.80			.80	.88

07 25 Weather Barriers

07 25 10 – Weather Barriers or Wraps

07 25 10.10 Weather Barriers

		Crew	Daily Output	Labor-Hours	Unit	Material	Labor	Equipment	Total	Total Incl O&P
0010	**WEATHER BARRIERS**									
0400	Asphalt felt paper, 15#	1 Carp	37	.216	Sq.	5.40	7.10		12.50	17.85
0401	Per square foot	"	3700	.002	S.F.	.05	.07		.12	.18
0450	Housewrap, exterior, spun bonded polypropylene									
0470	Small roll	1 Carp	3800	.002	S.F.	.15	.07		.22	.28
0480	Large roll	"	4000	.002	"	.14	.07		.21	.26
2100	Asphalt felt roof deck vapor barrier, class 1 metal decks	1 Rofc	37	.216	Sq.	22	6		28	35
2200	For all other decks	"	37	.216		16.50	6		22.50	29
2800	Asphalt felt, 50% recycled content, 15 lb., 4 sq. per roll	1 Carp	36	.222		5.40	7.30		12.70	18.20
2810	30 lb., 2 sq. per roll	"	36	.222		10.80	7.30		18.10	24
3000	Building wrap, spunbonded polyethylene	2 Carp	8000	.002	S.F.	.15	.07		.22	.28

07 26 Vapor Retarders

07 26 10 – Above-Grade Vapor Retarders

07 26 10.10 Building Paper

			Crew	Daily Output	Labor-Hours	Unit	Material	Labor	Equipment	Total	Total Incl O&P
0011	**BUILDING PAPER** Aluminum and kraft laminated, foil 1 side		1 Carp	3700	.002	S.F.	.13	.07		.20	.26
0101	Foil 2 sides	G		3700	.002		.14	.07		.21	.27
0601	Polyethylene vapor barrier, standard, 2 mil	G		3700	.002		.02	.07		.09	.14
0701	4 mil	G		3700	.002		.03	.07		.10	.15
0901	6 mil	G		3700	.002		.04	.07		.11	.16
1201	10 mil	G		3700	.002		.09	.07		.16	.22
1801	Reinf. waterproof, 2 mil polyethylene backing, 1 side			3700	.002		.06	.07		.13	.19
1901	2 sides			3700	.002		.08	.07		.15	.21

07 27 Air Barriers

07 27 26 – Fluid-Applied Membrane Air Barriers

07 27 26.10 Fluid Applied Membrane Air Barrier		Crew	Daily Output	Labor-Hours	Unit	Material	2015 Bare Costs Labor	Equipment	Total	Total Incl O&P
0010	**FLUID APPLIED MEMBRANE AIR BARRIER**									
0100	Spray applied vapor barrier, 25 S.F./gallon	1 Pord	1375	.006	S.F.	.01	.17		.18	.29

07 31 Shingles and Shakes

07 31 13 – Asphalt Shingles

07 31 13.10 Asphalt Roof Shingles		Crew	Daily Output	Labor-Hours	Unit	Material	Labor	Equipment	Total	Total Incl O&P
0010	**ASPHALT ROOF SHINGLES**									
0100	Standard strip shingles									
0150	Inorganic, class A, 25 year	1 Rofc	5.50	1.455	Sq.	79.50	40		119.50	162
0155	Pneumatic nailed		7	1.143		79.50	31.50		111	146
0200	30 year		5	1.600		94	44		138	185
0205	Pneumatic nailed		6.25	1.280		94	35.50		129.50	169
0250	Standard laminated multi-layered shingles									
0300	Class A, 240-260 lb./square	1 Rofc	4.50	1.778	Sq.	110	49		159	212
0305	Pneumatic nailed		5.63	1.422		110	39.50		149.50	194
0350	Class A, 250-270 lb./square		4	2		110	55.50		165.50	223
0355	Pneumatic nailed		5	1.600		110	44		154	203
0400	Premium, laminated multi-layered shingles									
0450	Class A, 260-300 lb./square	1 Rofc	3.50	2.286	Sq.	153	63		216	285
0455	Pneumatic nailed		4.37	1.831		153	50.50		203.50	262
0500	Class A, 300-385 lb./square		3	2.667		230	73.50		303.50	390
0505	Pneumatic nailed		3.75	2.133		230	59		289	360
0800	#15 felt underlayment		64	.125		5.40	3.46		8.86	12.35
0825	#30 felt underlayment		58	.138		10.60	3.81		14.41	18.70
0850	Self adhering polyethylene and rubberized asphalt underlayment		22	.364		75	10.05		85.05	101
0900	Ridge shingles		330	.024	L.F.	2.10	.67		2.77	3.55
0905	Pneumatic nailed		412.50	.019	"	2.10	.54		2.64	3.30
1000	For steep roofs (7 to 12 pitch or greater), add						50%			

07 31 26 – Slate Shingles

07 31 26.10 Slate Roof Shingles		Crew	Daily Output	Labor-Hours	Unit	Material	Labor	Equipment	Total	Total Incl O&P
0010	**SLATE ROOF SHINGLES** R073126-20									
0100	Buckingham Virginia black, 3/16" - 1/4" thick [G]	1 Rots	1.75	4.571	Sq.	470	126		596	750
0200	1/4" thick [G]		1.75	4.571		470	126		596	750
0900	Pennsylvania black, Bangor, #1 clear [G]		1.75	4.571		490	126		616	775
1200	Vermont, unfading, green, mottled green [G]		1.75	4.571		475	126		601	760
1300	Semi-weathering green & gray [G]		1.75	4.571		345	126		471	615
1400	Purple [G]		1.75	4.571		425	126		551	700
1500	Black or gray [G]		1.75	4.571		455	126		581	735
2700	Ridge shingles, slate		200	.040	L.F.	10	1.11		11.11	13.05

07 31 29 – Wood Shingles and Shakes

07 31 29.13 Wood Shingles

		Crew	Daily Output	Labor-Hours	Unit	Material	Labor	Equipment	Total	Total Incl O&P
0010	**WOOD SHINGLES**									
0012	16" No. 1 red cedar shingles, 5" exposure, on roof	1 Carp	2.50	3.200	Sq.	287	105		392	490
0015	Pneumatic nailed		3.25	2.462		287	81		368	450
0200	7-1/2" exposure, on walls		2.05	3.902		191	128		319	425
0205	Pneumatic nailed		2.67	2.996		191	98.50		289.50	375
0300	18" No. 1 red cedar perfections, 5-1/2" exposure, on roof		2.75	2.909		246	95.50		341.50	430
0305	Pneumatic nailed		3.57	2.241		246	73.50		319.50	395
0500	7-1/2" exposure, on walls		2.25	3.556		181	117		298	395
0505	Pneumatic nailed		2.92	2.740		181	90		271	350

07 31 Shingles and Shakes

07 31 29 – Wood Shingles and Shakes

07 31 29.13 Wood Shingles

		Crew	Daily Output	Labor-Hours	Unit	Material	2015 Bare Costs Labor	2015 Bare Costs Equipment	Total	Total Incl O&P
0600	Resquared, and rebutted, 5-1/2" exposure, on roof	1 Carp	3	2.667	Sq.	279	87.50		366.50	450
0605	Pneumatic nailed		3.90	2.051		279	67.50		346.50	420
0900	7-1/2" exposure, on walls		2.45	3.265		205	107		312	405
0905	Pneumatic nailed		3.18	2.516		205	82.50		287.50	365
1000	Add to above for fire retardant shingles					56			56	61.50
1060	Preformed ridge shingles	1 Carp	400	.020	L.F.	5	.66		5.66	6.60
2000	White cedar shingles, 16" long, extras, 5" exposure, on roof		2.40	3.333	Sq.	192	110		302	395
2005	Pneumatic nailed		3.12	2.564		192	84		276	350
2050	5" exposure on walls		2	4		192	131		323	430
2055	Pneumatic nailed		2.60	3.077		192	101		293	380
2100	7-1/2" exposure, on walls		2	4		137	131		268	370
2105	Pneumatic nailed		2.60	3.077		137	101		238	320
2150	"B" grade, 5" exposure on walls		2	4		165	131		296	405
2155	Pneumatic nailed		2.60	3.077		165	101		266	350
2300	For 15# organic felt underlayment on roof, 1 layer, add		64	.125		5.40	4.11		9.51	12.85
2400	2 layers, add		32	.250		10.80	8.20		19	25.50
2600	For steep roofs (7/12 pitch or greater), add to above						50%			
2700	Panelized systems, No.1 cedar shingles on 5/16" CDX plywood									
2800	On walls, 8' strips, 7" or 14" exposure	2 Carp	700	.023	S.F.	6	.75		6.75	7.85
3000	Ridge shakes or shingle wood	1 Carp	280	.029	L.F.	4.50	.94		5.44	6.55

07 31 29.16 Wood Shakes

		Crew	Daily Output	Labor-Hours	Unit	Material	Labor	Equipment	Total	Total Incl O&P
0010	**WOOD SHAKES**									
1100	Hand-split red cedar shakes, 1/2" thick x 24" long, 10" exp. on roof	1 Carp	2.50	3.200	Sq.	272	105		377	475
1105	Pneumatic nailed		3.25	2.462		272	81		353	435
1110	3/4" thick x 24" long, 10" exp. on roof		2.25	3.556		272	117		389	495
1115	Pneumatic nailed		2.92	2.740		272	90		362	450
1200	1/2" thick, 18" long, 8-1/2" exp. on roof		2	4		180	131		311	420
1205	Pneumatic nailed		2.60	3.077		180	101		281	370
1210	3/4" thick x 18" long, 8 1/2" exp. on roof		1.80	4.444		180	146		326	445
1215	Pneumatic nailed		2.34	3.419		180	112		292	385
1255	10" exp. on walls		2	4		174	131		305	410
1260	10" exposure on walls, pneumatic nailed		2.60	3.077		174	101		275	360
1700	Add to above for fire retardant shakes, 24" long					56			56	61.50
1800	18" long					56			56	61.50
1810	Ridge shakes	1 Carp	350	.023	L.F.	5	.75		5.75	6.75

07 32 Roof Tiles

07 32 13 – Clay Roof Tiles

07 32 13.10 Clay Tiles

		Crew	Daily Output	Labor-Hours	Unit	Material	Labor	Equipment	Total	Total Incl O&P
0010	**CLAY TILES**, including accessories									
0300	Flat shingle, interlocking, 15", 166 pcs/sq, fireflashed blend	3 Rots	6	4	Sq.	505	111		616	760
0500	Terra cotta red		6	4		510	111		621	765
0600	Roman pan and top, 18", 102 pcs/sq, fireflashed blend		5.50	4.364		555	121		676	835
0640	Terra cotta red	1 Rots	2.40	3.333		555	92		647	780
1100	Barrel mission tile, 18", 166 pcs/sq, fireflashed blend	3 Rots	5.50	4.364		410	121		531	675
1140	Terra cotta red		5.50	4.364		410	121		531	675
1700	Scalloped edge flat shingle, 14", 145 pcs/sq, fireflashed blend		6	4		1,125	111		1,236	1,450
1800	Terra cotta red		6	4		1,100	111		1,211	1,425
3010	#15 felt underlayment	1 Rofc	64	.125		5.40	3.46		8.86	12.35
3020	#30 felt underlayment		58	.138		10.60	3.81		14.41	18.70
3040	Polyethylene and rubberized asph. underlayment		22	.364		75	10.05		85.05	101

07 32 Roof Tiles

07 32 16 – Concrete Roof Tiles

07 32 16.10 Concrete Tiles		Crew	Daily Output	Labor-Hours	Unit	Material	2015 Bare Costs Labor	Equipment	Total	Total Incl O&P
0010	**CONCRETE TILES**									
0020	Corrugated, 13" x 16-1/2", 90 per sq., 950 lb. per sq.									
0050	Earthtone colors, nailed to wood deck	1 Rots	1.35	5.926	Sq.	106	164		270	420
0150	Blues		1.35	5.926		106	164		270	420
0200	Greens		1.35	5.926		106	164		270	420
0250	Premium colors		1.35	5.926		106	164		270	420
0500	Shakes, 13" x 16-1/2", 90 per sq., 950 lb. per sq.									
0600	All colors, nailed to wood deck	1 Rots	1.50	5.333	Sq.	108	147		255	390
1500	Accessory pieces, ridge & hip, 10" x 16-1/2", 8 lb. each	"	120	.067	Ea.	3.20	1.84		5.04	6.90
1700	Rake, 6-1/2" x 16-3/4", 9 lb. each					3.20			3.20	3.52
1800	Mansard hip, 10" x 16-1/2", 9.2 lb. each					3.20			3.20	3.52
1900	Hip starter, 10" x 16-1/2", 10.5 lb. each					9.90			9.90	10.90
2000	3 or 4 way apex, 10" each side, 11.5 lb. each					11.40			11.40	12.55

07 33 Natural Roof Coverings

07 33 63 – Vegetated Roofing

07 33 63.10 Green Roof Systems			Crew	Daily Output	Labor-Hours	Unit	Material	2015 Bare Costs Labor	Equipment	Total	Total Incl O&P
0010	**GREEN ROOF SYSTEMS**										
0020	Soil mixture for green roof 30% sand, 55% gravel, 15% soil										
0100	Hoist and spread soil mixture 4 inch depth up to five stories tall roof	G	B-13B	4000	.014	S.F.	.25	.38	.28	.91	1.23
0150	6 inch depth	G		2667	.021		.38	.57	.42	1.37	1.85
0200	8 inch depth	G		2000	.028		.50	.76	.56	1.82	2.45
0250	10 inch depth	G		1600	.035		.63	.96	.70	2.29	3.07
0300	12 inch depth	G		1335	.042		.76	1.15	.84	2.75	3.67
0310	Alt. man-made soil mix, hoist & spread, 4" deep up to 5 stories tall roof	G		4000	.014		1.86	.38	.28	2.52	3
0350	Mobilization 55 ton crane to site	G	1 Eqhv	3.60	2.222	Ea.		79		79	131
0355	Hoisting cost to five stories per day (Avg. 28 picks per day)	G	B-13B	1	56	Day		1,525	1,125	2,650	3,800
0360	Mobilization or demobilization, 100 ton crane to site driver & escort	G	A-3E	2.50	6.400	Ea.		205	62.50	267.50	410
0365	Hoisting cost six to ten stories per day (Avg. 21 picks per day)	G	B-13C	1	56	Day		1,525	1,675	3,200	4,375
0370	Hoist and spread soil mixture 4 inch depth six to ten stories tall roof	G		4000	.014	S.F.	.25	.38	.42	1.05	1.38
0375	6 inch depth	G		2667	.021		.38	.57	.63	1.58	2.07
0380	8 inch depth	G		2000	.028		.50	.76	.83	2.09	2.75
0385	10 inch depth	G		1600	.035		.63	.96	1.04	2.63	3.44
0390	12 inch depth	G		1335	.042		.76	1.15	1.25	3.16	4.11
0400	Green roof edging treated lumber 4" x 4" no hoisting included	G	2 Carp	400	.040	L.F.	1.34	1.31		2.65	3.68
0410	4" x 6"	G		400	.040		1.91	1.31		3.22	4.31
0420	4" x 8"	G		360	.044		3.81	1.46		5.27	6.65
0430	4" x 6" double stacked	G		300	.053		3.83	1.75		5.58	7.15
0500	Green roof edging redwood lumber 4" x 4" no hoisting included	G		400	.040		6.45	1.31		7.76	9.25
0510	4" x 6"	G		400	.040		12.60	1.31		13.91	16.05
0520	4" x 8"	G		360	.044		23.50	1.46		24.96	28.50
0530	4" x 6" double stacked	G		300	.053		25	1.75		26.75	30.50
0550	Components, not including membrane or insulation:										
0560	Fluid applied rubber membrane, reinforced, 215 mil thick	G	G-5	350	.114	S.F.	.28	2.89	.55	3.72	6.25
0570	Root barrier	G	2 Rofc	775	.021		.60	.57		1.17	1.71
0580	Moisture retention barrier and reservoir	G	"	900	.018		2.45	.49		2.94	3.61
0600	Planting sedum, light soil, potted, 2-1/4" diameter, two per S.F.	G	1 Clab	420	.019		4.90	.47		5.37	6.20
0610	one per S.F.	G	"	840	.010		2.45	.23		2.68	3.09
0630	Planting sedum mat per S.F. including shipping (4000 S.F. min)	G	4 Clab	4000	.008		5.90	.20		6.10	6.80
0640	Installation sedum mat system (no soil required) per S.F. (4000 S.F. min)	G	"	4000	.008		8.20	.20		8.40	9.35

07 33 Natural Roof Coverings

07 33 63 – Vegetated Roofing

07 33 63.10 Green Roof Systems	Crew	Daily Output	Labor-Hours	Unit	Material	2015 Bare Costs Labor	Equipment	Total	Total Incl O&P
0645 Note: pricing of sedum mats shipped in full truck loads (4000-5000 S.F.)									

07 41 Roof Panels

07 41 13 – Metal Roof Panels

07 41 13.10 Aluminum Roof Panels

0010 **ALUMINUM ROOF PANELS**									
0020 Corrugated or ribbed, .0155" thick, natural	G-3	1200	.027	S.F.	.99	.82		1.81	2.46
0300 Painted	"	1200	.027	"	1.44	.82		2.26	2.95

07 41 33 – Plastic Roof Panels

07 41 33.10 Fiberglass Panels

0010 **FIBERGLASS PANELS**									
0012 Corrugated panels, roofing, 8 oz. per S.F.	G-3	1000	.032	S.F.	2.10	.98		3.08	3.95
0100 12 oz. per S.F.		1000	.032		4.18	.98		5.16	6.25
0300 Corrugated siding, 6 oz. per S.F.		880	.036		2.10	1.12		3.22	4.18
0400 8 oz. per S.F.		880	.036		2.10	1.12		3.22	4.18
0500 Fire retardant		880	.036		3.80	1.12		4.92	6.05
0600 12 oz. siding, textured		880	.036		3.75	1.12		4.87	6
0700 Fire retardant		880	.036		4.35	1.12		5.47	6.65
0900 Flat panels, 6 oz. per S.F., clear or colors		880	.036		2.40	1.12		3.52	4.51
1100 Fire retardant, class A		880	.036		3.45	1.12		4.57	5.65
1300 8 oz. per S.F., clear or colors		880	.036		2.52	1.12		3.64	4.64

07 42 Wall Panels

07 42 13 – Metal Wall Panels

07 42 13.20 Aluminum Siding

0011 **ALUMINUM SIDING**									
6040 .024 thick smooth white single 8" wide	2 Carp	515	.031	S.F.	2.55	1.02		3.57	4.52
6060 Double 4" pattern		515	.031		2.61	1.02		3.63	4.58
6080 Double 5" pattern		550	.029		2.55	.96		3.51	4.41
6120 Embossed white, 8" wide		515	.031		2.54	1.02		3.56	4.50
6140 Double 4" pattern		515	.031		2.61	1.02		3.63	4.58
6160 Double 5" pattern		550	.029		2.59	.96		3.55	4.45
6170 Vertical, embossed white, 12" wide		590	.027		2.59	.89		3.48	4.35
6320 .019 thick, insulated, smooth white, 8" wide		515	.031		2.29	1.02		3.31	4.23
6340 Double 4" pattern		515	.031		2.27	1.02		3.29	4.21
6360 Double 5" pattern		550	.029		2.27	.96		3.23	4.10
6400 Embossed white, 8" wide		515	.031		2.64	1.02		3.66	4.61
6420 Double 4" pattern		515	.031		2.67	1.02		3.69	4.65
6440 Double 5" pattern		550	.029		2.67	.96		3.63	4.54
6500 Shake finish 10" wide white		550	.029		2.86	.96		3.82	4.75
6600 Vertical pattern, 12" wide, white		590	.027		2.18	.89		3.07	3.90
6640 For colors add					.14			.14	.15
6700 Accessories, white									
6720 Starter strip 2-1/8"	2 Carp	610	.026	L.F.	.45	.86		1.31	1.95
6740 Sill trim		450	.036		.61	1.17		1.78	2.63
6760 Inside corner		610	.026		1.59	.86		2.45	3.20
6780 Outside corner post		610	.026		3.38	.86		4.24	5.15
6800 Door & window trim		440	.036		.58	1.19		1.77	2.65
6820 For colors add					.13			.13	.14

07 42 Wall Panels

07 42 13 – Metal Wall Panels

07 42 13.20 Aluminum Siding		Crew	Daily Output	Labor-Hours	Unit	Material	2015 Bare Costs Labor	Equipment	Total	Total Incl O&P
6900	Soffit & fascia 1' overhang solid	2 Carp	110	.145	L.F.	4.35	4.78		9.13	12.80
6920	Vented		110	.145		4.38	4.78		9.16	12.80
6940	2' overhang solid		100	.160		6.45	5.25		11.70	15.85
6960	Vented		100	.160		6.45	5.25		11.70	15.85

07 42 13.30 Steel Siding										
0010	**STEEL SIDING**									
0020	Beveled, vinyl coated, 8" wide	1 Carp	265	.030	S.F.	1.85	.99		2.84	3.71
0050	10" wide	"	275	.029		1.88	.96		2.84	3.67
0081	Galv., corrugated or ribbed, on steel frame, 30 ga.	G-3	775	.041		1.20	1.27		2.47	3.44
0101	28 ga.		775	.041		1.28	1.27		2.55	3.53
0301	26 ga.		775	.041		1.80	1.27		3.07	4.10
0401	24 ga.		775	.041		1.85	1.27		3.12	4.16
0601	22 ga.		775	.041		2.15	1.27		3.42	4.49
0701	Colored, corrugated/ribbed, on steel frame, 10 yr. finish, 28 ga.		775	.041		1.95	1.27		3.22	4.27
0901	26 ga.		775	.041		2.09	1.27		3.36	4.42
1001	24 ga.		775	.041		2.37	1.27		3.64	4.73

07 46 Siding

07 46 23 – Wood Siding

07 46 23.10 Wood Board Siding		Crew	Daily Output	Labor-Hours	Unit	Material	Labor	Equipment	Total	Total Incl O&P
0010	**WOOD BOARD SIDING**									
2000	Board & batten, cedar, "B" grade, 1" x 10"	1 Carp	375	.021	S.F.	5.15	.70		5.85	6.90
2200	Redwood, clear, vertical grain, 1" x 10"		375	.021		5.30	.70		6	7.05
2400	White pine, #2 & better, 1" x 10"		375	.021		3.18	.70		3.88	4.68
2410	Board & batten siding, white pine #2, 1" x 12"		420	.019		3.18	.63		3.81	4.55
3200	Wood, cedar bevel, A grade, 1/2" x 6"		295	.027		4.14	.89		5.03	6.05
3300	1/2" x 8"		330	.024		5.90	.80		6.70	7.85
3500	3/4" x 10", clear grade		375	.021		6.30	.70		7	8.10
3600	"B" grade		375	.021		3.79	.70		4.49	5.35
3800	Cedar, rough sawn, 1" x 4", A grade, natural		220	.036		6.55	1.19		7.74	9.25
3900	Stained		220	.036		6.70	1.19		7.89	9.35
4100	1" x 12", board & batten, #3 & Btr., natural		420	.019		4.48	.63		5.11	6
4200	Stained		420	.019		4.81	.63		5.44	6.35
4400	1" x 8" channel siding, #3 & Btr., natural		330	.024		4.57	.80		5.37	6.35
4500	Stained		330	.024		4.92	.80		5.72	6.75
4700	Redwood, clear, beveled, vertical grain, 1/2" x 4"		220	.036		4.11	1.19		5.30	6.55
4750	1/2" x 6"		295	.027		4.49	.89		5.38	6.45
4800	1/2" x 8"		330	.024		4.89	.80		5.69	6.75
5000	3/4" x 10"		375	.021		4.50	.70		5.20	6.15
5200	Channel siding, 1" x 10", B grade		375	.021		4.12	.70		4.82	5.70
5250	Redwood, T&G boards, B grade, 1" x 4"		220	.036		2.54	1.19		3.73	4.80
5270	1" x 8"		330	.024		4.10	.80		4.90	5.85
5400	White pine, rough sawn, 1" x 8", natural		330	.024		2.19	.80		2.99	3.75
5500	Stained		330	.024		2.19	.80		2.99	3.75
5600	Tongue and groove, 1" x 8"		330	.024		2.19	.80		2.99	3.75

07 46 29 – Plywood Siding

07 46 29.10 Plywood Siding Options		Crew	Daily Output	Labor-Hours	Unit	Material	Labor	Equipment	Total	Total Incl O&P
0010	**PLYWOOD SIDING OPTIONS**									
0900	Plywood, medium density overlaid, 3/8" thick	2 Carp	750	.021	S.F.	1.25	.70		1.95	2.56
1000	1/2" thick		700	.023		1.44	.75		2.19	2.84

07 46 Siding

07 46 29 – Plywood Siding

07 46 29.10 Plywood Siding Options		Crew	Daily Output	Labor-Hours	Unit	Material	2015 Bare Costs Labor	Equipment	Total	Total Incl O&P
1100	3/4" thick	2 Carp	650	.025	S.F.	1.85	.81		2.66	3.40
1600	Texture 1-11, cedar, 5/8" thick, natural		675	.024		2.57	.78		3.35	4.14
1700	Factory stained		675	.024		2.80	.78		3.58	4.39
1900	Texture 1-11, fir, 5/8" thick, natural		675	.024		1.57	.78		2.35	3.04
2000	Factory stained		675	.024		1.80	.78		2.58	3.29
2050	Texture 1-11, S.Y.P., 5/8" thick, natural		675	.024		1.37	.78		2.15	2.82
2100	Factory stained		675	.024		1.44	.78		2.22	2.89
2200	Rough sawn cedar, 3/8" thick, natural		675	.024		1.25	.78		2.03	2.69
2300	Factory stained		675	.024		1.50	.78		2.28	2.96
2500	Rough sawn fir, 3/8" thick, natural		675	.024		.84	.78		1.62	2.23
2600	Factory stained		675	.024		1.04	.78		1.82	2.45
2800	Redwood, textured siding, 5/8" thick		675	.024		1.92	.78		2.70	3.42

07 46 33 – Plastic Siding

07 46 33.10 Vinyl Siding

		Crew	Daily Output	Labor-Hours	Unit	Material	2015 Bare Costs Labor	Equipment	Total	Total Incl O&P
0010	**VINYL SIDING**									
3995	Clapboard profile, woodgrain texture, .048 thick, double 4	2 Carp	495	.032	S.F.	1.04	1.06		2.10	2.93
4000	Double 5		550	.029		1.04	.96		2	2.75
4005	Single 8		495	.032		1.26	1.06		2.32	3.17
4010	Single 10		550	.029		1.51	.96		2.47	3.27
4015	.044 thick, double 4		495	.032		.99	1.06		2.05	2.87
4020	Double 5		550	.029		.99	.96		1.95	2.69
4025	.042 thick, double 4		495	.032		.92	1.06		1.98	2.80
4030	Double 5		550	.029		.92	.96		1.88	2.62
4035	Cross sawn texture, .040 thick, double 4		495	.032		.67	1.06		1.73	2.52
4040	Double 5		550	.029		.67	.96		1.63	2.34
4045	Smooth texture, .042 thick, double 4		495	.032		.75	1.06		1.81	2.61
4050	Double 5		550	.029		.75	.96		1.71	2.43
4055	Single 8		495	.032		.75	1.06		1.81	2.61
4060	Cedar texture, .044 thick, double 4		495	.032		1.01	1.06		2.07	2.90
4065	Double 6		600	.027		1.01	.88		1.89	2.59
4070	Dutch lap profile, woodgrain texture, .048 thick, double 5		550	.029		1.04	.96		2	2.75
4075	.044 thick, double 4.5		525	.030		1.01	1		2.01	2.80
4080	.042 thick, double 4.5		525	.030		.84	1		1.84	2.61
4085	.040 thick, double 4.5		525	.030		.67	1		1.67	2.42
4100	Shake profile, 10" wide		400	.040		3.45	1.31		4.76	6
4105	Vertical pattern, .046 thick, double 5		550	.029		1.42	.96		2.38	3.17
4110	.044 thick, triple 3		550	.029		1.56	.96		2.52	3.32
4115	.040 thick, triple 4		550	.029		1.50	.96		2.46	3.26
4120	.040 thick, triple 2.66		550	.029		1.81	.96		2.77	3.60
4125	Insulation, fan folded extruded polystyrene, 1/4"		2000	.008		.28	.26		.54	.75
4130	3/8"		2000	.008		.31	.26		.57	.78
4135	Accessories, J channel, 5/8" pocket		700	.023	L.F.	.46	.75		1.21	1.76
4140	3/4" pocket		695	.023		.50	.76		1.26	1.82
4145	1-1/4" pocket		680	.024		.77	.77		1.54	2.14
4150	Flexible, 3/4" pocket		600	.027		2.33	.88		3.21	4.03
4155	Under sill finish trim		500	.032		.50	1.05		1.55	2.31
4160	Vinyl starter strip		700	.023		.58	.75		1.33	1.90
4165	Aluminum starter strip		700	.023		.28	.75		1.03	1.57
4170	Window casing, 2-1/2" wide, 3/4" pocket		510	.031		1.57	1.03		2.60	3.45
4175	Outside corner, woodgrain finish, 4" face, 3/4" pocket		700	.023		1.92	.75		2.67	3.38
4180	5/8" pocket		700	.023		1.92	.75		2.67	3.38
4185	Smooth finish, 4" face, 3/4" pocket		700	.023		1.92	.75		2.67	3.38

For customer support on your Residential Cost Data, call 877.713.2617.

07 46 Siding

07 46 33 – Plastic Siding

07 46 33.10 Vinyl Siding

		Crew	Daily Output	Labor-Hours	Unit	Material	2015 Bare Costs Labor	Equipment	Total	Total Incl O&P
4190	7/8" pocket	2 Carp	690	.023	L.F.	1.89	.76		2.65	3.36
4195	1-1/4" pocket		700	.023		1.33	.75		2.08	2.73
4200	Soffit and fascia, 1' overhang, solid		120	.133		4.29	4.38		8.67	12.05
4205	Vented		120	.133		4.29	4.38		8.67	12.05
4207	18" overhang, solid		110	.145		5	4.78		9.78	13.50
4208	Vented		110	.145		5	4.78		9.78	13.50
4210	2' overhang, solid		100	.160		5.70	5.25		10.95	15.10
4215	Vented		100	.160		5.70	5.25		10.95	15.10
4217	3' overhang, solid		100	.160		7.10	5.25		12.35	16.65
4218	Vented		100	.160		7.10	5.25		12.35	16.65
4220	Colors for siding and soffits, add				S.F.	.14			.14	.15
4225	Colors for accessories and trim, add				L.F.	.30			.30	.33

07 46 33.20 Polypropylene Siding

		Crew	Daily Output	Labor-Hours	Unit	Material	Labor	Equipment	Total	Total Incl O&P
0010	**POLYPROPYLENE SIDING**									
4090	Shingle profile, random grooves, double 7	2 Carp	400	.040	S.F.	2.99	1.31		4.30	5.50
4092	Cornerpost for above	1 Carp	365	.022	L.F.	12	.72		12.72	14.40
4095	Triple 5	2 Carp	400	.040	S.F.	2.99	1.31		4.30	5.50
4097	Cornerpost for above	1 Carp	365	.022	L.F.	11.05	.72		11.77	13.35
5000	Staggered butt, double 7"	2 Carp	400	.040	S.F.	3.45	1.31		4.76	6
5002	Cornerpost for above	1 Carp	365	.022	L.F.	12	.72		12.72	14.40
5010	Half round, double 6-1/4"	2 Carp	360	.044	S.F.	3.40	1.46		4.86	6.20
5020	Shake profile, staggered butt, double 9"	"	510	.031	"	3.45	1.03		4.48	5.55
5022	Cornerpost for above	1 Carp	365	.022	L.F.	9	.72		9.72	11.10
5030	Straight butt, double 7"	2 Carp	400	.040	S.F.	3.40	1.31		4.71	5.95
5032	Cornerpost for above	1 Carp	365	.022	L.F.	11.90	.72		12.62	14.25
6000	Accessories, J channel, 5/8" pocket	2 Carp	700	.023		.46	.75		1.21	1.76
6010	3/4" pocket		695	.023		.50	.76		1.26	1.82
6020	1-1/4" pocket		680	.024		.77	.77		1.54	2.14
6030	Aluminum starter strip		700	.023		.28	.75		1.03	1.57

07 46 46 – Fiber Cement Siding

07 46 46.10 Fiber Cement Siding

		Crew	Daily Output	Labor-Hours	Unit	Material	Labor	Equipment	Total	Total Incl O&P
0010	**FIBER CEMENT SIDING**									
0020	Lap siding, 5/16" thick, 6" wide, 4-3/4" exposure, smooth texture	2 Carp	415	.039	S.F.	1.18	1.27		2.45	3.43
0025	Woodgrain texture		415	.039		1.18	1.27		2.45	3.43
0030	7-1/2" wide, 6-1/4" exposure, smooth texture		425	.038		1.46	1.24		2.70	3.69
0035	Woodgrain texture		425	.038		1.46	1.24		2.70	3.69
0040	8" wide, 6-3/4" exposure, smooth texture		425	.038		1.42	1.24		2.66	3.64
0045	Roughsawn texture		425	.038		1.42	1.24		2.66	3.64
0050	9-1/2" wide, 8-1/4" exposure, smooth texture		440	.036		1.22	1.19		2.41	3.35
0055	Woodgrain texture		440	.036		1.22	1.19		2.41	3.35
0060	12" wide, 10-3/8" exposure, smooth texture		455	.035		2.01	1.16		3.17	4.15
0065	Woodgrain texture		455	.035		2.01	1.16		3.17	4.15
0070	Panel siding, 5/16" thick, smooth texture		750	.021		1.25	.70		1.95	2.56
0075	Stucco texture		750	.021		1.25	.70		1.95	2.56
0080	Grooved woodgrain texture		750	.021		1.25	.70		1.95	2.56
0085	V - grooved woodgrain texture		750	.021		1.25	.70		1.95	2.56
0088	Shingle siding, 48" x 15-1/4" panels, 7" exposure		700	.023		3.94	.75		4.69	5.60
0090	Wood starter strip		400	.040	L.F.	.42	1.31		1.73	2.67
0200	Fiber cement siding, accessories, fascia, 5/4" x 3-1/2"		275	.058		.79	1.91		2.70	4.08
0210	5/4 x 5-1/2"		250	.064		1.93	2.10		4.03	5.65
0220	5/4 x 7-1/2"		225	.071		2.67	2.34		5.01	6.85
0230	5/4" x 9-1/2"		210	.076		3.20	2.50		5.70	7.70

07 46 Siding

07 46 73 – Soffit

07 46 73.10 Soffit Options

		Crew	Daily Output	Labor-Hours	Unit	Material	2015 Bare Costs Labor	2015 Bare Costs Equipment	Total	Total Incl O&P
0010	**SOFFIT OPTIONS**									
0012	Aluminum, residential, .020" thick	1 Carp	210	.038	S.F.	2	1.25		3.25	4.30
0100	Baked enamel on steel, 16 or 18 ga.		105	.076		5.90	2.50		8.40	10.70
0300	Polyvinyl chloride, white, solid		230	.035		2.10	1.14		3.24	4.23
0400	Perforated		230	.035		2.10	1.14		3.24	4.23
0500	For colors, add					.15			.15	.17

07 51 Built-Up Bituminous Roofing

07 51 13 – Built-Up Asphalt Roofing

07 51 13.10 Built-Up Roofing Components

		Crew	Daily Output	Labor-Hours	Unit	Material	Labor	Equipment	Total	Total Incl O&P
0010	**BUILT-UP ROOFING COMPONENTS**									
0012	Asphalt saturated felt, #30, 2 square per roll	1 Rofc	58	.138	Sq.	10.60	3.81		14.41	18.70
0200	#15, 4 sq. per roll, plain or perforated, not mopped		58	.138		5.40	3.81		9.21	13
0250	Perforated		58	.138		5.40	3.81		9.21	13
0300	Roll roofing, smooth, #65		15	.533		9.65	14.75		24.40	37.50
0500	#90		12	.667		37	18.45		55.45	74.50
0520	Mineralized		12	.667		35	18.45		53.45	72.50
0540	D.C. (Double coverage), 19" selvage edge		10	.800		52.50	22		74.50	98.50
0580	Adhesive (lap cement)				Gal.	8.85			8.85	9.75

07 51 13.20 Built-Up Roofing Systems

		Crew	Daily Output	Labor-Hours	Unit	Material	Labor	Equipment	Total	Total Incl O&P
0010	**BUILT-UP ROOFING SYSTEMS**									
0120	Asphalt flood coat with gravel/slag surfacing, not including									
0140	Insulation, flashing or wood nailers									
0200	Asphalt base sheet, 3 plies #15 asphalt felt, mopped	G-1	22	2.545	Sq.	100	66	26	192	261
0350	On nailable decks		21	2.667		102	69	27	198	270
0500	4 plies #15 asphalt felt, mopped		20	2.800		136	72.50	28.50	237	315
0550	On nailable decks		19	2.947		120	76.50	30	226.50	305
2000	Asphalt flood coat, smooth surface									
2200	Asphalt base sheet & 3 plies #15 asphalt felt, mopped	G-1	24	2.333	Sq.	105	60.50	23.50	189	253
2400	On nailable decks		23	2.435		96.50	63	24.50	184	250
2600	4 plies #15 asphalt felt, mopped		24	2.333		123	60.50	23.50	207	273
2700	On nailable decks		23	2.435		115	63	24.50	202.50	270
4500	Coal tar pitch with gravel/slag surfacing									
4600	4 plies #15 tarred felt, mopped	G-1	21	2.667	Sq.	233	69	27	329	415
4800	3 plies glass fiber felt (type IV), mopped	"	19	2.947	"	193	76.50	30	299.50	385

07 51 13.30 Cants

		Crew	Daily Output	Labor-Hours	Unit	Material	Labor	Equipment	Total	Total Incl O&P
0010	**CANTS**									
0012	Lumber, treated, 4" x 4" cut diagonally	1 Rofc	325	.025	L.F.	1.75	.68		2.43	3.19
0300	Mineral or fiber, trapezoidal, 1" x 4" x 48"		325	.025		.25	.68		.93	1.54
0400	1-1/2" x 5-5/8" x 48"		325	.025		.44	.68		1.12	1.74

07 52 Modified Bituminous Membrane Roofing

07 52 13 – Atactic-Polypropylene-Modified Bituminous Membrane Roofing

07 52 13.10 APP Modified Bituminous Membrane

		Crew	Daily Output	Labor-Hours	Unit	Material	2015 Bare Costs Labor	Equipment	Total	Total Incl O&P
0010	**APP MODIFIED BITUMINOUS MEMBRANE** R075213-30									
0020	Base sheet, #15 glass fiber felt, nailed to deck	1 Rofc	58	.138	Sq.	10.65	3.81		14.46	18.75
0030	Spot mopped to deck	G-1	295	.190		15.20	4.93	1.93	22.06	28
0040	Fully mopped to deck	"	192	.292		20.50	7.55	2.96	31.01	40.50
0050	#15 organic felt, nailed to deck	1 Rofc	58	.138		6.40	3.81		10.21	14.10
0060	Spot mopped to deck	G-1	295	.190		10.95	4.93	1.93	17.81	23.50
0070	Fully mopped to deck	"	192	.292		16.50	7.55	2.96	27.01	35.50
2100	APP mod., smooth surf. cap sheet, poly. reinf., torched, 160 mils	G-5	2100	.019	S.F.	.68	.48	.09	1.25	1.74
2150	170 mils		2100	.019		.70	.48	.09	1.27	1.76
2200	Granule surface cap sheet, poly. reinf., torched, 180 mils		2000	.020		.89	.51	.10	1.50	2.02
2250	Smooth surface flashing, torched, 160 mils		1260	.032		.68	.80	.15	1.63	2.40
2300	170 mils		1260	.032		.70	.80	.15	1.65	2.42
2350	Granule surface flashing, torched, 180 mils		1260	.032		.89	.80	.15	1.84	2.63
2400	Fibrated aluminum coating	1 Rofc	3800	.002		.08	.06		.14	.20
2450	Seam heat welding	"	205	.039	L.F.	.08	1.08		1.16	2.08

07 52 16 – Styrene-Butadiene-Styrene Modified Bituminous Membrane Roofing

07 52 16.10 SBS Modified Bituminous Membrane

		Crew	Daily Output	Labor-Hours	Unit	Material	Labor	Equipment	Total	Total Incl O&P
0010	**SBS MODIFIED BITUMINOUS MEMBRANE**									
0080	Mod bit rfng, SBS mod, gran surf cap sheet, poly reinf									
0650	120 to 149 mils thick	G-1	2000	.028	S.F.	1.25	.73	.28	2.26	3.03
0750	150 to 160 mils	"	2000	.028		1.72	.73	.28	2.73	3.54
1150	For reflective granules, add									
1600	Smooth surface cap sheet, mopped, 145 mils	G-1	2100	.027		.78	.69	.27	1.74	2.44
1620	Lightweight base sheet, fiberglass reinforced, 35 to 47 mil		2100	.027		.27	.69	.27	1.23	1.88
1625	Heavyweight base/ply sheet, reinforced, 87 to 120 mil thick		2100	.027		.89	.69	.27	1.85	2.56
1650	Granulated walkpad, 180 – 220 mils	1 Rofc	400	.020		1.82	.55		2.37	3.02
1700	Smooth surface flashing, 145 mils	G-1	1260	.044		.78	1.15	.45	2.38	3.49
1800	150 mils		1260	.044		.49	1.15	.45	2.09	3.17
1900	Granular surface flashing, 150 mils		1260	.044		.65	1.15	.45	2.25	3.35
2000	160 mils		1260	.044		.71	1.15	.45	2.31	3.41
2010	Elastomeric asphalt primer	1 Rofc	2600	.003		.17	.09		.26	.35
2015	Roofing asphalt, 30 lb. per square	G-1	19000	.003		.14	.08	.03	.25	.32
2020	Cold process adhesive, 20 – 30 mils thick	1 Rofc	750	.011		.24	.30		.54	.80
2025	Self adhering vapor retarder, 30 to 45 mils thick	G-5	2150	.019		1.02	.47	.09	1.58	2.09
2050	Seam heat welding	1 Rofc	205	.039	L.F.	.08	1.08		1.16	2.08

07 57 Coated Foamed Roofing

07 57 13 – Sprayed Polyurethane Foam Roofing

07 57 13.10 Sprayed Polyurethane Foam Roofing (S.P.F.)

		Crew	Daily Output	Labor-Hours	Unit	Material	Labor	Equipment	Total	Total Incl O&P
0010	**SPRAYED POLYURETHANE FOAM ROOFING (S.P.F.)**									
0100	Primer for metal substrate (when required)	G-2A	3000	.008	S.F.	.44	.19	.24	.87	1.09
0200	Primer for non-metal substrate (when required)		3000	.008		.17	.19	.24	.60	.80
0300	Closed cell spray, polyurethane foam, 3 lb. per C.F. density, 1", R6.7		15000	.002		.66	.04	.05	.75	.85
0400	2", R13.4		13125	.002		1.32	.04	.05	1.41	1.59
0500	3", R18.6		11485	.002		1.98	.05	.06	2.09	2.34
0550	4", R24.8		10080	.002		2.64	.06	.07	2.77	3.09
0700	Spray-on silicone coating		2500	.010		1.13	.23	.29	1.65	1.98
0800	Warranty 5-20 year manufacturer's								.15	.15
0900	Warranty 20 year, no dollar limit								.20	.20

07 58 Roll Roofing

07 58 10 – Asphalt Roll Roofing

07 58 10.10 Roll Roofing

		Crew	Daily Output	Labor-Hours	Unit	Material	2015 Bare Costs Labor	Equipment	Total	Total Incl O&P
0010	**ROLL ROOFING**									
0100	Asphalt, mineral surface									
0200	1 ply #15 organic felt, 1 ply mineral surfaced									
0300	Selvage roofing, lap 19", nailed & mopped	G-1	27	2.074	Sq.	74.50	54	21	149.50	205
0400	3 plies glass fiber felt (type IV), 1 ply mineral surfaced									
0500	Selvage roofing, lapped 19", mopped	G-1	25	2.240	Sq.	126	58	23	207	270
0600	Coated glass fiber base sheet, 2 plies of glass fiber									
0700	Felt (type IV), 1 ply mineral surfaced selvage									
0800	Roofing, lapped 19", mopped	G-1	25	2.240	Sq.	133	58	23	214	278
0900	On nailable decks	"	24	2.333	"	122	60.50	23.50	206	272
1000	3 plies glass fiber felt (type III), 1 ply mineral surfaced									
1100	Selvage roofing, lapped 19", mopped	G-1	25	2.240	Sq.	126	58	23	207	270

07 61 Sheet Metal Roofing

07 61 13 – Standing Seam Sheet Metal Roofing

07 61 13.10 Standing Seam Sheet Metal Roofing, Field Fab.

		Crew	Daily Output	Labor-Hours	Unit	Material	2015 Bare Costs Labor	Equipment	Total	Total Incl O&P
0010	**STANDING SEAM SHEET METAL ROOFING, FIELD FABRICATED**									
0400	Copper standing seam roofing, over 10 squares, 16 oz., 125 lb. per sq.	1 Shee	1.30	6.154	Sq.	960	227		1,187	1,425
0600	18 oz., 140 lb. per sq.	"	1.20	6.667		1,075	246		1,321	1,575
1200	For abnormal conditions or small areas, add					25%	100%			
1300	For lead-coated copper, add					25%				

07 61 16 – Batten Seam Sheet Metal Roofing

07 61 16.10 Batten Seam Sheet Metal Roofing, Field Fabricated

		Crew	Daily Output	Labor-Hours	Unit	Material	2015 Bare Costs Labor	Equipment	Total	Total Incl O&P
0010	**BATTEN SEAM SHEET METAL ROOFING, FIELD FABRICATED**									
0012	Copper batten seam roofing, over 10 sq., 16 oz., 130 lb. per sq.	1 Shee	1.10	7.273	Sq.	1,225	269		1,494	1,775
0100	Zinc/copper alloy batten seam roofing, .020 thick		1.20	6.667		1,225	246		1,471	1,750
0200	Copper roofing, batten seam, over 10 sq., 18 oz., 145 lb. per sq.		1	8		1,350	296		1,646	2,000
0800	Zinc, copper alloy roofing, batten seam, .027" thick		1.15	6.957		1,550	257		1,807	2,125
0900	.032" thick		1.10	7.273		1,975	269		2,244	2,625
1000	.040" thick		1.05	7.619		2,475	282		2,757	3,200

07 61 19 – Flat Seam Sheet Metal Roofing

07 61 19.10 Flat Seam Sheet Metal Roofing, Field Fabricated

		Crew	Daily Output	Labor-Hours	Unit	Material	2015 Bare Costs Labor	Equipment	Total	Total Incl O&P
0010	**FLAT SEAM SHEET METAL ROOFING, FIELD FABRICATED**									
0900	Copper flat seam roofing, over 10 squares, 16 oz., 115 lb./sq.	1 Shee	1.20	6.667	Sq.	890	246		1,136	1,400
0950	18 oz., 130 lb./sq.		1.15	6.957		995	257		1,252	1,525
1000	20 oz., 145 lb./sq.		1.10	7.273		1,100	269		1,369	1,650
1008	Zinc flat seam roofing, .020" thick		1.20	6.667		1,050	246		1,296	1,550
1010	.027" thick		1.15	6.957		1,325	257		1,582	1,875
1020	.032" thick		1.12	7.143		1,675	264		1,939	2,300
1030	.040" thick		1.05	7.619		2,100	282		2,382	2,800
1100	Lead flat seam roofing, 5 lb. per S.F.		1.30	6.154		1,150	227		1,377	1,650

07 62 Sheet Metal Flashing and Trim

07 62 10 – Sheet Metal Trim

07 62 10.10 Sheet Metal Cladding

		Crew	Daily Output	Labor-Hours	Unit	Material	2015 Bare Costs Labor	Equipment	Total	Total Incl O&P
0010	**SHEET METAL CLADDING**									
0100	Aluminum, up to 6 bends, .032" thick, window casing	1 Carp	180	.044	S.F.	1.40	1.46		2.86	3.99
0200	Window sill		72	.111	L.F.	1.40	3.65		5.05	7.70
0300	Door casing		180	.044	S.F.	1.40	1.46		2.86	3.99
0400	Fascia		250	.032		1.40	1.05		2.45	3.30
0500	Rake trim		225	.036		1.40	1.17		2.57	3.50
0700	.024" thick, window casing		180	.044		1.22	1.46		2.68	3.79
0800	Window sill		72	.111	L.F.	1.22	3.65		4.87	7.50
0900	Door casing		180	.044	S.F.	1.22	1.46		2.68	3.79
1000	Fascia		250	.032		1.22	1.05		2.27	3.10
1100	Rake trim		225	.036		1.22	1.17		2.39	3.30
1200	Vinyl coated aluminum, up to 6 bends, window casing		180	.044		1.35	1.46		2.81	3.94
1300	Window sill		72	.111	L.F.	1.35	3.65		5	7.65
1400	Door casing		180	.044	S.F.	1.35	1.46		2.81	3.94
1500	Fascia		250	.032		1.35	1.05		2.40	3.25
1600	Rake trim		225	.036		1.35	1.17		2.52	3.45

07 65 Flexible Flashing

07 65 10 – Sheet Metal Flashing

07 65 10.10 Sheet Metal Flashing and Counter Flashing

		Crew	Daily Output	Labor-Hours	Unit	Material	Labor	Equipment	Total	Total Incl O&P
0010	**SHEET METAL FLASHING AND COUNTER FLASHING**									
0011	Including up to 4 bends									
0020	Aluminum, mill finish, .013" thick	1 Rofc	145	.055	S.F.	.79	1.53		2.32	3.69
0030	.016" thick		145	.055		.91	1.53		2.44	3.82
0060	.019" thick		145	.055		1.15	1.53		2.68	4.09
0100	.032" thick		145	.055		1.35	1.53		2.88	4.31
0200	.040" thick		145	.055		2.31	1.53		3.84	5.35
0300	.050" thick		145	.055		2.43	1.53		3.96	5.50
0325	Mill finish 5" x 7" step flashing, .016" thick		1920	.004	Ea.	.15	.12		.27	.38
0350	Mill finish 12" x 12" step flashing, .016" thick		1600	.005	"	.50	.14		.64	.81
0400	Painted finish, add				S.F.	.29			.29	.32
1600	Copper, 16 oz., sheets, under 1000 lb.	1 Rofc	115	.070		7.65	1.92		9.57	12
1900	20 oz. sheets, under 1000 lb.		110	.073		9.40	2.01		11.41	14.05
2200	24 oz. sheets, under 1000 lb.		105	.076		13.30	2.11		15.41	18.55
2500	32 oz. sheets, under 1000 lb.		100	.080		17.80	2.21		20.01	23.50
2700	W shape for valleys, 16 oz., 24" wide		100	.080	L.F.	15.40	2.21		17.61	21
5800	Lead, 2.5 lb. per S.F., up to 12" wide		135	.059	S.F.	4.99	1.64		6.63	8.55
5900	Over 12" wide		135	.059		3.99	1.64		5.63	7.40
8900	Stainless steel sheets, 32 ga.,		155	.052		3	1.43		4.43	5.95
9000	28 ga.		155	.052		3.99	1.43		5.42	7
9100	26 ga.		155	.052		4.25	1.43		5.68	7.30
9200	24 ga.		155	.052		4.75	1.43		6.18	7.90
9290	For mechanically keyed flashing, add					40%				
9320	Steel sheets, galvanized, 20 ga.	1 Rofc	130	.062	S.F.	1.25	1.70		2.95	4.52
9322	22 ga.		135	.059		1.21	1.64		2.85	4.36
9324	24 ga.		140	.057		.92	1.58		2.50	3.93
9326	26 ga.		148	.054		.80	1.49		2.29	3.64
9328	28 ga.		155	.052		.69	1.43		2.12	3.39
9340	30 ga.		160	.050		.58	1.38		1.96	3.19
9400	Terne coated stainless steel, .015" thick, 28 ga		155	.052		7.35	1.43		8.78	10.75
9500	.018" thick, 26 ga		155	.052		8.30	1.43		9.73	11.75

07 65 Flexible Flashing

07 65 10 – Sheet Metal Flashing

07 65 10.10 Sheet Metal Flashing and Counter Flashing

		Crew	Daily Output	Labor-Hours	Unit	Material	2015 Bare Costs Labor	Equipment	Total	Total Incl O&P
9600	Zinc and copper alloy (brass), .020" thick	1 Rofc	155	.052	S.F.	9.30	1.43		10.73	12.90
9700	.027" thick		155	.052		11	1.43		12.43	14.75
9800	.032" thick		155	.052		14	1.43		15.43	18.05
9900	.040" thick		155	.052		18.50	1.43		19.93	23

07 65 13 – Laminated Sheet Flashing

07 65 13.10 Laminated Sheet Flashing

		Crew	Daily Output	Labor-Hours	Unit	Material	Labor	Equipment	Total	Incl O&P
0010	**LAMINATED SHEET FLASHING**, Including up to 4 bends									
8550	Shower pan, 3 ply copper and fabric, 3 oz.	1 Rofc	155	.052	S.F.	3.50	1.43		4.93	6.50
8600	7 oz.	"	155	.052	"	4.20	1.43		5.63	7.25

07 65 19 – Plastic Sheet Flashing

07 65 19.10 Plastic Sheet Flashing and Counter Flashing

		Crew	Daily Output	Labor-Hours	Unit	Material	Labor	Equipment	Total	Incl O&P
0010	**PLASTIC SHEET FLASHING AND COUNTER FLASHING**									
7300	Polyvinyl chloride, black, 10 mil	1 Rofc	285	.028	S.F.	.21	.78		.99	1.66
7400	20 mil		285	.028		.28	.78		1.06	1.74
7600	30 mil		285	.028		.35	.78		1.13	1.82
7700	60 mil		285	.028		.85	.78		1.63	2.37
8060	PVC tape, 5" x 45 mils, for joint covers, 100 L.F./roll				Ea.	170			170	187

07 65 23 – Rubber Sheet Flashing

07 65 23.10 Rubber Sheet Flashing and Counterflashing

		Crew	Daily Output	Labor-Hours	Unit	Material	Labor	Equipment	Total	Incl O&P
0010	**RUBBER SHEET FLASHING AND COUNTERFLASHING**									
4810	EPDM 90 mils, 1" diameter pipe flashing	1 Rofc	32	.250	Ea.	19.35	6.90		26.25	34.50
4820	2" diameter		30	.267		19.25	7.35		26.60	34.50
4830	3" diameter		28	.286		20.50	7.90		28.40	37.50
4840	4" diameter		24	.333		23.50	9.20		32.70	43
4850	6" diameter		22	.364		23.50	10.05		33.55	44.50
8100	Rubber, butyl, 1/32" thick		285	.028	S.F.	1.92	.78		2.70	3.54
8200	1/16" thick		285	.028		2.60	.78		3.38	4.29
8300	Neoprene, cured, 1/16" thick		285	.028		2.23	.78		3.01	3.88
8400	1/8" thick		285	.028		6.05	.78		6.83	8.10

07 71 Roof Specialties

07 71 19 – Manufactured Gravel Stops and Fasciae

07 71 19.10 Gravel Stop

		Crew	Daily Output	Labor-Hours	Unit	Material	Labor	Equipment	Total	Incl O&P
0010	**GRAVEL STOP**									
0020	Aluminum, .050" thick, 4" face height, mill finish	1 Shee	145	.055	L.F.	6.30	2.04		8.34	10.30
0080	Duranodic finish		145	.055		6.75	2.04		8.79	10.85
0100	Painted		145	.055		6.95	2.04		8.99	11.05
1200	Copper, 16 oz., 3" face height		145	.055		24	2.04		26.04	30
1300	6" face height		135	.059		33	2.19		35.19	40
1350	Galv steel, 24 ga., 4" leg, plain, with continuous cleat, 4" face		145	.055		6.25	2.04		8.29	10.30
1360	6" face height		145	.055		6.50	2.04		8.54	10.55
1500	Polyvinyl chloride, 6" face height		135	.059		5.20	2.19		7.39	9.35
1800	Stainless steel, 24 ga., 6" face height		135	.059		15.30	2.19		17.49	20.50

07 71 19.30 Fascia

		Crew	Daily Output	Labor-Hours	Unit	Material	Labor	Equipment	Total	Incl O&P
0010	**FASCIA**									
0100	Aluminum, reverse board and batten, .032" thick, colored, no furring incl	1 Shee	145	.055	S.F.	6.75	2.04		8.79	10.85
0200	Residential type, aluminum	1 Carp	200	.040	L.F.	1.95	1.31		3.26	4.36
0220	Vinyl	"	200	.040	"	2.10	1.31		3.41	4.52
0300	Steel, galv and enameled, stock, no furring, long panels	1 Shee	145	.055	S.F.	4.25	2.04		6.29	8.10

07 71 Roof Specialties

07 71 19 – Manufactured Gravel Stops and Fasciae

07 71 19.30 Fascia		Crew	Daily Output	Labor-Hours	Unit	Material	2015 Bare Costs Labor	Equipment	Total	Total Incl O&P
0600	Short panels	1 Shee	115	.070	S.F.	5.80	2.57		8.37	10.70

07 71 23 – Manufactured Gutters and Downspouts

07 71 23.10 Downspouts

		Crew	Daily Output	Labor-Hours	Unit	Material	Labor	Equipment	Total	Total Incl O&P
0010	**DOWNSPOUTS**									
0020	Aluminum, embossed, .020" thick, 2" x 3"	1 Shee	190	.042	L.F.	.98	1.56		2.54	3.68
0100	Enameled		190	.042		1.37	1.56		2.93	4.11
0300	.024" thick, 2' x 3"		180	.044		2.05	1.64		3.69	5
0400	3" x 4"		140	.057		2.33	2.11		4.44	6.10
0600	Round, corrugated aluminum, 3" diameter, .020" thick		190	.042		1.95	1.56		3.51	4.75
0700	4" diameter, .025" thick		140	.057		2.53	2.11		4.64	6.30
0900	Wire strainer, round, 2" diameter		155	.052	Ea.	2.67	1.91		4.58	6.10
1000	4" diameter		155	.052		3.67	1.91		5.58	7.20
1200	Rectangular, perforated, 2" x 3"		145	.055		2.30	2.04		4.34	5.95
1300	3" x 4"		145	.055		3.32	2.04		5.36	7.05
1500	Copper, round, 16 oz., stock, 2" diameter		190	.042	L.F.	7.60	1.56		9.16	10.95
1600	3" diameter		190	.042		7.50	1.56		9.06	10.85
1800	4" diameter		145	.055		9.70	2.04		11.74	14.05
1900	5" diameter		130	.062		17.05	2.27		19.32	22.50
2100	Rectangular, corrugated copper, stock, 2" x 3"		190	.042		8.55	1.56		10.11	12
2200	3" x 4"		145	.055		10.40	2.04		12.44	14.85
2400	Rectangular, plain copper, stock, 2" x 3"		190	.042		11.05	1.56		12.61	14.75
2500	3" x 4"		145	.055		13.95	2.04		15.99	18.75
2700	Wire strainers, rectangular, 2" x 3"		145	.055	Ea.	17.05	2.04		19.09	22
2800	3" x 4"		145	.055		17.80	2.04		19.84	23
3000	Round, 2" diameter		145	.055		6.40	2.04		8.44	10.45
3100	3" diameter		145	.055		7.10	2.04		9.14	11.20
3300	4" diameter		145	.055		12.10	2.04		14.14	16.70
3400	5" diameter		115	.070		22.50	2.57		25.07	29
3600	Lead-coated copper, round, stock, 2" diameter		190	.042	L.F.	22	1.56		23.56	27
3700	3" diameter		190	.042		22	1.56		23.56	27
3900	4" diameter		145	.055		23	2.04		25.04	29
4000	5" diameter, corrugated		130	.062		22	2.27		24.27	28.50
4200	6" diameter, corrugated		105	.076		30	2.82		32.82	37.50
4300	Rectangular, corrugated, stock, 2" x 3"		190	.042		15.40	1.56		16.96	19.55
4500	Plain, stock, 2" x 3"		190	.042		24	1.56		25.56	29
4600	3" x 4"		145	.055		33	2.04		35.04	40
4800	Steel, galvanized, round, corrugated, 2" or 3" diameter, 28 ga.		190	.042		2.08	1.56		3.64	4.89
4900	4" diameter, 28 ga.		145	.055		2.52	2.04		4.56	6.15
5700	Rectangular, corrugated, 28 ga., 2" x 3"		190	.042		1.94	1.56		3.50	4.73
5800	3" x 4"		145	.055		1.82	2.04		3.86	5.40
6000	Rectangular, plain, 28 ga., galvanized, 2" x 3"		190	.042		3.65	1.56		5.21	6.60
6100	3" x 4"		145	.055		4.06	2.04		6.10	7.85
6300	Epoxy painted, 24 ga., corrugated, 2" x 3"		190	.042		2.25	1.56		3.81	5.10
6400	3" x 4"		145	.055		2.80	2.04		4.84	6.50
6600	Wire strainers, rectangular, 2" x 3"		145	.055	Ea.	17.05	2.04		19.09	22
6700	3" x 4"		145	.055		17.80	2.04		19.84	23
6900	Round strainers, 2" or 3" diameter		145	.055		3.86	2.04		5.90	7.65
7000	4" diameter		145	.055		5.80	2.04		7.84	9.75
8200	Vinyl, rectangular, 2" x 3"		210	.038	L.F.	2.08	1.41		3.49	4.64
8300	Round, 2-1/2"		220	.036	"	1.32	1.34		2.66	3.69

07 71 Roof Specialties

07 71 23 – Manufactured Gutters and Downspouts

07 71 23.20 Downspout Elbows

		Crew	Daily Output	Labor-Hours	Unit	Material	2015 Bare Costs Labor	Equipment	Total	Total Incl O&P
0010	**DOWNSPOUT ELBOWS**									
0020	Aluminum, embossed, 2" x 3", .020" thick	1 Shee	100	.080	Ea.	.95	2.96		3.91	6
0100	Enameled		100	.080		1.75	2.96		4.71	6.85
0200	Embossed, 3" x 4", .025" thick		100	.080		4.51	2.96		7.47	9.90
0300	Enameled		100	.080		3.75	2.96		6.71	9.05
0400	Embossed, corrugated, 3" diameter, .020" thick		100	.080		2.74	2.96		5.70	7.95
0500	4" diameter, .025" thick		100	.080		5.80	2.96		8.76	11.30
0600	Copper, 16 oz., 2" diameter		100	.080		9.45	2.96		12.41	15.35
0700	3" diameter		100	.080		9	2.96		11.96	14.85
0800	4" diameter		100	.080		13.90	2.96		16.86	20
1000	2" x 3" corrugated		100	.080		9.35	2.96		12.31	15.25
1100	3" x 4" corrugated		100	.080		13.25	2.96		16.21	19.50
1300	Vinyl, 2-1/2" diameter, 45 or 75 degree bend		100	.080		3.88	2.96		6.84	9.20
1400	Tee Y junction		75	.107		12.90	3.94		16.84	21

07 71 23.30 Gutters

		Crew	Daily Output	Labor-Hours	Unit	Material	2015 Bare Costs Labor	Equipment	Total	Total Incl O&P
0010	**GUTTERS**									
0012	Aluminum, stock units, 5" K type, .027" thick, plain	1 Shee	125	.064	L.F.	2.72	2.36		5.08	6.95
0100	Enameled		125	.064		2.66	2.36		5.02	6.85
0300	5" K type type, .032" thick, plain		125	.064		3.23	2.36		5.59	7.50
0400	Enameled		125	.064		3.02	2.36		5.38	7.25
0700	Copper, half round, 16 oz., stock units, 4" wide		125	.064		8.90	2.36		11.26	13.75
0900	5" wide		125	.064		7.25	2.36		9.61	11.90
1000	6" wide		118	.068		11.85	2.51		14.36	17.20
1200	K type, 16 oz., stock, 5" wide		125	.064		8.05	2.36		10.41	12.80
1300	6" wide		125	.064		8.50	2.36		10.86	13.30
1500	Lead coated copper, 16 oz., half round, stock, 4" wide		125	.064		13.80	2.36		16.16	19.15
1600	6" wide		118	.068		21.50	2.51		24.01	28
1800	K type, stock, 5" wide		125	.064		18.35	2.36		20.71	24
1900	6" wide		125	.064		17.90	2.36		20.26	23.50
2100	Copper clad stainless steel, K type, 5" wide		125	.064		7.50	2.36		9.86	12.20
2200	6" wide		125	.064		9.45	2.36		11.81	14.35
2400	Steel, galv, half round or box, 28 ga., 5" wide, plain		125	.064		2.10	2.36		4.46	6.25
2500	Enameled		125	.064		2.10	2.36		4.46	6.25
2700	26 ga., stock, 5" wide		125	.064		2.18	2.36		4.54	6.35
2800	6" wide		125	.064		2.71	2.36		5.07	6.95
3000	Vinyl, O.G., 4" wide	1 Carp	115	.070		1.25	2.29		3.54	5.20
3100	5" wide		115	.070		1.55	2.29		3.84	5.55
3200	4" half round, stock units		115	.070		1.30	2.29		3.59	5.25
3250	Joint connectors				Ea.	2.90			2.90	3.19
3300	Wood, clear treated cedar, fir or hemlock, 3" x 4"	1 Carp	100	.080	L.F.	10	2.63		12.63	15.40
3400	4" x 5"	"	100	.080	"	16.75	2.63		19.38	23
5000	Accessories, end cap, K type, aluminum 5"	1 Shee	625	.013	Ea.	.66	.47		1.13	1.52
5010	6"		625	.013		1.44	.47		1.91	2.37
5020	Copper, 5"		625	.013		3.22	.47		3.69	4.33
5030	6"		625	.013		3.54	.47		4.01	4.68
5040	Lead coated copper, 5"		625	.013		12.50	.47		12.97	14.55
5050	6"		625	.013		13.35	.47		13.82	15.50
5060	Copper clad stainless steel, 5"		625	.013		3	.47		3.47	4.09
5070	6"		625	.013		3.60	.47		4.07	4.75
5080	Galvanized steel, 5"		625	.013		1.28	.47		1.75	2.20
5090	6"		625	.013		2.18	.47		2.65	3.19
5100	Vinyl, 4"	1 Carp	625	.013		13.15	.42		13.57	15.20

07 71 Roof Specialties

07 71 23 – Manufactured Gutters and Downspouts

07 71 23.30 Gutters

		Crew	Daily Output	Labor-Hours	Unit	Material	2015 Bare Costs Labor	Equipment	Total	Total Incl O&P
5110	5"	1 Carp	625	.013	Ea.	13.15	.42		13.57	15.20
5120	Half round, copper, 4"	1 Shee	625	.013		4.53	.47		5	5.75
5130	5"		625	.013		4.53	.47		5	5.75
5140	6"		625	.013		7.65	.47		8.12	9.25
5150	Lead coated copper, 5"		625	.013		14.15	.47		14.62	16.35
5160	6"		625	.013		21	.47		21.47	24.50
5170	Copper clad stainless steel, 5"		625	.013		4.50	.47		4.97	5.75
5180	6"		625	.013		4.50	.47		4.97	5.75
5190	Galvanized steel, 5"		625	.013		2.25	.47		2.72	3.27
5200	6"		625	.013		2.80	.47		3.27	3.87
5210	Outlet, aluminum, 2" x 3"		420	.019		.62	.70		1.32	1.85
5220	3" x 4"		420	.019		1.05	.70		1.75	2.33
5230	2-3/8" round		420	.019		.56	.70		1.26	1.79
5240	Copper, 2" x 3"		420	.019		6.50	.70		7.20	8.30
5250	3" x 4"		420	.019		7.85	.70		8.55	9.80
5260	2-3/8" round		420	.019		4.55	.70		5.25	6.15
5270	Lead coated copper, 2" x 3"		420	.019		25.50	.70		26.20	29
5280	3" x 4"		420	.019		28.50	.70		29.20	32.50
5290	2-3/8" round		420	.019		25.50	.70		26.20	29
5300	Copper clad stainless steel, 2" x 3"		420	.019		6.50	.70		7.20	8.30
5310	3" x 4"		420	.019		7.85	.70		8.55	9.80
5320	2-3/8" round		420	.019		4.55	.70		5.25	6.15
5330	Galvanized steel, 2" x 3"		420	.019		3.18	.70		3.88	4.67
5340	3" x 4"		420	.019		4.90	.70		5.60	6.55
5350	2-3/8" round		420	.019		3.98	.70		4.68	5.55
5360	K type mitres, aluminum		65	.123		3	4.55		7.55	10.90
5370	Copper		65	.123		17.40	4.55		21.95	26.50
5380	Lead coated copper		65	.123		54.50	4.55		59.05	67.50
5390	Copper clad stainless steel		65	.123		27.50	4.55		32.05	38
5400	Galvanized steel		65	.123		22.50	4.55		27.05	32.50
5420	Half round mitres, copper		65	.123		63.50	4.55		68.05	77.50
5430	Lead coated copper		65	.123		89.50	4.55		94.05	106
5440	Copper clad stainless steel		65	.123		56	4.55		60.55	69.50
5450	Galvanized steel		65	.123		25.50	4.55		30.05	35.50
5460	Vinyl mitres and outlets		65	.123		9.75	4.55		14.30	18.35
5470	Sealant		940	.009	L.F.	.01	.31		.32	.53
5480	Soldering		96	.083	"	.19	3.08		3.27	5.35

07 71 23.35 Gutter Guard

		Crew	Daily Output	Labor-Hours	Unit	Material	Labor	Equipment	Total	Total Incl O&P
0010	**GUTTER GUARD**									
0020	6" wide strip, aluminum mesh	1 Carp	500	.016	L.F.	2.42	.53		2.95	3.54
0100	Vinyl mesh	"	500	.016	"	2.61	.53		3.14	3.75

07 71 43 – Drip Edge

07 71 43.10 Drip Edge, Rake Edge, Ice Belts

		Crew	Daily Output	Labor-Hours	Unit	Material	Labor	Equipment	Total	Total Incl O&P
0010	**DRIP EDGE, RAKE EDGE, ICE BELTS**									
0020	Aluminum, .016" thick, 5" wide, mill finish	1 Carp	400	.020	L.F.	.55	.66		1.21	1.71
0100	White finish		400	.020		.63	.66		1.29	1.79
0200	8" wide, mill finish		400	.020		1.45	.66		2.11	2.70
0300	Ice belt, 28" wide, mill finish		100	.080		7.65	2.63		10.28	12.85
0310	Vented, mill finish		400	.020		1.99	.66		2.65	3.29
0320	Painted finish		400	.020		2.24	.66		2.90	3.56
0400	Galvanized, 5" wide		400	.020		.53	.66		1.19	1.68
0500	8" wide, mill finish		400	.020		.80	.66		1.46	1.98

07 71 Roof Specialties

07 71 43 – Drip Edge

07 71 43.10 Drip Edge, Rake Edge, Ice Belts		Crew	Daily Output	Labor-Hours	Unit	Material	2015 Bare Costs Labor	Equipment	Total	Total Incl O&P
0510	Rake edge, aluminum, 1-1/2" x 1-1/2"	1 Carp	400	.020	L.F.	.30	.66		.96	1.43
0520	3-1/2" x 1-1/2"	↓	400	.020	↓	.42	.66		1.08	1.56

07 72 Roof Accessories

07 72 23 – Relief Vents

07 72 23.20 Vents

		Crew	Daily Output	Labor-Hours	Unit	Material	Labor	Equipment	Total	Total Incl O&P
0010	**VENTS**									
0100	Soffit or eave, aluminum, mill finish, strips, 2-1/2" wide	1 Carp	200	.040	L.F.	.43	1.31		1.74	2.68
0200	3" wide		200	.040		.41	1.31		1.72	2.66
0300	Enamel finish, 3" wide		200	.040	↓	.50	1.31		1.81	2.76
0400	Mill finish, rectangular, 4" x 16"		72	.111	Ea.	1.45	3.65		5.10	7.75
0500	8" x 16"	↓	72	.111		2.26	3.65		5.91	8.65
2420	Roof ventilator	Q-9	16	1		52	33.50		85.50	113
2500	Vent, roof vent	1 Rofc	24	.333	↓	26	9.20		35.20	45.50

07 72 26 – Ridge Vents

07 72 26.10 Ridge Vents and Accessories

		Crew	Daily Output	Labor-Hours	Unit	Material	Labor	Equipment	Total	Total Incl O&P
0010	**RIDGE VENTS AND ACCESSORIES**									
0100	Aluminum strips, mill finish	1 Rofc	160	.050	L.F.	2.75	1.38		4.13	5.60
0150	Painted finish		160	.050	"	4.05	1.38		5.43	7
0200	Connectors		48	.167	Ea.	4.77	4.61		9.38	13.75
0300	End caps		48	.167	"	2.03	4.61		6.64	10.75
0400	Galvanized strips		160	.050	L.F.	3.58	1.38		4.96	6.50
0430	Molded polyethylene, shingles not included		160	.050	"	2.74	1.38		4.12	5.55
0440	End plugs		48	.167	Ea.	2.03	4.61		6.64	10.75
0450	Flexible roll, shingles not included	↓	160	.050	L.F.	2.38	1.38		3.76	5.15
2300	Ridge vent strip, mill finish	1 Shee	155	.052	"	3.85	1.91		5.76	7.40

07 72 53 – Snow Guards

07 72 53.10 Snow Guard Options

		Crew	Daily Output	Labor-Hours	Unit	Material	Labor	Equipment	Total	Total Incl O&P
0010	**SNOW GUARD OPTIONS**									
0100	Slate & asphalt shingle roofs, fastened with nails	1 Rofc	160	.050	Ea.	12.55	1.38		13.93	16.35
0200	Standing seam metal roofs, fastened with set screws		48	.167		16.95	4.61		21.56	27
0300	Surface mount for metal roofs, fastened with solder		48	.167	↓	6.05	4.61		10.66	15.15
0400	Double rail pipe type, including pipe	↓	130	.062	L.F.	32	1.70		33.70	38

07 72 80 – Vents

07 72 80.30 Vent Options

		Crew	Daily Output	Labor-Hours	Unit	Material	Labor	Equipment	Total	Total Incl O&P
0010	**VENT OPTIONS**									
0800	Polystyrene baffles, 12" wide for 16" O.C. rafter spacing	1 Carp	90	.089	Ea.	.42	2.92		3.34	5.35
0900	For 24" O.C. rafter spacing	"	110	.073	"	.77	2.39		3.16	4.86

07 76 Roof Pavers

07 76 16 – Roof Decking Pavers

07 76 16.10 Roof Pavers and Supports

		Crew	Daily Output	Labor-Hours	Unit	Material	2015 Bare Costs Labor	Equipment	Total	Total Incl O&P
0010	**ROOF PAVERS AND SUPPORTS**									
1000	Roof decking pavers, concrete blocks, 2" thick, natural	1 Clab	115	.070	S.F.	3.37	1.70		5.07	6.55
1100	Colors		115	.070	"	3.73	1.70		5.43	6.95
1200	Support pedestal, bottom cap		960	.008	Ea.	3	.20		3.20	3.64
1300	Top cap		960	.008		4.80	.20		5	5.65
1400	Leveling shims, 1/16"		1920	.004		1.20	.10		1.30	1.49
1500	1/8"		1920	.004		1.20	.10		1.30	1.49
1600	Buffer pad		960	.008		2.50	.20		2.70	3.09
1700	PVC legs (4" SDR 35)		2880	.003	Inch	.12	.07		.19	.24
2000	Alternate pricing method, system in place		101	.079	S.F.	7	1.94		8.94	10.95

07 91 Preformed Joint Seals

07 91 13 – Compression Seals

07 91 13.10 Compression Seals

		Crew	Daily Output	Labor-Hours	Unit	Material	Labor	Equipment	Total	Total Incl O&P
0010	**COMPRESSION SEALS**									
4900	O-ring type cord, 1/4"	1 Bric	472	.017	L.F.	.37	.55		.92	1.32
4910	1/2"		440	.018		.95	.59		1.54	2.03
4920	3/4"		424	.019		1.80	.61		2.41	3
4930	1"		408	.020		3.40	.63		4.03	4.80
4940	1-1/4"		384	.021		6.30	.67		6.97	8.05
4950	1-1/2"		368	.022		7.95	.70		8.65	9.90
4960	1-3/4"		352	.023		13.25	.73		13.98	15.80
4970	2"		344	.023		18.50	.75		19.25	22

07 91 16 – Joint Gaskets

07 91 16.10 Joint Gaskets

		Crew	Daily Output	Labor-Hours	Unit	Material	Labor	Equipment	Total	Total Incl O&P
0010	**JOINT GASKETS**									
4400	Joint gaskets, neoprene, closed cell w/adh, 1/8" x 3/8"	1 Bric	240	.033	L.F.	.28	1.08		1.36	2.10
4500	1/4" x 3/4"		215	.037		.59	1.20		1.79	2.65
4700	1/2" x 1"		200	.040		1.30	1.29		2.59	3.58
4800	3/4" x 1-1/2"		165	.048		1.45	1.57		3.02	4.21

07 91 23 – Backer Rods

07 91 23.10 Backer Rods

		Crew	Daily Output	Labor-Hours	Unit	Material	Labor	Equipment	Total	Total Incl O&P
0010	**BACKER RODS**									
0030	Backer rod, polyethylene, 1/4" diameter	1 Bric	4.60	1.739	C.L.F.	2.13	56		58.13	96
0050	1/2" diameter		4.60	1.739		3.40	56		59.40	97
0070	3/4" diameter		4.60	1.739		5.55	56		61.55	99.50
0090	1" diameter		4.60	1.739		8.85	56		64.85	103

07 91 26 – Joint Fillers

07 91 26.10 Joint Fillers

		Crew	Daily Output	Labor-Hours	Unit	Material	Labor	Equipment	Total	Total Incl O&P
0010	**JOINT FILLERS**									
4360	Butyl rubber filler, 1/4" x 1/4"	1 Bric	290	.028	L.F.	.22	.89		1.11	1.74
4365	1/2" x 1/2"		250	.032		.89	1.03		1.92	2.70
4370	1/2" x 3/4"		210	.038		1.34	1.23		2.57	3.52
4375	3/4" x 3/4"		230	.035		2.01	1.12		3.13	4.08
4380	1" x 1"		180	.044		2.68	1.44		4.12	5.35
4390	For coloring, add					12%				
4980	Polyethylene joint backing, 1/4" x 2"	1 Bric	2.08	3.846	C.L.F.	12	124		136	220
4990	1/4" x 6"		1.28	6.250	"	28	202		230	365
5600	Silicone, room temp vulcanizing foam seal, 1/4" x 1/2"		1312	.006	L.F.	.33	.20		.53	.70

07 91 Preformed Joint Seals

07 91 26 – Joint Fillers

07 91 26.10 Joint Fillers

		Crew	Daily Output	Labor-Hours	Unit	Material	2015 Bare Costs Labor	Equipment	Total	Total Incl O&P
5610	1/2" x 1/2"	1 Bric	656	.012	L.F.	.67	.39		1.06	1.39
5620	1/2" x 3/4"		442	.018		1	.58		1.58	2.07
5630	3/4" x 3/4"		328	.024		1.50	.79		2.29	2.96
5640	1/8" x 1"		1312	.006		.33	.20		.53	.70
5650	1/8" x 3"		442	.018		1	.58		1.58	2.07
5670	1/4" x 3"		295	.027		2	.88		2.88	3.66
5680	1/4" x 6"		148	.054		3.99	1.75		5.74	7.30
5690	1/2" x 6"		82	.098		8	3.15		11.15	14.05
5700	1/2" x 9"		52.50	.152		12	4.92		16.92	21.50
5710	1/2" x 12"		33	.242		15.95	7.85		23.80	30.50

07 92 Joint Sealants

07 92 13 – Elastomeric Joint Sealants

07 92 13.20 Caulking and Sealant Options

		Crew	Daily Output	Labor-Hours	Unit	Material	Labor	Equipment	Total	Total Incl O&P
0010	**CAULKING AND SEALANT OPTIONS**									
0050	Latex acrylic based, bulk				Gal.	26.50			26.50	29.50
0055	Bulk in place 1/4" x 1/4" bead	1 Bric	300	.027	L.F.	.08	.86		.94	1.53
0060	1/4" x 3/8"		294	.027		.14	.88		1.02	1.62
0065	1/4" x 1/2"		288	.028		.19	.90		1.09	1.71
0075	3/8" x 3/8"		284	.028		.21	.91		1.12	1.75
0080	3/8" x 1/2"		280	.029		.28	.92		1.20	1.85
0085	3/8" x 5/8"		276	.029		.35	.94		1.29	1.95
0095	3/8" x 3/4"		272	.029		.42	.95		1.37	2.04
0100	1/2" x 1/2"		275	.029		.37	.94		1.31	1.98
0105	1/2" x 5/8"		269	.030		.47	.96		1.43	2.11
0110	1/2" x 3/4"		263	.030		.56	.98		1.54	2.26
0115	1/2" x 7/8"		256	.031		.65	1.01		1.66	2.40
0120	1/2" x 1"		250	.032		.75	1.03		1.78	2.54
0125	3/4" x 3/4"		244	.033		.84	1.06		1.90	2.70
0130	3/4" x 1"		225	.036		1.12	1.15		2.27	3.14
0135	1" x 1"		200	.040		1.50	1.29		2.79	3.80
0190	Cartridges				Gal.	32			32	35.50
0200	11 fl. oz. cartridge				Ea.	2.76			2.76	3.04
0500	1/4" x 1/2"	1 Bric	288	.028	L.F.	.23	.90		1.13	1.75
0600	1/2" x 1/2"		275	.029		.45	.94		1.39	2.07
0800	3/4" x 3/4"		244	.033		1.01	1.06		2.07	2.89
0900	3/4" x 1"		225	.036		1.35	1.15		2.50	3.40
1000	1" x 1"		200	.040		1.69	1.29		2.98	4.01
1400	Butyl based, bulk				Gal.	35			35	38.50
1500	Cartridges				"	38.50			38.50	42
1700	1/4" x 1/2", 154 L.F./gal.	1 Bric	288	.028	L.F.	.23	.90		1.13	1.75
1800	1/2" x 1/2", 77 L.F./gal.	"	275	.029	"	.45	.94		1.39	2.07
2300	Polysulfide compounds, 1 component, bulk				Gal.	72			72	79
2400	Cartridges				"	116			116	128
2600	1 or 2 component, in place, 1/4" x 1/4", 308 L.F./gal.	1 Bric	300	.027	L.F.	.23	.86		1.09	1.70
2700	1/2" x 1/4", 154 L.F./gal.		288	.028		.47	.90		1.37	2.01
2900	3/4" x 3/8", 68 L.F./gal.		272	.029		1.06	.95		2.01	2.74
3000	1" x 1/2", 38 L.F./gal.		250	.032		1.89	1.03		2.92	3.80
3200	Polyurethane, 1 or 2 component				Gal.	49			49	54
3300	Cartridges				"	64			64	70.50
3500	Bulk, in place, 1/4" x 1/4"	1 Bric	300	.027	L.F.	.16	.86		1.02	1.62

07 92 Joint Sealants

07 92 13 – Elastomeric Joint Sealants

07 92 13.20 Caulking and Sealant Options		Crew	Daily Output	Labor-Hours	Unit	Material	2015 Bare Costs Labor	Equipment	Total	Total Incl O&P
3655	1/2" x 1/4"	1 Bric	288	.028	L.F.	.32	.90		1.22	1.85
3800	3/4" x 3/8"		272	.029		.72	.95		1.67	2.38
3900	1" x 1/2"		250	.032		1.28	1.03		2.31	3.13
4100	Silicone rubber, bulk				Gal.	49			49	54
4200	Cartridges				"	51			51	56

07 92 19 – Acoustical Joint Sealants

07 92 19.10 Acoustical Sealant

		Crew	Daily Output	Labor-Hours	Unit	Material	Labor	Equipment	Total	Total Incl O&P
0010	**ACOUSTICAL SEALANT**									
0020	Acoustical sealant, elastomeric, cartridges				Ea.	8.50			8.50	9.35
0025	In place, 1/4" x 1/4"	1 Bric	300	.027	L.F.	.35	.86		1.21	1.82
0030	1/4" x 1/2"		288	.028		.69	.90		1.59	2.26
0035	1/2" x 1/2"		275	.029		1.39	.94		2.33	3.09
0040	1/2" x 3/4"		263	.030		2.08	.98		3.06	3.93
0045	3/4" x 3/4"		244	.033		3.12	1.06		4.18	5.20
0050	1" x 1"		200	.040		5.55	1.29		6.84	8.25

Division 8 – Openings

Estimating Tips

08 10 00 Doors and Frames

All exterior doors should be addressed for their energy conservation (insulation and seals).

- Most metal doors and frames look alike, but there may be significant differences among them. When estimating these items, be sure to choose the line item that most closely compares to the specification or door schedule requirements regarding:
    - type of metal
    - metal gauge
    - door core material
    - fire rating
    - finish

- Wood and plastic doors vary considerably in price. The primary determinant is the veneer material. Lauan, birch, and oak are the most common veneers. Other variables include the following:
    - hollow or solid core
    - fire rating
    - flush or raised panel
    - finish

- Door pricing includes bore for cylindrical lockset and mortise for hinges.

08 30 00 Specialty Doors and Frames

- There are many varieties of special doors, and they are usually priced per each. Add frames, hardware, or operators required for a complete installation.

08 40 00 Entrances, Storefronts, and Curtain Walls

- Glazed curtain walls consist of the metal tube framing and the glazing material. The cost data in this subdivision is presented for the metal tube framing alone or the composite wall. If your estimate requires a detailed takeoff of the framing, be sure to add the glazing cost and any tints.

08 50 00 Windows

- Most metal windows are delivered preglazed. However, some metal windows are priced without glass. Refer to 08 80 00 Glazing for glass pricing. The grade C indicates commercial grade windows, usually ASTM C-35.

- All wood windows and vinyl are priced preglazed. The glazing is insulating glass. Add the cost of screens and grills if required, and not already included.

08 70 00 Hardware

- Hardware costs add considerably to the cost of a door. The most efficient method to determine the hardware requirements for a project is to review the door and hardware schedule together. One type of door may have different hardware, depending on the door usage.

- Door hinges are priced by the pair, with most doors requiring 1-1/2 pairs per door. The hinge prices do not include installation labor, because it is included in door installation. Hinges are classified according to the frequency of use, base material, and finish.

08 80 00 Glazing

- Different openings require different types of glass. The most common types are:
    - float
    - tempered
    - insulating
    - impact-resistant
    - ballistic-resistant

- Most exterior windows are glazed with insulating glass. Entrance doors and window walls, where the glass is less than 18" from the floor, are generally glazed with tempered glass. Interior windows and some residential windows are glazed with float glass.

- Coastal communities require the use of impact-resistant glass, dependant on wind speed.

- The insulation or 'u' value is a strong consideration, along with solar heat gain, to determine total energy efficiency.

Reference Numbers

Reference numbers are shown in shaded boxes at the beginning of some major classifications. These numbers refer to related items in the Reference Section. The reference information may be an estimating procedure, an alternate pricing method, or technical information.

Note: Not all subdivisions listed here necessarily appear in this publication. ■

No part of this publication may be reproduced, stored in a retrieval system, or transmitted in any form or by any means without prior written permission of RSMeans.

08 01 Operation and Maintenance of Openings

08 01 53 – Operation and Maintenance of Plastic Windows

08 01 53.81 Solid Vinyl Replacement Windows

			Crew	Daily Output	Labor-Hours	Unit	Material	2015 Bare Costs Labor	Equipment	Total	Total Incl O&P
0010	**SOLID VINYL REPLACEMENT WINDOWS**	R085313-20									
0020	Double hung, insulated glass, up to 83 united inches	G	2 Carp	8	2	Ea.	340	65.50		405.50	485
0040	84 to 93	G		8	2		380	65.50		445.50	525
0060	94 to 101	G		6	2.667		380	87.50		467.50	560
0080	102 to 111	G		6	2.667		395	87.50		482.50	580
0100	112 to 120	G		6	2.667		430	87.50		517.50	620
0120	For each united inch over 120, add	G		800	.020	Inch	5.50	.66		6.16	7.15
0140	Casement windows, one operating sash, 42 to 60 united inches	G		8	2	Ea.	230	65.50		295.50	365
0160	61 to 70	G		8	2		260	65.50		325.50	395
0180	71 to 80	G		8	2		280	65.50		345.50	420
0200	81 to 96	G		8	2		300	65.50		365.50	440
0220	Two operating sash, 58 to 78 united inches	G		8	2		455	65.50		520.50	610
0240	79 to 88	G		8	2		485	65.50		550.50	645
0260	89 to 98	G		8	2		525	65.50		590.50	690
0280	99 to 108	G		6	2.667		550	87.50		637.50	750
0300	109 to 121	G		6	2.667		590	87.50		677.50	795
0320	Two operating, one fixed sash, 73 to 108 united inches	G		8	2		715	65.50		780.50	900
0340	109 to 118	G		8	2		755	65.50		820.50	940
0360	119 to 128	G		6	2.667		775	87.50		862.50	995
0380	129 to 138	G		6	2.667		870	87.50		957.50	1,100
0400	139 to 156	G		6	2.667		945	87.50		1,032.50	1,200
0420	Four operating sash, 98 to 118 united inches	G		8	2		1,025	65.50		1,090.50	1,225
0440	119 to 128	G		8	2		1,100	65.50		1,165.50	1,300
0460	129 to 138	G		6	2.667		1,150	87.50		1,237.50	1,425
0480	139 to 148	G		6	2.667		1,225	87.50		1,312.50	1,475
0500	149 to 168	G		6	2.667		1,300	87.50		1,387.50	1,575
0520	169 to 178	G		6	2.667		1,400	87.50		1,487.50	1,675
0560	Fixed picture window, up to 63 united inches	G		8	2		180	65.50		245.50	310
0580	64 to 83	G		8	2		206	65.50		271.50	335
0600	84 to 101	G		8	2		257	65.50		322.50	395
0620	For each united inch over 101, add	G		900	.018	Inch	3.10	.58		3.68	4.39
0800	Cellulose fiber insulation, poured into sash balance cavity	G	1 Carp	36	.222	C.F.	.69	7.30		7.99	13
0820	Silicone caulking at perimeter	G	"	800	.010	L.F.	.17	.33		.50	.73

08 05 Common Work Results for Openings

08 05 05 – Selective Demolition for Openings

08 05 05.10 Selective Demolition Doors

			Crew	Daily Output	Labor-Hours	Unit	Material	2015 Bare Costs Labor	Equipment	Total	Total Incl O&P
0010	**SELECTIVE DEMOLITION DOORS**	R024119-10									
0200	Doors, exterior, 1-3/4" thick, single, 3' x 7' high		1 Clab	16	.500	Ea.		12.25		12.25	20.50
0210	3' x 8' high			10	.800			19.55		19.55	33
0215	Double, 3' x 8' high			6	1.333			32.50		32.50	54.50
0220	Double, 6' x 7' high			12	.667			16.30		16.30	27.50
0500	Interior, 1-3/8" thick, single, 3' x 7' high			20	.400			9.80		9.80	16.40
0520	Double, 6' x 7' high			16	.500			12.25		12.25	20.50
0700	Bi-folding, 3' x 6'-8" high			20	.400			9.80		9.80	16.40
0720	6' x 6'-8" high			18	.444			10.85		10.85	18.25
0900	Bi-passing, 3' x 6'-8" high			16	.500			12.25		12.25	20.50
0940	6' x 6'-8" high			14	.571			13.95		13.95	23.50
1500	Remove and reset, hollow core		1 Carp	8	1			33		33	55
1520	Solid			6	1.333			44		44	73.50
2000	Frames, including trim, metal			8	1			33		33	55

08 05 Common Work Results for Openings

08 05 05 – Selective Demolition for Openings

	08 05 05.10 Selective Demolition Doors	Crew	Daily Output	Labor-Hours	Unit	Material	2015 Bare Costs Labor	2015 Bare Costs Equipment	Total	Total Incl O&P
2200	Wood	2 Carp	32	.500	Ea.		16.45		16.45	27.50
2201	Alternate pricing method	1 Carp	200	.040	L.F.		1.31		1.31	2.21
3000	Special doors, counter doors	2 Carp	6	2.667	Ea.		87.50		87.50	147
3300	Glass, sliding, including frames		12	1.333			44		44	73.50
3400	Overhead, commercial, 12' x 12' high		4	4			131		131	221
3500	Residential, 9' x 7' high		8	2			65.50		65.50	110
3540	16' x 7' high		7	2.286			75		75	126
3600	Remove and reset, small		4	4			131		131	221
3620	Large		2.50	6.400			210		210	355
3660	Remove and reset elec. garage door opener	1 Carp	8	1			33		33	55
4000	Residential lockset, exterior		28	.286			9.40		9.40	15.75
4010	Residential lockset, exterior w/deadbolt		26	.308			10.10		10.10	16.95
4020	Residential lockset, interior		30	.267			8.75		8.75	14.70
4200	Deadbolt lock		32	.250			8.20		8.20	13.80
4224	Pocket door		8	1			33		33	55
5590	Remove mail slot	1 Clab	45	.178			4.35		4.35	7.30
5600	Remove door sidelight	1 Carp	6	1.333			44		44	73.50

	08 05 05.20 Selective Demolition of Windows	Crew	Daily Output	Labor-Hours	Unit	Material	Labor	Equipment	Total	Total Incl O&P
0010	**SELECTIVE DEMOLITION OF WINDOWS** R024119-10									
0200	Aluminum, including trim, to 12 S.F.	1 Clab	16	.500	Ea.		12.25		12.25	20.50
0240	To 25 S.F.		11	.727			17.80		17.80	30
0280	To 50 S.F.		5	1.600			39		39	65.50
0320	Storm windows/screens, to 12 S.F.		27	.296			7.25		7.25	12.15
0360	To 25 S.F.		21	.381			9.30		9.30	15.65
0400	To 50 S.F.		16	.500			12.25		12.25	20.50
0500	Screens, incl. aluminum frame, small		20	.400			9.80		9.80	16.40
0510	Large		16	.500			12.25		12.25	20.50
0600	Glass, up to 10 SF per window		200	.040	S.F.		.98		.98	1.64
0620	Over 10 SF per window		150	.053	"		1.30		1.30	2.19
2000	Wood, including trim, to 12 S.F.		22	.364	Ea.		8.90		8.90	14.95
2020	To 25 S.F.		18	.444			10.85		10.85	18.25
2060	To 50 S.F.		13	.615			15.05		15.05	25.50
2065	To 180 S.F.		8	1			24.50		24.50	41
4300	Remove bay/bow window	2 Carp	6	2.667			87.50		87.50	147
4410	Remove skylight, plstc domes, flush/curb mtd	G-3	395	.081	S.F.		2.49		2.49	4.16
5020	Remove and reset window, up to a 2'x2' widow	1 Carp	6	1.333	Ea.		44		44	73.50
5040	Up to a 3'x3' window		4	2			65.50		65.50	110
5080	Up to a 4'x5' window		2	4			131		131	221
6000	Screening only	1 Clab	4000	.002	S.F.		.05		.05	.08
9100	Window awning, residential	"	80	.100	L.F.		2.45		2.45	4.11

08 11 Metal Doors and Frames

08 11 63 – Metal Screen and Storm Doors and Frames

08 11 63.23 Aluminum Screen and Storm Doors and Frames

		Crew	Daily Output	Labor-Hours	Unit	Material	2015 Bare Costs Labor	Equipment	Total	Total Incl O&P
0010	**ALUMINUM SCREEN AND STORM DOORS AND FRAMES**									
0020	Combination storm and screen									
0420	Clear anodic coating, 2'-8" wide	2 Carp	14	1.143	Ea.	205	37.50		242.50	289
0440	3'-0" wide	"	14	1.143	"	178	37.50		215.50	258
0500	For 7' door height, add					8%				
1020	Mill finish, 2'-8" wide	2 Carp	14	1.143	Ea.	235	37.50		272.50	320
1040	3'-0" wide	"	14	1.143		258	37.50		295.50	345
1100	For 7'-0" door, add					8%				
1520	White painted, 2'-8" wide	2 Carp	14	1.143		286	37.50		323.50	380
1540	3'-0" wide		14	1.143		310	37.50		347.50	405
1541	Storm door, painted, alum., insul., 6'-8" x 2'-6" wide		14	1.143		255	37.50		292.50	345
1545	2'-8" wide		14	1.143		274	37.50		311.50	365
1600	For 7'-0" door, add					8%				
1800	Aluminum screen door, 6'-8" x 2'-8" wide	2 Carp	14	1.143		148	37.50		185.50	226
1810	3'-0" wide	"	14	1.143		255	37.50		292.50	345
2000	Wood door & screen, see Section 08 14 33.20									

08 12 Metal Frames

08 12 13 – Hollow Metal Frames

08 12 13.13 Standard Hollow Metal Frames

			Crew	Daily Output	Labor-Hours	Unit	Material	Labor	Equipment	Total	Total Incl O&P
0010	**STANDARD HOLLOW METAL FRAMES**										
0020	16 ga., up to 5-3/4" jamb depth										
0025	3'-0" x 6'-8" single	G	2 Carp	16	1	Ea.	146	33		179	215
0028	3'-6" wide, single	G		16	1		153	33		186	224
0030	4'-0" wide, single	G		16	1		152	33		185	222
0040	6'-0" wide, double	G		14	1.143		204	37.50		241.50	287
0045	8'-0" wide, double	G		14	1.143		213	37.50		250.50	297
0100	3'-0" x 7'-0" single	G		16	1		151	33		184	221
0110	3'-6" wide, single	G		16	1		159	33		192	229
0112	4'-0" wide, single	G		16	1		159	33		192	229
0140	6'-0" wide, double	G		14	1.143		194	37.50		231.50	277
0145	8'-0" wide, double	G		14	1.143		229	37.50		266.50	315
1000	16 ga., up to 4-7/8" deep, 3'-0" x 7'-0" single	G		16	1		166	33		199	238
1140	6'-0" wide, double	G		14	1.143		188	37.50		225.50	270
1200	16 ga., 8-3/4" deep, 3'-0" x 7'-0" single	G		16	1		199	33		232	274
1240	6'-0" wide, double	G		14	1.143		234	37.50		271.50	320
2800	14 ga., up to 3-7/8" deep, 3'-0" x 7'-0" single	G		16	1		181	33		214	254
2840	6'-0" wide, double	G		14	1.143		217	37.50		254.50	300
3000	14 ga., up to 5-3/4" deep, 3'-0" x 6'-8" single	G		16	1		155	33		188	226
3002	3'-6" wide, single	G		16	1		198	33		231	272
3005	4'-0" wide, single	G		16	1		203	33		236	279
3600	up to 5-3/4" jamb depth, 4'-0" x 7'-0" single	G		15	1.067		185	35		220	262
3620	6'-0" wide, double	G		12	1.333		235	44		279	335
3640	8'-0" wide, double	G		12	1.333		246	44		290	345
3700	8'-0" high, 4'-0" wide, single	G		15	1.067		235	35		270	320
3740	8'-0" wide, double	G		12	1.333		290	44		334	395
4000	6-3/4" deep, 4'-0" x 7'-0" single	G		15	1.067		222	35		257	305
4020	6'-0" wide, double	G		12	1.333		276	44		320	380
4040	8'-0", wide double	G		12	1.333		284	44		328	385
4100	8'-0" high, 4'-0" wide, single	G		15	1.067		276	35		311	365
4140	8'-0" wide, double	G		12	1.333		320	44		364	430

08 12 Metal Frames

08 12 13 – Hollow Metal Frames

08 12 13.13 Standard Hollow Metal Frames		Crew	Daily Output	Labor-Hours	Unit	Material	2015 Bare Costs Labor	Equipment	Total	Total Incl O&P
4400	8-3/4" deep, 4'-0" x 7'-0", single	G	2 Carp	15	1.067	Ea.	252	35		287	335
4440	8'-0" wide, double	G		12	1.333		315	44		359	420
4500	4'-0" x 8'-0", single	G		15	1.067		283	35		318	370
4540	8'-0" wide, double	G		12	1.333		350	44		394	460
4900	For welded frames, add						63			63	69.50
5400	14 ga., "B" label, up to 5-3/4" deep, 4'-0" x 7'-0" single	G	2 Carp	15	1.067		210	35		245	290
5440	8'-0" wide, double	G		12	1.333		272	44		316	375
5800	6-3/4" deep, 7'-0" high, 4'-0" wide, single	G		15	1.067		218	35		253	299
5840	8'-0" wide, double	G		12	1.333		385	44		429	500
6200	8-3/4" deep, 4'-0" x 7'-0" single	G		15	1.067		291	35		326	380
6240	8'-0" wide, double	G		12	1.333		380	44		424	490
6300	For "A" label use same price as "B" label										
6400	For baked enamel finish, add						30%	15%			
6500	For galvanizing, add						20%				
6600	For hospital stop, add					Ea.	295			295	325
6620	For hospital stop, stainless steel add					"	380			380	420
7900	Transom lite frames, fixed, add		2 Carp	155	.103	S.F.	54	3.39		57.39	65
8000	Movable, add		"	130	.123	"	68	4.04		72.04	82

08 13 Metal Doors

08 13 13 – Hollow Metal Doors

08 13 13.15 Metal Fire Doors

			Crew	Daily Output	Labor-Hours	Unit	Material	Labor	Equipment	Total	Total Incl O&P
0010	**METAL FIRE DOORS**	R081313-20									
0015	Steel, flush, "B" label, 90 minute										
0020	Full panel, 20 ga., 2'-0" x 6'-8"		2 Carp	20	.800	Ea.	420	26.50		446.50	505
0040	2'-8" x 6'-8"			18	.889		435	29		464	530
0060	3'-0" x 6'-8"			17	.941		435	31		466	530
0080	3'-0" x 7'-0"			17	.941		455	31		486	550
0140	18 ga., 3'-0" x 6'-8"			16	1		495	33		528	600
0160	2'-8" x 7'-0"			17	.941		520	31		551	620
0180	3'-0" x 7'-0"			16	1		505	33		538	610
0200	4'-0" x 7'-0"			15	1.067		650	35		685	775
0220	For "A" label, 3 hour, 18 ga., use same price as "B" label										
0240	For vision lite, add					Ea.	156			156	172
0520	Flush, "B" label 90 min., egress core, 20 ga., 2'-0" x 6'-8"		2 Carp	18	.889		655	29		684	770
0540	2'-8" x 6'-8"			17	.941		665	31		696	780
0560	3'-0" x 6'-8"			16	1		665	33		698	785
0580	3'-0" x 7'-0"			16	1		685	33		718	805
0640	Flush, "A" label 3 hour, egress core, 18 ga., 3'-0" x 6'-8"			15	1.067		720	35		755	855
0660	2'-8" x 7'-0"			16	1		750	33		783	875
0680	3'-0" x 7'-0"			15	1.067		740	35		775	875
0700	4'-0" x 7'-0"			14	1.143		885	37.50		922.50	1,025

08 13 13.20 Residential Steel Doors

			Crew	Daily Output	Labor-Hours	Unit	Material	Labor	Equipment	Total	Total Incl O&P
0010	**RESIDENTIAL STEEL DOORS**										
0020	Prehung, insulated, exterior										
0030	Embossed, full panel, 2'-8" x 6'-8"	G	2 Carp	17	.941	Ea.	315	31		346	395
0040	3'-0" x 6'-8"	G		15	1.067		270	35		305	355
0060	3'-0" x 7'-0"	G		15	1.067		330	35		365	425
0070	5'-4" x 6'-8", double	G		8	2		630	65.50		695.50	805
0220	Half glass, 2'-8" x 6'-8"	G		17	.941		320	31		351	400
0240	3'-0" x 6'-8"	G		16	1		320	33		353	405

08 13 Metal Doors

08 13 13 – Hollow Metal Doors

08 13 13.20 Residential Steel Doors

		Crew	Daily Output	Labor-Hours	Unit	Material	2015 Bare Costs Labor	Equipment	Total	Total Incl O&P
0260	3'-0" x 7'-0" G	2 Carp	16	1	Ea.	370	33		403	460
0270	5'-4" x 6'-8", double G		8	2		650	65.50		715.50	825
1320	Flush face, full panel, 2'-8" x 6'-8" G		16	1		264	33		297	345
1340	3'-0" x 6'-8" G		15	1.067		264	35		299	350
1360	3'-0" x 7'-0" G		15	1.067		296	35		331	385
1380	5'-4" x 6'-8", double G		8	2		560	65.50		625.50	725
1420	Half glass, 2'-8" x 6'-8" G		17	.941		325	31		356	405
1440	3'-0" x 6'-8" G		16	1		325	33		358	410
1460	3'-0" x 7'-0" G		16	1		360	33		393	450
1480	5'-4" x 6'-8", double G		8	2		640	65.50		705.50	815
1500	Sidelight, full lite, 1'-0" x 6'-8" with grille G					252			252	277
1510	1'-0" x 6'-8", low e G					268			268	295
1520	1'-0" x 6'-8", half lite G					280			280	310
1530	1'-0" x 6'-8", half lite, low e G					284			284	310
2300	Interior, residential, closet, bi-fold, 2'-0" x 6'-8" G	2 Carp	16	1		161	33		194	232
2330	3'-0" wide G		16	1		200	33		233	275
2360	4'-0" wide G		15	1.067		260	35		295	345
2400	5'-0" wide G		14	1.143		315	37.50		352.50	415
2420	6'-0" wide G		13	1.231		335	40.50		375.50	440
2510	Bi-passing closet, incl. hardware, no frame or trim incl.									
2511	Mirrored, metal frame, 4'-0" x 6'-8"	2 Carp	10	1.600	Opng.	208	52.50		260.50	315
2512	5'-0" wide		10	1.600		238	52.50		290.50	350
2513	6'-0" wide		10	1.600		272	52.50		324.50	385
2514	7'-0" wide		9	1.778		269	58.50		327.50	395
2515	8'-0" wide		9	1.778		440	58.50		498.50	580
2611	Mirrored, metal, 4'-0" x 8'-0"		10	1.600		315	52.50		367.50	440
2612	5'-0" wide		10	1.600		330	52.50		382.50	450
2613	6'-0" wide		10	1.600		385	52.50		437.50	510
2614	7'-0" wide		9	1.778		400	58.50		458.50	540
2615	8'-0" wide		9	1.778		440	58.50		498.50	585

08 14 Wood Doors

08 14 13 – Carved Wood Doors

08 14 13.10 Types of Wood Doors, Carved

		Crew	Daily Output	Labor-Hours	Unit	Material	Labor	Equipment	Total	Total Incl O&P
0010	**TYPES OF WOOD DOORS, CARVED**									
3000	Solid wood, 1-3/4" thick stile and rail									
3020	Mahogany, 3'-0" x 7'-0", six panel	2 Carp	14	1.143	Ea.	1,125	37.50		1,162.50	1,325
3030	With two lites		10	1.600		1,850	52.50		1,902.50	2,125
3040	3'-6" x 8'-0", six panel		10	1.600		1,500	52.50		1,552.50	1,750
3050	With two lites		8	2		2,500	65.50		2,565.50	2,850
3100	Pine, 3'-0" x 7'-0", six panel		14	1.143		525	37.50		562.50	645
3110	With two lites		10	1.600		825	52.50		877.50	1,000
3120	3'-6" x 8'-0", six panel		10	1.600		920	52.50		972.50	1,100
3130	With two lites		8	2		1,825	65.50		1,890.50	2,125
3200	Red oak, 3'-0" x 7'-0", six panel		14	1.143		1,750	37.50		1,787.50	1,975
3210	With two lites		10	1.600		2,300	52.50		2,352.50	2,625
3220	3'-6" x 8'-0", six panel		10	1.600		2,600	52.50		2,652.50	2,950
3230	With two lites		8	2		3,300	65.50		3,365.50	3,725
4000	Hand carved door, mahogany									
4020	3'-0" x 7'-0", simple design	2 Carp	14	1.143	Ea.	1,750	37.50		1,787.50	2,000
4030	Intricate design		11	1.455		3,700	48		3,748	4,150

08 14 Wood Doors

08 14 13 – Carved Wood Doors

08 14 13.10 Types of Wood Doors, Carved		Crew	Daily Output	Labor-Hours	Unit	Material	2015 Bare Costs Labor	Equipment	Total	Total Incl O&P
4040	3'-6" x 8'-0", simple design	2 Carp	10	1.600	Ea.	3,000	52.50		3,052.50	3,400
4050	Intricate design	↓	8	2		3,700	65.50		3,765.50	4,175
4400	For custom finish, add					475			475	525
4600	Side light, mahogany, 7'-0" x 1'-6" wide, 4 lites	2 Carp	18	.889		1,100	29		1,129	1,250
4610	6 lites		14	1.143		2,625	37.50		2,662.50	2,975
4620	8'-0" x 1'-6" wide, 4 lites		14	1.143		1,800	37.50		1,837.50	2,050
4630	6 lites		10	1.600		2,100	52.50		2,152.50	2,400
4640	Side light, oak, 7'-0" x 1'-6" wide, 4 lites		18	.889		1,200	29		1,229	1,375
4650	6 lites		14	1.143		2,100	37.50		2,137.50	2,375
4660	8'-0" x 1'-6" wide, 4 lites		14	1.143		1,100	37.50		1,137.50	1,275
4670	6 lites	↓	10	1.600	↓	2,100	52.50		2,152.50	2,400

08 14 16 – Flush Wood Doors

08 14 16.09 Smooth Wood Doors

		Crew	Daily Output	Labor-Hours	Unit	Material	Labor	Equipment	Total	Total Incl O&P
0010	**SMOOTH WOOD DOORS**									
0015	Flush, interior, hollow core									
0025	Lauan face, 1-3/8", 3'-0" x 6'-8"	2 Carp	17	.941	Ea.	54.50	31		85.50	112
0030	4'-0" x 6'-8"		16	1		126	33		159	194
0140	Birch face, 1-3/8", 2'-6" x 6'-8"		17	.941		85	31		116	145
0180	3'-0" x 6'-8"		17	.941		95.50	31		126.50	157
0200	4'-0" x 6'-8"		16	1		158	33		191	229
0202	1-3/4", 2'-0" x 6'-8"		17	.941		50.50	31		81.50	108
0204	2'-4" x 7'-0"		16	1		96.50	33		129.50	161
0206	2'-6" x 7'-0"		16	1		100	33		133	165
0208	2'-8" x 7'-0"		16	1		105	33		138	170
0210	3'-0" x 7'-0"		16	1		110	33		143	176
0212	3'-4" x 7'-0"		15	1.067	↓	193	35		228	271
0214	Pair of 3'-0" x 7'-0"	↓	9	1.778	Pr.	213	58.50		271.50	330
0480	For prefinishing, clear, add				Ea.	46			46	50.50
0500	For prefinishing, stain, add				"	57			57	62.50
0620	For dutch door with shelf, add					140%				
1320	M.D. overlay on hardboard, 1-3/8", 2'-0" x 6'-8"	2 Carp	17	.941	Ea.	115	31		146	179
1340	2'-6" x 6'-8"		17	.941		115	31		146	179
1380	3'-0" x 6'-8"		17	.941		126	31		157	191
1400	4'-0" x 6'-8"		16	1		178	33		211	251
1720	H.P. plastic laminate, 1-3/8", 2'-0" x 6'-8"		16	1		260	33		293	340
1740	2'-6" x 6'-8"		16	1		260	33		293	340
1780	3'-0" x 6'-8"		15	1.067		290	35		325	380
1785	Door, plastic laminate, 3'-0" x 6'-8"					290			290	320
1800	4'-0" x 6'-8"	2 Carp	14	1.143		385	37.50		422.50	485
2020	Particle core, lauan face, 1-3/8", 2'-6" x 6'-8"		15	1.067		92	35		127	160
2040	3'-0" x 6'-8"		14	1.143		94	37.50		131.50	166
2120	Birch face, 1-3/8", 2'-6" x 6'-8"		15	1.067		103	35		138	172
2140	3'-0" x 6'-8"		14	1.143		113	37.50		150.50	187
3320	M.D. overlay on hardboard, 1-3/8", 2'-6" x 6'-8"		14	1.143		106	37.50		143.50	180
3340	3'-0" x 6'-8"		13	1.231		115	40.50		155.50	195
4000	Exterior, flush, solid core, birch, 1-3/4" x 2'-6" x 7'-0"		15	1.067		173	35		208	249
4020	2'-8" wide		15	1.067		159	35		194	234
4040	3'-0" wide	↓	14	1.143		209	37.50		246.50	293
4045	3'-0" x 8'-0"	1 Carp	8	1		430	33		463	530
4100	Oak faced 1-3/4" x 2'-6" x 7'-0"	2 Carp	15	1.067		215	35		250	296
4120	2'-8" wide		15	1.067		225	35		260	305
4140	3'-0" wide	↓	14	1.143		230	37.50		267.50	315

08 14 Wood Doors

08 14 16 – Flush Wood Doors

08 14 16.09 Smooth Wood Doors

		Crew	Daily Output	Labor-Hours	Unit	Material	2015 Bare Costs Labor	Equipment	Total	Total Incl O&P
4160	Walnut faced, 1-3/4" x 3'-0" x 6'-8"	1 Carp	17	.471	Ea.	295	15.45		310.45	350
4180	3'-6" wide	"	17	.471		365	15.45		380.45	425
4200	Walnut faced, 1-3/4" x 2'-6" x 7'-0"	2 Carp	15	1.067		310	35		345	400
4220	2'-8" wide		15	1.067		315	35		350	405
4240	3'-0" wide		14	1.143		320	37.50		357.50	415
4250	3'-6" wide	1 Carp	14	.571		420	18.75		438.75	490
4260	3'-0" x 8'-0"		8	1		375	33		408	465
4270	3'-6" wide		8	1		450	33		483	550
4285	Cherry faced, flush, sc, 1-3/4" x 3'-0" x 8'-0" wide		8	1		465	33		498	565

08 14 16.10 Wood Doors Decorator

		Crew	Daily Output	Labor-Hours	Unit	Material	Labor	Equipment	Total	Total Incl O&P
0010	**WOOD DOORS DECORATOR**									
1800	Exterior, flush, solid wood core, birch 1-3/4" x 2'-6" x 7'-0"	2 Carp	15	1.067	Ea.	320	35		355	410
1820	2'-8" wide		15	1.067		325	35		360	420
1840	3'-0" wide		14	1.143		335	37.50		372.50	435
1900	Oak faced, 1-3/4" x 2'-6" x 7'-0"		15	1.067		320	35		355	410
1920	2'-8" wide		15	1.067		330	35		365	425
1940	3'-0" wide		14	1.143		340	37.50		377.50	440
2100	Walnut faced, 1-3/4" x 2'-6" x 7'-0"		15	1.067		385	35		420	485
2120	2'-8" wide		15	1.067		400	35		435	500
2140	3'-0" wide		14	1.143		430	37.50		467.50	540

08 14 33 – Stile and Rail Wood Doors

08 14 33.10 Wood Doors Paneled

		Crew	Daily Output	Labor-Hours	Unit	Material	Labor	Equipment	Total	Total Incl O&P
0010	**WOOD DOORS PANELED**									
0020	Interior, six panel, hollow core, 1-3/8" thick									
0040	Molded hardboard, 2'-0" x 6'-8"	2 Carp	17	.941	Ea.	62	31		93	120
0060	2'-6" x 6'-8"		17	.941		64	31		95	123
0070	2'-8" x 6'-8"		17	.941		67	31		98	126
0080	3'-0" x 6'-8"		17	.941		72	31		103	131
0140	Embossed print, molded hardboard, 2'-0" x 6'-8"		17	.941		64	31		95	123
0160	2'-6" x 6'-8"		17	.941		64	31		95	123
0180	3'-0" x 6'-8"		17	.941		72	31		103	131
0540	Six panel, solid, 1-3/8" thick, pine, 2'-0" x 6'-8"		15	1.067		155	35		190	230
0560	2'-6" x 6'-8"		14	1.143		170	37.50		207.50	250
0580	3'-0" x 6'-8"		13	1.231		145	40.50		185.50	228
1020	Two panel, bored rail, solid, 1-3/8" thick, pine, 1'-6" x 6'-8"		16	1		270	33		303	350
1040	2'-0" x 6'-8"		15	1.067		355	35		390	450
1060	2'-6" x 6'-8"		14	1.143		400	37.50		437.50	505
1340	Two panel, solid, 1-3/8" thick, fir, 2'-0" x 6'-8"		15	1.067		160	35		195	235
1360	2'-6" x 6'-8"		14	1.143		210	37.50		247.50	294
1380	3'-0" x 6'-8"		13	1.231		415	40.50		455.50	525
1740	Five panel, solid, 1-3/8" thick, fir, 2'-0" x 6'-8"		15	1.067		280	35		315	370
1760	2'-6" x 6'-8"		14	1.143		420	37.50		457.50	525
1780	3'-0" x 6'-8"		13	1.231		420	40.50		460.50	530
4190	Exterior, Knotty pine, paneled, 1-3/4", 3'-0" x 6'-8"		16	1		780	33		813	915
4195	Double 1-3/4", 3'-0" x 6'-8"		16	1		1,550	33		1,583	1,775
4200	Ash, paneled, 1-3/4", 3'-0" x 6'-8"		16	1		890	33		923	1,025
4205	Double 1-3/4", 3'-0" x 6'-8"		16	1		1,775	33		1,808	2,000
4210	Cherry, paneled, 1-3/4", 3'-0" x 6'-8"		16	1		890	33		923	1,025
4215	Double 1-3/4", 3'-0" x 6'-8"		16	1		1,775	33		1,808	2,000
4230	Ash, paneled, 1-3/4", 3'-0" x 8'-0"		16	1		950	33		983	1,100
4235	Double 1-3/4", 3'-0" x 8'-0"		16	1		1,900	33		1,933	2,150
4240	Hard Maple, paneled, 1-3/4", 3'-0" x 8'-0"		16	1		950	33		983	1,100

08 14 Wood Doors

08 14 33 – Stile and Rail Wood Doors

08 14 33.10 Wood Doors Paneled		Crew	Daily Output	Labor-Hours	Unit	Material	2015 Bare Costs Labor	2015 Bare Costs Equipment	Total	Total Incl O&P
4245	Double 1-3/4", 3'-0" x 8'-0"	2 Carp	16	1	Ea.	1,900	33		1,933	2,150
4250	Cherry, paneled, 1-3/4", 3'-0" x 8'-0"		16	1		1,000	33		1,033	1,150
4255	Double 1-3/4", 3'-0" x 8'-0"		16	1		2,000	33		2,033	2,250
08 14 33.20 Wood Doors Residential										
0010	**WOOD DOORS RESIDENTIAL**									
0200	Exterior, combination storm & screen, pine									
0260	2'-8" wide	2 Carp	10	1.600	Ea.	300	52.50		352.50	420
0280	3'-0" wide		9	1.778		310	58.50		368.50	440
0300	7'-1" x 3'-0" wide		9	1.778		335	58.50		393.50	470
0400	Full lite, 6'-9" x 2'-6" wide		11	1.455		320	48		368	430
0420	2'-8" wide		10	1.600		320	52.50		372.50	440
0440	3'-0" wide		9	1.778		325	58.50		383.50	455
0500	7'-1" x 3'-0" wide		9	1.778		355	58.50		413.50	490
0604	Door, screen, plain full		12	1.333		330	44		374	440
0614	Divided		12	1.333		370	44		414	485
0634	Decor full		12	1.333		455	44		499	575
0700	Dutch door, pine, 1-3/4" x 2'-8" x 6'-8", 6 panel		12	1.333		790	44		834	940
0720	Half glass		10	1.600		900	52.50		952.50	1,075
0800	3'-0" wide, 6 panel		12	1.333		790	44		834	940
0820	Half glass		10	1.600		900	52.50		952.50	1,075
1000	Entrance door, colonial, 1-3/4" x 6'-8" x 2'-8" wide		16	1		500	33		533	605
1020	6 panel pine, 3'-0" wide		15	1.067		445	35		480	550
1100	8 panel pine, 2'-8" wide		16	1		580	33		613	690
1120	3'-0" wide		15	1.067		570	35		605	685
1200	For tempered safety glass lites, (min of 2) add					79			79	87
1300	Flush, birch, solid core, 1-3/4" x 6'-8" x 2'-8" wide	2 Carp	16	1		113	33		146	180
1320	3'-0" wide		15	1.067		119	35		154	189
1350	7'-0" x 2'-8" wide		16	1		122	33		155	189
1360	3'-0" wide		15	1.067		143	35		178	216
1420	6'-8" x 3'-0" wide, fir		16	1		470	33		503	570
1720	Carved mahogany 3'-0" x 6'-8"		15	1.067		1,375	35		1,410	1,575
1760	Mahogany, 3'-0" x 6'-8"		15	1.067		750	35		785	885
1920	For fixed wood louvers, add					80			80	88
1930	For dutch door with shelf, add					140%				
2700	Interior, closet, bi-fold, w/hardware, no frame or trim incl.									
2720	Flush, birch, 2'-6" x 6'-8"	2 Carp	13	1.231	Ea.	67.50	40.50		108	143
2740	3'-0" wide		13	1.231		71.50	40.50		112	147
2760	4'-0" wide		12	1.333		109	44		153	194
2780	5'-0" wide		11	1.455		105	48		153	196
2800	6'-0" wide		10	1.600		128	52.50		180.50	229
2804	Flush lauan 2'-0" x 6'-8"		14	1.143		52.50	37.50		90	121
2810	8'-0" wide		9	1.778		195	58.50		253.50	315
2817	6'-0" wide		9	1.778		113	58.50		171.50	222
2820	Flush, hardboard, primed, 6'-8" x 2'-6" wide		13	1.231		56.50	40.50		97	130
2840	3'-0" wide		13	1.231		64	40.50		104.50	138
2860	4'-0" wide		12	1.333		94	44		138	177
2880	5'-0" wide		11	1.455		101	48		149	191
2900	6'-0" wide		10	1.600		139	52.50		191.50	241
2920	Hardboard, primed 7'-0" x 4'-0", wide		12	1.333		182	44		226	274
2930	6'-0" wide		10	1.600		177	52.50		229.50	283
3000	Raised panel pine, 6'-6" or 6'-8" x 2'-6" wide		13	1.231		198	40.50		238.50	286
3020	3'-0" wide		13	1.231		278	40.50		318.50	375

08 14 Wood Doors

08 14 33 – Stile and Rail Wood Doors

08 14 33.20 Wood Doors Residential		Crew	Daily Output	Labor-Hours	Unit	Material	2015 Bare Costs Labor	Equipment	Total	Total Incl O&P
3040	4'-0" wide	2 Carp	12	1.333	Ea.	305	44		349	410
3060	5'-0" wide		11	1.455		365	48		413	480
3080	6'-0" wide		10	1.600		400	52.50		452.50	530
3180	Louvered, pine, 6'-6" or 6'-8" x 1'-6" wide		13	1.231		180	40.50		220.50	267
3190	2'-0" wide		14	1.143		158	37.50		195.50	236
3200	Louvered, pine 6'-6" or 6'-8" x 2'-6" wide		13	1.231		144	40.50		184.50	226
3220	3'-0" wide		13	1.231		208	40.50		248.50	296
3225	Door, interior louvered bi-fold, pine, 3'-0" x 6'-8"					208			208	228
3240	4'-0" wide	2 Carp	12	1.333		235	44		279	330
3260	5'-0" wide		11	1.455		262	48		310	370
3280	6'-0" wide		10	1.600		289	52.50		341.50	410
3290	8'-0" wide		10	1.600		445	52.50		497.50	580
3300	7'-0" x 3'-0" wide		12	1.333		267	44		311	370
3320	6'-0" wide		10	1.600		305	52.50		357.50	425
4400	Bi-passing closet, incl. hardware and frame, no trim incl.									
4420	Flush, lauan, 6'-8" x 4'-0" wide	2 Carp	12	1.333	Opng.	176	44		220	268
4440	5'-0" wide		11	1.455		187	48		235	285
4460	6'-0" wide		10	1.600		209	52.50		261.50	320
4600	Flush, birch, 6'-8" x 4'-0" wide		12	1.333		223	44		267	320
4620	5'-0" wide		11	1.455		212	48		260	315
4640	6'-0" wide		10	1.600		260	52.50		312.50	375
4800	Louvered, pine, 6'-8" x 4'-0" wide		12	1.333		410	44		454	530
4820	5'-0" wide		11	1.455		395	48		443	515
4840	6'-0" wide		10	1.600		535	52.50		587.50	675
4900	Mirrored, 6'-8" x 4'-0" wide		12	1.333	Ea.	300	44		344	405
5000	Paneled, pine, 6'-8" x 4'-0" wide		12	1.333	Opng.	515	44		559	640
5020	5'-0" wide		11	1.455		410	48		458	530
5040	6'-0" wide		10	1.600		610	52.50		662.50	760
5042	8'-0" wide		12	1.333		1,025	44		1,069	1,200
5061	Hardboard, 6'-8" x 4'-0" wide		10	1.600		190	52.50		242.50	297
5062	5'-0" wide		10	1.600		196	52.50		248.50	305
5063	6'-0" wide		10	1.600		217	52.50		269.50	325
6100	Folding accordion, closet, including track and frame									
6200	Rigid PVC	2 Carp	10	1.600	Ea.	55.50	52.50		108	149
7310	Passage doors, flush, no frame included									
7320	Hardboard, hollow core, 1-3/8" x 6'-8" x 1'-6" wide	2 Carp	18	.889	Ea.	40.50	29		69.50	93.50
7330	2'-0" wide		18	.889		43	29		72	96.50
7340	2'-6" wide		18	.889		47.50	29		76.50	102
7350	2'-8" wide		18	.889		49	29		78	103
7360	3'-0" wide		17	.941		52	31		83	110
7420	Lauan, hollow core, 1-3/8" x 6'-8" x 1'-6" wide		18	.889		35	29		64	87.50
7440	2'-0" wide		18	.889		35	29		64	87.50
7450	2'-4" wide		18	.889		38.50	29		67.50	91.50
7460	2'-6" wide		18	.889		38.50	29		67.50	91.50
7480	2'-8" wide		18	.889		40	29		69	93
7500	3'-0" wide		17	.941		42.50	31		73.50	98.50
7540	2'-6" wide		16	1		49.50	33		82.50	110
7560	2'-8" wide		16	1		50.50	33		83.50	111
7580	3'-0" wide		16	1		52.50	33		85.50	113
7595	Pair of 3'-0" wide		9	1.778	Pr.	105	58.50		163.50	214
7700	Birch, hollow core, 1-3/8" x 6'-8" x 1'-6" wide		18	.889	Ea.	41	29		70	94
7720	2'-0" wide		18	.889		42.50	29		71.50	95.50
7740	2'-6" wide		18	.889		50	29		79	104

08 14 Wood Doors

08 14 33 – Stile and Rail Wood Doors

08 14 33.20 Wood Doors Residential		Crew	Daily Output	Labor-Hours	Unit	Material	2015 Bare Costs Labor	Equipment	Total	Total Incl O&P
7760	2'-8" wide	2 Carp	18	.889	Ea.	50.50	29		79.50	105
7780	3'-0" wide		17	.941		52	31		83	110
7790	2'-6" ash/oak door with hinges		18	.889		90	29		119	148
7910	2'-8" wide		16	1		64.50	33		97.50	126
7920	3'-0" wide		16	1		68.50	33		101.50	131
7940	Pair of 3'-0" wide		9	1.778	Pr.	146	58.50		204.50	259
8000	Pine louvered, 1-3/8" x 6'-8" x 1'-6" wide		19	.842	Ea.	120	27.50		147.50	179
8020	2'-0" wide		18	.889		131	29		160	193
8040	2'-6" wide		18	.889		150	29		179	214
8060	2'-8" wide		18	.889		161	29		190	226
8080	3'-0" wide		17	.941		175	31		206	244
8300	Pine paneled, 1-3/8" x 6'-8" x 1'-6" wide		19	.842		128	27.50		155.50	188
8320	2'-0" wide		18	.889		153	29		182	217
8330	2'-4" wide		18	.889		182	29		211	249
8340	2'-6" wide		18	.889		182	29		211	249
8360	2'-8" wide		18	.889		185	29		214	253
8380	3'-0" wide		17	.941		204	31		235	276
8450	French door, pine, 15 lites, 1-3/8" x 6'-8" x 2'-6" wide		18	.889		195	29		224	264
8470	2'-8" wide		18	.889		251	29		280	325
8490	3'-0" wide		17	.941		276	31		307	355
8804	Pocket door, 6 panel pine, 2'-6" x 6'-8" with frame		10.50	1.524		315	50		365	430
8814	2'-8" x 6'-8"		10.50	1.524		325	50		375	440
8824	3'-0" x 6'-8"		10.50	1.524		355	50		405	475
9000	Passage doors, flush, no frame, birch, solid core, 1-3/8" x 2'-4" x 7'-0"		16	1		118	33		151	185
9020	2'-8" wide		16	1		125	33		158	193
9040	3'-0" wide		16	1		135	33		168	204
9060	3'-4" wide		15	1.067		156	35		191	231
9080	Pair of 3'-0" wide		9	1.778	Pr.	267	58.50		325.50	390
9100	Lauan, solid core, 1-3/8" x 7'-0" x 2'-4" wide		16	1	Ea.	108	33		141	174
9120	2'-8" wide		16	1		116	33		149	182
9140	3'-0" wide		16	1		123	33		156	190
9160	3'-4" wide		15	1.067		129	35		164	201
9180	Pair of 3'-0" wide		9	1.778	Pr.	223	58.50		281.50	345
9200	Hardboard, solid core, 1-3/8" x 7'-0" x 2'-4" wide		16	1	Ea.	134	33		167	203
9220	2'-8" wide		16	1		139	33		172	208
9240	3'-0" wide		16	1		143	33		176	212
9260	3'-4" wide		15	1.067		158	35		193	233

08 14 35 – Torrified Doors

08 14 35.10 Torrified Exterior Doors

		Crew	Daily Output	Labor-Hours	Unit	Material	Labor	Equipment	Total	Total Incl O&P
0010	**TORRIFIED EXTERIOR DOORS**									
0020	Wood doors made from torrified wood, exterior.									
0030	All doors require a finish be applied, all glass is insulated									
0040	All doors require pilot holes for all fasteners									
0100	6 panel, Paint grade poplar, 1-3/4" x 3'-0" x 6'-8"	2 Carp	12	1.333	Ea.	1,075	44		1,119	1,250
0120	Half glass 3'-0" x 6'-8"	"	12	1.333		1,175	44		1,219	1,375
0200	Side lite, full glass, 1-3/4" x 1'-2" x 6'-8"					905			905	995
0220	Side lite, half glass, 1-3/4" x 1'-2" x 6'-8"					905			905	995
0300	Raised Face, 2 Panel, Paint grade poplar, 1-3/4" x 3'-0" x 7'-0"	2 Carp	12	1.333		1,275	44		1,319	1,475
0320	Side lite, raised face, half glass, 1-3/4" x 1'-2" x 7'-0"					1,050			1,050	1,175
0500	6 panel, Fir, 1-3/4" x 3'-0" x 6'-8"	2 Carp	12	1.333		1,550	44		1,594	1,775
0520	Half glass 3'-0" x 6'-8"	"	12	1.333		1,650	44		1,694	1,900
0600	Side lite, full glass, 1-3/4" x 1'-2" x 6'-8"					1,150			1,150	1,275

08 14 Wood Doors

08 14 35 – Torrified Doors

08 14 35.10 Torrified Exterior Doors		Crew	Daily Output	Labor-Hours	Unit	Material	2015 Bare Costs Labor	Equipment	Total	Total Incl O&P
0620	Side lite, half glass, 1-3/4" x 1'-2" x 6'-8"				Ea.	1,450			1,450	1,600
0700	6 panel, Mahogany, 1-3/4" x 3'-0"x 6'-8"	2 Carp	12	1.333		1,625	44		1,669	1,850
0800	Side lite, full glass, 1-3/4" x 1'-2" x 6'-8"					1,225			1,225	1,350
0820	Side lite, half glass, 1-3/4" x 1'-2" x 6'-8"					1,200			1,200	1,325

08 14 40 – Interior Cafe Doors

08 14 40.10 Cafe Style Doors										
0010	**CAFE STYLE DOORS**									
6520	Interior cafe doors, 2'-6" opening, stock, panel pine	2 Carp	16	1	Ea.	218	33		251	294
6540	3'-0" opening	"	16	1	"	240	33		273	320
6550	Louvered pine									
6560	2'-6" opening	2 Carp	16	1	Ea.	180	33		213	253
8000	3'-0" opening		16	1		193	33		226	267
8010	2'-6" opening, hardwood		16	1		264	33		297	345
8020	3'-0" opening		16	1		281	33		314	365

08 16 Composite Doors

08 16 13 – Fiberglass Doors

08 16 13.10 Entrance Doors, Fiberous Glass			Crew	Daily Output	Labor-Hours	Unit	Material	Labor	Equipment	Total	Total Incl O&P
0010	**ENTRANCE DOORS, FIBEROUS GLASS**										
0020	Exterior, fiberglass, door, 2'-8" wide x 6'-8" high	G	2 Carp	15	1.067	Ea.	270	35		305	355
0040	3'-0" wide x 6'-8" high	G		15	1.067		270	35		305	355
0060	3'-0" wide x 7'-0" high	G		15	1.067		460	35		495	565
0080	3'-0" wide x 6'-8" high, with two lites	G		15	1.067		315	35		350	405
0100	3'-0" wide x 8'-0" high, with two lites	G		15	1.067		525	35		560	640
0110	Half glass, 3'-0" wide x 6'-8" high	G		15	1.067		435	35		470	540
0120	3'-0" wide x 6'-8" high, low e	G		15	1.067		465	35		500	570
0130	3'-0" wide x 8'-0" high	G		15	1.067		585	35		620	705
0140	3'-0" wide x 8'-0" high, low e	G		15	1.067		655	35		690	730
0150	Side lights, 1'-0" wide x 6'-8" high	G					266			266	293
0160	1'-0" wide x 6'-8" high, low e	G					281			281	310
0180	1'-0" wide x 6'-8" high, full glass	G					310			310	340
0190	1'-0" wide x 6'-8" high, low e	G					340			340	370

08 16 14 – French Doors

08 16 14.10 Exterior Doors With Glass Lites		Crew	Daily Output	Labor-Hours	Unit	Material	Labor	Equipment	Total	Total Incl O&P
0010	**EXTERIOR DOORS WITH GLASS LITES**									
0020	French, Fir, 1-3/4", 3'-0"wide x 6'-8" high	2 Carp	12	1.333	Ea.	600	44		644	735
0025	Double		12	1.333		1,200	44		1,244	1,400
0030	Maple, 1-3/4", 3'-0"wide x 6'-8" high		12	1.333		675	44		719	820
0035	Double		12	1.333		1,350	44		1,394	1,550
0040	Cherry, 1-3/4", 3'-0"wide x 6'-8" high		12	1.333		790	44		834	940
0045	Double		12	1.333		1,575	44		1,619	1,800
0100	Mahogany, 1-3/4", 3'-0"wide x 8'-0" high		10	1.600		800	52.50		852.50	970
0105	Double		10	1.600		1,600	52.50		1,652.50	1,850
0110	Fir, 1-3/4", 3'-0"wide x 8'-0" high		10	1.600		1,200	52.50		1,252.50	1,425
0115	Double		10	1.600		2,400	52.50		2,452.50	2,750
0120	Oak, 1-3/4", 3'-0"wide x 8'-0" high		10	1.600		1,825	52.50		1,877.50	2,100
0125	Double		10	1.600		3,650	52.50		3,702.50	4,100

08 17 Integrated Door Opening Assemblies

08 17 23 – Integrated Wood Door Opening Assemblies

08 17 23.10 Pre-Hung Doors		Crew	Daily Output	Labor-Hours	Unit	Material	2015 Bare Costs Labor	Equipment	Total	Total Incl O&P
0010	**PRE-HUNG DOORS**									
0300	Exterior, wood, comb. storm & screen, 6'-9" x 2'-6" wide	2 Carp	15	1.067	Ea.	296	35		331	385
0320	2'-8" wide		15	1.067		296	35		331	385
0340	3'-0" wide		15	1.067		305	35		340	395
0360	For 7'-0" high door, add					30			30	33
1600	Entrance door, flush, birch, solid core									
1620	4-5/8" solid jamb, 1-3/4" x 6'-8" x 2'-8" wide	2 Carp	16	1	Ea.	289	33		322	375
1640	3'-0" wide		16	1		375	33		408	470
1642	5-5/8" jamb		16	1		325	33		358	415
1680	For 7'-0" high door, add					25			25	27.50
2000	Entrance door, colonial, 6 panel pine									
2020	4-5/8" solid jamb, 1-3/4" x 6'-8" x 2'-8" wide	2 Carp	16	1	Ea.	640	33		673	760
2040	3'-0" wide	"	16	1		675	33		708	795
2060	For 7'-0" high door, add					54			54	59
2200	For 5-5/8" solid jamb, add					41.50			41.50	46
2230	French style, exterior, 1 lite, 1-3/4" x 3'-0" x 6'-8"	1 Carp	14	.571		590	18.75		608.75	680
2235	9 lites	"	14	.571		660	18.75		678.75	755
2245	15 lites	2 Carp	14	1.143		640	37.50		677.50	770
2250	Double, 15 lites, 2'-0" x 6'-8", 4'-0" opening		7	2.286	Pr.	1,200	75		1,275	1,450
2260	2'-6" x 6'-8", 5'-0" opening		7	2.286		1,325	75		1,400	1,575
2280	3'-0" x 6'-8", 6'-0" opening		7	2.286		1,350	75		1,425	1,600
2430	3'-0" x 7'-0", 15 lites		14	1.143	Ea.	950	37.50		987.50	1,125
2432	Two 3'-0" x 7'-0"		7	2.286	Pr.	1,975	75		2,050	2,275
2435	3'-0" x 8'-0"		14	1.143	Ea.	1,000	37.50		1,037.50	1,175
2437	Two, 3'-0" x 8'-0"		7	2.286	Pr.	2,075	75		2,150	2,425
2500	Exterior, metal face, insulated, incl. jamb, brickmold and									
2520	Threshold, flush, 2'-8" x 6'-8"	2 Carp	16	1	Ea.	289	33		322	375
2550	3'-0" x 6'-8"		16	1		289	33		322	375
3500	Embossed, 6 panel, 2'-8" x 6'-8"		16	1		320	33		353	410
3550	3'-0" x 6'-8"		16	1		325	33		358	410
3600	2 narrow lites, 2'-8" x 6'-8"		16	1		292	33		325	375
3650	3'-0" x 6'-8"		16	1		296	33		329	380
3700	Half glass, 2'-8" x 6'-8"		16	1		315	33		348	405
3750	3'-0" x 6'-8"		16	1		340	33		373	425
3800	2 top lites, 2'-8" x 6'-8"		16	1		300	33		333	385
3850	3'-0" x 6'-8"		16	1		315	33		348	400
4000	Interior, passage door, 4-5/8" solid jamb									
4370	Pine, louvered, 2'-8" x 6'-8"	2 Carp	17	.941	Ea.	193	31		224	264
4380	3'-0"		17	.941		203	31		234	275
4400	Lauan, flush, solid core, 1-3/8" x 6'-8" x 2'-6" wide		17	.941		186	31		217	257
4420	2'-8" wide		17	.941		186	31		217	257
4440	3'-0" wide		16	1		201	33		234	276
4600	Hollow core, 1-3/8" x 6'-8" x 2'-6" wide		17	.941		125	31		156	190
4620	2'-8" wide		17	.941		125	31		156	189
4640	3'-0" wide		16	1		140	33		173	209
4700	For 7'-0" high door, add					35.50			35.50	39
5000	Birch, flush, solid core, 1-3/8" x 6'-8" x 2'-6" wide	2 Carp	17	.941		275	31		306	355
5020	2'-8" wide		17	.941		201	31		232	273
5040	3'-0" wide		16	1		300	33		333	390
5200	Hollow core, 1-3/8" x 6'-8" x 2'-6" wide		17	.941		222	31		253	296
5220	2'-8" wide		17	.941		266	31		297	345
5240	3'-0" wide		16	1		230	33		263	310
5280	For 7'-0" high door, add					30.50			30.50	33.50

For customer support on your Residential Cost Data, call 877.713.2617.

08 17 Integrated Door Opening Assemblies

08 17 23 – Integrated Wood Door Opening Assemblies

08 17 23.10 Pre-Hung Doors		Crew	Daily Output	Labor-Hours	Unit	Material	2015 Bare Costs Labor	Equipment	Total	Total Incl O&P
5500	Hardboard paneled, 1-3/8" x 6'-8" x 2'-6" wide	2 Carp	17	.941	Ea.	145	31		176	211
5520	2'-8" wide		17	.941		153	31		184	220
5540	3'-0" wide		16	1		150	33		183	220
6000	Pine paneled, 1-3/8" x 6'-8" x 2'-6" wide		17	.941		255	31		286	335
6020	2'-8" wide		17	.941		274	31		305	350
6040	3'-0" wide		16	1		282	33		315	365
7200	Prehung, bifold, mirrored, 6'-8" x 5'-0"	1 Carp	9	.889		380	29		409	470
7220	6'-8" x 6'-0"		9	.889		380	29		409	470
7240	6'-8" x 8'-0"		6	1.333		635	44		679	770
7600	Oak, 6 panel, 1-3/4" x 6'-8" x 3'-0"		17	.471		895	15.45		910.45	1,000
8500	Pocket door frame with lauan, flush, hollow core, 1-3/8" x 3'-0" x 6'-8"		17	.471		223	15.45		238.45	272

08 31 Access Doors and Panels

08 31 13 – Access Doors and Frames

08 31 13.20 Bulkhead/Cellar Doors

0010	BULKHEAD/CELLAR DOORS									
0020	Steel, not incl. sides, 44" x 62"	1 Carp	5.50	1.455	Ea.	560	48		608	695
0100	52" x 73"		5.10	1.569		790	51.50		841.50	955
0500	With sides and foundation plates, 57" x 45" x 24"		4.70	1.702		845	56		901	1,025
0600	42" x 49" x 51"		4.30	1.860		915	61		976	1,100

08 31 13.40 Kennel Doors

0010	KENNEL DOORS									
0020	2 way, swinging type, 13" x 19" opening	2 Carp	11	1.455	Opng.	90	48		138	179
0100	17" x 29" opening		11	1.455		110	48		158	201
0200	9" x 9" opening, electronic with accessories		11	1.455		144	48		192	238

08 32 Sliding Glass Doors

08 32 13 – Sliding Aluminum-Framed Glass Doors

08 32 13.10 Sliding Aluminum Doors

0010	SLIDING ALUMINUM DOORS									
0350	Aluminum, 5/8" tempered insulated glass, 6' wide									
0400	Premium	2 Carp	4	4	Ea.	1,575	131		1,706	1,950
0450	Economy		4	4		815	131		946	1,125
0500	8' wide, premium		3	5.333		1,650	175		1,825	2,125
0550	Economy		3	5.333		1,425	175		1,600	1,850
0600	12' wide, premium		2.50	6.400		2,975	210		3,185	3,625
0650	Economy		2.50	6.400		1,550	210		1,760	2,050
4000	Aluminum, baked on enamel, temp glass, 6'-8" x 10'-0" wide		4	4		1,075	131		1,206	1,400
4020	Insulating glass, 6'-8" x 6'-0" wide		4	4		935	131		1,066	1,250
4040	8'-0" wide		3	5.333		1,100	175		1,275	1,500
4060	10'-0" wide		2	8		1,350	263		1,613	1,925
4080	Anodized, temp glass, 6'-8" x 6'-0" wide		4	4		455	131		586	720
4100	8'-0" wide		3	5.333		575	175		750	930
4120	10'-0" wide		2	8		645	263		908	1,150

08 32 Sliding Glass Doors

08 32 19 – Sliding Wood-Framed Glass Doors

08 32 19.10 Sliding Wood Doors

		Crew	Daily Output	Labor-Hours	Unit	Material	2015 Bare Costs Labor	2015 Bare Costs Equipment	Total	Total Incl O&P
0010	**SLIDING WOOD DOORS**									
0020	Wood, tempered insul. glass, 6' wide, premium	2 Carp	4	4	Ea.	1,425	131		1,556	1,800
0100	Economy		4	4		1,175	131		1,306	1,500
0150	8' wide, wood, premium		3	5.333		1,825	175		2,000	2,300
0200	Economy		3	5.333		1,500	175		1,675	1,950
0235	10' wide, wood, premium		2.50	6.400		2,600	210		2,810	3,200
0240	Economy		2.50	6.400		2,200	210		2,410	2,775
0250	12' wide, wood, premium		2.50	6.400		3,000	210		3,210	3,650
0300	Economy		2.50	6.400		2,450	210		2,660	3,050

08 32 19.15 Sliding Glass Vinyl-Clad Wood Doors

			Crew	Daily Output	Labor-Hours	Unit	Material	Labor	Equipment	Total	Total Incl O&P
0010	**SLIDING GLASS VINYL-CLAD WOOD DOORS**										
0020	Glass, sliding vinyl clad, insul. glass, 6'-0" x 6'-8"	G	2 Carp	4	4	Opng.	1,500	131		1,631	1,900
0025	6'-0" x 6'-10" high	G		4	4		1,650	131		1,781	2,050
0030	6'-0" x 8'-0" high	G		4	4		1,975	131		2,106	2,400
0050	5'-0" x 6'-8" high	G		4	4		1,500	131		1,631	1,875
0100	8'-0" x 6'-10" high	G		4	4		2,025	131		2,156	2,450
0104	8'-0" x 6'-8"	G		4	4		1,925	131		2,056	2,325
0150	8'-0" x 8'-0" high	G		4	4		2,225	131		2,356	2,675
0500	4 leaf, 9'-0" x 6'-10" high	G		3	5.333		3,250	175		3,425	3,875
0550	9'-0" x 8'-0" high	G		3	5.333		3,775	175		3,950	4,450
0600	12'-0" x 6'-10" high	G		3	5.333		3,875	175		4,050	4,575

08 36 Panel Doors

08 36 13 – Sectional Doors

08 36 13.20 Residential Garage Doors

		Crew	Daily Output	Labor-Hours	Unit	Material	Labor	Equipment	Total	Total Incl O&P
0010	**RESIDENTIAL GARAGE DOORS**									
0050	Hinged, wood, custom, double door, 9' x 7'	2 Carp	4	4	Ea.	800	131		931	1,100
0070	16' x 7'		3	5.333		1,200	175		1,375	1,625
0200	Overhead, sectional, incl. hardware, fiberglass, 9' x 7', standard		5.28	3.030		945	99.50		1,044.50	1,225
0220	Deluxe		5.28	3.030		1,150	99.50		1,249.50	1,425
0300	16' x 7', standard		6	2.667		1,575	87.50		1,662.50	1,875
0320	Deluxe		6	2.667		2,150	87.50		2,237.50	2,500
0500	Hardboard, 9' x 7', standard		8	2		625	65.50		690.50	800
0520	Deluxe		8	2		815	65.50		880.50	1,000
0600	16' x 7', standard		6	2.667		1,225	87.50		1,312.50	1,500
0620	Deluxe		6	2.667		1,425	87.50		1,512.50	1,725
0700	Metal, 9' x 7', standard		5.28	3.030		740	99.50		839.50	980
0720	Deluxe		8	2		915	65.50		980.50	1,100
0800	16' x 7', standard		3	5.333		940	175		1,115	1,325
0820	Deluxe		6	2.667		1,400	87.50		1,487.50	1,700
0900	Wood, 9' x 7', standard		8	2		970	65.50		1,035.50	1,175
0920	Deluxe		8	2		2,150	65.50		2,215.50	2,450
1000	16' x 7', standard		6	2.667		1,625	87.50		1,712.50	1,950
1020	Deluxe		6	2.667		3,025	87.50		3,112.50	3,475
1800	Door hardware, sectional	1 Carp	4	2		350	65.50		415.50	495
1810	Door tracks only		4	2		163	65.50		228.50	289
1820	One side only		7	1.143		120	37.50		157.50	195
3000	Swing-up, including hardware, fiberglass, 9' x 7', standard	2 Carp	8	2		1,000	65.50		1,065.50	1,200
3020	Deluxe		8	2		1,100	65.50		1,165.50	1,300
3100	16' x 7', standard		6	2.667		1,250	87.50		1,337.50	1,525

For customer support on your Residential Cost Data, call 877.713.2617.

08 36 Panel Doors

08 36 13 – Sectional Doors

08 36 13.20 Residential Garage Doors

		Crew	Daily Output	Labor-Hours	Unit	Material	2015 Bare Costs Labor	Equipment	Total	Total Incl O&P
3120	Deluxe	2 Carp	6	2.667	Ea.	1,600	87.50		1,687.50	1,900
3200	Hardboard, 9' x 7', standard		8	2		550	65.50		615.50	715
3220	Deluxe		8	2		650	65.50		715.50	825
3300	16' x 7', standard		6	2.667		670	87.50		757.50	880
3320	Deluxe		6	2.667		850	87.50		937.50	1,075
3400	Metal, 9' x 7', standard		8	2		600	65.50		665.50	770
3420	Deluxe		8	2		965	65.50		1,030.50	1,150
3500	16' x 7', standard		6	2.667		800	87.50		887.50	1,025
3520	Deluxe		6	2.667		1,100	87.50		1,187.50	1,350
3600	Wood, 9' x 7', standard		8	2		700	65.50		765.50	880
3620	Deluxe		8	2		1,125	65.50		1,190.50	1,325
3700	16' x 7', standard		6	2.667		900	87.50		987.50	1,125
3720	Deluxe		6	2.667		2,100	87.50		2,187.50	2,450
3900	Door hardware only, swing up	1 Carp	4	2		168	65.50		233.50	295
3920	One side only		7	1.143		90	37.50		127.50	162
4000	For electric operator, economy, add		8	1		425	33		458	525
4100	Deluxe, including remote control		8	1		610	33		643	730
4500	For transmitter/receiver control, add to operator				Total	110			110	121
4600	Transmitters, additional				"	60			60	66
6000	Replace section, on sectional door, fiberglass, 9' x 7'	1 Carp	4	2	Ea.	640	65.50		705.50	810
6020	16' x 7'		3.50	2.286		745	75		820	945
6200	Hardboard, 9' x 7'		4	2		115	65.50		180.50	236
6220	16' x 7'		3.50	2.286		218	75		293	365
6300	Metal, 9' x 7'		4	2		184	65.50		249.50	315
6320	16' x 7'		3.50	2.286		300	75		375	455
6500	Wood, 9' x 7'		4	2		116	65.50		181.50	238
6520	16' x 7'		3.50	2.286		225	75		300	375
7010	Garage doors, row of lites					115			115	127

08 51 Metal Windows

08 51 13 – Aluminum Windows

08 51 13.20 Aluminum Windows

		Crew	Daily Output	Labor-Hours	Unit	Material	2015 Bare Costs Labor	Equipment	Total	Total Incl O&P
0010	**ALUMINUM WINDOWS**, incl. frame and glazing, commercial grade									
1000	Stock units, casement, 3'-1" x 3'-2" opening	2 Sswk	10	1.600	Ea.	375	56.50		431.50	515
1040	Insulating glass	"	10	1.600		515	56.50		571.50	670
1050	Add for storms					120			120	132
1600	Projected, with screen, 3'-1" x 3'-2" opening	2 Sswk	10	1.600		355	56.50		411.50	495
1650	Insulating glass	"	10	1.600		385	56.50		441.50	525
1700	Add for storms					117			117	129
2000	4'-5" x 5'-3" opening	2 Sswk	8	2		400	70.50		470.50	575
2050	Insulating glass	"	8	2		470	70.50		540.50	650
2100	Add for storms					126			126	139
2500	Enamel finish windows, 3'-1" x 3'-2"	2 Sswk	10	1.600		360	56.50		416.50	500
2550	Insulating glass		10	1.600		310	56.50		366.50	445
2600	4'-5" x 5'-3"		8	2		405	70.50		475.50	580
2700	Insulating glass		8	2		410	70.50		480.50	585
3000	Single hung, 2' x 3' opening, enameled, standard glazed		10	1.600		208	56.50		264.50	335
3100	Insulating glass		10	1.600		252	56.50		308.50	385
3300	2'-8" x 6'-8" opening, standard glazed		8	2		365	70.50		435.50	535
3400	Insulating glass		8	2		475	70.50		545.50	655
3700	3'-4" x 5'-0" opening, standard glazed		9	1.778		300	63		363	450

08 51 Metal Windows

08 51 13 – Aluminum Windows

08 51 13.20 Aluminum Windows

		Crew	Daily Output	Labor-Hours	Unit	Material	2015 Bare Costs Labor	2015 Bare Costs Equipment	Total	Total Incl O&P
3800	Insulating glass	2 Sswk	9	1.778	Ea.	335	63		398	485
4000	Sliding aluminum, 3' x 2' opening, standard glazed		10	1.600		217	56.50		273.50	345
4100	Insulating glass		10	1.600		232	56.50		288.50	360
4300	5' x 3' opening, standard glazed		9	1.778		330	63		393	485
4400	Insulating glass		9	1.778		385	63		448	545
4600	8' x 4' opening, standard glazed		6	2.667		350	94		444	560
4700	Insulating glass		6	2.667		565	94		659	795
5000	9' x 5' opening, standard glazed		4	4		530	141		671	850
5100	Insulating glass		4	4		850	141		991	1,200
5500	Sliding, with thermal barrier and screen, 6' x 4', 2 track		8	2		725	70.50		795.50	930
5700	4 track		8	2		910	70.50		980.50	1,125
6000	For above units with bronze finish, add					15%				
6200	For installation in concrete openings, add					8%				

08 51 23 – Steel Windows

08 51 23.40 Basement Utility Windows

		Crew	Daily Output	Labor-Hours	Unit	Material	2015 Bare Costs Labor	2015 Bare Costs Equipment	Total	Total Incl O&P
0010	**BASEMENT UTILITY WINDOWS**									
0015	1'-3" x 2'-8"	1 Carp	16	.500	Ea.	128	16.45		144.45	168
1100	1'-7" x 2'-8"		16	.500		140	16.45		156.45	182
1200	1'-11" x 2'-8"		14	.571		143	18.75		161.75	190

08 51 66 – Metal Window Screens

08 51 66.10 Screens

		Crew	Daily Output	Labor-Hours	Unit	Material	2015 Bare Costs Labor	2015 Bare Costs Equipment	Total	Total Incl O&P
0010	**SCREENS**									
0020	For metal sash, aluminum or bronze mesh, flat screen	2 Sswk	1200	.013	S.F.	4.30	.47		4.77	5.60
0500	Wicket screen, inside window	"	1000	.016	"	6.65	.56		7.21	8.35
0600	Residential, aluminum mesh and frame, 2' x 3'	2 Carp	32	.500	Ea.	16	16.45		32.45	45
0610	Rescreen		50	.320		13	10.50		23.50	32
0620	3' x 5'		32	.500		52.50	16.45		68.95	85.50
0630	Rescreen		45	.356		33	11.70		44.70	56
0640	4' x 8'		25	.640		82.50	21		103.50	127
0650	Rescreen		40	.400		49	13.15		62.15	76
0660	Patio door		25	.640		200	21		221	256
0680	Rescreening		1600	.010	S.F.	2.50	.33		2.83	3.30
1000	Screens for solar louvers	2 Sswk	160	.100	"	24.50	3.53		28.03	33.50

08 52 Wood Windows

08 52 10 – Plain Wood Windows

08 52 10.10 Wood Windows

		Crew	Daily Output	Labor-Hours	Unit	Material	2015 Bare Costs Labor	2015 Bare Costs Equipment	Total	Total Incl O&P
0010	**WOOD WINDOWS**, including frame, screens and grilles									
0020	Residential, stock units									
0050	Awning type, double insulated glass, 2'-10" x 1'-9" opening	2 Carp	12	1.333	Opng.	228	44		272	325
0100	2'-10" x 6'-0" opening	1 Carp	8	1		545	33		578	655
0200	4'-0" x 3'-6" single pane		10	.800		365	26.50		391.50	450
0300	6' x 5' single pane		8	1	Ea.	500	33		533	605
1000	Casement, 2'-0" x 3'-4" high	2 Carp	20	.800		262	26.50		288.50	330
1020	2'-0" x 4'-0"		18	.889		279	29		308	355
1040	2'-0" x 5'-0"		17	.941		305	31		336	390
1060	2'-0" x 6'-0"		16	1		305	33		338	390
1080	4'-0" x 3'-4"		15	1.067		585	35		620	700
1100	4'-0" x 4'-0"		15	1.067		655	35		690	780
1120	4'-0" x 5'-0"		14	1.143		740	37.50		777.50	880

08 52 Wood Windows

08 52 10 – Plain Wood Windows

08 52 10.10 Wood Windows		Crew	Daily Output	Labor-Hours	Unit	Material	2015 Bare Costs Labor	Equipment	Total	Total Incl O&P
1140	4'-0" x 6'-0"	2 Carp	12	1.333	Ea.	835	44		879	990
1600	Casement units, 8' x 5', with screens, double insulated glass		2.50	6.400	Opng.	1,425	210		1,635	1,900
1700	Low E glass		2.50	6.400		1,600	210		1,810	2,100
2300	Casements, including screens, 2'-0" x 3'-4", dbl. insulated glass		11	1.455		269	48		317	375
2400	Low E glass		11	1.455		269	48		317	375
2600	2 lite, 4'-0" x 4'-0", double insulated glass		9	1.778		495	58.50		553.50	640
2700	Low E glass		9	1.778		505	58.50		563.50	655
2900	3 lite, 5'-2" x 5'-0", double insulated glass		7	2.286		750	75		825	950
3000	Low E glass		7	2.286		790	75		865	995
3200	4 lite, 7'-0" x 5'-0", double insulated glass		6	2.667		1,075	87.50		1,162.50	1,325
3300	Low E glass		6	2.667		1,125	87.50		1,212.50	1,400
3500	5 lite, 8'-6" x 5'-0", double insulated glass		5	3.200		1,425	105		1,530	1,725
3600	Low E glass		5	3.200		1,500	105		1,605	1,825
3800	For removable wood grilles, diamond pattern, add				Leaf	38			38	42
3900	Rectangular pattern, add				"	38			38	42
4000	Bow, fixed lites, 8' x 5', double insulated glass	2 Carp	3	5.333	Opng.	1,425	175		1,600	1,850
4100	Low E glass	"	3	5.333	"	1,875	175		2,050	2,375
4150	6'-0" x 5'-0"	1 Carp	8	1	Ea.	1,250	33		1,283	1,425
4300	Fixed lites, 9'-9" x 5'-0", double insulated glass	2 Carp	2	8	Opng.	975	263		1,238	1,525
4400	Low E glass	"	2	8	"	1,075	263		1,338	1,625
5000	Bow, casement, 8'-1" x 4'-8" high	3 Carp	8	3	Ea.	1,575	98.50		1,673.50	1,900
5020	9'-6" x 4'-8"		8	3		1,700	98.50		1,798.50	2,025
5040	8'-1" x 5'-1"		8	3		1,900	98.50		1,998.50	2,275
5060	9'-6" x 5'-1"		6	4		1,925	131		2,056	2,325
5080	8'-1" x 6'-0"		6	4		1,900	131		2,031	2,325
5100	9'-6" x 6'-0"		6	4		2,000	131		2,131	2,425
5800	Skylights, hatches, vents, and sky roofs, see Section 08 62 13.00									

08 52 10.20 Awning Window		Crew	Daily Output	Labor-Hours	Unit	Material	Labor	Equipment	Total	Total Incl O&P
0010	**AWNING WINDOW**, Including frame, screens and grilles									
0100	34" x 22", insulated glass	1 Carp	10	.800	Ea.	270	26.50		296.50	340
0200	Low E glass		10	.800		271	26.50		297.50	340
0300	40" x 28", insulated glass		9	.889		315	29		344	395
0400	Low E Glass		9	.889		340	29		369	425
0500	48" x 36", insulated glass		8	1		465	33		498	570
0600	Low E glass		8	1		490	33		523	595

08 52 10.30 Wood Windows		Crew	Daily Output	Labor-Hours	Unit	Material	Labor	Equipment	Total	Total Incl O&P
0010	**WOOD WINDOWS**, double hung									
0020	Including frame, double insulated glass, screens and grilles									
0040	Double hung, 2'-2" x 3'-4" high	2 Carp	15	1.067	Ea.	210	35		245	290
0060	2'-2" x 4'-4"		14	1.143		229	37.50		266.50	315
0080	2'-6" x 3'-4"		13	1.231		219	40.50		259.50	310
0100	2'-6" x 4'-0"		12	1.333		229	44		273	325
0120	2'-6" x 4'-8"		12	1.333		250	44		294	350
0140	2'-10" x 3'-4"		10	1.600		224	52.50		276.50	335
0160	2'-10" x 4'-0"		10	1.600		247	52.50		299.50	360
0180	3'-7" x 3'-4"		9	1.778		254	58.50		312.50	375
0200	3'-7" x 5'-4"		9	1.778		284	58.50		342.50	410
0220	3'-10" x 5'-4"		8	2		510	65.50		575.50	670
3800	Triple glazing for above, add					25%				

08 52 Wood Windows

08 52 10 – Plain Wood Windows

08 52 10.40 Casement Window		Crew	Daily Output	Labor-Hours	Unit	Material	2015 Bare Costs Labor	Equipment	Total	Total Incl O&P
0010	**CASEMENT WINDOW**, including frame, screen and grilles R085216-10									
0100	2'-0" x 3'-0" H, dbl. insulated glass [G]	1 Carp	10	.800	Ea.	268	26.50		294.50	340
0150	Low E glass [G]		10	.800		259	26.50		285.50	330
0200	2'-0" x 4'-6" high, double insulated glass [G]		9	.889		360	29		389	445
0250	Low E glass [G]		9	.889		370	29		399	455
0260	Casement 4'-2" x 4'-2" double insulated glass [G]		11	.727		875	24		899	1,000
0270	4'-0" x 4'-0" Low E glass [G]		11	.727		535	24		559	630
0290	6'-4" x 5'-7" Low E glass [G]		9	.889		1,125	29		1,154	1,275
0300	2'-4" x 6'-0" high, double insulated glass [G]		8	1		440	33		473	540
0350	Low E glass [G]		8	1		475	33		508	580
0522	Vinyl clad, premium, double insulated glass, 2'-0" x 3'-0" [G]		10	.800		271	26.50		297.50	340
0524	2'-0" x 4'-0" [G]		9	.889		315	29		344	400
0525	2'-0" x 5'-0" [G]		8	1		360	33		393	455
0528	2'-0" x 6'-0" [G]		8	1		380	33		413	475
0600	3'-0" x 5'-0" [G]		8	1		665	33		698	785
0700	4'-0" x 3'-0" [G]		8	1		730	33		763	860
0710	4'-0" x 4'-0" [G]		8	1		625	33		658	740
0720	4'-8" x 4'-0" [G]		8	1		690	33		723	815
0730	4'-8" x 5'-0" [G]		6	1.333		790	44		834	940
0740	4'-8" x 6'-0" [G]		6	1.333		880	44		924	1,050
0750	6'-0" x 4'-0" [G]		6	1.333		805	44		849	960
0800	6'-0" x 5'-0" [G]		6	1.333		895	44		939	1,050
0900	5'-6" x 5'-6" [G]	2 Carp	15	1.067		1,425	35		1,460	1,625
2000	Bay, casement units, 8' x 5', w/screens, dbl. insul. glass		2.50	6.400	Opng.	1,600	210		1,810	2,100
2100	Low E glass		2.50	6.400	"	1,675	210		1,885	2,200
3020	Vinyl clad, premium, double insulated glass, multiple leaf units									
3080	Single unit, 1'-6" x 5'-0" [G]	2 Carp	20	.800	Ea.	315	26.50		341.50	390
3100	2'-0" x 2'-0" [G]		20	.800		215	26.50		241.50	280
3140	2'-0" x 2'-6" [G]		20	.800		271	26.50		297.50	340
3220	2'-0" x 3'-6" [G]		20	.800		273	26.50		299.50	345
3260	2'-0" x 4'-0" [G]		19	.842		315	27.50		342.50	395
3300	2'-0" x 4'-6" [G]		19	.842		315	27.50		342.50	395
3340	2'-0" x 5'-0" [G]		18	.889		360	29		389	450
3460	2'-4" x 3'-0" [G]		20	.800		275	26.50		301.50	345
3500	2'-4" x 4'-0" [G]		19	.842		340	27.50		367.50	420
3540	2'-4" x 5'-0" [G]		18	.889		405	29		434	495
3700	Double unit, 2'-8" x 5'-0" [G]		18	.889		570	29		599	675
3740	2'-8" x 6'-0" [G]		17	.941		660	31		691	775
3840	3'-0" x 4'-6" [G]		18	.889		515	29		544	620
3860	3'-0" x 5'-0" [G]		17	.941		665	31		696	780
3880	3'-0" x 6'-0" [G]		17	.941		715	31		746	835
3980	3'-4" x 2'-6" [G]		19	.842		435	27.50		462.50	525
4000	3'-4" x 3'-0" [G]		12	1.333		435	44		479	555
4030	3'-4" x 4'-0" [G]		18	.889		530	29		559	635
4050	3'-4" x 5'-0" [G]		12	1.333		660	44		704	800
4100	3'-4" x 6'-0" [G]		11	1.455		715	48		763	865
4200	3'-6" x 3'-0" [G]		18	.889		520	29		549	620
4340	4'-0" x 3'-0" [G]		18	.889		490	29		519	585
4380	4'-0" x 3'-6" [G]		17	.941		530	31		561	630
4420	4'-0" x 4'-0" [G]		16	1		625	33		658	740
4460	4'-0" x 4'-4" [G]		16	1		625	33		658	740
4540	4'-0" x 5'-0" [G]		16	1		675	33		708	800

08 52 Wood Windows

08 52 10 – Plain Wood Windows

08 52 10.40 Casement Window

		Crew	Daily Output	Labor-Hours	Unit	Material	2015 Bare Costs Labor	Equipment	Total	Total Incl O&P
4580	4'-0" x 6'-0" G	2 Carp	15	1.067	Ea.	765	35		800	900
4740	4'-8" x 3'-0" G		18	.889		555	29		584	660
4780	4'-8" x 3'-6" G		17	.941		595	31		626	700
4820	4'-8" x 4'-0" G		16	1		690	33		723	815
4860	4'-8" x 5'-0" G		15	1.067		790	35		825	925
4900	4'-8" x 6'-0" G		15	1.067		880	35		915	1,025
5060	5'-0" x 5'-0" G		15	1.067		1,025	35		1,060	1,175
5100	Triple unit, 5'-6" x 3'-0" G		17	.941		700	31		731	820
5140	5'-6" x 3'-6" G		16	1		760	33		793	890
5180	5'-6" x 4'-6" G		15	1.067		855	35		890	1,000
5220	5'-6" x 5'-6" G		15	1.067		1,125	35		1,160	1,275
5300	6'-0" x 4'-6" G		15	1.067		835	35		870	975
5850	5'-0" x 3'-0" G		12	1.333		590	44		634	725
5900	5'-0" x 4'-0" G		11	1.455		910	48		958	1,075
6100	5'-0" x 5'-6" G		10	1.600		1,050	52.50		1,102.50	1,250
6150	5'-0" x 6'-0" G		10	1.600		1,175	52.50		1,227.50	1,400
6200	6'-0" x 3'-0" G		12	1.333		1,150	44		1,194	1,350
6250	6'-0" x 3'-4" G		12	1.333		740	44		784	890
6300	6'-0" x 4'-0" G		11	1.455		825	48		873	990
6350	6'-0" x 5'-0" G		10	1.600		900	52.50		952.50	1,075
6400	6'-0" x 6'-0" G		10	1.600		1,150	52.50		1,202.50	1,350
6500	Quadruple unit, 7'-0" x 4'-0" G		9	1.778		1,100	58.50		1,158.50	1,300
6700	8'-0" x 4'-6" G		9	1.778		1,350	58.50		1,408.50	1,600
6950	6'-8" x 4'-0" G		10	1.600		1,075	52.50		1,127.50	1,275
7000	6'-8" x 6'-0" G		10	1.600		1,425	52.50		1,477.50	1,675
8190	For installation, add per leaf						15%			
8200	For multiple leaf units, deduct for stationary sash									
8220	2' high				Ea.	23			23	25.50
8240	4'-6" high					26			26	28.50
8260	6' high					34.50			34.50	38

08 52 10.50 Double Hung

		Crew	Daily Output	Labor-Hours	Unit	Material	Labor	Equipment	Total	Total Incl O&P
0010	**DOUBLE HUNG**, Including frame, screens and grilles									
0100	2'-0" x 3'-0" high, low E insul. glass G	1 Carp	10	.800	Ea.	210	26.50		236.50	275
0200	3'-0" x 4'-0" high, double insulated glass G		9	.889		279	29		308	355
0300	4'-0" x 4'-6" high, low E insulated glass G		8	1		320	33		353	410

08 52 10.55 Picture Window

		Crew	Daily Output	Labor-Hours	Unit	Material	Labor	Equipment	Total	Total Incl O&P
0010	**PICTURE WINDOW**, Including frame and grilles									
0100	3'-6" x 4'-0" high, dbl. insulated glass	2 Carp	12	1.333	Ea.	420	44		464	540
0150	Low E glass		12	1.333		435	44		479	550
0200	4'-0" x 4'-6" high, double insulated glass		11	1.455		550	48		598	685
0250	Low E glass		11	1.455		530	48		578	665
0300	5'-0" x 4'-0" high, double insulated glass		11	1.455		580	48		628	720
0350	Low E glass		11	1.455		605	48		653	745
0400	6'-0" x 4'-6" high, double insulated glass		10	1.600		625	52.50		677.50	780
0450	Low E glass		10	1.600		635	52.50		687.50	790

08 52 10.65 Wood Sash

		Crew	Daily Output	Labor-Hours	Unit	Material	Labor	Equipment	Total	Total Incl O&P
0010	**WOOD SASH**, Including glazing but not trim									
0050	Custom, 5'-0" x 4'-0", 1" dbl. glazed, 3/16" thick lites	2 Carp	3.20	5	Ea.	230	164		394	530
0100	1/4" thick lites		5	3.200		245	105		350	445
0200	1" thick, triple glazed		5	3.200		415	105		520	630
0300	7'-0" x 4'-6" high, 1" double glazed, 3/16" thick lites		4.30	3.721		420	122		542	665
0400	1/4" thick lites		4.30	3.721		475	122		597	725

For customer support on your Residential Cost Data, call 877.713.2617.

08 52 Wood Windows

08 52 10 – Plain Wood Windows

08 52 10.65 Wood Sash		Crew	Daily Output	Labor-Hours	Unit	Material	2015 Bare Costs Labor	Equipment	Total	Total Incl O&P
0500	1" thick, triple glazed	2 Carp	4.30	3.721	Ea.	540	122		662	800
0600	8'-6" x 5'-0" high, 1" double glazed, 3/16" thick lites		3.50	4.571		565	150		715	875
0700	1/4" thick lites		3.50	4.571		620	150		770	930
0800	1" thick, triple glazed		3.50	4.571		625	150		775	935
0900	Window frames only, based on perimeter length				L.F.	4.02			4.02	4.42

08 52 10.70 Sliding Windows

0010	**SLIDING WINDOWS**									
0100	3'-0" x 3'-0" high, double insulated	G	1 Carp	10	.800	Ea.	284	26.50		310.50	355
0120	Low E glass	G		10	.800		310	26.50		336.50	385
0200	4'-0" x 3'-6" high, double insulated	G		9	.889		355	29		384	440
0220	Low E glass	G		9	.889		360	29		389	450
0300	6'-0" x 5'-0" high, double insulated	G		8	1		490	33		523	590
0320	Low E glass	G		8	1		530	33		563	635
6000	Sliding, insulating glass, including screens,										
6100	3'-0" x 3'-0"		2 Carp	6.50	2.462	Ea.	325	81		406	495
6200	4'-0" x 3'-6"			6.30	2.540		330	83.50		413.50	505
6300	5'-0" x 4'-0"			6	2.667		410	87.50		497.50	595

08 52 13 – Metal-Clad Wood Windows

08 52 13.10 Awning Windows, Metal-Clad

		Crew	Daily Output	Labor-Hours	Unit	Material	Labor	Equipment	Total	Total Incl O&P
0010	**AWNING WINDOWS, METAL-CLAD**									
2000	Metal clad, awning deluxe, double insulated glass, 34" x 22"	1 Carp	9	.889	Ea.	247	29		276	320
2050	36" x 25"		9	.889		272	29		301	350
2100	40" x 22"		9	.889		291	29		320	370
2150	40" x 30"		9	.889		340	29		369	425
2200	48" x 28"		8	1		345	33		378	435
2250	60" x 36"		8	1		370	33		403	465

08 52 13.20 Casement Windows, Metal-Clad

			Crew	Daily Output	Labor-Hours	Unit	Material	Labor	Equipment	Total	Total Incl O&P
0010	**CASEMENT WINDOWS, METAL-CLAD**										
0100	Metal clad, deluxe, dbl. insul. glass, 2'-0" x 3'-0" high	G	1 Carp	10	.800	Ea.	279	26.50		305.50	350
0120	2'-0" x 4'-0" high	G		9	.889		315	29		344	395
0130	2'-0" x 5'-0" high	G		8	1		325	33		358	415
0140	2'-0" x 6'-0" high	G		8	1		365	33		398	455
0150	Casement window, metal clad, double insul. glass, 3'-6" x 3'-6"	G		8.90	.899		480	29.50		509.50	580
0300	Metal clad, casement, bldrs mdl, 6'-0" x 4'-0", dbl. insltd gls, 3 panels		2 Carp	10	1.600		1,200	52.50		1,252.50	1,425
0310	9'-0" x 4'-0", 4 panels			8	2		1,550	65.50		1,615.50	1,800
0320	10'-0" x 5'-0", 5 panels			7	2.286		2,100	75		2,175	2,450
0330	12'-0" x 6'-0", 6 panels			6	2.667		2,700	87.50		2,787.50	3,100

08 52 13.30 Double-Hung Windows, Metal-clad

			Crew	Daily Output	Labor-Hours	Unit	Material	Labor	Equipment	Total	Total Incl O&P
0010	**DOUBLE-HUNG WINDOWS, METAL-CLAD**										
0100	Metal clad, deluxe, dbl. insul. glass, 2'-6" x 3'-0" high	G	1 Carp	10	.800	Ea.	272	26.50		298.50	345
0120	3'-0" x 3'-6" high	G		10	.800		315	26.50		341.50	390
0140	3'-0" x 4'-0" high	G		9	.889		325	29		354	410
0160	3'-0" x 4'-6" high	G		9	.889		345	29		374	430
0180	3'-0" x 5'-0" high	G		8	1		370	33		403	465
0200	3'-6" x 6'-0" high	G		8	1		450	33		483	550

08 52 13.35 Picture and Sliding Windows Metal-Clad

		Crew	Daily Output	Labor-Hours	Unit	Material	Labor	Equipment	Total	Total Incl O&P
0010	**PICTURE AND SLIDING WINDOWS METAL-CLAD**									
2000	Metal clad, dlx picture, dbl. insul. glass, 4'-0" x 4'-0" high	2 Carp	12	1.333	Ea.	375	44		419	485
2100	4'-0" x 6'-0" high		11	1.455		545	48		593	680
2200	5'-0" x 6'-0" high		10	1.600		610	52.50		662.50	760
2300	6'-0" x 6'-0" high		10	1.600		695	52.50		747.50	855

08 52 Wood Windows

08 52 13 – Metal-Clad Wood Windows

08 52 13.35 Picture and Sliding Windows Metal-Clad		Crew	Daily Output	Labor-Hours	Unit	Material	2015 Bare Costs Labor	Equipment	Total	Total Incl O&P
2400	Metal clad, dlx sliding, double insulated glass, 3'-0" x 3'-0" high [G]	1 Carp	10	.800	Ea.	330	26.50		356.50	405
2420	4'-0" x 3'-6" high [G]		9	.889		400	29		429	490
2440	5'-0" x 4'-0" high [G]		9	.889		480	29		509	575
2460	6'-0" x 5'-0" high [G]	↓	8	1	↓	730	33		763	855

08 52 13.40 Bow and Bay Windows, Metal-Clad

		Crew	Daily Output	Labor-Hours	Unit	Material	Labor	Equipment	Total	Total Incl O&P
0010	**BOW AND BAY WINDOWS, METAL-CLAD**									
0100	Metal clad, deluxe, dbl. insul. glass, 8'-0" x 5'-0" high, 4 panels	2 Carp	10	1.600	Ea.	1,675	52.50		1,727.50	1,925
0120	10'-0" x 5'-0" high, 5 panels		8	2		1,800	65.50		1,865.50	2,075
0140	10'-0" x 6'-0" high, 5 panels		7	2.286		2,100	75		2,175	2,450
0160	12'-0" x 6'-0" high, 6 panels		6	2.667		2,925	87.50		3,012.50	3,375
0400	Double hung, bldrs. model, bay, 8' x 4' high, dbl. insulated glass		10	1.600		1,325	52.50		1,377.50	1,550
0440	Low E glass		10	1.600		1,425	52.50		1,477.50	1,675
0460	9'-0" x 5'-0" high, double insulated glass		6	2.667		1,425	87.50		1,512.50	1,725
0480	Low E glass		6	2.667		1,500	87.50		1,587.50	1,800
0500	Metal clad, deluxe, dbl. insul. glass, 7'-0" x 4'-0" high		10	1.600		1,275	52.50		1,327.50	1,500
0520	8'-0" x 4'-0" high		8	2		1,300	65.50		1,365.50	1,550
0540	8'-0" x 5'-0" high		7	2.286		1,350	75		1,425	1,625
0560	9'-0" x 5'-0" high	↓	6	2.667	↓	1,450	87.50		1,537.50	1,725

08 52 16 – Plastic-Clad Wood Windows

08 52 16.10 Bow Window

		Crew	Daily Output	Labor-Hours	Unit	Material	Labor	Equipment	Total	Total Incl O&P
0010	**BOW WINDOW** Including frames, screens, and grilles									
0020	End panels operable									
1000	Bow type, casement, wood, bldrs. mdl., 8' x 5' dbl. insltd glass, 4 panel	2 Carp	10	1.600	Ea.	1,500	52.50		1,552.50	1,750
1050	Low E glass		10	1.600		1,325	52.50		1,377.50	1,550
1100	10'-0" x 5'-0", double insulated glass, 6 panels		6	2.667		1,350	87.50		1,437.50	1,650
1200	Low E glass, 6 panels		6	2.667		1,450	87.50		1,537.50	1,750
1300	Vinyl clad, bldrs. model, double insulated glass, 6'-0" x 4'-0", 3 panel		10	1.600		1,025	52.50		1,077.50	1,225
1340	9'-0" x 4'-0", 4 panel		8	2		1,350	65.50		1,415.50	1,600
1380	10'-0" x 6'-0", 5 panels		7	2.286		2,250	75		2,325	2,600
1420	12'-0" x 6'-0", 6 panels		6	2.667		2,925	87.50		3,012.50	3,375
2000	Bay window, 8' x 5', dbl. insul glass		10	1.600		1,875	52.50		1,927.50	2,175
2050	Low E glass		10	1.600		2,275	52.50		2,327.50	2,600
2100	12'-0" x 6'-0", double insulated glass, 6 panels		6	2.667		2,350	87.50		2,437.50	2,725
2200	Low E glass		6	2.667		2,375	87.50		2,462.50	2,775
2280	6'-0" x 4'-0"		11	1.455		1,250	48		1,298	1,450
2300	Vinyl clad, premium, double insulated glass, 8'-0" x 5'-0"		10	1.600		1,750	52.50		1,802.50	2,025
2340	10'-0" x 5'-0"		8	2		2,300	65.50		2,365.50	2,625
2380	10'-0" x 6'-0"		7	2.286		2,625	75		2,700	3,000
2420	12'-0" x 6'-0"		6	2.667		3,200	87.50		3,287.50	3,675
2430	14'-0" x 3'-0"		7	2.286		2,950	75		3,025	3,375
2440	14'-0" x 6'-0"		5	3.200		5,000	105		5,105	5,675
3300	Vinyl clad, premium, double insulated glass, 7'-0" x 4'-6"		10	1.600		1,375	52.50		1,427.50	1,600
3340	8'-0" x 4'-6"		8	2		1,400	65.50		1,465.50	1,625
3380	8'-0" x 5'-0"		7	2.286		1,450	75		1,525	1,725
3420	9'-0" x 5'-0"	↓	6	2.667	↓	1,500	87.50		1,587.50	1,800

08 52 16.15 Awning Window Vinyl-Clad

		Crew	Daily Output	Labor-Hours	Unit	Material	Labor	Equipment	Total	Total Incl O&P
0010	**AWNING WINDOW VINYL-CLAD** Including frames, screens, and grilles									
0200	Vinyl clad, premium, double insulated glass, 24" x 17"	1 Carp	12	.667	Ea.	210	22		232	268
0210	24" x 28"		11	.727		254	24		278	320
0250	36" x 17"		9	.889		261	29		290	335
0280	36" x 28"		9	.889		305	29		334	385
0300	36" x 36"	↓	9	.889	↓	340	29		369	425

08 52 Wood Windows

08 52 16 – Plastic-Clad Wood Windows

08 52 16.15 Awning Window Vinyl-Clad		Crew	Daily Output	Labor-Hours	Unit	Material	2015 Bare Costs Labor	Equipment	Total	Total Incl O&P
0320	36" x 40"	1 Carp	9	.889	Ea.	395	29		424	480
0340	40" x 22"		10	.800		288	26.50		314.50	360
0360	48" x 28"		8	1		370	33		403	460
0380	60" x 36"		8	1		500	33		533	605
08 52 16.20 Half Round, Vinyl Clad										
0010	**HALF ROUND, VINYL CLAD**, double insulated glass, incl. grille									
0800	14" height x 24" base	2 Carp	9	1.778	Ea.	415	58.50		473.50	555
1040	15" height x 25" base		8	2		385	65.50		450.50	530
1060	16" height x 28" base		7	2.286		405	75		480	570
1080	17" height x 29" base		7	2.286		420	75		495	585
2000	19" height x 33" base	1 Carp	6	1.333		465	44		509	585
2100	20" height x 35" base		6	1.333		465	44		509	585
2200	21" height x 37" base		6	1.333		480	44		524	605
2250	23" height x 41" base	2 Carp	6	2.667		520	87.50		607.50	715
2300	26" height x 48" base		6	2.667		535	87.50		622.50	730
2350	30" height x 56" base		6	2.667		615	87.50		702.50	820
3000	36" height x 67" base	1 Carp	4	2		1,050	65.50		1,115.50	1,250
3040	38" height x 71" base	2 Carp	5	3.200		975	105		1,080	1,250
3050	40" height x 75" base	"	5	3.200		1,275	105		1,380	1,575
5000	Elliptical, 71" x 16"	1 Carp	11	.727		975	24		999	1,125
5100	95" x 21"	"	10	.800		1,375	26.50		1,401.50	1,575
08 52 16.30 Palladian Windows										
0010	**PALLADIAN WINDOWS**									
0020	Vinyl clad, double insulated glass, including frame and grilles									
0040	3'-2" x 2'-6" high	2 Carp	11	1.455	Ea.	1,250	48		1,298	1,450
0060	3'-2" x 4'-10"		11	1.455		1,750	48		1,798	1,975
0080	3'-2" x 6'-4"		10	1.600		1,700	52.50		1,752.50	1,975
0100	4'-0" x 4'-0"		10	1.600		1,525	52.50		1,577.50	1,775
0120	4'-0" x 5'-4"	3 Carp	10	2.400		1,875	79		1,954	2,175
0140	4'-0" x 6'-0"		9	2.667		1,950	87.50		2,037.50	2,300
0160	4'-0" x 7'-4"		9	2.667		2,125	87.50		2,212.50	2,475
0180	5'-5" x 4'-10"		9	2.667		2,275	87.50		2,362.50	2,650
0200	5'-5" x 6'-10"		9	2.667		2,575	87.50		2,662.50	3,000
0220	5'-5" x 7'-9"		9	2.667		2,800	87.50		2,887.50	3,225
0240	6'-0" x 7'-11"		8	3		3,475	98.50		3,573.50	4,000
0260	8'-0" x 6'-0"		8	3		3,075	98.50		3,173.50	3,550
08 52 16.35 Double-Hung Window										
0010	**DOUBLE-HUNG WINDOW** Including frames, screens, and grilles									
0300	Vinyl clad, premium, double insulated glass, 2'-6" x 3'-0" [G]	1 Carp	10	.800	Ea.	310	26.50		336.50	385
0305	2'-6" x 4'-0" [G]		10	.800		360	26.50		386.50	440
0400	3'-0" x 3'-6" [G]		10	.800		335	26.50		361.50	415
0500	3'-0" x 4'-0" [G]		9	.889		395	29		424	485
0600	3'-0" x 4'-6" [G]		9	.889		410	29		439	500
0700	3'-0" x 5'-0" [G]		8	1		445	33		478	540
0790	3'-4" x 5'-0" [G]		8	1		455	33		488	555
0800	3'-6" x 6'-0" [G]		8	1		490	33		523	595
0820	4'-0" x 5'-0" [G]		7	1.143		560	37.50		597.50	680
0830	4'-0" x 6'-0" [G]		7	1.143		700	37.50		737.50	830
08 52 16.40 Transom Windows										
0010	**TRANSOM WINDOWS**									
0050	Vinyl clad, premium, double insulated glass, 32" x 8"	1 Carp	16	.500	Ea.	188	16.45		204.45	234
0100	36" x 8"		16	.500		199	16.45		215.45	247

For customer support on your Residential Cost Data, call 877.713.2617.

08 52 Wood Windows

08 52 16 – Plastic-Clad Wood Windows

08 52 16.40 Transom Windows

		Crew	Daily Output	Labor-Hours	Unit	Material	2015 Bare Costs Labor	Equipment	Total	Total Incl O&P
0110	36" x 12"	1 Carp	16	.500	Ea.	212	16.45		228.45	261
1000	Vinyl clad, premium, dbl. insul. glass, 4'-0" x 4'-0"	2 Carp	12	1.333		510	44		554	635
1100	4'-0" x 6'-0"		11	1.455		935	48		983	1,100
1200	5'-0" x 6'-0"		10	1.600		1,050	52.50		1,102.50	1,250
1300	6'-0" x 6'-0"		10	1.600		1,050	52.50		1,102.50	1,250

08 52 16.45 Trapezoid Windows

		Crew	Daily Output	Labor-Hours	Unit	Material	Labor	Equipment	Total	Total Incl O&P
0010	**TRAPEZOID WINDOWS**									
0100	Vinyl clad, including frame and exterior trim									
0900	20" base x 44" leg x 53" leg	2 Carp	13	1.231	Ea.	390	40.50		430.50	500
1000	24" base x 90" leg x 102" leg		8	2		670	65.50		735.50	850
3000	36" base x 40" leg x 22" leg		12	1.333		430	44		474	545
3010	36" base x 44" leg x 25" leg		13	1.231		450	40.50		490.50	565
3050	36" base x 26" leg x 48" leg		9	1.778		455	58.50		513.50	600
3100	36" base x 42" legs, 50" peak		9	1.778		545	58.50		603.50	700
3200	36" base x 60" leg x 81" leg		11	1.455		710	48		758	865
4320	44" base x 23" leg x 56" leg		11	1.455		560	48		608	700
4350	44" base x 59" leg x 92" leg		10	1.600		840	52.50		892.50	1,025
4500	46" base x 15" leg x 46" leg		8	2		425	65.50		490.50	575
4550	46" base x 16" leg x 48" leg		8	2		450	65.50		515.50	605
4600	46" base x 50" leg x 80" leg		7	2.286		680	75		755	875
6600	66" base x 12" leg x 42" leg		8	2		575	65.50		640.50	745
6650	66" base x 12" legs, 28" peak		9	1.778		455	58.50		513.50	600
6700	68" base x 23" legs, 31" peak		8	2		575	65.50		640.50	745

08 52 16.70 Vinyl Clad, Premium, Dbl. Insulated Glass

			Crew	Daily Output	Labor-Hours	Unit	Material	Labor	Equipment	Total	Total Incl O&P
0010	**VINYL CLAD, PREMIUM, DBL. INSULATED GLASS**										
1000	Sliding, 3'-0" x 3'-0"	G	1 Carp	10	.800	Ea.	605	26.50		631.50	710
1020	4'-0" x 1'-11"	G		11	.727		585	24		609	680
1040	4'-0" x 3'-0"	G		10	.800		700	26.50		726.50	810
1050	4'-0" x 3'-6"	G		9	.889		685	29		714	805
1090	4'-0" x 5'-0"	G		9	.889		915	29		944	1,050
1100	5'-0" x 4'-0"	G		9	.889		900	29		929	1,050
1120	5'-0" x 5'-0"	G		8	1		1,000	33		1,033	1,150
1140	6'-0" x 4'-0"	G		8	1		1,025	33		1,058	1,175
1150	6'-0" x 5'-0"	G		8	1		1,125	33		1,158	1,300

08 52 50 – Window Accessories

08 52 50.10 Window Grille or Muntin

		Crew	Daily Output	Labor-Hours	Unit	Material	Labor	Equipment	Total	Total Incl O&P
0010	**WINDOW GRILLE OR MUNTIN**, snap in type									
0020	Standard pattern interior grilles									
2000	Wood, awning window, glass size 28" x 16" high	1 Carp	30	.267	Ea.	28	8.75		36.75	45.50
2060	44" x 24" high		32	.250		40	8.20		48.20	58
2100	Casement, glass size, 20" x 36" high		30	.267		32	8.75		40.75	49.50
2180	20" x 56" high		32	.250		43	8.20		51.20	61.50
2200	Double hung, glass size, 16" x 24" high		24	.333	Set	51	10.95		61.95	75
2280	32" x 32" high		34	.235	"	131	7.75		138.75	157
2500	Picture, glass size, 48" x 48" high		30	.267	Ea.	120	8.75		128.75	147
2580	60" x 68" high		28	.286	"	183	9.40		192.40	218
2600	Sliding, glass size, 14" x 36" high		24	.333	Set	35.50	10.95		46.45	57.50
2680	36" x 36" high		22	.364	"	43.50	11.95		55.45	67.50

08 52 Wood Windows

08 52 66 – Wood Window Screens

08 52 66.10 Wood Screens		Crew	Daily Output	Labor-Hours	Unit	Material	2015 Bare Costs Labor	Equipment	Total	Total Incl O&P
0010	**WOOD SCREENS**									
0020	Over 3 S.F., 3/4" frames	2 Carp	375	.043	S.F.	4.89	1.40		6.29	7.75
0100	1-1/8" frames	"	375	.043	"	8.20	1.40		9.60	11.35

08 52 69 – Wood Storm Windows

08 52 69.10 Storm Windows			Crew	Daily Output	Labor-Hours	Unit	Material	Labor	Equipment	Total	Total Incl O&P
0010	**STORM WINDOWS**, aluminum residential										
0300	Basement, mill finish, incl. fiberglass screen										
0320	1'-10" x 1'-0" high	G	2 Carp	30	.533	Ea.	35	17.50		52.50	68
0340	2'-9" x 1'-6" high	G		30	.533		38	17.50		55.50	71.50
0360	3'-4" x 2'-0" high	G		30	.533		45	17.50		62.50	79
1600	Double-hung, combination, storm & screen										
1700	Custom, clear anodic coating, 2'-0" x 3'-5" high		2 Carp	30	.533	Ea.	95	17.50		112.50	135
1720	2'-6" x 5'-0" high			28	.571		115	18.75		133.75	159
1740	4'-0" x 6'-0" high			25	.640		230	21		251	289
1800	White painted, 2'-0" x 3'-5" high			30	.533		110	17.50		127.50	151
1820	2'-6" x 5'-0" high			28	.571		170	18.75		188.75	219
1840	4'-0" x 6'-0" high			25	.640		280	21		301	345
2000	Clear anodic coating, 2'-0" x 3'-5" high	G		30	.533		95	17.50		112.50	135
2020	2'-6" x 5'-0" high	G		28	.571		117	18.75		135.75	160
2040	4'-0" x 6'-0" high	G		25	.640		130	21		151	179
2400	White painted, 2'-0" x 3'-5" high	G		30	.533		90	17.50		107.50	129
2420	2'-6" x 5'-0" high	G		28	.571		95	18.75		113.75	137
2440	4'-0" x 6'-0" high	G		25	.640		110	21		131	157
2600	Mill finish, 2'-0" x 3'-5" high	G		30	.533		85	17.50		102.50	123
2620	2'-6" x 5'-0" high	G		28	.571		90	18.75		108.75	131
2640	4'-0" x 6'-8" high	G		25	.640		110	21		131	157
4000	Picture window, storm, 1 lite, white or bronze finish										
4020	4'-6" x 4'-6" high		2 Carp	25	.640	Ea.	130	21		151	179
4040	5'-8" x 4'-6" high			20	.800		148	26.50		174.50	207
4400	Mill finish, 4'-6" x 4'-6" high			25	.640		130	21		151	179
4420	5'-8" x 4'-6" high			20	.800		150	26.50		176.50	209
4600	3 lite, white or bronze finish										
4620	4'-6" x 4'-6" high		2 Carp	25	.640	Ea.	160	21		181	212
4640	5'-8" x 4'-6" high			20	.800		170	26.50		196.50	231
4800	Mill finish, 4'-6" x 4'-6" high			25	.640		150	21		171	201
4820	5'-8" x 4'-6" high			20	.800		160	26.50		186.50	220
6000	Sliding window, storm, 2 lite, white or bronze finish										
6020	3'-4" x 2'-7" high		2 Carp	28	.571	Ea.	135	18.75		153.75	181
6040	4'-4" x 3'-3" high			25	.640		150	21		171	201
6060	5'-4" x 6'-0" high			20	.800		240	26.50		266.50	310
9000	Magnetic interior storm window										
9100	3/16" plate glass		1 Glaz	107	.075	S.F.	6.10	2.39		8.49	10.70

08 53 Plastic Windows

08 53 13 – Vinyl Windows

08 53 13.10 Solid Vinyl Windows

			Crew	Daily Output	Labor-Hours	Unit	Material	2015 Bare Costs Labor	Equipment	Total	Total Incl O&P
0010	**SOLID VINYL WINDOWS**										
0020	Double hung, including frame and screen, 2'-0" x 2'-6"		2 Carp	15	1.067	Ea.	199	35		234	278
0040	2'-0" x 3'-6"			14	1.143		212	37.50		249.50	296
0060	2'-6" x 4'-6"			13	1.231		250	40.50		290.50	345
0080	3'-0" x 4'-0"			10	1.600		216	52.50		268.50	325
0100	3'-0" x 4'-6"			9	1.778		278	58.50		336.50	405
0120	3'-6" x 4'-6"			8	2		300	65.50		365.50	440
0140	3'-6" x 6'-0"			7	2.286		355	75		430	515

08 53 13.20 Vinyl Single Hung Windows

			Crew	Daily Output	Labor-Hours	Unit	Material	2015 Bare Costs Labor	Equipment	Total	Total Incl O&P
0010	**VINYL SINGLE HUNG WINDOWS**, insulated glass										
0100	Grids, low E, J fin, ext. jambs, 21" x 53"	G	2 Carp	18	.889	Ea.	198	29		227	267
0110	21" x 57"	G		17	.941		200	31		231	272
0120	21" x 65"	G		16	1		205	33		238	281
0130	25" x 41"	G		20	.800		190	26.50		216.50	253
0140	25" x 49"	G		18	.889		200	29		229	269
0150	25" x 57"	G		17	.941		205	31		236	278
0160	25" x 65"	G		16	1		240	33		273	320
0170	29" x 41"	G		18	.889		195	29		224	264
0180	29" x 53"	G		18	.889		205	29		234	275
0190	29" x 57"	G		17	.941		210	31		241	283
0200	29" x 65"	G		16	1		215	33		248	292
0210	33" x 41"	G		20	.800		200	26.50		226.50	264
0220	33" x 53"	G		18	.889		215	29		244	286
0230	33" x 57"	G		17	.941		215	31		246	289
0240	33" x 65"	G		16	1		220	33		253	297
0250	37" x 41"	G		20	.800		225	26.50		251.50	292
0260	37" x 53"	G		18	.889		235	29		264	310
0270	37" x 57"	G		17	.941		240	31		271	315
0280	37" x 65"	G		16	1		256	33		289	335
0500	Vinyl clad, premium, double insulated glass, circle, 24" diameter		1 Carp	6	1.333		460	44		504	580
1000	2'-4" diameter			6	1.333		540	44		584	670
1500	2'-11" diameter			6	1.333		640	44		684	775

08 53 13.30 Vinyl Double Hung Windows

			Crew	Daily Output	Labor-Hours	Unit	Material	2015 Bare Costs Labor	Equipment	Total	Total Incl O&P
0010	**VINYL DOUBLE HUNG WINDOWS**, insulated glass										
0100	Grids, low E, J fin, ext. jambs, 21" x 53"	G	2 Carp	18	.889	Ea.	215	29		244	286
0102	21" x 37"	G		18	.889		223	29		252	294
0104	21" x 41"	G		18	.889		232	29		261	305
0106	21" x 49"	G		18	.889		247	29		276	320
0110	21" x 57"	G		17	.941		263	31		294	340
0120	21" x 65"	G		16	1		279	33		312	360
0128	25" x 37"	G		20	.800		234	26.50		260.50	300
0130	25" x 41"	G		20	.800		242	26.50		268.50	310
0140	25" x 49"	G		18	.889		258	29		287	335
0145	25" x 53"	G		18	.889		265	29		294	340
0150	25" x 57"	G		17	.941		274	31		305	350
0160	25" x 65"	G		16	1		290	33		323	375
0162	25" x 69"	G		16	1		297	33		330	380
0164	25" x 77"	G		16	1		325	33		358	410
0168	29" x 37"	G		18	.889		242	29		271	315
0170	29" x 41"	G		18	.889		250	29		279	325
0172	29" x 49"	G		18	.889		267	29		296	345
0180	29" x 53"	G		18	.889		271	29		300	345

08 53 Plastic Windows

08 53 13 – Vinyl Windows

08 53 13.30 Vinyl Double Hung Windows

		Crew	Daily Output	Labor-Hours	Unit	Material	2015 Bare Costs Labor	Equipment	Total	Total Incl O&P
0190	29" x 57" G	2 Carp	17	.941	Ea.	280	31		311	360
0200	29" x 65" G		16	1		300	33		333	385
0202	29" x 69" G		16	1		310	33		343	395
0205	29" x 77" G		16	1		335	33		368	420
0208	33" x 37" G		20	.800		255	26.50		281.50	325
0210	33" x 41" G		20	.800		263	26.50		289.50	335
0215	33" x 49" G		20	.800		280	26.50		306.50	355
0220	33" x 53" G		18	.889		285	29		314	365
0230	33" x 57" G		17	.941		297	31		328	375
0240	33" x 65" G		16	1		315	33		348	400
0242	33" x 69" G		16	1		330	33		363	415
0246	33" x 77" G		16	1		350	33		383	440
0250	37" x 41" G		20	.800		289	26.50		315.50	365
0255	37" x 49" G		20	.800		282	26.50		308.50	355
0260	37" x 53" G		18	.889		315	29		344	395
0270	37" x 57" G		17	.941		325	31		356	410
0280	37" x 65" G		16	1		350	33		383	440
0282	37" x 69" G		16	1		360	33		393	450
0286	37" x 77" G		16	1		380	33		413	470
0300	Solid vinyl, average quality, double insulated glass, 2'-0" x 3'-0" G	1 Carp	10	.800		291	26.50		317.50	365
0310	3'-0" x 4'-0" G		9	.889		207	29		236	276
0320	4'-0" x 4'-6" G		8	1		335	33		368	420
0330	Premium, double insulated glass, 2'-6" x 3'-0" G		10	.800		271	26.50		297.50	340
0340	3'-0" x 3'-6" G		9	.889		292	29		321	370
0350	3'-0" x 4'-0" G		9	.889		315	29		344	400
0360	3'-0" x 4'-6" G		9	.889		330	29		359	415
0370	3'-0" x 5'-0" G		8	1		355	33		388	445
0380	3'-6" x 6"-0" G		8	1		380	33		413	475

08 53 13.40 Vinyl Casement Windows

		Crew	Daily Output	Labor-Hours	Unit	Material	Labor	Equipment	Total	Total Incl O&P
0010	**VINYL CASEMENT WINDOWS**, insulated glass									
0015	Grids, low E, J fin, extension jambs, screens									
0100	One lite, 21" x 41" G	2 Carp	20	.800	Ea.	295	26.50		321.50	370
0110	21" x 47" G		20	.800		320	26.50		346.50	400
0120	21" x 53" G		20	.800		345	26.50		371.50	425
0128	24" x 35" G		19	.842		283	27.50		310.50	355
0130	24" x 41" G		19	.842		310	27.50		337.50	385
0140	24" x 47" G		19	.842		335	27.50		362.50	410
0150	24" x 53" G		19	.842		360	27.50		387.50	440
0158	28" x 35" G		19	.842		300	27.50		327.50	375
0160	28" x 41" G		19	.842		330	27.50		357.50	410
0170	28" x 47" G		19	.842		350	27.50		377.50	430
0180	28" x 53" G		19	.842		385	27.50		412.50	470
0184	28" x 59" G		19	.842		395	27.50		422.50	480
0188	Two lites, 33" x 35" G		18	.889		470	29		499	565
0190	33" x 41" G		18	.889		500	29		529	600
0200	33" x 47" G		18	.889		540	29		569	640
0210	33" x 53" G		18	.889		575	29		604	680
0212	33" x 59" G		18	.889		610	29		639	720
0215	33" x 72" G		18	.889		635	29		664	745
0220	41" x 41" G		18	.889		545	29		574	650
0230	41" x 47" G		18	.889		585	29		614	690
0240	41" x 53" G		17	.941		620	31		651	730

08 53 Plastic Windows

08 53 13 – Vinyl Windows

08 53 13.40 Vinyl Casement Windows

			Crew	Daily Output	Labor-Hours	Unit	Material	2015 Bare Costs Labor	Equipment	Total	Total Incl O&P
0242	41" x 59"	G	2 Carp	17	.941	Ea.	650	31		681	765
0246	41" x 72"	G		17	.941		685	31		716	805
0250	47" x 41"	G		17	.941		555	31		586	660
0260	47" x 47"	G		17	.941		590	31		621	700
0270	47" x 53"	G		17	.941		625	31		656	740
0272	47" x 59"	G		17	.941		680	31		711	800
0280	56" x 41"	G		15	1.067		595	35		630	715
0290	56" x 47"	G		15	1.067		625	35		660	750
0300	56" x 53"	G		15	1.067		680	35		715	810
0302	56" x 59"	G		15	1.067		710	35		745	840
0310	56" x 72"	G		15	1.067		770	35		805	910
0340	Solid vinyl, premium, double insulated glass, 2'-0" x 3'-0" high	G	1 Carp	10	.800		270	26.50		296.50	340
0360	2'-0" x 4'-0" high	G		9	.889		299	29		328	380
0380	2'-0" x 5'-0" high	G		8	1		335	33		368	425

08 53 13.50 Vinyl Picture Windows

		Crew	Daily Output	Labor-Hours	Unit	Material	Labor	Equipment	Total	Total Incl O&P
0010	**VINYL PICTURE WINDOWS**, insulated glass									
0100	Grids, low E, J fin, ext. jambs, 33" x 47"	2 Carp	12	1.333	Ea.	280	44		324	385
0110	35" x 71"		12	1.333		375	44		419	490
0120	47" x 35"		12	1.333		300	44		344	405
0130	47" x 41"		12	1.333		390	44		434	505
0140	47" x 47"		12	1.333		345	44		389	450
0150	47" x 53"		11	1.455		370	48		418	485
0160	71" x 35"		11	1.455		390	48		438	510
0170	71" x 41"		11	1.455		410	48		458	530
0180	71" x 47"		11	1.455		440	48		488	565

08 53 13.60 Vinyl Half Round Windows

		Crew	Daily Output	Labor-Hours	Unit	Material	Labor	Equipment	Total	Total Incl O&P
0010	**VINYL HALF ROUND WINDOWS**, Including grille, j fin low E, ext. jambs									
0100	10" height x 20" base	2 Carp	8	2	Ea.	475	65.50		540.50	630
0110	15" height x 30" base		8	2		440	65.50		505.50	595
0120	17" height x 34" base		7	2.286		320	75		395	475
0130	19" height x 38" base		7	2.286		345	75		420	505
0140	19" height x 33" base		7	2.286		555	75		630	740
0150	24" height x 48" base	1 Carp	6	1.333		365	44		409	475
0160	25" height x 50" base	"	6	1.333		755	44		799	905
0170	30" height x 60" base	2 Carp	6	2.667		660	87.50		747.50	870

08 54 Composite Windows

08 54 13 – Fiberglass Windows

08 54 13.10 Fiberglass Single Hung Windows

			Crew	Daily Output	Labor-Hours	Unit	Material	Labor	Equipment	Total	Total Incl O&P
0010	**FIBERGLASS SINGLE HUNG WINDOWS**										
0100	Grids, low E, 18" x 24"	G	2 Carp	18	.889	Ea.	335	29		364	420
0110	18" x 40"	G		17	.941		340	31		371	425
0130	24" x 40"	G		20	.800		360	26.50		386.50	440
0230	36" x 36"	G		17	.941		370	31		401	455
0250	36" x 48"	G		20	.800		405	26.50		431.50	490
0260	36" x 60"	G		18	.889		445	29		474	540
0280	36" x 72"	G		16	1		470	33		503	570
0290	48" x 40"	G		16	1		470	33		503	570

08 54 Composite Windows

08 54 13 – Fiberglass Windows

08 54 13.30 Fiberglass Slider Windows		Crew	Daily Output	Labor-Hours	Unit	Material	2015 Bare Costs Labor	Equipment	Total	Total Incl O&P
0010	**FIBERGLASS SLIDER WINDOWS**									
0100	Grids, low E, 36" x 24"	G	2 Carp	20	.800	Ea.	340	26.50		366.50	420
0110	36" x 36"	G	"	20	.800	"	390	26.50		416.50	475

08 54 13.50 Fiberglass Bay Windows										
0010	**FIBERGLASS BAY WINDOWS**									
0150	48" x 36"	G	2 Carp	11	1.455	Ea.	1,100	48		1,148	1,275

08 61 Roof Windows

08 61 13 – Metal Roof Windows

08 61 13.10 Roof Windows		Crew	Daily Output	Labor-Hours	Unit	Material	Labor	Equipment	Total	Total Incl O&P
0010	**ROOF WINDOWS**, fixed high perf tmpd glazing, metallic framed									
0020	46" x 21-1/2", Flashed for shingled roof	1 Carp	8	1	Ea.	270	33		303	350
0100	46" x 28"		8	1		300	33		333	385
0125	57" x 44"		6	1.333		370	44		414	480
0130	72" x 28"		7	1.143		370	37.50		407.50	470
0150	Fixed, laminated tempered glazing, 46" x 21-1/2"		8	1		455	33		488	555
0175	46" x 28"		8	1		505	33		538	610
0200	57" x 44"		6	1.333		475	44		519	595
0500	Vented flashing set for shingled roof, 46" x 21-1/2"		7	1.143		455	37.50		492.50	565
0525	46" x 28"		6	1.333		505	44		549	630
0550	57" x 44"		5	1.600		635	52.50		687.50	785
0560	72" x 28"		5	1.600		635	52.50		687.50	785
0575	Flashing set for low pitched roof, 46" x 21-1/2"		7	1.143		525	37.50		562.50	645
0600	46" x 28"		7	1.143		575	37.50		612.50	700
0625	57" x 44"		5	1.600		715	52.50		767.50	875
0650	Flashing set for curb 46" x 21-1/2"		7	1.143		605	37.50		642.50	730
0675	46" x 28"		7	1.143		660	37.50		697.50	790
0700	57" x 44"		5	1.600		810	52.50		862.50	980

08 61 16 – Wood Roof Windows

08 61 16.16 Roof Windows, Wood Framed		Crew	Daily Output	Labor-Hours	Unit	Material	Labor	Equipment	Total	Total Incl O&P
0010	**ROOF WINDOWS, WOOD FRAMED**									
5600	Roof window incl. frame, flashing, double insulated glass & screens,									
5610	complete unit, 22" x 38"	2 Carp	3	5.333	Ea.	715	175		890	1,075
5650	2'-5" x 3'-8"		3.20	5		875	164		1,039	1,225
5700	3'-5" x 4'-9"		3.40	4.706		995	155		1,150	1,350

08 62 Unit Skylights

08 62 13 – Domed Unit Skylights

08 62 13.10 Domed Skylights			Crew	Daily Output	Labor-Hours	Unit	Material	Labor	Equipment	Total	Total Incl O&P
0010	**DOMED SKYLIGHTS**										
0020	Skylight, fixed dome type, 22" x 22"	G	G-3	12	2.667	Ea.	216	82		298	375
0030	22" x 46"	G		10	3.200		269	98		367	460
0040	30" x 30"	G		12	2.667		286	82		368	450
0050	30" x 46"	G		10	3.200		380	98		478	585
0110	Fixed, double glazed, 22" x 27"	G		12	2.667		275	82		357	440
0120	22" x 46"	G		10	3.200		293	98		391	485
0130	44" x 46"	G		10	3.200		430	98		528	640
0210	Operable, double glazed, 22" x 27"	G		12	2.667		405	82		487	580

For customer support on your Residential Cost Data, call 877.713.2617.

08 62 Unit Skylights

08 62 13 – Domed Unit Skylights

08 62 13.10 Domed Skylights		Crew	Daily Output	Labor-Hours	Unit	Material	2015 Bare Costs Labor	Equipment	Total	Total Incl O&P
0220	22" x 46" G	G-3	10	3.200	Ea.	465	98		563	680
0230	44" x 46" G		10	3.200		875	98		973	1,125

08 62 13.20 Skylights

		Crew	Daily Output	Labor-Hours	Unit	Material	Labor	Equipment	Total	Total Incl O&P
0010	**SKYLIGHTS**, flush or curb mounted									
2120	Ventilating insulated plexiglass dome with									
2130	curb mounting, 36" x 36" G	G-3	12	2.667	Ea.	480	82		562	665
2150	52" x 52" G		12	2.667		670	82		752	870
2160	28" x 52" G		10	3.200		490	98		588	705
2170	36" x 52" G		10	3.200		545	98		643	760
2180	For electric opening system, add G					315			315	345
2210	Operating skylight, with thermopane glass, 24" x 48" G	G-3	10	3.200		585	98		683	810
2220	32" x 48"	"	9	3.556		615	109		724	860
2310	Non venting insulated plexiglass dome skylight with									
2320	Flush mount 22" x 46" G	G-3	15.23	2.101	Ea.	330	64.50		394.50	475
2330	30" x 30" G		16	2		305	61.50		366.50	440
2340	46" x 46" G		13.91	2.301		565	70.50		635.50	740
2350	Curb mount 22" x 46" G		15.23	2.101		365	64.50		429.50	510
2360	30" x 30" G		16	2		405	61.50		466.50	555
2370	46" x 46" G		13.91	2.301		625	70.50		695.50	810
2381	Non-insulated flush mount 22" x 46"		15.23	2.101		225	64.50		289.50	355
2382	30" x 30"		16	2		202	61.50		263.50	325
2383	46" x 46"		13.91	2.301		380	70.50		450.50	540
2384	Curb mount 22" x 46"		15.23	2.101		189	64.50		253.50	315
2385	30" x 30"		16	2		188	61.50		249.50	310
4000	Skylight, solar tube kit, incl dome, flashing, diffuser, 1 pipe, 10" dia. G	1 Carp	2	4		235	131		366	480
4010	14" diam. G		2	4		325	131		456	580
4020	21" diam. G		2	4		410	131		541	670
4030	Accessories for, 1' long x 9" dia pipe G		24	.333		47	10.95		57.95	70
4040	2' long x 9" dia pipe G		24	.333		37	10.95		47.95	59
4050	4' long x 9" dia pipe G		20	.400		70	13.15		83.15	99
4060	1' long x 13" dia pipe G		24	.333		65	10.95		75.95	90
4070	2' long x 13" dia pipe G		24	.333		51	10.95		61.95	74.50
4080	4' long x 13" dia pipe G		20	.400		97	13.15		110.15	129
4090	6.5" turret ext for 21" dia pipe G		16	.500		101	16.45		117.45	139
4100	12' long x 21" dia flexible pipe G		12	.667		82	22		104	127
4110	45 degree elbow, 10" G		16	.500		144	16.45		160.45	186
4120	14" G		16	.500		76	16.45		92.45	111
4130	Interior decorative ring, 9" G		20	.400		35	13.15		48.15	60.50
4140	13" G		20	.400		55	13.15		68.15	82.50

08 71 Door Hardware

08 71 20 – Hardware

08 71 20.15 Hardware

		Crew	Daily Output	Labor-Hours	Unit	Material	Labor	Equipment	Total	Total Incl O&P
0010	**HARDWARE**									
0020	Average percentage for hardware, total job cost									
0500	Total hardware for building, average distribution				Job	85%	15%			
1000	Door hardware, apartment, interior	1 Carp	4	2	Door	455	65.50		520.50	610
1300	Average, door hardware, motel/hotel interior, with access card		4	2	"	580	65.50		645.50	750
2100	Pocket door		6	1.333	Ea.	100	44		144	184
4000	Door knocker, bright brass		32	.250		75	8.20		83.20	96.50
4100	Mail slot, bright brass, 2" x 11"		25	.320		60	10.50		70.50	83.50

08 71 Door Hardware

08 71 20 – Hardware

08 71 20.15 Hardware		Crew	Daily Output	Labor-Hours	Unit	Material	2015 Bare Costs Labor	Equipment	Total	Total Incl O&P
4200	Peep hole, add to price of door				Ea.	11			11	12.10
08 71 20.40 Lockset										
0010	**LOCKSET**, Standard duty									
0020	Non-keyed, passage, w/sect. trim	1 Carp	12	.667	Ea.	70	22		92	114
0100	Privacy		12	.667		75	22		97	120
0400	Keyed, single cylinder function		10	.800		108	26.50		134.50	162
0500	Lever handled, keyed, single cylinder function		10	.800		128	26.50		154.50	185
08 71 20.41 Dead Locks										
0010	**DEAD LOCKS**									
1203	Deadlock night latch	1 Carp	7.70	1.039	Ea.	45	34		79	107
1420	Deadbolt lock, single cylinder		10	.800		48.50	26.50		75	97.50
1440	Double cylinder		10	.800		63	26.50		89.50	114
08 71 20.42 Mortise Locksets										
0010	**MORTISE LOCKSETS**, Comm., wrought knobs & full escutcheon trim									
0015	Assumes mortise is cut									
0020	Non-keyed, passage, grade 3	1 Carp	9	.889	Ea.	149	29		178	213
0030	Grade 1		8	1		405	33		438	500
0040	Privacy set Grade 3		9	.889		169	29		198	235
0050	Grade 1		8	1		455	33		488	555
0100	Keyed, office/entrance/apartment, Grade 2		8	1		197	33		230	272
0110	Grade 1		7	1.143		515	37.50		552.50	635
0120	Single cylinder, typical, Grade 3		8	1		189	33		222	263
0130	Grade 1		7	1.143		500	37.50		537.50	615
0200	Hotel, room, Grade 3		7	1.143		190	37.50		227.50	272
0210	Grade 1 (see also Section 08 71 20.15)		6	1.333		510	44		554	635
0300	Double cylinder, Grade 3		8	1		225	33		258	305
0310	Grade 1		7	1.143		515	37.50		552.50	635
1000	Wrought knobs and sectional trim, non-keyed, passage, Grade 3		10	.800		130	26.50		156.50	187
1010	Grade 1		9	.889		405	29		434	495
1040	Privacy, Grade 3		10	.800		145	26.50		171.50	204
1050	Grade 1		9	.889		455	29		484	550
1100	Keyed, entrance, office/apartment, Grade 3		9	.889		220	29		249	291
1103	Install lockset		6.92	1.156		220	38		258	305
1110	Grade 1		8	1		520	33		553	625
1120	Single cylinder, Grade 3		9	.889		225	29		254	297
1130	Grade 1		8	1		500	33		533	605
2000	Cast knobs and full escutcheon trim									
2010	Non-keyed, passage, Grade 3	1 Carp	9	.889	Ea.	275	29		304	355
2020	Grade 1		8	1		380	33		413	475
2040	Privacy, Grade 3		9	.889		320	29		349	400
2050	Grade 1		8	1		440	33		473	540
2120	Keyed, single cylinder, Grade 3		8	1		330	33		363	420
2123	Mortise lock		6.15	1.301		330	42.50		372.50	435
2130	Grade 1		7	1.143		525	37.50		562.50	645
08 71 20.50 Door Stops										
0010	**DOOR STOPS**									
0020	Holder & bumper, floor or wall	1 Carp	32	.250	Ea.	34.50	8.20		42.70	51.50
1300	Wall bumper, 4" diameter, with rubber pad, aluminum		32	.250		11.70	8.20		19.90	26.50
1600	Door bumper, floor type, aluminum		32	.250		8.10	8.20		16.30	22.50
1620	Brass		32	.250		10	8.20		18.20	25
1630	Bronze		32	.250		18.20	8.20		26.40	34
1900	Plunger type, door mounted		32	.250		28	8.20		36.20	44.50

08 71 Door Hardware

08 71 20 – Hardware

08 71 20.50 Door Stops

		Crew	Daily Output	Labor-Hours	Unit	Material	2015 Bare Costs Labor	Equipment	Total	Total Incl O&P
2520	Wall type, aluminum	1 Carp	32	.250	Ea.	35	8.20		43.20	52.50
2540	Plunger type, aluminum		32	.250		28	8.20		36.20	44.50
2560	Brass		32	.250		43	8.20		51.20	61
3020	Floor mounted, US3		3	2.667		315	87.50		402.50	495

08 71 20.60 Entrance Locks

0010	**ENTRANCE LOCKS**									
0015	Cylinder, grip handle deadlocking latch	1 Carp	9	.889	Ea.	175	29		204	242
0020	Deadbolt		8	1		175	33		208	248
0100	Push and pull plate, dead bolt		8	1		225	33		258	305
0900	For handicapped lever, add					150			150	165

08 71 20.65 Thresholds

0010	**THRESHOLDS**									
0011	Threshold 3' long saddles aluminum	1 Carp	48	.167	L.F.	9.40	5.50		14.90	19.55
0100	Aluminum, 8" wide, 1/2" thick		12	.667	Ea.	49	22		71	90.50
0500	Bronze		60	.133	L.F.	42	4.38		46.38	54
0600	Bronze, panic threshold, 5" wide, 1/2" thick		12	.667	Ea.	158	22		180	210
0700	Rubber, 1/2" thick, 5-1/2" wide		20	.400		40	13.15		53.15	66
0800	2-3/4" wide		20	.400		45	13.15		58.15	71.50
1950	ADA Compliant Thresholds									
2000	Threshold, wood oak 3-1/2" wide x 24" long	1 Carp	12	.667	Ea.	9	22		31	47
2010	3-1/2" wide x 36" long		12	.667		13.50	22		35.50	52
2020	3-1/2" wide x 48" long		12	.667		18	22		40	57
2030	4-1/2" wide x 24" long		12	.667		12	22		34	50
2040	4-1/2" wide x 36" long		12	.667		17	22		39	55.50
2050	4-1/2" wide x 48" long		12	.667		23	22		45	62.50
2060	6-1/2" wide x 24" long		12	.667		16.50	22		38.50	55
2070	6-1/2" wide x 36" long		12	.667		25	22		47	64.50
2080	6-1/2" wide x 48" long		12	.667		33	22		55	73.50
2090	Threshold, wood cherry 3-1/2" wide x 24" long		12	.667		13.50	22		35.50	52
2100	3-1/2" wide x 36" long		12	.667		21	22		43	60
2110	3-1/2" wide x 48" long		12	.667		27	22		49	66.50
2120	4-1/2" wide x 24" long		12	.667		17.50	22		39.50	56.50
2130	4-1/2" wide x 36" long		12	.667		25.50	22		47.50	65
2140	4-1/2" wide x 48" long		12	.667		34	22		56	74.50
2150	6-1/2" wide x 24" long		12	.667		25.50	22		47.50	65
2160	6-1/2" wide x 36" long		12	.667		36.50	22		58.50	77
2170	6-1/2" wide x 48" long		12	.667		49	22		71	91
2180	Threshold, wood walnut 3-1/2" wide x 24" long		12	.667		15.50	22		37.50	54
2190	3-1/2" wide x 36" long		12	.667		23	22		45	62.50
2200	3-1/2" wide x 48" long		12	.667		30	22		52	70
2210	4-1/2" wide x 24" long		12	.667		19.50	22		41.50	58.50
2220	4-1/2" wide x 36" long		12	.667		28.50	22		50.50	68.50
2230	4-1/2" wide x 48" long		12	.667		38	22		60	79
2240	6-1/2" wide x 24" long		12	.667		28	22		50	68
2250	6-1/2" wide x 36" long		12	.667		41.50	22		63.50	82.50
2260	6-1/2" wide x 48" long		12	.667		55	22		77	97.50
2300	Threshold, aluminum 4" wide x 36" long		12	.667		33	22		55	73.50
2310	4" wide x 48" long		12	.667		41	22		63	82
2320	4" wide x 72" long		12	.667		66	22		88	110
2330	5" wide x 36" long		12	.667		45	22		67	86.50
2340	5" wide x 48" long		12	.667		56	22		78	98.50
2350	5" wide x 72" long		12	.667		89	22		111	135

08 71 Door Hardware

08 71 20 – Hardware

08 71 20.65 Thresholds

		Crew	Daily Output	Labor-Hours	Unit	Material	2015 Bare Costs Labor	Equipment	Total	Total Incl O&P
2360	6" wide x 36" long	1 Carp	12	.667	Ea.	53	22		75	95.50
2370	6" wide x 48" long		12	.667		68	22		90	112
2380	6" wide x 72" long		12	.667		106	22		128	154
2390	7" wide x 36" long		12	.667		69	22		91	113
2400	7" wide x 48" long		12	.667		90	22		112	136
2410	7" wide x 72" long		12	.667		137	22		159	188
2500	Threshold, ramp, aluminum or rubber 24" x 24"		12	.667		190	22		212	246

08 71 20.75 Door Hardware Accessories

		Crew	Daily Output	Labor-Hours	Unit	Material	Labor	Equipment	Total	Total Incl O&P
0010	**DOOR HARDWARE ACCESSORIES**									
1000	Knockers, brass, standard	1 Carp	16	.500	Ea.	46.50	16.45		62.95	78.50
1100	Deluxe		10	.800		121	26.50		147.50	177
4500	Rubber door silencers		540	.015		.44	.49		.93	1.30

08 71 20.90 Hinges

		Crew	Daily Output	Labor-Hours	Unit	Material	Labor	Equipment	Total	Total Incl O&P
0010	**HINGES**									
0012	Full mortise, avg. freq., steel base, USP, 4-1/2" x 4-1/2"				Pr.	36			36	40
0100	5" x 5", USP					57.50			57.50	63
0200	6" x 6", USP					119			119	131
0400	Brass base, 4-1/2" x 4-1/2", US10					58.50			58.50	64.50
0500	5" x 5", US10					86			86	94.50
0600	6" x 6", US10					162			162	178
0800	Stainless steel base, 4-1/2" x 4-1/2", US32					74			74	81.50
0900	For non removable pin, add (security item)				Ea.	4.88			4.88	5.35
0910	For floating pin, driven tips, add					3.30			3.30	3.63
0930	For hospital type tip on pin, add					13.95			13.95	15.35
0940	For steeple type tip on pin, add					18.95			18.95	21
0950	Full mortise, high frequency, steel base, 3-1/2" x 3-1/2", US26D				Pr.	30			30	33
1000	4-1/2" x 4-1/2", USP					63			63	69
1100	5" x 5", USP					50			50	55
1200	6" x 6", USP					134			134	148
1400	Brass base, 3-1/2" x 3-1/2", US4					51			51	56
1430	4-1/2" x 4-1/2", US10					73.50			73.50	81
1500	5" x 5", US10					122			122	134
1600	6" x 6", US10					165			165	181
1800	Stainless steel base, 4-1/2" x 4-1/2", US32					103			103	114
1810	5" x 4-1/2", US32					137			137	151
1930	For hospital type tip on pin, add				Ea.	13.15			13.15	14.45
1950	Full mortise, low frequency, steel base, 3-1/2" x 3-1/2", US26D				Pr.	23.50			23.50	26
2000	4-1/2" x 4-1/2", USP					22			22	24.50
2100	5" x 5", USP					47.50			47.50	52
2200	6" x 6", USP					89			89	98
2300	4-1/2" x 4-1/2", US3					16.85			16.85	18.50
2310	5" x 5", US3					41			41	45
2400	Brass bass, 4-1/2" x 4-1/2", US10					53			53	58.50
2500	5" x 5", US10					76.50			76.50	84
2800	Stainless steel base, 4-1/2" x 4-1/2", US32					74.50			74.50	82
8000	Install hinge	1 Carp	34	.235			7.75		7.75	13

08 71 20.91 Special Hinges

		Crew	Daily Output	Labor-Hours	Unit	Material	Labor	Equipment	Total	Total Incl O&P
0010	**SPECIAL HINGES**									
8000	Continuous hinges									
8010	Steel, piano, 2" x 72"	1 Carp	20	.400	Ea.	22	13.15		35.15	46
8020	Brass, piano, 1-1/16" x 30"		30	.267		8	8.75		16.75	23.50
8030	Acrylic, piano, 1-3/4" x 12"		40	.200		15	6.55		21.55	27.50

08 71 Door Hardware

08 71 20 – Hardware

08 71 20.92 Mortised Hinges

		Crew	Daily Output	Labor-Hours	Unit	Material	2015 Bare Costs Labor	Equipment	Total	Total Incl O&P
0010	**MORTISED HINGES**									
0200	Average frequency, steel plated, ball bearing, 3-1/2" x 3-1/2"				Pr.	27.50			27.50	30.50
0300	Bronze, ball bearing					32.50			32.50	35.50
0900	High frequency, steel plated, ball bearing					73.50			73.50	81
1100	Bronze, ball bearing					73.50			73.50	81
1300	Average frequency, steel plated, ball bearing, 4-1/2" x 4-1/2"					35			35	38.50
1500	Bronze, ball bearing, to 36" wide					36.50			36.50	40.50
1700	Low frequency, steel, plated, plain bearing					19.40			19.40	21.50
1900	Bronze, plain bearing					27			27	30

08 71 20.95 Kick Plates

		Crew	Daily Output	Labor-Hours	Unit	Material	Labor	Equipment	Total	Total Incl O&P
0010	**KICK PLATES**									
0020	Stainless steel, .050, 16 ga., 8" x 28", US32	1 Carp	15	.533	Ea.	38	17.50		55.50	71.50
0080	Mop/Kick, 4" x 28"		15	.533		34	17.50		51.50	67
0090	4" x 30"		15	.533		36	17.50		53.50	69
0100	4" x 34"		15	.533		41	17.50		58.50	74.50
0110	6" x 28"		15	.533		43	17.50		60.50	77
0120	6" x 30"		15	.533		47	17.50		64.50	81
0130	6" x 34"		15	.533		53	17.50		70.50	88

08 71 21 – Astragals

08 71 21.10 Exterior Mouldings, Astragals

		Crew	Daily Output	Labor-Hours	Unit	Material	Labor	Equipment	Total	Total Incl O&P
0010	**EXTERIOR MOULDINGS, ASTRAGALS**									
4170	Astragal for double doors, aluminum	1 Carp	4	2	Opng.	33	65.50		98.50	147
4174	Bronze	"	4	2	"	47	65.50		112.50	162

08 71 25 – Weatherstripping

08 71 25.10 Mechanical Seals, Weatherstripping

		Crew	Daily Output	Labor-Hours	Unit	Material	Labor	Equipment	Total	Total Incl O&P
0010	**MECHANICAL SEALS, WEATHERSTRIPPING**									
1000	Doors, wood frame, interlocking, for 3' x 7' door, zinc	1 Carp	3	2.667	Opng.	44	87.50		131.50	196
1100	Bronze		3	2.667		56	87.50		143.50	209
1300	6' x 7' opening, zinc		2	4		54	131		185	281
1400	Bronze		2	4		65	131		196	293
1500	Vinyl V strip		6.40	1.250	Ea.	12.50	41		53.50	83
1700	Wood frame, spring type, bronze									
1800	3' x 7' door	1 Carp	7.60	1.053	Opng.	23	34.50		57.50	83.50
1900	6' x 7' door		7	1.143		29.50	37.50		67	95.50
1920	Felt, 3' x 7' door		14	.571		3.90	18.75		22.65	36
1930	6' x 7' door		13	.615		4.45	20		24.45	39
1950	Rubber, 3' x 7' door		7.60	1.053		9.15	34.50		43.65	68
1960	6' x 7' door		7	1.143		11.15	37.50		48.65	75.50
2200	Metal frame, spring type, bronze									
2300	3' x 7' door	1 Carp	3	2.667	Opng.	46.50	87.50		134	198
2400	6' x 7' door	"	2.50	3.200	"	52	105		157	233
2500	For stainless steel, spring type, add					133%				
2700	Metal frame, extruded sections, 3' x 7' door, aluminum	1 Carp	3	2.667	Opng.	28	87.50		115.50	178
2800	Bronze		3	2.667		82	87.50		169.50	237
3100	6' x 7' door, aluminum		1.50	5.333		35	175		210	335
3200	Bronze		1.50	5.333		137	175		312	445
3500	Threshold weatherstripping									
3650	Door sweep, flush mounted, aluminum	1 Carp	25	.320	Ea.	19	10.50		29.50	38.50
3700	Vinyl		25	.320		18	10.50		28.50	37.50
5000	Garage door bottom weatherstrip, 12' aluminum, clear		14	.571		25	18.75		43.75	59
5010	Bronze		14	.571		90	18.75		108.75	131

08 71 Door Hardware

08 71 25 – Weatherstripping

08 71 25.10 Mechanical Seals, Weatherstripping	Crew	Daily Output	Labor-Hours	Unit	Material	2015 Bare Costs Labor	Equipment	Total	Total Incl O&P
5050 Bottom protection, Rubber	1 Carp	14	.571	Ea.	37	18.75		55.75	72
5100 Threshold		14	.571		72	18.75		90.75	111

08 75 Window Hardware

08 75 10 – Window Handles and Latches

08 75 10.10 Handles and Latches

	Crew	Daily Output	Labor-Hours	Unit	Material	Labor	Equipment	Total	Total Incl O&P
0010 **HANDLES AND LATCHES**									
1000 Handles, surface mounted, aluminum	1 Carp	24	.333	Ea.	4.40	10.95		15.35	23
1020 Brass		24	.333		5.70	10.95		16.65	24.50
1040 Chrome		24	.333		7	10.95		17.95	26
1500 Recessed, aluminum		12	.667		2.51	22		24.51	40
1520 Brass		12	.667		3.95	22		25.95	41.50
1540 Chrome		12	.667		3.51	22		25.51	41
2000 Latches, aluminum		20	.400		3.36	13.15		16.51	25.50
2020 Brass		20	.400		4.55	13.15		17.70	27
2040 Chrome		20	.400		3.28	13.15		16.43	25.50

08 75 30 – Weatherstripping

08 75 30.10 Mechanical Weather Seals

	Crew	Daily Output	Labor-Hours	Unit	Material	Labor	Equipment	Total	Total Incl O&P
0010 **MECHANICAL WEATHER SEALS**, Window, double hung, 3' X 5'									
0020 Zinc	1 Carp	7.20	1.111	Opng.	20	36.50		56.50	83.50
0100 Bronze		7.20	1.111		40	36.50		76.50	106
0200 Vinyl V strip		7	1.143		9.50	37.50		47	73.50
0500 As above but heavy duty, zinc		4.60	1.739		20	57		77	118
0600 Bronze		4.60	1.739		70	57		127	173

08 79 Hardware Accessories

08 79 20 – Door Accessories

08 79 20.10 Door Hardware Accessories

	Crew	Daily Output	Labor-Hours	Unit	Material	Labor	Equipment	Total	Total Incl O&P
0010 **DOOR HARDWARE ACCESSORIES**									
0140 Door bolt, surface, 4"	1 Carp	32	.250	Ea.	11.70	8.20		19.90	26.50
0160 Door latch	"	12	.667	"	8.55	22		30.55	46.50
0200 Sliding closet door									
0220 Track and hanger, single	1 Carp	10	.800	Ea.	58.50	26.50		85	109
0240 Double		8	1		80	33		113	143
0260 Door guide, single		48	.167		30	5.50		35.50	42
0280 Double		48	.167		40	5.50		45.50	53
0600 Deadbolt and lock cover plate, brass or stainless steel		30	.267		28	8.75		36.75	45.50
0620 Hole cover plate, brass or chrome		35	.229		8	7.50		15.50	21.50
2240 Mortise lockset, passage, lever handle		9	.889		160	29		189	225
4000 Security chain, standard		18	.444		10	14.60		24.60	35.50
4100 Deluxe		18	.444		45	14.60		59.60	74

08 81 Glass Glazing

08 81 10 – Float Glass

08 81 10.10 Various Types and Thickness of Float Glass		Crew	Daily Output	Labor-Hours	Unit	Material	2015 Bare Costs Labor	Equipment	Total	Total Incl O&P
0010	**VARIOUS TYPES AND THICKNESS OF FLOAT GLASS**									
0020	3/16" Plain	2 Glaz	130	.123	S.F.	5.05	3.94		8.99	12.10
0200	Tempered, clear		130	.123		6.95	3.94		10.89	14.20
0300	Tinted		130	.123		8	3.94		11.94	15.35
0600	1/4" thick, clear, plain		120	.133		5.95	4.27		10.22	13.65
0700	Tinted		120	.133		9.10	4.27		13.37	17.10
0800	Tempered, clear		120	.133		8.85	4.27		13.12	16.85
0900	Tinted		120	.133		10.90	4.27		15.17	19.10
1600	3/8" thick, clear, plain		75	.213		10.20	6.85		17.05	22.50
1700	Tinted		75	.213		15.70	6.85		22.55	28.50
1800	Tempered, clear		75	.213		16.75	6.85		23.60	30
1900	Tinted		75	.213		18.85	6.85		25.70	32
2200	1/2" thick, clear, plain		55	.291		17.30	9.30		26.60	34.50
2300	Tinted		55	.291		27.50	9.30		36.80	45.50
2400	Tempered, clear		55	.291		25	9.30		34.30	43
2500	Tinted		55	.291		26	9.30		35.30	44
2800	5/8" thick, clear, plain		45	.356		27.50	11.40		38.90	49
2900	Tempered, clear		45	.356		31.50	11.40		42.90	53.50

08 81 25 – Glazing Variables

08 81 25.10 Applications of Glazing		Crew	Daily Output	Labor-Hours	Unit	Material	2015 Bare Costs Labor	Equipment	Total	Total Incl O&P
0010	**APPLICATIONS OF GLAZING**									
0600	For glass replacement, add				S.F.		100%			
0700	For gasket settings, add				L.F.	5.75			5.75	6.35
0900	For sloped glazing, add				S.F.		26%			
2000	Fabrication, polished edges, 1/4" thick				Inch	.55			.55	.61
2100	1/2" thick					1.30			1.30	1.43
2500	Mitered edges, 1/4" thick					1.30			1.30	1.43
2600	1/2" thick					2.15			2.15	2.37

08 81 30 – Insulating Glass

08 81 30.10 Reduce Heat Transfer Glass			Crew	Daily Output	Labor-Hours	Unit	Material	2015 Bare Costs Labor	Equipment	Total	Total Incl O&P
0010	**REDUCE HEAT TRANSFER GLASS**										
0015	2 lites 1/8" float, 1/2" thk under 15 S.F.										
0100	Tinted	G	2 Glaz	95	.168	S.F.	14.05	5.40		19.45	24.50
0280	Double glazed, 5/8" thk unit, 3/16" float, 15-30 S.F., clear			90	.178		13.70	5.70		19.40	24.50
0400	1" thk, dbl. glazed, 1/4" float, 30-70 S.F., clear	G		75	.213		16.80	6.85		23.65	30
0500	Tinted	G		75	.213		23.50	6.85		30.35	37.50
2000	Both lites, light & heat reflective	G		85	.188		31.50	6		37.50	44.50
2500	Heat reflective, film inside, 1" thick unit, clear	G		85	.188		27.50	6		33.50	40.50
2600	Tinted	G		85	.188		28.50	6		34.50	41.50
3000	Film on weatherside, clear, 1/2" thick unit	G		95	.168		19.60	5.40		25	30.50
3100	5/8" thick unit	G		90	.178		20	5.70		25.70	31.50
3200	1" thick unit	G		85	.188		27	6		33	40
5000	Spectrally selective film, on ext, blocks solar gain/allows 70% of light	G		95	.168		14.25	5.40		19.65	24.50

08 81 40 – Plate Glass

08 81 40.10 Plate Glass		Crew	Daily Output	Labor-Hours	Unit	Material	2015 Bare Costs Labor	Equipment	Total	Total Incl O&P
0010	**PLATE GLASS** Twin ground, polished,									
0015	3/16" thick, material				S.F.	5.30			5.30	5.80
0020	3/16" thick	2 Glaz	100	.160		5.30	5.10		10.40	14.30
0100	1/4" thick		94	.170		7.25	5.45		12.70	17
0200	3/8" thick		60	.267		12.55	8.55		21.10	28
0300	1/2" thick		40	.400		24.50	12.80		37.30	48.50

08 81 Glass Glazing
08 81 55 – Window Glass

08 81 55.10 Sheet Glass		Crew	Daily Output	Labor-Hours	Unit	Material	2015 Bare Costs Labor	2015 Bare Costs Equipment	Total	Total Incl O&P
0010	SHEET GLASS (window), clear float, stops, putty bed									
0015	1/8" thick, clear float	2 Glaz	480	.033	S.F.	3.65	1.07		4.72	5.80
0500	3/16" thick, clear		480	.033		5.90	1.07		6.97	8.20
0600	Tinted		480	.033		7.45	1.07		8.52	9.95
0700	Tempered		480	.033		9.20	1.07		10.27	11.85

08 83 Mirrors
08 83 13 – Mirrored Glass Glazing

08 83 13.10 Mirrors		Crew	Daily Output	Labor-Hours	Unit	Material	Labor	Equipment	Total	Total Incl O&P
0010	MIRRORS, No frames, wall type, 1/4" plate glass, polished edge									
0100	Up to 5 S.F.	2 Glaz	125	.128	S.F.	9.50	4.10		13.60	17.25
0200	Over 5 S.F.		160	.100		9.25	3.20		12.45	15.50
0500	Door type, 1/4" plate glass, up to 12 S.F.		160	.100		8.70	3.20		11.90	14.90
1000	Float glass, up to 10 S.F., 1/8" thick		160	.100		5.90	3.20		9.10	11.80
1100	3/16" thick		150	.107		7.30	3.41		10.71	13.70
1500	12" x 12" wall tiles, square edge, clear		195	.082		2.22	2.63		4.85	6.80
1600	Veined		195	.082		5.85	2.63		8.48	10.75
2010	Bathroom, unframed, laminated		160	.100		13.90	3.20		17.10	20.50

08 87 Glazing Surface Films
08 87 23 – Safety and Security Films

08 87 23.16 Security Films		Crew	Daily Output	Labor-Hours	Unit	Material	Labor	Equipment	Total	Total Incl O&P
0010	SECURITY FILMS, clear, 32000 psi tensile strength, adhered to glass									
0100	.002" thick, daylight installation	H-2	950	.025	S.F.	2.70	.74		3.44	4.21
0150	.004" thick, daylight installation		800	.030		3.35	.88		4.23	5.15
0200	.006" thick, daylight installation		700	.034		3.60	1.01		4.61	5.65
0210	Install for anchorage		600	.040		4	1.18		5.18	6.35
0400	.007" thick, daylight installation		600	.040		4.45	1.18		5.63	6.85
0410	Install for anchorage		500	.048		4.94	1.42		6.36	7.80
0500	.008" thick, daylight installation		500	.048		5	1.42		6.42	7.85
0510	Install for anchorage		500	.048		5.55	1.42		6.97	8.45
0600	.015" thick, daylight installation		400	.060		8.80	1.77		10.57	12.65
0610	Install for anchorage		400	.060		5.55	1.77		7.32	9.05
0900	Security film anchorage, mechanical attachment and cover plate	H-3	370	.043	L.F.	9.90	1.22		11.12	12.95
0950	Security film anchorage, wet glaze structural caulking	1 Glaz	225	.036	"	.99	1.14		2.13	2.98
1000	Adhered security film removal	1 Clab	275	.029	S.F.		.71		.71	1.19

08 91 Louvers

08 91 19 – Fixed Louvers

08 91 19.10 Aluminum Louvers

		Crew	Daily Output	Labor-Hours	Unit	Material	2015 Bare Costs Labor	Equipment	Total	Total Incl O&P
0010	**ALUMINUM LOUVERS**									
0020	Aluminum with screen, residential, 8" x 8"	1 Carp	38	.211	Ea.	20	6.90		26.90	33.50
0100	12" x 12"		38	.211		16	6.90		22.90	29
0200	12" x 18"		35	.229		20	7.50		27.50	34.50
0250	14" x 24"		30	.267		29	8.75		37.75	46.50
0300	18" x 24"		27	.296		32	9.75		41.75	51.50
0500	24" x 30"		24	.333		60	10.95		70.95	84.50
0700	Triangle, adjustable, small		20	.400		55	13.15		68.15	82.50
0800	Large		15	.533		76	17.50		93.50	113
2100	Midget, aluminum, 3/4" deep, 1" diameter		85	.094		.75	3.09		3.84	6.05
2150	3" diameter		60	.133		2.47	4.38		6.85	10.05
2200	4" diameter		50	.160		4.95	5.25		10.20	14.25
2250	6" diameter		30	.267		4.10	8.75		12.85	19.20

08 95 Vents

08 95 13 – Soffit Vents

08 95 13.10 Wall Louvers

		Crew	Daily Output	Labor-Hours	Unit	Material	Labor	Equipment	Total	Total Incl O&P
0010	**WALL LOUVERS**									
2400	Under eaves vent, aluminum, mill finish, 16" x 4"	1 Carp	48	.167	Ea.	1.90	5.50		7.40	11.30
2500	16" x 8"	"	48	.167	"	2.18	5.50		7.68	11.60

08 95 16 – Wall Vents

08 95 16.10 Louvers

		Crew	Daily Output	Labor-Hours	Unit	Material	Labor	Equipment	Total	Total Incl O&P
0010	**LOUVERS**									
0020	Redwood, 2'-0" diameter, full circle	1 Carp	16	.500	Ea.	190	16.45		206.45	237
0100	Half circle		16	.500		180	16.45		196.45	226
0200	Octagonal		16	.500		142	16.45		158.45	184
0300	Triangular, 5/12 pitch, 5'-0" at base		16	.500		200	16.45		216.45	248
1000	Rectangular, 1'-4" x 1'-3"		16	.500		19.50	16.45		35.95	49
1100	Rectangular, 1'-4" x 1'-8"		16	.500		28	16.45		44.45	58.50
1200	1'-4" x 2'-2"		15	.533		31	17.50		48.50	64
1300	1'-9" x 2'-2"		15	.533		37.50	17.50		55	71
1400	2'-3" x 2'-2"		14	.571		49.50	18.75		68.25	86
1700	2'-4" x 2'-11"		13	.615		49.50	20		69.50	88.50
2000	Aluminum, 12" x 16"		25	.320		20	10.50		30.50	39.50
2010	16" x 20"		25	.320		27.50	10.50		38	48
2020	24" x 30"		25	.320		59	10.50		69.50	82.50
2100	6' triangle		12	.667		168	22		190	222
3100	Round, 2'-2" diameter		16	.500		124	16.45		140.45	164
7000	Vinyl gable vent, 8" x 8"		38	.211		14	6.90		20.90	27
7020	12" x 12"		38	.211		27	6.90		33.90	41
7080	12" x 18"		35	.229		35	7.50		42.50	51
7200	18" x 24"		30	.267		45	8.75		53.75	64

Division 9 – Finishes

Estimating Tips

General
- Room Finish Schedule: A complete set of plans should contain a room finish schedule. If one is not available, it would be well worth the time and effort to obtain one.

09 20 00 Plaster and Gypsum Board
- Lath is estimated by the square yard plus a 5% allowance for waste. Furring, channels, and accessories are measured by the linear foot. An extra foot should be allowed for each accessory miter or stop.
- Plaster is also estimated by the square yard. Deductions for openings vary by preference, from zero deduction to 50% of all openings over 2 feet in width. The estimator should allow one extra square foot for each linear foot of horizontal interior or exterior angle located below the ceiling level. Also, double the areas of small radius work.
- Drywall accessories, studs, track, and acoustical caulking are all measured by the linear foot. Drywall taping is figured by the square foot. Gypsum wallboard is estimated by the square foot. No material deductions should be made for door or window openings under 32 S.F.

09 60 00 Flooring
- Tile and terrazzo areas are taken off on a square foot basis. Trim and base materials are measured by the linear foot. Accent tiles are listed per each. Two basic methods of installation are used. Mud set is approximately 30% more expensive than thin set. In terrazzo work, be sure to include the linear footage of embedded decorative strips, grounds, machine rubbing, and power cleanup.
- Wood flooring is available in strip, parquet, or block configuration. The latter two types are set in adhesives with quantities estimated by the square foot. The laying pattern will influence labor costs and material waste. In addition to the material and labor for laying wood floors, the estimator must make allowances for sanding and finishing these areas, unless the flooring is prefinished.
- Sheet flooring is measured by the square yard. Roll widths vary, so consideration should be given to use the most economical width, as waste must be figured into the total quantity. Consider also the installation methods available, direct glue down or stretched.

09 70 00 Wall Finishes
- Wall coverings are estimated by the square foot. The area to be covered is measured, length by height of wall above baseboards, to calculate the square footage of each wall. This figure is divided by the number of square feet in the single roll which is being used. Deduct, in full, the areas of openings such as doors and windows. Where a pattern match is required allow 25%–30% waste.

09 80 00 Acoustic Treatment
- Acoustical systems fall into several categories. The takeoff of these materials should be by the square foot of area with a 5% allowance for waste. Do not forget about scaffolding, if applicable, when estimating these systems.

09 90 00 Painting and Coating
- A major portion of the work in painting involves surface preparation. Be sure to include cleaning, sanding, filling, and masking costs in the estimate.
- Protection of adjacent surfaces is not included in painting costs. When considering the method of paint application, an important factor is the amount of protection and masking required. These must be estimated separately and may be the determining factor in choosing the method of application.

Reference Numbers
Reference numbers are shown in shaded boxes at the beginning of some major classifications. These numbers refer to related items in the Reference Section. The reference information may be an estimating procedure, an alternate pricing method, or technical information.

Note: Not all subdivisions listed here necessarily appear in this publication. ■

No part of this publication may be reproduced, stored in a retrieval system, or transmitted in any form or by any means without prior written permission of RSMeans.

09 01 Maintenance of Finishes

09 01 70 – Maintenance of Wall Finishes

09 01 70.10 Gypsum Wallboard Repairs

		Crew	Daily Output	Labor-Hours	Unit	Material	2015 Bare Costs Labor	Equipment	Total	Total Incl O&P
0010	**GYPSUM WALLBOARD REPAIRS**									
0100	Fill and sand, pin/nail holes	1 Carp	960	.008	Ea.		.27		.27	.46
0110	Screw head pops		480	.017			.55		.55	.92
0120	Dents, up to 2" square		48	.167		.01	5.50		5.51	9.20
0130	2" to 4" square		24	.333		.03	10.95		10.98	18.45
0140	Cut square, patch, sand and finish, holes, up to 2" square		12	.667		.03	22		22.03	37
0150	2" to 4" square		11	.727		.09	24		24.09	40
0160	4" to 8" square		10	.800		.23	26.50		26.73	44.50
0170	8" to 12" square		8	1		.46	33		33.46	55.50
0180	12" to 32" square		6	1.333		1.55	44		45.55	75
0210	16" by 48"		5	1.600		2.65	52.50		55.15	91
0220	32" by 48"		4	2		4.09	65.50		69.59	115
0230	48" square		3.50	2.286		5.75	75		80.75	132
0240	60" square		3.20	2.500		9.50	82		91.50	148
0500	Skim coat surface with joint compound		1600	.005	S.F.	.03	.16		.19	.32
0510	Prepare, retape and refinish joints		60	.133	L.F.	.64	4.38		5.02	8.05

09 05 Common Work Results for Finishes

09 05 05 – Selective Demolition for Finishes

09 05 05.10 Selective Demolition, Ceilings

		Crew	Daily Output	Labor-Hours	Unit	Material	2015 Bare Costs Labor	Equipment	Total	Total Incl O&P
0010	**SELECTIVE DEMOLITION, CEILINGS** R024119-10									
0200	Ceiling, drywall, furred and nailed or screwed	2 Clab	800	.020	S.F.		.49		.49	.82
1000	Plaster, lime and horse hair, on wood lath, incl. lath		700	.023			.56		.56	.94
1200	Suspended ceiling, mineral fiber, 2' x 2' or 2' x 4'		1500	.011			.26		.26	.44
1250	On suspension system, incl. system		1200	.013			.33		.33	.55
1500	Tile, wood fiber, 12" x 12", glued		900	.018			.43		.43	.73
1540	Stapled		1500	.011			.26		.26	.44
2000	Wood, tongue and groove, 1" x 4"		1000	.016			.39		.39	.66
2040	1" x 8"		1100	.015			.36		.36	.60
2400	Plywood or wood fiberboard, 4' x 8' sheets		1200	.013			.33		.33	.55

09 05 05.20 Selective Demolition, Flooring

		Crew	Daily Output	Labor-Hours	Unit	Material	2015 Bare Costs Labor	Equipment	Total	Total Incl O&P
0010	**SELECTIVE DEMOLITION, FLOORING** R024119-10									
0200	Brick with mortar	2 Clab	475	.034	S.F.		.82		.82	1.38
0400	Carpet, bonded, including surface scraping		2000	.008			.20		.20	.33
0480	Tackless		9000	.002			.04		.04	.07
0550	Carpet tile, releasable adhesive		5000	.003			.08		.08	.13
0560	Permanent adhesive		1850	.009			.21		.21	.36
0800	Resilient, sheet goods		1400	.011			.28		.28	.47
0850	Vinyl or rubber cove base	1 Clab	1000	.008	L.F.		.20		.20	.33
0860	Vinyl or rubber cove base, molded corner	"	1000	.008	Ea.		.20		.20	.33
0870	For glued and caulked installation, add to labor						50%			
0900	Vinyl composition tile, 12" x 12"	2 Clab	1000	.016	S.F.		.39		.39	.66
2000	Tile, ceramic, thin set		675	.024			.58		.58	.97
2020	Mud set		625	.026			.63		.63	1.05
3000	Wood, block, on end	1 Carp	400	.020			.66		.66	1.10
3200	Parquet		450	.018			.58		.58	.98
3400	Strip flooring, interior, 2-1/4" x 25/32" thick		325	.025			.81		.81	1.36
3500	Exterior, porch flooring, 1" x 4"		220	.036			1.19		1.19	2.01
3800	Subfloor, tongue and groove, 1" x 6"		325	.025			.81		.81	1.36
3820	1" x 8"		430	.019			.61		.61	1.03
3840	1" x 10"		520	.015			.51		.51	.85

09 05 Common Work Results for Finishes

09 05 05 – Selective Demolition for Finishes

09 05 05.20 Selective Demolition, Flooring		Crew	Daily Output	Labor-Hours	Unit	Material	2015 Bare Costs Labor	Equipment	Total	Total Incl O&P
4000	Plywood, nailed	1 Carp	600	.013	S.F.		.44		.44	.74
4100	Glued and nailed		400	.020			.66		.66	1.10
4200	Hardboard, 1/4" thick		760	.011			.35		.35	.58

09 05 05.30 Selective Demolition, Walls and Partitions

		Crew	Daily Output	Labor-Hours	Unit	Material	Labor	Equipment	Total	Total Incl O&P
0010	**SELECTIVE DEMOLITION, WALLS AND PARTITIONS** R024119-10									
0020	Walls, concrete, reinforced	B-39	120	.400	C.F.		9.90	1.94	11.84	18.80
0025	Plain	"	160	.300	"		7.45	1.46	8.91	14.10
1000	Drywall, nailed or screwed	1 Clab	1000	.008	S.F.		.20		.20	.33
1010	2 layers		400	.020			.49		.49	.82
1500	Fiberboard, nailed		900	.009			.22		.22	.36
1568	Plenum barrier, sheet lead		300	.027			.65		.65	1.09
2200	Metal or wood studs, finish 2 sides, fiberboard	B-1	520	.046			1.16		1.16	1.95
2250	Lath and plaster		260	.092			2.32		2.32	3.89
2300	Plasterboard (drywall)		520	.046			1.16		1.16	1.95
2350	Plywood		450	.053			1.34		1.34	2.25
2800	Paneling, 4' x 8' sheets	1 Clab	475	.017			.41		.41	.69
3000	Plaster, lime and horsehair, on wood lath		400	.020			.49		.49	.82
3020	On metal lath		335	.024			.58		.58	.98
3450	Plaster, interior gypsum, acoustic, or cement		60	.133	S.Y.		3.26		3.26	5.45
3500	Stucco, on masonry		145	.055			1.35		1.35	2.26
3510	Commercial 3-coat		80	.100			2.45		2.45	4.11
3520	Interior stucco		25	.320			7.80		7.80	13.15
3760	Tile, ceramic, on walls, thin set		300	.027	S.F.		.65		.65	1.09
3765	Mud set		250	.032	"		.78		.78	1.31

09 22 Supports for Plaster and Gypsum Board

09 22 03 – Fastening Methods for Finishes

09 22 03.20 Drilling Plaster/Drywall

		Crew	Daily Output	Labor-Hours	Unit	Material	Labor	Equipment	Total	Total Incl O&P
0010	**DRILLING PLASTER/DRYWALL**									
1100	Drilling & layout for drywall/plaster walls, up to 1" deep, no anchor									
1200	Holes, 1/4" diameter	1 Carp	150	.053	Ea.	.01	1.75		1.76	2.95
1300	3/8" diameter		140	.057		.01	1.88		1.89	3.16
1400	1/2" diameter		130	.062		.01	2.02		2.03	3.40
1500	3/4" diameter		120	.067		.01	2.19		2.20	3.70
1600	1" diameter		110	.073		.02	2.39		2.41	4.03
1700	1-1/4" diameter		100	.080		.03	2.63		2.66	4.45
1800	1-1/2" diameter		90	.089		.05	2.92		2.97	4.95
1900	For ceiling installations, add						40%			

09 22 13 – Metal Furring

09 22 13.13 Metal Channel Furring

		Crew	Daily Output	Labor-Hours	Unit	Material	Labor	Equipment	Total	Total Incl O&P
0010	**METAL CHANNEL FURRING**									
0030	Beams and columns, 7/8" channels, galvanized, 12" O.C.	1 Lath	155	.052	S.F.	.40	1.55		1.95	2.96
0050	16" O.C.		170	.047		.33	1.42		1.75	2.65
0070	24" O.C.		185	.043		.22	1.30		1.52	2.34
0100	Ceilings, on steel, 7/8" channels, galvanized, 12" O.C.		210	.038		.37	1.15		1.52	2.25
0300	16" O.C.		290	.028		.33	.83		1.16	1.70
0400	24" O.C.		420	.019		.22	.57		.79	1.17
0600	1-5/8" channels, galvanized, 12" O.C.		190	.042		.49	1.27		1.76	2.59
0700	16" O.C.		260	.031		.44	.93		1.37	1.98
0900	24" O.C.		390	.021		.29	.62		.91	1.32

For customer support on your Residential Cost Data, call 877.713.2617.

09 22 Supports for Plaster and Gypsum Board

09 22 13 – Metal Furring

09 22 13.13 Metal Channel Furring

		Crew	Daily Output	Labor-Hours	Unit	Material	2015 Bare Costs Labor	Equipment	Total	Total Incl O&P
0930	7/8" channels with sound isolation clips, 12" O.C.	1 Lath	120	.067	S.F.	1.74	2.01		3.75	5.15
0940	16" O.C.		100	.080		1.31	2.41		3.72	5.35
0950	24" O.C.		165	.048		.87	1.46		2.33	3.32
0960	1-5/8" channels, galvanized, 12" O.C.		110	.073		1.86	2.19		4.05	5.60
0970	16" O.C.		100	.080		1.40	2.41		3.81	5.45
0980	24" O.C.		155	.052		.93	1.55		2.48	3.54
1000	Walls, 7/8" channels, galvanized, 12" O.C.		235	.034		.37	1.02		1.39	2.05
1200	16" O.C.		265	.030		.33	.91		1.24	1.83
1300	24" O.C.		350	.023		.22	.69		.91	1.35
1500	1-5/8" channels, galvanized, 12" O.C.		210	.038		.49	1.15		1.64	2.39
1600	16" O.C.		240	.033		.44	1		1.44	2.10
1800	24" O.C.		305	.026		.29	.79		1.08	1.59
1920	7/8" channels with sound isolation clips, 12" O.C.		125	.064		1.74	1.93		3.67	5.05
1940	16" O.C.		100	.080		1.31	2.41		3.72	5.35
1950	24" O.C.		150	.053		.87	1.61		2.48	3.55
1960	1-5/8" channels, galvanized, 12" O.C.		115	.070		1.86	2.09		3.95	5.45
1970	16" O.C.		95	.084		1.40	2.53		3.93	5.65
1980	24" O.C.		140	.057		.93	1.72		2.65	3.81

09 22 16 – Non-Structural Metal Framing

09 22 16.13 Non-Structural Metal Stud Framing

		Crew	Daily Output	Labor-Hours	Unit	Material	2015 Bare Costs Labor	Equipment	Total	Total Incl O&P
0010	**NON-STRUCTURAL METAL STUD FRAMING**									
1600	Non-load bearing, galv., 8' high, 25 ga. 1-5/8" wide, 16" O.C.	1 Carp	619	.013	S.F.	.27	.42		.69	1
1610	24" O.C.		950	.008		.20	.28		.48	.68
1620	2-1/2" wide, 16" O.C.		613	.013		.33	.43		.76	1.08
1630	24" O.C.		938	.009		.24	.28		.52	.74
1640	3-5/8" wide, 16" O.C.		600	.013		.39	.44		.83	1.17
1650	24" O.C.		925	.009		.29	.28		.57	.80
1660	4" wide, 16" O.C.		594	.013		.43	.44		.87	1.21
1670	24" O.C.		925	.009		.32	.28		.60	.84
1680	6" wide, 16" O.C.		588	.014		.52	.45		.97	1.32
1690	24" O.C.		906	.009		.39	.29		.68	.92
1700	20 ga. studs, 1-5/8" wide, 16" O.C.		494	.016		.34	.53		.87	1.26
1710	24" O.C.		763	.010		.25	.34		.59	.86
1720	2-1/2" wide, 16" O.C.		488	.016		.42	.54		.96	1.36
1730	24" O.C.		750	.011		.31	.35		.66	.94
1740	3-5/8" wide, 16" O.C.		481	.017		.48	.55		1.03	1.45
1750	24" O.C.		738	.011		.36	.36		.72	1
1760	4" wide, 16" O.C.		475	.017		.57	.55		1.12	1.56
1770	24" O.C.		738	.011		.43	.36		.79	1.07
1780	6" wide, 16" O.C.		469	.017		.66	.56		1.22	1.67
1790	24" O.C.		725	.011		.50	.36		.86	1.16
2000	Non-load bearing, galv., 10' high, 25 ga. 1-5/8" wide, 16" O.C.		495	.016		.25	.53		.78	1.17
2100	24" O.C.		760	.011		.19	.35		.54	.78
2200	2-1/2" wide, 16" O.C.		490	.016		.31	.54		.85	1.24
2250	24" O.C.		750	.011		.23	.35		.58	.84
2300	3-5/8" wide, 16" O.C.		480	.017		.37	.55		.92	1.32
2350	24" O.C.		740	.011		.27	.36		.63	.90
2400	4" wide, 16" O.C.		475	.017		.41	.55		.96	1.38
2450	24" O.C.		740	.011		.30	.36		.66	.93
2500	6" wide, 16" O.C.		470	.017		.49	.56		1.05	1.48
2550	24" O.C.		725	.011		.36	.36		.72	1.01
2600	20 ga. studs, 1-5/8" wide, 16" O.C.		395	.020		.32	.67		.99	1.47

09 22 Supports for Plaster and Gypsum Board

09 22 16 – Non-Structural Metal Framing

09 22 16.13 Non-Structural Metal Stud Framing		Crew	Daily Output	Labor-Hours	Unit	Material	2015 Bare Costs Labor	2015 Bare Costs Equipment	Total	Total Incl O&P
2650	24" O.C.	1 Carp	610	.013	S.F.	.23	.43		.66	.98
2700	2-1/2" wide, 16" O.C.		390	.021		.39	.67		1.06	1.56
2750	24" O.C.		600	.013		.29	.44		.73	1.06
2800	3-5/8" wide, 16" OC		385	.021		.46	.68		1.14	1.65
2850	24" O.C.		590	.014		.34	.45		.79	1.12
2900	4" wide, 16" O.C.		380	.021		.54	.69		1.23	1.75
2950	24" O.C.		590	.014		.40	.45		.85	1.19
3000	6" wide, 16" O.C.		375	.021		.63	.70		1.33	1.87
3050	24" O.C.		580	.014		.46	.45		.91	1.27
3060	Non-load bearing, galv., 12' high, 25 ga. 1-5/8" wide, 16" O.C.		413	.019		.24	.64		.88	1.34
3070	24" O.C.		633	.013		.18	.42		.60	.89
3080	2-1/2" wide, 16" O.C.		408	.020		.30	.64		.94	1.41
3090	24" O.C.		625	.013		.22	.42		.64	.95
3100	3-5/8" wide, 16" O.C.		400	.020		.35	.66		1.01	1.49
3110	24" O.C.		617	.013		.26	.43		.69	1
3120	4" wide, 16" O.C.		396	.020		.39	.66		1.05	1.54
3130	24" O.C.		617	.013		.28	.43		.71	1.03
3140	6" wide, 16" O.C.		392	.020		.47	.67		1.14	1.65
3150	24" O.C.		604	.013		.34	.44		.78	1.11
3160	20 ga. studs, 1-5/8" wide, 16" O.C.		329	.024		.30	.80		1.10	1.67
3170	24" O.C.		508	.016		.22	.52		.74	1.11
3180	2-1/2" wide, 16" O.C.		325	.025		.38	.81		1.19	1.77
3190	24" O.C.		500	.016		.27	.53		.80	1.18
3200	3-5/8" wide, 16" O.C.		321	.025		.44	.82		1.26	1.85
3210	24" O.C.		492	.016		.32	.53		.85	1.25
3220	4" wide, 16" O.C.		317	.025		.52	.83		1.35	1.96
3230	24" O.C.		492	.016		.38	.53		.91	1.31
3240	6" wide, 16" O.C.		313	.026		.60	.84		1.44	2.07
3250	24" O.C.		483	.017		.44	.54		.98	1.39
5000	Load bearing studs, see Section 05 41 13.30									

09 22 26 – Suspension Systems

09 22 26.13 Ceiling Suspension Systems

		Crew	Daily Output	Labor-Hours	Unit	Material	Labor	Equipment	Total	Total Incl O&P
0010	**CEILING SUSPENSION SYSTEMS** for gypsum board or plaster									
8000	Suspended ceilings, including carriers									
8200	1-1/2" carriers, 24" O.C. with:									
8300	7/8" channels, 16" O.C.	1 Lath	275	.029	S.F.	.54	.88		1.42	2
8320	24" O.C.		310	.026		.42	.78		1.20	1.72
8400	1-5/8" channels, 16" O.C.		205	.039		.64	1.17		1.81	2.61
8420	24" O.C.		250	.032		.50	.96		1.46	2.11
8600	2" carriers, 24" O.C. with:									
8700	7/8" channels, 16" O.C.	1 Lath	250	.032	S.F.	.59	.96		1.55	2.21
8720	24" O.C.		285	.028		.48	.84		1.32	1.89
8800	1-5/8" channels, 16" O.C.		190	.042		.70	1.27		1.97	2.82
8820	24" O.C.		225	.036		.55	1.07		1.62	2.34

09 22 36 – Lath

09 22 36.13 Gypsum Lath

		Crew	Daily Output	Labor-Hours	Unit	Material	Labor	Equipment	Total	Total Incl O&P
0011	**GYPSUM LATH** Plain or perforated, nailed, 3/8" thick	1 Lath	765	.010	S.F.	.34	.31		.65	.88
0101	1/2" thick, nailed		720	.011		.27	.33		.60	.84
0301	Clipped to steel studs, 3/8" thick		675	.012		.34	.36		.70	.95
0401	1/2" thick		630	.013		.27	.38		.65	.92
0601	Firestop gypsum base, to steel studs, 3/8" thick		630	.013		.27	.38		.65	.92
0701	1/2" thick		585	.014		.31	.41		.72	1

09 22 Supports for Plaster and Gypsum Board

09 22 36 – Lath

09 22 36.13 Gypsum Lath

		Crew	Daily Output	Labor-Hours	Unit	Material	2015 Bare Costs Labor	Equipment	Total	Total Incl O&P
0901	Foil back, to steel studs, 3/8" thick	1 Lath	675	.012	S.F.	.38	.36		.74	1
1001	1/2" thick		630	.013		.44	.38		.82	1.10
1501	For ceiling installations, add		1950	.004			.12		.12	.20
1601	For columns and beams, add		1550	.005			.16		.16	.25

09 22 36.23 Metal Lath

		Crew	Daily Output	Labor-Hours	Unit	Material	Labor	Equipment	Total	Total Incl O&P
0010	**METAL LATH** R092000-50									
3601	2.5 lb. diamond painted, on wood framing, on walls	1 Lath	765	.010	S.F.	.36	.31		.67	.91
3701	On ceilings		675	.012		.36	.36		.72	.98
4201	3.4 lb. diamond painted, wired to steel framing, on walls		675	.012		.42	.36		.78	1.05
4301	On ceilings		540	.015		.42	.45		.87	1.19
5101	Rib lath, painted, wired to steel, on walls, 2.75 lb.		675	.012		.34	.36		.70	.95
5201	3.4 lb.		630	.013		.41	.38		.79	1.07
5701	Suspended ceiling system, incl. 3.4 lb. diamond lath, painted		135	.059		.47	1.78		2.25	3.40
5801	Galvanized		135	.059		.47	1.78		2.25	3.40

09 22 36.83 Accessories, Plaster

		Crew	Daily Output	Labor-Hours	Unit	Material	Labor	Equipment	Total	Total Incl O&P
0010	**ACCESSORIES, PLASTER**									
0020	Casing bead, expanded flange, galvanized	1 Lath	2.70	2.963	C.L.F.	55	89		144	205
0200	Foundation weep screed, galvanized	"	2.70	2.963		52	89		141	201
0900	Channels, cold rolled, 16 ga., 3/4" deep, galvanized					37			37	40.50
1620	Corner bead, expanded bullnose, 3/4" radius, #10, galvanized	1 Lath	2.60	3.077		24.50	92.50		117	177
1650	#1, galvanized		2.55	3.137		48.50	94.50		143	205
1670	Expanded wing, 2-3/4" wide, #1, galvanized		2.65	3.019		37	91		128	188
1700	Inside corner (corner rite), 3" x 3", painted		2.60	3.077		20.50	92.50		113	173
1750	Strip-ex, 4" wide, painted		2.55	3.137		24	94.50		118.50	179
1800	Expansion joint, 3/4" grounds, limited expansion, galv., 1 piece		2.70	2.963		75	89		164	227
2100	Extreme expansion, galvanized, 2 piece		2.60	3.077		140	92.50		232.50	305

09 23 Gypsum Plastering

09 23 13 – Acoustical Gypsum Plastering

09 23 13.10 Perlite or Vermiculite Plaster

		Crew	Daily Output	Labor-Hours	Unit	Material	Labor	Equipment	Total	Total Incl O&P
0010	**PERLITE OR VERMICULITE PLASTER** R092000-50									
0020	In 100 lb. bags, under 200 bags				Bag	17.55			17.55	19.35
0301	2 coats, no lath included, on walls	J-1	830	.048	S.F.	.65	1.38	.17	2.20	3.18
0401	On ceilings		710	.056		.65	1.62	.20	2.47	3.59
0901	3 coats, no lath included, on walls		665	.060		.71	1.73	.21	2.65	3.84
1001	On ceilings		565	.071		.71	2.03	.25	2.99	4.38
1700	For irregular or curved surfaces, add to above				S.Y.		30%			
1800	For columns and beams, add to above						50%			
1900	For soffits, add to ceiling prices						40%			

09 23 20 – Gypsum Plaster

09 23 20.10 Gypsum Plaster On Walls and Ceilings

		Crew	Daily Output	Labor-Hours	Unit	Material	Labor	Equipment	Total	Total Incl O&P
0010	**GYPSUM PLASTER ON WALLS AND CEILINGS** R092000-50									
0020	80# bag, less than 1 ton				Bag	15.95			15.95	17.55
0302	2 coats, no lath included, on walls	J-1	750	.053	S.F.	.40	1.53	.19	2.12	3.16
0402	On ceilings		660	.061		.40	1.74	.21	2.35	3.52
0903	3 coats, no lath included, on walls		620	.065		.58	1.85	.23	2.66	3.92
1002	On ceilings		560	.071		.98	2.05	.25	3.28	4.72
1600	For irregular or curved surfaces, add						30%			
1800	For columns & beams, add						50%			

09 24 Cement Plastering

09 24 23 – Cement Stucco

09 24 23.40 Stucco

		Crew	Daily Output	Labor-Hours	Unit	Material	2015 Bare Costs Labor	2015 Bare Costs Equipment	Total	Total Incl O&P
0010	STUCCO R092000-50									
0011	3 coats 1" thick, float finish, with mesh, on wood frame	J-2	470	.102	S.F.	.94	2.95	.30	4.19	6.20
0101	On masonry construction	J-1	495	.081		.26	2.32	.28	2.86	4.39
0151	2 coats, 3/4" thick, float finish, no lath incl.	"	980	.041		.26	1.17	.14	1.57	2.36
0301	For trowel finish, add	1 Plas	1530	.005			.16		.16	.26
0600	For coloring add	J-1	685	.058	S.Y.	.40	1.68	.20	2.28	3.42
0700	For special texture add		200	.200	"	1.41	5.75	.70	7.86	11.70
1001	Exterior stucco, with bonding agent, 3 coats, on walls		1800	.022	S.F.	.37	.64	.08	1.09	1.54
1201	Ceilings		1620	.025		.37	.71	.09	1.17	1.66
1301	Beams		720	.056		.37	1.59	.20	2.16	3.22
1501	Columns		900	.044		.37	1.28	.16	1.81	2.66
1601	Mesh, painted, nailed to wood, 1.8 lb.	1 Lath	540	.015		.69	.45		1.14	1.48
1801	3.6 lb.		495	.016		.41	.49		.90	1.24
1901	Wired to steel, painted, 1.8 lb.		477	.017		.69	.50		1.19	1.58
2101	3.6 lb.		450	.018		.41	.54		.95	1.31

09 25 Other Plastering

09 25 23 – Lime Based Plastering

09 25 23.10 Venetian Plaster

		Crew	Daily Output	Labor-Hours	Unit	Material	Labor	Equipment	Total	Total Incl O&P
0010	VENETIAN PLASTER									
0100	Walls, 1 coat primer, roller applied	1 Plas	950	.008	S.F.	.16	.25		.41	.59
0210	For pigment, light colors add per ea.				Ea.	3			3	3.30
0220	For pigment, dark colors add per ea.				"	9			9	9.90
0300	For sealer/wax coat incl. burnishing, add	1 Plas	300	.027	S.F.	.41	.80		1.21	1.76

09 26 Veneer Plastering

09 26 13 – Gypsum Veneer Plastering

09 26 13.20 Blueboard

		Crew	Daily Output	Labor-Hours	Unit	Material	Labor	Equipment	Total	Total Incl O&P
0010	BLUEBOARD For use with thin coat									
0100	plaster application see Section 09 26 13.80									
1000	3/8" thick, on walls or ceilings, standard, no finish included	2 Carp	1900	.008	S.F.	.34	.28		.62	.83
1100	With thin coat plaster finish		875	.018		.45	.60		1.05	1.51
1400	On beams, columns, or soffits, standard, no finish included		675	.024		.39	.78		1.17	1.74
1450	With thin coat plaster finish		475	.034		.50	1.11		1.61	2.41
3000	1/2" thick, on walls or ceilings, standard, no finish included		1900	.008		.33	.28		.61	.82
3100	With thin coat plaster finish		875	.018		.44	.60		1.04	1.50
3300	Fire resistant, no finish included		1900	.008		.33	.28		.61	.82
3400	With thin coat plaster finish		875	.018		.44	.60		1.04	1.50
3450	On beams, columns, or soffits, standard, no finish included		675	.024		.38	.78		1.16	1.73
3500	With thin coat plaster finish		475	.034		.49	1.11		1.60	2.40
3700	Fire resistant, no finish included		675	.024		.38	.78		1.16	1.73
3800	With thin coat plaster finish		475	.034		.49	1.11		1.60	2.40
5000	5/8" thick, on walls or ceilings, fire resistant, no finish included		1900	.008		.34	.28		.62	.83
5100	With thin coat plaster finish		875	.018		.45	.60		1.05	1.51
5500	On beams, columns, or soffits, no finish included		675	.024		.39	.78		1.17	1.74
5600	With thin coat plaster finish		475	.034		.50	1.11		1.61	2.41
6000	For high ceilings, over 8' high, add		3060	.005			.17		.17	.29
6500	For over 3 stories high, add per story		6100	.003			.09		.09	.14

09 26 Veneer Plastering

09 26 13 – Gypsum Veneer Plastering

09 26 13.80 Thin Coat Plaster

		Crew	Daily Output	Labor-Hours	Unit	Material	2015 Bare Costs Labor	2015 Bare Costs Equipment	Total	Total Incl O&P
0010	**THIN COAT PLASTER** R092000-50									
0012	1 coat veneer, not incl. lath	J-1	3600	.011	S.F.	.11	.32	.04	.47	.68
1000	In 50 lb. bags				Bag	15.25			15.25	16.80

09 28 Backing Boards and Underlayments

09 28 13 – Cementitious Backing Boards

09 28 13.10 Cementitious Backerboard

		Crew	Daily Output	Labor-Hours	Unit	Material	2015 Bare Costs Labor	2015 Bare Costs Equipment	Total	Total Incl O&P
0010	**CEMENTITIOUS BACKERBOARD**									
0070	Cementitious backerboard, on floor, 3' x 4' x 1/2" sheets	2 Carp	525	.030	S.F.	.78	1		1.78	2.54
0080	3' x 5' x 1/2" sheets		525	.030		.76	1		1.76	2.52
0090	3' x 6' x 1/2" sheets		525	.030		.74	1		1.74	2.49
0100	3' x 4' x 5/8" sheets		525	.030		.99	1		1.99	2.77
0110	3' x 5' x 5/8" sheets		525	.030		.99	1		1.99	2.77
0120	3' x 6' x 5/8" sheets		525	.030		.96	1		1.96	2.74
0150	On wall, 3' x 4' x 1/2" sheets		350	.046		.78	1.50		2.28	3.38
0160	3' x 5' x 1/2" sheets		350	.046		.76	1.50		2.26	3.36
0170	3' x 6' x 1/2" sheets		350	.046		.74	1.50		2.24	3.33
0180	3' x 4' x 5/8" sheets		350	.046		.99	1.50		2.49	3.61
0190	3' x 5' x 5/8" sheets		350	.046		.99	1.50		2.49	3.61
0200	3' x 6' x 5/8" sheets		350	.046		.96	1.50		2.46	3.58
0250	On counter, 3' x 4' x 1/2" sheets		180	.089		.78	2.92		3.70	5.75
0260	3' x 5' x 1/2" sheets		180	.089		.76	2.92		3.68	5.75
0270	3' x 6' x 1/2" sheets		180	.089		.74	2.92		3.66	5.70
0300	3' x 4' x 5/8" sheets		180	.089		.99	2.92		3.91	6
0310	3' x 5' x 5/8" sheets		180	.089		.99	2.92		3.91	6
0320	3' x 6' x 5/8" sheets		180	.089		.96	2.92		3.88	5.95

09 29 Gypsum Board

09 29 10 – Gypsum Board Panels

09 29 10.20 Taping and Finishing

		Crew	Daily Output	Labor-Hours	Unit	Material	2015 Bare Costs Labor	2015 Bare Costs Equipment	Total	Total Incl O&P
0010	**TAPING AND FINISHING**									
3600	For taping and finishing joints, add	2 Carp	2000	.008	S.F.	.05	.26		.31	.49
4500	For thin coat plaster instead of taping, add	J-1	3600	.011	"	.11	.32	.04	.47	.68

09 29 10.30 Gypsum Board

		Crew	Daily Output	Labor-Hours	Unit	Material	2015 Bare Costs Labor	2015 Bare Costs Equipment	Total	Total Incl O&P
0010	**GYPSUM BOARD** on walls & ceilings R092910-10									
0100	Nailed or screwed to studs unless otherwise noted									
0110	1/4" thick, on walls or ceilings, standard, no finish included	2 Carp	1330	.012	S.F.	.35	.40		.75	1.05
0115	1/4" thick, on walls or ceilings, flexible, no finish included		1050	.015		.50	.50		1	1.39
0117	1/4" thick, on columns or soffits, flexible, no finish included		1050	.015		.50	.50		1	1.39
0130	1/4" thick, standard, no finish included, less than 800 S.F.		510	.031		.35	1.03		1.38	2.12
0150	3/8" thick, on walls, standard, no finish included		2000	.008		.34	.26		.60	.81
0200	On ceilings, standard, no finish included		1800	.009		.34	.29		.63	.86
0250	On beams, columns, or soffits, no finish included		675	.024		.34	.78		1.12	1.68
0300	1/2" thick, on walls, standard, no finish included		2000	.008		.30	.26		.56	.77
0350	Taped and finished (level 4 finish)		965	.017		.35	.54		.89	1.29
0390	With compound skim coat (level 5 finish)		775	.021		.40	.68		1.08	1.58
0400	Fire resistant, no finish included		2000	.008		.35	.26		.61	.83
0450	Taped and finished (level 4 finish)		965	.017		.40	.54		.94	1.35

09 29 Gypsum Board

09 29 10 – Gypsum Board Panels

09 29 10.30 Gypsum Board		Crew	Daily Output	Labor-Hours	Unit	Material	2015 Bare Costs Labor	2015 Bare Costs Equipment	Total	Total Incl O&P
0490	With compound skim coat (level 5 finish)	2 Carp	775	.021	S.F.	.45	.68		1.13	1.64
0500	Water resistant, no finish included		2000	.008		.39	.26		.65	.87
0550	Taped and finished (level 4 finish)		965	.017		.44	.54		.98	1.39
0590	With compound skim coat (level 5 finish)		775	.021		.49	.68		1.17	1.68
0600	Prefinished, vinyl, clipped to studs		900	.018		.48	.58		1.06	1.51
0700	Mold resistant, no finish included		2000	.008		.44	.26		.70	.92
0710	Taped and finished (level 4 finish)		965	.017		.49	.54		1.03	1.45
0720	With compound skim coat (level 5 finish)		775	.021		.54	.68		1.22	1.73
1000	On ceilings, standard, no finish included		1800	.009		.30	.29		.59	.82
1050	Taped and finished (level 4 finish)		765	.021		.35	.69		1.04	1.53
1090	With compound skim coat (level 5 finish)		610	.026		.40	.86		1.26	1.89
1100	Fire resistant, no finish included		1800	.009		.35	.29		.64	.88
1150	Taped and finished (level 4 finish)		765	.021		.40	.69		1.09	1.59
1195	With compound skim coat (level 5 finish)		610	.026		.45	.86		1.31	1.95
1200	Water resistant, no finish included		1800	.009		.39	.29		.68	.92
1250	Taped and finished (level 4 finish)		765	.021		.44	.69		1.13	1.63
1290	With compound skim coat (level 5 finish)		610	.026		.49	.86		1.35	1.99
1310	Mold resistant, no finish included		1800	.009		.44	.29		.73	.97
1320	Taped and finished (level 4 finish)		765	.021		.49	.69		1.18	1.69
1330	With compound skim coat (level 5 finish)		610	.026		.54	.86		1.40	2.04
1350	Sag resistant, no finish included		1600	.010		.34	.33		.67	.92
1360	Taped and finished (level 4 finish)		765	.021		.39	.69		1.08	1.58
1370	With compound skim coat (level 5 finish)		610	.026		.44	.86		1.30	1.93
1500	On beams, columns, or soffits, standard, no finish included		675	.024		.35	.78		1.13	1.69
1550	Taped and finished (level 4 finish)		540	.030		.35	.97		1.32	2.01
1590	With compound skim coat (level 5 finish)		475	.034		.40	1.11		1.51	2.30
1600	Fire resistant, no finish included		675	.024		.35	.78		1.13	1.70
1650	Taped and finished (level 4 finish)		540	.030		.40	.97		1.37	2.07
1690	With compound skim coat (level 5 finish)		475	.034		.45	1.11		1.56	2.36
1700	Water resistant, no finish included		675	.024		.45	.78		1.23	1.80
1750	Taped and finished (level 4 finish)		540	.030		.44	.97		1.41	2.11
1790	With compound skim coat (level 5 finish)		475	.034		.49	1.11		1.60	2.40
1800	Mold resistant, no finish included		675	.024		.51	.78		1.29	1.87
1810	Taped and finished (level 4 finish)		540	.030		.49	.97		1.46	2.17
1820	With compound skim coat (level 5 finish)		475	.034		.54	1.11		1.65	2.45
1850	Sag resistant, no finish included		675	.024		.39	.78		1.17	1.74
1860	Taped and finished (level 4 finish)		540	.030		.39	.97		1.36	2.06
1870	With compound skim coat (level 5 finish)		475	.034		.44	1.11		1.55	2.34
2000	5/8" thick, on walls, standard, no finish included		2000	.008		.33	.26		.59	.80
2050	Taped and finished (level 4 finish)		965	.017		.38	.54		.92	1.32
2090	With compound skim coat (level 5 finish)		775	.021		.43	.68		1.11	1.61
2100	Fire resistant, no finish included		2000	.008		.34	.26		.60	.81
2150	Taped and finished (level 4 finish)		965	.017		.39	.54		.93	1.34
2195	With compound skim coat (level 5 finish)		775	.021		.44	.68		1.12	1.62
2200	Water resistant, no finish included		2000	.008		.42	.26		.68	.90
2250	Taped and finished (level 4 finish)		965	.017		.47	.54		1.01	1.42
2290	With compound skim coat (level 5 finish)		775	.021		.52	.68		1.20	1.71
2300	Prefinished, vinyl, clipped to studs		900	.018		.76	.58		1.34	1.82
2510	Mold resistant, no finish included		2000	.008		.46	.26		.72	.95
2520	Taped and finished (level 4 finish)		965	.017		.51	.54		1.05	1.47
2530	With compound skim coat (level 5 finish)		775	.021		.56	.68		1.24	1.76
3000	On ceilings, standard, no finish included		1800	.009		.33	.29		.62	.85
3050	Taped and finished (level 4 finish)		765	.021		.38	.69		1.07	1.56

09 29 Gypsum Board

09 29 10 – Gypsum Board Panels

09 29 10.30 Gypsum Board		Crew	Daily Output	Labor-Hours	Unit	Material	2015 Bare Costs Labor	Equipment	Total	Total Incl O&P
3090	With compound skim coat (level 5 finish)	2 Carp	615	.026	S.F.	.43	.85		1.28	1.91
3100	Fire resistant, no finish included		1800	.009		.34	.29		.63	.86
3150	Taped and finished (level 4 finish)		765	.021		.39	.69		1.08	1.58
3190	With compound skim coat (level 5 finish)		615	.026		.44	.85		1.29	1.92
3200	Water resistant, no finish included		1800	.009		.42	.29		.71	.95
3250	Taped and finished (level 4 finish)		765	.021		.47	.69		1.16	1.66
3290	With compound skim coat (level 5 finish)		615	.026		.52	.85		1.37	2.01
3300	Mold resistant, no finish included		1800	.009		.46	.29		.75	1
3310	Taped and finished (level 4 finish)		765	.021		.51	.69		1.20	1.71
3320	With compound skim coat (level 5 finish)		615	.026		.56	.85		1.41	2.06
3500	On beams, columns, or soffits, no finish included		675	.024		.38	.78		1.16	1.73
3550	Taped and finished (level 4 finish)		475	.034		.43	1.11		1.54	2.34
3590	With compound skim coat (level 5 finish)		380	.042		.49	1.38		1.87	2.86
3600	Fire resistant, no finish included		675	.024		.39	.78		1.17	1.74
3650	Taped and finished (level 4 finish)		475	.034		.45	1.11		1.56	2.35
3690	With compound skim coat (level 5 finish)		380	.042		.44	1.38		1.82	2.80
3700	Water resistant, no finish included		675	.024		.48	.78		1.26	1.84
3750	Taped and finished (level 4 finish)		475	.034		.52	1.11		1.63	2.43
3790	With compound skim coat (level 5 finish)		380	.042		.54	1.38		1.92	2.91
3800	Mold resistant, no finish included		675	.024		.53	.78		1.31	1.89
3810	Taped and finished (level 4 finish)		475	.034		.56	1.11		1.67	2.48
3820	With compound skim coat (level 5 finish)		380	.042		.58	1.38		1.96	2.96
4000	Fireproofing, beams or columns, 2 layers, 1/2" thick, incl finish		330	.048		.79	1.59		2.38	3.54
4010	Mold resistant		330	.048		.97	1.59		2.56	3.74
4050	5/8" thick		300	.053		.77	1.75		2.52	3.79
4060	Mold resistant		300	.053		1.01	1.75		2.76	4.06
4100	3 layers, 1/2" thick		225	.071		1.19	2.34		3.53	5.25
4110	Mold resistant		225	.071		1.46	2.34		3.80	5.55
4150	5/8" thick		210	.076		1.16	2.50		3.66	5.50
4160	Mold resistant		210	.076		1.52	2.50		4.02	5.85
5200	For work over 8' high, add		3060	.005			.17		.17	.29
5270	For textured spray, add	2 Lath	1600	.010		.04	.30		.34	.53
5350	For finishing inner corners, add	2 Carp	950	.017	L.F.	.10	.55		.65	1.04
5355	For finishing outer corners, add	"	1250	.013		.22	.42		.64	.96
5500	For acoustical sealant, add per bead	1 Carp	500	.016		.04	.53		.57	.93
5550	Sealant, 1 quart tube				Ea.	7.05			7.05	7.80
6000	Gypsum sound dampening panels									
6010	1/2" thick on walls, multi-layer, light weight, no finish included	2 Carp	1500	.011	S.F.	1.87	.35		2.22	2.65
6015	Taped and finished (level 4 finish)		725	.022		1.92	.73		2.65	3.33
6020	With compound skim coat (level 5 finish)		580	.028		1.97	.91		2.88	3.69
6025	5/8" thick on walls, for wood studs, no finish included		1500	.011		2.17	.35		2.52	2.98
6030	Taped and finished (level 4 finish)		725	.022		2.22	.73		2.95	3.66
6035	With compound skim coat (level 5 finish)		580	.028		2.27	.91		3.18	4.02
6040	For metal stud, no finish included		1500	.011		2.06	.35		2.41	2.86
6045	Taped and finished (level 4 finish)		725	.022		2.11	.73		2.84	3.54
6050	With compound skim coat (level 5 finish)		580	.028		2.16	.91		3.07	3.90
6055	Abuse resist, no finish included		1500	.011		3.75	.35		4.10	4.72
6060	Taped and finished (level 4 finish)		725	.022		3.80	.73		4.53	5.40
6065	With compound skim coat (level 5 finish)		580	.028		3.85	.91		4.76	5.75
6070	Shear rated, no finish included		1500	.011		4.30	.35		4.65	5.30
6075	Taped and finished (level 4 finish)		725	.022		4.35	.73		5.08	6
6080	With compound skim coat (level 5 finish)		580	.028		4.40	.91		5.31	6.35
6085	For SCIF applications, no finish included		1500	.011		4.72	.35		5.07	5.80

09 29 Gypsum Board

09 29 10 – Gypsum Board Panels

09 29 10.30 Gypsum Board		Crew	Daily Output	Labor-Hours	Unit	Material	2015 Bare Costs Labor	Equipment	Total	Total Incl O&P
6090	Taped and finished (level 4 finish)	2 Carp	725	.022	S.F.	4.77	.73		5.50	6.45
6095	With compound skim coat (level 5 finish)		580	.028		4.82	.91		5.73	6.80
6100	1-3/8" thick on walls, THX Certified, no finish included		1500	.011		8.40	.35		8.75	9.85
6105	Taped and finished (level 4 finish)		725	.022		8.45	.73		9.18	10.50
6110	With compound skim coat (level 5 finish)		580	.028		8.50	.91		9.41	10.85
6115	5/8" thick on walls, score & snap installation, no finish included		2000	.008		1.69	.26		1.95	2.30
6120	Taped and finished (level 4 finish)		965	.017		1.74	.54		2.28	2.82
6125	With compound skim coat (level 5 finish)		775	.021		1.79	.68		2.47	3.11
7020	5/8" thick on ceilings, for wood joists, no finish included		1200	.013		2.17	.44		2.61	3.13
7025	Taped and finished (level 4 finish)		510	.031		2.22	1.03		3.25	4.17
7030	With compound skim coat (level 5 finish)		410	.039		2.27	1.28		3.55	4.65
7035	For metal joists, no finish included		1200	.013		2.06	.44		2.50	3.01
7040	Taped and finished (level 4 finish)		510	.031		2.11	1.03		3.14	4.05
7045	With compound skim coat (level 5 finish)		410	.039		2.16	1.28		3.44	4.53
7050	Abuse resist, no finish included		1200	.013		3.75	.44		4.19	4.87
7055	Taped and finished (level 4 finish)		510	.031		3.80	1.03		4.83	5.90
7060	With compound skim coat (level 5 finish)		410	.039		3.85	1.28		5.13	6.40
7065	Shear rated, no finish included		1200	.013		4.30	.44		4.74	5.45
7070	Taped and finished (level 4 finish)		510	.031		4.35	1.03		5.38	6.50
7075	With compound skim coat (level 5 finish)		410	.039		4.40	1.28		5.68	7
7080	For SCIF applications, no finish included		1200	.013		4.72	.44		5.16	5.95
7085	Taped and finished (level 4 finish)		510	.031		4.77	1.03		5.80	7
7090	With compound skim coat (level 5 finish)		410	.039		4.82	1.28		6.10	7.45
8010	5/8" thick on ceilings, score & snap installation, no finish included		1600	.010		1.69	.33		2.02	2.41
8015	Taped and finished (level 4 finish)		680	.024		1.74	.77		2.51	3.21
8020	With compound skim coat (level 5 finish)		545	.029		1.79	.96		2.75	3.59

09 29 15 – Gypsum Board Accessories

09 29 15.10 Accessories, Drywall

		Crew	Daily Output	Labor-Hours	Unit	Material	Labor	Equipment	Total	Total Incl O&P
0011	**ACCESSORIES, DRYWALL** Casing bead, galvanized steel	1 Carp	290	.028	L.F.	.24	.91		1.15	1.78
0101	Vinyl		290	.028		.22	.91		1.13	1.76
0401	Corner bead, galvanized steel, 1-1/4" x 1-1/4"		350	.023		.16	.75		.91	1.44
0411	1-1/4" x 1-1/4", 10' long		35	.229	Ea.	1.61	7.50		9.11	14.35
0601	Vinyl corner bead		400	.020	L.F.	.20	.66		.86	1.32
0901	Furring channel, galv. steel, 7/8" deep, standard		260	.031		.34	1.01		1.35	2.07
1001	Resilient		260	.031		.25	1.01		1.26	1.98
1101	J trim, galvanized steel, 1/2" wide		300	.027			.88		.88	1.47
1121	5/8" wide		300	.027		.31	.88		1.19	1.81
1160	Screws #6 x 1" A				M	10.05			10.05	11.05
1170	#6 x 1-5/8" A				"	15.10			15.10	16.60
1501	Z stud, galvanized steel, 1-1/2" wide	1 Carp	260	.031	L.F.	.39	1.01		1.40	2.13

09 30 Tiling

09 30 13 – Ceramic Tiling

09 30 13.10 Ceramic Tile

		Crew	Daily Output	Labor-Hours	Unit	Material	2015 Bare Costs Labor	Equipment	Total	Total Incl O&P
0010	**CERAMIC TILE**									
0020	Backsplash, thinset, average grade tiles	1 Tilf	50	.160	S.F.	2.46	4.79		7.25	10.45
0022	Custom grade tiles		50	.160		4.93	4.79		9.72	13.15
0024	Luxury grade tiles		50	.160		9.85	4.79		14.64	18.60
0026	Economy grade tiles		50	.160		2.25	4.79		7.04	10.25
0050	Base, using 1' x 4" high pc. with 1" x 1" tiles, mud set	D-7	82	.195	L.F.	5.20	5.20		10.40	14.15
0100	Thin set	"	128	.125		4.84	3.34		8.18	10.70
0300	For 6" high base, 1" x 1" tile face, add					.77			.77	.85
0400	For 2" x 2" tile face, add to above					.42			.42	.46
0600	Cove base, 4-1/4" x 4-1/4" high, mud set	D-7	91	.176		3.99	4.70		8.69	12
0700	Thin set		128	.125		3.87	3.34		7.21	9.65
0900	6" x 4-1/4" high, mud set		100	.160		4.54	4.28		8.82	11.90
1000	Thin set		137	.117		4.42	3.12		7.54	9.90
1200	Sanitary cove base, 6" x 4-1/4" high, mud set		93	.172		4.36	4.60		8.96	12.25
1300	Thin set		124	.129		4.24	3.45		7.69	10.25
1500	6" x 6" high, mud set		84	.190		5.35	5.10		10.45	14.15
1600	Thin set		117	.137		5.20	3.66		8.86	11.65
1800	Bathroom accessories, average (soap dish, tooth brush holder)		82	.195	Ea.	12	5.20		17.20	21.50
1900	Bathtub, 5', rec. 4-1/4" x 4-1/4" tile wainscot, adhesive set 6' high		2.90	5.517		156	147		303	410
2100	7' high wainscot		2.50	6.400		179	171		350	475
2200	8' high wainscot		2.20	7.273		190	194		384	525
2400	Bullnose trim, 4-1/4" x 4-1/4", mud set		82	.195	L.F.	3.92	5.20		9.12	12.75
2500	Thin set		128	.125		3.84	3.34		7.18	9.60
2700	2" x 6" bullnose trim, mud set		84	.190		4.05	5.10		9.15	12.70
2800	Thin set		124	.129		3.99	3.45		7.44	10
3000	Floors, natural clay, random or uniform, thin set, color group 1		183	.087	S.F.	4.15	2.34		6.49	8.35
3100	Color group 2		183	.087		5.85	2.34		8.19	10.25
3255	Floors, glazed, thin set, 6" x 6", color group 1		300	.053		4.45	1.43		5.88	7.20
3260	8" x 8" tile		300	.053		4.45	1.43		5.88	7.20
3270	12" x 12" tile		290	.055		6.25	1.47		7.72	9.30
3280	16" x 16" tile		280	.057		6.70	1.53		8.23	9.85
3281	18" x 18" tile		270	.059		8.65	1.58		10.23	12.05
3282	20" x 20" tile		260	.062		9.90	1.65		11.55	13.55
3283	24" x 24" tile		250	.064		11.25	1.71		12.96	15.10
3285	Border, 6" x 12" tile		200	.080		12.55	2.14		14.69	17.25
3290	3" x 12" tile		200	.080		40	2.14		42.14	47.50
3300	Porcelain type, 1 color, color group 2, 1" x 1"		183	.087		5.20	2.34		7.54	9.50
3310	2" x 2" or 2" x 1", thin set		190	.084		6.10	2.25		8.35	10.35
3350	For random blend, 2 colors, add					1			1	1.10
3360	4 colors, add					1.50			1.50	1.65
4300	Specialty tile, 4-1/4" x 4-1/4" x 1/2", decorator finish	D-7	183	.087		10.40	2.34		12.74	15.25
4500	Add for epoxy grout, 1/16" joint, 1" x 1" tile		800	.020		.67	.53		1.20	1.61
4600	2" x 2" tile		820	.020		.62	.52		1.14	1.52
4610	Add for epoxy grout, 1/8" joint, 8" x 8" x 3/8" tile, add		900	.018		1.44	.48		1.92	2.36
4800	Pregrouted sheets, walls, 4-1/4" x 4-1/4", 6" x 4-1/4"									
4810	and 8-1/2" x 4-1/4", 4 S.F. sheets, silicone grout	D-7	240	.067	S.F.	5.10	1.78		6.88	8.55
5100	Floors, unglazed, 2 S.F. sheets,									
5110	Urethane adhesive	D-7	180	.089	S.F.	5.10	2.38		7.48	9.45
5400	Walls, interior, thin set, 4-1/4" x 4-1/4" tile		190	.084		2.26	2.25		4.51	6.10
5500	6" x 4-1/4" tile		190	.084		2.92	2.25		5.17	6.85
5700	8-1/2" x 4-1/4" tile		190	.084		4.86	2.25		7.11	9
5800	6" x 6" tile		175	.091		3.28	2.44		5.72	7.55
5810	8" x 8" tile		170	.094		4.44	2.52		6.96	8.95

09 30 Tiling

09 30 13 – Ceramic Tiling

09 30 13.10 Ceramic Tile

		Crew	Daily Output	Labor-Hours	Unit	Material	2015 Bare Costs Labor	Equipment	Total	Total Incl O&P
5820	12" x 12" tile	D-7	160	.100	S.F.	4.35	2.67		7.02	9.10
5830	16" x 16" tile		150	.107		4.77	2.85		7.62	9.85
6000	Decorated wall tile, 4-1/4" x 4-1/4", color group 1		270	.059		3.18	1.58		4.76	6.05
6100	Color group 4		180	.089		49.50	2.38		51.88	58.50
6600	Crystalline glazed, 4-1/4" x 4-1/4", mud set, plain		100	.160		4.36	4.28		8.64	11.70
6700	4-1/4" x 4-1/4", scored tile		100	.160		5.80	4.28		10.08	13.30
6900	6" x 6" plain		93	.172		6.70	4.60		11.30	14.80
7000	For epoxy grout, 1/16" joints, 4-1/4" tile, add		800	.020		.41	.53		.94	1.32
7200	For tile set in dry mortar, add		1735	.009			.25		.25	.40
7300	For tile set in Portland cement mortar, add		290	.055		.16	1.47		1.63	2.57
9300	Ceramic tiles, recycled glass, standard colors, 2" x 2" thru 6" x 6" G		190	.084		21	2.25		23.25	27
9310	6" x 6" G		175	.091		21.50	2.44		23.94	27.50
9320	8" x 8" G		170	.094		22.50	2.52		25.02	29
9330	12" x 12" G		160	.100		22.50	2.67		25.17	29.50
9340	Earthtones, 2" x 2" to 4" x 8" G		190	.084		25	2.25		27.25	31
9350	6" x 6" G		175	.091		25	2.44		27.44	31.50
9360	8" x 8" G		170	.094		26	2.52		28.52	32.50
9370	12" x 12" G		160	.100		26	2.67		28.67	33
9380	Deep colors, 2" x 2" to 4" x 8" G		190	.084		29.50	2.25		31.75	36
9390	6" x 6" G		175	.091		29.50	2.44		31.94	36.50
9400	8" x 8" G		170	.094		31	2.52		33.52	38
9410	12" x 12" G		160	.100		31	2.67		33.67	38.50

09 30 16 – Quarry Tiling

09 30 16.10 Quarry Tile

		Crew	Daily Output	Labor-Hours	Unit	Material	Labor	Equipment	Total	Total Incl O&P
0010	**QUARRY TILE**									
0100	Base, cove or sanitary, mud set, to 5" high, 1/2" thick	D-7	110	.145	L.F.	5.35	3.89		9.24	12.20
0300	Bullnose trim, red, mud set, 6" x 6" x 1/2" thick		120	.133		4.39	3.56		7.95	10.60
0400	4" x 4" x 1/2" thick		110	.145		4.50	3.89		8.39	11.25
0600	4" x 8" x 1/2" thick, using 8" as edge		130	.123		4.50	3.29		7.79	10.30
0700	Floors, mud set, 1,000 S.F. lots, red, 4" x 4" x 1/2" thick		120	.133	S.F.	8.05	3.56		11.61	14.60
0900	6" x 6" x 1/2" thick		140	.114		7.55	3.06		10.61	13.25
1000	4" x 8" x 1/2" thick		130	.123		5.55	3.29		8.84	11.45
1300	For waxed coating, add					.75			.75	.83
1500	For non-standard colors, add					.46			.46	.51
1600	For abrasive surface, add					.52			.52	.57
1800	Brown tile, imported, 6" x 6" x 3/4"	D-7	120	.133		7.95	3.56		11.51	14.50
1900	8" x 8" x 1"		110	.145		8.70	3.89		12.59	15.85
2100	For thin set mortar application, deduct		700	.023			.61		.61	.99
2700	Stair tread, 6" x 6" x 3/4", plain		50	.320		7.20	8.55		15.75	22
2800	Abrasive		47	.340		6.05	9.10		15.15	21.50
3000	Wainscot, 6" x 6" x 1/2", thin set, red		105	.152		4.68	4.07		8.75	11.75
3100	Non-standard colors		105	.152		5.20	4.07		9.27	12.30
3300	Window sill, 6" wide, 3/4" thick		90	.178	L.F.	9.20	4.75		13.95	17.85
3400	Corners		80	.200	Ea.	6.65	5.35		12	15.95

09 30 23 – Glass Mosaic Tiling

09 30 23.10 Glass Mosaics

		Crew	Daily Output	Labor-Hours	Unit	Material	Labor	Equipment	Total	Total Incl O&P
0010	**GLASS MOSAICS** 3/4" tile on 12" sheets, standard grout									
1020	1" tile on 12" sheets, opalescent finish	D-7	73	.219	S.F.	16.85	5.85		22.70	28
1040	1" x 2" tile on 12" sheet, blend		73	.219		18.15	5.85		24	29.50
1060	2" tile on 12" sheet, blend		73	.219		16.50	5.85		22.35	27.50
1080	5/8" x random tile, linear, on 12" sheet, blend		73	.219		26	5.85		31.85	38
1600	Dots on 12" sheet		73	.219		26	5.85		31.85	38

For customer support on your Residential Cost Data, call 877.713.2617.

09 30 Tiling

09 30 23 – Glass Mosaic Tiling

09 30 23.10 Glass Mosaics		Crew	Daily Output	Labor-Hours	Unit	Material	2015 Bare Costs Labor	2015 Bare Costs Equipment	Total	Total Incl O&P
1700	For glass mosaic tiles set in dry mortar, add	D-7	290	.055	S.F.	.45	1.47		1.92	2.89
1720	For glass mosaic tile set in Portland cement mortar, add		290	.055		.01	1.47		1.48	2.40
1730	For polyblend sanded tile grout		96.15	.166	Lb.	2.19	4.45		6.64	9.60

09 30 29 – Metal Tiling

09 30 29.10 Metal Tile

		Crew	Daily Output	Labor-Hours	Unit	Material	Labor	Equipment	Total	Total Incl O&P
0010	**METAL TILE** 4' x 4' sheet, 24 ga., tile pattern, nailed									
0200	Stainless steel	2 Carp	512	.031	S.F.	28	1.03		29.03	32.50
0400	Aluminized steel	"	512	.031	"	15.10	1.03		16.13	18.30

09 34 Waterproofing-Membrane Tiling

09 34 13 – Waterproofing-Membrane Ceramic Tiling

09 34 13.10 Ceramic Tile Waterproofing Membrane

		Crew	Daily Output	Labor-Hours	Unit	Material	Labor	Equipment	Total	Total Incl O&P
0010	**CERAMIC TILE WATERPROOFING MEMBRANE**									
0020	On floors, including thinset									
0030	Fleece laminated polyethylene grid, 1/8" thick	D-7	250	.064	S.F.	2.26	1.71		3.97	5.25
0040	5/16" thick	"	250	.064	"	2.58	1.71		4.29	5.60
0050	On walls, including thinset									
0060	Fleece laminated polyethylene sheet, 8 mil thick	D-7	480	.033	S.F.	2.26	.89		3.15	3.93
0070	Accessories, including thinset									
0080	Joint and corner sheet, 4 mils thick, 5" wide	1 Tilf	240	.033	L.F.	1.33	1		2.33	3.09
0090	7-1/4" wide		180	.044		1.69	1.33		3.02	4.02
0100	10" wide		120	.067		2.06	2		4.06	5.50
0110	Pre-formed corners, inside		32	.250	Ea.	6.90	7.50		14.40	19.75
0120	Outside		32	.250		7.65	7.50		15.15	20.50
0130	2" flanged floor drain with 6" stainless steel grate		16	.500		370	15		385	435
0140	EPS, sloped shower floor		480	.017	S.F.	4.95	.50		5.45	6.25
0150	Curb		32	.250	L.F.	14	7.50		21.50	27.50

09 51 Acoustical Ceilings

09 51 23 – Acoustical Tile Ceilings

09 51 23.10 Suspended Acoustic Ceiling Tiles

		Crew	Daily Output	Labor-Hours	Unit	Material	Labor	Equipment	Total	Total Incl O&P
0010	**SUSPENDED ACOUSTIC CEILING TILES**, not including									
0100	suspension system									
0300	Fiberglass boards, film faced, 2' x 2' or 2' x 4', 5/8" thick	1 Carp	625	.013	S.F.	1.24	.42		1.66	2.07
0400	3/4" thick		600	.013		2.63	.44		3.07	3.63
0500	3" thick, thermal, R11		450	.018		2.42	.58		3	3.64
0600	Glass cloth faced fiberglass, 3/4" thick		500	.016		2.65	.53		3.18	3.80
0700	1" thick		485	.016		3.13	.54		3.67	4.35
0820	1-1/2" thick, nubby face		475	.017		2.53	.55		3.08	3.71
1110	Mineral fiber tile, lay-in, 2' x 2' or 2' x 4', 5/8" thick, fine texture		625	.013		.91	.42		1.33	1.71
1115	Rough textured		625	.013		.85	.42		1.27	1.65
1125	3/4" thick, fine textured		600	.013		1.94	.44		2.38	2.87
1130	Rough textured		600	.013		1.56	.44		2	2.46
1135	Fissured		600	.013		1.96	.44		2.40	2.90
1150	Tegular, 5/8" thick, fine textured		470	.017		1.03	.56		1.59	2.07
1155	Rough textured		470	.017		1.14	.56		1.70	2.19
1165	3/4" thick, fine textured		450	.018		2.14	.58		2.72	3.33
1170	Rough textured		450	.018		1.43	.58		2.01	2.55
1175	Fissured		450	.018		2.16	.58		2.74	3.36

09 51 Acoustical Ceilings

09 51 23 – Acoustical Tile Ceilings

09 51 23.10 Suspended Acoustic Ceiling Tiles

		Crew	Daily Output	Labor-Hours	Unit	Material	2015 Bare Costs Labor	Equipment	Total	Total Incl O&P
1185	For plastic film face, add				S.F.	.75			.75	.83
1190	For fire rating, add					.44			.44	.48
1600	Solid alum. planks, 3-1/4"x12', open reveal	1 Carp	500	.016		2.35	.53		2.88	3.47
1650	Closed reveal		500	.016		3	.53		3.53	4.18
1700	7-1/4"x12', open reveal		500	.016		4	.53		4.53	5.30
1750	Closed reveal		500	.016		5.10	.53		5.63	6.50
1775	Metal, open cell, 2'x2', 6" cell		500	.016		8	.53		8.53	9.70
1800	8" cell		500	.016		8.85	.53		9.38	10.65
1825	2'x4', 6" cell		500	.016		5.10	.53		5.63	6.50
1850	8" cell		500	.016		5.10	.53		5.63	6.50
1870	Translucent lay-in panels, 2'x2'		500	.016		23	.53		23.53	26
1890	2'x6'		500	.016		17.20	.53		17.73	19.80
3750	Wood fiber in cementitious binder, 2' x 2' or 4', painted, 1" thick		600	.013		2	.44		2.44	2.94
3760	2" thick		550	.015		3.50	.48		3.98	4.65
3770	2-1/2" thick		500	.016		4.37	.53		4.90	5.70
3780	3" thick		450	.018		5.25	.58		5.83	6.80

09 51 23.30 Suspended Ceilings, Complete

		Crew	Daily Output	Labor-Hours	Unit	Material	Labor	Equipment	Total	Total Incl O&P
0010	**SUSPENDED CEILINGS, COMPLETE**, including standard									
0100	suspension system but not incl. 1-1/2" carrier channels									
0600	Fiberglass ceiling board, 2' x 4' x 5/8", plain faced	1 Carp	500	.016	S.F.	1.97	.53		2.50	3.05
0700	Offices, 2' x 4' x 3/4"		380	.021		3.36	.69		4.05	4.86
1800	Tile, Z bar suspension, 5/8" mineral fiber tile		150	.053		2.32	1.75		4.07	5.50
1900	3/4" mineral fiber tile		150	.053		2.48	1.75		4.23	5.65

09 51 53 – Direct-Applied Acoustical Ceilings

09 51 53.10 Ceiling Tile

		Crew	Daily Output	Labor-Hours	Unit	Material	Labor	Equipment	Total	Total Incl O&P
0010	**CEILING TILE**, stapled or cemented									
0100	12" x 12" or 12" x 24", not including furring									
0600	Mineral fiber, vinyl coated, 5/8" thick	1 Carp	300	.027	S.F.	2.18	.88		3.06	3.87
0700	3/4" thick		300	.027		2.40	.88		3.28	4.11
0900	Fire rated, 3/4" thick, plain faced		300	.027		1.27	.88		2.15	2.87
1000	Plastic coated face		300	.027		1.84	.88		2.72	3.49
1200	Aluminum faced, 5/8" thick, plain		300	.027		1.66	.88		2.54	3.30
3300	For flameproofing, add					.10			.10	.11
3400	For sculptured 3 dimensional, add					.31			.31	.34
3900	For ceiling primer, add					.13			.13	.14
4000	For ceiling cement, add					.39			.39	.43

09 53 Acoustical Ceiling Suspension Assemblies

09 53 23 – Metal Acoustical Ceiling Suspension Assemblies

09 53 23.30 Ceiling Suspension Systems

		Crew	Daily Output	Labor-Hours	Unit	Material	Labor	Equipment	Total	Total Incl O&P
0010	**CEILING SUSPENSION SYSTEMS** for boards and tile									
0050	Class A suspension system, 15/16" T bar, 2' x 4' grid	1 Carp	800	.010	S.F.	.73	.33		1.06	1.36
0300	2' x 2' grid	"	650	.012		.95	.40		1.35	1.72
0350	For 9/16" grid, add					.16			.16	.18
0360	For fire rated grid, add					.09			.09	.10
0370	For colored grid, add					.21			.21	.23
0400	Concealed Z bar suspension system, 12" module	1 Carp	520	.015		.84	.51		1.35	1.77
0600	1-1/2" carrier channels, 4' O.C., add		470	.017		.11	.56		.67	1.06
0650	1-1/2" x 3-1/2" channels		470	.017		.30	.56		.86	1.27
0700	Carrier channels for ceilings with									

09 53 Acoustical Ceiling Suspension Assemblies

09 53 23 – Metal Acoustical Ceiling Suspension Assemblies

09 53 23.30 Ceiling Suspension Systems		Crew	Daily Output	Labor-Hours	Unit	Material	2015 Bare Costs Labor	Equipment	Total	Total Incl O&P
0900	recessed lighting fixtures, add	1 Carp	460	.017	S.F.	.20	.57		.77	1.18
5000	Wire hangers, #12 wire	"	300	.027	Ea.	.06	.88		.94	1.54

09 61 Flooring Treatment

09 61 19 – Concrete Floor Staining

09 61 19.40 Floors, Interior

		Crew	Daily Output	Labor-Hours	Unit	Material	Labor	Equipment	Total	Total Incl O&P
0010	**FLOORS, INTERIOR**									
0300	Acid stain and sealer									
0310	Stain, one coat	1 Pord	650	.012	S.F.	.12	.35		.47	.71
0320	Two coats		570	.014		.23	.40		.63	.92
0330	Acrylic sealer, one coat		2600	.003		.23	.09		.32	.40
0340	Two coats		1400	.006		.46	.16		.62	.78

09 62 Specialty Flooring

09 62 19 – Laminate Flooring

09 62 19.10 Floating Floor

		Crew	Daily Output	Labor-Hours	Unit	Material	Labor	Equipment	Total	Total Incl O&P
0010	**FLOATING FLOOR**									
8300	Floating floor, laminate, wood pattern strip, complete	1 Clab	133	.060	S.F.	4.35	1.47		5.82	7.25
8310	Components, T & G wood composite strips					3.91			3.91	4.30
8320	Film					.14			.14	.15
8330	Foam					.25			.25	.28
8340	Adhesive					.65			.65	.72
8350	Installation kit					.17			.17	.19
8360	Trim, 2" wide x 3' long				L.F.	4.30			4.30	4.73
8370	Reducer moulding				"	5.70			5.70	6.25

09 62 23 – Bamboo Flooring

09 62 23.10 Flooring, Bamboo

			Crew	Daily Output	Labor-Hours	Unit	Material	Labor	Equipment	Total	Total Incl O&P
0010	**FLOORING, BAMBOO**										
8600	Flooring, wood, bamboo strips, unfinished, 5/8" x 4" x 3'	G	1 Carp	255	.031	S.F.	4.60	1.03		5.63	6.80
8610	5/8" x 4" x 4'	G		275	.029		4.77	.96		5.73	6.85
8620	5/8" x 4" x 6'	G		295	.027		5.25	.89		6.14	7.25
8630	Finished, 5/8" x 4" x 3'	G		255	.031		5.05	1.03		6.08	7.30
8640	5/8" x 4" x 4'	G		275	.029		5.30	.96		6.26	7.45
8650	5/8" x 4" x 6'	G		295	.027		4.61	.89		5.50	6.55
8660	Stair treads, unfinished, 1-1/16" x 11-1/2" x 4'	G		18	.444	Ea.	44	14.60		58.60	73
8670	Finished, 1-1/16" x 11-1/2" x 4'	G		18	.444		78	14.60		92.60	111
8680	Stair risers, unfinished, 5/8" x 7-1/2" x 4'	G		18	.444		16.30	14.60		30.90	42.50
8690	Finished, 5/8" x 7-1/2" x 4'	G		18	.444		31	14.60		45.60	58.50
8700	Stair nosing, unfinished, 6' long	G		16	.500		36	16.45		52.45	67
8710	Finished, 6' long	G		16	.500		42.50	16.45		58.95	74.50

09 63 Masonry Flooring

09 63 13 – Brick Flooring

09 63 13.10 Miscellaneous Brick Flooring

		Crew	Daily Output	Labor-Hours	Unit	Material	2015 Bare Costs Labor	Equipment	Total	Total Incl O&P
0010	**MISCELLANEOUS BRICK FLOORING**									
0020	Acid-proof shales, red, 8" x 3-3/4" x 1-1/4" thick	D-7	.43	37.209	M	695	995		1,690	2,375
0050	2-1/4" thick	D-1	.40	40		965	1,175		2,140	3,025
0200	Acid-proof clay brick, 8" x 3-3/4" x 2-1/4" thick G	"	.40	40		935	1,175		2,110	3,000
0260	Cast ceramic, pressed, 4" x 8" x 1/2", unglazed	D-7	100	.160	S.F.	6.35	4.28		10.63	13.85
0270	Glazed		100	.160		8.45	4.28		12.73	16.20
0280	Hand molded flooring, 4" x 8" x 3/4", unglazed		95	.168		8.35	4.50		12.85	16.50
0290	Glazed		95	.168		10.50	4.50		15	18.85
0300	8" hexagonal, 3/4" thick, unglazed		85	.188		9.20	5.05		14.25	18.25
0310	Glazed		85	.188		16.60	5.05		21.65	26.50
0450	Acid-proof joints, 1/4" wide	D-1	65	.246		1.46	7.25		8.71	13.70
0500	Pavers, 8" x 4", 1" to 1-1/4" thick, red	D-7	95	.168		3.69	4.50		8.19	11.35
0510	Ironspot	"	95	.168		5.20	4.50		9.70	13.05
0540	1-3/8" to 1-3/4" thick, red	D-1	95	.168		3.56	4.97		8.53	12.20
0560	Ironspot		95	.168		5.15	4.97		10.12	14
0580	2-1/4" thick, red		90	.178		3.62	5.25		8.87	12.75
0590	Ironspot		90	.178		5.60	5.25		10.85	14.95
0700	Paver, adobe brick, 6" x 12", 1/2" joint G		42	.381		1.30	11.25		12.55	20
0710	Mexican red, 12" x 12" G	1 Tilf	48	.167		1.66	4.99		6.65	9.95
0720	Saltillo, 12" x 12" G	"	48	.167		1.40	4.99		6.39	9.65
0800	For sidewalks and patios with pavers, see Section 32 14 16.10									
0870	For epoxy joints, add	D-1	600	.027	S.F.	2.81	.79		3.60	4.40
0880	For Furan underlayment, add	"	600	.027		2.33	.79		3.12	3.87
0890	For waxed surface, steam cleaned, add	A-1H	1000	.008		.20	.20	.08	.48	.63

09 63 40 – Stone Flooring

09 63 40.10 Marble

		Crew	Daily Output	Labor-Hours	Unit	Material	Labor	Equipment	Total	Total Incl O&P
0010	**MARBLE**									
0020	Thin gauge tile, 12" x 6", 3/8", white Carara	D-7	60	.267	S.F.	14.40	7.15		21.55	27.50
0100	Travertine		60	.267		11.60	7.15		18.75	24.50
0200	12" x 12" x 3/8", thin set, floors		60	.267		10.15	7.15		17.30	22.50
0300	On walls		52	.308		9.85	8.20		18.05	24
1000	Marble threshold, 4" wide x 36" long x 5/8" thick, white		60	.267	Ea.	10.15	7.15		17.30	22.50

09 63 40.20 Slate Tile

		Crew	Daily Output	Labor-Hours	Unit	Material	Labor	Equipment	Total	Total Incl O&P
0010	**SLATE TILE**									
0020	Vermont, 6" x 6" x 1/4" thick, thin set	D-7	180	.089	S.F.	7.50	2.38		9.88	12.10

09 64 Wood Flooring

09 64 23 – Wood Parquet Flooring

09 64 23.10 Wood Parquet

		Crew	Daily Output	Labor-Hours	Unit	Material	Labor	Equipment	Total	Total Incl O&P
0010	**WOOD PARQUET** flooring									
5200	Parquetry, 5/16" thk, no finish, oak, plain pattern	1 Carp	160	.050	S.F.	5.25	1.64		6.89	8.50
5300	Intricate pattern		100	.080		9.60	2.63		12.23	14.95
5500	Teak, plain pattern		160	.050		5.85	1.64		7.49	9.20
5600	Intricate pattern		100	.080		10.05	2.63		12.68	15.45
5650	13/16" thick, select grade oak, plain pattern		160	.050		9.75	1.64		11.39	13.45
5700	Intricate pattern		100	.080		16.40	2.63		19.03	22.50
5800	Custom parquetry, including finish, plain pattern		100	.080		16.80	2.63		19.43	23
5900	Intricate pattern		50	.160		24	5.25		29.25	35
6700	Parquetry, prefinished white oak, 5/16" thick, plain pattern		160	.050		7.90	1.64		9.54	11.45
6800	Intricate pattern		100	.080		8.45	2.63		11.08	13.65

09 64 Wood Flooring

09 64 23 – Wood Parquet Flooring

09 64 23.10 Wood Parquet		Crew	Daily Output	Labor-Hours	Unit	Material	2015 Bare Costs Labor	2015 Bare Costs Equipment	Total	Total Incl O&P
7000	Walnut or teak, parquetry, plain pattern	1 Carp	160	.050	S.F.	8	1.64		9.64	11.55
7100	Intricate pattern	↓	100	.080	↓	11.50	2.63		14.13	17.05
7200	Acrylic wood parquet blocks, 12" x 12" x 5/16",									
7210	Irradiated, set in epoxy	1 Carp	160	.050	S.F.	10	1.64		11.64	13.75

09 64 29 – Wood Strip and Plank Flooring

09 64 29.10 Wood		Crew	Daily Output	Labor-Hours	Unit	Material	Labor	Equipment	Total	Total Incl O&P
0010	**WOOD**									
0020	Fir, vertical grain, 1" x 4", not incl. finish, grade B & better	1 Carp	255	.031	S.F.	2.79	1.03		3.82	4.80
0100	C grade & better		255	.031		2.63	1.03		3.66	4.62
0300	Flat grain, 1" x 4", not incl. finish, B & better		255	.031		3.19	1.03		4.22	5.25
0400	C & better		255	.031		3.07	1.03		4.10	5.10
4000	Maple, strip, 25/32" x 2-1/4", not incl. finish, select		170	.047		4.95	1.55		6.50	8.05
4100	#2 & better		170	.047		4.29	1.55		5.84	7.30
4300	33/32" x 3-1/4", not incl. finish, #1 grade		170	.047		4.56	1.55		6.11	7.60
4400	#2 & better	↓	170	.047	↓	4.06	1.55		5.61	7.05
4600	Oak, white or red, 25/32" x 2-1/4", not incl. finish									
4700	#1 common	1 Carp	170	.047	S.F.	3.19	1.55		4.74	6.10
4900	Select quartered, 2-1/4" wide		170	.047		3.89	1.55		5.44	6.90
5000	Clear		170	.047		4.01	1.55		5.56	7
6100	Prefinished, white oak, prime grade, 2-1/4" wide		170	.047		4.69	1.55		6.24	7.75
6200	3-1/4" wide		185	.043		5.10	1.42		6.52	8
6400	Ranch plank		145	.055		7.15	1.81		8.96	10.90
6500	Hardwood blocks, 9" x 9", 25/32" thick		160	.050		6	1.64		7.64	9.35
7400	Yellow pine, 3/4" x 3-1/8", T & G, C & better, not incl. finish	↓	200	.040		1.49	1.31		2.80	3.85
7500	Refinish wood floor, sand, 2 coats poly, wax, soft wood	1 Clab	400	.020		.90	.49		1.39	1.81
7600	Hard wood		130	.062		1.34	1.50		2.84	4
7800	Sanding and finishing, 2 coats polyurethane	↓	295	.027	↓	.90	.66		1.56	2.10
7900	Subfloor and underlayment, see Section 06 16									
8015	Transition molding, 2 1/4" wide, 5' long	1 Carp	19.20	.417	Ea.	10.85	13.70		24.55	35

09 65 Resilient Flooring

09 65 10 – Resilient Tile Underlayment

09 65 10.10 Latex Underlayment		Crew	Daily Output	Labor-Hours	Unit	Material	Labor	Equipment	Total	Total Incl O&P
0010	**LATEX UNDERLAYMENT**									
3600	Latex underlayment, 1/8" thk., cementitious for resilient flooring	1 Tilf	160	.050	S.F.	1.19	1.50		2.69	3.74
4000	Liquid, fortified				Gal.	33			33	36

09 65 13 – Resilient Base and Accessories

09 65 13.13 Resilient Base		Crew	Daily Output	Labor-Hours	Unit	Material	Labor	Equipment	Total	Total Incl O&P
0010	**RESILIENT BASE**									
0690	1/8" vinyl base, 2 1/2" H, straight or cove, standard colors	1 Tilf	315	.025	L.F.	.67	.76		1.43	1.97
0700	4" high		315	.025		1.32	.76		2.08	2.68
0710	6" high		315	.025	↓	1.40	.76		2.16	2.77
0720	Corners, 2 1/2" high		315	.025	Ea.	2.02	.76		2.78	3.45
0730	4" high		315	.025		2.14	.76		2.90	3.58
0740	6" high		315	.025	↓	2.45	.76		3.21	3.93
0800	1/8" rubber base, 2 1/2" H, straight or cove, standard colors		315	.025	L.F.	1.13	.76		1.89	2.47
1100	4" high		315	.025		1.02	.76		1.78	2.35
1110	6" high		315	.025	↓	1.77	.76		2.53	3.18
1150	Corners, 2 1/2" high		315	.025	Ea.	2	.76		2.76	3.43
1153	4" high	↓	315	.025		2.05	.76		2.81	3.49

09 65 Resilient Flooring

09 65 13 – Resilient Base and Accessories

09 65 13.13 Resilient Base		Crew	Daily Output	Labor-Hours	Unit	Material	2015 Bare Costs Labor	Equipment	Total	Total Incl O&P
1155	6" high	1 Tilf	315	.025	Ea.	2.51	.76		3.27	3.99
1450	For premium color/finish add					50%				
1500	Millwork profile	1 Tilf	315	.025	L.F.	5.85	.76		6.61	7.70

09 65 16 – Resilient Sheet Flooring

09 65 16.10 Rubber and Vinyl Sheet Flooring

			Crew	Daily Output	Labor-Hours	Unit	Material	Labor	Equipment	Total	Total Incl O&P
0010	**RUBBER AND VINYL SHEET FLOORING**										
5500	Linoleum, sheet goods	G	1 Tilf	360	.022	S.F.	3.59	.67		4.26	5.05
5900	Rubber, sheet goods, 36" wide, 1/8" thick			120	.067		7.75	2		9.75	11.80
5950	3/16" thick			100	.080		10.50	2.40		12.90	15.45
6000	1/4" thick			90	.089		12.45	2.66		15.11	17.95
8000	Vinyl sheet goods, backed, .065" thick, plain pattern/colors			250	.032		4.09	.96		5.05	6.05
8050	Intricate pattern/colors			200	.040		4.57	1.20		5.77	7
8100	.080" thick, plain pattern/colors			230	.035		4.10	1.04		5.14	6.20
8150	Intricate pattern/colors			200	.040		5.90	1.20		7.10	8.45
8200	.125" thick, plain pattern/colors			230	.035		4.25	1.04		5.29	6.35
8250	intricate pattern/colors			200	.040		7.50	1.20		8.70	10.20
8700	Adhesive cement, 1 gallon per 200 to 300 S.F.					Gal.	27			27	29.50
8800	Asphalt primer, 1 gallon per 300 S.F.						14			14	15.40
8900	Emulsion, 1 gallon per 140 S.F.						18			18	19.80

09 65 19 – Resilient Tile Flooring

09 65 19.10 Miscellaneous Resilient Tile Flooring

			Crew	Daily Output	Labor-Hours	Unit	Material	Labor	Equipment	Total	Total Incl O&P
0010	**MISCELLANEOUS RESILIENT TILE FLOORING**										
2200	Cork tile, standard finish, 1/8" thick	G	1 Tilf	315	.025	S.F.	6.95	.76		7.71	8.90
2250	3/16" thick	G		315	.025		6.65	.76		7.41	8.55
2300	5/16" thick	G		315	.025		8.30	.76		9.06	10.35
2350	1/2" thick	G		315	.025		10.95	.76		11.71	13.25
2500	Urethane finish, 1/8" thick	G		315	.025		8.10	.76		8.86	10.15
2550	3/16" thick	G		315	.025		8.25	.76		9.01	10.30
2600	5/16" thick	G		315	.025		8.85	.76		9.61	11
2650	1/2" thick	G		315	.025		12.15	.76		12.91	14.60

09 65 19.19 Vinyl Composition Tile Flooring

		Crew	Daily Output	Labor-Hours	Unit	Material	Labor	Equipment	Total	Total Incl O&P
0010	**VINYL COMPOSITION TILE FLOORING**									
7000	Vinyl composition tile, 12" x 12", 1/16" thick	1 Tilf	500	.016	S.F.	1.18	.48		1.66	2.08
7050	Embossed		500	.016		2.21	.48		2.69	3.21
7100	Marbleized		500	.016		2.21	.48		2.69	3.21
7150	Solid		500	.016		2.85	.48		3.33	3.92
7200	3/32" thick, embossed		500	.016		1.51	.48		1.99	2.44
7250	Marbleized		500	.016		2.54	.48		3.02	3.57
7300	Solid		500	.016		2.36	.48		2.84	3.38
7350	1/8" thick, marbleized		500	.016		2.40	.48		2.88	3.42
7400	Solid		500	.016		1.53	.48		2.01	2.46
7450	Conductive		500	.016		6.05	.48		6.53	7.45

09 65 19.23 Vinyl Tile Flooring

		Crew	Daily Output	Labor-Hours	Unit	Material	Labor	Equipment	Total	Total Incl O&P
0010	**VINYL TILE FLOORING**									
7500	Vinyl tile, 12" x 12", 3/32" thick,	1 Tilf	500	.016	S.F.	3.59	.48		4.07	4.73
7550	3/32" thick, premium colors/patterns		500	.016		7.25	.48		7.73	8.80
7600	1/8" thick, standard colors/patterns		500	.016		5.65	.48		6.13	7
7650	Solid colors		500	.016		3.25	.48		3.73	4.36
7700	Marbleized or Travertine pattern		500	.016		5.90	.48		6.38	7.30
7750	Florentine pattern		500	.016		6.30	.48		6.78	7.75
7800	Premium colors/patterns		500	.016		6.20	.48		6.68	7.65

For customer support on your Residential Cost Data, call 877.713.2617.

09 65 Resilient Flooring

09 65 19 – Resilient Tile Flooring

09 65 19.33 Rubber Tile Flooring

		Crew	Daily Output	Labor-Hours	Unit	Material	2015 Bare Costs Labor	Equipment	Total	Total Incl O&P
0010	**RUBBER TILE FLOORING**									
6050	Rubber tile, marbleized colors, 12" x 12", 1/8" thick	1 Tilf	400	.020	S.F.	5.70	.60		6.30	7.20
6100	3/16" thick		400	.020		9.25	.60		9.85	11.15
6300	Special tile, plain colors, 1/8" thick		400	.020		7.70	.60		8.30	9.45
6350	3/16" thick	↓	400	.020	↓	9.05	.60		9.65	10.90

09 65 33 – Conductive Resilient Flooring

09 65 33.10 Conductive Rubber and Vinyl Flooring

		Crew	Daily Output	Labor-Hours	Unit	Material	Labor	Equipment	Total	Total Incl O&P
0010	**CONDUCTIVE RUBBER AND VINYL FLOORING**									
1700	Conductive flooring, rubber tile, 1/8" thick	1 Tilf	315	.025	S.F.	6.95	.76		7.71	8.90
1800	Homogeneous vinyl tile, 1/8" thick	"	315	.025	"	6.40	.76		7.16	8.30

09 66 Terrazzo Flooring

09 66 13 – Portland Cement Terrazzo Flooring

09 66 13.10 Portland Cement Terrazzo

		Crew	Daily Output	Labor-Hours	Unit	Material	Labor	Equipment	Total	Total Incl O&P
0010	**PORTLAND CEMENT TERRAZZO**, cast-in-place									
4300	Stone chips, onyx gemstone, per 50 lb. bag				Bag	16.80			16.80	18.50

09 66 16 – Terrazzo Floor Tile

09 66 16.13 Portland Cement Terrazzo Floor Tile

		Crew	Daily Output	Labor-Hours	Unit	Material	Labor	Equipment	Total	Total Incl O&P
0010	**PORTLAND CEMENT TERRAZZO FLOOR TILE**									
1200	Floor tiles, non-slip, 1" thick, 12" x 12"	D-1	60	.267	S.F.	20	7.85		27.85	35
1300	1-1/4" thick, 12" x 12"		60	.267		20.50	7.85		28.35	36
1500	16" x 16"		50	.320		22.50	9.45		31.95	40.50
1600	1-1/2" thick, 16" x 16"	↓	45	.356	↓	20.50	10.50		31	40

09 66 16.30 Terrazzo, Precast

		Crew	Daily Output	Labor-Hours	Unit	Material	Labor	Equipment	Total	Total Incl O&P
0010	**TERRAZZO, PRECAST**									
0020	Base, 6" high, straight	1 Mstz	70	.114	L.F.	11.95	3.43		15.38	18.70
0100	Cove		60	.133		12.80	4.01		16.81	20.50
0300	8" high, straight		60	.133		11.45	4.01		15.46	19.10
0400	Cove	↓	50	.160		16.85	4.81		21.66	26.50
0600	For white cement, add					.45			.45	.50
0700	For 16 ga. zinc toe strip, add					1.72			1.72	1.89
0900	Curbs, 4" x 4" high	1 Mstz	40	.200		32.50	6		38.50	45.50
1000	8" x 8" high		30	.267	↓	38	8		46	55
4800	Wainscot, 12" x 12" x 1" tiles		12	.667	S.F.	6.85	20		26.85	40
4900	16" x 16" x 1-1/2" tiles	↓	8	1	"	14.25	30		44.25	64

09 68 Carpeting

09 68 05 – Carpet Accessories

09 68 05.11 Flooring Transition Strip

		Crew	Daily Output	Labor-Hours	Unit	Material	Labor	Equipment	Total	Total Incl O&P
0010	**FLOORING TRANSITION STRIP**									
0107	Clamp down brass divider, 12' strip, vinyl to carpet	1 Tilf	31.25	.256	Ea.	13.30	7.65		20.95	27
0117	Vinyl to hard surface	"	31.25	.256	"	13.30	7.65		20.95	27

09 68 10 – Carpet Pad

09 68 10.10 Commercial Grade Carpet Pad

		Crew	Daily Output	Labor-Hours	Unit	Material	Labor	Equipment	Total	Total Incl O&P
0010	**COMMERCIAL GRADE CARPET PAD**									
9001	Sponge rubber pad, 20 oz./sq. yd.	1 Tilf	1350	.006	S.F.	.50	.18		.68	.84
9101	40-62 oz./sq. yd.	↓	1350	.006		.97	.18		1.15	1.36

09 68 Carpeting

09 68 10 – Carpet Pad

09 68 10.10 Commercial Grade Carpet Pad		Crew	Daily Output	Labor-Hours	Unit	Material	2015 Bare Costs Labor	Equipment	Total	Total Incl O&P
9201	Felt pad, 20-32 oz./sq. yd.	1 Tilf	1350	.006	S.F.	.59	.18		.77	.94
9301	Maximum		1350	.006		1	.18		1.18	1.39
9401	Bonded urethane pad, 2.7 density		1350	.006		.66	.18		.84	1.02
9501	13.0 density		1350	.006		.89	.18		1.07	1.27
9601	Prime urethane pad, 2.7 density		1350	.006		.35	.18		.53	.68
9701	13.0 density		1350	.006		.55	.18		.73	.90

09 68 13 – Tile Carpeting

09 68 13.10 Carpet Tile		Crew	Daily Output	Labor-Hours	Unit	Material	Labor	Equipment	Total	Total Incl O&P
0010	**CARPET TILE**									
0100	Tufted nylon, 18" x 18", hard back, 20 oz.	1 Tilf	80	.100	S.Y.	24	3		27	31.50
0110	26 oz.		80	.100		32	3		35	40
0200	Cushion back, 20 oz.		80	.100		28	3		31	35.50
0210	26 oz.		80	.100		43	3		46	52

09 68 16 – Sheet Carpeting

09 68 16.10 Sheet Carpet		Crew	Daily Output	Labor-Hours	Unit	Material	Labor	Equipment	Total	Total Incl O&P
0010	**SHEET CARPET**									
0701	Nylon, level loop, 26 oz., light to medium traffic	1 Tilf	445	.018	S.F.	2.77	.54		3.31	3.91
0901	32 oz., medium traffic		445	.018		4.74	.54		5.28	6.05
1101	40 oz., medium to heavy traffic		445	.018		6.60	.54		7.14	8.15
2101	Nylon, plush, 20 oz., light traffic		445	.018		1.91	.54		2.45	2.97
2801	24 oz., light to medium traffic		445	.018		2.07	.54		2.61	3.15
2901	30 oz., medium traffic		445	.018		3.28	.54		3.82	4.48
3001	36 oz., medium traffic		445	.018		3.84	.54		4.38	5.10
3101	42 oz., medium to heavy traffic		370	.022		5	.65		5.65	6.55
3201	46 oz., medium to heavy traffic		370	.022		5.45	.65		6.10	7.05
3301	54 oz., heavy traffic		370	.022		6.10	.65		6.75	7.75
3501	Olefin, 15 oz., light traffic		445	.018		1.67	.54		2.21	2.71
3651	22 oz., light traffic		445	.018		1.70	.54		2.24	2.74
4501	50 oz., medium to heavy traffic, level loop		445	.018		11.15	.54		11.69	13.15
4701	32 oz., medium to heavy traffic, patterned		400	.020		10.10	.60		10.70	12.10
4901	48 oz., heavy traffic, patterned		400	.020		11.15	.60		11.75	13.25
5000	For less than full roll (approx. 1500 S.F.), add					25%				
5100	For small rooms, less than 12' wide, add						25%			
5200	For large open areas (no cuts), deduct						25%			
5600	For bound carpet baseboard, add	1 Tilf	300	.027	L.F.	3	.80		3.80	4.59
5610	For stairs, not incl. price of carpet, add	"	30	.267	Riser		8		8	12.95
8950	For tackless, stretched installation, add padding from 09 68 10.10 to above									
9850	For brand-named specific fiber, add				S.Y.	25%				

09 68 20 – Athletic Carpet

09 68 20.10 Indoor Athletic Carpet		Crew	Daily Output	Labor-Hours	Unit	Material	Labor	Equipment	Total	Total Incl O&P
0010	**INDOOR ATHLETIC CARPET**									
3700	Polyethylene, in rolls, no base incl., landscape surfaces	1 Tilf	275	.029	S.F.	4.09	.87		4.96	5.90
3800	Nylon action surface, 1/8" thick		275	.029		3.76	.87		4.63	5.55
3900	1/4" thick		275	.029		5.45	.87		6.32	7.35
4000	3/8" thick		275	.029		6.80	.87		7.67	8.90

09 72 Wall Coverings

09 72 19 – Textile Wall Covering

09 72 19.10 Textile Wall Covering

		Crew	Daily Output	Labor-Hours	Unit	Material	2015 Bare Costs Labor	Equipment	Total	Total Incl O&P
0010	TEXTILE WALL COVERING, including sizing; add 10-30% waste @ takeoff									
0020	Silk	1 Pape	640	.013	S.F.	4.15	.36		4.51	5.15
0030	Cotton		640	.013		6.65	.36		7.01	7.90
0040	Linen		640	.013		1.85	.36		2.21	2.64
0050	Blend		640	.013		2.97	.36		3.33	3.87

09 72 20 – Natural Fiber Wall Covering

09 72 20.10 Natural Fiber Wall Covering

		Crew	Daily Output	Labor-Hours	Unit	Material	Labor	Equipment	Total	Total Incl O&P
0010	NATURAL FIBER WALL COVERING, including sizing; add 10-30% waste @ takeoff									
0015	Bamboo	1 Pape	640	.013	S.F.	2.26	.36		2.62	3.09
0030	Burlap		640	.013		2.03	.36		2.39	2.83
0045	Jute		640	.013		1.29	.36		1.65	2.02
0060	Sisal		640	.013		1.46	.36		1.82	2.21

09 72 23 – Wallpapering

09 72 23.10 Wallpaper

			Crew	Daily Output	Labor-Hours	Unit	Material	Labor	Equipment	Total	Total Incl O&P
0010	WALLPAPER including sizing; add 10-30 percent waste @ takeoff	R097223-10									
0050	Aluminum foil		1 Pape	275	.029	S.F.	1.04	.85		1.89	2.53
0100	Copper sheets, .025" thick, vinyl backing			240	.033		5.55	.97		6.52	7.70
0300	Phenolic backing			240	.033		7.20	.97		8.17	9.55
0600	Cork tiles, light or dark, 12" x 12" x 3/16"			240	.033		4.51	.97		5.48	6.55
0700	5/16" thick			235	.034		3.13	.99		4.12	5.05
0900	1/4" basketweave			240	.033		3.50	.97		4.47	5.45
1000	1/2" natural, non-directional pattern			240	.033		6.90	.97		7.87	9.20
1100	3/4" natural, non-directional pattern			240	.033		11.35	.97		12.32	14.10
1200	Granular surface, 12" x 36", 1/2" thick			385	.021		1.31	.60		1.91	2.43
1300	1" thick			370	.022		1.68	.63		2.31	2.88
1500	Polyurethane coated, 12" x 12" x 3/16" thick			240	.033		4.08	.97		5.05	6.10
1600	5/16" thick			235	.034		6.60	.99		7.59	8.90
1800	Cork wallpaper, paperbacked, natural			480	.017		2.06	.48		2.54	3.07
1900	Colors			480	.017		2.87	.48		3.35	3.96
2100	Flexible wood veneer, 1/32" thick, plain woods			100	.080		2.44	2.32		4.76	6.50
2200	Exotic woods			95	.084		3.69	2.45		6.14	8.10
2400	Gypsum-based, fabric-backed, fire resistant										
2500	for masonry walls, 21 oz./S.Y.		1 Pape	800	.010	S.F.	.85	.29		1.14	1.42
2600	Average			720	.011		1.27	.32		1.59	1.93
2700	Small quantities			640	.013		.76	.36		1.12	1.44
2750	Acrylic, modified, semi-rigid PVC, .028" thick		2 Carp	330	.048		1.37	1.59		2.96	4.18
2800	.040" thick		"	320	.050		1.81	1.64		3.45	4.75
3000	Vinyl wall covering, fabric-backed, lightweight, type 1 (12-15 oz./S.Y.)		1 Pape	640	.013		.97	.36		1.33	1.67
3300	Medium weight, type 2 (20-24 oz./S.Y.)			480	.017		.90	.48		1.38	1.79
3400	Heavy weight, type 3 (28 oz./S.Y.)			435	.018		1.37	.53		1.90	2.39
3600	Adhesive, 5 gal. lots (18 S.Y./gal.)					Gal.	14.20			14.20	15.60
3700	Wallpaper, average workmanship, solid pattern, low cost paper		1 Pape	640	.013	S.F.	.61	.36		.97	1.27
3900	basic patterns (matching required), avg. cost paper			535	.015		1.11	.43		1.54	1.94
4000	Paper at $85 per double roll, quality workmanship			435	.018		2.20	.53		2.73	3.30
4100	Linen wall covering, paper backed										
4150	Flame treatment					S.F.	1.01			1.01	1.11
4180	Stain resistance treatment						1.84			1.84	2.02
4200	Grass cloths with lining paper	G	1 Pape	400	.020		.91	.58		1.49	1.96
4300	Premium texture/color	G	"	350	.023		2.91	.66		3.57	4.29

09 91 Painting

09 91 03 – Paint Restoration

09 91 03.20 Sanding		Crew	Daily Output	Labor-Hours	Unit	Material	2015 Bare Costs Labor	2015 Bare Costs Equipment	Total	Total Incl O&P
0010	SANDING and puttying interior trim, compared to									
0100	Painting 1 coat, on quality work				L.F.		100%			
0300	Medium work						50%			
0400	Industrial grade						25%			
0500	Surface protection, placement and removal									
0510	Basic drop cloths	1 Pord	6400	.001	S.F.		.04		.04	.06
0520	Masking with paper		800	.010		.07	.29		.36	.55
0530	Volume cover up (using plastic sheathing, or building paper)		16000	.001			.01		.01	.02

09 91 03.30 Exterior Surface Preparation		Crew	Daily Output	Labor-Hours	Unit	Material	Labor	Equipment	Total	Total Incl O&P
0010	EXTERIOR SURFACE PREPARATION									
0015	Doors, per side, not incl. frames or trim									
0020	Scrape & sand									
0030	Wood, flush	1 Pord	616	.013	S.F.		.37		.37	.61
0040	Wood, detail		496	.016			.46		.46	.76
0050	Wood, louvered		280	.029			.82		.82	1.35
0060	Wood, overhead		616	.013			.37		.37	.61
0070	Wire brush									
0080	Metal, flush	1 Pord	640	.013	S.F.		.36		.36	.59
0090	Metal, detail		520	.015			.44		.44	.73
0100	Metal, louvered		360	.022			.64		.64	1.05
0110	Metal or fibr., overhead		640	.013			.36		.36	.59
0120	Metal, roll up		560	.014			.41		.41	.67
0130	Metal, bulkhead		640	.013			.36		.36	.59
0140	Power wash, based on 2500 lb. operating pressure									
0150	Metal, flush	A-1H	2240	.004	S.F.		.09	.03	.12	.19
0160	Metal, detail		2120	.004			.09	.04	.13	.19
0170	Metal, louvered		2000	.004			.10	.04	.14	.20
0180	Metal or fibr., overhead		2400	.003			.08	.03	.11	.17
0190	Metal, roll up		2400	.003			.08	.03	.11	.17
0200	Metal, bulkhead		2200	.004			.09	.03	.12	.19
0400	Windows, per side, not incl. trim									
0410	Scrape & sand									
0420	Wood, 1-2 lite	1 Pord	320	.025	S.F.		.72		.72	1.18
0430	Wood, 3-6 lite		280	.029			.82		.82	1.35
0440	Wood, 7-10 lite		240	.033			.95		.95	1.57
0450	Wood, 12 lite		200	.040			1.15		1.15	1.89
0460	Wood, Bay/Bow		320	.025			.72		.72	1.18
0470	Wire brush									
0480	Metal, 1-2 lite	1 Pord	480	.017	S.F.		.48		.48	.79
0490	Metal, 3-6 lite		400	.020			.57		.57	.94
0500	Metal, Bay/Bow		480	.017			.48		.48	.79
0510	Power wash, based on 2500 lb. operating pressure									
0520	1-2 lite	A-1H	4400	.002	S.F.		.04	.02	.06	.09
0530	3-6 lite		4320	.002			.05	.02	.07	.10
0540	7-10 lite		4240	.002			.05	.02	.07	.10
0550	12 lite		4160	.002			.05	.02	.07	.10
0560	Bay/Bow		4400	.002			.04	.02	.06	.09
0600	Siding, scrape and sand, light=10-30%, med.=30-70%									
0610	Heavy=70-100% of surface to sand									
0650	Texture 1-11, light	1 Pord	480	.017	S.F.		.48		.48	.79
0660	Med.		440	.018			.52		.52	.86
0670	Heavy		360	.022			.64		.64	1.05

09 91 Painting

09 91 03 – Paint Restoration

09 91 03.30 Exterior Surface Preparation

		Crew	Daily Output	Labor-Hours	Unit	Material	2015 Bare Costs Labor	2015 Bare Costs Equipment	Total	Total Incl O&P
0680	Wood shingles, shakes, light	1 Pord	440	.018	S.F.		.52		.52	.86
0690	Med.		360	.022			.64		.64	1.05
0700	Heavy		280	.029			.82		.82	1.35
0710	Clapboard, light		520	.015			.44		.44	.73
0720	Med.		480	.017			.48		.48	.79
0730	Heavy	▼	400	.020	▼		.57		.57	.94
0740	Wire brush									
0750	Aluminum, light	1 Pord	600	.013	S.F.		.38		.38	.63
0760	Med.		520	.015			.44		.44	.73
0770	Heavy	▼	440	.018	▼		.52		.52	.86
0780	Pressure wash, based on 2500 lb. operating pressure									
0790	Stucco	A-1H	3080	.003	S.F.		.06	.02	.08	.14
0800	Aluminum or vinyl		3200	.003			.06	.02	.08	.13
0810	Siding, masonry, brick & block	▼	2400	.003	▼		.08	.03	.11	.17
1300	Miscellaneous, wire brush									
1310	Metal, pedestrian gate	1 Pord	100	.080	S.F.		2.29		2.29	3.78

09 91 03.40 Interior Surface Preparation

		Crew	Daily Output	Labor-Hours	Unit	Material	2015 Bare Costs Labor	2015 Bare Costs Equipment	Total	Total Incl O&P
0010	**INTERIOR SURFACE PREPARATION**									
0020	Doors, per side, not incl. frames or trim									
0030	Scrape & sand									
0040	Wood, flush	1 Pord	616	.013	S.F.		.37		.37	.61
0050	Wood, detail		496	.016			.46		.46	.76
0060	Wood, louvered	▼	280	.029	▼		.82		.82	1.35
0070	Wire brush									
0080	Metal, flush	1 Pord	640	.013	S.F.		.36		.36	.59
0090	Metal, detail		520	.015			.44		.44	.73
0100	Metal, louvered	▼	360	.022			.64		.64	1.05
0110	Hand wash									
0120	Wood, flush	1 Pord	2160	.004	S.F.		.11		.11	.17
0130	Wood, detailed		2000	.004			.11		.11	.19
0140	Wood, louvered		1360	.006			.17		.17	.28
0150	Metal, flush		2160	.004			.11		.11	.17
0160	Metal, detail		2000	.004			.11		.11	.19
0170	Metal, louvered	▼	1360	.006			.17		.17	.28
0400	Windows, per side, not incl. trim									
0410	Scrape & sand									
0420	Wood, 1-2 lite	1 Pord	360	.022	S.F.		.64		.64	1.05
0430	Wood, 3-6 lite		320	.025			.72		.72	1.18
0440	Wood, 7-10 lite		280	.029			.82		.82	1.35
0450	Wood, 12 lite		240	.033			.95		.95	1.57
0460	Wood, Bay/Bow	▼	360	.022	▼		.64		.64	1.05
0470	Wire brush									
0480	Metal, 1-2 lite	1 Pord	520	.015	S.F.		.44		.44	.73
0490	Metal, 3-6 lite		440	.018			.52		.52	.86
0500	Metal, Bay/Bow	▼	520	.015			.44		.44	.73
0600	Walls, sanding, light=10-30%, medium - 30-70%,									
0610	heavy=70-100% of surface to sand									
0650	Walls, sand									
0660	Gypsum board or plaster, light	1 Pord	3077	.003	S.F.		.07		.07	.12
0670	Gypsum board or plaster, medium		2160	.004			.11		.11	.17
0680	Gypsum board or plaster, heavy		923	.009			.25		.25	.41
0690	Wood, T&G, light		2400	.003			.10		.10	.16

09 91 Painting

09 91 03 – Paint Restoration

09 91 03.40 Interior Surface Preparation

		Crew	Daily Output	Labor-Hours	Unit	Material	2015 Bare Costs Labor	Equipment	Total	Total Incl O&P
0700	Wood, T&G, med.	1 Pord	1600	.005	S.F.		.14		.14	.24
0710	Wood, T&G, heavy	↓	800	.010	↓		.29		.29	.47
0720	Walls, wash									
0730	Gypsum board or plaster	1 Pord	3200	.003	S.F.		.07		.07	.12
0740	Wood, T&G		3200	.003			.07		.07	.12
0750	Masonry, brick & block, smooth		2800	.003			.08		.08	.14
0760	Masonry, brick & block, coarse	↓	2000	.004			.11		.11	.19
8000	For chemical washing, see Section 04 01 30									

09 91 03.41 Scrape After Fire Damage

		Crew	Daily Output	Labor-Hours	Unit	Material	Labor	Equipment	Total	Total Incl O&P
0010	**SCRAPE AFTER FIRE DAMAGE**									
0050	Boards, 1" x 4"	1 Pord	336	.024	L.F.		.68		.68	1.12
0060	1" x 6"		260	.031			.88		.88	1.45
0070	1" x 8"		207	.039			1.11		1.11	1.82
0080	1" x 10"		174	.046			1.32		1.32	2.17
0500	Framing, 2" x 4"		265	.030			.86		.86	1.43
0510	2" x 6"		221	.036			1.04		1.04	1.71
0520	2" x 8"		190	.042			1.21		1.21	1.99
0530	2" x 10"		165	.048			1.39		1.39	2.29
0540	2" x 12"		144	.056			1.59		1.59	2.62
1000	Heavy framing, 3" x 4"		226	.035			1.01		1.01	1.67
1010	4" x 4"		210	.038			1.09		1.09	1.80
1020	4" x 6"		191	.042			1.20		1.20	1.98
1030	4" x 8"		165	.048			1.39		1.39	2.29
1040	4" x 10"		144	.056			1.59		1.59	2.62
1060	4" x 12"		131	.061	↓		1.75		1.75	2.88
2900	For sealing, light damage		825	.010	S.F.	.14	.28		.42	.61
2920	Heavy damage	↓	460	.017	"	.31	.50		.81	1.16

09 91 13 – Exterior Painting

09 91 13.30 Fences

		Crew	Daily Output	Labor-Hours	Unit	Material	Labor	Equipment	Total	Total Incl O&P
0010	**FENCES** R099100-20									
0100	Chain link or wire metal, one side, water base									
0110	Roll & brush, first coat	1 Pord	960	.008	S.F.	.08	.24		.32	.48
0120	Second coat		1280	.006		.07	.18		.25	.38
0130	Spray, first coat		2275	.004		.08	.10		.18	.26
0140	Second coat	↓	2600	.003	↓	.08	.09		.17	.24
0150	Picket, water base									
0160	Roll & brush, first coat	1 Pord	865	.009	S.F.	.08	.27		.35	.53
0170	Second coat		1050	.008		.08	.22		.30	.45
0180	Spray, first coat		2275	.004		.08	.10		.18	.26
0190	Second coat	↓	2600	.003	↓	.08	.09		.17	.24
0200	Stockade, water base									
0210	Roll & brush, first coat	1 Pord	1040	.008	S.F.	.08	.22		.30	.45
0220	Second coat		1200	.007		.08	.19		.27	.40
0230	Spray, first coat		2275	.004		.08	.10		.18	.26
0240	Second coat	↓	2600	.003	↓	.08	.09		.17	.24

09 91 13.42 Miscellaneous, Exterior

		Crew	Daily Output	Labor-Hours	Unit	Material	Labor	Equipment	Total	Total Incl O&P
0010	**MISCELLANEOUS, EXTERIOR** R099100-20									
0100	Railing, ext., decorative wood, incl. cap & baluster									
0110	Newels & spindles @ 12" O.C.									
0120	Brushwork, stain, sand, seal & varnish									
0130	First coat	1 Pord	90	.089	L.F.	.81	2.55		3.36	5.10
0140	Second coat	"	120	.067	"	.81	1.91		2.72	4.04

09 91 Painting

09 91 13 – Exterior Painting

09 91 13.42 Miscellaneous, Exterior

		Crew	Daily Output	Labor-Hours	Unit	Material	2015 Bare Costs Labor	Equipment	Total	Total Incl O&P
0150	Rough sawn wood, 42" high, 2" x 2" verticals, 6" O.C.									
0160	Brushwork, stain, each coat	1 Pord	90	.089	L.F.	.26	2.55		2.81	4.48
0170	Wrought iron, 1" rail, 1/2" sq. verticals									
0180	Brushwork, zinc chromate, 60" high, bars 6" O.C.									
0190	Primer	1 Pord	130	.062	L.F.	.86	1.76		2.62	3.85
0200	Finish coat		130	.062		1.13	1.76		2.89	4.14
0210	Additional coat		190	.042		1.32	1.21		2.53	3.44
0220	Shutters or blinds, single panel, 2' x 4', paint all sides									
0230	Brushwork, primer	1 Pord	20	.400	Ea.	.66	11.45		12.11	19.60
0240	Finish coat, exterior latex		20	.400		.62	11.45		12.07	19.60
0250	Primer & 1 coat, exterior latex		13	.615		1.14	17.65		18.79	30.50
0260	Spray, primer		35	.229		.96	6.55		7.51	11.85
0270	Finish coat, exterior latex		35	.229		1.33	6.55		7.88	12.25
0280	Primer & 1 coat, exterior latex		20	.400		1.04	11.45		12.49	20
0290	For louvered shutters, add					10%				
0300	Stair stringers, exterior, metal									
0310	Roll & brush, zinc chromate, to 14", each coat	1 Pord	320	.025	L.F.	.38	.72		1.10	1.59
0320	Rough sawn wood, 4" x 12"									
0330	Roll & brush, exterior latex, each coat	1 Pord	215	.037	L.F.	.09	1.07		1.16	1.86
0340	Trellis/lattice, 2" x 2" @ 3" O.C. with 2" x 8" supports									
0350	Spray, latex, per side, each coat	1 Pord	475	.017	S.F.	.09	.48		.57	.89
0450	Decking, ext., sealer, alkyd, brushwork, sealer coat		1140	.007		.10	.20		.30	.44
0460	1st coat		1140	.007		.10	.20		.30	.44
0470	2nd coat		1300	.006		.07	.18		.25	.37
0500	Paint, alkyd, brushwork, primer coat		1140	.007		.11	.20		.31	.46
0510	1st coat		1140	.007		.13	.20		.33	.47
0520	2nd coat		1300	.006		.09	.18		.27	.39
0600	Sand paint, alkyd, brushwork, 1 coat		150	.053		.14	1.53		1.67	2.67

09 91 13.60 Siding Exterior

		Crew	Daily Output	Labor-Hours	Unit	Material	2015 Bare Costs Labor	Equipment	Total	Total Incl O&P
0010	**SIDING EXTERIOR**, Alkyd (oil base)									
0450	Steel siding, oil base, paint 1 coat, brushwork	2 Pord	2015	.008	S.F.	.10	.23		.33	.48
0500	Spray		4550	.004		.15	.10		.25	.34
0800	Paint 2 coats, brushwork		1300	.012		.20	.35		.55	.80
1000	Spray		2750	.006		.17	.17		.34	.46
1200	Stucco, rough, oil base, paint 2 coats, brushwork		1300	.012		.20	.35		.55	.80
1400	Roller		1625	.010		.21	.28		.49	.69
1600	Spray		2925	.005		.22	.16		.38	.51
1800	Texture 1-11 or clapboard, oil base, primer coat, brushwork		1300	.012		.15	.35		.50	.74
2000	Spray		4550	.004		.15	.10		.25	.33
2100	Paint 1 coat, brushwork		1300	.012		.15	.35		.50	.74
2200	Spray		4550	.004		.15	.10		.25	.33
2400	Paint 2 coats, brushwork		810	.020		.30	.57		.87	1.25
2600	Spray		2600	.006		.33	.18		.51	.65
3000	Stain 1 coat, brushwork		1520	.011		.09	.30		.39	.59
3200	Spray		5320	.003		.10	.09		.19	.25
3400	Stain 2 coats, brushwork		950	.017		.17	.48		.65	.98
4000	Spray		3050	.005		.19	.15		.34	.46
4200	Wood shingles, oil base primer coat, brushwork		1300	.012		.14	.35		.49	.73
4400	Spray		3900	.004		.13	.12		.25	.33
4600	Paint 1 coat, brushwork		1300	.012		.12	.35		.47	.72
4800	Spray		3900	.004		.15	.12		.27	.36
5000	Paint 2 coats, brushwork		810	.020		.25	.57		.82	1.20

09 91 Painting

09 91 13 – Exterior Painting

09 91 13.60 Siding Exterior

		Crew	Daily Output	Labor-Hours	Unit	Material	2015 Bare Costs Labor	2015 Bare Costs Equipment	Total	Total Incl O&P
5200	Spray	2 Pord	2275	.007	S.F.	.23	.20		.43	.59
5800	Stain 1 coat, brushwork		1500	.011		.09	.31		.40	.59
6000	Spray		3900	.004		.09	.12		.21	.28
6500	Stain 2 coats, brushwork		950	.017		.17	.48		.65	.98
7000	Spray		2660	.006		.24	.17		.41	.54
8000	For latex paint, deduct					10%				
8100	For work over 12' H, from pipe scaffolding, add						15%			
8200	For work over 12' H, from extension ladder, add						25%			
8300	For work over 12' H, from swing staging, add						35%			

09 91 13.62 Siding, Misc.

		Crew	Daily Output	Labor-Hours	Unit	Material	Labor	Equipment	Total	Total Incl O&P
0010	**SIDING, MISC.**, latex paint R099100-10									
0100	Aluminum siding									
0110	Brushwork, primer	2 Pord	2275	.007	S.F.	.06	.20		.26	.40
0120	Finish coat, exterior latex		2275	.007		.06	.20		.26	.39
0130	Primer & 1 coat exterior latex		1300	.012		.13	.35		.48	.72
0140	Primer & 2 coats exterior latex		975	.016		.19	.47		.66	.97
0150	Mineral fiber shingles									
0160	Brushwork, primer	2 Pord	1495	.011	S.F.	.15	.31		.46	.67
0170	Finish coat, industrial enamel		1495	.011		.18	.31		.49	.71
0180	Primer & 1 coat enamel		810	.020		.33	.57		.90	1.29
0190	Primer & 2 coats enamel		540	.030		.51	.85		1.36	1.97
0200	Roll, primer		1625	.010		.17	.28		.45	.64
0210	Finish coat, industrial enamel		1625	.010		.20	.28		.48	.68
0220	Primer & 1 coat enamel		975	.016		.36	.47		.83	1.17
0230	Primer & 2 coats enamel		650	.025		.56	.71		1.27	1.78
0240	Spray, primer		3900	.004		.13	.12		.25	.33
0250	Finish coat, industrial enamel		3900	.004		.16	.12		.28	.37
0260	Primer & 1 coat enamel		2275	.007		.29	.20		.49	.65
0270	Primer & 2 coats enamel		1625	.010		.46	.28		.74	.96
0280	Waterproof sealer, first coat		4485	.004		.09	.10		.19	.26
0290	Second coat		5235	.003		.08	.09		.17	.23
0300	Rough wood incl. shingles, shakes or rough sawn siding									
0310	Brushwork, primer	2 Pord	1280	.013	S.F.	.13	.36		.49	.74
0320	Finish coat, exterior latex		1280	.013		.10	.36		.46	.70
0330	Primer & 1 coat exterior latex		960	.017		.24	.48		.72	1.05
0340	Primer & 2 coats exterior latex		700	.023		.34	.65		.99	1.45
0350	Roll, primer		2925	.005		.18	.16		.34	.46
0360	Finish coat, exterior latex		2925	.005		.12	.16		.28	.39
0370	Primer & 1 coat exterior latex		1790	.009		.30	.26		.56	.75
0380	Primer & 2 coats exterior latex		1300	.012		.42	.35		.77	1.04
0390	Spray, primer		3900	.004		.15	.12		.27	.35
0400	Finish coat, exterior latex		3900	.004		.09	.12		.21	.29
0410	Primer & 1 coat exterior latex		2600	.006		.24	.18		.42	.56
0420	Primer & 2 coats exterior latex		2080	.008		.34	.22		.56	.73
0430	Waterproof sealer, first coat		4485	.004		.16	.10		.26	.34
0440	Second coat		4485	.004		.09	.10		.19	.26
0450	Smooth wood incl. butt, T&G, beveled, drop or B&B siding									
0460	Brushwork, primer	2 Pord	2325	.007	S.F.	.10	.20		.30	.43
0470	Finish coat, exterior latex		1280	.013		.10	.36		.46	.70
0480	Primer & 1 coat exterior latex		800	.020		.20	.57		.77	1.16
0490	Primer & 2 coats exterior latex		630	.025		.30	.73		1.03	1.53
0500	Roll, primer		2275	.007		.11	.20		.31	.45

09 91 Painting

09 91 13 – Exterior Painting

09 91 13.62 Siding, Misc.

		Crew	Daily Output	Labor-Hours	Unit	Material	2015 Bare Costs Labor	Equipment	Total	Total Incl O&P
0510	Finish coat, exterior latex	2 Pord	2275	.007	S.F.	.11	.20		.31	.45
0520	Primer & 1 coat exterior latex		1300	.012		.22	.35		.57	.82
0530	Primer & 2 coats exterior latex		975	.016		.33	.47		.80	1.13
0540	Spray, primer		4550	.004		.08	.10		.18	.26
0550	Finish coat, exterior latex		4550	.004		.09	.10		.19	.27
0560	Primer & 1 coat exterior latex		2600	.006		.18	.18		.36	.49
0570	Primer & 2 coats exterior latex		1950	.008		.27	.24		.51	.69
0580	Waterproof sealer, first coat		5230	.003		.09	.09		.18	.23
0590	Second coat		5980	.003		.09	.08		.17	.22
0600	For oil base paint, add					10%				

09 91 13.70 Doors and Windows, Exterior

		Crew	Daily Output	Labor-Hours	Unit	Material	Labor	Equipment	Total	Total Incl O&P
0010	**DOORS AND WINDOWS, EXTERIOR** R099100-10									
0100	Door frames & trim, only									
0110	Brushwork, primer R099100-20	1 Pord	512	.016	L.F.	.06	.45		.51	.81
0120	Finish coat, exterior latex		512	.016		.08	.45		.53	.83
0130	Primer & 1 coat, exterior latex		300	.027		.14	.76		.90	1.41
0140	Primer & 2 coats, exterior latex		265	.030		.22	.86		1.08	1.67
0150	Doors, flush, both sides, incl. frame & trim									
0160	Roll & brush, primer	1 Pord	10	.800	Ea.	4.55	23		27.55	43
0170	Finish coat, exterior latex		10	.800		5.95	23		28.95	44.50
0180	Primer & 1 coat, exterior latex		7	1.143		10.50	32.50		43	65.50
0190	Primer & 2 coats, exterior latex		5	1.600		16.40	46		62.40	93.50
0200	Brushwork, stain, sealer & 2 coats polyurethane		4	2		28.50	57.50		86	126
0210	Doors, French, both sides, 10-15 lite, incl. frame & trim									
0220	Brushwork, primer	1 Pord	6	1.333	Ea.	2.27	38		40.27	65.50
0230	Finish coat, exterior latex		6	1.333		2.97	38		40.97	66.50
0240	Primer & 1 coat, exterior latex		3	2.667		5.25	76.50		81.75	132
0250	Primer & 2 coats, exterior latex		2	4		8.05	115		123.05	198
0260	Brushwork, stain, sealer & 2 coats polyurethane		2.50	3.200		10.30	91.50		101.80	162
0270	Doors, louvered, both sides, incl. frame & trim									
0280	Brushwork, primer	1 Pord	7	1.143	Ea.	4.55	32.50		37.05	59
0290	Finish coat, exterior latex		7	1.143		5.95	32.50		38.45	60.50
0300	Primer & 1 coat, exterior latex		4	2		10.50	57.50		68	106
0310	Primer & 2 coats, exterior latex		3	2.667		16.10	76.50		92.60	144
0320	Brushwork, stain, sealer & 2 coats polyurethane		4.50	1.778		28.50	51		79.50	116
0330	Doors, panel, both sides, incl. frame & trim									
0340	Roll & brush, primer	1 Pord	6	1.333	Ea.	4.55	38		42.55	68
0350	Finish coat, exterior latex		6	1.333		5.95	38		43.95	69.50
0360	Primer & 1 coat, exterior latex		3	2.667		10.50	76.50		87	138
0370	Primer & 2 coats, exterior latex		2.50	3.200		16.10	91.50		107.60	169
0380	Brushwork, stain, sealer & 2 coats polyurethane		3	2.667		28.50	76.50		105	158
0400	Windows, per ext. side, based on 15 S.F.									
0410	1 to 6 lite									
0420	Brushwork, primer	1 Pord	13	.615	Ea.	.90	17.65		18.55	30
0430	Finish coat, exterior latex		13	.615		1.17	17.65		18.82	30.50
0440	Primer & 1 coat, exterior latex		8	1		2.07	28.50		30.57	49.50
0450	Primer & 2 coats, exterior latex		6	1.333		3.17	38		41.17	66.50
0460	Stain, sealer & 1 coat varnish		7	1.143		4.06	32.50		36.56	58.50
0470	7 to 10 lite									
0480	Brushwork, primer	1 Pord	11	.727	Ea.	.90	21		21.90	35.50
0490	Finish coat, exterior latex		11	.727		1.17	21		22.17	36
0500	Primer & 1 coat, exterior latex		7	1.143		2.07	32.50		34.57	56.50

09 91 Painting

09 91 13 – Exterior Painting

09 91 13.70 Doors and Windows, Exterior		Crew	Daily Output	Labor-Hours	Unit	Material	2015 Bare Costs Labor	Equipment	Total	Total Incl O&P
0510	Primer & 2 coats, exterior latex	1 Pord	5	1.600	Ea.	3.17	46		49.17	79
0520	Stain, sealer & 1 coat varnish	↓	6	1.333	↓	4.06	38		42.06	67.50
0530	12 lite									
0540	Brushwork, primer	1 Pord	10	.800	Ea.	.90	23		23.90	39
0550	Finish coat, exterior latex		10	.800		1.17	23		24.17	39.50
0560	Primer & 1 coat, exterior latex		6	1.333		2.07	38		40.07	65.50
0570	Primer & 2 coats, exterior latex		5	1.600		3.17	46		49.17	79
0580	Stain, sealer & 1 coat varnish	↓	6	1.333	↓	4.15	38		42.15	67.50
0590	For oil base paint, add					10%				

09 91 13.80 Trim, Exterior		Crew	Daily Output	Labor-Hours	Unit	Material	Labor	Equipment	Total	Total Incl O&P
0010	**TRIM, EXTERIOR** R099100-10									
0100	Door frames & trim (see Doors, interior or exterior)									
0110	Fascia, latex paint, one coat coverage									
0120	1" x 4", brushwork	1 Pord	640	.013	L.F.	.02	.36		.38	.62
0130	Roll		1280	.006		.03	.18		.21	.33
0140	Spray		2080	.004		.02	.11		.13	.20
0150	1" x 6" to 1" x 10", brushwork		640	.013		.08	.36		.44	.68
0160	Roll		1230	.007		.09	.19		.28	.41
0170	Spray		2100	.004		.07	.11		.18	.25
0180	1" x 12", brushwork		640	.013		.08	.36		.44	.68
0190	Roll		1050	.008		.09	.22		.31	.46
0200	Spray	↓	2200	.004	↓	.07	.10		.17	.24
0210	Gutters & downspouts, metal, zinc chromate paint									
0220	Brushwork, gutters, 5", first coat	1 Pord	640	.013	L.F.	.40	.36		.76	1.03
0230	Second coat		960	.008		.38	.24		.62	.80
0240	Third coat		1280	.006		.30	.18		.48	.64
0250	Downspouts, 4", first coat		640	.013		.40	.36		.76	1.03
0260	Second coat		960	.008		.38	.24		.62	.80
0270	Third coat		1280	.006		.30	.18		.48	.64
0280	Gutters & downspouts, wood									
0290	Brushwork, gutters, 5", primer	1 Pord	640	.013	L.F.	.06	.36		.42	.66
0300	Finish coat, exterior latex		640	.013		.07	.36		.43	.67
0310	Primer & 1 coat exterior latex		400	.020		.14	.57		.71	1.09
0320	Primer & 2 coats exterior latex		325	.025		.22	.71		.93	1.40
0330	Downspouts, 4", primer		640	.013		.06	.36		.42	.66
0340	Finish coat, exterior latex		640	.013		.07	.36		.43	.67
0350	Primer & 1 coat exterior latex		400	.020		.14	.57		.71	1.09
0360	Primer & 2 coats exterior latex		325	.025		.11	.71		.82	1.28
0370	Molding, exterior, up to 14" wide									
0380	Brushwork, primer	1 Pord	640	.013	L.F.	.07	.36		.43	.67
0390	Finish coat, exterior latex		640	.013		.08	.36		.44	.68
0400	Primer & 1 coat exterior latex		400	.020		.17	.57		.74	1.12
0410	Primer & 2 coats exterior latex		315	.025		.17	.73		.90	1.38
0420	Stain & fill		1050	.008		.10	.22		.32	.47
0430	Shellac		1850	.004		.13	.12		.25	.34
0440	Varnish	↓	1275	.006	↓	.11	.18		.29	.42

09 91 13.90 Walls, Masonry (CMU), Exterior		Crew	Daily Output	Labor-Hours	Unit	Material	Labor	Equipment	Total	Total Incl O&P
0010	**WALLS, MASONRY (CMU), EXTERIOR**									
0360	Concrete masonry units (CMU), smooth surface									
0370	Brushwork, latex, first coat	1 Pord	640	.013	S.F.	.07	.36		.43	.67
0380	Second coat		960	.008		.06	.24		.30	.45
0390	Waterproof sealer, first coat	↓	736	.011	↓	.25	.31		.56	.78

09 91 Painting

09 91 13 – Exterior Painting

09 91 13.90 Walls, Masonry (CMU), Exterior

		Crew	Daily Output	Labor-Hours	Unit	Material	2015 Bare Costs Labor	2015 Bare Costs Equipment	Total	Total Incl O&P
0400	Second coat	1 Pord	1104	.007	S.F.	.25	.21		.46	.61
0410	Roll, latex, paint, first coat		1465	.005		.09	.16		.25	.35
0420	Second coat		1790	.004		.07	.13		.20	.28
0430	Waterproof sealer, first coat		1680	.005		.25	.14		.39	.49
0440	Second coat		2060	.004		.25	.11		.36	.45
0450	Spray, latex, paint, first coat		1950	.004		.07	.12		.19	.26
0460	Second coat		2600	.003		.05	.09		.14	.21
0470	Waterproof sealer, first coat		2245	.004		.25	.10		.35	.44
0480	Second coat	↓	2990	.003	↓	.25	.08		.33	.40
0490	Concrete masonry unit (CMU), porous									
0500	Brushwork, latex, first coat	1 Pord	640	.013	S.F.	.14	.36		.50	.75
0510	Second coat		960	.008		.07	.24		.31	.47
0520	Waterproof sealer, first coat		736	.011		.25	.31		.56	.78
0530	Second coat		1104	.007		.25	.21		.46	.61
0540	Roll latex, first coat		1465	.005		.11	.16		.27	.38
0550	Second coat		1790	.004		.07	.13		.20	.29
0560	Waterproof sealer, first coat		1680	.005		.25	.14		.39	.49
0570	Second coat		2060	.004		.25	.11		.36	.45
0580	Spray latex, first coat		1950	.004		.08	.12		.20	.28
0590	Second coat		2600	.003		.05	.09		.14	.21
0600	Waterproof sealer, first coat		2245	.004		.25	.10		.35	.44
0610	Second coat	↓	2990	.003	↓	.25	.08		.33	.40

09 91 23 – Interior Painting

09 91 23.20 Cabinets and Casework

		Crew	Daily Output	Labor-Hours	Unit	Material	Labor	Equipment	Total	Total Incl O&P
0010	**CABINETS AND CASEWORK**									
1000	Primer coat, oil base, brushwork	1 Pord	650	.012	S.F.	.07	.35		.42	.66
2000	Paint, oil base, brushwork, 1 coat		650	.012		.11	.35		.46	.70
2500	2 coats		400	.020		.21	.57		.78	1.17
3000	Stain, brushwork, wipe off		650	.012		.09	.35		.44	.67
4000	Shellac, 1 coat, brushwork		650	.012		.11	.35		.46	.70
4500	Varnish, 3 coats, brushwork, sand after 1st coat	↓	325	.025	↓	.28	.71		.99	1.47
5000	For latex paint, deduct					10%				

09 91 23.33 Doors and Windows, Interior Alkyd (Oil Base)

		Crew	Daily Output	Labor-Hours	Unit	Material	Labor	Equipment	Total	Total Incl O&P
0010	**DOORS AND WINDOWS, INTERIOR ALKYD (OIL BASE)**									
0500	Flush door & frame, 3' x 7', oil, primer, brushwork	1 Pord	10	.800	Ea.	3.45	23		26.45	42
1000	Paint, 1 coat		10	.800		4.14	23		27.14	42.50
1200	2 coats		6	1.333		5	38		43	68.50
1400	Stain, brushwork, wipe off		18	.444		1.81	12.75		14.56	23
1600	Shellac, 1 coat, brushwork		25	.320		2.25	9.15		11.40	17.60
1800	Varnish, 3 coats, brushwork, sand after 1st coat		9	.889		5.85	25.50		31.35	48.50
2000	Panel door & frame, 3' x 7', oil, primer, brushwork		6	1.333		2.66	38		40.66	66
2200	Paint, 1 coat		6	1.333		4.14	38		42.14	67.50
2400	2 coats		3	2.667		10.95	76.50		87.45	138
2600	Stain, brushwork, panel door, 3' x 7', not incl. frame		16	.500		1.81	14.35		16.16	25.50
2800	Shellac, 1 coat, brushwork		22	.364		2.25	10.40		12.65	19.65
3000	Varnish, 3 coats, brushwork, sand after 1st coat	↓	7.50	1.067	↓	5.85	30.50		36.35	57
3020	French door, incl. 3' x 7', 6 lites, frame & trim									
3022	Paint, 1 coat, over existing paint	1 Pord	5	1.600	Ea.	8.30	46		54.30	84.50
3024	2 coats, over existing paint		5	1.600		16.10	46		62.10	93
3026	Primer & 1 coat		3.50	2.286		13.60	65.50		79.10	123
3028	Primer & 2 coats		3	2.667		22	76.50		98.50	150
3032	Varnish or polyurethane, 1 coat		5	1.600		7.85	46		53.85	84

09 91 Painting

09 91 23 - Interior Painting

09 91 23.33 Doors and Windows, Interior Alkyd (Oil Base)		Crew	Daily Output	Labor-Hours	Unit	Material	2015 Bare Costs Labor	2015 Bare Costs Equipment	Total	Total Incl O&P
3034	2 coats, sanding between	1 Pord	3	2.667	Ea.	15.70	76.50		92.20	143
4400	Windows, including frame and trim, per side									
4600	Colonial type, 6/6 lites, 2' x 3', oil, primer, brushwork	1 Pord	14	.571	Ea.	.42	16.35		16.77	27.50
5800	Paint, 1 coat		14	.571		.65	16.35		17	27.50
6000	2 coats		9	.889		1.27	25.50		26.77	43.50
6200	3' x 5' opening, 6/6 lites, primer coat, brushwork		12	.667		1.05	19.10		20.15	32.50
6400	Paint, 1 coat		12	.667		1.64	19.10		20.74	33.50
6600	2 coats		7	1.143		3.18	32.50		35.68	57.50
6800	4' x 8' opening, 6/6 lites, primer coat, brushwork		8	1		2.24	28.50		30.74	49.50
7000	Paint, 1 coat		8	1		3.49	28.50		31.99	51
7200	2 coats		5	1.600		6.80	46		52.80	83
8000	Single lite type, 2' x 3', oil base, primer coat, brushwork		33	.242		.42	6.95		7.37	11.90
8200	Paint, 1 coat		33	.242		.65	6.95		7.60	12.15
8400	2 coats		20	.400		1.27	11.45		12.72	20.50
8600	3' x 5' opening, primer coat, brushwork		20	.400		1.05	11.45		12.50	20
8800	Paint, 1 coat		20	.400		1.64	11.45		13.09	20.50
8900	2 coats		13	.615		3.18	17.65		20.83	32.50
9200	4' x 8' opening, primer coat, brushwork		14	.571		2.24	16.35		18.59	29.50
9400	Paint, 1 coat		14	.571		3.49	16.35		19.84	31
9600	2 coats		8	1		6.80	28.50		35.30	54.50

09 91 23.35 Doors and Windows, Interior Latex		Crew	Daily Output	Labor-Hours	Unit	Material	Labor	Equipment	Total	Total Incl O&P
0010	**DOORS & WINDOWS, INTERIOR LATEX** R099100-10									
0100	Doors, flush, both sides, incl. frame & trim									
0110	Roll & brush, primer	1 Pord	10	.800	Ea.	4.13	23		27.13	42.50
0120	Finish coat, latex		10	.800		5.30	23		28.30	44
0130	Primer & 1 coat latex		7	1.143		9.40	32.50		41.90	64.50
0140	Primer & 2 coats latex		5	1.600		14.40	46		60.40	91.50
0160	Spray, both sides, primer		20	.400		4.35	11.45		15.80	23.50
0170	Finish coat, latex		20	.400		5.55	11.45		17	25
0180	Primer & 1 coat latex		11	.727		9.95	21		30.95	45.50
0190	Primer & 2 coats latex		8	1		15.25	28.50		43.75	64
0200	Doors, French, both sides, 10-15 lite, incl. frame & trim									
0210	Roll & brush, primer	1 Pord	6	1.333	Ea.	2.06	38		40.06	65.50
0220	Finish coat, latex		6	1.333		2.65	38		40.65	66
0230	Primer & 1 coat latex		3	2.667		4.71	76.50		81.21	131
0240	Primer & 2 coats latex		2	4		7.20	115		122.20	197
0260	Doors, louvered, both sides, incl. frame & trim									
0270	Roll & brush, primer	1 Pord	7	1.143	Ea.	4.13	32.50		36.63	58.50
0280	Finish coat, latex		7	1.143		5.30	32.50		37.80	60
0290	Primer & 1 coat, latex		4	2		9.20	57.50		66.70	105
0300	Primer & 2 coats, latex		3	2.667		14.70	76.50		91.20	142
0320	Spray, both sides, primer		20	.400		4.35	11.45		15.80	23.50
0330	Finish coat, latex		20	.400		5.55	11.45		17	25
0340	Primer & 1 coat, latex		11	.727		9.95	21		30.95	45.50
0350	Primer & 2 coats, latex		8	1		15.60	28.50		44.10	64
0360	Doors, panel, both sides, incl. frame & trim									
0370	Roll & brush, primer	1 Pord	6	1.333	Ea.	4.35	38		42.35	68
0380	Finish coat, latex		6	1.333		5.30	38		43.30	69
0390	Primer & 1 coat, latex		3	2.667		9.40	76.50		85.90	136
0400	Primer & 2 coats, latex		2.50	3.200		14.70	91.50		106.20	167
0420	Spray, both sides, primer		10	.800		4.35	23		27.35	43
0430	Finish coat, latex		10	.800		5.55	23		28.55	44

09 91 Painting

09 91 23 – Interior Painting

09 91 23.35 Doors and Windows, Interior Latex		Crew	Daily Output	Labor-Hours	Unit	Material	2015 Bare Costs Labor	Equipment	Total	Total Incl O&P
0440	Primer & 1 coat, latex	1 Pord	5	1.600	Ea.	9.95	46		55.95	86.50
0450	Primer & 2 coats, latex	↓	4	2	↓	15.60	57.50		73.10	112
0460	Windows, per interior side, based on 15 S.F.									
0470	1 to 6 lite									
0480	Brushwork, primer	1 Pord	13	.615	Ea.	.81	17.65		18.46	30
0490	Finish coat, enamel		13	.615		1.04	17.65		18.69	30
0500	Primer & 1 coat enamel		8	1		1.86	28.50		30.36	49
0510	Primer & 2 coats enamel	↓	6	1.333	↓	2.90	38		40.90	66
0530	7 to 10 lite									
0540	Brushwork, primer	1 Pord	11	.727	Ea.	.81	21		21.81	35.50
0550	Finish coat, enamel		11	.727		1.04	21		22.04	35.50
0560	Primer & 1 coat enamel		7	1.143		1.86	32.50		34.36	56
0570	Primer & 2 coats enamel		5	1.600		2.90	46		48.90	78.50
0590	12 lite									
0600	Brushwork, primer	1 Pord	10	.800	Ea.	.81	23		23.81	39
0610	Finish coat, enamel		10	.800		1.04	23		24.04	39
0620	Primer & 1 coat enamel		6	1.333		1.86	38		39.86	65
0630	Primer & 2 coats enamel	↓	5	1.600	↓	2.90	46		48.90	78.50
0650	For oil base paint, add					10%				

09 91 23.39 Doors and Windows, Interior Latex, Zero Voc

			Crew	Daily Output	Labor-Hours	Unit	Material	Labor	Equipment	Total	Total Incl O&P
0010	DOORS & WINDOWS, INTERIOR LATEX, ZERO VOC										
0100	Doors flush, both sides, incl. frame & trim										
0110	Roll & brush, primer	G	1 Pord	10	.800	Ea.	4.90	23		27.90	43.50
0120	Finish coat, latex	G		10	.800		5.70	23		28.70	44.50
0130	Primer & 1 coat latex	G		7	1.143		10.60	32.50		43.10	65.50
0140	Primer & 2 coats latex	G		5	1.600		16	46		62	93
0160	Spray, both sides, primer	G		20	.400		5.15	11.45		16.60	24.50
0170	Finish coat, latex	G		20	.400		6	11.45		17.45	25.50
0180	Primer & 1 coat latex	G		11	.727		11.25	21		32.25	47
0190	Primer & 2 coats latex	G	↓	8	1		16.95	28.50		45.45	65.50
0200	Doors, French, both sides, 10-15 lite, incl. frame & trim										
0210	Roll & brush, primer	G	1 Pord	6	1.333	Ea.	2.45	38		40.45	65.50
0220	Finish coat, latex	G		6	1.333		2.86	38		40.86	66
0230	Primer & 1 coat latex	G		3	2.667		5.30	76.50		81.80	132
0240	Primer & 2 coats latex	G	↓	2	4	↓	8	115		123	198
0360	Doors, panel, both sides, incl. frame & trim										
0370	Roll & brush, primer	G	1 Pord	6	1.333	Ea.	5.15	38		43.15	68.50
0380	Finish coat, latex	G		6	1.333		5.70	38		43.70	69.50
0390	Primer & 1 coat, latex	G		3	2.667		10.60	76.50		87.10	138
0400	Primer & 2 coats, latex	G		2.50	3.200		16.35	91.50		107.85	169
0420	Spray, both sides, primer	G		10	.800		5.15	23		28.15	43.50
0430	Finish coat, latex	G		10	.800		6	23		29	44.50
0440	Primer & 1 coat, latex	G		5	1.600		11.25	46		57.25	88
0450	Primer & 2 coats, latex	G	↓	4	2	↓	17.30	57.50		74.80	114
0460	Windows, per interior side, based on 15 S.F.										
0470	1 to 6 lite										
0480	Brushwork, primer	G	1 Pord	13	.615	Ea.	.97	17.65		18.62	30
0490	Finish coat, enamel	G		13	.615		1.13	17.65		18.78	30
0500	Primer & 1 coat enamel	G		8	1		2.09	28.50		30.59	49.50
0510	Primer & 2 coats enamel	G	↓	6	1.333	↓	3.22	38		41.22	66.50

09 91 Painting

09 91 23 – Interior Painting

09 91 23.40 Floors, Interior

		Crew	Daily Output	Labor-Hours	Unit	Material	2015 Bare Costs Labor	2015 Bare Costs Equipment	Total	Total Incl O&P
0010	**FLOORS, INTERIOR**									
0100	Concrete paint, latex									
0110	Brushwork									
0120	1st coat	1 Pord	975	.008	S.F.	.15	.24		.39	.55
0130	2nd coat		1150	.007		.10	.20		.30	.44
0140	3rd coat		1300	.006		.08	.18		.26	.38
0150	Roll									
0160	1st coat	1 Pord	2600	.003	S.F.	.20	.09		.29	.37
0170	2nd coat		3250	.002		.12	.07		.19	.25
0180	3rd coat		3900	.002		.09	.06		.15	.20
0190	Spray									
0200	1st coat	1 Pord	2600	.003	S.F.	.17	.09		.26	.34
0210	2nd coat		3250	.002		.09	.07		.16	.22
0220	3rd coat		3900	.002		.07	.06		.13	.18

09 91 23.52 Miscellaneous, Interior

		Crew	Daily Output	Labor-Hours	Unit	Material	2015 Bare Costs Labor	2015 Bare Costs Equipment	Total	Total Incl O&P
0010	**MISCELLANEOUS, INTERIOR**									
2400	Floors, conc./wood, oil base, primer/sealer coat, brushwork	2 Pord	1950	.008	S.F.	.09	.24		.33	.49
2450	Roller		5200	.003		.10	.09		.19	.26
2600	Spray		6000	.003		.10	.08		.18	.24
2650	Paint 1 coat, brushwork		1950	.008		.10	.24		.34	.50
2800	Roller		5200	.003		.10	.09		.19	.26
2850	Spray		6000	.003		.11	.08		.19	.25
3000	Stain, wood floor, brushwork, 1 coat		4550	.004		.09	.10		.19	.26
3200	Roller		5200	.003		.09	.09		.18	.25
3250	Spray		6000	.003		.09	.08		.17	.23
3400	Varnish, wood floor, brushwork		4550	.004		.09	.10		.19	.27
3450	Roller		5200	.003		.10	.09		.19	.26
3600	Spray		6000	.003		.10	.08		.18	.24
3800	Grilles, per side, oil base, primer coat, brushwork	1 Pord	520	.015		.14	.44		.58	.88
3850	Spray		1140	.007		.15	.20		.35	.49
3880	Paint 1 coat, brushwork		520	.015		.22	.44		.66	.97
3900	Spray		1140	.007		.24	.20		.44	.60
3920	Paint 2 coats, brushwork		325	.025		.42	.71		1.13	1.63
3940	Spray		650	.012		.48	.35		.83	1.11
4500	Louvers, one side, primer, brushwork		524	.015		.09	.44		.53	.82
4520	Paint one coat, brushwork		520	.015		.10	.44		.54	.83
4530	Spray		1140	.007		.11	.20		.31	.45
4540	Paint two coats, brushwork		325	.025		.18	.71		.89	1.36
4550	Spray		650	.012		.20	.35		.55	.80
4560	Paint three coats, brushwork		270	.030		.27	.85		1.12	1.70
4570	Spray		500	.016		.30	.46		.76	1.09
5000	Pipe, 1" - 4" diameter, primer or sealer coat, oil base, brushwork	2 Pord	1250	.013	L.F.	.10	.37		.47	.71
5100	Spray		2165	.007		.09	.21		.30	.45
5200	Paint 1 coat, brushwork		1250	.013		.11	.37		.48	.72
5300	Spray		2165	.007		.10	.21		.31	.46
5350	Paint 2 coats, brushwork		775	.021		.19	.59		.78	1.18
5400	Spray		1240	.013		.22	.37		.59	.85
5450	5" - 8" diameter, primer or sealer coat, brushwork		620	.026		.20	.74		.94	1.44
5500	Spray		1085	.015		.32	.42		.74	1.05
5550	Paint 1 coat, brushwork		620	.026		.29	.74		1.03	1.54
5600	Spray		1085	.015		.33	.42		.75	1.06
5650	Paint 2 coats, brushwork		385	.042		.39	1.19		1.58	2.39

09 91 Painting

09 91 23 – Interior Painting

09 91 23.52 Miscellaneous, Interior

		Crew	Daily Output	Labor-Hours	Unit	Material	2015 Bare Costs Labor	Equipment	Total	Total Incl O&P
5700	Spray	2 Pord	620	.026	L.F.	.43	.74		1.17	1.69
6600	Radiators, per side, primer, brushwork	1 Pord	520	.015	S.F.	.09	.44		.53	.83
6620	Paint, one coat		520	.015		.09	.44		.53	.83
6640	Two coats		340	.024		.18	.67		.85	1.31
6660	Three coats		283	.028		.27	.81		1.08	1.63
7000	Trim, wood, incl. puttying, under 6" wide									
7200	Primer coat, oil base, brushwork	1 Pord	650	.012	L.F.	.03	.35		.38	.62
7250	Paint, 1 coat, brushwork		650	.012		.05	.35		.40	.64
7400	2 coats		400	.020		.11	.57		.68	1.06
7450	3 coats		325	.025		.16	.71		.87	1.33
7500	Over 6" wide, primer coat, brushwork		650	.012		.07	.35		.42	.66
7550	Paint, 1 coat, brushwork		650	.012		.11	.35		.46	.70
7600	2 coats		400	.020		.21	.57		.78	1.17
7650	3 coats		325	.025		.31	.71		1.02	1.51
8000	Cornice, simple design, primer coat, oil base, brushwork		650	.012	S.F.	.07	.35		.42	.66
8250	Paint, 1 coat		650	.012		.11	.35		.46	.70
8300	2 coats		400	.020		.21	.57		.78	1.17
8350	Ornate design, primer coat		350	.023		.07	.65		.72	1.16
8400	Paint, 1 coat		350	.023		.11	.65		.76	1.20
8450	2 coats		400	.020		.21	.57		.78	1.17
8600	Balustrades, primer coat, oil base, brushwork		520	.015		.07	.44		.51	.81
8650	Paint, 1 coat		520	.015		.11	.44		.55	.85
8700	2 coats		325	.025		.21	.71		.92	1.39
8900	Trusses and wood frames, primer coat, oil base, brushwork		800	.010		.07	.29		.36	.55
8950	Spray		1200	.007		.07	.19		.26	.39
9000	Paint 1 coat, brushwork		750	.011		.11	.31		.42	.62
9200	Spray		1200	.007		.12	.19		.31	.44
9220	Paint 2 coats, brushwork		500	.016		.21	.46		.67	.99
9240	Spray		600	.013		.24	.38		.62	.89
9260	Stain, brushwork, wipe off		600	.013		.09	.38		.47	.72
9280	Varnish, 3 coats, brushwork		275	.029		.28	.83		1.11	1.68
9350	For latex paint, deduct					10%				

09 91 23.72 Walls and Ceilings, Interior

		Crew	Daily Output	Labor-Hours	Unit	Material	2015 Bare Costs Labor	Equipment	Total	Total Incl O&P
0010	**WALLS AND CEILINGS, INTERIOR**									
0100	Concrete, drywall or plaster, latex, primer or sealer coat									
0200	Smooth finish, brushwork	1 Pord	1150	.007	S.F.	.06	.20		.26	.40
0240	Roller		1350	.006		.06	.17		.23	.35
0280	Spray		2750	.003		.05	.08		.13	.20
0300	Sand finish, brushwork		975	.008		.06	.24		.30	.46
0340	Roller		1150	.007		.06	.20		.26	.40
0380	Spray		2275	.004		.05	.10		.15	.23
0400	Paint 1 coat, smooth finish, brushwork		1200	.007		.07	.19		.26	.38
0440	Roller		1300	.006		.07	.18		.25	.36
0480	Spray		2275	.004		.05	.10		.15	.23
0500	Sand finish, brushwork		1050	.008		.06	.22		.28	.43
0540	Roller		1600	.005		.07	.14		.21	.31
0580	Spray		2100	.004		.02	.11		.13	.20
0800	Paint 2 coats, smooth finish, brushwork		680	.012		.13	.34		.47	.71
0840	Roller		800	.010		.13	.29		.42	.62
0880	Spray		1625	.005		.12	.14		.26	.37
0900	Sand finish, brushwork		605	.013		.13	.38		.51	.77
0940	Roller		1020	.008		.13	.22		.35	.52

09 91 Painting

09 91 23 – Interior Painting

09 91 23.72 Walls and Ceilings, Interior		Crew	Daily Output	Labor-Hours	Unit	Material	2015 Bare Costs Labor	Equipment	Total	Total Incl O&P
0980	Spray	1 Pord	1700	.005	S.F.	.12	.13		.25	.36
1200	Paint 3 coats, smooth finish, brushwork		510	.016		.20	.45		.65	.96
1240	Roller		650	.012		.20	.35		.55	.80
1280	Spray		1625	.005		.19	.14		.33	.43
1300	Sand finish, brushwork		454	.018		.33	.50		.83	1.19
1340	Roller		680	.012		.35	.34		.69	.94
1380	Spray		1133	.007		.30	.20		.50	.66
1600	Glaze coating, 2 coats, spray, clear		1200	.007		.49	.19		.68	.85
1640	Multicolor		1200	.007		.99	.19		1.18	1.40
1660	Painting walls, complete, including surface prep, primer &									
1670	2 coats finish, on drywall or plaster, with roller	1 Pord	325	.025	S.F.	.20	.71		.91	1.38
1700	For oil base paint, add					10%				
1800	For ceiling installations, add						25%			
2000	Masonry or concrete block, primer/sealer, latex paint									
2100	Primer, smooth finish, brushwork	1 Pord	1000	.008	S.F.	.11	.23		.34	.50
2110	Roller		1150	.007		.11	.20		.31	.45
2180	Spray		2400	.003		.10	.10		.20	.27
2200	Sand finish, brushwork		850	.009		.11	.27		.38	.56
2210	Roller		975	.008		.11	.24		.35	.51
2280	Spray		2050	.004		.10	.11		.21	.29
2400	Finish coat, smooth finish, brush		1100	.007		.08	.21		.29	.43
2410	Roller		1300	.006		.08	.18		.26	.38
2480	Spray		2400	.003		.07	.10		.17	.24
2500	Sand finish, brushwork		950	.008		.08	.24		.32	.49
2510	Roller		1090	.007		.08	.21		.29	.44
2580	Spray		2040	.004		.07	.11		.18	.27
2800	Primer plus one finish coat, smooth brush		525	.015		.30	.44		.74	1.05
2810	Roller		615	.013		.19	.37		.56	.82
2880	Spray		1200	.007		.17	.19		.36	.49
2900	Sand finish, brushwork		450	.018		.19	.51		.70	1.05
2910	Roller		515	.016		.19	.44		.63	.94
2980	Spray		1025	.008		.17	.22		.39	.55
3200	Primer plus 2 finish coats, smooth, brush		355	.023		.27	.65		.92	1.35
3210	Roller		415	.019		.27	.55		.82	1.20
3280	Spray		800	.010		.23	.29		.52	.73
3300	Sand finish, brushwork		305	.026		.27	.75		1.02	1.53
3310	Roller		350	.023		.27	.65		.92	1.37
3380	Spray		675	.012		.23	.34		.57	.82
3600	Glaze coating, 3 coats, spray, clear		900	.009		.70	.25		.95	1.19
3620	Multicolor		900	.009		1.15	.25		1.40	1.69
4000	Block filler, 1 coat, brushwork		425	.019		.12	.54		.66	1.02
4100	Silicone, water repellent, 2 coats, spray		2000	.004		.33	.11		.44	.56
4120	For oil base paint, add					10%				
8200	For work 8' - 15' H, add						10%			
8300	For work over 15' H, add						20%			
8400	For light textured surfaces, add						10%			
8410	Heavy textured, add						25%			
09 91 23.74 Walls and Ceilings, Interior, Zero VOC Latex										
0010	**WALLS AND CEILINGS, INTERIOR, ZERO VOC LATEX**									
0100	Concrete, dry wall or plaster, latex, primer or sealer coat									
0200	Smooth finish, brushwork	G	1 Pord	1150	.007	S.F.	.06	.20		.26	.40
0240	Roller	G		1350	.006		.06	.17		.23	.35

For customer support on your Residential Cost Data, call 877.713.2617.

09 91 Painting

09 91 23 – Interior Painting

09 91 23.74 Walls and Ceilings, Interior, Zero VOC Latex

			Crew	Daily Output	Labor-Hours	Unit	Material	2015 Bare Costs Labor	Equipment	Total	Total Incl O&P
0280	Spray	G	1 Pord	2750	.003	S.F.	.05	.08		.13	.19
0300	Sand finish, brushwork	G		975	.008		.06	.24		.30	.46
0340	Roller	G		1150	.007		.07	.20		.27	.40
0380	Spray	G		2275	.004		.05	.10		.15	.23
0400	Paint 1 coat, smooth finish, brushwork	G		1200	.007		.08	.19		.27	.40
0440	Roller	G		1300	.006		.08	.18		.26	.38
0480	Spray	G		2275	.004		.07	.10		.17	.24
0500	Sand finish, brushwork	G		1050	.008		.07	.22		.29	.44
0540	Roller	G		1600	.005		.08	.14		.22	.33
0580	Spray	G		2100	.004		.07	.11		.18	.25
0800	Paint 2 coats, smooth finish, brushwork	G		680	.012		.15	.34		.49	.73
0840	Roller	G		800	.010		.16	.29		.45	.64
0880	Spray	G		1625	.005		.13	.14		.27	.38
0900	Sand finish, brushwork	G		605	.013		.15	.38		.53	.78
0940	Roller	G		1020	.008		.16	.22		.38	.54
0980	Spray	G		1700	.005		.13	.13		.26	.37
1200	Paint 3 coats, smooth finish, brushwork	G		510	.016		.22	.45		.67	.98
1240	Roller	G		650	.012		.23	.35		.58	.84
1280	Spray	G		1625	.005		.20	.14		.34	.45
1800	For ceiling installations, add	G						25%			
8200	For work 8' - 15' H, add							10%			
8300	For work over 15' H, add							20%			

09 91 23.75 Dry Fall Painting

		Crew	Daily Output	Labor-Hours	Unit	Material	Labor	Equipment	Total	Total Incl O&P
0010	**DRY FALL PAINTING**									
0100	Sprayed on walls, gypsum board or plaster									
0220	One coat	1 Pord	2600	.003	S.F.	.06	.09		.15	.21
0250	Two coats		1560	.005		.11	.15		.26	.36
0280	Concrete or textured plaster, one coat		1560	.005		.06	.15		.21	.30
0310	Two coats		1300	.006		.11	.18		.29	.41
0340	Concrete block, one coat		1560	.005		.06	.15		.21	.30
0370	Two coats		1300	.006		.11	.18		.29	.41
0400	Wood, one coat		877	.009		.06	.26		.32	.49
0430	Two coats		650	.012		.11	.35		.46	.70
0440	On ceilings, gypsum board or plaster									
0470	One coat	1 Pord	1560	.005	S.F.	.06	.15		.21	.30
0500	Two coats		1300	.006		.11	.18		.29	.41
0530	Concrete or textured plaster, one coat		1560	.005		.06	.15		.21	.30
0560	Two coats		1300	.006		.11	.18		.29	.41
0570	Structural steel, bar joists or metal deck, one coat		1560	.005		.06	.15		.21	.30
0580	Two coats		1040	.008		.11	.22		.33	.48

09 93 Staining and Transparent Finishing

09 93 23 – Interior Staining and Finishing

09 93 23.10 Varnish

		Crew	Daily Output	Labor-Hours	Unit	Material	Labor	Equipment	Total	Total Incl O&P
0010	**VARNISH**									
0012	1 coat + sealer, on wood trim, brush, no sanding included	1 Pord	400	.020	S.F.	.07	.57		.64	1.02
0020	1 coat + sealer, on wood trim, brush, no sanding included, no VOC		400	.020		.21	.57		.78	1.17
0100	Hardwood floors, 2 coats, no sanding included, roller		1890	.004		.15	.12		.27	.37

09 96 High-Performance Coatings

09 96 56 – Epoxy Coatings

09 96 56.20 Wall Coatings		Crew	Daily Output	Labor-Hours	Unit	Material	2015 Bare Costs Labor	Equipment	Total	Total Incl O&P
0010	**WALL COATINGS**									
0100	Acrylic glazed coatings, matte	1 Pord	525	.015	S.F.	.31	.44		.75	1.06
0200	Gloss		305	.026		.65	.75		1.40	1.96
0300	Epoxy coatings, solvent based		525	.015		.40	.44		.84	1.16
0400	Water based		170	.047		1.20	1.35		2.55	3.54
0600	Exposed aggregate, troweled on, 1/16" to 1/4", solvent based		235	.034		.62	.98		1.60	2.29
0700	Water based (epoxy or polyacrylate)		130	.062		1.33	1.76		3.09	4.36
0900	1/2" to 5/8" aggregate, solvent based		130	.062		1.20	1.76		2.96	4.22
1000	Water based		80	.100		2.08	2.87		4.95	7
1200	1" aggregate size, solvent based		90	.089		2.12	2.55		4.67	6.55
1300	Water based		55	.145		3.23	4.17		7.40	10.40
1500	Exposed aggregate, sprayed on, 1/8" aggregate, solvent based		295	.027		.57	.78		1.35	1.91
1600	Water based		145	.055		1.05	1.58		2.63	3.76

Division Notes

	CREW	DAILY OUTPUT	LABOR-HOURS	UNIT	BARE COSTS				TOTAL INCL O&P
					MAT.	LABOR	EQUIP.	TOTAL	

Estimating Tips
General
- The items in this division are usually priced per square foot or each.
- Many items in Division 10 require some type of support system or special anchors that are not usually furnished with the item. The required anchors must be added to the estimate in the appropriate division.
- Some items in Division 10, such as lockers, may require assembly before installation. Verify the amount of assembly required. Assembly can often exceed installation time.

10 20 00 Interior Specialties
- Support angles and blocking are not included in the installation of toilet compartments, shower/dressing compartments, or cubicles. Appropriate line items from Divisions 5 or 6 may need to be added to support the installations.
- Toilet partitions are priced by the stall. A stall consists of a side wall, pilaster, and door with hardware. Toilet tissue holders and grab bars are extra.
- The required acoustical rating of a folding partition can have a significant impact on costs. Verify the sound transmission coefficient rating of the panel priced to the specification requirements.
- Grab bar installation does not include supplemental blocking or backing to support the required load. When grab bars are installed at an existing facility, provisions must be made to attach the grab bars to solid structure.

Reference Numbers
Reference numbers are shown in shaded boxes at the beginning of some major classifications. These numbers refer to related items in the Reference Section. The reference information may be an estimating procedure, an alternate pricing method, or technical information.

Note: Not all subdivisions listed here necessarily appear in this publication. ■

Did you know?
RSMeans Online gives you the same access to RSMeans' data with 24/7 access:
- Quickly locate costs in the searchable database.
- Build cost lists, estimates, and reports in minutes.
- Adjust costs to any location in the U.S. and Canada with the click of a button.

Start your free trial today at **www.rsmeansonline.com**

No part of this publication may be reproduced, stored in a retrieval system, or transmitted in any form or by any means without prior written permission of RSMeans.

10 28 Toilet, Bath, and Laundry Accessories

10 28 13 – Toilet Accessories

10 28 13.13 Commercial Toilet Accessories

		Crew	Daily Output	Labor-Hours	Unit	Material	2015 Bare Costs Labor	Equipment	Total	Total Incl O&P
0010	**COMMERCIAL TOILET ACCESSORIES**									
0200	Curtain rod, stainless steel, 5' long, 1" diameter	1 Carp	13	.615	Ea.	28.50	20		48.50	65
0300	1-1/4" diameter		13	.615		29	20		49	65.50
0800	Grab bar, straight, 1-1/4" diameter, stainless steel, 18" long		24	.333		29	10.95		39.95	50.50
1100	36" long		20	.400		38.50	13.15		51.65	64.50
1105	42" long		20	.400		46	13.15		59.15	72.50
1120	Corner, 36" long		20	.400		85.50	13.15		98.65	116
3000	Mirror, with stainless steel 3/4" square frame, 18" x 24"		20	.400		49	13.15		62.15	76
3100	36" x 24"		15	.533		109	17.50		126.50	150
3300	72" x 24"		6	1.333		289	44		333	395
4300	Robe hook, single, regular		36	.222		18.95	7.30		26.25	33.50
4400	Heavy duty, concealed mounting		36	.222		19.60	7.30		26.90	34
6400	Towel bar, stainless steel, 18" long		23	.348		41.50	11.45		52.95	65
6500	30" long		21	.381		111	12.50		123.50	144
7400	Tumbler holder, for tumbler only		30	.267		17.80	8.75		26.55	34.50
7410	Tumbler holder, recessed		20	.400		9.80	13.15		22.95	33
7500	Soap, tumbler & toothbrush		30	.267		19.60	8.75		28.35	36
7510	Tumbler & toothbrush holder		20	.400		13.60	13.15		26.75	37

10 28 16 – Bath Accessories

10 28 16.20 Medicine Cabinets

		Crew	Daily Output	Labor-Hours	Unit	Material	Labor	Equipment	Total	Total Incl O&P
0010	**MEDICINE CABINETS**									
0020	With mirror, sst frame, 16" x 22", unlighted	1 Carp	14	.571	Ea.	98	18.75		116.75	140
0100	Wood frame		14	.571		128	18.75		146.75	173
0300	Sliding mirror doors, 20" x 16" x 4-3/4", unlighted		7	1.143		124	37.50		161.50	199
0400	24" x 19" x 8-1/2", lighted		5	1.600		179	52.50		231.50	285
0600	Triple door, 30" x 32", unlighted, plywood body		7	1.143		325	37.50		362.50	425
0700	Steel body		7	1.143		375	37.50		412.50	475
0900	Oak door, wood body, beveled mirror, single door		7	1.143		199	37.50		236.50	282
1000	Double door		6	1.333		380	44		424	495

10 28 19 – Tub and Shower Enclosures

10 28 19.10 Partitions, Shower

		Crew	Daily Output	Labor-Hours	Unit	Material	Labor	Equipment	Total	Total Incl O&P
0010	**PARTITIONS, SHOWER** floor mounted, no plumbing									
0400	Cabinet, one piece, fiberglass, 32" x 32"	2 Carp	5	3.200	Ea.	530	105		635	760
0420	36" x 36"		5	3.200		570	105		675	800
0440	36" x 48"		5	3.200		1,375	105		1,480	1,675
0460	Acrylic, 32" x 32"		5	3.200		310	105		415	515
0480	36" x 36"		5	3.200		1,025	105		1,130	1,300
0500	36" x 48"		5	3.200		1,500	105		1,605	1,850
0520	Shower door for above, clear plastic, 24" wide	1 Carp	8	1		186	33		219	259
0540	28" wide		8	1		206	33		239	282
0560	Tempered glass, 24" wide		8	1		200	33		233	275
0580	28" wide		8	1		226	33		259	305
2400	Glass stalls, with doors, no receptors, chrome on brass	2 Shee	3	5.333		1,650	197		1,847	2,150
2700	Anodized aluminum	"	4	4		1,150	148		1,298	1,525
3200	Receptors, precast terrazzo, 32" x 32"	2 Marb	14	1.143		360	35		395	455
3300	48" x 34"		9.50	1.684		465	51.50		516.50	595
3500	Plastic, simulated terrazzo receptor, 32" x 32"		14	1.143		148	35		183	221
3600	32" x 48"		12	1.333		215	40.50		255.50	305
3800	Precast concrete, colors, 32" x 32"		14	1.143		186	35		221	263
3900	48" x 48"		8	2		251	61		312	380
4100	Shower doors, economy plastic, 24" wide	1 Shee	9	.889		144	33		177	213
4200	Tempered glass door, economy		8	1		258	37		295	345

10 28 Toilet, Bath, and Laundry Accessories

10 28 19 – Tub and Shower Enclosures

10 28 19.10 Partitions, Shower		Crew	Daily Output	Labor-Hours	Unit	Material	2015 Bare Costs Labor	Equipment	Total	Total Incl O&P
4400	Folding, tempered glass, aluminum frame	1 Shee	6	1.333	Ea.	390	49.50		439.50	510
4700	Deluxe, tempered glass, chrome on brass frame, 42" to 44"		8	1		385	37		422	480
4800	39" to 48" wide		1	8		625	296		921	1,175
4850	On anodized aluminum frame, obscure glass		2	4		540	148		688	835
4900	Clear glass		1	8		625	296		921	1,175
5100	Shower enclosure, tempered glass, anodized alum. frame									
5120	2 panel & door, corner unit, 32" x 32"	1 Shee	2	4	Ea.	1,025	148		1,173	1,400
5140	Neo-angle corner unit, 16" x 24" x 16"	"	2	4		1,075	148		1,223	1,450
5200	Shower surround, 3 wall, polypropylene, 32" x 32"	1 Carp	4	2		455	65.50		520.50	610
5220	PVC, 32" x 32"		4	2		380	65.50		445.50	530
5240	Fiberglass		4	2		420	65.50		485.50	570
5250	2 wall, polypropylene, 32" x 32"		4	2		315	65.50		380.50	460
5270	PVC		4	2		395	65.50		460.50	545
5290	Fiberglass		4	2		400	65.50		465.50	550
5300	Tub doors, tempered glass & frame, obscure glass	1 Shee	8	1		229	37		266	315
5400	Clear glass		6	1.333		535	49.50		584.50	670
5600	Chrome plated, brass frame, obscure glass		8	1		305	37		342	395
5700	Clear glass		6	1.333		745	49.50		794.50	895
5900	Tub/shower enclosure, temp. glass, alum. frame, obscure glass		2	4		410	148		558	695
6200	Clear glass		1.50	5.333		855	197		1,052	1,275
6500	On chrome-plated brass frame, obscure glass		2	4		565	148		713	870
6600	Clear glass		1.50	5.333		1,200	197		1,397	1,650
6800	Tub surround, 3 wall, polypropylene	1 Carp	4	2		256	65.50		321.50	390
6900	PVC		4	2		390	65.50		455.50	540
7000	Fiberglass, obscure glass		4	2		400	65.50		465.50	550
7100	Clear glass		3	2.667		680	87.50		767.50	895

10 28 23 – Laundry Accessories

10 28 23.13 Built-In Ironing Boards

		Crew	Daily Output	Labor-Hours	Unit	Material	Labor	Equipment	Total	Total Incl O&P
0010	**BUILT-IN IRONING BOARDS**									
0020	Including cabinet, board & light, 42"	1 Carp	2	4	Ea.	395	131		526	655

10 31 Manufactured Fireplaces

10 31 13 – Manufactured Fireplace Chimneys

10 31 13.10 Fireplace Chimneys

		Crew	Daily Output	Labor-Hours	Unit	Material	Labor	Equipment	Total	Total Incl O&P
0010	**FIREPLACE CHIMNEYS**									
0500	Chimney dbl. wall, all stainless, over 8'-6", 7" diam., add to fireplace	1 Carp	33	.242	V.L.F.	83.50	7.95		91.45	105
0600	10" diameter, add to fireplace		32	.250		132	8.20		140.20	159
0700	12" diameter, add to fireplace		31	.258		158	8.50		166.50	188
0800	14" diameter, add to fireplace		30	.267		227	8.75		235.75	265
1000	Simulated brick chimney top, 4' high, 16" x 16"		10	.800	Ea.	425	26.50		451.50	515
1100	24" x 24"		7	1.143	"	540	37.50		577.50	660

10 31 13.20 Chimney Accessories

		Crew	Daily Output	Labor-Hours	Unit	Material	Labor	Equipment	Total	Total Incl O&P
0010	**CHIMNEY ACCESSORIES**									
0020	Chimney screens, galv., 13" x 13" flue	1 Bric	8	1	Ea.	58.50	32.50		91	118
0050	24" x 24" flue		5	1.600		124	51.50		175.50	223
0200	Stainless steel, 13" x 13" flue		8	1		97.50	32.50		130	161
0250	20" x 20" flue		5	1.600		152	51.50		203.50	253
2400	Squirrel and bird screens, galvanized, 8" x 8" flue		16	.500		53	16.15		69.15	85.50
2450	13" x 13" flue		12	.667		55	21.50		76.50	96.50

10 31 Manufactured Fireplaces

10 31 16 – Manufactured Fireplace Forms

10 31 16.10 Fireplace Forms

		Crew	Daily Output	Labor-Hours	Unit	Material	2015 Bare Costs Labor	Equipment	Total	Total Incl O&P
0010	**FIREPLACE FORMS**									
1800	Fireplace forms, no accessories, 32" opening	1 Bric	3	2.667	Ea.	670	86		756	885
1900	36" opening		2.50	3.200		855	103		958	1,100
2000	40" opening		2	4		1,125	129		1,254	1,475
2100	78" opening		1.50	5.333		1,650	172		1,822	2,100

10 31 23 – Prefabricated Fireplaces

10 31 23.10 Fireplace, Prefabricated

		Crew	Daily Output	Labor-Hours	Unit	Material	Labor	Equipment	Total	Total Incl O&P
0010	**FIREPLACE, PREFABRICATED**, free standing or wall hung									
0100	With hood & screen, painted	1 Carp	1.30	6.154	Ea.	1,375	202		1,577	1,875
0150	Average		1	8		1,625	263		1,888	2,225
0200	Stainless steel		.90	8.889		3,050	292		3,342	3,875
1500	Simulated logs, gas fired, 40,000 BTU, 2' long, manual safety pilot		7	1.143	Set	425	37.50		462.50	530
1600	Adjustable flame remote pilot		6	1.333		1,100	44		1,144	1,300
1700	Electric, 1,500 BTU, 1'-6" long, incandescent flame		7	1.143		197	37.50		234.50	279
1800	1,500 BTU, LED flame		6	1.333		300	44		344	410
2000	Fireplace, built-in, 36" hearth, radiant		1.30	6.154	Ea.	660	202		862	1,075
2100	Recirculating, small fan		1	8		880	263		1,143	1,400
2150	Large fan		.90	8.889		1,900	292		2,192	2,600
2200	42" hearth, radiant		1.20	6.667		890	219		1,109	1,350
2300	Recirculating, small fan		.90	8.889		1,175	292		1,467	1,800
2350	Large fan		.80	10		1,300	330		1,630	1,975
2400	48" hearth, radiant		1.10	7.273		2,075	239		2,314	2,675
2500	Recirculating, small fan		.80	10		2,350	330		2,680	3,150
2550	Large fan		.70	11.429		2,375	375		2,750	3,225
3000	See through, including doors		.80	10		2,225	330		2,555	3,000
3200	Corner (2 wall)		1	8		3,250	263		3,513	4,025

10 32 Fireplace Specialties

10 32 13 – Fireplace Dampers

10 32 13.10 Dampers

		Crew	Daily Output	Labor-Hours	Unit	Material	Labor	Equipment	Total	Total Incl O&P
0010	**DAMPERS**									
0800	Damper, rotary control, steel, 30" opening	1 Bric	6	1.333	Ea.	119	43		162	203
0850	Cast iron, 30" opening		6	1.333		125	43		168	209
1200	Steel plate, poker control, 60" opening		8	1		320	32.50		352.50	410
1250	84" opening, special order		5	1.600		585	51.50		636.50	730
1400	"Universal" type, chain operated, 32" x 20" opening		8	1		250	32.50		282.50	330
1450	48" x 24" opening		5	1.600		375	51.50		426.50	495

10 32 23 – Fireplace Doors

10 32 23.10 Doors

		Crew	Daily Output	Labor-Hours	Unit	Material	Labor	Equipment	Total	Total Incl O&P
0010	**DOORS**									
0400	Cleanout doors and frames, cast iron, 8" x 8"	1 Bric	12	.667	Ea.	41	21.50		62.50	81
0450	12" x 12"		10	.800		108	26		134	162
0500	18" x 24"		8	1		150	32.50		182.50	219
0550	Cast iron frame, steel door, 24" x 30"		5	1.600		315	51.50		366.50	430
1600	Dutch Oven door and frame, cast iron, 12" x 15" opening		13	.615		131	19.90		150.90	178
1650	Copper plated, 12" x 15" opening		13	.615		257	19.90		276.90	315

10 35 Stoves

10 35 13 – Heating Stoves

10 35 13.10 Woodburning Stoves

		Crew	Daily Output	Labor-Hours	Unit	Material	2015 Bare Costs Labor	2015 Bare Costs Equipment	Total	Total Incl O&P
0010	**WOODBURNING STOVES**									
0015	Cast iron, °1500sf	2 Carp	1.30	12.308	Ea.	1,175	405		1,580	1,975
0020	1500-2000sf		1	16		2,025	525		2,550	3,100
0030	2,000 sf		.80	20		2,775	655		3,430	4,150
0050	For gas log lighter, add					45			45	49.50

10 44 Fire Protection Specialties

10 44 16 – Fire Extinguishers

10 44 16.13 Portable Fire Extinguishers

		Crew	Daily Output	Labor-Hours	Unit	Material	2015 Bare Costs Labor	2015 Bare Costs Equipment	Total	Total Incl O&P
0010	**PORTABLE FIRE EXTINGUISHERS**									
0140	CO_2, with hose and "H" horn, 10 lb.				Ea.	275			275	305
1000	Dry chemical, pressurized									
1040	Standard type, portable, painted, 2-1/2 lb.				Ea.	37.50			37.50	41
1080	10 lb.					81			81	89.50
1100	20 lb.					136			136	150
1120	30 lb.					425			425	470
2000	ABC all purpose type, portable, 2-1/2 lb.					21			21	23.50
2080	9-1/2 lb.					44			44	48

10 55 Postal Specialties

10 55 23 – Mail Boxes

10 55 23.10 Mail Boxes

		Crew	Daily Output	Labor-Hours	Unit	Material	2015 Bare Costs Labor	2015 Bare Costs Equipment	Total	Total Incl O&P
0011	**MAIL BOXES**									
1900	Letter slot, residential	1 Carp	20	.400	Ea.	80	13.15		93.15	110
2400	Residential, galv. steel, small 20" x 7" x 9"	1 Clab	16	.500		200	12.25		212.25	241
2410	With galv. steel post, 54" long		6	1.333		250	32.50		282.50	330
2420	Large, 24" x 12" x 15"		16	.500		200	12.25		212.25	241
2430	With galv. steel post, 54" long		6	1.333		225	32.50		257.50	305
2440	Decorative, polyethylene, 22" x 10" x 10"		16	.500		60	12.25		72.25	86.50
2450	With alum. post, decorative, 54" long		6	1.333		180	32.50		212.50	253

10 56 Storage Assemblies

10 56 13 – Metal Storage Shelving

10 56 13.10 Shelving

		Crew	Daily Output	Labor-Hours	Unit	Material	2015 Bare Costs Labor	2015 Bare Costs Equipment	Total	Total Incl O&P
0010	**SHELVING**									
0020	Metal, industrial, cross-braced, 3' wide, 12" deep	1 Sswk	175	.046	SF Shlf	7.65	1.61		9.26	11.45
0100	24" deep		330	.024		5.05	.86		5.91	7.15
2200	Wide span, 1600 lb. capacity per shelf, 6' wide, 24" deep		380	.021		7	.74		7.74	9.10
2400	36" deep		440	.018		6.85	.64		7.49	8.70
3000	Residential, vinyl covered wire, wardrobe, 12" deep	1 Carp	195	.041	L.F.	14.20	1.35		15.55	17.85
3100	16" deep		195	.041		10.30	1.35		11.65	13.55
3200	Standard, 6" deep		195	.041		3.85	1.35		5.20	6.50
3300	9" deep		195	.041		5.80	1.35		7.15	8.60
3400	12" deep		195	.041		7.70	1.35		9.05	10.70
3500	16" deep		195	.041		17.70	1.35		19.05	21.50
3600	20" deep		195	.041		20.50	1.35		21.85	25
3700	Support bracket		80	.100	Ea.	6.40	3.29		9.69	12.55

10 57 Wardrobe and Closet Specialties

10 57 23 – Closet and Utility Shelving

10 57 23.19 Wood Closet and Utility Shelving		Crew	Daily Output	Labor-Hours	Unit	Material	2015 Bare Costs Labor	Equipment	Total	Total Incl O&P
0010	**WOOD CLOSET AND UTILITY SHELVING**									
0020	Pine, clear grade, no edge band, 1" x 8"	1 Carp	115	.070	L.F.	3.49	2.29		5.78	7.70
0100	1" x 10"		110	.073		4.34	2.39		6.73	8.80
0200	1" x 12"		105	.076		5.25	2.50		7.75	9.95
0450	1" x 18"		95	.084		7.85	2.77		10.62	13.30
0460	1" x 24"		85	.094		10.45	3.09		13.54	16.70
0600	Plywood, 3/4" thick with lumber edge, 12" wide		75	.107		1.87	3.50		5.37	7.95
0700	24" wide		70	.114		3.30	3.75		7.05	9.95
0900	Bookcase, clear grade pine, shelves 12" O.C., 8" deep, per S.F. shelf		70	.114	S.F.	11.35	3.75		15.10	18.75
1000	12" deep shelves		65	.123	"	17	4.04		21.04	25.50
1200	Adjustable closet rod and shelf, 12" wide, 3' long		20	.400	Ea.	11.15	13.15		24.30	34.50
1300	8' long		15	.533	"	21.50	17.50		39	53
1500	Prefinished shelves with supports, stock, 8" wide		75	.107	L.F.	4.58	3.50		8.08	10.95
1600	10" wide		70	.114	"	6.50	3.75		10.25	13.45

10 73 Protective Covers

10 73 16 – Canopies

10 73 16.10 Canopies, Residential

		Crew	Daily Output	Labor-Hours	Unit	Material	2015 Bare Costs Labor	Equipment	Total	Total Incl O&P
0010	**CANOPIES, RESIDENTIAL** Prefabricated									
0500	Carport, free standing, baked enamel, alum., .032", 40 psf									
0520	16' x 8', 4 posts	2 Carp	3	5.333	Ea.	4,725	175		4,900	5,500
0600	20' x 10', 6 posts		2	8		4,825	263		5,088	5,775
0605	30' x 10', 8 posts		2	8		7,250	263		7,513	8,425
1000	Door canopies, extruded alum., .032", 42" projection, 4' wide	1 Carp	8	1		201	33		234	276
1020	6' wide	"	6	1.333		284	44		328	385
1040	8' wide	2 Carp	9	1.778		370	58.50		428.50	510
1060	10' wide		7	2.286		390	75		465	555
1080	12' wide		5	3.200		620	105		725	855
1200	54" projection, 4' wide	1 Carp	8	1		248	33		281	325
1220	6' wide	"	6	1.333		290	44		334	395
1240	8' wide	2 Carp	9	1.778		350	58.50		408.50	485
1260	10' wide		7	2.286		515	75		590	695
1280	12' wide		5	3.200		555	105		660	790
1300	Painted, add					20%				
1310	Bronze anodized, add					50%				
3000	Window awnings, aluminum, window 3' high, 4' wide	1 Carp	10	.800		126	26.50		152.50	183
3020	6' wide	"	8	1		180	33		213	253
3040	9' wide	2 Carp	9	1.778		305	58.50		363.50	435
3060	12' wide	"	5	3.200		305	105		410	510
3100	Window, 4' high, 4' wide	1 Carp	10	.800		159	26.50		185.50	219
3120	6' wide	"	8	1		229	33		262	305
3140	9' wide	2 Carp	9	1.778		375	58.50		433.50	510
3160	12' wide	"	5	3.200		375	105		480	585
3200	Window, 6' high, 4' wide	1 Carp	10	.800		295	26.50		321.50	370
3220	6' wide	"	8	1		340	33		373	430
3240	9' wide	2 Carp	9	1.778		900	58.50		958.50	1,100
3260	12' wide	"	5	3.200		1,375	105		1,480	1,700
3400	Roll-up aluminum, 2'-6" wide	1 Carp	14	.571		190	18.75		208.75	241
3420	3' wide		12	.667		195	22		217	251
3440	4' wide		10	.800		244	26.50		270.50	315
3460	6' wide		8	1		290	33		323	375

10 73 Protective Covers

10 73 16 – Canopies

10 73 16.10 Canopies, Residential

		Crew	Daily Output	Labor-Hours	Unit	Material	2015 Bare Costs Labor	2015 Bare Costs Equipment	Total	Total Incl O&P
3480	9' wide	2 Carp	9	1.778	Ea.	400	58.50		458.50	540
3500	12' wide	"	5	3.200	↓	535	105		640	765
3600	Window awnings, canvas, 24" drop, 3' wide	1 Carp	30	.267	L.F.	50	8.75		58.75	69.50
3620	4' wide		40	.200		43	6.55		49.55	58.50
3700	30" drop, 3' wide		30	.267		51.50	8.75		60.25	71.50
3720	4' wide		40	.200		44.50	6.55		51.05	60
3740	5' wide		45	.178		40	5.85		45.85	54
3760	6' wide		48	.167		37	5.50		42.50	49.50
3780	8' wide		48	.167		34.50	5.50		40	47
3800	10' wide	↓	50	.160	↓	42	5.25		47.25	55

10 74 Manufactured Exterior Specialties

10 74 23 – Cupolas

10 74 23.10 Wood Cupolas

		Crew	Daily Output	Labor-Hours	Unit	Material	Labor	Equipment	Total	Incl O&P
0010	**WOOD CUPOLAS**									
0020	Stock units, pine, painted, 18" sq., 28" high, alum. roof	1 Carp	4.10	1.951	Ea.	184	64		248	310
0100	Copper roof		3.80	2.105		255	69		324	395
0300	23" square, 33" high, aluminum roof		3.70	2.162		360	71		431	515
0400	Copper roof		3.30	2.424		470	79.50		549.50	650
0600	30" square, 37" high, aluminum roof		3.70	2.162		560	71		631	735
0700	Copper roof		3.30	2.424		670	79.50		749.50	870
0900	Hexagonal, 31" wide, 46" high, copper roof		4	2		930	65.50		995.50	1,125
1000	36" wide, 50" high, copper roof	↓	3.50	2.286		1,400	75		1,475	1,675
1200	For deluxe stock units, add to above					25%				
1400	For custom built units, add to above					50%	50%			

10 74 33 – Weathervanes

10 74 33.10 Residential Weathervanes

		Crew	Daily Output	Labor-Hours	Unit	Material	Labor	Equipment	Total	Incl O&P
0010	**RESIDENTIAL WEATHERVANES**									
0020	Residential types, 18" to 24"	1 Carp	8	1	Ea.	151	33		184	221
0100	24" to 48"	"	2	4	"	1,700	131		1,831	2,100

10 74 46 – Window Wells

10 74 46.10 Area Window Wells

		Crew	Daily Output	Labor-Hours	Unit	Material	Labor	Equipment	Total	Incl O&P
0010	**AREA WINDOW WELLS**, Galvanized steel									
0020	20 ga., 3'-2" wide, 1' deep	1 Sswk	29	.276	Ea.	17.45	9.75		27.20	37.50
0100	2' deep		23	.348		31	12.30		43.30	57.50
0300	16 ga., 3'-2" wide, 1' deep		29	.276		23	9.75		32.75	44
0400	3' deep		23	.348		47	12.30		59.30	74.50
0600	Welded grating for above, 15 lb., painted		45	.178		87.50	6.30		93.80	108
0700	Galvanized		45	.178		118	6.30		124.30	142
0900	Translucent plastic cap for above	↓	60	.133	↓	19	4.71		23.71	30

10 75 Flagpoles

10 75 16 – Ground-Set Flagpoles

10 75 16.10 Flagpoles		Crew	Daily Output	Labor-Hours	Unit	Material	2015 Bare Costs Labor	Equipment	Total	Total Incl O&P
0010	**FLAGPOLES**, ground set									
0050	Not including base or foundation									
0100	Aluminum, tapered, ground set 20' high	K-1	2	8	Ea.	1,050	242	152	1,444	1,725
0200	25' high		1.70	9.412		1,100	285	178	1,563	1,900
0300	30' high		1.50	10.667		1,300	325	202	1,827	2,175
0500	40' high		1.20	13.333		2,775	405	253	3,433	4,000

Estimating Tips
General

- The items in this division are usually priced per square foot or each. Many of these items are purchased by the owner for installation by the contractor. Check the specifications for responsibilities and include time for receiving, storage, installation, and mechanical and electrical hookups in the appropriate divisions.
- Many items in Division 11 require some type of support system that is not usually furnished with the item. Examples of these systems include blocking for the attachment of casework and support angles for ceiling-hung projection screens. The required blocking or supports must be added to the estimate in the appropriate division.
- Some items in Division 11 may require assembly or electrical hookups. Verify the amount of assembly required or the need for a hard electrical connection and add the appropriate costs.

Reference Numbers
Reference numbers are shown in shaded boxes at the beginning of some major classifications. These numbers refer to related items in the Reference Section. The reference information may be an estimating procedure, an alternate pricing method, or technical information.

Note: Not all subdivisions listed here necessarily appear in this publication. ∎

Division 11 – Equipment

Estimate with precision with RSMeans Online...
Quick, intuitive and easy-to-use, RSMeans Online gives you instant access to hundreds of thousands of material, labor, and equipment costs from RSMeans' comprehensive database, delivering the information you need to build competitive estimates every time.

Start your free trial today at **www.rsmeansonline.com**

RSMeansOnline

No part of this publication may be reproduced, stored in a retrieval system, or transmitted in any form or by any means without prior written permission of RSMeans.

11 30 Residential Equipment

11 30 13 – Residential Appliances

11 30 13.15 Cooking Equipment

		Crew	Daily Output	Labor-Hours	Unit	Material	2015 Bare Costs Labor	2015 Bare Costs Equipment	Total	Total Incl O&P
0010	**COOKING EQUIPMENT**									
0020	Cooking range, 30" free standing, 1 oven, minimum	2 Clab	10	1.600	Ea.	440	39		479	550
0050	Maximum		4	4		2,025	98		2,123	2,400
0150	2 oven, minimum		10	1.600		1,225	39		1,264	1,400
0200	Maximum	▼	10	1.600		2,550	39		2,589	2,875
0350	Built-in, 30" wide, 1 oven, minimum	1 Elec	6	1.333		680	49		729	830
0400	Maximum	2 Carp	2	8		1,350	263		1,613	1,950
0500	2 oven, conventional, minimum		4	4		1,325	131		1,456	1,675
0550	1 conventional, 1 microwave, maximum	▼	2	8		1,800	263		2,063	2,425
0700	Free-standing, 1 oven, 21" wide range, minimum	2 Clab	10	1.600		455	39		494	565
0750	21" wide, maximum	"	4	4		450	98		548	660
0900	Countertop cooktops, 4 burner, standard, minimum	1 Elec	6	1.333		296	49		345	405
0950	Maximum		3	2.667		1,300	97.50		1,397.50	1,600
1050	As above, but with grill and griddle attachment, minimum		6	1.333		1,250	49		1,299	1,450
1100	Maximum		3	2.667		3,700	97.50		3,797.50	4,225
1250	Microwave oven, minimum		4	2		120	73.50		193.50	252
1300	Maximum		2	4		465	147		612	750
5380	Oven, built in, standard		4	2		635	73.50		708.50	815
5390	Deluxe	▼	2	4		2,275	147		2,422	2,750

11 30 13.16 Refrigeration Equipment

		Crew	Daily Output	Labor-Hours	Unit	Material	Labor	Equipment	Total	Total Incl O&P
0010	**REFRIGERATION EQUIPMENT**									
2000	Deep freeze, 15 to 23 C.F., minimum	2 Clab	10	1.600	Ea.	610	39		649	735
2050	Maximum		5	3.200		820	78		898	1,025
2200	30 C.F., minimum		8	2		705	49		754	855
2250	Maximum	▼	3	5.333		910	130		1,040	1,225
5200	Icemaker, automatic, 20 lb. per day	1 Plum	7	1.143		895	44.50		939.50	1,050
5350	51 lb. per day	"	2	4		1,575	155		1,730	2,000
5450	Refrigerator, no frost, 6 C.F.	2 Clab	15	1.067		244	26		270	315
5500	Refrigerator, no frost, 10 C.F. to 12 C.F., minimum		10	1.600		425	39		464	530
5600	Maximum		6	2.667		450	65		515	605
5750	14 C.F. to 16 C.F., minimum		9	1.778		505	43.50		548.50	630
5800	Maximum		5	3.200		620	78		698	815
5950	18 C.F. to 20 C.F., minimum		8	2		630	49		679	770
6000	Maximum		4	4		1,425	98		1,523	1,750
6150	21 C.F. to 29 C.F., minimum		7	2.286		965	56		1,021	1,175
6200	Maximum	▼	3	5.333		3,375	130		3,505	3,925
6790	Energy-star qualified, 18 C.F., minimum [G]	2 Carp	4	4		510	131		641	780
6795	Maximum [G]		2	8		1,175	263		1,438	1,725
6797	21.7 C.F., minimum [G]		4	4		1,025	131		1,156	1,350
6799	Maximum [G]	▼	4	4		2,100	131		2,231	2,550

11 30 13.17 Kitchen Cleaning Equipment

		Crew	Daily Output	Labor-Hours	Unit	Material	Labor	Equipment	Total	Total Incl O&P
0010	**KITCHEN CLEANING EQUIPMENT**									
2750	Dishwasher, built-in, 2 cycles, minimum	L-1	4	2.500	Ea.	238	96		334	420
2800	Maximum		2	5		435	192		627	795
2950	4 or more cycles, minimum		4	2.500		375	96		471	575
2960	Average		4	2.500		500	96		596	710
3000	Maximum		2	5		1,100	192		1,292	1,525
3100	Energy-star qualified, minimum [G]		4	2.500		370	96		466	565
3110	Maximum [G]	▼	2	5	▼	1,500	192		1,692	1,975

11 30 Residential Equipment

11 30 13 – Residential Appliances

11 30 13.18 Waste Disposal Equipment

		Crew	Daily Output	Labor-Hours	Unit	Material	2015 Bare Costs Labor	Equipment	Total	Total Incl O&P
0010	**WASTE DISPOSAL EQUIPMENT**									
1750	Compactor, residential size, 4 to 1 compaction, minimum	1 Carp	5	1.600	Ea.	645	52.50		697.50	800
1800	Maximum	"	3	2.667		1,025	87.50		1,112.50	1,275
3300	Garbage disposal, sink type, minimum	L-1	10	1		87.50	38.50		126	159
3350	Maximum	"	10	1		220	38.50		258.50	305

11 30 13.19 Kitchen Ventilation Equipment

		Crew	Daily Output	Labor-Hours	Unit	Material	Labor	Equipment	Total	Total Incl O&P
0010	**KITCHEN VENTILATION EQUIPMENT**									
4150	Hood for range, 2 speed, vented, 30" wide, minimum	L-3	5	2	Ea.	70.50	67		137.50	190
4200	Maximum		3	3.333		780	112		892	1,050
4300	42" wide, minimum		5	2		168	67		235	297
4330	Custom		5	2		1,650	67		1,717	1,925
4350	Maximum		3	3.333		2,025	112		2,137	2,400
4500	For ventless hood, 2 speed, add					18.65			18.65	20.50
4650	For vented 1 speed, deduct from maximum					50			50	55

11 30 13.24 Washers

		Crew	Daily Output	Labor-Hours	Unit	Material	Labor	Equipment	Total	Total Incl O&P
0010	**WASHERS**									
6650	Washing machine, automatic, minimum	1 Plum	3	2.667	Ea.	485	103		588	700
6700	Maximum		1	8		1,525	310		1,835	2,175
6750	Energy star qualified, front loading, minimum G		3	2.667		655	103		758	890
6760	Maximum G		1	8		1,600	310		1,910	2,250
6764	Top loading, minimum G		3	2.667		450	103		553	665
6766	Maximum G		3	2.667		1,250	103		1,353	1,550

11 30 13.25 Dryers

		Crew	Daily Output	Labor-Hours	Unit	Material	Labor	Equipment	Total	Total Incl O&P
0010	**DRYERS**									
6770	Electric, front loading, energy-star qualified, minimum G	L-2	3	5.333	Ea.	385	153		538	680
6780	Maximum G	"	2	8		1,925	229		2,154	2,475
7450	Vent kits for dryers	1 Carp	10	.800		39	26.50		65.50	86.50

11 30 15 – Miscellaneous Residential Appliances

11 30 15.13 Sump Pumps

		Crew	Daily Output	Labor-Hours	Unit	Material	Labor	Equipment	Total	Total Incl O&P
0010	**SUMP PUMPS**									
6400	Cellar drainer, pedestal, 1/3 H.P., molded PVC base	1 Plum	3	2.667	Ea.	135	103		238	320
6450	Solid brass	"	2	4	"	289	155		444	575
6460	Sump pump, see also Section 22 14 29.16									

11 30 15.23 Water Heaters

		Crew	Daily Output	Labor-Hours	Unit	Material	Labor	Equipment	Total	Total Incl O&P
0010	**WATER HEATERS**									
6900	Electric, glass lined, 30 gallon, minimum	L-1	5	2	Ea.	430	76.50		506.50	595
6950	Maximum		3	3.333		595	128		723	865
7100	80 gallon, minimum		2	5		1,225	192		1,417	1,675
7150	Maximum		1	10		1,700	385		2,085	2,475
7180	Gas, glass lined, 30 gallon, minimum	2 Plum	5	3.200		805	124		929	1,100
7220	Maximum		3	5.333		1,125	207		1,332	1,575
7260	50 gallon, minimum		2.50	6.400		845	248		1,093	1,325
7300	Maximum		1.50	10.667		1,175	415		1,590	1,950
7310	Water heater, see also Section 22 33 30.13									

11 30 15.43 Air Quality

		Crew	Daily Output	Labor-Hours	Unit	Material	Labor	Equipment	Total	Total Incl O&P
0010	**AIR QUALITY**									
2450	Dehumidifier, portable, automatic, 15 pint	1 Elec	4	2	Ea.	152	73.50		225.50	287
2550	40 pint		3.75	2.133		209	78		287	360
3550	Heater, electric, built-in, 1250 watt, ceiling type, minimum		4	2		107	73.50		180.50	238
3600	Maximum		3	2.667		175	97.50		272.50	350

11 30 Residential Equipment

11 30 15 – Miscellaneous Residential Appliances

11 30 15.43 Air Quality		Crew	Daily Output	Labor-Hours	Unit	Material	2015 Bare Costs Labor	Equipment	Total	Total Incl O&P
3700	Wall type, minimum	1 Elec	4	2	Ea.	172	73.50		245.50	310
3750	Maximum		3	2.667		185	97.50		282.50	365
3900	1500 watt wall type, with blower		4	2		172	73.50		245.50	310
3950	3000 watt		3	2.667		350	97.50		447.50	545
4850	Humidifier, portable, 8 gallons per day					133			133	146
5000	15 gallons per day					211			211	232

11 30 33 – Retractable Stairs

11 30 33.10 Disappearing Stairway

		Crew	Daily Output	Labor-Hours	Unit	Material	Labor	Equipment	Total	Total Incl O&P
0010	**DISAPPEARING STAIRWAY** No trim included									
0020	One piece, yellow pine, 8'-0" ceiling	2 Carp	4	4	Ea.	257	131		388	505
0030	9'-0" ceiling		4	4		263	131		394	510
0040	10'-0" ceiling		3	5.333		274	175		449	595
0050	11'-0" ceiling		3	5.333		430	175		605	770
0060	12'-0" ceiling		3	5.333		495	175		670	840
0100	Custom grade, pine, 8'-6" ceiling, minimum	1 Carp	4	2		177	65.50		242.50	305
0150	Average		3.50	2.286		253	75		328	405
0200	Maximum		3	2.667		325	87.50		412.50	505
0500	Heavy duty, pivoted, from 7'-7" to 12'-10" floor to floor		3	2.667		740	87.50		827.50	960
0600	16'-0" ceiling		2	4		1,525	131		1,656	1,900
0800	Economy folding, pine, 8'-6" ceiling		4	2		176	65.50		241.50	305
0900	9'-6" ceiling		4	2		196	65.50		261.50	325
1100	Automatic electric, aluminum, floor to floor height, 8' to 9'	2 Carp	1	16		8,550	525		9,075	10,300

11 32 Unit Kitchens

11 32 13 – Metal Unit Kitchens

11 32 13.10 Commercial Unit Kitchens

		Crew	Daily Output	Labor-Hours	Unit	Material	Labor	Equipment	Total	Total Incl O&P
0010	**COMMERCIAL UNIT KITCHENS**									
1500	Combination range, refrigerator and sink, 30" wide, minimum	L-1	2	5	Ea.	1,100	192		1,292	1,550
1550	Maximum		1	10		1,475	385		1,860	2,250
1570	60" wide, average		1.40	7.143		1,525	274		1,799	2,150
1590	72" wide, average		1.20	8.333		1,625	320		1,945	2,300

11 41 Foodservice Storage Equipment

11 41 13 – Refrigerated Food Storage Cases

11 41 13.30 Wine Cellar

		Crew	Daily Output	Labor-Hours	Unit	Material	Labor	Equipment	Total	Total Incl O&P
0010	**WINE CELLAR**, refrigerated, Redwood interior, carpeted, walk-in type									
0020	6'-8" high, including racks									
0200	80" W x 48" D for 900 bottles	2 Carp	1.50	10.667	Ea.	4,300	350		4,650	5,325
0250	80" W x 72" D for 1300 bottles		1.33	12.030		5,225	395		5,620	6,425
0300	80" W x 94" D for 1900 bottles		1.17	13.675		6,350	450		6,800	7,725

11 81 Facility Maintenance Equipment

11 81 19 – Vacuum Cleaning Systems

11 81 19.10 Vacuum Cleaning		Crew	Daily Output	Labor-Hours	Unit	Material	2015 Bare Costs Labor	Equipment	Total	Total Incl O&P
0010	**VACUUM CLEANING**									
0020	Central, 3 inlet, residential	1 Skwk	.90	8.889	Total	1,075	296		1,371	1,675
0400	5 inlet system, residential		.50	16		1,500	535		2,035	2,550
0600	7 inlet system, commercial		.40	20		1,700	665		2,365	3,000
0800	9 inlet system, residential	↓	.30	26.667	↓	3,750	890		4,640	5,625
4010	Rule of thumb: First 1200 S.F., installed								1,425	1,575
4020	For each additional S.F., add				S.F.				.26	.26

Division Notes

	CREW	DAILY OUTPUT	LABOR-HOURS	UNIT	BARE COSTS				TOTAL INCL O&P
					MAT.	LABOR	EQUIP.	TOTAL	

Estimating Tips
General

- The items in this division are usually priced per square foot or each. Most of these items are purchased by the owner and installed by the contractor. Do not assume the items in Division 12 will be purchased and installed by the contractor. Check the specifications for responsibilities and include receiving, storage, installation, and mechanical and electrical hookups in the appropriate divisions.

- Some items in this division require some type of support system that is not usually furnished with the item. Examples of these systems include blocking for the attachment of casework and heavy drapery rods. The required blocking must be added to the estimate in the appropriate division.

Reference Numbers
Reference numbers are shown in shaded boxes at the beginning of some major classifications. These numbers refer to related items in the Reference Section. The reference information may be an estimating procedure, an alternate pricing method, or technical information.

Note: Not all subdivisions listed here necessarily appear in this publication. ■

Did you know?
RSMeans Online gives you the same access to RSMeans' data with 24/7 access:
- Quickly locate costs in the searchable database.
- Build cost lists, estimates, and reports in minutes.
- Adjust costs to any location in the U.S. and Canada with the click of a button.

Start your free trial today at **www.rsmeansonline.com**

RSMeansOnline

12 21 Window Blinds

12 21 13 – Horizontal Louver Blinds

12 21 13.13 Metal Horizontal Louver Blinds	Crew	Daily Output	Labor-Hours	Unit	Material	2015 Bare Costs Labor	Equipment	Total	Total Incl O&P
0010 **METAL HORIZONTAL LOUVER BLINDS**									
0020 Horizontal, 1" aluminum slats, solid color, stock	1 Carp	590	.014	S.F.	4.90	.45		5.35	6.15
0070 Horizontal, 1" aluminum slats, custom color	"	590	.014	"	5.40	.45		5.85	6.70
12 21 13.33 Vinyl Horizontal Louver Blinds									
0010 **VINYL HORIZONTAL LOUVER BLINDS**									
0015 1" composite, 48" wide, 48" high	1 Carp	30	.267	Ea.	34	8.75		42.75	52
0020 72" high		30	.267		41.50	8.75		50.25	60.50
0030 60" wide, 60" high		30	.267		49.50	8.75		58.25	69
0040 72" high		30	.267		49.50	8.75		58.25	69
0050 72" wide x 72" high		30	.267		58	8.75		66.75	78.50

12 22 Curtains and Drapes

12 22 16 – Drapery Track and Accessories

12 22 16.10 Drapery Hardware	Crew	Daily Output	Labor-Hours	Unit	Material	2015 Bare Costs Labor	Equipment	Total	Total Incl O&P
0010 **DRAPERY HARDWARE**									
0030 Standard traverse, per foot, minimum	1 Carp	59	.136	L.F.	7	4.45		11.45	15.20
0100 Maximum		51	.157	"	10.20	5.15		15.35	19.90
0200 Decorative traverse, 28"-48", minimum		22	.364	Ea.	24.50	11.95		36.45	47
0220 Maximum		21	.381		50	12.50		62.50	76
0300 48"-84", minimum		20	.400		22.50	13.15		35.65	46.50
0320 Maximum		19	.421		64	13.85		77.85	93
0400 66"-120", minimum		18	.444		43	14.60		57.60	71.50
0420 Maximum		17	.471		97.50	15.45		112.95	133
0500 84"-156", minimum		16	.500		62.50	16.45		78.95	96.50
0520 Maximum		15	.533		133	17.50		150.50	176
0600 130"-240", minimum		14	.571		33.50	18.75		52.25	68.50
0620 Maximum		13	.615		184	20		204	236
0700 Slide rings, each, minimum					1.06			1.06	1.17
0720 Maximum					2.05			2.05	2.26
4000 Traverse rods, adjustable, 28" to 48"	1 Carp	22	.364		22.50	11.95		34.45	45
4020 48" to 84"		20	.400		29	13.15		42.15	53.50
4040 66" to 120"		18	.444		35	14.60		49.60	63
4060 84" to 156"		16	.500		40	16.45		56.45	71
4080 100" to 180"		14	.571		46.50	18.75		65.25	82.50
4100 228" to 312"		13	.615		64.50	20		84.50	105
4500 Curtain rod, 28" to 48", single		22	.364		5.50	11.95		17.45	26
4510 Double		22	.364		9.35	11.95		21.30	30.50
4520 48" to 86", single		20	.400		9.40	13.15		22.55	32.50
4530 Double		20	.400		15.65	13.15		28.80	39
4540 66" to 120", single		18	.444		15.70	14.60		30.30	42
4550 Double		18	.444		24.50	14.60		39.10	51.50
4600 Valance, pinch pleated fabric, 12" deep, up to 54" long, minimum					39			39	43
4610 Maximum					98			98	108
4620 Up to 77" long, minimum					60.50			60.50	66.50
4630 Maximum					158			158	174
5000 Stationary rods, first 2'					8.25			8.25	9.10

12 23 Interior Shutters

12 23 10 – Wood Interior Shutters

12 23 10.10 Wood Interior Shutters

		Crew	Daily Output	Labor-Hours	Unit	Material	2015 Bare Costs Labor	2015 Bare Costs Equipment	Total	Total Incl O&P
0010	**WOOD INTERIOR SHUTTERS**, louvered									
0200	Two panel, 27" wide, 36" high	1 Carp	5	1.600	Set	150	52.50		202.50	253
0300	33" wide, 36" high		5	1.600		194	52.50		246.50	300
0500	47" wide, 36" high		5	1.600		260	52.50		312.50	375
1000	Four panel, 27" wide, 36" high		5	1.600		220	52.50		272.50	330
1100	33" wide, 36" high		5	1.600		282	52.50		334.50	400
1300	47" wide, 36" high		5	1.600		375	52.50		427.50	505

12 23 10.13 Wood Panels

		Crew	Daily Output	Labor-Hours	Unit	Material	Labor	Equipment	Total	Total Incl O&P
0010	**WOOD PANELS**									
3000	Wood folding panels with movable louvers, 7" x 20" each	1 Carp	17	.471	Pr.	79.50	15.45		94.95	114
3300	8" x 28" each		17	.471		79.50	15.45		94.95	114
3450	9" x 36" each		17	.471		91.50	15.45		106.95	127
3600	10" x 40" each		17	.471		100	15.45		115.45	136
4000	Fixed louver type, stock units, 8" x 20" each		17	.471		94	15.45		109.45	129
4150	10" x 28" each		17	.471		79.50	15.45		94.95	114
4300	12" x 36" each		17	.471		94	15.45		109.45	129
4450	18" x 40" each		17	.471		134	15.45		149.45	174
5000	Insert panel type, stock, 7" x 20" each		17	.471		21	15.45		36.45	49
5150	8" x 28" each		17	.471		38	15.45		53.45	68
5300	9" x 36" each		17	.471		48.50	15.45		63.95	79
5450	10" x 40" each		17	.471		52	15.45		67.45	83
5600	Raised panel type, stock, 10" x 24" each		17	.471		247	15.45		262.45	298
5650	12" x 26" each		17	.471		247	15.45		262.45	298
5700	14" x 30" each		17	.471		273	15.45		288.45	325
5750	16" x 36" each		17	.471		300	15.45		315.45	360
6000	For custom built pine, add					22%				
6500	For custom built hardwood blinds, add					42%				

12 24 Window Shades

12 24 13 – Roller Window Shades

12 24 13.10 Shades

		Crew	Daily Output	Labor-Hours	Unit	Material	Labor	Equipment	Total	Total Incl O&P
0010	**SHADES**									
0020	Basswood, roll-up, stain finish, 3/8" slats	1 Carp	300	.027	S.F.	14.95	.88		15.83	17.85
5011	Insulative shades	G	125	.064		11.80	2.10		13.90	16.50
6011	Solar screening, fiberglass	G	85	.094		6.75	3.09		9.84	12.60
8011	Interior insulative shutter									
8111	Stock unit, 15" x 60"	G 1 Carp	17	.471	Pr.	11.90	15.45		27.35	39

12 32 Manufactured Wood Casework

12 32 16 – Manufactured Plastic-Laminate-Clad Casework

12 32 16.20 Plastic Laminate Casework Doors

		Crew	Daily Output	Labor-Hours	Unit	Material	Labor	Equipment	Total	Total Incl O&P
0010	**PLASTIC LAMINATE CASEWORK DOORS**									
1000	For casework frames, see Section 12 32 23.15									
1100	For casework hardware, see Section 12 32 23.35									
6000	Plastic laminate on particle board									
6100	12" wide, 18" high	1 Carp	25	.320	Ea.	24	10.50		34.50	44
6120	24" high		24	.333		32	10.95		42.95	53.50
6140	30" high		23	.348		40	11.45		51.45	63
6160	36" high		21	.381		48	12.50		60.50	73.50

12 32 Manufactured Wood Casework

12 32 16 – Manufactured Plastic-Laminate-Clad Casework

12 32 16.20 Plastic Laminate Casework Doors

		Crew	Daily Output	Labor-Hours	Unit	Material	2015 Bare Costs Labor	Equipment	Total	Total Incl O&P
6200	48" high	1 Carp	16	.500	Ea.	64	16.45		80.45	97.50
6250	60" high		13	.615		79.50	20		99.50	122
6300	72" high		12	.667		95.50	22		117.50	142
6320	15" wide, 18" high		24.50	.327		32	10.75		42.75	53
6340	24" high		23.50	.340		42.50	11.20		53.70	66
6360	30" high		22.50	.356		53	11.70		64.70	78
6380	36" high		20.50	.390		64	12.80		76.80	91.50
6400	48" high		15.50	.516		85	16.95		101.95	122
6450	60" high		12.50	.640		106	21		127	153
6480	72" high		11.50	.696		128	23		151	179
6500	18" wide, 18" high		24	.333		36	10.95		46.95	58
6550	24" high		23	.348		48	11.45		59.45	71.50
6600	30" high		22	.364		60	11.95		71.95	86
6650	36" high		20	.400		71.50	13.15		84.65	101
6700	48" high		15	.533		95.50	17.50		113	135
6750	60" high		12	.667		120	22		142	169
6800	72" high		11	.727		143	24		167	198

12 32 16.25 Plastic Laminate Drawer Fronts

		Crew	Daily Output	Labor-Hours	Unit	Material	Labor	Equipment	Total	Total Incl O&P
0010	**PLASTIC LAMINATE DRAWER FRONTS**									
2800	Plastic laminate on particle board front									
3000	4" high, 12" wide	1 Carp	17	.471	Ea.	4.51	15.45		19.96	31
3200	18" wide		16	.500		6.75	16.45		23.20	35
3600	24" wide		15	.533		9.05	17.50		26.55	39.50
3800	6" high, 12" wide		16	.500		6.85	16.45		23.30	35
4000	18" wide		15	.533		10.85	17.50		28.35	41.50
4500	24" wide		14	.571		13.70	18.75		32.45	46.50
4800	9" high, 12" wide		15	.533		10.25	17.50		27.75	41
5000	18" wide		14	.571		15.40	18.75		34.15	48.50
5200	24" wide		13	.615		20.50	20		40.50	56.50

12 32 23 – Hardwood Casework

12 32 23.10 Manufactured Wood Casework, Stock Units

		Crew	Daily Output	Labor-Hours	Unit	Material	Labor	Equipment	Total	Total Incl O&P
0010	**MANUFACTURED WOOD CASEWORK, STOCK UNITS**									
0700	Kitchen base cabinets, hardwood, not incl. counter tops,									
0710	24" deep, 35" high, prefinished									
0800	One top drawer, one door below, 12" wide	2 Carp	24.80	.645	Ea.	265	21		286	330
0820	15" wide		24	.667		276	22		298	340
0840	18" wide		23.30	.687		300	22.50		322.50	370
0860	21" wide		22.70	.705		305	23		328	380
0880	24" wide		22.30	.717		365	23.50		388.50	440
1000	Four drawers, 12" wide		24.80	.645		279	21		300	340
1020	15" wide		24	.667		283	22		305	345
1040	18" wide		23.30	.687		310	22.50		332.50	385
1060	24" wide		22.30	.717		345	23.50		368.50	420
1200	Two top drawers, two doors below, 27" wide		22	.727		390	24		414	470
1220	30" wide		21.40	.748		425	24.50		449.50	510
1240	33" wide		20.90	.766		440	25		465	525
1260	36" wide		20.30	.788		455	26		481	545
1280	42" wide		19.80	.808		480	26.50		506.50	575
1300	48" wide		18.90	.847		515	28		543	615
1500	Range or sink base, two doors below, 30" wide		21.40	.748		350	24.50		374.50	425
1520	33" wide		20.90	.766		375	25		400	455
1540	36" wide		20.30	.788		395	26		421	480

12 32 Manufactured Wood Casework

12 32 23 – Hardwood Casework

12 32 23.10 Manufactured Wood Casework, Stock Units

	Description	Crew	Daily Output	Labor-Hours	Unit	Material	2015 Bare Costs Labor	Equipment	Total	Total Incl O&P
1560	42" wide	2 Carp	19.80	.808	Ea.	415	26.50		441.50	500
1580	48" wide		18.90	.847		435	28		463	525
1800	For sink front units, deduct					161			161	177
2000	Corner base cabinets, 36" wide, standard	2 Carp	18	.889		625	29		654	740
2100	Lazy Susan with revolving door	"	16.50	.970		840	32		872	980
4000	Kitchen wall cabinets, hardwood, 12" deep with two doors									
4050	12" high, 30" wide	2 Carp	24.80	.645	Ea.	237	21		258	297
4100	36" wide		24	.667		282	22		304	345
4400	15" high, 30" wide		24	.667		241	22		263	300
4420	33" wide		23.30	.687		298	22.50		320.50	370
4440	36" wide		22.70	.705		290	23		313	360
4450	42" wide		22.70	.705		325	23		348	400
4700	24" high, 30" wide		23.30	.687		325	22.50		347.50	395
4720	36" wide		22.70	.705		355	23		378	430
4740	42" wide		22.30	.717		405	23.50		428.50	490
5000	30" high, one door, 12" wide		22	.727		216	24		240	278
5020	15" wide		21.40	.748		241	24.50		265.50	305
5040	18" wide		20.90	.766		265	25		290	335
5060	24" wide		20.30	.788		310	26		336	385
5300	Two doors, 27" wide		19.80	.808		340	26.50		366.50	420
5320	30" wide		19.30	.829		355	27		382	440
5340	36" wide		18.80	.851		405	28		433	495
5360	42" wide		18.50	.865		445	28.50		473.50	540
5380	48" wide		18.40	.870		500	28.50		528.50	600
6000	Corner wall, 30" high, 24" wide		18	.889		355	29		384	440
6050	30" wide		17.20	.930		380	30.50		410.50	465
6100	36" wide		16.50	.970		430	32		462	525
6500	Revolving Lazy Susan		15.20	1.053		480	34.50		514.50	585
7000	Broom cabinet, 84" high, 24" deep, 18" wide		10	1.600		650	52.50		702.50	805
7500	Oven cabinets, 84" high, 24" deep, 27" wide		8	2		1,000	65.50		1,065.50	1,200
7750	Valance board trim		396	.040	L.F.	13	1.33		14.33	16.55
7780	Toe kick trim	1 Carp	256	.031	"	2.79	1.03		3.82	4.79
7790	Base cabinet corner filler		16	.500	Ea.	41.50	16.45		57.95	73
7800	Cabinet filler, 3" x 24"		20	.400		18.05	13.15		31.20	42
7810	3" x 30"		20	.400		22.50	13.15		35.65	47
7820	3" x 42"		18	.444		31.50	14.60		46.10	59
7830	3" x 80"		16	.500		60	16.45		76.45	93.50
7850	Cabinet panel		50	.160	S.F.	8.65	5.25		13.90	18.35
9000	For deluxe models of all cabinets, add					40%				
9500	For custom built in place, add					25%	10%			
9558	Rule of thumb, kitchen cabinets not including									
9560	appliances & counter top, minimum	2 Carp	30	.533	L.F.	176	17.50		193.50	223
9600	Maximum	"	25	.640	"	395	21		416	470

12 32 23.15 Manufactured Wood Casework Frames

	Description	Crew	Daily Output	Labor-Hours	Unit	Material	Labor	Equipment	Total	Total Incl O&P
0010	**MANUFACTURED WOOD CASEWORK FRAMES**									
0050	Base cabinets, counter storage, 36" high									
0100	One bay, 18" wide	1 Carp	2.70	2.963	Ea.	171	97.50		268.50	350
0400	Two bay, 36" wide		2.20	3.636		261	119		380	490
1100	Three bay, 54" wide		1.50	5.333		310	175		485	635
2800	Bookcases, one bay, 7' high, 18" wide		2.40	3.333		201	110		311	405
3500	Two bay, 36" wide		1.60	5		292	164		456	595
4100	Three bay, 54" wide		1.20	6.667		485	219		704	900

12 32 Manufactured Wood Casework

12 32 23 – Hardwood Casework

12 32 23.15 Manufactured Wood Casework Frames		Crew	Daily Output	Labor-Hours	Unit	Material	2015 Bare Costs Labor	Equipment	Total	Total Incl O&P
6100	Wall mounted cabinet, one bay, 24" high, 18" wide	1 Carp	3.60	2.222	Ea.	110	73		183	244
6800	Two bay, 36" wide		2.20	3.636		161	119		280	380
7400	Three bay, 54" wide		1.70	4.706		201	155		356	480
8400	30" high, one bay, 18" wide		3.60	2.222		120	73		193	255
9000	Two bay, 36" wide		2.15	3.721		160	122		282	380
9400	Three bay, 54" wide		1.60	5		199	164		363	495
9800	Wardrobe, 7' high, single, 24" wide		2.70	2.963		222	97.50		319.50	405
9950	Partition, adjustable shelves & drawers, 48" wide		1.40	5.714		425	188		613	780

12 32 23.20 Manufactured Hardwood Casework Doors		Crew	Daily Output	Labor-Hours	Unit	Material	Labor	Equipment	Total	Total Incl O&P
0010	**MANUFACTURED HARDWOOD CASEWORK DOORS**									
2000	Glass panel, hardwood frame									
2200	12" wide, 18" high	1 Carp	34	.235	Ea.	27	7.75		34.75	42.50
2400	24" high		33	.242		36	7.95		43.95	53
2600	30" high		32	.250		45	8.20		53.20	63.50
2800	36" high		30	.267		54	8.75		62.75	74
3000	48" high		23	.348		72	11.45		83.45	98
3200	60" high		17	.471		90	15.45		105.45	125
3400	72" high		15	.533		108	17.50		125.50	149
3600	15" wide, 18" high		33	.242		34	7.95		41.95	50.50
3800	24" high		32	.250		45	8.20		53.20	63.50
4000	30" high		30	.267		56.50	8.75		65.25	76.50
4250	36" high		28	.286		67.50	9.40		76.90	90.50
4300	48" high		22	.364		90	11.95		101.95	119
4350	60" high		16	.500		113	16.45		129.45	152
4400	72" high		14	.571		135	18.75		153.75	181
4450	18" wide, 18" high		32	.250		40.50	8.20		48.70	58.50
4500	24" high		30	.267		54	8.75		62.75	74
4550	30" high		29	.276		67.50	9.05		76.55	89.50
4600	36" high		27	.296		81	9.75		90.75	105
4650	48" high		21	.381		108	12.50		120.50	140
4700	60" high		15	.533		135	17.50		152.50	179
4750	72" high		13	.615		162	20		182	212
5000	Hardwood, raised panel									
5100	12" wide, 18" high	1 Carp	16	.500	Ea.	28.50	16.45		44.95	59
5150	24" high		15.50	.516		38	16.95		54.95	70.50
5200	30" high		15	.533		47.50	17.50		65	82
5250	36" high		14	.571		57	18.75		75.75	94
5300	48" high		11	.727		76	24		100	124
5320	60" high		8	1		95	33		128	160
5340	72" high		7	1.143		114	37.50		151.50	188
5360	15" wide, 18" high		15.50	.516		35.50	16.95		52.45	67.50
5380	24" high		15	.533		47.50	17.50		65	82
5400	30" high		14.50	.552		59.50	18.10		77.60	96
5420	36" high		13.50	.593		71.50	19.45		90.95	111
5440	48" high		10.50	.762		95	25		120	147
5460	60" high		7.50	1.067		119	35		154	190
5480	72" high		6.50	1.231		143	40.50		183.50	225
5500	18" wide, 18" high		15	.533		43	17.50		60.50	76.50
5550	24" high		14.50	.552		57	18.10		75.10	93
5600	30" high		14	.571		71.50	18.75		90.25	110
5650	36" high		13	.615		85.50	20		105.50	128
5700	48" high		10	.800		114	26.50		140.50	169

12 32 Manufactured Wood Casework

12 32 23 – Hardwood Casework

12 32 23.20 Manufactured Hardwood Casework Doors

		Crew	Daily Output	Labor-Hours	Unit	Material	2015 Bare Costs Labor	Equipment	Total	Total Incl O&P
5750	60" high	1 Carp	7	1.143	Ea.	143	37.50		180.50	220
5800	72" high		6	1.333		171	44		215	262

12 32 23.25 Manufactured Wood Casework Drawer Fronts

		Crew	Daily Output	Labor-Hours	Unit	Material	Labor	Equipment	Total	Total Incl O&P
0010	**MANUFACTURED WOOD CASEWORK DRAWER FRONTS**									
0100	Solid hardwood front									
1000	4" high, 12" wide	1 Carp	17	.471	Ea.	4.33	15.45		19.78	31
1200	18" wide		16	.500		6.50	16.45		22.95	34.50
1400	24" wide		15	.533		8.65	17.50		26.15	39
1600	6" high, 12" wide		16	.500		6.50	16.45		22.95	34.50
1800	18" wide		15	.533		9.75	17.50		27.25	40.50
2000	24" wide		14	.571		13	18.75		31.75	46
2200	9" high, 12" wide		15	.533		9.75	17.50		27.25	40.50
2400	18" wide		14	.571		14.60	18.75		33.35	47.50
2600	24" wide		13	.615		19.50	20		39.50	55.50

12 32 23.30 Manufactured Wood Casework Vanities

		Crew	Daily Output	Labor-Hours	Unit	Material	Labor	Equipment	Total	Total Incl O&P
0010	**MANUFACTURED WOOD CASEWORK VANITIES**									
8000	Vanity bases, 2 doors, 30" high, 21" deep, 24" wide	2 Carp	20	.800	Ea.	310	26.50		336.50	385
8050	30" wide		16	1		370	33		403	460
8100	36" wide		13.33	1.200		360	39.50		399.50	460
8150	48" wide		11.43	1.400		470	46		516	595
9000	For deluxe models of all vanities, add to above					40%				
9500	For custom built in place, add to above					25%	10%			

12 32 23.35 Manufactured Wood Casework Hardware

		Crew	Daily Output	Labor-Hours	Unit	Material	Labor	Equipment	Total	Total Incl O&P
0010	**MANUFACTURED WOOD CASEWORK HARDWARE**									
1000	Catches, minimum	1 Carp	235	.034	Ea.	1.22	1.12		2.34	3.22
1020	Average		119.40	.067		3.96	2.20		6.16	8.05
1040	Maximum		80	.100		7.55	3.29		10.84	13.80
2000	Door/drawer pulls, handles									
2200	Handles and pulls, projecting, metal, minimum	1 Carp	48	.167	Ea.	5	5.50		10.50	14.70
2220	Average		42	.190		7.75	6.25		14	19.05
2240	Maximum		36	.222		10.60	7.30		17.90	24
2300	Wood, minimum		48	.167		5.25	5.50		10.75	15
2320	Average		42	.190		7.05	6.25		13.30	18.25
2340	Maximum		36	.222		9.65	7.30		16.95	23
2600	Flush, metal, minimum		48	.167		5.25	5.50		10.75	15
2620	Average		42	.190		7.05	6.25		13.30	18.25
2640	Maximum		36	.222		9.65	7.30		16.95	23
3000	Drawer tracks/glides, minimum		48	.167	Pr.	8.95	5.50		14.45	19
3020	Average		32	.250		15.15	8.20		23.35	30.50
3040	Maximum		24	.333		26	10.95		36.95	47
4000	Cabinet hinges, minimum		160	.050		3.02	1.64		4.66	6.10
4020	Average		95.24	.084		6.95	2.76		9.71	12.30
4040	Maximum		68	.118		11.45	3.86		15.31	19.10

12 34 Manufactured Plastic Casework

12 34 16 – Manufactured Solid-Plastic Casework

12 34 16.10 Outdoor Casework		Crew	Daily Output	Labor-Hours	Unit	Material	2015 Bare Costs Labor	Equipment	Total	Total Incl O&P
0010	**OUTDOOR CASEWORK**									
0020	Cabinet, base, sink/range, 36"	2 Carp	20.30	.788	Ea.	1,875	26		1,901	2,100
0100	Base, 36"		20.30	.788		2,100	26		2,126	2,350
0200	Filler strip, 1" x 30"		158	.101		32.50	3.33		35.83	41

12 36 Countertops

12 36 16 – Metal Countertops

12 36 16.10 Stainless Steel Countertops

		Crew	Daily Output	Labor-Hours	Unit	Material	Labor	Equipment	Total	Total Incl O&P
0010	**STAINLESS STEEL COUNTERTOPS**									
3200	Stainless steel, custom	1 Carp	24	.333	S.F.	153	10.95		163.95	186

12 36 19 – Wood Countertops

12 36 19.10 Maple Countertops

0010	**MAPLE COUNTERTOPS**									
2900	Solid, laminated, 1-1/2" thick, no splash	1 Carp	28	.286	L.F.	75.50	9.40		84.90	99
3000	With square splash		28	.286	"	90	9.40		99.40	114
3400	Recessed cutting block with trim, 16" x 20" x 1"		8	1	Ea.	92	33		125	156
3411	Replace cutting block only		16	.500	"	69	16.45		85.45	103

12 36 23 – Plastic Countertops

12 36 23.13 Plastic-Laminate-Clad Countertops

0010	**PLASTIC-LAMINATE-CLAD COUNTERTOPS**									
0020	Stock, 24" wide w/backsplash, minimum	1 Carp	30	.267	L.F.	17	8.75		25.75	33.50
0100	Maximum		25	.320		34.50	10.50		45	55.50
0300	Custom plastic, 7/8" thick, aluminum molding, no splash		30	.267		30	8.75		38.75	47.50
0400	Cove splash		30	.267		29	8.75		37.75	46.50
0600	1-1/4" thick, no splash		28	.286		35.50	9.40		44.90	55
0700	Square splash		28	.286		42.50	9.40		51.90	62.50
0900	Square edge, plastic face, 7/8" thick, no splash		30	.267		33	8.75		41.75	51
1000	With splash		30	.267		39.50	8.75		48.25	57.50
1200	For stainless channel edge, 7/8" thick, add					3.12			3.12	3.43
1300	1-1/4" thick, add					3.72			3.72	4.09
1500	For solid color suede finish, add					4.08			4.08	4.49
1700	For end splash, add				Ea.	18.35			18.35	20
1901	For cut outs, standard, add, minimum	1 Carp	32	.250			8.20		8.20	13.80
2000	Maximum		8	1		6.10	33		39.10	62
2010	Cut out in backsplash for elec. wall outlet		38	.211			6.90		6.90	11.60
2020	Cut out for sink		20	.400			13.15		13.15	22
2030	Cut out for stove top		18	.444			14.60		14.60	24.50
2100	Postformed, including backsplash and front edge		30	.267	L.F.	10.20	8.75		18.95	26
2110	Mitred, add		12	.667	Ea.		22		22	37
2200	Built-in place, 25" wide, plastic laminate		25	.320	L.F.	41.50	10.50		52	63

12 36 33 – Tile Countertops

12 36 33.10 Ceramic Tile Countertops

0010	**CERAMIC TILE COUNTERTOPS**									
2300	Ceramic tile mosaic	1 Carp	25	.320	L.F.	33.50	10.50		44	54.50

12 36 40 – Stone Countertops

12 36 40.10 Natural Stone Countertops

0010	**NATURAL STONE COUNTERTOPS**									
2500	Marble, stock, with splash, 1/2" thick, minimum	1 Bric	17	.471	L.F.	42	15.20		57.20	71.50
2700	3/4" thick, maximum		13	.615		105	19.90		124.90	149

12 36 Countertops

12 36 40 – Stone Countertops

	12 36 40.10 Natural Stone Countertops	Crew	Daily Output	Labor-Hours	Unit	Material	2015 Bare Costs Labor	Equipment	Total	Total Incl O&P
2800	Granite, average, 1-1/4" thick, 24" wide, no splash	1 Bric	13.01	.615	L.F.	138	19.85		157.85	184

12 36 61 – Simulated Stone Countertops

12 36 61.16 Solid Surface Countertops

		Crew	Daily Output	Labor-Hours	Unit	Material	Labor	Equipment	Total	Total Incl O&P
0010	**SOLID SURFACE COUNTERTOPS**, Acrylic polymer									
2000	Pricing for order of 1 – 50 L.F.									
2100	25" wide, solid colors	2 Carp	20	.800	L.F.	73.50	26.50		100	125
2200	Patterned colors		20	.800		93.50	26.50		120	147
2300	Premium patterned colors		20	.800		117	26.50		143.50	173
2400	With silicone attached 4" backsplash, solid colors		19	.842		81	27.50		108.50	136
2500	Patterned colors		19	.842		102	27.50		129.50	160
2600	Premium patterned colors		19	.842		128	27.50		155.50	187
2700	With hard seam attached 4" backsplash, solid colors		15	1.067		81	35		116	148
2800	Patterned colors		15	1.067		102	35		137	172
2900	Premium patterned colors		15	1.067		128	35		163	199
3800	Sinks, pricing for order of 1 – 50 units									
3900	Single bowl, hard seamed, solid colors, 13" x 17"	1 Carp	2	4	Ea.	500	131		631	770
4000	10" x 15"		4.55	1.758		230	58		288	350
4100	Cutouts for sinks		5.25	1.524			50		50	84

12 36 61.17 Solid Surface Vanity Tops

		Crew	Daily Output	Labor-Hours	Unit	Material	Labor	Equipment	Total	Total Incl O&P
0010	**SOLID SURFACE VANITY TOPS**									
0015	Solid surface, center bowl, 17" x 19"	1 Carp	12	.667	Ea.	190	22		212	246
0020	19" x 25"		12	.667		194	22		216	251
0030	19" x 31"		12	.667		227	22		249	287
0040	19" x 37"		12	.667		264	22		286	330
0050	22" x 25"		10	.800		345	26.50		371.50	425
0060	22" x 31"		10	.800		405	26.50		431.50	490
0070	22" x 37"		10	.800		470	26.50		496.50	565
0080	22" x 43"		10	.800		535	26.50		561.50	635
0090	22" x 49"		10	.800		595	26.50		621.50	700
0110	22" x 55"		8	1		675	33		708	800
0120	22" x 61"		8	1		770	33		803	905
0220	Double bowl, 22" x 61"		8	1		870	33		903	1,025
0230	Double bowl, 22" x 73"		8	1		950	33		983	1,100
0240	For aggregate colors, add					35%				
0250	For faucets and fittings, see Section 22 41 39.10									

12 36 61.19 Quartz Agglomerate Countertops

		Crew	Daily Output	Labor-Hours	Unit	Material	Labor	Equipment	Total	Total Incl O&P
0010	**QUARTZ AGGLOMERATE COUNTERTOPS**									
0100	25" wide, 4" backsplash, color group A, minimum	2 Carp	15	1.067	L.F.	64.50	35		99.50	130
0110	Maximum		15	1.067		90	35		125	158
0120	Color group B, minimum		15	1.067		66.50	35		101.50	133
0130	Maximum		15	1.067		94.50	35		129.50	163
0140	Color group C, minimum		15	1.067		78	35		113	145
0150	Maximum		15	1.067		107	35		142	176
0160	Color group D, minimum		15	1.067		84.50	35		119.50	152
0170	Maximum		15	1.067		115	35		150	185

Division Notes

		CREW	DAILY OUTPUT	LABOR-HOURS	UNIT	BARE COSTS				TOTAL INCL O&P
						MAT.	LABOR	EQUIP.	TOTAL	

Division 13 – Special Construction

Estimating Tips
General

- The items and systems in this division are usually estimated, purchased, supplied, and installed as a unit by one or more subcontractors. The estimator must ensure that all parties are operating from the same set of specifications and assumptions, and that all necessary items are estimated and will be provided. Many times the complex items and systems are covered, but the more common ones, such as excavation or a crane, are overlooked for the very reason that everyone assumes nobody could miss them. The estimator should be the central focus and be able to ensure that all systems are complete.

- Another area where problems can develop in this division is at the interface between systems. The estimator must ensure, for instance, that anchor bolts, nuts, and washers are estimated and included for the air-supported structures and pre-engineered buildings to be bolted to their foundations. Utility supply is a common area where essential items or pieces of equipment can be missed or overlooked, because each subcontractor may feel it is another's responsibility. The estimator should also be aware of certain items which may be supplied as part of a package but installed by others, and ensure that the installing contractor's estimate includes the cost of installation. Conversely, the estimator must also ensure that items are not costed by two different subcontractors, resulting in an inflated overall estimate.

13 30 00 Special Structures

- The foundations and floor slab, as well as rough mechanical and electrical, should be estimated, as this work is required for the assembly and erection of the structure. Generally, as noted in the book, the pre-engineered building comes as a shell. Pricing is based on the size and structural design parameters stated in the reference section. Additional features, such as windows and doors with their related structural framing, must also be included by the estimator. Here again, the estimator must have a clear understanding of the scope of each portion of the work and all the necessary interfaces.

Reference Numbers

Reference numbers are shown in shaded boxes at the beginning of some major classifications. These numbers refer to related items in the Reference Section. The reference information may be an estimating procedure, an alternate pricing method, or technical information.

Note: Not all subdivisions listed here necessarily appear in this publication. ■

Did you know?
RSMeans Online gives you the same access to RSMeans' data with 24/7 access:
- Quickly locate costs in the searchable database.
- Build cost lists, estimates, and reports in minutes.
- Adjust costs to any location in the U.S. and Canada with the click of a button.

Start your free trial today at **www.rsmeansonline.com**

RSMeansOnline

13 11 Swimming Pools

13 11 13 – Below-Grade Swimming Pools

13 11 13.50 Swimming Pools

		Crew	Daily Output	Labor-Hours	Unit	Material	2015 Bare Costs Labor	Equipment	Total	Total Incl O&P
0010	**SWIMMING POOLS** Residential in-ground, vinyl lined, concrete									
0020	Swimming pools, resi in-ground, vyl lined, conc sides, W/ equip, sand bot	B-52	300	.187	SF Surf	23.50	5.10	1.98	30.58	36.50
0100	Metal or polystyrene sides [R131113-20]	B-14	410	.117		19.55	3.08	.89	23.52	27.50
0200	Add for vermiculite bottom					1.49			1.49	1.64
0500	Gunite bottom and sides, white plaster finish									
0600	12' x 30' pool	B-52	145	.386	SF Surf	43.50	10.60	4.10	58.20	70.50
0720	16' x 32' pool		155	.361		39	9.90	3.83	52.73	64
0750	20' x 40' pool		250	.224		35	6.15	2.38	43.53	51.50
0810	Concrete bottom and sides, tile finish									
0820	12' x 30' pool	B-52	80	.700	SF Surf	44	19.20	7.45	70.65	89
0830	16' x 32' pool		95	.589		36.50	16.20	6.25	58.95	74
0840	20' x 40' pool		130	.431		29	11.80	4.57	45.37	57
1600	For water heating system, see Section 23 52 28.10									
1700	Filtration and deck equipment only, as % of total				Total				20%	20%
1800	Deck equipment, rule of thumb, 20' x 40' pool				SF Pool				1.18	1.30
3000	Painting pools, preparation + 3 coats, 20' x 40' pool, epoxy	2 Pord	.33	48.485	Total	1,775	1,400		3,175	4,250
3100	Rubber base paint, 18 gallons	"	.33	48.485	"	1,225	1,400		2,625	3,650

13 11 46 – Swimming Pool Accessories

13 11 46.50 Swimming Pool Equipment

		Crew	Daily Output	Labor-Hours	Unit	Material	Labor	Equipment	Total	Total Incl O&P
0010	**SWIMMING POOL EQUIPMENT**									
0020	Diving stand, stainless steel, 3 meter	2 Carp	.40	40	Ea.	15,100	1,325		16,425	18,800
0300	1 meter		2.70	5.926		9,175	195		9,370	10,400
0600	Diving boards, 16' long, aluminum		2.70	5.926		3,950	195		4,145	4,675
0700	Fiberglass		2.70	5.926		3,250	195		3,445	3,900
0800	14' long, aluminum		2.70	5.926		3,575	195		3,770	4,250
0850	Fiberglass		2.70	5.926		3,225	195		3,420	3,850
1200	Ladders, heavy duty, stainless steel, 2 tread		7	2.286		810	75		885	1,025
1500	4 tread		6	2.667		1,075	87.50		1,162.50	1,325
2100	Lights, underwater, 12 volt, with transformer, 300 watt	1 Elec	1	8		330	293		623	845
2200	110 volt, 500 watt, standard	"	1	8		294	293		587	805
3000	Pool covers, reinforced vinyl	3 Clab	1800	.013	S.F.	1.13	.33		1.46	1.79
3100	Vinyl, for winter, 400 SF max pool surface		3200	.008		.26	.18		.44	.60
3200	With water tubes, 400SF max pool surface		3000	.008		.29	.20		.49	.65
3300	Slides, tubular, fiberglass, aluminum handrails & ladder, 5'-0", straight	2 Carp	1.60	10	Ea.	3,575	330		3,905	4,500
3320	8'-0", curved	"	3	5.333	"	7,225	175		7,400	8,250

13 12 Fountains

13 12 13 – Exterior Fountains

13 12 13.10 Yard Fountains

		Crew	Daily Output	Labor-Hours	Unit	Material	Labor	Equipment	Total	Total Incl O&P
0010	**YARD FOUNTAINS**									
0100	Outdoor fountain, 48" high with bowl and figures	2 Clab	2	8	Ea.	570	196		766	955

13 17 Tubs and Pools

13 17 13 – Hot Tubs

13 17 13.10 Redwood Hot Tub System

		Crew	Daily Output	Labor-Hours	Unit	Material	Labor	Equipment	Total	Total Incl O&P
0010	**REDWOOD HOT TUB SYSTEM**									
7050	4' diameter x 4' deep	Q-1	1	16	Ea.	3,225	560		3,785	4,475
7150	6' diameter x 4' deep		.80	20		4,950	700		5,650	6,575
7200	8' diameter x 4' deep		.80	20		7,250	700		7,950	9,125

13 17 33 – Whirlpool Tubs

13 17 33.10 Whirlpool Bath

		Crew	Daily Output	Labor-Hours	Unit	Material	Labor	Equipment	Total	Total Incl O&P
0010	**WHIRLPOOL BATH**									
6000	Whirlpool, bath with vented overflow, molded fiberglass									
6100	66" x 36" x 24"	Q-1	1	16	Ea.	3,475	560		4,035	4,750
6400	72" x 36" x 21"		1	16		1,800	560		2,360	2,925
6500	60" x 34" x 21"		1	16		1,800	560		2,360	2,900
6600	72" x 42" x 23"		1	16		2,175	560		2,735	3,325

13 24 Special Activity Rooms

13 24 16 – Saunas

13 24 16.50 Saunas and Heaters

		Crew	Daily Output	Labor-Hours	Unit	Material	Labor	Equipment	Total	Total Incl O&P
0010	**SAUNAS AND HEATERS**									
0020	Prefabricated, incl. heater & controls, 7' high, 6' x 4', C/C	L-7	2.20	11.818	Ea.	5,175	330		5,505	6,250
0050	6' x 4', C/P		2	13		4,700	365		5,065	5,750
0400	6' x 5', C/C		2	13		5,875	365		6,240	7,075
0450	6' x 5', C/P		2	13		5,325	365		5,690	6,450
0600	6' x 6', C/C		1.80	14.444		6,225	405		6,630	7,525
0650	6' x 6', C/P		1.80	14.444		5,675	405		6,080	6,900
0800	6' x 9', C/C		1.60	16.250		7,975	455		8,430	9,550
0850	6' x 9', C/P		1.60	16.250		7,225	455		7,680	8,700
1000	8' x 12', C/C		1.10	23.636		11,700	660		12,360	14,000
1050	8' x 12', C/P		1.10	23.636		10,500	660		11,160	12,700
1200	8' x 8', C/C		1.40	18.571		9,175	520		9,695	11,000
1250	8' x 8', C/P		1.40	18.571		8,450	520		8,970	10,200
1400	8' x 10', C/C		1.20	21.667		10,200	605		10,805	12,200
1450	8' x 10', C/P		1.20	21.667		9,250	605		9,855	11,200
1600	10' x 12', C/C		1	26		12,200	725		12,925	14,700
1650	10' x 12', C/P		1	26		11,000	725		11,725	13,300
1700	Door only, cedar, 2'x6', with 1'x4' tempered insulated glass window	2 Carp	3.40	4.706		750	155		905	1,075
1800	Prehung, incl. jambs, pulls & hardware	"	12	1.333		745	44		789	895
2500	Heaters only (incl. above), wall mounted, to 200 C.F.					685			685	755
2750	To 300 C.F.					930			930	1,025
3000	Floor standing, to 720 C.F., 10,000 watts, w/controls	1 Elec	3	2.667		2,950	97.50		3,047.50	3,400
3250	To 1,000 C.F., 16,000 watts	"	3	2.667		3,825	97.50		3,922.50	4,350

13 24 26 – Steam Baths

13 24 26.50 Steam Baths and Components

		Crew	Daily Output	Labor-Hours	Unit	Material	Labor	Equipment	Total	Total Incl O&P
0010	**STEAM BATHS AND COMPONENTS**									
0020	Heater, timer & head, single, to 140 C.F.	1 Plum	1.20	6.667	Ea.	2,200	258		2,458	2,825
0500	To 300 C.F.	"	1.10	7.273		2,425	282		2,707	3,150
2700	Conversion unit for residential tub, including door					3,550			3,550	3,925

13 34 Fabricated Engineered Structures

13 34 13 – Glazed Structures

13 34 13.13 Greenhouses

		Crew	Daily Output	Labor-Hours	Unit	Material	2015 Bare Costs Labor	2015 Bare Costs Equipment	Total	Total Incl O&P
0010	**GREENHOUSES**, Shell only, stock units, not incl. 2' stub walls,									
0020	foundation, floors, heat or compartments									
0300	Residential type, free standing, 8'-6" long x 7'-6" wide	2 Carp	59	.271	SF Flr.	20	8.90		28.90	37
0400	10'-6" wide		85	.188		37	6.20		43.20	51
0600	13'-6" wide		108	.148		39	4.87		43.87	50.50
0700	17'-0" wide		160	.100		43.50	3.29		46.79	53.50
0900	Lean-to type, 3'-10" wide		34	.471		41.50	15.45		56.95	72
1000	6'-10" wide		58	.276		50	9.05		59.05	70
1100	Wall mounted to existing window, 3' x 3'	1 Carp	4	2	Ea.	1,575	65.50		1,640.50	1,850
1120	4' x 5'	"	3	2.667	"	2,025	87.50		2,112.50	2,375
1200	Deluxe quality, free standing, 7'-6" wide	2 Carp	55	.291	SF Flr.	81.50	9.55		91.05	106
1220	10'-6" wide		81	.198		61	6.50		67.50	78
1240	13'-6" wide		104	.154		55	5.05		60.05	69
1260	17'-0" wide		150	.107		56.50	3.50		60	68.50
1400	Lean-to type, 3'-10" wide		31	.516		105	16.95		121.95	145
1420	6'-10" wide		55	.291		66.50	9.55		76.05	89
1440	8'-0" wide		97	.165		56	5.40		61.40	70.50

13 34 13.19 Swimming Pool Enclosures

		Crew	Daily Output	Labor-Hours	Unit	Material	Labor	Equipment	Total	Total Incl O&P
0010	**SWIMMING POOL ENCLOSURES** Translucent, free standing									
0020	not including foundations, heat or light									
0200	Economy	2 Carp	200	.080	SF Hor.	37	2.63		39.63	45.50
0600	Deluxe	"	70	.229	"	93	7.50		100.50	115

13 34 63 – Natural Fiber Construction

13 34 63.50 Straw Bale Construction

			Crew	Daily Output	Labor-Hours	Unit	Material	Labor	Equipment	Total	Total Incl O&P
0010	**STRAW BALE CONSTRUCTION**										
2020	Straw bales in walls w/modified post and beam frame	G	2 Carp	320	.050	S.F.	6.15	1.64		7.79	9.50

Estimating Tips
General
- Many products in Division 14 will require some type of support or blocking for installation not included with the item itself. Examples are supports for conveyors or tube systems, attachment points for lifts, and footings for hoists or cranes. Add these supports in the appropriate division.

14 10 00 Dumbwaiters
14 20 00 Elevators
- Dumbwaiters and elevators are estimated and purchased in a method similar to buying a car. The manufacturer has a base unit with standard features. Added to this base unit price will be whatever options the owner or specifications require. Increased load capacity, additional vertical travel, additional stops, higher speed, and cab finish options are items to be considered. When developing an estimate for dumbwaiters and elevators, remember that some items needed by the installers may have to be included as part of the general contract.

Examples are:
- shaftway
- rail support brackets
- machine room
- electrical supply
- sill angles
- electrical connections
- pits
- roof penthouses
- pit ladders

Check the job specifications and drawings before pricing.
- Installation of elevators and handicapped lifts in historic structures can require significant additional costs. The associated structural requirements may involve cutting into and repairing finishes, moldings, flooring, etc. The estimator must account for these special conditions.

14 30 00 Escalators and Moving Walks
- Escalators and moving walks are specialty items installed by specialty contractors. There are numerous options associated with these items. For specific options, contact a manufacturer or contractor. In a method similar to estimating dumbwaiters and elevators, you should verify the extent of general contract work and add items as necessary.

14 40 00 Lifts
14 90 00 Other Conveying Equipment
- Products such as correspondence lifts, chutes, and pneumatic tube systems, as well as other items specified in this subdivision, may require trained installers. The general contractor might not have any choice as to who will perform the installation, or when it will be performed. Long lead times are often required for these products, making early decisions in scheduling necessary.

Reference Numbers
Reference numbers are shown in shaded boxes at the beginning of some major classifications. These numbers refer to related items in the Reference Section. The reference information may be an estimating procedure, an alternate pricing method, or technical information.

Note: Not all subdivisions listed here necessarily appear in this publication. ■

Estimate with precision with RSMeans Online...
Quick, intuitive and easy-to-use, RSMeans Online gives you instant access to hundreds of thousands of material, labor, and equipment costs from RSMeans' comprehensive database, delivering the information you need to build competitive estimates every time.

Start your free trial today at **www.rsmeansonline.com**

RSMeansOnline

No part of this publication may be reproduced, stored in a retrieval system, or transmitted in any form or by any means without prior written permission of RSMeans.

14 21 Electric Traction Elevators

14 21 33 – Electric Traction Residential Elevators

14 21 33.20 Residential Elevators	Crew	Daily Output	Labor-Hours	Unit	Material	2015 Bare Costs Labor	Equipment	Total	Total Incl O&P
0010 **RESIDENTIAL ELEVATORS**									
7000 Residential, cab type, 1 floor, 2 stop, economy model	2 Elev	.20	80	Ea.	11,400	4,100		15,500	19,300
7100 Custom model		.10	160		19,300	8,200		27,500	34,600
7200 2 floor, 3 stop, economy model		.12	133		17,000	6,825		23,825	29,900
7300 Custom model	↓	.06	266	↓	27,700	13,700		41,400	53,000

14 42 Wheelchair Lifts

14 42 13 – Inclined Wheelchair Lifts

14 42 13.10 Inclined Wheelchair Lifts and Stairclimbers

	Crew	Daily Output	Labor-Hours	Unit	Material	Labor	Equipment	Total	Total Incl O&P
0010 **INCLINED WHEELCHAIR LIFTS AND STAIRCLIMBERS**									
7700 Stair climber (chair lift), single seat, minimum	2 Elev	1	16	Ea.	5,325	820		6,145	7,225
7800 Maximum	"	.20	80	"	7,350	4,100		11,450	14,800

Estimating Tips

Pipe for fire protection and all uses is located in Subdivisions 21 11 13 and 22 11 13.

The labor adjustment factors listed in Subdivision 22 01 02.20 also apply to Division 21.

Many, but not all, areas in the U.S. require backflow protection in the fire system. It is advisable to check local building codes for specific requirements.

For your reference, the following is a list of the most applicable Fire Codes and Standards which may be purchased from the NFPA, 1 Batterymarch Park, Quincy, MA 02169-7471.

- NFPA 1: Uniform Fire Code
- NFPA 10: Portable Fire Extinguishers
- NFPA 11: Low-, Medium-, and High-Expansion Foam
- NFPA 12: Carbon Dioxide Extinguishing Systems (Also companion 12A)
- NFPA 13: Installation of Sprinkler Systems (Also companion 13D, 13E, and 13R)
- NFPA 14: Installation of Standpipe and Hose Systems
- NFPA 15: Water Spray Fixed Systems for Fire Protection
- NFPA 16: Installation of Foam-Water Sprinkler and Foam-Water Spray Systems
- NFPA 17: Dry Chemical Extinguishing Systems (Also companion 17A)
- NFPA 18: Wetting Agents
- NFPA 20: Installation of Stationary Pumps for Fire Protection
- NFPA 22: Water Tanks for Private Fire Protection
- NFPA 24: Installation of Private Fire Service Mains and their Appurtenances
- NFPA 25: Inspection, Testing and Maintenance of Water-Based Fire Protection

Reference Numbers

Reference numbers are shown in shaded boxes at the beginning of some major classifications. These numbers refer to related items in the Reference Section. The reference information may be an estimating procedure, an alternate pricing method, or technical information.

Note: Not all subdivisions listed here necessarily appear in this publication. ■

Did you know?

RSMeans Online gives you the same access to RSMeans' data with 24/7 access:
- Quickly locate costs in the searchable database.
- Build cost lists, estimates, and reports in minutes.
- Adjust costs to any location in the U.S. and Canada with the click of a button.

Start your free trial today at **www.rsmeansonline.com**

21 05 Common Work Results for Fire Suppression

21 05 23 – General-Duty Valves for Water-Based Fire-Suppression Piping

21 05 23.50 General-Duty Valves

		Crew	Daily Output	Labor-Hours	Unit	Material	2015 Bare Costs Labor	2015 Bare Costs Equipment	Total	Total Incl O&P
0010	**GENERAL-DUTY VALVES**, for water-based fire suppression									
6200	Valves and components									
6210	Alarm, includes									
6220	retard chamber, trim, gauges, alarm line strainer									
6260	3" size	Q-12	3	5.333	Ea.	1,625	178		1,803	2,100
6280	4" size	"	2	8		1,700	267		1,967	2,325
6300	6" size	Q-13	4	8	↓	1,900	267		2,167	2,550
6500	Check, swing, C.I. body, brass fittings, auto. ball drip									
6520	4" size	Q-12	3	5.333	Ea.	350	178		528	680
6800	Check, wafer, butterfly type, C.I. body, bronze fittings									
6820	4" size	Q-12	4	4	Ea.	1,000	133		1,133	1,325
8800	Flow control valve, includes trim and gauges, 2" size		2	8		4,700	267		4,967	5,625
8820	3" size	↓	1.50	10.667		5,175	355		5,530	6,275
8840	4" size	Q-13	2.80	11.429		5,775	380		6,155	6,975
8860	6" size	"	2	16		6,675	535		7,210	8,225
9200	Pressure operated relief valve, brass body	1 Spri	18	.444		580	16.45		596.45	660
9600	Waterflow indicator, with recycling retard and									
9610	two single pole retard switches, 2" thru 6" pipe size	1 Spri	8	1	Ea.	137	37		174	213

21 11 Facility Fire-Suppression Water-Service Piping

21 11 13 – Facility Water Distribution Piping

21 11 13.16 Pipe, Plastic

		Crew	Daily Output	Labor-Hours	Unit	Material	2015 Bare Costs Labor	2015 Bare Costs Equipment	Total	Total Incl O&P
0010	**PIPE, PLASTIC**									
0020	CPVC, fire suppression, (C-UL-S. FM, NFPA 13, 13D & 13R)									
0030	Socket joint, no couplings or hangers									
0100	SDR 13.5, (ASTM F442)									
0120	3/4" diameter	Q-12	420	.038	L.F.	1.67	1.27		2.94	3.94
0130	1" diameter		340	.047		2.58	1.57		4.15	5.45
0140	1-1/4" diameter		260	.062		4.10	2.05		6.15	7.90
0150	1-1/2" diameter		190	.084		5.65	2.81		8.46	10.90
0160	2" diameter		140	.114		9	3.81		12.81	16.20
0170	2-1/2" diameter		130	.123		16.65	4.10		20.75	25
0180	3" diameter	↓	120	.133	↓	25.50	4.45		29.95	35.50

21 11 13.18 Pipe Fittings, Plastic

		Crew	Daily Output	Labor-Hours	Unit	Material	2015 Bare Costs Labor	2015 Bare Costs Equipment	Total	Total Incl O&P
0010	**PIPE FITTINGS, PLASTIC**									
0020	CPVC, fire suppression, (C-UL-S. FM, NFPA 13, 13D & 13R)									
0030	Socket joint									
0100	90° Elbow									
0120	3/4"	1 Plum	26	.308	Ea.	2.10	11.90		14	22
0130	1"		22.70	.352		4.62	13.65		18.27	27.50
0140	1-1/4"		20.20	.396		5.85	15.35		21.20	32
0150	1-1/2"		18.20	.440		8.30	17.05		25.35	37
0160	2"	Q-1	33.10	.483		10.30	16.85		27.15	39.50
0170	2-1/2"		24.20	.661		19.80	23		42.80	60
0180	3"	↓	20.80	.769	↓	27	27		54	74
0200	45° Elbow									
0210	3/4"	1 Plum	26	.308	Ea.	2.89	11.90		14.79	23
0220	1"		22.70	.352		3.39	13.65		17.04	26
0230	1-1/4"		20.20	.396		4.90	15.35		20.25	31
0240	1-1/2"		18.20	.440		6.85	17.05		23.90	35.50
0250	2"	Q-1	33.10	.483		8.50	16.85		25.35	37.50

21 11 Facility Fire-Suppression Water-Service Piping

21 11 13 – Facility Water Distribution Piping

21 11 13.18 Pipe Fittings, Plastic		Crew	Daily Output	Labor-Hours	Unit	Material	2015 Bare Costs Labor	Equipment	Total	Total Incl O&P
0260	2-1/2"	Q-1	24.20	.661	Ea.	15.30	23		38.30	55
0270	3"	↓	20.80	.769	↓	22	27		49	68.50
0300	Tee									
0310	3/4"	1 Plum	17.30	.462	Ea.	2.89	17.90		20.79	32.50
0320	1"		15.20	.526		5.70	20.50		26.20	40
0330	1-1/4"		13.50	.593		8.55	23		31.55	47.50
0340	1-1/2"	↓	12.10	.661		12.60	25.50		38.10	56.50
0350	2"	Q-1	20	.800		18.65	28		46.65	66.50
0360	2-1/2"		16.20	.988		30.50	34.50		65	90.50
0370	3"	↓	13.90	1.151		47.50	40		87.50	118
0400	Tee, reducing x any size									
0420	1"	1 Plum	15.20	.526	Ea.	4.84	20.50		25.34	39
0430	1-1/4"		13.50	.593		8.85	23		31.85	48
0440	1-1/2"	↓	12.10	.661		10.75	25.50		36.25	54.50
0450	2"	Q-1	20	.800		19.85	28		47.85	68
0460	2-1/2"		16.20	.988		23.50	34.50		58	83
0470	3"	↓	13.90	1.151		27	40		67	96
0500	Coupling									
0510	3/4"	1 Plum	26	.308	Ea.	2.02	11.90		13.92	22
0520	1"		22.70	.352		2.67	13.65		16.32	25.50
0530	1-1/4"		20.20	.396		3.89	15.35		19.24	30
0540	1-1/2"	↓	18.20	.440		5.55	17.05		22.60	34
0550	2"	Q-1	33.10	.483		7.50	16.85		24.35	36.50
0560	2-1/2"		24.20	.661		11.45	23		34.45	50.50
0570	3"	↓	20.80	.769		14.90	27		41.90	61
0600	Coupling, reducing									
0610	1" x 3/4"	1 Plum	22.70	.352	Ea.	2.67	13.65		16.32	25.50
0620	1-1/4" x 1"		20.20	.396		4.04	15.35		19.39	30
0630	1-1/2" x 3/4"		18.20	.440		6.05	17.05		23.10	34.50
0640	1-1/2" x 1"		18.20	.440		5.85	17.05		22.90	34.50
0650	1-1/2" x 1-1/4"	↓	18.20	.440		5.55	17.05		22.60	34
0660	2" x 1"	Q-1	33.10	.483		7.80	16.85		24.65	36.50
0670	2" x 1-1/2"	"	33.10	.483	↓	7.50	16.85		24.35	36.50
0700	Cross									
0720	3/4"	1 Plum	13	.615	Ea.	4.54	24		28.54	44.50
0730	1"		11.30	.708		5.70	27.50		33.20	52
0740	1-1/4"		10.10	.792		7.85	30.50		38.35	59
0750	1-1/2"	↓	9.10	.879		10.90	34		44.90	68
0760	2"	Q-1	16.60	.964		17.75	33.50		51.25	75
0770	2-1/2"	"	12.10	1.322	↓	39	46		85	119
0800	Cap									
0820	3/4"	1 Plum	52	.154	Ea.	1.12	5.95		7.07	11.10
0830	1"		45	.178		1.74	6.90		8.64	13.25
0840	1-1/4"		40	.200		2.81	7.75		10.56	15.90
0850	1-1/2"	↓	36.40	.220		3.89	8.50		12.39	18.35
0860	2"	Q-1	66	.242		5.85	8.45		14.30	20.50
0870	2-1/2"		48.40	.331		8.45	11.55		20	28.50
0880	3"	↓	41.60	.385		13.60	13.40		27	37
0900	Adapter, sprinkler head, female w/metal thd. insert, (s x FNPT)									
0920	3/4" x 1/2"	1 Plum	52	.154	Ea.	5.45	5.95		11.40	15.85
0930	1" x 1/2"		45	.178		5.75	6.90		12.65	17.70
0940	1" x 3/4"	↓	45	.178		9.05	6.90		15.95	21.50

21 11 Facility Fire-Suppression Water-Service Piping

21 11 16 – Facility Fire Hydrants

21 11 16.50 Fire Hydrants for Buildings

		Crew	Daily Output	Labor-Hours	Unit	Material	2015 Bare Costs Labor	Equipment	Total	Total Incl O&P
0010	**FIRE HYDRANTS FOR BUILDINGS**									
3750	Hydrants, wall, w/caps, single, flush, polished brass									
3800	2-1/2" x 2-1/2"	Q-12	5	3.200	Ea.	213	107		320	410
3840	2-1/2" x 3"		5	3.200		430	107		537	645
3860	3" x 3"	↓	4.80	3.333		350	111		461	570
3900	For polished chrome, add				↓	20%				
3950	Double, flush, polished brass									
4000	2-1/2" x 2-1/2" x 4"	Q-12	5	3.200	Ea.	570	107		677	805
4040	2-1/2" x 2-1/2" x 6"		4.60	3.478		825	116		941	1,100
4080	3" x 3" x 4"		4.90	3.265		1,225	109		1,334	1,525
4120	3" x 3" x 6"	↓	4.50	3.556		1,250	119		1,369	1,575
4200	For polished chrome, add					10%				
4350	Double, projecting, polished brass									
4400	2-1/2" x 2-1/2" x 4"	Q-12	5	3.200	Ea.	254	107		361	455
4450	2-1/2" x 2-1/2" x 6"	"	4.60	3.478	"	520	116		636	760
4460	Valve control, dbl. flush/projecting hydrant, cap &									
4470	chain, extension rod & cplg., escutcheon, polished brass	Q-12	8	2	Ea.	300	66.50		366.50	440
4480	Four-way square, flush, polished brass									
4540	2-1/2"(4) x 6"	Q-12	3.60	4.444	Ea.	3,075	148		3,223	3,625

21 11 19 – Fire-Department Connections

21 11 19.50 Connections for the Fire-Department

		Crew	Daily Output	Labor-Hours	Unit	Material	2015 Bare Costs Labor	Equipment	Total	Total Incl O&P
0010	**CONNECTIONS FOR THE FIRE-DEPARTMENT**									
6000	Roof manifold, horiz., brass, without valves & caps									
6040	2-1/2" x 2-1/2" x 4"	Q-12	4.80	3.333	Ea.	168	111		279	370
6060	2-1/2" x 2-1/2" x 6"		4.60	3.478		184	116		300	395
6080	2-1/2" x 2-1/2" x 2-1/2" x 4"		4.60	3.478		271	116		387	490
6090	2-1/2" x 2-1/2" x 2-1/2" x 6"	↓	4.60	3.478	↓	280	116		396	500
7000	Sprinkler line tester, cast brass					24			24	26.50
7140	Standpipe connections, wall, w/plugs & chains									
7160	Single, flush, brass, 2-1/2" x 2-1/2"	Q-12	5	3.200	Ea.	159	107		266	350
7180	2-1/2" x 3"	"	5	3.200	"	164	107		271	355
7240	For polished chrome, add					15%				
7280	Double, flush, polished brass									
7300	2-1/2" x 2-1/2" x 4"	Q-12	5	3.200	Ea.	520	107		627	745
7330	2-1/2" x 2-1/2" x 6"		4.60	3.478		725	116		841	985
7340	3" x 3" x 4"		4.90	3.265		945	109		1,054	1,225
7370	3" x 3" x 6"	↓	4.50	3.556	↓	1,125	119		1,244	1,450
7400	For polished chrome, add					15%				
7440	For sill cock combination, add				Ea.	90.50			90.50	99.50
7580	Double projecting, polished brass									
7600	2-1/2" x 2-1/2" x 4"	Q-12	5	3.200	Ea.	475	107		582	695
7630	2-1/2" x 2-1/2" x 6"	"	4.60	3.478	"	800	116		916	1,075
7680	For polished chrome, add					15%				
7900	Three way, flush, polished brass									
7920	2-1/2" (3) x 4"	Q-12	4.80	3.333	Ea.	1,650	111		1,761	2,000
7930	2-1/2" (3) x 6"	"	4.80	3.333		1,650	111		1,761	2,000
8000	For polished chrome, add				↓	9%				
8020	Three way, projecting, polished brass									
8040	2-1/2"(3) x 4"	Q-12	4.80	3.333	Ea.	1,575	111		1,686	1,900
8070	2-1/2" (3) x 6"	"	4.60	3.478		1,575	116		1,691	1,925
8100	For polished chrome, add				↓	12%				
8200	Four way, square, flush, polished brass,									

21 11 Facility Fire-Suppression Water-Service Piping

21 11 19 – Fire-Department Connections

21 11 19.50 Connections for the Fire-Department

		Crew	Daily Output	Labor-Hours	Unit	Material	2015 Bare Costs Labor	Equipment	Total	Total Incl O&P
8240	2-1/2"(4) x 6"	Q-12	3.60	4.444	Ea.	1,250	148		1,398	1,625
8300	For polished chrome, add				"	10%				
8550	Wall, vertical, flush, cast brass									
8600	Two way, 2-1/2" x 2-1/2" x 4"	Q-12	5	3.200	Ea.	395	107		502	610
8660	Four way, 2-1/2"(4) x 6"		3.80	4.211		1,300	140		1,440	1,650
8680	Six way, 2-1/2"(6) x 6"		3.40	4.706		1,550	157		1,707	1,975
8700	For polished chrome, add					10%				
8800	Sidewalk siamese unit, polished brass, two way									
8820	2-1/2" x 2-1/2" x 4"	Q-12	2.50	6.400	Ea.	635	213		848	1,050
8850	2-1/2" x 2-1/2" x 6"		2	8		780	267		1,047	1,300
8860	3" x 3" x 4"		2.50	6.400		900	213		1,113	1,350
8890	3" x 3" x 6"		2	8		1,225	267		1,492	1,800
8940	For polished chrome, add					12%				
9100	Sidewalk siamese unit, polished brass, three way									
9120	2-1/2" x 2-1/2" x 2-1/2" x 6"	Q-12	2	8	Ea.	1,000	267		1,267	1,550
9160	For polished chrome, add				"	15%				

21 12 Fire-Suppression Standpipes

21 12 19 – Fire-Suppression Hose Racks

21 12 19.50 Fire Hose Racks

		Crew	Daily Output	Labor-Hours	Unit	Material	Labor	Equipment	Total	Total Incl O&P
0010	**FIRE HOSE RACKS**									
2600	Hose rack, swinging, for 1-1/2" diameter hose,									
2620	Enameled steel, 50' & 75' lengths of hose	Q-12	20	.800	Ea.	61	26.50		87.50	111
2640	100' and 125' lengths of hose		20	.800		61	26.50		87.50	111
2680	Chrome plated, 50' and 75' lengths of hose		20	.800		98.50	26.50		125	152
2700	100' and 125' lengths of hose		20	.800		104	26.50		130.50	159
2780	For hose rack nipple, 1-1/2" polished brass, add					27			27	29.50
2820	2-1/2" polished brass, add					49			49	54
2840	1-1/2" polished chrome, add					39.50			39.50	43.50
2860	2-1/2" polished chrome, add					52			52	57.50

21 12 23 – Fire-Suppression Hose Valves

21 12 23.70 Fire Hose Valves

		Crew	Daily Output	Labor-Hours	Unit	Material	Labor	Equipment	Total	Total Incl O&P
0010	**FIRE HOSE VALVES**									
0020	Angle, combination pressure adjust/restricting, rough brass									
0030	1-1/2"	1 Spri	12	.667	Ea.	85.50	24.50		110	135
0040	2-1/2"	"	7	1.143	"	175	42.50		217.50	262
0042	Nonpressure adjustable/restricting, rough brass									
0044	1-1/2"	1 Spri	12	.667	Ea.	53.50	24.50		78	99.50
0046	2-1/2"	"	7	1.143	"	90	42.50		132.50	169
0050	For polished brass, add					30%				
0060	For polished chrome, add					40%				
1000	Ball drip, automatic, rough brass, 1/2"	1 Spri	20	.400	Ea.	16.15	14.80		30.95	42.50
1010	3/4"	"	20	.400	"	18	14.80		32.80	44.50
1100	Ball, 175 lb., sprinkler system, FM/UL, threaded, bronze									
1120	Slow close									
1150	1" size	1 Spri	19	.421	Ea.	241	15.60		256.60	291
1160	1-1/4" size		15	.533		260	19.75		279.75	320
1170	1-1/2" size		13	.615		440	23		463	525
1180	2" size		11	.727		415	27		442	505
1190	2-1/2" size	Q-12	15	1.067		560	35.50		595.50	680

21 12 Fire-Suppression Standpipes

21 12 23 – Fire-Suppression Hose Valves

21 12 23.70 Fire Hose Valves

		Crew	Daily Output	Labor-Hours	Unit	Material	2015 Bare Costs Labor	2015 Bare Costs Equipment	Total	Total Incl O&P
1230	For supervisory switch kit, all sizes									
1240	One circuit, add	1 Spri	48	.167	Ea.	165	6.20		171.20	192
1280	Quarter turn for trim									
1300	1/2" size	1 Spri	22	.364	Ea.	32	13.45		45.45	57.50
1310	3/4" size		20	.400		34.50	14.80		49.30	62
1320	1" size		19	.421		38	15.60		53.60	68
1330	1-1/4" size		15	.533		62.50	19.75		82.25	101
1340	1-1/2" size		13	.615		78.50	23		101.50	124
1350	2" size		11	.727		93.50	27		120.50	148
1400	Caps, polished brass with chain, 3/4"					42			42	46
1420	1"					53.50			53.50	59
1440	1-1/2"					14.35			14.35	15.75
1460	2-1/2"					21.50			21.50	24
1480	3"					33.50			33.50	37
1900	Escutcheon plate, for angle valves, polished brass, 1-1/2"					15.20			15.20	16.70
1920	2-1/2"					24.50			24.50	27
1940	3"					31			31	34
1980	For polished chrome, add					15%				
3000	Gate, hose, wheel handle, N.R.S., rough brass, 1-1/2"	1 Spri	12	.667		135	24.50		159.50	189
3040	2-1/2", 300 lb.	"	7	1.143		189	42.50		231.50	278
3080	For polished brass, add					40%				
3090	For polished chrome, add					50%				
5000	Pressure restricting, adjustable rough brass, 1-1/2"	1 Spri	12	.667		107	24.50		131.50	158
5020	2-1/2"	"	7	1.143		160	42.50		202.50	246
5080	For polished brass, add					30%				
5090	For polished chrome, add					45%				

21 13 Fire-Suppression Sprinkler Systems

21 13 13 – Wet-Pipe Sprinkler Systems

21 13 13.50 Wet-Pipe Sprinkler System Components

		Crew	Daily Output	Labor-Hours	Unit	Material	Labor	Equipment	Total	Total Incl O&P
0010	**WET-PIPE SPRINKLER SYSTEM COMPONENTS**									
1100	Alarm, electric pressure switch (circuit closer)	1 Spri	26	.308	Ea.	88.50	11.40		99.90	116
1140	For explosion proof, max 20 PSI, contacts close or open		26	.308		535	11.40		546.40	605
1220	Water motor, complete with gong		4	2		415	74		489	580
1900	Flexible sprinkler head connectors									
1910	Braided stainless steel hose with mounting bracket									
1920	1/2" and 3/4" outlet size									
1940	40" length	1 Spri	30	.267	Ea.	70.50	9.90		80.40	94
1960	60" length	"	22	.364	"	81	13.45		94.45	112
1982	May replace hard-pipe armovers									
1984	For wet, pre-action, deluge or dry pipe systems									
2000	Release, emergency, manual, for hydraulic or pneumatic system	1 Spri	12	.667	Ea.	218	24.50		242.50	280
2060	Release, thermostatic, for hydraulic or pneumatic release line		20	.400		675	14.80		689.80	765
2200	Sprinkler cabinets, 6 head capacity		16	.500		79	18.55		97.55	117
2260	12 head capacity		16	.500		83.50	18.55		102.05	123
2340	Sprinkler head escutcheons, standard, brass tone, 1" size		40	.200		2.85	7.40		10.25	15.40
2360	Chrome, 1" size		40	.200		3.04	7.40		10.44	15.60
2400	Recessed type, bright brass		40	.200		10.10	7.40		17.50	23.50
2440	Chrome or white enamel		40	.200		3.35	7.40		10.75	15.95
2600	Sprinkler heads, not including supply piping									
3700	Standard spray, pendent or upright, brass, 135°F to 286°F									

21 13 Fire-Suppression Sprinkler Systems

21 13 13 – Wet-Pipe Sprinkler Systems

21 13 13.50 Wet-Pipe Sprinkler System Components

		Crew	Daily Output	Labor-Hours	Unit	Material	2015 Bare Costs Labor	Equipment	Total	Total Incl O&P
3720	1/2" NPT, 3/8" orifice	1 Spri	16	.500	Ea.	15.40	18.55		33.95	47.50
3730	1/2" NPT, 7/16" orifice		16	.500		15.20	18.55		33.75	47.50
3732	1/2" NPT, 7/16" orifice, chrome		16	.500		16	18.55		34.55	48
3740	1/2" NPT, 1/2" orifice		16	.500		9.90	18.55		28.45	41.50
3760	1/2" NPT, 17/32" orifice		16	.500		12.90	18.55		31.45	44.50
3780	3/4" NPT, 17/32" orifice		16	.500		11.90	18.55		30.45	43.50
3840	For chrome, add					3.38			3.38	3.72
4200	Sidewall, vertical brass, 135°F to 286°F									
4240	1/2" NPT, 1/2" orifice	1 Spri	16	.500	Ea.	25	18.55		43.55	58
4280	3/4" NPT, 17/32" orifice	"	16	.500		70	18.55		88.55	107
4360	For satin chrome, add					2.75			2.75	3.03
4500	Sidewall, horizontal, brass, 135°F to 286°F									
4520	1/2" NPT, 1/2" orifice	1 Spri	16	.500	Ea.	25	18.55		43.55	58
5600	Concealed, complete with cover plate									
5620	1/2" NPT, 1/2" orifice, 135°F to 212°F	1 Spri	9	.889	Ea.	24	33		57	80.50
6025	Residential sprinkler components, (one and two family)									
6026	Water motor alarm, with strainer	1 Spri	4	2	Ea.	415	74		489	580
6027	Fast response, glass bulb, 135°F to 155°F									
6028	1/2" NPT, pendent, brass	1 Spri	16	.500	Ea.	27	18.55		45.55	60.50
6029	1/2" NPT, sidewall, brass		16	.500		27	18.55		45.55	60.50
6030	1/2" NPT, pendent, brass, extended coverage		16	.500		24	18.55		42.55	56.50
6031	1/2" NPT, sidewall, brass, extended coverage		16	.500		22	18.55		40.55	54.50
6032	3/4" NPT sidewall, brass, extended coverage		16	.500		24	18.55		42.55	57
6033	For chrome, add					15%				
6034	For polyester/teflon coating add					20%				
6100	Sprinkler head wrenches, standard head				Ea.	24.50			24.50	26.50
6120	Recessed head					37.50			37.50	41
6160	Tamper switch, (valve supervisory switch)	1 Spri	16	.500		102	18.55		120.55	143

21 13 16 – Dry-Pipe Sprinkler Systems

21 13 16.50 Dry-Pipe Sprinkler System Components

		Crew	Daily Output	Labor-Hours	Unit	Material	Labor	Equipment	Total	Total Incl O&P
0010	**DRY-PIPE SPRINKLER SYSTEM COMPONENTS**									
0600	Accelerator	1 Spri	8	1	Ea.	755	37		792	890
0800	Air compressor for dry pipe system, automatic, complete									
0820	30 gal. system capacity, 3/4 HP	1 Spri	1.30	6.154	Ea.	765	228		993	1,225
0860	30 gal. system capacity, 1 HP		1.30	6.154		790	228		1,018	1,250
0960	Air pressure maintenance control		24	.333		345	12.35		357.35	400
1600	Dehydrator package, incl. valves and nipples		12	.667		710	24.50		734.50	825
2600	Sprinkler heads, not including supply piping									
2640	Dry, pendent, 1/2" orifice, 3/4" or 1" NPT									
2660	1/2" to 6" length	1 Spri	14	.571	Ea.	124	21		145	171
2670	6-1/4" to 8" length		14	.571		129	21		150	177
2680	8-1/4" to 12" length		14	.571		135	21		156	183
2800	For each inch or fraction, add					2.84			2.84	3.12
6330	Valves and components									
6340	Alarm test/shut off valve, 1/2"	1 Spri	20	.400	Ea.	21.50	14.80		36.30	48.50
8000	Dry pipe air check valve, 3" size	Q-12	2	8		1,775	267		2,042	2,400
8200	Dry pipe valve, incl. trim and gauges, 3" size		2	8		2,575	267		2,842	3,300
8220	4" size		1	16		2,850	535		3,385	4,025
8240	6" size	Q-13	2	16		3,350	535		3,885	4,550
8280	For accelerator trim with gauges, add	1 Spri	8	1		239	37		276	325

Division Notes

		CREW	DAILY OUTPUT	LABOR-HOURS	UNIT	BARE COSTS				TOTAL INCL O&P
						MAT.	LABOR	EQUIP.	TOTAL	

Estimating Tips
22 10 00 Plumbing Piping and Pumps
This subdivision is primarily basic pipe and related materials. The pipe may be used by any of the mechanical disciplines, i.e., plumbing, fire protection, heating, and air conditioning.

Note: CPVC plastic piping approved for fire protection is located in 21 11 13.

- The labor adjustment factors listed in Subdivision 22 01 02.20 apply throughout Divisions 21, 22, and 23. CAUTION: the correct percentage may vary for the same items. For example, the percentage add for the basic pipe installation should be based on the maximum height that the craftsman must install for that particular section. If the pipe is to be located 14' above the floor but it is suspended on threaded rod from beams, the bottom flange of which is 18' high (4' rods), then the height is actually 18' and the add is 20%. The pipe coverer, however, does not have to go above the 14', and so the add should be 10%.
- Most pipe is priced first as straight pipe with a joint (coupling, weld, etc.) every 10' and a hanger usually every 10'. There are exceptions with hanger spacing such as for cast iron pipe (5') and plastic pipe (3 per 10'). Following each type of pipe there are several lines listing sizes and the amount to be subtracted to delete couplings and hangers. This is for pipe that is to be buried or supported together on trapeze hangers. The reason that the couplings are deleted is that these runs are usually long, and frequently longer lengths of pipe are used. By deleting the couplings, the estimator is expected to look up and add back the correct reduced number of couplings.
- When preparing an estimate, it may be necessary to approximate the fittings. Fittings usually run between 25% and 50% of the cost of the pipe. The lower percentage is for simpler runs, and the higher number is for complex areas, such as mechanical rooms.
- For historic restoration projects, the systems must be as invisible as possible, and pathways must be sought for pipes, conduit, and ductwork. While installations in accessible spaces (such as basements and attics) are relatively straightforward to estimate, labor costs may be more difficult to determine when delivery systems must be concealed.

22 40 00 Plumbing Fixtures
- Plumbing fixture costs usually require two lines: the fixture itself and its "rough-in, supply, and waste."
- In the Assemblies Section (Plumbing D2010) for the desired fixture, the System Components Group at the center of the page shows the fixture on the first line. The rest of the list (fittings, pipe, tubing, etc.) will total up to what we refer to in the Unit Price section as "Rough-in, supply, waste, and vent." Note that for most fixtures we allow a nominal 5' of tubing to reach from the fixture to a main or riser.
- Remember that gas- and oil-fired units need venting.

Reference Numbers
Reference numbers are shown in shaded boxes at the beginning of some major classifications. These numbers refer to related items in the Reference Section. The reference information may be an estimating procedure, an alternate pricing method, or technical information.

Note: Not all subdivisions listed here necessarily appear in this publication. ■

Estimate with precision with RSMeans Online...
Quick, intuitive and easy-to-use, RSMeans Online gives you instant access to hundreds of thousands of material, labor, and equipment costs from RSMeans' comprehensive database, delivering the information you need to build competitive estimates every time.

Start your free trial today at **www.rsmeansonline.com**

RSMeansOnline

22 05 Common Work Results for Plumbing

22 05 05 – Selective Demolition for Plumbing

22 05 05.10 Plumbing Demolition

		Crew	Daily Output	Labor-Hours	Unit	Material	2015 Bare Costs Labor	2015 Bare Costs Equipment	Total	Total Incl O&P
0010	**PLUMBING DEMOLITION**									
1020	Fixtures, including 10' piping									
1101	Bathtubs, cast iron	1 Clab	4	2	Ea.		49		49	82
1121	Fiberglass		6	1.333			32.50		32.50	54.50
1141	Steel	↓	5	1.600			39		39	65.50
1200	Lavatory, wall hung	1 Plum	10	.800			31		31	51
1221	Counter top	1 Clab	16	.500			12.25		12.25	20.50
1301	Sink, single compartment		16	.500			12.25		12.25	20.50
1321	Double	↓	10	.800			19.55		19.55	33
1400	Water closet, floor mounted	1 Plum	8	1			39		39	64
1421	Wall mounted	1 Clab	7	1.143			28		28	47
2001	Piping, metal, to 1-1/2" diameter		200	.040	L.F.		.98		.98	1.64
2051	2" thru 3-1/2" diameter		150	.053			1.30		1.30	2.19
2101	4" thru 6" diameter	↓	100	.080			1.96		1.96	3.28
2160	Plastic pipe with fittings, up thru 1-1/2" diameter	1 Plum	250	.032			1.24		1.24	2.05
2162	2" thru 3" diameter	"	200	.040			1.55		1.55	2.56
2164	4" thru 6" diameter	Q-1	200	.080			2.79		2.79	4.60
2166	8" thru 14" diameter		150	.107			3.72		3.72	6.15
2168	16" diameter	↓	100	.160	↓		5.60		5.60	9.20
3000	Submersible sump pump	1 Plum	24	.333	Ea.		12.90		12.90	21.50
6000	Remove and reset fixtures, easy access		6	1.333			51.50		51.50	85.50
6100	Difficult access	↓	4	2	↓		77.50		77.50	128

22 05 23 – General-Duty Valves for Plumbing Piping

22 05 23.20 Valves, Bronze

		Crew	Daily Output	Labor-Hours	Unit	Material	Labor	Equipment	Total	Total Incl O&P
0010	**VALVES, BRONZE**									
1750	Check, swing, class 150, regrinding disc, threaded									
1860	3/4"	1 Plum	20	.400	Ea.	100	15.50		115.50	136
1870	1"	"	19	.421	"	144	16.30		160.30	185
2850	Gate, N.R.S., soldered, 125 psi									
2940	3/4"	1 Plum	20	.400	Ea.	63.50	15.50		79	95.50
2950	1"		19	.421		76.50	16.30		92.80	111
5600	Relief, pressure & temperature, self-closing, ASME, threaded									
5640	3/4"	1 Plum	28	.286	Ea.	253	11.05		264.05	296
5650	1"		24	.333		405	12.90		417.90	465
5660	1-1/4"	↓	20	.400	↓	895	15.50		910.50	1,000
6400	Pressure, water, ASME, threaded									
6440	3/4"	1 Plum	28	.286	Ea.	116	11.05		127.05	145
6450	1"	"	24	.333	"	260	12.90		272.90	310
6900	Reducing, water pressure									
6920	300 psi to 25-75 psi, threaded or sweat									
6940	1/2"	1 Plum	24	.333	Ea.	395	12.90		407.90	455
6950	3/4"		20	.400		405	15.50		420.50	475
6960	1"	↓	19	.421	↓	630	16.30		646.30	715
8350	Tempering, water, sweat connections									
8400	1/2"	1 Plum	24	.333	Ea.	98	12.90		110.90	130
8440	3/4"	"	20	.400	"	126	15.50		141.50	165
8650	Threaded connections									
8700	1/2"	1 Plum	24	.333	Ea.	126	12.90		138.90	161
8740	3/4"	"	20	.400	"	770	15.50		785.50	870

22 05 Common Work Results for Plumbing

22 05 29 – Hangers and Supports for Plumbing Piping and Equipment

22 05 29.10 Hangers & Supp. for Plumb'g/HVAC Pipe/Equip.

		Crew	Daily Output	Labor-Hours	Unit	Material	2015 Bare Costs Labor	Equipment	Total	Total Incl O&P
0010	**HANGERS AND SUPPORTS FOR PLUMB'G/HVAC PIPE/EQUIP.**									
8000	Pipe clamp, plastic, 1/2" CTS	1 Plum	80	.100	Ea.	.23	3.88		4.11	6.65
8010	3/4" CTS		73	.110		.24	4.25		4.49	7.25
8020	1" CTS		68	.118		.54	4.56		5.10	8.10
8080	Economy clamp, 1/4" CTS		175	.046		.05	1.77		1.82	2.98
8090	3/8" CTS		168	.048		.07	1.85		1.92	3.13
8100	1/2" CTS		160	.050		.07	1.94		2.01	3.28
8110	3/4" CTS		145	.055		.15	2.14		2.29	3.70
8200	Half clamp, 1/2" CTS		80	.100		.08	3.88		3.96	6.50
8210	3/4" CTS		73	.110		.12	4.25		4.37	7.15
8300	Suspension clamp, 1/2" CTS		80	.100		.23	3.88		4.11	6.65
8310	3/4" CTS		73	.110		.24	4.25		4.49	7.25
8320	1" CTS		68	.118		.54	4.56		5.10	8.10
8400	Insulator, 1/2" CTS		80	.100		.37	3.88		4.25	6.80
8410	3/4" CTS		73	.110		.39	4.25		4.64	7.45
8420	1" CTS		68	.118		.40	4.56		4.96	7.95
8500	J hook clamp with nail 1/2" CTS		240	.033		.13	1.29		1.42	2.27
8501	3/4" CTS		240	.033		.13	1.29		1.42	2.27

22 05 48 – Vibration and Seismic Controls for Plumbing Piping and Equipment

22 05 48.10 Seismic Bracing Supports

		Crew	Daily Output	Labor-Hours	Unit	Material	Labor	Equipment	Total	Total Incl O&P
0010	**SEISMIC BRACING SUPPORTS**									
0020	Clamps									
0030	C-clamp, for mounting on steel beam									
0040	3/8" threaded rod	1 Skwk	160	.050	Ea.	2.05	1.67		3.72	5.05
0050	1/2" threaded rod		160	.050		2.20	1.67		3.87	5.25
0060	5/8" threaded rod		160	.050		3.70	1.67		5.37	6.90
0070	3/4" threaded rod		160	.050		4.55	1.67		6.22	7.80
0100	Brackets									
0110	Beam side or wall malleable iron									
0120	3/8" threaded rod	1 Skwk	48	.167	Ea.	3.11	5.55		8.66	12.75
0130	1/2" threaded rod		48	.167		4.39	5.55		9.94	14.20
0140	5/8" threaded rod		48	.167		8.20	5.55		13.75	18.40
0150	3/4" threaded rod		48	.167		11.30	5.55		16.85	22
0160	7/8" threaded rod		48	.167		11.70	5.55		17.25	22
0170	For concrete installation, add						30%			
0180	Wall, welded steel									
0190	0 size 12" wide 18" deep	1 Skwk	34	.235	Ea.	172	7.85		179.85	202
0200	1 size 18" wide 24" deep		34	.235		211	7.85		218.85	245
0210	2 size 24" wide 30" deep		34	.235		290	7.85		297.85	335
0300	Rod, carbon steel									
0310	Continuous thread									
0320	1/4" thread	1 Skwk	144	.056	L.F.	1.70	1.85		3.55	4.99
0330	3/8" thread		144	.056		1.81	1.85		3.66	5.10
0340	1/2" thread		144	.056		2.86	1.85		4.71	6.25
0350	5/8" thread		144	.056		4.05	1.85		5.90	7.60
0360	3/4" thread		144	.056		7.15	1.85		9	10.95
0370	7/8" thread		144	.056		8.95	1.85		10.80	12.95
0380	For galvanized, add					30%				
0400	Channel, steel									
0410	3/4" x 1-1/2"	1 Skwk	80	.100	L.F.	4.10	3.34		7.44	10.10
0420	1-1/2" x 1-1/2"		70	.114		5.40	3.81		9.21	12.35
0430	1-7/8" x 1-1/2"		60	.133		23	4.45		27.45	33

22 05 Common Work Results for Plumbing

22 05 48 – Vibration and Seismic Controls for Plumbing Piping and Equipment

22 05 48.10 Seismic Bracing Supports		Crew	Daily Output	Labor-Hours	Unit	Material	2015 Bare Costs Labor	Equipment	Total	Total Incl O&P
0440	3" x 1-1/2"	1 Skwk	50	.160	L.F.	39.50	5.35		44.85	52.50
0450	Spring nuts									
0460	3/8"	1 Skwk	100	.080	Ea.	1.20	2.67		3.87	5.80
0470	1/2"	"	80	.100	"	1.67	3.34		5.01	7.45
0500	Welding, field									
0510	Cleaning and welding plates, bars, or rods									
0520	To existing beams, columns, or trusses									
0530	1" weld	1 Skwk	144	.056	Ea.	.23	1.85		2.08	3.37
0540	2" weld		72	.111		.39	3.71		4.10	6.70
0550	3" weld		54	.148		.61	4.94		5.55	9
0560	4" weld		36	.222		.83	7.40		8.23	13.40
0570	5" weld		30	.267		1.05	8.90		9.95	16.15
0580	6" weld	↓	24	.333	↓	1.18	11.10		12.28	20
0600	Vibration absorbers									
0610	Hangers, neoprene flex									
0620	10-120 lb. capacity	1 Skwk	8	1	Ea.	26	33.50		59.50	84.50
0630	75-550 lb. capacity		8	1		37	33.50		70.50	96.50
0640	250-1100 lb. capacity		6	1.333		72	44.50		116.50	154
0650	1000-4000 lb. capacity		6	1.333		134	44.50		178.50	223

22 05 76 – Facility Drainage Piping Cleanouts

22 05 76.10 Cleanouts

		Crew	Daily Output	Labor-Hours	Unit	Material	Labor	Equipment	Total	Total Incl O&P
0010	**CLEANOUTS**									
0060	Floor type									
0080	Round or square, scoriated nickel bronze top									
0100	2" pipe size	1 Plum	10	.800	Ea.	197	31		228	268
0120	3" pipe size		8	1		295	39		334	390
0140	4" pipe size	↓	6	1.333	↓	295	51.50		346.50	410

22 05 76.20 Cleanout Tees

		Crew	Daily Output	Labor-Hours	Unit	Material	Labor	Equipment	Total	Total Incl O&P
0010	**CLEANOUT TEES**									
0100	Cast iron, B&S, with countersunk plug									
0220	3" pipe size	1 Plum	3.60	2.222	Ea.	292	86		378	460
0240	4" pipe size	"	3.30	2.424	"	365	94		459	555
0500	For round smooth access cover, same price									
4000	Plastic, tees and adapters. Add plugs									
4010	ABS, DWV									
4020	Cleanout tee, 1-1/2" pipe size	1 Plum	15	.533	Ea.	23	20.50		43.50	59

22 07 Plumbing Insulation

22 07 16 – Plumbing Equipment Insulation

22 07 16.10 Insulation for Plumbing Equipment

			Crew	Daily Output	Labor-Hours	Unit	Material	Labor	Equipment	Total	Total Incl O&P
0010	**INSULATION FOR PLUMBING EQUIPMENT**										
2900	Domestic water heater wrap kit										
2920	1-1/2" with vinyl jacket, 20-60 gal.	G	1 Plum	8	1	Ea.	18.60	39		57.60	84.50

22 07 19 – Plumbing Piping Insulation

22 07 19.10 Piping Insulation

		Crew	Daily Output	Labor-Hours	Unit	Material	Labor	Equipment	Total	Total Incl O&P
0010	**PIPING INSULATION**									
0230	Insulated protectors, (ADA)									
0235	For exposed piping under sinks or lavatories									
0240	Vinyl coated foam, velcro tabs									
0245	P Trap, 1-1/4" or 1-1/2"	1 Plum	32	.250	Ea.	22.50	9.70		32.20	40.50

22 07 Plumbing Insulation

22 07 19 – Plumbing Piping Insulation

22 07 19.10 Piping Insulation		Crew	Daily Output	Labor-Hours	Unit	Material	2015 Bare Costs Labor	Equipment	Total	Total Incl O&P
0260	Valve and supply cover									
0265	1/2", 3/8", and 7/16" pipe size	1 Plum	32	.250	Ea.	22	9.70		31.70	40
0285	1-1/4" pipe size	"	32	.250	"	13.65	9.70		23.35	31
0600	Pipe covering (price copper tube one size less than IPS)									
6600	Fiberglass, with all service jacket									
6840	1" wall, 1/2" iron pipe size	G	Q-14	240	.067	L.F.	.83	2.07		2.90	4.43
6860	3/4" iron pipe size	G		230	.070		.90	2.16		3.06	4.66
6870	1" iron pipe size	G		220	.073		.97	2.26		3.23	4.91
6900	2" iron pipe size	G		200	.080		1.22	2.49		3.71	5.55
7879	Rubber tubing, flexible closed cell foam									
8100	1/2" wall, 1/4" iron pipe size	G	1 Asbe	90	.089	L.F.	.54	3.07		3.61	5.80
8130	1/2" iron pipe size	G		89	.090		.67	3.11		3.78	6
8140	3/4" iron pipe size	G		89	.090		.75	3.11		3.86	6.10
8150	1" iron pipe size	G		88	.091		.82	3.14		3.96	6.25
8170	1-1/2" iron pipe size	G		87	.092		1.15	3.18		4.33	6.65
8180	2" iron pipe size	G		86	.093		1.47	3.21		4.68	7.05
8300	3/4" wall, 1/4" iron pipe size	G		90	.089		.85	3.07		3.92	6.15
8330	1/2" iron pipe size	G		89	.090		1.10	3.11		4.21	6.45
8340	3/4" iron pipe size	G		89	.090		1.35	3.11		4.46	6.75
8350	1" iron pipe size	G		88	.091		1.54	3.14		4.68	7.05
8380	2" iron pipe size	G		86	.093		2.68	3.21		5.89	8.40
8444	1" wall, 1/2" iron pipe size	G		86	.093		2.05	3.21		5.26	7.70
8445	3/4" iron pipe size	G		84	.095		2.48	3.29		5.77	8.35
8446	1" iron pipe size	G		84	.095		2.89	3.29		6.18	8.80
8447	1-1/4" iron pipe size	G		82	.098		3.23	3.37		6.60	9.25
8448	1-1/2" iron pipe size	G		82	.098		3.76	3.37		7.13	9.85
8449	2" iron pipe size	G		80	.100		4.95	3.46		8.41	11.30
8450	2-1/2" iron pipe size	G		80	.100		6.45	3.46		9.91	12.95
8456	Rubber insulation tape, 1/8" x 2" x 30'	G				Ea.	12.05			12.05	13.25

22 11 Facility Water Distribution

22 11 13 – Facility Water Distribution Piping

22 11 13.23 Pipe/Tube, Copper

		Crew	Daily Output	Labor-Hours	Unit	Material	Labor	Equipment	Total	Total Incl O&P
0010	**PIPE/TUBE, COPPER**, Solder joints									
1000	Type K tubing, couplings & clevis hanger assemblies 10' O.C.									
1180	3/4" diameter	1 Plum	74	.108	L.F.	7.80	4.19		11.99	15.45
1200	1" diameter	"	66	.121	"	10.35	4.70		15.05	19.15
2000	Type L tubing, couplings & clevis hanger assemblies 10' O.C.									
2140	1/2" diameter	1 Plum	81	.099	L.F.	3.70	3.83		7.53	10.35
2160	5/8" diameter		79	.101		5.35	3.92		9.27	12.35
2180	3/4" diameter		76	.105		5.20	4.08		9.28	12.45
2200	1" diameter		68	.118		7.85	4.56		12.41	16.10
2220	1-1/4" diameter		58	.138		10.35	5.35		15.70	20
3000	Type M tubing, couplings & clevis hanger assemblies 10' O.C.									
3140	1/2" diameter	1 Plum	84	.095	L.F.	3.05	3.69		6.74	9.45
3180	3/4" diameter		78	.103		4.13	3.97		8.10	11.10
3200	1" diameter		70	.114		6.60	4.43		11.03	14.55
3220	1-1/4" diameter		60	.133		8.80	5.15		13.95	18.20
3240	1-1/2" diameter		54	.148		11.55	5.75		17.30	22
3260	2" diameter		44	.182		17.35	7.05		24.40	31
4000	Type DWV tubing, couplings & clevis hanger assemblies 10' O.C.									

22 11 Facility Water Distribution

22 11 13 – Facility Water Distribution Piping

22 11 13.23 Pipe/Tube, Copper

		Crew	Daily Output	Labor-Hours	Unit	Material	2015 Bare Costs Labor	Equipment	Total	Total Incl O&P
4100	1-1/4" diameter	1 Plum	60	.133	L.F.	9.15	5.15		14.30	18.60
4120	1-1/2" diameter		54	.148		11.15	5.75		16.90	22
4140	2" diameter	▼	44	.182		14.60	7.05		21.65	28
4160	3" diameter	Q-1	58	.276		26.50	9.60		36.10	45
4180	4" diameter	"	40	.400	▼	44	13.95		57.95	71

22 11 13.25 Pipe/Tube Fittings, Copper

		Crew	Daily Output	Labor-Hours	Unit	Material	2015 Bare Costs Labor	Equipment	Total	Total Incl O&P
0010	**PIPE/TUBE FITTINGS, COPPER**, Wrought unless otherwise noted									
0040	Solder joints, copper x copper									
0070	90° elbow, 1/4"	1 Plum	22	.364	Ea.	3.28	14.10		17.38	27
0100	1/2"		20	.400		1.10	15.50		16.60	26.50
0120	3/4"		19	.421		2.46	16.30		18.76	29.50
0250	45° elbow, 1/4"		22	.364		5.90	14.10		20	30
0280	1/2"		20	.400		2.01	15.50		17.51	27.50
0290	5/8"		19	.421		9.55	16.30		25.85	37.50
0300	3/4"		19	.421		3.52	16.30		19.82	31
0310	1"		16	.500		8.85	19.40		28.25	41.50
0320	1-1/4"		15	.533		11.85	20.50		32.35	47
0450	Tee, 1/4"		14	.571		6.55	22		28.55	43.50
0480	1/2"		13	.615		1.87	24		25.87	41.50
0490	5/8"		12	.667		12.05	26		38.05	56
0500	3/4"		12	.667		4.51	26		30.51	47.50
0510	1"		10	.800		13.95	31		44.95	66.50
0520	1-1/4"		9	.889		18.85	34.50		53.35	77.50
0612	Tee, reducing on the outlet, 1/4"		15	.533		16.45	20.50		36.95	52
0613	3/8"		15	.533		15.40	20.50		35.90	51
0614	1/2"		14	.571		10.05	22		32.05	47.50
0615	5/8"		13	.615		32	24		56	75
0616	3/4"		12	.667		6.45	26		32.45	49.50
0617	1"		11	.727		18.50	28		46.50	67
0618	1-1/4"		10	.800		20.50	31		51.50	73.50
0619	1-1/2"		9	.889		21.50	34.50		56	81
0620	2"	▼	8	1		39	39		78	107
0621	2-1/2"	Q-1	9	1.778		107	62		169	219
0622	3"		8	2		110	70		180	236
0623	4"		6	2.667		203	93		296	375
0624	5"	▼	5	3.200		1,125	112		1,237	1,425
0625	6"	Q-2	7	3.429		1,550	115		1,665	1,900
0626	8"	"	6	4		5,950	134		6,084	6,775
0630	Tee, reducing on the run, 1/4"	1 Plum	15	.533		13.90	20.50		34.40	49.50
0631	3/8"		15	.533		19.60	20.50		40.10	55.50
0632	1/2"		14	.571		11.35	22		33.35	49
0633	5/8"		13	.615		19.40	24		43.40	61
0634	3/4"		12	.667		15.15	26		41.15	59
0635	1"		11	.727		16.65	28		44.65	65
0636	1-1/4"		10	.800		26	31		57	80
0637	1-1/2"		9	.889		46	34.50		80.50	108
0638	2"	▼	8	1		59	39		98	129
0639	2-1/2"	Q-1	9	1.778		152	62		214	269
0640	3"		8	2		202	70		272	335
0641	4"		6	2.667		425	93		518	620
0642	5"	▼	5	3.200		1,075	112		1,187	1,350
0643	6"	Q-2	7	3.429	▼	1,625	115		1,740	2,000

22 11 Facility Water Distribution

22 11 13 – Facility Water Distribution Piping

22 11 13.25 Pipe/Tube Fittings, Copper

		Crew	Daily Output	Labor-Hours	Unit	Material	2015 Bare Costs Labor	Equipment	Total	Total Incl O&P
0644	8"	Q-2	6	4	Ea.	5,950	134		6,084	6,775
0650	Coupling, 1/4"	1 Plum	24	.333		.75	12.90		13.65	22.50
0680	1/2"		22	.364		.83	14.10		14.93	24.50
0690	5/8"		21	.381		2.50	14.75		17.25	27.50
0700	3/4"		21	.381		1.67	14.75		16.42	26.50
0710	1"		18	.444		5.55	17.20		22.75	34.50
0715	1-1/4"		17	.471		5.85	18.25		24.10	36.50
2000	DWV, solder joints, copper x copper									
2030	90° Elbow, 1-1/4"	1 Plum	13	.615	Ea.	10.45	24		34.45	51
2050	1-1/2"		12	.667		14.55	26		40.55	58.50
2070	2"		10	.800		28	31		59	82
2090	3"	Q-1	10	1.600		56.50	56		112.50	155
2100	4"	"	9	1.778		335	62		397	465
2250	Tee, Sanitary, 1-1/4"	1 Plum	9	.889		20.50	34.50		55	80
2270	1-1/2"		8	1		26	39		65	92.50
2290	2"		7	1.143		30	44.50		74.50	106
2310	3"	Q-1	7	2.286		138	79.50		217.50	283
2330	4"	"	6	2.667		330	93		423	515
2400	Coupling, 1-1/4"	1 Plum	14	.571		4.91	22		26.91	42
2420	1-1/2"		13	.615		6.10	24		30.10	46
2440	2"		11	.727		8.45	28		36.45	56
2460	3"	Q-1	11	1.455		15.40	50.50		65.90	100
2480	4"	"	10	1.600		49	56		105	146

22 11 13.44 Pipe, Steel

		Crew	Daily Output	Labor-Hours	Unit	Material	Labor	Equipment	Total	Total Incl O&P
0010	**PIPE, STEEL**									
0050	Schedule 40, threaded, with couplings, and clevis hanger									
0060	assemblies sized for covering, 10' O.C.									
0540	Black, 1/4" diameter	1 Plum	66	.121	L.F.	6.55	4.70		11.25	14.95
0560	1/2" diameter		63	.127		3.32	4.92		8.24	11.75
0570	3/4" diameter		61	.131		3.91	5.10		9.01	12.70
0580	1" diameter		53	.151		5.05	5.85		10.90	15.20
0590	1-1/4" diameter	Q-1	89	.180		6.20	6.25		12.45	17.15
0600	1-1/2" diameter		80	.200		7.10	7		14.10	19.30
0610	2" diameter		64	.250		9	8.70		17.70	24.50

22 11 13.45 Pipe Fittings, Steel, Threaded

		Crew	Daily Output	Labor-Hours	Unit	Material	Labor	Equipment	Total	Total Incl O&P
0010	**PIPE FITTINGS, STEEL, THREADED**									
5000	Malleable iron, 150 lb.									
5020	Black									
5040	90° elbow, straight									
5090	3/4"	1 Plum	14	.571	Ea.	3.99	22		25.99	41
5100	1"	"	13	.615		6.95	24		30.95	47
5120	1-1/2"	Q-1	20	.800		15.05	28		43.05	62.50
5130	2"	"	18	.889		26	31		57	79.50
5450	Tee, straight									
5500	3/4"	1 Plum	9	.889	Ea.	6.35	34.50		40.85	64
5510	1"	"	8	1		10.85	39		49.85	76
5520	1-1/4"	Q-1	14	1.143		17.60	40		57.60	85.50
5530	1-1/2"		13	1.231		22	43		65	95
5540	2"		11	1.455		37.50	50.50		88	125
5650	Coupling									
5700	3/4"	1 Plum	18	.444	Ea.	5.35	17.20		22.55	34.50
5710	1"	"	15	.533		8	20.50		28.50	43

For customer support on your Residential Cost Data, call 877.713.2617.

22 11 Facility Water Distribution

22 11 13 – Facility Water Distribution Piping

22 11 13.45 Pipe Fittings, Steel, Threaded		Crew	Daily Output	Labor-Hours	Unit	Material	2015 Bare Costs Labor	Equipment	Total	Total Incl O&P
5720	1-1/4"	Q-1	26	.615	Ea.	10.40	21.50		31.90	47
5730	1-1/2"		24	.667		14.05	23.50		37.55	54
5740	2"	↓	21	.762	↓	21	26.50		47.50	67

22 11 13.74 Pipe, Plastic

		Crew	Daily Output	Labor-Hours	Unit	Material	Labor	Equipment	Total	Total Incl O&P
0010	**PIPE, PLASTIC**									
1800	PVC, couplings 10' O.C., clevis hanger assemblies, 3 per 10'									
1820	Schedule 40									
1860	1/2" diameter	1 Plum	54	.148	L.F.	4.88	5.75		10.63	14.80
1870	3/4" diameter		51	.157		5.20	6.10		11.30	15.80
1880	1" diameter		46	.174		5.85	6.75		12.60	17.55
1890	1-1/4" diameter		42	.190		6.55	7.40		13.95	19.40
1900	1-1/2" diameter	↓	36	.222		6.85	8.60		15.45	22
1910	2" diameter	Q-1	59	.271		8.05	9.45		17.50	24.50
1920	2-1/2" diameter		56	.286		10.35	9.95		20.30	28
1930	3" diameter		53	.302		12.60	10.55		23.15	31
1940	4" diameter	↓	48	.333		16.10	11.65		27.75	37
4100	DWV type, schedule 40, couplings 10' O.C., clevis hanger assy's, 3 per 10'									
4210	ABS, schedule 40, foam core type									
4212	Plain end black									
4214	1-1/2" diameter	1 Plum	39	.205	L.F.	5.20	7.95		13.15	18.85
4216	2" diameter	Q-1	62	.258		5.60	9		14.60	21
4218	3" diameter		56	.286		8	9.95		17.95	25
4220	4" diameter		51	.314		10.20	10.95		21.15	29.50
4222	6" diameter	↓	42	.381	↓	18	13.30		31.30	42
4240	To delete coupling & hangers, subtract									
4244	1-1/2" diam. to 6" diam.					43%	48%			
4400	PVC									
4410	1-1/4" diameter	1 Plum	42	.190	L.F.	5.35	7.40		12.75	18.10
4420	1-1/2" diameter	"	36	.222		5.30	8.60		13.90	20
4460	2" diameter	Q-1	59	.271		5.65	9.45		15.10	22
4470	3" diameter		53	.302		8.05	10.55		18.60	26
4480	4" diameter	↓	48	.333	↓	10	11.65		21.65	30
5300	CPVC, socket joint, couplings 10' O.C., clevis hanger assemblies, 3 per 10'									
5302	Schedule 40									
5304	1/2" diameter	1 Plum	54	.148	L.F.	5.90	5.75		11.65	15.95
5305	3/4" diameter		51	.157		6.75	6.10		12.85	17.50
5306	1" diameter		46	.174		8.10	6.75		14.85	20
5307	1-1/4" diameter		42	.190		9.65	7.40		17.05	23
5308	1-1/2" diameter		36	.222		10.85	8.60		19.45	26
5309	2" diameter	Q-1	59	.271	↓	12.85	9.45		22.30	29.50
5360	CPVC, threaded, couplings 10' O.C., clevis hanger assemblies, 3 per 10'									
5380	Schedule 40									
5460	1/2" diameter	1 Plum	54	.148	L.F.	6.70	5.75		12.45	16.85
5470	3/4" diameter		51	.157		8.15	6.10		14.25	19
5480	1" diameter		46	.174		9.55	6.75		16.30	21.50
5490	1-1/4" diameter		42	.190		10.80	7.40		18.20	24
5500	1-1/2" diameter	↓	36	.222		11.80	8.60		20.40	27
5510	2" diameter	Q-1	59	.271	↓	14	9.45		23.45	31
6500	Residential installation, plastic pipe									
6510	Couplings 10' O.C., strap hangers 3 per 10'									
6520	PVC, Schedule 40									
6530	1/2" diameter	1 Plum	138	.058	L.F.	1.02	2.25		3.27	4.84

22 11 Facility Water Distribution

22 11 13 – Facility Water Distribution Piping

22 11 13.74 Pipe, Plastic

		Crew	Daily Output	Labor-Hours	Unit	Material	2015 Bare Costs Labor	Equipment	Total	Total Incl O&P
6540	3/4" diameter	1 Plum	128	.063	L.F.	1.23	2.42		3.65	5.35
6550	1" diameter		119	.067		1.81	2.61		4.42	6.30
6560	1-1/4" diameter		111	.072		2.20	2.79		4.99	7.05
6570	1-1/2" diameter		104	.077		2.44	2.98		5.42	7.60
6580	2" diameter	Q-1	197	.081		3.10	2.83		5.93	8.10
6590	2-1/2" diameter		162	.099		5.10	3.45		8.55	11.30
6600	4" diameter		123	.130		8.60	4.54		13.14	16.95
6700	PVC, DWV, Schedule 40									
6720	1-1/4" diameter	1 Plum	100	.080	L.F.	1.84	3.10		4.94	7.10
6730	1-1/2" diameter	"	94	.085		1.77	3.30		5.07	7.40
6740	2" diameter	Q-1	178	.090		2.09	3.14		5.23	7.45
6760	4" diameter	"	110	.145		5.70	5.05		10.75	14.65
7280	PEX, flexible, no couplings or hangers									
7282	Note: For labor costs add 25% to the couplings and fittings labor total.									
7285	For fittings see section 23 83 16.10 7000									
7300	Non-barrier type, hot/cold tubing rolls									
7310	1/4" diameter x 100'				L.F.	.49			.49	.54
7350	3/8" diameter x 100'					.55			.55	.61
7360	1/2" diameter x 100'					.61			.61	.67
7370	1/2" diameter x 500'					.61			.61	.67
7380	1/2" diameter x 1000'					.61			.61	.67
7400	3/4" diameter x 100'					1.11			1.11	1.22
7410	3/4" diameter x 500'					1.11			1.11	1.22
7420	3/4" diameter x 1000'					1.11			1.11	1.22
7460	1" diameter x 100'					1.90			1.90	2.09
7470	1" diameter x 300'					1.90			1.90	2.09
7480	1" diameter x 500'					1.90			1.90	2.09
7500	1-1/4" diameter x 100'					3.23			3.23	3.55
7510	1-1/4" diameter x 300'					3.23			3.23	3.55
7540	1-1/2" diameter x 100'					4.39			4.39	4.83
7550	1-1/2" diameter x 300'					4.39			4.39	4.83
7596	Most sizes available in red or blue									
7700	Non-barrier type, hot/cold tubing straight lengths									
7710	1/2" diameter x 20'				L.F.	.60			.60	.66
7750	3/4" diameter x 20'					1.10			1.10	1.21
7760	1" diameter x 20'					1.90			1.90	2.09
7770	1-1/4" diameter x 20'					3.23			3.23	3.55
7780	1-1/2" diameter x 20'					4.39			4.39	4.83
7790	2" diameter					8.60			8.60	9.45
7796	Most sizes available in red or blue									

22 11 13.76 Pipe Fittings, Plastic

		Crew	Daily Output	Labor-Hours	Unit	Material	2015 Bare Costs Labor	Equipment	Total	Total Incl O&P
0010	**PIPE FITTINGS, PLASTIC**									
2700	PVC (white), schedule 40, socket joints									
2760	90° elbow, 1/2"	1 Plum	33.30	.240	Ea.	.48	9.30		9.78	15.90
2770	3/4"		28.60	.280		.54	10.85		11.39	18.50
2780	1"		25	.320		.97	12.40		13.37	21.50
2790	1-1/4"		22.20	.360		1.71	13.95		15.66	25
2800	1-1/2"		20	.400		1.85	15.50		17.35	27.50
2810	2"	Q-1	36.40	.440		2.89	15.35		18.24	28.50
2820	2-1/2"		26.70	.599		8.80	21		29.80	44
2830	3"		22.90	.699		10.50	24.50		35	51.50
2840	4"		18.20	.879		18.80	30.50		49.30	71

22 11 Facility Water Distribution

22 11 13 – Facility Water Distribution Piping

22 11 13.76 Pipe Fittings, Plastic

		Crew	Daily Output	Labor-Hours	Unit	Material	2015 Bare Costs Labor	Equipment	Total	Total Incl O&P
3180	Tee, 1/2"	1 Plum	22.20	.360	Ea.	.60	13.95		14.55	23.50
3190	3/4"		19	.421		.69	16.30		16.99	28
3200	1"		16.70	.479		1.29	18.55		19.84	32
3210	1-1/4"		14.80	.541		2.01	21		23.01	36.50
3220	1-1/2"		13.30	.602		2.45	23.50		25.95	41
3230	2"	Q-1	24.20	.661		3.56	23		26.56	42
3240	2-1/2"		17.80	.899		11.75	31.50		43.25	64.50
3250	3"		15.20	1.053		15.45	36.50		51.95	77.50
3260	4"		12.10	1.322		28	46		74	107
3380	Coupling, 1/2"	1 Plum	33.30	.240		.32	9.30		9.62	15.70
3390	3/4"		28.60	.280		.44	10.85		11.29	18.40
3400	1"		25	.320		.77	12.40		13.17	21.50
3410	1-1/4"		22.20	.360		1.06	13.95		15.01	24
3420	1-1/2"		20	.400		1.13	15.50		16.63	26.50
3430	2"	Q-1	36.40	.440		1.72	15.35		17.07	27.50
3440	2-1/2"		26.70	.599		3.81	21		24.81	38.50
3450	3"		22.90	.699		5.95	24.50		30.45	46.50
3460	4"		18.20	.879		8.65	30.50		39.15	60
4500	DWV, ABS, non pressure, socket joints									
4540	1/4 Bend, 1-1/4"	1 Plum	20.20	.396	Ea.	2.57	15.35		17.92	28.50
4560	1-1/2"	"	18.20	.440		1.92	17.05		18.97	30
4570	2"	Q-1	33.10	.483		3.05	16.85		19.90	31.50
4650	1/8 Bend, same as 1/4 Bend									
4800	Tee, sanitary									
4820	1-1/4"	1 Plum	13.50	.593	Ea.	3.42	23		26.42	42
4830	1-1/2"	"	12.10	.661		2.92	25.50		28.42	45.50
4840	2"	Q-1	20	.800		4.52	28		32.52	51
5000	DWV, PVC, schedule 40, socket joints									
5040	1/4 bend, 1-1/4"	1 Plum	20.20	.396	Ea.	5.20	15.35		20.55	31
5060	1-1/2"	"	18.20	.440		1.48	17.05		18.53	29.50
5070	2"	Q-1	33.10	.483		2.33	16.85		19.18	30.50
5080	3"		20.80	.769		6.85	27		33.85	52
5090	4"		16.50	.970		13.05	34		47.05	70.50
5110	1/4 bend, long sweep, 1-1/2"	1 Plum	18.20	.440		3.42	17.05		20.47	32
5112	2"	Q-1	33.10	.483		3.83	16.85		20.68	32
5114	3"		20.80	.769		8.75	27		35.75	54
5116	4"		16.50	.970		16.65	34		50.65	74.50
5250	Tee, sanitary 1-1/4"	1 Plum	13.50	.593		5.20	23		28.20	44
5254	1-1/2"	"	12.10	.661		2.58	25.50		28.08	45.50
5255	2"	Q-1	20	.800		3.80	28		31.80	50
5256	3"		13.90	1.151		10	40		50	77
5257	4"		11	1.455		17.75	50.50		68.25	103
5259	6"		6.70	2.388		74	83.50		157.50	219
5261	8"	Q-2	6.20	3.871		173	130		303	405
5264	2" x 1-1/2"	Q-1	22	.727		3.35	25.50		28.85	45.50
5266	3" x 1-1/2"		15.50	1.032		7.05	36		43.05	67.50
5268	4" x 3"		12.10	1.322		21.50	46		67.50	99.50
5271	6" x 4"		6.90	2.319		71.50	81		152.50	212
5314	Combination Y & 1/8 bend, 1-1/2"	1 Plum	12.10	.661		6.25	25.50		31.75	49.50
5315	2"	Q-1	20	.800		6.70	28		34.70	53.50
5317	3"		13.90	1.151		16.60	40		56.60	84.50
5318	4"		11	1.455		33	50.50		83.50	120
5324	Combination Y & 1/8 bend, reducing									

22 11 Facility Water Distribution

22 11 13 – Facility Water Distribution Piping

22 11 13.76 Pipe Fittings, Plastic		Crew	Daily Output	Labor-Hours	Unit	Material	2015 Bare Costs Labor	Equipment	Total	Total Incl O&P
5325	2" x 2" x 1-1/2"	Q-1	22	.727	Ea.	8.80	25.50		34.30	51.50
5327	3" x 3" x 1-1/2"		15.50	1.032		15.65	36		51.65	76.50
5328	3" x 3" x 2"		15.30	1.046		11.35	36.50		47.85	72.50
5329	4" x 4" x 2"		12.20	1.311		17.85	45.50		63.35	95
5331	Wye, 1-1/4"	1 Plum	13.50	.593		6.65	23		29.65	45.50
5332	1-1/2"	"	12.10	.661		4.87	25.50		30.37	48
5333	2"	Q-1	20	.800		4.63	28		32.63	51
5334	3"		13.90	1.151		12.50	40		52.50	80
5335	4"		11	1.455		22.50	50.50		73	109
5336	6"		6.70	2.388		68.50	83.50		152	213
5337	8"	Q-2	6.20	3.871		124	130		254	350
5341	2" x 1-1/2"	Q-1	22	.727		5.70	25.50		31.20	48.50
5342	3" x 1-1/2"		15.50	1.032		8.20	36		44.20	68.50
5343	4" x 3"		12.10	1.322		18.40	46		64.40	96
5344	6" x 4"		6.90	2.319		50.50	81		131.50	189
5345	8" x 6"	Q-2	6.40	3.750		93.50	126		219.50	310
5347	Double wye, 1-1/2"	1 Plum	9.10	.879		10.55	34		44.55	67.50
5348	2"	Q-1	16.60	.964		12.20	33.50		45.70	69
5349	3"		10.40	1.538		25	53.50		78.50	116
5350	4"		8.25	1.939		50.50	67.50		118	168
5353	Double wye, reducing									
5354	2" x 2" x 1-1/2" x 1-1/2"	Q-1	16.80	.952	Ea.	10.75	33		43.75	67
5355	3" x 3" x 2" x 2"		10.60	1.509		18.45	52.50		70.95	108
5356	4" x 4" x 3" x 3"		8.45	1.893		40	66		106	153
5357	6" x 6" x 4" x 4"		7.25	2.207		138	77		215	279
5374	Coupling, 1-1/4"	1 Plum	20.20	.396		3.15	15.35		18.50	29
5376	1-1/2"	"	18.20	.440		.72	17.05		17.77	29
5378	2"	Q-1	33.10	.483		.95	16.85		17.80	29
5380	3"		20.80	.769		3.29	27		30.29	48
5390	4"		16.50	.970		5.65	34		39.65	62
5410	Reducer bushing, 2" x 1-1/4"		36.50	.438		2.82	15.30		18.12	28
5412	3" x 1-1/2"		27.30	.586		5.85	20.50		26.35	40
5414	4" x 2"		18.20	.879		10.20	30.50		40.70	61.50
5416	6" x 4"		11.10	1.441		27	50.50		77.50	113
5418	8" x 6"	Q-2	10.20	2.353		46	79		125	181
5500	CPVC, Schedule 80, threaded joints									
5540	90° Elbow, 1/4"	1 Plum	32	.250	Ea.	12.15	9.70		21.85	29.50
5560	1/2"		30.30	.264		7.05	10.25		17.30	24.50
5570	3/4"		26	.308		10.55	11.90		22.45	31.50
5580	1"		22.70	.352		14.80	13.65		28.45	39
5590	1-1/4"		20.20	.396		28.50	15.35		43.85	57
5600	1-1/2"		18.20	.440		30.50	17.05		47.55	62
5610	2"	Q-1	33.10	.483		41	16.85		57.85	73
5730	Coupling, 1/4"	1 Plum	32	.250		15.45	9.70		25.15	33
5732	1/2"		30.30	.264		12.75	10.25		23	31
5734	3/4"		26	.308		20.50	11.90		32.40	42
5736	1"		22.70	.352		23.50	13.65		37.15	48.50
5738	1-1/4"		20.20	.396		25	15.35		40.35	52.50
5740	1-1/2"		18.20	.440		26.50	17.05		43.55	57.50
5742	2"	Q-1	33.10	.483		31.50	16.85		48.35	62.50
5900	CPVC, Schedule 80, socket joints									
5904	90° Elbow, 1/4"	1 Plum	32	.250	Ea.	11.55	9.70		21.25	28.50
5906	1/2"		30.30	.264		4.53	10.25		14.78	22

22 11 Facility Water Distribution

22 11 13 – Facility Water Distribution Piping

22 11 13.76 Pipe Fittings, Plastic

		Crew	Daily Output	Labor-Hours	Unit	Material	2015 Bare Costs Labor	Equipment	Total	Total Incl O&P
5908	3/4"	1 Plum	26	.308	Ea.	5.80	11.90		17.70	26
5910	1"		22.70	.352		9.20	13.65		22.85	32.50
5912	1-1/4"		20.20	.396		19.85	15.35		35.20	47.50
5914	1-1/2"		18.20	.440		22	17.05		39.05	52.50
5916	2"	Q-1	33.10	.483		26.50	16.85		43.35	57.50
5930	45° Elbow, 1/4"	1 Plum	32	.250		17.20	9.70		26.90	35
5932	1/2"		30.30	.264		5.55	10.25		15.80	23
5934	3/4"		26	.308		7.95	11.90		19.85	28.50
5936	1"		22.70	.352		12.75	13.65		26.40	36.50
5938	1-1/4"		20.20	.396		25	15.35		40.35	53
5940	1-1/2"		18.20	.440		25.50	17.05		42.55	56
5942	2"	Q-1	33.10	.483		29	16.85		45.85	60
5990	Coupling, 1/4"	1 Plum	32	.250		12.30	9.70		22	29.50
5992	1/2"		30.30	.264		4.79	10.25		15.04	22
5994	3/4"		26	.308		6.70	11.90		18.60	27
5996	1"		22.70	.352		9	13.65		22.65	32.50
5998	1-1/4"		20.20	.396		13.50	15.35		28.85	40.50
6000	1-1/2"		18.20	.440		17	17.05		34.05	46.50
6002	2"	Q-1	33.10	.483		19.75	16.85		36.60	49.50

22 11 19 – Domestic Water Piping Specialties

22 11 19.38 Water Supply Meters

		Crew	Daily Output	Labor-Hours	Unit	Material	Labor	Equipment	Total	Total Incl O&P
0010	**WATER SUPPLY METERS**									
2000	Domestic/commercial, bronze									
2020	Threaded									
2060	5/8" diameter, to 20 GPM	1 Plum	16	.500	Ea.	50	19.40		69.40	87
2080	3/4" diameter, to 30 GPM		14	.571		91	22		113	137
2100	1" diameter, to 50 GPM		12	.667		138	26		164	195

22 11 19.42 Backflow Preventers

		Crew	Daily Output	Labor-Hours	Unit	Material	Labor	Equipment	Total	Total Incl O&P
0010	**BACKFLOW PREVENTERS**, Includes valves									
0020	and four test cocks, corrosion resistant, automatic operation									
4000	Reduced pressure principle									
4100	Threaded, bronze, valves are ball									
4120	3/4" pipe size	1 Plum	16	.500	Ea.	445	19.40		464.40	520

22 11 19.50 Vacuum Breakers

		Crew	Daily Output	Labor-Hours	Unit	Material	Labor	Equipment	Total	Total Incl O&P
0010	**VACUUM BREAKERS**									
0013	See also backflow preventers Section 22 11 19.42									
1000	Anti-siphon continuous pressure type									
1010	Max. 150 PSI - 210°F									
1020	Bronze body									
1030	1/2" size	1 Stpi	24	.333	Ea.	187	13.15		200.15	228
1040	3/4" size		20	.400		187	15.80		202.80	232
1050	1" size		19	.421		194	16.60		210.60	241
1060	1-1/4" size		15	.533		380	21		401	455
1070	1-1/2" size		13	.615		470	24.50		494.50	555
1080	2" size		11	.727		485	28.50		513.50	580
1200	Max. 125 PSI with atmospheric vent									
1210	Brass, in-line construction									
1220	1/4" size	1 Stpi	24	.333	Ea.	117	13.15		130.15	151
1230	3/8" size	"	24	.333		117	13.15		130.15	151
1260	For polished chrome finish, add					13%				
2000	Anti-siphon, non-continuous pressure type									
2010	Hot or cold water 125 PSI - 210°F									

22 11 Facility Water Distribution

22 11 19 – Domestic Water Piping Specialties

22 11 19.50 Vacuum Breakers

		Crew	Daily Output	Labor-Hours	Unit	Material	2015 Bare Costs Labor	2015 Bare Costs Equipment	Total	Total Incl O&P
2020	Bronze body									
2030	1/4" size	1 Stpi	24	.333	Ea.	64	13.15		77.15	92
2040	3/8" size		24	.333		64	13.15		77.15	92
2050	1/2" size		24	.333		72.50	13.15		85.65	102
2060	3/4" size		20	.400		86.50	15.80		102.30	121
2070	1" size		19	.421		134	16.60		150.60	175
2080	1-1/4" size		15	.533		235	21		256	293
2090	1-1/2" size		13	.615		276	24.50		300.50	345
2100	2" size		11	.727		430	28.50		458.50	520
2110	2-1/2" size		8	1		1,225	39.50		1,264.50	1,425
2120	3" size	▼	6	1.333	▼	1,625	52.50		1,677.50	1,875
2150	For polished chrome finish, add					50%				

22 11 19.54 Water Hammer Arresters/Shock Absorbers

		Crew	Daily Output	Labor-Hours	Unit	Material	Labor	Equipment	Total	Total Incl O&P
0010	**WATER HAMMER ARRESTERS/SHOCK ABSORBERS**									
0490	Copper									
0500	3/4" male I.P.S. For 1 to 11 fixtures	1 Plum	12	.667	Ea.	28	26		54	73.50

22 13 Facility Sanitary Sewerage

22 13 16 – Sanitary Waste and Vent Piping

22 13 16.20 Pipe, Cast Iron

		Crew	Daily Output	Labor-Hours	Unit	Material	Labor	Equipment	Total	Total Incl O&P
0010	**PIPE, CAST IRON**, Soil, on clevis hanger assemblies, 5' O.C. R221113-50									
0020	Single hub, service wt., lead & oakum joints 10' O.C.									
2120	2" diameter	Q-1	63	.254	L.F.	11.35	8.85		20.20	27
2140	3" diameter		60	.267		15	9.30		24.30	32
2160	4" diameter	▼	55	.291	▼	18.55	10.15		28.70	37.50
4000	No hub, couplings 10' O.C.									
4100	1-1/2" diameter	Q-1	71	.225	L.F.	11.05	7.85		18.90	25
4120	2" diameter		67	.239		11.25	8.35		19.60	26
4140	3" diameter		64	.250		15.20	8.70		23.90	31
4160	4" diameter	▼	58	.276		19.10	9.60		28.70	37

22 13 16.30 Pipe Fittings, Cast Iron

		Crew	Daily Output	Labor-Hours	Unit	Material	Labor	Equipment	Total	Total Incl O&P
0010	**PIPE FITTINGS, CAST IRON**, Soil									
0040	Hub and spigot, service weight, lead & oakum joints									
0080	1/4 bend, 2"	Q-1	16	1	Ea.	19.90	35		54.90	79.50
0120	3"		14	1.143		26.50	40		66.50	95
0140	4"		13	1.231		41.50	43		84.50	117
0340	1/8 bend, 2"		16	1		14.15	35		49.15	73
0350	3"		14	1.143		22	40		62	90.50
0360	4"		13	1.231		33.50	43		76.50	108
0500	Sanitary tee, 2"		10	1.600		27.50	56		83.50	123
0540	3"		9	1.778		45	62		107	152
0620	4"	▼	8	2	▼	55	70		125	176
5990	No hub									
6000	Cplg. & labor required at joints not incl. in fitting									
6010	price. Add 1 coupling per joint for installed price									
6020	1/4 Bend, 1-1/2"				Ea.	10.20			10.20	11.25
6060	2"					11.15			11.15	12.25
6080	3"					16.25			16.25	17.85
6120	4"					23			23	25.50
6184	1/4 Bend, long sweep, 1-1/2"				▼	25.50			25.50	28

22 13 Facility Sanitary Sewerage

22 13 16 – Sanitary Waste and Vent Piping

22 13 16.30 Pipe Fittings, Cast Iron		Crew	Daily Output	Labor-Hours	Unit	Material	2015 Bare Costs Labor	Equipment	Total	Total Incl O&P
6186	2"				Ea.	24.50			24.50	26.50
6188	3"					29.50			29.50	32.50
6189	4"					47			47	51.50
6190	5"					90.50			90.50	99.50
6191	6"					103			103	114
6192	8"					282			282	310
6193	10"					570			570	625
6200	1/8 Bend, 1-1/2"					8.60			8.60	9.45
6210	2"					9.60			9.60	10.55
6212	3"					12.80			12.80	14.10
6214	4"					16.80			16.80	18.50
6380	Sanitary Tee, tapped, 1-1/2"					20.50			20.50	22.50
6382	2" x 1-1/2"					17.95			17.95	19.70
6384	2"					19.25			19.25	21
6386	3" x 2"					28.50			28.50	31.50
6388	3"					49.50			49.50	54.50
6390	4" x 1-1/2"					25.50			25.50	28
6392	4" x 2"					29			29	32
6393	4"					29			29	32
6394	6" x 1-1/2"					66			66	73
6396	6" x 2"					67.50			67.50	74
6459	Sanitary Tee, 1-1/2"					14.35			14.35	15.80
6460	2"					15.35			15.35	16.85
6470	3"					18.90			18.90	21
6472	4"					36			36	39.50
8000	Coupling, standard (by CISPI Mfrs.)									
8020	1-1/2"	Q-1	48	.333	Ea.	13.95	11.65		25.60	34.50
8040	2"		44	.364		15.25	12.70		27.95	38
8080	3"		38	.421		17.10	14.70		31.80	43
8120	4"		33	.485		20	16.90		36.90	50

22 13 16.50 Shower Drains

		Crew	Daily Output	Labor-Hours	Unit	Material	Labor	Equipment	Total	Total Incl O&P
0010	**SHOWER DRAINS**									
2780	Shower, with strainer, uniform diam. trap, bronze top									
2800	2" and 3" pipe size	Q-1	8	2	Ea.	480	70		550	645
2820	4" pipe size	"	7	2.286		485	79.50		564.50	660
2840	For galvanized body, add					189			189	208

22 13 16.60 Traps

		Crew	Daily Output	Labor-Hours	Unit	Material	Labor	Equipment	Total	Total Incl O&P
0010	**TRAPS**									
0030	Cast iron, service weight									
0050	Running P trap, without vent									
1100	2"	Q-1	16	1	Ea.	143	35		178	215
1150	4"	"	13	1.231		143	43		186	228
1160	6"	Q-2	17	1.412		635	47.50		682.50	775
3000	P trap, B&S, 2" pipe size	Q-1	16	1		34	35		69	95
3040	3" pipe size	"	14	1.143		50.50	40		90.50	122
4700	Copper, drainage, drum trap									
4840	3" x 6" swivel, 1-1/2" pipe size	1 Plum	16	.500	Ea.	160	19.40		179.40	208
5100	P trap, standard pattern									
5200	1-1/4" pipe size	1 Plum	18	.444	Ea.	78.50	17.20		95.70	115
5240	1-1/2" pipe size		17	.471		72	18.25		90.25	109
5260	2" pipe size		15	.533		111	20.50		131.50	156
5280	3" pipe size		11	.727		281	28		309	355

22 13 Facility Sanitary Sewerage

22 13 16 – Sanitary Waste and Vent Piping

22 13 16.60 Traps		Crew	Daily Output	Labor-Hours	Unit	Material	2015 Bare Costs Labor	Equipment	Total	Total Incl O&P
6710	ABS DWV P trap, solvent weld joint									
6720	1-1/2" pipe size	1 Plum	18	.444	Ea.	6.30	17.20		23.50	35.50
6722	2" pipe size		17	.471		8.30	18.25		26.55	39
6724	3" pipe size		15	.533		32.50	20.50		53	70
6726	4" pipe size		14	.571		65	22		87	108
6732	PVC DWV P trap, solvent weld joint									
6733	1-1/2" pipe size	1 Plum	18	.444	Ea.	4.95	17.20		22.15	34
6734	2" pipe size		17	.471		6.70	18.25		24.95	37.50
6735	3" pipe size		15	.533		22.50	20.50		43	59
6736	4" pipe size		14	.571		51.50	22		73.50	93.50
6860	PVC DWV hub x hub, basin trap, 1-1/4" pipe size		18	.444		40	17.20		57.20	72.50
6870	Sink P trap, 1-1/2" pipe size		18	.444		9.65	17.20		26.85	39
6880	Tubular S trap, 1-1/2" pipe size		17	.471		23.50	18.25		41.75	55.50
6890	PVC sch. 40 DWV, drum trap									
6900	1-1/2" pipe size	1 Plum	16	.500	Ea.	23.50	19.40		42.90	58
6910	P trap, 1-1/2" pipe size		18	.444		5.85	17.20		23.05	35
6920	2" pipe size		17	.471		7.90	18.25		26.15	38.50
6930	3" pipe size		15	.533		27	20.50		47.50	63.50
6940	4" pipe size		14	.571		61	22		83	104
6950	P trap w/clean out, 1-1/2" pipe size		18	.444		9.65	17.20		26.85	39
6960	2" pipe size		17	.471		16.40	18.25		34.65	48

22 13 16.80 Vent Flashing and Caps

		Crew	Daily Output	Labor-Hours	Unit	Material	Labor	Equipment	Total	Total Incl O&P
0010	**VENT FLASHING AND CAPS**									
0120	Vent caps									
0140	Cast iron									
0160	1-1/4" - 1-1/2" pipe	1 Plum	23	.348	Ea.	34	13.50		47.50	59.50
0170	2" - 2-1/8" pipe		22	.364		39.50	14.10		53.60	67
0180	2-1/2" - 3-5/8" pipe		21	.381		45	14.75		59.75	74
0190	4" - 4-1/8" pipe		19	.421		55	16.30		71.30	87.50
0200	5" - 6" pipe		17	.471		83.50	18.25		101.75	122
0300	PVC									
0320	1-1/4" - 1-1/2" pipe	1 Plum	24	.333	Ea.	9.20	12.90		22.10	31.50
0330	2" - 2-1/8" pipe	"	23	.348	"	9.30	13.50		22.80	32
0900	Vent flashing									
1350	Copper with neoprene ring									
1400	1-1/4" pipe	1 Plum	20	.400	Ea.	61.50	15.50		77	93
1430	1-1/2" pipe		20	.400		61.50	15.50		77	93
1440	2" pipe		18	.444		61.50	17.20		78.70	96
1450	3" pipe		17	.471		74.50	18.25		92.75	112
1460	4" pipe		16	.500		74.50	19.40		93.90	114
2980	Neoprene, one piece									
3000	1-1/4" pipe	1 Plum	24	.333	Ea.	5.05	12.90		17.95	27
3030	1-1/2" pipe		24	.333		5.10	12.90		18	27
3040	2" pipe		23	.348		5.10	13.50		18.60	27.50
3050	3" pipe		21	.381		5.85	14.75		20.60	31
3060	4" pipe		20	.400		8.65	15.50		24.15	35

22 13 19 – Sanitary Waste Piping Specialties

22 13 19.13 Sanitary Drains

		Crew	Daily Output	Labor-Hours	Unit	Material	Labor	Equipment	Total	Total Incl O&P
0010	**SANITARY DRAINS**									
2000	Floor, medium duty, C.I., deep flange, 7" diam. top									
2040	2" and 3" pipe size	Q-1	12	1.333	Ea.	208	46.50		254.50	305
2080	For galvanized body, add					96.50			96.50	106

22 13 Facility Sanitary Sewerage

22 13 19 – Sanitary Waste Piping Specialties

22 13 19.13 Sanitary Drains		Crew	Daily Output	Labor-Hours	Unit	Material	2015 Bare Costs Labor	Equipment	Total	Total Incl O&P
2120	With polished bronze top				Ea.	315			315	345

22 14 Facility Storm Drainage

22 14 26 – Facility Storm Drains

22 14 26.13 Roof Drains

		Crew	Daily Output	Labor-Hours	Unit	Material	Labor	Equipment	Total	Total Incl O&P
0010	**ROOF DRAINS**									
3860	Roof, flat metal deck, C.I. body, 12" C.I. dome									
3890	3" pipe size	Q-1	14	1.143	Ea.	395	40		435	500

22 14 29 – Sump Pumps

22 14 29.16 Submersible Sump Pumps

		Crew	Daily Output	Labor-Hours	Unit	Material	Labor	Equipment	Total	Total Incl O&P
0010	**SUBMERSIBLE SUMP PUMPS**									
7000	Sump pump, automatic									
7100	Plastic, 1-1/4" discharge, 1/4 HP	1 Plum	6.40	1.250	Ea.	138	48.50		186.50	232
7500	Cast iron, 1-1/4" discharge, 1/4 HP	"	6	1.333	"	194	51.50		245.50	300

22 31 Domestic Water Softeners

22 31 13 – Residential Domestic Water Softeners

22 31 13.10 Residential Water Softeners

		Crew	Daily Output	Labor-Hours	Unit	Material	Labor	Equipment	Total	Total Incl O&P
0010	**RESIDENTIAL WATER SOFTENERS**									
7350	Water softener, automatic, to 30 grains per gallon	2 Plum	5	3.200	Ea.	405	124		529	650
7400	To 100 grains per gallon	"	4	4	"	660	155		815	980

22 33 Electric Domestic Water Heaters

22 33 30 – Residential, Electric Domestic Water Heaters

22 33 30.13 Residential, Small-Capacity Elec. Water Heaters

		Crew	Daily Output	Labor-Hours	Unit	Material	Labor	Equipment	Total	Total Incl O&P
0010	**RESIDENTIAL, SMALL-CAPACITY ELECTRIC DOMESTIC WATER HEATERS**									
1000	Residential, electric, glass lined tank, 5 yr., 10 gal., single element	1 Plum	2.30	3.478	Ea.	330	135		465	580
1060	30 gallon, double element		2.20	3.636		475	141		616	760
1080	40 gallon, double element		2	4		800	155		955	1,125
1100	52 gallon, double element		2	4		895	155		1,050	1,250
1120	66 gallon, double element		1.80	4.444		1,200	172		1,372	1,600
1140	80 gallon, double element		1.60	5		1,350	194		1,544	1,800

22 34 Fuel-Fired Domestic Water Heaters

22 34 13 – Instantaneous, Tankless, Gas Domestic Water Heaters

22 34 13.10 Instantaneous, Tankless, Gas Water Heaters

			Crew	Daily Output	Labor-Hours	Unit	Material	Labor	Equipment	Total	Total Incl O&P
0010	**INSTANTANEOUS, TANKLESS, GAS WATER HEATERS**										
9410	Natural gas/propane, 3.2 GPM	G	1 Plum	2	4	Ea.	370	155		525	660
9420	6.4 GPM	G		1.90	4.211		625	163		788	955
9430	8.4 GPM	G		1.80	4.444		730	172		902	1,075
9440	9.5 GPM	G		1.60	5		930	194		1,124	1,350

22 34 Fuel-Fired Domestic Water Heaters

22 34 30 – Residential Gas Domestic Water Heaters

22 34 30.13 Residential, Atmos, Gas Domestic Wtr Heaters

		Crew	Daily Output	Labor-Hours	Unit	Material	2015 Bare Costs Labor	2015 Bare Costs Equipment	Total	Total Incl O&P
0010	**RESIDENTIAL, ATMOSPHERIC, GAS DOMESTIC WATER HEATERS**									
2000	Gas fired, foam lined tank, 10 yr., vent not incl.									
2040	30 gallon	1 Plum	2	4	Ea.	895	155		1,050	1,250
2100	75 gallon	"	1.50	5.333	"	1,350	207		1,557	1,850

22 34 46 – Oil-Fired Domestic Water Heaters

22 34 46.10 Residential Oil-Fired Water Heaters

		Crew	Daily Output	Labor-Hours	Unit	Material	Labor	Equipment	Total	Incl O&P
0010	**RESIDENTIAL OIL-FIRED WATER HEATERS**									
3000	Oil fired, glass lined tank, 5 yr., vent not included, 30 gallon	1 Plum	2	4	Ea.	1,175	155		1,330	1,525
3040	50 gallon	"	1.80	4.444	"	1,375	172		1,547	1,800

22 41 Residential Plumbing Fixtures

22 41 13 – Residential Water Closets, Urinals, and Bidets

22 41 13.13 Water Closets

		Crew	Daily Output	Labor-Hours	Unit	Material	Labor	Equipment	Total	Incl O&P
0010	**WATER CLOSETS**									
0150	Tank type, vitreous china, incl. seat, supply pipe w/stop, 1.6 gpf or noted									
0200	Wall hung									
0400	Two piece, close coupled	Q-1	5.30	3.019	Ea.	630	105		735	870
0960	For rough-in, supply, waste, vent and carrier	"	2.73	5.861	"	1,025	204		1,229	1,450
0999	Floor mounted									
1020	One piece, low profile	Q-1	5.30	3.019	Ea.	1,125	105		1,230	1,400
1100	Two piece, close coupled		5.30	3.019		237	105		342	435
1102	Economy		5.30	3.019		132	105		237	320
1110	Two piece, close coupled, dual flush		5.30	3.019		310	105		415	520
1140	Two piece, close coupled, 1.28 gpf, ADA [G]		5.30	3.019		310	105		415	520
1960	For color, add					30%				
1980	For rough-in, supply, waste and vent	Q-1	3.05	5.246	Ea.	330	183		513	665

22 41 16 – Residential Lavatories and Sinks

22 41 16.13 Lavatories

		Crew	Daily Output	Labor-Hours	Unit	Material	Labor	Equipment	Total	Incl O&P
0010	**LAVATORIES**, With trim, white unless noted otherwise									
0500	Vanity top, porcelain enamel on cast iron									
0600	20" x 18"	Q-1	6.40	2.500	Ea.	335	87		422	515
0640	33" x 19" oval		6.40	2.500		475	87		562	670
0720	19" round		6.40	2.500		435	87		522	625
0860	For color, add					25%				
1000	Cultured marble, 19" x 17", single bowl	Q-1	6.40	2.500	Ea.	175	87		262	335
1120	25" x 22", single bowl		6.40	2.500		215	87		302	380
1160	37" x 22", single bowl		6.40	2.500		245	87		332	415
1580	For color, same price									
1900	Stainless steel, self-rimming, 25" x 22", single bowl, ledge	Q-1	6.40	2.500	Ea.	365	87		452	550
1960	17" x 22", single bowl		6.40	2.500		355	87		442	535
2600	Steel, enameled, 20" x 17", single bowl		5.80	2.759		161	96		257	335
2900	Vitreous china, 20" x 16", single bowl		5.40	2.963		260	103		363	455
3200	22" x 13", single bowl		5.40	2.963		267	103		370	465
3580	Rough-in, supply, waste and vent for all above lavatories		2.30	6.957		231	243		474	655
4000	Wall hung									
4040	Porcelain enamel on cast iron, 16" x 14", single bowl	Q-1	8	2	Ea.	520	70		590	685
4180	20" x 18", single bowl	"	8	2	"	277	70		347	420
4580	For color, add					30%				
6000	Vitreous china, 18" x 15", single bowl with backsplash	Q-1	7	2.286	Ea.	231	79.50		310.50	385

22 41 Residential Plumbing Fixtures

22 41 16 – Residential Lavatories and Sinks

22 41 16.13 Lavatories

		Crew	Daily Output	Labor-Hours	Unit	Material	2015 Bare Costs Labor	2015 Bare Costs Equipment	Total	Total Incl O&P
6060	19" x 17", single bowl	Q-1	7	2.286	Ea.	207	79.50		286.50	360
6960	Rough-in, supply, waste and vent for above lavatories	↓	1.66	9.639	↓	455	335		790	1,050
7000	Pedestal type									
7600	Vitreous china, 27" x 21", white	Q-1	6.60	2.424	Ea.	700	84.50		784.50	910
7610	27" x 21", colored		6.60	2.424		880	84.50		964.50	1,100
7620	27" x 21", premium color		6.60	2.424		995	84.50		1,079.50	1,250
7660	26" x 20", white		6.60	2.424		795	84.50		879.50	1,025
7670	26" x 20", colored		6.60	2.424		1,000	84.50		1,084.50	1,250
7680	26" x 20", premium color		6.60	2.424		1,125	84.50		1,209.50	1,400
7700	24" x 20", white		6.60	2.424		405	84.50		489.50	585
7710	24" x 20", colored		6.60	2.424		475	84.50		559.50	665
7720	24" x 20", premium color		6.60	2.424		495	84.50		579.50	685
7760	21" x 18", white		6.60	2.424		291	84.50		375.50	460
7770	21" x 18", colored		6.60	2.424		291	84.50		375.50	460
7990	Rough-in, supply, waste and vent for pedestal lavatories	↓	1.66	9.639	↓	455	335		790	1,050

22 41 16.16 Sinks

		Crew	Daily Output	Labor-Hours	Unit	Material	2015 Bare Costs Labor	2015 Bare Costs Equipment	Total	Total Incl O&P
0010	**SINKS**, With faucets and drain									
2000	Kitchen, counter top style, P.E. on C.I., 24" x 21" single bowl	Q-1	5.60	2.857	Ea.	285	99.50		384.50	480
2100	31" x 22" single bowl		5.60	2.857		615	99.50		714.50	845
2200	32" x 21" double bowl		4.80	3.333		355	116		471	580
3000	Stainless steel, self rimming, 19" x 18" single bowl		5.60	2.857		590	99.50		689.50	815
3100	25" x 22" single bowl		5.60	2.857		660	99.50		759.50	890
3200	33" x 22" double bowl		4.80	3.333		960	116		1,076	1,250
3300	43" x 22" double bowl		4.80	3.333		1,125	116		1,241	1,425
4000	Steel, enameled, with ledge, 24" x 21" single bowl		5.60	2.857		510	99.50		609.50	725
4100	32" x 21" double bowl	↓	4.80	3.333		495	116		611	735
4960	For color sinks except stainless steel, add					10%				
4980	For rough-in, supply, waste and vent, counter top sinks	Q-1	2.14	7.477	↓	260	261		521	715
5000	Kitchen, raised deck, P.E. on C.I.									
5100	32" x 21", dual level, double bowl	Q-1	2.60	6.154	Ea.	420	215		635	815
5790	For rough-in, supply, waste & vent, sinks	"	1.85	8.649	"	260	300		560	785

22 41 19 – Residential Bathtubs

22 41 19.10 Baths

		Crew	Daily Output	Labor-Hours	Unit	Material	2015 Bare Costs Labor	2015 Bare Costs Equipment	Total	Total Incl O&P
0010	**BATHS**									
0100	Tubs, recessed porcelain enamel on cast iron, with trim									
0180	48" x 42"	Q-1	4	4	Ea.	2,625	140		2,765	3,125
0220	72" x 36"	"	3	5.333	"	2,725	186		2,911	3,300
0300	Mat bottom									
0380	5' long	Q-1	4.40	3.636	Ea.	1,125	127		1,252	1,450
0480	Above floor drain, 5' long		4	4		785	140		925	1,100
0560	Corner 48" x 44"		4.40	3.636		2,625	127		2,752	3,100
2000	Enameled formed steel, 4'-6" long	↓	5.80	2.759	↓	495	96		591	705
4600	Module tub & showerwall surround, molded fiberglass									
4610	5' long x 34" wide x 76" high	Q-1	4	4	Ea.	840	140		980	1,150
9600	Rough-in, supply, waste and vent, for all above tubs, add	"	2.07	7.729	"	345	270		615	825

22 41 23 – Residential Showers

22 41 23.20 Showers

		Crew	Daily Output	Labor-Hours	Unit	Material	2015 Bare Costs Labor	2015 Bare Costs Equipment	Total	Total Incl O&P
0010	**SHOWERS**									
1500	Stall, with drain only. Add for valve and door/curtain									
1520	32" square	Q-1	5	3.200	Ea.	1,175	112		1,287	1,450
1530	36" square	↓	4.80	3.333	↓	2,925	116		3,041	3,400

22 41 Residential Plumbing Fixtures

22 41 23 – Residential Showers

22 41 23.20 Showers

		Crew	Daily Output	Labor-Hours	Unit	Material	2015 Bare Costs Labor	Equipment	Total	Total Incl O&P
1540	Terrazzo receptor, 32" square	Q-1	5	3.200	Ea.	1,350	112		1,462	1,650
1560	36" square		4.80	3.333		1,475	116		1,591	1,825
1580	36" corner angle		4.80	3.333		1,725	116		1,841	2,100
3000	Fiberglass, one piece, with 3 walls, 32" x 32" square		5.50	2.909		490	101		591	705
3100	36" x 36" square		5.50	2.909		505	101		606	720
4200	Rough-in, supply, waste and vent for above showers		2.05	7.805		345	272		617	825

22 41 36 – Residential Laundry Trays

22 41 36.10 Laundry Sinks

		Crew	Daily Output	Labor-Hours	Unit	Material	Labor	Equipment	Total	Total Incl O&P
0010	**LAUNDRY SINKS**, With trim									
0020	Porcelain enamel on cast iron, black iron frame									
0050	24" x 21", single compartment	Q-1	6	2.667	Ea.	580	93		673	790
0100	26" x 21", single compartment	"	6	2.667	"	610	93		703	825
3000	Plastic, on wall hanger or legs									
3020	18" x 23", single compartment	Q-1	6.50	2.462	Ea.	135	86		221	291
3100	20" x 24", single compartment		6.50	2.462		154	86		240	310
3200	36" x 23", double compartment		5.50	2.909		185	101		286	370
3300	40" x 24", double compartment		5.50	2.909		278	101		379	470
5000	Stainless steel, counter top, 22" x 17" single compartment		6	2.667		64	93		157	224
5200	33" x 22", double compartment		5	3.200		79	112		191	271
9600	Rough-in, supply, waste and vent, for all laundry sinks		2.14	7.477		260	261		521	715

22 41 39 – Residential Faucets, Supplies and Trim

22 41 39.10 Faucets and Fittings

		Crew	Daily Output	Labor-Hours	Unit	Material	Labor	Equipment	Total	Total Incl O&P
0010	**FAUCETS AND FITTINGS**									
0150	Bath, faucets, diverter spout combination, sweat	1 Plum	8	1	Ea.	86.50	39		125.50	159
0200	For integral stops, IPS unions, add					109			109	120
0420	Bath, press-bal mix valve w/diverter, spout, shower head, arm/flange	1 Plum	8	1		168	39		207	249
0500	Drain, central lift, 1-1/2" IPS male		20	.400		71	15.50		86.50	104
0600	Trip lever, 1-1/2" IPS male		20	.400		45	15.50		60.50	75
1000	Kitchen sink faucets, top mount, cast spout		10	.800		61.50	31		92.50	119
1100	For spray, add		24	.333		16.15	12.90		29.05	39.50
1300	Single control lever handle									
1310	With pull out spray									
1320	Polished chrome	1 Plum	10	.800	Ea.	196	31		227	266
2000	Laundry faucets, shelf type, IPS or copper unions		12	.667		49.50	26		75.50	97
2100	Lavatory faucet, centerset, without drain		10	.800		44.50	31		75.50	100
2120	With pop-up drain		6.66	1.201		62.50	46.50		109	146
2210	Porcelain cross handles and pop-up drain									
2220	Polished chrome	1 Plum	6.66	1.201	Ea.	189	46.50		235.50	285
2230	Polished brass	"	6.66	1.201	"	298	46.50		344.50	405
2260	Single lever handle and pop-up drain									
2280	Satin nickel	1 Plum	6.66	1.201	Ea.	277	46.50		323.50	380
2290	Polished chrome		6.66	1.201		198	46.50		244.50	295
2800	Self-closing, center set		10	.800		131	31		162	195
4000	Shower by-pass valve with union		18	.444		68.50	17.20		85.70	104
4200	Shower thermostatic mixing valve, concealed, with shower head trim kit		8	1		345	39		384	445
4220	Shower pressure balancing mixing valve,									
4230	With shower head, arm, flange and diverter tub spout									
4240	Chrome	1 Plum	6.14	1.303	Ea.	360	50.50		410.50	480
4250	Satin nickel		6.14	1.303		555	50.50		605.50	695
4260	Polished graphite		6.14	1.303		555	50.50		605.50	695
5000	Sillcock, compact, brass, IPS or copper to hose		24	.333		9.70	12.90		22.60	32

22 41 Residential Plumbing Fixtures

22 41 39 – Residential Faucets, Supplies and Trim

22 41 39.70 Washer/Dryer Accessories

		Crew	Daily Output	Labor-Hours	Unit	Material	2015 Bare Costs Labor	2015 Bare Costs Equipment	Total	Total Incl O&P
0010	**WASHER/DRYER ACCESSORIES**									
1020	Valves ball type single lever									
1030	1/2" diam., IPS	1 Plum	21	.381	Ea.	54.50	14.75		69.25	84.50
1040	1/2" diam., solder	"	21	.381	"	54.50	14.75		69.25	84.50
1050	Recessed box, 16 ga., two hose valves and drain									
1060	1/2" size, 1-1/2" drain	1 Plum	18	.444	Ea.	116	17.20		133.20	157
1070	1/2" size, 2" drain	"	17	.471	"	108	18.25		126.25	149
1080	With grounding electric receptacle									
1090	1/2" size, 1-1/2" drain	1 Plum	18	.444	Ea.	127	17.20		144.20	169
1100	1/2" size, 2" drain	"	17	.471	"	138	18.25		156.25	181
1110	With grounding and dryer receptacle									
1120	1/2" size, 1-1/2" drain	1 Plum	18	.444	Ea.	157	17.20		174.20	202
1130	1/2" size, 2" drain	"	17	.471	"	159	18.25		177.25	205
1140	Recessed box 16 ga., ball valves with single lever and drain									
1150	1/2" size, 1-1/2" drain	1 Plum	19	.421	Ea.	219	16.30		235.30	268
1160	1/2" size, 2" drain	"	18	.444	"	197	17.20		214.20	246
1170	With grounding electric receptacle									
1180	1/2" size, 1-1/2" drain	1 Plum	19	.421	Ea.	190	16.30		206.30	236
1190	1/2" size, 2" drain	"	18	.444	"	211	17.20		228.20	261
1200	With grounding and dryer receptacles									
1210	1/2" size, 1-1/2" drain	1 Plum	19	.421	Ea.	208	16.30		224.30	255
1220	1/2" size, 2" drain	"	18	.444	"	230	17.20		247.20	283
1300	Recessed box, 20 ga., two hose valves and drain (economy type)									
1310	1/2" size, 1-1/2" drain	1 Plum	19	.421	Ea.	90.50	16.30		106.80	127
1320	1/2" size, 2" drain		18	.444		86	17.20		103.20	123
1330	Box with drain only		24	.333		52.50	12.90		65.40	79
1340	1/2" size, 1-1/2" ABS/PVC drain		19	.421		100	16.30		116.30	137
1350	1/2" size, 2" ABS/PVC drain		18	.444		108	17.20		125.20	147
1352	Box with drain and 15 A receptacle		24	.333		90.50	12.90		103.40	121
1360	1/2" size, 2" drain ABS/PVC, 15 A receptacle		24	.333		127	12.90		139.90	162
1400	Wall mounted									
1410	1/2" size, 1-1/2" plastic drain	1 Plum	19	.421	Ea.	24.50	16.30		40.80	54
1420	1/2" size, 2" plastic drain	"	18	.444	"	21	17.20		38.20	51.50
1500	Dryer vent kit									
1510	8' flex duct, clamps and outside hood	1 Plum	20	.400	Ea.	15.30	15.50		30.80	42.50
1980	Rough-in, supply, waste, and vent for washer boxes		3.46	2.310		244	89.50		333.50	415
9605	Washing machine valve assembly, hot & cold water supply, recessed		8	1		70	39		109	141
9610	Washing machine valve assembly, hot & cold water supply, mounted		8	1		54.50	39		93.50	124

22 42 Commercial Plumbing Fixtures

22 42 13 – Commercial Water Closets, Urinals, and Bidets

22 42 13.13 Water Closets

		Crew	Daily Output	Labor-Hours	Unit	Material	2015 Bare Costs Labor	2015 Bare Costs Equipment	Total	Total Incl O&P
0010	**WATER CLOSETS**									
3000	Bowl only, with flush valve, seat, 1.6 gpf unless noted									
3100	Wall hung	Q-1	5.80	2.759	Ea.	945	96		1,041	1,200
3200	For rough-in, supply, waste and vent, single WC		2.56	6.250		1,075	218		1,293	1,525
3300	Floor mounted		5.80	2.759		315	96		411	505
3370	For rough-in, supply, waste and vent, single WC		2.84	5.634		370	197		567	730
3390	Floor mounted children's size, 10-3/4" high									
3392	With automatic flush sensor, 1.6 gpf	Q-1	6.20	2.581	Ea.	615	90		705	830

22 42 Commercial Plumbing Fixtures

22 42 13 – Commercial Water Closets, Urinals, and Bidets

22 42 13.13 Water Closets		Crew	Daily Output	Labor-Hours	Unit	Material	2015 Bare Costs Labor	Equipment	Total	Total Incl O&P
3396	With automatic flush sensor, 1.28 gpf	Q-1	6.20	2.581	Ea.	620	90		710	830
3400	For rough-in, supply, waste and vent, single WC		2.84	5.634		370	197		567	730

22 42 16 – Commercial Lavatories and Sinks

22 42 16.13 Lavatories

0010	LAVATORIES, With trim, white unless noted otherwise
0020	Commercial lavatories same as residential. See Section 22 41 16

22 42 16.40 Service Sinks

		Crew	Daily Output	Labor-Hours	Unit	Material	Labor	Equipment	Total	Incl O&P
0010	SERVICE SINKS									
6650	Service, floor, corner, P.E. on C.I., 28" x 28"	Q-1	4.40	3.636	Ea.	1,000	127		1,127	1,300
6750	Vinyl coated rim guard, add					67.50			67.50	74
6760	Mop sink, molded stone, 24" x 36"	1 Plum	3.33	2.402		273	93		366	455
6770	Mop sink, molded stone, 24" x 36", w/rim 3 sides	"	3.33	2.402		278	93		371	460
6790	For rough-in, supply, waste & vent, floor service sinks	Q-1	1.64	9.756		705	340		1,045	1,325

22 42 39 – Commercial Faucets, Supplies, and Trim

22 42 39.10 Faucets and Fittings

		Crew	Daily Output	Labor-Hours	Unit	Material	Labor	Equipment	Total	Incl O&P
0010	FAUCETS AND FITTINGS									
2790	Faucets for lavatories									
2800	Self-closing, center set	1 Plum	10	.800	Ea.	131	31		162	195
2810	Automatic sensor and operator, with faucet head		6.15	1.301		450	50.50		500.50	580
3000	Service sink faucet, cast spout, pail hook, hose end		14	.571		80	22		102	125

22 42 39.30 Carriers and Supports

		Crew	Daily Output	Labor-Hours	Unit	Material	Labor	Equipment	Total	Incl O&P
0010	CARRIERS AND SUPPORTS, For plumbing fixtures									
0600	Plate type with studs, top back plate	1 Plum	7	1.143	Ea.	95	44.50		139.50	178
3000	Lavatory, concealed arm									
3050	Floor mounted, single									
3100	High back fixture	1 Plum	6	1.333	Ea.	525	51.50		576.50	665
3200	Flat slab fixture	"	6	1.333	"	455	51.50		506.50	585
8200	Water closet, residential									
8220	Vertical centerline, floor mount									
8240	Single, 3" caulk, 2" or 3" vent	1 Plum	6	1.333	Ea.	610	51.50		661.50	755
8260	4" caulk, 2" or 4" vent	"	6	1.333	"	785	51.50		836.50	945

22 51 Swimming Pool Plumbing Systems

22 51 19 – Swimming Pool Water Treatment Equipment

22 51 19.50 Swimming Pool Filtration Equipment

		Crew	Daily Output	Labor-Hours	Unit	Material	Labor	Equipment	Total	Incl O&P
0010	SWIMMING POOL FILTRATION EQUIPMENT									
0900	Filter system, sand or diatomite type, incl. pump, 6,000 gal./hr.	2 Plum	1.80	8.889	Total	1,975	345		2,320	2,750
1020	Add for chlorination system, 800 S.F. pool	"	3	5.333	Ea.	181	207		388	540

Division Notes

	CREW	DAILY OUTPUT	LABOR-HOURS	UNIT	BARE COSTS				TOTAL INCL O&P
					MAT.	LABOR	EQUIP.	TOTAL	

Estimating Tips

The labor adjustment factors listed in Subdivision 22 01 02.20 also apply to Division 23.

23 10 00 Facility Fuel Systems

- The prices in this subdivision for above- and below-ground storage tanks do not include foundations or hold-down slabs, unless noted. The estimator should refer to Divisions 3 and 31 for foundation system pricing. In addition to the foundations, required tank accessories, such as tank gauges, leak detection devices, and additional manholes and piping, must be added to the tank prices.

23 50 00 Central Heating Equipment

- When estimating the cost of an HVAC system, check to see who is responsible for providing and installing the temperature control system. It is possible to overlook controls, assuming that they would be included in the electrical estimate.
- When looking up a boiler, be careful on specified capacity. Some manufacturers rate their products on output while others use input.
- Include HVAC insulation for pipe, boiler, and duct (wrap and liner).
- Be careful when looking up mechanical items to get the correct pressure rating and connection type (thread, weld, flange).

23 70 00 Central HVAC Equipment

- Combination heating and cooling units are sized by the air conditioning requirements. (See Reference No. R236000-20 for preliminary sizing guide.)
- A ton of air conditioning is nominally 400 CFM.
- Rectangular duct is taken off by the linear foot for each size, but its cost is usually estimated by the pound. Remember that SMACNA standards now base duct on internal pressure.
- Prefabricated duct is estimated and purchased like pipe: straight sections and fittings.
- Note that cranes or other lifting equipment are not included on any lines in Division 23. For example, if a crane is required to lift a heavy piece of pipe into place high above a gym floor, or to put a rooftop unit on the roof of a four-story building, etc., it must be added. Due to the potential for extreme variation—from nothing additional required to a major crane or helicopter—we feel that including a nominal amount for "lifting contingency" would be useless and detract from the accuracy of the estimate. When using equipment rental cost data from RSMeans, do not forget to include the cost of the operator(s).

Reference Numbers

Reference numbers are shown in shaded boxes at the beginning of some major classifications. These numbers refer to related items in the Reference Section. The reference information may be an estimating procedure, an alternate pricing method, or technical information.

Note: Not all subdivisions listed here necessarily appear in this publication. ∎

*Note: **Trade Service**, in part, has been used as a reference source for some of the material prices used in Division 23.*

Did you know?
RSMeans Online gives you the same access to RSMeans' data with 24/7 access:
- Quickly locate costs in the searchable database.
- Build cost lists, estimates, and reports in minutes.
- Adjust costs to any location in the U.S. and Canada with the click of a button.

Start your free trial today at **www.rsmeansonline.com**

RSMeansOnline

23 05 Common Work Results for HVAC

23 05 05 – Selective Demolition for HVAC

23 05 05.10 HVAC Demolition

		Crew	Daily Output	Labor-Hours	Unit	Material	2015 Bare Costs Labor	2015 Bare Costs Equipment	Total	Total Incl O&P
0010	**HVAC DEMOLITION**									
0100	Air conditioner, split unit, 3 ton	Q-5	2	8	Ea.		284		284	470
0150	Package unit, 3 ton	Q-6	3	8	"		273		273	450
0260	Baseboard, hydronic fin tube, 1/2"	Q-5	117	.137	L.F.		4.85		4.85	8
0298	Boilers									
0300	Electric, up thru 148 kW	Q-19	2	12	Ea.		430		430	710
0310	150 thru 518 kW	"	1	24			860		860	1,425
0320	550 thru 2000 kW	Q-21	.40	80			2,950		2,950	4,850
0330	2070 kW and up	"	.30	106			3,925		3,925	6,450
0340	Gas and/or oil, up thru 150 MBH	Q-7	2.20	14.545			515		515	850
0350	160 thru 2000 MBH		.80	40			1,425		1,425	2,350
0360	2100 thru 4500 MBH		.50	64			2,275		2,275	3,750
0370	4600 thru 7000 MBH		.30	106			3,775		3,775	6,250
0390	12,200 thru 25,000 MBH		.12	266			9,475		9,475	15,600
1000	Ductwork, 4" high, 8" wide	1 Clab	200	.040	L.F.		.98		.98	1.64
1100	6" high, 8" wide		165	.048			1.19		1.19	1.99
1200	10" high, 12" wide		125	.064			1.56		1.56	2.63
1300	12"-14" high, 16"-18" wide		85	.094			2.30		2.30	3.86
1500	30" high, 36" wide		56	.143			3.49		3.49	5.85
2200	Furnace, electric	Q-20	2	10	Ea.		340		340	565
2300	Gas or oil, under 120 MBH	Q-9	4	4			133		133	222
2340	Over 120 MBH	"	3	5.333			177		177	296
2800	Heat pump, package unit, 3 ton	Q-5	2.40	6.667			237		237	390
2840	Split unit, 3 ton		2	8			284		284	470
2950	Tank, steel, oil, 275 gal., above ground		10	1.600			57		57	93.50
2960	Remove and reset		3	5.333			189		189	310
5090	Remove refrigerant from system	1 Stpi	40	.200	Lb.		7.90		7.90	13

23 07 HVAC Insulation

23 07 13 – Duct Insulation

23 07 13.10 Duct Thermal Insulation

			Crew	Daily Output	Labor-Hours	Unit	Material	2015 Bare Costs Labor	2015 Bare Costs Equipment	Total	Total Incl O&P
0010	**DUCT THERMAL INSULATION**										
3000	Ductwork										
3020	Blanket type, fiberglass, flexible										
3030	Fire rated for grease and hazardous exhaust ducts										
3060	1-1/2" thick		Q-14	300	.053	S.F.	4.54	1.66		6.20	7.80
3090	Fire rated for plenums										
3100	1/2" x24" x 25'		Q-14	7.20	2.222	Roll	167	69		236	300
3110	1/2" x24" x 25'			360	.044	S.F.	3.35	1.38		4.73	6.05
3120	1/2" x 48" x 25'			3.80	4.211	Roll	335	131		466	590
3126	1/2" x 48" x 25'			380	.042	S.F.	3.35	1.31		4.66	5.90
3140	FSK vapor barrier wrap, .75 lb. density										
3160	1" thick	G	Q-14	350	.046	S.F.	.18	1.42		1.60	2.61
3170	1-1/2" thick	G	"	320	.050	"	.22	1.56		1.78	2.88
3210	Vinyl jacket, same as FSK										

23 09 Instrumentation and Control for HVAC

23 09 53 – Pneumatic and Electric Control System for HVAC

23 09 53.10 Control Components

		Crew	Daily Output	Labor-Hours	Unit	Material	2015 Bare Costs Labor	Equipment	Total	Total Incl O&P
0010	**CONTROL COMPONENTS**									
5000	Thermostats									
5030	Manual	1 Stpi	8	1	Ea.	51	39.50		90.50	121
5040	1 set back, electric, timed [G]	↓	8	1		37	39.50		76.50	106
5050	2 set back, electric, timed [G]	↓	8	1	↓	186	39.50		225.50	270

23 13 Facility Fuel-Storage Tanks

23 13 13 – Facility Underground Fuel-Oil, Storage Tanks

23 13 13.09 Single-Wall Steel Fuel-Oil Tanks

		Crew	Daily Output	Labor-Hours	Unit	Material	Labor	Equipment	Total	Total Incl O&P
0010	**SINGLE-WALL STEEL FUEL-OIL TANKS**									
5000	Tanks, steel ugnd., sti-p3, not incl. hold-down bars									
5500	Excavation, pad, pumps and piping not included									
5510	Single wall, 500 gallon capacity, 7 ga. shell	Q-5	2.70	5.926	Ea.	2,050	210		2,260	2,600
5520	1,000 gallon capacity, 7 ga. shell	"	2.50	6.400		3,825	227		4,052	4,575
5530	2,000 gallon capacity, 1/4" thick shell	Q-7	4.60	6.957		6,200	247		6,447	7,225
5535	2,500 gallon capacity, 7 ga. shell	Q-5	3	5.333		6,850	189		7,039	7,825
5610	25,000 gallon capacity, 3/8" thick shell	Q-7	1.30	24.615		37,900	875		38,775	43,200
5630	40,000 gallon capacity, 3/8" thick shell		.90	35.556		42,000	1,250		43,250	48,300
5640	50,000 gallon capacity, 3/8" thick shell	↓	.80	40	↓	46,600	1,425		48,025	54,000

23 13 13.23 Glass-Fiber-Reinfcd-Plastic, Fuel-Oil, Storage

		Crew	Daily Output	Labor-Hours	Unit	Material	Labor	Equipment	Total	Total Incl O&P
0010	**GLASS-FIBER-REINFCD-PLASTIC, UNDERGRND. FUEL-OIL, STORAGE**									
0210	Fiberglass, underground, single wall, U.L. listed, not including									
0220	manway or hold-down strap									
0230	1,000 gallon capacity	Q-5	2.46	6.504	Ea.	4,125	231		4,356	4,900
0240	2,000 gallon capacity	Q-7	4.57	7.002		5,950	249		6,199	6,950
0245	3,000 gallon capacity		3.90	8.205		7,125	291		7,416	8,325
0255	5,000 gallon capacity	↓	3.20	10	↓	9,200	355		9,555	10,700
0500	For manway, fittings and hold-downs, add					20%	15%			
2210	Fiberglass, underground, single wall, U.L. listed, including									
2220	hold-down straps, no manways									
2230	1,000 gallon capacity	Q-5	1.88	8.511	Ea.	4,575	300		4,875	5,525
2240	2,000 gallon capacity	Q-7	3.55	9.014	"	6,400	320		6,720	7,575

23 13 23 – Facility Aboveground Fuel-Oil, Storage Tanks

23 13 23.16 Steel

		Crew	Daily Output	Labor-Hours	Unit	Material	Labor	Equipment	Total	Total Incl O&P
3001	**STEEL**, storage, above ground, including supports, coating									
3020	fittings, not including foundation, pumps or piping									
3040	Single wall, 275 gallon	Q-5	5	3.200	Ea.	490	114		604	725
3060	550 gallon	"	2.70	5.926		3,750	210		3,960	4,475
3080	1,000 gallon	Q-7	5	6.400		4,025	227		4,252	4,800
3320	Double wall, 500 gallon capacity	Q-5	2.40	6.667		2,625	237		2,862	3,275
3330	2000 gallon capacity	Q-7	4.15	7.711		9,975	274		10,249	11,500
3340	4000 gallon capacity		3.60	8.889		17,800	315		18,115	20,100
3350	6000 gallon capacity		2.40	13.333		21,000	475		21,475	23,900
3360	8000 gallon capacity		2	16		27,000	570		27,570	30,600
3370	10000 gallon capacity		1.80	17.778		30,200	630		30,830	34,300
3380	15000 gallon capacity		1.50	21.333		45,900	755		46,655	52,000
3390	20000 gallon capacity		1.30	24.615		52,500	875		53,375	59,000
3400	25000 gallon capacity		1.15	27.826		63,500	990		64,490	71,500
3410	30000 gallon capacity	↓	1	32	↓	69,500	1,125		70,625	78,500

23 13 Facility Fuel-Storage Tanks

23 13 23 – Facility Aboveground Fuel-Oil, Storage Tanks

23 13 23.26 Horizontal, Conc., Abvgrd Fuel-Oil, Stor. Tanks	Crew	Daily Output	Labor-Hours	Unit	Material	2015 Bare Costs Labor	Equipment	Total	Total Incl O&P
0010 **HORIZONTAL, CONCRETE, ABOVEGROUND FUEL-OIL, STORAGE TANKS**									
0050 Concrete, storage, above ground, including pad & pump									
0100 500 gallon	F-3	2	20	Ea.	10,000	600	325	10,925	12,400
0200 1,000 gallon	"	2	20	"	14,000	600	325	14,925	16,800

23 21 Hydronic Piping and Pumps

23 21 20 – Hydronic HVAC Piping Specialties

23 21 20.46 Expansion Tanks

	Crew	Daily Output	Labor-Hours	Unit	Material	Labor	Equipment	Total	Total Incl O&P
0010 **EXPANSION TANKS**									
1507 Underground fuel-oil storage tanks, see Section 23 13 13									
2000 Steel, liquid expansion, ASME, painted, 15 gallon capacity	Q-5	17	.941	Ea.	640	33.50		673.50	755
2040 30 gallon capacity		12	1.333		715	47.50		762.50	865
3000 Steel ASME expansion, rubber diaphragm, 19 gal. cap. accept.		12	1.333		2,425	47.50		2,472.50	2,750
3020 31 gallon capacity	↓	8	2	↓	2,700	71		2,771	3,100

23 21 23 – Hydronic Pumps

23 21 23.13 In-Line Centrifugal Hydronic Pumps

	Crew	Daily Output	Labor-Hours	Unit	Material	Labor	Equipment	Total	Total Incl O&P
0010 **IN-LINE CENTRIFUGAL HYDRONIC PUMPS**									
0600 Bronze, sweat connections, 1/40 HP, in line									
0640 3/4" size	Q-1	16	1	Ea.	218	35		253	298
1000 Flange connection, 3/4" to 1-1/2" size									
1040 1/12 HP	Q-1	6	2.667	Ea.	565	93		658	775
1060 1/8 HP		6	2.667		970	93		1,063	1,225
2101 Pumps, circulating, 3/4" to 1-1/2" size, 1/3 HP	↓	6	2.667	↓	790	93		883	1,025

23 23 Refrigerant Piping

23 23 16 – Refrigerant Piping Specialties

23 23 16.16 Refrigerant Line Sets

	Crew	Daily Output	Labor-Hours	Unit	Material	Labor	Equipment	Total	Total Incl O&P
0010 **REFRIGERANT LINE SETS**									
0100 Copper tube									
0110 1/2" insulation, both tubes									
0120 Combination 1/4" and 1/2" tubes									
0130 10' set	Q-5	42	.381	Ea.	70.50	13.50		84	100
0140 20' set		40	.400		110	14.20		124.20	145
0150 30' set		37	.432		148	15.35		163.35	188
0160 40' set		35	.457		193	16.25		209.25	240
0170 50' set		32	.500		229	17.75		246.75	282
0180 100' set	↓	22	.727		520	26		546	620
0300 Combination 1/4" and 3/4" tubes									
0310 10' set	Q-5	40	.400	Ea.	78.50	14.20		92.70	110
0320 20' set		38	.421		139	14.95		153.95	178
0330 30' set		35	.457		208	16.25		224.25	256
0340 40' set		33	.485		272	17.20		289.20	330
0350 50' set		30	.533		345	18.95		363.95	405
0380 100' set	↓	20	.800	↓	780	28.50		808.50	905
0500 Combination 3/8" & 3/4" tubes									
0510 10' set	Q-5	28	.571	Ea.	90.50	20.50		111	133
0520 20' set		36	.444		141	15.80		156.80	181
0530 30' set	↓	34	.471	↓	191	16.70		207.70	239

23 23 Refrigerant Piping

23 23 16 – Refrigerant Piping Specialties

23 23 16.16 Refrigerant Line Sets

		Crew	Daily Output	Labor-Hours	Unit	Material	2015 Bare Costs Labor	2015 Bare Costs Equipment	Total	Total Incl O&P
0540	40' set	Q-5	31	.516	Ea.	252	18.30		270.30	305
0550	50' set		28	.571		296	20.50		316.50	360
0580	100' set		18	.889		875	31.50		906.50	1,000
0700	Combination 3/8" & 1-1/8" tubes									
0710	10' set	Q-5	36	.444	Ea.	158	15.80		173.80	200
0720	20' set		33	.485		265	17.20		282.20	320
0730	30' set		31	.516		355	18.30		373.30	420
0740	40' set		28	.571		535	20.50		555.50	620
0750	50' set		26	.615		615	22		637	715
0900	Combination 1/2" & 3/4" tubes									
0910	10' set	Q-5	37	.432	Ea.	95.50	15.35		110.85	131
0920	20' set		35	.457		169	16.25		185.25	213
0930	30' set		33	.485		255	17.20		272.20	310
0940	40' set		30	.533		335	18.95		353.95	400
0950	50' set		27	.593		420	21		441	500
0980	100' set		17	.941		950	33.50		983.50	1,100
2100	Combination 1/2" & 1-1/8" tubes									
2110	10' set	Q-5	35	.457	Ea.	163	16.25		179.25	206
2120	20' set		31	.516		279	18.30		297.30	335
2130	30' set		29	.552		420	19.60		439.60	495
2140	40' set		25	.640		565	22.50		587.50	660
2150	50' set		14	1.143		625	40.50		665.50	750
2300	For 1" thick insulation add					30%	15%			

23 31 HVAC Ducts and Casings

23 31 13 – Metal Ducts

23 31 13.13 Rectangular Metal Ducts

		Crew	Daily Output	Labor-Hours	Unit	Material	2015 Bare Costs Labor	2015 Bare Costs Equipment	Total	Total Incl O&P
0010	**RECTANGULAR METAL DUCTS**									
0020	Fabricated rectangular, includes fittings, joints, supports,									
0021	allowance for flexible connections and field sketches.									
0030	Does not include "as-built dwgs." or insulation.									
0031	NOTE: Fabrication and installation are combined									
0040	as LABOR cost. Approx. 25% fittings assumed.									
0042	Fabrication/Inst. is to commercial quality standards									
0043	(SMACNA or equiv.) for structure, sealing, leak testing, etc.									
0100	Aluminum, alloy 3003-H14, under 100 lb.	Q-10	75	.320	Lb.	3.20	11.05		14.25	22
0110	100 to 500 lb.		80	.300		1.88	10.35		12.23	19.30
0120	500 to 1,000 lb.		95	.253		1.82	8.70		10.52	16.55
0140	1,000 to 2,000 lb.		120	.200		1.77	6.90		8.67	13.45
0500	Galvanized steel, under 200 lb.		235	.102		.65	3.52		4.17	6.60
0520	200 to 500 lb.		245	.098		.64	3.38		4.02	6.35
0540	500 to 1,000 lb.		255	.094		.62	3.25		3.87	6.10

23 33 Air Duct Accessories

23 33 13 – Dampers

23 33 13.13 Volume-Control Dampers		Crew	Daily Output	Labor-Hours	Unit	Material	2015 Bare Costs Labor	Equipment	Total	Total Incl O&P
0010	**VOLUME-CONTROL DAMPERS**									
6000	12" x 12"	1 Shee	21	.381	Ea.	29.50	14.10		43.60	56
8000	Multi-blade dampers, parallel blade									
8100	8" x 8"	1 Shee	24	.333	Ea.	82	12.30		94.30	111

23 33 13.16 Fire Dampers

0010	**FIRE DAMPERS**									
3000	Fire damper, curtain type, 1-1/2 hr. rated, vertical, 6" x 6"	1 Shee	24	.333	Ea.	24	12.30		36.30	47
3020	8" x 6"	"	22	.364	"	24	13.45		37.45	49

23 33 46 – Flexible Ducts

23 33 46.10 Flexible Air Ducts

0010	**FLEXIBLE AIR DUCTS**									
1300	Flexible, coated fiberglass fabric on corr. resist. metal helix									
1400	pressure to 12" (WG) UL-181									
1500	Noninsulated, 3" diameter	Q-9	400	.040	L.F.	1.12	1.33		2.45	3.45
1540	5" diameter		320	.050		1.30	1.66		2.96	4.20
1560	6" diameter		280	.057		1.50	1.90		3.40	4.82
1580	7" diameter		240	.067		1.53	2.22		3.75	5.40
1900	Insulated, 1" thick, PE jacket, 3" diameter	G	380	.042		2.60	1.40		4	5.20
1910	4" diameter	G	340	.047		2.60	1.56		4.16	5.45
1920	5" diameter	G	300	.053		2.60	1.77		4.37	5.80
1940	6" diameter	G	260	.062		2.94	2.05		4.99	6.65
1960	7" diameter	G	220	.073		3.20	2.42		5.62	7.55
1980	8" diameter	G	180	.089		3.49	2.96		6.45	8.75
2040	12" diameter	G	100	.160		4.90	5.30		10.20	14.30

23 33 53 – Duct Liners

23 33 53.10 Duct Liner Board

0010	**DUCT LINER BOARD**									
3490	Board type, fiberglass liner, 3 lb. density									
3940	Board type, non-fibrous foam									
3950	Temperature, bacteria and fungi resistant									
3960	1" thick	G Q-14	150	.107	S.F.	2.38	3.32		5.70	8.25
3970	1-1/2" thick	G	130	.123		3.28	3.83		7.11	10.10
3980	2" thick	G	120	.133		3.98	4.15		8.13	11.45

23 34 HVAC Fans

23 34 23 – HVAC Power Ventilators

23 34 23.10 HVAC Power Circulators and Ventilators

0010	**HVAC POWER CIRCULATORS AND VENTILATORS**									
6650	Residential, bath exhaust, grille, back draft damper									
6660	50 CFM	Q-20	24	.833	Ea.	63.50	28.50		92	117
6670	110 CFM		22	.909		98	31		129	160
6900	Kitchen exhaust, grille, complete, 160 CFM		22	.909		108	31		139	171
6910	180 CFM		20	1		90.50	34		124.50	156
6920	270 CFM		18	1.111		171	37.50		208.50	251
6930	350 CFM		16	1.250		129	42.50		171.50	213
6940	Residential roof jacks and wall caps									
6944	Wall cap with back draft damper									
6946	3" & 4" diam. round duct	1 Shee	11	.727	Ea.	26	27		53	73.50
6948	6" diam. round duct	"	11	.727	"	64	27		91	116
6958	Roof jack with bird screen and back draft damper									

23 34 HVAC Fans

23 34 23 – HVAC Power Ventilators

23 34 23.10 HVAC Power Circulators and Ventilators

		Crew	Daily Output	Labor-Hours	Unit	Material	2015 Bare Costs Labor	Equipment	Total	Total Incl O&P
6960	3" & 4" diam. round duct	1 Shee	11	.727	Ea.	26	27		53	73.50
6962	3-1/4" x 10" rectangular duct	"	10	.800	"	48.50	29.50		78	103
8020	Attic, roof type									
8030	Aluminum dome, damper & curb									
8080	12" diameter, 1000 CFM (gravity)	1 Elec	10	.800	Ea.	575	29.50		604.50	680
8090	16" diameter, 1500 CFM (gravity)		9	.889		690	32.50		722.50	815
8100	20" diameter, 2500 CFM (gravity)	↓	8	1	↓	850	36.50		886.50	995
8160	Plastic, ABS dome									
8180	1050 CFM	1 Elec	14	.571	Ea.	168	21		189	220
8200	1600 CFM	"	12	.667	"	252	24.50		276.50	320
8240	Attic, wall type, with shutter, one speed									
8250	12" diameter, 1000 CFM	1 Elec	14	.571	Ea.	395	21		416	470
8260	14" diameter, 1500 CFM		12	.667		430	24.50		454.50	515
8270	16" diameter, 2000 CFM	↓	9	.889	↓	485	32.50		517.50	590
8290	Whole house, wall type, with shutter, one speed									
8300	30" diameter, 4800 CFM	1 Elec	7	1.143	Ea.	1,050	42		1,092	1,225
8310	36" diameter, 7000 CFM		6	1.333		1,125	49		1,174	1,325
8320	42" diameter, 10,000 CFM		5	1.600		1,275	58.50		1,333.50	1,500
8330	48" diameter, 16,000 CFM		4	2		1,575	73.50		1,648.50	1,875
8340	For two speed, add				↓	95			95	105
8350	Whole house, lay-down type, with shutter, one speed									
8360	30" diameter, 4500 CFM	1 Elec	8	1	Ea.	1,100	36.50		1,136.50	1,275
8370	36" diameter, 6500 CFM		7	1.143		1,200	42		1,242	1,375
8380	42" diameter, 9000 CFM		6	1.333		1,300	49		1,349	1,525
8390	48" diameter, 12,000 CFM	↓	5	1.600		1,500	58.50		1,558.50	1,725
8440	For two speed, add					71.50			71.50	78.50
8450	For 12 hour timer switch, add	1 Elec	32	.250	↓	71.50	9.15		80.65	93.50

23 37 Air Outlets and Inlets

23 37 13 – Diffusers, Registers, and Grilles

23 37 13.10 Diffusers

		Crew	Daily Output	Labor-Hours	Unit	Material	2015 Bare Costs Labor	Equipment	Total	Total Incl O&P
0010	**DIFFUSERS**, Aluminum, opposed blade damper unless noted									
0100	Ceiling, linear, also for sidewall									
0120	2" wide	1 Shee	32	.250	L.F.	16.20	9.25		25.45	33
0160	4" wide		26	.308	"	21.50	11.35		32.85	42.50
0500	Perforated, 24" x 24" lay-in panel size, 6" x 6"		16	.500	Ea.	151	18.50		169.50	197
0520	8" x 8"		15	.533		159	19.70		178.70	208
0530	9" x 9"		14	.571		161	21		182	212
0590	16" x 16"		11	.727		189	27		216	252
1000	Rectangular, 1 to 4 way blow, 6" x 6"		16	.500		50	18.50		68.50	86
1010	8" x 8"		15	.533		58	19.70		77.70	96.50
1014	9" x 9"		15	.533		66	19.70		85.70	106
1016	10" x 10"		15	.533		80	19.70		99.70	121
1020	12" x 6"		15	.533		71.50	19.70		91.20	112
1040	12" x 9"		14	.571		75.50	21		96.50	119
1060	12" x 12"		12	.667		84.50	24.50		109	134
1070	14" x 6"		13	.615		77.50	22.50		100	124
1074	14" x 14"		12	.667		128	24.50		152.50	181
1150	18" x 18"		9	.889		138	33		171	206
1170	24" x 12"		10	.800		163	29.50		192.50	230
1180	24" x 24"		7	1.143		270	42		312	365

23 37 Air Outlets and Inlets

23 37 13 – Diffusers, Registers, and Grilles

23 37 13.10 Diffusers

		Crew	Daily Output	Labor-Hours	Unit	Material	2015 Bare Costs Labor	Equipment	Total	Total Incl O&P
1500	Round, butterfly damper, steel, diffuser size, 6" diameter	1 Shee	18	.444	Ea.	9	16.40		25.40	37.50
1520	8" diameter		16	.500		9.55	18.50		28.05	41.50
2000	T bar mounting, 24" x 24" lay-in frame, 6" x 6"		16	.500		61	18.50		79.50	98
2020	8" x 8"		14	.571		62	21		83	103
2040	12" x 12"		12	.667		74	24.50		98.50	122
2060	16" x 16"		11	.727		94.50	27		121.50	149
2080	18" x 18"		10	.800		106	29.50		135.50	167
6000	For steel diffusers instead of aluminum, deduct					10%				

23 37 13.30 Grilles

		Crew	Daily Output	Labor-Hours	Unit	Material	Labor	Equipment	Total	Total Incl O&P
0010	**GRILLES**									
0020	Aluminum, unless noted otherwise									
1000	Air return, steel, 6" x 6"	1 Shee	26	.308	Ea.	18.70	11.35		30.05	39.50
1020	10" x 6"		24	.333		18.70	12.30		31	41
1080	16" x 8"		22	.364		26.50	13.45		39.95	51.50
1100	12" x 12"		22	.364		26.50	13.45		39.95	51.50
1120	24" x 12"		18	.444		36	16.40		52.40	67
1180	16" x 16"		22	.364		34	13.45		47.45	60

23 37 13.60 Registers

		Crew	Daily Output	Labor-Hours	Unit	Material	Labor	Equipment	Total	Total Incl O&P
0010	**REGISTERS**									
0980	Air supply									
3000	Baseboard, hand adj. damper, enameled steel									
3012	8" x 6"	1 Shee	26	.308	Ea.	4.85	11.35		16.20	24.50
3020	10" x 6"		24	.333		5.30	12.30		17.60	26.50
3040	12" x 5"		23	.348		5.60	12.85		18.45	27.50
3060	12" x 6"		23	.348		5.75	12.85		18.60	28
4000	Floor, toe operated damper, enameled steel									
4020	4" x 8"	1 Shee	32	.250	Ea.	9	9.25		18.25	25.50
4040	4" x 12"	"	26	.308	"	10.55	11.35		21.90	30.50
4300	Spiral pipe supply register									
4310	Steel, with air scoop									
4320	4" x 12", for 8" thru 13" diameter duct	1 Shee	25	.320	Ea.	74.50	11.80		86.30	101
4330	4" x 18", for 8" thru 13" diameter duct		18	.444		87	16.40		103.40	123
4340	6" x 12", for 14" thru 21" diameter duct		19	.421		78.50	15.55		94.05	113
4350	6" x 16", for 14" thru 21" diameter duct		18	.444		89	16.40		105.40	126
4360	6" x 20", for 14" thru 21" diameter duct		17	.471		98.50	17.40		115.90	137
4370	6" x 24", for 14" thru 21" diameter duct		16	.500		116	18.50		134.50	158
4380	8" x 16", for 22" thru 31" diameter duct		19	.421		94	15.55		109.55	129
4390	8" x 18", for 22" thru 31" diameter duct		18	.444		98.50	16.40		114.90	136
4400	8" x 24", for 22" thru 31" diameter duct		15	.533		122	19.70		141.70	167

23 41 Particulate Air Filtration

23 41 13 – Panel Air Filters

23 41 13.10 Panel Type Air Filters

		Crew	Daily Output	Labor-Hours	Unit	Material	Labor	Equipment	Total	Total Incl O&P
0010	**PANEL TYPE AIR FILTERS**									
2950	Mechanical media filtration units									
3000	High efficiency type, with frame, non-supported	G			MCFM	35			35	38.50
3100	Supported type	G			"	45			45	49.50
5500	Throwaway glass or paper media type				Ea.	3.42			3.42	3.76

23 41 Particulate Air Filtration

23 41 16 – Renewable-Media Air Filters

23 41 16.10 Disposable Media Air Filters	Crew	Daily Output	Labor-Hours	Unit	Material	2015 Bare Costs Labor	Equipment	Total	Total Incl O&P
0010 **DISPOSABLE MEDIA AIR FILTERS**									
5000 Renewable disposable roll				C.S.F.	1.54			1.54	1.70

23 41 19 – Washable Air Filters

23 41 19.10 Permanent Air Filters

	Crew	Daily Output	Labor-Hours	Unit	Material	Labor	Equipment	Total	Total Incl O&P
0010 **PERMANENT AIR FILTERS**									
4500 Permanent washable [G]				MCFM	20			20	22

23 41 23 – Extended Surface Filters

23 41 23.10 Expanded Surface Filters

	Crew	Daily Output	Labor-Hours	Unit	Material	Labor	Equipment	Total	Total Incl O&P
0010 **EXPANDED SURFACE FILTERS**									
4000 Medium efficiency, extended surface [G]				MCFM	5.50			5.50	6.05

23 42 Gas-Phase Air Filtration

23 42 13 – Activated-Carbon Air Filtration

23 42 13.10 Charcoal Type Air Filtration

	Crew	Daily Output	Labor-Hours	Unit	Material	Labor	Equipment	Total	Total Incl O&P
0010 **CHARCOAL TYPE AIR FILTRATION**									
0050 Activated charcoal type, full flow				MCFM	600			600	660
0060 Full flow, impregnated media 12" deep					225			225	248
0070 HEPA filter & frame for field erection					350			350	385
0080 HEPA filter-diffuser, ceiling install.					300			300	330

23 43 Electronic Air Cleaners

23 43 13 – Washable Electronic Air Cleaners

23 43 13.10 Electronic Air Cleaners

	Crew	Daily Output	Labor-Hours	Unit	Material	Labor	Equipment	Total	Total Incl O&P
0010 **ELECTRONIC AIR CLEANERS**									
2000 Electronic air cleaner, duct mounted									
2150 1000 CFM	1 Shee	4	2	Ea.	420	74		494	590
2200 1200 CFM		3.80	2.105		505	78		583	685
2250 1400 CFM		3.60	2.222		520	82		602	710

23 51 Breechings, Chimneys, and Stacks

23 51 23 – Gas Vents

23 51 23.10 Gas Chimney Vents

	Crew	Daily Output	Labor-Hours	Unit	Material	Labor	Equipment	Total	Total Incl O&P
0010 **GAS CHIMNEY VENTS**, Prefab metal, U.L. listed									
0020 Gas, double wall, galvanized steel									
0080 3" diameter	Q-9	72	.222	V.L.F.	5.55	7.40		12.95	18.45
0100 4" diameter	"	68	.235	"	7.20	7.80		15	21

23 52 Heating Boilers

23 52 13 – Electric Boilers

23 52 13.10 Electric Boilers, ASME

		Crew	Daily Output	Labor-Hours	Unit	Material	2015 Bare Costs Labor	2015 Bare Costs Equipment	Total	Total Incl O&P
0010	**ELECTRIC BOILERS, ASME**, Standard controls and trim									
1000	Steam, 6 KW, 20.5 MBH	Q-19	1.20	20	Ea.	3,950	720		4,670	5,525
1160	60 KW, 205 MBH		1	24		6,650	860		7,510	8,750
2000	Hot water, 7.5 KW, 25.6 MBH		1.30	18.462		4,975	660		5,635	6,575
2040	30 KW, 102 MBH		1.20	20		5,325	720		6,045	7,025
2060	45 KW, 164 MBH		1.20	20		5,425	720		6,145	7,125

23 52 23 – Cast-Iron Boilers

23 52 23.20 Gas-Fired Boilers

		Crew	Daily Output	Labor-Hours	Unit	Material	Labor	Equipment	Total	Total Incl O&P
0010	**GAS-FIRED BOILERS**, Natural or propane, standard controls, packaged									
1000	Cast iron, with insulated jacket									
3000	Hot water, gross output, 80 MBH	Q-7	1.46	21.918	Ea.	1,975	780		2,755	3,450
3020	100 MBH	"	1.35	23.704		2,550	840		3,390	4,225
7000	For tankless water heater, add					10%				
7050	For additional zone valves up to 312 MBH add					194			194	214

23 52 23.30 Gas/Oil Fired Boilers

		Crew	Daily Output	Labor-Hours	Unit	Material	Labor	Equipment	Total	Total Incl O&P
0010	**GAS/OIL FIRED BOILERS**, Combination with burners and controls, packaged									
1000	Cast iron with insulated jacket									
2000	Steam, gross output, 720 MBH	Q-7	.43	74.074	Ea.	14,700	2,625		17,325	20,600
2900	Hot water, gross output									
2910	200 MBH	Q-6	.62	39.024	Ea.	10,500	1,325		11,825	13,800
2920	300 MBH		.49	49.080		10,500	1,675		12,175	14,400
2930	400 MBH		.41	57.971		12,300	1,975		14,275	16,800
2940	500 MBH		.36	67.039		13,300	2,300		15,600	18,400
3000	584 MBH	Q-7	.44	72.072		14,700	2,550		17,250	20,300

23 52 23.40 Oil-Fired Boilers

		Crew	Daily Output	Labor-Hours	Unit	Material	Labor	Equipment	Total	Total Incl O&P
0010	**OIL-FIRED BOILERS**, Standard controls, flame retention burner, packaged									
1000	Cast iron, with insulated flush jacket									
2000	Steam, gross output, 109 MBH	Q-7	1.20	26.667	Ea.	2,325	945		3,270	4,100
2060	207 MBH	"	.90	35.556	"	3,175	1,250		4,425	5,575
3000	Hot water, same price as steam									

23 52 26 – Steel Boilers

23 52 26.40 Oil-Fired Boilers

		Crew	Daily Output	Labor-Hours	Unit	Material	Labor	Equipment	Total	Total Incl O&P
0010	**OIL-FIRED BOILERS**, Standard controls, flame retention burner									
5000	Steel, with insulated flush jacket									
7000	Hot water, gross output, 103 MBH	Q-6	1.60	15	Ea.	1,775	515		2,290	2,800
7020	122 MBH		1.45	16.506		1,875	565		2,440	3,000
7060	168 MBH		1.30	18.405		2,100	630		2,730	3,375
7080	225 MBH		1.22	19.704		2,575	675		3,250	3,925

23 52 28 – Swimming Pool Boilers

23 52 28.10 Swimming Pool Heaters

		Crew	Daily Output	Labor-Hours	Unit	Material	Labor	Equipment	Total	Total Incl O&P
0010	**SWIMMING POOL HEATERS**, Not including wiring, external									
0020	piping, base or pad,									
0160	Gas fired, input, 155 MBH	Q-6	1.50	16	Ea.	2,000	545		2,545	3,100
0200	199 MBH		1	24		2,125	820		2,945	3,700
0280	500 MBH		.40	60		8,900	2,050		10,950	13,200
2000	Electric, 12 KW, 4,800 gallon pool	Q-19	3	8		2,075	287		2,362	2,750
2020	15 KW, 7,200 gallon pool		2.80	8.571		2,100	310		2,410	2,825
2040	24 KW, 9,600 gallon pool		2.40	10		2,425	360		2,785	3,275
2100	57 KW, 24,000 gallon pool		1.20	20		3,575	720		4,295	5,100

23 52 Heating Boilers

23 52 88 – Burners

23 52 88.10 Replacement Type Burners

		Crew	Daily Output	Labor-Hours	Unit	Material	2015 Bare Costs Labor	2015 Bare Costs Equipment	Total	Total Incl O&P
0010	**REPLACEMENT TYPE BURNERS**									
0990	Residential, conversion, gas fired, LP or natural									
1000	Gun type, atmospheric input 50 to 225 MBH	Q-1	2.50	6.400	Ea.	720	223		943	1,150
1020	100 to 400 MBH	"	2	8		1,200	279		1,479	1,775
1025	Burner, gas, 100 – 400 MBH [G]					1,200			1,200	1,325
1040	300 to 1000 MBH	Q-1	1.70	9.412		3,625	330		3,955	4,525

23 54 Furnaces

23 54 13 – Electric-Resistance Furnaces

23 54 13.10 Electric Furnaces

		Crew	Daily Output	Labor-Hours	Unit	Material	Labor	Equipment	Total	Total Incl O&P
0010	**ELECTRIC FURNACES**, Hot air, blowers, std. controls									
0011	not including gas, oil or flue piping									
1000	Electric, UL listed									
1100	34.1 MBH	Q-20	4.40	4.545	Ea.	455	154		609	755

23 54 16 – Fuel-Fired Furnaces

23 54 16.13 Gas-Fired Furnaces

		Crew	Daily Output	Labor-Hours	Unit	Material	Labor	Equipment	Total	Total Incl O&P
0010	**GAS-FIRED FURNACES**									
3000	Gas, AGA certified, upflow, direct drive models									
3020	45 MBH input	Q-9	4	4	Ea.	535	133		668	810
3040	60 MBH input		3.80	4.211		535	140		675	825
3060	75 MBH input		3.60	4.444		575	148		723	880
3100	100 MBH input		3.20	5		625	166		791	965
3120	125 MBH input		3	5.333		655	177		832	1,025
3130	150 MBH input		2.80	5.714		670	190		860	1,050
3140	200 MBH input		2.60	6.154		2,800	205		3,005	3,425
4000	For starter plenum, add		16	1		89.50	33.50		123	154

23 54 16.16 Oil-Fired Furnaces

		Crew	Daily Output	Labor-Hours	Unit	Material	Labor	Equipment	Total	Total Incl O&P
0010	**OIL-FIRED FURNACES**									
6000	Oil, UL listed, atomizing gun type burner									
6020	56 MBH output	Q-9	3.60	4.444	Ea.	1,725	148		1,873	2,150
6030	84 MBH output		3.50	4.571		1,850	152		2,002	2,300
6040	95 MBH output		3.40	4.706		1,875	156		2,031	2,325
6060	134 MBH output		3.20	5		2,175	166		2,341	2,675
6080	151 MBH output		3	5.333		2,250	177		2,427	2,775
6100	200 MBH input		2.60	6.154		2,450	205		2,655	3,050

23 54 16.21 Solid Fuel-Fired Furnaces

		Crew	Daily Output	Labor-Hours	Unit	Material	Labor	Equipment	Total	Total Incl O&P
0010	**SOLID FUEL-FIRED FURNACES**									
6020	Wood fired furnaces									
6030	Includes hot water coil, thermostat, and auto draft control									
6040	24" long firebox [G]	Q-9	4	4	Ea.	4,300	133		4,433	4,950
6050	30" long firebox [G]		3.60	4.444		5,025	148		5,173	5,775
6060	With fireplace glass doors [G]		3.20	5		6,350	166		6,516	7,275
6200	Wood/oil fired furnaces, includes two thermostats									
6210	Includes hot water coil and auto draft control									
6240	24" long firebox [G]	Q-9	3.40	4.706	Ea.	5,150	156		5,306	5,925
6250	30" long firebox [G]		3	5.333		5,775	177		5,952	6,650
6260	With fireplace glass doors [G]		2.80	5.714		7,075	190		7,265	8,100
6400	Wood/gas fired furnaces, includes two thermostats									
6410	Includes hot water coil and auto draft control									

23 54 Furnaces

23 54 16 – Fuel-Fired Furnaces

23 54 16.21 Solid Fuel-Fired Furnaces

			Crew	Daily Output	Labor-Hours	Unit	Material	2015 Bare Costs Labor	Equipment	Total	Total Incl O&P
6440		24" long firebox	G	Q-9	2.80	5.714	Ea.	5,925	190		6,115	6,850
6450		30" long firebox	G		2.40	6.667		6,625	222		6,847	7,675
6460		With fireplace glass doors	G		2	8		7,900	266		8,166	9,150
6600	Wood/oil/gas fired furnaces, optional accessories											
6610		Hot air plenum		Q-9	16	1	Ea.	108	33.50		141.50	175
6620		Safety heat dump			24	.667		94	22		116	140
6630		Auto air intake			18	.889		148	29.50		177.50	213
6640		Cold air return package			14	1.143		159	38		197	239
6650		Wood fork						51			51	56
6700	Wood fired outdoor furnace											
6740		24" long firebox	G	Q-9	3.80	4.211	Ea.	4,500	140		4,640	5,175
6760	Wood fired outdoor furnace, optional accessories											
6770		Chimney section, stainless steel, 6" ID x 3' lg.	G	Q-9	36	.444	Ea.	165	14.80		179.80	207
6780		Chimney cap, stainless steel	G	"	40	.400	"	95	13.30		108.30	127
6800	Wood fired hot water furnace											
6820		Includes 200 gal. hot water storage, thermostat, and auto draft control										
6840		30" long firebox	G	Q-9	2.10	7.619	Ea.	8,275	253		8,528	9,525
6850	Water to air heat exchanger											
6870		Includes mounting kit and blower relay										
6880		140 MBH, 18.75" W x 18.75" L		Q-9	7.50	2.133	Ea.	370	71		441	525
6890		200 MBH, 24" W x 24" L		"	7	2.286	"	525	76		601	705
6900	Water to water heat exchanger											
6940		100 MBH		Q-9	6.50	2.462	Ea.	370	82		452	545
6960		290 MBH		"	6	2.667	"	535	88.50		623.50	740
7000	Optional accessories											
7010		Large volume circulation pump, (2 Included)		Q-9	14	1.143	Ea.	242	38		280	330
7020		Air bleed fittings, (package)			24	.667		50.50	22		72.50	92.50
7030		Domestic water preheater			6	2.667		219	88.50		307.50	390
7040		Smoke pipe kit			4	4		88	133		221	320

23 54 24 – Furnace Components for Cooling

23 54 24.10 Furnace Components and Combinations

			Crew	Daily Output	Labor-Hours	Unit	Material	Labor	Equipment	Total	Total Incl O&P
0010	**FURNACE COMPONENTS AND COMBINATIONS**										
0080	Coils, A.C. evaporator, for gas or oil furnaces										
0090		Add-on, with holding charge									
0100		Upflow									
0120		1-1/2 ton cooling		Q-5	4	4	Ea.	220	142		362	475
0130		2 ton cooling			3.70	4.324		241	154		395	520
0140		3 ton cooling			3.30	4.848		375	172		547	700
0150		4 ton cooling			3	5.333		465	189		654	825
0160		5 ton cooling			2.70	5.926		505	210		715	900
0300		Downflow									
0330		2-1/2 ton cooling		Q-5	3	5.333	Ea.	295	189		484	635
0340		3-1/2 ton cooling			2.60	6.154		435	218		653	835
0350		5 ton cooling			2.20	7.273		505	258		763	980
0600		Horizontal									
0630		2 ton cooling		Q-5	3.90	4.103	Ea.	320	146		466	595
0640		3 ton cooling			3.50	4.571		370	162		532	675
0650		4 ton cooling			3.20	5		460	178		638	800
0660		5 ton cooling			2.90	5.517		460	196		656	830
2000		Cased evaporator coils for air handlers									
2100		1-1/2 ton cooling		Q-5	4.40	3.636	Ea.	267	129		396	505
2110		2 ton cooling			4.10	3.902		300	139		439	560

23 54 Furnaces

23 54 24 – Furnace Components for Cooling

23 54 24.10 Furnace Components and Combinations		Crew	Daily Output	Labor-Hours	Unit	Material	2015 Bare Costs Labor	Equipment	Total	Total Incl O&P
2120	2-1/2 ton cooling	Q-5	3.90	4.103	Ea.	320	146		466	590
2130	3 ton cooling		3.70	4.324		360	154		514	650
2140	3-1/2 ton cooling		3.50	4.571		485	162		647	805
2150	4 ton cooling		3.20	5		515	178		693	865
2160	5 ton cooling		2.90	5.517		525	196		721	905
3010	Air handler, modular									
3100	With cased evaporator cooling coil									
3120	1-1/2 ton cooling	Q-5	3.80	4.211	Ea.	850	149		999	1,175
3130	2 ton cooling		3.50	4.571		940	162		1,102	1,300
3140	2-1/2 ton cooling		3.30	4.848		995	172		1,167	1,375
3150	3 ton cooling		3.10	5.161		1,150	183		1,333	1,550
3160	3-1/2 ton cooling		2.90	5.517		1,325	196		1,521	1,775
3170	4 ton cooling		2.50	6.400		1,425	227		1,652	1,925
3180	5 ton cooling		2.10	7.619		1,700	270		1,970	2,300
3500	With no cooling coil									
3520	1-1/2 ton coil size	Q-5	12	1.333	Ea.	620	47.50		667.50	760
3530	2 ton coil size		10	1.600		620	57		677	780
3540	2-1/2 ton coil size		10	1.600		725	57		782	890
3554	3 ton coil size		9	1.778		825	63		888	1,000
3560	3-1/2 ton coil size		9	1.778		845	63		908	1,025
3570	4 ton coil size		8.50	1.882		1,050	67		1,117	1,275
3580	5 ton coil size		8	2		1,075	71		1,146	1,300
4000	With heater									
4120	5 kW, 17.1 MBH	Q-5	16	1	Ea.	670	35.50		705.50	795
4130	7.5 kW, 25.6 MBH		15.60	1.026		680	36.50		716.50	805
4140	10 kW, 34.2 MBH		15.20	1.053		845	37.50		882.50	990
4150	12.5 KW, 42.7 MBH		14.80	1.081		950	38.50		988.50	1,125
4160	15 KW, 51.2 MBH		14.40	1.111		1,025	39.50		1,064.50	1,200
4170	25 KW, 85.4 MBH		14	1.143		1,375	40.50		1,415.50	1,575
4180	30 KW, 102 MBH		13	1.231		1,600	43.50		1,643.50	1,825

23 62 Packaged Compressor and Condenser Units

23 62 13 – Packaged Air-Cooled Refrigerant Compressor and Condenser Units

23 62 13.10 Packaged Air-Cooled Refrig. Condensing Units

0010	PACKAGED AIR-COOLED REFRIGERANT CONDENSING UNITS									
0020	Condensing unit									
0030	Air cooled, compressor, standard controls									
0050	1.5 ton	Q-5	2.50	6.400	Ea.	1,250	227		1,477	1,750
0100	2 ton		2.10	7.619		1,350	270		1,620	1,925
0200	2.5 ton		1.70	9.412		1,425	335		1,760	2,125
0300	3 ton		1.30	12.308		1,500	435		1,935	2,375
0350	3.5 ton		1.10	14.545		1,725	515		2,240	2,750
0400	4 ton		.90	17.778		1,975	630		2,605	3,225
0500	5 ton		.60	26.667		2,375	945		3,320	4,150

23 74 Packaged Outdoor HVAC Equipment

23 74 33 – Dedicated Outdoor-Air Units

23 74 33.10 Rooftop Air Conditioners		Crew	Daily Output	Labor-Hours	Unit	Material	2015 Bare Costs Labor	2015 Bare Costs Equipment	Total	Total Incl O&P
0010	**ROOFTOP AIR CONDITIONERS**, Standard controls, curb, economizer									
1000	Single zone, electric cool, gas heat									
1140	5 ton cooling, 112 MBH heating	Q-5	.56	28.521	Ea.	5,125	1,000		6,125	7,300
1150	7.5 ton cooling, 170 MBH heating		.50	32.258		6,825	1,150		7,975	9,425
1156	8.5 ton cooling, 170 MBH heating	↓	.46	34.783		8,125	1,225		9,350	11,000
1160	10 ton cooling, 200 MBH heating	Q-6	.67	35.982	↓	9,425	1,225		10,650	12,400

23 81 Decentralized Unitary HVAC Equipment

23 81 13 – Packaged Terminal Air-Conditioners

23 81 13.10 Packaged Cabinet Type Air-Conditioners

		Crew	Daily Output	Labor-Hours	Unit	Material	Labor	Equipment	Total	Total Incl O&P
0010	**PACKAGED CABINET TYPE AIR-CONDITIONERS**, Cabinet, wall sleeve,									
0100	louver, electric heat, thermostat, manual changeover, 208 V									
0200	6,000 BTUH cooling, 8800 BTU heat	Q-5	6	2.667	Ea.	775	94.50		869.50	1,000
0220	9,000 BTUH cooling, 13,900 BTU heat		5	3.200		1,200	114		1,314	1,500
0240	12,000 BTUH cooling, 13,900 BTU heat		4	4		1,350	142		1,492	1,725
0260	15,000 BTUH cooling, 13,900 BTU heat	↓	3	5.333		1,450	189		1,639	1,900

23 81 19 – Self-Contained Air-Conditioners

23 81 19.10 Window Unit Air Conditioners

		Crew	Daily Output	Labor-Hours	Unit	Material	Labor	Equipment	Total	Total Incl O&P
0010	**WINDOW UNIT AIR CONDITIONERS**									
4000	Portable/window, 15 amp 125 V grounded receptacle required									
4060	5000 BTUH	1 Carp	8	1	Ea.	295	33		328	380
4340	6000 BTUH		8	1		340	33		373	430
4480	8000 BTUH		6	1.333		460	44		504	585
4500	10,000 BTUH	↓	6	1.333		590	44		634	725
4520	12,000 BTUH	L-2	8	2	↓	725	57.50		782.50	890
4600	Window/thru-the-wall, 15 amp 230 V grounded receptacle required									
4780	18,000 BTUH	L-2	6	2.667	Ea.	970	76.50		1,046.50	1,200
4940	25,000 BTUH		4	4		1,175	115		1,290	1,500
4960	29,000 BTUH	↓	4	4		1,325	115		1,440	1,675

23 81 19.20 Self-Contained Single Package

		Crew	Daily Output	Labor-Hours	Unit	Material	Labor	Equipment	Total	Total Incl O&P
0010	**SELF-CONTAINED SINGLE PACKAGE**									
0100	Air cooled, for free blow or duct, not incl. remote condenser									
0110	Constant volume									
0200	3 ton cooling	Q-5	1	16	Ea.	3,750	570		4,320	5,025
0210	4 ton cooling	"	.80	20	"	4,075	710		4,785	5,650
1000	Water cooled for free blow or duct, not including tower									
1010	Constant volume									
1100	3 ton cooling	Q-6	1	24	Ea.	3,725	820		4,545	5,450
1300	For hot water or steam heat coils, add				"	12%	10%			

23 81 26 – Split-System Air-Conditioners

23 81 26.10 Split Ductless Systems

		Crew	Daily Output	Labor-Hours	Unit	Material	Labor	Equipment	Total	Total Incl O&P
0010	**SPLIT DUCTLESS SYSTEMS**									
0100	Cooling only, single zone									
0110	Wall mount									
0120	3/4 ton cooling	Q-5	2	8	Ea.	1,100	284		1,384	1,700
0130	1 ton cooling		1.80	8.889		1,225	315		1,540	1,875
0140	1-1/2 ton cooling		1.60	10		1,975	355		2,330	2,750
0150	2 ton cooling	↓	1.40	11.429	↓	2,250	405		2,655	3,150
1000	Ceiling mount									
1020	2 ton cooling	Q-5	1.40	11.429	Ea.	2,000	405		2,405	2,875

23 81 Decentralized Unitary HVAC Equipment

23 81 26 – Split-System Air-Conditioners

23 81 26.10 Split Ductless Systems		Crew	Daily Output	Labor-Hours	Unit	Material	2015 Bare Costs Labor	Equipment	Total	Total Incl O&P
1030	3 ton cooling	Q-5	1.20	13.333	Ea.	2,550	475		3,025	3,575
3000	Multizone									
3010	Wall mount									
3020	2 @ 3/4 ton cooling	Q-5	1.80	8.889	Ea.	3,375	315		3,690	4,225
5000	Cooling/Heating									
5010	Wall mount									
5110	1 ton cooling	Q-5	1.70	9.412	Ea.	1,300	335		1,635	1,975
5120	1-1/2 ton cooling	"	1.50	10.667	"	1,975	380		2,355	2,800
7000	Accessories for all split ductless systems									
7010	Add for ambient frost control	Q-5	8	2	Ea.	126	71		197	256
7020	Add for tube/wiring kit, (Line sets)									
7030	15' kit	Q-5	32	.500	Ea.	103	17.75		120.75	143
7036	25' kit		28	.571		166	20.50		186.50	216
7040	35' kit		24	.667		174	23.50		197.50	230
7050	50' kit		20	.800		261	28.50		289.50	335

23 81 43 – Air-Source Unitary Heat Pumps

23 81 43.10 Air-Source Heat Pumps

		Crew	Daily Output	Labor-Hours	Unit	Material	Labor	Equipment	Total	Total Incl O&P
0010	**AIR-SOURCE HEAT PUMPS**, Not including interconnecting tubing									
1000	Air to air, split system, not including curbs, pads, fan coil and ductwork									
1012	Outside condensing unit only, for fan coil see Section 23 82 19.10									
1020	2 ton cooling, 8.5 MBH heat @ 0°F	Q-5	2	8	Ea.	2,400	284		2,684	3,125
1054	4 ton cooling, 24 MBH heat @ 0°F	"	.80	20	"	3,250	710		3,960	4,750
1500	Single package, not including curbs, pads, or plenums									
1520	2 ton cooling, 6.5 MBH heat @ 0°F	Q-5	1.50	10.667	Ea.	3,100	380		3,480	4,050
1580	4 ton cooling, 13 MBH heat @ 0°F	"	.96	16.667	"	4,200	590		4,790	5,600

23 81 46 – Water-Source Unitary Heat Pumps

23 81 46.10 Water Source Heat Pumps

		Crew	Daily Output	Labor-Hours	Unit	Material	Labor	Equipment	Total	Total Incl O&P
0010	**WATER SOURCE HEAT PUMPS**, Not incl. connecting tubing or water source									
2000	Water source to air, single package									
2100	1 ton cooling, 13 MBH heat @ 75°F	Q-5	2	8	Ea.	2,075	284		2,359	2,775
2200	4 ton cooling, 31 MBH heat @ 75°F	"	1.20	13.333	"	3,475	475		3,950	4,600

23 82 Convection Heating and Cooling Units

23 82 19 – Fan Coil Units

23 82 19.10 Fan Coil Air Conditioning

		Crew	Daily Output	Labor-Hours	Unit	Material	Labor	Equipment	Total	Total Incl O&P
0010	**FAN COIL AIR CONDITIONING**									
0030	Fan coil AC, cabinet mounted, filters and controls									
0100	Chilled water, 1/2 ton cooling	Q-5	8	2	Ea.	815	71		886	1,000
0110	3/4 ton cooling		7	2.286		850	81		931	1,075
0120	1 ton cooling		6	2.667		950	94.50		1,044.50	1,200

23 82 29 – Radiators

23 82 29.10 Hydronic Heating

		Crew	Daily Output	Labor-Hours	Unit	Material	Labor	Equipment	Total	Total Incl O&P
0010	**HYDRONIC HEATING**, Terminal units, not incl. main supply pipe									
1000	Radiation									
1100	Panel, baseboard, C.I., including supports, no covers	Q-5	46	.348	L.F.	38	12.35		50.35	62
3000	Radiators, cast iron									
3100	Free standing or wall hung, 6 tube, 25" high	Q-5	96	.167	Section	44	5.90		49.90	58.50
3200	4 tube, 19" high	"	96	.167	"	31.50	5.90		37.40	44.50
3250	Adj. brackets, 2 per wall radiator up to 30 sections	1 Stpi	32	.250	Ea.	53	9.85		62.85	75

23 82 Convection Heating and Cooling Units

23 82 36 – Finned-Tube Radiation Heaters

23 82 36.10 Finned Tube Radiation

		Crew	Daily Output	Labor-Hours	Unit	Material	2015 Bare Costs Labor	Equipment	Total	Total Incl O&P
0010	**FINNED TUBE RADIATION**, Terminal units, not incl. main supply pipe									
1310	Baseboard, pkgd, 1/2" copper tube, alum. fin, 7" high	Q-5	60	.267	L.F.	11.20	9.45		20.65	28
1320	3/4" copper tube, alum. fin, 7" high		58	.276		7.35	9.80		17.15	24.50
1340	1" copper tube, alum. fin, 8-7/8" high		56	.286		18.85	10.15		29	37.50
1360	1-1/4" copper tube, alum. fin, 8-7/8" high	↓	54	.296	↓	28	10.50		38.50	48
1500	Note: fin tube may also require corners, caps, etc.									

23 83 Radiant Heating Units

23 83 16 – Radiant-Heating Hydronic Piping

23 83 16.10 Radiant Floor Heating

		Crew	Daily Output	Labor-Hours	Unit	Material	2015 Bare Costs Labor	Equipment	Total	Total Incl O&P
0010	**RADIANT FLOOR HEATING**									
0100	Tubing, PEX (cross-linked polyethylene)									
0110	Oxygen barrier type for systems with ferrous materials									
0120	1/2"	Q-5	800	.020	L.F.	.96	.71		1.67	2.23
0130	3/4"		535	.030		1.35	1.06		2.41	3.24
0140	1"	↓	400	.040	↓	2.11	1.42		3.53	4.66
0200	Non barrier type for ferrous free systems									
0210	1/2"	Q-5	800	.020	L.F.	.61	.71		1.32	1.84
0220	3/4"		535	.030		1.11	1.06		2.17	2.97
0230	1"	↓	400	.040	↓	1.90	1.42		3.32	4.43
1000	Manifolds									
1110	Brass									
1120	With supply and return valves, flow meter, thermometer,									
1122	auto air vent and drain/fill valve.									
1130	1", 2 circuit	Q-5	14	1.143	Ea.	257	40.50		297.50	350
1140	1", 3 circuit		13.50	1.185		294	42		336	395
1150	1", 4 circuit		13	1.231		320	43.50		363.50	420
1154	1", 5 circuit		12.50	1.280		365	45.50		410.50	480
1158	1", 6 circuit		12	1.333		415	47.50		462.50	535
1162	1", 7 circuit		11.50	1.391		455	49.50		504.50	580
1166	1", 8 circuit		11	1.455		500	51.50		551.50	635
1172	1", 9 circuit		10.50	1.524		540	54		594	685
1174	1", 10 circuit		10	1.600		580	57		637	735
1178	1", 11 circuit		9.50	1.684		605	60		665	765
1182	1", 12 circuit	↓	9	1.778	↓	670	63		733	840
1610	Copper manifold header, (cut to size)									
1620	1" header, 12 – 1/2" sweat outlets	Q-5	3.33	4.805	Ea.	91	171		262	380
1630	1-1/4" header, 12 – 1/2" sweat outlets		3.20	5		106	178		284	410
1640	1-1/4" header, 12 – 3/4" sweat outlets		3	5.333		114	189		303	435
1650	1-1/2" header, 12 – 3/4" sweat outlets		3.10	5.161		137	183		320	450
1660	2" header, 12 – 3/4" sweat outlets	↓	2.90	5.517	↓	201	196		397	545
3000	Valves									
3110	Thermostatic zone valve actuator with end switch	Q-5	40	.400	Ea.	38.50	14.20		52.70	66
3114	Thermostatic zone valve actuator	"	36	.444	"	81	15.80		96.80	115
3120	Motorized straight zone valve with operator complete									
3130	3/4"	Q-5	35	.457	Ea.	130	16.25		146.25	170
3140	1"		32	.500		141	17.75		158.75	185
3150	1-1/4"	↓	29.60	.541	↓	179	19.20		198.20	229
3500	4 Way mixing valve, manual, brass									
3530	1"	Q-5	13.30	1.203	Ea.	180	42.50		222.50	269

23 83 Radiant Heating Units

23 83 16 – Radiant-Heating Hydronic Piping

23 83 16.10 Radiant Floor Heating

		Crew	Daily Output	Labor-Hours	Unit	Material	2015 Bare Costs Labor	Equipment	Total	Total Incl O&P
3540	1-1/4"	Q-5	11.40	1.404	Ea.	195	50		245	296
3550	1-1/2"		11	1.455		249	51.50		300.50	360
3560	2"		10.60	1.509		350	53.50		403.50	480
3800	Mixing valve motor, 4 way for valves, 1" and 1-1/4"		34	.471		310	16.70		326.70	370
3810	Mixing valve motor, 4 way for valves, 1-1/2" and 2"		30	.533		355	18.95		373.95	420
5000	Radiant floor heating, zone control panel									
5120	4 Zone actuator valve control, expandable	Q-5	20	.800	Ea.	163	28.50		191.50	226
5130	6 Zone actuator valve control, expandable		18	.889		226	31.50		257.50	300
6070	Thermal track, straight panel for long continuous runs, 5.333 S.F.		40	.400		26.50	14.20		40.70	53
6080	Thermal track, utility panel, for direction reverse at run end, 5.333 S.F.		40	.400		26.50	14.20		40.70	53
6090	Combination panel, for direction reverse plus straight run, 5.333 S.F.		40	.400		26.50	14.20		40.70	53
7000	PEX tubing fittings									
7100	Compression type									
7116	Coupling									
7120	1/2" x 1/2"	1 Stpi	27	.296	Ea.	6.70	11.70		18.40	26.50
7124	3/4" x 3/4"	"	23	.348	"	10.60	13.70		24.30	34
7130	Adapter									
7132	1/2" x female sweat 1/2"	1 Stpi	27	.296	Ea.	4.36	11.70		16.06	24
7134	1/2" x female sweat 3/4"		26	.308		4.88	12.15		17.03	25.50
7136	5/8" x female sweat 3/4"		24	.333		7	13.15		20.15	29
7140	Elbow									
7142	1/2" x female sweat 1/2"	1 Stpi	27	.296	Ea.	6.65	11.70		18.35	26.50
7144	1/2" x female sweat 3/4"		26	.308		7.80	12.15		19.95	28.50
7146	5/8" x female sweat 3/4"		24	.333		8.75	13.15		21.90	31
7200	Insert type									
7206	PEX x male NPT									
7210	1/2" x 1/2"	1 Stpi	29	.276	Ea.	2.59	10.90		13.49	21
7220	3/4" x 3/4"		27	.296		3.81	11.70		15.51	23.50
7230	1" x 1"		26	.308		6.45	12.15		18.60	27
7300	PEX coupling									
7310	1/2" x 1/2"	1 Stpi	30	.267	Ea.	1.76	10.50		12.26	19.30
7320	3/4" x 3/4"		29	.276		2.15	10.90		13.05	20.50
7330	1" x 1"		28	.286		5.85	11.25		17.10	25
7400	PEX stainless crimp ring									
7410	1/2" x 1/2"	1 Stpi	86	.093	Ea.	.37	3.67		4.04	6.45
7420	3/4" x 3/4"		84	.095		.51	3.76		4.27	6.75
7430	1" x 1"		82	.098		.73	3.85		4.58	7.15

23 83 33 – Electric Radiant Heaters

23 83 33.10 Electric Heating

		Crew	Daily Output	Labor-Hours	Unit	Material	2015 Bare Costs Labor	Equipment	Total	Total Incl O&P
0010	**ELECTRIC HEATING**, not incl. conduit or feed wiring									
1100	Rule of thumb: Baseboard units, including control	1 Elec	4.40	1.818	kW	104	66.50		170.50	223
1300	Baseboard heaters, 2' long, 350 watt		8	1	Ea.	28.50	36.50		65	91.50
1400	3' long, 750 watt		8	1		33	36.50		69.50	96.50
1600	4' long, 1000 watt		6.70	1.194		39	44		83	115
1800	5' long, 935 watt		5.70	1.404		48	51.50		99.50	137
2000	6' long, 1500 watt		5	1.600		54.50	58.50		113	156
2400	8' long, 2000 watt		4	2		66	73.50		139.50	193
2800	10' long, 1875 watt		3.30	2.424		190	89		279	355
2950	Wall heaters with fan, 120 to 277 volt									
3170	1000 watt	1 Elec	6	1.333	Ea.	134	49		183	227
3180	1250 watt		5	1.600		134	58.50		192.50	243
3190	1500 watt		4	2		134	73.50		207.50	267

23 83 Radiant Heating Units

23 83 33 – Electric Radiant Heaters

23 83 33.10 Electric Heating		Crew	Daily Output	Labor-Hours	Unit	Material	2015 Bare Costs Labor	Equipment	Total	Total Incl O&P
3600	Thermostats, integral	1 Elec	16	.500	Ea.	30	18.35		48.35	63
3800	Line voltage, 1 pole		8	1		19.20	36.50		55.70	81
5000	Radiant heating ceiling panels, 2' x 4', 500 watt		16	.500		305	18.35		323.35	365
5050	750 watt		16	.500		310	18.35		328.35	375
5300	Infrared quartz heaters, 120 volts, 1000 watts		6.70	1.194		284	44		328	380
5350	1500 watt		5	1.600		284	58.50		342.50	405
5400	240 volts, 1500 watt		5	1.600		284	58.50		342.50	405
5450	2000 watt		4	2		284	73.50		357.50	430
5500	3000 watt	↓	3	2.667	↓	325	97.50		422.50	515

Estimating Tips
26 05 00 Common Work Results for Electrical

- Conduit should be taken off in three main categories—power distribution, branch power, and branch lighting—so the estimator can concentrate on systems and components, therefore making it easier to ensure all items have been accounted for.
- For cost modifications for elevated conduit installation, add the percentages to labor according to the height of installation, and only to the quantities exceeding the different height levels, not to the total conduit quantities.
- Remember that aluminum wiring of equal ampacity is larger in diameter than copper and may require larger conduit.
- If more than three wires at a time are being pulled, deduct percentages from the labor hours of that grouping of wires.
- When taking off grounding systems, identify separately the type and size of wire, and list each unique type of ground connection.
- The estimator should take the weights of materials into consideration when completing a takeoff. Topics to consider include: How will the materials be supported? What methods of support are available? How high will the support structure have to reach? Will the final support structure be able to withstand the total burden? Is the support material included or separate from the fixture, equipment, and material specified?
- Do not overlook the costs for equipment used in the installation. If scaffolding or highlifts are available in the field, contractors may use them in lieu of the proposed ladders and rolling staging.

26 20 00 Low-Voltage Electrical Transmission

- Supports and concrete pads may be shown on drawings for the larger equipment, or the support system may be only a piece of plywood for the back of a panelboard. In either case, it must be included in the costs.

26 40 00 Electrical and Cathodic Protection

- When taking off cathodic protections systems, identify the type and size of cable, and list each unique type of anode connection.

26 50 00 Lighting

- Fixtures should be taken off room by room, using the fixture schedule, specifications, and the ceiling plan. For large concentrations of lighting fixtures in the same area, deduct the percentages from labor hours.

Reference Numbers
Reference numbers are shown in shaded boxes at the beginning of some major classifications. These numbers refer to related items in the Reference Section. The reference information may be an estimating procedure, an alternate pricing method, or technical information.

Note: Not all subdivisions listed here necessarily appear in this publication. ■

*Note: **Trade Service**, in part, has been used as a reference source for some of the material prices used in Division 26.*

Did you know?
RSMeans Online gives you the same access to RSMeans' data with 24/7 access:
- Quickly locate costs in the searchable database.
- Build cost lists, estimates, and reports in minutes.
- Adjust costs to any location in the U.S. and Canada with the click of a button.

Start your free trial today at **www.rsmeansonline.com**

RSMeansOnline

26 05 Common Work Results for Electrical

26 05 05 – Selective Demolition for Electrical

26 05 05.10 Electrical Demolition

		Crew	Daily Output	Labor-Hours	Unit	Material	2015 Bare Costs Labor	2015 Bare Costs Equipment	Total	Total Incl O&P
0010	**ELECTRICAL DEMOLITION**									
0020	Conduit to 15' high, including fittings & hangers									
0100	Rigid galvanized steel, 1/2" to 1" diameter	1 Elec	242	.033	L.F.		1.21		1.21	1.99
0120	1-1/4" to 2"	"	200	.040	"		1.47		1.47	2.40
0270	Armored cable, (BX) avg. 50' runs									
0280	#14, 2 wire	1 Elec	690	.012	L.F.		.42		.42	.70
0290	#14, 3 wire		571	.014			.51		.51	.84
0300	#12, 2 wire		605	.013			.48		.48	.79
0310	#12, 3 wire		514	.016			.57		.57	.93
0320	#10, 2 wire		514	.016			.57		.57	.93
0330	#10, 3 wire		425	.019			.69		.69	1.13
0340	#8, 3 wire		342	.023			.86		.86	1.40
0350	Non metallic sheathed cable (Romex)									
0360	#14, 2 wire	1 Elec	720	.011	L.F.		.41		.41	.67
0370	#14, 3 wire		657	.012			.45		.45	.73
0380	#12, 2 wire		629	.013			.47		.47	.76
0390	#10, 3 wire		450	.018			.65		.65	1.07
0400	Wiremold raceway, including fittings & hangers									
0420	No. 3000	1 Elec	250	.032	L.F.		1.17		1.17	1.92
0440	No. 4000		217	.037			1.35		1.35	2.21
0460	No. 6000		166	.048			1.77		1.77	2.89
0462	Plugmold with receptacle		114	.070			2.57		2.57	4.21
0465	Telephone/power pole		12	.667	Ea.		24.50		24.50	40
0470	Non-metallic, straight section		480	.017	L.F.		.61		.61	1
0500	Channels, steel, including fittings & hangers									
0520	3/4" x 1-1/2"	1 Elec	308	.026	L.F.		.95		.95	1.56
0540	1-1/2" x 1-1/2"		269	.030			1.09		1.09	1.79
0560	1-1/2" x 1-7/8"		229	.035			1.28		1.28	2.10
1210	Panel boards, incl. removal of all breakers,									
1220	conduit terminations & wire connections									
1230	3 wire, 120/240 V, 100A, to 20 circuits	1 Elec	2.60	3.077	Ea.		113		113	185
1240	200 amps, to 42 circuits	2 Elec	2.60	6.154			226		226	370
1260	4 wire, 120/208 V, 125A, to 20 circuits	1 Elec	2.40	3.333			122		122	200
1270	200 amps, to 42 circuits	2 Elec	2.40	6.667			244		244	400
1720	Junction boxes, 4" sq. & oct.	1 Elec	80	.100			3.67		3.67	6
1740	Handy box		107	.075			2.74		2.74	4.49
1760	Switch box		107	.075			2.74		2.74	4.49
1780	Receptacle & switch plates		257	.031			1.14		1.14	1.87
1800	Wire, THW-THWN-THHN, removed from									
1810	in place conduit, to 15' high									
1830	#14	1 Elec	65	.123	C.L.F.		4.51		4.51	7.40
1840	#12		55	.145			5.35		5.35	8.75
1850	#10		45.50	.176			6.45		6.45	10.55
2000	Interior fluorescent fixtures, incl. supports									
2010	& whips, to 15' high									
2100	Recessed drop-in 2' x 2', 2 lamp	2 Elec	35	.457	Ea.		16.75		16.75	27.50
2140	2' x 4', 4 lamp	"	30	.533	"		19.55		19.55	32
2180	Surface mount, acrylic lens & hinged frame									
2220	2' x 2', 2 lamp	2 Elec	44	.364	Ea.		13.35		13.35	22
2260	2' x 4', 4 lamp	"	33	.485	"		17.75		17.75	29
2300	Strip fixtures, surface mount									
2320	4' long, 1 lamp	2 Elec	53	.302	Ea.		11.05		11.05	18.15
2380	8' long, 2 lamp	"	40	.400	"		14.65		14.65	24

26 05 Common Work Results for Electrical

26 05 05 – Selective Demolition for Electrical

26 05 05.10 Electrical Demolition		Crew	Daily Output	Labor-Hours	Unit	Material	2015 Bare Costs Labor	Equipment	Total	Total Incl O&P
2460	Interior incandescent, surface, ceiling									
2470	or wall mount, to 12' high									
2480	Metal cylinder type, 75 Watt	2 Elec	62	.258	Ea.		9.45		9.45	15.50
2600	Exterior fixtures, incandescent, wall mount									
2620	100 Watt	2 Elec	50	.320	Ea.		11.75		11.75	19.20
3000	Ceiling fan, tear out and remove	1 Elec	24	.333	"		12.20		12.20	20

26 05 19 – Low-Voltage Electrical Power Conductors and Cables

26 05 19.20 Armored Cable

		Crew	Daily Output	Labor-Hours	Unit	Material	Labor	Equipment	Total	Total Incl O&P
0010	**ARMORED CABLE**									
0051	600 volt, copper (BX), #14, 2 conductor, solid	1 Elec	240	.033	L.F.	.46	1.22		1.68	2.50
0101	3 conductor, solid		200	.040		.71	1.47		2.18	3.18
0151	#12, 2 conductor, solid		210	.038		.46	1.40		1.86	2.80
0201	3 conductor, solid		180	.044		.75	1.63		2.38	3.50
0251	#10, 2 conductor, solid		180	.044		.85	1.63		2.48	3.60
0301	3 conductor, solid		150	.053		1.18	1.95		3.13	4.50
0351	3 conductor, stranded		120	.067		2.22	2.44		4.66	6.45

26 05 19.55 Non-Metallic Sheathed Cable

		Crew	Daily Output	Labor-Hours	Unit	Material	Labor	Equipment	Total	Total Incl O&P
0010	**NON-METALLIC SHEATHED CABLE** 600 volt									
0100	Copper with ground wire, (Romex)									
0151	#14, 2 wire	1 Elec	250	.032	L.F.	.24	1.17		1.41	2.18
0201	3 wire		230	.035		.34	1.27		1.61	2.47
0251	#12, 2 wire		220	.036		.36	1.33		1.69	2.58
0301	3 wire		200	.040		.52	1.47		1.99	2.97
0351	#10, 2 wire		200	.040		.56	1.47		2.03	3.02
0401	3 wire		140	.057		.82	2.09		2.91	4.34
0451	#8, 3 conductor		130	.062		1.33	2.26		3.59	5.15
0501	#6, 3 wire		120	.067		2.15	2.44		4.59	6.35
0550	SE type SER aluminum cable, 3 RHW and									
0601	1 bare neutral, 3 #8 & 1 #8	1 Elec	150	.053	L.F.	1.58	1.95		3.53	4.94
0651	3 #6 & 1 #6	"	130	.062		1.79	2.26		4.05	5.65
0701	3 #4 & 1 #6	2 Elec	220	.073		1.65	2.67		4.32	6.20
0751	3 #2 & 1 #4		200	.080		2.96	2.93		5.89	8.05
0801	3 #1/0 & 1 #2		180	.089		4.49	3.26		7.75	10.30
0851	3 #2/0 & 1 #1		160	.100		5.30	3.67		8.97	11.80
0901	3 #4/0 & 1 #2/0		140	.114		7.55	4.19		11.74	15.15
2401	SEU service entrance cable, copper 2 conductors, #8 + #8 neutral	1 Elec	150	.053		1.15	1.95		3.10	4.46
2601	#6 + #8 neutral		130	.062		1.70	2.26		3.96	5.55
2801	#6 + #6 neutral		130	.062		1.93	2.26		4.19	5.80
3001	#4 + #6 neutral	2 Elec	220	.073		2.66	2.67		5.33	7.30
3201	#4 + #4 neutral		220	.073		3	2.67		5.67	7.65
3401	#3 + #5 neutral		210	.076		3.72	2.79		6.51	8.65
6500	Service entrance cap for copper SEU									
6600	100 amp	1 Elec	12	.667	Ea.	8.75	24.50		33.25	49.50
6700	150 amp		10	.800		16.10	29.50		45.60	65.50
6800	200 amp		8	1		24.50	36.50		61	87

26 05 19.90 Wire

		Crew	Daily Output	Labor-Hours	Unit	Material	Labor	Equipment	Total	Total Incl O&P
0010	**WIRE**									
0021	600 volt, copper type THW, solid, #14	1 Elec	1300	.006	L.F.	.08	.23		.31	.46
0031	#12		1100	.007		.12	.27		.39	.57
0041	#10		1000	.008		.19	.29		.48	.69
0161	#6		650	.012		.64	.45		1.09	1.44
0181	#4	2 Elec	1060	.015		1	.55		1.55	2.01

26 05 Common Work Results for Electrical

26 05 19 – Low-Voltage Electrical Power Conductors and Cables

26 05 19.90 Wire

		Crew	Daily Output	Labor-Hours	Unit	Material	2015 Bare Costs Labor	Equipment	Total	Total Incl O&P
0201	#3	2 Elec	1000	.016	L.F.	1.26	.59		1.85	2.34
0221	#2		900	.018		1.58	.65		2.23	2.81
0241	#1		800	.020		2	.73		2.73	3.40
0261	1/0		660	.024		2.50	.89		3.39	4.21
0281	2/0		580	.028		3.13	1.01		4.14	5.10
0301	3/0		500	.032		3.93	1.17		5.10	6.25
0351	4/0		440	.036		5	1.33		6.33	7.70

26 05 26 – Grounding and Bonding for Electrical Systems

26 05 26.80 Grounding

		Crew	Daily Output	Labor-Hours	Unit	Material	Labor	Equipment	Total	Total Incl O&P
0010	**GROUNDING**									
0030	Rod, copper clad, 8' long, 1/2" diameter	1 Elec	5.50	1.455	Ea.	20	53.50		73.50	110
0050	3/4" diameter		5.30	1.509		33.50	55.50		89	128
0080	10' long, 1/2" diameter		4.80	1.667		22	61		83	125
0100	3/4" diameter		4.40	1.818		38.50	66.50		105	152
0261	Wire, ground bare armored, #8-1 conductor		200	.040	L.F.	.76	1.47		2.23	3.24
0271	#6-1 conductor		180	.044	"	.91	1.63		2.54	3.68
0390	Bare copper wire, stranded, #8		11	.727	C.L.F.	29.50	26.50		56	76
0401	Bare copper, #6 wire		1000	.008	L.F.	.52	.29		.81	1.05
0601	#2 stranded	2 Elec	1000	.016	"	1.38	.59		1.97	2.48
1800	Water pipe ground clamps, heavy duty									
2000	Bronze, 1/2" to 1" diameter	1 Elec	8	1	Ea.	24	36.50		60.50	86

26 05 33 – Raceway and Boxes for Electrical Systems

26 05 33.13 Conduit

		Crew	Daily Output	Labor-Hours	Unit	Material	Labor	Equipment	Total	Total Incl O&P
0010	**CONDUIT** To 15' high, includes 2 terminations, 2 elbows,									
0020	11 beam clamps, and 11 couplings per 100 L.F.									
1750	Rigid galvanized steel, 1/2" diameter	1 Elec	90	.089	L.F.	2.49	3.26		5.75	8.10
1770	3/4" diameter		80	.100		2.74	3.67		6.41	9
1800	1" diameter		65	.123		3.95	4.51		8.46	11.75
1830	1-1/4" diameter		60	.133		5.15	4.89		10.04	13.65
1850	1-1/2" diameter		55	.145		6.10	5.35		11.45	15.50
1870	2" diameter		45	.178		7.90	6.50		14.40	19.40
5000	Electric metallic tubing (EMT), 1/2" diameter		170	.047		.67	1.72		2.39	3.57
5020	3/4" diameter		130	.062		.94	2.26		3.20	4.74
5040	1" diameter		115	.070		1.60	2.55		4.15	5.95
5060	1-1/4" diameter		100	.080		2.61	2.93		5.54	7.65
5080	1-1/2" diameter		90	.089		3.35	3.26		6.61	9.05
9100	PVC, schedule 40, 1/2" diameter		190	.042		.90	1.54		2.44	3.52
9110	3/4" diameter		145	.055		.98	2.02		3	4.39
9120	1" diameter		125	.064		2.11	2.35		4.46	6.15
9130	1-1/4" diameter		110	.073		2.63	2.67		5.30	7.25
9140	1-1/2" diameter		100	.080		2.99	2.93		5.92	8.10
9150	2" diameter		90	.089		3.90	3.26		7.16	9.65

26 05 33.16 Outlet Boxes

		Crew	Daily Output	Labor-Hours	Unit	Material	Labor	Equipment	Total	Total Incl O&P
0010	**OUTLET BOXES**									
0021	Pressed steel, octagon, 4"	1 Elec	18	.444	Ea.	2.61	16.30		18.91	29.50
0060	Covers, blank		64	.125		1.10	4.58		5.68	8.70
0100	Extension rings		40	.200		4.33	7.35		11.68	16.75
0151	Square 4"		18	.444		2.42	16.30		18.72	29
0200	Extension rings		40	.200		4.37	7.35		11.72	16.80
0250	Covers, blank		64	.125		1.18	4.58		5.76	8.80
0300	Plaster rings		64	.125		2.40	4.58		6.98	10.15

26 05 Common Work Results for Electrical

26 05 33 – Raceway and Boxes for Electrical Systems

26 05 33.16 Outlet Boxes		Crew	Daily Output	Labor-Hours	Unit	Material	2015 Bare Costs Labor	Equipment	Total	Total Incl O&P
0651	Switchbox	1 Elec	24	.333	Ea.	4.18	12.20		16.38	24.50
1100	Concrete, floor, 1 gang		5.30	1.509		89.50	55.50		145	189

26 05 33.17 Outlet Boxes, Plastic										
0010	**OUTLET BOXES, PLASTIC**									
0051	4" diameter, round, with 2 mounting nails	1 Elec	23	.348	Ea.	2.39	12.75		15.14	23.50
0101	Bar hanger mounted		23	.348		5.10	12.75		17.85	26.50
0201	Square with 2 mounting nails		23	.348		4.63	12.75		17.38	26
0300	Plaster ring		64	.125		1.98	4.58		6.56	9.70
0401	Switch box with 2 mounting nails, 1 gang		27	.296		4.39	10.85		15.24	22.50
0501	2 gang		23	.348		3.33	12.75		16.08	24.50
0601	3 gang		18	.444		4.93	16.30		21.23	32

26 05 33.18 Pull Boxes										
0010	**PULL BOXES**									
0100	Steel, pull box, NEMA 1, type SC, 6" W x 6" H x 4" D	1 Elec	8	1	Ea.	9.65	36.50		46.15	70.50
0200	8" W x 8" H x 4" D		8	1		14.50	36.50		51	76
0300	10" W x 12" H x 6" D		5.30	1.509		24.50	55.50		80	118

26 05 33.25 Conduit Fittings for Rigid Galvanized Steel										
0010	**CONDUIT FITTINGS FOR RIGID GALVANIZED STEEL**									
2280	LB, LR or LL fittings & covers, 1/2" diameter	1 Elec	16	.500	Ea.	8.60	18.35		26.95	39.50
2290	3/4" diameter		13	.615		10.65	22.50		33.15	49
2300	1" diameter		11	.727		15.55	26.50		42.05	60.50
2330	1-1/4" diameter		8	1		24.50	36.50		61	86.50
2350	1-1/2" diameter		6	1.333		32	49		81	116
2370	2" diameter		5	1.600		52.50	58.50		111	154
5280	Service entrance cap, 1/2" diameter		16	.500		5.65	18.35		24	36
5300	3/4" diameter		13	.615		6.50	22.50		29	44
5320	1" diameter		10	.800		7.85	29.50		37.35	56.50
5340	1-1/4" diameter		8	1		10.55	36.50		47.05	71.50
5360	1-1/2" diameter		6.50	1.231		21	45		66	97
5380	2" diameter		5.50	1.455		31.50	53.50		85	122

26 05 33.35 Flexible Metallic Conduit										
0010	**FLEXIBLE METALLIC CONDUIT**									
0050	Steel, 3/8" diameter	1 Elec	200	.040	L.F.	.43	1.47		1.90	2.87
0100	1/2" diameter		200	.040		.48	1.47		1.95	2.93
0200	3/4" diameter		160	.050		.67	1.83		2.50	3.74
0250	1" diameter		100	.080		1.21	2.93		4.14	6.15
0300	1-1/4" diameter		70	.114		1.57	4.19		5.76	8.60
0350	1-1/2" diameter		50	.160		2.55	5.85		8.40	12.40
0370	2" diameter		40	.200		3.11	7.35		10.46	15.40

26 05 39 – Underfloor Raceways for Electrical Systems

26 05 39.30 Conduit In Concrete Slab										
0010	**CONDUIT IN CONCRETE SLAB** Including terminations,									
0020	fittings and supports									
3230	PVC, schedule 40, 1/2" diameter	1 Elec	270	.030	L.F.	.57	1.09		1.66	2.41
3250	3/4" diameter		230	.035		.66	1.27		1.93	2.81
3270	1" diameter		200	.040		.87	1.47		2.34	3.35
3300	1-1/4" diameter		170	.047		1.17	1.72		2.89	4.12
3330	1-1/2" diameter		140	.057		1.42	2.09		3.51	4.99
3350	2" diameter		120	.067		1.78	2.44		4.22	5.95
4350	Rigid galvanized steel, 1/2" diameter		200	.040		2.48	1.47		3.95	5.15
4400	3/4" diameter		170	.047		2.62	1.72		4.34	5.70

For customer support on your Residential Cost Data, call 877.713.2617.

26 05 Common Work Results for Electrical

26 05 39 – Underfloor Raceways for Electrical Systems

26 05 39.30 Conduit In Concrete Slab		Crew	Daily Output	Labor-Hours	Unit	Material	2015 Bare Costs Labor	Equipment	Total	Total Incl O&P
4450	1" diameter	1 Elec	130	.062	L.F.	3.60	2.26		5.86	7.65
4500	1-1/4" diameter		110	.073		4.94	2.67		7.61	9.80
4600	1-1/2" diameter		100	.080		5.50	2.93		8.43	10.85
4800	2" diameter		90	.089		6.80	3.26		10.06	12.85

26 05 39.40 Conduit In Trench

		Crew	Daily Output	Labor-Hours	Unit	Material	Labor	Equipment	Total	Total Incl O&P
0010	**CONDUIT IN TRENCH** Includes terminations and fittings									
0020	Does not include excavation or backfill, see Section 31 23 16.00									
0200	Rigid galvanized steel, 2" diameter	1 Elec	150	.053	L.F.	6.40	1.95		8.35	10.25
0400	2-1/2" diameter	"	100	.080		12.45	2.93		15.38	18.50
0600	3" diameter	2 Elec	160	.100		14.55	3.67		18.22	22
0800	3-1/2" diameter	"	140	.114		19.20	4.19		23.39	28

26 05 80 – Wiring Connections

26 05 80.10 Motor Connections

		Crew	Daily Output	Labor-Hours	Unit	Material	Labor	Equipment	Total	Total Incl O&P
0010	**MOTOR CONNECTIONS**									
0020	Flexible conduit and fittings, 115 volt, 1 phase, up to 1 HP motor	1 Elec	8	1	Ea.	6.35	36.50		42.85	67

26 05 90 – Residential Applications

26 05 90.10 Residential Wiring

		Crew	Daily Output	Labor-Hours	Unit	Material	Labor	Equipment	Total	Total Incl O&P
0010	**RESIDENTIAL WIRING**									
0020	20' avg. runs and #14/2 wiring incl. unless otherwise noted									
1000	Service & panel, includes 24' SE-AL cable, service eye, meter,									
1010	Socket, panel board, main bkr., ground rod, 15 or 20 amp									
1020	1-pole circuit breakers, and misc. hardware									
1100	100 amp, with 10 branch breakers	1 Elec	1.19	6.723	Ea.	560	246		806	1,025
1110	With PVC conduit and wire		.92	8.696		605	320		925	1,175
1120	With RGS conduit and wire		.73	10.959		765	400		1,165	1,500
1150	150 amp, with 14 branch breakers		1.03	7.767		860	285		1,145	1,400
1170	With PVC conduit and wire		.82	9.756		955	360		1,315	1,625
1180	With RGS conduit and wire		.67	11.940		1,275	440		1,715	2,125
1200	200 amp, with 18 branch breakers	2 Elec	1.80	8.889		1,150	325		1,475	1,800
1220	With PVC conduit and wire		1.46	10.959		1,250	400		1,650	2,025
1230	With RGS conduit and wire		1.24	12.903		1,650	475		2,125	2,575
1800	Lightning surge suppressor	1 Elec	32	.250		50.50	9.15		59.65	71
2000	Switch devices									
2100	Single pole, 15 amp, Ivory, with a 1-gang box, cover plate,									
2110	Type NM (Romex) cable	1 Elec	17.10	.468	Ea.	12.35	17.15		29.50	41.50
2120	Type MC (BX) cable		14.30	.559		21.50	20.50		42	57
2130	EMT & wire		5.71	1.401		30	51.50		81.50	117
2150	3-way, #14/3, type NM cable		14.55	.550		14.85	20		34.85	49.50
2170	Type MC cable		12.31	.650		27	24		51	68.50
2180	EMT & wire		5	1.600		32	58.50		90.50	131
2200	4-way, #14/3, type NM cable		14.55	.550		21	20		41	56
2220	Type MC cable		12.31	.650		33	24		57	75.50
2230	EMT & wire		5	1.600		38.50	58.50		97	138
2250	S.P., 20 amp, #12/2, type NM cable		13.33	.600		22	22		44	60
2270	Type MC cable		11.43	.700		28.50	25.50		54	73.50
2280	EMT & wire		4.85	1.649		41	60.50		101.50	144
2290	S.P. rotary dimmer, 600W, no wiring		17	.471		29.50	17.25		46.75	61
2300	S.P. rotary dimmer, 600W, type NM cable		14.55	.550		34.50	20		54.50	71
2320	Type MC cable		12.31	.650		43.50	24		67.50	87
2330	EMT & wire		5	1.600		53.50	58.50		112	155
2350	3-way rotary dimmer, type NM cable		13.33	.600		28.50	22		50.50	67.50

26 05 Common Work Results for Electrical

26 05 90 − Residential Applications

26 05 90.10 Residential Wiring

		Crew	Daily Output	Labor-Hours	Unit	Material	2015 Bare Costs Labor	Equipment	Total	Total Incl O&P
2370	Type MC cable	1 Elec	11.43	.700	Ea.	37.50	25.50		63	83
2380	EMT & wire		4.85	1.649		47.50	60.50		108	152
2400	Interval timer wall switch, 20 amp, 1-30 min., #12/2									
2410	Type NM cable	1 Elec	14.55	.550	Ea.	56	20		76	94.50
2420	Type MC cable		12.31	.650		60	24		84	105
2430	EMT & wire		5	1.600		75	58.50		133.50	179
2500	Decorator style									
2510	S.P., 15 amp, type NM cable	1 Elec	17.10	.468	Ea.	16.25	17.15		33.40	46
2520	Type MC cable		14.30	.559		25	20.50		45.50	61
2530	EMT & wire		5.71	1.401		34	51.50		85.50	121
2550	3-way, #14/3, type NM cable		14.55	.550		18.75	20		38.75	53.50
2570	Type MC cable		12.31	.650		30.50	24		54.50	72.50
2580	EMT & wire		5	1.600		36	58.50		94.50	136
2600	4-way, #14/3, type NM cable		14.55	.550		25	20		45	60.50
2620	Type MC cable		12.31	.650		37	24		61	79.50
2630	EMT & wire		5	1.600		42	58.50		100.50	143
2650	S.P., 20 amp, #12/2, type NM cable		13.33	.600		26	22		48	64.50
2670	Type MC cable		11.43	.700		32.50	25.50		58	77.50
2680	EMT & wire		4.85	1.649		45	60.50		105.50	149
2700	S.P., slide dimmer, type NM cable		17.10	.468		28.50	17.15		45.65	59.50
2720	Type MC cable		14.30	.559		37.50	20.50		58	75
2730	EMT & wire		5.71	1.401		47.50	51.50		99	137
2750	S.P., touch dimmer, type NM cable		17.10	.468		32	17.15		49.15	63.50
2770	Type MC cable		14.30	.559		41	20.50		61.50	79
2780	EMT & wire		5.71	1.401		51.50	51.50		103	141
2800	3-way touch dimmer, type NM cable		13.33	.600		52	22		74	93
2820	Type MC cable		11.43	.700		60.50	25.50		86	109
2830	EMT & wire		4.85	1.649		71	60.50		131.50	177
3000	Combination devices									
3100	S.P. switch/15 amp recpt., Ivory, 1-gang box, plate									
3110	Type NM cable	1 Elec	11.43	.700	Ea.	24.50	25.50		50	69
3120	Type MC cable		10	.800		33.50	29.50		63	84.50
3130	EMT & wire		4.40	1.818		43.50	66.50		110	157
3150	S.P. switch/pilot light, type NM cable		11.43	.700		24.50	25.50		50	69
3170	Type MC cable		10	.800		33.50	29.50		63	85
3180	EMT & wire		4.43	1.806		43.50	66		109.50	156
3190	2-S.P. switches, 2-#14/2, no wiring		14	.571		6.90	21		27.90	42
3200	2-S.P. switches, 2-#14/2, type NM cables		10	.800		27.50	29.50		57	78
3220	Type MC cable		8.89	.900		41	33		74	99
3230	EMT & wire		4.10	1.951		46.50	71.50		118	168
3250	3-way switch/15 amp recpt., #14/3, type NM cable		10	.800		32	29.50		61.50	83
3270	Type MC cable		8.89	.900		44	33		77	102
3280	EMT & wire		4.10	1.951		49	71.50		120.50	171
3300	2-3 way switches, 2-#14/3, type NM cables		8.89	.900		41.50	33		74.50	99.50
3320	Type MC cable		8	1		60.50	36.50		97	127
3330	EMT & wire		4	2		56.50	73.50		130	182
3350	S.P. switch/20 amp recpt., #12/2, type NM cable		10	.800		32	29.50		61.50	83
3370	Type MC cable		8.89	.900		36	33		69	93.50
3380	EMT & wire		4.10	1.951		51	71.50		122.50	173
3400	Decorator style									
3410	S.P. switch/15 amp recpt., type NM cable	1 Elec	11.43	.700	Ea.	28.50	25.50		54	73
3420	Type MC cable		10	.800		37	29.50		66.50	89
3430	EMT & wire		4.40	1.818		47.50	66.50		114	161

For customer support on your Residential Cost Data, call 877.713.2617.

26 05 Common Work Results for Electrical

26 05 90 – Residential Applications

26 05 90.10 Residential Wiring		Crew	Daily Output	Labor-Hours	Unit	Material	2015 Bare Costs Labor	Equipment	Total	Total Incl O&P
3450	S.P. switch/pilot light, type NM cable	1 Elec	11.43	.700	Ea.	28.50	25.50		54	73
3470	Type MC cable		10	.800		37.50	29.50		67	89
3480	EMT & wire		4.40	1.818		47.50	66.50		114	162
3500	2-S.P. switches, 2-#14/2, type NM cables		10	.800		31.50	29.50		61	82.50
3520	Type MC cable		8.89	.900		44.50	33		77.50	103
3530	EMT & wire		4.10	1.951		50.50	71.50		122	173
3550	3-way/15 amp recpt., #14/3, type NM cable		10	.800		36	29.50		65.50	87.50
3570	Type MC cable		8.89	.900		47.50	33		80.50	107
3580	EMT & wire		4.10	1.951		53	71.50		124.50	175
3650	2-3 way switches, 2-#14/3, type NM cables		8.89	.900		45.50	33		78.50	104
3670	Type MC cable		8	1		64.50	36.50		101	131
3680	EMT & wire		4	2		60.50	73.50		134	187
3700	S.P. switch/20 amp recpt., #12/2, type NM cable		10	.800		35.50	29.50		65	87.50
3720	Type MC cable		8.89	.900		39.50	33		72.50	97.50
3730	EMT & wire		4.10	1.951		55	71.50		126.50	178
4000	Receptacle devices									
4010	Duplex outlet, 15 amp recpt., Ivory, 1-gang box, plate									
4015	Type NM cable	1 Elec	14.55	.550	Ea.	10.80	20		30.80	45
4020	Type MC cable		12.31	.650		19.75	24		43.75	60.50
4030	EMT & wire		5.33	1.501		28.50	55		83.50	122
4050	With #12/2, type NM cable		12.31	.650		13.30	24		37.30	53.50
4070	Type MC cable		10.67	.750		19.85	27.50		47.35	67
4080	EMT & wire		4.71	1.699		32.50	62.50		95	138
4100	20 amp recpt., #12/2, type NM cable		12.31	.650		19.95	24		43.95	61
4120	Type MC cable		10.67	.750		26.50	27.50		54	74
4130	EMT & wire		4.71	1.699		39	62.50		101.50	145
4140	For GFI see Section 26 05 90.10 line 4300 below									
4150	Decorator style, 15 amp recpt., type NM cable	1 Elec	14.55	.550	Ea.	14.70	20		34.70	49
4170	Type MC cable		12.31	.650		23.50	24		47.50	65
4180	EMT & wire		5.33	1.501		32.50	55		87.50	126
4200	With #12/2, type NM cable		12.31	.650		17.20	24		41.20	58
4220	Type MC cable		10.67	.750		23.50	27.50		51	71
4230	EMT & wire		4.71	1.699		36.50	62.50		99	142
4250	20 amp recpt. #12/2, type NM cable		12.31	.650		24	24		48	65
4270	Type MC cable		10.67	.750		30.50	27.50		58	78.50
4280	EMT & wire		4.71	1.699		43	62.50		105.50	150
4300	GFI, 15 amp recpt., type NM cable		12.31	.650		41.50	24		65.50	84.50
4320	Type MC cable		10.67	.750		50.50	27.50		78	101
4330	EMT & wire		4.71	1.699		59	62.50		121.50	167
4350	GFI with #12/2, type NM cable		10.67	.750		44	27.50		71.50	93.50
4370	Type MC cable		9.20	.870		50.50	32		82.50	108
4380	EMT & wire		4.21	1.900		63	69.50		132.50	184
4400	20 amp recpt., #12/2 type NM cable		10.67	.750		52	27.50		79.50	103
4420	Type MC cable		9.20	.870		58.50	32		90.50	117
4430	EMT & wire		4.21	1.900		71.50	69.50		141	193
4500	Weather-proof cover for above receptacles, add		32	.250		3.46	9.15		12.61	18.80
4550	Air conditioner outlet, 20 amp-240 volt recpt.									
4560	30' of #12/2, 2 pole circuit breaker									
4570	Type NM cable	1 Elec	10	.800	Ea.	63	29.50		92.50	118
4580	Type MC cable		9	.889		70.50	32.50		103	131
4590	EMT & wire		4	2		82.50	73.50		156	211
4600	Decorator style, type NM cable		10	.800		68	29.50		97.50	123
4620	Type MC cable		9	.889		75.50	32.50		108	137

26 05 Common Work Results for Electrical

26 05 90 – Residential Applications

26 05 90.10 Residential Wiring		Crew	Daily Output	Labor-Hours	Unit	Material	2015 Bare Costs Labor	Equipment	Total	Total Incl O&P
4630	EMT & wire	1 Elec	4	2	Ea.	87	73.50		160.50	216
4650	Dryer outlet, 30 amp-240 volt recpt., 20' of #10/3									
4660	2 pole circuit breaker									
4670	Type NM cable	1 Elec	6.41	1.248	Ea.	61.50	45.50		107	143
4680	Type MC cable		5.71	1.401		64.50	51.50		116	155
4690	EMT & wire		3.48	2.299		76	84.50		160.50	222
4700	Range outlet, 50 amp-240 volt recpt., 30' of #8/3									
4710	Type NM cable	1 Elec	4.21	1.900	Ea.	90.50	69.50		160	214
4720	Type MC cable		4	2		124	73.50		197.50	257
4730	EMT & wire		2.96	2.703		108	99		207	281
4750	Central vacuum outlet, Type NM cable		6.40	1.250		58.50	46		104.50	140
4770	Type MC cable		5.71	1.401		69.50	51.50		121	161
4780	EMT & wire		3.48	2.299		83.50	84.50		168	230
4800	30 amp-110 volt locking recpt., #10/2 circ. bkr.									
4810	Type NM cable	1 Elec	6.20	1.290	Ea.	67	47.50		114.50	151
4820	Type MC cable		5.40	1.481		80	54.50		134.50	177
4830	EMT & wire		3.20	2.500		94	91.50		185.50	254
4900	Low voltage outlets									
4910	Telephone recpt., 20' of 4/C phone wire	1 Elec	26	.308	Ea.	9.30	11.30		20.60	29
4920	TV recpt., 20' of RG59U coax wire, F type connector	"	16	.500	"	18.75	18.35		37.10	50.50
4950	Door bell chime, transformer, 2 buttons, 60' of bellwire									
4970	Economy model	1 Elec	11.50	.696	Ea.	51.50	25.50		77	99
4980	Custom model		11.50	.696		99	25.50		124.50	151
4990	Luxury model, 3 buttons		9.50	.842		281	31		312	360
6000	Lighting outlets									
6050	Wire only (for fixture), type NM cable	1 Elec	32	.250	Ea.	6.90	9.15		16.05	22.50
6070	Type MC cable		24	.333		12.30	12.20		24.50	33.50
6080	EMT & wire		10	.800		20	29.50		49.50	70
6100	Box (4"), and wire (for fixture), type NM cable		25	.320		14.45	11.75		26.20	35
6120	Type MC cable		20	.400		19.90	14.65		34.55	46
6130	EMT & wire		11	.727		27.50	26.50		54	74
6200	Fixtures (use with lines 6050 or 6100 above)									
6210	Canopy style, economy grade	1 Elec	40	.200	Ea.	32	7.35		39.35	47
6220	Custom grade		40	.200		53.50	7.35		60.85	71
6250	Dining room chandelier, economy grade		19	.421		79.50	15.45		94.95	113
6260	Custom grade		19	.421		315	15.45		330.45	370
6270	Luxury grade		15	.533		715	19.55		734.55	815
6310	Kitchen fixture (fluorescent), economy grade		30	.267		71.50	9.75		81.25	95
6320	Custom grade		25	.320		219	11.75		230.75	260
6350	Outdoor, wall mounted, economy grade		30	.267		30	9.75		39.75	49
6360	Custom grade		30	.267		120	9.75		129.75	148
6370	Luxury grade		25	.320		248	11.75		259.75	292
6410	Outdoor PAR floodlights, 1 lamp, 150 watt		20	.400		32.50	14.65		47.15	59.50
6420	2 lamp, 150 watt each		20	.400		53.50	14.65		68.15	83
6425	Motion sensing, 2 lamp, 150 watt each		20	.400		87	14.65		101.65	120
6430	For infrared security sensor, add		32	.250		132	9.15		141.15	160
6450	Outdoor, quartz-halogen, 300 watt flood		20	.400		39	14.65		53.65	66.50
6600	Recessed downlight, round, pre-wired, 50 or 75 watt trim		30	.267		80	9.75		89.75	104
6610	With shower light trim		30	.267		89	9.75		98.75	114
6620	With wall washer trim		28	.286		99.50	10.45		109.95	126
6630	With eye-ball trim		28	.286		99.50	10.45		109.95	126
6700	Porcelain lamp holder		40	.200		2.96	7.35		10.31	15.25
6710	With pull switch		40	.200		6.55	7.35		13.90	19.25

For customer support on your Residential Cost Data, call 877.713.2617.

26 05 Common Work Results for Electrical

26 05 90 – Residential Applications

26 05 90.10 Residential Wiring

		Crew	Daily Output	Labor-Hours	Unit	Material	2015 Bare Costs Labor	Equipment	Total	Total Incl O&P
6750	Fluorescent strip, 2-20 watt tube, wrap around diffuser, 24"	1 Elec	24	.333	Ea.	49.50	12.20		61.70	74.50
6760	1-34 watt tube, 48"		24	.333		87	12.20		99.20	116
6770	2-34 watt tubes, 48"		20	.400		103	14.65		117.65	137
6800	Bathroom heat lamp, 1-250 watt		28	.286		44	10.45		54.45	65
6810	2-250 watt lamps		28	.286		70	10.45		80.45	94
6820	For timer switch, see Section 26 05 90.10 line 2400									
6900	Outdoor post lamp, incl. post, fixture, 35' of #14/2									
6910	Type NMC cable	1 Elec	3.50	2.286	Ea.	258	84		342	420
6920	Photo-eye, add		27	.296		32	10.85		42.85	53.50
6950	Clock dial time switch, 24 hr., w/enclosure, type NM cable		11.43	.700		73.50	25.50		99	123
6970	Type MC cable		11	.727		82.50	26.50		109	134
6980	EMT & wire		4.85	1.649		91	60.50		151.50	199
7000	Alarm systems									
7050	Smoke detectors, box, #14/3, type NM cable	1 Elec	14.55	.550	Ea.	34.50	20		54.50	71
7070	Type MC cable		12.31	.650		44	24		68	87.50
7080	EMT & wire		5	1.600		49.50	58.50		108	151
7090	For relay output to security system, add					12			12	13.20
8000	Residential equipment									
8050	Disposal hook-up, incl. switch, outlet box, 3' of flex									
8060	20 amp-1 pole circ. bkr., and 25' of #12/2									
8070	Type NM cable	1 Elec	10	.800	Ea.	32.50	29.50		62	83.50
8080	Type MC cable		8	1		39.50	36.50		76	104
8090	EMT & wire		5	1.600		54.50	58.50		113	156
8100	Trash compactor or dishwasher hook-up, incl. outlet box,									
8110	3' of flex, 15 amp-1 pole circ. bkr., and 25' of #14/2									
8130	Type MC cable	1 Elec	8	1	Ea.	34	36.50		70.50	97.50
8140	EMT & wire	"	5	1.600	"	46	58.50		104.50	147
8150	Hot water sink dispensor hook-up, use line 8100									
8200	Vent/exhaust fan hook-up, type NM cable	1 Elec	32	.250	Ea.	6.90	9.15		16.05	22.50
8220	Type MC cable		24	.333		12.30	12.20		24.50	33.50
8230	EMT & wire		10	.800		20	29.50		49.50	70
8250	Bathroom vent fan, 50 CFM (use with above hook-up)									
8260	Economy model	1 Elec	15	.533	Ea.	23	19.55		42.55	57.50
8270	Low noise model		15	.533		41	19.55		60.55	77
8280	Custom model		12	.667		127	24.50		151.50	180
8300	Bathroom or kitchen vent fan, 110 CFM									
8310	Economy model	1 Elec	15	.533	Ea.	67.50	19.55		87.05	107
8320	Low noise model	"	15	.533	"	92	19.55		111.55	133
8350	Paddle fan, variable speed (w/o lights)									
8360	Economy model (AC motor)	1 Elec	10	.800	Ea.	109	29.50		138.50	168
8362	With light kit		10	.800		150	29.50		179.50	213
8370	Custom model (AC motor)		10	.800		227	29.50		256.50	298
8372	With light kit		10	.800		268	29.50		297.50	345
8380	Luxury model (DC motor)		8	1		330	36.50		366.50	425
8382	With light kit		8	1		375	36.50		411.50	470
8390	Remote speed switch for above, add		12	.667		37.50	24.50		62	81
8500	Whole house exhaust fan, ceiling mount, 36", variable speed									
8510	Remote switch, incl. shutters, 20 amp-1 pole circ. bkr.									
8520	30' of #12/2, type NM cable	1 Elec	4	2	Ea.	1,300	73.50		1,373.50	1,550
8530	Type MC cable		3.50	2.286		1,300	84		1,384	1,550
8540	EMT & wire		3	2.667		1,325	97.50		1,422.50	1,600
8600	Whirlpool tub hook-up, incl. timer switch, outlet box									
8610	3' of flex, 20 amp-1 pole GFI circ. bkr.									

26 05 Common Work Results for Electrical

26 05 90 – Residential Applications

26 05 90.10 Residential Wiring

		Crew	Daily Output	Labor-Hours	Unit	Material	2015 Bare Costs Labor	Equipment	Total	Total Incl O&P
8620	30' of #12/2, type NM cable	1 Elec	5	1.600	Ea.	142	58.50		200.50	252
8630	Type MC cable		4.20	1.905		145	70		215	274
8640	EMT & wire		3.40	2.353		158	86		244	315
8650	Hot water heater hook-up, incl. 1-2 pole circ. bkr., box;									
8660	3' of flex, 20' of #10/2, type NM cable	1 Elec	5	1.600	Ea.	33	58.50		91.50	133
8670	Type MC cable		4.20	1.905		43	70		113	162
8680	EMT & wire		3.40	2.353		48.50	86		134.50	194
9000	Heating/air conditioning									
9050	Furnace/boiler hook-up, incl. firestat, local on-off switch									
9060	Emergency switch, and 40' of type NM cable	1 Elec	4	2	Ea.	52.50	73.50		126	178
9070	Type MC cable		3.50	2.286		65.50	84		149.50	210
9080	EMT & wire		1.50	5.333		83	195		278	410
9100	Air conditioner hook-up, incl. local 60 amp disc. switch									
9110	3' sealtite, 40 amp, 2 pole circuit breaker									
9130	40' of #8/2, type NM cable	1 Elec	3.50	2.286	Ea.	163	84		247	315
9140	Type MC cable		3	2.667		216	97.50		313.50	400
9150	EMT & wire		1.30	6.154		204	226		430	595
9200	Heat pump hook-up, 1-40 & 1-100 amp 2 pole circ. bkr.									
9210	Local disconnect switch, 3' sealtite									
9220	40' of #8/2 & 30' of #3/2									
9230	Type NM cable	1 Elec	1.30	6.154	Ea.	520	226		746	940
9240	Type MC cable		1.08	7.407		535	271		806	1,025
9250	EMT & wire		.94	8.511		585	310		895	1,150
9500	Thermostat hook-up, using low voltage wire									
9520	Heating only, 25' of #18-3	1 Elec	24	.333	Ea.	8.80	12.20		21	29.50
9530	Heating/cooling, 25' of #18-4	"	20	.400	"	11.45	14.65		26.10	36.50

26 24 Switchboards and Panelboards

26 24 16 – Panelboards

26 24 16.10 Load Centers

		Crew	Daily Output	Labor-Hours	Unit	Material	2015 Bare Costs Labor	Equipment	Total	Total Incl O&P
0010	**LOAD CENTERS** (residential type)									
0100	3 wire, 120/240 V, 1 phase, including 1 pole plug-in breakers									
0200	100 amp main lugs, indoor, 8 circuits	1 Elec	1.40	5.714	Ea.	178	209		387	540
0300	12 circuits		1.20	6.667		249	244		493	675
0400	Rainproof, 8 circuits		1.40	5.714		214	209		423	580
0500	12 circuits		1.20	6.667		300	244		544	730
0600	200 amp main lugs, indoor, 16 circuits	R-1A	1.80	8.889		330	271		601	815
0700	20 circuits		1.50	10.667		415	325		740	995
0800	24 circuits		1.30	12.308		545	375		920	1,225
1200	Rainproof, 16 circuits		1.80	8.889		395	271		666	885
1300	20 circuits		1.50	10.667		475	325		800	1,050
1400	24 circuits		1.30	12.308		710	375		1,085	1,400

26 24 16.20 Panelboard and Load Center Circuit Breakers

		Crew	Daily Output	Labor-Hours	Unit	Material	2015 Bare Costs Labor	Equipment	Total	Total Incl O&P
0010	**PANELBOARD AND LOAD CENTER CIRCUIT BREAKERS**									
2000	Plug-in panel or load center, 120/240 volt, to 60 amp, 1 pole	1 Elec	12	.667	Ea.	11.45	24.50		35.95	52.50
2004	Circuit breaker, 120/240 volt, 20 A, 1 pole with NM cable		6.50	1.231		18.75	45		63.75	94.50
2006	30 A, 1 pole with NM cable		6.50	1.231		18.75	45		63.75	94.50
2010	2 pole		9	.889		26	32.50		58.50	82
2014	50 A, 2 pole with NM cable		5.50	1.455		33	53.50		86.50	124
2020	3 pole		7.50	1.067		89.50	39		128.50	162
2030	100 amp, 2 pole		6	1.333		120	49		169	212

26 24 Switchboards and Panelboards

26 24 16 – Panelboards

26 24 16.20 Panelboard and Load Center Circuit Breakers		Crew	Daily Output	Labor-Hours	Unit	Material	2015 Bare Costs Labor	Equipment	Total	Total Incl O&P
2040	3 pole	1 Elec	4.50	1.778	Ea.	133	65		198	253
2050	150 to 200 amp, 2 pole		3	2.667		218	97.50		315.50	400
2060	Plug-in tandem, 120/240 V, 2-15 A, 1 pole		11	.727		24.50	26.50		51	70.50
2070	1-15 A & 1-20 A		11	.727		24.50	26.50		51	70.50
2080	2-20 A		11	.727		24.50	26.50		51	70.50
2300	Ground fault, 240 volt, 30 amp, 1 pole		7	1.143		102	42		144	181
2310	2 pole		6	1.333		182	49		231	280

26 27 Low-Voltage Distribution Equipment

26 27 13 – Electricity Metering

26 27 13.10 Meter Centers and Sockets

		Crew	Daily Output	Labor-Hours	Unit	Material	2015 Bare Costs Labor	Equipment	Total	Total Incl O&P
0010	**METER CENTERS AND SOCKETS**									
0100	Sockets, single position, 4 terminal, 100 amp	1 Elec	3.20	2.500	Ea.	43	91.50		134.50	198
0200	150 amp		2.30	3.478		48	127		175	262
0300	200 amp		1.90	4.211		92	154		246	355
0500	Double position, 4 terminal, 100 amp		2.80	2.857		204	105		309	395
0600	150 amp		2.10	3.810		240	140		380	495
0700	200 amp		1.70	4.706		450	172		622	780
1100	Meter centers and sockets, three phase, single pos, 7 terminal, 100 amp		2.80	2.857		164	105		269	355
1200	200 amp		2.10	3.810		295	140		435	555
1400	400 amp		1.70	4.706		720	172		892	1,075
2590	Basic meter device									
2600	1P 3W 120/240 V 4 jaw 125A sockets, 3 meter	2 Elec	1	16	Ea.	620	585		1,205	1,650
2620	5 meter		.80	20		930	735		1,665	2,225
2640	7 meter		.56	28.571		1,375	1,050		2,425	3,225
2660	10 meter		.48	33.333		1,875	1,225		3,100	4,050
2680	Rainproof 1P 3W 120/240 V 4 jaw 125A sockets									
2690	3 meter	2 Elec	1	16	Ea.	620	585		1,205	1,650
2710	6 meter		.60	26.667		1,075	975		2,050	2,775
2730	8 meter		.52	30.769		1,500	1,125		2,625	3,500
2750	1P 3W 120/240 V 4 jaw sockets									
2760	with 125A circuit breaker, 3 meter	2 Elec	1	16	Ea.	1,175	585		1,760	2,225
2780	5 meter		.80	20		1,825	735		2,560	3,225
2800	7 meter		.56	28.571		2,625	1,050		3,675	4,625
2820	10 meter		.48	33.333		3,675	1,225		4,900	6,050
2830	Rainproof 1P 3W 120/240 V 4 jaw sockets									
2840	with 125A circuit breaker, 3 meter	2 Elec	1	16	Ea.	1,175	585		1,760	2,225
2870	6 meter		.60	26.667		2,150	975		3,125	3,975
2890	8 meter		.52	30.769		2,925	1,125		4,050	5,075
3250	1P 3W 120/240 V 4 jaw sockets									
3260	with 200A circuit breaker, 3 meter	2 Elec	1	16	Ea.	1,750	585		2,335	2,875
3290	6 meter		.60	26.667		3,500	975		4,475	5,450
3310	8 meter		.56	28.571		4,725	1,050		5,775	6,925
3330	Rainproof 1P 3W 120/240 V 4 jaw sockets									
3350	with 200A circuit breaker, 3 meter	2 Elec	1	16	Ea.	1,750	585		2,335	2,875
3380	6 meter		.60	26.667		3,500	975		4,475	5,450
3400	8 meter		.52	30.769		4,725	1,125		5,850	7,050

26 27 Low-Voltage Distribution Equipment

26 27 23 – Indoor Service Poles

26 27 23.40 Surface Raceway

		Crew	Daily Output	Labor-Hours	Unit	Material	2015 Bare Costs Labor	Equipment	Total	Total Incl O&P
0010	**SURFACE RACEWAY**									
0090	Metal, straight section									
0100	No. 500	1 Elec	100	.080	L.F.	1.04	2.93		3.97	5.95
0110	No. 700		100	.080		1.17	2.93		4.10	6.10
0200	No. 1000		90	.089		1.67	3.26		4.93	7.20
0400	No. 1500, small pancake		90	.089		2.15	3.26		5.41	7.70
0600	No. 2000, base & cover, blank		90	.089		2.19	3.26		5.45	7.75
0800	No. 3000, base & cover, blank		75	.107		4.18	3.91		8.09	11
2400	Fittings, elbows, No. 500		40	.200	Ea.	1.90	7.35		9.25	14.10
2800	Elbow cover, No. 2000		40	.200		3.57	7.35		10.92	15.95
2880	Tee, No. 500		42	.190		3.66	7		10.66	15.50
2900	No. 2000		27	.296		11.85	10.85		22.70	31
3000	Switch box, No. 500		16	.500		11	18.35		29.35	42
3400	Telephone outlet, No. 1500		16	.500		14.10	18.35		32.45	45.50
3600	Junction box, No. 1500		16	.500		9.65	18.35		28	40.50
3800	Plugmold wired sections, No. 2000									
4000	1 circuit, 6 outlets, 3 ft. long	1 Elec	8	1	Ea.	36	36.50		72.50	99.50
4100	2 circuits, 8 outlets, 6 ft. long	"	5.30	1.509	"	53	55.50		108.50	149

26 27 26 – Wiring Devices

26 27 26.10 Low Voltage Switching

		Crew	Daily Output	Labor-Hours	Unit	Material	Labor	Equipment	Total	Total Incl O&P
0010	**LOW VOLTAGE SWITCHING**									
3600	Relays, 120 V or 277 V standard	1 Elec	12	.667	Ea.	41.50	24.50		66	86
3800	Flush switch, standard		40	.200		11.50	7.35		18.85	24.50
4000	Interchangeable		40	.200		15.05	7.35		22.40	28.50
4100	Surface switch, standard		40	.200		8.15	7.35		15.50	21
4200	Transformer 115 V to 25 V		12	.667		130	24.50		154.50	183
4400	Master control, 12 circuit, manual		4	2		126	73.50		199.50	259
4500	25 circuit, motorized		4	2		140	73.50		213.50	274
4600	Rectifier, silicon		12	.667		45.50	24.50		70	90
4800	Switchplates, 1 gang, 1, 2 or 3 switch, plastic		80	.100		5	3.67		8.67	11.50
5000	Stainless steel		80	.100		11.35	3.67		15.02	18.50
5400	2 gang, 3 switch, stainless steel		53	.151		23	5.55		28.55	34
5500	4 switch, plastic		53	.151		10.35	5.55		15.90	20.50
5600	2 gang, 4 switch, stainless steel		53	.151		21.50	5.55		27.05	32.50
5700	6 switch, stainless steel		53	.151		43	5.55		48.55	56.50
5800	3 gang, 9 switch, stainless steel		32	.250		64.50	9.15		73.65	86

26 27 26.20 Wiring Devices Elements

		Crew	Daily Output	Labor-Hours	Unit	Material	Labor	Equipment	Total	Total Incl O&P
0010	**WIRING DEVICES ELEMENTS**									
0200	Toggle switch, quiet type, single pole, 15 amp	1 Elec	40	.200	Ea.	6.55	7.35		13.90	19.25
0600	3 way, 15 amp		23	.348		5.05	12.75		17.80	26.50
0900	4 way, 15 amp		15	.533		9.10	19.55		28.65	42
1650	Dimmer switch, 120 volt, incandescent, 600 watt, 1 pole [G]		16	.500		21	18.35		39.35	53
2460	Receptacle, duplex, 120 volt, grounded, 15 amp		40	.200		1.26	7.35		8.61	13.40
2470	20 amp		27	.296		7.90	10.85		18.75	26.50
2490	Dryer, 30 amp		15	.533		4.39	19.55		23.94	37
2500	Range, 50 amp		11	.727		12.15	26.50		38.65	57
2600	Wall plates, stainless steel, 1 gang		80	.100		2.56	3.67		6.23	8.80
2800	2 gang		53	.151		4.33	5.55		9.88	13.80
3200	Lampholder, keyless		26	.308		12.65	11.30		23.95	32.50
3400	Pullchain with receptacle		22	.364		20.50	13.35		33.85	44.50

26 27 Low-Voltage Distribution Equipment

26 27 73 – Door Chimes

26 27 73.10 Doorbell System		Crew	Daily Output	Labor-Hours	Unit	Material	2015 Bare Costs Labor	Equipment	Total	Total Incl O&P
0010	DOORBELL SYSTEM, incl. transformer, button & signal									
1000	Door chimes, 2 notes	1 Elec	16	.500	Ea.	22.50	18.35		40.85	55
1020	with ambient light		12	.667		109	24.50		133.50	160
1100	Tube type, 3 tube system		12	.667		187	24.50		211.50	246
1180	4 tube system		10	.800		385	29.50		414.50	470
1900	For transformer & button, add		5	1.600		15.85	58.50		74.35	113
3000	For push button only		24	.333		2.58	12.20		14.78	23

26 28 Low-Voltage Circuit Protective Devices

26 28 16 – Enclosed Switches and Circuit Breakers

26 28 16.10 Circuit Breakers

		Crew	Daily Output	Labor-Hours	Unit	Material	Labor	Equipment	Total	Total Incl O&P
0010	CIRCUIT BREAKERS (in enclosure)									
0100	Enclosed (NEMA 1), 600 volt, 3 pole, 30 amp	1 Elec	3.20	2.500	Ea.	500	91.50		591.50	700
0200	60 amp		2.80	2.857		615	105		720	850
0400	100 amp		2.30	3.478		705	127		832	985

26 28 16.20 Safety Switches

		Crew	Daily Output	Labor-Hours	Unit	Material	Labor	Equipment	Total	Total Incl O&P
0010	SAFETY SWITCHES									
0100	General duty 240 volt, 3 pole NEMA 1, fusible, 30 amp	1 Elec	3.20	2.500	Ea.	73.50	91.50		165	231
0200	60 amp		2.30	3.478		124	127		251	345
0300	100 amp		1.90	4.211		213	154		367	485
0400	200 amp		1.30	6.154		455	226		681	870
0500	400 amp	2 Elec	1.80	8.889		1,150	325		1,475	1,800
9010	Disc. switch, 600 V 3 pole fusible, 30 amp, to 10 HP motor	1 Elec	3.20	2.500		355	91.50		446.50	540
9050	60 amp, to 30 HP motor		2.30	3.478		815	127		942	1,100
9070	100 amp, to 60 HP motor		1.90	4.211		815	154		969	1,150

26 28 16.40 Time Switches

		Crew	Daily Output	Labor-Hours	Unit	Material	Labor	Equipment	Total	Total Incl O&P
0010	TIME SWITCHES									
0100	Single pole, single throw, 24 hour dial	1 Elec	4	2	Ea.	145	73.50		218.50	280
0200	24 hour dial with reserve power		3.60	2.222		615	81.50		696.50	810
0300	Astronomic dial		3.60	2.222		209	81.50		290.50	365
0400	Astronomic dial with reserve power		3.30	2.424		665	89		754	875
0500	7 day calendar dial		3.30	2.424		150	89		239	310
0600	7 day calendar dial with reserve power		3.20	2.500		335	91.50		426.50	520
0700	Photo cell 2000 watt		8	1		28.50	36.50		65	91.50

26 32 Packaged Generator Assemblies

26 32 13 – Engine Generators

26 32 13.16 Gas-Engine-Driven Generator Sets

		Crew	Daily Output	Labor-Hours	Unit	Material	Labor	Equipment	Total	Total Incl O&P
0010	GAS-ENGINE-DRIVEN GENERATOR SETS									
0020	Gas or gasoline operated, includes battery,									
0050	charger, & muffler									
0200	3 phase 4 wire, 277/480 volt, 7.5 kW	R-3	.83	24.096	Ea.	7,350	885	166	8,401	9,700
0300	11.5 kW		.71	28.169		10,400	1,025	194	11,619	13,400
0400	20 kW		.63	31.746		12,300	1,175	219	13,694	15,600
0500	35 kW		.55	36.364		14,600	1,325	251	16,176	18,600

26 33 Battery Equipment

26 33 43 – Battery Chargers

26 33 43.55 Electric Vehicle Charging

		Crew	Daily Output	Labor-Hours	Unit	Material	2015 Bare Costs Labor	Equipment	Total	Total Incl O&P
0010	**ELECTRIC VEHICLE CHARGING**									
0020	Level 2, wall mounted									
2100	Light duty, hard wired	[G] 1 Elec	20.48	.391	Ea.	1,125	14.30		1,139.30	1,250
2110	plug in	[G] "	30.72	.260	"	1,125	9.55		1,134.55	1,250

26 36 Transfer Switches

26 36 23 – Automatic Transfer Switches

26 36 23.10 Automatic Transfer Switch Devices

		Crew	Daily Output	Labor-Hours	Unit	Material	Labor	Equipment	Total	Total Incl O&P
0010	**AUTOMATIC TRANSFER SWITCH DEVICES**									
0100	Switches, enclosed 480 volt, 3 pole, 30 amp	1 Elec	2.30	3.478	Ea.	2,650	127		2,777	3,125
0200	60 amp	"	1.90	4.211	"	2,650	154		2,804	3,175

26 41 Facility Lightning Protection

26 41 13 – Lightning Protection for Structures

26 41 13.13 Lightning Protection for Buildings

		Crew	Daily Output	Labor-Hours	Unit	Material	Labor	Equipment	Total	Total Incl O&P
0010	**LIGHTNING PROTECTION FOR BUILDINGS**									
0200	Air terminals & base, copper									
0400	3/8" diameter x 10" (to 75' high)	1 Elec	8	1	Ea.	22	36.50		58.50	84
1000	Aluminum, 1/2" diameter x 12" (to 75' high)		8	1		15.10	36.50		51.60	76.50
1020	1/2" diameter x 24"		7.30	1.096		17.25	40		57.25	85
1040	1/2" diameter x 60"		6.70	1.194		25.50	44		69.50	99.50
2000	Cable, copper, 220 lb. per thousand ft. (to 75' high)		320	.025	L.F.	3.03	.92		3.95	4.83
2500	Aluminum, 101 lb. per thousand ft. (to 75' high)		280	.029	"	.95	1.05		2	2.77
3000	Arrester, 175 volt AC to ground		8	1	Ea.	103	36.50		139.50	173

26 51 Interior Lighting

26 51 13 – Interior Lighting Fixtures, Lamps, and Ballasts

26 51 13.50 Interior Lighting Fixtures

		Crew	Daily Output	Labor-Hours	Unit	Material	Labor	Equipment	Total	Total Incl O&P
0010	**INTERIOR LIGHTING FIXTURES** Including lamps, mounting									
0030	hardware and connections									
0100	Fluorescent, C.W. lamps, troffer, recess mounted in grid, RS									
0130	Grid ceiling mount									
0200	Acrylic lens, 1'W x 4'L, two 40 watt	1 Elec	5.70	1.404	Ea.	47.50	51.50		99	137
0300	2'W x 2'L, two U40 watt		5.70	1.404		51	51.50		102.50	141
0600	2'W x 4'L, four 40 watt		4.70	1.702		57.50	62.50		120	166
1000	Surface mounted, RS									
1030	Acrylic lens with hinged & latched door frame									
1100	1'W x 4'L, two 40 watt	1 Elec	7	1.143	Ea.	64	42		106	139
1200	2'W x 2'L, two U40 watt		7	1.143		68.50	42		110.50	144
1500	2'W x 4'L, four 40 watt		5.30	1.509		81	55.50		136.50	180
2100	Strip fixture									
2200	4' long, one 40 watt, RS	1 Elec	8.50	.941	Ea.	27.50	34.50		62	86.50
2300	4' long, two 40 watt, RS	"	8	1		38.50	36.50		75	102
2600	8' long, one 75 watt, SL	2 Elec	13.40	1.194		48	44		92	125
2700	8' long, two 75 watt, SL	"	12.40	1.290		58	47.50		105.50	141
4450	Incandescent, high hat can, round alzak reflector, prewired									
4470	100 watt	1 Elec	8	1	Ea.	71.50	36.50		108	139

26 51 Interior Lighting

26 51 13 – Interior Lighting Fixtures, Lamps, and Ballasts

26 51 13.50 Interior Lighting Fixtures

		Crew	Daily Output	Labor-Hours	Unit	Material	2015 Bare Costs Labor	Equipment	Total	Total Incl O&P
4480	150 watt	1 Elec	8	1	Ea.	104	36.50		140.50	174
5200	Ceiling, surface mounted, opal glass drum									
5300	8", one 60 watt lamp	1 Elec	10	.800	Ea.	44.50	29.50		74	97
5400	10", two 60 watt lamps		8	1		50	36.50		86.50	115
5500	12", four 60 watt lamps		6.70	1.194		70.50	44		114.50	149
6900	Mirror light, fluorescent, RS, acrylic enclosure, two 40 watt		8	1		128	36.50		164.50	200
6910	One 40 watt		8	1		106	36.50		142.50	176
6920	One 20 watt		12	.667		81.50	24.50		106	130

26 51 13.55 Interior LED Fixtures

			Crew	Daily Output	Labor-Hours	Unit	Material	Labor	Equipment	Total	Total Incl O&P
0010	**INTERIOR LED FIXTURES** Incl. lamps, and mounting hardware										
0100	Downlight, recess mounted, 7.5" diameter, 25 watt	G	1 Elec	8	1	Ea.	335	36.50		371.50	425
0120	10" diameter, 36 watt	G		8	1		360	36.50		396.50	455
0160	cylinder, 10 watts	G		8	1		102	36.50		138.50	172
0180	20 watts	G		8	1		585	36.50		621.50	705
1000	Troffer, recess mounted, 2' x 4', 3200 Lumens	G		5.30	1.509		138	55.50		193.50	242
1010	4800 Lumens	G		5	1.600		179	58.50		237.50	292
1020	6400 Lumens	G		4.70	1.702		198	62.50		260.50	320
1100	Troffer retrofit lamp, 38 watt	G		21	.381		238	13.95		251.95	284
1110	60 watt	G		20	.400		340	14.65		354.65	400
1120	100 watt	G		18	.444		510	16.30		526.30	585
1200	Troffer, volumetric recess mounted, 2' x 2'	G		5.70	1.404		251	51.50		302.50	360
2000	Strip, surface mounted, one light bar 4' long, 3500K	G		8.50	.941		299	34.50		333.50	385
2010	5000K	G		8	1		299	36.50		335.50	390
2020	Two light bar 4' long, 5000K	G		7	1.143		470	42		512	585
3000	Linear, suspended mounted, one light bar 4' long, 37 watt	G		6.70	1.194		195	44		239	287
3010	One light bar 8' long, 74 watt	G	2 Elec	12.20	1.311		360	48		408	480
3020	Two light bar 4' long, 74 watt	G	1 Elec	5.70	1.404		390	51.50		441.50	510
3030	Two light bar 8' long, 148 watt	G	2 Elec	8.80	1.818		450	66.50		516.50	605
4000	High bay, surface mounted, round, 150 watts	G		5.41	2.959		605	108		713	845
4010	2 bars, 164 watts	G		5.41	2.959		570	108		678	805
4020	3 bars, 246 watts	G		5.01	3.197		730	117		847	995
4030	4 bars, 328 watts	G		4.60	3.478		895	127		1,022	1,200
4040	5 bars, 410 watts	G	3 Elec	4.20	5.716		1,050	209		1,259	1,500
4050	6 bars, 492 watts	G		3.80	6.324		1,200	232		1,432	1,675
4060	7 bars, 574 watts	G		3.39	7.075		1,350	259		1,609	1,925
4070	8 bars, 656 watts	G		2.99	8.029		1,500	294		1,794	2,125
5000	track, lighthead, 6 watt	G	1 Elec	32	.250		54.50	9.15		63.65	75
5010	9 watt	G	"	32	.250		61.50	9.15		70.65	82.50
6000	Garage, surface mount, 103 watts	G	2 Elec	6.50	2.462		970	90		1,060	1,225
6100	pendent mount, 80 watts	G		6.50	2.462		565	90		655	770
6200	95 watts	G		6.50	2.462		635	90		725	850
6300	125 watts	G		6.50	2.462		690	90		780	910

26 51 13.70 Residential Fixtures

		Crew	Daily Output	Labor-Hours	Unit	Material	Labor	Equipment	Total	Total Incl O&P
0010	**RESIDENTIAL FIXTURES**									
0400	Fluorescent, interior, surface, circline, 32 watt & 40 watt	1 Elec	20	.400	Ea.	130	14.65		144.65	167
0500	2' x 2', two U-tube 32 watt T8		8	1		165	36.50		201.50	241
0700	Shallow under cabinet, two 20 watt		16	.500		62	18.35		80.35	98.50
0900	Wall mounted, 4'L, two 32 watt T8, with baffle		10	.800		133	29.50		162.50	194
2000	Incandescent, exterior lantern, wall mounted, 60 watt		16	.500		55.50	18.35		73.85	91
2100	Post light, 150W, with 7' post		4	2		243	73.50		316.50	385
2500	Lamp holder, weatherproof with 150W PAR		16	.500		32	18.35		50.35	65
2550	With reflector and guard		12	.667		56	24.50		80.50	102

26 51 Interior Lighting

26 51 13 – Interior Lighting Fixtures, Lamps, and Ballasts

26 51 13.70 Residential Fixtures

		Crew	Daily Output	Labor-Hours	Unit	Material	2015 Bare Costs Labor	Equipment	Total	Total Incl O&P
2600	Interior pendent, globe with shade, 150 watt	1 Elec	20	.400	Ea.	201	14.65		215.65	245

26 55 Special Purpose Lighting

26 55 59 – Display Lighting

26 55 59.10 Track Lighting

		Crew	Daily Output	Labor-Hours	Unit	Material	Labor	Equipment	Total	Total Incl O&P
0010	**TRACK LIGHTING**									
0100	8' section	2 Elec	10.60	1.509	Ea.	69	55.50		124.50	167
0300	3 circuits, 4' section	1 Elec	6.70	1.194		83	44		127	163
0400	8' section	2 Elec	10.60	1.509		109	55.50		164.50	211
0500	12' section	"	8.80	1.818		159	66.50		225.50	284
1000	Feed kit, surface mounting	1 Elec	16	.500		13	18.35		31.35	44.50
1100	End cover		24	.333		5.55	12.20		17.75	26
1200	Feed kit, stem mounting, 1 circuit		16	.500		39	18.35		57.35	72.50
1300	3 circuit		16	.500		39	18.35		57.35	72.50
2000	Electrical joiner, for continuous runs, 1 circuit		32	.250		21	9.15		30.15	38
2100	3 circuit		32	.250		59	9.15		68.15	80
2200	Fixtures, spotlight, 75W PAR halogen		16	.500		65.50	18.35		83.85	102
2210	50W MR16 halogen		16	.500		171	18.35		189.35	218
3000	Wall washer, 250 watt tungsten halogen		16	.500		132	18.35		150.35	175
3100	Low voltage, 25/50 watt, 1 circuit		16	.500		133	18.35		151.35	176
3120	3 circuit		16	.500		198	18.35		216.35	248

26 56 Exterior Lighting

26 56 13 – Lighting Poles and Standards

26 56 13.10 Lighting Poles

		Crew	Daily Output	Labor-Hours	Unit	Material	Labor	Equipment	Total	Total Incl O&P
0010	**LIGHTING POLES**									
6420	Wood pole, 4-1/2" x 5-1/8", 8' high	1 Elec	6	1.333	Ea.	330	49		379	445
6440	12' high		5.70	1.404		475	51.50		526.50	610
6460	20' high		4	2		670	73.50		743.50	860

26 56 23 – Area Lighting

26 56 23.10 Exterior Fixtures

		Crew	Daily Output	Labor-Hours	Unit	Material	Labor	Equipment	Total	Total Incl O&P
0010	**EXTERIOR FIXTURES** With lamps									
0400	Quartz, 500 watt	1 Elec	5.30	1.509	Ea.	50	55.50		105.50	146
1100	Wall pack, low pressure sodium, 35 watt		4	2		227	73.50		300.50	370
1150	55 watt		4	2		270	73.50		343.50	415

26 56 26 – Landscape Lighting

26 56 26.20 Landscape Fixtures

		Crew	Daily Output	Labor-Hours	Unit	Material	Labor	Equipment	Total	Total Incl O&P
0010	**LANDSCAPE FIXTURES**									
7380	Landscape recessed uplight, incl. housing, ballast, transformer									
7390	& reflector, not incl. conduit, wire, trench									
7420	Incandescent, 250 watt	1 Elec	5	1.600	Ea.	590	58.50		648.50	745
7440	Quartz, 250 watt	"	5	1.600	"	560	58.50		618.50	710

26 56 33 – Walkway Lighting

26 56 33.10 Walkway Luminaire

		Crew	Daily Output	Labor-Hours	Unit	Material	Labor	Equipment	Total	Total Incl O&P
0010	**WALKWAY LUMINAIRE**									
6500	Bollard light, lamp & ballast, 42" high with polycarbonate lens									
7200	Incandescent, 150 watt	1 Elec	3	2.667	Ea.	595	97.50		692.50	815

26 61 Lighting Systems and Accessories

26 61 23 – Lamps Applications

26 61 23.10 Lamps		Crew	Daily Output	Labor-Hours	Unit	Material	2015 Bare Costs Labor	Equipment	Total	Total Incl O&P
0010	**LAMPS**									
0081	Fluorescent, rapid start, cool white, 2' long, 20 watt	1 Elec	100	.080	Ea.	3.09	2.93		6.02	8.20
0101	4' long, 40 watt		90	.089		2.36	3.26		5.62	7.95
1351	High pressure sodium, 70 watt		30	.267		31	9.75		40.75	50.50
1371	150 watt		30	.267		33.50	9.75		43.25	52.50

26 61 23.55 LED Lamps			Crew	Daily Output	Labor-Hours	Unit	Material	2015 Bare Costs Labor	Equipment	Total	Total Incl O&P
0010	**LED LAMPS**										
0100	LED lamp, interior, shape A60, equal to 60 watt	G	1 Elec	160	.050	Ea.	22.50	1.83		24.33	27.50
0200	Globe frosted A60, equal to 60 watt	G		160	.050		22.50	1.83		24.33	27.50
0300	Globe earth, equal to 100 watt	G		160	.050		74	1.83		75.83	84.50
1100	MR16, 3 watt, replacement of halogen lamp 25 watt	G		130	.062		21.50	2.26		23.76	27
1200	6 watt replacement of halogen lamp 45 watt	G		130	.062		47	2.26		49.26	55
2100	10 watt, PAR20, equal to 60 watt	G		130	.062		47.50	2.26		49.76	56
2200	15 watt, PAR30, equal to 100 watt	G		130	.062		79	2.26		81.26	90.50

Estimating Tips
27 20 00 Data Communications
27 30 00 Voice Communications
27 40 00 Audio-Video Communications

When estimating material costs for special systems, it is always prudent to obtain manufacturers' quotations for equipment prices and special installation requirements which will affect the total costs.

Reference Numbers
Reference numbers are shown in shaded boxes at the beginning of some major classifications. These numbers refer to related items in the Reference Section. The reference information may be an estimating procedure, an alternate pricing method, or technical information.

Note: Not all subdivisions listed here necessarily appear in this publication. ■

*Note: **Trade Service**, in part, has been used as a reference source for some of the material prices used in Division 27.*

Estimate with precision with RSMeans Online...
Quick, intuitive and easy-to-use, RSMeans Online gives you instant access to hundreds of thousands of material, labor, and equipment costs from RSMeans' comprehensive database, delivering the information you need to build competitive estimates every time.

Start your free trial today at **www.rsmeansonline.com**

RSMeansOnline

27 41 Audio-Video Systems

27 41 33 – Master Antenna Television Systems

27 41 33.10 T.V. Systems		Crew	Daily Output	Labor-Hours	Unit	Material	2015 Bare Costs Labor	Equipment	Total	Total Incl O&P
0010	**T.V. SYSTEMS**, not including rough-in wires, cables & conduits									
0100	Master TV antenna system									
0200	VHF reception & distribution, 12 outlets	1 Elec	6	1.333	Outlet	133	49		182	226
0800	VHF & UHF reception & distribution, 12 outlets		6	1.333	"	213	49		262	315
5000	Antenna, small		6	1.333	Ea.	47	49		96	132
5100	Large		4	2		198	73.50		271.50	340
5110	Rotor unit		8	1		77.50	36.50		114	146
5120	Single booster		8	1		38	36.50		74.50	102
5130	Antenna pole, 10'		3.20	2.500		28	91.50		119.50	181
6100	Satellite TV system	2 Elec	1	16		2,150	585		2,735	3,300
6110	Dish, mesh, 10' diam.	"	2.40	6.667		1,175	244		1,419	1,700
6111	Two way RF/IF tapeoff	1 Elec	36	.222		3.56	8.15		11.71	17.25
6112	Two way RF/IF splitter		24	.333		14.75	12.20		26.95	36
6113	Line amplifier		24	.333		14.75	12.20		26.95	36
6114	Line splitters		36	.222		4.69	8.15		12.84	18.50
6115	Line multi switches		8	1		680	36.50		716.50	810
6120	Motor unit		2.40	3.333		370	122		492	610
7000	Home theater, widescreen, 42", high definition, TV		10.24	.781		510	28.50		538.50	605
7050	Flat wall mount bracket		10.24	.781		129	28.50		157.50	189
7100	7 channel home theater receiver		10.24	.781		380	28.50		408.50	465
7200	Home theater speakers		10.24	.781	Set	191	28.50		219.50	257
7300	Home theater programmable remote		10.24	.781	Ea.	259	28.50		287.50	330
8000	Main video splitter		4	2		2,450	73.50		2,523.50	2,825
8010	Video distribution units		4	2		124	73.50		197.50	256

Estimating Tips

- When estimating material costs for electronic safety and security systems, it is always prudent to obtain manufacturers' quotations for equipment prices and special installation requirements that affect the total cost.

- Fire alarm systems consist of control panels, annunciator panels, battery with rack, charger, and fire alarm actuating and indicating devices. Some fire alarm systems include speakers, telephone lines, door closer controls, and other components. Be careful not to overlook the costs related to installation for these items. Also be aware of costs for integrated automation instrumentation and terminal devices, control equipment, control wiring, and programming.

- Security equipment includes items such as CCTV, access control, and other detection and identification systems to perform alert and alarm functions. Be sure to consider the costs related to installation for this security equipment, such as for integrated automation instrumentation and terminal devices, control equipment, control wiring, and programming.

Reference Numbers

Reference numbers are shown in shaded boxes at the beginning of some major classifications. These numbers refer to related items in the Reference Section. The reference information may be an estimating procedure, an alternate pricing method, or technical information.

Note: Not all subdivisions listed here necessarily appear in this publication. ■

Did you know?
RSMeans Online gives you the same access to RSMeans' data with 24/7 access:
- Quickly locate costs in the searchable database.
- Build cost lists, estimates, and reports in minutes.
- Adjust costs to any location in the U.S. and Canada with the click of a button.

Start your free trial today at **www.rsmeansonline.com**

RSMeansOnline

No part of this publication may be reproduced, stored in a retrieval system, or transmitted in any form or by any means without prior written permission of RSMeans.

28 16 Intrusion Detection

28 16 16 – Intrusion Detection Systems Infrastructure

28 16 16.50 Intrusion Detection		Crew	Daily Output	Labor-Hours	Unit	Material	2015 Bare Costs Labor	Equipment	Total	Total Incl O&P
0010	**INTRUSION DETECTION**, not including wires & conduits									
0100	Burglar alarm, battery operated, mechanical trigger	1 Elec	4	2	Ea.	278	73.50		351.50	425
0200	Electrical trigger		4	2		330	73.50		403.50	485
0400	For outside key control, add		8	1		84	36.50		120.50	153
0600	For remote signaling circuitry, add		8	1		125	36.50		161.50	197
0800	Card reader, flush type, standard		2.70	2.963		930	109		1,039	1,200
1000	Multi-code		2.70	2.963		1,200	109		1,309	1,500
1200	Door switches, hinge switch		5.30	1.509		58.50	55.50		114	155
1400	Magnetic switch		5.30	1.509		69	55.50		124.50	167
2800	Ultrasonic motion detector, 12 volt		2.30	3.478		230	127		357	460
3000	Infrared photoelectric detector		4	2		189	73.50		262.50	330
3200	Passive infrared detector		4	2		283	73.50		356.50	430
3420	Switchmats, 30" x 5'		5.30	1.509		84.50	55.50		140	184
3440	30" x 25'		4	2		203	73.50		276.50	345
3460	Police connect panel		4	2		244	73.50		317.50	390
3480	Telephone dialer		5.30	1.509		385	55.50		440.50	515
3500	Alarm bell		4	2		104	73.50		177.50	235
3520	Siren		4	2		146	73.50		219.50	281

28 31 Fire Detection and Alarm

28 31 23 – Fire Detection and Alarm Annunciation Panels and Fire Stations

28 31 23.50 Alarm Panels and Devices

		Crew	Daily Output	Labor-Hours	Unit	Material	Labor	Equipment	Total	Incl O&P
0010	**ALARM PANELS AND DEVICES**, not including wires & conduits									
5600	Strobe and horn	1 Elec	5.30	1.509	Ea.	152	55.50		207.50	258
5800	Fire alarm horn		6.70	1.194		61	44		105	139
6600	Drill switch		8	1		370	36.50		406.50	470
6800	Master box		2.70	2.963		6,400	109		6,509	7,225
7800	Remote annunciator, 8 zone lamp		1.80	4.444		209	163		372	495
8000	12 zone lamp	2 Elec	2.60	6.154		335	226		561	740
8200	16 zone lamp	"	2.20	7.273		420	267		687	895

28 31 46 – Smoke Detection Sensors

28 31 46.50 Smoke Detectors

		Crew	Daily Output	Labor-Hours	Unit	Material	Labor	Equipment	Total	Incl O&P
0010	**SMOKE DETECTORS**									
5200	Smoke detector, ceiling type	1 Elec	6.20	1.290	Ea.	110	47.50		157.50	199

Division 31 – Earthwork

Estimating Tips
31 05 00 Common Work Results for Earthwork

- Estimating the actual cost of performing earthwork requires careful consideration of the variables involved. This includes items such as type of soil, whether water will be encountered, dewatering, whether banks need bracing, disposal of excavated earth, and length of haul to fill or spoil sites, etc. If the project has large quantities of cut or fill, consider raising or lowering the site to reduce costs, while paying close attention to the effect on site drainage and utilities.

- If the project has large quantities of fill, creating a borrow pit on the site can significantly lower the costs.

- It is very important to consider what time of year the project is scheduled for completion. Bad weather can create large cost overruns from dewatering, site repair, and lost productivity from cold weather.

Reference Numbers

Reference numbers are shown in shaded boxes at the beginning of some major classifications. These numbers refer to related items in the Reference Section. The reference information may be an estimating procedure, an alternate pricing method, or technical information.

Note: Not all subdivisions listed here necessarily appear in this publication. ■

Estimate with precision with RSMeans Online...

Quick, intuitive and easy-to-use, RSMeans Online gives you instant access to hundreds of thousands of material, labor, and equipment costs from RSMeans' comprehensive database, delivering the information you need to build competitive estimates every time.

Start your free trial today at **www.rsmeansonline.com**

RSMeansOnline

No part of this publication may be reproduced, stored in a retrieval system, or transmitted in any form or by any means without prior written permission of RSMeans.

31 05 Common Work Results for Earthwork

31 05 13 – Soils for Earthwork

31 05 13.10 Borrow

		Crew	Daily Output	Labor-Hours	Unit	Material	2015 Bare Costs Labor	Equipment	Total	Total Incl O&P
0010	**BORROW**									
0020	Spread, 200 H.P. dozer, no compaction, 2 mi. RT haul									
0200	Common borrow	B-15	600	.047	C.Y.	12.40	1.39	4.62	18.41	21

31 05 16 – Aggregates for Earthwork

31 05 16.10 Borrow

		Crew	Daily Output	Labor-Hours	Unit	Material	Labor	Equipment	Total	Total Incl O&P
0010	**BORROW**									
0020	Spread, with 200 H.P. dozer, no compaction, 2 mi. RT haul									
0100	Bank run gravel	B-15	600	.047	L.C.Y.	22	1.39	4.62	28.01	32
0300	Crushed stone (1.40 tons per CY), 1-1/2"		600	.047		23.50	1.39	4.62	29.51	33.50
0320	3/4"		600	.047		23.50	1.39	4.62	29.51	33.50
0340	1/2"		600	.047		26.50	1.39	4.62	32.51	36.50
0360	3/8"		600	.047		27.50	1.39	4.62	33.51	38
0400	Sand, washed, concrete		600	.047		41	1.39	4.62	47.01	52.50
0500	Dead or bank sand		600	.047		17.85	1.39	4.62	23.86	27

31 11 Clearing and Grubbing

31 11 10 – Clearing and Grubbing Land

31 11 10.10 Clear and Grub Site

		Crew	Daily Output	Labor-Hours	Unit	Material	Labor	Equipment	Total	Total Incl O&P
0010	**CLEAR AND GRUB SITE**									
0020	Cut & chip light trees to 6" diam.	B-7	1	48	Acre		1,275	1,675	2,950	3,950
0150	Grub stumps and remove	B-30	2	12			365	1,200	1,565	1,925
0200	Cut & chip medium, trees to 12" diam.	B-7	.70	68.571			1,825	2,375	4,200	5,675
0250	Grub stumps and remove	B-30	1	24			735	2,400	3,135	3,875
0300	Cut & chip heavy, trees to 24" diam.	B-7	.30	160			4,250	5,575	9,825	13,200
0350	Grub stumps and remove	B-30	.50	48			1,475	4,825	6,300	7,725
0400	If burning is allowed, deduct cut & chip								40%	40%

31 13 Selective Tree and Shrub Removal and Trimming

31 13 13 – Selective Tree and Shrub Removal

31 13 13.10 Selective Clearing

		Crew	Daily Output	Labor-Hours	Unit	Material	Labor	Equipment	Total	Total Incl O&P
0010	**SELECTIVE CLEARING**									
0020	Clearing brush with brush saw	A-1C	.25	32	Acre		780	125	905	1,450
0100	By hand	1 Clab	.12	66.667			1,625		1,625	2,725
0300	With dozer, ball and chain, light clearing	B-11A	2	8			237	695	932	1,150
0400	Medium clearing	"	1.50	10.667			315	925	1,240	1,550

31 14 Earth Stripping and Stockpiling

31 14 13 – Soil Stripping and Stockpiling

31 14 13.23 Topsoil Stripping and Stockpiling

		Crew	Daily Output	Labor-Hours	Unit	Material	Labor	Equipment	Total	Total Incl O&P
0010	**TOPSOIL STRIPPING AND STOCKPILING**									
1400	Loam or topsoil, remove and stockpile on site									
1420	6" deep, 200' haul	B-10B	865	.009	C.Y.		.32	1.60	1.92	2.30
1430	300' haul		520	.015			.54	2.67	3.21	3.82
1440	500' haul		225	.036			1.24	6.15	7.39	8.85
1450	Alternate method: 6" deep, 200' haul		5090	.002	S.Y.		.05	.27	.32	.39
1460	500' haul		1325	.006	"		.21	1.05	1.26	1.50
1500	Loam or topsoil, remove/stockpile on site									

31 14 Earth Stripping and Stockpiling

31 14 13 – Soil Stripping and Stockpiling

31 14 13.23 Topsoil Stripping and Stockpiling

		Crew	Daily Output	Labor-Hours	Unit	Material	2015 Bare Costs Labor	Equipment	Total	Total Incl O&P
1510	By hand, 6" deep, 50' haul, less than 100 S.Y.	B-1	100	.240	S.Y.		6.05		6.05	10.10
1520	By skid steer, 6" deep, 100' haul, 101-500 S.Y.	B-62	500	.048			1.32	.35	1.67	2.58
1530	100' haul, 501-900 S.Y.	"	900	.027			.73	.19	.92	1.43
1540	200' haul, 901-1100 S.Y.	B-63	1000	.040			.98	.17	1.15	1.83
1550	By dozer, 200' haul, 1101-4000 S.Y.	B-10B	4000	.002			.07	.35	.42	.50

31 22 Grading

31 22 16 – Fine Grading

31 22 16.10 Finish Grading

		Crew	Daily Output	Labor-Hours	Unit	Material	Labor	Equipment	Total	Total Incl O&P
0010	**FINISH GRADING**									
0012	Finish grading area to be paved with grader, small area	B-11L	400	.040	S.Y.		1.19	1.84	3.03	4
0100	Large area		2000	.008			.24	.37	.61	.80
0200	Grade subgrade for base course, roadways		3500	.005			.14	.21	.35	.46
1020	For large parking lots	B-32C	5000	.010			.29	.47	.76	1
1050	For small irregular areas	"	2000	.024			.72	1.18	1.90	2.50
1100	Fine grade for slab on grade, machine	B-11L	1040	.015			.46	.71	1.17	1.54
1150	Hand grading	B-18	700	.034			.86	.07	.93	1.52
1200	Fine grade granular base for sidewalks and bikeways	B-62	1200	.020			.55	.14	.69	1.08
2550	Hand grade select gravel	2 Clab	60	.267	C.S.F.		6.50		6.50	10.95
3000	Hand grade select gravel, including compaction, 4" deep	B-18	555	.043	S.Y.		1.09	.08	1.17	1.91
3100	6" deep		400	.060			1.51	.12	1.63	2.66
3120	8" deep		300	.080			2.01	.16	2.17	3.54
3300	Finishing grading slopes, gentle	B-11L	8900	.002			.05	.08	.13	.18
3310	Steep slopes	"	7100	.002			.07	.10	.17	.22

31 23 Excavation and Fill

31 23 16 – Excavation

31 23 16.13 Excavating, Trench

		Crew	Daily Output	Labor-Hours	Unit	Material	Labor	Equipment	Total	Total Incl O&P
0010	**EXCAVATING, TRENCH**									
0011	Or continuous footing									
0050	1' to 4' deep, 3/8 C.Y. excavator	B-11C	150	.107	B.C.Y.		3.17	2.43	5.60	7.90
0060	1/2 C.Y. excavator	B-11M	200	.080			2.37	1.96	4.33	6.10
0090	4' to 6' deep, 1/2 C.Y. excavator	"	200	.080			2.37	1.96	4.33	6.10
0100	5/8 C.Y. excavator	B-12Q	250	.064			1.92	2.36	4.28	5.80
0300	1/2 C.Y. excavator, truck mounted	B-12J	200	.080			2.40	4.41	6.81	8.85
1352	4' to 6' deep, 1/2 C.Y. excavator w/trench box	B-13H	188	.085			2.56	5.10	7.66	9.90
1354	5/8 C.Y. excavator	"	235	.068			2.05	4.10	6.15	7.90
1400	By hand with pick and shovel 2' to 6' deep, light soil	1 Clab	8	1			24.50		24.50	41
1500	Heavy soil	"	4	2			49		49	82
5020	Loam & Sandy clay with no sheeting or dewatering included									
5050	1' to 4' deep, 3/8 C.Y. tractor loader/backhoe	B-11C	162	.099	B.C.Y.		2.93	2.25	5.18	7.35
5060	1/2 C.Y. excavator	B-11M	216	.074			2.20	1.81	4.01	5.65
5080	4' to 6' deep, 1/2 C.Y. excavator	"	216	.074			2.20	1.81	4.01	5.65
5090	5/8 C.Y. excavator	B-12Q	276	.058			1.74	2.14	3.88	5.25
5130	1/2 C.Y. excavator, truck mounted	B-12J	216	.074			2.23	4.08	6.31	8.20
5352	4' to 6' deep, 1/2 C.Y. excavator w/trench box	B-13H	205	.078			2.35	4.70	7.05	9.05
5354	5/8 C.Y. excavator	"	257	.062			1.87	3.75	5.62	7.25
6020	Sand & gravel with no sheeting or dewatering included									
6050	1' to 4' deep, 3/8 C.Y. excavator	B-11C	165	.097	B.C.Y.		2.88	2.21	5.09	7.25

31 23 Excavation and Fill

31 23 16 – Excavation

31 23 16.13 Excavating, Trench

		Crew	Daily Output	Labor-Hours	Unit	Material	2015 Bare Costs Labor	2015 Bare Costs Equipment	Total	Total Incl O&P
6060	1/2 C.Y. excavator	B-11M	220	.073	B.C.Y.		2.16	1.78	3.94	5.55
6080	4' to 6' deep, 1/2 C.Y. excavator	"	220	.073			2.16	1.78	3.94	5.55
6090	5/8 C.Y. excavator	B-12Q	275	.058			1.75	2.14	3.89	5.25
6130	1/2 C.Y. excavator, truck mounted	B-12J	220	.073			2.19	4.01	6.20	8.05
6352	4' to 6' deep, 1/2 C.Y. excavator w/trench box	B-13H	209	.077			2.30	4.61	6.91	8.90
6354	5/8 C.Y. excavator	"	261	.061			1.84	3.69	5.53	7.15
7020	Dense hard clay with no sheeting or dewatering included									
7050	1' to 4' deep, 3/8 C.Y. excavator	B-11C	132	.121	B.C.Y.		3.60	2.76	6.36	9.05
7060	1/2 C.Y. excavator	B-11M	176	.091			2.70	2.23	4.93	6.95
7080	4' to 6' deep, 1/2 C.Y. excavator	"	176	.091			2.70	2.23	4.93	6.95
7090	5/8 C.Y. excavator	B-12Q	220	.073			2.19	2.68	4.87	6.60
7130	1/2 C.Y. excavator, truck mounted	B-12J	176	.091			2.73	5	7.73	10.05

31 23 16.14 Excavating, Utility Trench

		Crew	Daily Output	Labor-Hours	Unit	Material	Labor	Equipment	Total	Total Incl O&P
0010	**EXCAVATING, UTILITY TRENCH**									
0011	Common earth									
0050	Trenching with chain trencher, 12 H.P., operator walking									
0100	4" wide trench, 12" deep	B-53	800	.010	L.F.		.24	.09	.33	.50
1000	Backfill by hand including compaction, add									
1050	4" wide trench, 12" deep	A-1G	800	.010	L.F.		.24	.07	.31	.49

31 23 16.16 Structural Excavation for Minor Structures

		Crew	Daily Output	Labor-Hours	Unit	Material	Labor	Equipment	Total	Total Incl O&P
0010	**STRUCTURAL EXCAVATION FOR MINOR STRUCTURES**									
0015	Hand, pits to 6' deep, sandy soil	1 Clab	8	1	B.C.Y.		24.50		24.50	41
0100	Heavy soil or clay		4	2			49		49	82
1100	Hand loading trucks from stock pile, sandy soil		12	.667			16.30		16.30	27.50
1300	Heavy soil or clay		8	1			24.50		24.50	41
1500	For wet or muck hand excavation, add to above								50%	50%

31 23 16.42 Excavating, Bulk Bank Measure

		Crew	Daily Output	Labor-Hours	Unit	Material	Labor	Equipment	Total	Total Incl O&P
0010	**EXCAVATING, BULK BANK MEASURE**									
0011	Common earth piled									
0020	For loading onto trucks, add								15%	15%
0200	Excavator, hydraulic, crawler mtd., 1 C.Y. cap. = 100 C.Y./hr.	B-12A	800	.020	B.C.Y.		.60	1.02	1.62	2.12
0310	Wheel mounted, 1/2 C.Y. cap. = 40 C.Y./hr.	B-12E	320	.050			1.50	1.40	2.90	4.04
1200	Front end loader, track mtd., 1-1/2 C.Y. cap. = 70 C.Y./hr.	B-10N	560	.014			.50	.93	1.43	1.86
1500	Wheel mounted, 3/4 C.Y. cap. = 45 C.Y./hr.	B-10R	360	.022			.78	.83	1.61	2.20
5000	Excavating, bulk bank measure, sandy clay & loam piled									
5020	For loading onto trucks, add								15%	15%
5100	Excavator, hydraulic, crawler mtd., 1 C.Y. cap. = 120 C.Y./hr.	B-12A	960	.017	B.C.Y.		.50	.85	1.35	1.76
5610	Wheel mounted, 1/2 C.Y. cap. = 44 C.Y./hr.	B-12E	352	.045	"		1.37	1.27	2.64	3.67
8000	For hauling excavated material, see Section 31 23 23.20									

31 23 23 – Fill

31 23 23.13 Backfill

		Crew	Daily Output	Labor-Hours	Unit	Material	Labor	Equipment	Total	Total Incl O&P
0010	**BACKFILL**									
0015	By hand, no compaction, light soil	1 Clab	14	.571	L.C.Y.		13.95		13.95	23.50
0100	Heavy soil		11	.727	"		17.80		17.80	30
0300	Compaction in 6" layers, hand tamp, add to above		20.60	.388	E.C.Y.		9.50		9.50	15.95
0500	Air tamp, add	B-9D	190	.211			5.25	1.40	6.65	10.35
0600	Vibrating plate, add	A-1D	60	.133			3.26	.60	3.86	6.10
0800	Compaction in 12" layers, hand tamp, add to above	1 Clab	34	.235			5.75		5.75	9.65
1300	Dozer backfilling, bulk, up to 300' haul, no compaction	B-10B	1200	.007	L.C.Y.		.23	1.16	1.39	1.66
1400	Air tamped, add	B-11B	80	.200	E.C.Y.		5.80	3.77	9.57	13.80

31 23 Excavation and Fill

31 23 23 – Fill

31 23 23.16 Fill By Borrow and Utility Bedding

		Crew	Daily Output	Labor-Hours	Unit	Material	2015 Bare Costs Labor	2015 Bare Costs Equipment	Total	Total Incl O&P
0010	**FILL BY BORROW AND UTILITY BEDDING**									
0049	Utility bedding, for pipe & conduit, not incl. compaction									
0050	Crushed or screened bank run gravel	B-6	150	.160	L.C.Y.	25.50	4.40	2.43	32.33	38
0100	Crushed stone 3/4" to 1/2"		150	.160		23.50	4.40	2.43	30.33	36
0200	Sand, dead or bank	↓	150	.160	↓	17.85	4.40	2.43	24.68	29.50
0500	Compacting bedding in trench	A-1D	90	.089	E.C.Y.		2.17	.40	2.57	4.09
0600	If material source exceeds 2 miles, add for extra mileage.									
0610	See Section 31 23 23.20 for hauling mileage add.									

31 23 23.17 General Fill

		Crew	Daily Output	Labor-Hours	Unit	Material	Labor	Equipment	Total	Total Incl O&P
0010	**GENERAL FILL**									
0011	Spread dumped material, no compaction									
0020	By dozer, no compaction	B-10B	1000	.008	L.C.Y.		.28	1.39	1.67	1.99
0100	By hand	1 Clab	12	.667	"		16.30		16.30	27.50
0500	Gravel fill, compacted, under floor slabs, 4" deep	B-37	10000	.005	S.F.	.42	.13	.02	.57	.69
0600	6" deep		8600	.006		.63	.15	.02	.80	.96
0700	9" deep		7200	.007		1.05	.18	.02	1.25	1.46
0800	12" deep		6000	.008		1.47	.21	.03	1.71	2
1000	Alternate pricing method, 4" deep		120	.400	E.C.Y.	31.50	10.50	1.31	43.31	53.50
1100	6" deep		160	.300		31.50	7.90	.98	40.38	49
1200	9" deep		200	.240		31.50	6.30	.79	38.59	46
1300	12" deep	↓	220	.218		31.50	5.75	.72	37.97	45

31 23 23.20 Hauling

		Crew	Daily Output	Labor-Hours	Unit	Material	Labor	Equipment	Total	Total Incl O&P
0010	**HAULING**									
0011	Excavated or borrow, loose cubic yards									
0012	no loading equipment, including hauling, waiting, loading/dumping									
0013	time per cycle (wait, load, travel, unload or dump & return)									
0014	8 C.Y. truck, 15 MPH ave, cycle 0.5 miles, 10 min. wait/Ld./Uld.	B-34A	320	.025	L.C.Y.		.71	1.29	2	2.59
0016	cycle 1 mile		272	.029			.84	1.51	2.35	3.04
0018	cycle 2 miles		208	.038			1.09	1.98	3.07	3.98
0020	cycle 4 miles		144	.056			1.58	2.86	4.44	5.75
0022	cycle 6 miles		112	.071			2.03	3.67	5.70	7.40
0024	cycle 8 miles		88	.091			2.59	4.67	7.26	9.40
0026	20 MPH ave, cycle 0.5 mile		336	.024			.68	1.22	1.90	2.47
0028	cycle 1 mile		296	.027			.77	1.39	2.16	2.80
0030	cycle 2 miles		240	.033			.95	1.71	2.66	3.45
0032	cycle 4 miles		176	.045			1.29	2.34	3.63	4.71
0034	cycle 6 miles		136	.059			1.67	3.02	4.69	6.10
0036	cycle 8 miles		112	.071			2.03	3.67	5.70	7.40
0044	25 MPH ave, cycle 4 miles		192	.042			1.19	2.14	3.33	4.32
0046	cycle 6 miles		160	.050			1.42	2.57	3.99	5.20
0048	cycle 8 miles		128	.063			1.78	3.21	4.99	6.45
0050	30 MPH ave, cycle 4 miles		216	.037			1.05	1.90	2.95	3.83
0052	cycle 6 miles		176	.045			1.29	2.34	3.63	4.71
0054	cycle 8 miles		144	.056			1.58	2.86	4.44	5.75
0114	15 MPH ave, cycle 0.5 mile, 15 min. wait/Ld./Uld.		224	.036			1.02	1.84	2.86	3.70
0116	cycle 1 mile		200	.040			1.14	2.06	3.20	4.14
0118	cycle 2 miles		168	.048			1.35	2.45	3.80	4.93
0120	cycle 4 miles		120	.067			1.90	3.43	5.33	6.90
0122	cycle 6 miles		96	.083			2.37	4.28	6.65	8.65
0124	cycle 8 miles		80	.100			2.85	5.15	8	10.35
0126	20 MPH ave, cycle 0.5 mile		232	.034			.98	1.77	2.75	3.57
0128	cycle 1 mile		208	.038			1.09	1.98	3.07	3.98

31 23 Excavation and Fill

31 23 23 – Fill

31 23 23.20 Hauling

		Crew	Daily Output	Labor-Hours	Unit	Material	2015 Bare Costs Labor	Equipment	Total	Total Incl O&P
0130	cycle 2 miles	B-34A	184	.043	L.C.Y.		1.24	2.23	3.47	4.50
0132	cycle 4 miles		144	.056			1.58	2.86	4.44	5.75
0134	cycle 6 miles		112	.071			2.03	3.67	5.70	7.40
0136	cycle 8 miles		96	.083			2.37	4.28	6.65	8.65
0144	25 MPH ave, cycle 4 miles		152	.053			1.50	2.71	4.21	5.45
0146	cycle 6 miles		128	.063			1.78	3.21	4.99	6.45
0148	cycle 8 miles		112	.071			2.03	3.67	5.70	7.40
0150	30 MPH ave, cycle 4 miles		168	.048			1.35	2.45	3.80	4.93
0152	cycle 6 miles		144	.056			1.58	2.86	4.44	5.75
0154	cycle 8 miles		120	.067			1.90	3.43	5.33	6.90
0214	15 MPH ave, cycle 0.5 mile, 20 min wait/Ld./Uld.		176	.045			1.29	2.34	3.63	4.71
0216	cycle 1 mile		160	.050			1.42	2.57	3.99	5.20
0218	cycle 2 miles		136	.059			1.67	3.02	4.69	6.10
0220	cycle 4 miles		104	.077			2.19	3.95	6.14	7.95
0222	cycle 6 miles		88	.091			2.59	4.67	7.26	9.40
0224	cycle 8 miles		72	.111			3.16	5.70	8.86	11.50
0226	20 MPH ave, cycle 0.5 mile		176	.045			1.29	2.34	3.63	4.71
0228	cycle 1 mile		168	.048			1.35	2.45	3.80	4.93
0230	cycle 2 miles		144	.056			1.58	2.86	4.44	5.75
0232	cycle 4 miles		120	.067			1.90	3.43	5.33	6.90
0234	cycle 6 miles		96	.083			2.37	4.28	6.65	8.65
0236	cycle 8 miles		88	.091			2.59	4.67	7.26	9.40
0244	25 MPH ave, cycle 4 miles		128	.063			1.78	3.21	4.99	6.45
0246	cycle 6 miles		112	.071			2.03	3.67	5.70	7.40
0248	cycle 8 miles		96	.083			2.37	4.28	6.65	8.65
0250	30 MPH ave, cycle 4 miles		136	.059			1.67	3.02	4.69	6.10
0252	cycle 6 miles		120	.067			1.90	3.43	5.33	6.90
0254	cycle 8 miles		104	.077			2.19	3.95	6.14	7.95
0314	15 MPH ave, cycle 0.5 mile, 25 min wait/Ld./Uld.		144	.056			1.58	2.86	4.44	5.75
0316	cycle 1 mile		128	.063			1.78	3.21	4.99	6.45
0318	cycle 2 miles		112	.071			2.03	3.67	5.70	7.40
0320	cycle 4 miles		96	.083			2.37	4.28	6.65	8.65
0322	cycle 6 miles		80	.100			2.85	5.15	8	10.35
0324	cycle 8 miles		64	.125			3.56	6.45	10.01	12.95
0326	20 MPH ave, cycle 0.5 mile		144	.056			1.58	2.86	4.44	5.75
0328	cycle 1 mile		136	.059			1.67	3.02	4.69	6.10
0330	cycle 2 miles		120	.067			1.90	3.43	5.33	6.90
0332	cycle 4 miles		104	.077			2.19	3.95	6.14	7.95
0334	cycle 6 miles		88	.091			2.59	4.67	7.26	9.40
0336	cycle 8 miles		80	.100			2.85	5.15	8	10.35
0344	25 MPH ave, cycle 4 miles		112	.071			2.03	3.67	5.70	7.40
0346	cycle 6 miles		96	.083			2.37	4.28	6.65	8.65
0348	cycle 8 miles		88	.091			2.59	4.67	7.26	9.40
0350	30 MPH ave, cycle 4 miles		112	.071			2.03	3.67	5.70	7.40
0352	cycle 6 miles		104	.077			2.19	3.95	6.14	7.95
0354	cycle 8 miles		96	.083			2.37	4.28	6.65	8.65
0414	15 MPH ave, cycle 0.5 mile, 30 min wait/Ld./Uld.		120	.067			1.90	3.43	5.33	6.90
0416	cycle 1 mile		112	.071			2.03	3.67	5.70	7.40
0418	cycle 2 miles		96	.083			2.37	4.28	6.65	8.65
0420	cycle 4 miles		80	.100			2.85	5.15	8	10.35
0422	cycle 6 miles		72	.111			3.16	5.70	8.86	11.50
0424	cycle 8 miles		64	.125			3.56	6.45	10.01	12.95
0426	20 MPH ave, cycle 0.5 mile		120	.067			1.90	3.43	5.33	6.90

31 23 Excavation and Fill

31 23 23 – Fill

31 23 23.20 Hauling		Crew	Daily Output	Labor-Hours	Unit	Material	2015 Bare Costs Labor	Equipment	Total	Total Incl O&P
0428	cycle 1 mile	B-34A	112	.071	L.C.Y.		2.03	3.67	5.70	7.40
0430	cycle 2 miles		104	.077			2.19	3.95	6.14	7.95
0432	cycle 4 miles		88	.091			2.59	4.67	7.26	9.40
0434	cycle 6 miles		80	.100			2.85	5.15	8	10.35
0436	cycle 8 miles		72	.111			3.16	5.70	8.86	11.50
0444	25 MPH ave, cycle 4 miles		96	.083			2.37	4.28	6.65	8.65
0446	cycle 6 miles		88	.091			2.59	4.67	7.26	9.40
0448	cycle 8 miles		80	.100			2.85	5.15	8	10.35
0450	30 MPH ave, cycle 4 miles		96	.083			2.37	4.28	6.65	8.65
0452	cycle 6 miles		88	.091			2.59	4.67	7.26	9.40
0454	cycle 8 miles		80	.100			2.85	5.15	8	10.35
0514	15 MPH ave, cycle 0.5 mile, 35 min wait/Ld./Uld.		104	.077			2.19	3.95	6.14	7.95
0516	cycle 1 mile		96	.083			2.37	4.28	6.65	8.65
0518	cycle 2 miles		88	.091			2.59	4.67	7.26	9.40
0520	cycle 4 miles		72	.111			3.16	5.70	8.86	11.50
0522	cycle 6 miles		64	.125			3.56	6.45	10.01	12.95
0524	cycle 8 miles		56	.143			4.06	7.35	11.41	14.80
0526	20 MPH ave, cycle 0.5 mile		104	.077			2.19	3.95	6.14	7.95
0528	cycle 1 mile		96	.083			2.37	4.28	6.65	8.65
0530	cycle 2 miles		96	.083			2.37	4.28	6.65	8.65
0532	cycle 4 miles		80	.100			2.85	5.15	8	10.35
0534	cycle 6 miles		72	.111			3.16	5.70	8.86	11.50
0536	cycle 8 miles		64	.125			3.56	6.45	10.01	12.95
0544	25 MPH ave, cycle 4 miles		88	.091			2.59	4.67	7.26	9.40
0546	cycle 6 miles		80	.100			2.85	5.15	8	10.35
0548	cycle 8 miles		72	.111			3.16	5.70	8.86	11.50
0550	30 MPH ave, cycle 4 miles		88	.091			2.59	4.67	7.26	9.40
0552	cycle 6 miles		80	.100			2.85	5.15	8	10.35
0554	cycle 8 miles		72	.111			3.16	5.70	8.86	11.50
1014	12 C.Y. truck, cycle 0.5 mile, 15 MPH ave, 15 min. wait/Ld./Uld.	B-34B	336	.024			.68	2.06	2.74	3.38
1016	cycle 1 mile		300	.027			.76	2.30	3.06	3.78
1018	cycle 2 miles		252	.032			.90	2.74	3.64	4.51
1020	cycle 4 miles		180	.044			1.26	3.84	5.10	6.30
1022	cycle 6 miles		144	.056			1.58	4.80	6.38	7.90
1024	cycle 8 miles		120	.067			1.90	5.75	7.65	9.50
1025	cycle 10 miles		96	.083			2.37	7.20	9.57	11.80
1026	20 MPH ave, cycle 0.5 mile		348	.023			.65	1.99	2.64	3.26
1028	cycle 1 mile		312	.026			.73	2.21	2.94	3.65
1030	cycle 2 miles		276	.029			.82	2.50	3.32	4.11
1032	cycle 4 miles		216	.037			1.05	3.20	4.25	5.25
1034	cycle 6 miles		168	.048			1.35	4.11	5.46	6.75
1036	cycle 8 miles		144	.056			1.58	4.80	6.38	7.90
1038	cycle 10 miles		120	.067			1.90	5.75	7.65	9.50
1040	25 MPH ave, cycle 4 miles		228	.035			1	3.03	4.03	4.98
1042	cycle 6 miles		192	.042			1.19	3.60	4.79	5.90
1044	cycle 8 miles		168	.048			1.35	4.11	5.46	6.75
1046	cycle 10 miles		144	.056			1.58	4.80	6.38	7.90
1050	30 MPH ave, cycle 4 miles		252	.032			.90	2.74	3.64	4.51
1052	cycle 6 miles		216	.037			1.05	3.20	4.25	5.25
1054	cycle 8 miles		180	.044			1.26	3.84	5.10	6.30
1056	cycle 10 miles		156	.051			1.46	4.43	5.89	7.30
1060	35 MPH ave, cycle 4 miles		264	.030			.86	2.62	3.48	4.30
1062	cycle 6 miles		228	.035			1	3.03	4.03	4.98

For customer support on your Residential Cost Data, call 877.713.2617.

31 23 Excavation and Fill

31 23 23 – Fill

31 23 23.20 Hauling		Crew	Daily Output	Labor-Hours	Unit	Material	2015 Bare Costs Labor	2015 Bare Costs Equipment	Total	Total Incl O&P
1064	cycle 8 miles	B-34B	204	.039	L.C.Y.		1.12	3.39	4.51	5.55
1066	cycle 10 miles		180	.044			1.26	3.84	5.10	6.30
1068	cycle 20 miles		120	.067			1.90	5.75	7.65	9.50
1069	cycle 30 miles		84	.095			2.71	8.25	10.96	13.55
1070	cycle 40 miles		72	.111			3.16	9.60	12.76	15.75
1072	40 MPH ave, cycle 6 miles		240	.033			.95	2.88	3.83	4.74
1074	cycle 8 miles		216	.037			1.05	3.20	4.25	5.25
1076	cycle 10 miles		192	.042			1.19	3.60	4.79	5.90
1078	cycle 20 miles		120	.067			1.90	5.75	7.65	9.50
1080	cycle 30 miles		96	.083			2.37	7.20	9.57	11.80
1082	cycle 40 miles		72	.111			3.16	9.60	12.76	15.75
1084	cycle 50 miles		60	.133			3.79	11.50	15.29	18.90
1094	45 MPH ave, cycle 8 miles		216	.037			1.05	3.20	4.25	5.25
1096	cycle 10 miles		204	.039			1.12	3.39	4.51	5.55
1098	cycle 20 miles		132	.061			1.72	5.25	6.97	8.60
1100	cycle 30 miles		108	.074			2.11	6.40	8.51	10.55
1102	cycle 40 miles		84	.095			2.71	8.25	10.96	13.55
1104	cycle 50 miles		72	.111			3.16	9.60	12.76	15.75
1106	50 MPH ave, cycle 10 miles		216	.037			1.05	3.20	4.25	5.25
1108	cycle 20 miles		144	.056			1.58	4.80	6.38	7.90
1110	cycle 30 miles		108	.074			2.11	6.40	8.51	10.55
1112	cycle 40 miles		84	.095			2.71	8.25	10.96	13.55
1114	cycle 50 miles		72	.111			3.16	9.60	12.76	15.75
1214	15 MPH ave, cycle 0.5 mile, 20 min. wait/Ld./Uld.		264	.030			.86	2.62	3.48	4.30
1216	cycle 1 mile		240	.033			.95	2.88	3.83	4.74
1218	cycle 2 miles		204	.039			1.12	3.39	4.51	5.55
1220	cycle 4 miles		156	.051			1.46	4.43	5.89	7.30
1222	cycle 6 miles		132	.061			1.72	5.25	6.97	8.60
1224	cycle 8 miles		108	.074			2.11	6.40	8.51	10.55
1225	cycle 10 miles		96	.083			2.37	7.20	9.57	11.80
1226	20 MPH ave, cycle 0.5 mile		264	.030			.86	2.62	3.48	4.30
1228	cycle 1 mile		252	.032			.90	2.74	3.64	4.51
1230	cycle 2 miles		216	.037			1.05	3.20	4.25	5.25
1232	cycle 4 miles		180	.044			1.26	3.84	5.10	6.30
1234	cycle 6 miles		144	.056			1.58	4.80	6.38	7.90
1236	cycle 8 miles		132	.061			1.72	5.25	6.97	8.60
1238	cycle 10 miles		108	.074			2.11	6.40	8.51	10.55
1240	25 MPH ave, cycle 4 miles		192	.042			1.19	3.60	4.79	5.90
1242	cycle 6 miles		168	.048			1.35	4.11	5.46	6.75
1244	cycle 8 miles		144	.056			1.58	4.80	6.38	7.90
1246	cycle 10 miles		132	.061			1.72	5.25	6.97	8.60
1250	30 MPH ave, cycle 4 miles		204	.039			1.12	3.39	4.51	5.55
1252	cycle 6 miles		180	.044			1.26	3.84	5.10	6.30
1254	cycle 8 miles		156	.051			1.46	4.43	5.89	7.30
1256	cycle 10 miles		144	.056			1.58	4.80	6.38	7.90
1260	35 MPH ave, cycle 4 miles		216	.037			1.05	3.20	4.25	5.25
1262	cycle 6 miles		192	.042			1.19	3.60	4.79	5.90
1264	cycle 8 miles		168	.048			1.35	4.11	5.46	6.75
1266	cycle 10 miles		156	.051			1.46	4.43	5.89	7.30
1268	cycle 20 miles		108	.074			2.11	6.40	8.51	10.55
1269	cycle 30 miles		72	.111			3.16	9.60	12.76	15.75
1270	cycle 40 miles		60	.133			3.79	11.50	15.29	18.90
1272	40 MPH ave, cycle 6 miles		192	.042			1.19	3.60	4.79	5.90

31 23 Excavation and Fill

31 23 23 – Fill

31 23 23.20 Hauling

		Crew	Daily Output	Labor-Hours	Unit	Material	2015 Bare Costs Labor	2015 Bare Costs Equipment	Total	Total Incl O&P
1274	cycle 8 miles	B-34B	180	.044	L.C.Y.		1.26	3.84	5.10	6.30
1276	cycle 10 miles		156	.051			1.46	4.43	5.89	7.30
1278	cycle 20 miles		108	.074			2.11	6.40	8.51	10.55
1280	cycle 30 miles		84	.095			2.71	8.25	10.96	13.55
1282	cycle 40 miles		72	.111			3.16	9.60	12.76	15.75
1284	cycle 50 miles		60	.133			3.79	11.50	15.29	18.90
1294	45 MPH ave, cycle 8 miles		180	.044			1.26	3.84	5.10	6.30
1296	cycle 10 miles		168	.048			1.35	4.11	5.46	6.75
1298	cycle 20 miles		120	.067			1.90	5.75	7.65	9.50
1300	cycle 30 miles		96	.083			2.37	7.20	9.57	11.80
1302	cycle 40 miles		72	.111			3.16	9.60	12.76	15.75
1304	cycle 50 miles		60	.133			3.79	11.50	15.29	18.90
1306	50 MPH ave, cycle 10 miles		180	.044			1.26	3.84	5.10	6.30
1308	cycle 20 miles		132	.061			1.72	5.25	6.97	8.60
1310	cycle 30 miles		96	.083			2.37	7.20	9.57	11.80
1312	cycle 40 miles		84	.095			2.71	8.25	10.96	13.55
1314	cycle 50 miles		72	.111			3.16	9.60	12.76	15.75
1414	15 MPH ave, cycle 0.5 mile, 25 min. wait/Ld./Uld.		204	.039			1.12	3.39	4.51	5.55
1416	cycle 1 mile		192	.042			1.19	3.60	4.79	5.90
1418	cycle 2 miles		168	.048			1.35	4.11	5.46	6.75
1420	cycle 4 miles		132	.061			1.72	5.25	6.97	8.60
1422	cycle 6 miles		120	.067			1.90	5.75	7.65	9.50
1424	cycle 8 miles		96	.083			2.37	7.20	9.57	11.80
1425	cycle 10 miles		84	.095			2.71	8.25	10.96	13.55
1426	20 MPH ave, cycle 0.5 mile		216	.037			1.05	3.20	4.25	5.25
1428	cycle 1 mile		204	.039			1.12	3.39	4.51	5.55
1430	cycle 2 miles		180	.044			1.26	3.84	5.10	6.30
1432	cycle 4 miles		156	.051			1.46	4.43	5.89	7.30
1434	cycle 6 miles		132	.061			1.72	5.25	6.97	8.60
1436	cycle 8 miles		120	.067			1.90	5.75	7.65	9.50
1438	cycle 10 miles		96	.083			2.37	7.20	9.57	11.80
1440	25 MPH ave, cycle 4 miles		168	.048			1.35	4.11	5.46	6.75
1442	cycle 6 miles		144	.056			1.58	4.80	6.38	7.90
1444	cycle 8 miles		132	.061			1.72	5.25	6.97	8.60
1446	cycle 10 miles		108	.074			2.11	6.40	8.51	10.55
1450	30 MPH ave, cycle 4 miles		168	.048			1.35	4.11	5.46	6.75
1452	cycle 6 miles		156	.051			1.46	4.43	5.89	7.30
1454	cycle 8 miles		132	.061			1.72	5.25	6.97	8.60
1456	cycle 10 miles		120	.067			1.90	5.75	7.65	9.50
1460	35 MPH ave, cycle 4 miles		180	.044			1.26	3.84	5.10	6.30
1462	cycle 6 miles		156	.051			1.46	4.43	5.89	7.30
1464	cycle 8 miles		144	.056			1.58	4.80	6.38	7.90
1466	cycle 10 miles		132	.061			1.72	5.25	6.97	8.60
1468	cycle 20 miles		96	.083			2.37	7.20	9.57	11.80
1469	cycle 30 miles		72	.111			3.16	9.60	12.76	15.75
1470	cycle 40 miles		60	.133			3.79	11.50	15.29	18.90
1472	40 MPH ave, cycle 6 miles		168	.048			1.35	4.11	5.46	6.75
1474	cycle 8 miles		156	.051			1.46	4.43	5.89	7.30
1476	cycle 10 miles		144	.056			1.58	4.80	6.38	7.90
1478	cycle 20 miles		96	.083			2.37	7.20	9.57	11.80
1480	cycle 30 miles		84	.095			2.71	8.25	10.96	13.55
1482	cycle 40 miles		60	.133			3.79	11.50	15.29	18.90
1484	cycle 50 miles		60	.133			3.79	11.50	15.29	18.90

31 23 Excavation and Fill

31 23 23 – Fill

31 23 23.20 Hauling

		Crew	Daily Output	Labor-Hours	Unit	Material	2015 Bare Costs Labor	2015 Bare Costs Equipment	Total	Total Incl O&P
1494	45 MPH ave, cycle 8 miles	B-34B	156	.051	L.C.Y.		1.46	4.43	5.89	7.30
1496	cycle 10 miles		144	.056			1.58	4.80	6.38	7.90
1498	cycle 20 miles		108	.074			2.11	6.40	8.51	10.55
1500	cycle 30 miles		84	.095			2.71	8.25	10.96	13.55
1502	cycle 40 miles		72	.111			3.16	9.60	12.76	15.75
1504	cycle 50 miles		60	.133			3.79	11.50	15.29	18.90
1506	50 MPH ave, cycle 10 miles		156	.051			1.46	4.43	5.89	7.30
1508	cycle 20 miles		120	.067			1.90	5.75	7.65	9.50
1510	cycle 30 miles		96	.083			2.37	7.20	9.57	11.80
1512	cycle 40 miles		72	.111			3.16	9.60	12.76	15.75
1514	cycle 50 miles		60	.133			3.79	11.50	15.29	18.90
1614	15 MPH, cycle 0.5 mile, 30 min. wait/Ld./Uld.		180	.044			1.26	3.84	5.10	6.30
1616	cycle 1 mile		168	.048			1.35	4.11	5.46	6.75
1618	cycle 2 miles		144	.056			1.58	4.80	6.38	7.90
1620	cycle 4 miles		120	.067			1.90	5.75	7.65	9.50
1622	cycle 6 miles		108	.074			2.11	6.40	8.51	10.55
1624	cycle 8 miles		84	.095			2.71	8.25	10.96	13.55
1625	cycle 10 miles		84	.095			2.71	8.25	10.96	13.55
1626	20 MPH ave, cycle 0.5 mile		180	.044			1.26	3.84	5.10	6.30
1628	cycle 1 mile		168	.048			1.35	4.11	5.46	6.75
1630	cycle 2 miles		156	.051			1.46	4.43	5.89	7.30
1632	cycle 4 miles		132	.061			1.72	5.25	6.97	8.60
1634	cycle 6 miles		120	.067			1.90	5.75	7.65	9.50
1636	cycle 8 miles		108	.074			2.11	6.40	8.51	10.55
1638	cycle 10 miles		96	.083			2.37	7.20	9.57	11.80
1640	25 MPH ave, cycle 4 miles		144	.056			1.58	4.80	6.38	7.90
1642	cycle 6 miles		132	.061			1.72	5.25	6.97	8.60
1644	cycle 8 miles		108	.074			2.11	6.40	8.51	10.55
1646	cycle 10 miles		108	.074			2.11	6.40	8.51	10.55
1650	30 MPH ave, cycle 4 miles		144	.056			1.58	4.80	6.38	7.90
1652	cycle 6 miles		132	.061			1.72	5.25	6.97	8.60
1654	cycle 8 miles		120	.067			1.90	5.75	7.65	9.50
1656	cycle 10 miles		108	.074			2.11	6.40	8.51	10.55
1660	35 MPH ave, cycle 4 miles		156	.051			1.46	4.43	5.89	7.30
1662	cycle 6 miles		144	.056			1.58	4.80	6.38	7.90
1664	cycle 8 miles		132	.061			1.72	5.25	6.97	8.60
1666	cycle 10 miles		120	.067			1.90	5.75	7.65	9.50
1668	cycle 20 miles		84	.095			2.71	8.25	10.96	13.55
1669	cycle 30 miles		72	.111			3.16	9.60	12.76	15.75
1670	cycle 40 miles		60	.133			3.79	11.50	15.29	18.90
1672	40 MPH, cycle 6 miles		144	.056			1.58	4.80	6.38	7.90
1674	cycle 8 miles		132	.061			1.72	5.25	6.97	8.60
1676	cycle 10 miles		120	.067			1.90	5.75	7.65	9.50
1678	cycle 20 miles		96	.083			2.37	7.20	9.57	11.80
1680	cycle 30 miles		72	.111			3.16	9.60	12.76	15.75
1682	cycle 40 miles		60	.133			3.79	11.50	15.29	18.90
1684	cycle 50 miles		48	.167			4.74	14.40	19.14	23.50
1694	45 MPH ave, cycle 8 miles		144	.056			1.58	4.80	6.38	7.90
1696	cycle 10 miles		132	.061			1.72	5.25	6.97	8.60
1698	cycle 20 miles		96	.083			2.37	7.20	9.57	11.80
1700	cycle 30 miles		84	.095			2.71	8.25	10.96	13.55
1702	cycle 40 miles		60	.133			3.79	11.50	15.29	18.90
1704	cycle 50 miles		60	.133			3.79	11.50	15.29	18.90

31 23 Excavation and Fill

31 23 23 – Fill

31 23 23.20 Hauling		Crew	Daily Output	Labor-Hours	Unit	Material	2015 Bare Costs Labor	2015 Bare Costs Equipment	Total	Total Incl O&P
1706	50 MPH ave, cycle 10 miles	B-34B	132	.061	L.C.Y.		1.72	5.25	6.97	8.60
1708	cycle 20 miles		108	.074			2.11	6.40	8.51	10.55
1710	cycle 30 miles		84	.095			2.71	8.25	10.96	13.55
1712	cycle 40 miles		72	.111			3.16	9.60	12.76	15.75
1714	cycle 50 miles		60	.133			3.79	11.50	15.29	18.90
2000	Hauling, 8 C.Y. truck, small project cost per hour	B-34A	8	1	Hr.		28.50	51.50	80	104
2100	12 C.Y. Truck	B-34B	8	1			28.50	86.50	115	142
2150	16.5 C.Y. Truck	B-34C	8	1			28.50	91	119.50	147
2175	18 C.Y. 8 wheel Truck	B-34I	8	1			28.50	109	137.50	166
2200	20 C.Y. Truck	B-34D	8	1			28.50	92.50	121	149
9014	18 C.Y. truck, 8 wheels,15 min. wait/Ld./Uld.,15 MPH, cycle 0.5 mi.	B-34I	504	.016	L.C.Y.		.45	1.72	2.17	2.65
9016	cycle 1 mile		450	.018			.51	1.93	2.44	2.96
9018	cycle 2 miles		378	.021			.60	2.30	2.90	3.52
9020	cycle 4 miles		270	.030			.84	3.22	4.06	4.93
9022	cycle 6 miles		216	.037			1.05	4.02	5.07	6.15
9024	cycle 8 miles		180	.044			1.26	4.83	6.09	7.40
9025	cycle 10 miles		144	.056			1.58	6.05	7.63	9.25
9026	20 MPH ave, cycle 0.5 mile		522	.015			.44	1.67	2.11	2.55
9028	cycle 1 mile		468	.017			.49	1.86	2.35	2.84
9030	cycle 2 miles		414	.019			.55	2.10	2.65	3.22
9032	cycle 4 miles		324	.025			.70	2.68	3.38	4.11
9034	cycle 6 miles		252	.032			.90	3.45	4.35	5.30
9036	cycle 8 miles		216	.037			1.05	4.02	5.07	6.15
9038	cycle 10 miles		180	.044			1.26	4.83	6.09	7.40
9040	25 MPH ave, cycle 4 miles		342	.023			.67	2.54	3.21	3.89
9042	cycle 6 miles		288	.028			.79	3.02	3.81	4.63
9044	cycle 8 miles		252	.032			.90	3.45	4.35	5.30
9046	cycle 10 miles		216	.037			1.05	4.02	5.07	6.15
9050	30 MPH ave, cycle 4 miles		378	.021			.60	2.30	2.90	3.52
9052	cycle 6 miles		324	.025			.70	2.68	3.38	4.11
9054	cycle 8 miles		270	.030			.84	3.22	4.06	4.93
9056	cycle 10 miles		234	.034			.97	3.71	4.68	5.70
9060	35 MPH ave, cycle 4 miles		396	.020			.57	2.19	2.76	3.36
9062	cycle 6 miles		342	.023			.67	2.54	3.21	3.89
9064	cycle 8 miles		288	.028			.79	3.02	3.81	4.63
9066	cycle 10 miles		270	.030			.84	3.22	4.06	4.93
9068	cycle 20 miles		162	.049			1.40	5.35	6.75	8.20
9070	cycle 30 miles		126	.063			1.81	6.90	8.71	10.60
9072	cycle 40 miles		90	.089			2.53	9.65	12.18	14.80
9074	40 MPH ave, cycle 6 miles		360	.022			.63	2.41	3.04	3.70
9076	cycle 8 miles		324	.025			.70	2.68	3.38	4.11
9078	cycle 10 miles		288	.028			.79	3.02	3.81	4.63
9080	cycle 20 miles		180	.044			1.26	4.83	6.09	7.40
9082	cycle 30 miles		144	.056			1.58	6.05	7.63	9.25
9084	cycle 40 miles		108	.074			2.11	8.05	10.16	12.35
9086	cycle 50 miles		90	.089			2.53	9.65	12.18	14.80
9094	45 MPH ave, cycle 8 miles		324	.025			.70	2.68	3.38	4.11
9096	cycle 10 miles		306	.026			.74	2.84	3.58	4.35
9098	cycle 20 miles		198	.040			1.15	4.39	5.54	6.75
9100	cycle 30 miles		144	.056			1.58	6.05	7.63	9.25
9102	cycle 40 miles		126	.063			1.81	6.90	8.71	10.60
9104	cycle 50 miles		108	.074			2.11	8.05	10.16	12.35
9106	50 MPH ave, cycle 10 miles		324	.025			.70	2.68	3.38	4.11

31 23 Excavation and Fill

31 23 23 – Fill

31 23 23.20 Hauling

		Crew	Daily Output	Labor-Hours	Unit	Material	2015 Bare Costs Labor	2015 Bare Costs Equipment	Total	Total Incl O&P
9108	cycle 20 miles	B-34I	216	.037	L.C.Y.		1.05	4.02	5.07	6.15
9110	cycle 30 miles		162	.049			1.40	5.35	6.75	8.20
9112	cycle 40 miles		126	.063			1.81	6.90	8.71	10.60
9114	cycle 50 miles		108	.074			2.11	8.05	10.16	12.35
9214	20 min. wait/Ld./Uld.,15 MPH, cycle 0.5 mi.		396	.020			.57	2.19	2.76	3.36
9216	cycle 1 mile		360	.022			.63	2.41	3.04	3.70
9218	cycle 2 miles		306	.026			.74	2.84	3.58	4.35
9220	cycle 4 miles		234	.034			.97	3.71	4.68	5.70
9222	cycle 6 miles		198	.040			1.15	4.39	5.54	6.75
9224	cycle 8 miles		162	.049			1.40	5.35	6.75	8.20
9225	cycle 10 miles		144	.056			1.58	6.05	7.63	9.25
9226	20 MPH ave, cycle 0.5 mile		396	.020			.57	2.19	2.76	3.36
9228	cycle 1 mile		378	.021			.60	2.30	2.90	3.52
9230	cycle 2 miles		324	.025			.70	2.68	3.38	4.11
9232	cycle 4 miles		270	.030			.84	3.22	4.06	4.93
9234	cycle 6 miles		216	.037			1.05	4.02	5.07	6.15
9236	cycle 8 miles		198	.040			1.15	4.39	5.54	6.75
9238	cycle 10 miles		162	.049			1.40	5.35	6.75	8.20
9240	25 MPH ave, cycle 4 miles		288	.028			.79	3.02	3.81	4.63
9242	cycle 6 miles		252	.032			.90	3.45	4.35	5.30
9244	cycle 8 miles		216	.037			1.05	4.02	5.07	6.15
9246	cycle 10 miles		198	.040			1.15	4.39	5.54	6.75
9250	30 MPH ave, cycle 4 miles		306	.026			.74	2.84	3.58	4.35
9252	cycle 6 miles		270	.030			.84	3.22	4.06	4.93
9254	cycle 8 miles		234	.034			.97	3.71	4.68	5.70
9256	cycle 10 miles		216	.037			1.05	4.02	5.07	6.15
9260	35 MPH ave, cycle 4 miles		324	.025			.70	2.68	3.38	4.11
9262	cycle 6 miles		288	.028			.79	3.02	3.81	4.63
9264	cycle 8 miles		252	.032			.90	3.45	4.35	5.30
9266	cycle 10 miles		234	.034			.97	3.71	4.68	5.70
9268	cycle 20 miles		162	.049			1.40	5.35	6.75	8.20
9270	cycle 30 miles		108	.074			2.11	8.05	10.16	12.35
9272	cycle 40 miles		90	.089			2.53	9.65	12.18	14.80
9274	40 MPH ave, cycle 6 miles		288	.028			.79	3.02	3.81	4.63
9276	cycle 8 miles		270	.030			.84	3.22	4.06	4.93
9278	cycle 10 miles		234	.034			.97	3.71	4.68	5.70
9280	cycle 20 miles		162	.049			1.40	5.35	6.75	8.20
9282	cycle 30 miles		126	.063			1.81	6.90	8.71	10.60
9284	cycle 40 miles		108	.074			2.11	8.05	10.16	12.35
9286	cycle 50 miles		90	.089			2.53	9.65	12.18	14.80
9294	45 MPH ave, cycle 8 miles		270	.030			.84	3.22	4.06	4.93
9296	cycle 10 miles		252	.032			.90	3.45	4.35	5.30
9298	cycle 20 miles		180	.044			1.26	4.83	6.09	7.40
9300	cycle 30 miles		144	.056			1.58	6.05	7.63	9.25
9302	cycle 40 miles		108	.074			2.11	8.05	10.16	12.35
9304	cycle 50 miles		90	.089			2.53	9.65	12.18	14.80
9306	50 MPH ave, cycle 10 miles		270	.030			.84	3.22	4.06	4.93
9308	cycle 20 miles		198	.040			1.15	4.39	5.54	6.75
9310	cycle 30 miles		144	.056			1.58	6.05	7.63	9.25
9312	cycle 40 miles		126	.063			1.81	6.90	8.71	10.60
9314	cycle 50 miles		108	.074			2.11	8.05	10.16	12.35
9414	25 min. wait/Ld./Uld.,15 MPH, cycle 0.5 mi.		306	.026			.74	2.84	3.58	4.35
9416	cycle 1 mile		288	.028			.79	3.02	3.81	4.63

31 23 Excavation and Fill

31 23 23 – Fill

31 23 23.20 Hauling		Crew	Daily Output	Labor-Hours	Unit	Material	Labor	Equipment	Total	Total Incl O&P
9418	cycle 2 miles	B-34I	252	.032	L.C.Y.		.90	3.45	4.35	5.30
9420	cycle 4 miles		198	.040			1.15	4.39	5.54	6.75
9422	cycle 6 miles		180	.044			1.26	4.83	6.09	7.40
9424	cycle 8 miles		144	.056			1.58	6.05	7.63	9.25
9425	cycle 10 miles		126	.063			1.81	6.90	8.71	10.60
9426	20 MPH ave, cycle 0.5 mile		324	.025			.70	2.68	3.38	4.11
9428	cycle 1 mile		306	.026			.74	2.84	3.58	4.35
9430	cycle 2 miles		270	.030			.84	3.22	4.06	4.93
9432	cycle 4 miles		234	.034			.97	3.71	4.68	5.70
9434	cycle 6 miles		198	.040			1.15	4.39	5.54	6.75
9436	cycle 8 miles		180	.044			1.26	4.83	6.09	7.40
9438	cycle 10 miles		144	.056			1.58	6.05	7.63	9.25
9440	25 MPH ave, cycle 4 miles		252	.032			.90	3.45	4.35	5.30
9442	cycle 6 miles		216	.037			1.05	4.02	5.07	6.15
9444	cycle 8 miles		198	.040			1.15	4.39	5.54	6.75
9446	cycle 10 miles		180	.044			1.26	4.83	6.09	7.40
9450	30 MPH ave, cycle 4 miles		252	.032			.90	3.45	4.35	5.30
9452	cycle 6 miles		234	.034			.97	3.71	4.68	5.70
9454	cycle 8 miles		198	.040			1.15	4.39	5.54	6.75
9456	cycle 10 miles		180	.044			1.26	4.83	6.09	7.40
9460	35 MPH ave, cycle 4 miles		270	.030			.84	3.22	4.06	4.93
9462	cycle 6 miles		234	.034			.97	3.71	4.68	5.70
9464	cycle 8 miles		216	.037			1.05	4.02	5.07	6.15
9466	cycle 10 miles		198	.040			1.15	4.39	5.54	6.75
9468	cycle 20 miles		144	.056			1.58	6.05	7.63	9.25
9470	cycle 30 miles		108	.074			2.11	8.05	10.16	12.35
9472	cycle 40 miles		90	.089			2.53	9.65	12.18	14.80
9474	40 MPH ave, cycle 6 miles		252	.032			.90	3.45	4.35	5.30
9476	cycle 8 miles		234	.034			.97	3.71	4.68	5.70
9478	cycle 10 miles		216	.037			1.05	4.02	5.07	6.15
9480	cycle 20 miles		144	.056			1.58	6.05	7.63	9.25
9482	cycle 30 miles		126	.063			1.81	6.90	8.71	10.60
9484	cycle 40 miles		90	.089			2.53	9.65	12.18	14.80
9486	cycle 50 miles		90	.089			2.53	9.65	12.18	14.80
9494	45 MPH ave, cycle 8 miles		234	.034			.97	3.71	4.68	5.70
9496	cycle 10 miles		216	.037			1.05	4.02	5.07	6.15
9498	cycle 20 miles		162	.049			1.40	5.35	6.75	8.20
9500	cycle 30 miles		126	.063			1.81	6.90	8.71	10.60
9502	cycle 40 miles		108	.074			2.11	8.05	10.16	12.35
9504	cycle 50 miles		90	.089			2.53	9.65	12.18	14.80
9506	50 MPH ave, cycle 10 miles		234	.034			.97	3.71	4.68	5.70
9508	cycle 20 miles		180	.044			1.26	4.83	6.09	7.40
9510	cycle 30 miles		144	.056			1.58	6.05	7.63	9.25
9512	cycle 40 miles		108	.074			2.11	8.05	10.16	12.35
9514	cycle 50 miles		90	.089			2.53	9.65	12.18	14.80
9614	30 min. wait/Ld./Uld.,15 MPH, cycle 0.5 mi.		270	.030			.84	3.22	4.06	4.93
9616	cycle 1 mile		252	.032			.90	3.45	4.35	5.30
9618	cycle 2 miles		216	.037			1.05	4.02	5.07	6.15
9620	cycle 4 miles		180	.044			1.26	4.83	6.09	7.40
9622	cycle 6 miles		162	.049			1.40	5.35	6.75	8.20
9624	cycle 8 miles		126	.063			1.81	6.90	8.71	10.60
9625	cycle 10 miles		126	.063			1.81	6.90	8.71	10.60
9626	20 MPH ave, cycle 0.5 mile		270	.030			.84	3.22	4.06	4.93

31 23 Excavation and Fill

31 23 23 – Fill

31 23 23.20 Hauling

		Crew	Daily Output	Labor-Hours	Unit	Material	2015 Bare Costs Labor	2015 Bare Costs Equipment	Total	Total Incl O&P
9628	cycle 1 mile	B-34I	252	.032	L.C.Y.		.90	3.45	4.35	5.30
9630	cycle 2 miles		234	.034			.97	3.71	4.68	5.70
9632	cycle 4 miles		198	.040			1.15	4.39	5.54	6.75
9634	cycle 6 miles		180	.044			1.26	4.83	6.09	7.40
9636	cycle 8 miles		162	.049			1.40	5.35	6.75	8.20
9638	cycle 10 miles		144	.056			1.58	6.05	7.63	9.25
9640	25 MPH ave, cycle 4 miles		216	.037			1.05	4.02	5.07	6.15
9642	cycle 6 miles		198	.040			1.15	4.39	5.54	6.75
9644	cycle 8 miles		180	.044			1.26	4.83	6.09	7.40
9646	cycle 10 miles		162	.049			1.40	5.35	6.75	8.20
9650	30 MPH ave, cycle 4 miles		216	.037			1.05	4.02	5.07	6.15
9652	cycle 6 miles		198	.040			1.15	4.39	5.54	6.75
9654	cycle 8 miles		180	.044			1.26	4.83	6.09	7.40
9656	cycle 10 miles		162	.049			1.40	5.35	6.75	8.20
9660	35 MPH ave, cycle 4 miles		234	.034			.97	3.71	4.68	5.70
9662	cycle 6 miles		216	.037			1.05	4.02	5.07	6.15
9664	cycle 8 miles		198	.040			1.15	4.39	5.54	6.75
9666	cycle 10 miles		180	.044			1.26	4.83	6.09	7.40
9668	cycle 20 miles		126	.063			1.81	6.90	8.71	10.60
9670	cycle 30 miles		108	.074			2.11	8.05	10.16	12.35
9672	cycle 40 miles		90	.089			2.53	9.65	12.18	14.80
9674	40 MPH ave, cycle 6 miles		216	.037			1.05	4.02	5.07	6.15
9676	cycle 8 miles		198	.040			1.15	4.39	5.54	6.75
9678	cycle 10 miles		180	.044			1.26	4.83	6.09	7.40
9680	cycle 20 miles		144	.056			1.58	6.05	7.63	9.25
9682	cycle 30 miles		108	.074			2.11	8.05	10.16	12.35
9684	cycle 40 miles		90	.089			2.53	9.65	12.18	14.80
9686	cycle 50 miles		72	.111			3.16	12.05	15.21	18.50
9694	45 MPH ave, cycle 8 miles		216	.037			1.05	4.02	5.07	6.15
9696	cycle 10 miles		198	.040			1.15	4.39	5.54	6.75
9698	cycle 20 miles		144	.056			1.58	6.05	7.63	9.25
9700	cycle 30 miles		126	.063			1.81	6.90	8.71	10.60
9702	cycle 40 miles		108	.074			2.11	8.05	10.16	12.35
9704	cycle 50 miles		90	.089			2.53	9.65	12.18	14.80
9706	50 MPH ave, cycle 10 miles		198	.040			1.15	4.39	5.54	6.75
9708	cycle 20 miles		162	.049			1.40	5.35	6.75	8.20
9710	cycle 30 miles		126	.063			1.81	6.90	8.71	10.60
9712	cycle 40 miles		108	.074			2.11	8.05	10.16	12.35
9714	cycle 50 miles		90	.089			2.53	9.65	12.18	14.80

31 23 23.24 Compaction, Structural

		Crew	Daily Output	Labor-Hours	Unit	Material	Labor	Equipment	Total	Total Incl O&P
0010	**COMPACTION, STRUCTURAL**									
0020	Steel wheel tandem roller, 5 tons	B-10E	8	1	Hr.		35	19.70	54.70	79.50
0050	Air tamp, 6" to 8" lifts, common fill	B-9	250	.160	E.C.Y.		3.98	.93	4.91	7.70
0060	Select fill	"	300	.133			3.31	.78	4.09	6.40
0600	Vibratory plate, 8" lifts, common fill	A-1D	200	.040			.98	.18	1.16	1.84
0700	Select fill	"	216	.037			.91	.17	1.08	1.70

31 25 Erosion and Sedimentation Controls

31 25 14 – Stabilization Measures for Erosion and Sedimentation Control

31 25 14.16 Rolled Erosion Control Mats and Blankets

	31 25 14.16 Rolled Erosion Control Mats and Blankets		Crew	Daily Output	Labor-Hours	Unit	Material	2015 Bare Costs Labor	2015 Bare Costs Equipment	Total	Total Incl O&P
0010	**ROLLED EROSION CONTROL MATS AND BLANKETS**										
0020	Jute mesh, 100 S.Y. per roll, 4' wide, stapled	G	B-80A	2400	.010	S.Y.	1.03	.24	.13	1.40	1.68
0100	Plastic netting, stapled, 2" x 1" mesh, 20 mil	G	B-1	2500	.010		.24	.24		.48	.66
0200	Polypropylene mesh, stapled, 6.5 oz./S.Y.	G		2500	.010		2.30	.24		2.54	2.93
0300	Tobacco netting, or jute mesh #2, stapled	G	↓	2500	.010	↓	.19	.24		.43	.61
1000	Silt fence, install and maintain, remove	G	B-62	1300	.018	L.F.	.72	.51	.13	1.36	1.79
1100	Allow 25% per month for maintenance; 6-month max life										

31 31 Soil Treatment

31 31 16 – Termite Control

31 31 16.13 Chemical Termite Control

	31 31 16.13 Chemical Termite Control	Crew	Daily Output	Labor-Hours	Unit	Material	2015 Bare Costs Labor	2015 Bare Costs Equipment	Total	Total Incl O&P
0010	**CHEMICAL TERMITE CONTROL**									
0020	Slab and walls, residential	1 Skwk	1200	.007	SF Flr.	.32	.22		.54	.72
0400	Insecticides for termite control, minimum		14.20	.563	Gal.	68.50	18.80		87.30	107
0500	Maximum	↓	11	.727	"	117	24.50		141.50	170

Division Notes

		CREW	DAILY OUTPUT	LABOR-HOURS	UNIT	BARE COSTS				TOTAL INCL O&P
						MAT.	LABOR	EQUIP.	TOTAL	

Estimating Tips

32 01 00 Operations and Maintenance of Exterior Improvements

- Recycling of asphalt pavement is becoming very popular and is an alternative to removal and replacement. It can be a good value engineering proposal if removed pavement can be recycled, either at the project site or at another site that is reasonably close to the project site. Sections on repair of flexible and rigid pavement are included.

32 10 00 Bases, Ballasts, and Paving

- When estimating paving, keep in mind the project schedule. Also note that prices for asphalt and concrete are generally higher in the cold seasons. Lines for pavement markings, including tactile warning systems and fence lines, are included.

32 90 00 Planting

- The timing of planting and guarantee specifications often dictate the costs for establishing tree and shrub growth and a stand of grass or ground cover. Establish the work performance schedule to coincide with the local planting season. Maintenance and growth guarantees can add from 20%–100% to the total landscaping cost and can be contractually cumbersome. The cost to replace trees and shrubs can be as high as 5% of the total cost, depending on the planting zone, soil conditions, and time of year.

Reference Numbers

Reference numbers are shown in shaded boxes at the beginning of some major classifications. These numbers refer to related items in the Reference Section. The reference information may be an estimating procedure, an alternate pricing method, or technical information.

Note: Not all subdivisions listed here necessarily appear in this publication. ■

Did you know?
RSMeans Online gives you the same access to RSMeans' data with 24/7 access:
- Quickly locate costs in the searchable database.
- Build cost lists, estimates, and reports in minutes.
- Adjust costs to any location in the U.S. and Canada with the click of a button.

Start your free trial today at **www.rsmeansonline.com**

RSMeansOnline

No part of this publication may be reproduced, stored in a retrieval system, or transmitted in any form or by any means without prior written permission of RSMeans.

32 01 Operation and Maintenance of Exterior Improvements

32 01 13 – Flexible Paving Surface Treatment

32 01 13.66 Fog Seal

		Crew	Daily Output	Labor-Hours	Unit	Material	2015 Bare Costs Labor	2015 Bare Costs Equipment	Total	Total Incl O&P
0010	**FOG SEAL**									
0012	Sealcoating, 2 coat coal tar pitch emulsion over 10,000 S.Y.	B-45	5000	.003	S.Y.	.90	.08	.06	1.04	1.20
0030	1000 to 10,000 S.Y.	"	3000	.005		.90	.14	.10	1.14	1.33
0100	Under 1000 S.Y.	B-1	1050	.023		.90	.57		1.47	1.95
0300	Petroleum resistant, over 10,000 S.Y.	B-45	5000	.003		1.30	.08	.06	1.44	1.64
0320	1000 to 10,000 S.Y.	"	3000	.005		1.30	.14	.10	1.54	1.77
0400	Under 1000 S.Y.	B-1	1050	.023		1.30	.57		1.87	2.39

32 06 Schedules for Exterior Improvements

32 06 10 – Schedules for Bases, Ballasts, and Paving

32 06 10.10 Sidewalks, Driveways and Patios

		Crew	Daily Output	Labor-Hours	Unit	Material	Labor	Equipment	Total	Total Incl O&P
0010	**SIDEWALKS, DRIVEWAYS AND PATIOS** No base									
0021	Asphaltic concrete, 2" thick	B-37	6480	.007	S.F.	.82	.19	.02	1.03	1.27
0101	2-1/2" thick	"	5950	.008	"	1.04	.21	.03	1.28	1.54
0300	Concrete, 3000 psi, CIP, 6 x 6 - W1.4 x W1.4 mesh,									
0310	broomed finish, no base, 4" thick	B-24	600	.040	S.F.	1.69	1.18		2.87	3.82
0350	5" thick		545	.044		2.26	1.30		3.56	4.65
0400	6" thick		510	.047		2.64	1.39		4.03	5.20
0450	For bank run gravel base, 4" thick, add	B-18	2500	.010		.52	.24	.02	.78	.99
0520	8" thick, add	"	1600	.015		1.04	.38	.03	1.45	1.81
1000	Crushed stone, 1" thick, white marble	2 Clab	1700	.009		.44	.23		.67	.88
1050	Bluestone	"	1700	.009		.13	.23		.36	.53
1700	Redwood, prefabricated, 4' x 4' sections	2 Carp	316	.051		4.77	1.66		6.43	8.05
1750	Redwood planks, 1" thick, on sleepers	"	240	.067		4.77	2.19		6.96	8.95
2250	Stone dust, 4" thick	B-62	900	.027	S.Y.	3.33	.73	.19	4.25	5.10

32 06 10.20 Steps

		Crew	Daily Output	Labor-Hours	Unit	Material	Labor	Equipment	Total	Total Incl O&P
0010	**STEPS**									
0011	Incl. excav., borrow & concrete base as required									
0100	Brick steps	B-24	35	.686	LF Riser	15.35	20.50		35.85	50.50
0200	Railroad ties	2 Clab	25	.640		3.59	15.65		19.24	30.50
0300	Bluestone treads, 12" x 2" or 12" x 1-1/2"	B-24	30	.800		33.50	23.50		57	76
0600	Precast concrete, see Section 03 41 23.50									
4025	Steel edge strips, incl. stakes, 1/4" x 5"	B-1	390	.062	L.F.	3.92	1.55		5.47	6.90
4050	Edging, landscape timber or railroad ties, 6" x 8"	2 Carp	170	.094	"	2.79	3.09		5.88	8.25

32 11 Base Courses

32 11 23 – Aggregate Base Courses

32 11 23.23 Base Course Drainage Layers

		Crew	Daily Output	Labor-Hours	Unit	Material	Labor	Equipment	Total	Total Incl O&P
0010	**BASE COURSE DRAINAGE LAYERS**									
0011	For roadways and large areas									
0051	3/4" stone compacted to 3" deep	B-36	36000	.001	S.F.	.26	.03	.05	.34	.38
0101	6" deep		35100	.001		.51	.03	.05	.59	.67
0201	9" deep		25875	.002		.75	.05	.06	.86	.97
0305	12" deep		21150	.002		1.49	.05	.08	1.62	1.81
0306	Crushed 1-1/2" stone base, compacted to 4" deep		47000	.001		.05	.02	.04	.11	.14
0307	6" deep		35100	.001		.74	.03	.05	.82	.93
0308	8" deep		27000	.001		.99	.04	.06	1.09	1.23
0309	12" deep		16200	.002		1.49	.07	.10	1.66	1.86
0350	Bank run gravel, spread and compacted									

32 11 Base Courses

32 11 23 – Aggregate Base Courses

32 11 23.23 Base Course Drainage Layers

		Crew	Daily Output	Labor-Hours	Unit	Material	2015 Bare Costs Labor	Equipment	Total	Total Incl O&P
0371	6" deep	B-32	54000	.001	S.F.	.48	.02	.04	.54	.61
0391	9" deep	↓	39600	.001		.70	.03	.06	.79	.88
0401	12" deep		32400	.001	↓	.96	.03	.07	1.06	1.19
6900	For small and irregular areas, add						50%	50%		

32 11 26 – Asphaltic Base Courses

32 11 26.19 Bituminous-Stabilized Base Courses

		Crew	Daily Output	Labor-Hours	Unit	Material	Labor	Equipment	Total	Total Incl O&P
0010	**BITUMINOUS-STABILIZED BASE COURSES**									
0020	And large paved areas									
0700	Liquid application to gravel base, asphalt emulsion	B-45	6000	.003	Gal.	4.29	.07	.05	4.41	4.90
0800	Prime and seal, cut back asphalt		6000	.003	"	5.05	.07	.05	5.17	5.75
1000	Macadam penetration crushed stone, 2 gal. per S.Y., 4" thick	↓	6000	.003	S.Y.	8.60	.07	.05	8.72	9.65
1100	6" thick, 3 gal. per S.Y.		4000	.004		12.85	.11	.08	13.04	14.40
1200	8" thick, 4 gal. per S.Y.	↓	3000	.005	↓	17.15	.14	.10	17.39	19.25
8900	For small and irregular areas, add						50%	50%		

32 12 Flexible Paving

32 12 16 – Asphalt Paving

32 12 16.14 Paving Asphaltic Concrete

		Crew	Daily Output	Labor-Hours	Unit	Material	Labor	Equipment	Total	Total Incl O&P
0010	**PAVING ASPHALTIC CONCRETE**									
0020	6" stone base, 2" binder course, 1" topping	B-25C	9000	.005	S.F.	1.86	.15	.26	2.27	2.59
0025	2" binder course, 2" topping		9000	.005		2.28	.15	.26	2.69	3.05
0030	3" binder course, 2" topping		9000	.005		2.71	.15	.26	3.12	3.52
0035	4" binder course, 2" topping		9000	.005		3.13	.15	.26	3.54	3.98
0040	1.5" binder course, 1" topping		9000	.005		1.65	.15	.26	2.06	2.36
0042	3" binder course, 1" topping		9000	.005		2.29	.15	.26	2.70	3.06
0045	3" binder course, 3" topping		9000	.005		3.13	.15	.26	3.54	3.98
0050	4" binder course, 3" topping		9000	.005		3.55	.15	.26	3.96	4.44
0055	4" binder course, 4" topping		9000	.005		3.96	.15	.26	4.37	4.90
0300	Binder course, 1-1/2" thick		35000	.001		.64	.04	.07	.75	.83
0400	2" thick		25000	.002		.82	.05	.09	.96	1.10
0500	3" thick		15000	.003		1.27	.09	.16	1.52	1.72
0600	4" thick		10800	.004		1.67	.13	.22	2.02	2.28
0800	Sand finish course, 3/4" thick		41000	.001		.32	.03	.06	.41	.47
0900	1" thick	↓	34000	.001		.40	.04	.07	.51	.59
1000	Fill pot holes, hot mix, 2" thick	B-16	4200	.008		.85	.20	.16	1.21	1.45
1100	4" thick		3500	.009		1.25	.24	.20	1.69	1.99
1120	6" thick	↓	3100	.010		1.67	.27	.22	2.16	2.54
1140	Cold patch, 2" thick	B-51	3000	.016		.99	.41	.08	1.48	1.86
1160	4" thick		2700	.018		1.89	.45	.09	2.43	2.93
1180	6" thick	↓	1900	.025	↓	2.94	.64	.13	3.71	4.45

32 13 Rigid Paving

32 13 13 – Concrete Paving

32 13 13.23 Concrete Paving Surface Treatment

		Crew	Daily Output	Labor-Hours	Unit	Material	2015 Bare Costs Labor	Equipment	Total	Total Incl O&P
0010	**CONCRETE PAVING SURFACE TREATMENT**									
0015	Including joints, finishing and curing									
0021	Fixed form, 12' pass, unreinforced, 6" thick	B-26	18000	.005	S.F.	3.18	.14	.20	3.52	3.95
0101	8" thick	"	13500	.007		4.29	.18	.26	4.73	5.30
0701	Finishing, broom finish small areas	2 Cefi	1215	.013	↓		.41		.41	.67

32 14 Unit Paving

32 14 13 – Precast Concrete Unit Paving

32 14 13.18 Precast Concrete Plantable Pavers

		Crew	Daily Output	Labor-Hours	Unit	Material	Labor	Equipment	Total	Total Incl O&P
0010	**PRECAST CONCRETE PLANTABLE PAVERS** (50% grass)									
0300	3/4" crushed stone base for plantable pavers, 6 inch depth	B-62	1000	.024	S.Y.	4.61	.66	.17	5.44	6.35
0400	8 inch depth		900	.027		6.15	.73	.19	7.07	8.25
0500	10 inch depth		800	.030		7.70	.82	.22	8.74	10.05
0600	12 inch depth	↓	700	.034	↓	9.20	.94	.25	10.39	12
0700	Hydro seeding plantable pavers	B-81A	20	.800	M.S.F.	12.45	21	21.50	54.95	72
0800	Apply fertilizer and seed to plantable pavers	1 Clab	8	1	"	50.50	24.50		75	96.50

32 14 16 – Brick Unit Paving

32 14 16.10 Brick Paving

		Crew	Daily Output	Labor-Hours	Unit	Material	Labor	Equipment	Total	Total Incl O&P
0010	**BRICK PAVING**									
0012	4" x 8" x 1-1/2", without joints (4.5 brick/S.F.)	D-1	110	.145	S.F.	2.21	4.29		6.50	9.60
0100	Grouted, 3/8" joint (3.9 brick/S.F.)		90	.178		1.95	5.25		7.20	10.90
0200	4" x 8" x 2-1/4", without joints (4.5 bricks/S.F.)		110	.145		2.25	4.29		6.54	9.65
0300	Grouted, 3/8" joint (3.9 brick/S.F.)		90	.178		1.95	5.25		7.20	10.90
0455	Pervious brick paving, 4" x 8" x 3-1/4", without joints (4.5 bricks/S.F.)	↓	110	.145		3.38	4.29		7.67	10.85
0500	Bedding, asphalt, 3/4" thick	B-25	5130	.017		.70	.47	.53	1.70	2.14
0540	Course washed sand bed, 1" thick	B-18	5000	.005		.32	.12	.01	.45	.56
0580	Mortar, 1" thick	D-1	300	.053		.74	1.57		2.31	3.43
0620	2" thick		200	.080		1.48	2.36		3.84	5.55
1500	Brick on 1" thick sand bed laid flat, 4.5 per S.F.		100	.160		3.18	4.72		7.90	11.35
2000	Brick pavers, laid on edge, 7.2 per S.F.	↓	70	.229	↓	3.86	6.75		10.61	15.50

32 14 23 – Asphalt Unit Paving

32 14 23.10 Asphalt Blocks

		Crew	Daily Output	Labor-Hours	Unit	Material	Labor	Equipment	Total	Total Incl O&P
0010	**ASPHALT BLOCKS**									
0020	Rectangular, 6" x 12" x 1-1/4", w/bed & neopr. adhesive	D-1	135	.119	S.F.	9.30	3.49		12.79	16.05
0100	3" thick		130	.123		13	3.63		16.63	20.50
0300	Hexagonal tile, 8" wide, 1-1/4" thick		135	.119		9.30	3.49		12.79	16.05
0400	2" thick		130	.123		13	3.63		16.63	20.50
0500	Square, 8" x 8", 1-1/4" thick		135	.119		9.30	3.49		12.79	16.05
0600	2" thick		130	.123	↓	13	3.63		16.63	20.50

32 14 40 – Stone Paving

32 14 40.10 Stone Pavers

		Crew	Daily Output	Labor-Hours	Unit	Material	Labor	Equipment	Total	Total Incl O&P
0010	**STONE PAVERS**									
1100	Flagging, bluestone, irregular, 1" thick,	D-1	81	.198	S.F.	6.40	5.80		12.20	16.75
1150	Snapped random rectangular, 1" thick		92	.174		9.70	5.15		14.85	19.25
1200	1-1/2" thick		85	.188		11.65	5.55		17.20	22
1250	2" thick		83	.193		13.60	5.70		19.30	24.50
1300	Slate, natural cleft, irregular, 3/4" thick		92	.174		9.20	5.15		14.35	18.65
1310	1" thick		85	.188		10.70	5.55		16.25	21
1351	Random rectangular, gauged, 1/2" thick		105	.152	↓	19.85	4.49		24.34	29.50

32 14 Unit Paving

32 14 40 – Stone Paving

32 14 40.10 Stone Pavers		Crew	Daily Output	Labor-Hours	Unit	Material	2015 Bare Costs Labor	2015 Bare Costs Equipment	Total	Total Incl O&P
1400	Random rectangular, butt joint, gauged, 1/4" thick	D-1	150	.107	S.F.	21.50	3.14		24.64	29
1450	For sand rubbed finish, add					9.30			9.30	10.20
1500	For interior setting, add								25%	25%
1550	Granite blocks, 3-1/2" x 3-1/2" x 3-1/2"	D-1	92	.174	S.F.	12	5.15		17.15	22

32 16 Curbs, Gutters, Sidewalks, and Driveways

32 16 13 – Curbs and Gutters

32 16 13.13 Cast-in-Place Concrete Curbs and Gutters

		Crew	Daily Output	Labor-Hours	Unit	Material	Labor	Equipment	Total	Total Incl O&P
0010	**CAST-IN-PLACE CONCRETE CURBS AND GUTTERS**									
0290	Forms only, no concrete									
0300	Concrete, wood forms, 6" x 18", straight	C-2	500	.096	L.F.	2.99	2.78		5.77	7.95
0400	6" x 18", radius	"	200	.240		3.11	6.95		10.06	15.10
0404	Concrete, wood forms, 6" x 18", straight & concrete	C-2A	500	.096		5.85	3.03		8.88	11.45
0406	6" x 18", radius	"	200	.240		5.95	7.55		13.50	19.20

32 16 13.23 Precast Concrete Curbs and Gutters

0010	**PRECAST CONCRETE CURBS AND GUTTERS**									
0550	Precast, 6" x 18", straight	B-29	700	.069	L.F.	10.30	1.83	1.26	13.39	15.80
0600	6" x 18", radius	"	325	.148	"	10.80	3.94	2.71	17.45	21.50

32 16 13.33 Asphalt Curbs

0010	**ASPHALT CURBS**									
0012	Curbs, asphaltic, machine formed, 8" wide, 6" high, 40 L.F./ton	B-27	1000	.032	L.F.	1.82	.80	.30	2.92	3.67
0100	8" wide, 8" high, 30 L.F. per ton		900	.036		2.43	.89	.33	3.65	4.52
0150	Asphaltic berm, 12" W, 3"-6" H, 35 L.F./ton, before pavement		700	.046		.04	1.14	.42	1.60	2.42
0200	12" W, 1-1/2" to 4" H, 60 L.F. per ton, laid with pavement	B-2	1050	.038		.02	.95		.97	1.62

32 16 13.43 Stone Curbs

0010	**STONE CURBS**									
1000	Granite, split face, straight, 5" x 16"	D-13	275	.175	L.F.	11.65	5.45	1.72	18.82	24
1100	6" x 18"	"	250	.192		15.35	6	1.89	23.24	29
1300	Radius curbing, 6" x 18", over 10' radius	B-29	260	.185		18.75	4.92	3.39	27.06	32.50
1400	Corners, 2' radius	"	80	.600	Ea.	63	16	11.05	90.05	109
1600	Edging, 4-1/2" x 12", straight	d-13	300	.160	L.F.	5.85	5	1.58	12.43	16.50
1800	Curb inlets, (guttermouth) straight	B-29	41	1.171	Ea.	140	31	21.50	192.50	230
2000	Indian granite (belgian block)									
2100	Jumbo, 10-1/2" x 7-1/2" x 4", grey	D-1	150	.107	L.F.	6.15	3.14		9.29	12
2150	Pink		150	.107		7.75	3.14		10.89	13.75
2200	Regular, 9" x 4-1/2" x 4-1/2", grey		160	.100		4.67	2.95		7.62	10.05
2250	Pink		160	.100		6.15	2.95		9.10	11.65
2300	Cubes, 4" x 4" x 4", grey		175	.091		3.96	2.70		6.66	8.85
2350	Pink		175	.091		3.81	2.70		6.51	8.70
2400	6" x 6" x 6", pink		155	.103		12.60	3.04		15.64	18.90
2500	Alternate pricing method for indian granite									
2550	Jumbo, 10-1/2" x 7-1/2" x 4" (30 lb.), grey				Ton	350			350	385
2600	Pink					450			450	495
2650	Regular, 9" x 4-1/2" x 4-1/2" (20 lb.), grey					330			330	365
2700	Pink					430			430	475
2750	Cubes, 4" x 4" x 4" (5 lb.), grey					480			480	530
2800	Pink					490			490	540
2850	6" x 6" x 6" (25 lb.), pink					490			490	540
2900	For pallets, add					22			22	24

32 31 Fences and Gates

32 31 13 – Chain Link Fences and Gates

32 31 13.15 Chain Link Fence	Crew	Daily Output	Labor-Hours	Unit	Material	2015 Bare Costs Labor	Equipment	Total	Total Incl O&P
0010 **CHAIN LINK FENCE**									
0020 1-5/8" post 10' O.C., 1-3/8" top rail, 2" corner post galv. stl., 3' high	B-1	185	.130	L.F.	9.10	3.26		12.36	15.50
0050 4' high		170	.141		9.20	3.55		12.75	16.10
0100 6' high		115	.209	↓	11.35	5.25		16.60	21.50
0150 Add for gate 3' wide, 1-3/8" frame 3' high		12	2	Ea.	89	50		139	183
0170 4' high		10	2.400		98	60.50		158.50	209
0190 6' high		10	2.400		118	60.50		178.50	230
0200 Add for gate 4' wide, 1-3/8" frame 3' high		9	2.667		93.50	67		160.50	215
0220 4' high		9	2.667		102	67		169	224
0240 6' high		8	3	↓	124	75.50		199.50	263
0350 Aluminized steel, 9 ga. wire, 3' high		185	.130	L.F.	7.45	3.26		10.71	13.65
0380 4' high		170	.141		7.85	3.55		11.40	14.55
0400 6' high		115	.209	↓	10.20	5.25		15.45	20
0450 Add for gate 3' wide, 1-3/8" frame 3' high		12	2	Ea.	125	50		175	222
0470 4' high		10	2.400		134	60.50		194.50	249
0490 6' high		10	2.400		169	60.50		229.50	287
0500 Add for gate 4' wide, 1-3/8" frame 3' high		10	2.400		130	60.50		190.50	244
0520 4' high		9	2.667		140	67		207	266
0540 6' high		8	3	↓	178	75.50		253.50	325
0620 Vinyl covered 9 ga. wire, 3' high		185	.130	L.F.	7.25	3.26		10.51	13.45
0640 4' high		170	.141		7.10	3.55		10.65	13.75
0660 6' high		115	.209	↓	9.50	5.25		14.75	19.25
0720 Add for gate 3' wide, 1-3/8" frame 3' high		12	2	Ea.	102	50		152	198
0740 4' high		10	2.400		105	60.50		165.50	216
0760 6' high		10	2.400		129	60.50		189.50	243
0780 Add for gate 4' wide, 1-3/8" frame 3' high		10	2.400		103	60.50		163.50	214
0800 4' high		9	2.667		108	67		175	230
0820 6' high		8	3	↓	139	75.50		214.50	280
0860 Tennis courts, 11 ga. wire, 2 1/2" post 10' O.C., 1-5/8" top rail									
0900 2-1/2" corner post, 10' high	B-1	95	.253	L.F.	7.55	6.35		13.90	19
0920 12' high		80	.300	"	7.80	7.55		15.35	21.50
1000 Add for gate 3' wide, 1-5/8" frame 10' high		10	2.400	Ea.	229	60.50		289.50	355
1040 Aluminized, 11 ga. wire 10' high		95	.253	L.F.	9.10	6.35		15.45	20.50
1100 12' high		80	.300	"	10	7.55		17.55	23.50
1140 Add for gate 3' wide, 1-5/8" frame, 10' high		10	2.400	Ea.	139	60.50		199.50	254
1250 Vinyl covered 11 ga. wire, 10' high		95	.253	L.F.	7.90	6.35		14.25	19.30
1300 12' high		80	.300	"	9.40	7.55		16.95	23
1400 Add for gate 3' wide, 1-3/8" frame, 10' high	↓	10	2.400	Ea.	227	60.50		287.50	350

32 31 23 – Plastic Fences and Gates

32 31 23.10 Fence, Vinyl

	Crew	Daily Output	Labor-Hours	Unit	Material	Labor	Equipment	Total	Total Incl O&P
0010 **FENCE, VINYL**									
0011 White, steel reinforced, stainless steel fasteners									
0020 Picket, 4" x 4" posts @ 6'-0" OC, 3' high	B-1	140	.171	L.F.	22.50	4.31		26.81	32
0030 4' high		130	.185		24.50	4.64		29.14	35
0040 5' high		120	.200		27.50	5		32.50	38.50
0100 Board (semi-privacy), 5" x 5" posts @ 7'-6" OC, 5' high		130	.185		25	4.64		29.64	35.50
0120 6' high		125	.192		28.50	4.82		33.32	39
0200 Basketweave, 5" x 5" posts @ 7'-6" OC, 5' high		160	.150		25.50	3.77		29.27	34.50
0220 6' high		150	.160		28.50	4.02		32.52	38
0300 Privacy, 5" x 5" posts @ 7'-6" OC, 5' high		130	.185		24.50	4.64		29.14	35
0320 6' high		150	.160	↓	28	4.02		32.02	38
0350 Gate, 5' high	↓	9	2.667	Ea.	310	67		377	450

32 31 Fences and Gates

32 31 23 – Plastic Fences and Gates

32 31 23.10 Fence, Vinyl		Crew	Daily Output	Labor-Hours	Unit	Material	2015 Bare Costs Labor	Equipment	Total	Total Incl O&P
0360	6' high	B-1	9	2.667	Ea.	355	67		422	500
0400	For posts set in concrete, add		25	.960		8.45	24		32.45	50
0500	Post and rail fence, 2 rail		150	.160	L.F.	5.90	4.02		9.92	13.25
0510	3 rail		150	.160		7.60	4.02		11.62	15.10
0515	4 rail		150	.160		9.85	4.02		13.87	17.60

32 31 26 – Wire Fences and Gates

32 31 26.10 Fences, Misc. Metal

		Crew	Daily Output	Labor-Hours	Unit	Material	Labor	Equipment	Total	Incl O&P
0010	**FENCES, MISC. METAL**									
0012	Chicken wire, posts @ 4', 1" mesh, 4' high	B-80C	410	.059	L.F.	3.30	1.50	.62	5.42	6.80
0100	2" mesh, 6' high		350	.069		3.82	1.75	.73	6.30	7.90
0200	Galv. steel, 12 ga., 2" x 4" mesh, posts 5' O.C., 3' high		300	.080		2.78	2.04	.85	5.67	7.40
0300	5' high		300	.080		3.38	2.04	.85	6.27	8.05
0400	14 ga., 1" x 2" mesh, 3' high		300	.080		3.42	2.04	.85	6.31	8.10
0500	5' high		300	.080		4.57	2.04	.85	7.46	9.40
1000	Kennel fencing, 1-1/2" mesh, 6' long, 3'-6" wide, 6'-2" high	2 Clab	4	4	Ea.	510	98		608	725
1050	12' long		4	4		720	98		818	955
1200	Top covers, 1-1/2" mesh, 6' long		15	1.067		136	26		162	194
1250	12' long		12	1.333		191	32.50		223.50	265

32 31 29 – Wood Fences and Gates

32 31 29.10 Fence, Wood

		Crew	Daily Output	Labor-Hours	Unit	Material	Labor	Equipment	Total	Incl O&P
0010	**FENCE, WOOD**									
0011	Basket weave, 3/8" x 4" boards, 2" x 4"									
0020	stringers on spreaders, 4" x 4" posts									
0050	No. 1 cedar, 6' high	B-80C	160	.150	L.F.	26	3.83	1.59	31.42	36.50
0070	Treated pine, 6' high		150	.160		34.50	4.09	1.69	40.28	46.50
0090	Vertical weave 6' high		145	.166		22.50	4.23	1.75	28.48	34
0200	Board fence, 1" x 4" boards, 2" x 4" rails, 4" x 4" post									
0220	Preservative treated, 2 rail, 3' high	B-80C	145	.166	L.F.	9.30	4.23	1.75	15.28	19.20
0240	4' high		135	.178		10.90	4.54	1.88	17.32	21.50
0260	3 rail, 5' high		130	.185		11.85	4.72	1.95	18.52	23
0300	6' high		125	.192		13.35	4.91	2.03	20.29	25
0320	No. 2 grade western cedar, 2 rail, 3' high		145	.166		10.20	4.23	1.75	16.18	20
0340	4' high		135	.178		11.25	4.54	1.88	17.67	22
0360	3 rail, 5' high		130	.185		12.60	4.72	1.95	19.27	24
0400	6' high		125	.192		13.45	4.91	2.03	20.39	25
0420	No. 1 grade cedar, 2 rail, 3' high		145	.166		12.30	4.23	1.75	18.28	22.50
0440	4' high		135	.178		13.95	4.54	1.88	20.37	25
0460	3 rail, 5' high		130	.185		17	4.72	1.95	23.67	28.50
0500	6' high		125	.192		18.75	4.91	2.03	25.69	31
0540	Shadow box, 1" x 6" board, 2" x 4" rail, 4" x 4"post									
0560	Pine, pressure treated, 3 rail, 6' high	B-80C	150	.160	L.F.	17.20	4.09	1.69	22.98	27.50
0600	Gate, 3'-6" wide		8	3	Ea.	107	76.50	31.50	215	281
0620	No. 1 cedar, 3 rail, 4' high		130	.185	L.F.	18	4.72	1.95	24.67	30
0640	6' high		125	.192		23	4.91	2.03	29.94	35.50
0860	Open rail fence, split rails, 2 rail 3' high, no. 1 cedar		160	.150		11.75	3.83	1.59	17.17	21
0870	No. 2 cedar		160	.150		10.75	3.83	1.59	16.17	20
0880	3 rail, 4' high, no. 1 cedar		150	.160		12.10	4.09	1.69	17.88	22
0890	No. 2 cedar		150	.160		8.55	4.09	1.69	14.33	18.05
0920	Rustic rails, 2 rail 3' high, no. 1 cedar		160	.150		9.35	3.83	1.59	14.77	18.40
0930	No. 2 cedar		160	.150		8.75	3.83	1.59	14.17	17.80
0940	3 rail, 4' high		150	.160		8.95	4.09	1.69	14.73	18.50
0950	No. 2 cedar		150	.160		7.30	4.09	1.69	13.08	16.70

32 31 Fences and Gates

32 31 29 – Wood Fences and Gates

32 31 29.10 Fence, Wood

		Crew	Daily Output	Labor-Hours	Unit	Material	2015 Bare Costs Labor	Equipment	Total	Total Incl O&P
0960	Picket fence, gothic, pressure treated pine									
1000	2 rail, 3' high	B-80C	140	.171	L.F.	7.15	4.38	1.81	13.34	17.20
1020	3 rail, 4' high		130	.185	"	8.15	4.72	1.95	14.82	19
1040	Gate, 3'-6" wide		9	2.667	Ea.	68	68	28	164	220
1060	No. 2 cedar, 2 rail, 3' high		140	.171	L.F.	8.25	4.38	1.81	14.44	18.35
1100	3 rail, 4' high		130	.185	"	8.35	4.72	1.95	15.02	19.20
1120	Gate, 3'-6" wide		9	2.667	Ea.	72.50	68	28	168.50	225
1140	No. 1 cedar, 2 rail 3' high		140	.171	L.F.	12.55	4.38	1.81	18.74	23
1160	3 rail, 4' high		130	.185		16.35	4.72	1.95	23.02	28
1200	Rustic picket, molded pine, 2 rail, 3' high		140	.171		7.75	4.38	1.81	13.94	17.80
1220	No. 1 cedar, 2 rail, 3' high		140	.171		8.95	4.38	1.81	15.14	19.15
1240	Stockade fence, no. 1 cedar, 3-1/4" rails, 6' high		160	.150		12.05	3.83	1.59	17.47	21.50
1260	8' high		155	.155		16.60	3.96	1.64	22.20	26.50
1300	No. 2 cedar, treated wood rails, 6' high		160	.150		12.35	3.83	1.59	17.77	21.50
1320	Gate, 3'-6" wide		8	3	Ea.	82	76.50	31.50	190	253
1360	Treated pine, treated rails, 6' high		160	.150	L.F.	12.80	3.83	1.59	18.22	22
1400	8' high		150	.160	"	19	4.09	1.69	24.78	29.50

32 31 29.20 Fence, Wood Rail

		Crew	Daily Output	Labor-Hours	Unit	Material	Labor	Equipment	Total	Total Incl O&P
0010	**FENCE, WOOD RAIL**									
0012	Picket, No. 2 cedar, Gothic, 2 rail, 3' high	B-1	160	.150	L.F.	7.60	3.77		11.37	14.70
0050	Gate, 3'-6" wide	B-80C	9	2.667	Ea.	77	68	28	173	230
0400	3 rail, 4' high		150	.160	L.F.	8.50	4.09	1.69	14.28	18
0500	Gate, 3'-6" wide		9	2.667	Ea.	95	68	28	191	249
1200	Stockade, No. 2 cedar, treated wood rails, 6' high		160	.150	L.F.	8.90	3.83	1.59	14.32	17.95
1250	Gate, 3' wide		9	2.667	Ea.	96	68	28	192	251
1300	No. 1 cedar, 3-1/4" cedar rails, 6' high		160	.150	L.F.	20.50	3.83	1.59	25.92	30.50
1500	Gate, 3' wide		9	2.667	Ea.	217	68	28	313	385
2700	Prefabricated redwood or cedar, 4' high		160	.150	L.F.	15.80	3.83	1.59	21.22	25.50
2800	6' high		150	.160		23	4.09	1.69	28.78	34
3300	Board, shadow box, 1" x 6", treated pine, 6' high		160	.150		12.45	3.83	1.59	17.87	22
3400	No. 1 cedar, 6' high		150	.160		24.50	4.09	1.69	30.28	35.50
3900	Basket weave, No. 1 cedar, 6' high		160	.150		34	3.83	1.59	39.42	45
4200	Gate, 3'-6" wide	B-1	9	2.667	Ea.	176	67		243	305
5000	Fence rail, redwood, 2" x 4", merch. grade 8'	"	2400	.010	L.F.	2.42	.25		2.67	3.08

32 32 Retaining Walls

32 32 13 – Cast-in-Place Concrete Retaining Walls

32 32 13.10 Retaining Walls, Cast Concrete

		Crew	Daily Output	Labor-Hours	Unit	Material	Labor	Equipment	Total	Total Incl O&P
0010	**RETAINING WALLS, CAST CONCRETE**									
1800	Concrete gravity wall with vertical face including excavation & backfill									
1850	No reinforcing									
1900	6' high, level embankment	C-17C	36	2.306	L.F.	74.50	78	16.90	169.40	232
2000	33° slope embankment	"	32	2.594	"	86.50	87.50	19.05	193.05	264
2800	Reinforced concrete cantilever, incl. excavation, backfill & reinf.									
2900	6' high, 33° slope embankment	C-17C	35	2.371	L.F.	68	80	17.40	165.40	229

32 32 23 – Segmental Retaining Walls

32 32 23.13 Segmental Conc. Unit Masonry Retaining Walls

0010	**SEGMENTAL CONC. UNIT MASONRY RETAINING WALLS**									
7100	Segmental Retaining Wall system, incl. pins, and void fill									
7120	base and backfill not included									

32 32 Retaining Walls

32 32 23 – Segmental Retaining Walls

32 32 23.13 Segmental Conc. Unit Masonry Retaining Walls		Crew	Daily Output	Labor-Hours	Unit	Material	2015 Bare Costs Labor	Equipment	Total	Total Incl O&P
7140	Large unit, 8" high x 18" wide x 20" deep, 3 plane split	B-62	300	.080	S.F.	13.50	2.20	.58	16.28	19.15
7150	Straight split		300	.080		13.60	2.20	.58	16.38	19.25
7160	Medium, lt. wt., 8" high x 18" wide x 12" deep, 3 plane split		400	.060		10.50	1.65	.43	12.58	14.80
7170	Straight split		400	.060		10.40	1.65	.43	12.48	14.70
7180	Small unit, 4" x 18" x 10" deep, 3 plane split		400	.060		13.40	1.65	.43	15.48	18
7190	Straight split		400	.060		13.20	1.65	.43	15.28	17.75
7200	Cap unit, 3 plane split		300	.080		13.80	2.20	.58	16.58	19.50
7210	Cap unit, straight split	↓	300	.080		13.80	2.20	.58	16.58	19.50
7250	Geo-grid soil reinforcement 4' x 50'	2 Clab	22500	.001		.76	.02		.78	.87
7255	Geo-grid soil reinforcement 6' x 150'	"	22500	.001	↓	.60	.02		.62	.69
8000	For higher walls, add components as necessary									

32 32 26 – Metal Crib Retaining Walls

32 32 26.10 Metal Bin Retaining Walls

		Crew	Daily Output	Labor-Hours	Unit	Material	Labor	Equipment	Total	Total Incl O&P
0010	**METAL BIN RETAINING WALLS**									
0011	Aluminized steel bin, excavation									
0020	and backfill not included, 10' wide									
0100	4' high, 5.5' deep	B-13	650	.074	S.F.	27	1.97	1.13	30.10	34
0200	8' high, 5.5' deep		615	.078		31	2.08	1.20	34.28	39
0300	10' high, 7.7' deep		580	.083		34.50	2.21	1.27	37.98	43
0400	12' high, 7.7' deep		530	.091		37	2.41	1.39	40.80	46.50
0500	16' high, 7.7' deep	↓	515	.093	↓	39.50	2.48	1.43	43.41	48.50

32 32 29 – Timber Retaining Walls

32 32 29.10 Landscape Timber Retaining Walls

		Crew	Daily Output	Labor-Hours	Unit	Material	Labor	Equipment	Total	Total Incl O&P
0010	**LANDSCAPE TIMBER RETAINING WALLS**									
0100	Treated timbers, 6" x 6"	1 Clab	265	.030	L.F.	2.01	.74		2.75	3.45
0110	6" x 8"	"	200	.040	"	2.65	.98		3.63	4.56
0120	Drilling holes in timbers for fastening, 1/2"	1 Carp	450	.018	Inch		.58		.58	.98
0130	5/8"	"	450	.018	"		.58		.58	.98
0140	Reinforcing rods for fastening, 1/2"	1 Clab	312	.026	L.F.	.36	.63		.99	1.44
0150	5/8"	"	312	.026	"	.56	.63		1.19	1.66
0160	Reinforcing fabric	2 Clab	2500	.006	S.Y.	1.90	.16		2.06	2.35
0170	Gravel backfill		28	.571	C.Y.	20	13.95		33.95	45.50
0180	Perforated pipe, 4" diameter with silt sock		1200	.013	L.F.	1.25	.33		1.58	1.93
0190	Galvanized 60d common nails	1 Clab	625	.013	Ea.	.16	.31		.47	.71
0200	20d common nails	"	3800	.002	"	.04	.05		.09	.13

32 32 53 – Stone Retaining Walls

32 32 53.10 Retaining Walls, Stone

		Crew	Daily Output	Labor-Hours	Unit	Material	Labor	Equipment	Total	Total Incl O&P
0010	**RETAINING WALLS, STONE**									
0015	Including excavation, concrete footing and									
0020	stone 3' below grade. Price is exposed face area.									
0200	Decorative random stone, to 6' high, 1'-6" thick, dry set	D-1	35	.457	S.F.	58.50	13.50		72	87
0300	Mortar set		40	.400		60.50	11.80		72.30	86
0500	Cut stone, to 6' high, 1'-6" thick, dry set		35	.457		60.50	13.50		74	89.50
0600	Mortar set		40	.400		61.50	11.80		73.30	87
0800	Random stone, 6' to 10' high, 2' thick, dry set		45	.356		66.50	10.50		77	90.50
0900	Mortar set		50	.320		69	9.45		78.45	91.50
1100	Cut stone, 6' to 10' high, 2' thick, dry set		45	.356		67	10.50		77.50	91
1200	Mortar set	↓	50	.320	↓	69	9.45		78.45	92

32 33 Site Furnishings

32 33 33 – Site Manufactured Planters

32 33 33.10 Planters

		Crew	Daily Output	Labor-Hours	Unit	Material	2015 Bare Costs Labor	Equipment	Total	Total Incl O&P
0010	**PLANTERS**									
0012	Concrete, sandblasted, precast, 48" diameter, 24" high	2 Clab	15	1.067	Ea.	630	26		656	740
0300	Fiberglass, circular, 36" diameter, 24" high		15	1.067		725	26		751	840
1200	Wood, square, 48" side, 24" high		15	1.067		1,400	26		1,426	1,600
1300	Circular, 48" diameter, 30" high		10	1.600		985	39		1,024	1,150
1600	Planter/bench, 72"		5	3.200		3,275	78		3,353	3,725

32 33 43 – Site Seating and Tables

32 33 43.13 Site Seating

		Crew	Daily Output	Labor-Hours	Unit	Material	Labor	Equipment	Total	Total Incl O&P
0010	**SITE SEATING**									
0012	Seating, benches, park, precast conc., w/backs, wood rails, 4' long	2 Clab	5	3.200	Ea.	595	78		673	785
0100	8' long		4	4		955	98		1,053	1,225
0500	Steel barstock pedestals w/backs, 2" x 3" wood rails, 4' long		10	1.600		1,125	39		1,164	1,300
0510	8' long		7	2.286		1,425	56		1,481	1,675
0800	Cast iron pedestals, back & arms, wood slats, 4' long		8	2		385	49		434	500
0820	8' long		5	3.200		1,075	78		1,153	1,300
1700	Steel frame, fir seat, 10' long		10	1.600		355	39		394	460

32 84 Planting Irrigation

32 84 23 – Underground Sprinklers

32 84 23.10 Sprinkler Irrigation System

		Crew	Daily Output	Labor-Hours	Unit	Material	Labor	Equipment	Total	Total Incl O&P
0010	**SPRINKLER IRRIGATION SYSTEM**									
0011	For lawns									
0800	Residential system, custom, 1" supply	B-20	2000	.012	S.F.	.26	.30		.56	.80
0900	1-1/2" supply	"	1800	.013	"	.49	.33		.82	1.10

32 91 Planting Preparation

32 91 13 – Soil Preparation

32 91 13.16 Mulching

		Crew	Daily Output	Labor-Hours	Unit	Material	Labor	Equipment	Total	Total Incl O&P
0010	**MULCHING**									
0100	Aged barks, 3" deep, hand spread	1 Clab	100	.080	S.Y.	3.55	1.96		5.51	7.20
0150	Skid steer loader	B-63	13.50	2.963	M.S.F.	395	72.50	12.85	480.35	570
0200	Hay, 1" deep, hand spread	1 Clab	475	.017	S.Y.	.52	.41		.93	1.26
0250	Power mulcher, small	B-64	180	.089	M.S.F.	58	2.32	2.21	62.53	70
0350	Large	B-65	530	.030	"	58	.79	1.07	59.86	66
0400	Humus peat, 1" deep, hand spread	1 Clab	700	.011	S.Y.	2.42	.28		2.70	3.13
0450	Push spreader	"	2500	.003	"	2.42	.08		2.50	2.79
0550	Tractor spreader	B-66	700	.011	M.S.F.	269	.38	.38	269.76	297
0600	Oat straw, 1" deep, hand spread	1 Clab	475	.017	S.Y.	.60	.41		1.01	1.35
0650	Power mulcher, small	B-64	180	.089	M.S.F.	66.50	2.32	2.21	71.03	80
0700	Large	B-65	530	.030	"	66.50	.79	1.07	68.36	76
0750	Add for asphaltic emulsion	B-45	1770	.009	Gal.	5.80	.24	.17	6.21	7
0800	Peat moss, 1" deep, hand spread	1 Clab	900	.009	S.Y.	2.80	.22		3.02	3.44
0850	Push spreader	"	2500	.003	"	2.80	.08		2.88	3.21
0950	Tractor spreader	B-66	700	.011	M.S.F.	310	.38	.38	310.76	340
1000	Polyethylene film, 6 mil	2 Clab	2000	.008	S.Y.	.46	.20		.66	.84
1100	Redwood nuggets, 3" deep, hand spread	1 Clab	150	.053	"	2.77	1.30		4.07	5.25
1150	Skid steer loader	B-63	13.50	2.963	M.S.F.	310	72.50	12.85	395.35	475
1200	Stone mulch, hand spread, ceramic chips, economy	1 Clab	125	.064	S.Y.	6.85	1.56		8.41	10.20
1250	Deluxe	"	95	.084	"	10.60	2.06		12.66	15.10

32 91 Planting Preparation

32 91 13 – Soil Preparation

32 91 13.16 Mulching

		Crew	Daily Output	Labor-Hours	Unit	Material	2015 Bare Costs Labor	Equipment	Total	Total Incl O&P
1300	Granite chips	B-1	10	2.400	C.Y.	64	60.50		124.50	172
1400	Marble chips		10	2.400		152	60.50		212.50	268
1600	Pea gravel		28	.857		108	21.50		129.50	154
1700	Quartz		10	2.400		188	60.50		248.50	310
1800	Tar paper, 15 lb. felt	1 Clab	800	.010	S.Y.	.49	.24		.73	.94
1900	Wood chips, 2" deep, hand spread	"	220	.036	"	1.60	.89		2.49	3.25
1950	Skid steer loader	B-63	20.30	1.970	M.S.F.	178	48	8.55	234.55	286

32 91 13.26 Planting Beds

		Crew	Daily Output	Labor-Hours	Unit	Material	Labor	Equipment	Total	Total Incl O&P
0010	**PLANTING BEDS**									
0100	Backfill planting pit, by hand, on site topsoil	2 Clab	18	.889	C.Y.		21.50		21.50	36.50
0200	Prepared planting mix, by hand	"	24	.667			16.30		16.30	27.50
0300	Skid steer loader, on site topsoil	B-62	340	.071			1.94	.51	2.45	3.80
0400	Prepared planting mix	"	410	.059			1.61	.42	2.03	3.16
1000	Excavate planting pit, by hand, sandy soil	2 Clab	16	1			24.50		24.50	41
1100	Heavy soil or clay	"	8	2			49		49	82
1200	1/2 C.Y. backhoe, sandy soil	B-11C	150	.107			3.17	2.43	5.60	7.90
1300	Heavy soil or clay	"	115	.139			4.13	3.17	7.30	10.40
2000	Mix planting soil, incl. loam, manure, peat, by hand	2 Clab	60	.267		43	6.50		49.50	58
2100	Skid steer loader	B-62	150	.160		43	4.40	1.16	48.56	55.50
3000	Pile sod, skid steer loader	"	2800	.009	S.Y.		.24	.06	.30	.46
3100	By hand	2 Clab	400	.040			.98		.98	1.64
4000	Remove sod, F.E. loader	B-10S	2000	.004			.14	.19	.33	.44
4100	Sod cutter	B-12K	3200	.005			.15	.31	.46	.59
4200	By hand	2 Clab	240	.067			1.63		1.63	2.74

32 91 19 – Landscape Grading

32 91 19.13 Topsoil Placement and Grading

		Crew	Daily Output	Labor-Hours	Unit	Material	Labor	Equipment	Total	Total Incl O&P
0010	**TOPSOIL PLACEMENT AND GRADING**									
0300	Fine grade, base course for paving, see Section 32 11 23.23									
0701	Furnish and place, truck dumped, unscreened, 4" deep	B-10S	12000	.001	S.F.	.41	.02	.03	.46	.52
0801	6" deep	"	7400	.001	"	.49	.04	.05	.58	.66
0900	Fine grading and seeding, incl. lime, fertilizer & seed,									
1001	With equipment	B-14	9000	.005	S.F.	.07	.14	.04	.25	.35

32 92 Turf and Grasses

32 92 19 – Seeding

32 92 19.13 Mechanical Seeding

		Crew	Daily Output	Labor-Hours	Unit	Material	Labor	Equipment	Total	Total Incl O&P
0010	**MECHANICAL SEEDING**									
0020	Mechanical seeding, 215 lb./acre	B-66	1.50	5.333	Acre	560	179	175	914	1,100
0100	44 lb./M.S.Y.	"	2500	.003	S.Y.	.18	.11	.11	.40	.50
0101	$2.00/lb., 44 lb./M.S.Y.	1 Clab	13950	.001	S.F.	.02	.01		.03	.04
0300	Fine grading and seeding incl. lime, fertilizer & seed,									
0310	with equipment	B-14	1000	.048	S.Y.	.44	1.26	.36	2.06	2.99
0400	Fertilizer hand push spreader, 35 lb. per M.S.F.	1 Clab	200	.040	M.S.F.	9.80	.98		10.78	12.45
0600	Limestone hand push spreader, 50 lb. per M.S.F.		180	.044		5.40	1.09		6.49	7.70
0800	Grass seed hand push spreader, 4.5 lb. per M.S.F.		180	.044		20	1.09		21.09	24
1000	Hydro or air seeding for large areas, incl. seed and fertilizer	B-81	8900	.002	S.Y.	.43	.05	.08	.56	.64
1100	With wood fiber mulch added	"	8900	.002	"	1.76	.05	.08	1.89	2.11
1300	Seed only, over 100 lb., field seed, minimum				Lb.	1.75			1.75	1.93
1400	Maximum					1.75			1.75	1.93
1500	Lawn seed, minimum					1.41			1.41	1.55

32 92 Turf and Grasses

32 92 19 – Seeding

32 92 19.13 Mechanical Seeding

		Crew	Daily Output	Labor-Hours	Unit	Material	2015 Bare Costs Labor	Equipment	Total	Total Incl O&P
1600	Maximum				Lb.	2.46			2.46	2.71
1800	Aerial operations, seeding only, field seed	B-58	50	.480	Acre	570	13.20	63	646.20	715
1900	Lawn seed		50	.480		460	13.20	63	536.20	595
2100	Seed and liquid fertilizer, field seed		50	.480		690	13.20	63	766.20	850
2200	Lawn seed		50	.480		580	13.20	63	656.20	725

32 92 23 – Sodding

32 92 23.10 Sodding Systems

		Crew	Daily Output	Labor-Hours	Unit	Material	Labor	Equipment	Total	Total Incl O&P
0010	**SODDING SYSTEMS**									
0020	Sodding, 1" deep, bluegrass sod, on level ground, over 8 M.S.F.	B-63	22	1.818	M.S.F.	243	44.50	7.90	295.40	350
0200	4 M.S.F.		17	2.353		254	57.50	10.20	321.70	385
0300	1000 S.F.		13.50	2.963		295	72.50	12.85	380.35	460
0500	Sloped ground, over 8 M.S.F.		6	6.667		243	163	29	435	575
0600	4 M.S.F.		5	8		254	196	34.50	484.50	645
0700	1000 S.F.		4	10		295	245	43.50	583.50	785
1000	Bent grass sod, on level ground, over 6 M.S.F.		20	2		252	49	8.70	309.70	370
1100	3 M.S.F.		18	2.222		261	54.50	9.65	325.15	390
1200	Sodding 1000 S.F. or less		14	2.857		286	70	12.40	368.40	445
1500	Sloped ground, over 6 M.S.F.		15	2.667		252	65	11.55	328.55	400
1600	3 M.S.F.		13.50	2.963		261	72.50	12.85	346.35	425
1700	1000 S.F.		12	3.333		286	81.50	14.45	381.95	470

32 93 Plants

32 93 13 – Ground Covers

32 93 13.10 Ground Cover Plants

		Crew	Daily Output	Labor-Hours	Unit	Material	Labor	Equipment	Total	Total Incl O&P
0010	**GROUND COVER PLANTS**									
0012	Plants, pachysandra, in prepared beds	B-1	15	1.600	C	62.50	40		102.50	137
0200	Vinca minor, 1 yr., bare root, in prepared beds		12	2	"	111	50		161	207
0600	Stone chips, in 50 lb. bags, Georgia marble		520	.046	Bag	3.76	1.16		4.92	6.10
0700	Onyx gemstone		260	.092		16.90	2.32		19.22	22.50
0800	Quartz		260	.092		16.80	2.32		19.12	22.50
0900	Pea gravel, truckload lots		28	.857	Ton	25.50	21.50		47	64

32 93 33 – Shrubs

32 93 33.10 Shrubs and Trees

		Crew	Daily Output	Labor-Hours	Unit	Material	Labor	Equipment	Total	Total Incl O&P
0010	**SHRUBS AND TREES**									
0011	Evergreen, in prepared beds, B & B									
0100	Arborvitae pyramidal, 4'-5'	B-17	30	1.067	Ea.	100	29.50	26	155.50	189
0150	Globe, 12"-15"	B-1	96	.250		21.50	6.30		27.80	34
0300	Cedar, blue, 8'-10'	B-17	18	1.778		231	49.50	43	323.50	385
0500	Hemlock, Canadian, 2-1/2'-3'	B-1	36	.667		31	16.75		47.75	62.50
0550	Holly, Savannah, 8'-10' H		9.68	2.479		257	62.50		319.50	385
0600	Juniper, andorra, 18"-24"		80	.300		36.50	7.55		44.05	52.50
0620	Wiltoni, 15"-18"		80	.300		26	7.55		33.55	41
0640	Skyrocket, 4-1/2'-5'	B-17	55	.582		107	16.15	14.10	137.25	160
0660	Blue pfitzer, 2'-2-1/2'	B-1	44	.545		38.50	13.70		52.20	65.50
0680	Ketleerie, 2-1/2'-3'		50	.480		52	12.05		64.05	77
0700	Pine, black, 2-1/2'-3'		50	.480		59.50	12.05		71.55	85
0720	Mugo, 18"-24"		60	.400		59	10.05		69.05	82
0740	White, 4'-5'	B-17	75	.427		51	11.85	10.35	73.20	87
0800	Spruce, blue, 18"-24"	B-1	60	.400		66.50	10.05		76.55	90.50
0840	Norway, 4'-5'	B-17	75	.427		83	11.85	10.35	105.20	122

32 93 Plants

32 93 33 – Shrubs

32 93 33.10 Shrubs and Trees

		Crew	Daily Output	Labor-Hours	Unit	Material	2015 Bare Costs Labor	Equipment	Total	Total Incl O&P
0900	Yew, denisforma, 12"-15"	B-1	60	.400	Ea.	35	10.05		45.05	55.50
1000	Capitata, 18"-24"		30	.800		32.50	20		52.50	69.50
1100	Hicksi, 2'-2-1/2'		30	.800		78.50	20		98.50	120

32 93 33.20 Shrubs

		Crew	Daily Output	Labor-Hours	Unit	Material	Labor	Equipment	Total	Total Incl O&P
0010	**SHRUBS**									
0011	Broadleaf Evergreen, planted in prepared beds									
0100	Andromeda, 15"-18", container	B-1	96	.250	Ea.	32.50	6.30		38.80	46.50
0200	Azalea, 15" - 18", container		96	.250		29.50	6.30		35.80	43
0300	Barberry, 9"-12", container		130	.185		17.60	4.64		22.24	27
0400	Boxwood, 15"-18", B&B		96	.250		43	6.30		49.30	57.50
0500	Euonymus, emerald gaiety, 12" to 15", container		115	.209		24	5.25		29.25	35
0600	Holly, 15"-18", B & B		96	.250		36	6.30		42.30	50.50
0900	Mount laurel, 18" - 24", B & B		80	.300		70.50	7.55		78.05	90
1000	Paxistema, 9 – 12" high		130	.185		21	4.64		25.64	31.50
1100	Rhododendron, 18"-24", container		48	.500		37.50	12.55		50.05	62.50
1200	Rosemary, 1 gal. container		600	.040		17.70	1		18.70	21
2000	Deciduous, planted in prepared beds, amelanchier, 2'-3', B & B		57	.421		119	10.60		129.60	149
2100	Azalea, 15"-18", B & B		96	.250		29	6.30		35.30	42.50
2300	Bayberry, 2'-3', B & B		57	.421		24	10.60		34.60	44.50
2600	Cotoneaster, 15"-18", B & B		80	.300		26.50	7.55		34.05	41.50
2800	Dogwood, 3'-4', B & B	B-17	40	.800		32	22	19.40	73.40	93.50
2900	Euonymus, alatus compacta, 15" to 18", container	B-1	80	.300		26	7.55		33.55	41.50
3200	Forsythia, 2'-3', container	"	60	.400		18.45	10.05		28.50	37.50
3300	Hibiscus, 3'-4', B & B	B-17	75	.427		48	11.85	10.35	70.20	84
3400	Honeysuckle, 3'-4', B & B	B-1	60	.400		26.50	10.05		36.55	46
3500	Hydrangea, 2'-3', B & B	"	57	.421		30	10.60		40.60	51
3600	Lilac, 3'-4', B & B	B-17	40	.800		28	22	19.40	69.40	89
3900	Privet, bare root, 18"-24"	B-1	80	.300		14.50	7.55		22.05	28.50
4100	Quince, 2'-3', B & B	"	57	.421		28	10.60		38.60	49
4200	Russian olive, 3'-4', B & B	B-17	75	.427		29	11.85	10.35	51.20	62.50
4400	Spirea, 3'-4', B & B	B-1	70	.343		20.50	8.60		29.10	37
4500	Viburnum, 3'-4', B & B	B-17	40	.800		25	22	19.40	66.40	86

32 93 43 – Trees

32 93 43.20 Trees

			Crew	Daily Output	Labor-Hours	Unit	Material	Labor	Equipment	Total	Total Incl O&P
0010	**TREES**										
0011	Deciduous, in prep. beds, balled & burlapped (B&B)										
0100	Ash, 2" caliper	G	B-17	8	4	Ea.	186	111	97	394	495
0200	Beech, 5'-6'	G		50	.640		187	17.75	15.50	220.25	253
0300	Birch, 6'-8', 3 stems	G		20	1.600		168	44.50	39	251.50	300
0500	Crabapple, 6'-8'	G		20	1.600		139	44.50	39	222.50	269
0600	Dogwood, 4'-5'	G		40	.800		131	22	19.40	172.40	203
0700	Eastern redbud 4'-5'	G		40	.800		149	22	19.40	190.40	223
0800	Elm, 8'-10'	G		20	1.600		258	44.50	39	341.50	400
0900	Ginkgo, 6'-7'	G		24	1.333		148	37	32.50	217.50	260
1000	Hawthorn, 8'-10', 1" caliper	G		20	1.600		162	44.50	39	245.50	295
1100	Honeylocust, 10'-12', 1-1/2" caliper	G		10	3.200		206	88.50	77.50	372	460
1300	Larch, 8'	G		32	1		127	27.50	24	178.50	213
1400	Linden, 8'-10', 1" caliper	G		20	1.600		144	44.50	39	227.50	275
1500	Magnolia, 4'-5'	G		20	1.600		101	44.50	39	184.50	228
1600	Maple, red, 8'-10', 1-1/2" caliper	G		10	3.200		202	88.50	77.50	368	455
1700	Mountain ash, 8'-10', 1" caliper	G		16	2		176	55.50	48.50	280	340
1800	Oak, 2-1/2"-3" caliper	G		6	5.333		325	148	129	602	750

32 93 Plants

32 93 43 – Trees

32 93 43.20 Trees		Crew	Daily Output	Labor-Hours	Unit	Material	2015 Bare Costs Labor	Equipment	Total	Total Incl O&P
2100	Planetree, 9'-11', 1-1/4" caliper G	B-17	10	3.200	Ea.	236	88.50	77.50	402	495
2200	Plum, 6'-8', 1" caliper G		20	1.600		82	44.50	39	165.50	207
2300	Poplar, 9'-11', 1-1/4" caliper G		10	3.200		144	88.50	77.50	310	390
2500	Sumac, 2'-3' G		75	.427		44	11.85	10.35	66.20	79.50
2700	Tulip, 5'-6' G		40	.800		47	22	19.40	88.40	110
2800	Willow, 6'-8', 1" caliper G		20	1.600		96	44.50	39	179.50	223

32 94 Planting Accessories

32 94 13 – Landscape Edging

32 94 13.20 Edging

		Crew	Daily Output	Labor-Hours	Unit	Material	Labor	Equipment	Total	Total Incl O&P
0010	**EDGING**									
0050	Aluminum alloy, including stakes, 1/8" x 4", mill finish	B-1	390	.062	L.F.	2.30	1.55		3.85	5.15
0051	Black paint		390	.062		2.67	1.55		4.22	5.55
0052	Black anodized		390	.062		3.08	1.55		4.63	6
0100	Brick, set horizontally, 1-1/2 bricks per L.F.	D-1	370	.043		1.22	1.27		2.49	3.48
0150	Set vertically, 3 bricks per L.F.	"	135	.119		3.68	3.49		7.17	9.90
0200	Corrugated aluminum, roll, 4" wide	1 Carp	650	.012		1.93	.40		2.33	2.80
0250	6" wide	"	550	.015		2.41	.48		2.89	3.45
0600	Railroad ties, 6" x 8"	2 Carp	170	.094		2.79	3.09		5.88	8.25
0650	7" x 9"		136	.118		3.10	3.86		6.96	9.90
0750	Redwood 2" x 4"		330	.048		2.25	1.59		3.84	5.15
0800	Steel edge strips, incl. stakes, 1/4" x 5"	B-1	390	.062		3.92	1.55		5.47	6.90
0850	3/16" x 4"	"	390	.062		3.10	1.55		4.65	6

32 94 50 – Tree Guying

32 94 50.10 Tree Guying Systems

		Crew	Daily Output	Labor-Hours	Unit	Material	Labor	Equipment	Total	Total Incl O&P
0010	**TREE GUYING SYSTEMS**									
0015	Tree guying Including stakes, guy wire and wrap									
0100	Less than 3" caliper, 2 stakes	2 Clab	35	.457	Ea.	14.25	11.20		25.45	34.50
0200	3" to 4" caliper, 3 stakes	"	21	.762	"	19.25	18.65		37.90	52.50
1000	Including arrowhead anchor, cable, turnbuckles and wrap									
1100	Less than 3" caliper, 3" anchors	2 Clab	20	.800	Ea.	44.50	19.55		64.05	82
1200	3" to 6" caliper, 4" anchors		15	1.067		38.50	26		64.50	86
1300	6" caliper, 6" anchors		12	1.333		44.50	32.50		77	104
1400	8" caliper, 8" anchors		9	1.778		129	43.50		172.50	215

32 96 Transplanting

32 96 23 – Plant and Bulb Transplanting

32 96 23.23 Planting

		Crew	Daily Output	Labor-Hours	Unit	Material	Labor	Equipment	Total	Total Incl O&P
0010	**PLANTING**									
0012	Moving shrubs on site, 12" ball	B-62	28	.857	Ea.		23.50	6.20	29.70	46.50
0100	24" ball	"	22	1.091	"		30	7.90	37.90	58.50

32 96 23.43 Moving Trees

		Crew	Daily Output	Labor-Hours	Unit	Material	Labor	Equipment	Total	Total Incl O&P
0010	**MOVING TREES**, On site									
0300	Moving trees on site, 36" ball	B-6	3.75	6.400	Ea.		176	97	273	400
0400	60" ball	"	1	24	"		660	365	1,025	1,500

Estimating Tips
33 10 00 Water Utilities
33 30 00 Sanitary Sewerage Utilities
33 40 00 Storm Drainage Utilities

- Never assume that the water, sewer, and drainage lines will go in at the early stages of the project. Consider the site access needs before dividing the site in half with open trenches, loose pipe, and machinery obstructions. Always inspect the site to establish that the site drawings are complete. Check off all existing utilities on your drawings as you locate them. Be especially careful with underground utilities because appurtenances are sometimes buried during regrading or repaving operations. If you find any discrepancies, mark up the site plan for further research. Differing site conditions can be very costly if discovered later in the project.

- See also Section 33 01 00 for restoration of pipe where removal/replacement may be undesirable. Use of new types of piping materials can reduce the overall project cost. Owners/design engineers should consider the installing contractor as a valuable source of current information on utility products and local conditions that could lead to significant cost savings.

Reference Numbers
Reference numbers are shown in shaded boxes at the beginning of some major classifications. These numbers refer to related items in the Reference Section. The reference information may be an estimating procedure, an alternate pricing method, or technical information.

Note: Not all subdivisions listed here necessarily appear in this publication. ■

---

*Note: **Trade Service**, in part, has been used as a reference source for some of the material prices used in Division 33.*

Estimate with precision with RSMeans Online...
Quick, intuitive and easy-to-use, RSMeans Online gives you instant access to hundreds of thousands of material, labor and equipment costs from RSMeans' comprehensive database, delivering the information you need to build competitive estimates every time.

Start your free trial today at www.rsmeansonline.com

RSMeansOnline

33 05 Common Work Results for Utilities

33 05 16 – Utility Structures

33 05 16.13 Precast Concrete Utility Boxes

		Crew	Daily Output	Labor-Hours	Unit	Material	2015 Bare Costs Labor	Equipment	Total	Total Incl O&P
0010	**PRECAST CONCRETE UTILITY BOXES**, 6" thick									
0050	5' x 10' x 6' high, I.D.	B-13	2	24	Ea.	3,750	640	370	4,760	5,600
0350	Hand hole, precast concrete, 1-1/2" thick									
0400	1'-0" x 2'-0" x 1'-9", I.D., light duty	B-1	4	6	Ea.	400	151		551	695
0450	4'-6" x 3'-2" x 2'-0", O.D., heavy duty	B-6	3	8	"	1,475	220	121	1,816	2,100

33 05 23 – Trenchless Utility Installation

33 05 23.19 Microtunneling

		Crew	Daily Output	Labor-Hours	Unit	Material	Labor	Equipment	Total	Total Incl O&P
0010	**MICROTUNNELING**									
0011	Not including excavation, backfill, shoring,									
0020	or dewatering, average 50'/day, slurry method									
0100	24" to 48" outside diameter, minimum				L.F.				875	965
0110	Adverse conditions, add				%				50%	50%
1000	Rent microtunneling machine, average monthly lease				Month				97,500	107,000
1010	Operating technician				Day				630	690
1100	Mobilization and demobilization, minimum				Job				41,200	45,900
1110	Maximum				"				445,500	490,500

33 11 Water Utility Distribution Piping

33 11 13 – Public Water Utility Distribution Piping

33 11 13.15 Water Supply, Ductile Iron Pipe

		Crew	Daily Output	Labor-Hours	Unit	Material	Labor	Equipment	Total	Total Incl O&P
0010	**WATER SUPPLY, DUCTILE IRON PIPE**									
0020	Not including excavation or backfill									
2000	Pipe, class 50 water piping, 18' lengths									
2020	Mechanical joint, 4" diameter	B-21A	200	.200	L.F.	30.50	6.25	2.37	39.12	46.50
2040	6" diameter		160	.250		32	7.80	2.96	42.76	51.50
3000	Push-on joint, 4" diameter		400	.100		21	3.13	1.18	25.31	29.50
3020	6" diameter		333.33	.120		17	3.75	1.42	22.17	26.50
8000	Piping, fittings, mechanical joint, AWWA C110									
8006	90° bend, 4" diameter	B-20A	16	2	Ea.	155	60.50		215.50	270
8020	6" diameter		12.80	2.500		229	75.50		304.50	375
8200	Wye or tee, 4" diameter		10.67	2.999		355	90.50		445.50	540
8220	6" diameter		8.53	3.751		535	113		648	780
8398	45° bends, 4" diameter		16	2		255	60.50		315.50	380
8400	6" diameter		12.80	2.500		199	75.50		274.50	345
8405	8" diameter		10.67	2.999		291	90.50		381.50	470
8450	Decreaser, 6" x 4" diameter		14.22	2.250		207	68		275	340
8460	8" x 6" diameter		11.64	2.749		310	83		393	480
8550	Piping, butterfly valves, cast iron									
8560	4" diameter	B-20	6	4	Ea.	470	100		570	685
8700	Joint restraint, ductile iron mechanical joints									
8710	4" diameter	B-20A	32	1	Ea.	29	30		59	82
8720	6" diameter		25.60	1.250		31.50	37.50		69	97
8730	8" diameter		21.33	1.500		30	45		75	108
8740	10" diameter		18.28	1.751		38.50	53		91.50	131
8750	12" diameter		16.84	1.900		50	57.50		107.50	151
8760	14" diameter		16	2		53	60.50		113.50	159
8770	16" diameter		11.64	2.749		58	83		141	202
8780	18" diameter		11.03	2.901		73	87.50		160.50	225
8785	20" diameter		9.14	3.501		78	106		184	262
8790	24" diameter		7.53	4.250		104	128		232	330

33 11 Water Utility Distribution Piping

33 11 13 – Public Water Utility Distribution Piping

33 11 13.15 Water Supply, Ductile Iron Pipe

		Crew	Daily Output	Labor-Hours	Unit	Material	2015 Bare Costs Labor	2015 Bare Costs Equipment	Total	Total Incl O&P
9600	Steel sleeve with tap, 4" diameter	B-20	3	8	Ea.	435	201		636	815
9620	6" diameter	"	2	12		475	300		775	1,025

33 11 13.25 Water Supply, Polyvinyl Chloride Pipe

		Crew	Daily Output	Labor-Hours	Unit	Material	Labor	Equipment	Total	Total Incl O&P
0010	**WATER SUPPLY, POLYVINYL CHLORIDE PIPE**									
2100	PVC pipe, Class 150, 1-1/2" diameter	Q-1A	750	.013	L.F.	.45	.52		.97	1.36
2120	2" diameter		686	.015		.68	.57		1.25	1.69
2140	2-1/2" diameter		500	.020		1.30	.78		2.08	2.72
2160	3" diameter	B-20	430	.056		1.40	1.40		2.80	3.89
8700	PVC pipe, joint restraint									
8710	4" diameter	B-20A	32	1	Ea.	45.50	30		75.50	100
8720	6" diameter		25.60	1.250		56	37.50		93.50	125
8730	8" diameter		21.33	1.500		74.50	45		119.50	157
8740	10" diameter		18.28	1.751		130	53		183	231
8750	12" diameter		16.84	1.900		145	57.50		202.50	255
8760	14" diameter		16	2		204	60.50		264.50	325
8770	16" diameter		11.64	2.749		265	83		348	430
8780	18" diameter		11.03	2.901		340	87.50		427.50	520
8785	20" diameter		9.14	3.501		420	106		526	635
8790	24" diameter		7.53	4.250		490	128		618	755

33 12 Water Utility Distribution Equipment

33 12 13 – Water Service Connections

33 12 13.15 Tapping, Crosses and Sleeves

		Crew	Daily Output	Labor-Hours	Unit	Material	Labor	Equipment	Total	Total Incl O&P
0010	**TAPPING, CROSSES AND SLEEVES**									
4000	Drill and tap pressurized main (labor only)									
4100	6" main, 1" to 2" service	Q-1	3	5.333	Ea.		186		186	305
4150	8" main, 1" to 2" service	"	2.75	5.818	"		203		203	335
4500	Tap and insert gate valve									
4600	8" main, 4" branch	B-21	3.20	8.750	Ea.		233	43	276	440
4650	6" branch		2.70	10.370			276	51	327	515
4700	10" Main, 4" branch		2.70	10.370			276	51	327	515
4750	6" branch		2.35	11.915			315	58.50	373.50	595
4800	12" main, 6" branch		2.35	11.915			315	58.50	373.50	595

33 21 Water Supply Wells

33 21 13 – Public Water Supply Wells

33 21 13.10 Wells and Accessories

		Crew	Daily Output	Labor-Hours	Unit	Material	Labor	Equipment	Total	Total Incl O&P
0010	**WELLS & ACCESSORIES**									
0011	Domestic									
0100	Drilled, 4" to 6" diameter	B-23	120	.333	L.F.		8.30	23.50	31.80	40
1500	Pumps, installed in wells to 100' deep, 4" submersible									
1520	3/4 H.P.	Q-1	2.66	6.015	Ea.	400	210		610	785
1600	1 H.P.	"	2.29	6.987	"	425	244		669	870

33 31 Sanitary Utility Sewerage Piping

33 31 13 – Public Sanitary Utility Sewerage Piping

33 31 13.15 Sewage Collection, Concrete Pipe

		Crew	Daily Output	Labor-Hours	Unit	Material	2015 Bare Costs Labor	Equipment	Total	Total Incl O&P
0010	**SEWAGE COLLECTION, CONCRETE PIPE**									
0020	See Section 33 41 13.60 for sewage/drainage collection, concrete pipe									

33 31 13.25 Sewage Collection, Polyvinyl Chloride Pipe

		Crew	Daily Output	Labor-Hours	Unit	Material	Labor	Equipment	Total	Total Incl O&P
0010	**SEWAGE COLLECTION, POLYVINYL CHLORIDE PIPE**									
0020	Not including excavation or backfill									
2000	20' lengths, SDR 35, B&S, 4" diameter	B-20	375	.064	L.F.	1.45	1.61		3.06	4.30
2040	6" diameter		350	.069		3.28	1.72		5	6.50
2080	13' lengths, SDR 35, B&S, 8" diameter		335	.072		6.90	1.80		8.70	10.60
2120	10" diameter	B-21	330	.085		11.40	2.26	.42	14.08	16.75
4000	Piping, DWV PVC, no exc./bkfill., 10' L, Sch 40, 4" diameter	B-20	375	.064		3.84	1.61		5.45	6.95
4010	6" diameter		350	.069		7.65	1.72		9.37	11.30
4020	8" diameter		335	.072		12	1.80		13.80	16.20
4030	Piping, fittings, 1/4 bend DWV PVC, 4" diameter		19	1.263	Ea.	13.05	31.50		44.55	68
4040	Piping, fittings, 1/4 bend DWV PVC, 6" diameter		12	2		65	50		115	156
4050	Piping, fittings, 1/4 bend DWV PVC, 8" diameter		11	2.182		65	55		120	164
4060	Piping, fittings, 1/8 bend DWV PVC, 4" diameter		19	1.263		10.40	31.50		41.90	65
4070	Piping, fittings, 1/8 bend DWV PVC, 6" diameter		12	2		42	50		92	131
4080	Piping, fittings, 1/8 bend DWV PVC, 8" diameter		11	2.182		158	55		213	265
4090	Piping, fittings, Tee DWV PVC, 4" diameter		12	2		17.75	50		67.75	104
4100	Piping, fittings, Tee DWV PVC, 6" diameter		10	2.400		74	60.50		134.50	183
4110	Piping, fittings, Tee DWV PVC, 8" diameter		9	2.667		173	67		240	300

33 36 Utility Septic Tanks

33 36 13 – Utility Septic Tank and Effluent Wet Wells

33 36 13.13 Concrete Utility Septic Tank

		Crew	Daily Output	Labor-Hours	Unit	Material	Labor	Equipment	Total	Total Incl O&P
0010	**CONCRETE UTILITY SEPTIC TANK**									
0011	Not including excavation or piping									
0015	Septic tanks, precast, 1,000 gallon	B-21	8	3.500	Ea.	1,000	93	17.20	1,110.20	1,275
0060	1,500 gallon		7	4		1,325	106	19.70	1,450.70	1,675
0100	2,000 gallon		5	5.600		1,925	149	27.50	2,101.50	2,375
0900	Concrete riser 24" X 12" with standard lid	1 Clab	6	1.333		83	32.50		115.50	146
0905	24" X 12" with heavy duty lid		6	1.333		105	32.50		137.50	171
0910	24" X 8" with standard lid		6	1.333		78	32.50		110.50	141
0915	24" X 8" with heavy duty lid		6	1.333		100	32.50		132.50	165
0917	24" X 12" extension		12	.667		56	16.30		72.30	89
0920	24" X 8" extension		12	.667		51	16.30		67.30	83.50
0950	HDPE riser 20" X 12" with standard lid		8	1		122	24.50		146.50	175
0951	HDPE 20" X 12" with heavy duty lid		8	1		136	24.50		160.50	191
0955	HDPE 24" X 12" with standard lid		8	1		117	24.50		141.50	170
0956	HDPE 24" X 12" with heavy duty lid		8	1		140	24.50		164.50	195
0960	HDPE 20" X 6" extension		48	.167		27	4.08		31.08	36.50
0962	HDPE 20" X 12" extension		48	.167		48	4.08		52.08	59.50
0965	HDPE 24" X 6" extension		48	.167		28.50	4.08		32.58	38
0967	HDPE 24" X 12" extension		48	.167		46.50	4.08		50.58	58
1150	Leaching field chambers, 13' x 3'-7" x 1'-4", standard	B-13	16	3		485	80	46	611	715
1420	Leaching pit, precast concrete, 6', dia, 3' deep	B-21	4.70	5.957		885	159	29.50	1,073.50	1,275

33 36 13.19 Polyethylene Utility Septic Tank

		Crew	Daily Output	Labor-Hours	Unit	Material	Labor	Equipment	Total	Total Incl O&P
0010	**POLYETHYLENE UTILITY SEPTIC TANK**									
0015	High density polyethylene, 1,000 gallon	B-21	8	3.500	Ea.	1,550	93	17.20	1,660.20	1,875
0020	1,250 gallon		8	3.500		1,200	93	17.20	1,310.20	1,500

33 36 Utility Septic Tanks

33 36 13 – Utility Septic Tank and Effluent Wet Wells

33 36 13.19 Polyethylene Utility Septic Tank	Crew	Daily Output	Labor-Hours	Unit	Material	2015 Bare Costs Labor	Equipment	Total	Total Incl O&P
0025 1,500 gallon	B-21	7	4	Ea.	1,475	106	19.70	1,600.70	1,825

33 36 19 – Utility Septic Tank Effluent Filter

33 36 19.13 Utility Septic Tank Effluent Tube Filter

	Crew	Daily Output	Labor-Hours	Unit	Material	Labor	Equipment	Total	Total Incl O&P
0010 **UTILITY SEPTIC TANK EFFLUENT TUBE FILTER**									
3000 Effluent filter, 4" diameter	1 Skwk	8	1	Ea.	42	33.50		75.50	102
3020 6" diameter	"	7	1.143	"	249	38		287	335

33 36 33 – Utility Septic Tank Drainage Field

33 36 33.13 Utility Septic Tank Tile Drainage Field

	Crew	Daily Output	Labor-Hours	Unit	Material	Labor	Equipment	Total	Total Incl O&P
0010 **UTILITY SEPTIC TANK TILE DRAINAGE FIELD**									
0015 Distribution box, concrete, 5 outlets	2 Clab	20	.800	Ea.	80.50	19.55		100.05	122
0020 7 outlets		16	1		80.50	24.50		105	130
0025 9 outlets		8	2		490	49		539	620
0115 Distribution boxes, HDPE, 5 outlets		20	.800		64.50	19.55		84.05	104
0117 6 outlets		15	1.067		65	26		91	116
0118 7 outlets		15	1.067		65	26		91	116
0120 8 outlets		10	1.600		68	39		107	141
0240 Distribution boxes, Outlet Flow Leveler	1 Clab	50	.160		2.25	3.91		6.16	9.05
0300 Precast concrete, galley, 4' x 4' x 4'	B-21	16	1.750		250	46.50	8.60	305.10	360
0350 HDPE infiltration chamber 12" H X 15" W	2 Clab	300	.053	L.F.	7.30	1.30		8.60	10.25
0351 12" H X 15" W End Cap	1 Clab	32	.250	Ea.	19.75	6.10		25.85	32
0355 chamber 12" H X 22" W	2 Clab	300	.053	L.F.	6.70	1.30		8	9.55
0356 12" H X 22" W End Cap	1 Clab	32	.250	Ea.	16.85	6.10		22.95	29
0360 chamber 13" H X 34" W	2 Clab	300	.053	L.F.	15	1.30		16.30	18.70
0361 13" H X 34" W End Cap	1 Clab	32	.250	Ea.	53.50	6.10		59.60	69
0365 chamber 16" H X 34" W	2 Clab	300	.053	L.F.	10.70	1.30		12	14
0366 16" H X 34" W End Cap	1 Clab	32	.250	Ea.	8	6.10		14.10	19.05
0370 chamber 8" H X 16" W	2 Clab	300	.053	L.F.	9.65	1.30		10.95	12.80
0371 8" H X 16" W End Cap	1 Clab	32	.250	Ea.	10.80	6.10		16.90	22

33 36 50 – Drainage Field Systems

33 36 50.10 Drainage Field Excavation and Fill

	Crew	Daily Output	Labor-Hours	Unit	Material	Labor	Equipment	Total	Total Incl O&P
0010 **DRAINAGE FIELD EXCAVATION AND FILL**									
2200 Septic tank & drainage field excavation with 3/4 c.y backhoe	B-12F	145	.110	C.Y.		3.32	4.52	7.84	10.45
2400 4' trench for disposal field, 3/4 C.Y. backhoe	"	335	.048	L.F.		1.44	1.95	3.39	4.54
2600 Gravel fill, run of bank	B-6	150	.160	C.Y.	20	4.40	2.43	26.83	32
2800 Crushed stone, 3/4"	"	150	.160	"	36.50	4.40	2.43	43.33	50

33 41 Storm Utility Drainage Piping

33 41 13 – Public Storm Utility Drainage Piping

33 41 13.60 Sewage/Drainage Collection, Concrete Pipe

	Crew	Daily Output	Labor-Hours	Unit	Material	Labor	Equipment	Total	Total Incl O&P
0010 **SEWAGE/DRAINAGE COLLECTION, CONCRETE PIPE**									
0020 Not including excavation or backfill									
1020 8" diameter	B-14	224	.214	L.F.	6.60	5.65	1.63	13.88	18.50
1030 10" diameter	"	216	.222	"	7.30	5.85	1.69	14.84	19.70
3780 Concrete slotted pipe, class 4 mortar joint									
3800 12" diameter	B-21	168	.167	L.F.	28	4.44	.82	33.26	39.50
3840 18" diameter	"	152	.184	"	32	4.90	.91	37.81	44
3900 Concrete slotted pipe, Class 4 O-ring joint									
3940 12" diameter	B-21	168	.167	L.F.	28	4.44	.82	33.26	39.50
3960 18" diameter	"	152	.184	"	32	4.90	.91	37.81	44

33 44 Storm Utility Water Drains

33 44 13 – Utility Area Drains

33 44 13.13 Catchbasins

		Crew	Daily Output	Labor-Hours	Unit	Material	2015 Bare Costs Labor	Equipment	Total	Total Incl O&P
0010	**CATCHBASINS**									
0011	Not including footing & excavation									
1600	Frames & grates, C.I., 24" square, 500 lb.	B-6	7.80	3.077	Ea.	340	84.50	46.50	471	570

33 46 Subdrainage

33 46 16 – Subdrainage Piping

33 46 16.25 Piping, Subdrainage, Corrugated Metal

		Crew	Daily Output	Labor-Hours	Unit	Material	Labor	Equipment	Total	Total Incl O&P
0010	**PIPING, SUBDRAINAGE, CORRUGATED METAL**									
0021	Not including excavation and backfill									
2010	Aluminum, perforated									
2020	6" diameter, 18 ga.	B-20	380	.063	L.F.	6.50	1.59		8.09	9.80
2200	8" diameter, 16 ga.	"	370	.065		8.55	1.63		10.18	12.15
2220	10" diameter, 16 ga.	B-21	360	.078		10.70	2.07	.38	13.15	15.65
3000	Uncoated galvanized, perforated									
3020	6" diameter, 18 ga.	B-20	380	.063	L.F.	6.05	1.59		7.64	9.30
3200	8" diameter, 16 ga.	"	370	.065		8.30	1.63		9.93	11.90
3220	10" diameter, 16 ga.	B-21	360	.078		8.80	2.07	.38	11.25	13.60
3240	12" diameter, 16 ga.	"	285	.098		9.80	2.62	.48	12.90	15.65
4000	Steel, perforated, asphalt coated									
4020	6" diameter 18 ga.	B-20	380	.063	L.F.	6.50	1.59		8.09	9.80
4030	8" diameter 18 ga.	"	370	.065		8.55	1.63		10.18	12.15
4040	10" diameter 16 ga.	B-21	360	.078		10.05	2.07	.38	12.50	14.95
4050	12" diameter 16 ga.		285	.098		11.05	2.62	.48	14.15	17.05
4060	18" diameter 16 ga.		205	.137		17.10	3.64	.67	21.41	25.50

33 49 Storm Drainage Structures

33 49 13 – Storm Drainage Manholes, Frames, and Covers

33 49 13.10 Storm Drainage Manholes, Frames and Covers

		Crew	Daily Output	Labor-Hours	Unit	Material	Labor	Equipment	Total	Total Incl O&P
0010	**STORM DRAINAGE MANHOLES, FRAMES & COVERS**									
0020	Excludes footing, excavation, backfill (See line items for frame & cover)									
0050	Brick, 4' inside diameter, 4' deep	D-1	1	16	Ea.	520	470		990	1,350
1110	Precast, 4' I.D., 4' deep	B-22	4.10	7.317	"	725	199	50.50	974.50	1,200

33 51 Natural-Gas Distribution

33 51 13 – Natural-Gas Piping

33 51 13.10 Piping, Gas Service and Distribution, P.E.

		Crew	Daily Output	Labor-Hours	Unit	Material	Labor	Equipment	Total	Total Incl O&P
0010	**PIPING, GAS SERVICE AND DISTRIBUTION, POLYETHYLENE**									
0020	Not including excavation or backfill									
1000	60 psi coils, compression coupling @ 100', 1/2" diameter, SDR 11	B-20A	608	.053	L.F.	.45	1.59		2.04	3.14
1010	1" diameter, SDR 11		544	.059		1.05	1.77		2.82	4.11
1040	1-1/4" diameter, SDR 11		544	.059		1.62	1.77		3.39	4.73
1100	2" diameter, SDR 11		488	.066		2.82	1.98		4.80	6.40
1160	3" diameter, SDR 11		408	.078		6.45	2.37		8.82	11.05
1500	60 PSI 40' joints with coupling, 3" diameter, SDR 11	B-21A	408	.098		10.05	3.06	1.16	14.27	17.45
1540	4" diameter, SDR 11		352	.114		14.25	3.55	1.35	19.15	23
1600	6" diameter, SDR 11		328	.122		33.50	3.81	1.44	38.75	45
1640	8" diameter, SDR 11		272	.147		51.50	4.60	1.74	57.84	66

33 51 Natural-Gas Distribution

33 51 13 – Natural-Gas Piping

33 51 13.20 Piping, Gas Service and Distribution, Steel		Crew	Daily Output	Labor-Hours	Unit	Material	2015 Bare Costs Labor	Equipment	Total	Total Incl O&P
0010	**PIPING, GAS SERVICE & DISTRIBUTION, STEEL**									
0020	Not including excavation or backfill, tar coated and wrapped									
4000	Pipe schedule 40, plain end									
4040	1" diameter	Q-4	300	.107	L.F.	4.75	3.93	.19	8.87	11.95
4080	2" diameter	"	280	.114	"	7.45	4.21	.21	11.87	15.40

Division Notes

Estimating Tips

- When estimating costs for the installation of electrical power generation equipment, factors to review include access to the job site, access and setting at the installation site, required connections, uncrating pads, anchors, leveling, final assembly of the components, and temporary protection from physical damage, including from exposure to the environment.

- Be aware of the cost of equipment supports, concrete pads, and vibration isolators, and cross-reference to other trades' specifications. Also, review site and structural drawings for items that must be included in the estimates.

- It is important to include items that are not documented in the plans and specifications but must be priced. These items include, but are not limited to, testing, dust protection, roof penetration, core drilling concrete floors and walls, patching, cleanup, and final adjustments. Add a contingency or allowance for utility company fees for power hookups, if needed.

- The project size and scope of electrical power generation equipment will have a significant impact on cost. The intent of RSMeans cost data is to provide a benchmark cost so that owners, engineers, and electrical contractors will have a comfortable number with which to start a project. Additionally, there are many websites available to use for research and to obtain a vendor's quote to finalize costs.

Reference Numbers

Reference numbers are shown in shaded boxes at the beginning of some major classifications. These numbers refer to related items in the Reference Section. The reference information may be an estimating procedure, an alternate pricing method, or technical information.

Note: Not all subdivisions listed here necessarily appear in this publication. ■

Estimate with precision with RSMeans Online...

Quick, intuitive and easy-to-use, RSMeans Online gives you instant access to hundreds of thousands of material, labor, and equipment costs from RSMeans' comprehensive database, delivering the information you need to build competitive estimates every time.

Start your free trial today at **www.rsmeansonline.com**

RSMeansOnline

No part of this publication may be reproduced, stored in a retrieval system, or transmitted in any form or by any means without prior written permission of RSMeans.

48 15 Wind Energy Electrical Power Generation Equipment

48 15 13 – Wind Turbines

48 15 13.50 Wind Turbines and Components		Crew	Daily Output	Labor-Hours	Unit	Material	2015 Bare Costs Labor	Equipment	Total	Total Incl O&P
0010	**WIND TURBINES & COMPONENTS**									
0500	Complete system, grid connected									
1000	20 kW, 31' dia, incl. labor & material	G			System				49,900	49,900
1010	Enhanced	G							55,000	55,000
1500	10 kW, 23' dia, incl. labor & material	G							74,000	74,000
2000	2.4 kW, 12' dia, incl. labor & material	G			↓				18,000	18,000
2900	Component system									
3000	Turbine, 400 watt, 3' dia	G	1 Elec	3.41	2.346	Ea.	650	86		736	855
3100	600 watt, 3' dia	G		2.56	3.125		880	115		995	1,150
3200	1000 watt, 9' dia	G	↓	2.05	3.902	↓	3,650	143		3,793	4,250
3400	Mounting hardware										
3500	30' guyed tower kit	G	2 Clab	5.12	3.125	Ea.	415	76.50		491.50	590
3505	3' galvanized helical earth screw	G	1 Clab	8	1		49.50	24.50		74	95.50
3510	Attic mount kit	G	1 Rofc	2.56	3.125		154	86.50		240.50	330
3520	Roof mount kit	G	1 Clab	3.41	2.346	↓	136	57.50		193.50	247
8900	Equipment										
9100	DC to AC inverter for, 48 V, 4,000 watt	G	1 Elec	2	4	Ea.	2,325	147		2,472	2,825

Reference Section

All the reference information is in one section, making it easy to find what you need to know ... and easy to use the book on a daily basis. This section is visually identified by a vertical black bar on the page edges.

In this Reference Section, we've included Equipment Rental Costs, a listing of rental and operating costs; Crew Listings, a full listing of all crews, equipment, and their costs; Location Factors for adjusting costs to the region you are in; Reference Tables, where you will find explanations, estimating information and procedures, or technical data; an explanation of all the Abbreviations in the book; and sample Estimating Forms.

Table of Contents

Construction Equipment Rental Costs	661
Crew Listings	673
Location Factors	707
Reference Tables	713
R01 General Requirements	713
R02 Existing Conditions	721
R04 Masonry	722
R05 Metals	724
R06 Wood, Plastics & Composites	725

Reference Tables (cont.)

R07 Thermal & Moisture Protection	725
R08 Openings	726
R09 Finishes	728
R13 Special Construction	731
R22 Plumbing	732
R32 Exterior Improvements	732
R33 Utilities	733
Abbreviations	734
Estimating Forms	738

Equipment Rental Costs

Estimating Tips
- This section contains the average costs to rent and operate hundreds of pieces of construction equipment. This is useful information when estimating the time and material requirements of any particular operation in order to establish a unit or total cost. Equipment costs include not only rental, but also operating costs for equipment under normal use.

Rental Costs
- Equipment rental rates are obtained from the following industry sources throughout North America: contractors, suppliers, dealers, manufacturers, and distributors.
- Rental rates vary throughout the country, with larger cities generally having lower rates. Lease plans for new equipment are available for periods in excess of six months, with a percentage of payments applying toward purchase.
- Monthly rental rates vary from 2% to 5% of the purchase price of the equipment depending on the anticipated life of the equipment and its wearing parts.
- Weekly rental rates are about 1/3 the monthly rates, and daily rental rates are about 1/3 the weekly rate.
- Rental rates can also be treated as reimbursement costs for contractor-owned equipment. Owned equipment costs include depreciation, loan payments, interest, taxes, insurance, storage, and major repairs.

Operating Costs
- The operating costs include parts and labor for routine servicing, such as repair and replacement of pumps, filters and worn lines. Normal operating expendables, such as fuel, lubricants, tires and electricity (where applicable), are also included.
- Extraordinary operating expendables with highly variable wear patterns, such as diamond bits and blades, are excluded. These costs can be found as material costs in the Unit Price section.
- The hourly operating costs listed do not include the operator's wages.

Equipment Cost/Day
- Any power equipment required by a crew is shown in the Crew Listings with a daily cost.
- The daily cost of equipment needed by a crew is based on dividing the weekly rental rate by 5 (number of working days in the week), and then adding the hourly operating cost times 8 (the number of hours in a day). This "Equipment Cost/Day" is shown in the far right column of the Equipment Rental pages.
- If equipment is needed for only one or two days, it is best to develop your own cost by including components for daily rent and hourly operating cost. This is important when the listed Crew for a task does not contain the equipment needed, such as a crane for lifting mechanical heating/cooling equipment up onto a roof.
- If the quantity of work is less than the crew's Daily Output shown for a Unit Price line item that includes a bare unit equipment cost, it is recommended to estimate one day's rental cost and operating cost for equipment shown in the Crew Listing for that line item.

Mobilization/ Demobilization
- The cost to move construction equipment from an equipment yard or rental company to the job site and back again is not included in equipment rental costs listed in the Reference Section, nor in the bare equipment cost of any unit price line item, nor in any equipment costs shown in the Crew listings.
- Mobilization (to the site) and demobilization (from the site) costs can be found in the Unit Price Section.
- If a piece of equipment is already at the job site, it is not appropriate to utilize mobilization/demobilization costs again in an estimate. ∎

01 54 | Construction Aids

01 54 33 | Equipment Rental

			UNIT	HOURLY OPER. COST	RENT PER DAY	RENT PER WEEK	RENT PER MONTH	EQUIPMENT COST/DAY	
10	0010	**CONCRETE EQUIPMENT RENTAL** without operators R015433-10							10
	0200	Bucket, concrete lightweight, 1/2 C.Y.	Ea.	.80	23.50	70	210	20.40	
	0300	1 C.Y.		.90	27.50	83	249	23.80	
	0400	1-1/2 C.Y.		1.10	36.50	110	330	30.80	
	0500	2 C.Y.		1.25	45	135	405	37	
	0580	8 C.Y.		6.10	257	770	2,300	202.80	
	0600	Cart, concrete, self-propelled, operator walking, 10 C.F.		3.25	56.50	170	510	60	
	0700	Operator riding, 18 C.F.		5.35	95	285	855	99.80	
	0800	Conveyer for concrete, portable, gas, 16" wide, 26' long		12.60	123	370	1,100	174.80	
	0900	46' long		13.00	148	445	1,325	193	
	1000	56' long		13.10	157	470	1,400	198.80	
	1100	Core drill, electric, 2-1/2 H.P., 1" to 8" bit diameter		1.77	68.50	205	615	55.15	
	1150	11 H.P., 8" to 18" cores		5.95	113	340	1,025	115.60	
	1200	Finisher, concrete floor, gas, riding trowel, 96" wide		12.75	145	435	1,300	189	
	1300	Gas, walk-behind, 3 blade, 36" trowel		2.15	20	60	180	29.20	
	1400	4 blade, 48" trowel		4.20	27.50	83	249	50.20	
	1500	Float, hand-operated (Bull float) 48" wide		.08	13.35	40	120	8.65	
	1570	Curb builder, 14 H.P., gas, single screw		14.65	248	745	2,225	266.20	
	1590	Double screw		15.35	290	870	2,600	296.80	
	1600	Floor grinder, concrete and terrazzo, electric, 22" path		2.57	158	475	1,425	115.55	
	1700	Edger, concrete, electric, 7" path		1.04	51.50	155	465	39.30	
	1750	Vacuum pick-up system for floor grinders, wet/dry		1.49	81.50	245	735	60.90	
	1800	Mixer, powered, mortar and concrete, gas, 6 C.F., 18 H.P.		8.65	118	355	1,075	140.20	
	1900	10 C.F., 25 H.P.		10.80	143	430	1,300	172.40	
	2000	16 C.F.		11.15	165	495	1,475	188.20	
	2100	Concrete, stationary, tilt drum, 2 C.Y.		6.90	232	695	2,075	194.20	
	2120	Pump, concrete, truck mounted 4" line 80' boom		25.00	865	2,600	7,800	720	
	2140	5" line, 110' boom		32.40	1,150	3,435	10,300	946.20	
	2160	Mud jack, 50 C.F. per hr.		7.30	125	375	1,125	133.40	
	2180	225 C.F. per hr.		9.50	145	435	1,300	163	
	2190	Shotcrete pump rig, 12 C.Y./hr.		14.85	220	660	1,975	250.80	
	2200	35 C.Y./hr.		17.50	235	705	2,125	281	
	2600	Saw, concrete, manual, gas, 18 H.P.		6.75	45	135	405	81	
	2650	Self-propelled, gas, 30 H.P.		13.30	102	305	915	167.40	
	2675	V-groove crack chaser, manual, gas, 6 H.P.		2.35	17.35	52	156	29.20	
	2700	Vibrators, concrete, electric, 60 cycle, 2 H.P.		.46	8.65	26	78	8.90	
	2800	3 H.P.		.60	11.65	35	105	11.80	
	2900	Gas engine, 5 H.P.		2.00	16	48	144	25.60	
	3000	8 H.P.		2.75	15.35	46	138	31.20	
	3050	Vibrating screed, gas engine, 8 H.P.		2.91	71.50	215	645	66.30	
	3120	Concrete transit mixer, 6 x 4, 250 H.P., 8 C.Y., rear discharge		61.55	570	1,715	5,150	835.40	
	3200	Front discharge		72.45	700	2,095	6,275	998.60	
	3300	6 x 6, 285 H.P., 12 C.Y., rear discharge		71.50	660	1,980	5,950	968	
	3400	Front discharge		74.00	705	2,120	6,350	1,016	
20	0010	**EARTHWORK EQUIPMENT RENTAL** without operators R015433-10							20
	0040	Aggregate spreader, push type 8' to 12' wide	Ea.	3.05	25.50	76	228	39.60	
	0045	Tailgate type, 8' wide		2.90	32.50	98	294	42.80	
	0055	Earth auger, truck-mounted, for fence & sign posts, utility poles		19.15	440	1,320	3,950	417.20	
	0060	For borings and monitoring wells		46.65	675	2,030	6,100	779.20	
	0070	Portable, trailer mounted		3.30	32.50	98	294	46	
	0075	Truck-mounted, for caissons, water wells		100.15	2,900	8,705	26,100	2,542	
	0080	Horizontal boring machine, 12" to 36" diameter, 45 H.P.		24.20	192	575	1,725	308.60	
	0090	12" to 48" diameter, 65 H.P.		34.35	335	1,000	3,000	474.80	
	0095	Auger, for fence posts, gas engine, hand held		.60	6	18	54	8.40	
	0100	Excavator, diesel hydraulic, crawler mounted, 1/2 C.Y. cap.		24.65	420	1,255	3,775	448.20	
	0120	5/8 C.Y. capacity		32.55	550	1,645	4,925	589.40	
	0140	3/4 C.Y. capacity		35.60	615	1,850	5,550	654.80	
	0150	1 C.Y. capacity		49.70	690	2,075	6,225	812.60	

01 54 | Construction Aids

01 54 33 | Equipment Rental

		UNIT	HOURLY OPER. COST	RENT PER DAY	RENT PER WEEK	RENT PER MONTH	EQUIPMENT COST/DAY
0200	1-1/2 C.Y. capacity	Ea.	59.80	920	2,760	8,275	1,030
0300	2 C.Y. capacity		67.35	1,050	3,180	9,550	1,175
0320	2-1/2 C.Y. capacity		102.55	1,300	3,925	11,800	1,605
0325	3-1/2 C.Y. capacity		144.45	2,150	6,420	19,300	2,440
0330	4-1/2 C.Y. capacity		173.40	2,625	7,880	23,600	2,963
0335	6 C.Y. capacity		219.80	2,900	8,680	26,000	3,494
0340	7 C.Y. capacity		222.05	3,025	9,070	27,200	3,590
0342	Excavator attachments, bucket thumbs		3.20	245	735	2,200	172.60
0345	Grapples		2.75	193	580	1,750	138
0346	Hydraulic hammer for boom mounting, 4000 ft lb.		12.40	350	1,045	3,125	308.20
0347	5000 ft lb.		14.40	425	1,280	3,850	371.20
0348	8000 ft lb.		21.30	625	1,875	5,625	545.40
0349	12,000 ft lb.		23.35	750	2,245	6,725	635.80
0350	Gradall type, truck mounted, 3 ton @ 15' radius, 5/8 C.Y.		43.65	890	2,665	8,000	882.20
0370	1 C.Y. capacity		47.85	1,025	3,095	9,275	1,002
0400	Backhoe-loader, 40 to 45 H.P., 5/8 C.Y. capacity		14.90	240	720	2,150	263.20
0450	45 H.P. to 60 H.P., 3/4 C.Y. capacity		23.40	295	885	2,650	364.20
0460	80 H.P., 1-1/4 C.Y. capacity		25.60	310	935	2,800	391.80
0470	112 H.P., 1-1/2 C.Y. capacity		40.95	645	1,930	5,800	713.60
0482	Backhoe-loader attachment, compactor, 20,000 lb.		5.80	142	425	1,275	131.40
0485	Hydraulic hammer, 750 ft lb.		3.30	96.50	290	870	84.40
0486	Hydraulic hammer, 1200 ft lb.		6.30	217	650	1,950	180.40
0500	Brush chipper, gas engine, 6" cutter head, 35 H.P.		11.10	105	315	945	151.80
0550	Diesel engine, 12" cutter head, 130 H.P.		27.55	285	855	2,575	391.40
0600	15" cutter head, 165 H.P.		33.10	335	1,005	3,025	465.80
0750	Bucket, clamshell, general purpose, 3/8 C.Y.		1.30	38.50	115	345	33.40
0800	1/2 C.Y.		1.40	45	135	405	38.20
0850	3/4 C.Y.		1.55	55	165	495	45.40
0900	1 C.Y.		1.60	58.50	175	525	47.80
0950	1-1/2 C.Y.		2.55	81.50	245	735	69.40
1000	2 C.Y.		2.70	90	270	810	75.60
1010	Bucket, dragline, medium duty, 1/2 C.Y.		.75	23.50	70	210	20
1020	3/4 C.Y.		.75	24.50	73	219	20.60
1030	1 C.Y.		.80	26	78	234	22
1040	1-1/2 C.Y.		1.20	40	120	360	33.60
1050	2 C.Y.		1.30	45	135	405	37.40
1070	3 C.Y.		1.95	61.50	185	555	52.60
1200	Compactor, manually guided 2-drum vibratory smooth roller, 7.5 H.P.		7.25	203	610	1,825	180
1250	Rammer/tamper, gas, 8"		2.75	46.50	140	420	50
1260	15"		3.05	53.50	160	480	56.40
1300	Vibratory plate, gas, 18" plate, 3000 lb. blow		2.70	24.50	73	219	36.20
1350	21" plate, 5000 lb. blow		3.40	32.50	98	294	46.80
1370	Curb builder/extruder, 14 H.P., gas, single screw		14.65	248	745	2,225	266.20
1390	Double screw		15.35	290	870	2,600	296.80
1500	Disc harrow attachment, for tractor		.44	73.50	221	665	47.70
1810	Feller buncher, shearing & accumulating trees, 100 H.P.		48.30	755	2,260	6,775	838.40
1860	Grader, self-propelled, 25,000 lb.		39.50	640	1,925	5,775	701
1910	30,000 lb.		43.15	650	1,955	5,875	736.90
1920	40,000 lb.		64.90	1,100	3,275	9,825	1,174
1930	55,000 lb.		84.15	1,700	5,130	15,400	1,699
1950	Hammer, pavement breaker, self-propelled, diesel, 1000 to 1250 lb.		29.90	350	1,055	3,175	450.20
2000	1300 to 1500 lb.		44.85	705	2,110	6,325	780.80
2050	Pile driving hammer, steam or air, 4150 ft lb. @ 225 bpm		10.40	475	1,420	4,250	367.20
2100	8750 ft lb. @ 145 bpm		12.40	660	1,975	5,925	494.20
2150	15,000 ft lb. @ 60 bpm		14.00	790	2,370	7,100	586
2200	24,450 ft lb. @ 111 bpm		15.00	875	2,630	7,900	646
2250	Leads, 60' high for pile driving hammers up to 20,000 ft lb.		3.30	80.50	242	725	74.80
2300	90' high for hammers over 20,000 ft lb.		4.95	141	424	1,275	124.40

For customer support on your Residential Cost Data, call 877.713.2617.

01 54 | Construction Aids

01 54 33 | Equipment Rental

			UNIT	HOURLY OPER. COST	RENT PER DAY	RENT PER WEEK	RENT PER MONTH	EQUIPMENT COST/DAY
2350		Diesel type hammer, 22,400 ft lb.	Ea.	22.55	460	1,380	4,150	456.40
2400		41,300 ft lb.		32.00	540	1,625	4,875	581
2450		141,000 ft lb.		51.65	960	2,875	8,625	988.20
2500		Vib. elec. hammer/extractor, 200 kW diesel generator, 34 H.P.		54.90	635	1,900	5,700	819.20
2550		80 H.P.		100.65	925	2,775	8,325	1,360
2600		150 H.P.		190.90	1,775	5,300	15,900	2,587
2800		Log chipper, up to 22" diameter, 600 H.P.		63.30	640	1,915	5,750	889.40
2850		Logger, for skidding & stacking logs, 150 H.P.		52.75	815	2,445	7,325	911
2860		Mulcher, diesel powered, trailer mounted		24.05	212	635	1,900	319.40
2900		Rake, spring tooth, with tractor		18.23	345	1,042	3,125	354.25
3000		Roller, vibratory, tandem, smooth drum, 20 H.P.		8.80	145	435	1,300	157.40
3050		35 H.P.		11.10	247	740	2,225	236.80
3100		Towed type vibratory compactor, smooth drum, 50 H.P.		26.00	315	940	2,825	396
3150		Sheepsfoot, 50 H.P.		27.35	345	1,035	3,100	425.80
3170		Landfill compactor, 220 H.P.		91.00	1,475	4,440	13,300	1,616
3200		Pneumatic tire roller, 80 H.P.		16.80	350	1,050	3,150	344.40
3250		120 H.P.		25.20	600	1,800	5,400	561.60
3300		Sheepsfoot vibratory roller, 240 H.P.		68.95	1,100	3,270	9,800	1,206
3320		340 H.P.		100.85	1,575	4,730	14,200	1,753
3350		Smooth drum vibratory roller, 75 H.P.		24.85	585	1,755	5,275	549.80
3400		125 H.P.		32.55	710	2,135	6,400	687.40
3410		Rotary mower, brush, 60", with tractor		22.60	315	950	2,850	370.80
3420		Rototiller, walk-behind, gas, 5 H.P.		1.96	51.50	155	465	46.70
3422		8 H.P.		3.08	80	240	720	72.65
3440		Scrapers, towed type, 7 C.Y. capacity		5.80	113	340	1,025	114.40
3450		10 C.Y. capacity		6.60	157	470	1,400	146.80
3500		15 C.Y. capacity		7.10	180	540	1,625	164.80
3525		Self-propelled, single engine, 14 C.Y. capacity		114.76	1,600	4,830	14,500	1,884
3550		Dual engine, 21 C.Y. capacity		173.80	2,175	6,490	19,500	2,688
3600		31 C.Y. capacity		231.70	3,075	9,195	27,600	3,693
3640		44 C.Y. capacity		287.80	3,950	11,885	35,700	4,679
3650		Elevating type, single engine, 11 C.Y. capacity		71.45	985	2,955	8,875	1,163
3700		22 C.Y. capacity		139.55	2,350	7,025	21,100	2,521
3710		Screening plant 110 H.P. w/5' x 10' screen		25.35	400	1,200	3,600	442.80
3720		5' x 16' screen		29.65	505	1,515	4,550	540.20
3850		Shovel, crawler-mounted, front-loading, 7 C.Y. capacity		258.85	3,200	9,610	28,800	3,993
3855		12 C.Y. capacity		386.45	4,400	13,220	39,700	5,736
3860		Shovel/backhoe bucket, 1/2 C.Y.		2.50	66.50	200	600	60
3870		3/4 C.Y.		2.55	75	225	675	65.40
3880		1 C.Y.		2.65	83.50	250	750	71.20
3890		1-1/2 C.Y.		2.75	96.50	290	870	80
3910		3 C.Y.		3.15	130	390	1,175	103.20
3950		Stump chipper, 18" deep, 30 H.P.		7.82	205	615	1,850	185.55
4110		Dozer, crawler, torque converter, diesel 80 H.P.		29.05	400	1,200	3,600	472.40
4150		105 H.P.		35.60	530	1,590	4,775	602.80
4200		140 H.P.		51.25	795	2,380	7,150	886
4260		200 H.P.		77.25	1,275	3,845	11,500	1,387
4310		300 H.P.		100.65	1,825	5,460	16,400	1,897
4360		410 H.P.		132.60	2,250	6,720	20,200	2,405
4370		500 H.P.		171.15	3,225	9,650	29,000	3,299
4380		700 H.P.		280.90	4,550	13,660	41,000	4,979
4400		Loader, crawler, torque conv., diesel, 1-1/2 C.Y., 80 H.P.		31.25	455	1,360	4,075	522
4450		1-1/2 to 1-3/4 C.Y., 95 H.P.		34.60	590	1,765	5,300	629.80
4510		1-3/4 to 2-1/4 C.Y., 130 H.P.		53.80	875	2,630	7,900	956.40
4530		2-1/2 to 3-1/4 C.Y., 190 H.P.		65.95	1,100	3,300	9,900	1,188
4560		3-1/2 to 5 C.Y., 275 H.P.		87.80	1,425	4,295	12,900	1,561
4610		Front end loader, 4WD, articulated frame, diesel, 1 to 1-1/4 C.Y., 70 H.P.		19.75	235	705	2,125	299
4620		1-1/2 to 1-3/4 C.Y., 95 H.P.		24.80	300	900	2,700	378.40

01 54 | Construction Aids

01 54 33 | Equipment Rental

		UNIT	HOURLY OPER. COST	RENT PER DAY	RENT PER WEEK	RENT PER MONTH	EQUIPMENT COST/DAY	
4650	1-3/4 to 2 C.Y., 130 H.P.	Ea.	29.15	380	1,140	3,425	461.20	20
4710	2-1/2 to 3-1/2 C.Y., 145 H.P.		33.10	425	1,275	3,825	519.80	
4730	3 to 4-1/2 C.Y., 185 H.P.		42.40	550	1,650	4,950	669.20	
4760	5-1/4 to 5-3/4 C.Y., 270 H.P.		67.50	905	2,710	8,125	1,082	
4810	7 to 9 C.Y., 475 H.P.		114.05	1,700	5,125	15,400	1,937	
4870	9 - 11 C.Y., 620 H.P.		160.55	2,775	8,290	24,900	2,942	
4880	Skid steer loader, wheeled, 10 C.F., 30 H.P. gas		10.45	150	450	1,350	173.60	
4890	1 C.Y., 78 H.P., diesel		20.20	262	785	2,350	318.60	
4892	Skid-steer attachment, auger		.48	80.50	242	725	52.25	
4893	Backhoe		.66	111	332	995	71.70	
4894	Broom		.74	124	372	1,125	80.30	
4895	Forks		.22	36	108	325	23.35	
4896	Grapple		.56	92.50	278	835	60.10	
4897	Concrete hammer		1.06	177	531	1,600	114.70	
4898	Tree spade		1.03	172	515	1,550	111.25	
4899	Trencher		.54	90	270	810	58.30	
4900	Trencher, chain, boom type, gas, operator walking, 12 H.P.		5.00	46.50	140	420	68	
4910	Operator riding, 40 H.P.		20.15	290	870	2,600	335.20	
5000	Wheel type, diesel, 4' deep, 12" wide		85.95	845	2,535	7,600	1,195	
5100	6' deep, 20" wide		92.75	1,925	5,740	17,200	1,890	
5150	Chain type, diesel, 5' deep, 8" wide		38.35	560	1,675	5,025	641.80	
5200	Diesel, 8' deep, 16" wide		155.95	3,600	10,770	32,300	3,402	
5202	Rock trencher, wheel type, 6" wide x 18" deep		21.50	335	1,000	3,000	372	
5206	Chain type, 18" wide x 7' deep		112.00	2,800	8,395	25,200	2,575	
5210	Tree spade, self-propelled		14.38	267	800	2,400	275.05	
5250	Truck, dump, 2-axle, 12 ton, 8 C.Y. payload, 220 H.P.		34.40	227	680	2,050	411.20	
5300	Three axle dump, 16 ton, 12 C.Y. payload, 400 H.P.		61.00	340	1,015	3,050	691	
5310	Four axle dump, 25 ton, 18 C.Y. payload, 450 H.P.		71.75	490	1,475	4,425	869	
5350	Dump trailer only, rear dump, 16-1/2 C.Y.		5.45	138	415	1,250	126.60	
5400	20 C.Y.		5.90	157	470	1,400	141.20	
5450	Flatbed, single axle, 1-1/2 ton rating		25.55	68.50	205	615	245.40	
5500	3 ton rating		30.65	96.50	290	870	303.20	
5550	Off highway rear dump, 25 ton capacity		73.90	1,275	3,800	11,400	1,351	
5600	35 ton capacity		82.50	1,425	4,260	12,800	1,512	
5610	50 ton capacity		102.65	1,700	5,080	15,200	1,837	
5620	65 ton capacity		105.65	1,700	5,090	15,300	1,863	
5630	100 ton capacity		151.40	2,850	8,570	25,700	2,925	
6000	Vibratory plow, 25 H.P., walking		8.45	60	180	540	103.60	
0010	**GENERAL EQUIPMENT RENTAL** without operators R015433-10							40
0150	Aerial lift, scissor type, to 15' high, 1000 lb. cap., electric	Ea.	3.05	51.50	155	465	55.40	
0160	To 25' high, 2000 lb. capacity		3.45	66.50	200	600	67.60	
0170	Telescoping boom to 40' high, 500 lb. capacity, diesel		13.55	320	965	2,900	301.40	
0180	To 45' high, 500 lb. capacity		14.75	350	1,055	3,175	329	
0190	To 60' high, 600 lb. capacity		17.20	495	1,490	4,475	435.60	
0195	Air compressor, portable, 6.5 CFM, electric		.67	12.65	38	114	12.95	
0196	Gasoline		.81	19	57	171	17.90	
0200	Towed type, gas engine, 60 CFM		13.55	48.50	145	435	137.40	
0300	160 CFM		15.80	50	150	450	156.40	
0400	Diesel engine, rotary screw, 250 CFM		17.05	108	325	975	201.40	
0500	365 CFM		23.10	132	395	1,175	263.80	
0550	450 CFM		29.35	165	495	1,475	333.80	
0600	600 CFM		51.70	228	685	2,050	550.60	
0700	750 CFM		51.90	237	710	2,125	557.20	
0800	For silenced models, small sizes, add to rent		3%	5%	5%	5%		
0900	Large sizes, add to rent		5%	7%	7%	7%		
0930	Air tools, breaker, pavement, 60 lb.	Ea.	.50	9.65	29	87	9.80	
0940	80 lb.		.50	10.35	31	93	10.20	
0950	Drills, hand (jackhammer) 65 lb.		.60	17.35	52	156	15.20	

01 54 | Construction Aids

01 54 33 | Equipment Rental

		UNIT	HOURLY OPER. COST	RENT PER DAY	RENT PER WEEK	RENT PER MONTH	EQUIPMENT COST/DAY
0960	Track or wagon, swing boom, 4" drifter	Ea.	62.50	880	2,640	7,925	1,028
0970	5" drifter		74.45	1,075	3,235	9,700	1,243
0975	Track mounted quarry drill, 6" diameter drill		128.10	1,600	4,765	14,300	1,978
0980	Dust control per drill		1.02	23.50	71	213	22.35
0990	Hammer, chipping, 12 lb.		.55	26	78	234	20
1000	Hose, air with couplings, 50' long, 3/4" diameter		.03	5	15	45	3.25
1100	1" diameter		.04	6.35	19	57	4.10
1200	1-1/2" diameter		.05	9	27	81	5.80
1300	2" diameter		.07	12	36	108	7.75
1400	2-1/2" diameter		.11	19	57	171	12.30
1410	3" diameter		.14	23	69	207	14.90
1450	Drill, steel, 7/8" x 2'		.05	8.65	26	78	5.60
1460	7/8" x 6'		.05	9	27	81	5.80
1520	Moil points		.02	3.33	10	30	2.15
1525	Pneumatic nailer w/accessories		.58	38.50	115	345	27.65
1530	Sheeting driver for 60 lb. breaker		.04	6	18	54	3.90
1540	For 90 lb. breaker		.12	8	24	72	5.75
1550	Spade, 25 lb.		.45	6.65	20	60	7.60
1560	Tamper, single, 35 lb.		.55	36.50	109	325	26.20
1570	Triple, 140 lb.		.82	54.50	164	490	39.35
1580	Wrenches, impact, air powered, up to 3/4" bolt		.40	12.65	38	114	10.80
1590	Up to 1-1/4" bolt		.50	23.50	70	210	18
1600	Barricades, barrels, reflectorized, 1 to 99 barrels		.03	4.60	13.80	41.50	3
1610	100 to 200 barrels		.02	3.53	10.60	32	2.30
1620	Barrels with flashers, 1 to 99 barrels		.03	5.25	15.80	47.50	3.40
1630	100 to 200 barrels		.03	4.20	12.60	38	2.75
1640	Barrels with steady burn type C lights		.04	7	21	63	4.50
1650	Illuminated board, trailer mounted, with generator		3.50	130	390	1,175	106
1670	Portable barricade, stock, with flashers, 1 to 6 units		.03	5.25	15.80	47.50	3.40
1680	25 to 50 units		.03	4.90	14.70	44	3.20
1685	Butt fusion machine, wheeled, 1.5 HP electric, 2" - 8" diameter pipe		2.63	167	500	1,500	121.05
1690	Tracked, 20 HP diesel, 4"-12" diameter pipe		11.14	490	1,465	4,400	382.10
1695	83 HP diesel, 8" - 24" diameter pipe		30.46	975	2,930	8,800	829.70
1700	Carts, brick, hand powered, 1000 lb. capacity		.50	83.50	251	755	54.20
1800	Gas engine, 1500 lb., 7-1/2' lift		4.17	115	345	1,025	102.35
1822	Dehumidifier, medium, 6 lb./hr., 150 CFM		1.00	62	186	560	45.20
1824	Large, 18 lb./hr., 600 CFM		2.02	126	378	1,125	91.75
1830	Distributor, asphalt, trailer mounted, 2000 gal., 38 H.P. diesel		9.95	325	980	2,950	275.60
1840	3000 gal., 38 H.P. diesel		11.45	355	1,070	3,200	305.60
1850	Drill, rotary hammer, electric		1.02	26	78	234	23.75
1860	Carbide bit, 1-1/2" diameter, add to electric rotary hammer		.02	3.61	10.84	32.50	2.35
1865	Rotary, crawler, 250 H.P.		148.50	2,050	6,185	18,600	2,425
1870	Emulsion sprayer, 65 gal., 5 H.P. gas engine		2.91	97	291	875	81.50
1880	200 gal., 5 H.P. engine		7.85	162	485	1,450	159.80
1900	Floor auto-scrubbing machine, walk-behind, 28" path		4.98	325	970	2,900	233.85
1930	Floodlight, mercury vapor, or quartz, on tripod, 1000 watt		.44	20.50	62	186	15.90
1940	2000 watt		.83	41	123	370	31.25
1950	Floodlights, trailer mounted with generator, 1 - 300 watt light		3.65	71.50	215	645	72.20
1960	2 - 1000 watt lights		4.85	96.50	290	870	96.80
2000	4 - 300 watt lights		4.55	93.50	280	840	92.40
2005	Foam spray rig, incl. box trailer, compressor, generator, proportioner		35.03	490	1,465	4,400	573.25
2020	Forklift, straight mast, 12' lift, 5000 lb., 2 wheel drive, gas		26.35	202	605	1,825	331.80
2040	21' lift, 5000 lb., 4 wheel drive, diesel		20.90	240	720	2,150	311.20
2050	For rough terrain, 42' lift, 35' reach, 9000 lb., 110 H.P.		29.55	485	1,450	4,350	526.40
2060	For plant, 4 ton capacity, 80 H.P., 2 wheel drive, gas		16.15	93.50	280	840	185.20
2080	10 ton capacity, 120 H.P., 2 wheel drive, diesel		24.10	162	485	1,450	289.80
2100	Generator, electric, gas engine, 1.5 kW to 3 kW		3.75	11.35	34	102	36.80
2200	5 kW		4.80	15.35	46	138	47.60

01 54 | Construction Aids

01 54 33 | Equipment Rental

		UNIT	HOURLY OPER. COST	RENT PER DAY	RENT PER WEEK	RENT PER MONTH	EQUIPMENT COST/DAY	
2300	10 kW	Ea.	9.10	36.50	110	330	94.80	40
2400	25 kW		10.60	85	255	765	135.80	
2500	Diesel engine, 20 kW		12.10	66.50	200	600	136.80	
2600	50 kW		23.35	103	310	930	248.80	
2700	100 kW		43.30	128	385	1,150	423.40	
2800	250 kW		85.15	235	705	2,125	822.20	
2850	Hammer, hydraulic, for mounting on boom, to 500 ft lb.		2.55	75	225	675	65.40	
2860	1000 ft lb.		4.35	127	380	1,150	110.80	
2900	Heaters, space, oil or electric, 50 MBH		2.01	7.65	23	69	20.70	
3000	100 MBH		3.76	10.65	32	96	36.50	
3100	300 MBH		11.03	38.50	115	345	111.25	
3150	500 MBH		18.15	45	135	405	172.20	
3200	Hose, water, suction with coupling, 20' long, 2" diameter		.02	3	9	27	1.95	
3210	3" diameter		.03	4.33	13	39	2.85	
3220	4" diameter		.03	5	15	45	3.25	
3230	6" diameter		.11	17.65	53	159	11.50	
3240	8" diameter		.27	44.50	133	400	28.75	
3250	Discharge hose with coupling, 50' long, 2" diameter		.01	1.33	4	12	.90	
3260	3" diameter		.01	2.33	7	21	1.50	
3270	4" diameter		.02	3.67	11	33	2.35	
3280	6" diameter		.06	9.35	28	84	6.10	
3290	8" diameter		.18	30	90	270	19.45	
3295	Insulation blower		.78	6	18	54	9.85	
3300	Ladders, extension type, 16' to 36' long		.14	22.50	68	204	14.70	
3400	40' to 60' long		.19	31	93	279	20.10	
3405	Lance for cutting concrete		2.35	71.50	215	645	61.80	
3407	Lawn mower, rotary, 22", 5 H.P.		1.80	49	147	440	43.80	
3408	48" self propelled		2.98	75	225	675	68.85	
3410	Level, electronic, automatic, with tripod and leveling rod		1.20	75.50	226	680	54.80	
3430	Laser type, for pipe and sewer line and grade		.73	48.50	145	435	34.85	
3440	Rotating beam for interior control		.78	52	156	470	37.45	
3460	Builder's optical transit, with tripod and rod		.09	14.65	44	132	9.50	
3500	Light towers, towable, with diesel generator, 2000 watt		4.55	93.50	280	840	92.40	
3600	4000 watt		4.85	96.50	290	870	96.80	
3700	Mixer, powered, plaster and mortar, 6 C.F., 7 H.P.		2.70	20	60	180	33.60	
3800	10 C.F., 9 H.P.		2.80	31.50	94	282	41.20	
3850	Nailer, pneumatic		.58	38.50	115	345	27.65	
3900	Paint sprayers complete, 8 CFM		1.04	69.50	208	625	49.90	
4000	17 CFM		1.90	127	380	1,150	91.20	
4020	Pavers, bituminous, rubber tires, 8' wide, 50 H.P., diesel		31.15	490	1,465	4,400	542.20	
4030	10' wide, 150 H.P.		106.10	1,800	5,370	16,100	1,923	
4050	Crawler, 8' wide, 100 H.P., diesel		88.65	1,800	5,385	16,200	1,786	
4060	10' wide, 150 H.P.		113.15	2,200	6,610	19,800	2,227	
4070	Concrete paver, 12' to 24' wide, 250 H.P.		101.40	1,550	4,675	14,000	1,746	
4080	Placer-spreader-trimmer, 24' wide, 300 H.P.		150.25	2,650	7,960	23,900	2,794	
4100	Pump, centrifugal gas pump, 1-1/2" diam., 65 GPM		3.90	48.50	145	435	60.20	
4200	2" diameter, 130 GPM		5.35	53.50	160	480	74.80	
4300	3" diameter, 250 GPM		5.65	55	165	495	78.20	
4400	6" diameter, 1500 GPM		30.45	172	515	1,550	346.60	
4500	Submersible electric pump, 1-1/4" diameter, 55 GPM		.39	16.35	49	147	12.90	
4600	1-1/2" diameter, 83 GPM		.43	18.65	56	168	14.65	
4700	2" diameter, 120 GPM		1.50	23.50	70	210	26	
4800	3" diameter, 300 GPM		2.69	41.50	125	375	46.50	
4900	4" diameter, 560 GPM		12.34	158	475	1,425	193.70	
5000	6" diameter, 1590 GPM		18.40	212	635	1,900	274.20	
5100	Diaphragm pump, gas, single, 1-1/2" diameter		1.18	49.50	148	445	39.05	
5200	2" diameter		4.25	61.50	185	555	71	
5300	3" diameter		4.25	61.50	185	555	71	

01 54 | Construction Aids

01 54 33 | Equipment Rental

		UNIT	HOURLY OPER. COST	RENT PER DAY	RENT PER WEEK	RENT PER MONTH	EQUIPMENT COST/DAY	
5400	Double, 4" diameter	Ea.	6.40	105	315	945	114.20	
5450	Pressure washer 5 GPM, 3000 psi		4.90	51.50	155	465	70.20	
5460	7 GPM, 3000 psi		6.45	60	180	540	87.60	
5500	Trash pump, self-priming, gas, 2" diameter		4.60	21	63	189	49.40	
5600	Diesel, 4" diameter		9.15	88.50	265	795	126.20	
5650	Diesel, 6" diameter		25.25	147	440	1,325	290	
5655	Grout Pump		26.35	268	805	2,425	371.80	
5700	Salamanders, L.P. gas fired, 100,000 Btu		3.96	13.65	41	123	39.90	
5705	50,000 Btu		2.21	10.35	31	93	23.90	
5720	Sandblaster, portable, open top, 3 C.F. capacity		.55	26.50	80	240	20.40	
5730	6 C.F. capacity		.95	40	120	360	31.60	
5740	Accessories for above		.13	21.50	65	195	14.05	
5750	Sander, floor		.70	14.35	43	129	14.20	
5760	Edger		.50	14.35	43	129	12.60	
5800	Saw, chain, gas engine, 18" long		2.30	21.50	64	192	31.20	
5900	Hydraulic powered, 36" long		.75	65	195	585	45	
5950	60" long		.75	66.50	200	600	46	
6000	Masonry, table mounted, 14" diameter, 5 H.P.		1.32	56.50	170	510	44.55	
6050	Portable cut-off, 8 H.P.		2.50	33.50	100	300	40	
6100	Circular, hand held, electric, 7-1/4" diameter		.23	4.67	14	42	4.65	
6200	12" diameter		.23	8	24	72	6.65	
6250	Wall saw, w/hydraulic power, 10 H.P.		9.70	61.50	185	555	114.60	
6275	Shot blaster, walk-behind, 20" wide		4.75	293	880	2,650	214	
6280	Sidewalk broom, walk-behind		2.52	78.50	235	705	67.15	
6300	Steam cleaner, 100 gallons per hour		3.70	76.50	230	690	75.60	
6310	200 gallons per hour		5.35	95	285	855	99.80	
6340	Tar Kettle/Pot, 400 gallons		15.60	75	225	675	169.80	
6350	Torch, cutting, acetylene-oxygen, 150' hose, excludes gases		.30	15	45	135	11.40	
6360	Hourly operating cost includes tips and gas		19.00				152	
6410	Toilet, portable chemical		.13	21	63	189	13.65	
6420	Recycle flush type		.15	25	75	225	16.20	
6430	Toilet, fresh water flush, garden hose,		.18	30.50	91	273	19.65	
6440	Hoisted, non-flush, for high rise		.15	24.50	74	222	16	
6465	Tractor, farm with attachment		21.50	297	890	2,675	350	
6480	Trailers, platform, flush deck, 2 axle, 3 ton capacity		1.45	20	60	180	23.60	
6500	25 ton capacity		5.45	117	350	1,050	113.60	
6600	40 ton capacity		7.00	163	490	1,475	154	
6700	3 axle, 50 ton capacity		7.55	180	540	1,625	168.40	
6800	75 ton capacity		9.40	235	705	2,125	216.20	
6810	Trailer mounted cable reel for high voltage line work		5.45	260	779	2,325	199.40	
6820	Trailer mounted cable tensioning rig		10.85	515	1,550	4,650	396.80	
6830	Cable pulling rig		73.05	2,900	8,680	26,000	2,320	
6900	Water tank trailer, engine driven discharge, 5000 gallons		7.00	142	425	1,275	141	
6925	10,000 gallons		9.50	197	590	1,775	194	
6950	Water truck, off highway, 6000 gallons		88.15	770	2,315	6,950	1,168	
7010	Tram car for high voltage line work, powered, 2 conductor		6.60	141	423	1,275	137.40	
7020	Transit (builder's level) with tripod		.09	14.65	44	132	9.50	
7030	Trench box, 3000 lb., 6' x 8'		.56	93.50	280	840	60.50	
7040	7200 lb., 6' x 20'		.75	125	375	1,125	81	
7050	8000 lb., 8' x 16'		1.08	180	540	1,625	116.65	
7060	9500 lb., 8' x 20'		1.21	201	603	1,800	130.30	
7065	11,000 lb., 8' x 24'		1.27	211	633	1,900	136.75	
7070	12,000 lb., 10' x 20'		1.50	249	748	2,250	161.60	
7100	Truck, pickup, 3/4 ton, 2 wheel drive		13.65	58.50	175	525	144.20	
7200	4 wheel drive		13.95	73.50	220	660	155.60	
7250	Crew carrier, 9 passenger		19.30	86.50	260	780	206.40	
7290	Flat bed truck, 20,000 lb. GVW		21.70	125	375	1,125	248.60	
7300	Tractor, 4 x 2, 220 H.P.		30.20	197	590	1,775	359.60	

01 54 | Construction Aids

01 54 33 | Equipment Rental

		UNIT	HOURLY OPER. COST	RENT PER DAY	RENT PER WEEK	RENT PER MONTH	EQUIPMENT COST/DAY	
7410	330 H.P.	Ea.	44.75	270	810	2,425	520	40
7500	6 x 4, 380 H.P.		51.35	315	945	2,825	599.80	
7600	450 H.P.		62.30	380	1,145	3,425	727.40	
7610	Tractor, with A frame, boom and winch, 225 H.P.		33.00	272	815	2,450	427	
7620	Vacuum truck, hazardous material, 2500 gallons		12.05	290	870	2,600	270.40	
7625	5,000 gallons		14.26	405	1,220	3,650	358.10	
7650	Vacuum, HEPA, 16 gallon, wet/dry		.85	18	54	162	17.60	
7655	55 gallon, wet/dry		.80	27	81	243	22.60	
7660	Water tank, portable		.16	26.50	80	240	17.30	
7690	Sewer/catch basin vacuum, 14 C.Y., 1500 gallons		18.53	610	1,830	5,500	514.25	
7700	Welder, electric, 200 amp		3.88	16.35	49	147	40.85	
7800	300 amp		5.71	20	60	180	57.70	
7900	Gas engine, 200 amp		14.65	23.50	70	210	131.20	
8000	300 amp		16.38	24.50	74	222	145.85	
8100	Wheelbarrow, any size		.08	13	39	117	8.45	
8200	Wrecking ball, 4000 lb.		2.35	68.50	205	615	59.80	
0010	**HIGHWAY EQUIPMENT RENTAL** without operators							50
0050	Asphalt batch plant, portable drum mixer, 100 ton/hr.	R015433 -10 Ea.	78.06	1,425	4,270	12,800	1,478	
0060	200 ton/hr.		89.18	1,500	4,530	13,600	1,619	
0070	300 ton/hr.		106.34	1,775	5,325	16,000	1,916	
0100	Backhoe attachment, long stick, up to 185 H.P., 10.5' long		.35	23.50	70	210	16.80	
0140	Up to 250 H.P., 12' long		.38	25	75	225	18.05	
0180	Over 250 H.P., 15' long		.53	35	105	315	25.25	
0200	Special dipper arm, up to 100 H.P., 32' long		1.08	71.50	215	645	51.65	
0240	Over 100 H.P., 33' long		1.34	89.50	268	805	64.30	
0280	Catch basin/sewer cleaning truck, 3 ton, 9 C.Y., 1000 gal.		42.70	385	1,160	3,475	573.60	
0300	Concrete batch plant, portable, electric, 200 C.Y./hr.		22.87	515	1,550	4,650	492.95	
0520	Grader/dozer attachment, ripper/scarifier, rear mounted, up to 135 H.P.		2.90	56.50	170	510	57.20	
0540	Up to 180 H.P.		3.90	88.50	265	795	84.20	
0580	Up to 250 H.P.		5.00	117	350	1,050	110	
0700	Pvmt. removal bucket, for hyd. excavator, up to 90 H.P.		1.90	53.50	160	480	47.20	
0740	Up to 200 H.P.		2.15	73.50	220	660	61.20	
0780	Over 200 H.P.		2.25	85	255	765	69	
0900	Aggregate spreader, self-propelled, 187 H.P.		53.00	685	2,050	6,150	834	
1000	Chemical spreader, 3 C.Y.		3.30	43.50	130	390	52.40	
1900	Hammermill, traveling, 250 H.P.		78.83	2,075	6,240	18,700	1,879	
2000	Horizontal borer, 3" diameter, 13 H.P. gas driven		6.35	55	165	495	83.80	
2150	Horizontal directional drill, 20,000 lb. thrust, 78 H.P. diesel		31.15	675	2,025	6,075	654.20	
2160	30,000 lb. thrust, 115 H.P.		38.90	1,025	3,100	9,300	931.20	
2170	50,000 lb. thrust, 170 H.P.		55.65	1,325	3,960	11,900	1,237	
2190	Mud trailer for HDD, 1500 gallons, 175 H.P., gas		31.90	152	455	1,375	346.20	
2200	Hydromulcher, diesel, 3000 gallon, for truck mounting		23.40	245	735	2,200	334.20	
2300	Gas, 600 gallon		8.60	98.50	295	885	127.80	
2400	Joint & crack cleaner, walk behind, 25 H.P.		3.95	50	150	450	61.60	
2500	Filler, trailer mounted, 400 gallons, 20 H.P.		9.50	210	630	1,900	202	
3000	Paint striper, self-propelled, 40 gallon, 22 H.P.		7.30	155	465	1,400	151.40	
3100	120 gallon, 120 H.P.		23.00	395	1,185	3,550	421	
3200	Post drivers, 6" I-Beam frame, for truck mounting		19.05	380	1,135	3,400	379.40	
3400	Road sweeper, self-propelled, 8' wide, 90 H.P.		39.80	575	1,720	5,150	662.40	
3450	Road sweeper, vacuum assisted, 4 C.Y., 220 gallons		75.75	635	1,900	5,700	986	
4000	Road mixer, self-propelled, 130 H.P.		46.85	765	2,295	6,875	833.80	
4100	310 H.P.		83.25	2,100	6,315	18,900	1,929	
4220	Cold mix paver, incl. pug mill and bitumen tank, 165 H.P.		97.00	2,275	6,855	20,600	2,147	
4240	Pavement brush, towed		3.20	93.50	280	840	81.60	
4250	Paver, asphalt, wheel or crawler, 130 H.P., diesel		96.65	2,275	6,795	20,400	2,132	
4300	Paver, road widener, gas 1' to 6', 67 H.P.		48.50	895	2,680	8,050	924	
4400	Diesel, 2' to 14', 88 H.P.		62.70	1,050	3,135	9,400	1,129	
4600	Slipform pavers, curb and gutter, 2 track, 75 H.P.		57.20	900	2,700	8,100	997.60	

For customer support on your Residential Cost Data, call 877.713.2617.

01 54 | Construction Aids

01 54 33 | Equipment Rental

			UNIT	HOURLY OPER. COST	RENT PER DAY	RENT PER WEEK	RENT PER MONTH	EQUIPMENT COST/DAY	
50	4700	4 track, 165 H.P.	Ea.	41.50	740	2,220	6,650	776	50
	4800	Median barrier, 215 H.P.		60.50	1,050	3,150	9,450	1,114	
	4901	Trailer, low bed, 75 ton capacity		10.10	235	705	2,125	221.80	
	5000	Road planer, walk behind, 10" cutting width, 10 H.P.		3.70	33.50	100	300	49.60	
	5100	Self-propelled, 12" cutting width, 64 H.P.		10.40	113	340	1,025	151.20	
	5120	Traffic line remover, metal ball blaster, truck mounted, 115 H.P.		47.95	730	2,190	6,575	821.60	
	5140	Grinder, truck mounted, 115 H.P.		54.60	790	2,375	7,125	911.80	
	5160	Walk-behind, 11 H.P.		4.20	53.50	160	480	65.60	
	5200	Pavement profiler, 4' to 6' wide, 450 H.P.		254.50	3,325	9,960	29,900	4,028	
	5300	8' to 10' wide, 750 H.P.		399.20	4,350	13,055	39,200	5,805	
	5400	Roadway plate, steel, 1" x 8' x 20'		.08	13.35	40	120	8.65	
	5600	Stabilizer, self-propelled, 150 H.P.		49.30	600	1,795	5,375	753.40	
	5700	310 H.P.		99.05	1,675	5,030	15,100	1,798	
	5800	Striper, truck mounted, 120 gallon paint, 460 H.P.		64.20	485	1,455	4,375	804.60	
	5900	Thermal paint heating kettle, 115 gallons		8.20	25.50	77	231	81	
	6000	Tar kettle, 330 gallon, trailer mounted		12.25	56.50	170	510	132	
	7000	Tunnel locomotive, diesel, 8 to 12 ton		31.85	585	1,750	5,250	604.80	
	7005	Electric, 10 ton		26.00	665	1,995	5,975	607	
	7010	Muck cars, 1/2 C.Y. capacity		2.00	24.50	73	219	30.60	
	7020	1 C.Y. capacity		2.25	32.50	97	291	37.40	
	7030	2 C.Y. capacity		2.35	36.50	110	330	40.80	
	7040	Side dump, 2 C.Y. capacity		2.60	45	135	405	47.80	
	7050	3 C.Y. capacity		3.45	50	150	450	57.60	
	7060	5 C.Y. capacity		4.95	63.50	190	570	77.60	
	7100	Ventilating blower for tunnel, 7-1/2 H.P.		2.05	51.50	155	465	47.40	
	7110	10 H.P.		2.28	51.50	155	465	49.25	
	7120	20 H.P.		3.48	67.50	202	605	68.25	
	7140	40 H.P.		5.76	96.50	290	870	104.10	
	7160	60 H.P.		8.77	152	455	1,375	161.15	
	7175	75 H.P.		11.66	207	620	1,850	217.30	
	7180	200 H.P.		23.65	305	910	2,725	371.20	
	7800	Windrow loader, elevating		55.00	1,300	3,935	11,800	1,227	
60	0010	**LIFTING AND HOISTING EQUIPMENT RENTAL** without operators R015433-10	Ea.						60
	0120	Aerial lift truck, 2 person, to 80'		26.30	705	2,120	6,350	634.40	
	0140	Boom work platform, 40' snorkel		12.00	275	825	2,475	261	
	0150	Crane, flatbed mounted, 3 ton capacity R015433-15		16.35	188	565	1,700	243.80	
	0200	Crane, climbing, 106' jib, 6000 lb. capacity, 410 fpm R312316-45		39.35	1,650	4,960	14,900	1,307	
	0300	101' jib, 10,250 lb. capacity, 270 fpm		45.95	2,100	6,280	18,800	1,624	
	0500	Tower, static, 130' high, 106' jib 6200 lb. capacity at 400 fpm		43.20	1,900	5,730	17,200	1,492	
	0600	Crawler mounted, lattice boom, 1/2 C.Y., 15 tons at 12' radius		36.98	625	1,870	5,600	669.85	
	0700	3/4 C.Y., 20 tons at 12' radius		49.31	780	2,340	7,025	862.50	
	0800	1 C.Y., 25 tons at 12' radius		65.75	1,050	3,120	9,350	1,150	
	0900	1-1/2 C.Y., 40 tons at 12' radius		65.75	1,050	3,150	9,450	1,156	
	1000	2 C.Y., 50 tons at 12' radius		69.70	1,225	3,680	11,000	1,294	
	1100	3 C.Y., 75 tons at 12' radius		74.70	1,425	4,305	12,900	1,459	
	1200	100 ton capacity, 60' boom		84.60	1,650	4,950	14,900	1,667	
	1300	165 ton capacity, 60' boom		107.75	1,925	5,785	17,400	2,019	
	1400	200 ton capacity, 70' boom		130.55	2,400	7,210	21,600	2,486	
	1500	350 ton capacity, 80' boom		182.90	3,625	10,845	32,500	3,632	
	1600	Truck mounted, lattice boom, 6 x 4, 20 tons at 10' radius		37.11	1,075	3,260	9,775	948.90	
	1700	25 tons at 10' radius		40.09	1,175	3,550	10,700	1,031	
	1800	8 x 4, 30 tons at 10' radius		43.61	1,250	3,780	11,300	1,105	
	1900	40 tons at 12' radius		46.70	1,325	3,950	11,900	1,164	
	2000	60 tons at 15' radius		53.07	1,400	4,180	12,500	1,261	
	2050	82 tons at 15' radius		59.80	1,475	4,460	13,400	1,370	
	2100	90 tons at 15' radius		67.36	1,625	4,860	14,600	1,511	
	2200	115 tons at 15' radius		76.11	1,800	5,430	16,300	1,695	
	2300	150 tons at 18' radius		83.65	1,900	5,720	17,200	1,813	

01 54 | Construction Aids

01 54 33 | Equipment Rental

		UNIT	HOURLY OPER. COST	RENT PER DAY	RENT PER WEEK	RENT PER MONTH	EQUIPMENT COST/DAY	
2350	165 tons at 18' radius	Ea.	90.05	2,025	6,060	18,200	1,932	60
2400	Truck mounted, hydraulic, 12 ton capacity		42.60	520	1,565	4,700	653.80	
2500	25 ton capacity		44.95	630	1,885	5,650	736.60	
2550	33 ton capacity		45.50	645	1,930	5,800	750	
2560	40 ton capacity		58.85	750	2,250	6,750	920.80	
2600	55 ton capacity		76.80	855	2,570	7,700	1,128	
2700	80 ton capacity		100.45	1,375	4,105	12,300	1,625	
2720	100 ton capacity		94.20	1,425	4,250	12,800	1,604	
2740	120 ton capacity		100.05	1,525	4,575	13,700	1,715	
2760	150 ton capacity		127.60	2,000	5,995	18,000	2,220	
2800	Self-propelled, 4 x 4, with telescoping boom, 5 ton		17.55	225	675	2,025	275.40	
2900	12-1/2 ton capacity		32.30	360	1,075	3,225	473.40	
3000	15 ton capacity		33.00	375	1,130	3,400	490	
3050	20 ton capacity		35.95	445	1,335	4,000	554.60	
3100	25 ton capacity		37.55	500	1,495	4,475	599.40	
3150	40 ton capacity		46.00	560	1,675	5,025	703	
3200	Derricks, guy, 20 ton capacity, 60' boom, 75' mast		27.55	405	1,214	3,650	463.20	
3300	100' boom, 115' mast		43.21	695	2,090	6,275	763.70	
3400	Stiffleg, 20 ton capacity, 70' boom, 37' mast		30.11	525	1,580	4,750	556.90	
3500	100' boom, 47' mast		46.29	845	2,530	7,600	876.30	
3550	Helicopter, small, lift to 1250 lb. maximum, w/pilot		103.70	3,275	9,800	29,400	2,790	
3600	Hoists, chain type, overhead, manual, 3/4 ton		.10	.33	1	3	1	
3900	10 ton		.70	6	18	54	9.20	
4000	Hoist and tower, 5000 lb. cap., portable electric, 40' high		4.95	233	699	2,100	179.40	
4100	For each added 10' section, add		.11	18.35	55	165	11.90	
4200	Hoist and single tubular tower, 5000 lb. electric, 100' high		6.70	325	976	2,925	248.80	
4300	For each added 6'-6" section, add		.19	31.50	95	285	20.50	
4400	Hoist and double tubular tower, 5000 lb., 100' high		7.19	360	1,075	3,225	272.50	
4500	For each added 6'-6" section, add		.21	35	105	315	22.70	
4550	Hoist and tower, mast type, 6000 lb., 100' high		7.76	370	1,115	3,350	285.10	
4570	For each added 10' section, add		.13	21.50	65	195	14.05	
4600	Hoist and tower, personnel, electric, 2000 lb., 100' @ 125 fpm		16.31	990	2,970	8,900	724.50	
4700	3000 lb., 100' @ 200 fpm		18.62	1,125	3,360	10,100	820.95	
4800	3000 lb., 150' @ 300 fpm		20.62	1,250	3,760	11,300	916.95	
4900	4000 lb., 100' @ 300 fpm		21.38	1,275	3,840	11,500	939.05	
5000	6000 lb., 100' @ 275 fpm		23.06	1,350	4,030	12,100	990.50	
5100	For added heights up to 500', add	L.F.	.01	1.67	5	15	1.10	
5200	Jacks, hydraulic, 20 ton	Ea.	.05	2	6	18	1.60	
5500	100 ton		.40	11.65	35	105	10.20	
6100	Jacks, hydraulic, climbing w/50' jackrods, control console, 30 ton cap.		2.01	134	402	1,200	96.50	
6150	For each added 10' jackrod section, add		.05	3.33	10	30	2.40	
6300	50 ton capacity		3.23	215	646	1,950	155.05	
6350	For each added 10' jackrod section, add		.06	4	12	36	2.90	
6500	125 ton capacity		8.45	565	1,690	5,075	405.60	
6550	For each added 10' jackrod section, add		.58	38.50	115	345	27.65	
6600	Cable jack, 10 ton capacity with 200' cable		1.68	112	336	1,000	80.65	
6650	For each added 50' of cable, add		.20	13.35	40	120	9.60	
0010	**WELLPOINT EQUIPMENT RENTAL** without operators		R015433-10					70
0020	Based on 2 months rental							
0100	Combination jetting & wellpoint pump, 60 H.P. diesel	Ea.	18.40	330	996	3,000	346.40	
0200	High pressure gas jet pump, 200 H.P., 300 psi	"	44.06	284	851	2,550	522.70	
0300	Discharge pipe, 8" diameter	L.F.	.01	.54	1.62	4.86	.40	
0350	12" diameter		.01	.79	2.38	7.15	.55	
0400	Header pipe, flows up to 150 GPM, 4" diameter		.01	.49	1.47	4.41	.35	
0500	400 GPM, 6" diameter		.01	.57	1.72	5.15	.40	
0600	800 GPM, 8" diameter		.01	.79	2.38	7.15	.55	
0700	1500 GPM, 10" diameter		.01	.84	2.51	7.55	.60	

01 54 | Construction Aids

01 54 33 | Equipment Rental

			UNIT	HOURLY OPER. COST	RENT PER DAY	RENT PER WEEK	RENT PER MONTH	EQUIPMENT COST/DAY	
70	0800	2500 GPM, 12" diameter	L.F.	.02	1.58	4.74	14.20	1.10	70
	0900	4500 GPM, 16" diameter		.03	2.02	6.07	18.20	1.45	
	0950	For quick coupling aluminum and plastic pipe, add		.03	2.09	6.28	18.85	1.50	
	1100	Wellpoint, 25' long, with fittings & riser pipe, 1-1/2" or 2" diameter	Ea.	.06	4.18	12.54	37.50	3	
	1200	Wellpoint pump, diesel powered, 4" suction, 20 H.P.		7.83	191	574	1,725	177.45	
	1300	6" suction, 30 H.P.		10.70	237	712	2,125	228	
	1400	8" suction, 40 H.P.		14.45	325	976	2,925	310.80	
	1500	10" suction, 75 H.P.		22.27	380	1,141	3,425	406.35	
	1600	12" suction, 100 H.P.		31.72	605	1,810	5,425	615.75	
	1700	12" suction, 175 H.P.		47.41	670	2,010	6,025	781.30	
80	0010	**MARINE EQUIPMENT RENTAL** without operators							80
	0200	Barge, 400 Ton, 30' wide x 90' long	Ea.	17.50	1,050	3,180	9,550	776	
	0240	800 Ton, 45' wide x 90' long		21.25	1,300	3,870	11,600	944	
	2000	Tugboat, diesel, 100 H.P.		38.20	217	650	1,950	435.60	
	2040	250 H.P.		79.25	395	1,180	3,550	870	
	2080	380 H.P.		156.80	1,175	3,535	10,600	1,961	
	3000	Small work boat, gas, 16-foot, 50 H.P.		17.15	60	180	540	173.20	
	4000	Large, diesel, 48-foot, 200 H.P.		89.40	1,250	3,740	11,200	1,463	

R015433-10

Crews

Crew No.	Bare Costs Hr.	Bare Costs Daily	Incl. Subs O&P Hr.	Incl. Subs O&P Daily	Cost Per Labor-Hour Bare Costs	Cost Per Labor-Hour Incl. O&P
Crew A-1	Hr.	Daily	Hr.	Daily	Bare Costs	Incl. O&P
1 Building Laborer	$24.45	$195.60	$41.05	$328.40	$24.45	$41.05
1 Concrete Saw, Gas Manual		81.00		89.10	10.13	11.14
8 L.H., Daily Totals		$276.60		$417.50	$34.58	$52.19
Crew A-1A	Hr.	Daily	Hr.	Daily	Bare Costs	Incl. O&P
1 Skilled Worker	$33.35	$266.80	$56.20	$449.60	$33.35	$56.20
1 Shot Blaster, 20"		214.00		235.40	26.75	29.43
8 L.H., Daily Totals		$480.80		$685.00	$60.10	$85.63
Crew A-1B	Hr.	Daily	Hr.	Daily	Bare Costs	Incl. O&P
1 Building Laborer	$24.45	$195.60	$41.05	$328.40	$24.45	$41.05
1 Concrete Saw		167.40		184.14	20.93	23.02
8 L.H., Daily Totals		$363.00		$512.54	$45.38	$64.07
Crew A-1C	Hr.	Daily	Hr.	Daily	Bare Costs	Incl. O&P
1 Building Laborer	$24.45	$195.60	$41.05	$328.40	$24.45	$41.05
1 Chain Saw, Gas, 18"		31.20		34.32	3.90	4.29
8 L.H., Daily Totals		$226.80		$362.72	$28.35	$45.34
Crew A-1D	Hr.	Daily	Hr.	Daily	Bare Costs	Incl. O&P
1 Building Laborer	$24.45	$195.60	$41.05	$328.40	$24.45	$41.05
1 Vibrating Plate, Gas, 18"		36.20		39.82	4.53	4.98
8 L.H., Daily Totals		$231.80		$368.22	$28.98	$46.03
Crew A-1E	Hr.	Daily	Hr.	Daily	Bare Costs	Incl. O&P
1 Building Laborer	$24.45	$195.60	$41.05	$328.40	$24.45	$41.05
1 Vibrating Plate, Gas, 21"		46.80		51.48	5.85	6.43
8 L.H., Daily Totals		$242.40		$379.88	$30.30	$47.48
Crew A-1F	Hr.	Daily	Hr.	Daily	Bare Costs	Incl. O&P
1 Building Laborer	$24.45	$195.60	$41.05	$328.40	$24.45	$41.05
1 Rammer/Tamper, Gas, 8"		50.00		55.00	6.25	6.88
8 L.H., Daily Totals		$245.60		$383.40	$30.70	$47.92
Crew A-1G	Hr.	Daily	Hr.	Daily	Bare Costs	Incl. O&P
1 Building Laborer	$24.45	$195.60	$41.05	$328.40	$24.45	$41.05
1 Rammer/Tamper, Gas, 15"		56.40		62.04	7.05	7.75
8 L.H., Daily Totals		$252.00		$390.44	$31.50	$48.81
Crew A-1H	Hr.	Daily	Hr.	Daily	Bare Costs	Incl. O&P
1 Building Laborer	$24.45	$195.60	$41.05	$328.40	$24.45	$41.05
1 Exterior Steam Cleaner		75.60		83.16	9.45	10.40
8 L.H., Daily Totals		$271.20		$411.56	$33.90	$51.45
Crew A-1J	Hr.	Daily	Hr.	Daily	Bare Costs	Incl. O&P
1 Building Laborer	$24.45	$195.60	$41.05	$328.40	$24.45	$41.05
1 Cultivator, Walk-Behind, 5 H.P.		46.70		51.37	5.84	6.42
8 L.H., Daily Totals		$242.30		$379.77	$30.29	$47.47
Crew A-1K	Hr.	Daily	Hr.	Daily	Bare Costs	Incl. O&P
1 Building Laborer	$24.45	$195.60	$41.05	$328.40	$24.45	$41.05
1 Cultivator, Walk-Behind, 8 H.P.		72.65		79.92	9.08	9.99
8 L.H., Daily Totals		$268.25		$408.32	$33.53	$51.04
Crew A-1M	Hr.	Daily	Hr.	Daily	Bare Costs	Incl. O&P
1 Building Laborer	$24.45	$195.60	$41.05	$328.40	$24.45	$41.05
1 Snow Blower, Walk-Behind		67.15		73.86	8.39	9.23
8 L.H., Daily Totals		$262.75		$402.26	$32.84	$50.28

Crew No.	Bare Costs Hr.	Bare Costs Daily	Incl. Subs O&P Hr.	Incl. Subs O&P Daily	Cost Per Labor-Hour Bare Costs	Cost Per Labor-Hour Incl. O&P
Crew A-2	Hr.	Daily	Hr.	Daily	Bare Costs	Incl. O&P
2 Laborers	$24.45	$391.20	$41.05	$656.80	$25.55	$42.65
1 Truck Driver (light)	27.75	222.00	45.85	366.80		
1 Flatbed Truck, Gas, 1.5 Ton		245.40		269.94	10.23	11.25
24 L.H., Daily Totals		$858.60		$1293.54	$35.77	$53.90
Crew A-2A	Hr.	Daily	Hr.	Daily	Bare Costs	Incl. O&P
2 Laborers	$24.45	$391.20	$41.05	$656.80	$25.55	$42.65
1 Truck Driver (light)	27.75	222.00	45.85	366.80		
1 Flatbed Truck, Gas, 1.5 Ton		245.40		269.94		
1 Concrete Saw		167.40		184.14	17.20	18.92
24 L.H., Daily Totals		$1026.00		$1477.68	$42.75	$61.57
Crew A-2B	Hr.	Daily	Hr.	Daily	Bare Costs	Incl. O&P
1 Truck Driver (light)	$27.75	$222.00	$45.85	$366.80	$27.75	$45.85
1 Flatbed Truck, Gas, 1.5 Ton		245.40		269.94	30.68	33.74
8 L.H., Daily Totals		$467.40		$636.74	$58.42	$79.59
Crew A-3A	Hr.	Daily	Hr.	Daily	Bare Costs	Incl. O&P
1 Equip. Oper. (light)	$33.55	$268.40	$55.60	$444.80	$33.55	$55.60
1 Pickup Truck, 4 x 4, 3/4 Ton		155.60		171.16	19.45	21.40
8 L.H., Daily Totals		$424.00		$615.96	$53.00	$77.00
Crew A-3B	Hr.	Daily	Hr.	Daily	Bare Costs	Incl. O&P
1 Equip. Oper. (medium)	$34.90	$279.20	$57.85	$462.80	$31.68	$52.42
1 Truck Driver (heavy)	28.45	227.60	47.00	376.00		
1 Dump Truck, 12 C.Y., 400 H.P.		691.00		760.10		
1 F.E. Loader, W.M., 2.5 C.Y.		519.80		571.78	75.67	83.24
16 L.H., Daily Totals		$1717.60		$2170.68	$107.35	$135.67
Crew A-3C	Hr.	Daily	Hr.	Daily	Bare Costs	Incl. O&P
1 Equip. Oper. (light)	$33.55	$268.40	$55.60	$444.80	$33.55	$55.60
1 Loader, Skid Steer, 78 H.P.		318.60		350.46	39.83	43.81
8 L.H., Daily Totals		$587.00		$795.26	$73.38	$99.41
Crew A-3D	Hr.	Daily	Hr.	Daily	Bare Costs	Incl. O&P
1 Truck Driver (light)	$27.75	$222.00	$45.85	$366.80	$27.75	$45.85
1 Pickup Truck, 4 x 4, 3/4 Ton		155.60		171.16		
1 Flatbed Trailer, 25 Ton		113.60		124.96	33.65	37.02
8 L.H., Daily Totals		$491.20		$662.92	$61.40	$82.86
Crew A-3E	Hr.	Daily	Hr.	Daily	Bare Costs	Incl. O&P
1 Equip. Oper. (crane)	$35.65	$285.20	$59.05	$472.40	$32.05	$53.02
1 Truck Driver (heavy)	28.45	227.60	47.00	376.00		
1 Pickup Truck, 4 x 4, 3/4 Ton		155.60		171.16	9.72	10.70
16 L.H., Daily Totals		$668.40		$1019.56	$41.77	$63.72
Crew A-3F	Hr.	Daily	Hr.	Daily	Bare Costs	Incl. O&P
1 Equip. Oper. (crane)	$35.65	$285.20	$59.05	$472.40	$32.05	$53.02
1 Truck Driver (heavy)	28.45	227.60	47.00	376.00		
1 Pickup Truck, 4 x 4, 3/4 Ton		155.60		171.16		
1 Truck Tractor, 6x4, 380 H.P.		599.80		659.78		
1 Lowbed Trailer, 75 Ton		221.80		243.98	61.08	67.18
16 L.H., Daily Totals		$1490.00		$1923.32	$93.13	$120.21

Crews

Crew No.		Bare Costs		Incl. Subs O&P		Cost Per Labor-Hour	
Crew A-3G	Hr.	Daily	Hr.	Daily	Bare Costs	Incl. O&P
1 Equip. Oper. (crane)	$35.65	$285.20	$59.05	$472.40	$32.05	$53.02
1 Truck Driver (heavy)	28.45	227.60	47.00	376.00		
1 Pickup Truck, 4 x 4, 3/4 Ton		155.60		171.16		
1 Truck Tractor, 6x4, 450 H.P.		727.40		800.14		
1 Lowbed Trailer, 75 Ton		221.80		243.98	69.05	75.95
16 L.H., Daily Totals		$1617.60		$2063.68	$101.10	$128.98
Crew A-3H	Hr.	Daily	Hr.	Daily	Bare Costs	Incl. O&P
1 Equip. Oper. (crane)	$35.65	$285.20	$59.05	$472.40	$35.65	$59.05
1 Hyd. Crane, 12 Ton (Daily)		860.80		946.88	107.60	118.36
8 L.H., Daily Totals		$1146.00		$1419.28	$143.25	$177.41
Crew A-3I	Hr.	Daily	Hr.	Daily	Bare Costs	Incl. O&P
1 Equip. Oper. (crane)	$35.65	$285.20	$59.05	$472.40	$35.65	$59.05
1 Hyd. Crane, 25 Ton (Daily)		989.60		1088.56	123.70	136.07
8 L.H., Daily Totals		$1274.80		$1560.96	$159.35	$195.12
Crew A-3J	Hr.	Daily	Hr.	Daily	Bare Costs	Incl. O&P
1 Equip. Oper. (crane)	$35.65	$285.20	$59.05	$472.40	$35.65	$59.05
1 Hyd. Crane, 40 Ton (Daily)		1221.00		1343.10	152.63	167.89
8 L.H., Daily Totals		$1506.20		$1815.50	$188.28	$226.94
Crew A-3K	Hr.	Daily	Hr.	Daily	Bare Costs	Incl. O&P
1 Equip. Oper. (crane)	$35.65	$285.20	$59.05	$472.40	$33.42	$55.38
1 Equip. Oper. (oiler)	31.20	249.60	51.70	413.60		
1 Hyd. Crane, 55 Ton (Daily)		1469.00		1615.90		
1 P/U Truck, 3/4 Ton (Daily)		167.20		183.92	102.26	112.49
16 L.H., Daily Totals		$2171.00		$2685.82	$135.69	$167.86
Crew A-3L	Hr.	Daily	Hr.	Daily	Bare Costs	Incl. O&P
1 Equip. Oper. (crane)	$35.65	$285.20	$59.05	$472.40	$33.42	$55.38
1 Equip. Oper. (oiler)	31.20	249.60	51.70	413.60		
1 Hyd. Crane, 80 Ton (Daily)		2174.00		2391.40		
1 P/U Truck, 3/4 Ton (Daily)		167.20		183.92	146.32	160.96
16 L.H., Daily Totals		$2876.00		$3461.32	$179.75	$216.33
Crew A-3M	Hr.	Daily	Hr.	Daily	Bare Costs	Incl. O&P
1 Equip. Oper. (crane)	$35.65	$285.20	$59.05	$472.40	$33.42	$55.38
1 Equip. Oper. (oiler)	31.20	249.60	51.70	413.60		
1 Hyd. Crane, 100 Ton (Daily)		2169.00		2385.90		
1 P/U Truck, 3/4 Ton (Daily)		167.20		183.92	146.01	160.61
16 L.H., Daily Totals		$2871.00		$3455.82	$179.44	$215.99
Crew A-3N	Hr.	Daily	Hr.	Daily	Bare Costs	Incl. O&P
1 Equip. Oper. (crane)	$35.65	$285.20	$59.05	$472.40	$35.65	$59.05
1 Tower Cane (monthly)		1128.00		1240.80	141.00	155.10
8 L.H., Daily Totals		$1413.20		$1713.20	$176.65	$214.15
Crew A-3P	Hr.	Daily	Hr.	Daily	Bare Costs	Incl. O&P
1 Equip. Oper. (light)	$33.55	$268.40	$55.60	$444.80	$33.55	$55.60
1 A.T. Forklift, 42' lift		526.40		579.04	65.80	72.38
8 L.H., Daily Totals		$794.80		$1023.84	$99.35	$127.98
Crew A-3Q	Hr.	Daily	Hr.	Daily	Bare Costs	Incl. O&P
1 Equip. Oper. (light)	$33.55	$268.40	$55.60	$444.80	$33.55	$55.60
1 Pickup Truck, 4 x 4, 3/4 Ton		155.60		171.16		
1 Flatbed trailer, 3 Ton		23.60		25.96	22.40	24.64
8 L.H., Daily Totals		$447.60		$641.92	$55.95	$80.24

Crew No.		Bare Costs		Incl. Subs O&P		Cost Per Labor-Hour	
Crew A-4	Hr.	Daily	Hr.	Daily	Bare Costs	Incl. O&P
2 Carpenters	$32.85	$525.60	$55.15	$882.40	$31.45	$52.50
1 Painter, Ordinary	28.65	229.20	47.20	377.60		
24 L.H., Daily Totals		$754.80		$1260.00	$31.45	$52.50
Crew A-5	Hr.	Daily	Hr.	Daily	Bare Costs	Incl. O&P
2 Laborers	$24.45	$391.20	$41.05	$656.80	$24.82	$41.58
.25 Truck Driver (light)	27.75	55.50	45.85	91.70		
.25 Flatbed Truck, Gas, 1.5 Ton		61.35		67.48	3.41	3.75
18 L.H., Daily Totals		$508.05		$815.99	$28.23	$45.33
Crew A-6	Hr.	Daily	Hr.	Daily	Bare Costs	Incl. O&P
1 Instrument Man	$33.35	$266.80	$56.20	$449.60	$32.67	$54.70
1 Rodman/Chainman	32.00	256.00	53.20	425.60		
1 Level, Electronic		54.80		60.28	3.42	3.77
16 L.H., Daily Totals		$577.60		$935.48	$36.10	$58.47
Crew A-7	Hr.	Daily	Hr.	Daily	Bare Costs	Incl. O&P
1 Chief of Party	$40.20	$321.60	$67.15	$537.20	$35.18	$58.85
1 Instrument Man	33.35	266.80	56.20	449.60		
1 Rodman/Chainman	32.00	256.00	53.20	425.60		
1 Level, Electronic		54.80		60.28	2.28	2.51
24 L.H., Daily Totals		$899.20		$1472.68	$37.47	$61.36
Crew A-8	Hr.	Daily	Hr.	Daily	Bare Costs	Incl. O&P
1 Chief of Party	$40.20	$321.60	$67.15	$537.20	$34.39	$57.44
1 Instrument Man	33.35	266.80	56.20	449.60		
2 Rodmen/Chainmen	32.00	512.00	53.20	851.20		
1 Level, Electronic		54.80		60.28	1.71	1.88
32 L.H., Daily Totals		$1155.20		$1898.28	$36.10	$59.32
Crew A-9	Hr.	Daily	Hr.	Daily	Bare Costs	Incl. O&P
1 Asbestos Foreman	$35.05	$280.40	$59.50	$476.00	$34.61	$58.76
7 Asbestos Workers	34.55	1934.80	58.65	3284.40		
64 L.H., Daily Totals		$2215.20		$3760.40	$34.61	$58.76
Crew A-10A	Hr.	Daily	Hr.	Daily	Bare Costs	Incl. O&P
1 Asbestos Foreman	$35.05	$280.40	$59.50	$476.00	$34.72	$58.93
2 Asbestos Workers	34.55	552.80	58.65	938.40		
24 L.H., Daily Totals		$833.20		$1414.40	$34.72	$58.93
Crew A-10B	Hr.	Daily	Hr.	Daily	Bare Costs	Incl. O&P
1 Asbestos Foreman	$35.05	$280.40	$59.50	$476.00	$34.67	$58.86
3 Asbestos Workers	34.55	829.20	58.65	1407.60		
32 L.H., Daily Totals		$1109.60		$1883.60	$34.67	$58.86
Crew A-10C	Hr.	Daily	Hr.	Daily	Bare Costs	Incl. O&P
3 Asbestos Workers	$34.55	$829.20	$58.65	$1407.60	$34.55	$58.65
1 Flatbed Truck, Gas, 1.5 Ton		245.40		269.94	10.23	11.25
24 L.H., Daily Totals		$1074.60		$1677.54	$44.77	$69.90
Crew A-10D	Hr.	Daily	Hr.	Daily	Bare Costs	Incl. O&P
2 Asbestos Workers	$34.55	$552.80	$58.65	$938.40	$33.99	$57.01
1 Equip. Oper. (crane)	35.65	285.20	59.05	472.40		
1 Equip. Oper. (oiler)	31.20	249.60	51.70	413.60		
1 Hydraulic Crane, 33 Ton		750.00		825.00	23.44	25.78
32 L.H., Daily Totals		$1837.60		$2649.40	$57.42	$82.79

674

For customer support on your Residential Cost Data, call 877.713.2617.

Crews

Crew No.	Bare Costs		Incl. Subs O&P		Cost Per Labor-Hour	
Crew A-11	Hr.	Daily	Hr.	Daily	Bare Costs	Incl. O&P
1 Asbestos Foreman	$35.05	$280.40	$59.50	$476.00	$34.61	$58.76
7 Asbestos Workers	34.55	1934.80	58.65	3284.40		
2 Chip. Hammers, 12 Lb., Elec.		40.00		44.00	0.63	0.69
64 L.H., Daily Totals		$2255.20		$3804.40	$35.24	$59.44
Crew A-12	Hr.	Daily	Hr.	Daily	Bare Costs	Incl. O&P
1 Asbestos Foreman	$35.05	$280.40	$59.50	$476.00	$34.61	$58.76
7 Asbestos Workers	34.55	1934.80	58.65	3284.40		
1 Trk-Mtd Vac, 14 CY, 1500 Gal.		514.25		565.67		
1 Flatbed Truck, 20,000 GVW		248.60		273.46	11.92	13.11
64 L.H., Daily Totals		$2978.05		$4599.53	$46.53	$71.87
Crew A-13	Hr.	Daily	Hr.	Daily	Bare Costs	Incl. O&P
1 Equip. Oper. (light)	$33.55	$268.40	$55.60	$444.80	$33.55	$55.60
1 Trk-Mtd Vac, 14 CY, 1500 Gal.		514.25		565.67		
1 Flatbed Truck, 20,000 GVW		248.60		273.46	95.36	104.89
8 L.H., Daily Totals		$1031.25		$1283.93	$128.91	$160.49
Crew B-1	Hr.	Daily	Hr.	Daily	Bare Costs	Incl. O&P
1 Labor Foreman (outside)	$26.45	$211.60	$44.40	$355.20	$25.12	$42.17
2 Laborers	24.45	391.20	41.05	656.80		
24 L.H., Daily Totals		$602.80		$1012.00	$25.12	$42.17
Crew B-1A	Hr.	Daily	Hr.	Daily	Bare Costs	Incl. O&P
1 Labor Foreman (outside)	$26.45	$211.60	$44.40	$355.20	$25.12	$42.17
2 Laborers	24.45	391.20	41.05	656.80		
2 Cutting Torches		22.80		25.08		
2 Sets of Gases		304.00		334.40	13.62	14.98
24 L.H., Daily Totals		$929.60		$1371.48	$38.73	$57.15
Crew B-1B	Hr.	Daily	Hr.	Daily	Bare Costs	Incl. O&P
1 Labor Foreman (outside)	$26.45	$211.60	$44.40	$355.20	$27.75	$46.39
2 Laborers	24.45	391.20	41.05	656.80		
1 Equip. Oper. (crane)	35.65	285.20	59.05	472.40		
2 Cutting Torches		22.80		25.08		
2 Sets of Gases		304.00		334.40		
1 Hyd. Crane, 12 Ton		653.80		719.18	30.64	33.71
32 L.H., Daily Totals		$1868.60		$2563.06	$58.39	$80.10
Crew B-1C	Hr.	Daily	Hr.	Daily	Bare Costs	Incl. O&P
1 Labor Foreman (outside)	$26.45	$211.60	$44.40	$355.20	$25.12	$42.17
2 Laborers	24.45	391.20	41.05	656.80		
1 Aerial Lift Truck, 60' Boom		435.60		479.16	18.15	19.97
24 L.H., Daily Totals		$1038.40		$1491.16	$43.27	$62.13
Crew B-1D	Hr.	Daily	Hr.	Daily	Bare Costs	Incl. O&P
2 Laborers	$24.45	$391.20	$41.05	$656.80	$24.45	$41.05
1 Small Work Boat, Gas, 50 H.P.		173.20		190.52		
1 Pressure Washer, 7 GPM		87.60		96.36	16.30	17.93
16 L.H., Daily Totals		$652.00		$943.68	$40.75	$58.98
Crew B-1E	Hr.	Daily	Hr.	Daily	Bare Costs	Incl. O&P
1 Labor Foreman (outside)	$26.45	$211.60	$44.40	$355.20	$24.95	$41.89
3 Laborers	24.45	586.80	41.05	985.20		
1 Work Boat, Diesel, 200 H.P.		1463.00		1609.30		
2 Pressure Washer, 7 GPM		175.20		192.72	51.19	56.31
32 L.H., Daily Totals		$2436.60		$3142.42	$76.14	$98.20

Crew No.	Bare Costs		Incl. Subs O&P		Cost Per Labor-Hour	
Crew B-1F	Hr.	Daily	Hr.	Daily	Bare Costs	Incl. O&P
2 Skilled Workers	$33.35	$533.60	$56.20	$899.20	$30.38	$51.15
1 Laborer	24.45	195.60	41.05	328.40		
1 Small Work Boat, Gas, 50 H.P.		173.20		190.52		
1 Pressure Washer, 7 GPM		87.60		96.36	10.87	11.95
24 L.H., Daily Totals		$990.00		$1514.48	$41.25	$63.10
Crew B-1G	Hr.	Daily	Hr.	Daily	Bare Costs	Incl. O&P
2 Laborers	$24.45	$391.20	$41.05	$656.80	$24.45	$41.05
1 Small Work Boat, Gas, 50 H.P.		173.20		190.52	10.82	11.91
16 L.H., Daily Totals		$564.40		$847.32	$35.27	$52.96
Crew B-1H	Hr.	Daily	Hr.	Daily	Bare Costs	Incl. O&P
2 Skilled Workers	$33.35	$533.60	$56.20	$899.20	$30.38	$51.15
1 Laborer	24.45	195.60	41.05	328.40		
1 Small Work Boat, Gas, 50 H.P.		173.20		190.52	7.22	7.94
24 L.H., Daily Totals		$902.40		$1418.12	$37.60	$59.09
Crew B-1J	Hr.	Daily	Hr.	Daily	Bare Costs	Incl. O&P
1 Labor Foreman (inside)	$24.95	$199.60	$41.90	$335.20	$24.70	$41.48
1 Laborer	24.45	195.60	41.05	328.40		
16 L.H., Daily Totals		$395.20		$663.60	$24.70	$41.48
Crew B-1K	Hr.	Daily	Hr.	Daily	Bare Costs	Incl. O&P
1 Carpenter Foreman (inside)	$33.35	$266.80	$56.00	$448.00	$33.10	$55.58
1 Carpenter	32.85	262.80	55.15	441.20		
16 L.H., Daily Totals		$529.60		$889.20	$33.10	$55.58
Crew B-2	Hr.	Daily	Hr.	Daily	Bare Costs	Incl. O&P
1 Labor Foreman (outside)	$26.45	$211.60	$44.40	$355.20	$24.85	$41.72
4 Laborers	24.45	782.40	41.05	1313.60		
40 L.H., Daily Totals		$994.00		$1668.80	$24.85	$41.72
Crew B-2A	Hr.	Daily	Hr.	Daily	Bare Costs	Incl. O&P
1 Labor Foreman (outside)	$26.45	$211.60	$44.40	$355.20	$25.12	$42.17
2 Laborers	24.45	391.20	41.05	656.80		
1 Aerial Lift Truck, 60' Boom		435.60		479.16	18.15	19.97
24 L.H., Daily Totals		$1038.40		$1491.16	$43.27	$62.13
Crew B-3	Hr.	Daily	Hr.	Daily	Bare Costs	Incl. O&P
1 Labor Foreman (outside)	$26.45	$211.60	$44.40	$355.20	$27.86	$46.39
2 Laborers	24.45	391.20	41.05	656.80		
1 Equip. Oper. (medium)	34.90	279.20	57.85	462.80		
2 Truck Drivers (heavy)	28.45	455.20	47.00	752.00		
1 Crawler Loader, 3 C.Y.		1188.00		1306.80		
2 Dump Trucks, 12 C.Y., 400 H.P.		1382.00		1520.20	53.54	58.90
48 L.H., Daily Totals		$3907.20		$5053.80	$81.40	$105.29
Crew B-3A	Hr.	Daily	Hr.	Daily	Bare Costs	Incl. O&P
4 Laborers	$24.45	$782.40	$41.05	$1313.60	$26.54	$44.41
1 Equip. Oper. (medium)	34.90	279.20	57.85	462.80		
1 Hyd. Excavator, 1.5 C.Y.		1030.00		1133.00	25.75	28.32
40 L.H., Daily Totals		$2091.60		$2909.40	$52.29	$72.73
Crew B-3B	Hr.	Daily	Hr.	Daily	Bare Costs	Incl. O&P
2 Laborers	$24.45	$391.20	$41.05	$656.80	$28.06	$46.74
1 Equip. Oper. (medium)	34.90	279.20	57.85	462.80		
1 Truck Driver (heavy)	28.45	227.60	47.00	376.00		
1 Backhoe Loader, 80 H.P.		391.80		430.98		
1 Dump Truck, 12 C.Y., 400 H.P.		691.00		760.10	33.84	37.22
32 L.H., Daily Totals		$1980.80		$2686.68	$61.90	$83.96

Crews

Crew No.		Bare Costs		Incl. Subs O&P		Cost Per Labor-Hour	
Crew B-3C	Hr.	Daily	Hr.	Daily	Bare Costs	Incl. O&P
3 Laborers	$24.45	$586.80	$41.05	$985.20	$27.06	$45.25
1 Equip. Oper. (medium)	34.90	279.20	57.85	462.80		
1 Crawler Loader, 4 C.Y.		1561.00		1717.10	48.78	53.66
32 L.H., Daily Totals		$2427.00		$3165.10	$75.84	$98.91
Crew B-4	Hr.	Daily	Hr.	Daily	Bare Costs	Incl. O&P
1 Labor Foreman (outside)	$26.45	$211.60	$44.40	$355.20	$25.45	$42.60
4 Laborers	24.45	782.40	41.05	1313.60		
1 Truck Driver (heavy)	28.45	227.60	47.00	376.00		
1 Truck Tractor, 220 H.P.		359.60		395.56		
1 Flatbed Trailer, 40 Ton		154.00		169.40	10.70	11.77
48 L.H., Daily Totals		$1735.20		$2609.76	$36.15	$54.37
Crew B-5	Hr.	Daily	Hr.	Daily	Bare Costs	Incl. O&P
1 Labor Foreman (outside)	$26.45	$211.60	$44.40	$355.20	$26.94	$45.08
3 Laborers	24.45	586.80	41.05	985.20		
1 Equip. Oper. (medium)	34.90	279.20	57.85	462.80		
1 Air Compressor, 250 cfm		201.40		221.54		
2 Breakers, Pavement, 60 lb.		19.60		21.56		
2 -50' Air Hoses, 1.5"		11.60		12.76		
1 Crawler Loader, 3 C.Y.		1188.00		1306.80	35.52	39.07
40 L.H., Daily Totals		$2498.20		$3365.86	$62.45	$84.15
Crew B-5A	Hr.	Daily	Hr.	Daily	Bare Costs	Incl. O&P
1 Labor Foreman (outside)	$26.45	$211.60	$44.40	$355.20	$27.78	$46.33
6 Laborers	24.45	1173.60	41.05	1970.40		
2 Equip. Oper. (medium)	34.90	558.40	57.85	925.60		
1 Equip. Oper. (light)	33.55	268.40	55.60	444.80		
2 Truck Drivers (heavy)	28.45	455.20	47.00	752.00		
1 Air Compressor, 365 cfm		263.80		290.18		
2 Breakers, Pavement, 60 lb.		19.60		21.56		
8 -50' Air Hoses, 1"		32.80		36.08		
2 Dump Trucks, 8 C.Y., 220 H.P.		822.40		904.64	11.86	13.05
96 L.H., Daily Totals		$3805.80		$5700.46	$39.64	$59.38
Crew B-5B	Hr.	Daily	Hr.	Daily	Bare Costs	Incl. O&P
1 Powderman	$33.35	$266.80	$56.20	$449.60	$31.42	$52.15
2 Equip. Oper. (medium)	34.90	558.40	57.85	925.60		
3 Truck Drivers (heavy)	28.45	682.80	47.00	1128.00		
1 F.E. Loader, W.M.,2.5 C.Y.		519.80		571.78		
3 Dump Trucks, 12 C.Y., 400 H.P.		2073.00		2280.30		
1 Air Compressor, 365 CFM		263.80		290.18	59.51	65.46
48 L.H., Daily Totals		$4364.60		$5645.46	$90.93	$117.61
Crew B-5C	Hr.	Daily	Hr.	Daily	Bare Costs	Incl. O&P
3 Laborers	$24.45	$586.80	$41.05	$985.20	$29.00	$48.22
1 Equip. Oper. (medium)	34.90	279.20	57.85	462.80		
2 Truck Drivers (heavy)	28.45	455.20	47.00	752.00		
1 Equip. Oper. (crane)	35.65	285.20	59.05	472.40		
1 Equip. Oper. (oiler)	31.20	249.60	51.70	413.60		
2 Dump Trucks, 12 C.Y., 400 H.P.		1382.00		1520.20		
1 Crawler Loader, 4 C.Y.		1561.00		1717.10		
1 S.P. Crane, 4x4, 25 Ton		599.40		659.34	55.35	60.88
64 L.H., Daily Totals		$5398.40		$6982.64	$84.35	$109.10

Crew No.		Bare Costs		Incl. Subs O&P		Cost Per Labor-Hour	
Crew B-5D	Hr.	Daily	Hr.	Daily	Bare Costs	Incl. O&P
1 Labor Foreman (outside)	$26.45	$211.60	$44.40	$355.20	$27.19	$45.40
3 Laborers	24.45	586.80	41.05	985.20		
1 Equip. Oper. (medium)	34.90	279.20	57.85	462.80		
1 Truck Driver (heavy)	28.45	227.60	47.00	376.00		
1 Air Compressor, 250 cfm		201.40		221.54		
2 Breakers, Pavement, 60 lb.		19.60		21.56		
2 -50' Air Hoses, 1.5"		11.60		12.76		
1 Crawler Loader, 3 C.Y.		1188.00		1306.80		
1 Dump Truck, 12 C.Y., 400 H.P.		691.00		760.10	43.99	48.39
48 L.H., Daily Totals		$3416.80		$4501.96	$71.18	$93.79
Crew B-6	Hr.	Daily	Hr.	Daily	Bare Costs	Incl. O&P
2 Laborers	$24.45	$391.20	$41.05	$656.80	$27.48	$45.90
1 Equip. Oper. (light)	33.55	268.40	55.60	444.80		
1 Backhoe Loader, 48 H.P.		364.20		400.62	15.18	16.69
24 L.H., Daily Totals		$1023.80		$1502.22	$42.66	$62.59
Crew B-6B	Hr.	Daily	Hr.	Daily	Bare Costs	Incl. O&P
2 Labor Foremen (outside)	$26.45	$423.20	$44.40	$710.40	$25.12	$42.17
4 Laborers	24.45	782.40	41.05	1313.60		
1 S.P. Crane, 4x4, 5 Ton		275.40		302.94		
1 Flatbed Truck, Gas, 1.5 Ton		245.40		269.94		
1 Butt Fusion Mach., 4"-12" diam.		382.10		420.31	18.81	20.69
48 L.H., Daily Totals		$2108.50		$3017.19	$43.93	$62.86
Crew B-6C	Hr.	Daily	Hr.	Daily	Bare Costs	Incl. O&P
2 Labor Foremen (outside)	$26.45	$423.20	$44.40	$710.40	$25.12	$42.17
4 Laborers	24.45	782.40	41.05	1313.60		
1 S.P. Crane, 4x4, 12 Ton		473.40		520.74		
1 Flatbed Truck, Gas, 3 Ton		303.20		333.52		
1 Butt Fusion Mach., 8"-24" diam.		829.70		912.67	33.46	36.81
48 L.H., Daily Totals		$2811.90		$3790.93	$58.58	$78.98
Crew B-7	Hr.	Daily	Hr.	Daily	Bare Costs	Incl. O&P
1 Labor Foreman (outside)	$26.45	$211.60	$44.40	$355.20	$26.52	$44.41
4 Laborers	24.45	782.40	41.05	1313.60		
1 Equip. Oper. (medium)	34.90	279.20	57.85	462.80		
1 Brush Chipper, 12", 130 H.P.		391.40		430.54		
1 Crawler Loader, 3 C.Y.		1188.00		1306.80		
2 Chain Saws, Gas, 36" Long		90.00		99.00	34.78	38.25
48 L.H., Daily Totals		$2942.60		$3967.94	$61.30	$82.67
Crew B-7A	Hr.	Daily	Hr.	Daily	Bare Costs	Incl. O&P
2 Laborers	$24.45	$391.20	$41.05	$656.80	$27.48	$45.90
1 Equip. Oper. (light)	33.55	268.40	55.60	444.80		
1 Rake w/Tractor		354.25		389.68		
2 Chain Saw, Gas, 18"		62.40		68.64	17.36	19.10
24 L.H., Daily Totals		$1076.25		$1559.92	$44.84	$65.00
Crew B-7B	Hr.	Daily	Hr.	Daily	Bare Costs	Incl. O&P
1 Labor Foreman (outside)	$26.45	$211.60	$44.40	$355.20	$26.80	$44.78
4 Laborers	24.45	782.40	41.05	1313.60		
1 Equip. Oper. (medium)	34.90	279.20	57.85	462.80		
1 Truck Driver (heavy)	28.45	227.60	47.00	376.00		
1 Brush Chipper, 12", 130 H.P.		391.40		430.54		
1 Crawler Loader, 3 C.Y.		1188.00		1306.80		
2 Chain Saws, Gas, 36" Long		90.00		99.00		
1 Dump Truck, 8 C.Y., 220 H.P.		411.20		452.32	37.15	40.87
56 L.H., Daily Totals		$3581.40		$4796.26	$63.95	$85.65

Crews

Crew No.	Bare Costs		Incl. Subs O&P		Cost Per Labor-Hour	
Crew B-7C	**Hr.**	**Daily**	**Hr.**	**Daily**	**Bare Costs**	**Incl. O&P**
1 Labor Foreman (outside)	$26.45	$211.60	$44.40	$355.20	$26.80	$44.78
4 Laborers	24.45	782.40	41.05	1313.60		
1 Equip. Oper. (medium)	34.90	279.20	57.85	462.80		
1 Truck Driver (heavy)	28.45	227.60	47.00	376.00		
1 Brush Chipper, 12", 130 H.P.		391.40		430.54		
1 Crawler Loader, 3 C.Y.		1188.00		1306.80		
2 Chain Saws, Gas, 36" Long		90.00		99.00		
1 Dump Truck, 12 C.Y., 400 H.P.		691.00		760.10	42.15	46.37
56 L.H., Daily Totals		$3861.20		$5104.04	$68.95	$91.14
Crew B-8	**Hr.**	**Daily**	**Hr.**	**Daily**	**Bare Costs**	**Incl. O&P**
1 Labor Foreman (outside)	$26.45	$211.60	$44.40	$355.20	$28.86	$48.03
2 Laborers	24.45	391.20	41.05	656.80		
2 Equip. Oper. (medium)	34.90	558.40	57.85	925.60		
2 Truck Drivers (heavy)	28.45	455.20	47.00	752.00		
1 Hyd. Crane, 25 Ton		736.60		810.26		
1 Crawler Loader, 3 C.Y.		1188.00		1306.80		
2 Dump Trucks, 12 C.Y., 400 H.P.		1382.00		1520.20	59.05	64.95
56 L.H., Daily Totals		$4923.00		$6326.86	$87.91	$112.98
Crew B-9	**Hr.**	**Daily**	**Hr.**	**Daily**	**Bare Costs**	**Incl. O&P**
1 Labor Foreman (outside)	$26.45	$211.60	$44.40	$355.20	$24.85	$41.72
4 Laborers	24.45	782.40	41.05	1313.60		
1 Air Compressor, 250 cfm		201.40		221.54		
2 Breakers, Pavement, 60 lb.		19.60		21.56		
2 -50' Air Hoses, 1.5"		11.60		12.76	5.82	6.40
40 L.H., Daily Totals		$1226.60		$1924.66	$30.66	$48.12
Crew B-9A	**Hr.**	**Daily**	**Hr.**	**Daily**	**Bare Costs**	**Incl. O&P**
2 Laborers	$24.45	$391.20	$41.05	$656.80	$25.78	$43.03
1 Truck Driver (heavy)	28.45	227.60	47.00	376.00		
1 Water Tank Trailer, 5000 Gal.		141.00		155.10		
1 Truck Tractor, 220 H.P.		359.60		395.56		
2 -50' Discharge Hoses, 3"		3.00		3.30	20.98	23.08
24 L.H., Daily Totals		$1122.40		$1586.76	$46.77	$66.11
Crew B-9B	**Hr.**	**Daily**	**Hr.**	**Daily**	**Bare Costs**	**Incl. O&P**
2 Laborers	$24.45	$391.20	$41.05	$656.80	$25.78	$43.03
1 Truck Driver (heavy)	28.45	227.60	47.00	376.00		
2 -50' Discharge Hoses, 3"		3.00		3.30		
1 Water Tank Trailer, 5000 Gal.		141.00		155.10		
1 Truck Tractor, 220 H.P.		359.60		395.56		
1 Pressure Washer		70.20		77.22	23.91	26.30
24 L.H., Daily Totals		$1192.60		$1663.98	$49.69	$69.33
Crew B-9D	**Hr.**	**Daily**	**Hr.**	**Daily**	**Bare Costs**	**Incl. O&P**
1 Labor Foreman (outside)	$26.45	$211.60	$44.40	$355.20	$24.85	$41.72
4 Common Laborers	24.45	782.40	41.05	1313.60		
1 Air Compressor, 250 cfm		201.40		221.54		
2 -50' Air Hoses, 1.5"		11.60		12.76		
2 Air Powered Tampers		52.40		57.64	6.63	7.30
40 L.H., Daily Totals		$1259.40		$1960.74	$31.48	$49.02
Crew B-10	**Hr.**	**Daily**	**Hr.**	**Daily**	**Bare Costs**	**Incl. O&P**
1 Equip. Oper. (medium)	34.90	279.20	57.85	462.80	34.90	57.85
8 L.H., Daily Totals		$279.20		$462.80	$34.90	$57.85
Crew B-10A	**Hr.**	**Daily**	**Hr.**	**Daily**	**Bare Costs**	**Incl. O&P**
1 Equip. Oper. (medium)	$34.90	$279.20	$57.85	$462.80	$34.90	$57.85
1 Roller, 2-Drum, W.B., 7.5 H.P.		180.00		198.00	22.50	24.75
8 L.H., Daily Totals		$459.20		$660.80	$57.40	$82.60

Crew No.	Bare Costs		Incl. Subs O&P		Cost Per Labor-Hour	
Crew B-10B	**Hr.**	**Daily**	**Hr.**	**Daily**	**Bare Costs**	**Incl. O&P**
1 Equip. Oper. (medium)	$34.90	$279.20	$57.85	$462.80	$34.90	$57.85
1 Dozer, 200 H.P.		1387.00		1525.70	173.38	190.71
8 L.H., Daily Totals		$1666.20		$1988.50	$208.28	$248.56
Crew B-10C	**Hr.**	**Daily**	**Hr.**	**Daily**	**Bare Costs**	**Incl. O&P**
1 Equip. Oper. (medium)	$34.90	$279.20	$57.85	$462.80	$34.90	$57.85
1 Dozer, 200 H.P.		1387.00		1525.70		
1 Vibratory Roller, Towed, 23 Ton		396.00		435.60	222.88	245.16
8 L.H., Daily Totals		$2062.20		$2424.10	$257.77	$303.01
Crew B-10D	**Hr.**	**Daily**	**Hr.**	**Daily**	**Bare Costs**	**Incl. O&P**
1 Equip. Oper. (medium)	$34.90	$279.20	$57.85	$462.80	$34.90	$57.85
1 Dozer, 200 H.P.		1387.00		1525.70		
1 Sheepsft. Roller, Towed		425.80		468.38	226.60	249.26
8 L.H., Daily Totals		$2092.00		$2456.88	$261.50	$307.11
Crew B-10E	**Hr.**	**Daily**	**Hr.**	**Daily**	**Bare Costs**	**Incl. O&P**
1 Equip. Oper. (medium)	$34.90	$279.20	$57.85	$462.80	$34.90	$57.85
1 Tandem Roller, 5 Ton		157.40		173.14	19.68	21.64
8 L.H., Daily Totals		$436.60		$635.94	$54.58	$79.49
Crew B-10F	**Hr.**	**Daily**	**Hr.**	**Daily**	**Bare Costs**	**Incl. O&P**
1 Equip. Oper. (medium)	$34.90	$279.20	$57.85	$462.80	$34.90	$57.85
1 Tandem Roller, 10 Ton		236.80		260.48	29.60	32.56
8 L.H., Daily Totals		$516.00		$723.28	$64.50	$90.41
Crew B-10G	**Hr.**	**Daily**	**Hr.**	**Daily**	**Bare Costs**	**Incl. O&P**
1 Equip. Oper. (medium)	$34.90	$279.20	$57.85	$462.80	$34.90	$57.85
1 Sheepsfoot Roller, 240 H.P.		1206.00		1326.60	150.75	165.82
8 L.H., Daily Totals		$1485.20		$1789.40	$185.65	$223.68
Crew B-10H	**Hr.**	**Daily**	**Hr.**	**Daily**	**Bare Costs**	**Incl. O&P**
1 Equip. Oper. (medium)	$34.90	$279.20	$57.85	$462.80	$34.90	$57.85
1 Diaphragm Water Pump, 2"		71.00		78.10		
1 -20' Suction Hose, 2"		1.95		2.15		
2 -50' Discharge Hoses, 2"		1.80		1.98	9.34	10.28
8 L.H., Daily Totals		$353.95		$545.02	$44.24	$68.13
Crew B-10I	**Hr.**	**Daily**	**Hr.**	**Daily**	**Bare Costs**	**Incl. O&P**
1 Equip. Oper. (medium)	$34.90	$279.20	$57.85	$462.80	$34.90	$57.85
1 Diaphragm Water Pump, 4"		114.20		125.62		
1 -20' Suction Hose, 4"		3.25		3.58		
2 -50' Discharge Hoses, 4"		4.70		5.17	15.27	16.80
8 L.H., Daily Totals		$401.35		$597.16	$50.17	$74.65
Crew B-10J	**Hr.**	**Daily**	**Hr.**	**Daily**	**Bare Costs**	**Incl. O&P**
1 Equip. Oper. (medium)	$34.90	$279.20	$57.85	$462.80	$34.90	$57.85
1 Centrifugal Water Pump, 3"		78.20		86.02		
1 -20' Suction Hose, 3"		2.85		3.13		
2 -50' Discharge Hoses, 3"		3.00		3.30	10.51	11.56
8 L.H., Daily Totals		$363.25		$555.26	$45.41	$69.41
Crew B-10K	**Hr.**	**Daily**	**Hr.**	**Daily**	**Bare Costs**	**Incl. O&P**
1 Equip. Oper. (medium)	$34.90	$279.20	$57.85	$462.80	$34.90	$57.85
1 Centr. Water Pump, 6"		346.60		381.26		
1 -20' Suction Hose, 6"		11.50		12.65		
2 -50' Discharge Hoses, 6"		12.20		13.42	46.29	50.92
8 L.H., Daily Totals		$649.50		$870.13	$81.19	$108.77

Crews

Crew No.	Bare Costs		Incl. Subs O&P		Cost Per Labor-Hour	
	Hr.	Daily	Hr.	Daily	Bare Costs	Incl. O&P
Crew B-10L						
1 Equip. Oper. (medium)	$34.90	$279.20	$57.85	$462.80	$34.90	$57.85
1 Dozer, 80 H.P.		472.40		519.64	59.05	64.95
8 L.H., Daily Totals		$751.60		$982.44	$93.95	$122.81
Crew B-10M	Hr.	Daily	Hr.	Daily	Bare Costs	Incl. O&P
1 Equip. Oper. (medium)	$34.90	$279.20	$57.85	$462.80	$34.90	$57.85
1 Dozer, 300 H.P.		1897.00		2086.70	237.13	260.84
8 L.H., Daily Totals		$2176.20		$2549.50	$272.02	$318.69
Crew B-10N	Hr.	Daily	Hr.	Daily	Bare Costs	Incl. O&P
1 Equip. Oper. (medium)	$34.90	$279.20	$57.85	$462.80	$34.90	$57.85
1 F.E. Loader, T.M., 1.5 C.Y		522.00		574.20	65.25	71.78
8 L.H., Daily Totals		$801.20		$1037.00	$100.15	$129.63
Crew B-10O	Hr.	Daily	Hr.	Daily	Bare Costs	Incl. O&P
1 Equip. Oper. (medium)	$34.90	$279.20	$57.85	$462.80	$34.90	$57.85
1 F.E. Loader, T.M., 2.25 C.Y.		956.40		1052.04	119.55	131.51
8 L.H., Daily Totals		$1235.60		$1514.84	$154.45	$189.35
Crew B-10P	Hr.	Daily	Hr.	Daily	Bare Costs	Incl. O&P
1 Equip. Oper. (medium)	$34.90	$279.20	$57.85	$462.80	$34.90	$57.85
1 Crawler Loader, 3 C.Y.		1188.00		1306.80	148.50	163.35
8 L.H., Daily Totals		$1467.20		$1769.60	$183.40	$221.20
Crew B-10Q	Hr.	Daily	Hr.	Daily	Bare Costs	Incl. O&P
1 Equip. Oper. (medium)	$34.90	$279.20	$57.85	$462.80	$34.90	$57.85
1 Crawler Loader, 4 C.Y.		1561.00		1717.10	195.13	214.64
8 L.H., Daily Totals		$1840.20		$2179.90	$230.03	$272.49
Crew B-10R	Hr.	Daily	Hr.	Daily	Bare Costs	Incl. O&P
1 Equip. Oper. (medium)	$34.90	$279.20	$57.85	$462.80	$34.90	$57.85
1 F.E. Loader, W.M., 1 C.Y.		299.00		328.90	37.38	41.11
8 L.H., Daily Totals		$578.20		$791.70	$72.28	$98.96
Crew B-10S	Hr.	Daily	Hr.	Daily	Bare Costs	Incl. O&P
1 Equip. Oper. (medium)	$34.90	$279.20	$57.85	$462.80	$34.90	$57.85
1 F.E. Loader, W.M., 1.5 C.Y.		378.40		416.24	47.30	52.03
8 L.H., Daily Totals		$657.60		$879.04	$82.20	$109.88
Crew B-10T	Hr.	Daily	Hr.	Daily	Bare Costs	Incl. O&P
1 Equip. Oper. (medium)	$34.90	$279.20	$57.85	$462.80	$34.90	$57.85
1 F.E. Loader, W.M., 2.5 C.Y.		519.80		571.78	64.97	71.47
8 L.H., Daily Totals		$799.00		$1034.58	$99.88	$129.32
Crew B-10U	Hr.	Daily	Hr.	Daily	Bare Costs	Incl. O&P
1 Equip. Oper. (medium)	$34.90	$279.20	$57.85	$462.80	$34.90	$57.85
1 F.E. Loader, W.M., 5.5 C.Y.		1082.00		1190.20	135.25	148.78
8 L.H., Daily Totals		$1361.20		$1653.00	$170.15	$206.63
Crew B-10V	Hr.	Daily	Hr.	Daily	Bare Costs	Incl. O&P
1 Equip. Oper. (medium)	$34.90	$279.20	$57.85	$462.80	$34.90	$57.85
1 Dozer, 700 H.P.		4979.00		5476.90	622.38	684.61
8 L.H., Daily Totals		$5258.20		$5939.70	$657.27	$742.46
Crew B-10W	Hr.	Daily	Hr.	Daily	Bare Costs	Incl. O&P
1 Equip. Oper. (medium)	$34.90	$279.20	$57.85	$462.80	$34.90	$57.85
1 Dozer, 105 H.P.		602.80		663.08	75.35	82.89
8 L.H., Daily Totals		$882.00		$1125.88	$110.25	$140.74

Crew No.	Bare Costs		Incl. Subs O&P		Cost Per Labor-Hour	
	Hr.	Daily	Hr.	Daily	Bare Costs	Incl. O&P
Crew B-10X						
1 Equip. Oper. (medium)	$34.90	$279.20	$57.85	$462.80	$34.90	$57.85
1 Dozer, 410 H.P.		2405.00		2645.50	300.63	330.69
8 L.H., Daily Totals		$2684.20		$3108.30	$335.52	$388.54
Crew B-10Y	Hr.	Daily	Hr.	Daily	Bare Costs	Incl. O&P
1 Equip. Oper. (medium)	$34.90	$279.20	$57.85	$462.80	$34.90	$57.85
1 Vibr. Roller, Towed, 12 Ton		549.80		604.78	68.72	75.60
8 L.H., Daily Totals		$829.00		$1067.58	$103.63	$133.45
Crew B-11A	Hr.	Daily	Hr.	Daily	Bare Costs	Incl. O&P
1 Equipment Oper. (med.)	$34.90	$279.20	$57.85	$462.80	$29.68	$49.45
1 Laborer	24.45	195.60	41.05	328.40		
1 Dozer, 200 H.P.		1387.00		1525.70	86.69	95.36
16 L.H., Daily Totals		$1861.80		$2316.90	$116.36	$144.81
Crew B-11B	Hr.	Daily	Hr.	Daily	Bare Costs	Incl. O&P
1 Equipment Oper. (light)	$33.55	$268.40	$55.60	$444.80	$29.00	$48.33
1 Laborer	24.45	195.60	41.05	328.40		
1 Air Powered Tamper		26.20		28.82		
1 Air Compressor, 365 cfm		263.80		290.18		
2 -50' Air Hoses, 1.5"		11.60		12.76	18.85	20.73
16 L.H., Daily Totals		$765.60		$1104.96	$47.85	$69.06
Crew B-11C	Hr.	Daily	Hr.	Daily	Bare Costs	Incl. O&P
1 Equipment Oper. (med.)	$34.90	$279.20	$57.85	$462.80	$29.68	$49.45
1 Laborer	24.45	195.60	41.05	328.40		
1 Backhoe Loader, 48 H.P.		364.20		400.62	22.76	25.04
16 L.H., Daily Totals		$839.00		$1191.82	$52.44	$74.49
Crew B-11K	Hr.	Daily	Hr.	Daily	Bare Costs	Incl. O&P
1 Equipment Oper. (med.)	$34.90	$279.20	$57.85	$462.80	$29.68	$49.45
1 Laborer	24.45	195.60	41.05	328.40		
1 Trencher, Chain Type, 8' D		3402.00		3742.20	212.63	233.89
16 L.H., Daily Totals		$3876.80		$4533.40	$242.30	$283.34
Crew B-11L	Hr.	Daily	Hr.	Daily	Bare Costs	Incl. O&P
1 Equipment Oper. (med.)	$34.90	$279.20	$57.85	$462.80	$29.68	$49.45
1 Laborer	24.45	195.60	41.05	328.40		
1 Grader, 30,000 Lbs.		736.20		809.82	46.01	50.61
16 L.H., Daily Totals		$1211.00		$1601.02	$75.69	$100.06
Crew B-11M	Hr.	Daily	Hr.	Daily	Bare Costs	Incl. O&P
1 Equipment Oper. (med.)	$34.90	$279.20	$57.85	$462.80	$29.68	$49.45
1 Laborer	24.45	195.60	41.05	328.40		
1 Backhoe Loader, 80 H.P.		391.80		430.98	24.49	26.94
16 L.H., Daily Totals		$866.60		$1222.18	$54.16	$76.39
Crew B-11W	Hr.	Daily	Hr.	Daily	Bare Costs	Incl. O&P
1 Equipment Operator (med.)	$34.90	$279.20	$57.85	$462.80	$28.65	$47.41
1 Common Laborer	24.45	195.60	41.05	328.40		
10 Truck Drivers (heavy)	28.45	2276.00	47.00	3760.00		
1 Dozer, 200 H.P.		1387.00		1525.70		
1 Vibratory Roller, Towed, 23 Ton		396.00		435.60		
10 Dump Trucks, 8 C.Y., 220 H.P.		4112.00		4523.20	61.41	67.55
96 L.H., Daily Totals		$8645.80		$11035.70	$90.06	$114.96

Crews

Crew No.	Bare Costs		Incl. Subs O&P		Cost Per Labor-Hour	
Crew B-11Y	**Hr.**	**Daily**	**Hr.**	**Daily**	**Bare Costs**	**Incl. O&P**
1 Labor Foreman (outside)	$26.45	$211.60	$44.40	$355.20	$28.16	$47.02
5 Common Laborers	24.45	978.00	41.05	1642.00		
3 Equipment Operators (med.)	34.90	837.60	57.85	1388.40		
1 Dozer, 80 H.P.		472.40		519.64		
2 Roller, 2-Drum, W.B., 7.5 H.P.		360.00		396.00		
4 Vibrating Plate, Gas, 21"		187.20		205.92	14.16	15.58
72 L.H., Daily Totals		$3046.80		$4507.16	$42.32	$62.60
Crew B-12A	**Hr.**	**Daily**	**Hr.**	**Daily**	**Bare Costs**	**Incl. O&P**
1 Equip. Oper. (crane)	$35.65	$285.20	$59.05	$472.40	$30.05	$50.05
1 Laborer	24.45	195.60	41.05	328.40		
1 Hyd. Excavator, 1 C.Y.		812.60		893.86	50.79	55.87
16 L.H., Daily Totals		$1293.40		$1694.66	$80.84	$105.92
Crew B-12B	**Hr.**	**Daily**	**Hr.**	**Daily**	**Bare Costs**	**Incl. O&P**
1 Equip. Oper. (crane)	$35.65	$285.20	$59.05	$472.40	$30.05	$50.05
1 Laborer	24.45	195.60	41.05	328.40		
1 Hyd. Excavator, 1.5 C.Y.		1030.00		1133.00	64.38	70.81
16 L.H., Daily Totals		$1510.80		$1933.80	$94.42	$120.86
Crew B-12C	**Hr.**	**Daily**	**Hr.**	**Daily**	**Bare Costs**	**Incl. O&P**
1 Equip. Oper. (crane)	$35.65	$285.20	$59.05	$472.40	$30.05	$50.05
1 Laborer	24.45	195.60	41.05	328.40		
1 Hyd. Excavator, 2 C.Y.		1175.00		1292.50	73.44	80.78
16 L.H., Daily Totals		$1655.80		$2093.30	$103.49	$130.83
Crew B-12D	**Hr.**	**Daily**	**Hr.**	**Daily**	**Bare Costs**	**Incl. O&P**
1 Equip. Oper. (crane)	$35.65	$285.20	$59.05	$472.40	$30.05	$50.05
1 Laborer	24.45	195.60	41.05	328.40		
1 Hyd. Excavator, 3.5 C.Y.		2440.00		2684.00	152.50	167.75
16 L.H., Daily Totals		$2920.80		$3484.80	$182.55	$217.80
Crew B-12E	**Hr.**	**Daily**	**Hr.**	**Daily**	**Bare Costs**	**Incl. O&P**
1 Equip. Oper. (crane)	$35.65	$285.20	$59.05	$472.40	$30.05	$50.05
1 Laborer	24.45	195.60	41.05	328.40		
1 Hyd. Excavator, .5 C.Y.		448.20		493.02	28.01	30.81
16 L.H., Daily Totals		$929.00		$1293.82	$58.06	$80.86
Crew B-12F	**Hr.**	**Daily**	**Hr.**	**Daily**	**Bare Costs**	**Incl. O&P**
1 Equip. Oper. (crane)	$35.65	$285.20	$59.05	$472.40	$30.05	$50.05
1 Laborer	24.45	195.60	41.05	328.40		
1 Hyd. Excavator, .75 C.Y.		654.80		720.28	40.92	45.02
16 L.H., Daily Totals		$1135.60		$1521.08	$70.97	$95.07
Crew B-12G	**Hr.**	**Daily**	**Hr.**	**Daily**	**Bare Costs**	**Incl. O&P**
1 Equip. Oper. (crane)	$35.65	$285.20	$59.05	$472.40	$30.05	$50.05
1 Laborer	24.45	195.60	41.05	328.40		
1 Crawler Crane, 15 Ton		669.85		736.84		
1 Clamshell Bucket, .5 C.Y.		38.20		42.02	44.25	48.68
16 L.H., Daily Totals		$1188.85		$1579.66	$74.30	$98.73
Crew B-12H	**Hr.**	**Daily**	**Hr.**	**Daily**	**Bare Costs**	**Incl. O&P**
1 Equip. Oper. (crane)	$35.65	$285.20	$59.05	$472.40	$30.05	$50.05
1 Laborer	24.45	195.60	41.05	328.40		
1 Crawler Crane, 25 Ton		1150.00		1265.00		
1 Clamshell Bucket, 1 C.Y.		47.80		52.58	74.86	82.35
16 L.H., Daily Totals		$1678.60		$2118.38	$104.91	$132.40
Crew B-12I	**Hr.**	**Daily**	**Hr.**	**Daily**	**Bare Costs**	**Incl. O&P**
1 Equip. Oper. (crane)	$35.65	$285.20	$59.05	$472.40	$30.05	$50.05
1 Laborer	24.45	195.60	41.05	328.40		
1 Crawler Crane, 20 Ton		862.50		948.75		
1 Dragline Bucket, .75 C.Y.		20.60		22.66	55.19	60.71
16 L.H., Daily Totals		$1363.90		$1772.21	$85.24	$110.76
Crew B-12J	**Hr.**	**Daily**	**Hr.**	**Daily**	**Bare Costs**	**Incl. O&P**
1 Equip. Oper. (crane)	$35.65	$285.20	$59.05	$472.40	$30.05	$50.05
1 Laborer	24.45	195.60	41.05	328.40		
1 Gradall, 5/8 C.Y.		882.20		970.42	55.14	60.65
16 L.H., Daily Totals		$1363.00		$1771.22	$85.19	$110.70
Crew B-12K	**Hr.**	**Daily**	**Hr.**	**Daily**	**Bare Costs**	**Incl. O&P**
1 Equip. Oper. (crane)	$35.65	$285.20	$59.05	$472.40	$30.05	$50.05
1 Laborer	24.45	195.60	41.05	328.40		
1 Gradall, 3 Ton, 1 C.Y.		1002.00		1102.20	62.63	68.89
16 L.H., Daily Totals		$1482.80		$1903.00	$92.67	$118.94
Crew B-12L	**Hr.**	**Daily**	**Hr.**	**Daily**	**Bare Costs**	**Incl. O&P**
1 Equip. Oper. (crane)	$35.65	$285.20	$59.05	$472.40	$30.05	$50.05
1 Laborer	24.45	195.60	41.05	328.40		
1 Crawler Crane, 15 Ton		669.85		736.84		
1 F.E. Attachment, .5 C.Y.		60.00		66.00	45.62	50.18
16 L.H., Daily Totals		$1210.65		$1603.64	$75.67	$100.23
Crew B-12M	**Hr.**	**Daily**	**Hr.**	**Daily**	**Bare Costs**	**Incl. O&P**
1 Equip. Oper. (crane)	$35.65	$285.20	$59.05	$472.40	$30.05	$50.05
1 Laborer	24.45	195.60	41.05	328.40		
1 Crawler Crane, 20 Ton		862.50		948.75		
1 F.E. Attachment, .75 C.Y.		65.40		71.94	57.99	63.79
16 L.H., Daily Totals		$1408.70		$1821.49	$88.04	$113.84
Crew B-12N	**Hr.**	**Daily**	**Hr.**	**Daily**	**Bare Costs**	**Incl. O&P**
1 Equip. Oper. (crane)	$35.65	$285.20	$59.05	$472.40	$30.05	$50.05
1 Laborer	24.45	195.60	41.05	328.40		
1 Crawler Crane, 25 Ton		1150.00		1265.00		
1 F.E. Attachment, 1 C.Y.		71.20		78.32	76.33	83.96
16 L.H., Daily Totals		$1702.00		$2144.12	$106.38	$134.01
Crew B-12O	**Hr.**	**Daily**	**Hr.**	**Daily**	**Bare Costs**	**Incl. O&P**
1 Equip. Oper. (crane)	$35.65	$285.20	$59.05	$472.40	$30.05	$50.05
1 Laborer	24.45	195.60	41.05	328.40		
1 Crawler Crane, 40 Ton		1156.00		1271.60		
1 F.E. Attachment, 1.5 C.Y.		80.00		88.00	77.25	84.97
16 L.H., Daily Totals		$1716.80		$2160.40	$107.30	$135.03
Crew B-12P	**Hr.**	**Daily**	**Hr.**	**Daily**	**Bare Costs**	**Incl. O&P**
1 Equip. Oper. (crane)	$35.65	$285.20	$59.05	$472.40	$30.05	$50.05
1 Laborer	24.45	195.60	41.05	328.40		
1 Crawler Crane, 40 Ton		1156.00		1271.60		
1 Dragline Bucket, 1.5 C.Y.		33.60		36.96	74.35	81.78
16 L.H., Daily Totals		$1670.40		$2109.36	$104.40	$131.84
Crew B-12Q	**Hr.**	**Daily**	**Hr.**	**Daily**	**Bare Costs**	**Incl. O&P**
1 Equip. Oper. (crane)	$35.65	$285.20	$59.05	$472.40	$30.05	$50.05
1 Laborer	24.45	195.60	41.05	328.40		
1 Hyd. Excavator, 5/8 C.Y.		589.40		648.34	36.84	40.52
16 L.H., Daily Totals		$1070.20		$1449.14	$66.89	$90.57

Crews

Crew No.	Bare Costs		Incl. Subs O&P		Cost Per Labor-Hour	

Crew B-12S	Hr.	Daily	Hr.	Daily	Bare Costs	Incl. O&P
1 Equip. Oper. (crane)	$35.65	$285.20	$59.05	$472.40	$30.05	$50.05
1 Laborer	24.45	195.60	41.05	328.40		
1 Hyd. Excavator, 2.5 C.Y.		1605.00		1765.50	100.31	110.34
16 L.H., Daily Totals		$2085.80		$2566.30	$130.36	$160.39

Crew B-12T	Hr.	Daily	Hr.	Daily	Bare Costs	Incl. O&P
1 Equip. Oper. (crane)	$35.65	$285.20	$59.05	$472.40	$30.05	$50.05
1 Laborer	24.45	195.60	41.05	328.40		
1 Crawler Crane, 75 Ton		1459.00		1604.90		
1 F.E. Attachment, 3 C.Y.		103.20		113.52	97.64	107.40
16 L.H., Daily Totals		$2043.00		$2519.22	$127.69	$157.45

Crew B-12V	Hr.	Daily	Hr.	Daily	Bare Costs	Incl. O&P
1 Equip. Oper. (crane)	$35.65	$285.20	$59.05	$472.40	$30.05	$50.05
1 Laborer	24.45	195.60	41.05	328.40		
1 Crawler Crane, 75 Ton		1459.00		1604.90		
1 Dragline Bucket, 3 C.Y.		52.60		57.86	94.47	103.92
16 L.H., Daily Totals		$1992.40		$2463.56	$124.53	$153.97

Crew B-12Y	Hr.	Daily	Hr.	Daily	Bare Costs	Incl. O&P
1 Equip. Oper. (crane)	$35.65	$285.20	$59.05	$472.40	$28.18	$47.05
2 Laborers	24.45	391.20	41.05	656.80		
1 Hyd. Excavator, 3.5 C.Y.		2440.00		2684.00	101.67	111.83
24 L.H., Daily Totals		$3116.40		$3813.20	$129.85	$158.88

Crew B-12Z	Hr.	Daily	Hr.	Daily	Bare Costs	Incl. O&P
1 Equip. Oper. (crane)	$35.65	$285.20	$59.05	$472.40	$28.18	$47.05
2 Laborers	24.45	391.20	41.05	656.80		
1 Hyd. Excavator, 2.5 C.Y.		1605.00		1765.50	66.88	73.56
24 L.H., Daily Totals		$2281.40		$2894.70	$95.06	$120.61

Crew B-13	Hr.	Daily	Hr.	Daily	Bare Costs	Incl. O&P
1 Labor Foreman (outside)	$26.45	$211.60	$44.40	$355.20	$26.65	$44.61
4 Laborers	24.45	782.40	41.05	1313.60		
1 Equip. Oper. (crane)	35.65	285.20	59.05	472.40		
1 Hyd. Crane, 25 Ton		736.60		810.26	15.35	16.88
48 L.H., Daily Totals		$2015.80		$2951.46	$42.00	$61.49

Crew B-13A	Hr.	Daily	Hr.	Daily	Bare Costs	Incl. O&P
1 Labor Foreman (outside)	$26.45	$211.60	$44.40	$355.20	$28.86	$48.03
2 Laborers	24.45	391.20	41.05	656.80		
2 Equipment Operators (med.)	34.90	558.40	57.85	925.60		
2 Truck Drivers (heavy)	28.45	455.20	47.00	752.00		
1 Crawler Crane, 75 Ton		1459.00		1604.90		
1 Crawler Loader, 4 C.Y.		1561.00		1717.10		
2 Dump Trucks, 8 C.Y., 220 H.P.		822.40		904.64	68.61	75.48
56 L.H., Daily Totals		$5458.80		$6916.24	$97.48	$123.50

Crew B-13B	Hr.	Daily	Hr.	Daily	Bare Costs	Incl. O&P
1 Labor Foreman (outside)	$26.45	$211.60	$44.40	$355.20	$27.30	$45.62
4 Laborers	24.45	782.40	41.05	1313.60		
1 Equip. Oper. (crane)	35.65	285.20	59.05	472.40		
1 Equip. Oper. (oiler)	31.20	249.60	51.70	413.60		
1 Hyd. Crane, 55 Ton		1128.00		1240.80	20.14	22.16
56 L.H., Daily Totals		$2656.80		$3795.60	$47.44	$67.78

Crew B-13C	Hr.	Daily	Hr.	Daily	Bare Costs	Incl. O&P
1 Labor Foreman (outside)	$26.45	$211.60	$44.40	$355.20	$27.30	$45.62
4 Laborers	24.45	782.40	41.05	1313.60		
1 Equip. Oper. (crane)	35.65	285.20	59.05	472.40		
1 Equip. Oper. (oiler)	31.20	249.60	51.70	413.60		
1 Crawler Crane, 100 Ton		1667.00		1833.70	29.77	32.74
56 L.H., Daily Totals		$3195.80		$4388.50	$57.07	$78.37

Crew B-13D	Hr.	Daily	Hr.	Daily	Bare Costs	Incl. O&P
1 Laborer	$24.45	$195.60	$41.05	$328.40	$30.05	$50.05
1 Equip. Oper. (crane)	35.65	285.20	59.05	472.40		
1 Hyd. Excavator, 1 C.Y.		812.60		893.86		
1 Trench Box		81.00		89.10	55.85	61.44
16 L.H., Daily Totals		$1374.40		$1783.76	$85.90	$111.49

Crew B-13E	Hr.	Daily	Hr.	Daily	Bare Costs	Incl. O&P
1 Laborer	$24.45	$195.60	$41.05	$328.40	$30.05	$50.05
1 Equip. Oper. (crane)	35.65	285.20	59.05	472.40		
1 Hyd. Excavator, 1.5 C.Y.		1030.00		1133.00		
1 Trench Box		81.00		89.10	69.44	76.38
16 L.H., Daily Totals		$1591.80		$2022.90	$99.49	$126.43

Crew B-13F	Hr.	Daily	Hr.	Daily	Bare Costs	Incl. O&P
1 Laborer	$24.45	$195.60	$41.05	$328.40	$30.05	$50.05
1 Equip. Oper. (crane)	35.65	285.20	59.05	472.40		
1 Hyd. Excavator, 3.5 C.Y.		2440.00		2684.00		
1 Trench Box		81.00		89.10	157.56	173.32
16 L.H., Daily Totals		$3001.80		$3573.90	$187.61	$223.37

Crew B-13G	Hr.	Daily	Hr.	Daily	Bare Costs	Incl. O&P
1 Laborer	$24.45	$195.60	$41.05	$328.40	$30.05	$50.05
1 Equip. Oper. (crane)	35.65	285.20	59.05	472.40		
1 Hyd. Excavator, .75 C.Y.		654.80		720.28		
1 Trench Box		81.00		89.10	45.99	50.59
16 L.H., Daily Totals		$1216.60		$1610.18	$76.04	$100.64

Crew B-13H	Hr.	Daily	Hr.	Daily	Bare Costs	Incl. O&P
1 Laborer	$24.45	$195.60	$41.05	$328.40	$30.05	$50.05
1 Equip. Oper. (crane)	35.65	285.20	59.05	472.40		
1 Gradall, 5/8 C.Y.		882.20		970.42		
1 Trench Box		81.00		89.10	60.20	66.22
16 L.H., Daily Totals		$1444.00		$1860.32	$90.25	$116.27

Crew B-13I	Hr.	Daily	Hr.	Daily	Bare Costs	Incl. O&P
1 Laborer	$24.45	$195.60	$41.05	$328.40	$30.05	$50.05
1 Equip. Oper. (crane)	35.65	285.20	59.05	472.40		
1 Gradall, 3 Ton, 1 C.Y.		1002.00		1102.20		
1 Trench Box		81.00		89.10	67.69	74.46
16 L.H., Daily Totals		$1563.80		$1992.10	$97.74	$124.51

Crew B-13J	Hr.	Daily	Hr.	Daily	Bare Costs	Incl. O&P
1 Laborer	$24.45	$195.60	$41.05	$328.40	$30.05	$50.05
1 Equip. Oper. (crane)	35.65	285.20	59.05	472.40		
1 Hyd. Excavator, 2.5 C.Y.		1605.00		1765.50		
1 Trench Box		81.00		89.10	105.38	115.91
16 L.H., Daily Totals		$2166.80		$2655.40	$135.43	$165.96

Crews

Crew No.	Bare Costs		Incl. Subs O&P		Cost Per Labor-Hour	
Crew B-13K	Hr.	Daily	Hr.	Daily	Bare Costs	Incl. O&P
2 Equip. Opers. (crane)	$35.65	$570.40	$59.05	$944.80	$35.65	$59.05
1 Hyd. Excavator, .75 C.Y.		654.80		720.28		
1 Hyd. Hammer, 4000 ft-lb		308.20		339.02		
1 Hyd. Excavator, .75 C.Y.		654.80		720.28	101.11	111.22
16 L.H., Daily Totals		$2188.20		$2724.38	$136.76	$170.27
Crew B-13L	Hr.	Daily	Hr.	Daily	Bare Costs	Incl. O&P
2 Equip. Opers. (crane)	$35.65	$570.40	$59.05	$944.80	$35.65	$59.05
1 Hyd. Excavator, 1.5 C.Y.		1030.00		1133.00		
1 Hyd. Hammer, 5000 ft-lb		371.20		408.32		
1 Hyd. Excavator, .75 C.Y.		654.80		720.28	128.50	141.35
16 L.H., Daily Totals		$2626.40		$3206.40	$164.15	$200.40
Crew B-13M	Hr.	Daily	Hr.	Daily	Bare Costs	Incl. O&P
2 Equip. Opers. (crane)	$35.65	$570.40	$59.05	$944.80	$35.65	$59.05
1 Hyd. Excavator, 2.5 C.Y.		1605.00		1765.50		
1 Hyd. Hammer, 8000 ft-lb		545.40		599.94		
1 Hyd. Excavator, 1.5 C.Y.		1030.00		1133.00	198.78	218.65
16 L.H., Daily Totals		$3750.80		$4443.24	$234.43	$277.70
Crew B-13N	Hr.	Daily	Hr.	Daily	Bare Costs	Incl. O&P
2 Equip. Opers. (crane)	$35.65	$570.40	$59.05	$944.80	$35.65	$59.05
1 Hyd. Excavator, 3.5 C.Y.		2440.00		2684.00		
1 Hyd. Hammer, 12,000 ft-lb		635.80		699.38		
1 Hyd. Excavator, 1.5 C.Y.		1030.00		1133.00	256.61	282.27
16 L.H., Daily Totals		$4676.20		$5461.18	$292.26	$341.32
Crew B-14	Hr.	Daily	Hr.	Daily	Bare Costs	Incl. O&P
1 Labor Foreman (outside)	$26.45	$211.60	$44.40	$355.20	$26.30	$44.03
4 Laborers	24.45	782.40	41.05	1313.60		
1 Equip. Oper. (light)	33.55	268.40	55.60	444.80		
1 Backhoe Loader, 48 H.P.		364.20		400.62	7.59	8.35
48 L.H., Daily Totals		$1626.60		$2514.22	$33.89	$52.38
Crew B-14A	Hr.	Daily	Hr.	Daily	Bare Costs	Incl. O&P
1 Equip. Oper. (crane)	$35.65	$285.20	$59.05	$472.40	$31.92	$53.05
.5 Laborer	24.45	97.80	41.05	164.20		
1 Hyd. Excavator, 4.5 C.Y.		2963.00		3259.30	246.92	271.61
12 L.H., Daily Totals		$3346.00		$3895.90	$278.83	$324.66
Crew B-14B	Hr.	Daily	Hr.	Daily	Bare Costs	Incl. O&P
1 Equip. Oper. (crane)	$35.65	$285.20	$59.05	$472.40	$31.92	$53.05
.5 Laborer	24.45	97.80	41.05	164.20		
1 Hyd. Excavator, 6 C.Y.		3494.00		3843.40	291.17	320.28
12 L.H., Daily Totals		$3877.00		$4480.00	$323.08	$373.33
Crew B-14C	Hr.	Daily	Hr.	Daily	Bare Costs	Incl. O&P
1 Equip. Oper. (crane)	$35.65	$285.20	$59.05	$472.40	$31.92	$53.05
.5 Laborer	24.45	97.80	41.05	164.20		
1 Hyd. Excavator, 7 C.Y.		3590.00		3949.00	299.17	329.08
12 L.H., Daily Totals		$3973.00		$4585.60	$331.08	$382.13
Crew B-14F	Hr.	Daily	Hr.	Daily	Bare Costs	Incl. O&P
1 Equip. Oper. (crane)	$35.65	$285.20	$59.05	$472.40	$31.92	$53.05
.5 Laborer	24.45	97.80	41.05	164.20		
1 Hyd. Shovel, 7 C.Y.		3993.00		4392.30	332.75	366.02
12 L.H., Daily Totals		$4376.00		$5028.90	$364.67	$419.07

Crew No.	Bare Costs		Incl. Subs O&P		Cost Per Labor-Hour	
Crew B-14G	Hr.	Daily	Hr.	Daily	Bare Costs	Incl. O&P
1 Equip. Oper. (crane)	$35.65	$285.20	$59.05	$472.40	$31.92	$53.05
.5 Laborer	24.45	97.80	41.05	164.20		
1 Hyd. Shovel, 12 C.Y.		5736.00		6309.60	478.00	525.80
12 L.H., Daily Totals		$6119.00		$6946.20	$509.92	$578.85
Crew B-14J	Hr.	Daily	Hr.	Daily	Bare Costs	Incl. O&P
1 Equip. Oper. (medium)	$34.90	$279.20	$57.85	$462.80	$31.42	$52.25
.5 Laborer	24.45	97.80	41.05	164.20		
1 F.E. Loader, 8 C.Y.		1937.00		2130.70	161.42	177.56
12 L.H., Daily Totals		$2314.00		$2757.70	$192.83	$229.81
Crew B-14K	Hr.	Daily	Hr.	Daily	Bare Costs	Incl. O&P
1 Equip. Oper. (medium)	$34.90	$279.20	$57.85	$462.80	$31.42	$52.25
.5 Laborer	24.45	97.80	41.05	164.20		
1 F.E. Loader, 10 C.Y.		2942.00		3236.20	245.17	269.68
12 L.H., Daily Totals		$3319.00		$3863.20	$276.58	$321.93
Crew B-15	Hr.	Daily	Hr.	Daily	Bare Costs	Incl. O&P
1 Equipment Oper. (med.)	$34.90	$279.20	$57.85	$462.80	$29.72	$49.25
.5 Laborer	24.45	97.80	41.05	164.20		
2 Truck Drivers (heavy)	28.45	455.20	47.00	752.00		
2 Dump Trucks, 12 C.Y., 400 H.P.		1382.00		1520.20		
1 Dozer, 200 H.P.		1387.00		1525.70	98.89	108.78
28 L.H., Daily Totals		$3601.20		$4424.90	$128.61	$158.03
Crew B-16	Hr.	Daily	Hr.	Daily	Bare Costs	Incl. O&P
1 Labor Foreman (outside)	$26.45	$211.60	$44.40	$355.20	$25.95	$43.38
2 Laborers	24.45	391.20	41.05	656.80		
1 Truck Driver (heavy)	28.45	227.60	47.00	376.00		
1 Dump Truck, 12 C.Y., 400 H.P.		691.00		760.10	21.59	23.75
32 L.H., Daily Totals		$1521.40		$2148.10	$47.54	$67.13
Crew B-17	Hr.	Daily	Hr.	Daily	Bare Costs	Incl. O&P
2 Laborers	$24.45	$391.20	$41.05	$656.80	$27.73	$46.17
1 Equip. Oper. (light)	33.55	268.40	55.60	444.80		
1 Truck Driver (heavy)	28.45	227.60	47.00	376.00		
1 Backhoe Loader, 48 H.P.		364.20		400.62		
1 Dump Truck, 8 C.Y., 220 H.P.		411.20		452.32	24.23	26.65
32 L.H., Daily Totals		$1662.60		$2330.54	$51.96	$72.83
Crew B-17A	Hr.	Daily	Hr.	Daily	Bare Costs	Incl. O&P
2 Labor Foremen (outside)	$26.45	$423.20	$44.40	$710.40	$26.83	$45.09
6 Laborers	24.45	1173.60	41.05	1970.40		
1 Skilled Worker Foreman (out)	35.35	282.80	59.55	476.40		
1 Skilled Worker	33.35	266.80	56.20	449.60		
80 L.H., Daily Totals		$2146.40		$3606.80	$26.83	$45.09
Crew B-17B	Hr.	Daily	Hr.	Daily	Bare Costs	Incl. O&P
2 Laborers	$24.45	$391.20	$41.05	$656.80	$27.73	$46.17
1 Equip. Oper. (light)	33.55	268.40	55.60	444.80		
1 Truck Driver (heavy)	28.45	227.60	47.00	376.00		
1 Backhoe Loader, 48 H.P.		364.20		400.62		
1 Dump Truck, 12 C.Y., 400 H.P.		691.00		760.10	32.98	36.27
32 L.H., Daily Totals		$1942.40		$2638.32	$60.70	$82.45
Crew B-18	Hr.	Daily	Hr.	Daily	Bare Costs	Incl. O&P
1 Labor Foreman (outside)	$26.45	$211.60	$44.40	$355.20	$25.12	$42.17
2 Laborers	24.45	391.20	41.05	656.80		
1 Vibrating Plate, Gas, 21"		46.80		51.48	1.95	2.15
24 L.H., Daily Totals		$649.60		$1063.48	$27.07	$44.31

Crews

Crew No.	Bare Costs		Incl. Subs O&P		Cost Per Labor-Hour	
Crew B-19	**Hr.**	**Daily**	**Hr.**	**Daily**	**Bare Costs**	**Incl. O&P**
1 Pile Driver Foreman (outside)	$34.25	$274.00	$59.00	$472.00	$31.91	$54.47
4 Pile Drivers	32.25	1032.00	55.55	1777.60		
1 Equip. Oper. (crane)	35.65	285.20	59.05	472.40		
1 Building Laborer	24.45	195.60	41.05	328.40		
1 Crawler Crane, 40 Ton		1156.00		1271.60		
1 Lead, 90' High		124.40		136.84		
1 Hammer, Diesel, 22k ft-lb		456.40		502.04	31.01	34.12
56 L.H., Daily Totals		$3523.60		$4960.88	$62.92	$88.59
Crew B-19A	**Hr.**	**Daily**	**Hr.**	**Daily**	**Bare Costs**	**Incl. O&P**
1 Pile Driver Foreman (outside)	$34.25	$274.00	$59.00	$472.00	$31.91	$54.47
4 Pile Drivers	32.25	1032.00	55.55	1777.60		
1 Equip. Oper. (crane)	35.65	285.20	59.05	472.40		
1 Common Laborer	24.45	195.60	41.05	328.40		
1 Crawler Crane, 75 Ton		1459.00		1604.90		
1 Lead, 90' high		124.40		136.84		
1 Hammer, Diesel, 41k ft-lb		581.00		639.10	38.65	42.52
56 L.H., Daily Totals		$3951.20		$5431.24	$70.56	$96.99
Crew B-19B	**Hr.**	**Daily**	**Hr.**	**Daily**	**Bare Costs**	**Incl. O&P**
1 Pile Driver Foreman (outside)	$34.25	$274.00	$59.00	$472.00	$31.91	$54.47
4 Pile Drivers	32.25	1032.00	55.55	1777.60		
1 Equip. Oper. (crane)	35.65	285.20	59.05	472.40		
1 Common Laborer	24.45	195.60	41.05	328.40		
1 Crawler Crane, 40 Ton		1156.00		1271.60		
1 Lead, 90' High		124.40		136.84		
1 Hammer, Diesel, 22k ft-lb		456.40		502.04		
1 Barge, 400 Ton		776.00		853.60	44.87	49.36
56 L.H., Daily Totals		$4299.60		$5814.48	$76.78	$103.83
Crew B-19C	**Hr.**	**Daily**	**Hr.**	**Daily**	**Bare Costs**	**Incl. O&P**
1 Pile Driver Foreman (outside)	$34.25	$274.00	$59.00	$472.00	$31.91	$54.47
4 Pile Drivers	32.25	1032.00	55.55	1777.60		
1 Equip. Oper. (crane)	35.65	285.20	59.05	472.40		
1 Common Laborer	24.45	195.60	41.05	328.40		
1 Crawler Crane, 75 Ton		1459.00		1604.90		
1 Lead, 90' High		124.40		136.84		
1 Hammer, Diesel, 41k ft-lb		581.00		639.10		
1 Barge, 400 Ton		776.00		853.60	52.51	57.76
56 L.H., Daily Totals		$4727.20		$6284.84	$84.41	$112.23
Crew B-20	**Hr.**	**Daily**	**Hr.**	**Daily**	**Bare Costs**	**Incl. O&P**
1 Labor Foreman (outside)	$26.45	$211.60	$44.40	$355.20	$25.12	$42.17
2 Laborers	24.45	391.20	41.05	656.80		
24 L.H., Daily Totals		$602.80		$1012.00	$25.12	$42.17
Crew B-20A	**Hr.**	**Daily**	**Hr.**	**Daily**	**Bare Costs**	**Incl. O&P**
1 Labor Foreman (outside)	$26.45	$211.60	$44.40	$355.20	$30.16	$50.14
1 Laborer	24.45	195.60	41.05	328.40		
1 Plumber	38.75	310.00	63.95	511.60		
1 Plumber Apprentice	31.00	248.00	51.15	409.20		
32 L.H., Daily Totals		$965.20		$1604.40	$30.16	$50.14
Crew B-21	**Hr.**	**Daily**	**Hr.**	**Daily**	**Bare Costs**	**Incl. O&P**
1 Labor Foreman (outside)	$26.45	$211.60	$44.40	$355.20	$26.62	$44.58
2 Laborers	24.45	391.20	41.05	656.80		
.5 Equip. Oper. (crane)	35.65	142.60	59.05	236.20		
.5 S.P. Crane, 4x4, 5 Ton		137.70		151.47	4.92	5.41
28 L.H., Daily Totals		$883.10		$1399.67	$31.54	$49.99

Crew No.	Bare Costs		Incl. Subs O&P		Cost Per Labor-Hour	
Crew B-21A	**Hr.**	**Daily**	**Hr.**	**Daily**	**Bare Costs**	**Incl. O&P**
1 Labor Foreman (outside)	$26.45	$211.60	$44.40	$355.20	$31.26	$51.92
1 Laborer	24.45	195.60	41.05	328.40		
1 Plumber	38.75	310.00	63.95	511.60		
1 Plumber Apprentice	31.00	248.00	51.15	409.20		
1 Equip. Oper. (crane)	35.65	285.20	59.05	472.40		
1 S.P. Crane, 4x4, 12 Ton		473.40		520.74	11.84	13.02
40 L.H., Daily Totals		$1723.80		$2597.54	$43.09	$64.94
Crew B-21B	**Hr.**	**Daily**	**Hr.**	**Daily**	**Bare Costs**	**Incl. O&P**
1 Labor Foreman (outside)	$26.45	$211.60	$44.40	$355.20	$27.09	$45.32
3 Laborers	24.45	586.80	41.05	985.20		
1 Equip. Oper. (crane)	35.65	285.20	59.05	472.40		
1 Hyd. Crane, 12 Ton		653.80		719.18	16.34	17.98
40 L.H., Daily Totals		$1737.40		$2531.98	$43.44	$63.30
Crew B-21C	**Hr.**	**Daily**	**Hr.**	**Daily**	**Bare Costs**	**Incl. O&P**
1 Labor Foreman (outside)	$26.45	$211.60	$44.40	$355.20	$27.30	$45.62
4 Laborers	24.45	782.40	41.05	1313.60		
1 Equip. Oper. (crane)	35.65	285.20	59.05	472.40		
1 Equip. Oper. (oiler)	31.20	249.60	51.70	413.60		
2 Cutting Torches		22.80		25.08		
2 Sets of Gases		304.00		334.40		
1 Lattice Boom Crane, 90 Ton		1511.00		1662.10	32.82	36.10
56 L.H., Daily Totals		$3366.60		$4576.38	$60.12	$81.72
Crew B-22	**Hr.**	**Daily**	**Hr.**	**Daily**	**Bare Costs**	**Incl. O&P**
1 Labor Foreman (outside)	$26.45	$211.60	$44.40	$355.20	$27.22	$45.54
2 Laborers	24.45	391.20	41.05	656.80		
.75 Equip. Oper. (crane)	35.65	213.90	59.05	354.30		
.75 S.P. Crane, 4x4, 5 Ton		206.55		227.21	6.88	7.57
30 L.H., Daily Totals		$1023.25		$1593.51	$34.11	$53.12
Crew B-22A	**Hr.**	**Daily**	**Hr.**	**Daily**	**Bare Costs**	**Incl. O&P**
1 Labor Foreman (outside)	$26.45	$211.60	$44.40	$355.20	$28.87	$48.35
1 Skilled Worker	33.35	266.80	56.20	449.60		
2 Laborers	24.45	391.20	41.05	656.80		
1 Equipment Operator, Crane	35.65	285.20	59.05	472.40		
1 S.P. Crane, 4x4, 5 Ton		275.40		302.94		
1 Butt Fusion Mach., 4"-12" diam.		382.10		420.31	16.44	18.08
40 L.H., Daily Totals		$1812.30		$2657.25	$45.31	$66.43
Crew B-22B	**Hr.**	**Daily**	**Hr.**	**Daily**	**Bare Costs**	**Incl. O&P**
1 Labor Foreman (outside)	$26.45	$211.60	$44.40	$355.20	$28.87	$48.35
1 Skilled Worker	33.35	266.80	56.20	449.60		
2 Laborers	24.45	391.20	41.05	656.80		
1 Equip. Oper. (crane)	35.65	285.20	59.05	472.40		
1 S.P. Crane, 4x4, 5 Ton		275.40		302.94		
1 Butt Fusion Mach., 8"-24" diam.		829.70		912.67	27.63	30.39
40 L.H., Daily Totals		$2259.90		$3149.61	$56.50	$78.74
Crew B-22C	**Hr.**	**Daily**	**Hr.**	**Daily**	**Bare Costs**	**Incl. O&P**
1 Skilled Worker	$33.35	$266.80	$56.20	$449.60	$28.90	$48.63
1 Laborer	24.45	195.60	41.05	328.40		
1 Butt Fusion Mach., 2"-8" diam.		121.05		133.16	7.57	8.32
16 L.H., Daily Totals		$583.45		$911.15	$36.47	$56.95

Crews

Crew No.	Bare Costs		Incl. Subs O&P		Cost Per Labor-Hour	
	Hr.	Daily	Hr.	Daily	Bare Costs	Incl. O&P
Crew B-23						
1 Labor Foreman (outside)	$26.45	$211.60	$44.40	$355.20	$24.85	$41.72
4 Laborers	24.45	782.40	41.05	1313.60		
1 Drill Rig, Truck-Mounted		2542.00		2796.20		
1 Flatbed Truck, Gas, 3 Ton		303.20		333.52	71.13	78.24
40 L.H., Daily Totals		$3839.20		$4798.52	$95.98	$119.96
Crew B-23A						
1 Labor Foreman (outside)	$26.45	$211.60	$44.40	$355.20	$28.60	$47.77
1 Laborer	24.45	195.60	41.05	328.40		
1 Equip. Oper. (medium)	34.90	279.20	57.85	462.80		
1 Drill Rig, Truck-Mounted		2542.00		2796.20		
1 Pickup Truck, 3/4 Ton		144.20		158.62	111.93	123.12
24 L.H., Daily Totals		$3372.60		$4101.22	$140.53	$170.88
Crew B-23B						
1 Labor Foreman (outside)	$26.45	$211.60	$44.40	$355.20	$28.60	$47.77
1 Laborer	24.45	195.60	41.05	328.40		
1 Equip. Oper. (medium)	34.90	279.20	57.85	462.80		
1 Drill Rig, Truck-Mounted		2542.00		2796.20		
1 Pickup Truck, 3/4 Ton		144.20		158.62		
1 Centr. Water Pump, 6"		346.60		381.26	126.37	139.00
24 L.H., Daily Totals		$3719.20		$4482.48	$154.97	$186.77
Crew B-24						
1 Cement Finisher	$31.50	$252.00	$51.05	$408.40	$29.60	$49.08
1 Laborer	24.45	195.60	41.05	328.40		
1 Carpenter	32.85	262.80	55.15	441.20		
24 L.H., Daily Totals		$710.40		$1178.00	$29.60	$49.08
Crew B-25						
1 Labor Foreman (outside)	$26.45	$211.60	$44.40	$355.20	$27.48	$45.94
7 Laborers	24.45	1369.20	41.05	2298.80		
3 Equip. Oper. (medium)	34.90	837.60	57.85	1388.40		
1 Asphalt Paver, 130 H.P.		2132.00		2345.20		
1 Tandem Roller, 10 Ton		236.80		260.48		
1 Roller, Pneum. Whl., 12 Ton		344.40		378.84	30.83	33.91
88 L.H., Daily Totals		$5131.60		$7026.92	$58.31	$79.85
Crew B-25B						
1 Labor Foreman (outside)	$26.45	$211.60	$44.40	$355.20	$28.10	$46.93
7 Laborers	24.45	1369.20	41.05	2298.80		
4 Equip. Oper. (medium)	34.90	1116.80	57.85	1851.20		
1 Asphalt Paver, 130 H.P.		2132.00		2345.20		
2 Tandem Rollers, 10 Ton		473.60		520.96		
1 Roller, Pneum. Whl., 12 Ton		344.40		378.84	30.73	33.80
96 L.H., Daily Totals		$5647.60		$7750.20	$58.83	$80.73
Crew B-25C						
1 Labor Foreman (outside)	$26.45	$211.60	$44.40	$355.20	$28.27	$47.21
3 Laborers	24.45	586.80	41.05	985.20		
2 Equip. Oper. (medium)	34.90	558.40	57.85	925.60		
1 Asphalt Paver, 130 H.P.		2132.00		2345.20		
1 Tandem Roller, 10 Ton		236.80		260.48	49.35	54.28
48 L.H., Daily Totals		$3725.60		$4871.68	$77.62	$101.49
Crew B-25D						
1 Labor Foreman (outside)	$26.45	$211.60	$44.40	$355.20	$28.40	$47.42
3 Laborers	24.45	586.80	41.05	985.20		
2.125 Equip. Oper. (medium)	34.90	593.30	57.85	983.45		
.125 Truck Driver (heavy)	28.45	28.45	47.00	47.00		
.125 Truck Tractor, 6x4, 380 H.P.		74.97		82.47		
.125 Dist. Tanker, 3000 Gallon		38.20		42.02		
1 Asphalt Paver, 130 H.P.		2132.00		2345.20		
1 Tandem Roller, 10 Ton		236.80		260.48	49.64	54.60
50 L.H., Daily Totals		$3902.13		$5101.02	$78.04	$102.02
Crew B-25E						
1 Labor Foreman (outside)	$26.45	$211.60	$44.40	$355.20	$28.53	$47.61
3 Laborers	24.45	586.80	41.05	985.20		
2.250 Equip. Oper. (medium)	34.90	628.20	57.85	1041.30		
.25 Truck Driver (heavy)	28.45	56.90	47.00	94.00		
.25 Truck Tractor, 6x4, 380 H.P.		149.95		164.94		
.25 Dist. Tanker, 3000 Gallon		76.40		84.04		
1 Asphalt Paver, 130 H.P.		2132.00		2345.20		
1 Tandem Roller, 10 Ton		236.80		260.48	49.91	54.90
52 L.H., Daily Totals		$4078.65		$5330.36	$78.44	$102.51
Crew B-26						
1 Labor Foreman (outside)	$26.45	$211.60	$44.40	$355.20	$28.15	$47.03
6 Laborers	24.45	1173.60	41.05	1970.40		
2 Equip. Oper. (medium)	34.90	558.40	57.85	925.60		
1 Rodman (reinf.)	35.20	281.60	59.90	479.20		
1 Cement Finisher	31.50	252.00	51.05	408.40		
1 Grader, 30,000 Lbs.		736.20		809.82		
1 Paving Mach. & Equip.		2794.00		3073.40	40.12	44.13
88 L.H., Daily Totals		$6007.40		$8022.02	$68.27	$91.16
Crew B-26A						
1 Labor Foreman (outside)	$26.45	$211.60	$44.40	$355.20	$28.15	$47.03
6 Laborers	24.45	1173.60	41.05	1970.40		
2 Equip. Oper. (medium)	34.90	558.40	57.85	925.60		
1 Rodman (reinf.)	35.20	281.60	59.90	479.20		
1 Cement Finisher	31.50	252.00	51.05	408.40		
1 Grader, 30,000 Lbs.		736.20		809.82		
1 Paving Mach. & Equip.		2794.00		3073.40		
1 Concrete Saw		167.40		184.14	42.02	46.22
88 L.H., Daily Totals		$6174.80		$8206.16	$70.17	$93.25
Crew B-26B						
1 Labor Foreman (outside)	$26.45	$211.60	$44.40	$355.20	$28.71	$47.93
6 Laborers	24.45	1173.60	41.05	1970.40		
3 Equip. Oper. (medium)	34.90	837.60	57.85	1388.40		
1 Rodman (reinf.)	35.20	281.60	59.90	479.20		
1 Cement Finisher	31.50	252.00	51.05	408.40		
1 Grader, 30,000 Lbs.		736.20		809.82		
1 Paving Mach. & Equip.		2794.00		3073.40		
1 Concrete Pump, 110' Boom		946.20		1040.82	46.63	51.29
96 L.H., Daily Totals		$7232.80		$9525.64	$75.34	$99.23
Crew B-26C						
1 Labor Foreman (outside)	$26.45	$211.60	$44.40	$355.20	$27.48	$45.95
6 Laborers	24.45	1173.60	41.05	1970.40		
1 Equip. Oper. (medium)	34.90	279.20	57.85	462.80		
1 Rodman (reinf.)	35.20	281.60	59.90	479.20		
1 Cement Finisher	31.50	252.00	51.05	408.40		
1 Paving Mach. & Equip.		2794.00		3073.40		
1 Concrete Saw		167.40		184.14	37.02	40.72
80 L.H., Daily Totals		$5159.40		$6933.54	$64.49	$86.67

Crews

Crew No.		Bare Costs		Incl. Subs O&P		Cost Per Labor-Hour	
Crew B-27	Hr.	Daily	Hr.	Daily	Bare Costs	Incl. O&P
1 Labor Foreman (outside)	$26.45	$211.60	$44.40	$355.20	$24.95	$41.89
3 Laborers	24.45	586.80	41.05	985.20		
1 Berm Machine		296.80		326.48	9.28	10.20
32 L.H., Daily Totals		$1095.20		$1666.88	$34.23	$52.09
Crew B-28	Hr.	Daily	Hr.	Daily	Bare Costs	Incl. O&P
2 Carpenters	$32.85	$525.60	$55.15	$882.40	$30.05	$50.45
1 Laborer	24.45	195.60	41.05	328.40		
24 L.H., Daily Totals		$721.20		$1210.80	$30.05	$50.45
Crew B-29	Hr.	Daily	Hr.	Daily	Bare Costs	Incl. O&P
1 Labor Foreman (outside)	$26.45	$211.60	$44.40	$355.20	$26.65	$44.61
4 Laborers	24.45	782.40	41.05	1313.60		
1 Equip. Oper. (crane)	35.65	285.20	59.05	472.40		
1 Gradall, 5/8 C.Y.		882.20		970.42	18.38	20.22
48 L.H., Daily Totals		$2161.40		$3111.62	$45.03	$64.83
Crew B-30	Hr.	Daily	Hr.	Daily	Bare Costs	Incl. O&P
1 Equip. Oper. (medium)	$34.90	$279.20	$57.85	$462.80	$30.60	$50.62
2 Truck Drivers (heavy)	28.45	455.20	47.00	752.00		
1 Hyd. Excavator, 1.5 C.Y.		1030.00		1133.00		
2 Dump Trucks, 12 C.Y., 400 H.P.		1382.00		1520.20	100.50	110.55
24 L.H., Daily Totals		$3146.40		$3868.00	$131.10	$161.17
Crew B-31	Hr.	Daily	Hr.	Daily	Bare Costs	Incl. O&P
1 Labor Foreman (outside)	$26.45	$211.60	$44.40	$355.20	$24.85	$41.72
4 Laborers	24.45	782.40	41.05	1313.60		
1 Air Compressor, 250 cfm		201.40		221.54		
1 Sheeting Driver		5.75		6.33		
2 -50' Air Hoses, 1.5"		11.60		12.76	5.47	6.02
40 L.H., Daily Totals		$1212.75		$1909.43	$30.32	$47.74
Crew B-32	Hr.	Daily	Hr.	Daily	Bare Costs	Incl. O&P
1 Laborer	$24.45	$195.60	$41.05	$328.40	$32.29	$53.65
3 Equip. Oper. (medium)	34.90	837.60	57.85	1388.40		
1 Grader, 30,000 Lbs.		736.20		809.82		
1 Tandem Roller, 10 Ton		236.80		260.48		
1 Dozer, 200 H.P.		1387.00		1525.70	73.75	81.13
32 L.H., Daily Totals		$3393.20		$4312.80	$106.04	$134.78
Crew B-32A	Hr.	Daily	Hr.	Daily	Bare Costs	Incl. O&P
1 Laborer	$24.45	$195.60	$41.05	$328.40	$31.42	$52.25
2 Equip. Oper. (medium)	34.90	558.40	57.85	925.60		
1 Grader, 30,000 Lbs.		736.20		809.82		
1 Roller, Vibratory, 25 Ton		687.40		756.14	59.32	65.25
24 L.H., Daily Totals		$2177.60		$2819.96	$90.73	$117.50
Crew B-32B	Hr.	Daily	Hr.	Daily	Bare Costs	Incl. O&P
1 Laborer	$24.45	$195.60	$41.05	$328.40	$31.42	$52.25
2 Equip. Oper. (medium)	34.90	558.40	57.85	925.60		
1 Dozer, 200 H.P.		1387.00		1525.70		
1 Roller, Vibratory, 25 Ton		687.40		756.14	86.43	95.08
24 L.H., Daily Totals		$2828.40		$3535.84	$117.85	$147.33
Crew B-32C	Hr.	Daily	Hr.	Daily	Bare Costs	Incl. O&P
1 Labor Foreman (outside)	$26.45	$211.60	$44.40	$355.20	$30.01	$50.01
2 Laborers	24.45	391.20	41.05	656.80		
3 Equip. Oper. (medium)	34.90	837.60	57.85	1388.40		
1 Grader, 30,000 Lbs.		736.20		809.82		
1 Tandem Roller, 10 Ton		236.80		260.48		
1 Dozer, 200 H.P.		1387.00		1525.70	49.17	54.08
48 L.H., Daily Totals		$3800.40		$4996.40	$79.17	$104.09
Crew B-33A	Hr.	Daily	Hr.	Daily	Bare Costs	Incl. O&P
1 Equip. Oper. (medium)	$34.90	$279.20	$57.85	$462.80	$34.90	$57.85
.25 Equip. Oper. (medium)	34.90	69.80	57.85	115.70		
1 Scraper, Towed, 7 C.Y.		114.40		125.84		
1.250 Dozers, 300 H.P.		2371.25		2608.38	248.57	273.42
10 L.H., Daily Totals		$2834.65		$3312.72	$283.46	$331.27
Crew B-33B	Hr.	Daily	Hr.	Daily	Bare Costs	Incl. O&P
1 Equip. Oper. (medium)	$34.90	$279.20	$57.85	$462.80	$34.90	$57.85
.25 Equip. Oper. (medium)	34.90	69.80	57.85	115.70		
1 Scraper, Towed, 10 C.Y.		146.80		161.48		
1.250 Dozers, 300 H.P.		2371.25		2608.38	251.81	276.99
10 L.H., Daily Totals		$2867.05		$3348.36	$286.70	$334.84
Crew B-33C	Hr.	Daily	Hr.	Daily	Bare Costs	Incl. O&P
1 Equip. Oper. (medium)	$34.90	$279.20	$57.85	$462.80	$34.90	$57.85
.25 Equip. Oper. (medium)	34.90	69.80	57.85	115.70		
1 Scraper, Towed, 15 C.Y.		164.80		181.28		
1.250 Dozers, 300 H.P.		2371.25		2608.38	253.60	278.97
10 L.H., Daily Totals		$2885.05		$3368.16	$288.51	$336.82
Crew B-33D	Hr.	Daily	Hr.	Daily	Bare Costs	Incl. O&P
1 Equip. Oper. (medium)	$34.90	$279.20	$57.85	$462.80	$34.90	$57.85
.25 Equip. Oper. (medium)	34.90	69.80	57.85	115.70		
1 S.P. Scraper, 14 C.Y.		1884.00		2072.40		
.25 Dozer, 300 H.P.		474.25		521.67	235.82	259.41
10 L.H., Daily Totals		$2707.25		$3172.57	$270.73	$317.26
Crew B-33E	Hr.	Daily	Hr.	Daily	Bare Costs	Incl. O&P
1 Equip. Oper. (medium)	$34.90	$279.20	$57.85	$462.80	$34.90	$57.85
.25 Equip. Oper. (medium)	34.90	69.80	57.85	115.70		
1 S.P. Scraper, 21 C.Y.		2688.00		2956.80		
.25 Dozer, 300 H.P.		474.25		521.67	316.23	347.85
10 L.H., Daily Totals		$3511.25		$4056.97	$351.13	$405.70
Crew B-33F	Hr.	Daily	Hr.	Daily	Bare Costs	Incl. O&P
1 Equip. Oper. (medium)	$34.90	$279.20	$57.85	$462.80	$34.90	$57.85
.25 Equip. Oper. (medium)	34.90	69.80	57.85	115.70		
1 Elev. Scraper, 11 C.Y.		1163.00		1279.30		
.25 Dozer, 300 H.P.		474.25		521.67	163.72	180.10
10 L.H., Daily Totals		$1986.25		$2379.47	$198.63	$237.95
Crew B-33G	Hr.	Daily	Hr.	Daily	Bare Costs	Incl. O&P
1 Equip. Oper. (medium)	$34.90	$279.20	$57.85	$462.80	$34.90	$57.85
.25 Equip. Oper. (medium)	34.90	69.80	57.85	115.70		
1 Elev. Scraper, 22 C.Y.		2521.00		2773.10		
.25 Dozer, 300 H.P.		474.25		521.67	299.52	329.48
10 L.H., Daily Totals		$3344.25		$3873.28	$334.43	$387.33

Crews

Crew No.	Bare Costs Hr.	Bare Costs Daily	Incl. Subs O&P Hr.	Incl. Subs O&P Daily	Cost Per Labor-Hour Bare Costs	Cost Per Labor-Hour Incl. O&P
Crew B-33K						
1 Equipment Operator (med.)	$34.90	$279.20	$57.85	$462.80	$31.91	$53.05
.25 Equipment Operator (med.)	34.90	69.80	57.85	115.70		
.5 Laborer	24.45	97.80	41.05	164.20		
1 S.P. Scraper, 31 C.Y.		3693.00		4062.30		
.25 Dozer, 410 H.P.		601.25		661.38	306.73	337.41
14 L.H., Daily Totals		$4741.05		$5466.38	$338.65	$390.46
Crew B-34A					Bare Costs	Incl. O&P
1 Truck Driver (heavy)	$28.45	$227.60	$47.00	$376.00	$28.45	$47.00
1 Dump Truck, 8 C.Y., 220 H.P.		411.20		452.32	51.40	56.54
8 L.H., Daily Totals		$638.80		$828.32	$79.85	$103.54
Crew B-34B					Bare Costs	Incl. O&P
1 Truck Driver (heavy)	$28.45	$227.60	$47.00	$376.00	$28.45	$47.00
1 Dump Truck, 12 C.Y., 400 H.P.		691.00		760.10	86.38	95.01
8 L.H., Daily Totals		$918.60		$1136.10	$114.83	$142.01
Crew B-34C					Bare Costs	Incl. O&P
1 Truck Driver (heavy)	$28.45	$227.60	$47.00	$376.00	$28.45	$47.00
1 Truck Tractor, 6x4, 380 H.P.		599.80		659.78		
1 Dump Trailer, 16.5 C.Y.		126.60		139.26	90.80	99.88
8 L.H., Daily Totals		$954.00		$1175.04	$119.25	$146.88
Crew B-34D					Bare Costs	Incl. O&P
1 Truck Driver (heavy)	$28.45	$227.60	$47.00	$376.00	$28.45	$47.00
1 Truck Tractor, 6x4, 380 H.P.		599.80		659.78		
1 Dump Trailer, 20 C.Y.		141.20		155.32	92.63	101.89
8 L.H., Daily Totals		$968.60		$1191.10	$121.08	$148.89
Crew B-34E					Bare Costs	Incl. O&P
1 Truck Driver (heavy)	$28.45	$227.60	$47.00	$376.00	$28.45	$47.00
1 Dump Truck, Off Hwy., 25 Ton		1351.00		1486.10	168.88	185.76
8 L.H., Daily Totals		$1578.60		$1862.10	$197.32	$232.76
Crew B-34F					Bare Costs	Incl. O&P
1 Truck Driver (heavy)	$28.45	$227.60	$47.00	$376.00	$28.45	$47.00
1 Dump Truck, Off Hwy., 35 Ton		1512.00		1663.20	189.00	207.90
8 L.H., Daily Totals		$1739.60		$2039.20	$217.45	$254.90
Crew B-34G					Bare Costs	Incl. O&P
1 Truck Driver (heavy)	$28.45	$227.60	$47.00	$376.00	$28.45	$47.00
1 Dump Truck, Off Hwy., 50 Ton		1837.00		2020.70	229.63	252.59
8 L.H., Daily Totals		$2064.60		$2396.70	$258.07	$299.59
Crew B-34H					Bare Costs	Incl. O&P
1 Truck Driver (heavy)	$28.45	$227.60	$47.00	$376.00	$28.45	$47.00
1 Dump Truck, Off Hwy., 65 Ton		1863.00		2049.30	232.88	256.16
8 L.H., Daily Totals		$2090.60		$2425.30	$261.32	$303.16
Crew B-34I					Bare Costs	Incl. O&P
1 Truck Driver (heavy)	$28.45	$227.60	$47.00	$376.00	$28.45	$47.00
1 Dump Truck, 18 C.Y., 450 H.P.		869.00		955.90	108.63	119.49
8 L.H., Daily Totals		$1096.60		$1331.90	$137.07	$166.49
Crew B-34J					Bare Costs	Incl. O&P
1 Truck Driver (heavy)	$28.45	$227.60	$47.00	$376.00	$28.45	$47.00
1 Dump Truck, Off Hwy., 100 Ton		2925.00		3217.50	365.63	402.19
8 L.H., Daily Totals		$3152.60		$3593.50	$394.07	$449.19

Crew No.	Bare Costs Hr.	Bare Costs Daily	Incl. Subs O&P Hr.	Incl. Subs O&P Daily	Cost Per Labor-Hour Bare Costs	Cost Per Labor-Hour Incl. O&P
Crew B-34K						
1 Truck Driver (heavy)	$28.45	$227.60	$47.00	$376.00	$28.45	$47.00
1 Truck Tractor, 6x4, 450 H.P.		727.40		800.14		
1 Lowbed Trailer, 75 Ton		221.80		243.98	118.65	130.51
8 L.H., Daily Totals		$1176.80		$1420.12	$147.10	$177.51
Crew B-34L					Bare Costs	Incl. O&P
1 Equip. Oper. (light)	$33.55	$268.40	$55.60	$444.80	$33.55	$55.60
1 Flatbed Truck, Gas, 1.5 Ton		245.40		269.94	30.68	33.74
8 L.H., Daily Totals		$513.80		$714.74	$64.22	$89.34
Crew B-34M					Bare Costs	Incl. O&P
1 Equip. Oper. (light)	$33.55	$268.40	$55.60	$444.80	$33.55	$55.60
1 Flatbed Truck, Gas, 3 Ton		303.20		333.52	37.90	41.69
8 L.H., Daily Totals		$571.60		$778.32	$71.45	$97.29
Crew B-34N					Bare Costs	Incl. O&P
1 Truck Driver (heavy)	$28.45	$227.60	$47.00	$376.00	$31.68	$52.42
1 Equip. Oper. (medium)	34.90	279.20	57.85	462.80		
1 Truck Tractor, 6x4, 380 H.P.		599.80		659.78		
1 Flatbed Trailer, 40 Ton		154.00		169.40	47.11	51.82
16 L.H., Daily Totals		$1260.60		$1667.98	$78.79	$104.25
Crew B-34P					Bare Costs	Incl. O&P
1 Pipe Fitter	$39.45	$315.60	$65.10	$520.80	$34.03	$56.27
1 Truck Driver (light)	27.75	222.00	45.85	366.80		
1 Equip. Oper. (medium)	34.90	279.20	57.85	462.80		
1 Flatbed Truck, Gas, 3 Ton		303.20		333.52		
1 Backhoe Loader, 48 H.P.		364.20		400.62	27.81	30.59
24 L.H., Daily Totals		$1484.20		$2084.54	$61.84	$86.86
Crew B-34Q					Bare Costs	Incl. O&P
1 Pipe Fitter	$39.45	$315.60	$65.10	$520.80	$34.28	$56.67
1 Truck Driver (light)	27.75	222.00	45.85	366.80		
1 Equip. Oper. (crane)	35.65	285.20	59.05	472.40		
1 Flatbed Trailer, 25 Ton		113.60		124.96		
1 Dump Truck, 8 C.Y., 220 H.P.		411.20		452.32		
1 Hyd. Crane, 25 Ton		736.60		810.26	52.56	57.81
24 L.H., Daily Totals		$2084.20		$2747.54	$86.84	$114.48
Crew B-34R					Bare Costs	Incl. O&P
1 Pipe Fitter	$39.45	$315.60	$65.10	$520.80	$34.28	$56.67
1 Truck Driver (light)	27.75	222.00	45.85	366.80		
1 Equip. Oper. (crane)	35.65	285.20	59.05	472.40		
1 Flatbed Trailer, 25 Ton		113.60		124.96		
1 Dump Truck, 8 C.Y., 220 H.P.		411.20		452.32		
1 Hyd. Crane, 25 Ton		736.60		810.26		
1 Hyd. Excavator, 1 C.Y.		812.60		893.86	86.42	95.06
24 L.H., Daily Totals		$2896.80		$3641.40	$120.70	$151.72
Crew B-34S					Bare Costs	Incl. O&P
2 Pipe Fitters	$39.45	$631.20	$65.10	$1041.60	$35.75	$59.06
1 Truck Driver (heavy)	28.45	227.60	47.00	376.00		
1 Equip. Oper. (crane)	35.65	285.20	59.05	472.40		
1 Flatbed Trailer, 40 Ton		154.00		169.40		
1 Truck Tractor, 6x4, 380 H.P.		599.80		659.78		
1 Hyd. Crane, 80 Ton		1625.00		1787.50		
1 Hyd. Excavator, 2 C.Y.		1175.00		1292.50	111.06	122.16
32 L.H., Daily Totals		$4697.80		$5799.18	$146.81	$181.22

Crews

Crew No.	Bare Costs		Incl. Subs O&P		Cost Per Labor-Hour	
Crew B-34T	Hr.	Daily	Hr.	Daily	Bare Costs	Incl. O&P
2 Pipe Fitters	$39.45	$631.20	$65.10	$1041.60	$35.75	$59.06
1 Truck Driver (heavy)	28.45	227.60	47.00	376.00		
1 Equip. Oper. (crane)	35.65	285.20	59.05	472.40		
1 Flatbed Trailer, 40 Ton		154.00		169.40		
1 Truck Tractor, 6x4, 380 H.P.		599.80		659.78		
1 Hyd. Crane, 80 Ton		1625.00		1787.50	74.34	81.77
32 L.H., Daily Totals		$3522.80		$4506.68	$110.09	$140.83
Crew B-34U	Hr.	Daily	Hr.	Daily	Bare Costs	Incl. O&P
1 Truck Driver (heavy)	$28.45	$227.60	$47.00	$376.00	$31.00	$51.30
1 Equip. Oper. (light)	33.55	268.40	55.60	444.80		
1 Truck Tractor, 220 H.P.		359.60		395.56		
1 Flatbed Trailer, 25 Ton		113.60		124.96	29.57	32.53
16 L.H., Daily Totals		$969.20		$1341.32	$60.58	$83.83
Crew B-34V	Hr.	Daily	Hr.	Daily	Bare Costs	Incl. O&P
1 Truck Driver (heavy)	$28.45	$227.60	$47.00	$376.00	$32.55	$53.88
1 Equip. Oper. (crane)	35.65	285.20	59.05	472.40		
1 Equip. Oper. (light)	33.55	268.40	55.60	444.80		
1 Truck Tractor, 6x4, 450 H.P.		727.40		800.14		
1 Equipment Trailer, 50 Ton		168.40		185.24		
1 Pickup Truck, 4 x 4, 3/4 Ton		155.60		171.16	43.81	48.19
24 L.H., Daily Totals		$1832.60		$2449.74	$76.36	$102.07
Crew B-34W	Hr.	Daily	Hr.	Daily	Bare Costs	Incl. O&P
5 Truck Drivers (heavy)	$28.45	$1138.00	$47.00	$1880.00	$30.41	$50.36
2 Equip. Opers. (crane)	35.65	570.40	59.05	944.80		
1 Equip. Oper. (mechanic)	35.65	285.20	59.05	472.40		
1 Laborer	24.45	195.60	41.05	328.40		
4 Truck Tractor, 6x4, 380 H.P.		2399.20		2639.12		
2 Equipment Trailer, 50 Ton		336.80		370.48		
2 Flatbed Trailer, 40 Ton		308.00		338.80		
1 Pickup Truck, 4 x 4, 3/4 Ton		155.60		171.16		
1 S.P. Crane, 4x4, 20 Ton		554.60		610.06	52.14	57.36
72 L.H., Daily Totals		$5943.40		$7755.22	$82.55	$107.71
Crew B-35	Hr.	Daily	Hr.	Daily	Bare Costs	Incl. O&P
1 Labor Foreman (outside)	$26.45	$211.60	$44.40	$355.20	$31.73	$52.93
1 Skilled Worker	33.35	266.80	56.20	449.60		
1 Welder (plumber)	38.75	310.00	63.95	511.60		
1 Laborer	24.45	195.60	41.05	328.40		
1 Equip. Oper. (crane)	35.65	285.20	59.05	472.40		
1 Welder, Electric, 300 amp		57.70		63.47		
1 Hyd. Excavator, .75 C.Y.		654.80		720.28	17.81	19.59
40 L.H., Daily Totals		$1981.70		$2900.95	$49.54	$72.52
Crew B-35A	Hr.	Daily	Hr.	Daily	Bare Costs	Incl. O&P
1 Labor Foreman (outside)	$26.45	$211.60	$44.40	$355.20	$30.61	$51.06
2 Laborers	24.45	391.20	41.05	656.80		
1 Skilled Worker	33.35	266.80	56.20	449.60		
1 Welder (plumber)	38.75	310.00	63.95	511.60		
1 Equip. Oper. (crane)	35.65	285.20	59.05	472.40		
1 Equip. Oper. (oiler)	31.20	249.60	51.70	413.60		
1 Welder, Gas Engine, 300 amp		145.85		160.44		
1 Crawler Crane, 75 Ton		1459.00		1604.90	28.66	31.52
56 L.H., Daily Totals		$3319.25		$4624.53	$59.27	$82.58

Crew No.	Bare Costs		Incl. Subs O&P		Cost Per Labor-Hour	
Crew B-36	Hr.	Daily	Hr.	Daily	Bare Costs	Incl. O&P
1 Labor Foreman (outside)	$26.45	$211.60	$44.40	$355.20	$29.03	$48.44
2 Laborers	24.45	391.20	41.05	656.80		
2 Equip. Oper. (medium)	34.90	558.40	57.85	925.60		
1 Dozer, 200 H.P.		1387.00		1525.70		
1 Aggregate Spreader		39.60		43.56		
1 Tandem Roller, 10 Ton		236.80		260.48	41.59	45.74
40 L.H., Daily Totals		$2824.60		$3767.34	$70.61	$94.18
Crew B-36A	Hr.	Daily	Hr.	Daily	Bare Costs	Incl. O&P
1 Labor Foreman (outside)	$26.45	$211.60	$44.40	$355.20	$30.71	$51.13
2 Laborers	24.45	391.20	41.05	656.80		
4 Equip. Oper. (medium)	34.90	1116.80	57.85	1851.20		
1 Dozer, 200 H.P.		1387.00		1525.70		
1 Aggregate Spreader		39.60		43.56		
1 Tandem Roller, 10 Ton		236.80		260.48		
1 Roller, Pneum. Whl., 12 Ton		344.40		378.84	35.85	39.44
56 L.H., Daily Totals		$3727.40		$5071.78	$66.56	$90.57
Crew B-36B	Hr.	Daily	Hr.	Daily	Bare Costs	Incl. O&P
1 Labor Foreman (outside)	$26.45	$211.60	$44.40	$355.20	$30.43	$50.61
2 Laborers	24.45	391.20	41.05	656.80		
4 Equip. Oper. (medium)	34.90	1116.80	57.85	1851.20		
1 Truck Driver (heavy)	28.45	227.60	47.00	376.00		
1 Grader, 30,000 Lbs.		736.20		809.82		
1 F.E. Loader, Crl, 1.5 C.Y.		629.80		692.78		
1 Dozer, 300 H.P.		1897.00		2086.70		
1 Roller, Vibratory, 25 Ton		687.40		756.14		
1 Truck Tractor, 6x4, 450 H.P.		727.40		800.14		
1 Water Tank Trailer, 5000 Gal.		141.00		155.10	75.29	82.82
64 L.H., Daily Totals		$6766.00		$8539.88	$105.72	$133.44
Crew B-36C	Hr.	Daily	Hr.	Daily	Bare Costs	Incl. O&P
1 Labor Foreman (outside)	$26.45	$211.60	$44.40	$355.20	$31.92	$52.99
3 Equip. Oper. (medium)	34.90	837.60	57.85	1388.40		
1 Truck Driver (heavy)	28.45	227.60	47.00	376.00		
1 Grader, 30,000 Lbs.		736.20		809.82		
1 Dozer, 300 H.P.		1897.00		2086.70		
1 Roller, Vibratory, 25 Ton		687.40		756.14		
1 Truck Tractor, 6x4, 450 H.P.		727.40		800.14		
1 Water Tank Trailer, 5000 Gal.		141.00		155.10	104.72	115.20
40 L.H., Daily Totals		$5465.80		$6727.50	$136.65	$168.19
Crew B-36E	Hr.	Daily	Hr.	Daily	Bare Costs	Incl. O&P
1 Labor Foreman (outside)	$26.45	$211.60	$44.40	$355.20	$32.42	$53.80
4 Equip. Oper. (medium)	34.90	1116.80	57.85	1851.20		
1 Truck Driver (heavy)	28.45	227.60	47.00	376.00		
1 Grader, 30,000 Lbs.		736.20		809.82		
1 Dozer, 300 H.P.		1897.00		2086.70		
1 Roller, Vibratory, 25 Ton		687.40		756.14		
1 Truck Tractor, 6x4, 380 H.P.		599.80		659.78		
1 Dist. Tanker, 3000 Gallon		305.60		336.16	88.04	96.85
48 L.H., Daily Totals		$5782.00		$7231.00	$120.46	$150.65
Crew B-37	Hr.	Daily	Hr.	Daily	Bare Costs	Incl. O&P
1 Labor Foreman (outside)	$26.45	$211.60	$44.40	$355.20	$26.30	$44.03
4 Laborers	24.45	782.40	41.05	1313.60		
1 Equip. Oper. (light)	33.55	268.40	55.60	444.80		
1 Tandem Roller, 5 Ton		157.40		173.14	3.28	3.61
48 L.H., Daily Totals		$1419.80		$2286.74	$29.58	$47.64

Crews

Crew No.	Bare Costs		Incl. Subs O&P		Cost Per Labor-Hour	
Crew B-37A	Hr.	Daily	Hr.	Daily	Bare Costs	Incl. O&P
2 Laborers	$24.45	$391.20	$41.05	$656.80	$25.55	$42.65
1 Truck Driver (light)	27.75	222.00	45.85	366.80		
1 Flatbed Truck, Gas, 1.5 Ton		245.40		269.94		
1 Tar Kettle, T.M.		132.00		145.20	15.73	17.30
24 L.H., Daily Totals		$990.60		$1438.74	$41.27	$59.95
Crew B-37B	Hr.	Daily	Hr.	Daily	Bare Costs	Incl. O&P
3 Laborers	$24.45	$586.80	$41.05	$985.20	$25.27	$42.25
1 Truck Driver (light)	27.75	222.00	45.85	366.80		
1 Flatbed Truck, Gas, 1.5 Ton		245.40		269.94		
1 Tar Kettle, T.M.		132.00		145.20	11.79	12.97
32 L.H., Daily Totals		$1186.20		$1767.14	$37.07	$55.22
Crew B-37C	Hr.	Daily	Hr.	Daily	Bare Costs	Incl. O&P
2 Laborers	$24.45	$391.20	$41.05	$656.80	$26.10	$43.45
2 Truck Drivers (light)	27.75	444.00	45.85	733.60		
2 Flatbed Trucks, Gas, 1.5 Ton		490.80		539.88		
1 Tar Kettle, T.M.		132.00		145.20	19.46	21.41
32 L.H., Daily Totals		$1458.00		$2075.48	$45.56	$64.86
Crew B-37D	Hr.	Daily	Hr.	Daily	Bare Costs	Incl. O&P
1 Laborer	$24.45	$195.60	$41.05	$328.40	$26.10	$43.45
1 Truck Driver (light)	27.75	222.00	45.85	366.80		
1 Pickup Truck, 3/4 Ton		144.20		158.62	9.01	9.91
16 L.H., Daily Totals		$561.80		$853.82	$35.11	$53.36
Crew B-37E	Hr.	Daily	Hr.	Daily	Bare Costs	Incl. O&P
3 Laborers	$24.45	$586.80	$41.05	$985.20	$28.19	$46.90
1 Equip. Oper. (light)	33.55	268.40	55.60	444.80		
1 Equip. Oper. (medium)	34.90	279.20	57.85	462.80		
2 Truck Drivers (light)	27.75	444.00	45.85	733.60		
4 Barrels w/ Flasher		13.60		14.96		
1 Concrete Saw		167.40		184.14		
1 Rotary Hammer Drill		23.75		26.13		
1 Hammer Drill Bit		2.35		2.59		
1 Loader, Skid Steer, 30 H.P.		173.60		190.96		
1 Conc. Hammer Attach.		114.70		126.17		
1 Vibrating Plate, Gas, 18"		36.20		39.82		
2 Flatbed Trucks, Gas, 1.5 Ton		490.80		539.88	18.26	20.08
56 L.H., Daily Totals		$2600.80		$3751.04	$46.44	$66.98
Crew B-37F	Hr.	Daily	Hr.	Daily	Bare Costs	Incl. O&P
3 Laborers	$24.45	$586.80	$41.05	$985.20	$25.27	$42.25
1 Truck Driver (light)	27.75	222.00	45.85	366.80		
4 Barrels w/ Flasher		13.60		14.96		
1 Concrete Mixer, 10 C.F.		172.40		189.64		
1 Air Compressor, 60 cfm		137.40		151.14		
1 -50' Air Hose, 3/4"		3.25		3.58		
1 Spade (Chipper)		7.60		8.36		
1 Flatbed Truck, Gas, 1.5 Ton		245.40		269.94	18.11	19.93
32 L.H., Daily Totals		$1388.45		$1989.62	$43.39	$62.18
Crew B-37G	Hr.	Daily	Hr.	Daily	Bare Costs	Incl. O&P
1 Labor Foreman (outside)	$26.45	$211.60	$44.40	$355.20	$26.30	$44.03
4 Laborers	24.45	782.40	41.05	1313.60		
1 Equip. Oper. (light)	33.55	268.40	55.60	444.80		
1 Berm Machine		296.80		326.48		
1 Tandem Roller, 5 Ton		157.40		173.14	9.46	10.41
48 L.H., Daily Totals		$1716.60		$2613.22	$35.76	$54.44

Crew No.	Bare Costs		Incl. Subs O&P		Cost Per Labor-Hour	
Crew B-37H	Hr.	Daily	Hr.	Daily	Bare Costs	Incl. O&P
1 Labor Foreman (outside)	$26.45	$211.60	$44.40	$355.20	$26.30	$44.03
4 Laborers	24.45	782.40	41.05	1313.60		
1 Equip. Oper. (light)	33.55	268.40	55.60	444.80		
1 Tandem Roller, 5 Ton		157.40		173.14		
1 Flatbed Trucks, Gas, 1.5 Ton		245.40		269.94		
1 Tar Kettle, T.M.		132.00		145.20	11.14	12.26
48 L.H., Daily Totals		$1797.20		$2701.88	$37.44	$56.29
Crew B-37I	Hr.	Daily	Hr.	Daily	Bare Costs	Incl. O&P
3 Laborers	$24.45	$586.80	$41.05	$985.20	$28.19	$46.90
1 Equip. Oper. (light)	33.55	268.40	55.60	444.80		
1 Equip. Oper. (medium)	34.90	279.20	57.85	462.80		
2 Truck Drivers (light)	27.75	444.00	45.85	733.60		
4 Barrels w/ Flasher		13.60		14.96		
1 Concrete Saw		167.40		184.14		
1 Rotary Hammer Drill		23.75		26.13		
1 Hammer Drill Bit		2.35		2.59		
1 Air Compressor, 60 cfm		137.40		151.14		
1 -50' Air Hose, 3/4"		3.25		3.58		
1 Spade (Chipper)		7.60		8.36		
1 Loader, Skid Steer, 30 H.P.		173.60		190.96		
1 Conc. Hammer Attach.		114.70		126.17		
1 Concrete Mixer, 10 C.F.		172.40		189.64		
1 Vibrating Plate, Gas, 18"		36.20		39.82		
2 Flatbed Trucks, Gas, 1.5 Ton		490.80		539.88	23.98	26.38
56 L.H., Daily Totals		$2921.45		$4103.76	$52.17	$73.28
Crew B-37J	Hr.	Daily	Hr.	Daily	Bare Costs	Incl. O&P
1 Labor Foreman (outside)	$26.45	$211.60	$44.40	$355.20	$26.30	$44.03
4 Laborers	24.45	782.40	41.05	1313.60		
1 Equip. Oper. (light)	33.55	268.40	55.60	444.80		
1 Air Compressor, 60 cfm		137.40		151.14		
1 -50' Air Hose, 3/4"		3.25		3.58		
2 Concrete Mixer, 10 C.F.		344.80		379.28		
2 Flatbed Trucks, Gas, 1.5 Ton		490.80		539.88		
1 Shot Blaster, 20"		214.00		235.40	24.80	27.28
48 L.H., Daily Totals		$2452.65		$3422.88	$51.10	$71.31
Crew B-37K	Hr.	Daily	Hr.	Daily	Bare Costs	Incl. O&P
1 Labor Foreman (outside)	$26.45	$211.60	$44.40	$355.20	$26.30	$44.03
4 Laborers	24.45	782.40	41.05	1313.60		
1 Equip. Oper. (light)	33.55	268.40	55.60	444.80		
1 Air Compressor, 60 cfm		137.40		151.14		
1 -50' Air Hose, 3/4"		3.25		3.58		
2 Flatbed Trucks, Gas, 1.5 Ton		490.80		539.88		
1 Shot Blaster, 20"		214.00		235.40	17.61	19.37
48 L.H., Daily Totals		$2107.85		$3043.59	$43.91	$63.41
Crew B-38	Hr.	Daily	Hr.	Daily	Bare Costs	Incl. O&P
2 Laborers	$24.45	$391.20	$41.05	$656.80	$27.48	$45.90
1 Equip. Oper. (light)	33.55	268.40	55.60	444.80		
1 Backhoe Loader, 48 H.P.		364.20		400.62		
1 Hyd. Hammer, (1200 lb.)		180.40		198.44	22.69	24.96
24 L.H., Daily Totals		$1204.20		$1700.66	$50.17	$70.86
Crew B-39	Hr.	Daily	Hr.	Daily	Bare Costs	Incl. O&P
1 Labor Foreman (outside)	$26.45	$211.60	$44.40	$355.20	$24.78	$41.61
5 Laborers	24.45	978.00	41.05	1642.00		
1 Air Compressor, 250 cfm		201.40		221.54		
2 Breakers, Pavement, 60 lb.		19.60		21.56		
2 -50' Air Hoses, 1.5"		11.60		12.76	4.85	5.33
48 L.H., Daily Totals		$1422.20		$2253.06	$29.63	$46.94

Crews

Crew No.	Bare Costs		Incl. Subs O&P		Cost Per Labor-Hour	
	Hr.	Daily	Hr.	Daily	Bare Costs	Incl. O&P
Crew B-40						
1 Pile Driver Foreman (outside)	$34.25	$274.00	$59.00	$472.00	$31.91	$54.47
4 Pile Drivers	32.25	1032.00	55.55	1777.60		
1 Building Laborer	24.45	195.60	41.05	328.40		
1 Equip. Oper. (crane)	35.65	285.20	59.05	472.40		
1 Crawler Crane, 40 Ton		1156.00		1271.60		
1 Vibratory Hammer & Gen.		2587.00		2845.70	66.84	73.52
56 L.H., Daily Totals		$5529.80		$7167.70	$98.75	$127.99
Crew B-40B	Hr.	Daily	Hr.	Daily	Bare Costs	Incl. O&P
1 Labor Foreman (outside)	$26.45	$211.60	$44.40	$355.20	$27.77	$46.38
3 Laborers	24.45	586.80	41.05	985.20		
1 Equip. Oper. (crane)	35.65	285.20	59.05	472.40		
1 Equip. Oper. (oiler)	31.20	249.60	51.70	413.60		
1 Lattice Boom Crane, 40 Ton		1164.00		1280.40	24.25	26.68
48 L.H., Daily Totals		$2497.20		$3506.80	$52.02	$73.06
Crew B-41	Hr.	Daily	Hr.	Daily	Bare Costs	Incl. O&P
1 Labor Foreman (outside)	$26.45	$211.60	$44.40	$355.20	$25.63	$42.96
4 Laborers	24.45	782.40	41.05	1313.60		
.25 Equip. Oper. (crane)	35.65	71.30	59.05	118.10		
.25 Equip. Oper. (oiler)	31.20	62.40	51.70	103.40		
.25 Crawler Crane, 40 Ton		289.00		317.90	6.57	7.22
44 L.H., Daily Totals		$1416.70		$2208.20	$32.20	$50.19
Crew B-42	Hr.	Daily	Hr.	Daily	Bare Costs	Incl. O&P
1 Labor Foreman (outside)	$26.45	$211.60	$44.40	$355.20	$28.38	$47.37
4 Laborers	24.45	782.40	41.05	1313.60		
1 Equip. Oper. (crane)	35.65	285.20	59.05	472.40		
1 Welder	38.75	310.00	63.95	511.60		
1 Hyd. Crane, 25 Ton		736.60		810.26		
1 Welder, Gas Engine, 300 amp		145.85		160.44		
1 Horz. Boring Csg. Mch.		474.80		522.28	24.24	26.66
56 L.H., Daily Totals		$2946.45		$4145.77	$52.62	$74.03
Crew B-43	Hr.	Daily	Hr.	Daily	Bare Costs	Incl. O&P
1 Labor Foreman (outside)	$26.45	$211.60	$44.40	$355.20	$24.85	$41.72
4 Laborers	24.45	782.40	41.05	1313.60		
1 Drill Rig, Truck-Mounted		2542.00		2796.20	63.55	69.91
40 L.H., Daily Totals		$3536.00		$4465.00	$88.40	$111.63
Crew B-44	Hr.	Daily	Hr.	Daily	Bare Costs	Incl. O&P
1 Pile Driver Foreman (outside)	$34.25	$274.00	$59.00	$472.00	$30.98	$52.79
4 Pile Drivers	32.25	1032.00	55.55	1777.60		
1 Equip. Oper. (crane)	35.65	285.20	59.05	472.40		
2 Laborers	24.45	391.20	41.05	656.80		
1 Crawler Crane, 40 Ton		1156.00		1271.60		
1 Lead, 60' High		74.80		82.28		
1 Hammer, Diesel, 15K ft.-lbs.		586.00		644.60	28.39	31.23
64 L.H., Daily Totals		$3799.20		$5377.28	$59.36	$84.02
Crew B-45	Hr.	Daily	Hr.	Daily	Bare Costs	Incl. O&P
1 Building Laborer	$24.45	$195.60	$41.05	$328.40	$26.45	$44.02
1 Truck Driver (heavy)	28.45	227.60	47.00	376.00		
1 Dist. Tanker, 3000 Gallon		305.60		336.16	19.10	21.01
16 L.H., Daily Totals		$728.80		$1040.56	$45.55	$65.03

Crew No.	Bare Costs		Incl. Subs O&P		Cost Per Labor-Hour	
	Hr.	Daily	Hr.	Daily	Bare Costs	Incl. O&P
Crew B-46						
1 Pile Driver Foreman (outside)	$34.25	$274.00	$59.00	$472.00	$28.68	$48.88
2 Pile Drivers	32.25	516.00	55.55	888.80		
3 Laborers	24.45	586.80	41.05	985.20		
1 Chain Saw, Gas, 36" Long		45.00		49.50	0.94	1.03
48 L.H., Daily Totals		$1421.80		$2395.50	$29.62	$49.91
Crew B-47	Hr.	Daily	Hr.	Daily	Bare Costs	Incl. O&P
1 Blast Foreman (outside)	$26.45	$211.60	$44.40	$355.20	$25.45	$42.73
1 Driller	24.45	195.60	41.05	328.40		
1 Air Track Drill, 4"		1028.00		1130.80		
1 Air Compressor, 600 cfm		550.60		605.66		
2 -50' Air Hoses, 3"		29.80		32.78	100.53	110.58
16 L.H., Daily Totals		$2015.60		$2452.84	$125.97	$153.30
Crew B-47A	Hr.	Daily	Hr.	Daily	Bare Costs	Incl. O&P
1 Drilling Foreman (outside)	$26.45	$211.60	$44.40	$355.20	$31.10	$51.72
1 Equip. Oper. (heavy)	35.65	285.20	59.05	472.40		
1 Equip. Oper. (oiler)	31.20	249.60	51.70	413.60		
1 Air Track Drill, 5"		1243.00		1367.30	51.79	56.97
24 L.H., Daily Totals		$1989.40		$2608.50	$82.89	$108.69
Crew B-47C	Hr.	Daily	Hr.	Daily	Bare Costs	Incl. O&P
1 Laborer	$24.45	$195.60	$41.05	$328.40	$29.00	$48.33
1 Equip. Oper. (light)	33.55	268.40	55.60	444.80		
1 Air Compressor, 750 cfm		557.20		612.92		
2 -50' Air Hoses, 3"		29.80		32.78		
1 Air Track Drill, 4"		1028.00		1130.80	100.94	111.03
16 L.H., Daily Totals		$2079.00		$2549.70	$129.94	$159.36
Crew B-47E	Hr.	Daily	Hr.	Daily	Bare Costs	Incl. O&P
1 Labor Foreman (outside)	$26.45	$211.60	$44.40	$355.20	$24.95	$41.89
3 Laborers	24.45	586.80	41.05	985.20		
1 Flatbed Truck, Gas, 3 Ton		303.20		333.52	9.47	10.42
32 L.H., Daily Totals		$1101.60		$1673.92	$34.42	$52.31
Crew B-47G	Hr.	Daily	Hr.	Daily	Bare Costs	Incl. O&P
1 Labor Foreman (outside)	$26.45	$211.60	$44.40	$355.20	$25.12	$42.17
2 Laborers	24.45	391.20	41.05	656.80		
1 Air Track Drill, 4"		1028.00		1130.80		
1 Air Compressor, 600 cfm		550.60		605.66		
2 -50' Air Hoses, 3"		29.80		32.78		
1 Gunite Pump Rig		371.80		408.98	82.51	90.76
24 L.H., Daily Totals		$2583.00		$3190.22	$107.63	$132.93
Crew B-47H	Hr.	Daily	Hr.	Daily	Bare Costs	Incl. O&P
1 Skilled Worker Foreman (out)	$35.35	$282.80	$59.55	$476.40	$33.85	$57.04
3 Skilled Workers	33.35	800.40	56.20	1348.80		
1 Flatbed Truck, Gas, 3 Ton		303.20		333.52	9.47	10.42
32 L.H., Daily Totals		$1386.40		$2158.72	$43.33	$67.46
Crew B-48	Hr.	Daily	Hr.	Daily	Bare Costs	Incl. O&P
1 Labor Foreman (outside)	$26.45	$211.60	$44.40	$355.20	$26.65	$44.61
4 Laborers	24.45	782.40	41.05	1313.60		
1 Equip. Oper. (crane)	35.65	285.20	59.05	472.40		
1 Centr. Water Pump, 6"		346.60		381.26		
1 -20' Suction Hose, 6"		11.50		12.65		
1 -50' Discharge Hose, 6"		6.10		6.71		
1 Drill Rig, Truck-Mounted		2542.00		2796.20	60.55	66.60
48 L.H., Daily Totals		$4185.40		$5338.02	$87.20	$111.21

Crews

Crew No.	Bare Costs		Incl. Subs O&P		Cost Per Labor-Hour	
Crew B-49	**Hr.**	**Daily**	**Hr.**	**Daily**	**Bare Costs**	**Incl. O&P**
1 Labor Foreman (outside)	$26.45	$211.60	$44.40	$355.20	$27.65	$46.64
5 Laborers	24.45	978.00	41.05	1642.00		
1 Equip. Oper. (crane)	35.65	285.20	59.05	472.40		
2 Pile Drivers	32.25	516.00	55.55	888.80		
1 Hyd. Crane, 25 Ton		736.60		810.26		
1 Centr. Water Pump, 6"		346.60		381.26		
1 -20' Suction Hose, 6"		11.50		12.65		
1 -50' Discharge Hose, 6"		6.10		6.71		
1 Drill Rig, Truck-Mounted		2542.00		2796.20	50.59	55.65
72 L.H., Daily Totals		$5633.60		$7365.48	$78.24	$102.30
Crew B-50	**Hr.**	**Daily**	**Hr.**	**Daily**	**Bare Costs**	**Incl. O&P**
1 Pile Driver Foreman (outside)	$34.25	$274.00	$59.00	$472.00	$29.67	$50.51
6 Pile Drivers	32.25	1548.00	55.55	2666.40		
1 Equip. Oper. (crane)	35.65	285.20	59.05	472.40		
5 Laborers	24.45	978.00	41.05	1642.00		
1 Crawler Crane, 40 Ton		1156.00		1271.60		
1 Lead, 60' High		74.80		82.28		
1 Hammer, Diesel, 15K ft.-lbs.		586.00		644.60		
1 Air Compressor, 600 cfm		550.60		605.66		
2 -50' Air Hoses, 3"		29.80		32.78		
1 Chain Saw, Gas, 36" Long		45.00		49.50	23.48	25.83
104 L.H., Daily Totals		$5527.40		$7939.22	$53.15	$76.34
Crew B-51	**Hr.**	**Daily**	**Hr.**	**Daily**	**Bare Costs**	**Incl. O&P**
1 Labor Foreman (outside)	$26.45	$211.60	$44.40	$355.20	$25.33	$42.41
4 Laborers	24.45	782.40	41.05	1313.60		
1 Truck Driver (light)	27.75	222.00	45.85	366.80		
1 Flatbed Truck, Gas, 1.5 Ton		245.40		269.94	5.11	5.62
48 L.H., Daily Totals		$1461.40		$2305.54	$30.45	$48.03
Crew B-52	**Hr.**	**Daily**	**Hr.**	**Daily**	**Bare Costs**	**Incl. O&P**
1 Labor Foreman (outside)	$26.45	$211.60	$44.40	$355.20	$27.45	$46.09
1 Carpenter	32.85	262.80	55.15	441.20		
4 Laborers	24.45	782.40	41.05	1313.60		
.5 Rodman (reinf.)	35.20	140.80	59.90	239.60		
.5 Equip. Oper. (medium)	34.90	139.60	57.85	231.40		
.5 Crawler Loader, 3 C.Y.		594.00		653.40	10.61	11.67
56 L.H., Daily Totals		$2131.20		$3234.40	$38.06	$57.76
Crew B-53	**Hr.**	**Daily**	**Hr.**	**Daily**	**Bare Costs**	**Incl. O&P**
1 Building Laborer	$24.45	$195.60	$41.05	$328.40	$24.45	$41.05
1 Trencher, Chain, 12 H.P.		68.00		74.80	8.50	9.35
8 L.H., Daily Totals		$263.60		$403.20	$32.95	$50.40
Crew B-54	**Hr.**	**Daily**	**Hr.**	**Daily**	**Bare Costs**	**Incl. O&P**
1 Equip. Oper. (light)	$33.55	$268.40	$55.60	$444.80	$33.55	$55.60
1 Trencher, Chain, 40 H.P.		335.20		368.72	41.90	46.09
8 L.H., Daily Totals		$603.60		$813.52	$75.45	$101.69
Crew B-54A	**Hr.**	**Daily**	**Hr.**	**Daily**	**Bare Costs**	**Incl. O&P**
.17 Labor Foreman (outside)	$26.45	$35.97	$44.40	$60.38	$33.67	$55.90
1 Equipment Operator (med.)	34.90	279.20	57.85	462.80		
1 Wheel Trencher, 67 H.P.		1195.00		1314.50	127.67	140.44
9.36 L.H., Daily Totals		$1510.17		$1837.68	$161.34	$196.33
Crew B-54B	**Hr.**	**Daily**	**Hr.**	**Daily**	**Bare Costs**	**Incl. O&P**
.25 Labor Foreman (outside)	$26.45	$52.90	$44.40	$88.80	$33.21	$55.16
1 Equipment Operator (med.)	34.90	279.20	57.85	462.80		
1 Wheel Trencher, 150 H.P.		1890.00		2079.00	189.00	207.90
10 L.H., Daily Totals		$2222.10		$2630.60	$222.21	$263.06

Crew No.	Bare Costs		Incl. Subs O&P		Cost Per Labor-Hour	
Crew B-54D	**Hr.**	**Daily**	**Hr.**	**Daily**	**Bare Costs**	**Incl. O&P**
1 Laborer	$24.45	$195.60	$41.05	$328.40	$29.68	$49.45
1 Equipment Operator (med.)	34.90	279.20	57.85	462.80		
1 Rock Trencher, 6" Width		372.00		409.20	23.25	25.57
16 L.H., Daily Totals		$846.80		$1200.40	$52.92	$75.03
Crew B-54E	**Hr.**	**Daily**	**Hr.**	**Daily**	**Bare Costs**	**Incl. O&P**
1 Laborer	$24.45	$195.60	$41.05	$328.40	$29.68	$49.45
1 Equipment Operator (med.)	34.90	279.20	57.85	462.80		
1 Rock Trencher, 18" Width		2575.00		2832.50	160.94	177.03
16 L.H., Daily Totals		$3049.80		$3623.70	$190.61	$226.48
Crew B-55	**Hr.**	**Daily**	**Hr.**	**Daily**	**Bare Costs**	**Incl. O&P**
1 Laborer	$24.45	$195.60	$41.05	$328.40	$26.10	$43.45
1 Truck Driver (light)	27.75	222.00	45.85	366.80		
1 Truck-Mounted Earth Auger		779.20		857.12		
1 Flatbed Truck, Gas, 3 Ton		303.20		333.52	67.65	74.42
16 L.H., Daily Totals		$1500.00		$1885.84	$93.75	$117.86
Crew B-56	**Hr.**	**Daily**	**Hr.**	**Daily**	**Bare Costs**	**Incl. O&P**
2 Laborers	$24.45	$391.20	$41.05	$656.80	$24.45	$41.05
1 Air Track Drill, 4"		1028.00		1130.80		
1 Air Compressor, 600 cfm		550.60		605.66		
1 -50' Air Hose, 3"		14.90		16.39	99.59	109.55
16 L.H., Daily Totals		$1984.70		$2409.65	$124.04	$150.60
Crew B-57	**Hr.**	**Daily**	**Hr.**	**Daily**	**Bare Costs**	**Incl. O&P**
1 Labor Foreman (outside)	$26.45	$211.60	$44.40	$355.20	$27.09	$45.32
3 Laborers	24.45	586.80	41.05	985.20		
1 Equip. Oper. (crane)	35.65	285.20	59.05	472.40		
1 Barge, 400 Ton		776.00		853.60		
1 Crawler Crane, 25 Ton		1150.00		1265.00		
1 Clamshell Bucket, 1 C.Y.		47.80		52.58		
1 Centr. Water Pump, 6"		346.60		381.26		
1 -20' Suction Hose, 6"		11.50		12.65		
20 -50' Discharge Hoses, 6"		122.00		134.20	61.35	67.48
40 L.H., Daily Totals		$3537.50		$4512.09	$88.44	$112.80
Crew B-58	**Hr.**	**Daily**	**Hr.**	**Daily**	**Bare Costs**	**Incl. O&P**
2 Laborers	$24.45	$391.20	$41.05	$656.80	$27.48	$45.90
1 Equip. Oper. (light)	33.55	268.40	55.60	444.80		
1 Backhoe Loader, 48 H.P.		364.20		400.62		
1 Small Helicopter, w/ Pilot		2790.00		3069.00	131.43	144.57
24 L.H., Daily Totals		$3813.80		$4571.22	$158.91	$190.47
Crew B-59	**Hr.**	**Daily**	**Hr.**	**Daily**	**Bare Costs**	**Incl. O&P**
1 Truck Driver (heavy)	$28.45	$227.60	$47.00	$376.00	$28.45	$47.00
1 Truck Tractor, 220 H.P.		359.60		395.56		
1 Water Tank Trailer, 5000 Gal.		141.00		155.10	62.58	68.83
8 L.H., Daily Totals		$728.20		$926.66	$91.03	$115.83
Crew B-60	**Hr.**	**Daily**	**Hr.**	**Daily**	**Bare Costs**	**Incl. O&P**
1 Labor Foreman (outside)	$26.45	$211.60	$44.40	$355.20	$28.17	$47.03
3 Laborers	24.45	586.80	41.05	985.20		
1 Equip. Oper. (crane)	35.65	285.20	59.05	472.40		
1 Equip. Oper. (light)	33.55	268.40	55.60	444.80		
1 Crawler Crane, 40 Ton		1156.00		1271.60		
1 Lead, 60' High		74.80		82.28		
1 Hammer, Diesel, 15K ft.-lbs.		586.00		644.60		
1 Backhoe Loader, 48 H.P.		364.20		400.62	45.44	49.98
48 L.H., Daily Totals		$3533.00		$4656.70	$73.60	$97.01

Crews

Crew No.	Bare Costs			Incl. Subs O&P		Cost Per Labor-Hour	
Crew B-61	Hr.	Daily	Hr.	Daily	Bare Costs	Incl. O&P
1 Labor Foreman (outside)	$26.45	$211.60	$44.40	$355.20	$24.85	$41.72
4 Laborers	24.45	782.40	41.05	1313.60		
1 Cement Mixer, 2 C.Y.		194.20		213.62		
1 Air Compressor, 160 cfm		156.40		172.04	8.77	9.64
40 L.H., Daily Totals		$1344.60		$2054.46	$33.62	$51.36
Crew B-62	Hr.	Daily	Hr.	Daily	Bare Costs	Incl. O&P
2 Laborers	$24.45	$391.20	$41.05	$656.80	$27.48	$45.90
1 Equip. Oper. (light)	33.55	268.40	55.60	444.80		
1 Loader, Skid Steer, 30 H.P.		173.60		190.96	7.23	7.96
24 L.H., Daily Totals		$833.20		$1292.56	$34.72	$53.86
Crew B-62A	Hr.	Daily	Hr.	Daily	Bare Costs	Incl. O&P
2 Laborers	$24.45	$391.20	$41.05	$656.80	$27.48	$45.90
1 Equip. Oper. (light)	33.55	268.40	55.60	444.80		
1 Loader, Skid Steer, 30 H.P.		173.60		190.96		
1 Trencher Attachment		58.30		64.13	9.66	10.63
24 L.H., Daily Totals		$891.50		$1356.69	$37.15	$56.53
Crew B-63	Hr.	Daily	Hr.	Daily	Bare Costs	Incl. O&P
5 Laborers	$24.45	$978.00	$41.05	$1642.00	$24.45	$41.05
1 Loader, Skid Steer, 30 H.P.		173.60		190.96	4.34	4.77
40 L.H., Daily Totals		$1151.60		$1832.96	$28.79	$45.82
Crew B-63B	Hr.	Daily	Hr.	Daily	Bare Costs	Incl. O&P
1 Labor Foreman (inside)	$24.95	$199.60	$41.90	$335.20	$26.85	$44.90
2 Laborers	24.45	391.20	41.05	656.80		
1 Equip. Oper. (light)	33.55	268.40	55.60	444.80		
1 Loader, Skid Steer, 78 H.P.		318.60		350.46	9.96	10.95
32 L.H., Daily Totals		$1177.80		$1787.26	$36.81	$55.85
Crew B-64	Hr.	Daily	Hr.	Daily	Bare Costs	Incl. O&P
1 Laborer	$24.45	$195.60	$41.05	$328.40	$26.10	$43.45
1 Truck Driver (light)	27.75	222.00	45.85	366.80		
1 Power Mulcher (small)		151.80		166.98		
1 Flatbed Truck, Gas, 1.5 Ton		245.40		269.94	24.82	27.31
16 L.H., Daily Totals		$814.80		$1132.12	$50.92	$70.76
Crew B-65	Hr.	Daily	Hr.	Daily	Bare Costs	Incl. O&P
1 Laborer	$24.45	$195.60	$41.05	$328.40	$26.10	$43.45
1 Truck Driver (light)	27.75	222.00	45.85	366.80		
1 Power Mulcher (Large)		319.40		351.34		
1 Flatbed Truck, Gas, 1.5 Ton		245.40		269.94	35.30	38.83
16 L.H., Daily Totals		$982.40		$1316.48	$61.40	$82.28
Crew B-66	Hr.	Daily	Hr.	Daily	Bare Costs	Incl. O&P
1 Equip. Oper. (light)	$33.55	$268.40	$55.60	$444.80	$33.55	$55.60
1 Loader-Backhoe, 40 H.P.		263.20		289.52	32.90	36.19
8 L.H., Daily Totals		$531.60		$734.32	$66.45	$91.79
Crew B-67	Hr.	Daily	Hr.	Daily	Bare Costs	Incl. O&P
1 Millwright	$34.40	$275.20	$55.45	$443.60	$33.98	$55.52
1 Equip. Oper. (light)	33.55	268.40	55.60	444.80		
1 Forklift, R/T, 4,000 Lb.		311.20		342.32	19.45	21.40
16 L.H., Daily Totals		$854.80		$1230.72	$53.42	$76.92
Crew B-67B	Hr.	Daily	Hr.	Daily	Bare Costs	Incl. O&P
1 Millwright Foreman (inside)	$34.90	$279.20	$56.25	$450.00	$34.65	$55.85
1 Millwright	34.40	275.20	55.45	443.60		
16 L.H., Daily Totals		$554.40		$893.60	$34.65	$55.85

Crew No.	Bare Costs			Incl. Subs O&P		Cost Per Labor-Hour	
Crew B-68	Hr.	Daily	Hr.	Daily	Bare Costs	Incl. O&P
2 Millwrights	$34.40	$550.40	$55.45	$887.20	$34.12	$55.50
1 Equip. Oper. (light)	33.55	268.40	55.60	444.80		
1 Forklift, R/T, 4,000 Lb.		311.20		342.32	12.97	14.26
24 L.H., Daily Totals		$1130.00		$1674.32	$47.08	$69.76
Crew B-68A	Hr.	Daily	Hr.	Daily	Bare Costs	Incl. O&P
1 Millwright Foreman (inside)	$34.90	$279.20	$56.25	$450.00	$34.57	$55.72
2 Millwrights	34.40	550.40	55.45	887.20		
1 Forklift, 8,000 Lb.		185.20		203.72	7.72	8.49
24 L.H., Daily Totals		$1014.80		$1540.92	$42.28	$64.20
Crew B-68B	Hr.	Daily	Hr.	Daily	Bare Costs	Incl. O&P
1 Millwright Foreman (inside)	$34.90	$279.20	$56.25	$450.00	$36.36	$59.31
2 Millwrights	34.40	550.40	55.45	887.20		
2 Electricians	36.65	586.40	60.05	960.80		
2 Plumbers	38.75	620.00	63.95	1023.20		
1 Forklift, 5,000 Lb.		331.80		364.98	5.92	6.52
56 L.H., Daily Totals		$2367.80		$3686.18	$42.28	$65.82
Crew B-68C	Hr.	Daily	Hr.	Daily	Bare Costs	Incl. O&P
1 Millwright Foreman (inside)	$34.90	$279.20	$56.25	$450.00	$36.17	$58.92
1 Millwright	34.40	275.20	55.45	443.60		
1 Electrician	36.65	293.20	60.05	480.40		
1 Plumber	38.75	310.00	63.95	511.60		
1 Forklift, 5,000 Lb.		331.80		364.98	10.37	11.41
32 L.H., Daily Totals		$1489.40		$2250.58	$46.54	$70.33
Crew B-68D	Hr.	Daily	Hr.	Daily	Bare Costs	Incl. O&P
1 Labor Foreman (inside)	$24.95	$199.60	$41.90	$335.20	$27.65	$46.18
1 Laborer	24.45	195.60	41.05	328.40		
1 Equip. Oper. (light)	33.55	268.40	55.60	444.80		
1 Forklift, 5,000 Lb.		331.80		364.98	13.82	15.21
24 L.H., Daily Totals		$995.40		$1473.38	$41.48	$61.39
Crew B-68E	Hr.	Daily	Hr.	Daily	Bare Costs	Incl. O&P
1 Struc. Steel Foreman (inside)	$35.80	$286.40	$67.20	$537.60	$35.40	$66.44
3 Struc. Steel Workers	35.30	847.20	66.25	1590.00		
1 Welder	35.30	282.40	66.25	530.00		
1 Forklift, 8,000 Lb.		185.20		203.72	4.63	5.09
40 L.H., Daily Totals		$1601.20		$2861.32	$40.03	$71.53
Crew B-68F	Hr.	Daily	Hr.	Daily	Bare Costs	Incl. O&P
1 Skilled Worker Foreman (out)	$35.35	$282.80	$59.55	$476.40	$34.02	$57.32
2 Skilled Workers	33.35	533.60	56.20	899.20		
1 Forklift, 5,000 Lb.		331.80		364.98	13.82	15.21
24 L.H., Daily Totals		$1148.20		$1740.58	$47.84	$72.52
Crew B-68G	Hr.	Daily	Hr.	Daily	Bare Costs	Incl. O&P
2 Structural Steel Workers	$35.30	$564.80	$66.25	$1060.00	$35.30	$66.25
1 Forklift, 5,000 Lb.		331.80		364.98	20.74	22.81
16 L.H., Daily Totals		$896.60		$1424.98	$56.04	$89.06
Crew B-69	Hr.	Daily	Hr.	Daily	Bare Costs	Incl. O&P
1 Labor Foreman (outside)	$26.45	$211.60	$44.40	$355.20	$27.77	$46.38
3 Laborers	24.45	586.80	41.05	985.20		
1 Equip. Oper. (crane)	35.65	285.20	59.05	472.40		
1 Equip. Oper. (oiler)	31.20	249.60	51.70	413.60		
1 Hyd. Crane, 80 Ton		1625.00		1787.50	33.85	37.24
48 L.H., Daily Totals		$2958.20		$4013.90	$61.63	$83.62

Crews

Crew No.	Bare Costs		Incl. Subs O&P		Cost Per Labor-Hour	
Crew B-69A	Hr.	Daily	Hr.	Daily	Bare Costs	Incl. O&P
1 Labor Foreman (outside)	$26.45	$211.60	$44.40	$355.20	$27.70	$46.08
3 Laborers	24.45	586.80	41.05	985.20		
1 Equip. Oper. (medium)	34.90	279.20	57.85	462.80		
1 Concrete Finisher	31.50	252.00	51.05	408.40		
1 Curb/Gutter Paver, 2-Track		997.60		1097.36	20.78	22.86
48 L.H., Daily Totals		$2327.20		$3308.96	$48.48	$68.94
Crew B-69B	Hr.	Daily	Hr.	Daily	Bare Costs	Incl. O&P
1 Labor Foreman (outside)	$26.45	$211.60	$44.40	$355.20	$27.70	$46.08
3 Laborers	24.45	586.80	41.05	985.20		
1 Equip. Oper. (medium)	34.90	279.20	57.85	462.80		
1 Cement Finisher	31.50	252.00	51.05	408.40		
1 Curb/Gutter Paver, 4-Track		776.00		853.60	16.17	17.78
48 L.H., Daily Totals		$2105.60		$3065.20	$43.87	$63.86
Crew B-70	Hr.	Daily	Hr.	Daily	Bare Costs	Incl. O&P
1 Labor Foreman (outside)	$26.45	$211.60	$44.40	$355.20	$29.21	$48.73
3 Laborers	24.45	586.80	41.05	985.20		
3 Equip. Oper. (medium)	34.90	837.60	57.85	1388.40		
1 Grader, 30,000 Lbs.		736.20		809.82		
1 Ripper, Beam & 1 Shank		84.20		92.62		
1 Road Sweeper, S.P., 8' wide		662.40		728.64		
1 F.E. Loader, W.M., 1.5 C.Y.		378.40		416.24	33.24	36.56
56 L.H., Daily Totals		$3497.20		$4776.12	$62.45	$85.29
Crew B-71	Hr.	Daily	Hr.	Daily	Bare Costs	Incl. O&P
1 Labor Foreman (outside)	$26.45	$211.60	$44.40	$355.20	$29.21	$48.73
3 Laborers	24.45	586.80	41.05	985.20		
3 Equip. Oper. (medium)	34.90	837.60	57.85	1388.40		
1 Pvmt. Profiler, 750 H.P.		5805.00		6385.50		
1 Road Sweeper, S.P., 8' wide		662.40		728.64		
1 F.E. Loader, W.M., 1.5 C.Y.		378.40		416.24	122.25	134.47
56 L.H., Daily Totals		$8481.80		$10259.18	$151.46	$183.20
Crew B-72	Hr.	Daily	Hr.	Daily	Bare Costs	Incl. O&P
1 Labor Foreman (outside)	$26.45	$211.60	$44.40	$355.20	$29.93	$49.87
3 Laborers	24.45	586.80	41.05	985.20		
4 Equip. Oper. (medium)	34.90	1116.80	57.85	1851.20		
1 Pvmt. Profiler, 750 H.P.		5805.00		6385.50		
1 Hammermill, 250 H.P.		1879.00		2066.90		
1 Windrow Loader		1227.00		1349.70		
1 Mix Paver 165 H.P.		2147.00		2361.70		
1 Roller, Pneum. Whl., 12 Ton		344.40		378.84	178.16	195.98
64 L.H., Daily Totals		$13317.60		$15734.24	$208.09	$245.85
Crew B-73	Hr.	Daily	Hr.	Daily	Bare Costs	Incl. O&P
1 Labor Foreman (outside)	$26.45	$211.60	$44.40	$355.20	$31.23	$51.97
2 Laborers	24.45	391.20	41.05	656.80		
5 Equip. Oper. (medium)	34.90	1396.00	57.85	2314.00		
1 Road Mixer, 310 H.P.		1929.00		2121.90		
1 Tandem Roller, 10 Ton		236.80		260.48		
1 Hammermill, 250 H.P.		1879.00		2066.90		
1 Grader, 30,000 Lbs.		736.20		809.82		
.5 F.E. Loader, W.M., 1.5 C.Y.		189.20		208.12		
.5 Truck Tractor, 220 H.P.		179.80		197.78		
.5 Water Tank Trailer, 5000 Gal.		70.50		77.55	81.57	89.73
64 L.H., Daily Totals		$7219.30		$9068.55	$112.80	$141.70

Crew No.	Bare Costs		Incl. Subs O&P		Cost Per Labor-Hour	
Crew B-74	Hr.	Daily	Hr.	Daily	Bare Costs	Incl. O&P
1 Labor Foreman (outside)	$26.45	$211.60	$44.40	$355.20	$30.93	$51.36
1 Laborer	24.45	195.60	41.05	328.40		
4 Equip. Oper. (medium)	34.90	1116.80	57.85	1851.20		
2 Truck Drivers (heavy)	28.45	455.20	47.00	752.00		
1 Grader, 30,000 Lbs.		736.20		809.82		
1 Ripper, Beam & 1 Shank		84.20		92.62		
2 Stabilizers, 310 H.P.		3596.00		3955.60		
1 Flatbed Truck, Gas, 3 Ton		303.20		333.52		
1 Chem. Spreader, Towed		52.40		57.64		
1 Roller, Vibratory, 25 Ton		687.40		756.14		
1 Water Tank Trailer, 5000 Gal.		141.00		155.10		
1 Truck Tractor, 220 H.P.		359.60		395.56	93.13	102.44
64 L.H., Daily Totals		$7939.20		$9842.80	$124.05	$153.79
Crew B-75	Hr.	Daily	Hr.	Daily	Bare Costs	Incl. O&P
1 Labor Foreman (outside)	$26.45	$211.60	$44.40	$355.20	$31.28	$51.98
1 Laborer	24.45	195.60	41.05	328.40		
4 Equip. Oper. (medium)	34.90	1116.80	57.85	1851.20		
1 Truck Driver (heavy)	28.45	227.60	47.00	376.00		
1 Grader, 30,000 Lbs.		736.20		809.82		
1 Ripper, Beam & 1 Shank		84.20		92.62		
2 Stabilizers, 310 H.P.		3596.00		3955.60		
1 Dist. Tanker, 3000 Gallon		305.60		336.16		
1 Truck Tractor, 6x4, 380 H.P.		599.80		659.78		
1 Roller, Vibratory, 25 Ton		687.40		756.14	107.31	118.04
56 L.H., Daily Totals		$7760.80		$9520.92	$138.59	$170.02
Crew B-76	Hr.	Daily	Hr.	Daily	Bare Costs	Incl. O&P
1 Dock Builder Foreman (outside)	$34.25	$274.00	$59.00	$472.00	$33.11	$56.28
5 Dock Builders	32.25	1290.00	55.55	2222.00		
2 Equip. Oper. (crane)	35.65	570.40	59.05	944.80		
1 Equip. Oper. (oiler)	31.20	249.60	51.70	413.60		
1 Crawler Crane, 50 Ton		1294.00		1423.40		
1 Barge, 400 Ton		776.00		853.60		
1 Hammer, Diesel, 15K ft.-lbs.		586.00		644.60		
1 Lead, 60' High		74.80		82.28		
1 Air Compressor, 600 cfm		550.60		605.66		
2 -50' Air Hoses, 3"		29.80		32.78	45.99	50.59
72 L.H., Daily Totals		$5695.20		$7694.72	$79.10	$106.87
Crew B-76A	Hr.	Daily	Hr.	Daily	Bare Costs	Incl. O&P
1 Labor Foreman (outside)	$26.45	$211.60	$44.40	$355.20	$26.94	$45.05
5 Laborers	24.45	978.00	41.05	1642.00		
1 Equip. Oper. (crane)	35.65	285.20	59.05	472.40		
1 Equip. Oper. (oiler)	31.20	249.60	51.70	413.60		
1 Crawler Crane, 50 Ton		1294.00		1423.40		
1 Barge, 400 Ton		776.00		853.60	32.34	35.58
64 L.H., Daily Totals		$3794.40		$5160.20	$59.29	$80.63
Crew B-77	Hr.	Daily	Hr.	Daily	Bare Costs	Incl. O&P
1 Labor Foreman (outside)	$26.45	$211.60	$44.40	$355.20	$25.51	$42.68
3 Laborers	24.45	586.80	41.05	985.20		
1 Truck Driver (light)	27.75	222.00	45.85	366.80		
1 Crack Cleaner, 25 H.P.		61.60		67.76		
1 Crack Filler, Trailer Mtd.		202.00		222.20		
1 Flatbed Truck, Gas, 3 Ton		303.20		333.52	14.17	15.59
40 L.H., Daily Totals		$1587.20		$2330.68	$39.68	$58.27

Crews

Crew No.	Bare Costs		Incl. Subs O&P		Cost Per Labor-Hour	
Crew B-78	Hr.	Daily	Hr.	Daily	Bare Costs	Incl. O&P
1 Labor Foreman (outside)	$26.45	$211.60	$44.40	$355.20	$24.85	$41.72
4 Laborers	24.45	782.40	41.05	1313.60		
1 Paint Striper, S.P., 40 Gallon		151.40		166.54		
1 Flatbed Truck, Gas, 3 Ton		303.20		333.52		
1 Pickup Truck, 3/4 Ton		144.20		158.62	14.97	16.47
40 L.H., Daily Totals		$1592.80		$2327.48	$39.82	$58.19
Crew B-78B	Hr.	Daily	Hr.	Daily	Bare Costs	Incl. O&P
2 Laborers	$24.45	$391.20	$41.05	$656.80	$25.46	$42.67
.25 Equip. Oper. (light)	33.55	67.10	55.60	111.20		
1 Pickup Truck, 3/4 Ton		144.20		158.62		
1 Line Rem.,11 H.P.,Walk Behind		65.60		72.16		
.25 Road Sweeper, S.P., 8' wide		165.60		182.16	20.86	22.94
18 L.H., Daily Totals		$833.70		$1180.94	$46.32	$65.61
Crew B-78C	Hr.	Daily	Hr.	Daily	Bare Costs	Incl. O&P
1 Labor Foreman (outside)	$26.45	$211.60	$44.40	$355.20	$25.33	$42.41
4 Laborers	24.45	782.40	41.05	1313.60		
1 Truck Driver (light)	27.75	222.00	45.85	366.80		
1 Paint Striper, T.M., 120 Gal.		804.60		885.06		
1 Flatbed Truck, Gas, 3 Ton		303.20		333.52		
1 Pickup Truck, 3/4 Ton		144.20		158.62	26.08	28.69
48 L.H., Daily Totals		$2468.00		$3412.80	$51.42	$71.10
Crew B-78D	Hr.	Daily	Hr.	Daily	Bare Costs	Incl. O&P
2 Labor Foremen (outside)	$26.45	$423.20	$44.40	$710.40	$25.18	$42.20
7 Laborers	24.45	1369.20	41.05	2298.80		
1 Truck Driver (light)	27.75	222.00	45.85	366.80		
1 Paint Striper, T.M., 120 Gal.		804.60		885.06		
1 Flatbed Truck, Gas, 3 Ton		303.20		333.52		
3 Pickup Trucks, 3/4 Ton		432.60		475.86		
1 Air Compressor, 60 cfm		137.40		151.14		
1 -50' Air Hose, 3/4"		3.25		3.58		
1 Breakers, Pavement, 60 lb.		9.80		10.78	21.14	23.25
80 L.H., Daily Totals		$3705.25		$5235.94	$46.32	$65.45
Crew B-78E	Hr.	Daily	Hr.	Daily	Bare Costs	Incl. O&P
2 Labor Foremen (outside)	$26.45	$423.20	$44.40	$710.40	$25.06	$42.01
9 Laborers	24.45	1760.40	41.05	2955.60		
1 Truck Driver (light)	27.75	222.00	45.85	366.80		
1 Paint Striper, T.M., 120 Gal.		804.60		885.06		
1 Flatbed Truck, Gas, 3 Ton		303.20		333.52		
4 Pickup Trucks, 3/4 Ton		576.80		634.48		
2 Air Compressor, 60 cfm		274.80		302.28		
2 -50' Air Hose, 3/4"		6.50		7.15		
2 Breakers, Pavement, 60 lb.		19.60		21.56	20.68	22.75
96 L.H., Daily Totals		$4391.10		$6216.85	$45.74	$64.76
Crew B-78F	Hr.	Daily	Hr.	Daily	Bare Costs	Incl. O&P
2 Labor Foremen (outside)	$26.45	$423.20	$44.40	$710.40	$24.97	$41.87
11 Laborers	24.45	2151.60	41.05	3612.40		
1 Truck Driver (light)	27.75	222.00	45.85	366.80		
1 Paint Striper, T.M., 120 Gal.		804.60		885.06		
1 Flatbed Truck, Gas, 3 Ton		303.20		333.52		
7 Pickup Trucks, 3/4 Ton		1009.40		1110.34		
3 Air Compressor, 60 cfm		412.20		453.42		
3 -50' Air Hose, 3/4"		9.75		10.73		
3 Breakers, Pavement, 60 lb.		29.40		32.34	22.93	25.23
112 L.H., Daily Totals		$5365.35		$7515.01	$47.90	$67.10

Crew No.	Bare Costs		Incl. Subs O&P		Cost Per Labor-Hour	
Crew B-79	Hr.	Daily	Hr.	Daily	Bare Costs	Incl. O&P
1 Labor Foreman (outside)	$26.45	$211.60	$44.40	$355.20	$25.51	$42.68
3 Laborers	24.45	586.80	41.05	985.20		
1 Truck Driver (light)	27.75	222.00	45.85	366.80		
1 Paint Striper, T.M., 120 Gal.		804.60		885.06		
1 Heating Kettle, 115 Gallon		81.00		89.10		
1 Flatbed Truck, Gas, 3 Ton		303.20		333.52		
2 Pickup Trucks, 3/4 Ton		288.40		317.24	36.93	40.62
40 L.H., Daily Totals		$2497.60		$3332.12	$62.44	$83.30
Crew B-79B	Hr.	Daily	Hr.	Daily	Bare Costs	Incl. O&P
1 Laborer	$24.45	$195.60	$41.05	$328.40	$24.45	$41.05
1 Set of Gases		152.00		167.20	19.00	20.90
8 L.H., Daily Totals		$347.60		$495.60	$43.45	$61.95
Crew B-79C	Hr.	Daily	Hr.	Daily	Bare Costs	Incl. O&P
1 Labor Foreman (outside)	$26.45	$211.60	$44.40	$355.20	$25.21	$42.21
5 Laborers	24.45	978.00	41.05	1642.00		
1 Truck Driver (light)	27.75	222.00	45.85	366.80		
1 Paint Striper, T.M., 120 Gal.		804.60		885.06		
1 Heating Kettle, 115 Gallon		81.00		89.10		
1 Flatbed Truck, Gas, 3 Ton		303.20		333.52		
3 Pickup Trucks, 3/4 Ton		432.60		475.86		
1 Air Compressor, 60 cfm		137.40		151.14		
1 -50' Air Hose, 3/4"		3.25		3.58		
1 Breakers, Pavement, 60 lb.		9.80		10.78	31.64	34.80
56 L.H., Daily Totals		$3183.45		$4313.03	$56.85	$77.02
Crew B-79D	Hr.	Daily	Hr.	Daily	Bare Costs	Incl. O&P
2 Labor Foremen (outside)	$26.45	$423.20	$44.40	$710.40	$25.36	$42.49
5 Laborers	24.45	978.00	41.05	1642.00		
1 Truck Driver (light)	27.75	222.00	45.85	366.80		
1 Paint Striper, T.M., 120 Gal.		804.60		885.06		
1 Heating Kettle, 115 Gallon		81.00		89.10		
1 Flatbed Truck, Gas, 3 Ton		303.20		333.52		
4 Pickup Trucks, 3/4 Ton		576.80		634.48		
1 Air Compressor, 60 cfm		137.40		151.14		
1 -50' Air Hose, 3/4"		3.25		3.58		
1 Breakers, Pavement, 60 lb.		9.80		10.78	29.94	32.93
64 L.H., Daily Totals		$3539.25		$4826.85	$55.30	$75.42
Crew B-79E	Hr.	Daily	Hr.	Daily	Bare Costs	Incl. O&P
2 Labor Foremen (outside)	$26.45	$423.20	$44.40	$710.40	$25.18	$42.20
7 Laborers	24.45	1369.20	41.05	2298.80		
1 Truck Driver (light)	27.75	222.00	45.85	366.80		
1 Paint Striper, T.M., 120 Gal.		804.60		885.06		
1 Heating Kettle, 115 Gallon		81.00		89.10		
1 Flatbed Truck, Gas, 3 Ton		303.20		333.52		
5 Pickup Trucks, 3/4 Ton		721.00		793.10		
2 Air Compressors, 60 cfm		274.80		302.28		
2 -50' Air Hoses, 3/4"		6.50		7.15		
2 Breakers, Pavement, 60 lb.		19.60		21.56	27.63	30.40
80 L.H., Daily Totals		$4225.10		$5807.77	$52.81	$72.60
Crew B-80	Hr.	Daily	Hr.	Daily	Bare Costs	Incl. O&P
1 Labor Foreman (outside)	$26.45	$211.60	$44.40	$355.20	$25.12	$42.17
2 Laborers	24.45	391.20	41.05	656.80		
1 Flatbed Truck, Gas, 3 Ton		303.20		333.52		
1 Earth Auger, Truck-Mtd.		417.20		458.92	30.02	33.02
24 L.H., Daily Totals		$1323.20		$1804.44	$55.13	$75.19

Crews

Crew No.	Bare Costs			Incl. Subs O&P		Cost Per Labor-Hour	
	Hr.	Daily	Hr.	Daily		Bare Costs	Incl. O&P
Crew B-80A							
3 Laborers	$24.45	$586.80	$41.05	$985.20		$24.45	$41.05
1 Flatbed Truck, Gas, 3 Ton		303.20		333.52		12.63	13.90
24 L.H., Daily Totals		$890.00		$1318.72		$37.08	$54.95
Crew B-80B	Hr.	Daily	Hr.	Daily		Bare Costs	Incl. O&P
3 Laborers	$24.45	$586.80	$41.05	$985.20		$26.73	$44.69
1 Equip. Oper. (light)	33.55	268.40	55.60	444.80			
1 Crane, Flatbed Mounted, 3 Ton		243.80		268.18		7.62	8.38
32 L.H., Daily Totals		$1099.00		$1698.18		$34.34	$53.07
Crew B-80C	Hr.	Daily	Hr.	Daily		Bare Costs	Incl. O&P
2 Laborers	$24.45	$391.20	$41.05	$656.80		$25.55	$42.65
1 Truck Driver (light)	27.75	222.00	45.85	366.80			
1 Flatbed Truck, Gas, 1.5 Ton		245.40		269.94			
1 Manual Fence Post Auger, Gas		8.40		9.24		10.57	11.63
24 L.H., Daily Totals		$867.00		$1302.78		$36.13	$54.28
Crew B-81	Hr.	Daily	Hr.	Daily		Bare Costs	Incl. O&P
1 Laborer	$24.45	$195.60	$41.05	$328.40		$26.45	$44.02
1 Truck Driver (heavy)	28.45	227.60	47.00	376.00			
1 Hydromulcher, T.M., 3000 Gal.		334.20		367.62			
1 Truck Tractor, 220 H.P.		359.60		395.56		43.36	47.70
16 L.H., Daily Totals		$1117.00		$1467.58		$69.81	$91.72
Crew B-81A	Hr.	Daily	Hr.	Daily		Bare Costs	Incl. O&P
1 Laborer	$24.45	$195.60	$41.05	$328.40		$26.10	$43.45
1 Truck Driver (light)	27.75	222.00	45.85	366.80			
1 Hydromulcher, T.M., 600 Gal.		127.80		140.58			
1 Flatbed Truck, Gas, 3 Ton		303.20		333.52		26.94	29.63
16 L.H., Daily Totals		$848.60		$1169.30		$53.04	$73.08
Crew B-82	Hr.	Daily	Hr.	Daily		Bare Costs	Incl. O&P
1 Laborer	$24.45	$195.60	$41.05	$328.40		$29.00	$48.33
1 Equip. Oper. (light)	33.55	268.40	55.60	444.80			
1 Horiz. Borer, 6 H.P.		83.80		92.18		5.24	5.76
16 L.H., Daily Totals		$547.80		$865.38		$34.24	$54.09
Crew B-82A	Hr.	Daily	Hr.	Daily		Bare Costs	Incl. O&P
2 Laborers	$24.45	$391.20	$41.05	$656.80		$29.00	$48.33
2 Equip. Opers. (light)	33.55	536.80	55.60	889.60			
2 Dump Truck, 8 C.Y., 220 H.P.		822.40		904.64			
1 Flatbed Trailer, 25 Ton		113.60		124.96			
1 Horiz. Dir. Drill, 20k lb. Thrust		654.20		719.62			
1 Mud Trailer for HDD, 1500 Gal.		346.20		380.82			
1 Pickup Truck, 4 x 4, 3/4 Ton		155.60		171.16			
1 Flatbed trailer, 3 Ton		23.60		25.96			
1 Loader, Skid Steer, 78 H.P.		318.60		350.46		76.07	83.68
32 L.H., Daily Totals		$3362.20		$4224.02		$105.07	$132.00
Crew B-82B	Hr.	Daily	Hr.	Daily		Bare Costs	Incl. O&P
2 Laborers	$24.45	$391.20	$41.05	$656.80		$29.00	$48.33
2 Equip. Opers. (light)	33.55	536.80	55.60	889.60			
2 Dump Truck, 8 C.Y., 220 H.P.		822.40		904.64			
1 Flatbed Trailer, 25 Ton		113.60		124.96			
1 Horiz. Dir. Drill, 30k lb. Thrust		931.20		1024.32			
1 Mud Trailer for HDD, 1500 Gal.		346.20		380.82			
1 Pickup Truck, 4 x 4, 3/4 Ton		155.60		171.16			
1 Flatbed trailer, 3 Ton		23.60		25.96			
1 Loader, Skid Steer, 78 H.P.		318.60		350.46		84.72	93.20
32 L.H., Daily Totals		$3639.20		$4528.72		$113.72	$141.52
Crew B-82C	Hr.	Daily	Hr.	Daily		Bare Costs	Incl. O&P
2 Laborers	$24.45	$391.20	$41.05	$656.80		$29.00	$48.33
2 Equip. Opers. (light)	33.55	536.80	55.60	889.60			
2 Dump Truck, 8 C.Y., 220 H.P.		822.40		904.64			
1 Flatbed Trailer, 25 Ton		113.60		124.96			
1 Horiz. Dir. Drill, 50k lb. Thrust		1237.00		1360.70			
1 Mud Trailer for HDD, 1500 Gal.		346.20		380.82			
1 Pickup Truck, 4 x 4, 3/4 Ton		155.60		171.16			
1 Flatbed trailer, 3 Ton		23.60		25.96			
1 Loader, Skid Steer, 78 H.P.		318.60		350.46		94.28	103.71
32 L.H., Daily Totals		$3945.00		$4865.10		$123.28	$152.03
Crew B-82D	Hr.	Daily	Hr.	Daily		Bare Costs	Incl. O&P
1 Equip. Oper. (light)	$33.55	$268.40	$55.60	$444.80		$33.55	$55.60
1 Mud Trailer for HDD, 1500 Gal.		346.20		380.82		43.27	47.60
8 L.H., Daily Totals		$614.60		$825.62		$76.83	$103.20
Crew B-83	Hr.	Daily	Hr.	Daily		Bare Costs	Incl. O&P
1 Tugboat Captain	$34.90	$279.20	$57.85	$462.80		$29.68	$49.45
1 Tugboat Hand	24.45	195.60	41.05	328.40			
1 Tugboat, 250 H.P.		870.00		957.00		54.38	59.81
16 L.H., Daily Totals		$1344.80		$1748.20		$84.05	$109.26
Crew B-84	Hr.	Daily	Hr.	Daily		Bare Costs	Incl. O&P
1 Equip. Oper. (medium)	$34.90	$279.20	$57.85	$462.80		$34.90	$57.85
1 Rotary Mower/Tractor		370.80		407.88		46.35	50.98
8 L.H., Daily Totals		$650.00		$870.68		$81.25	$108.83
Crew B-85	Hr.	Daily	Hr.	Daily		Bare Costs	Incl. O&P
3 Laborers	$24.45	$586.80	$41.05	$985.20		$27.34	$45.60
1 Equip. Oper. (medium)	34.90	279.20	57.85	462.80			
1 Truck Driver (heavy)	28.45	227.60	47.00	376.00			
1 Aerial Lift Truck, 80'		634.40		697.84			
1 Brush Chipper, 12", 130 H.P.		391.40		430.54			
1 Pruning Saw, Rotary		6.65		7.32		25.81	28.39
40 L.H., Daily Totals		$2126.05		$2959.70		$53.15	$73.99
Crew B-86	Hr.	Daily	Hr.	Daily		Bare Costs	Incl. O&P
1 Equip. Oper. (medium)	$34.90	$279.20	$57.85	$462.80		$34.90	$57.85
1 Stump Chipper, S.P.		185.55		204.10		23.19	25.51
8 L.H., Daily Totals		$464.75		$666.90		$58.09	$83.36
Crew B-86A	Hr.	Daily	Hr.	Daily		Bare Costs	Incl. O&P
1 Equip. Oper. (medium)	$34.90	$279.20	$57.85	$462.80		$34.90	$57.85
1 Grader, 30,000 Lbs.		736.20		809.82		92.03	101.23
8 L.H., Daily Totals		$1015.40		$1272.62		$126.93	$159.08
Crew B-86B	Hr.	Daily	Hr.	Daily		Bare Costs	Incl. O&P
1 Equip. Oper. (medium)	$34.90	$279.20	$57.85	$462.80		$34.90	$57.85
1 Dozer, 200 H.P.		1387.00		1525.70		173.38	190.71
8 L.H., Daily Totals		$1666.20		$1988.50		$208.28	$248.56
Crew B-87	Hr.	Daily	Hr.	Daily		Bare Costs	Incl. O&P
1 Laborer	$24.45	$195.60	$41.05	$328.40		$32.81	$54.49
4 Equip. Oper. (medium)	34.90	1116.80	57.85	1851.20			
2 Feller Bunchers, 100 H.P.		1676.80		1844.48			
1 Log Chipper, 22" Tree		889.40		978.34			
1 Dozer, 105 H.P.		602.80		663.08			
1 Chain Saw, Gas, 36" Long		45.00		49.50		80.35	88.39
40 L.H., Daily Totals		$4526.40		$5715.00		$113.16	$142.88

Crews

Crew No.	Bare Costs		Incl. Subs O&P		Cost Per Labor-Hour	
Crew B-88	Hr.	Daily	Hr.	Daily	Bare Costs	Incl. O&P
1 Laborer	$24.45	$195.60	$41.05	$328.40	$33.41	$55.45
6 Equip. Oper. (medium)	34.90	1675.20	57.85	2776.80		
2 Feller Bunchers, 100 H.P.		1676.80		1844.48		
1 Log Chipper, 22" Tree		889.40		978.34		
2 Log Skidders, 50 H.P.		1822.00		2004.20		
1 Dozer, 105 H.P.		602.80		663.08		
1 Chain Saw, Gas, 36" Long		45.00		49.50	89.93	98.92
56 L.H., Daily Totals		$6906.80		$8644.80	$123.34	$154.37

Crew B-89	Hr.	Daily	Hr.	Daily	Bare Costs	Incl. O&P
1 Skilled Worker	$33.35	$266.80	$56.20	$449.60	$28.90	$48.63
1 Building Laborer	24.45	195.60	41.05	328.40		
1 Flatbed Truck, Gas, 3 Ton		303.20		333.52		
1 Concrete Saw		167.40		184.14		
1 Water Tank, 65 Gal.		17.30		19.03	30.49	33.54
16 L.H., Daily Totals		$950.30		$1314.69	$59.39	$82.17

Crew B-89A	Hr.	Daily	Hr.	Daily	Bare Costs	Incl. O&P
1 Skilled Worker	$33.35	$266.80	$56.20	$449.60	$28.90	$48.63
1 Laborer	24.45	195.60	41.05	328.40		
1 Core Drill (Large)		115.60		127.16	7.22	7.95
16 L.H., Daily Totals		$578.00		$905.16	$36.13	$56.57

Crew B-89B	Hr.	Daily	Hr.	Daily	Bare Costs	Incl. O&P
1 Equip. Oper. (light)	$33.55	$268.40	$55.60	$444.80	$30.65	$50.73
1 Truck Driver (light)	27.75	222.00	45.85	366.80		
1 Wall Saw, Hydraulic, 10 H.P.		114.60		126.06		
1 Generator, Diesel, 100 kW		423.40		465.74		
1 Water Tank, 65 Gal.		17.30		19.03		
1 Flatbed Truck, Gas, 3 Ton		303.20		333.52	53.66	59.02
16 L.H., Daily Totals		$1348.90		$1755.95	$84.31	$109.75

Crew B-90	Hr.	Daily	Hr.	Daily	Bare Costs	Incl. O&P
1 Labor Foreman (outside)	$26.45	$211.60	$44.40	$355.20	$27.98	$46.59
3 Laborers	24.45	586.80	41.05	985.20		
2 Equip. Oper. (light)	33.55	536.80	55.60	889.60		
2 Truck Drivers (heavy)	28.45	455.20	47.00	752.00		
1 Road Mixer, 310 H.P.		1929.00		2121.90		
1 Dist. Truck, 2000 Gal.		275.60		303.16	34.45	37.89
64 L.H., Daily Totals		$3995.00		$5407.06	$62.42	$84.49

Crew B-90A	Hr.	Daily	Hr.	Daily	Bare Costs	Incl. O&P
1 Labor Foreman (outside)	$26.45	$211.60	$44.40	$355.20	$30.71	$51.13
2 Laborers	24.45	391.20	41.05	656.80		
4 Equip. Oper. (medium)	34.90	1116.80	57.85	1851.20		
2 Graders, 30,000 Lbs.		1472.40		1619.64		
1 Tandem Roller, 10 Ton		236.80		260.48		
1 Roller, Pneum. Whl., 12 Ton		344.40		378.84	36.67	40.34
56 L.H., Daily Totals		$3773.20		$5122.16	$67.38	$91.47

Crew B-90B	Hr.	Daily	Hr.	Daily	Bare Costs	Incl. O&P
1 Labor Foreman (outside)	$26.45	$211.60	$44.40	$355.20	$30.01	$50.01
2 Laborers	24.45	391.20	41.05	656.80		
3 Equip. Oper. (medium)	34.90	837.60	57.85	1388.40		
1 Roller, Pneum. Whl., 12 Ton		344.40		378.84		
1 Road Mixer, 310 H.P.		1929.00		2121.90	47.36	52.10
48 L.H., Daily Totals		$3713.80		$4901.14	$77.37	$102.11

Crew B-90C	Hr.	Daily	Hr.	Daily	Bare Costs	Incl. O&P
1 Labor Foreman (outside)	$26.45	$211.60	$44.40	$355.20	$28.57	$47.56
4 Laborers	24.45	782.40	41.05	1313.60		
3 Equip. Oper. (medium)	34.90	837.60	57.85	1388.40		
3 Truck Drivers (heavy)	28.45	682.80	47.00	1128.00		
3 Road Mixers, 310 H.P.		5787.00		6365.70	65.76	72.34
88 L.H., Daily Totals		$8301.40		$10550.90	$94.33	$119.90

Crew B-90D	Hr.	Daily	Hr.	Daily	Bare Costs	Incl. O&P
1 Labor Foreman (outside)	$26.45	$211.60	$44.40	$355.20	$27.94	$46.56
6 Laborers	24.45	1173.60	41.05	1970.40		
3 Equip. Oper. (medium)	34.90	837.60	57.85	1388.40		
3 Truck Drivers (heavy)	28.45	682.80	47.00	1128.00		
3 Road Mixers, 310 H.P.		5787.00		6365.70	55.64	61.21
104 L.H., Daily Totals		$8692.60		$11207.70	$83.58	$107.77

Crew B-90E	Hr.	Daily	Hr.	Daily	Bare Costs	Incl. O&P
1 Labor Foreman (outside)	$26.45	$211.60	$44.40	$355.20	$28.60	$47.68
4 Laborers	24.45	782.40	41.05	1313.60		
3 Equip. Oper. (medium)	34.90	837.60	57.85	1388.40		
1 Truck Driver (heavy)	28.45	227.60	47.00	376.00		
1 Road Mixers, 310 H.P.		1929.00		2121.90	26.79	29.47
72 L.H., Daily Totals		$3988.20		$5555.10	$55.39	$77.15

Crew B-91	Hr.	Daily	Hr.	Daily	Bare Costs	Incl. O&P
1 Labor Foreman (outside)	$26.45	$211.60	$44.40	$355.20	$30.43	$50.61
2 Laborers	24.45	391.20	41.05	656.80		
4 Equip. Oper. (medium)	34.90	1116.80	57.85	1851.20		
1 Truck Driver (heavy)	28.45	227.60	47.00	376.00		
1 Dist. Tanker, 3000 Gallon		305.60		336.16		
1 Truck Tractor, 6x4, 380 H.P.		599.80		659.78		
1 Aggreg. Spreader, S.P.		834.00		917.40		
1 Roller, Pneum. Whl., 12 Ton		344.40		378.84		
1 Tandem Roller, 10 Ton		236.80		260.48	36.26	39.89
64 L.H., Daily Totals		$4267.80		$5791.86	$66.68	$90.50

Crew B-91B	Hr.	Daily	Hr.	Daily	Bare Costs	Incl. O&P
1 Laborer	$24.45	$195.60	$41.05	$328.40	$29.68	$49.45
1 Equipment Oper. (med.)	34.90	279.20	57.85	462.80		
1 Road Sweeper, Vac. Assist.		986.00		1084.60	61.63	67.79
16 L.H., Daily Totals		$1460.80		$1875.80	$91.30	$117.24

Crew B-91C	Hr.	Daily	Hr.	Daily	Bare Costs	Incl. O&P
1 Laborer	$24.45	$195.60	$41.05	$328.40	$26.10	$43.45
1 Truck Driver (light)	27.75	222.00	45.85	366.80		
1 Catch Basin Cleaning Truck		573.60		630.96	35.85	39.44
16 L.H., Daily Totals		$991.20		$1326.16	$61.95	$82.89

Crew B-91D	Hr.	Daily	Hr.	Daily	Bare Costs	Incl. O&P
1 Labor Foreman (outside)	$26.45	$211.60	$44.40	$355.20	$29.24	$48.68
5 Laborers	24.45	978.00	41.05	1642.00		
5 Equip. Oper. (medium)	34.90	1396.00	57.85	2314.00		
2 Truck Drivers (heavy)	28.45	455.20	47.00	752.00		
1 Aggreg. Spreader, S.P.		834.00		917.40		
2 Truck Tractor, 6x4, 380 H.P.		1199.60		1319.56		
2 Dist. Tanker, 3000 Gallon		611.20		672.32		
2 Pavement Brush, Towed		163.20		179.52		
2 Roller, Pneum. Whl., 12 Ton		688.80		757.68	33.62	36.99
104 L.H., Daily Totals		$6537.60		$8909.68	$62.86	$85.67

Crews

Crew No.	Bare Costs		Incl. Subs O&P		Cost Per Labor-Hour	
Crew B-92	Hr.	Daily	Hr.	Daily	Bare Costs	Incl. O&P
1 Labor Foreman (outside)	$26.45	$211.60	$44.40	$355.20	$24.95	$41.89
3 Laborers	24.45	586.80	41.05	985.20		
1 Crack Cleaner, 25 H.P.		61.60		67.76		
1 Air Compressor, 60 cfm		137.40		151.14		
1 Tar Kettle, T.M.		132.00		145.20		
1 Flatbed Truck, Gas, 3 Ton		303.20		333.52	19.82	21.80
32 L.H., Daily Totals		$1432.60		$2038.02	$44.77	$63.69
Crew B-93	Hr.	Daily	Hr.	Daily	Bare Costs	Incl. O&P
1 Equip. Oper. (medium)	$34.90	$279.20	$57.85	$462.80	$34.90	$57.85
1 Feller Buncher, 100 H.P.		838.40		922.24	104.80	115.28
8 L.H., Daily Totals		$1117.60		$1385.04	$139.70	$173.13
Crew B-94A	Hr.	Daily	Hr.	Daily	Bare Costs	Incl. O&P
1 Laborer	$24.45	$195.60	$41.05	$328.40	$24.45	$41.05
1 Diaphragm Water Pump, 2"		71.00		78.10		
1 -20' Suction Hose, 2"		1.95		2.15		
2 -50' Discharge Hoses, 2"		1.80		1.98	9.34	10.28
8 L.H., Daily Totals		$270.35		$410.63	$33.79	$51.33
Crew B-94B	Hr.	Daily	Hr.	Daily	Bare Costs	Incl. O&P
1 Laborer	$24.45	$195.60	$41.05	$328.40	$24.45	$41.05
1 Diaphragm Water Pump, 4"		114.20		125.62		
1 -20' Suction Hose, 4"		3.25		3.58		
2 -50' Discharge Hoses, 4"		4.70		5.17	15.27	16.80
8 L.H., Daily Totals		$317.75		$462.76	$39.72	$57.85
Crew B-94C	Hr.	Daily	Hr.	Daily	Bare Costs	Incl. O&P
1 Laborer	$24.45	$195.60	$41.05	$328.40	$24.45	$41.05
1 Centrifugal Water Pump, 3"		78.20		86.02		
1 -20' Suction Hose, 3"		2.85		3.13		
2 -50' Discharge Hoses, 3"		3.00		3.30	10.51	11.56
8 L.H., Daily Totals		$279.65		$420.86	$34.96	$52.61
Crew B-94D	Hr.	Daily	Hr.	Daily	Bare Costs	Incl. O&P
1 Laborer	$24.45	$195.60	$41.05	$328.40	$24.45	$41.05
1 Centr. Water Pump, 6"		346.60		381.26		
1 -20' Suction Hose, 6"		11.50		12.65		
2 -50' Discharge Hoses, 6"		12.20		13.42	46.29	50.92
8 L.H., Daily Totals		$565.90		$735.73	$70.74	$91.97
Crew C-1	Hr.	Daily	Hr.	Daily	Bare Costs	Incl. O&P
2 Carpenters	$32.85	$525.60	$55.15	$882.40	$28.64	$48.15
1 Carpenter Helper	24.40	195.20	41.25	330.00		
1 Laborer	24.45	195.60	41.05	328.40		
32 L.H., Daily Totals		$916.40		$1540.80	$28.64	$48.15
Crew C-2	Hr.	Daily	Hr.	Daily	Bare Costs	Incl. O&P
1 Carpenter Foreman (outside)	$34.85	$278.80	$58.50	$468.00	$28.97	$48.73
2 Carpenters	32.85	525.60	55.15	882.40		
2 Carpenter Helpers	24.40	390.40	41.25	660.00		
1 Laborer	24.45	195.60	41.05	328.40		
48 L.H., Daily Totals		$1390.40		$2338.80	$28.97	$48.73
Crew C-2A	Hr.	Daily	Hr.	Daily	Bare Costs	Incl. O&P
1 Carpenter Foreman (outside)	$34.85	$278.80	$58.50	$468.00	$31.56	$52.67
3 Carpenters	32.85	788.40	55.15	1323.60		
1 Cement Finisher	31.50	252.00	51.05	408.40		
1 Laborer	24.45	195.60	41.05	328.40		
48 L.H., Daily Totals		$1514.80		$2528.40	$31.56	$52.67

Crew No.	Bare Costs		Incl. Subs O&P		Cost Per Labor-Hour	
Crew C-3	Hr.	Daily	Hr.	Daily	Bare Costs	Incl. O&P
1 Rodman Foreman (outside)	$37.20	$297.60	$63.30	$506.40	$31.21	$52.72
3 Rodmen (reinf.)	35.20	844.80	59.90	1437.60		
1 Equip. Oper. (light)	33.55	268.40	55.60	444.80		
3 Laborers	24.45	586.80	41.05	985.20		
3 Stressing Equipment		30.60		33.66		
.5 Grouting Equipment		81.50		89.65	1.75	1.93
64 L.H., Daily Totals		$2109.70		$3497.31	$32.96	$54.65
Crew C-4	Hr.	Daily	Hr.	Daily	Bare Costs	Incl. O&P
1 Rodman Foreman (outside)	$37.20	$297.60	$63.30	$506.40	$33.01	$56.04
2 Rodmen (reinf.)	35.20	563.20	59.90	958.40		
1 Building Laborer	24.45	195.60	41.05	328.40		
3 Stressing Equipment		30.60		33.66	0.96	1.05
32 L.H., Daily Totals		$1087.00		$1826.86	$33.97	$57.09
Crew C-4A	Hr.	Daily	Hr.	Daily	Bare Costs	Incl. O&P
2 Rodmen (reinf.)	$35.20	$563.20	$59.90	$958.40	$35.20	$59.90
4 Stressing Equipment		40.80		44.88	2.55	2.81
16 L.H., Daily Totals		$604.00		$1003.28	$37.75	$62.70
Crew C-5	Hr.	Daily	Hr.	Daily	Bare Costs	Incl. O&P
1 Rodman Foreman (outside)	$37.20	$297.60	$63.30	$506.40	$32.02	$54.04
2 Rodmen (reinf.)	35.20	563.20	59.90	958.40		
1 Equip. Oper. (crane)	35.65	285.20	59.05	472.40		
2 Building Laborers	24.45	391.20	41.05	656.80		
1 Hyd. Crane, 25 Ton		736.60		810.26	15.35	16.88
48 L.H., Daily Totals		$2273.80		$3404.26	$47.37	$70.92
Crew C-6	Hr.	Daily	Hr.	Daily	Bare Costs	Incl. O&P
1 Labor Foreman (outside)	$26.45	$211.60	$44.40	$355.20	$25.96	$43.27
4 Laborers	24.45	782.40	41.05	1313.60		
1 Cement Finisher	31.50	252.00	51.05	408.40		
2 Gas Engine Vibrators		62.40		68.64	1.30	1.43
48 L.H., Daily Totals		$1308.40		$2145.84	$27.26	$44.70
Crew C-7	Hr.	Daily	Hr.	Daily	Bare Costs	Incl. O&P
1 Labor Foreman (outside)	$26.45	$211.60	$44.40	$355.20	$27.37	$45.58
5 Laborers	24.45	978.00	41.05	1642.00		
1 Cement Finisher	31.50	252.00	51.05	408.40		
1 Equip. Oper. (medium)	34.90	279.20	57.85	462.80		
1 Equip. Oper. (oiler)	31.20	249.60	51.70	413.60		
2 Gas Engine Vibrators		62.40		68.64		
1 Concrete Bucket, 1 C.Y.		23.80		26.18		
1 Hyd. Crane, 55 Ton		1128.00		1240.80	16.86	18.55
72 L.H., Daily Totals		$3184.60		$4617.62	$44.23	$64.13
Crew C-8	Hr.	Daily	Hr.	Daily	Bare Costs	Incl. O&P
1 Labor Foreman (outside)	$26.45	$211.60	$44.40	$355.20	$28.24	$46.79
3 Laborers	24.45	586.80	41.05	985.20		
2 Cement Finishers	31.50	504.00	51.05	816.80		
1 Equip. Oper. (medium)	34.90	279.20	57.85	462.80		
1 Concrete Pump (Small)		720.00		792.00	12.86	14.14
56 L.H., Daily Totals		$2301.60		$3412.00	$41.10	$60.93
Crew C-8A	Hr.	Daily	Hr.	Daily	Bare Costs	Incl. O&P
1 Labor Foreman (outside)	$26.45	$211.60	$44.40	$355.20	$27.13	$44.94
3 Laborers	24.45	586.80	41.05	985.20		
2 Cement Finishers	31.50	504.00	51.05	816.80		
48 L.H., Daily Totals		$1302.40		$2157.20	$27.13	$44.94

Crews

Crew No.		Bare Costs	Incl. Subs O&P		Cost Per Labor-Hour	
Crew C-8B	Hr.	Daily	Hr.	Daily	Bare Costs	Incl. O&P
1 Labor Foreman (outside)	$26.45	$211.60	$44.40	$355.20	$26.94	$45.08
3 Laborers	24.45	586.80	41.05	985.20		
1 Equip. Oper. (medium)	34.90	279.20	57.85	462.80		
1 Vibrating Power Screed		66.30		72.93		
1 Roller, Vibratory, 25 Ton		687.40		756.14		
1 Dozer, 200 H.P.		1387.00		1525.70	53.52	58.87
40 L.H., Daily Totals		$3218.30		$4157.97	$80.46	$103.95
Crew C-8C	Hr.	Daily	Hr.	Daily	Bare Costs	Incl. O&P
1 Labor Foreman (outside)	$26.45	$211.60	$44.40	$355.20	$27.70	$46.08
3 Laborers	24.45	586.80	41.05	985.20		
1 Cement Finisher	31.50	252.00	51.05	408.40		
1 Equip. Oper. (medium)	34.90	279.20	57.85	462.80		
1 Shotcrete Rig, 12 C.Y./hr		250.80		275.88		
1 Air Compressor, 160 cfm		156.40		172.04		
4 -50' Air Hoses, 1"		16.40		18.04		
4 -50' Air Hoses, 2"		31.00		34.10	9.47	10.42
48 L.H., Daily Totals		$1784.20		$2711.66	$37.17	$56.49
Crew C-8D	Hr.	Daily	Hr.	Daily	Bare Costs	Incl. O&P
1 Labor Foreman (outside)	$26.45	$211.60	$44.40	$355.20	$28.99	$48.02
1 Laborer	24.45	195.60	41.05	328.40		
1 Cement Finisher	31.50	252.00	51.05	408.40		
1 Equipment Oper. (light)	33.55	268.40	55.60	444.80		
1 Air Compressor, 250 cfm		201.40		221.54		
2 -50' Air Hoses, 1"		8.20		9.02	6.55	7.21
32 L.H., Daily Totals		$1137.20		$1767.36	$35.54	$55.23
Crew C-8E	Hr.	Daily	Hr.	Daily	Bare Costs	Incl. O&P
1 Labor Foreman (outside)	$26.45	$211.60	$44.40	$355.20	$27.48	$45.70
3 Laborers	24.45	586.80	41.05	985.20		
1 Cement Finisher	31.50	252.00	51.05	408.40		
1 Equipment Oper. (light)	33.55	268.40	55.60	444.80		
1 Shotcrete Rig, 35 C.Y./hr		281.00		309.10		
1 Air Compressor, 250 cfm		201.40		221.54		
4 -50' Air Hoses, 1"		16.40		18.04		
4 -50' Air Hoses, 2"		31.00		34.10	11.04	12.14
48 L.H., Daily Totals		$1848.60		$2776.38	$38.51	$57.84
Crew C-10	Hr.	Daily	Hr.	Daily	Bare Costs	Incl. O&P
1 Laborer	$24.45	$195.60	$41.05	$328.40	$29.15	$47.72
2 Cement Finishers	31.50	504.00	51.05	816.80		
24 L.H., Daily Totals		$699.60		$1145.20	$29.15	$47.72
Crew C-10B	Hr.	Daily	Hr.	Daily	Bare Costs	Incl. O&P
3 Laborers	$24.45	$586.80	$41.05	$985.20	$27.27	$45.05
2 Cement Finishers	31.50	504.00	51.05	816.80		
1 Concrete Mixer, 10 C.F.		172.40		189.64		
2 Trowels, 48" Walk-Behind		100.40		110.44	6.82	7.50
40 L.H., Daily Totals		$1363.60		$2102.08	$34.09	$52.55
Crew C-10C	Hr.	Daily	Hr.	Daily	Bare Costs	Incl. O&P
1 Laborer	$24.45	$195.60	$41.05	$328.40	$29.15	$47.72
2 Cement Finishers	31.50	504.00	51.05	816.80		
1 Trowel, 48" Walk-Behind		50.20		55.22	2.09	2.30
24 L.H., Daily Totals		$749.80		$1200.42	$31.24	$50.02

Crew No.		Bare Costs	Incl. Subs O&P		Cost Per Labor-Hour	
Crew C-10D	Hr.	Daily	Hr.	Daily	Bare Costs	Incl. O&P
1 Laborer	$24.45	$195.60	$41.05	$328.40	$29.15	$47.72
2 Cement Finishers	31.50	504.00	51.05	816.80		
1 Vibrating Power Screed		66.30		72.93		
1 Trowel, 48" Walk-Behind		50.20		55.22	4.85	5.34
24 L.H., Daily Totals		$816.10		$1273.35	$34.00	$53.06
Crew C-10E	Hr.	Daily	Hr.	Daily	Bare Costs	Incl. O&P
1 Laborer	$24.45	$195.60	$41.05	$328.40	$29.15	$47.72
2 Cement Finishers	31.50	504.00	51.05	816.80		
1 Vibrating Power Screed		66.30		72.93		
1 Cement Trowel, 96" Ride-On		189.00		207.90	10.64	11.70
24 L.H., Daily Totals		$954.90		$1426.03	$39.79	$59.42
Crew C-10F	Hr.	Daily	Hr.	Daily	Bare Costs	Incl. O&P
1 Laborer	$24.45	$195.60	$41.05	$328.40	$29.15	$47.72
2 Cement Finishers	31.50	504.00	51.05	816.80		
1 Aerial Lift Truck, 60' Boom		435.60		479.16	18.15	19.97
24 L.H., Daily Totals		$1135.20		$1624.36	$47.30	$67.68
Crew C-11	Hr.	Daily	Hr.	Daily	Bare Costs	Incl. O&P
1 Skilled Worker Foreman	$35.35	$282.80	$59.55	$476.40	$33.96	$57.09
5 Skilled Workers	33.35	1334.00	56.20	2248.00		
1 Equip. Oper. (crane)	35.65	285.20	59.05	472.40		
1 Lattice Boom Crane, 150 Ton		1813.00		1994.30	32.38	35.61
56 L.H., Daily Totals		$3715.00		$5191.10	$66.34	$92.70
Crew C-12	Hr.	Daily	Hr.	Daily	Bare Costs	Incl. O&P
1 Carpenter Foreman (outside)	$34.85	$278.80	$58.50	$468.00	$32.25	$54.01
3 Carpenters	32.85	788.40	55.15	1323.60		
1 Laborer	24.45	195.60	41.05	328.40		
1 Equip. Oper. (crane)	35.65	285.20	59.05	472.40		
1 Hyd. Crane, 12 Ton		653.80		719.18	13.62	14.98
48 L.H., Daily Totals		$2201.80		$3311.58	$45.87	$68.99
Crew C-13	Hr.	Daily	Hr.	Daily	Bare Costs	Incl. O&P
2 Struc. Steel Workers	$35.30	$564.80	$66.25	$1060.00	$34.48	$62.55
1 Carpenter	32.85	262.80	55.15	441.20		
1 Welder, Gas Engine, 300 amp		145.85		160.44	6.08	6.68
24 L.H., Daily Totals		$973.45		$1661.64	$40.56	$69.23
Crew C-14	Hr.	Daily	Hr.	Daily	Bare Costs	Incl. O&P
1 Carpenter Foreman (outside)	$34.85	$278.80	$58.50	$468.00	$29.29	$49.06
3 Carpenters	32.85	788.40	55.15	1323.60		
2 Carpenter Helpers	24.40	390.40	41.25	660.00		
4 Laborers	24.45	782.40	41.05	1313.60		
2 Rodmen (reinf.)	35.20	563.20	59.90	958.40		
2 Rodman Helpers	24.40	390.40	41.25	660.00		
2 Cement Finishers	31.50	504.00	51.05	816.80		
1 Equip. Oper. (crane)	35.65	285.20	59.05	472.40		
1 Hyd. Crane, 80 Ton		1625.00		1787.50	11.95	13.14
136 L.H., Daily Totals		$5607.80		$8460.30	$41.23	$62.21

Crews

Crew No.		Bare Costs	Incl. Subs O&P		Cost Per Labor-Hour	
Crew C-14A	Hr.	Daily	Hr.	Daily	Bare Costs	Incl. O&P
1 Carpenter Foreman (outside)	$34.85	$278.80	$58.50	$468.00	$32.66	$54.86
16 Carpenters	32.85	4204.80	55.15	7059.20		
4 Rodmen (reinf.)	35.20	1126.40	59.90	1916.80		
2 Laborers	24.45	391.20	41.05	656.80		
1 Cement Finisher	31.50	252.00	51.05	408.40		
1 Equip. Oper. (medium)	34.90	279.20	57.85	462.80		
1 Gas Engine Vibrator		31.20		34.32		
1 Concrete Pump (Small)		720.00		792.00	3.76	4.13
200 L.H., Daily Totals		$7283.60		$11798.32	$36.42	$58.99
Crew C-14B	Hr.	Daily	Hr.	Daily	Bare Costs	Incl. O&P
1 Carpenter Foreman (outside)	$34.85	$278.80	$58.50	$468.00	$32.62	$54.71
16 Carpenters	32.85	4204.80	55.15	7059.20		
4 Rodmen (reinf.)	35.20	1126.40	59.90	1916.80		
2 Laborers	24.45	391.20	41.05	656.80		
2 Cement Finishers	31.50	504.00	51.05	816.80		
1 Equip. Oper. (medium)	34.90	279.20	57.85	462.80		
1 Gas Engine Vibrator		31.20		34.32		
1 Concrete Pump (Small)		720.00		792.00	3.61	3.97
208 L.H., Daily Totals		$7535.60		$12206.72	$36.23	$58.69
Crew C-14C	Hr.	Daily	Hr.	Daily	Bare Costs	Incl. O&P
1 Carpenter Foreman (outside)	$34.85	$278.80	$58.50	$468.00	$30.83	$51.75
6 Carpenters	32.85	1576.80	55.15	2647.20		
2 Rodmen (reinf.)	35.20	563.20	59.90	958.40		
4 Laborers	24.45	782.40	41.05	1313.60		
1 Cement Finisher	31.50	252.00	51.05	408.40		
1 Gas Engine Vibrator		31.20		34.32	0.28	0.31
112 L.H., Daily Totals		$3484.40		$5829.92	$31.11	$52.05
Crew C-14D	Hr.	Daily	Hr.	Daily	Bare Costs	Incl. O&P
1 Carpenter Foreman (outside)	$34.85	$278.80	$58.50	$468.00	$32.47	$54.48
18 Carpenters	32.85	4730.40	55.15	7941.60		
2 Rodmen (reinf.)	35.20	563.20	59.90	958.40		
2 Laborers	24.45	391.20	41.05	656.80		
1 Cement Finisher	31.50	252.00	51.05	408.40		
1 Equip. Oper. (medium)	34.90	279.20	57.85	462.80		
1 Gas Engine Vibrator		31.20		34.32		
1 Concrete Pump (Small)		720.00		792.00	3.76	4.13
200 L.H., Daily Totals		$7246.00		$11722.32	$36.23	$58.61
Crew C-14E	Hr.	Daily	Hr.	Daily	Bare Costs	Incl. O&P
1 Carpenter Foreman (outside)	$34.85	$278.80	$58.50	$468.00	$31.47	$52.96
2 Carpenters	32.85	525.60	55.15	882.40		
4 Rodmen (reinf.)	35.20	1126.40	59.90	1916.80		
3 Laborers	24.45	586.80	41.05	985.20		
1 Cement Finisher	31.50	252.00	51.05	408.40		
1 Gas Engine Vibrator		31.20		34.32	0.35	0.39
88 L.H., Daily Totals		$2800.80		$4695.12	$31.83	$53.35
Crew C-14F	Hr.	Daily	Hr.	Daily	Bare Costs	Incl. O&P
1 Labor Foreman (outside)	$26.45	$211.60	$44.40	$355.20	$29.37	$48.09
2 Laborers	24.45	391.20	41.05	656.80		
6 Cement Finishers	31.50	1512.00	51.05	2450.40		
1 Gas Engine Vibrator		31.20		34.32	0.43	0.48
72 L.H., Daily Totals		$2146.00		$3496.72	$29.81	$48.57
Crew C-14G	Hr.	Daily	Hr.	Daily	Bare Costs	Incl. O&P
1 Labor Foreman (outside)	$26.45	$211.60	$44.40	$355.20	$28.76	$47.24
2 Laborers	24.45	391.20	41.05	656.80		
4 Cement Finishers	31.50	1008.00	51.05	1633.60		
1 Gas Engine Vibrator		31.20		34.32	0.56	0.61
56 L.H., Daily Totals		$1642.00		$2679.92	$29.32	$47.86
Crew C-14H	Hr.	Daily	Hr.	Daily	Bare Costs	Incl. O&P
1 Carpenter Foreman (outside)	$34.85	$278.80	$58.50	$468.00	$31.95	$53.47
2 Carpenters	32.85	525.60	55.15	882.40		
1 Rodman (reinf.)	35.20	281.60	59.90	479.20		
1 Laborer	24.45	195.60	41.05	328.40		
1 Cement Finisher	31.50	252.00	51.05	408.40		
1 Gas Engine Vibrator		31.20		34.32	0.65	0.71
48 L.H., Daily Totals		$1564.80		$2600.72	$32.60	$54.18
Crew C-14L	Hr.	Daily	Hr.	Daily	Bare Costs	Incl. O&P
1 Carpenter Foreman (outside)	$34.85	$278.80	$58.50	$468.00	$30.10	$50.39
6 Carpenters	32.85	1576.80	55.15	2647.20		
4 Laborers	24.45	782.40	41.05	1313.60		
1 Cement Finisher	31.50	252.00	51.05	408.40		
1 Gas Engine Vibrator		31.20		34.32	0.33	0.36
96 L.H., Daily Totals		$2921.20		$4871.52	$30.43	$50.74
Crew C-14M	Hr.	Daily	Hr.	Daily	Bare Costs	Incl. O&P
1 Carpenter Foreman (outside)	$34.85	$278.80	$58.50	$468.00	$31.38	$52.46
2 Carpenters	32.85	525.60	55.15	882.40		
1 Rodman (reinf.)	35.20	281.60	59.90	479.20		
2 Laborers	24.45	391.20	41.05	656.80		
1 Cement Finisher	31.50	252.00	51.05	408.40		
1 Equip. Oper. (medium)	34.90	279.20	57.85	462.80		
1 Gas Engine Vibrator		31.20		34.32		
1 Concrete Pump (Small)		720.00		792.00	11.74	12.91
64 L.H., Daily Totals		$2759.60		$4183.92	$43.12	$65.37
Crew C-15	Hr.	Daily	Hr.	Daily	Bare Costs	Incl. O&P
1 Carpenter Foreman (outside)	$34.85	$278.80	$58.50	$468.00	$30.23	$50.44
2 Carpenters	32.85	525.60	55.15	882.40		
3 Laborers	24.45	586.80	41.05	985.20		
2 Cement Finishers	31.50	504.00	51.05	816.80		
1 Rodman (reinf.)	35.20	281.60	59.90	479.20		
72 L.H., Daily Totals		$2176.80		$3631.60	$30.23	$50.44
Crew C-16	Hr.	Daily	Hr.	Daily	Bare Costs	Incl. O&P
1 Labor Foreman (outside)	$26.45	$211.60	$44.40	$355.20	$28.24	$46.79
3 Laborers	24.45	586.80	41.05	985.20		
2 Cement Finishers	31.50	504.00	51.05	816.80		
1 Equip. Oper. (medium)	34.90	279.20	57.85	462.80		
1 Gunite Pump Rig		371.80		408.98		
2 -50' Air Hoses, 3/4"		6.50		7.15		
2 -50' Air Hoses, 2"		15.50		17.05	7.03	7.74
56 L.H., Daily Totals		$1975.40		$3053.18	$35.27	$54.52
Crew C-16A	Hr.	Daily	Hr.	Daily	Bare Costs	Incl. O&P
1 Laborer	$24.45	$195.60	$41.05	$328.40	$30.59	$50.25
2 Cement Finishers	31.50	504.00	51.05	816.80		
1 Equip. Oper. (medium)	34.90	279.20	57.85	462.80		
1 Gunite Pump Rig		371.80		408.98		
2 -50' Air Hoses, 3/4"		6.50		7.15		
2 -50' Air Hoses, 2"		15.50		17.05		
1 Aerial Lift Truck, 60' Boom		435.60		479.16	25.92	28.51
32 L.H., Daily Totals		$1808.20		$2520.34	$56.51	$78.76

Crews

Crew No.	Bare Costs		Incl. Subs O&P		Cost Per Labor-Hour	
Crew C-17	Hr.	Daily	Hr.	Daily	Bare Costs	Incl. O&P
2 Skilled Worker Foremen (out)	$35.35	$565.60	$59.55	$952.80	$33.75	$56.87
8 Skilled Workers	33.35	2134.40	56.20	3596.80		
80 L.H., Daily Totals		$2700.00		$4549.60	$33.75	$56.87
Crew C-17A	Hr.	Daily	Hr.	Daily	Bare Costs	Incl. O&P
2 Skilled Worker Foremen (out)	$35.35	$565.60	$59.55	$952.80	$33.77	$56.90
8 Skilled Workers	33.35	2134.40	56.20	3596.80		
.125 Equip. Oper. (crane)	35.65	35.65	59.05	59.05		
.125 Hyd. Crane, 80 Ton		203.13		223.44	2.51	2.76
81 L.H., Daily Totals		$2938.78		$4832.09	$36.28	$59.66
Crew C-17B	Hr.	Daily	Hr.	Daily	Bare Costs	Incl. O&P
2 Skilled Worker Foremen (out)	$35.35	$565.60	$59.55	$952.80	$33.80	$56.92
8 Skilled Workers	33.35	2134.40	56.20	3596.80		
.25 Equip. Oper. (crane)	35.65	71.30	59.05	118.10		
.25 Hyd. Crane, 80 Ton		406.25		446.88		
.25 Trowel, 48" Walk-Behind		12.55		13.81	5.11	5.62
82 L.H., Daily Totals		$3190.10		$5128.38	$38.90	$62.54
Crew C-17C	Hr.	Daily	Hr.	Daily	Bare Costs	Incl. O&P
2 Skilled Worker Foremen (out)	$35.35	$565.60	$59.55	$952.80	$33.82	$56.95
8 Skilled Workers	33.35	2134.40	56.20	3596.80		
.375 Equip. Oper. (crane)	35.65	106.95	59.05	177.15		
.375 Hyd. Crane, 80 Ton		609.38		670.31	7.34	8.08
83 L.H., Daily Totals		$3416.32		$5397.06	$41.16	$65.02
Crew C-17D	Hr.	Daily	Hr.	Daily	Bare Costs	Incl. O&P
2 Skilled Worker Foremen (out)	$35.35	$565.60	$59.55	$952.80	$33.84	$56.97
8 Skilled Workers	33.35	2134.40	56.20	3596.80		
.5 Equip. Oper. (crane)	35.65	142.60	59.05	236.20		
.5 Hyd. Crane, 80 Ton		812.50		893.75	9.67	10.64
84 L.H., Daily Totals		$3655.10		$5679.55	$43.51	$67.61
Crew C-17E	Hr.	Daily	Hr.	Daily	Bare Costs	Incl. O&P
2 Skilled Worker Foremen (out)	$35.35	$565.60	$59.55	$952.80	$33.75	$56.87
8 Skilled Workers	33.35	2134.40	56.20	3596.80		
1 Hyd. Jack with Rods		96.50		106.15	1.21	1.33
80 L.H., Daily Totals		$2796.50		$4655.75	$34.96	$58.20
Crew C-18	Hr.	Daily	Hr.	Daily	Bare Costs	Incl. O&P
.125 Labor Foreman (outside)	$26.45	$26.45	$44.40	$44.40	$24.67	$41.42
1 Laborer	24.45	195.60	41.05	328.40		
1 Concrete Cart, 10 C.F.		60.00		66.00	6.67	7.33
9 L.H., Daily Totals		$282.05		$438.80	$31.34	$48.76
Crew C-19	Hr.	Daily	Hr.	Daily	Bare Costs	Incl. O&P
.125 Labor Foreman (outside)	$26.45	$26.45	$44.40	$44.40	$24.67	$41.42
1 Laborer	24.45	195.60	41.05	328.40		
1 Concrete Cart, 18 C.F.		99.80		109.78	11.09	12.20
9 L.H., Daily Totals		$321.85		$482.58	$35.76	$53.62
Crew C-20	Hr.	Daily	Hr.	Daily	Bare Costs	Incl. O&P
1 Labor Foreman (outside)	$26.45	$211.60	$44.40	$355.20	$26.89	$44.82
5 Laborers	24.45	978.00	41.05	1642.00		
1 Cement Finisher	31.50	252.00	51.05	408.40		
1 Equip. Oper. (medium)	34.90	279.20	57.85	462.80		
2 Gas Engine Vibrators		62.40		68.64		
1 Concrete Pump (Small)		720.00		792.00	12.23	13.45
64 L.H., Daily Totals		$2503.20		$3729.04	$39.11	$58.27

Crew No.	Bare Costs		Incl. Subs O&P		Cost Per Labor-Hour	
Crew C-21	Hr.	Daily	Hr.	Daily	Bare Costs	Incl. O&P
1 Labor Foreman (outside)	$26.45	$211.60	$44.40	$355.20	$26.89	$44.82
5 Laborers	24.45	978.00	41.05	1642.00		
1 Cement Finisher	31.50	252.00	51.05	408.40		
1 Equip. Oper. (medium)	34.90	279.20	57.85	462.80		
2 Gas Engine Vibrators		62.40		68.64		
1 Concrete Conveyer		198.80		218.68	4.08	4.49
64 L.H., Daily Totals		$1982.00		$3155.72	$30.97	$49.31
Crew C-22	Hr.	Daily	Hr.	Daily	Bare Costs	Incl. O&P
1 Rodman Foreman (outside)	$37.20	$297.60	$63.30	$506.40	$35.50	$60.33
4 Rodmen (reinf.)	35.20	1126.40	59.90	1916.80		
.125 Equip. Oper. (crane)	35.65	35.65	59.05	59.05		
.125 Equip. Oper. (oiler)	31.20	31.20	51.70	51.70		
.125 Hyd. Crane, 25 Ton		92.08		101.28	2.19	2.41
42 L.H., Daily Totals		$1582.93		$2635.23	$37.69	$62.74
Crew C-23	Hr.	Daily	Hr.	Daily	Bare Costs	Incl. O&P
2 Skilled Worker Foremen (out)	$35.35	$565.60	$59.55	$952.80	$33.77	$56.70
6 Skilled Workers	33.35	1600.80	56.20	2697.60		
1 Equip. Oper. (crane)	35.65	285.20	59.05	472.40		
1 Equip. Oper. (oiler)	31.20	249.60	51.70	413.60		
1 Lattice Boom Crane, 90 Ton		1511.00		1662.10	18.89	20.78
80 L.H., Daily Totals		$4212.20		$6198.50	$52.65	$77.48
Crew C-24	Hr.	Daily	Hr.	Daily	Bare Costs	Incl. O&P
2 Skilled Worker Foremen (out)	$35.35	$565.60	$59.55	$952.80	$33.77	$56.70
6 Skilled Workers	33.35	1600.80	56.20	2697.60		
1 Equip. Oper. (crane)	35.65	285.20	59.05	472.40		
1 Equip. Oper. (oiler)	31.20	249.60	51.70	413.60		
1 Lattice Boom Crane, 150 Ton		1813.00		1994.30	22.66	24.93
80 L.H., Daily Totals		$4514.20		$6530.70	$56.43	$81.63
Crew C-25	Hr.	Daily	Hr.	Daily	Bare Costs	Incl. O&P
2 Rodmen (reinf.)	$35.20	$563.20	$59.90	$958.40	$27.98	$49.13
2 Rodmen Helpers	20.75	332.00	38.35	613.60		
32 L.H., Daily Totals		$895.20		$1572.00	$27.98	$49.13
Crew C-27	Hr.	Daily	Hr.	Daily	Bare Costs	Incl. O&P
2 Cement Finishers	$31.50	$504.00	$51.05	$816.80	$31.50	$51.05
1 Concrete Saw		167.40		184.14	10.46	11.51
16 L.H., Daily Totals		$671.40		$1000.94	$41.96	$62.56
Crew C-28	Hr.	Daily	Hr.	Daily	Bare Costs	Incl. O&P
1 Cement Finisher	$31.50	$252.00	$51.05	$408.40	$31.50	$51.05
1 Portable Air Compressor, Gas		17.90		19.69	2.24	2.46
8 L.H., Daily Totals		$269.90		$428.09	$33.74	$53.51
Crew C-29	Hr.	Daily	Hr.	Daily	Bare Costs	Incl. O&P
1 Laborer	$24.45	$195.60	$41.05	$328.40	$24.45	$41.05
1 Pressure Washer		70.20		77.22	8.78	9.65
8 L.H., Daily Totals		$265.80		$405.62	$33.23	$50.70
Crew C-30	Hr.	Daily	Hr.	Daily	Bare Costs	Incl. O&P
1 Laborer	$24.45	$195.60	$41.05	$328.40	$24.45	$41.05
1 Concrete Mixer, 10 C.F.		172.40		189.64	21.55	23.70
8 L.H., Daily Totals		$368.00		$518.04	$46.00	$64.75

Crews

Crew No.	Bare Costs		Incl. Subs O&P		Cost Per Labor-Hour	
Crew C-31	Hr.	Daily	Hr.	Daily	Bare Costs	Incl. O&P
1 Cement Finisher	$31.50	$252.00	$51.05	$408.40	$31.50	$51.05
1 Grout Pump		371.80		408.98	46.48	51.12
8 L.H., Daily Totals		$623.80		$817.38	$77.97	$102.17
Crew C-32	Hr.	Daily	Hr.	Daily	Bare Costs	Incl. O&P
1 Cement Finisher	$31.50	$252.00	$51.05	$408.40	$27.98	$46.05
1 Laborer	24.45	195.60	41.05	328.40		
1 Crack Chaser Saw, Gas, 6 H.P.		29.20		32.12		
1 Vacuum Pick-Up System		60.90		66.99	5.63	6.19
16 L.H., Daily Totals		$537.70		$835.91	$33.61	$52.24
Crew D-1	Hr.	Daily	Hr.	Daily	Bare Costs	Incl. O&P
1 Bricklayer	$32.30	$258.40	$53.85	$430.80	$29.48	$49.15
1 Bricklayer Helper	26.65	213.20	44.45	355.60		
16 L.H., Daily Totals		$471.60		$786.40	$29.48	$49.15
Crew D-2	Hr.	Daily	Hr.	Daily	Bare Costs	Incl. O&P
3 Bricklayers	$32.30	$775.20	$53.85	$1292.40	$30.04	$50.09
2 Bricklayer Helpers	26.65	426.40	44.45	711.20		
40 L.H., Daily Totals		$1201.60		$2003.60	$30.04	$50.09
Crew D-3	Hr.	Daily	Hr.	Daily	Bare Costs	Incl. O&P
3 Bricklayers	$32.30	$775.20	$53.85	$1292.40	$30.17	$50.33
2 Bricklayer Helpers	26.65	426.40	44.45	711.20		
.25 Carpenter	32.85	65.70	55.15	110.30		
42 L.H., Daily Totals		$1267.30		$2113.90	$30.17	$50.33
Crew D-4	Hr.	Daily	Hr.	Daily	Bare Costs	Incl. O&P
1 Bricklayer	$32.30	$258.40	$53.85	$430.80	$27.34	$45.65
3 Bricklayer Helpers	26.65	639.60	44.45	1066.80		
1 Building Laborer	24.45	195.60	41.05	328.40		
1 Grout Pump, 50 C.F./hr.		133.40		146.74	3.34	3.67
40 L.H., Daily Totals		$1227.00		$1972.74	$30.68	$49.32
Crew D-5	Hr.	Daily	Hr.	Daily	Bare Costs	Incl. O&P
1 Block Mason Helper	26.65	213.20	44.45	355.60	26.65	44.45
8 L.H., Daily Totals		$213.20		$355.60	$26.65	$44.45
Crew D-6	Hr.	Daily	Hr.	Daily	Bare Costs	Incl. O&P
3 Bricklayers	$32.30	$775.20	$53.85	$1292.40	$29.48	$49.15
3 Bricklayer Helpers	26.65	639.60	44.45	1066.80		
48 L.H., Daily Totals		$1414.80		$2359.20	$29.48	$49.15
Crew D-7	Hr.	Daily	Hr.	Daily	Bare Costs	Incl. O&P
1 Tile Layer	$29.95	$239.60	$48.50	$388.00	$26.73	$43.27
1 Tile Layer Helper	23.50	188.00	38.05	304.40		
16 L.H., Daily Totals		$427.60		$692.40	$26.73	$43.27
Crew D-8	Hr.	Daily	Hr.	Daily	Bare Costs	Incl. O&P
3 Bricklayers	$32.30	$775.20	$53.85	$1292.40	$30.04	$50.09
2 Bricklayer Helpers	26.65	426.40	44.45	711.20		
40 L.H., Daily Totals		$1201.60		$2003.60	$30.04	$50.09
Crew D-9	Hr.	Daily	Hr.	Daily	Bare Costs	Incl. O&P
3 Bricklayers	$32.30	$775.20	$53.85	$1292.40	$29.48	$49.15
3 Bricklayer Helpers	26.65	639.60	44.45	1066.80		
48 L.H., Daily Totals		$1414.80		$2359.20	$29.48	$49.15

Crew No.	Bare Costs		Incl. Subs O&P		Cost Per Labor-Hour	
Crew D-10	Hr.	Daily	Hr.	Daily	Bare Costs	Incl. O&P
1 Bricklayer Foreman (outside)	$34.30	$274.40	$57.20	$457.60	$32.23	$53.64
1 Bricklayer	32.30	258.40	53.85	430.80		
1 Bricklayer Helper	26.65	213.20	44.45	355.60		
1 Equip. Oper. (crane)	35.65	285.20	59.05	472.40		
1 S.P. Crane, 4x4, 12 Ton		473.40		520.74	14.79	16.27
32 L.H., Daily Totals		$1504.60		$2237.14	$47.02	$69.91
Crew D-11	Hr.	Daily	Hr.	Daily	Bare Costs	Incl. O&P
2 Bricklayers	$32.30	$516.80	$53.85	$861.60	$30.42	$50.72
1 Bricklayer Helper	26.65	213.20	44.45	355.60		
24 L.H., Daily Totals		$730.00		$1217.20	$30.42	$50.72
Crew D-12	Hr.	Daily	Hr.	Daily	Bare Costs	Incl. O&P
2 Bricklayers	$32.30	$516.80	$53.85	$861.60	$29.48	$49.15
2 Bricklayer Helpers	26.65	426.40	44.45	711.20		
32 L.H., Daily Totals		$943.20		$1572.80	$29.48	$49.15
Crew D-13	Hr.	Daily	Hr.	Daily	Bare Costs	Incl. O&P
1 Bricklayer Foreman (outside)	$34.30	$274.40	$57.20	$457.60	$31.31	$52.14
2 Bricklayers	32.30	516.80	53.85	861.60		
2 Bricklayer Helpers	26.65	426.40	44.45	711.20		
1 Equip. Oper. (crane)	35.65	285.20	59.05	472.40		
1 S.P. Crane, 4x4, 12 Ton		473.40		520.74	9.86	10.85
48 L.H., Daily Totals		$1976.20		$3023.54	$41.17	$62.99
Crew E-1	Hr.	Daily	Hr.	Daily	Bare Costs	Incl. O&P
2 Struc. Steel Workers	$35.30	$564.80	$66.25	$1060.00	$35.30	$66.25
1 Welder, Gas Engine, 300 amp		145.85		160.44	9.12	10.03
16 L.H., Daily Totals		$710.65		$1220.43	$44.42	$76.28
Crew E-2	Hr.	Daily	Hr.	Daily	Bare Costs	Incl. O&P
1 Struc. Steel Foreman (outside)	$37.30	$298.40	$70.00	$560.00	$35.69	$65.67
4 Struc. Steel Workers	35.30	1129.60	66.25	2120.00		
1 Equip. Oper. (crane)	35.65	285.20	59.05	472.40		
1 Lattice Boom Crane, 90 Ton		1511.00		1662.10	31.48	34.63
48 L.H., Daily Totals		$3224.20		$4814.50	$67.17	$100.30
Crew E-3	Hr.	Daily	Hr.	Daily	Bare Costs	Incl. O&P
1 Struc. Steel Foreman (outside)	$37.30	$298.40	$70.00	$560.00	$35.97	$67.50
2 Struc. Steel Workers	35.30	564.80	66.25	1060.00		
1 Welder, Gas Engine, 300 amp		145.85		160.44	6.08	6.68
24 L.H., Daily Totals		$1009.05		$1780.43	$42.04	$74.18
Crew E-3A	Hr.	Daily	Hr.	Daily	Bare Costs	Incl. O&P
1 Struc. Steel Foreman (outside)	$37.30	$298.40	$70.00	$560.00	$35.97	$67.50
2 Struc. Steel Workers	35.30	564.80	66.25	1060.00		
1 Welder, Gas Engine, 300 amp		145.85		160.44		
1 Aerial Lift Truck, 40' Boom		301.40		331.54	18.64	20.50
24 L.H., Daily Totals		$1310.45		$2111.97	$54.60	$88.00
Crew E-4	Hr.	Daily	Hr.	Daily	Bare Costs	Incl. O&P
1 Struc. Steel Foreman (outside)	$37.30	$298.40	$70.00	$560.00	$35.80	$67.19
3 Struc. Steel Workers	35.30	847.20	66.25	1590.00		
1 Welder, Gas Engine, 300 amp		145.85		160.44	4.56	5.01
32 L.H., Daily Totals		$1291.45		$2310.43	$40.36	$72.20

Crews

Crew No.	Bare Costs		Incl. Subs O&P		Cost Per Labor-Hour	
Crew E-5	**Hr.**	**Daily**	**Hr.**	**Daily**	**Bare Costs**	**Incl. O&P**
1 Struc. Steel Foreman (outside)	$37.30	$298.40	$70.00	$560.00	$35.56	$65.87
7 Struc. Steel Workers	35.30	1976.80	66.25	3710.00		
1 Equip. Oper. (crane)	35.65	285.20	59.05	472.40		
1 Lattice Boom Crane, 90 Ton		1511.00		1662.10		
1 Welder, Gas Engine, 300 amp		145.85		160.44	23.01	25.31
72 L.H., Daily Totals		$4217.25		$6564.94	$58.57	$91.18
Crew E-6	**Hr.**	**Daily**	**Hr.**	**Daily**	**Bare Costs**	**Incl. O&P**
1 Struc. Steel Foreman (outside)	$37.30	$298.40	$70.00	$560.00	$35.34	$65.31
12 Struc. Steel Workers	35.30	3388.80	66.25	6360.00		
1 Equip. Oper. (crane)	35.65	285.20	59.05	472.40		
1 Equip. Oper. (light)	33.55	268.40	55.60	444.80		
1 Lattice Boom Crane, 90 Ton		1511.00		1662.10		
1 Welder, Gas Engine, 300 amp		145.85		160.44		
1 Air Compressor, 160 cfm		156.40		172.04		
2 Impact Wrenches		36.00		39.60	15.41	16.95
120 L.H., Daily Totals		$6090.05		$9871.38	$50.75	$82.26
Crew E-7	**Hr.**	**Daily**	**Hr.**	**Daily**	**Bare Costs**	**Incl. O&P**
1 Struc. Steel Foreman (outside)	$37.30	$298.40	$70.00	$560.00	$35.56	$65.87
7 Struc. Steel Workers	35.30	1976.80	66.25	3710.00		
1 Equip. Oper. (crane)	35.65	285.20	59.05	472.40		
1 Lattice Boom Crane, 90 Ton		1511.00		1662.10		
2 Welder, Gas Engine, 300 amp		291.70		320.87	25.04	27.54
72 L.H., Daily Totals		$4363.10		$6725.37	$60.60	$93.41
Crew E-8	**Hr.**	**Daily**	**Hr.**	**Daily**	**Bare Costs**	**Incl. O&P**
1 Struc. Steel Foreman (outside)	$37.30	$298.40	$70.00	$560.00	$35.51	$65.94
9 Struc. Steel Workers	35.30	2541.60	66.25	4770.00		
1 Equip. Oper. (crane)	35.65	285.20	59.05	472.40		
1 Lattice Boom Crane, 90 Ton		1511.00		1662.10		
4 Welder, Gas Engine, 300 amp		583.40		641.74	23.80	26.18
88 L.H., Daily Totals		$5219.60		$8106.24	$59.31	$92.12
Crew E-9	**Hr.**	**Daily**	**Hr.**	**Daily**	**Bare Costs**	**Incl. O&P**
2 Struc. Steel Foremen (outside)	$37.30	$596.80	$70.00	$1120.00	$35.33	$64.93
5 Struc. Steel Workers	35.30	1412.00	66.25	2650.00		
1 Welder Foreman (outside)	37.30	298.40	70.00	560.00		
5 Welders	35.30	1412.00	66.25	2650.00		
1 Equip. Oper. (crane)	35.65	285.20	59.05	472.40		
1 Equip. Oper. (oiler)	31.20	249.60	51.70	413.60		
1 Equip. Oper. (light)	33.55	268.40	55.60	444.80		
1 Lattice Boom Crane, 90 Ton		1511.00		1662.10		
5 Welder, Gas Engine, 300 amp		729.25		802.17	17.50	19.25
128 L.H., Daily Totals		$6762.65		$10775.08	$52.83	$84.18
Crew E-10	**Hr.**	**Daily**	**Hr.**	**Daily**	**Bare Costs**	**Incl. O&P**
1 Struc. Steel Foreman (outside)	$37.30	$298.40	$70.00	$560.00	$35.97	$67.50
2 Struc. Steel Workers	35.30	564.80	66.25	1060.00		
1 Welder, Gas Engine, 300 amp		145.85		160.44		
1 Flatbed Truck, Gas, 3 Ton		303.20		333.52	18.71	20.58
24 L.H., Daily Totals		$1312.25		$2113.95	$54.68	$88.08
Crew E-11	**Hr.**	**Daily**	**Hr.**	**Daily**	**Bare Costs**	**Incl. O&P**
2 Painters, Struc. Steel	$29.45	$471.20	$55.95	$895.20	$29.23	$52.14
1 Building Laborer	24.45	195.60	41.05	328.40		
1 Equip. Oper. (light)	33.55	268.40	55.60	444.80		
1 Air Compressor, 250 cfm		201.40		221.54		
1 Sandblaster, Portable, 3 C.F.		20.40		22.44		
1 Set Sand Blasting Accessories		14.05		15.46	7.37	8.11
32 L.H., Daily Totals		$1171.05		$1927.84	$36.60	$60.24

Crew No.	Bare Costs		Incl. Subs O&P		Cost Per Labor-Hour	
Crew E-11A	**Hr.**	**Daily**	**Hr.**	**Daily**	**Bare Costs**	**Incl. O&P**
2 Painters, Struc. Steel	$29.45	$471.20	$55.95	$895.20	$29.23	$52.14
1 Building Laborer	24.45	195.60	41.05	328.40		
1 Equip. Oper. (light)	33.55	268.40	55.60	444.80		
1 Air Compressor, 250 cfm		201.40		221.54		
1 Sandblaster, Portable, 3 C.F.		20.40		22.44		
1 Set Sand Blasting Accessories		14.05		15.46		
1 Aerial Lift Truck, 60' Boom		435.60		479.16	20.98	23.08
32 L.H., Daily Totals		$1606.65		$2406.99	$50.21	$75.22
Crew E-11B	**Hr.**	**Daily**	**Hr.**	**Daily**	**Bare Costs**	**Incl. O&P**
2 Painters, Struc. Steel	$29.45	$471.20	$55.95	$895.20	$27.78	$50.98
1 Building Laborer	24.45	195.60	41.05	328.40		
2 Paint Sprayer, 8 C.F.M.		99.80		109.78		
1 Aerial Lift Truck, 60' Boom		435.60		479.16	22.31	24.54
24 L.H., Daily Totals		$1202.20		$1812.54	$50.09	$75.52
Crew E-12	**Hr.**	**Daily**	**Hr.**	**Daily**	**Bare Costs**	**Incl. O&P**
1 Welder Foreman (outside)	$37.30	$298.40	$70.00	$560.00	$35.42	$62.80
1 Equip. Oper. (light)	33.55	268.40	55.60	444.80		
1 Welder, Gas Engine, 300 amp		145.85		160.44	9.12	10.03
16 L.H., Daily Totals		$712.65		$1165.23	$44.54	$72.83
Crew E-13	**Hr.**	**Daily**	**Hr.**	**Daily**	**Bare Costs**	**Incl. O&P**
1 Welder Foreman (outside)	$37.30	$298.40	$70.00	$560.00	$36.05	$65.10
.5 Equip. Oper. (light)	33.55	134.20	55.60	222.40		
1 Welder, Gas Engine, 300 amp		145.85		160.44	12.15	13.37
12 L.H., Daily Totals		$578.45		$942.84	$48.20	$78.57
Crew E-14	**Hr.**	**Daily**	**Hr.**	**Daily**	**Bare Costs**	**Incl. O&P**
1 Struc. Steel Worker	$35.30	$282.40	$66.25	$530.00	$35.30	$66.25
1 Welder, Gas Engine, 300 amp		145.85		160.44	18.23	20.05
8 L.H., Daily Totals		$428.25		$690.43	$53.53	$86.30
Crew E-16	**Hr.**	**Daily**	**Hr.**	**Daily**	**Bare Costs**	**Incl. O&P**
1 Welder Foreman (outside)	$37.30	$298.40	$70.00	$560.00	$36.30	$68.13
1 Welder	35.30	282.40	66.25	530.00		
1 Welder, Gas Engine, 300 amp		145.85		160.44	9.12	10.03
16 L.H., Daily Totals		$726.65		$1250.43	$45.42	$78.15
Crew E-17	**Hr.**	**Daily**	**Hr.**	**Daily**	**Bare Costs**	**Incl. O&P**
1 Struc. Steel Foreman (outside)	$37.30	$298.40	$70.00	$560.00	$36.30	$68.13
1 Structural Steel Worker	35.30	282.40	66.25	530.00		
16 L.H., Daily Totals		$580.80		$1090.00	$36.30	$68.13
Crew E-18	**Hr.**	**Daily**	**Hr.**	**Daily**	**Bare Costs**	**Incl. O&P**
1 Struc. Steel Foreman (outside)	$37.30	$298.40	$70.00	$560.00	$35.62	$65.32
3 Structural Steel Workers	35.30	847.20	66.25	1590.00		
1 Equipment Operator (med.)	34.90	279.20	57.85	462.80		
1 Lattice Boom Crane, 20 Ton		948.90		1043.79	23.72	26.09
40 L.H., Daily Totals		$2373.70		$3656.59	$59.34	$91.41
Crew E-19	**Hr.**	**Daily**	**Hr.**	**Daily**	**Bare Costs**	**Incl. O&P**
1 Struc. Steel Foreman (outside)	$37.30	$298.40	$70.00	$560.00	$35.38	$63.95
1 Structural Steel Worker	35.30	282.40	66.25	530.00		
1 Equip. Oper. (light)	33.55	268.40	55.60	444.80		
1 Lattice Boom Crane, 20 Ton		948.90		1043.79	39.54	43.49
24 L.H., Daily Totals		$1798.10		$2578.59	$74.92	$107.44

Crews

Crew No.	Bare Costs		Incl. Subs O&P		Cost Per Labor-Hour	
Crew E-20	**Hr.**	**Daily**	**Hr.**	**Daily**	**Bare Costs**	**Incl. O&P**
1 Struc. Steel Foreman (outside)	$37.30	$298.40	$70.00	$560.00	$35.08	$64.00
5 Structural Steel Workers	35.30	1412.00	66.25	2650.00		
1 Equip. Oper. (crane)	35.65	285.20	59.05	472.40		
1 Equip. Oper. (oiler)	31.20	249.60	51.70	413.60		
1 Lattice Boom Crane, 40 Ton		1164.00		1280.40	18.19	20.01
64 L.H., Daily Totals		$3409.20		$5376.40	$53.27	$84.01
Crew E-22	**Hr.**	**Daily**	**Hr.**	**Daily**	**Bare Costs**	**Incl. O&P**
1 Skilled Worker Foreman (out)	$35.35	$282.80	$59.55	$476.40	$34.02	$57.32
2 Skilled Workers	33.35	533.60	56.20	899.20		
24 L.H., Daily Totals		$816.40		$1375.60	$34.02	$57.32
Crew E-24	**Hr.**	**Daily**	**Hr.**	**Daily**	**Bare Costs**	**Incl. O&P**
3 Structural Steel Workers	$35.30	$847.20	$66.25	$1590.00	$35.20	$64.15
1 Equipment Operator (med.)	34.90	279.20	57.85	462.80		
1 Hyd. Crane, 25 Ton		736.60		810.26	23.02	25.32
32 L.H., Daily Totals		$1863.00		$2863.06	$58.22	$89.47
Crew E-25	**Hr.**	**Daily**	**Hr.**	**Daily**	**Bare Costs**	**Incl. O&P**
1 Welder	$35.30	$282.40	$66.25	$530.00	$35.30	$66.25
1 Cutting Torch		11.40		12.54	1.43	1.57
8 L.H., Daily Totals		$293.80		$542.54	$36.73	$67.82
Crew F-3	**Hr.**	**Daily**	**Hr.**	**Daily**	**Bare Costs**	**Incl. O&P**
2 Carpenters	$32.85	$525.60	$55.15	$882.40	$30.03	$50.37
2 Carpenter Helpers	24.40	390.40	41.25	660.00		
1 Equip. Oper. (crane)	35.65	285.20	59.05	472.40		
1 Hyd. Crane, 12 Ton		653.80		719.18	16.34	17.98
40 L.H., Daily Totals		$1855.00		$2733.98	$46.38	$68.35
Crew F-4	**Hr.**	**Daily**	**Hr.**	**Daily**	**Bare Costs**	**Incl. O&P**
2 Carpenters	$32.85	$525.60	$55.15	$882.40	$30.03	$50.37
2 Carpenter Helpers	24.40	390.40	41.25	660.00		
1 Equip. Oper. (crane)	35.65	285.20	59.05	472.40		
1 Hyd. Crane, 55 Ton		1128.00		1240.80	28.20	31.02
40 L.H., Daily Totals		$2329.20		$3255.60	$58.23	$81.39
Crew F-5	**Hr.**	**Daily**	**Hr.**	**Daily**	**Bare Costs**	**Incl. O&P**
2 Carpenters	$32.85	$525.60	$55.15	$882.40	$28.63	$48.20
2 Carpenter Helpers	24.40	390.40	41.25	660.00		
32 L.H., Daily Totals		$916.00		$1542.40	$28.63	$48.20
Crew F-6	**Hr.**	**Daily**	**Hr.**	**Daily**	**Bare Costs**	**Incl. O&P**
2 Carpenters	$32.85	$525.60	$55.15	$882.40	$30.05	$50.29
2 Building Laborers	24.45	391.20	41.05	656.80		
1 Equip. Oper. (crane)	35.65	285.20	59.05	472.40		
1 Hyd. Crane, 12 Ton		653.80		719.18	16.34	17.98
40 L.H., Daily Totals		$1855.80		$2730.78	$46.40	$68.27
Crew F-7	**Hr.**	**Daily**	**Hr.**	**Daily**	**Bare Costs**	**Incl. O&P**
2 Carpenters	$32.85	$525.60	$55.15	$882.40	$28.65	$48.10
2 Building Laborers	24.45	391.20	41.05	656.80		
32 L.H., Daily Totals		$916.80		$1539.20	$28.65	$48.10

Crew No.	Bare Costs		Incl. Subs O&P		Cost Per Labor-Hour	
Crew G-1	**Hr.**	**Daily**	**Hr.**	**Daily**	**Bare Costs**	**Incl. O&P**
1 Roofer Foreman (outside)	$29.65	$237.20	$54.75	$438.00	$25.96	$47.95
4 Roofers Composition	27.65	884.80	51.05	1633.60		
2 Roofer Helpers	20.75	332.00	38.35	613.60		
1 Application Equipment		193.00		212.30		
1 Tar Kettle/Pot		169.80		186.78		
1 Crew Truck		206.40		227.04	10.16	11.18
56 L.H., Daily Totals		$2023.20		$3311.32	$36.13	$59.13
Crew G-2	**Hr.**	**Daily**	**Hr.**	**Daily**	**Bare Costs**	**Incl. O&P**
1 Plasterer	$30.05	$240.40	$49.25	$394.00	$27.05	$44.67
1 Plasterer Helper	26.65	213.20	43.70	349.60		
1 Building Laborer	24.45	195.60	41.05	328.40		
1 Grout Pump, 50 C.F./hr.		133.40		146.74	5.56	6.11
24 L.H., Daily Totals		$782.60		$1218.74	$32.61	$50.78
Crew G-2A	**Hr.**	**Daily**	**Hr.**	**Daily**	**Bare Costs**	**Incl. O&P**
1 Roofer Composition	$27.65	$221.20	$51.05	$408.40	$24.28	$43.48
1 Roofer Helper	20.75	166.00	38.35	306.80		
1 Building Laborer	24.45	195.60	41.05	328.40		
1 Foam Spray Rig, Trailer-Mtd.		573.25		630.58		
1 Pickup Truck, 3/4 Ton		144.20		158.62	29.89	32.88
24 L.H., Daily Totals		$1300.25		$1832.80	$54.18	$76.37
Crew G-3	**Hr.**	**Daily**	**Hr.**	**Daily**	**Bare Costs**	**Incl. O&P**
2 Sheet Metal Workers	$36.95	$591.20	$61.65	$986.40	$30.70	$51.35
2 Building Laborers	24.45	391.20	41.05	656.80		
32 L.H., Daily Totals		$982.40		$1643.20	$30.70	$51.35
Crew G-4	**Hr.**	**Daily**	**Hr.**	**Daily**	**Bare Costs**	**Incl. O&P**
1 Labor Foreman (outside)	$26.45	$211.60	$44.40	$355.20	$25.12	$42.17
2 Building Laborers	24.45	391.20	41.05	656.80		
1 Flatbed Truck, Gas, 1.5 Ton		245.40		269.94		
1 Air Compressor, 160 cfm		156.40		172.04	16.74	18.42
24 L.H., Daily Totals		$1004.60		$1453.98	$41.86	$60.58
Crew G-5	**Hr.**	**Daily**	**Hr.**	**Daily**	**Bare Costs**	**Incl. O&P**
1 Roofer Foreman (outside)	$29.65	$237.20	$54.75	$438.00	$25.29	$46.71
2 Roofers Composition	27.65	442.40	51.05	816.80		
2 Roofer Helpers	20.75	332.00	38.35	613.60		
1 Application Equipment		193.00		212.30	4.83	5.31
40 L.H., Daily Totals		$1204.60		$2080.70	$30.11	$52.02
Crew G-6A	**Hr.**	**Daily**	**Hr.**	**Daily**	**Bare Costs**	**Incl. O&P**
2 Roofers Composition	$27.65	$442.40	$51.05	$816.80	$27.65	$51.05
1 Small Compressor, Electric		12.95		14.24		
2 Pneumatic Nailers		55.30		60.83	4.27	4.69
16 L.H., Daily Totals		$510.65		$891.88	$31.92	$55.74
Crew G-7	**Hr.**	**Daily**	**Hr.**	**Daily**	**Bare Costs**	**Incl. O&P**
1 Carpenter	$32.85	$262.80	$55.15	$441.20	$32.85	$55.15
1 Small Compressor, Electric		12.95		14.24		
1 Pneumatic Nailer		27.65		30.41	5.08	5.58
8 L.H., Daily Totals		$303.40		$485.86	$37.92	$60.73
Crew H-1	**Hr.**	**Daily**	**Hr.**	**Daily**	**Bare Costs**	**Incl. O&P**
2 Glaziers	$32.00	$512.00	$53.20	$851.20	$33.65	$59.73
2 Struc. Steel Workers	35.30	564.80	66.25	1060.00		
32 L.H., Daily Totals		$1076.80		$1911.20	$33.65	$59.73

Crews

Crew No.	Bare Costs		Incl. Subs O&P		Cost Per Labor-Hour	
Crew H-2	Hr.	Daily	Hr.	Daily	Bare Costs	Incl. O&P
2 Glaziers	$32.00	$512.00	$53.20	$851.20	$29.48	$49.15
1 Building Laborer	24.45	195.60	41.05	328.40		
24 L.H., Daily Totals		$707.60		$1179.60	$29.48	$49.15
Crew H-3	Hr.	Daily	Hr.	Daily	Bare Costs	Incl. O&P
1 Glazier	$32.00	$256.00	$53.20	$425.60	$28.20	$47.23
1 Helper	24.40	195.20	41.25	330.00		
16 L.H., Daily Totals		$451.20		$755.60	$28.20	$47.23
Crew H-4	Hr.	Daily	Hr.	Daily	Bare Costs	Incl. O&P
1 Carpenter	$32.85	$262.80	$55.15	$441.20	$30.23	$50.57
1 Carpenter Helper	24.40	195.20	41.25	330.00		
.5 Electrician	36.65	146.60	60.05	240.20		
20 L.H., Daily Totals		$604.60		$1011.40	$30.23	$50.57
Crew J-1	Hr.	Daily	Hr.	Daily	Bare Costs	Incl. O&P
3 Plasterers	$30.05	$721.20	$49.25	$1182.00	$28.69	$47.03
2 Plasterer Helpers	26.65	426.40	43.70	699.20		
1 Mixing Machine, 6 C.F.		140.20		154.22	3.50	3.86
40 L.H., Daily Totals		$1287.80		$2035.42	$32.20	$50.89
Crew J-2	Hr.	Daily	Hr.	Daily	Bare Costs	Incl. O&P
3 Plasterers	$30.05	$721.20	$49.25	$1182.00	$28.93	$47.29
2 Plasterer Helpers	26.65	426.40	43.70	699.20		
1 Lather	30.10	240.80	48.60	388.80		
1 Mixing Machine, 6 C.F.		140.20		154.22	2.92	3.21
48 L.H., Daily Totals		$1528.60		$2424.22	$31.85	$50.50
Crew J-3	Hr.	Daily	Hr.	Daily	Bare Costs	Incl. O&P
1 Terrazzo Worker	$30.05	$240.40	$48.65	$389.20	$27.45	$44.45
1 Terrazzo Helper	24.85	198.80	40.25	322.00		
1 Floor Grinder, 22" Path		115.55		127.11		
1 Terrazzo Mixer		188.20		207.02	18.98	20.88
16 L.H., Daily Totals		$742.95		$1045.33	$46.43	$65.33
Crew J-4	Hr.	Daily	Hr.	Daily	Bare Costs	Incl. O&P
2 Cement Finishers	$31.50	$504.00	$51.05	$816.80	$29.15	$47.72
1 Laborer	24.45	195.60	41.05	328.40		
1 Floor Grinder, 22" Path		115.55		127.11		
1 Floor Edger, 7" Path		39.30		43.23		
1 Vacuum Pick-Up System		60.90		66.99	8.99	9.89
24 L.H., Daily Totals		$915.35		$1382.53	$38.14	$57.61
Crew J-4A	Hr.	Daily	Hr.	Daily	Bare Costs	Incl. O&P
2 Cement Finishers	$31.50	$504.00	$51.05	$816.80	$27.98	$46.05
2 Laborers	24.45	391.20	41.05	656.80		
1 Floor Grinder, 22" Path		115.55		127.11		
1 Floor Edger, 7" Path		39.30		43.23		
1 Vacuum Pick-Up System		60.90		66.99		
1 Floor Auto Scrubber		233.85		257.24	14.05	15.46
32 L.H., Daily Totals		$1344.80		$1968.16	$42.02	$61.51
Crew J-4B	Hr.	Daily	Hr.	Daily	Bare Costs	Incl. O&P
1 Laborer	$24.45	$195.60	$41.05	$328.40	$24.45	$41.05
1 Floor Auto Scrubber		233.85		257.24	29.23	32.15
8 L.H., Daily Totals		$429.45		$585.63	$53.68	$73.20
Crew J-6	Hr.	Daily	Hr.	Daily	Bare Costs	Incl. O&P
2 Painters	$28.65	$458.40	$47.20	$755.20	$28.82	$47.76
1 Building Laborer	24.45	195.60	41.05	328.40		
1 Equip. Oper. (light)	33.55	268.40	55.60	444.80		
1 Air Compressor, 250 cfm		201.40		221.54		
1 Sandblaster, Portable, 3 C.F.		20.40		22.44		
1 Set Sand Blasting Accessories		14.05		15.46	7.37	8.11
32 L.H., Daily Totals		$1158.25		$1787.84	$36.20	$55.87
Crew J-7	Hr.	Daily	Hr.	Daily	Bare Costs	Incl. O&P
2 Painters	$28.65	$458.40	$47.20	$755.20	$28.65	$47.20
1 Floor Belt Sander		14.20		15.62		
1 Floor Sanding Edger		12.60		13.86	1.68	1.84
16 L.H., Daily Totals		$485.20		$784.68	$30.32	$49.04
Crew K-1	Hr.	Daily	Hr.	Daily	Bare Costs	Incl. O&P
1 Carpenter	$32.85	$262.80	$55.15	$441.20	$30.30	$50.50
1 Truck Driver (light)	27.75	222.00	45.85	366.80		
1 Flatbed Truck, Gas, 3 Ton		303.20		333.52	18.95	20.84
16 L.H., Daily Totals		$788.00		$1141.52	$49.25	$71.34
Crew K-2	Hr.	Daily	Hr.	Daily	Bare Costs	Incl. O&P
1 Struc. Steel Foreman (outside)	$37.30	$298.40	$70.00	$560.00	$33.45	$60.70
1 Struc. Steel Worker	35.30	282.40	66.25	530.00		
1 Truck Driver (light)	27.75	222.00	45.85	366.80		
1 Flatbed Truck, Gas, 3 Ton		303.20		333.52	12.63	13.90
24 L.H., Daily Totals		$1106.00		$1790.32	$46.08	$74.60
Crew L-1	Hr.	Daily	Hr.	Daily	Bare Costs	Incl. O&P
.25 Electrician	$36.65	$73.30	$60.05	$120.10	$38.33	$63.17
1 Plumber	38.75	310.00	63.95	511.60		
10 L.H., Daily Totals		$383.30		$631.70	$38.33	$63.17
Crew L-2	Hr.	Daily	Hr.	Daily	Bare Costs	Incl. O&P
1 Carpenter	$32.85	$262.80	$55.15	$441.20	$28.63	$48.20
1 Carpenter Helper	24.40	195.20	41.25	330.00		
16 L.H., Daily Totals		$458.00		$771.20	$28.63	$48.20
Crew L-3	Hr.	Daily	Hr.	Daily	Bare Costs	Incl. O&P
1 Carpenter	$32.85	$262.80	$55.15	$441.20	$33.61	$56.13
.25 Electrician	36.65	73.30	60.05	120.10		
10 L.H., Daily Totals		$336.10		$561.30	$33.61	$56.13
Crew L-3A	Hr.	Daily	Hr.	Daily	Bare Costs	Incl. O&P
1 Carpenter Foreman (outside)	$34.85	$278.80	$58.50	$468.00	$35.55	$59.55
.5 Sheet Metal Worker	36.95	147.80	61.65	246.60		
12 L.H., Daily Totals		$426.60		$714.60	$35.55	$59.55
Crew L-4	Hr.	Daily	Hr.	Daily	Bare Costs	Incl. O&P
1 Skilled Worker	$33.35	$266.80	$56.20	$449.60	$28.88	$48.73
1 Helper	24.40	195.20	41.25	330.00		
16 L.H., Daily Totals		$462.00		$779.60	$28.88	$48.73
Crew L-5	Hr.	Daily	Hr.	Daily	Bare Costs	Incl. O&P
1 Struc. Steel Foreman (outside)	$37.30	$298.40	$70.00	$560.00	$35.64	$65.76
5 Struc. Steel Workers	35.30	1412.00	66.25	2650.00		
1 Equip. Oper. (crane)	35.65	285.20	59.05	472.40		
1 Hyd. Crane, 25 Ton		736.60		810.26	13.15	14.47
56 L.H., Daily Totals		$2732.20		$4492.66	$48.79	$80.23

Crews

Crew No.		Bare Costs		Incl. Subs O&P		Cost Per Labor-Hour	
Crew L-5A	Hr.	Daily	Hr.	Daily	Bare Costs	Incl. O&P
1 Struc. Steel Foreman (outside)	$37.30	$298.40	$70.00	$560.00	$35.89	$65.39
2 Structural Steel Workers	35.30	564.80	66.25	1060.00		
1 Equip. Oper. (crane)	35.65	285.20	59.05	472.40		
1 S.P. Crane, 4x4, 25 Ton		599.40		659.34	18.73	20.60
32 L.H., Daily Totals		$1747.80		$2751.74	$54.62	$85.99
Crew L-5B	Hr.	Daily	Hr.	Daily	Bare Costs	Incl. O&P
1 Struc. Steel Foreman (outside)	$37.30	$298.40	$70.00	$560.00	$35.58	$61.43
2 Structural Steel Workers	35.30	564.80	66.25	1060.00		
2 Electricians	36.65	586.40	60.05	960.80		
2 Steamfitters/Pipefitters	39.45	631.20	65.10	1041.60		
1 Equip. Oper. (crane)	35.65	285.20	59.05	472.40		
1 Common Laborer	24.45	195.60	41.05	328.40		
1 Hyd. Crane, 80 Ton		1625.00		1787.50	22.57	24.83
72 L.H., Daily Totals		$4186.60		$6210.70	$58.15	$86.26
Crew L-6	Hr.	Daily	Hr.	Daily	Bare Costs	Incl. O&P
1 Plumber	$38.75	$310.00	$63.95	$511.60	$38.05	$62.65
.5 Electrician	36.65	146.60	60.05	240.20		
12 L.H., Daily Totals		$456.60		$751.80	$38.05	$62.65
Crew L-7	Hr.	Daily	Hr.	Daily	Bare Costs	Incl. O&P
1 Carpenter	$32.85	$262.80	$55.15	$441.20	$27.94	$46.97
2 Carpenter Helpers	24.40	390.40	41.25	660.00		
.25 Electrician	36.65	73.30	60.05	120.10		
26 L.H., Daily Totals		$726.50		$1221.30	$27.94	$46.97
Crew L-8	Hr.	Daily	Hr.	Daily	Bare Costs	Incl. O&P
1 Carpenter	$32.85	$262.80	$55.15	$441.20	$30.65	$51.35
1 Carpenter Helper	24.40	195.20	41.25	330.00		
.5 Plumber	38.75	155.00	63.95	255.80		
20 L.H., Daily Totals		$613.00		$1027.00	$30.65	$51.35
Crew L-9	Hr.	Daily	Hr.	Daily	Bare Costs	Incl. O&P
1 Skilled Worker Foreman	$35.35	$282.80	$59.55	$476.40	$30.18	$50.73
1 Skilled Worker	33.35	266.80	56.20	449.60		
2 Helpers	24.40	390.40	41.25	660.00		
.5 Electrician	36.65	146.60	60.05	240.20		
36 L.H., Daily Totals		$1086.60		$1826.20	$30.18	$50.73
Crew L-10	Hr.	Daily	Hr.	Daily	Bare Costs	Incl. O&P
1 Struc. Steel Foreman (outside)	$37.30	$298.40	$70.00	$560.00	$36.08	$65.10
1 Structural Steel Worker	35.30	282.40	66.25	530.00		
1 Equip. Oper. (crane)	35.65	285.20	59.05	472.40		
1 Hyd. Crane, 12 Ton		653.80		719.18	27.24	29.97
24 L.H., Daily Totals		$1519.80		$2281.58	$63.33	$95.07
Crew L-11	Hr.	Daily	Hr.	Daily	Bare Costs	Incl. O&P
2 Wreckers	$25.20	$403.20	$45.15	$722.40	$29.90	$51.24
1 Equip. Oper. (crane)	35.65	285.20	59.05	472.40		
1 Equip. Oper. (light)	33.55	268.40	55.60	444.80		
1 Hyd. Excavator, 2.5 C.Y.		1605.00		1765.50		
1 Loader, Skid Steer, 78 H.P.		318.60		350.46	60.11	66.12
32 L.H., Daily Totals		$2880.40		$3755.56	$90.01	$117.36
Crew M-1	Hr.	Daily	Hr.	Daily	Bare Costs	Incl. O&P
3 Elevator Constructors	$51.25	$1230.00	$83.70	$2008.80	$48.69	$79.51
1 Elevator Apprentice	41.00	328.00	66.95	535.60		
5 Hand Tools		46.00		50.60	1.44	1.58
32 L.H., Daily Totals		$1604.00		$2595.00	$50.13	$81.09

Crew No.		Bare Costs		Incl. Subs O&P		Cost Per Labor-Hour	
Crew M-3	Hr.	Daily	Hr.	Daily	Bare Costs	Incl. O&P
1 Electrician Foreman (outside)	$38.65	$309.20	$63.30	$506.40	$38.61	$63.40
1 Common Laborer	24.45	195.60	41.05	328.40		
.25 Equipment Operator (med.)	34.90	69.80	57.85	115.70		
1 Elevator Constructor	51.25	410.00	83.70	669.60		
1 Elevator Apprentice	41.00	328.00	66.95	535.60		
.25 S.P. Crane, 4x4, 20 Ton		138.65		152.51	4.08	4.49
34 L.H., Daily Totals		$1451.25		$2308.22	$42.68	$67.89
Crew M-4	Hr.	Daily	Hr.	Daily	Bare Costs	Incl. O&P
1 Electrician Foreman (outside)	$38.65	$309.20	$63.30	$506.40	$38.24	$62.82
1 Common Laborer	24.45	195.60	41.05	328.40		
.25 Equipment Operator, Crane	35.65	71.30	59.05	118.10		
.25 Equip. Oper. (oiler)	31.20	62.40	51.70	103.40		
1 Elevator Constructor	51.25	410.00	83.70	669.60		
1 Elevator Apprentice	41.00	328.00	66.95	535.60		
.25 S.P. Crane, 4x4, 40 Ton		175.75		193.32	4.88	5.37
36 L.H., Daily Totals		$1552.25		$2454.82	$43.12	$68.19
Crew Q-1	Hr.	Daily	Hr.	Daily	Bare Costs	Incl. O&P
1 Plumber	$38.75	$310.00	$63.95	$511.60	$34.88	$57.55
1 Plumber Apprentice	31.00	248.00	51.15	409.20		
16 L.H., Daily Totals		$558.00		$920.80	$34.88	$57.55
Crew Q-1A	Hr.	Daily	Hr.	Daily	Bare Costs	Incl. O&P
.25 Plumber Foreman (outside)	$40.75	$81.50	$67.25	$134.50	$39.15	$64.61
1 Plumber	38.75	310.00	63.95	511.60		
10 L.H., Daily Totals		$391.50		$646.10	$39.15	$64.61
Crew Q-1C	Hr.	Daily	Hr.	Daily	Bare Costs	Incl. O&P
1 Plumber	$38.75	$310.00	$63.95	$511.60	$34.88	$57.65
1 Plumber Apprentice	31.00	248.00	51.15	409.20		
1 Equip. Oper. (medium)	34.90	279.20	57.85	462.80		
1 Trencher, Chain Type, 8' D		3402.00		3742.20	141.75	155.93
24 L.H., Daily Totals		$4239.20		$5125.80	$176.63	$213.57
Crew Q-2	Hr.	Daily	Hr.	Daily	Bare Costs	Incl. O&P
1 Plumber	$38.75	$310.00	$63.95	$511.60	$33.58	$55.42
2 Plumber Apprentices	31.00	496.00	51.15	818.40		
24 L.H., Daily Totals		$806.00		$1330.00	$33.58	$55.42
Crew Q-3	Hr.	Daily	Hr.	Daily	Bare Costs	Incl. O&P
2 Plumbers	$38.75	$620.00	$63.95	$1023.20	$34.88	$57.55
2 Plumber Apprentices	31.00	496.00	51.15	818.40		
32 L.H., Daily Totals		$1116.00		$1841.60	$34.88	$57.55
Crew Q-4	Hr.	Daily	Hr.	Daily	Bare Costs	Incl. O&P
2 Plumbers	$38.75	$620.00	$63.95	$1023.20	$36.81	$60.75
1 Welder (plumber)	38.75	310.00	63.95	511.60		
1 Plumber Apprentice	31.00	248.00	51.15	409.20		
1 Welder, Electric, 300 amp		57.70		63.47	1.80	1.98
32 L.H., Daily Totals		$1235.70		$2007.47	$38.62	$62.73
Crew Q-5	Hr.	Daily	Hr.	Daily	Bare Costs	Incl. O&P
1 Steamfitter	$39.45	$315.60	$65.10	$520.80	$35.50	$58.58
1 Steamfitter Apprentice	31.55	252.40	52.05	416.40		
16 L.H., Daily Totals		$568.00		$937.20	$35.50	$58.58

Crews

Crew No.	Bare Costs		Incl. Subs O&P		Cost Per Labor-Hour	
Crew Q-6	Hr.	Daily	Hr.	Daily	Bare Costs	Incl. O&P
1 Steamfitter	$39.45	$315.60	$65.10	$520.80	$34.18	$56.40
2 Steamfitter Apprentices	31.55	504.80	52.05	832.80		
24 L.H., Daily Totals		$820.40		$1353.60	$34.18	$56.40
Crew Q-7	Hr.	Daily	Hr.	Daily	Bare Costs	Incl. O&P
2 Steamfitters	$39.45	$631.20	$65.10	$1041.60	$35.50	$58.58
2 Steamfitter Apprentices	31.55	504.80	52.05	832.80		
32 L.H., Daily Totals		$1136.00		$1874.40	$35.50	$58.58
Crew Q-8	Hr.	Daily	Hr.	Daily	Bare Costs	Incl. O&P
2 Steamfitters	$39.45	$631.20	$65.10	$1041.60	$37.48	$61.84
1 Welder (steamfitter)	39.45	315.60	65.10	520.80		
1 Steamfitter Apprentice	31.55	252.40	52.05	416.40		
1 Welder, Electric, 300 amp		57.70		63.47	1.80	1.98
32 L.H., Daily Totals		$1256.90		$2042.27	$39.28	$63.82
Crew Q-9	Hr.	Daily	Hr.	Daily	Bare Costs	Incl. O&P
1 Sheet Metal Worker	$36.95	$295.60	$61.65	$493.20	$33.25	$55.48
1 Sheet Metal Apprentice	29.55	236.40	49.30	394.40		
16 L.H., Daily Totals		$532.00		$887.60	$33.25	$55.48
Crew Q-10	Hr.	Daily	Hr.	Daily	Bare Costs	Incl. O&P
2 Sheet Metal Workers	$36.95	$591.20	$61.65	$986.40	$34.48	$57.53
1 Sheet Metal Apprentice	29.55	236.40	49.30	394.40		
24 L.H., Daily Totals		$827.60		$1380.80	$34.48	$57.53
Crew Q-11	Hr.	Daily	Hr.	Daily	Bare Costs	Incl. O&P
2 Sheet Metal Workers	$36.95	$591.20	$61.65	$986.40	$33.25	$55.48
2 Sheet Metal Apprentices	29.55	472.80	49.30	788.80		
32 L.H., Daily Totals		$1064.00		$1775.20	$33.25	$55.48
Crew Q-12	Hr.	Daily	Hr.	Daily	Bare Costs	Incl. O&P
1 Sprinkler Installer	$37.05	$296.40	$61.25	$490.00	$33.35	$55.13
1 Sprinkler Apprentice	29.65	237.20	49.00	392.00		
16 L.H., Daily Totals		$533.60		$882.00	$33.35	$55.13
Crew Q-13	Hr.	Daily	Hr.	Daily	Bare Costs	Incl. O&P
2 Sprinkler Installers	$37.05	$592.80	$61.25	$980.00	$33.35	$55.13
2 Sprinkler Apprentices	29.65	474.40	49.00	784.00		
32 L.H., Daily Totals		$1067.20		$1764.00	$33.35	$55.13
Crew Q-14	Hr.	Daily	Hr.	Daily	Bare Costs	Incl. O&P
1 Asbestos Worker	$34.55	$276.40	$58.65	$469.20	$31.10	$52.77
1 Asbestos Apprentice	27.65	221.20	46.90	375.20		
16 L.H., Daily Totals		$497.60		$844.40	$31.10	$52.77
Crew Q-15	Hr.	Daily	Hr.	Daily	Bare Costs	Incl. O&P
1 Plumber	$38.75	$310.00	$63.95	$511.60	$34.88	$57.55
1 Plumber Apprentice	31.00	248.00	51.15	409.20		
1 Welder, Electric, 300 amp		57.70		63.47	3.61	3.97
16 L.H., Daily Totals		$615.70		$984.27	$38.48	$61.52
Crew Q-16	Hr.	Daily	Hr.	Daily	Bare Costs	Incl. O&P
2 Plumbers	$38.75	$620.00	$63.95	$1023.20	$36.17	$59.68
1 Plumber Apprentice	31.00	248.00	51.15	409.20		
1 Welder, Electric, 300 amp		57.70		63.47	2.40	2.64
24 L.H., Daily Totals		$925.70		$1495.87	$38.57	$62.33

Crew No.	Bare Costs		Incl. Subs O&P		Cost Per Labor-Hour	
Crew Q-17	Hr.	Daily	Hr.	Daily	Bare Costs	Incl. O&P
1 Steamfitter	$39.45	$315.60	$65.10	$520.80	$35.50	$58.58
1 Steamfitter Apprentice	31.55	252.40	52.05	416.40		
1 Welder, Electric, 300 amp		57.70		63.47	3.61	3.97
16 L.H., Daily Totals		$625.70		$1000.67	$39.11	$62.54
Crew Q-17A	Hr.	Daily	Hr.	Daily	Bare Costs	Incl. O&P
1 Steamfitter	$39.45	$315.60	$65.10	$520.80	$35.55	$58.73
1 Steamfitter Apprentice	31.55	252.40	52.05	416.40		
1 Equip. Oper. (crane)	35.65	285.20	59.05	472.40		
1 Hyd. Crane, 12 Ton		653.80		719.18		
1 Welder, Electric, 300 amp		57.70		63.47	29.65	32.61
24 L.H., Daily Totals		$1564.70		$2192.25	$65.20	$91.34
Crew Q-18	Hr.	Daily	Hr.	Daily	Bare Costs	Incl. O&P
2 Steamfitters	$39.45	$631.20	$65.10	$1041.60	$36.82	$60.75
1 Steamfitter Apprentice	31.55	252.40	52.05	416.40		
1 Welder, Electric, 300 amp		57.70		63.47	2.40	2.64
24 L.H., Daily Totals		$941.30		$1521.47	$39.22	$63.39
Crew Q-19	Hr.	Daily	Hr.	Daily	Bare Costs	Incl. O&P
1 Steamfitter	$39.45	$315.60	$65.10	$520.80	$35.88	$59.07
1 Steamfitter Apprentice	31.55	252.40	52.05	416.40		
1 Electrician	36.65	293.20	60.05	480.40		
24 L.H., Daily Totals		$861.20		$1417.60	$35.88	$59.07
Crew Q-20	Hr.	Daily	Hr.	Daily	Bare Costs	Incl. O&P
1 Sheet Metal Worker	$36.95	$295.60	$61.65	$493.20	$33.93	$56.39
1 Sheet Metal Apprentice	29.55	236.40	49.30	394.40		
.5 Electrician	36.65	146.60	60.05	240.20		
20 L.H., Daily Totals		$678.60		$1127.80	$33.93	$56.39
Crew Q-21	Hr.	Daily	Hr.	Daily	Bare Costs	Incl. O&P
2 Steamfitters	$39.45	$631.20	$65.10	$1041.60	$36.77	$60.58
1 Steamfitter Apprentice	31.55	252.40	52.05	416.40		
1 Electrician	36.65	293.20	60.05	480.40		
32 L.H., Daily Totals		$1176.80		$1938.40	$36.77	$60.58
Crew Q-22	Hr.	Daily	Hr.	Daily	Bare Costs	Incl. O&P
1 Plumber	$38.75	$310.00	$63.95	$511.60	$34.88	$57.55
1 Plumber Apprentice	31.00	248.00	51.15	409.20		
1 Hyd. Crane, 12 Ton		653.80		719.18	40.86	44.95
16 L.H., Daily Totals		$1211.80		$1639.98	$75.74	$102.50
Crew Q-22A	Hr.	Daily	Hr.	Daily	Bare Costs	Incl. O&P
1 Plumber	$38.75	$310.00	$63.95	$511.60	$32.46	$53.80
1 Plumber Apprentice	31.00	248.00	51.15	409.20		
1 Laborer	24.45	195.60	41.05	328.40		
1 Equip. Oper. (crane)	35.65	285.20	59.05	472.40		
1 Hyd. Crane, 12 Ton		653.80		719.18	20.43	22.47
32 L.H., Daily Totals		$1692.60		$2440.78	$52.89	$76.27
Crew Q-23	Hr.	Daily	Hr.	Daily	Bare Costs	Incl. O&P
1 Plumber Foreman (outside)	$40.75	$326.00	$67.25	$538.00	$38.13	$63.02
1 Plumber	38.75	310.00	63.95	511.60		
1 Equip. Oper. (medium)	34.90	279.20	57.85	462.80		
1 Lattice Boom Crane, 20 Ton		948.90		1043.79	39.54	43.49
24 L.H., Daily Totals		$1864.10		$2556.19	$77.67	$106.51

Crews

Crew No.	Bare Costs		Incl. Subs O&P		Cost Per Labor-Hour	
Crew R-1	Hr.	Daily	Hr.	Daily	Bare Costs	Incl. O&P
1 Electrician Foreman	$37.15	$297.20	$60.85	$486.80	$32.65	$53.92
3 Electricians	36.65	879.60	60.05	1441.20		
2 Helpers	24.40	390.40	41.25	660.00		
48 L.H., Daily Totals		$1567.20		$2588.00	$32.65	$53.92
Crew R-1A	Hr.	Daily	Hr.	Daily	Bare Costs	Incl. O&P
1 Electrician	$36.65	$293.20	$60.05	$480.40	$30.52	$50.65
1 Helper	24.40	195.20	41.25	330.00		
16 L.H., Daily Totals		$488.40		$810.40	$30.52	$50.65
Crew R-2	Hr.	Daily	Hr.	Daily	Bare Costs	Incl. O&P
1 Electrician Foreman	$37.15	$297.20	$60.85	$486.80	$33.08	$54.65
3 Electricians	36.65	879.60	60.05	1441.20		
2 Helpers	24.40	390.40	41.25	660.00		
1 Equip. Oper. (crane)	35.65	285.20	59.05	472.40		
1 S.P. Crane, 4x4, 5 Ton		275.40		302.94	4.92	5.41
56 L.H., Daily Totals		$2127.80		$3363.34	$38.00	$60.06
Crew R-3	Hr.	Daily	Hr.	Daily	Bare Costs	Incl. O&P
1 Electrician Foreman	$37.15	$297.20	$60.85	$486.80	$36.65	$60.17
1 Electrician	36.65	293.20	60.05	480.40		
.5 Equip. Oper. (crane)	35.65	142.60	59.05	236.20		
.5 S.P. Crane, 4x4, 5 Ton		137.70		151.47	6.88	7.57
20 L.H., Daily Totals		$870.70		$1354.87	$43.53	$67.74
Crew R-4	Hr.	Daily	Hr.	Daily	Bare Costs	Incl. O&P
1 Struc. Steel Foreman (outside)	$37.30	$298.40	$70.00	$560.00	$35.97	$65.76
3 Struc. Steel Workers	35.30	847.20	66.25	1590.00		
1 Electrician	36.65	293.20	60.05	480.40		
1 Welder, Gas Engine, 300 amp		145.85		160.44	3.65	4.01
40 L.H., Daily Totals		$1584.65		$2790.84	$39.62	$69.77
Crew R-5	Hr.	Daily	Hr.	Daily	Bare Costs	Incl. O&P
1 Electrician Foreman	$37.15	$297.20	$60.85	$486.80	$32.24	$53.29
4 Electrician Linemen	36.65	1172.80	60.05	1921.60		
2 Electrician Operators	36.65	586.40	60.05	960.80		
4 Electrician Groundmen	24.40	780.80	41.25	1320.00		
1 Crew Truck		206.40		227.04		
1 Flatbed Truck, 20,000 GVW		248.60		273.46		
1 Pickup Truck, 3/4 Ton		144.20		158.62		
.2 Hyd. Crane, 55 Ton		225.60		248.16		
.2 Hyd. Crane, 12 Ton		130.76		143.84		
.2 Earth Auger, Truck-Mtd.		83.44		91.78		
1 Tractor w/Winch		427.00		469.70	16.66	18.32
88 L.H., Daily Totals		$4303.20		$6301.80	$48.90	$71.61

Crew No.	Bare Costs		Incl. Subs O&P		Cost Per Labor-Hour	
Crew R-6	Hr.	Daily	Hr.	Daily	Bare Costs	Incl. O&P
1 Electrician Foreman	$37.15	$297.20	$60.85	$486.80	$32.24	$53.29
4 Electrician Linemen	36.65	1172.80	60.05	1921.60		
2 Electrician Operators	36.65	586.40	60.05	960.80		
4 Electrician Groundmen	24.40	780.80	41.25	1320.00		
1 Crew Truck		206.40		227.04		
1 Flatbed Truck, 20,000 GVW		248.60		273.46		
1 Pickup Truck, 3/4 Ton		144.20		158.62		
.2 Hyd. Crane, 55 Ton		225.60		248.16		
.2 Hyd. Crane, 12 Ton		130.76		143.84		
.2 Earth Auger, Truck-Mtd.		83.44		91.78		
1 Tractor w/Winch		427.00		469.70		
3 Cable Trailers		598.20		658.02		
.5 Tensioning Rig		198.40		218.24		
.5 Cable Pulling Rig		1160.00		1276.00	38.89	42.78
88 L.H., Daily Totals		$6259.80		$8454.06	$71.13	$96.07
Crew R-7	Hr.	Daily	Hr.	Daily	Bare Costs	Incl. O&P
1 Electrician Foreman	$37.15	$297.20	$60.85	$486.80	$26.52	$44.52
5 Electrician Groundmen	24.40	976.00	41.25	1650.00		
1 Crew Truck		206.40		227.04	4.30	4.73
48 L.H., Daily Totals		$1479.60		$2363.84	$30.82	$49.25
Crew R-8	Hr.	Daily	Hr.	Daily	Bare Costs	Incl. O&P
1 Electrician Foreman	$37.15	$297.20	$60.85	$486.80	$32.65	$53.92
3 Electrician Linemen	36.65	879.60	60.05	1441.20		
2 Electrician Groundmen	24.40	390.40	41.25	660.00		
1 Pickup Truck, 3/4 Ton		144.20		158.62		
1 Crew Truck		206.40		227.04	7.30	8.03
48 L.H., Daily Totals		$1917.80		$2973.66	$39.95	$61.95
Crew R-9	Hr.	Daily	Hr.	Daily	Bare Costs	Incl. O&P
1 Electrician Foreman	$37.15	$297.20	$60.85	$486.80	$30.59	$50.75
1 Electrician Lineman	36.65	293.20	60.05	480.40		
2 Electrician Operators	36.65	586.40	60.05	960.80		
4 Electrician Groundmen	24.40	780.80	41.25	1320.00		
1 Pickup Truck, 3/4 Ton		144.20		158.62		
1 Crew Truck		206.40		227.04	5.48	6.03
64 L.H., Daily Totals		$2308.20		$3633.66	$36.07	$56.78
Crew R-10	Hr.	Daily	Hr.	Daily	Bare Costs	Incl. O&P
1 Electrician Foreman	$37.15	$297.20	$60.85	$486.80	$34.69	$57.05
4 Electrician Linemen	36.65	1172.80	60.05	1921.60		
1 Electrician Groundman	24.40	195.20	41.25	330.00		
1 Crew Truck		206.40		227.04		
3 Tram Cars		412.20		453.42	12.89	14.18
48 L.H., Daily Totals		$2283.80		$3418.86	$47.58	$71.23
Crew R-11	Hr.	Daily	Hr.	Daily	Bare Costs	Incl. O&P
1 Electrician Foreman	$37.15	$297.20	$60.85	$486.80	$34.84	$57.31
4 Electricians	36.65	1172.80	60.05	1921.60		
1 Equip. Oper. (crane)	35.65	285.20	59.05	472.40		
1 Common Laborer	24.45	195.60	41.05	328.40		
1 Crew Truck		206.40		227.04		
1 Hyd. Crane, 12 Ton		653.80		719.18	15.36	16.90
56 L.H., Daily Totals		$2811.00		$4155.42	$50.20	$74.20

Crews

Crew No.	Bare Costs		Incl. Subs O & P		Cost Per Labor-Hour	
Crew R-12	Hr.	Daily	Hr.	Daily	Bare Costs	Incl. O&P
1 Carpenter Foreman (inside)	$33.35	$266.80	$56.00	$448.00	$30.25	$51.35
4 Carpenters	32.85	1051.20	55.15	1764.80		
4 Common Laborers	24.45	782.40	41.05	1313.60		
1 Equip. Oper. (medium)	34.90	279.20	57.85	462.80		
1 Steel Worker	35.30	282.40	66.25	530.00		
1 Dozer, 200 H.P.		1387.00		1525.70		
1 Pickup Truck, 3/4 Ton		144.20		158.62	17.40	19.14
88 L.H., Daily Totals		$4193.20		$6203.52	$47.65	$70.49
Crew R-15	Hr.	Daily	Hr.	Daily	Bare Costs	Incl. O&P
1 Electrician Foreman	$37.15	$297.20	$60.85	$486.80	$36.22	$59.44
4 Electricians	36.65	1172.80	60.05	1921.60		
1 Equipment Oper. (light)	33.55	268.40	55.60	444.80		
1 Aerial Lift Truck, 40' Boom		301.40		331.54	6.28	6.91
48 L.H., Daily Totals		$2039.80		$3184.74	$42.50	$66.35
Crew R-15A	Hr.	Daily	Hr.	Daily	Bare Costs	Incl. O&P
1 Electrician Foreman	$37.15	$297.20	$60.85	$486.80	$32.15	$53.11
2 Electricians	36.65	586.40	60.05	960.80		
2 Common Laborers	24.45	391.20	41.05	656.80		
1 Equip. Oper. (light)	33.55	268.40	55.60	444.80		
1 Aerial Lift Truck, 40' Boom		301.40		331.54	6.28	6.91
48 L.H., Daily Totals		$1844.60		$2880.74	$38.43	$60.02
Crew R-18	Hr.	Daily	Hr.	Daily	Bare Costs	Incl. O&P
.25 Electrician Foreman	$37.15	$74.30	$60.85	$121.70	$29.15	$48.54
1 Electrician	36.65	293.20	60.05	480.40		
2 Helpers	24.40	390.40	41.25	660.00		
26 L.H., Daily Totals		$757.90		$1262.10	$29.15	$48.54
Crew R-19	Hr.	Daily	Hr.	Daily	Bare Costs	Incl. O&P
.5 Electrician Foreman	$37.15	$148.60	$60.85	$243.40	$36.75	$60.21
2 Electricians	36.65	586.40	60.05	960.80		
20 L.H., Daily Totals		$735.00		$1204.20	$36.75	$60.21
Crew R-21	Hr.	Daily	Hr.	Daily	Bare Costs	Incl. O&P
1 Electrician Foreman	$37.15	$297.20	$60.85	$486.80	$36.73	$60.19
3 Electricians	36.65	879.60	60.05	1441.20		
.1 Equip. Oper. (medium)	34.90	27.92	57.85	46.28		
.1 S.P. Crane, 4x4, 25 Ton		59.94		65.93	1.83	2.01
32.8 L.H., Daily Totals		$1264.66		$2040.21	$38.56	$62.20
Crew R-22	Hr.	Daily	Hr.	Daily	Bare Costs	Incl. O&P
.66 Electrician Foreman	$37.15	$196.15	$60.85	$321.29	$31.46	$52.09
2 Electricians	36.65	586.40	60.05	960.80		
2 Helpers	24.40	390.40	41.25	660.00		
37.28 L.H., Daily Totals		$1172.95		$1942.09	$31.46	$52.09
Crew R-30	Hr.	Daily	Hr.	Daily	Bare Costs	Incl. O&P
.25 Electrician Foreman (outside)	$38.65	$77.30	$63.30	$126.60	$29.30	$48.61
1 Electrician	36.65	293.20	60.05	480.40		
2 Laborers, (Semi-Skilled)	24.45	391.20	41.05	656.80		
26 L.H., Daily Totals		$761.70		$1263.80	$29.30	$48.61

Location Factors

Costs shown in *RSMeans Residential Cost Data* are based on national averages for materials and installation. To adjust these costs to a specific location, simply multiply the base cost by the factor for that city. The data is arranged alphabetically by state and postal zip code numbers. For a city not listed, use the factor for a nearby city with similar economic characteristics.

STATE	CITY	Residential
ALABAMA		
350-352	Birmingham	.88
354	Tuscaloosa	.78
355	Jasper	.72
356	Decatur	.76
357-358	Huntsville	.86
359	Gadsden	.75
360-361	Montgomery	.75
362	Anniston	.75
363	Dothan	.74
364	Evergreen	.71
365-366	Mobile	.80
367	Selma	.71
368	Phenix City	.73
369	Butler	.72
ALASKA		
995-996	Anchorage	1.25
997	Fairbanks	1.28
998	Juneau	1.25
999	Ketchikan	1.29
ARIZONA		
850,853	Phoenix	.86
851,852	Mesa/Tempe	.86
855	Globe	.83
856-857	Tucson	.85
859	Show Low	.87
860	Flagstaff	.86
863	Prescott	.82
864	Kingman	.83
865	Chambers	.83
ARKANSAS		
716	Pine Bluff	.78
717	Camden	.69
718	Texarkana	.73
719	Hot Springs	.70
720-722	Little Rock	.81
723	West Memphis	.77
724	Jonesboro	.75
725	Batesville	.73
726	Harrison	.73
727	Fayetteville	.69
728	Russellville	.75
729	Fort Smith	.81
CALIFORNIA		
900-902	Los Angeles	1.09
903-905	Inglewood	1.05
906-908	Long Beach	1.04
910-912	Pasadena	1.04
913-916	Van Nuys	1.07
917-918	Alhambra	1.08
919-921	San Diego	1.05
922	Palm Springs	1.04
923-924	San Bernardino	1.05
925	Riverside	1.08
926-927	Santa Ana	1.06
928	Anaheim	1.08
930	Oxnard	1.09
931	Santa Barbara	1.08
932-933	Bakersfield	1.07
934	San Luis Obispo	1.08
935	Mojave	1.05
936-938	Fresno	1.12
939	Salinas	1.15
940-941	San Francisco	1.25
942,956-958	Sacramento	1.12
943	Palo Alto	1.19
944	San Mateo	1.23
945	Vallejo	1.16
946	Oakland	1.22
947	Berkeley	1.24
948	Richmond	1.24
949	San Rafael	1.24
950	Santa Cruz	1.17
951	San Jose	1.23
952	Stockton	1.13
953	Modesto	1.12

STATE	CITY	Residential
CALIFORNIA (CONT'D)		
954	Santa Rosa	1.20
955	Eureka	1.14
959	Marysville	1.13
960	Redding	1.18
961	Susanville	1.17
COLORADO		
800-802	Denver	.90
803	Boulder	.91
804	Golden	.87
805	Fort Collins	.88
806	Greeley	.86
807	Fort Morgan	.89
808-809	Colorado Springs	.88
810	Pueblo	.90
811	Alamosa	.85
812	Salida	.88
813	Durango	.89
814	Montrose	.86
815	Grand Junction	.92
816	Glenwood Springs	.88
CONNECTICUT		
060	New Britain	1.11
061	Hartford	1.10
062	Willimantic	1.11
063	New London	1.11
064	Meriden	1.11
065	New Haven	1.11
066	Bridgeport	1.12
067	Waterbury	1.11
068	Norwalk	1.13
069	Stamford	1.13
D.C.		
200-205	Washington	.93
DELAWARE		
197	Newark	1.02
198	Wilmington	1.02
199	Dover	1.01
FLORIDA		
320,322	Jacksonville	.81
321	Daytona Beach	.86
323	Tallahassee	.78
324	Panama City	.76
325	Pensacola	.81
326,344	Gainesville	.80
327-328,347	Orlando	.86
329	Melbourne	.86
330-332,340	Miami	.84
333	Fort Lauderdale	.84
334,349	West Palm Beach	.85
335-336,346	Tampa	.88
337	St. Petersburg	.79
338	Lakeland	.85
339,341	Fort Myers	.84
342	Sarasota	.88
GEORGIA		
300-303,399	Atlanta	.87
304	Statesboro	.73
305	Gainesville	.77
306	Athens	.76
307	Dalton	.75
308-309	Augusta	.81
310-312	Macon	.77
313-314	Savannah	.78
315	Waycross	.77
316	Valdosta	.73
317,398	Albany	.75
318-319	Columbus	.78
HAWAII		
967	Hilo	1.21
968	Honolulu	1.25

Location Factors

STATE	CITY	Residential
STATES & POSS.		
969	Guam	.97
IDAHO		
832	Pocatello	.87
833	Twin Falls	.81
834	Idaho Falls	.86
835	Lewiston	1.00
836-837	Boise	.88
838	Coeur d'Alene	.98
ILLINOIS		
600-603	North Suburban	1.20
604	Joliet	1.22
605	South Suburban	1.20
606-608	Chicago	1.22
609	Kankakee	1.14
610-611	Rockford	1.12
612	Rock Island	.98
613	La Salle	1.11
614	Galesburg	1.03
615-616	Peoria	1.06
617	Bloomington	1.03
618-619	Champaign	1.04
620-622	East St. Louis	1.02
623	Quincy	1.01
624	Effingham	1.03
625	Decatur	1.02
626-627	Springfield	1.03
628	Centralia	1.02
629	Carbondale	1.01
INDIANA		
460	Anderson	.92
461-462	Indianapolis	.93
463-464	Gary	1.06
465-466	South Bend	.91
467-468	Fort Wayne	.89
469	Kokomo	.91
470	Lawrenceburg	.87
471	New Albany	.87
472	Columbus	.92
473	Muncie	.92
474	Bloomington	.94
475	Washington	.90
476-477	Evansville	.91
478	Terre Haute	.91
479	Lafayette	.93
IOWA		
500-503,509	Des Moines	.91
504	Mason City	.76
505	Fort Dodge	.74
506-507	Waterloo	.83
508	Creston	.83
510-511	Sioux City	.86
512	Sibley	.74
513	Spencer	.75
514	Carroll	.80
515	Council Bluffs	.86
516	Shenandoah	.81
520	Dubuque	.87
521	Decorah	.78
522-524	Cedar Rapids	.93
525	Ottumwa	.87
526	Burlington	.88
527-528	Davenport	.97
KANSAS		
660-662	Kansas City	.97
664-666	Topeka	.78
667	Fort Scott	.88
668	Emporia	.83
669	Belleville	.80
670-672	Wichita	.79
673	Independence	.88
674	Salina	.80
675	Hutchinson	.79
676	Hays	.82
677	Colby	.83
678	Dodge City	.81
679	Liberal	.80
KENTUCKY		
400-402	Louisville	.90
403-405	Lexington	.88

STATE	CITY	Residential
KENTUCKY (CONT'D)		
406	Frankfort	.87
407-409	Corbin	.81
410	Covington	.91
411-412	Ashland	.94
413-414	Campton	.87
415-416	Pikeville	.91
417-418	Hazard	.86
420	Paducah	.88
421-422	Bowling Green	.89
423	Owensboro	.91
424	Henderson	.91
425-426	Somerset	.85
427	Elizabethtown	.87
LOUISIANA		
700-701	New Orleans	.85
703	Thibodaux	.82
704	Hammond	.75
705	Lafayette	.82
706	Lake Charles	.83
707-708	Baton Rouge	.82
710-711	Shreveport	.79
712	Monroe	.75
713-714	Alexandria	.77
MAINE		
039	Kittery	.92
040-041	Portland	.96
042	Lewiston	.96
043	Augusta	.92
044	Bangor	.95
045	Bath	.92
046	Machias	.95
047	Houlton	.96
048	Rockland	.95
049	Waterville	.90
MARYLAND		
206	Waldorf	.89
207-208	College Park	.87
209	Silver Spring	.88
210-212	Baltimore	.90
214	Annapolis	.89
215	Cumberland	.88
216	Easton	.83
217	Hagerstown	.87
218	Salisbury	.82
219	Elkton	.88
MASSACHUSETTS		
010-011	Springfield	1.08
012	Pittsfield	1.07
013	Greenfield	1.06
014	Fitchburg	1.14
015-016	Worcester	1.16
017	Framingham	1.19
018	Lowell	1.19
019	Lawrence	1.20
020-022, 024	Boston	1.22
023	Brockton	1.18
025	Buzzards Bay	1.16
026	Hyannis	1.15
027	New Bedford	1.17
MICHIGAN		
480,483	Royal Oak	1.03
481	Ann Arbor	1.03
482	Detroit	1.06
484-485	Flint	.96
486	Saginaw	.93
487	Bay City	.94
488-489	Lansing	.95
490	Battle Creek	.93
491	Kalamazoo	.92
492	Jackson	.93
493,495	Grand Rapids	.91
494	Muskegon	.90
496	Traverse City	.85
497	Gaylord	.89
498-499	Iron mountain	.90
MINNESOTA		
550-551	Saint Paul	1.09
553-555	Minneapolis	1.12
556-558	Duluth	1.04

Location Factors

STATE		CITY	Residential
MINNESOTA (CONT'D)			
	559	Rochester	1.03
	560	Mankato	.99
	561	Windom	.93
	562	Willmar	.96
	563	St. Cloud	1.04
	564	Brainerd	.95
	565	Detroit Lakes	.94
	566	Bemidji	.94
	567	Thief River Falls	.93
MISSISSIPPI			
	386	Clarksdale	.74
	387	Greenville	.80
	388	Tupelo	.75
	389	Greenwood	.77
	390-392	Jackson	.81
	393	Meridian	.80
	394	Laurel	.77
	395	Biloxi	.79
	396	Mccomb	.74
	397	Columbus	.75
MISSOURI			
	630-631	St. Louis	1.02
	633	Bowling Green	.95
	634	Hannibal	.92
	635	Kirksville	.89
	636	Flat River	.95
	637	Cape Girardeau	.91
	638	Sikeston	.90
	639	Poplar Bluff	.89
	640-641	Kansas City	1.04
	644-645	St. Joseph	.97
	646	Chillicothe	.96
	647	Harrisonville	.98
	648	Joplin	.91
	650-651	Jefferson City	.94
	652	Columbia	.92
	653	Sedalia	.94
	654-655	Rolla	.96
	656-658	Springfield	.91
MONTANA			
	590-591	Billings	.90
	592	Wolf Point	.87
	593	Miles City	.88
	594	Great Falls	.91
	595	Havre	.84
	596	Helena	.87
	597	Butte	.87
	598	Missoula	.88
	599	Kalispell	.86
NEBRASKA			
	680-681	Omaha	.90
	683-685	Lincoln	.87
	686	Columbus	.87
	687	Norfolk	.89
	688	Grand Island	.88
	689	Hastings	.90
	690	Mccook	.83
	691	North Platte	.88
	692	Valentine	.84
	693	Alliance	.83
NEVADA			
	889-891	Las Vegas	1.03
	893	Ely	1.02
	894-895	Reno	.95
	897	Carson City	.96
	898	Elko	.93
NEW HAMPSHIRE			
	030	Nashua	.98
	031	Manchester	.97
	032-033	Concord	.98
	034	Keene	.88
	035	Littleton	.92
	036	Charleston	.86
	037	Claremont	.86
	038	Portsmouth	.97

STATE		CITY	Residential
NEW JERSEY			
	070-071	Newark	1.14
	072	Elizabeth	1.16
	073	Jersey City	1.12
	074-075	Paterson	1.13
	076	Hackensack	1.13
	077	Long Branch	1.13
	078	Dover	1.13
	079	Summit	1.14
	080,083	Vineland	1.11
	081	Camden	1.13
	082,084	Atlantic City	1.15
	085-086	Trenton	1.13
	087	Point Pleasant	1.12
	088-089	New Brunswick	1.15
NEW MEXICO			
	870-872	Albuquerque	.84
	873	Gallup	.83
	874	Farmington	.84
	875	Santa Fe	.85
	877	Las Vegas	.84
	878	Socorro	.83
	879	Truth/Consequences	.82
	880	Las Cruces	.81
	881	Clovis	.83
	882	Roswell	.83
	883	Carrizozo	.83
	884	Tucumcari	.84
NEW YORK			
	100-102	New York	1.35
	103	Staten Island	1.31
	104	Bronx	1.30
	105	Mount Vernon	1.16
	106	White Plains	1.19
	107	Yonkers	1.21
	108	New Rochelle	1.20
	109	Suffern	1.15
	110	Queens	1.31
	111	Long Island City	1.34
	112	Brooklyn	1.37
	113	Flushing	1.33
	114	Jamaica	1.32
	115,117,118	Hicksville	1.23
	116	Far Rockaway	1.32
	119	Riverhead	1.23
	120-122	Albany	1.00
	123	Schenectady	1.02
	124	Kingston	1.05
	125-126	Poughkeepsie	1.22
	127	Monticello	1.04
	128	Glens Falls	.93
	129	Plattsburgh	1.00
	130-132	Syracuse	.98
	133-135	Utica	.96
	136	Watertown	.94
	137-139	Binghamton	.99
	140-142	Buffalo	1.06
	143	Niagara Falls	1.03
	144-146	Rochester	1.00
	147	Jamestown	.93
	148-149	Elmira	.95
NORTH CAROLINA			
	270,272-274	Greensboro	.84
	271	Winston-Salem	.85
	275-276	Raleigh	.84
	277	Durham	.84
	278	Rocky Mount	.80
	279	Elizabeth City	.79
	280	Gastonia	.85
	281-282	Charlotte	.87
	283	Fayetteville	.90
	284	Wilmington	.86
	285	Kinston	.81
	286	Hickory	.80
	287-288	Asheville	.84
	289	Murphy	.83
NORTH DAKOTA			
	580-581	Fargo	.80
	582	Grand Forks	.74
	583	Devils Lake	.78
	584	Jamestown	.72
	585	Bismarck	.79

Location Factors

STATE		CITY	Residential
NORTH DAKOTA (CONT'D)			
586		Dickinson	.76
587		Minot	.84
588		Williston	.76
OHIO			
430-432		Columbus	.94
433		Marion	.91
434-436		Toledo	.99
437-438		Zanesville	.91
439		Steubenville	.95
440		Lorain	.96
441		Cleveland	1.00
442-443		Akron	.98
444-445		Youngstown	.95
446-447		Canton	.93
448-449		Mansfield	.91
450		Hamilton	.92
451-452		Cincinnati	.92
453-454		Dayton	.93
455		Springfield	.94
456		Chillicothe	.97
457		Athens	.90
458		Lima	.92
OKLAHOMA			
730-731		Oklahoma City	.83
734		Ardmore	.78
735		Lawton	.80
736		Clinton	.79
737		Enid	.76
738		Woodward	.79
739		Guymon	.78
740-741		Tulsa	.78
743		Miami	.84
744		Muskogee	.74
745		Mcalester	.75
746		Ponca City	.78
747		Durant	.77
748		Shawnee	.76
749		Poteau	.76
OREGON			
970-972		Portland	.99
973		Salem	.99
974		Eugene	.99
975		Medford	.97
976		Klamath Falls	.98
977		Bend	1.00
978		Pendleton	.98
979		Vale	.98
PENNSYLVANIA			
150-152		Pittsburgh	1.00
153		Washington	.96
154		Uniontown	.94
155		Bedford	.90
156		Greensburg	.96
157		Indiana	.93
158		Dubois	.92
159		Johnstown	.92
160		Butler	.93
161		New Castle	.92
162		Kittanning	.94
163		Oil City	.90
164-165		Erie	.94
166		Altoona	.88
167		Bradford	.91
168		State College	.90
169		Wellsboro	.91
170-171		Harrisburg	.95
172		Chambersburg	.88
173-174		York	.92
175-176		Lancaster	.91
177		Williamsport	.84
178		Sunbury	.93
179		Pottsville	.93
180		Lehigh Valley	1.01
181		Allentown	1.03
182		Hazleton	.92
183		Stroudsburg	.92
184-185		Scranton	.97
186-187		Wilkes-Barre	.94
188		Montrose	.91
189		Doylestown	1.06
PENNSYLVANIA (CONT'D)			
190-191		Philadelphia	1.16
193		Westchester	1.09
194		Norristown	1.09
195-196		Reading	.98
PUERTO RICO			
009		San Juan	.76
RHODE ISLAND			
028		Newport	1.09
029		Providence	1.10
SOUTH CAROLINA			
290-292		Columbia	.85
293		Spartanburg	.85
294		Charleston	.93
295		Florence	.81
296		Greenville	.84
297		Rock Hill	.80
298		Aiken	.96
299		Beaufort	.79
SOUTH DAKOTA			
570-571		Sioux Falls	.76
572		Watertown	.73
573		Mitchell	.74
574		Aberdeen	.75
575		Pierre	.78
576		Mobridge	.73
577		Rapid City	.78
TENNESSEE			
370-372		Nashville	.85
373-374		Chattanooga	.83
375,380-381		Memphis	.84
376		Johnson City	.71
377-379		Knoxville	.81
382		Mckenzie	.73
383		Jackson	.76
384		Columbia	.77
385		Cookeville	.70
TEXAS			
750		Mckinney	.82
751		Waxahachie	.83
752-753		Dallas	.85
754		Greenville	.83
755		Texarkana	.80
756		Longview	.80
757		Tyler	.84
758		Palestine	.80
759		Lufkin	.82
760-761		Fort Worth	.83
762		Denton	.84
763		Wichita Falls	.85
764		Eastland	.83
765		Temple	.82
766-767		Waco	.84
768		Brownwood	.80
769		San Angelo	.79
770-772		Houston	.84
773		Huntsville	.81
774		Wharton	.81
775		Galveston	.83
776-777		Beaumont	.85
778		Bryan	.80
779		Victoria	.82
780		Laredo	.82
781-782		San Antonio	.82
783-784		Corpus Christi	.86
785		Mc Allen	.85
786-787		Austin	.81
788		Del Rio	.82
789		Giddings	.81
790-791		Amarillo	.81
792		Childress	.83
793-794		Lubbock	.83
795-796		Abilene	.83
797		Midland	.86
798-799,885		El Paso	.80
UTAH			
840-841		Salt Lake City	.81
842,844		Ogden	.80
843		Logan	.80

Location Factors

STATE	CITY	Residential
UTAH (CONT'D)		
845	Price	.78
846-847	Provo	.81
VERMONT		
050	White River Jct.	.90
051	Bellows Falls	.96
052	Bennington	1.00
053	Brattleboro	.96
054	Burlington	.94
056	Montpelier	.93
057	Rutland	.94
058	St. Johnsbury	.91
059	Guildhall	.91
VIRGINIA		
220-221	Fairfax	1.02
222	Arlington	1.04
223	Alexandria	1.05
224-225	Fredericksburg	1.01
226	Winchester	1.00
227	Culpeper	1.01
228	Harrisonburg	.87
229	Charlottesville	.90
230-232	Richmond	.95
233-235	Norfolk	.96
236	Newport News	.96
237	Portsmouth	.90
238	Petersburg	.93
239	Farmville	.84
240-241	Roanoke	.95
242	Bristol	.83
243	Pulaski	.83
244	Staunton	.86
245	Lynchburg	.93
246	Grundy	.80
WASHINGTON		
980-981,987	Seattle	1.03
982	Everett	1.04
983-984	Tacoma	1.01
985	Olympia	.98
986	Vancouver	.96
988	Wenatchee	.93
989	Yakima	.98
990-992	Spokane	1.00
993	Richland	1.00
994	Clarkston	.94
WEST VIRGINIA		
247-248	Bluefield	.94
249	Lewisburg	.93
250-253	Charleston	.97
254	Martinsburg	.89
255-257	Huntington	.98
258-259	Beckley	.94
260	Wheeling	.95
261	Parkersburg	.93
262	Buckhannon	.95
263-264	Clarksburg	.95
265	Morgantown	.95
266	Gassaway	.94
267	Romney	.93
268	Petersburg	.94
WISCONSIN		
530,532	Milwaukee	1.07
531	Kenosha	1.03
534	Racine	1.02
535	Beloit	1.00
537	Madison	1.00
538	Lancaster	.97
539	Portage	.96
540	New Richmond	.98
541-543	Green Bay	1.01
544	Wausau	.96
545	Rhinelander	.95
546	La Crosse	.96
547	Eau Claire	.98
548	Superior	.96
549	Oshkosh	.96
WYOMING		
820	Cheyenne	.81
821	Yellowstone Nat. Pk.	.81
822	Wheatland	.79

STATE	CITY	Residential
WYOMING (CONT'D)		
823	Rawlins	.82
824	Worland	.81
825	Riverton	.81
826	Casper	.80
827	Newcastle	.81
828	Sheridan	.81
829-831	Rock Springs	.86
CANADIAN FACTORS (reflect Canadian currency)		
ALBERTA		
	Calgary	1.11
	Edmonton	1.10
	Fort McMurray	1.12
	Lethbridge	1.10
	Lloydminster	1.05
	Medicine Hat	1.06
	Red Deer	1.06
BRITISH COLUMBIA		
	Kamloops	1.04
	Prince George	1.03
	Vancouver	1.05
	Victoria	1.04
MANITOBA		
	Brandon	1.10
	Portage la Prairie	1.00
	Winnipeg	.98
NEW BRUNSWICK		
	Bathurst	.93
	Dalhousie	.93
	Fredericton	.98
	Moncton	.94
	Newcastle	.93
	St. John	1.02
NEWFOUNDLAND		
	Corner Brook	1.03
	St Johns	1.06
NORTHWEST TERRITORIES		
	Yellowknife	1.13
NOVA SCOTIA		
	Bridgewater	.96
	Dartmouth	1.04
	Halifax	1.00
	New Glasgow	1.03
	Sydney	1.02
	Truro	.96
	Yarmouth	1.03
ONTARIO		
	Barrie	1.12
	Brantford	1.13
	Cornwall	1.12
	Hamilton	1.09
	Kingston	1.12
	Kitchener	1.05
	London	1.08
	North Bay	1.19
	Oshawa	1.09
	Ottawa	1.09
	Owen Sound	1.11
	Peterborough	1.10
	Sarnia	1.13
	Sault Ste Marie	1.06
	St. Catharines	1.07
	Sudbury	1.04
	Thunder Bay	1.09
	Timmins	1.09
	Toronto	1.12
	Windsor	1.08
PRINCE EDWARD ISLAND		
	Charlottetown	.91
	Summerside	.98
QUEBEC		
	Cap-de-la-Madeleine	1.11
	Charlesbourg	1.11
	Chicoutimi	1.14
	Gatineau	1.10
	Granby	1.11

Location Factors

STATE	CITY	Residential
QUEBEC (CONT'D)		
	Hull	1.10
	Joliette	1.11
	Laval	1.11
	Montreal	1.11
	Quebec	1.10
	Rimouski	1.14
	Rouyn-Noranda	1.10
	Saint Hyacinthe	1.11
	Sherbrooke	1.11
	Sorel	1.11
	St Jerome	1.10
	Trois Rivieres	1.22
SASKATCHEWAN		
	Moose Jaw	.93
	Prince Albert	.92
	Regina	1.06
	Saskatoon	1.05
YUKON		
	Whitehorse	1.03

General Requirements — R0111 Summary of Work

R011105-05 Tips for Accurate Estimating

1. Use pre-printed or columnar forms for orderly sequence of dimensions and locations and for recording telephone quotations.
2. Use only the front side of each paper or form except for certain pre-printed summary forms.
3. Be consistent in listing dimensions: For example, length x width x height. This helps in rechecking to ensure that, the total length of partitions is appropriate for the building area.
4. Use printed (rather than measured) dimensions where given.
5. Add up multiple printed dimensions for a single entry where possible.
6. Measure all other dimensions carefully.
7. Use each set of dimensions to calculate multiple related quantities.
8. Convert foot and inch measurements to decimal feet when listing. Memorize decimal equivalents to .01 parts of a foot (1/8" equals approximately .01').
9. Do not "round off" quantities until the final summary.
10. Mark drawings with different colors as items are taken off.
11. Keep similar items together, different items separate.
12. Identify location and drawing numbers to aid in future checking for completeness.
13. Measure or list everything on the drawings or mentioned in the specifications.
14. It may be necessary to list items not called for to make the job complete.
15. Be alert for: Notes on plans such as N.T.S. (not to scale); changes in scale throughout the drawings; reduced size drawings; discrepancies between the specifications and the drawings.
16. Develop a consistent pattern of performing an estimate. For example:
    a. Start the quantity takeoff at the lower floor and move to the next higher floor.
    b. Proceed from the main section of the building to the wings.
    c. Proceed from south to north or vice versa, clockwise or counterclockwise.
    d. Take off floor plan quantities first, elevations next, then detail drawings.
17. List all gross dimensions that can be either used again for different quantities, or used as a rough check of other quantities for verification (exterior perimeter, gross floor area, individual floor areas, etc.).
18. Utilize design symmetry or repetition (repetitive floors, repetitive wings, symmetrical design around a center line, similar room layouts, etc.). Note: Extreme caution is needed here so as not to omit or duplicate an area.
19. Do not convert units until the final total is obtained. For instance, when estimating concrete work, keep all units to the nearest cubic foot, then summarize and convert to cubic yards.
20. When figuring alternatives, it is best to total all items involved in the basic system, then total all items involved in the alternates. Therefore you work with positive numbers in all cases. When adds and deducts are used, it is often confusing whether to add or subtract a portion of an item; especially on a complicated or involved alternate.

General Requirements R0111 Summary of Work

R011105-50 Metric Conversion Factors

Description: This table is primarily for converting customary U.S. units in the left hand column to SI metric units in the right hand column. In addition, conversion factors for some commonly encountered Canadian and non-SI metric units are included.

Category	If You Know		Multiply By		To Find
Length	Inches	x	25.4[a]	=	Millimeters
	Feet	x	0.3048[a]	=	Meters
	Yards	x	0.9144[a]	=	Meters
	Miles (statute)	x	1.609	=	Kilometers
Area	Square inches	x	645.2	=	Square millimeters
	Square feet	x	0.0929	=	Square meters
	Square yards	x	0.8361	=	Square meters
Volume (Capacity)	Cubic inches	x	16,387	=	Cubic millimeters
	Cubic feet	x	0.02832	=	Cubic meters
	Cubic yards	x	0.7646	=	Cubic meters
	Gallons (U.S. liquids)[b]	x	0.003785	=	Cubic meters[c]
	Gallons (Canadian liquid)[b]	x	0.004546	=	Cubic meters[c]
	Ounces (U.S. liquid)[b]	x	29.57	=	Milliliters[c, d]
	Quarts (U.S. liquid)[b]	x	0.9464	=	Liters[c, d]
	Gallons (U.S. liquid)[b]	x	3.785	=	Liters[c, d]
Force	Kilograms force[d]	x	9.807	=	Newtons
	Pounds force	x	4.448	=	Newtons
	Pounds force	x	0.4536	=	Kilograms force[d]
	Kips	x	4448	=	Newtons
	Kips	x	453.6	=	Kilograms force[d]
Pressure, Stress, Strength (Force per unit area)	Kilograms force per square centimeter[d]	x	0.09807	=	Megapascals
	Pounds force per square inch (psi)	x	0.006895	=	Megapascals
	Kips per square inch	x	6.895	=	Megapascals
	Pounds force per square inch (psi)	x	0.07031	=	Kilograms force per square centimeter[d]
	Pounds force per square foot	x	47.88	=	Pascals
	Pounds force per square foot	x	4.882	=	Kilograms force per square meter[d]
Bending Moment Or Torque	Inch-pounds force	x	0.01152	=	Meter-kilograms force[d]
	Inch-pounds force	x	0.1130	=	Newton-meters
	Foot-pounds force	x	0.1383	=	Meter-kilograms force[d]
	Foot-pounds force	x	1.356	=	Newton-meters
	Meter-kilograms force[d]	x	9.807	=	Newton-meters
Mass	Ounces (avoirdupois)	x	28.35	=	Grams
	Pounds (avoirdupois)	x	0.4536	=	Kilograms
	Tons (metric)	x	1000	=	Kilograms
	Tons, short (2000 pounds)	x	907.2	=	Kilograms
	Tons, short (2000 pounds)	x	0.9072	=	Megagrams[e]
Mass per Unit Volume	Pounds mass per cubic foot	x	16.02	=	Kilograms per cubic meter
	Pounds mass per cubic yard	x	0.5933	=	Kilograms per cubic meter
	Pounds mass per gallon (U.S. liquid)[b]	x	119.8	=	Kilograms per cubic meter
	Pounds mass per gallon (Canadian liquid)[b]	x	99.78	=	Kilograms per cubic meter
Temperature	Degrees Fahrenheit	(F-32)/1.8		=	Degrees Celsius
	Degrees Fahrenheit	(F+459.67)/1.8		=	Degrees Kelvin
	Degrees Celsius	C+273.15		=	Degrees Kelvin

[a] The factor given is exact
[b] One U.S. gallon = 0.8327 Canadian gallon
[c] 1 liter = 1000 milliliters = 1000 cubic centimeters
 1 cubic decimeter = 0.001 cubic meter
[d] Metric but not SI unit
[e] Called "tonne" in England and "metric ton" in other metric countries

General Requirements — R0111 Summary of Work

R011105-60 Weights and Measures

Measures of Length
1 Mile = 1760 Yards = 5280 Feet
1 Yard = 3 Feet = 36 inches
1 Foot = 12 Inches
1 Mil = 0.001 Inch
1 Fathom = 2 Yards = 6 Feet
1 Rod = 5.5 Yards = 16.5 Feet
1 Hand = 4 Inches
1 Span = 9 Inches
1 Micro-inch = One Millionth Inch or 0.000001 Inch
1 Micron = One Millionth Meter + 0.00003937 Inch

Surveyor's Measure
1 Mile = 8 Furlongs = 80 Chains
1 Furlong = 10 Chains = 220 Yards
1 Chain = 4 Rods = 22 Yards = 66 Feet = 100 Links
1 Link = 7.92 Inches

Square Measure
1 Square Mile = 640 Acres = 6400 Square Chains
1 Acre = 10 Square Chains = 4840 Square Yards = 43,560 Sq. Ft.
1 Square Chain = 16 Square Rods = 484 Square Yards = 4356 Sq. Ft.
1 Square Rod = 30.25 Square Yards = 272.25 Square Feet = 625 Square Lines
1 Square Yard = 9 Square Feet
1 Square Foot = 144 Square Inches
An Acre equals a Square 208.7 Feet per Side

Cubic Measure
1 Cubic Yard = 27 Cubic Feet
1 Cubic Foot = 1728 Cubic Inches
1 Cord of Wood = 4 x 4 x 8 Feet = 128 Cubic Feet
1 Perch of Masonry = 16½ x 1½ x 1 Foot = 24.75 Cubic Feet

Avoirdupois or Commercial Weight
1 Gross or Long Ton = 2240 Pounds
1 Net or Short ton = 2000 Pounds
1 Pound = 16 Ounces = 7000 Grains
1 Ounce = 16 Drachms = 437.5 Grains
1 Stone = 14 Pounds

Shipping Measure
For Measuring Internal Capacity of a Vessel:
  1 Register Ton = 100 Cubic Feet

For Measurement of Cargo:
  Approximately 40 Cubic Feet of Merchandise is considered a Shipping Ton, unless that bulk would weigh more than 2000 Pounds, in which case Freight Charge may be based upon weight.

40 Cubic Feet = 32.143 U.S. Bushels = 31.16 Imp. Bushels

Liquid Measure
1 Imperial Gallon = 1.2009 U.S. Gallon = 277.42 Cu. In.
1 Cubic Foot = 7.48 U.S. Gallons

R011110-10 Architectural Fees

Tabulated below are typical percentage fees by project size, for good professional architectural service. Fees may vary from those listed depending upon degree of design difficulty and economic conditions in any particular area.

Rates can be interpolated horizontally and vertically. Various portions of the same project requiring different rates should be adjusted proportionately. For alterations, add 50% to the fee for the first $500,000 of project cost and add 25% to the fee for project cost over $500,000.

Architectural fees tabulated below include Structural, Mechanical and Electrical Engineering Fees. They do not include the fees for special consultants such as kitchen planning, security, acoustical, interior design, etc.

Civil Engineering fees are included in the Architectural fee for project sites requiring minimal design such as city sites. However, separate Civil Engineering fees must be added when utility connections require design, drainage calculations are needed, stepped foundations are required, or provisions are required to protect adjacent wetlands.

Building Types	Total Project Size in Thousands of Dollars						
	100	250	500	1,000	5,000	10,000	50,000
Factories, garages, warehouses, repetitive housing	9.0%	8.0%	7.0%	6.2%	5.3%	4.9%	4.5%
Apartments, banks, schools, libraries, offices, municipal buildings	12.2	12.3	9.2	8.0	7.0	6.6	6.2
Churches, hospitals, homes, laboratories, museums, research	15.0	13.6	12.7	11.9	9.5	8.8	8.0
Memorials, monumental work, decorative furnishings	—	16.0	14.5	13.1	10.0	9.0	8.3

General Requirements — R0129 Payment Procedures

R012909-80 Sales Tax by State

State sales tax on materials is tabulated below (5 states have no sales tax). Many states allow local jurisdictions, such as a county or city, to levy additional sales tax.

Some projects may be sales tax exempt, particularly those constructed with public funds.

State	Tax (%)	State	Tax (%)	State	Tax (%)	State	Tax (%)
Alabama	4	Illinois	6.25	Montana	0	Rhode Island	7
Alaska	0	Indiana	7	Nebraska	5.5	South Carolina	6
Arizona	5.6	Iowa	6	Nevada	6.85	South Dakota	4
Arkansas	6	Kansas	6.3	New Hampshire	0	Tennessee	7
California	7.5	Kentucky	6	New Jersey	7	Texas	6.25
Colorado	2.9	Louisiana	4	New Mexico	5.13	Utah	4.7
Connecticut	6.35	Maine	5	New York	4	Vermont	6
Delaware	0	Maryland	6	North Carolina	4.75	Virginia	5
District of Columbia	6	Massachusetts	6.25	North Dakota	5	Washington	6.5
Florida	6	Michigan	6	Ohio	5.5	West Virginia	6
Georgia	4	Minnesota	6.88	Oklahoma	4.5	Wisconsin	5
Hawaii	4	Mississippi	7	Oregon	0	Wyoming	4
Idaho	6	Missouri	4.23	Pennsylvania	6	Average	5.04 %

Sales Tax by Province (Canada)

GST - a value-added tax, which the government imposes on most goods and services provided in or imported into Canada. PST - a retail sales tax, which three of the provinces impose on the price of most goods and some services. QST - a value-added tax, similar to the federal GST, which Quebec imposes. HST - Five provinces have combined their retail sales tax with the federal GST into one harmonized tax.

Province	PST (%)	QST (%)	GST (%)	HST (%)
Alberta	0	0	5	0
British Columbia	7	0	5	0
Manitoba	8	0	5	0
New Brunswick	0	0	0	13
Newfoundland	0	0	0	13
Northwest Territories	0	0	5	0
Nova Scotia	0	0	0	15
Ontario	0	0	0	13
Prince Edward Island	0	0	0	14
Quebec	0	9.975	5	0
Saskatchewan	5	0	5	0
Yukon	0	0	5	0

R012909-85 Unemployment Taxes and Social Security Taxes

State unemployment tax rates vary not only from state to state, but also with the experience rating of the contractor. The federal unemployment tax rate is 6.0% of the first $7,000 of wages. This is reduced by a credit of up to 5.4% for timely payment to the state. The minimum federal unemployment tax is 0.6% after all credits.

Social security (FICA) for 2015 is estimated at time of publication to be 7.65% of wages up to $117,000.

General Requirements — R0131 Project Management & Coordination

R013113-40 Builder's Risk Insurance

Builder's Risk Insurance is insurance on a building during construction. Premiums are paid by the owner or the contractor. Blasting, collapse and underground insurance would raise total insurance costs above those listed. Floater policy for materials delivered to the job runs $.75 to $1.25 per $100 value. Contractor equipment insurance runs $.50 to $1.50 per $100 value. Insurance for miscellaneous tools to $1,500 value runs from $3.00 to $7.50 per $100 value.

Tabulated below are New England Builder's Risk insurance rates in dollars per $100 value for $1,000 deductible. For $25,000 deductible, rates can be reduced 13% to 34%. On contracts over $1,000,000, rates may be lower than those tabulated. Policies are written annually for the total completed value in place. For "all risk" insurance (excluding flood, earthquake and certain other perils) add $.025 to total rates below.

Coverage	Frame Construction (Class 1) Range	Frame Construction (Class 1) Average	Brick Construction (Class 4) Range	Brick Construction (Class 4) Average	Fire Resistive (Class 6) Range	Fire Resistive (Class 6) Average
Fire Insurance	$.350 to $.850	$.600	$.158 to $.189	$.174	$.052 to $.080	$.070
Extended Coverage	.115 to .200	.158	.080 to .105	.101	.081 to .105	.100
Vandalism	.012 to .016	.014	.008 to .011	.011	.008 to .011	.010
Total Annual Rate	$.477 to $1.066	$.772	$.246 to $.305	$.286	$.141 to $.196	$.180

R013113-50 General Contractor's Overhead

There are two distinct types of overhead on a construction project: Project Overhead and Main Office Overhead. Project Overhead includes those costs at a construction site not directly associated with the installation of construction materials. Examples of Project Overhead costs include the following:

1. Superintendent
2. Construction office and storage trailers
3. Temporary sanitary facilities
4. Temporary utilities
5. Security fencing
6. Photographs
7. Clean up
8. Performance and payment bonds

The above Project Overhead items are also referred to as General Requirements and therefore are estimated in Division 1. Division 1 is the first division listed in the CSI MasterFormat but it is usually the last division estimated. The sum of the costs in Divisions 1 through 49 is referred to as the sum of the direct costs.

All construction projects also include indirect costs. The primary components of indirect costs are the contractor's Main Office Overhead and profit. The amount of the Main Office Overhead expense varies depending on the the following:

1. Owner's compensation
2. Project managers and estimator's wages
3. Clerical support wages
4. Office rent and utilities
5. Corporate legal and accounting costs
6. Advertising
7. Automobile expenses
8. Association dues
9. Travel and entertainment expenses

These costs are usually calculated as a percentage of annual sales volume. This percentage can range from 35% for a small contractor doing less than $500,000 to 5% for a large contractor with sales in excess of $100 million.

General Requirements — R0131 Project Management & Coordination

R013113-60 Workers' Compensation Insurance Rates by Trade

The table below tabulates the national averages for workers' compensation insurance rates by trade and type of building. The average "Insurance Rate" is multiplied by the "% of Building Cost" for each trade. This produces the "Workers' Compensation" cost by % of total labor cost, to be added for each trade by building type to determine the weighted average workers' compensation rate for the building types analyzed.

Trade	Insurance Rate (% Labor Cost) Range	Insurance Rate (% Labor Cost) Average	% of Building Cost Office Bldgs.	% of Building Cost Schools & Apts.	% of Building Cost Mfg.	Workers' Compensation Office Bldgs.	Workers' Compensation Schools & Apts.	Workers' Compensation Mfg.
Excavation, Grading, etc.	3.4 % to 21.2%	9.7%	4.8%	4.9%	4.5%	0.47%	0.48%	0.44%
Piles & Foundations	6.1 to 28.3	14.3	7.1	5.2	8.7	1.02	0.74	1.24
Concrete	4.2 to 35.8	12.9	5.0	14.8	3.7	0.65	1.91	0.48
Masonry	4.9 to 36.4	13.7	6.9	7.5	1.9	0.95	1.03	0.26
Structural Steel	6.1 to 101.1	31.7	10.7	3.9	17.6	3.39	1.24	5.58
Miscellaneous & Ornamental Metals	4.3 to 31.5	12.2	2.8	4.0	3.6	0.34	0.49	0.44
Carpentry & Millwork	5.0 to 36.8	14.9	3.7	4.0	0.5	0.55	0.60	0.07
Metal or Composition Siding	6.1 to 69.2	18.4	2.3	0.3	4.3	0.42	0.06	0.79
Roofing	6.1 to 100.3	31.7	2.3	2.6	3.1	0.73	0.82	0.98
Doors & Hardware	4.1 to 36.8	11.3	0.9	1.4	0.4	0.10	0.16	0.05
Sash & Glazing	4.8 to 30.7	13.3	3.5	4.0	1.0	0.47	0.53	0.13
Lath & Plaster	3.1 to 36.2	10.9	3.3	6.9	0.8	0.36	0.75	0.09
Tile, Marble & Floors	3 to 24.8	8.9	2.6	3.0	0.5	0.23	0.27	0.04
Acoustical Ceilings	3.8 to 33.8	8.4	2.4	0.2	0.3	0.20	0.02	0.03
Painting	4.5 to 36.1	11.7	1.5	1.6	1.6	0.18	0.19	0.19
Interior Partitions	5.0 to 36.8	14.9	3.9	4.3	4.4	0.58	0.64	0.66
Miscellaneous Items	2.5 to 132.3	13.4	5.2	3.7	9.7	0.70	0.50	1.30
Elevators	1.3 to 12.8	5.3	2.1	1.1	2.2	0.11	0.06	0.12
Sprinklers	2.9 to 18.5	7.3	0.5	—	2.0	0.04	—	0.15
Plumbing	2.2 to 19.3	7.0	4.9	7.2	5.2	0.34	0.50	0.36
Heat., Vent., Air Conditioning	3.8 to 18.1	8.8	13.5	11.0	12.9	1.19	0.97	1.14
Electrical	2.4 to 13.5	5.8	10.1	8.4	11.1	0.59	0.49	0.64
Total	1.3 % to 132.3%	—	100.0%	100.0%	100.0%	13.61%	12.45%	15.18%
				Overall Weighted Average	13.75%			

Workers' Compensation Insurance Rates by States

The table below lists the weighted average workers' compensation base rate for each state with a factor comparing this with the national average of 13.5%.

State	Weighted Average	Factor	State	Weighted Average	Factor	State	Weighted Average	Factor
Alabama	20.6%	153	Kentucky	12.5%	93	North Dakota	8.1%	60
Alaska	12.5	93	Louisiana	19.6	145	Ohio	8.1	60
Arizona	13.8	102	Maine	11.4	84	Oklahoma	10.3	76
Arkansas	7.7	57	Maryland	13.5	100	Oregon	11.3	84
California	27.1	201	Massachusetts	11.5	85	Pennsylvania	18.3	136
Colorado	7.3	54	Michigan	13.5	100	Rhode Island	14.7	109
Connecticut	25.9	192	Minnesota	22.8	169	South Carolina	16.9	125
Delaware	14.0	104	Mississippi	14.0	104	South Dakota	14.1	104
District of Columbia	10.4	77	Missouri	13.9	103	Tennessee	12.2	90
Florida	10.8	80	Montana	8.3	61	Texas	9.3	69
Georgia	29.5	219	Nebraska	17.0	126	Utah	8.8	65
Hawaii	8.7	64	Nevada	9.2	68	Vermont	13.3	99
Idaho	10.5	78	New Hampshire	18.8	139	Virginia	8.7	64
Illinois	26.9	199	New Jersey	15.7	116	Washington	9.5	70
Indiana	5.0	37	New Mexico	15.3	113	West Virginia	7.4	55
Iowa	15.5	115	New York	19.0	141	Wisconsin	13.2	98
Kansas	9.2	68	North Carolina	17.3	128	Wyoming	6.5	48
			Weighted Average for U.S. is	13.7% of payroll = 100%				

The weighted average skilled worker rate for 35 trades is 13.5%. For bidding purposes, apply the full value of workers' compensation directly to total labor costs, or if labor is 38%, materials 42% and overhead and profit 20% of total cost, carry 38/80 × 13.5% = 6.4% of cost (before overhead and profit) into overhead. Rates vary not only from state to state but also with the experience rating of the contractor.

Rates are the most current available at the time of publication.

General Requirements — R0154 Construction Aids

R015423-10 Steel Tubular Scaffolding

On new construction, tubular scaffolding is efficient up to 60' high or five stories. Above this it is usually better to use a hung scaffolding if construction permits. Swing scaffolding operations may interfere with tenants. In this case, the tubular is more practical at all heights.

In repairing or cleaning the front of an existing building the cost of tubular scaffolding per S.F. of building front increases as the height increases above the first tier. The first tier cost is relatively high due to leveling and alignment.

The minimum efficient crew for erecting and dismantling is three workers. They can set up and remove 18 frame sections per day up to 5 stories high. For 6 to 12 stories high, a crew of four is most efficient. Use two or more on top and two on the bottom for handing up or hoisting. They can also set up and remove 18 frame sections per day. At 7' horizontal spacing, this will run about 800 S.F. per day of erecting and dismantling. Time for placing and removing planks must be added to the above. A crew of three can place and remove 72 planks per day up to 5 stories. For over 5 stories, a crew of four can place and remove 80 planks per day.

The table below shows the number of pieces required to erect tubular steel scaffolding for 1000 S.F. of building frontage. This area is made up of a scaffolding system that is 12 frames (11 bays) long by 2 frames high.

For jobs under twenty-five frames, add 50% to rental cost. Rental rates will be lower for jobs over three months duration. Large quantities for long periods can reduce rental rates by 20%.

Description of Component	Number of Pieces for 1000 S.F. of Building Front	Unit
5' Wide Standard Frame, 6'-4" High	24	Ea.
Leveling Jack & Plate	24	
Cross Brace	44	
Side Arm Bracket, 21"	12	
Guardrail Post	12	
Guardrail, 7' section	22	
Stairway Section	2	
Stairway Starter Bar	1	
Stairway Inside Handrail	2	
Stairway Outside Handrail	2	
Walk-Thru Frame Guardrail	2	

Scaffolding is often used as falsework over 15' high during construction of cast-in-place concrete beams and slabs. Two foot wide scaffolding is generally used for heavy beam construction. The span between frames depends upon the load to be carried with a maximum span of 5'.

Heavy duty shoring frames with a capacity of 10,000#/leg can be spaced up to 10' O.C. depending upon form support design and loading.

Scaffolding used as horizontal shoring requires less than half the material required with conventional shoring.

On new construction, erection is done by carpenters.

Rolling towers supporting horizontal shores can reduce labor and speed the job. For maintenance work, catwalks with spans up to 70' can be supported by the rolling towers.

R015423-20 Pump Staging

Pump staging is generally not available for rent. The table below shows the number of pieces required to erect pump staging for 2400 S.F. of building frontage. This area is made up of a pump jack system that is 3 poles (2 bays) wide by 2 poles high.

Item	Number of Pieces for 2400 S.F. of Building Front	Unit
Aluminum pole section, 24' long	6	Ea.
Aluminum splice joint, 6' long	3	
Aluminum foldable brace	3	
Aluminum pump jack	3	
Aluminum support for workbench/back safety rail	3	
Aluminum scaffold plank/workbench, 14" wide x 24' long	4	
Safety net, 22' long	2	
Aluminum plank end safety rail	2	

The cost in place for this 2400 S.F. will depend on how many uses are realized during the life of the equipment.

General Requirements — R0154 Construction Aids

R015436-50 Mobilization

Costs to move rented construction equipment to a job site from an equipment dealer's or contractor's yard (mobilization), or to move the equipment off the job site (demobilization), are not included in the rental or operating rates, nor in the equipment cost on a unit price line or in a crew listing. These costs can be found consolidated in the Mobilization section of the data and elsewhere in particular site work sections. If a piece of equipment is already on the job site, it is not appropriate to include mob/demob costs in a new estimate that requires use of that equipment. The following table identifies approximate sizes of rented construction equipment that would be hauled on a towed trailer. Because this listing is not all-encompassing, the user can infer as to what size trailer might be required for a piece of equipment not listed.

3-ton Trailer	20-ton Trailer	40-ton Trailer	50-ton Trailer
20 H.P. Excavator	110 H.P. Excavator	200 H.P. Excavator	270 H.P. Excavator
50 H.P. Skid Steer	165 H.P. Dozer	300 H.P. Dozer	Small Crawler Crane
35 H.P. Roller	150 H.P. Roller	400 H.P. Scraper	500 H.P. Scraper
40 H.P. Trencher	Backhoe	450 H.P. Art. Dump Truck	500 H.P. Art. Dump Truck

Existing Conditions — R0241 Demolition

R024119-10 Demolition Defined

Whole Building Demolition - Demolition of the whole building with no concern for any particular building element, component, or material type being demolished. This type of demolition is accomplished with large pieces of construction equipment that break up the structure, load it into trucks and haul it to a disposal site, but disposal or dump fees are not included. Demolition of below-grade foundation elements, such as footings, foundation walls, grade beams, slabs on grade, etc., is not included. Certain mechanical equipment containing flammable liquids or ozone-depleting refrigerants, electric lighting elements, communication equipment components, and other building elements may contain hazardous waste, and must be removed, either selectively or carefully, as hazardous waste before the building can be demolished.

Foundation Demolition - Demolition of below-grade foundation footings, foundation walls, grade beams, and slabs on grade. This type of demolition is accomplished by hand or pneumatic hand tools, and does not include saw cutting, or handling, loading, hauling, or disposal of the debris.

Gutting - Removal of building interior finishes and electrical/mechanical systems down to the load-bearing and sub-floor elements of the rough building frame, with no concern for any particular building element, component, or material type being demolished. This type of demolition is accomplished by hand or pneumatic hand tools, and includes loading into trucks, but not hauling, disposal or dump fees, scaffolding, or shoring. Certain mechanical equipment containing flammable liquids or ozone-depleting refrigerants, electric lighting elements, communication equipment components, and other building elements may contain hazardous waste, and must be removed, either selectively or carefully, as hazardous waste, before the building is gutted.

Selective Demolition - Demolition of a selected building element, component, or finish, with some concern for surrounding or adjacent elements, components, or finishes (see the first Subdivision (s) at the beginning of appropriate Divisions). This type of demolition is accomplished by hand or pneumatic hand tools, and does not include handling, loading, storing, hauling, or disposal of the debris, scaffolding, or shoring. "Gutting" methods may be used in order to save time, but damage that is caused to surrounding or adjacent elements, components, or finishes may have to be repaired at a later time.

Careful Removal - Removal of a piece of service equipment, building element or component, or material type, with great concern for both the removed item and surrounding or adjacent elements, components or finishes. The purpose of careful removal may be to protect the removed item for later re-use, preserve a higher salvage value of the removed item, or replace an item while taking care to protect surrounding or adjacent elements, components, connections, or finishes from cosmetic and/or structural damage. An approximation of the time required to perform this type of removal is 1/3 to 1/2 the time it would take to install a new item of like kind. This type of removal is accomplished by hand or pneumatic hand tools, and does not include loading, hauling, or storing the removed item, scaffolding, shoring, or lifting equipment.

Cutout Demolition - Demolition of a small quantity of floor, wall, roof, or other assembly, with concern for the appearance and structural integrity of the surrounding materials. This type of demolition is accomplished by hand or pneumatic hand tools, and does not include saw cutting, handling, loading, hauling, or disposal of debris, scaffolding, or shoring.

Rubbish Handling - Work activities that involve handling, loading or hauling of debris. Generally, the cost of rubbish handling must be added to the cost of all types of demolition, with the exception of whole building demolition.

Minor Site Demolition - Demolition of site elements outside the footprint of a building. This type of demolition is accomplished by hand or pneumatic hand tools, or with larger pieces of construction equipment, and may include loading a removed item onto a truck (check the Crew for equipment used). It does not include saw cutting, hauling or disposal of debris, and, sometimes, handling or loading.

R024119-20 Dumpsters

Dumpster rental costs on construction sites are presented in two ways.

The cost per week rental includes the delivery of the dumpster; its pulling or emptying once per week, and its final removal. The assumption is made that the dumpster contractor could choose to empty a dumpster by simply bringing in an empty unit and removing the full one. These costs also include the disposal of the materials in the dumpster.

The Alternate Pricing can be used when actual planned conditions are not approximated by the weekly numbers. For example, these lines can be used when a dumpster is needed for 4 weeks and will need to be emptied 2 or 3 times per week. Conversely the Alternate Pricing lines can be used when a dumpster will be rented for several weeks or months but needs to be emptied only a few times over this period.

Masonry — R0401 Maintenance of Masonry

R040130-10 Cleaning Face Brick

On smooth brick a person can clean 70 S.F. an hour; on rough brick 50 S.F. per hour. Use one gallon muriatic acid to 20 gallons of water for 1000 S.F. Do not use acid solution until wall is at least seven days old, but a mild soap solution may be used after two days.

Time has been allowed for clean-up in brick prices.

Masonry — R0405 Common Work Results for Masonry

R040513-10 Cement Mortar (material only)

Type N - 1:1:6 mix by volume. Use everywhere above grade except as noted below. - 1:3 mix using conventional masonry cement which saves handling two separate bagged materials.

Type M - 1:1/4:3 mix by volume, or 1 part cement, 1/4 (10% by wt.) lime, 3 parts sand. Use for heavy loads and where earthquakes or hurricanes may occur. Also for reinforced brick, sewers, manholes and everywhere below grade.

Mix Proportions by Volume and Compressive Strength of Mortar

Where Used		Allowable Proportions by Volume				Compressive Strength @ 28 days
	Mortar Type	Portland Cement	Masonry Cement	Hydrated Lime	Masonry Sand	
Plain Masonry		1	1	—	6	
	M	1	—	1/4	3	2500 psi
		1/2	1	—	4	
	S	1	—	1/4 to 1/2	4	1800 psi
		—	1	—	3	
	N	1	—	1/2 to 1-1/4	6	750 psi
		—	1	—	3	
	O	1	—	1-1/4 to 2-1/2	9	350 psi
	K	1	—	2-1/2 to 4	12	75 psi
Reinforced Masonry	PM	1	1	—	6	2500 psi
	PL	1	—	1/4 to 1/2	4	2500 psi

Note: The total aggregate should be between 2.25 to 3 times the sum of the cement and lime used.

The labor cost to mix the mortar is included in the productivity and labor cost of unit price lines in unit cost sections for brickwork, blockwork and stonework.

The material cost of mixed mortar is included in the material cost of those same unit price lines and includes the cost of renting and operating a 10 C.F. mixer at the rate of 200 C.F. per day.

There are two types of mortar color used. One type is the inert additive type with about 100 lbs. per M brick as the typical quantity required. These colors are also available in smaller-batch-sized bags (1 lb. to 15 lb.) which can be placed directly into the mixer without measuring. The other type is premixed and replaces the masonry cement. Dark green color has the highest cost.

R040519-50 Masonry Reinforcing

Horizontal joint reinforcing helps prevent wall cracks where wall movement may occur and in many locations is required by code. Horizontal joint reinforcing is generally not considered to be structural reinforcing and an unreinforced wall may still contain joint reinforcing.

Reinforcing strips come in 10' and 12' lengths and in truss and ladder shapes, with and without drips. Field labor runs between 2.7 to 5.3 hours per 1000 L.F. for wall thicknesses up to 12".

The wire meets ASTM A82 for cold drawn steel wire and the typical size is 9 ga. sides and ties with 3/16" diameter also available. Typical finish is mill galvanized with zinc coating at .10 oz. per S.F. Class I (.40 oz. per S.F.) and Class III (.80 oz per S.F.) are also available, as is hot dipped galvanizing at 1.50 oz. per S.F.

Masonry — R0421 Clay Unit Masonry

R042110-10 Economy in Bricklaying

Have adequate supervision. Be sure bricklayers are always supplied with materials so there is no waiting. Place best bricklayers at corners and openings.

Use only screened sand for mortar. Otherwise, labor time will be wasted picking out pebbles. Use seamless metal tubs for mortar as they do not leak or catch the trowel. Locate stack and mortar for easy wheeling.

Have brick delivered for stacking. This makes for faster handling, reduces chipping and breakage, and requires less storage space. Many dealers will deliver select common in 2' x 3' x 4' pallets or face brick packaged. This affords quick handling with a crane or forklift and easy tonging in units of ten, which reduces waste.

Use wider bricks for one wythe wall construction. Keep scaffolding away from wall to allow mortar to fall clear and not stain wall.

On large jobs develop specialized crews for each type of masonry unit.

Consider designing for prefabricated panel construction on high rise projects.

Avoid excessive corners or openings. Each opening adds about 50% to labor cost for area of opening.

Bolting stone panels and using window frames as stops reduces labor costs and speeds up erection.

R042110-20 Common and Face Brick

Common building brick manufactured according to ASTM C62 and facing brick manufactured according to ASTM C216 are the two standard bricks available for general building use.

Building brick is made in three grades; SW, where high resistance to damage caused by cyclic freezing is required; MW, where moderate resistance to cyclic freezing is needed; and NW, where little resistance to cyclic freezing is needed. Facing brick is made in only the two grades SW and MW. Additionally, facing brick is available in three types; FBS, for general use; FBX, for general use where a higher degree of precision and lower permissible variation in size than FBS is needed; and FBA, for general use to produce characteristic architectural effects resulting from non-uniformity in size and texture of the units.

In figuring the material cost of brickwork, an allowance of 25% mortar waste and 3% brick breakage was included. If bricks are delivered palletized with 280 to 300 per pallet, or packaged, allow only 1-1/2% for breakage. Packaged or palletized delivery is practical when a job is big enough to have a crane or other equipment available to handle a package of brick. This is so on all industrial work but not always true on small commercial buildings.

The use of buff and gray face is increasing, and there is a continuing trend to the Norman, Roman, Jumbo and SCR brick.

Common red clay brick for backup is not used that often. Concrete block is the most usual backup material with occasional use of sand lime or cement brick. Building brick is commonly used in solid walls for strength and as a fire stop.

Brick panels built on the ground and then crane erected to the upper floors have proven to be economical. This allows the work to be done under cover and without scaffolding.

R042110-50 Brick, Block & Mortar Quantities

Type Brick	Nominal Size (incl. mortar) L H W	Modular Coursing	Number of Brick per S.F.	C.F. of Mortar per M Bricks, Waste Included 3/8" Joint	1/2" Joint	Bond Type	Description	Factor
Standard	8 x 2-2/3 x 4	3C=8"	6.75	10.3	12.9	Common	full header every fifth course	+20%
Economy	8 x 4 x 4	1C=4"	4.50	11.4	14.6		full header every sixth course	+16.7%
Engineer	8 x 3-1/5 x 4	5C=16"	5.63	10.6	13.6	English	full header every second course	+50%
Fire	9 x 2-1/2 x 4-1/2	2C=5"	6.40	550 # Fireclay	—	Flemish	alternate headers every course	+33.3%
Jumbo	12 x 4 x 6 or 8	1C=4"	3.00	23.8	30.8		every sixth course	+5.6%
Norman	12 x 2-2/3 x 4	3C=8"	4.50	14.0	17.9	Header = W x H exposed		+100%
Norwegian	12 x 3-1/5 x 4	5C=16"	3.75	14.6	18.6	Rowlock = H x W exposed		+100%
Roman	12 x 2 x 4	2C=4"	6.00	13.4	17.0	Rowlock stretcher = L x W exposed		+33.3%
SCR	12 x 2-2/3 x 6	3C=8"	4.50	21.8	28.0	Soldier = H x L exposed		—
Utility	12 x 4 x 4	1C=4"	3.00	15.4	19.6	Sailor = W x L exposed		-33.3%

Concrete Blocks Nominal Size	Approximate Weight per S.F. Standard	Lightweight	Blocks per 100 S.F.	Mortar per M block, waste included Partitions	Back up
2" x 8" x 16"	20 PSF	15 PSF	113	27 C.F.	36 C.F.
4"	30	20		41	51
6"	42	30		56	66
8"	55	38		72	82
10"	70	47		87	97
12"	85	55		102	112

Brick & Mortar Quantities
©Brick Industry Association. 2009 Feb. Technical Notes on Brick Construction 10:
  Dimensioning and Estimating Brick Masonry. Reston (VA): BIA. Table 1 Modular Brick Sizes and Table 4 Quantity Estimates for Brick Masonry.

Masonry — R0422 Concrete Unit Masonry

R042210-20 Concrete Block

The material cost of special block such as corner, jamb and head block can be figured at the same price as ordinary block of same size. Labor on specials is about the same as equal-sized regular block.

Bond beam and 16" high lintel blocks are more expensive than regular units of equal size. Lintel blocks are 8" long and either 8" or 16" high.

Use of motorized mortar spreader box will speed construction of continuous walls.

Hollow non-load-bearing units are made according to ASTM C129 and hollow load-bearing units according to ASTM C90.

Metals — R0531 Steel Decking

R053100-10 Decking Descriptions

General - All Deck Products

Steel deck is made by cold forming structural grade sheet steel into a repeating pattern of parallel ribs. The strength and stiffness of the panels are the result of the ribs and the material properties of the steel. Deck lengths can be varied to suit job conditions, but because of shipping considerations, are usually less than 40 feet. Standard deck width varies with the product used but full sheets are usually 12", 18", 24", 30", or 36". Deck is typically furnished in a standard width with the ends cut square. Any cutting for width, such as at openings or for angular fit, is done at the job site.

Deck is typically attached to the building frame with arc puddle welds, self-drilling screws, or powder or pneumatically driven pins. Sheet to sheet fastening is done with screws, button punching (crimping), or welds.

Composite Floor Deck

After installation and adequate fastening, floor deck serves several purposes. It (a) acts as a working platform, (b) stabilizes the frame, (c) serves as a concrete form for the slab, and (d) reinforces the slab to carry the design loads applied during the life of the building. Composite decks are distinguished by the presence of shear connector devices as part of the deck. These devices are designed to mechanically lock the concrete and deck together so that the concrete and the deck work together to carry subsequent floor loads. These shear connector devices can be rolled-in embossments, lugs, holes, or wires welded to the panels. The deck profile can also be used to interlock concrete and steel.

Composite deck finishes are either galvanized (zinc coated) or phosphatized/painted. Galvanized deck has a zinc coating on both the top and bottom surfaces. The phosphatized/painted deck has a bare (phosphatized) top surface that will come into contact with the concrete. This bare top surface can be expected to develop rust before the concrete is placed. The bottom side of the deck has a primer coat of paint.

Composite floor deck is normally installed so the panel ends do not overlap on the supporting beams. Shear lugs or panel profile shape often prevent a tight metal to metal fit if the panel ends overlap; the air gap caused by overlapping will prevent proper fusion with the structural steel supports when the panel end laps are shear stud welded.

Adequate end bearing of the deck must be obtained as shown on the drawings. If bearing is actually less in the field than shown on the drawings, further investigation is required.

Roof Deck

Roof deck is not designed to act compositely with other materials. Roof deck acts alone in transferring horizontal and vertical loads into the building frame. Roof deck rib openings are usually narrower than floor deck rib openings. This provides adequate support of rigid thermal insulation board.

Roof deck is typically installed to endlap approximately 2" over supports. However, it can be butted (or lapped more than 2") to solve field fit problems. Since designers frequently use the installed deck system as part of the horizontal bracing system (the deck as a diaphragm), any fastening substitution or change should be approved by the designer. Continuous perimeter support of the deck is necessary to limit edge deflection in the finished roof and may be required for diaphragm shear transfer.

Standard roof deck finishes are galvanized or primer painted. The standard factory applied paint for roof deck is a primer paint and is not intended to weather for extended periods of time. Field painting or touching up of abrasions and deterioration of the primer coat or other protective finishes is the responsibility of the contractor.

Cellular Deck

Cellular deck is made by attaching a bottom steel sheet to a roof deck or composite floor deck panel. Cellular deck can be used in the same manner as floor deck. Electrical, telephone, and data wires are easily run through the chase created between the deck panel and the bottom sheet.

When used as part of the electrical distribution system, the cellular deck must be installed so that the ribs line up and create a smooth cell transition at abutting ends. The joint that occurs at butting cell ends must be taped or otherwise sealed to prevent wet concrete from seeping into the cell. Cell interiors must be free of welding burrs, or other sharp intrusions, to prevent damage to wires.

When used as a roof deck, the bottom flat plate is usually left exposed to view. Care must be maintained during erection to keep good alignment and prevent damage.

Cellular deck is sometimes used with the flat plate on the top side to provide a flat working surface. Installation of the deck for this purpose requires special methods for attachment to the frame because the flat plate, now on the top, can prevent direct access to the deck material that is bearing on the structural steel. It may be advisable to treat the flat top surface to prevent slipping.

Cellular deck is always furnished galvanized or painted over galvanized.

Form Deck

Form deck can be any floor or roof deck product used as a concrete form. Connections to the frame are by the same methods used to anchor floor and roof deck. Welding washers are recommended when welding deck that is less than 20 gauge thickness.

Form deck is furnished galvanized, prime painted, or uncoated. Galvanized deck must be used for those roof deck systems where form deck is used to carry a lightweight insulating concrete fill.

Wood, Plastics & Comp. — R0611 Wood Framing

R061110-30 Lumber Product Material Prices

The price of forest products fluctuates widely from location to location and from season to season depending upon economic conditions. The bare material prices in the unit cost sections of the book show the National Average material prices in effect Jan. 1 of this book year. It must be noted that lumber prices in general may change significantly during the year.

Availability of certain items depends upon geographic location and must be checked prior to firm-price bidding.

Wood, Plastics & Comp. — R0616 Sheathing

R061636-20 Plywood

There are two types of plywood used in construction: interior, which is moisture-resistant but not waterproofed, and exterior, which is waterproofed.

The grade of the exterior surface of the plywood sheets is designated by the first letter: A, for smooth surface with patches allowed; B, for solid surface with patches and plugs allowed; C, which may be surface plugged or may have knot holes up to 1″ wide; and D, which is used only for interior type plywood and may have knot holes up to 2-1/2″ wide. "Structural Grade" is specifically designed for engineered applications such as box beams. All CC & DD grades have roof and floor spans marked on them.

Underlayment-grade plywood runs from 1/4″ to 1-1/4″ thick. Thicknesses 5/8″ and over have optional tongue and groove joints which eliminate the need for blocking the edges. Underlayment 19/32″ and over may be referred to as Sturd-i-Floor.

The price of plywood can fluctuate widely due to geographic and economic conditions.

Typical uses for various plywood grades are as follows:

AA-AD Interior — cupboards, shelving, paneling, furniture
BB Plyform — concrete form plywood
CDX — wall and roof sheathing
Structural — box beams, girders, stressed skin panels
AA-AC Exterior — fences, signs, siding, soffits, etc.
Underlayment — base for resilient floor coverings
Overlaid HDO — high density for concrete forms & highway signs
Overlaid MDO — medium density for painting, siding, soffits & signs
303 Siding — exterior siding, textured, striated, embossed, etc.

Thermal & Moist. Protec. — R0731 Shingles & Shakes

R073126-20 Roof Slate

16″, 18″ and 20″ are standard lengths, and slate usually comes in random widths. For standard 3/16″ thickness use 1-1/2″ copper nails. Allow for 3% breakage.

Thermal & Moist. Protec. — R0752 Modified Bituminous Membrane Roofing

R075213-30 Modified Bitumen Roofing

The cost of modified bitumen roofing is highly dependent on the type of installation that is planned. Installation is based on the type of modifier used in the bitumen. The two most popular modifiers are atactic polypropylene (APP) and styrene butadiene styrene (SBS). The modifiers are added to heated bitumen during the manufacturing process to change its characteristics. A polyethylene, polyester or fiberglass reinforcing sheet is then sandwiched between layers of this bitumen. When completed, the result is a pre-assembled, built-up roof that has increased elasticity and weatherablility. Some manufacturers include a surfacing material such as ceramic or mineral granules, metal particles or sand.

The preferred method of adhering SBS-modified bitumen roofing to the substrate is with hot-mopped asphalt (much the same as built-up roofing). This installation method requires a tar kettle/pot to heat the asphalt, as well as the labor, tools and equipment necessary to distribute and spread the hot asphalt.

The alternative method for applying APP and SBS modified bitumen is as follows. A skilled installer uses a torch to melt a small pool of bitumen off the membrane. This pool must form across the entire roll for proper adhesion. The installer must unroll the roofing at a pace slow enough to melt the bitumen, but fast enough to prevent damage to the rest of the membrane.

Modified bitumen roofing provides the advantages of both built-up and single-ply roofing. Labor costs are reduced over those of built-up roofing because only a single ply is necessary. The elasticity of single-ply roofing is attained with the reinforcing sheet and polymer modifiers. Modifieds have some self-healing characteristics and because of their multi-layer construction, they offer the reliability and safety of built-up roofing.

Openings — R0813 Metal Doors

R081313-20 Steel Door Selection Guide

Standard steel doors are classified into four levels, as recommended by the Steel Door Institute in the chart below. Each of the four levels offers a range of construction models and designs, to meet architectural requirements for preference and appearance, including full flush, seamless, and stile & rail. Recommended minimum gauge requirements are also included.

For complete standard steel door construction specifications and available sizes, refer to the Steel Door Institute Technical Data Series, ANSI A250.8-98 (SDI-100), and ANSI A250.4-94 Test Procedure and Acceptance Criteria for Physical Endurance of Steel Door and Hardware Reinforcements.

Level		Model	Construction	For Full Flush or Seamless		
				Min. Gauge	Thickness (in)	Thickness (mm)
I	Standard Duty	1	Full Flush	20	0.032	0.8
		2	Seamless			
II	Heavy Duty	1	Full Flush	18	0.042	1.0
		2	Seamless			
III	Extra Heavy Duty	1	Full Flush	16	0.053	1.3
		2	Seamless			
		3	*Stile & Rail			
IV	Maximum Duty	1	Full Flush	14	0.067	1.6
		2	Seamless			

*Stiles & rails are 16 gauge; flush panels, when specified, are 18 gauge.

Openings — R0852 Wood Windows

R085216-10 Window Estimates

To ensure a complete window estimate, be sure to include the material and labor costs for each window, as well as the material and labor costs for an interior wood trim set.

Openings — R0853 Plastic Windows

R085313-20 Replacement Windows

Replacement windows are typically measured per United Inch.

United Inches are calculated by rounding the width and height of the window opening up to the nearest inch, then adding the two figures.

The labor cost for replacement windows includes removal of sash, existing sash balance or weights, parting bead where necessary and installation of new window.

Debris hauling and dump fees are not included.

Openings — R0871 Door Hardware

R087110-10 Hardware Finishes

This table describes hardware finishes used throughout the industry. It also shows the base metal and the respective symbols in the three predominate systems of identification. Many of these are used in pricing descriptions in Division Eight.

US"	BMHA*	CDN^	Base	Description
US P	600	CP	Steel	Primed for Painting
US 1B	601	C1B	Steel	Bright Black Japanned
US 2C	602	C2C	Steel	Zinc Plated
US 2G	603	C2G	Steel	Zinc Plated
US 3	605	C3	Brass	Bright Brass, Clear Coated
US 4	606	C4	Brass	Satin Brass, Clear Coated
US 5	609	C5	Brass	Satin Brass, Blackened, Satin Relieved, Clear Coated
US 7	610	C7	Brass	Satin Brass, Blackened, Bright Relived, Clear Coated
US 9	611	C9	Bronze	Bright Bronze, Clear Coated
US 10	612	C10	Bronze	Satin Bronze, Clear Coated
US 10A	641	C10A	Steel	Antiqued Bronze, Oiled and Lacquered
US 10B	613	C10B	Bronze	Antiqued Bronze, Oiled
US 11	616	C11	Bronze	Satin Bronze, Blackened, Satin Relieved, Clear Coated
US 14	618	C14	Brass/Bronze	Bright Nickel Plated, Clear Coated
US 15	619	C15	Brass/Bronze	Satin Nickel, Clear Coated
US 15A	620	C15A	Brass/Bronze	Satin Nickel Plated, Blackened, Satin Relieved, Clear Coated
US 17A	621	C17A	Brass/Bronze	Nickel Plated, Blackened, Relieved, Clear Coated
US 19	622	C19	Brass/Bronze	Flat Black Coated
US 20	623	C20	Brass/Bronze	Statuary Bronze, Light
US 20A	624	C20A	Brass/Bronze	Statuary Bronze, Dark
US 26	625	C26	Brass/Bronze	Bright Chromium
US 26D	626	C26D	Brass/Bronze	Satin Chromium
US 20	627	C27	Aluminum	Satin Aluminum Clear
US 28	628	C28	Aluminum	Anodized Dull Aluminum
US 32	629	C32	Stainless Steel	Bright Stainless Steel
US 32D	630	C32D	Stainless Steel	Stainless Steel
US 3	632	C3	Steel	Bright Brass Plated, Clear Coated
US 4	633	C4	Steel	Satin Brass, Clear Coated
US 7	636	C7	Steel	Satin Brass Plated, Blackened, Bright Relieved, Clear Coated
US 9	637	C9	Steel	Bright Bronze Plated, Clear Coated
US 5	638	C5	Steel	Satin Brass Plated, Blackened, Bright Relieved, Clear Coated
US 10	639	C10	Steel	Satin Bronze Plated, Clear Coated
US 10B	640	C10B	Steel	Antique Bronze, Oiled
US 10A	641	C10A	Steel	Antiqued Bronze, Oiled and Lacquered
US 11	643	C11	Steel	Satin Bronze Plated, Blackened, Bright Relieved, Clear Coated
US 14	645	C14	Steel	Bright Nickel Plated, Clear Coated
US 15	646	C15	Steel	Satin Nickel Plated, Clear Coated
US 15A	647	C15A	Steel	Nickel Plated, Blackened, Bright Relieved, Clear Coated
US 17A	648	C17A	Steel	Nickel Plated, Blackened, Relieved, Clear Coated
US 20	649	C20	Steel	Statuary Bronze, Light
US 20A	650	C20A	Steel	Statuary Bronze, Dark
US 26	651	C26	Steel	Bright Chromium Plated
US 26D	652	C26D	Steel	Satin Chromium Plated

* - BMHA Builders Hardware Manufacturing Association
" - US Equivalent
^ - Canadian Equivalent
Japanning is imitating Asian lacquer work

Finishes

R0920 Plaster & Gypsum Board

R092000-50 Lath, Plaster and Gypsum Board

Gypsum board lath is available in 3/8" thick x 16" wide x 4' long sheets as a base material for multi-layer plaster applications. It is also available as a base for either multi-layer or veneer plaster applications in 1/2" and 5/8" thick–4' wide x 8', 10' or 12' long sheets. Fasteners are screws or blued ring shank nails for wood framing and screws for metal framing.

Metal lath is available in diamond mesh pattern with flat or self-furring profiles. Paper backing is available for applications where excessive plaster waste needs to be avoided. A slotted mesh ribbed lath should be used in areas where the span between structural supports is greater than normal. Most metal lath comes in 27" x 96" sheets. Diamond mesh weighs 1.75, 2.5 or 3.4 pounds per square yard, slotted mesh lath weighs 2.75 or 3.4 pounds per square yard. Metal lath can be nailed, screwed or tied in place.

Many **accessories** are available. Corner beads, flat reinforcing strips, casing beads, control and expansion joints, furring brackets and channels are some examples. Note that accessories are not included in plaster or stucco line items.

Plaster is defined as a material or combination of materials that when mixed with a suitable amount of water, forms a plastic mass or paste. When applied to a surface, the paste adheres to it and subsequently hardens, preserving in a rigid state the form or texture imposed during the period of elasticity.

Gypsum plaster is made from ground calcined gypsum. It is mixed with aggregates and water for use as a base coat plaster.

Vermiculite plaster is a fire-retardant plaster covering used on steel beams, concrete slabs and other heavy construction materials. Vermiculite is a group name for certain clay minerals, hydrous silicates or aluminum, magnesium and iron that have been expanded by heat.

Perlite plaster is a plaster using perlite as an aggregate instead of sand. Perlite is a volcanic glass that has been expanded by heat.

Gauging plaster is a mix of gypsum plaster and lime putty that when applied produces a quick drying finish coat.

Veneer plaster is a one or two component gypsum plaster used as a thin finish coat over special gypsum board.

Keenes cement is a white cementitious material manufactured from gypsum that has been burned at a high temperature and ground to a fine powder. Alum is added to accelerate the set. The resulting plaster is hard and strong and accepts and maintains a high polish, hence it is used as a finishing plaster.

Stucco is a Portland cement based plaster used primarily as an exterior finish.

Plaster is used on both interior and exterior surfaces. Generally it is applied in multiple-coat systems. A three-coat system uses the terms scratch, brown and finish to identify each coat. A two-coat system uses base and finish to describe each coat. Each type of plaster and application system has attributes that are chosen by the designer to best fit the intended use.

Gypsum Plaster Quantities for 100 S.Y.	2 Coat, 5/8" Thick		3 Coat, 3/4" Thick		
	Base	Finish	Scratch	Brown	Finish
	1:3 Mix	2:1 Mix	1:2 Mix	1:3 Mix	2:1 Mix
Gypsum plaster	1,300 lb.		1,350 lb.	650 lb.	
Sand	1.75 C.Y.		1.85 C.Y.	1.35 C.Y.	
Finish hydrated lime		340 lb.			340 lb.
Gauging plaster		170 lb.			170 lb.

Vermiculite or Perlite Plaster Quantities for 100 S.Y.	2 Coat, 5/8" Thick		3 Coat, 3/4" Thick		
	Base	Finish	Scratch	Brown	Finish
Gypsum plaster	1,250 lb.		1,450 lb.	800 lb.	
Vermiculite or perlite	7.8 bags		8.0 bags	3.3 bags	
Finish hydrated lime		340 lb.			340 lb.
Gauging plaster		170 lb.			170 lb.

Stucco–Three-Coat System Quantities for 100 S.Y.	On Wood Frame	On Masonry
Portland cement	29 bags	21 bags
Sand	2.6 C.Y.	2.0 C.Y.
Hydrated lime	180 lb.	120 lb.

Finishes

R0929 Gypsum Board

R092910-10 Levels of Gypsum Drywall Finish

In the past, contract documents often used phrases such as "industry standard" and "workmanlike finish" to specify the expected quality of gypsum board wall and ceiling installations. The vagueness of these descriptions led to unacceptable work and disputes.

In order to resolve this problem, four major trade associations concerned with the manufacture, erection, finish and decoration of gypsum board wall and ceiling systems have developed an industry-wide *Recommended Levels of Gypsum Board Finish*.

The finish of gypsum board walls and ceilings for specific final decoration is dependent on a number of factors. A primary consideration is the location of the surface and the degree of decorative treatment desired. Painted and unpainted surfaces in warehouses and other areas where appearance is normally not critical may simply require the taping of wallboard joints and 'spotting' of fastener heads. Blemish-free, smooth, monolithic surfaces often intended for painted and decorated walls and ceilings in habitated structures, ranging from single-family dwellings through monumental buildings, require additional finishing prior to the application of the final decoration.

Other factors to be considered in determining the level of finish of the gypsum board surface are (1) the type of angle of surface illumination (both natural and artificial lighting), and (2) the paint and method of application or the type and finish of wallcovering specified as the final decoration. Critical lighting conditions, gloss paints, and thin wall coverings require a higher level of gypsum board finish than do heavily textured surfaces which are subsequently painted or surfaces which are to be decorated with heavy grade wall coverings.

The following descriptions were developed jointly by the Association of the Wall and Ceiling Industries-International (AWCI), Ceiling & Interior Systems Construction Association (CISCA), Gypsum Association (GA), and Painting and Decorating Contractors of America (PDCA) as a guide.

Level 0: Used in temporary construction or wherever the final decoration has not been determined. Unfinished. No taping, finishing or corner beads are required. Also could be used where non-predecorated panels will be used in demountable-type partitions that are to be painted as a final finish.

Level 1: Frequently used in plenum areas above ceilings, in attics, in areas where the assembly would generally be concealed, or in building service corridors and other areas not normally open to public view. Some degree of sound and smoke control is provided; in some geographic areas, this level is referred to as "fire-taping," although this level of finish does not typically meet fire-resistant assembly requirements. Where a fire resistance rating is required for the gypsum board assembly, details of construction should be in accordance with reports of fire tests of assemblies that have met the requirements of the fire rating acceptable.

All joints and interior angles shall have tape embedded in joint compound. Accessories are optional at specifier discretion in corridors and other areas with pedestrian traffic. Tape and fastener heads need not be covered with joint compound. Surface shall be free of excess joint compound. Tool marks and ridges are acceptable.

Level 2: It may be specified for standard gypsum board surfaces in garages, warehouse storage, or other similar areas where surface appearance is not of primary importance.

All joints and interior angles shall have tape embedded in joint compound and shall be immediately wiped with a joint knife or trowel, leaving a thin coating of joint compound over all joints and interior angles. Fastener heads and accessories shall be covered with a coat of joint compound. Surface shall be free of excess joint compound. Tool marks and ridges are acceptable.

Level 3: Typically used in areas that are to receive heavy texture (spray or hand applied) finishes before final painting, or where commercial-grade (heavy duty) wall coverings are to be applied as the final decoration. This level of finish should not be used where smooth painted surfaces or where lighter weight wall coverings are specified. The prepared surface shall be coated with a drywall primer prior to the application of final finishes.

All joints and interior angles shall have tape embedded in joint compound and shall be immediately wiped with a joint knife or trowel, leaving a thin coating of joint compound over all joints and interior angles. One additional coat of joint compound shall be applied over all joints and interior angles. Fastener heads and accessories shall be covered with two separate coats of joint compound. All joint compounds shall be smooth and free of tool marks and ridges. The prepared surface shall be covered with a drywall primer prior to the application of the final decoration.

Level 4: This level should be used where residential grade (light duty) wall coverings, flat paints, or light textures are to be applied. The prepared surface shall be coated with a drywall primer prior to the application of final finishes. Release agents for wall coverings are specifically formulated to minimize damage if coverings are subsequently removed.

The weight, texture, and sheen level of the wall covering material selected should be taken into consideration when specifying wall coverings over this level of drywall treatment. Joints and fasteners must be sufficiently concealed if the wall covering material is lightweight, contains limited pattern, has a glossy finish, or has any combination of these features. In critical lighting areas, flat paints applied over light textures tend to reduce joint photographing. Gloss, semi-gloss, and enamel paints are not recommended over this level of finish.

All joints and interior angles shall have tape embedded in joint compound and shall be immediately wiped with a joint knife or trowel, leaving a thin coating of joint compound over all joints and interior angles. In addition, two separate coats of joint compound shall be applied over all flat joints and one separate coat of joint compound applied over interior angles. Fastener heads and accessories shall be covered with three separate coats of joint compound. All joint compounds shall be smooth and free of tool marks and ridges. The prepared surface shall be covered with a drywall primer like Sheetrock first coat prior to the application of the final decoration.

Level 5: The highest quality finish is the most effective method to provide a uniform surface and minimize the possibility of joint photographing and of fasteners showing through the final decoration. This level of finish is required where gloss, semi-gloss, or enamel is specified; when flat joints are specified over an untextured surface; or where critical lighting conditions occur. The prepared surface shall be coated with a drywall primer prior to the application of final decoration.

All joints and interior angles shall have tape embedded in joint compound and be immediately wiped with a joint knife or trowel, leaving a thin coating of joint compound over all joints and interior angles. Two separate coats of joint compound shall be applied over all flat joints and one separate coat of joint compound applied over interior angles. Fastener heads and accessories shall be covered with three separate coats of joint compound.

A thin skim coat of joint compound shall be trowel applied to the entire surface. Excess compound is immediately troweled off, leaving a film or skim coating of compound completely covering the paper. As an alternative to a skim coat, a material manufactured especially for this purpose may be applied such as Sheetrock Tuff-Hide primer surfacer. The surface must be smooth and free of tool marks and ridges. The prepared surface shall be covered with a drywall primer prior to the application of the final decoration.

Finishes — R0972 Wall Coverings

R097223-10 Wall Covering

The table below lists the quantities required for 100 S.F. of wall covering.

Description	Medium-Priced Paper	Expensive Paper
Paper	1.6 dbl. rolls	1.6 dbl. rolls
Wall sizing	0.25 gallon	0.25 gallon
Vinyl wall paste	0.6 gallon	0.6 gallon
Apply sizing	0.3 hour	0.3 hour
Apply paper	1.2 hours	1.5 hours

Most wallpapers now come in double rolls only.
To remove old paper, allow 1.3 hours per 100 S.F.

Finishes — R0991 Painting

R099100-10 Painting Estimating Techniques

Proper estimating methodology is needed to obtain an accurate painting estimate. There is no known reliable shortcut or square foot method. The following steps should be followed:

- List all surfaces to be painted, with an accurate quantity (area) of each. Items having similar surface condition, finish, application method and accessibility may be grouped together.
- List all the tasks required for each surface to be painted, including surface preparation, masking, and protection of adjacent surfaces. Surface preparation may include minor repairs, washing, sanding and puttying.
- Select the proper Means line for each task. Review and consider all adjustments to labor and materials for type of paint and location of work. Apply the height adjustment carefully. For instance, when applying the adjustment for work over 8' high to a wall that is 12' high, apply the adjustment only to the area between 8' and 12' high, and not to the entire wall.

When applying more than one percent (%) adjustment, apply each to the base cost of the data, rather than applying one percentage adjustment on top of the other.

When estimating the cost of painting walls and ceilings remember to add the brushwork for all cut-ins at inside corners and around windows and doors as a LF measure. One linear foot of cut-in with brush equals one square foot of painting.

All items for spray painting include the labor for roll-back.

Deduct for openings greater than 100 SF, or openings that extend from floor to ceiling and are greater than 5' wide. Do not deduct small openings.

The cost of brushes, rollers, ladders and spray equipment are considered to be part of a painting contractor's overhead, and should not be added to the estimate. The cost of rented equipment such as scaffolding and swing staging should be added to the estimate.

R099100-20 Painting

Item	Coat	One Gallon Covers			In 8 Hours a Laborer Covers			Labor-Hours per 100 S.F.		
		Brush	Roller	Spray	Brush	Roller	Spray	Brush	Roller	Spray
Paint wood siding	prime	250 S.F.	225 S.F.	290 S.F.	1150 S.F.	1300 S.F.	2275 S.F.	.695	.615	.351
	others	270	250	290	1300	1625	2600	.615	.492	.307
Paint exterior trim	prime	400	—	—	650	—	—	1.230	—	—
	1st	475	—	—	800	—	—	1.000	—	—
	2nd	520	—	—	975	—	—	.820	—	—
Paint shingle siding	prime	270	255	300	650	975	1950	1.230	.820	.410
	others	360	340	380	800	1150	2275	1.000	.695	.351
Stain shingle siding	1st	180	170	200	750	1125	2250	1.068	.711	.355
	2nd	270	250	290	900	1325	2600	.888	.603	.307
Paint brick masonry	prime	180	135	160	750	800	1800	1.066	1.000	.444
	1st	270	225	290	815	975	2275	.981	.820	.351
	2nd	340	305	360	815	1150	2925	.981	.695	.273
Paint interior plaster or drywall	prime	400	380	495	1150	2000	3250	.695	.400	.246
	others	450	425	495	1300	2300	4000	.615	.347	.200
Paint interior doors and windows	prime	400	—	—	650	—	—	1.230	—	—
	1st	425	—	—	800	—	—	1.000	—	—
	2nd	450	—	—	975	—	—	.820	—	—

Special Construction   R1311 Swimming Pools

R131113-20   Swimming Pools

Pool prices given per square foot of surface area include pool structure, filter and chlorination equipment, pumps, related piping, ladders/steps, maintenance kit, skimmer and vacuum system. Decks and electrical service to equipment are not included.

Residential in-ground pool construction can be divided into two categories: vinyl lined and gunite. Vinyl lined pool walls are constructed of different materials including wood, concrete, plastic or metal. The bottom is often graded with sand over which the vinyl liner is installed. Vermiculite or soil cement bottoms may be substituted for an added cost.

Gunite pool construction is used both in residential and municipal installations. These structures are steel reinforced for strength and finished with a white cement limestone plaster.

Municipal pools will have a higher cost because plumbing codes require more expensive materials, chlorination equipment and higher filtration rates.

Municipal pools greater than 1,800 S.F. require gutter systems to control waves. This gutter may be formed into the concrete wall. Often a vinyl/stainless steel gutter or gutter/wall system is specified, which will raise the pool cost.

Competition pools usually require tile bottoms and sides with contrasting lane striping, which will also raise the pool cost.

Plumbing — R2211 Facility Water Distribution

R221113-50 Pipe Material Considerations

1. Malleable fittings should be used for gas service.
2. Malleable fittings are used where there are stresses/strains due to expansion and vibration.
3. Cast fittings may be broken as an aid to disassembling of heating lines frozen by long use, temperature and minerals.
4. Cast iron pipe is extensively used for underground and submerged service.
5. Type M (light wall) copper tubing is available in hard temper only and is used for nonpressure and less severe applications than K and L.
6. Type L (medium wall) copper tubing, available hard or soft for interior service.
7. Type K (heavy wall) copper tubing, available in hard or soft temper for use where conditions are severe. For underground and interior service.
8. Hard drawn tubing requires fewer hangers or supports but should not be bent. Silver brazed fittings are recommended, however soft solder is normally used.
9. Type DMV (very light wall) copper tubing designed for drainage, waste and vent plus other non-critical pressure services.

Domestic/Imported Pipe and Fittings Cost

The prices shown in this publication for steel/cast iron pipe and steel, cast iron, malleable iron fittings are based on domestic production sold at the normal trade discounts. The above listed items of foreign manufacture may be available at prices of 1/3 to 1/2 those shown. Some imported items after minor machining or finishing operations are being sold as domestic to further complicate the system.

Caution: Most pipe prices in this book also include a coupling and pipe hangers which for the larger sizes can add significantly to the per foot cost and should be taken into account when comparing "book cost" with quoted supplier's cost.

Exterior Improvements — R3292 Turf & Grasses

R329219-50 Seeding

The type of grass is determined by light, shade and moisture content of soil plus intended use. Fertilizer should be disked 4" before seeding. For steep slopes disk five tons of mulch and lay two tons of hay or straw on surface per acre after seeding. Surface mulch can be staked, lightly disked or tar emulsion sprayed. Material for mulch can be wood chips, peat moss, partially rotted hay or straw, wood fibers and sprayed emulsions. Hemp seed blankets with fertilizer are also available. For spring seeding, watering is necessary. Late fall seeding may have to be reseeded in the spring. Hydraulic seeding, power mulching, and aerial seeding can be used on large areas.

Utilities

R3311 Water Utility Distribution Piping

R331113-80 Piping Designations

There are several systems currently in use to describe pipe and fittings. The following paragraphs will help to identify and clarify classifications of piping systems used for water distribution.

Piping may be classified by schedule. Piping schedules include 5S, 10S, 10, 20, 30, Standard, 40, 60, Extra Strong, 80, 100, 120, 140, 160 and Double Extra Strong. These schedules are dependent upon the pipe wall thickness. The wall thickness of a particular schedule may vary with pipe size.

Ductile iron pipe for water distribution is classified by Pressure Classes such as Class 150, 200, 250, 300 and 350. These classes are actually the rated water working pressure of the pipe in pounds per square inch (psi). The pipe in these pressure classes is designed to withstand the rated water working pressure plus a surge allowance of 100 psi.

The American Water Works Association (AWWA) provides standards for various types of **plastic pipe.** C-900 is the specification for polyvinyl chloride (PVC) piping used for water distribution in sizes ranging from 4" through 12". C-901 is the specification for polyethylene (PE) pressure pipe, tubing and fittings used for water distribution in sizes ranging from 1/2" through 3". C-905 is the specification for PVC piping sizes 14" and greater.

PVC pressure-rated pipe is identified using the standard dimensional ratio (SDR) method. This method is defined by the American Society for Testing and Materials (ASTM) Standard D 2241. This pipe is available in SDR numbers 64, 41, 32.5, 26, 21, 17, and 13.5. Pipe with an SDR of 64 will have the thinnest wall while pipe with an SDR of 13.5 will have the thickest wall. When the pressure rating (PR) of a pipe is given in psi, it is based on a line supplying water at 73 degrees F.

The National Sanitation Foundation (NSF) seal of approval is applied to products that can be used with potable water. These products have been tested to ANSI/NSF Standard 14.

Valves and strainers are classified by American National Standards Institute (ANSI) Classes. These Classes are 125, 150, 200, 250, 300, 400, 600, 900, 1500 and 2500. Within each class there is an operating pressure range dependent upon temperature. Design parameters should be compared to the appropriate material dependent, pressure-temperature rating chart for accurate valve selection.

Abbreviations

A	Area Square Feet; Ampere	Brk., brk	Brick	Csc	Cosecant
AAFES	Army and Air Force Exchange Service	brkt	Bracket	C.S.F.	Hundred Square Feet
ABS	Acrylonitrile Butadiene Stryrene; Asbestos Bonded Steel	Brng.	Bearing	CSI	Construction Specifications Institute
A.C., AC	Alternating Current; Air-Conditioning; Asbestos Cement; Plywood Grade A & C	Brs.	Brass	CT	Current Transformer
		Brz.	Bronze	CTS	Copper Tube Size
		Bsn.	Basin	Cu	Copper, Cubic
		Btr.	Better	Cu. Ft.	Cubic Foot
		Btu	British Thermal Unit	cw	Continuous Wave
ACI	American Concrete Institute	BTUH	BTU per Hour	C.W.	Cool White; Cold Water
ACR	Air Conditioning Refrigeration	Bu.	bushels	Cwt.	100 Pounds
ADA	Americans with Disabilities Act	BUR	Built-up Roofing	C.W.X.	Cool White Deluxe
AD	Plywood, Grade A & D	BX	Interlocked Armored Cable	C.Y.	Cubic Yard (27 cubic feet)
Addit.	Additional	°C	degree centigrade	C.Y./Hr.	Cubic Yard per Hour
Adj.	Adjustable	c	Conductivity, Copper Sweat	Cyl.	Cylinder
af	Audio-frequency	C	Hundred; Centigrade	d	Penny (nail size)
AFUE	Annual Fuel Utilization Efficiency	C/C	Center to Center, Cedar on Cedar	D	Deep; Depth; Discharge
AGA	American Gas Association	C-C	Center to Center	Dis., Disch.	Discharge
Agg.	Aggregate	Cab	Cabinet	Db	Decibel
A.H., Ah	Ampere Hours	Cair.	Air Tool Laborer	Dbl.	Double
A hr.	Ampere-hour	Cal.	caliper	DC	Direct Current
A.H.U., AHU	Air Handling Unit	Calc	Calculated	DDC	Direct Digital Control
A.I.A.	American Institute of Architects	Cap.	Capacity	Demob.	Demobilization
AIC	Ampere Interrupting Capacity	Carp.	Carpenter	d.f.t.	Dry Film Thickness
Allow.	Allowance	C.B.	Circuit Breaker	d.f.u.	Drainage Fixture Units
alt., alt	Alternate	C.C.A.	Chromate Copper Arsenate	D.H.	Double Hung
Alum.	Aluminum	C.C.F.	Hundred Cubic Feet	DHW	Domestic Hot Water
a.m.	Ante Meridiem	cd	Candela	DI	Ductile Iron
Amp.	Ampere	cd/sf	Candela per Square Foot	Diag.	Diagonal
Anod.	Anodized	CD	Grade of Plywood Face & Back	Diam., Dia	Diameter
ANSI	American National Standards Institute	CDX	Plywood, Grade C & D, exterior glue	Distrib.	Distribution
		Cefi.	Cement Finisher	Div.	Division
APA	American Plywood Association	Cem.	Cement	Dk.	Deck
Approx.	Approximate	CF	Hundred Feet	D.L.	Dead Load; Diesel
Apt.	Apartment	C.F.	Cubic Feet	DLH	Deep Long Span Bar Joist
Asb.	Asbestos	CFM	Cubic Feet per Minute	dlx	Deluxe
A.S.B.C.	American Standard Building Code	CFRP	Carbon Fiber Reinforced Plastic	Do.	Ditto
Asbe.	Asbestos Worker	c.g.	Center of Gravity	DOP	Dioctyl Phthalate Penetration Test (Air Filters)
ASCE.	American Society of Civil Engineers	CHW	Chilled Water; Commercial Hot Water		
A.S.H.R.A.E.	American Society of Heating, Refrig. & AC Engineers			Dp., dp	Depth
		C.I., CI	Cast Iron	D.P.S.T.	Double Pole, Single Throw
ASME	American Society of Mechanical Engineers	C.I.P., CIP	Cast in Place	Dr.	Drive
		Circ.	Circuit	DR	Dimension Ratio
ASTM	American Society for Testing and Materials	C.L.	Carload Lot	Drink.	Drinking
		CL	Chain Link	D.S.	Double Strength
Attchmt.	Attachment	Clab.	Common Laborer	D.S.A.	Double Strength A Grade
Avg., Ave.	Average	Clam	Common maintenance laborer	D.S.B.	Double Strength B Grade
AWG	American Wire Gauge	C.L.F.	Hundred Linear Feet	Dty.	Duty
AWWA	American Water Works Assoc.	CLF	Current Limiting Fuse	DWV	Drain Waste Vent
Bbl.	Barrel	CLP	Cross Linked Polyethylene	DX	Deluxe White, Direct Expansion
B&B, BB	Grade B and Better; Balled & Burlapped	cm	Centimeter	dyn	Dyne
		CMP	Corr. Metal Pipe	e	Eccentricity
B&S	Bell and Spigot	CMU	Concrete Masonry Unit	E	Equipment Only; East; emissivity
B.&W.	Black and White	CN	Change Notice	Ea.	Each
b.c.c.	Body-centered Cubic	Col.	Column	EB	Encased Burial
B.C.Y.	Bank Cubic Yards	CO_2	Carbon Dioxide	Econ.	Economy
BE	Bevel End	Comb.	Combination	E.C.Y	Embankment Cubic Yards
B.F.	Board Feet	comm.	Commercial, Communication	EDP	Electronic Data Processing
Bg. cem.	Bag of Cement	Compr.	Compressor	EIFS	Exterior Insulation Finish System
BHP	Boiler Horsepower; Brake Horsepower	Conc.	Concrete	E.D.R.	Equiv. Direct Radiation
		Cont., cont	Continuous; Continued, Container	Eq.	Equation
B.I.	Black Iron	Corr.	Corrugated	EL	elevation
bidir.	bidirectional	Cos	Cosine	Elec.	Electrician; Electrical
Bit., Bitum.	Bituminous	Cot	Cotangent	Elev.	Elevator; Elevating
Bit., Conc.	Bituminous Concrete	Cov.	Cover	EMT	Electrical Metallic Conduit; Thin Wall Conduit
Bk.	Backed	C/P	Cedar on Paneling		
Bkrs.	Breakers	CPA	Control Point Adjustment	Eng.	Engine, Engineered
Bldg., bldg	Building	Cplg.	Coupling	EPDM	Ethylene Propylene Diene Monomer
Blk.	Block	CPM	Critical Path Method		
Bm.	Beam	CPVC	Chlorinated Polyvinyl Chloride	EPS	Expanded Polystyrene
Boil.	Boilermaker	C.Pr.	Hundred Pair	Eqhv.	Equip. Oper., Heavy
bpm	Blows per Minute	CRC	Cold Rolled Channel	Eqlt.	Equip. Oper., Light
BR	Bedroom	Creos.	Creosote	Eqmd.	Equip. Oper., Medium
Brg.	Bearing	Crpt.	Carpet & Linoleum Layer	Eqmm.	Equip. Oper., Master Mechanic
Brhe.	Bricklayer Helper	CRT	Cathode-ray Tube	Eqol.	Equip. Oper., Oilers
Bric.	Bricklayer	CS	Carbon Steel, Constant Shear Bar Joist	Equip.	Equipment
				ERW	Electric Resistance Welded

Abbreviations

.S.	Energy Saver	H	High Henry	Lath.	Lather
.st.	Estimated	HC	High Capacity	Lav.	Lavatory
.su	Electrostatic Units	H.D., HD	Heavy Duty; High Density	lb.; #	Pound
.W.	Each Way	H.D.O.	High Density Overlaid	L.B., LB	Load Bearing; L Conduit Body
.WT	Entering Water Temperature	HDPE	High density polyethelene plastic	L. & E.	Labor & Equipment
.xcav.	Excavation	Hdr.	Header	lb./hr.	Pounds per Hour
.xcl	Excluding	Hdwe.	Hardware	lb./L.F.	Pounds per Linear Foot
.xp., exp	Expansion, Exposure	H.I.D., HID	High Intensity Discharge	lbf/sq.in.	Pound-force per Square Inch
.xt., ext	Exterior; Extension	Help.	Helper Average	L.C.L.	Less than Carload Lot
.xtru.	Extrusion	HEPA	High Efficiency Particulate Air Filter	L.C.Y.	Loose Cubic Yard
.	Fiber stress			Ld.	Load
.	Fahrenheit; Female; Fill	Hg	Mercury	LE	Lead Equivalent
.ab., fab	Fabricated; fabric	HIC	High Interrupting Capacity	LED	Light Emitting Diode
.BGS	Fiberglass	HM	Hollow Metal	L.F.	Linear Foot
.C.	Footcandles	HMWPE	high molecular weight polyethylene	L.F. Nose	Linear Foot of Stair Nosing
.c.c.	Face-centered Cubic			L.F. Rsr	Linear Foot of Stair Riser
.c.	Compressive Stress in Concrete; Extreme Compressive Stress	HO	High Output	Lg.	Long; Length; Large
		Horiz.	Horizontal	L & H	Light and Heat
.E.	Front End	H.P., HP	Horsepower; High Pressure	LH	Long Span Bar Joist
.EP	Fluorinated Ethylene Propylene (Teflon)	H.P.F.	High Power Factor	L.H.	Labor Hours
		Hr.	Hour	L.L., LL	Live Load
.G.	Flat Grain	Hrs./Day	Hours per Day	L.L.D.	Lamp Lumen Depreciation
.H.A.	Federal Housing Administration	HSC	High Short Circuit	lm	Lumen
.ig.	Figure	Ht.	Height	lm/sf	Lumen per Square Foot
.in.	Finished	Htg.	Heating	lm/W	Lumen per Watt
.IPS	Female Iron Pipe Size	Htrs.	Heaters	LOA	Length Over All
.ixt.	Fixture	HVAC	Heating, Ventilation & Air-Conditioning	log	Logarithm
FJP	Finger jointed and primed			L-O-L	Lateralolet
Fl. Oz.	Fluid Ounces	Hvy.	Heavy	long.	longitude
Flr.	Floor	HW	Hot Water	L.P., LP	Liquefied Petroleum; Low Pressure
FM	Frequency Modulation; Factory Mutual	Hyd.; Hydr.	Hydraulic	L.P.F.	Low Power Factor
		Hz	Hertz (cycles)	LR	Long Radius
Fmg.	Framing	I.	Moment of Inertia	L.S.	Lump Sum
FM/UL	Factory Mutual/Underwriters Labs	IBC	International Building Code	Lt.	Light
Fdn.	Foundation	I.C.	Interrupting Capacity	Lt. Ga.	Light Gauge
FNPT	Female National Pipe Thread	ID	Inside Diameter	L.T.L.	Less than Truckload Lot
Fori.	Foreman, Inside	I.D.	Inside Dimension; Identification	Lt. Wt.	Lightweight
Foro.	Foreman, Outside	I.F.	Inside Frosted	L.V.	Low Voltage
Fount.	Fountain	I.M.C.	Intermediate Metal Conduit	M	Thousand; Material; Male; Light Wall Copper Tubing
fpm	Feet per Minute	In.	Inch		
FPT	Female Pipe Thread	Incan.	Incandescent	M^2CA	Meters Squared Contact Area
Fr	Frame	Incl.	Included; Including	m/hr.; M.H.	Man-hour
F.R.	Fire Rating	Int.	Interior	mA	Milliampere
FRK	Foil Reinforced Kraft	Inst.	Installation	Mach.	Machine
FSK	Foil/scrim/kraft	Insul., insul	Insulation/Insulated	Mag. Str.	Magnetic Starter
FRP	Fiberglass Reinforced Plastic	I.P.	Iron Pipe	Maint.	Maintenance
FS	Forged Steel	I.P.S., IPS	Iron Pipe Size	Marb.	Marble Setter
FSC	Cast Body; Cast Switch Box	IPT	Iron Pipe Threaded	Mat; Mat'l.	Material
Ft., ft	Foot; Feet	I.W.	Indirect Waste	Max.	Maximum
Ftng.	Fitting	J	Joule	MBF	Thousand Board Feet
Ftg.	Footing	J.I.C.	Joint Industrial Council	MBH	Thousand BTU's per hr.
Ft lb.	Foot Pound	K	Thousand; Thousand Pounds; Heavy Wall Copper Tubing, Kelvin	MC	Metal Clad Cable
Furn.	Furniture			MCC	Motor Control Center
FVNR	Full Voltage Non-Reversing	K.A.H.	Thousand Amp. Hours	M.C.F.	Thousand Cubic Feet
FVR	Full Voltage Reversing	kcmil	Thousand Circular Mils	MCFM	Thousand Cubic Feet per Minute
FXM	Female by Male	KD	Knock Down	M.C.M.	Thousand Circular Mils
Fy.	Minimum Yield Stress of Steel	K.D.A.T.	Kiln Dried After Treatment	MCP	Motor Circuit Protector
g	Gram	kg	Kilogram	MD	Medium Duty
G	Gauss	kG	Kilogauss	MDF	Medium-density fibreboard
Ga.	Gauge	kgf	Kilogram Force	M.D.O.	Medium Density Overlaid
Gal., gal.	Gallon	kHz	Kilohertz	Med.	Medium
gpm, GPM	Gallon per Minute	Kip	1000 Pounds	MF	Thousand Feet
Galv., galv	Galvanized	KJ	Kiljoule	M.F.B.M.	Thousand Feet Board Measure
GC/MS	Gas Chromatograph/Mass Spectrometer	K.L.	Effective Length Factor	Mfg.	Manufacturing
		K.L.F.	Kips per Linear Foot	Mfrs.	Manufacturers
Gen.	General	Km	Kilometer	mg	Milligram
GFI	Ground Fault Interrupter	KO	Knock Out	MGD	Million Gallons per Day
GFRC	Glass Fiber Reinforced Concrete	K.S.F.	Kips per Square Foot	MGPH	Thousand Gallons per Hour
Glaz.	Glazier	K.S.I.	Kips per Square Inch	MH, M.H.	Manhole; Metal Halide; Man-Hour
GPD	Gallons per Day	kV	Kilovolt	MHz	Megahertz
gpf	Gallon per flush	kVA	Kilovolt Ampere	Mi.	Mile
GPH	Gallons per Hour	kVAR	Kilovar (Reactance)	MI	Malleable Iron; Mineral Insulated
GPM	Gallons per Minute	KW	Kilowatt	MIPS	Male Iron Pipe Size
GR	Grade	KWh	Kilowatt-hour	mj	Mechanical Joint
Gran.	Granular	L	Labor Only; Length; Long; Medium Wall Copper Tubing	m	Meter
Grnd.	Ground			mm	Millimeter
GVW	Gross Vehicle Weight	Lab.	Labor	Mill.	Millwright
GWB	Gypsum wall board	lat	Latitude	Min., min.	Minimum, minute

Abbreviations

Misc.	Miscellaneous	PDCA	Painting and Decorating Contractors of America	SC	Screw Cover
ml	Milliliter, Mainline			SCFM	Standard Cubic Feet per Minute
M.L.F.	Thousand Linear Feet	P.E., PE	Professional Engineer;	Scaf.	Scaffold
Mo.	Month		Porcelain Enamel;	Sch., Sched.	Schedule
Mobil.	Mobilization		Polyethylene; Plain End	S.C.R.	Modular Brick
Mog.	Mogul Base	P.E.C.I.	Porcelain Enamel on Cast Iron	S.D.	Sound Deadening
MPH	Miles per Hour	Perf.	Perforated	SDR	Standard Dimension Ratio
MPT	Male Pipe Thread	PEX	Cross linked polyethylene	S.E.	Surfaced Edge
MRGWB	Moisture Resistant Gypsum Wallboard	Ph.	Phase	Sel.	Select
		P.I.	Pressure Injected	SER, SEU	Service Entrance Cable
MRT	Mile Round Trip	Pile.	Pile Driver	S.F.	Square Foot
ms	Millisecond	Pkg.	Package	S.F.C.A.	Square Foot Contact Area
M.S.F.	Thousand Square Feet	Pl.	Plate	S.F. Flr.	Square Foot of Floor
Mstz.	Mosaic & Terrazzo Worker	Plah.	Plasterer Helper	S.F.G.	Square Foot of Ground
M.S.Y.	Thousand Square Yards	Plas.	Plasterer	S.F. Hor.	Square Foot Horizontal
Mtd., mtd., mtd	Mounted	plf	Pounds Per Linear Foot	SFR	Square Feet of Radiation
Mthe.	Mosaic & Terrazzo Helper	Pluh.	Plumbers Helper	S.F. Shlf.	Square Foot of Shelf
Mtng.	Mounting	Plum.	Plumber	S4S	Surface 4 Sides
Mult.	Multi; Multiply	Ply.	Plywood	Shee.	Sheet Metal Worker
M.V.A.	Million Volt Amperes	p.m.	Post Meridiem	Sin.	Sine
M.V.A.R.	Million Volt Amperes Reactance	Pntd.	Painted	Skwk.	Skilled Worker
MV	Megavolt	Pord.	Painter, Ordinary	SL	Saran Lined
MW	Megawatt	pp	Pages	S.L.	Slimline
MXM	Male by Male	PP, PPL	Polypropylene	Sldr.	Solder
MYD	Thousand Yards	P.P.M.	Parts per Million	SLH	Super Long Span Bar Joist
N	Natural; North	Pr.	Pair	S.N.	Solid Neutral
nA	Nanoampere	P.E.S.B.	Pre-engineered Steel Building	SO	Stranded with oil resistant inside insulation
NA	Not Available; Not Applicable	Prefab.	Prefabricated		
N.B.C.	National Building Code	Prefin.	Prefinished	S-O-L	Socketolet
NC	Normally Closed	Prop.	Propelled	sp	Standpipe
NEMA	National Electrical Manufacturers Assoc.	PSF, psf	Pounds per Square Foot	S.P.	Static Pressure; Single Pole; Self-Propelled
		PSI, psi	Pounds per Square Inch		
NEHB	Bolted Circuit Breaker to 600V.	PSIG	Pounds per Square Inch Gauge	Spri.	Sprinkler Installer
NFPA	National Fire Protection Association	PSP	Plastic Sewer Pipe	spwg	Static Pressure Water Gauge
NLB	Non-Load-Bearing	Pspr.	Painter, Spray	S.P.D.T.	Single Pole, Double Throw
NM	Non-Metallic Cable	Psst.	Painter, Structural Steel	SPF	Spruce Pine Fir; Sprayed Polyurethane Foam
nm	Nanometer	P.T.	Potential Transformer		
No.	Number	P. & T.	Pressure & Temperature	S.P.S.T.	Single Pole, Single Throw
NO	Normally Open	Ptd.	Painted	SPT	Standard Pipe Thread
N.O.C.	Not Otherwise Classified	Ptns.	Partitions	Sq.	Square; 100 Square Feet
Nose.	Nosing	Pu	Ultimate Load	Sq. Hd.	Square Head
NPT	National Pipe Thread	PVC	Polyvinyl Chloride	Sq. In.	Square Inch
NQOD	Combination Plug-on/Bolt on Circuit Breaker to 240V.	Pvmt.	Pavement	S.S.	Single Strength; Stainless Steel
		PRV	Pressure Relief Valve	S.S.B.	Single Strength B Grade
N.R.C., NRC	Noise Reduction Coefficient/ Nuclear Regulator Commission	Pwr.	Power	sst, ss	Stainless Steel
		Q	Quantity Heat Flow	Sswk.	Structural Steel Worker
N.R.S.	Non Rising Stem	Qt.	Quart	Sswl.	Structural Steel Welder
ns	Nanosecond	Quan., Qty.	Quantity	St.; Stl.	Steel
nW	Nanowatt	Q.C.	Quick Coupling	STC	Sound Transmission Coefficient
OB	Opposing Blade	r	Radius of Gyration	Std.	Standard
OC	On Center	R	Resistance	Stg.	Staging
OD	Outside Diameter	R.C.P.	Reinforced Concrete Pipe	STK	Select Tight Knot
O.D.	Outside Dimension	Rect.	Rectangle	STP	Standard Temperature & Pressure
ODS	Overhead Distribution System	recpt.	receptacle	Stpi.	Steamfitter, Pipefitter
O.G.	Ogee	Reg.	Regular	Str.	Strength; Starter; Straight
O.H.	Overhead	Reinf.	Reinforced	Strd.	Stranded
O&P	Overhead and Profit	Req'd.	Required	Struct.	Structural
Oper.	Operator	Res.	Resistant	Sty.	Story
Opng.	Opening	Resi.	Residential	Subj.	Subject
Orna.	Ornamental	RF	Radio Frequency	Subs.	Subcontractors
OSB	Oriented Strand Board	RFID	Radio-frequency identification	Surf.	Surface
OS&Y	Outside Screw and Yoke	Rgh.	Rough	Sw.	Switch
OSHA	Occupational Safety and Health Act	RGS	Rigid Galvanized Steel	Swbd.	Switchboard
		RHW	Rubber, Heat & Water Resistant; Residential Hot Water	S.Y.	Square Yard
Ovhd.	Overhead			Syn.	Synthetic
OWG	Oil, Water or Gas	rms	Root Mean Square	S.Y.P.	Southern Yellow Pine
Oz.	Ounce	Rnd.	Round	Sys.	System
P.	Pole; Applied Load; Projection	Rodm.	Rodman	t.	Thickness
p.	Page	Rofc.	Roofer, Composition	T	Temperature; Ton
Pape.	Paperhanger	Rofp.	Roofer, Precast	Tan	Tangent
P.A.P.R.	Powered Air Purifying Respirator	Rohe.	Roofer Helpers (Composition)	T.C.	Terra Cotta
PAR	Parabolic Reflector	Rots.	Roofer, Tile & Slate	T & C	Threaded and Coupled
P.B., PB	Push Button	R.O.W.	Right of Way	T.D.	Temperature Difference
Pc., Pcs.	Piece, Pieces	RPM	Revolutions per Minute	Tdd	Telecommunications Device for the Deaf
P.C.	Portland Cement; Power Connector	R.S.	Rapid Start		
P.C.F.	Pounds per Cubic Foot	Rsr	Riser	T.E.M.	Transmission Electron Microscopy
PCM	Phase Contrast Microscopy	RT	Round Trip	temp	Temperature, Tempered, Temporary
		S.	Suction; Single Entrance; South	TFFN	Nylon Jacketed Wire
		SBS	Styrene Butadiere Styrene		

Abbreviations

TFE	Tetrafluoroethylene (Teflon)	U.L., UL	Underwriters Laboratory	w/	With
T. & G.	Tongue & Groove; Tar & Gravel	Uld.	unloading	W.C., WC	Water Column; Water Closet
		Unfin.	Unfinished	W.F.	Wide Flange
Th., Thk.	Thick	UPS	Uninterruptible Power Supply	W.G.	Water Gauge
Thn.	Thin	URD	Underground Residential Distribution	Wldg.	Welding
Thrded	Threaded			W. Mile	Wire Mile
Tilf.	Tile Layer, Floor	US	United States	W-O-L	Weldolet
Tilh.	Tile Layer, Helper	USGBC	U.S. Green Building Council	W.R.	Water Resistant
THHN	Nylon Jacketed Wire	USP	United States Primed	Wrck.	Wrecker
THW.	Insulated Strand Wire	UTMCD	Uniform Traffic Manual For Control Devices	W.S.P.	Water, Steam, Petroleum
THWN	Nylon Jacketed Wire			WT., Wt.	Weight
TL., TL	Truckload	UTP	Unshielded Twisted Pair	WWF	Welded Wire Fabric
T.M.	Track Mounted	V	Volt	XFER	Transfer
Tot.	Total	VA	Volt Amperes	XFMR	Transformer
T-O-L	Threadolet	VAT	Vinyl Asbestos Tile	XHD	Extra Heavy Duty
tmpd	Tempered	V.C.T.	Vinyl Composition Tile	XHHW,	Cross-Linked Polyethylene Wire
TPO	Thermoplastic Polyolefin	VAV	Variable Air Volume	XLPE	Insulation
T.S.	Trigger Start	VC	Veneer Core	XLP	Cross-linked Polyethylene
Tr.	Trade	VDC	Volts Direct Current	Xport	Transport
Transf.	Transformer	Vent.	Ventilation	Y	Wye
Trhv.	Truck Driver, Heavy	Vert.	Vertical	yd	Yard
Trlr	Trailer	V.F.	Vinyl Faced	yr	Year
Trlt.	Truck Driver, Light	V.G.	Vertical Grain	Δ	Delta
TTY	Teletypewriter	VHF	Very High Frequency	%	Percent
TV	Television	VHO	Very High Output	~	Approximately
T.W.	Thermoplastic Water Resistant Wire	Vib.	Vibrating	Ø	Phase; diameter
		VLF	Vertical Linear Foot	@	At
UCI	Uniform Construction Index	VOC	Volitile Organic Compound	#	Pound; Number
UF	Underground Feeder	Vol.	Volume	<	Less Than
UGND	Underground Feeder	VRP	Vinyl Reinforced Polyester	>	Greater Than
UHF	Ultra High Frequency	W	Wire; Watt; Wide; West	Z	zone
U.I.	United Inch				

FORMS

RESIDENTIAL COST ESTIMATE

OWNER'S NAME: _____ APPRAISER: _____

RESIDENCE ADDRESS: _____ PROJECT: _____

CITY, STATE, ZIP CODE: _____ DATE: _____

CLASS OF CONSTRUCTION
- ☐ ECONOMY
- ☐ AVERAGE
- ☐ CUSTOM
- ☐ LUXURY

RESIDENCE TYPE
- ☐ 1 STORY
- ☐ 1-1/2 STORY
- ☐ 2 STORY
- ☐ 2-1/2 STORY
- ☐ 3 STORY
- ☐ BI-LEVEL
- ☐ TRI-LEVEL

CONFIGURATION
- ☐ DETACHED
- ☐ TOWN/ROW HOUSE
- ☐ SEMI-DETACHED

OCCUPANCY
- ☐ ONE STORY
- ☐ TWO FAMILY
- ☐ THREE FAMILY
- ☐ OTHER _____

EXTERIOR WALL SYSTEM
- ☐ WOOD SIDING—WOOD FRAME
- ☐ BRICK VENEER—WOOD FRAME
- ☐ STUCCO ON WOOD FRAME
- ☐ PAINTED CONCRETE BLOCK
- ☐ SOLID MASONRY (AVERAGE & CUSTOM)
- ☐ STONE VENEER—WOOD FRAME
- ☐ SOLID BRICK (LUXURY)
- ☐ SOLID STONE (LUXURY)

*LIVING AREA (Main Building)
- First Level _____ S.F.
- Second Level _____ S.F.
- Third Level _____ S.F.
- Total _____ S.F.

*LIVING AREA (Wing or Ell) ( )
- First Level _____ S.F.
- Second Level _____ S.F.
- Third Level _____ S.F.
- Total _____ S.F.

*LIVING AREA (Wing or Ell) ( )
- First Level _____ S.F.
- Second Level _____ S.F.
- Third Level _____ S.F.
- Total _____ S.F.

*Basement Area is not part of living area.

MAIN BUILDING	COSTS PER S.F. LIVING AREA
Cost per Square Foot of Living Area, from Page _____	$
Basement Addition: _____ % Finished, _____ % Unfinished	+
Roof Cover Adjustment: _____ Type, Page _____ (Add or Deduct)	( )
Central Air Conditioning: ☐ Separate Ducts ☐ Heating Ducts, Page _____	+
Heating System Adjustment: _____ Type, Page _____ (Add or Deduct)	( )
Main Building: Adjusted Cost per S.F. of Living Area	$

MAIN BUILDING TOTAL COST: $ _____ /S.F. (Cost per S.F. Living Area) × _____ S.F. (Living Area) × _____ (Town/Row House Multiplier, Use 1 for Detached) = $ _____ TOTAL COST

WING OR ELL ( ) _____ STORY	COSTS PER S.F. LIVING AREA
Cost per Square Foot of Living Area, from Page _____	$
Basement Addition: _____ % Finished, _____ % Unfinished	+
Roof Cover Adjustment: _____ Type, Page _____ (Add or Deduct)	( )
Central Air Conditioning: ☐ Separate Ducts ☐ Heating Ducts, Page _____	+
Heating System Adjustment: _____ Type, Page _____ (Add or Deduct)	( )
Wing or Ell ( ): Adjusted Cost per S.F. of Living Area	$

WING OR ELL ( ) TOTAL COST: $ _____ /S.F. (Cost per S.F. Living Area) × _____ S.F. (Living Area) = $ _____ TOTAL COST

WING OR ELL ( ) _____ STORY	COSTS PER S.F. LIVING AREA
Cost per Square Foot of Living Area, from Page _____	$
Basement Addition: _____ % Finished, _____ % Unfinished	+
Roof Cover Adjustment: _____ Type, Page _____ (Add or Deduct)	( )
Central Air Conditioning: ☐ Separate Ducts ☐ Heating Ducts, Page _____	+
Heating System Adjustment: _____ Type, Page _____ (Add or Deduct)	( )
Wing or Ell ( ): Adjusted Cost per S.F. of Living Area	$

WING OR ELL ( ) TOTAL COST: $ _____ /S.F. (Cost per S.F. Living Area) × _____ S.F. (Living Area) = $ _____ TOTAL COST

TOTAL THIS PAGE: _____

FORMS

RESIDENTIAL COST ESTIMATE

		QUANTITY	UNIT COST	
Total Page 1				$
Additional Bathrooms: _____ Full, _____ Half				
Finished Attic: _____ Ft. x _____ Ft.		S.F.		+
Breezeway: ☐ Open ☐ Enclosed _____ Ft. x _____ Ft.		S.F.		+
Covered Porch: ☐ Open ☐ Enclosed _____ Ft. x _____ Ft.		S.F.		+
Fireplace: ☐ Interior Chimney ☐ Exterior Chimney ☐ No. of Flues ☐ Additional Fireplaces				+
Appliances:				+
Kitchen Cabinets Adjustment: (+/−)				
☐ Garage ☐ Carport: _____ Car(s) Description _____ (+/−)				
Miscellaneous:				+

ADJUSTED TOTAL BUILDING COST $ _____

REPLACEMENT COST

ADJUSTED TOTAL BUILDING COST	$ _____
Site Improvements	
(A) Paving & Sidewalks	$ _____
(B) Landscaping	$ _____
(C) Fences	$ _____
(D) Swimming Pool	$ _____
(E) Miscellaneous	$ _____
TOTAL	$ _____
Location Factor	x _____
Location Replacement Cost	$ _____
Depreciation	−$ _____
LOCAL DEPRECIATED COST	$ _____

INSURANCE COST

ADJUSTED TOTAL BUILDING COST	$ _____
Insurance Exclusions	
(A) Footings, Site Work, Underground Piping	−$ _____
(B) Architects' Fees	−$ _____
Total Building Cost Less Exclusion	$ _____
Location Factor	x _____
LOCAL INSURABLE REPLACEMENT COST	$ _____

SKETCH AND ADDITIONAL CALCULATIONS

Index

A

Abandon catch basin 300
ABC extinguisher portable 523
Aboveground storage tank
   concrete 582
Abrasive floor tile 493
   tread 493
ABS DWV pipe 564
Absorber shock 569
A/C coils furnace 590
   packaged terminal 592
Accelerator sprinkler system 555
Access doors and panels 456
   door basement 318
Accessories anchor bolt 311
   bathroom 492, 520
   door 477
   drainage 437
   drywall 491
   duct 584
   fireplace 521
   gypsum board 491
   plaster 486
   toilet 520
Accordion door 452
Acid etch finish concrete 317
   proof floor 497
Acoustic ceiling board 494
   ceiling tile suspended 494
Acoustical ceiling 494, 495
   sealant 442, 490
Acrylic carpet 228
   wall coating 517
   wallcovering 502
   wood block 498
Adhesive cement 499
   roof 431
   wallpaper 502
Adobe brick 329
Aerial lift 665
   lift truck 670
   seeding 646
Aggregate exposed 317
   spreader 662, 669
   stone 646
Aid staging 295
Air cleaner electronic 587
   compressor 665
   compressor portable 665
   compressor sprinkler system .. 555
   conditioner cooling &
    heating 592
   conditioner fan coil 593
   conditioner packaged
    terminal 592
   conditioner portable 592
   conditioner receptacle 604
   conditioner removal 580
   conditioner rooftop 592
   conditioner self-contained .. 592
   conditioner thru-wall 592
   conditioner window 592
   conditioner wiring 607
   conditioning 268, 270
   conditioning ventilating .. 585, 592
   cooled condensing unit 591
   filter 586, 587
   filter mechanical media 587
   filter permanent washable .. 587
   filter roll type 587
   filter washable 587
   filtration charcoal type 587
   handler heater 591
   handler modular 591
   hose 666
   quality 529
   return grille 586
   spade 666
   supply register 586
Air-handling unit 593
Air-source heat pumps 593
Alarm fire 554
   panels and devices 618
   residential 606
   standpipe 554
   valve sprinkler 550
   water motor 554
Alteration fee 292
Alternative fastening 422
Aluminum ceiling tile 495
   chain link 640
   column 407
   coping 333
   diffuser perforated 585
   downspout 436
   drip edge 438
   ductwork 583
   edging 648
   flagpole 526
   flashing 211, 434
   foil 420, 423
   gravel stop 435
   grille 586
   gutter 437
   louver 480
   nail 361
   plank scaffold 295
   rivet 338
   roof 427
   roof panel 427
   screen/storm door 446
   service entrance cable 599
   sheet metal 436
   siding 166, 427
   siding accessories 427
   siding paint 507
   sliding door 456
   storm door 446
   threshold 474
   threshold ramp 475
   tile 494
   window 458
   window demolition 445
Anchor bolt 311, 324
   bolt accessories 311
   bolt sleeve 311
   brick 324
   buck 324
   chemical 319, 336
   epoxy 319, 336
   expansion 336
   framing 362
   hollow wall 336
   lead screw 336
   masonry 324
   metal nailing 336
   nailing 336
   nylon nailing 336
   partition 324
   plastic screw 337
   rafter 362
   rigid 324
   roof safety 295
   screw 336
   sill 362
   steel 324
   stone 324
   toggle bolt 336
Antenna T.V. 616
Anti-siphon device 568
Apartment hardware 472
Appliance 528
   residential 528, 606
Arch laminated 388
   radial 388
Architectural fee 292, 715
Area window well 525
Armored cable 599
Arrester lightning 611
Ashlar stone 331
Asphalt base sheet 431
   block 638
   block floor 638
   coating 417
   curb 639
   distributor 666
   felt 431
   flood coat 431
   paver 667
   paving 637
   plant portable 669
   primer 499
   roof shingle 424
   sheathing 386
   shingle 424
   sidewalk 636
   tile 229
Asphaltic cover board 421
   emulsion 644
Astragal 476
   molding 401
   one piece 476
Athletic carpet indoor 501
Attachment ripper 669
Attachments skid steer 665
Attic stair 530
   ventilation fan 585
Auger earth 662
   hole 300
Autoclave aerated concrete
   block 327
Automatic transfer switch .. 611
   washing machine 529
Auto-scrub machine 666
Awning canvas 525
   window 174, 460
   window metal-clad 463
   window vinyl-clad 464

B

Backer rod 440
Backerboard cementitious ... 488
Backfill 622, 623
   planting pit 645
   trench 622
Backflow preventer 568
Backhoe 622, 663
   bucket 664
   extension 669
Backsplash countertop 540
Backup block 327
Baffle roof 439
Baked enamel frame 447
Ball valve slow close 553
   wrecking 669
Baluster 231
   birch 231
   pine 231
   wood stair 404
Balustrade painting 514
Bamboo flooring 496
Band joist framing 345
Bank run gravel 620
Bar grab 520
   masonry reinforcing ... 325
   towel 520
   Zee 491
Barge construction 672
Bark mulch 644
   mulch redwood 644
Barrels flasher 666
   reflectorized 666
Barricade flasher 666
Barriers and enclosures .. 296
   dust 296
   slipform pavers 670
   weather 423
Base cabinet 536, 537
   carpet 501
   ceramic tile 492
   column 362
   course 636, 637
   course drainage layers .. 636
   gravel 636
   molding 391
   quarry tile 493
   resilient 498
   sheet 431
   sheet asphalt 431
   sink 536
   stabilizer 670
   stone 331
   terrazzo 500
   vanity 539
   wall 498
   wood 391
Baseboard demolition ... 360
   heat 593
   heater electric 595
   heating 279
   register 586
Basement bulkhead stair .. 318
   stair 231
Basement/utility window . 459
Baseplate scaffold 294
   shoring 294
Basic finish materials ... 482
   meter device 608
Basket weave fence vinyl .. 640
Basketweave fence 641
Bath accessory 237
   spa 545
   steam 545
   whirlpool 545
Bathroom ... 248, 250, 252, 254, 256,
    258, 260, 262, 264, 266, 574
   accessories 492, 520
   exhaust fan 584
   faucet 575
   fixture 573, 574
Bathtub enclosure 521
   removal 558
   residential 574
Batt insulation 168, 419
Bay window 178, 460
   wood window double hung . 464
Bead casing 491
   corner 486
   parting 402
Beam & girder framing .. 371
   bond 324
   bondcrete 487
   ceiling 407
   decorative 407
   drywall 488
   hanger 362
   laminated 388
   load-bearing stud wall box .. 342
   mantel 407
   plaster 486
   steel 340

740

For customer support on your Residential Cost Data, call 877.713.2617.

Index

wood 371, 383
Bed molding 394
Bedding brick 638
   pipe 623
Belgian block 639
Bell & spigot pipe 569
Bench 644
   park 644
   player 644
Berm pavement 639
   road 639
Bevel siding 428
Bi-fold door 226, 448, 451
Bin retaining walls metal 643
Bi-passing closet door 452
   door 226
Birch door 451, 452, 455
   molding 401
   paneling 403
   wood frame 408
Bituminous block 638
   coating 417
   concrete curbs 639
   paver 667
Bituminous-stabilized base
course 637
Blanket insulation 580
   insulation wall 419
Blaster shot 668
Blind 191
   exterior 409
   vinyl 409, 534
   window 534
   wood 535
Block asphalt 638
   autoclave concrete 327
   backup 327
   belgian 639
   bituminous 638
Block, brick and mortar 723
Block concrete 327-329, 724
   concrete bond beam 327
   concrete exterior 328
   filler 515
   floor 498
   glass 329
   insulation 327
   insulation insert concrete 327
   lightweight 329
   lintel 328
   partition 329
   profile 328
   reflective 329
   split rib 328
   wall 158
   wall removal 302
Blocking 370
   carpentry 370
   wood 374
Blower insulation 667
Blown-in cellulose 420
   fiberglass 420
   insulation 420
Blueboard 487
Bluegrass sod 646
Bluestone sidewalk 636
   sill 332
   step 636
Board & batten siding 428
   batten fence 642
   ceiling 494
   fence 641
   gypsum 488
   insulation 418
   insulation foam 418
   paneling 404

sheathing 385
siding 162
siding wood 428
valance 537
verge 400
Boat work 672
Boiler 588
   demolition 580
   electric 588
   electric steam 588
   gas-fired 588
   gas/oil combination 588
   hot water 588
   oil-fired 588
   removal 580
   steam 588
Bollard 339
   light 613
Bolt & hex nut steel 337
   anchor 311, 324
   steel 362
Bond beam 324
Bondcrete 487
Bookcase 524, 537
Boom truck 670
Borer horizontal 669
Boring and exploratory drilling .. 300
   cased 300
   machine horizontal 662
Borrow 620
Boundary and survey marker 300
Bow & bay metal-clad window .. 464
   window 178, 460, 464
Bowl toilet 573
Bowstring truss 388
Box & wiring device 601
   beam load-bearing stud wall .. 342
   distribution 653
   electrical pull 601
   fence shadow 641
   flow leveler distribution 653
   mail 523
   pull 601
   stair 404
   trench 668
   utility 650
Boxed headers/beams 346
   header/beam framing 346
Bracing 370
   framing 348
   let-in 370
   load-bearing stud wall 341
   metal 370
   metal joist 344
   roof rafter 348
   shoring 294
   support seismic 559
   wood 370
Bracket scaffold 294
   sidewalk 295
Brass hinge 475
   screw 361
Break ties concrete 317
Breaker circuit 610
   pavement 665
   vacuum 568
Brick adobe 329
   anchor 324
   bedding 638
Brick, block and mortar 723
Brick cart 666
   chimney simulated 521
   cleaning 322
   common building 327
   demolition 323, 482
   edging 648

face 327, 723
flooring miscellaneous 497
manhole 654
molding 398, 401
oversized 326
paving 638
removal 301
sidewalk 638
sill 332
stair 330
step 636
veneer 160, 325
veneer demolition 323
veneer thin 325
wall 160, 330
wash 322
Bricklaying 723
Bridging 371
   framing 348
   load-bearing stud wall 341
   metal joist 345
   roof rafter 348
Bronze swing check valve 558
   valve 558
Broom cabinet 537
   finish concrete 317
   sidewalk 668
Brownstone 331
Brush chipper 663
   clearing 620
   cutter 663, 664
Buck anchor 324
   rough 373
Bucket backhoe 664
   clamshell 663
   concrete 662
   dragline 663
   excavator 669
   pavement 669
Buggy concrete 316, 662
Builder's risk insurance 717
Building demo footing &
foundation 302
   demolition 301
   demolition selective 302
   excavation 102, 104
   greenhouse 546
   hardware 472
   insulation 420
   paper 423
   permit 293
   prefabricated 546
Built-in range 528
Built-up roof 431
   roofing 210
   roofing system 431
Bulk bank measure excavating .. 622
Bulkhead door 456
Bulkhead/cellar door 456
Bulldozer 622, 664
Bullfloat concrete 662
Bumper door 473
   wall 473
Buncher feller 663
Burlap curing 318
   rubbing concrete 317
Burner gas conversion 589
   gun type 589
   residential 589
Bush hammer 317
   hammer finish concrete 317
Butt fusion machine 666
Butterfly valve 650
Butyl caulking 441
   flashing 435
   rubber 440

C

Cabinet 234
   & casework paint/coating ... 510
   base 536, 537
   broom 537
   casework 538
   corner base 537
   corner wall 537
   demolition 360
   door 536, 538
   filler 537
   hardboard 536
   hardware 539
   hinge 539
   kitchen 536
   medicine 520
   outdoor 540
   oven 537
   shower 520
   stain 510
   varnish 510
   wall 537
Cable armored 599
   copper 599
   electric 599
   electrical 600
   jack 671
   pulling 668
   sheathed non-metallic 599
   sheathed Romex 599
   tensioning 668
   trailer 668
Cafe door 454
Canopy door 524
   framing 377
   residential 524
   wood 377
Cant roof 376, 431
Cantilever retaining wall 642
Canvas awning 525
Cap dowel 312
   post 362
   service entrance 599
   vent 571
Car tram 668
Carpet 228
   base 501
   felt pad 501
   floor 501
   nylon 501
   olefin 501
   pad 500
   pad commercial grade 500
   padding 500
   removal 482
   sheet 501
   stair 501
   tile 501
   tile removal 482
   urethane pad 501
   wool 501
Carrier ceiling 485
   channel 495
Carrier/support fixture 577
Cart brick 666
   concrete 316, 662
Cased boring 300
   evaporator coil 590
Casement window 172, 461, 464
   window metal-clad 463
   window vinyl 469
Casework 537
   cabinet 538
   custom 536
   demolition 360

Index

door manufactured hardwood . 538
door plastic laminate 535
frame wood 537
ground 381
painting 510
stain 510
varnish 510
Casing bead 491
molding 392
Cast-in-place concrete 314, 315
in place concrete curb 639
iron bench 644
iron damper 522
iron fitting 569
iron pipe 569
iron pipe fitting 569
iron radiator 593
iron trap 570
iron vent caps 571
trim lock 473
Caster scaffold 294
Catch basin 654
basin cleaner 669
basin precast 654
basin removal 300
basin vacuum 669
door 539
Caulking 442
polyurethane 441
sealant 441
Cavity wall insulation 420
wall reinforcing 325
Cedar closet 403
fence 641, 642
paneling 404
post 407
roof deck 384
roof plank 384
shake siding 164
shingle 424
siding 428
Ceiling 216, 220, 494
acoustical 494, 495
beam 407
board 494
board acoustic 494
board fiberglass 494
bondcrete 487
carrier 485
demolition 482
diffuser 585
drywall 489
framing 372
furring 380, 483, 485
heater 529
insulation 418, 420
lath 486
luminous 614
molding 394
painting 514, 515
plaster 486
stair 530
support 339
suspended 485, 495
suspension system 495
tile 495
Cellar door 456
wine 530
Cellulose blown-in 420
insulation 420
Cement adhesive 499
flashing 416
grout 324
masonry 324
masonry unit 327-329
mortar 493, 494, 722

parging 417
terrazzo Portland 500
Cementitious backerboard 488
Central vacuum 531
Centrifugal pump 667
Ceramic mulch 644
tile 229, 492
tile countertop 540
tile demolition 482
tile floor 492
tile waterproofing membrane .. 494
tiling 492, 494
Chain fall hoist 671
link fence 114, 640
link fence paint 505
saw 668
trencher 622, 665
Chair lifts 548
molding 401
rail demolition 360
Chamber infiltration 653
Chamfer strip 310
Channel carrier 495
furring 483, 491
siding 428
steel 486
Charcoal type air filter 587
Charge powder 338
Check valve 550
Chemical anchor 319, 336
dry extinguisher 523
spreader 669
termite control 633
toilet 668
Cherry threshold 474
Chime door 610
Chimney 326
accessories 521
brick 326
demolition 323
fireplace 521
flue 332
foundation 314
metal 521
screen 521
simulated brick 521
vent 587
Chip wood 645
Chipper brush 663
log 664
stump 664
Chipping hammer 666
Chloride accelerator concrete ... 316
Chlorination system 577
Circline fixture 612
Circuit wiring 280
Circuit breaker 610
Circular saw 668
Circulating pump 582
Cladding sheet metal 434
Clamp water pipe ground 600
Clamshell bucket 663
Clapboard painting 506
Clay tile 425
tile coping 333
Clean control joint 311
Cleaner catch basin 669
crack 669
steam 668
Cleaning brick 322
face brick 722
masonry 322
up 297
Cleanout door 522
floor 560
pipe 560

tee 560
Clear and grub site 620
grub 620
Clearing brush 620
selective 620
site 620
Climbing crane 670
hydraulic jack 671
Clip plywood 362
Clock timer 606
Closet cedar 403
door 226, 448, 452
pole 401
rod 524
water 573, 576
CMU 327
pointing 322
Coal tar pitch 431
Coat glaze 515
Coated roofing foamed 432
Coating bituminous 417
roof 416
rubber 418
silicone 418
spray 417
trowel 417
wall 517
water repellent 418
waterproofing 417
Cold mix paver 669
Collection concrete pipe sewage . 652
sewage/drainage 653
Colonial door 451, 455
wood frame 408
Column 407
accessories formwork 310
aluminum 407
base 362
bondcrete 487
brick 326
demolition 323
drywall 488
formwork 308
framing 372
laminated wood 388
lath 486
lightweight 339
plaster 486
plywood formwork 308
removal 323
round fiber tube formwork ... 308
steel 340
steel framed formwork 308
structural 339
wood 372, 383, 407
Combination device 603
storm door 451, 455
Commercial lavatory 577
toilet accessories 520
Common brick 327
building brick 327
nail 361
rafter roof framing 138
Communicating lockset 473
Compact fill 623
Compaction 632
soil 622
structural 632
Compactor earth 663
landfill 664
residential 529
Compensation workers' 293
Component control 581
furnace 590
Composite decking 375, 412
decking woodgrained 412

door 454
fabrication 412
insulation 422
rafter 376
railing 412
railing encased 412
Compression seals 440
Compressor air 665
Concrete acid etch finish 317
block 327-329, 724
block autoclave 327
block back-up 327
block bond beam 327
block decorative 328
block demolition 303
block exterior 328
block foundation 328
block grout 324
block insulation 420
block insulation insert 327
block lintel 328
block painting 515
block partition 329
block wall 158
break tie 317
broom finish 317
bucket 662
buggy 316, 662
bullfloat 662
burlap rubbing 317
bush hammer finish 317
cart 316, 662
cast-in-place 314, 315
chloride accelerator 316
conveyer 662
coping 333
curb 639
curing 318
cutout 302
demolition 302
distribution box 653
drilling 319
equipment rental 662
fiber reinforcing 316
float finish 317
floor edger 662
floor grinder 662
floor patching 308
floor staining 496
foundation 122
furring 380
grout 324
hand mix 315
hand trowel finish 317
impact drilling 319
lance 667
machine trowel finish 317
membrane curing 318
mixer 662
non-chloride accelerator ... 316
paver 667
paving surface treatment ... 638
pipe 653
placement footing 316
placement slab 316
placement wall 316
placing 316
plantable pavers precast ... 638
pump 662
ready mix 315
reinforcing 312
removal 300
retaining wall 642
retaining wall segmental ... 642
retarder 316
saw 662

Index

septic tank 652
shingle 426
short load 316
sidewalk 636
slab 126
slab conduit 601
small quantity 315
spreader 667
stair 315
stair finish 317
tank aboveground storage 582
tile 426
trowel 662
truck 662
truck holding 316
utility vault 650
vibrator 662
volumetric mixed 315
water reducer 316
wheeling 316
winter 316
Condensing unit air cooled 591
Conditioner portable air 592
Conditioning ventilating air 592
Conductive floor 500
 rubber & vinyl flooring 500
Conductor 599
 & grounding 599
Conduit & fitting flexible 602
 concrete slab 601
 electrical 600
 fitting 601
 flexible metallic 601
 Greenfield 601
 in slab PVC 601
 in trench electrical 602
 in trench steel 602
 PVC 600
 rigid in slab 601
 rigid steel 600
Connection fire-department 552
 motor 602
 split system tubing 593
 standpipe 552
Connector joist 362
 timber 361
Construction barge 672
 management fee 292
 temporary 296
 time 292
Containment area lead 304
Continuous hinge 475
Control component 581
 damper volume 584
 joint 311
 joint clean 311
 joint sawn 311
 pressure sprinkler 555
Cooking equipment 528
 range 528
Cooling 268, 270
Coping 331, 333
 aluminum 333
 clay tile 333
 concrete 333
 removal 323
 terra cotta 333
Copper cable 599
 downspout 436
 drum trap 570
 DWV tubing 561
 fitting 562
 flashing 434, 571
 gravel stop 435
 gutter 437
 pipe 561

rivet 338
roof 433
tube fittings 562
wall covering 502
wire 599
Core drill 662
Cork floor 499
 tile 499
 wall tile 502
Corner base cabinet 537
 bead 486
 wall cabinet 537
Cornerboard PVC 411
Cornice 330
 molding 394, 397
 painting 514
Corrugated metal pipe 654
 roof tile 425
 siding 427, 428
 steel pipe 654
Counter top 540
 top demolition 360
Countertop backsplash 540
 ceramic tile 540
 engineered stone 541
 granite 541
 laminated 540
 maple 540
 marble 540
 plastic-laminate-clad 540
 postformed 540
 sink 574
 solid surface 541
 stainless steel 540
Coupling plastic 567
Course drainage layers base 636
Cove base ceramic tile 492
 base terrazzo 500
 molding 394
Cover board asphaltic 421
 board gypsum 421
 pool 544
Covering wall 730
CPVC pipe 564
Crack cleaner 669
 filler 669
Crane climbing 670
 crawler 670
 crew daily 294
 hydraulic 671
 tower 670
 truck mounted 670
Crawler crane 670
 drill rotary 666
 shovel 664
Crew daily crane 294
 forklift 293
 survey 297
Crown molding 394, 401
Crushed stone 620
 stone sidewalk 636
Cubicle shower 575
Cultured stone 333
Cupola 525
Curb 639
 and gutter 639
 asphalt 639
 builder 662
 concrete 639
 edging 639
 extruder 663
 granite 639
 inlet 639
 precast 639
 removal 300
 roof 377

slipform paver 669
 terrazzo 500
Curing concrete 318
 concrete membrane 318
 concrete water 318
 paper 423
Curtain damper fire 584
 rod 520
Custom casework 536
Cut stone curb 639
 stone trim 331
Cutout counter 540
 demolition 302
 slab 302
Cutter brush 663, 664
Cutting block 540
 torch 668
Cylinder lockset 473

D

Daily crane crew 294
Damper fire curtain 584
 fireplace 522
 multi-blade 584
Dead lock 473
Deadbolt 474
Deciduous shrub 647
Deck framing 374, 410
 roof 384, 386
 slab form 340
 wood 244, 384
Decking 724
 composite 375, 412
 form 340
 log roof 383
 roof 340
 woodgrained composite 412
Decorative beam 407
Decorator device 603
 switch 603
Deep freeze 528
Dehumidifier 666
Dehydrator package sprinkler 555
Delivery charge 296
Demo footing & foundation
building 302
 thermal & moisture protection 416
Demolish decking 357
 remove pavement and curb 300
Demolition . 302, 323, 482, 558, 620, 721
 baseboard 360
 boiler 580
 brick 323, 482
 brick veneer 323
 cabinet 360
 casework 360
 ceiling 482
 ceramic tile 482
 chimney 323
 column 323
 concrete 302
 concrete block 303
 door 444
 drywall 483
 ductwork 580
 electrical 598
 fencing 301
 fireplace 323
 flooring 482
 framing 357
 furnace 580
 glass 445
 granite 323

gutter 416
gutting 304
hammer 663
HVAC 580
joist 357
masonry 302, 323
metal stud 483
millwork 360
millwork & trim 360
paneling 360
partition 483
pavement 300
plaster 482, 483
plenum 483
plumbing 558
plywood 482, 483
post 358
rafter 358
railing 360
roofing 416
selective building 302
site 300
tile 482
truss 360
walls and partitions 483
window 445
window & door 444
wood 482
wood framing 356
Dentil crown molding 395
Derrick crane guyed 671
 crane stiffleg 671
Detection intrusion 618
 system 618
Detector infrared 618
 motion 618
 smoke 618
Device anti-siphon 568
 combination 603
 decorator 603
 GFI 604
 receptacle 604
 residential 602
 wiring 609
Dewatering equipment 671
Diaphragm pump 667
Diesel hammer 664
 tugboat 672
Diffuser ceiling 585
 linear 585
 opposed blade damper 585
 perforated aluminum 585
 rectangular 585
 steel 586
 T-bar mount 586
Dimmer switch 602, 609
Directional drill horizontal 669
Disappearing stairs 530
Disc harrow 663
Discharge hose 667
Disconnect safety switch 610
Dishwasher residential 528
Disposal 302
 field 652
 garbage 529
Disposer garbage 236
Distribution box 653
 box concrete 653
 box flow leveler 653
 box HDPE 653
 pipe water 650, 651
Distributor asphalt 666
Diving board 544
 stand 544
Domed skylights 471

743

Index

Door & window interior
paint 511, 512
   accessories 477
   accordion 452
   and panel access 456
   and window materials 444
   bell residential 605
   bi-fold 448, 451
   bi-passing closet 452
   birch 451, 452, 455
   blind 409
   bulkhead 456
   bulkhead/cellar 456
   bumper 473
   cabinet 536, 538
   cafe 454
   canopy 524
   casing PVC 411
   catch 539
   chime 610
   cleanout 522
   closet 448, 452
   colonial 451, 455
   combination storm 451, 455
   composite 454
   decorator wood 450
   demolition 444
   double 447
   dutch 451
   dutch oven 522
   embossed metal 455
   entrance 451, 455
   exterior 182, 454
   exterior prehung 455
   fiberglass 454
   fireplace 522
   flush 449, 451
   frame 446
   frame exterior wood 408
   frame interior 408
   french 455
   french exterior 454
   garage 186, 457
   glass 457
   hand carved 448
   handle 539
   hardboard 451
   hardware 472
   hardware accessory 475
   interior pre-hung 455
   jamb molding 402
   kennel 456
   knob 473
   knocker 472, 475
   labeled 447
   louvered 453, 456
   mahogany 449
   manufactured hardwood
casework 538
   metal 446
   metal faced 455
   metal fire 447
   mirror 479
   molding 402
   moulded 450
   opener 458
   overhead 457
   panel 450, 451
   paneled 447
   passage 452, 455
   patio 457
   plastic laminate 449
   plastic laminate casework . 535
   porch 451
   prefinished 449
   pre-hung 454, 455
   removal 445
   residential 408, 447, 451
   residential garage 457
   residential steel 447
   residential storm 446
   rough buck 373
   sauna 545
   sectional 457
   sectional overhead 457
   shower 520
   sidelight 448, 454
   silencer 475
   sill 402, 408, 474
   sliding aluminum 456
   sliding glass 456, 457
   sliding glass vinyl-clad .. 457
   sliding vinyl clad 457
   sliding wood 457
   special 456
   stain 510
   steel 447, 726
   stop 473
   storm 190, 446
   switch burglar alarm 618
   threshold 408
   torrified 453
   varnish 510
   weatherstrip 476
   weatherstrip garage 476
   wood 224, 226, 448, 449
   wood panel 450
   wood storm 455
Doorbell system 610
Dormer framing 150, 152
   gable 380
   roofing 204, 206
Double hung bay wood
window 464
   hung window 465
   hung wood window 462
Double-hung window 170
Dowel cap 312
   reinforcing 312
   sleeve 312
Downspout 436, 437
   aluminum 436
   copper 436
   demolition 416
   elbow 437
   lead coated copper 436
   steel 436
   strainer 436
Dozer 664
Dragline bucket 663
Drain 570
   floor 571
   roof 572
   sanitary 571
   shower 570
Drainage accessories 437
   field 653
   pipe 652, 653
   trap 570
Drapery hardware 534
   ring 534
Drawer front plastic laminate . 536
   front wood 539
   track 539
   wood 539
Drill core 662
   main 651
   quarry 666
   rig 300
   steel 666
   tap main 651
   track 666
   wood 362, 643
Drilling concrete 319
   concrete impact 319
   plaster 483
   rig 662
   steel 337
Drinking fountain support ... 577
Drip edge 438
   edge aluminum 438
Driver post 669
   sheeting 666
Driveway 110, 636
   removal 300
Drum trap copper 570
Dry fall painting 516
   pipe sprinkler head 555
Dryer accessories washer 576
   receptacle 605
   vent 529, 576
   wiring 280
Dry-pipe sprinkler system ... 555
Drywall 214, 216, 488
   accessories 491
   column 488
   cutout 303
   demolition 483
   finish 729
   frame 446
   gypsum 488
   nail 361
   painting 514, 516
   prefinished 489
   screw 491
Duck tarpaulin 296
Duct accessories 584
   fire rated blanket 580
   flexible 584
   flexible insulated 584
   flexible non-insulated 584
   HVAC 583
   insulation 580
   liner 584
   liner non-fibrous 584
   mechanical 583
   thermal insulation 580
Ductile iron fitting 650
   iron fitting mech joint ... 650
   iron pipe 650
Ductless split system 592
Ductwork aluminum 583
   demolition 580
   fabric coated flexible 584
   fabricated 583
   galvanized 583
   metal 583
   rectangular 583
   rigid 583
Dump charges 304
   truck 665
Dumpster 303, 721
Dumptruck off highway 665
Duplex receptacle 280, 609
Dust barrier 296
Dutch door 451
   oven door 522
DWV pipe ABS 564
   PVC pipe 564, 652
   tubing copper 561

E

Earth auger 662
   compactor 663
   scraper 664
   vibrator 632
Earthwork 621
   equipment 622
   equipment rental 662
Eave vent 439
Edge drip 438
Edger concrete floor 662
Edging 648
   aluminum 648
   curb 639
Effluent filter septic system . 653
EIFS 423
Elbow downspout 437
Electric appliance 528
   baseboard heater 595
   boiler 588
   cable 599
   fixture 611-613
   furnace 589
   generator 666
   heater 529
   heating 279, 595
   lamp 614
   log 522
   metallic tubing 600
   pool heater 588
   service 278, 610
   stair 530
   switch 609, 610
   vehicle charging 611
   water heater residential .. 572
Electrical cable 600
   conduit 600
   conduit Greenfield 601
   demolition 598
   wire 599
Electronic air cleaner 587
Elevator 548
   residential 548
Embossed metal door 455
   print door 450
Employer liability 293
EMT 600
Emulsion adhesive 499
   asphaltic 644
   pavement 636
   penetration 637
   sprayer 666
Encased railing 412
Enclosure bathtub 521
   shower 521
   swimming pool 546
Engineered stone countertop . 541
Entrance cable service 599
   door 451, 455
   door fiberous glass 454
   frame 408
   lock 474
   weather cap 601
EPDM flashing 435
Epoxy anchor 319, 336
   grout 492
   wall coating 517
   welded wire 312
Equipment 662
   earthwork 622
   insurance 293
   pad 314
   rental 662, 720
   rental concrete 662
   rental earthwork 662
   rental general 665
   rental highway 669
   rental lifting 670
   rental marine 672
   rental wellpoint 671
   swimming pool 544

Index

Entry	Page
Erosion control synthetic	633
Escutcheon plate	554
sprinkler	554
Estimate electrical heating	595
Estimates window	726
Estimating	713
Evap/condes unit line sets	593
Evaporator coils cased	590
Evergreen shrub	646
tree	646
Excavating bulk bank measure	622
trench	621
utility trench	622
Excavation	621, 622
footing	102
foundation	104
hand	621, 622
planting pit	645
septic tank	653
structural	622
tractor	621
utility	106
Excavator bucket	669
hydraulic	662
Exhaust hood	529
Expansion anchor	336
joint	311, 486
shield	336
tank	582
tank steel	582
Exposed aggregate	317
aggregate coating	517
Extension backhoe	669
ladder	667
Exterior blind	409
concrete block	328
door	182, 454
door frame	408
fixture lighting	613
insulation	423
insulation finish system	423
light	281
lighting fixture	613
molding	395
paint doors & windows	508
plaster	487
prehung door	455
PVC molding	411
residential door	451
shutter	409
siding painting	506
surface preparation	503
trim	395
trim paint	509
wall framing	136
wood door frame	408
wood frame	408
Extinguisher chemical dry	523
fire	523
portable ABC	523
standard	523
Extruder curb	663

F

Fabric welded wire	312
Fabricated ductwork	583
Fabrication composite	412
Face brick	327, 723
brick cleaning	722
Facing stone	331
Fall painting dry	516
Fan attic ventilation	585
bathroom exhaust	584
house ventilation	585
kitchen exhaust	584
paddle	606
residential	606
ventilation	606
wiring	280, 606
Fascia aluminum	428
board	376
board demolition	357
metal	435
PVC	411
wood	398
Fastener timber	361
wood	361
Fastening alternative	422
Faucet & fitting	575
bathroom	575
laundry	575
lavatory	575
Fee architectural	292
Feller buncher	663
Felt asphalt	431
carpet pad	501
Fence	505
and gate	640
board	641
board batten	642
cedar	641, 642
chain link	640
metal	640
misc metal	641
open rail	641
picket paint	505
residential	640
vinyl	640
wire	641
wood	641
wood rail	642
Fencing demolition	301
Fertilizer	645
Fiber cement siding	430
reinforcing concrete	316
steel	313
synthetic	313
Fiberboard insulation	421
Fiberglass area wall cap	525
bay window	471
blown-in	420
ceiling board	494
door	454
insulation	418-420
panel	427
planter	644
single hung window	470
slider window	471
tank	581
window	470
wool	420
Field disposal	652
drainage	653
seeding	645
Fieldstone	330
Fill	622, 623
by borrow & utility bedding	623
gravel	620, 623
Filler block	515
cabinet	537
crack	669
joint	440
Film polyethylene	644
security	479
Filter air	587
mechanical media	586
swimming pool	577
Filtration equipment	544
Fine grade	645
Finish drywall	488
floor	498, 516
grading	621
nail	361
wall	502
Finishing floor concrete	317
wall concrete	317
Fir column	407
floor	498
molding	401
roof deck	384
roof plank	384
Fire alarm	554
call pullbox	618
damage repair	505
damper curtain type	584
door frame	447
door metal	447
escape stair	351
extinguisher	523
extinguisher portable	523
extinguishing system	550, 553
horn	618
hose gate valve	554
hose rack	553
hose valve	553
hydrant building	552
rated blanket duct	580
resistant drywall	489, 490
sprinkler head	555
Fire-department connection	552
Fireplace accessories	521
box	332
built-in	522
chimney	521
damper	522
demolition	323
door	522
form	522
free standing	522
mantel	406
mantel beam	407
masonry	238, 332
prefabricated	240, 521, 522
Firestop gypsum	485
wood	373
Fire-suppression pipe fitting	550
plastic pipe	550
Fitting cast iron	569
cast iron pipe	569
conduit	601
copper	562
copper pipe	562
ductile iron	650
DWV pipe	566
malleable iron	563
plastic	566
plastic pipe	550
PVC	565
steel	563
Fixed window	180
Fixture bathroom	573, 574
carrier/support	577
electric	611
fluorescent	611, 612
incandescent	611
interior light	611
landscape	613
lantern	612
mirror light	612
plumbing	573, 582
removal	558
residential	605, 612
Flagging	497, 638
Flagpole	526
aluminum	526
Flasher barrels	666
barricade	666
Flashing	434
aluminum	434
butyl	435
cement	416
copper	434, 571
EPDM	435
laminated sheet	435
masonry	434
neoprene	435, 571
plastic sheet	435
PVC	435
sheet metal	434
stainless	434
vent	571
Flat seam sheet metal roofing	433
Flatbed truck	665, 668
truck crane	670
Flexible conduit & fitting	602
duct	584
ductwork fabric coated	584
insulated duct	584
metallic conduit	601
non-insulated duct	584
sprinkler connector	554
Float finish concrete	317
glass	478
Floater equipment	293
Floating floor	496
pin	475
Floodlight trailer	666
tripod	666
Floor	497
acid proof	497
asphalt block	638
brick	497
carpet	501
ceramic tile	492
cleanout	560
concrete finishing	317
conductive	500
cork	499
drain	571
finish	516
flagging	638
framing	130, 132, 134
framing log	382
framing removal	303
heating radiant	594
insulation	418
marble	331
nail	361
oak	498
paint	513
paint & coating interior	513
parquet	497
plank	383
plywood	384
polyethylene	501
quarry tile	493
register	586
removal	302
resilient	499
rubber	499
sander	668
stain	513
subfloor	384
tile terrazzo	500
transition strip	500
underlayment	385
varnish	513, 516
vinyl	499
vinyl sheet	499
wood	497, 498
wood composition	497
Flooring	229

Index

bamboo	496	
conductive rubber & vinyl	500	
demolition	482	
masonry	497	
miscellaneous brick	497	
treatment	496	
wood strip	498	
Flue chimney	332	
chimney metal	587	
liner	326, 332	
screen	521	
tile	332	
Fluid applied membrane air barrier	424	
Fluorescent fixture	611, 612	
Flush door	449, 451	
wood door	450	
Foam board insulation	418	
core DWV ABS pipe	564	
insulation	420	
roofing	432	
spray rig	666	
Foamed coated roofing	432	
in place insulation	419	
Fog seal	636	
Foil aluminum	420, 423	
metallic	420	
Folding door shower	521	
Footing	118	
concrete placement	316	
excavation	102	
formwork continuous	308	
keyway	308	
keyway formwork	308	
reinforcing	312	
removal	302	
spread	314	
Forest stewardship council	370	
Forklift	666	
crew	293	
Form decking	340	
fireplace	522	
Formwork column	308	
column accessories	310	
column plywood	308	
column round fiber tube	308	
column steel framed	308	
continuous footing	308	
footing keyway	308	
grade beam	308	
insert concrete	311	
insulating concrete	309	
slab blockout	309	
slab bulkhead	308	
slab curb	309	
slab edge	309	
slab haunch	314	
slab on grade	308	
slab thickened edge	314	
slab turndown	314	
slab void	309	
sleeve	310	
snap ties	310	
spread footing	308	
stair	310	
stake	310	
wall	309	
wall accessories	310	
wall boxout	309	
wall brick shelf	309	
wall bulkhead	309	
wall plywood	309	
wall prefab plywood	309	
Foundation	118	
chimney	314	
concrete	122	

concrete block	328	
excavation	104	
mat	314	
mat concrete placement	316	
removal	301	
wall	328	
wood	124	
Fountain outdoor	544	
yard	544	
Frame baked enamel	447	
door	446	
drywall	446	
entrance	408	
exterior wood door	408	
fire door	447	
labeled	447	
metal	446	
metal butt	447	
scaffold	294	
steel	446	
welded	447	
window	463	
wood	537	
Framing anchor	362	
band joist	345	
beam & girder	371	
boxed headers/beams	346	
bracing	348	
bridging	348	
canopy	377	
ceiling	372	
columns	372	
deck	374, 410	
demolition	357	
dormer	150, 152	
floor	130, 132, 134	
heavy	383	
joist	372	
laminated	388	
ledger & sill	377	
load-bearing stud partition	342	
load-bearing stud wall	342	
log floor	382	
log roof	383	
metal joist	346	
metal roof parapet	348	
miscellaneous wood	373	
NLB partition	484	
open web joist wood	372	
partition	154	
porch	374	
removal	356	
roof	138, 140, 142, 144, 146, 148	
roof metal rafter	349	
roof metal truss	350	
roof rafters	376	
roof soffits	349	
sill & ledger	377	
sleepers	377	
soffit & canopy	377	
timber	383	
treated lumber	377	
wall	378	
web stiffeners	347	
wood	370, 371, 383	
Freeze deep	528	
French door	455	
exterior door	454	
Frieze PVC	411	
Front end loader	622	
plastic laminate drawer	536	
Fuel furnace solid	589	
Furnace A/C coils	590	
components	590	
demolition	580	
electric	589	

gas-fired	589	
hot air	589	
oil-fired	589	
wiring	280	
wood	589	
Furring and lathing	483	
ceiling	380, 483, 485	
channel	483, 491	
concrete	380	
masonry	380	
metal	483	
steel	483	
wall	380, 484	
wood	380	
Fusion machine butt	666	

G

Gable dormer	380	
dormer framing	150	
dormer roofing	204	
roof framing	138	
Galley septic	653	
Galvanized ductwork	583	
welded wire	312	
Gambrel roof framing	144	
Garage door	186, 457	
door residential	457	
door weatherstrip	476	
Garbage disposal	529	
Gas conversion burner	589	
fired boiler	588	
fired furnace	589	
furnace wood	589	
generator	610	
generator set	610	
heat air conditioner	592	
heating	268, 272	
log	522	
pipe	654, 655	
service steel piping	655	
vent	587	
water heater instantaneous	572	
water heater residential	573	
water heater tankless	572	
Gasket joint	440	
neoprene	440	
Gas/oil combination boiler	588	
Gasoline generator	610	
Gate metal	640	
General contractor's overhead	717	
equipment rental	665	
fill	623	
Generator electric	666	
emergency	610	
gas	610	
gasoline	610	
set	610	
Geo-grid	643	
GFI receptacle	604	
Girder wood	371, 383	
Glass	478, 479	
block	329	
demolition	445	
door	457	
door shower	521	
door sliding	457	
float	478	
heat reflective	478	
lined water heater	529	
mirror	479	
mosaic	493	
mosaic sheet	493	
plate	478	
reduce heat transfer	478	

sheet	479	
shower stall	520	
tempered	478	
tile	479	
tinted	478	
window	479	
Glaze coat	515	
Glazed ceramic tile	492	
wall coating	517	
Glazing	478	
application	478	
Glued laminated	388	
Grab bar	520	
Gradall	663	
Grade beam formwork	308	
fine	645	
Grader motorized	663	
Grading	620	
and seeding	645	
slope	621	
Granite building	330	
chip	645	
countertop	541	
curb	639	
demolition	323	
indian	639	
paver	330	
sidewalk	639	
Grass cloth wallpaper	502	
lawn	645	
seed	645	
Grating area wall	525	
area way	525	
Gravel base	636	
fill	620, 623	
pea	645	
stop	435	
Gravity retaining wall	642	
Green roof membrane	426	
roof soil mixture	426	
roof system	426	
Greenfield conduit	601	
Greenhouse	242, 546	
residential	546	
Grid spike	362	
Grille air return	586	
aluminum	586	
decorative wood	406	
painting	513	
window	466	
Grinder concrete floor	662	
Ground	381	
clamp water pipe	600	
cover plant	646	
fault	280	
rod	600	
wire	600	
Grounding	600	
& conductor	599	
Grout	324	
cement	324	
concrete block	324	
epoxy	492	
pump	668	
tile	492	
wall	324	
Guard gutter	438	
snow	439	
Guardrail scaffold	294	
Gun type burner	589	
Gunite pool	243	
Gutter	437	
aluminum	437	
copper	437	
demolition	416	
guard	438	

Index

lead coated copper 437
  stainless 437
  steel 437
  strainer 438
  vinyl 437
  wood 437
Gutting 304
  demolition 304
Guyed derrick crane 671
Gypsum block demolition 303
  board 488
  board accessories 491
  cover board 421
  drywall 488
  fabric wallcovering 502
  firestop 485
  lath 485
  lath nail 361
  plaster 486
  restoration 482
  sheathing 387
  sound dampening panel 490
  underlayment poured 319
  wallboard repair 482
  weatherproof 387

H

Half round molding 401
  round vinyl clad window 465
  round vinyl window 470
Hammer bush 317
  chipping 666
  demolition 663
  diesel 664
  drill rotary 666
  hydraulic 663, 667
  pile 663
  vibratory 664
Hammermill 669
Hand carved door 448
  clearing 620
  excavation 621, 622
  hole 650
  split shake 425
  trowel finish concrete 317
Handicap lever 474
  ramp 314
Handle door 539
  latch 477
Handrail oak 231
  wood 401, 406
Hand-split shake 164
Hanger beam 362
  joist 362
  pipe 559
  plastic pipe 559
Hardboard cabinet 536
  door 451
  molding 403
  overhead door 457
  paneling 402
  tempered 402
  underlayment 385
Hardware 472
  apartment 472
  cabinet 539
  door 472
  drapery 534
  finishes 727
  motel/hotel 472
  window 472
Hardwood floor 498
  grille 406
Harrow disc 663

Hauling 623
  cycle 623
Haunch slab 314
Hay 644
HDPE distribution box 653
  infiltration chamber 653
Head sprinkler 554
Header load-bearing stud wall . 342
  pipe wellpoint 671
  wood 378
Headers/beams boxed 346
Hearth 332
Heat baseboard 593
  electric baseboard 595
  pump 593
  pump air-source 593
  pump residential 607
  reflective glass 478
  temporary 308
Heater air handler 591
  electric 529
  gas residential water 573
  infrared quartz 596
  quartz 596
  sauna 545
  space 667
  swimming pool 588
  water 529, 607
  wiring 280
Heating 268, 270, 272, 561, 588
  electric 279, 595
  estimate electrical 595
  hot air 589
  hydronic 588, 593
  kettle 670
  panel radiant 596
Heavy duty shoring 294
  framing 383
  timber 383
Helicopter 671
Hemlock column 407
HEX bolt steel 337
Highway equipment rental 669
  paver 638
Hinge 475, 476
  brass 475
  cabinet 539
  continuous 475
  mortised 476
  paumelle 475
  residential 475
  stainless steel 475
  steel 475
Hip rafter 376
  roof framing 142
Hoist chain fall 671
  personnel 671
  tower 671
Holdown 362
Hollow core door 449, 453
  metal frame 446
  wall anchor 336
Home theater 616
Hood exhaust 529
  range 529
Hook robe 520
Horizontal aluminum siding ... 427
  borer 669
  boring machine 662
  directional drill 669
Horn fire 618
Hose air 666
  bibb sillcock 575
  discharge 667
  rack 553
  suction 667

water 667
Hospital tip pin 475
Hot air furnace 589
  air heating 589
  tub 545
  water boiler 588
  water heating 588, 593
Hot air heating 268, 270
Hotel lockset 473
Hot water heating 272
House ventilation fan 585
Housewrap 423
Humidifier 530
Humus peat 644
HVAC demolition 580
  duct 583
Hydrant building fire 552
Hydraulic crane 671
  excavator 662
  hammer 663, 667
  jack 671
  jack climbing 671
  seeding 645
Hydromulcher 669
Hydronic heating 588, 593

I

Icemaker 528
I-joists wood & composite 387
Impact wrench 666
Incandescent fixture 611, 612
  light 281
Indian granite 639
Indoor athletic carpet 501
Infiltration chamber 653
  chamber HDPE 653
Infrared detector 618
  quartz heater 596
Inlet curb 639
Insecticide 633
Insert concrete formwork 311
Instantaneous gas water heater . 572
Insulated panel 423
  protector ADA 560
Insulating concrete formwork . 309
  sheathing 384
Insulation 418, 420
  batt 419
  blanket 580
  blower 667
  blown-in 420
  board 211, 418
  building 420
  cavity wall 420
  ceiling 418, 420
  cellulose 420
  composite 422
  duct 580
  duct thermal 580
  exterior 423
  fiberglass 418-420
  finish system exterior 423
  floor 418
  foam 420
  foamed in place 419
  insert concrete block 327
  isocyanurate 418
  loose fill 420
  masonry 420
  mineral fiber 419
  pipe 560
  polystyrene 418, 420
  reflective 420
  removal 416

rigid 418
  roof 421
  roof deck 421
  sprayed-on 421
  vapor barrier 423
  vermiculite 420
  wall 418
  wall blanket 419
  water heater 560
Insurance 293, 717
  builder risk 293
  equipment 293
  public liability 293
Interior door 224, 226
  door frame 408
  LED fixture 612
  light 281
  light fixture 611
  paint doors & windows 510
  paint wall & ceiling 515
  partition framing 154
  pre-hung door 455
  residential door 451
  shutter 535
  wood door frame 408
Interval timer 603
Intrusion detection 618
  system 618
Ironing center 521
Ironspot brick 497
Irrigation system sprinkler .. 644
Isocyanurate insulation 418

J

J hook clamp w/nail 559
Jack cable 671
  hydraulic 671
  rafter 376
Jackhammer 665
Joint control 311
  expansion 311, 486
  filler 440
  gaskets 440
  push-on 650
  reinforcing 325
  restraint 650, 651
  sealer 440, 441
Joist connector 362
  demolition 357
  framing 372
  hanger 362
  metal framing 346
  removal 357
  sister 372
  web stiffener 347
  wood 372, 387
Jute mesh 633

K

Kennel door 456
  fence 641
Kettle heating 670
  tar 668, 670
Keyway footing 308
Kick plate 476
Kitchen 234
  appliance 528
  cabinet 536
  exhaust fan 584
  sink 574
  sink faucet 575
  sink residential 574

Index

unit commercial 530
Knob door 473
Knocker door 472, 475

L

Labeled door 447
   frame 447
Ladder extension 667
   swimming pool 544
Lag screw 338
   screw shield 336
Laminated beam 388
   countertop 540
   framing 388
   glued 388
   roof deck 384
   sheet flashing 435
   veneer members 388
   wood 388
Lamp electric 614
   LED 614
   post 352
Lampholder 609
Lance concrete 667
Landfill compactor 664
Landscape fixture 613
   light 613
   surface 501
Lantern fixture 612
Laser level 667
Latch handle 477
   set 473
Latex caulking 441
   underlayment 498
Lath gypsum 485
Lath, plaster and gypsum board . 728
Lattice molding 401
Lauan door 452
Laundry faucet 575
   sink 575
Lavatory 248
   commercial 577
   faucet 575
   pedestal type 574
   removal 558
   residential 573, 577
   sink 573
   support 577
   vanity top 573
   wall hung 573
Lawn grass 645
   mower 667
   seed 645
Lazy Susan 537
Leaching pit 652
Lead coated copper downspout .. 436
   coated copper gutter 437
   coated downspout 436
   containment area 304
   flashing 434
   paint encapsulation 305
   paint removal 305
   roof 433
   screw anchor 336
Leads pile 663
Lean-to type greenhouse 546
LED fixture interior 612
   lamps 614
Ledger & sill framing 377
Let-in bracing 370
Letter slot 523
Level laser 667
Leveling jack shoring 294
Lever handicap 474

Liability employer 293
   insurance 717
Lift 548
   aerial 665
   scissor 665
   truck aerial 670
Lifting equipment rental 670
Light bollard 613
   fixture interior 611
   landscape 613
   post 612
   support 339
   tower 667
   track 613
Lighting 280, 612, 614
   exterior fixture 613
   incandescent 612
   outlet 605
   pole 613
   residential 605
   strip 611
   track 613
Lightning arrester 611
   protection 611
   suppressor 602
Lightweight block 329
   column 339
Limestone 330, 645
   coping 333
Line remover traffic 670
   set 593
   sets refrigerant 582
Linear diffuser 585
Linen wallcoverings 502
Liner duct 584
   flue 326, 332
   non-fibrous duct 584
Lintel 331
   block 328
   concrete block 328
   precast concrete 318
   steel 339
Load center indoor 607
   center plug-in breaker 607
   center rainproof 607
   center residential 602
Load-bearing stud partition
   framing 342
   stud wall bracing 341
   stud wall bridging 341
   stud wall framing 342
   stud wall header 342
Loader front end 622
   skid steer 665
   tractor 664
   wheeled 664
   windrow 670
Loam 620
Lock entrance 474
Locking receptacle 605
Lockset communicating 473
   cylinder 473
   hotel 473
   mortise 473
Locomotive tunnel 670
Log chipper 664
   electric 522
   gas 522
   skidder 664
   structures 382
   wall 382
Loose fill insulation 420
Louver 480
   aluminum 480
   midget 480
   redwood 480

ventilation 480
wall 480
wood 406
Louvered blind 409
   door 224, 226, 453, 456
Lowbed trailer 670
Low-voltage silicon rectifier .. 609
   switching 609
   switchplate 609
   transformer 609
Lumber 371
   core paneling 403
   plastic 410
   product prices 725
   recycled plastic 410
   structural plastic 410
Luminaire walkway 613
Luminous ceiling 614

M

Macadam 637
   penetration 637
Machine auto-scrub 666
   trowel finish concrete ... 317
   welding 669
Mahogany door 449
Mail box 523
Malleable iron fitting 563
   iron pipe fitting 563
Management fee construction . 292
Manhole brick 654
   removal 301
Manifold roof fire valve 552
Man-made soil mix 426
Mansard roof framing 146
Mantel beam 407
   fireplace 406
Maple countertop 540
Marble 331
   chips 645
   coping 333
   countertop 540
   floor 331
   sill 332
   synthetic 497
   tile 497
Marine equipment rental 672
Marker boundary and survey .. 300
Mason scaffold 294
Masonry anchor 324
   brick 326
   cement 324
   cleaning 322
   cornice 330
   demolition 302, 323
   fireplace 238, 332
   flashing 434
   flooring 497
   furring 380
   insulation 420
   nail 361
   painting 515
   pointing 322
   reinforcing 325, 722
   removal 301, 323
   saw 668
   selective demolition 322
   sill 332
   step 636
   toothing 303
   wall 158, 160, 327, 643
   wall tie 324
Mat concrete placement
   foundation 316

foundation 314
Material door and window 444
Mechanical duct 583
   media filter 586
   seeding 645
Medicine cabinet 520
Membrane 211
   roofing 431
Metal bin retaining wall 643
   bracing 370
   butt frame 447
   chimney 521
   clad window .. 170, 172, 174, 176,
                      178, 180
   door 446
   door residential 447
   ductwork 583
   faced door 455
   fascia 435
   fence 640
   fireplace 240
   flue chimney 587
   frame 446
   framing parapet 348
   furring 483
   gate 640
   halide fixture 281
   joist bracing 344
   joist bridging 345
   joist framing 346
   nailing anchor 336
   overhead door 457
   rafter framing 349
   roof 427
   roof parapet framing 348
   roof window 471
   sheet 433
   shelf 523
   siding 166
   soffit 431
   stud demolition 483
   stud NLB 484
   support assemblies 483
   threshold 474
   tile 494
   truss framing 350
   window 458
Metal-clad double hung
   window 463
   window bow & bay 464
   window picture & sliding . 463
Metallic foil 420
Meter center rainproof 608
   device basic 608
   socket 608
   water supply 568
   water supply domestic 568
Metric conversion factors ... 714
Microtunneling 650
Microwave oven 528
Mill construction 383
Millwork 391, 537
   & trim demolition 360
   demolition 360
Mineral fiber ceiling .. 494, 495
   fiber insulation 419
   roof 433
Minor site demolition 301
Mirror 479
   door 479
   glass 479
   light fixture 612
   wall 479
Miscellaneous painting .. 505, 510
Mix planting pit 645
Mixer concrete 662

Index

mortar	662, 667
plaster	667
road	669
Mixing valve	575
Mobilization or demobilization	296
Modification to cost	292
Modified bitumen roof	432
bitumen roofing	725
bituminous membrane SBS	432
Modular air handler	591
Module tub-shower	574
Moil point	666
Molding base	391
bed	394
birch	401
brick	401
casing	392
ceiling	394
chair	401
cornice	394, 397
cove	394
crown	394
dentil crown	395
exterior	395
hardboard	403
pine	396
soffit	402
trim	401
window and door	402
wood transition	498
Monitor support	339
Monument survey	300
Mortar, brick and block	723
Mortar cement	493, 494, 722
mixer	662, 667
thinset	493
Mortise lockset	473
Mortised hinge	476
Mosaic glass	493
Moss peat	644
Motel/hotel hardware	472
Motion detector	618
Motor connections	602
support	339
Motorized grader	663
Moulded door	450
Mounting board plywood	386
Movable louver blind	535
Moving shrub	648
tree	648
Mower lawn	667
Muck car tunnel	670
Mud pump	662
trailer	669
Mulch bark	644
ceramic	644
stone	644
Mulcher power	664
Mulching	644
Multi-blade damper	584
Muntin	460
window	466

N

Nail	361
common	361
stake	310
Nailer pneumatic	666, 667
steel	373
wood	373
Nailing anchor	336
Natural fiber wall covering	502
Neoprene flashing	435, 571
gasket	440

Newel	231
wood stair	405
No hub pipe	569
hub pipe fitting	570
Non chloride accelerator concrete	316
Non removable pin	475
Nylon carpet	228, 501
nailing anchor	336

O

Oak door frame	408
floor	229, 498
molding	401
paneling	403
threshold	408, 474
Off highway dumptruck	665
Oil-fired boiler	588
fired furnace	589
furnace wood	589
heating	270, 272
Oil-fired boiler	588
water heater residential	573
Olefin carpet	501
One piece astragal	476
One-way vent	439
Onyx	500
Open rail fence	641
Opener door	458
OSB faced panel	381
subfloor	385
Outdoor cabinet	540
fountain	544
Outlet box plastic	601
box steel	600
lighting	605
Oven	528
cabinet	537
microwave	528
Overhaul	304
Overhead contractor	717
door	186, 457
Overlay face door	449
Oversized brick	326

P

P trap	570
trap running	570
Packaged terminal A/C	592
Pad carpet	500
commercial grade carpet	500
equipment	314
Padding	228
carpet	500
Paddle fan	606
Paint & coating	503
& coating interior floor	513
aluminum siding	507
chain link fence	505
door & windows exterior	508
door & windows interior	510-512
encapsulation lead	305
exterior miscellaneous	505
fence picket	505
floor	513
floor concrete	513
floor wood	513
interior miscellaneous	513
removal	305
removal lead	305
siding	507
sprayer	667

striper	669, 670
trim exterior	509
wall & ceilings interior	514, 515
wall masonry exterior	509
Paint/coating cabinet & casework	510
Painting	730
balustrade	514
casework	510
ceiling	514, 515
clapboard	506
concrete block	515
cornice	514
decking	506
drywall	514, 516
exterior siding	506
grille	513
masonry	515
miscellaneous	505, 510
pipe	513
plaster	514, 516
railing	505
shutter	506
siding	506
stair stringer	506
steel siding	506
stucco	506
swimming pool	544
trellis/lattice	506
trim	514
truss	514
wall	507, 515, 516
window	511
Palladian window	465
Panel and device alarm	618
door	450, 451
fiberglass	427
insulated	423
OSB faced	381
prefabricated	423
radiant heat	596
shearwall	381
sound dampening gypsum	490
structural	381
structural insulated	381
system	403
wood folding	535
Panelboard w/circuit breaker	607
Paneled door	224, 447
pine door	453
Paneling	402
birch	403
board	404
cedar	404
cutout	303
demolition	360
hardboard	402
plywood	403
redwood	404
wood	402
Panelized shingle	425
Paper building	423
sheathing	423
Paperhanging	502
Parapet metal framing	348
Parging cement	417
Park bench	644
Parquet floor	229, 497
wood	497
Particle board underlayment	385
core door	449
Parting bead	404
Partition	214, 218
anchor	324
block	329
concrete block	329

demolition	483
framing	154
framing load-bearing stud	342
framing NLB	484
shower	520
support	339
wood frame	373
Passage door	224, 452, 455
Patch roof	416
Patching concrete floor	308
Patio door	457
Paumelle hinge	475
Pavement	638
berm	639
breaker	665
bucket	669
demolition	300
emulsion	636
planer	670
profiler	670
widener	669
Paver asphalt	667
bituminous	667
cold mix	669
concrete	667
floor	497
highway	638
roof	440
shoulder	669
Paving	637
asphalt	637
brick	638
surface treatment concrete	638
Pea gravel	645
Peastone	646
Peat humus	644
moss	644
Pedestal type lavatory	574
Pegboard	402
Penetration macadam	637
Perforated aluminum pipe	654
pipe	654
Perlite insulation	418
plaster	486
Permit building	293
Personnel hoist	671
PEX pipe	565
tubing	565, 594
tubing fittings	595
Picket fence	642
fence vinyl	640
railing	352
Pickup truck	668
Pick-up vacuum	662
Picture window	180, 462
window vinyl	470
Pier brick	326
Pilaster wood column	407
Pile driving	720
hammer	663
leads	663
sod	645
Pin powder	338
Pine door	451
door frame	408
fireplace mantel	406
floor	498
molding	396
roof deck	384
shelving	524
siding	428
stair tread	404
Pipe & fittings	558, 562-564, 566, 568
and fittings	732
bedding	623
cast iron	569

749

Index

clamp plastic 559
cleanout 560
concrete 653
copper 561
corrugated metal 654
covering 561
covering fiberglass 561
CPVC 564
drainage 652, 653
ductile iron 650
DWV PVC 564, 652
fire-suppression plastic 550
fitting cast iron 569
fitting copper 562
fitting DWV 566
fitting fire-suppression 550
fitting no hub 569
fitting plastic 550, 565, 567
fitting soil 569
foam core DWV ABS 564
gas 654, 655
hanger 559
hanger plastic 559
insulation 560
no hub 569
painting 513
perforated aluminum 654
PEX 565
plastic 550, 564
polyethylene 654
PVC 564, 652
rail aluminum 351
rail galvanized 352
rail stainless 352
rail steel 352
rail wall 352
railing 351
removal 301, 558
removal plastic 558
residential PVC 564
sewage 652
sewage collection PVC 652
single hub 569
sleeve plastic 310
soil 569
steel 655
supply register spiral 586
Piping designations 733
Piping gas service polyethylene . . 654
gas service steel 655
Pit leaching 652
Pitch coal tar 431
emulsion tar 636
Placing concrete 316
Planer pavement 670
Plank floor 383
roof 384
scaffolding 295
Plant and bulb transplanting 648
bed preparation 645
ground cover 646
screening 664
Planter 644
fiberglass 644
Planting 648
Plaster 218, 220
accessories 486
beam 486
board 214, 216
ceiling 486
column 486
cutout 303
demolition 482, 483
drilling 483
ground 381
gypsum 486

mixer 667
painting 514, 516
perlite 486
soffit 486
thincoat 488
venetian 487
wall 486
Plasterboard 488
Plastic blind 191
clad window . . 170, 172, 174, 176,
178, 180
coupling 567
faced hardboard 403
fitting 566
laminate door 449
lumber 410
lumber structural 410
outlet box 601
pipe 550, 564
pipe clamp 559
pipe fire-suppression 550
pipe fitting 565, 567
pipe hanger 559
pipe removal 558
railing 411
screw anchor 337
sheet flashing 435
siding 166
skylight 472
trap 571
vent caps 571
window 468
Plastic-laminate-clad countertop . 540
Plate escutcheon 554
glass 478
roadway 670
shear 362
steel 340
stiffener 340
wall switch 609
Platform trailer 668
Plating zinc 361
Player bench 644
Plenum demolition 483
Plow vibrator 665
Plug in circuit breaker 607
in tandem circuit breaker 608
Plugmold raceway 609
Plumbing 568
demolition 558
fixture 573, 582
fixtures removal 558
Plywood 725
clip 362
demolition 482, 483
floor 384
joist 387
mounting board 386
paneling 403
sheathing roof & walls 386
shelving 524
siding 428
soffit 402
subfloor 385
underlayment 385
Pneumatic nailer 666, 667
Pocket door 472
door frame 408
Point moil 666
Pointing CMU 322
masonry 322
Pole closet 401
lighting 613
Police connect panel 618
Polyethylene film 644
floor 501

pipe 654
pool cover 544
septic tank 652
tarpaulin 296
Polypropylene siding 430
Polystyrene blind 409
insulation 418, 420
Polysulfide caulking 441
Polyurethane caulking 441
Polyvinyl soffit 402, 431
Pool accessories 544
cover 544
cover polyethylene 544
filtration swimming 577
heater electric 588
swimming 544
Porcelain tile 492
Porch door 451
framing 374
Portable air compressor 665
asphalt plant 669
fire extinguisher 523
Portland cement terrazzo 500
Post cap 362
cedar 407
demolition 358
driver 669
lamp 352
light 612
wood 372
Postformed countertop 540
Post-tensioned slab on grade . . . 313
Poured gypsum underlayment . . 319
insulation 168
Powder actuated tool 338
charge 338
pin 338
Power mulcher 664
trowel 662
wiring 610
Precast catch basin 654
concrete lintel 318
concrete plantable pavers 638
concrete stairs 318
concrete window sill 318
coping 333
curb 639
receptor 520
terrazzo 500
Prefabricated building 546
fireplace 240, 521
panels 423
wood stair 404
Prefinished door 449
drywall 489
floor 229, 498
hardboard paneling 402
shelving 524
Preformed roof panel 427
roofing & siding 428
Pre-hung door 454, 455
Preparation exterior surface 503
interior surface 504
plant bed 645
Pressure reducing valve water . . . 558
restricting valve 554
valve relief 558
wash 504
washer 668
Prestressing steel 313
Preventer backflow 568
Prices lumber products 725
Prime coat 637
Primer asphalt 499
Privacy fence vinyl 640
Profile block 328

Profiler pavement 670
Projected window aluminum . . . 458
Property line survey 300
Protection lightning 611
slope 633
temporary 297
termite 633
winter 308
Protector ADA insulated 560
P&T relief valve 558
PT/AC unit 592
Pull box 601
box electrical 601
door 539
Pump 651
centrifugal 667
circulating 582
concrete 662
diaphragm 667
grout 668
heat 593
mud 662
shotcrete 662
staging 294, 719
submersible 572, 651, 667
sump 529
trash 668
water 582, 651, 667
wellpoint 671
Purlin log roof 383
roof 383
Push-on joint 650
Puttying 503
PVC conduit 600
conduit in slab 601
cornerboards 411
door casing 411
fascia 411
fitting 565
flashing 435
frieze 411
gravel stop 435
molding exterior 411
pipe 564, 652
pipe residential 564
rake 412
soffit 412
trap 571
trim 411
waterstop 310

Q

Quarry drill 666
tile 493
Quarter round molding 401
Quartz 645
heater 596
Quoin 331

R

Raceway plugmold 609
surface 609
wiremold 609
Rack hose 553
Radial arch 388
Radiant floor heating 594
heating panel 596
Radiator cast iron 593
Rafter 376
anchor 362
composite 376
demolition 358

Index

framing metal 349
  metal bracing 348
  metal bridging 348
  roof framing 138
  wood 376
Rail aluminum pipe 351
  galvanized pipe 352
  stainless pipe 352
  steel pipe 352
  wall pipe 352
Railing composite 412
  demolition 360
  encased 412
  encased composite 412
  picket 352
  pipe 351
  plastic 411
  wood 401, 404, 406
  wood stair 405
Railroad tie 636, 648
  tie step 636
Rainproof meter center 608
Rake PVC 412
  tractor 664
Rammer/tamper 663
Ramp handicap 314
Ranch plank floor 498
Range circuit 280
  cooking 528
  hood 236, 529
  receptacle 605, 609
Ready mix concrete 315
Receptacle air conditioner 604
  device 604
  dryer 605
  duplex 609
  GFI 604
  locking 605
  range 605, 609
  telephone 605
  television 605
  weather-proof 604
Receptor precast 520
  shower 520
  terrazzo 520
Rectangular diffuser 585
  ductwork 583
Rectifier low-voltage silicon . . . 609
Recycled plastic lumber 410
Reduce heat transfer glass . . . 478
Redwood bark mulch 644
  cupola 525
  louver 480
  paneling 404
  siding 162, 428
  tub 545
  wine cellar 530
Refinish floor 498
Reflective block 329
  insulation 420
Reflectorized barrels 666
Refrigerant line sets 582
  removal 580
Refrigerated wine cellar 530
Refrigeration residential 528
Register air supply 586
  baseboard 586
Reinforcing concrete 312
  dowel 312
  footing 312
  joint 325
  masonry 325
  steel fiber 313
  synthetic fiber 313
  wall 312
Release emergency sprinkler 554

Relief valve P&T 558
  valve self-closing 558
  valve temperature 558
Removal air conditioner 580
  bathtub 558
  block wall 302
  boiler 580
  catch basin 300
  concrete 300
  curb 300
  driveway 300
  fixture 558
  floor 302
  foundation 301
  insulation 416
  lavatory 558
  masonry 301, 323
  paint 305
  pipe 301, 558
  plumbing fixtures 558
  refrigerant 580
  shingle 416
  sidewalk 301
  sink 558
  sod 645
  stone 301
  tree 620
  utility line 300
  water closet 558
  window 445
Rental equipment 662
Repair fire damage 505
Repellent water 515
Replacement windows 444
  windows 726
Residential alarm 606
  appliance 528, 606
  applications 602
  bathtubs 574
  burner 589
  canopy 524
  closet door 448
  device 602
  dishwasher 528
  door 408, 447, 451
  door bell 605
  elevator 548
  fan 606
  fence 640
  fixture 605, 612
  gas water heater 573
  greenhouse 546
  gutting 304
  heat pump 607
  hinge 475
  kitchen sinks 574
  lavatory 573, 577
  lighting 605
  load center 602
  oil-fired water heater 573
  overhead door 457
  PVC pipe 564
  refrigeration 528
  roof jack 585
  service 602
  sinks 574
  smoke detector 606
  sprinkler 555
  stair 404
  storm door 446
  switch 602
  wall cap 584
  wash bowl 573
  water heater 529, 573, 607
  water heater electric 572
  wiring 602, 606

wood window 459
Resilient base 498
  floor 499
  flooring 229
Resquared shingle 425
Restoration gypsum 482
  window 306
Retaining wall 643
  wall cast concrete 642
  wall concrete segmental 642
  wall segmental 643
  wall stone 643
  wall timber 643
Retarder concrete 316
  vapor 423
Retractable stair 530
Ribbed waterstop 310
Ridge board 376
  shingle 425
  shingle slate 424
  vent 439
Rig drill 300
Rigid anchor 324
  conduit in trench 602
  in slab conduit 601
  insulation 168, 418
Ring drapery 534
  split 362
  toothed 362
Ripper attachment 669
Riser beech 231
  oak 231
  pipe wellpoint 672
  wood stair 405
Rivet 338
  aluminum 338
  copper 338
  stainless 338
  steel 338
  tool 338
Road berm 639
  mixer 669
  sweeper 669
Roadway plate 670
Robe hook 520
Rock trencher 665
Rod backer 440
  closet 524
  curtain 520
  ground 600
Roll roof 431
  roofing 433
  type air filter 586
Roller sheepsfoot 664
  tandem 664
  vibratory 664
Romex copper 599
Roof adhesive 431
  aluminum 427
  baffle 439
  beam 388
  built-up 431
  cant 376, 431
  clay tile 425
  coating 416
  copper 433
  deck 384
  deck insulation 421
  deck laminated 384
  deck wood 384
  decking 340
  decking log 383
  drains 572
  fiberglass 427
  fire valve manifold 552

framing . . 138, 140, 142, 144, 146, 148
framing log 383
framing removal 303
hatch removal 417
insulation 421
jack residential 585
lead 433
membrane green 426
metal 427
metal rafter framing 349
metal truss framing 350
mineral 433
modified bitumen 432
nail 361
panel aluminum 427
panel preformed 427
patch 416
paver 440
paver and support 440
purlin 383
purlin log 383
rafter 376
rafter bracing 348
rafter bridging 348
rafter framing 376
roll 431
safety anchor 295
sheathing 386
shingle asphalt 424
slate 424, 725
soffit framing 349
soil mixture green 426
specialty prefab 436
system green 426
truss 387, 388
ventilator 439
window 471
window metal 471
zinc 433
Roofing 194, 196, 198, 200, 202
  & siding preformed 427
  demolition 416
  flat seam sheet metal 433
  gambrel 198
  hip roof 196
  Mansard 200
  membrane 431
  modified bitumen 432
  roll 433
  shed 202
  SPF 432
  system built-up 431
Rooftop air conditioner 592
Rope safety line 295
Rotary crawler drill 666
  hammer drill 666
Rototiller 664
Rough buck 373
  stone wall 330
Rough-in sink countertop 574
  sink raised deck 574
  sink service floor 577
  tub 574
Round diffuser 586
Rubber base 498
  coating 418
  floor 499
  floor tile 500
  sheet 499
  threshold 474
  tile 500
Rubbish handling 303
Run gravel bank 620

Index

S

Safety line rope	295
switch	610
switch disconnect	610
Salamander	668
Sales tax	292, 716
Sand fill	620
Sandblasting equipment	668
Sander floor	668
Sanding	503
floor	498
Sandstone	331
flagging	638
Sanitary base cove	492
drain	571
tee	570
Sash wood	462
Sauna	545
door	545
Saw chain	668
circular	668
concrete	662
masonry	668
Sawn control joint	311
SBS modified bituminous membrane	432
Scaffold aluminum plank	295
baseplate	294
bracket	294
caster	294
frame	294
guardrail	294
mason	294
stairway	294
wood plank	294
Scaffolding	719
plank	295
tubular	294
Scissor lift	665
Scrape after damage	505
Scraper earth	664
Screed, gas engine, 8HP vibrating	662
Screen chimney	521
fence	642
molding	401
security	459
squirrel and bird	521
window	459, 467
wood	467
Screening plant	664
Screen/storm door aluminum	446
Screw anchor	336
brass	361
drywall	491
lag	338
steel	361
wood	361
Seal compression	440
fog	636
pavement	636
Sealant	417
acoustical	442, 490
caulking	441
Sealcoat	636
Sealer joint	440, 441
Sectional door	457
overhead door	457
Security film	479
screen	459
Seeding	645, 732
Segmental concrete retaining wall	642
retaining wall	643
Seismic bracing supports	559
Selective clearing	620
demolition fencing	301
demolition masonry	322
Self-closing relief valve	558
Self-contained air conditioner	592
Septic galley	653
system	112, 652
system chamber	653
system effluent-filter	653
tank	652
tank concrete	652
tank polyethylene	652
Service electric	278, 610
entrance cable aluminum	599
entrance cap	599
residential	602
sink	577
sink faucet	577
Sewage collection concrete pipe	652
collection PVC pipe	652
pipe	652
Sewage/drainage collection	653
Shade	535
Shadow box fence	641
Shake siding	164
wood	425
Shear plate	362
wall	386
Shearwall panel	381
Sheathed non-metallic cable	599
Romex cable	599
Sheathing	385, 386
asphalt	386
board	385
gypsum	387
insulating	384
paper	423
roof	386
roof & walls plywood	386
wall	386
Shed dormer framing	152
dormer roofing	206
roof framing	148
Sheepsfoot roller	664
Sheet carpet	501
glass	479
glass mosaic	493
metal	433
metal aluminum	436
metal cladding	434
metal flashing	434
rock	214, 216
Sheeting driver	666
Shelf metal	523
Shellac door	510
Shelter temporary	308
Shelving	524
pine	524
plywood	524
prefinished	524
storage	523
wood	524
Shield expansion	336
lag screw	336
Shingle	196, 198, 200, 202, 424
asphalt	424
cedar	424
concrete	426
panelized	425
removal	416
ridge	425
siding	164
slate	424
stain	506
strip	424
wood	424
Shock absorber	569
Shoring baseplate	294
bracing	294
heavy duty	294
leveling jack	294
Short load concrete	316
Shot blaster	668
Shotcrete pump	662
Shoulder paver	669
Shovel crawler	664
Shower by-pass valve	575
cabinet	520
cubicle	575
door	520
drain	570
enclosure	521
glass door	521
partition	520
receptor	520
stall	574
surround	521
Shower/tub control set	575
Shower-tub valve spout set	575
Shrub and tree	646
broadleaf evergreen	647
deciduous	647
evergreen	646
moving	648
Shutter	191, 535
exterior	409
interior	535
wood	409
wood interior	535
Sidelight	408
door	448, 454
Sidewalk	497
asphalt	636
brick	638
broom	668
concrete	636
driveway and patio	636
removal	301
Sidewalk bracket	295
Siding aluminum	427
bevel	428
cedar	428
fiber cement	430
fiberglass	427
metal	166
nail	361
paint	507
painting	506
plywood	428
polypropylene	430
redwood	428
removal	417
shingle	164
stain	429, 506
steel	428
vinyl	429
wood	162, 428
wood board	428
Silencer door	475
Silicone	440
caulking	442
coating	418
water repellent	418
Sill	331
& ledger framing	377
anchor	362
door	402, 408, 474
masonry	332
precast concrete window	318
quarry tile	493
wood	377
Sillcock hose bibb	575
Silt fence	633
Simulated stone	333
Single hub pipe	569
hung aluminum window	458
zone rooftop unit	592
Sink	236
base	536
countertop	574
countertop rough-in	574
faucet kitchen	575
kitchen	574
laundry	575
lavatory	573
raised deck rough-in	574
removal	558
residential	574
service	577
service floor rough-in	577
slop	577
Siren	618
Sister joist	372
Site clear and grub	620
clearing	620
demolition	300
improvement	644
preparation	300
Sitework	102
Skidder log	664
Skid steer attachments	665
loader	665
Skim coat	214, 216
Skylight	208, 471
domed	471
plastic	472
removal	416
Skywindow	208
Slab blockout formwork	309
bulkhead formwork	308
concrete	126
concrete placement	316
curb formwork	309
cutout	302
edge formwork	309
haunch	314
haunch formwork	314
on grade	314
on grade formwork	308
on grade post-tensioned	313
on grade removal	301
stamping	317
textured	315
thickened edge	314
thickened edge formwork	314
turndown	314
turndown formwork	314
void formwork	309
Slate	331
roof	424, 725
shingle	424
sidewalk	638
sill	332
stair	331
tile	497
Slatwall	403
Sleeper	377
framing	377
wood	229
Sleeve anchor bolt	311
and tap	651
dowel	312
formwork	310
plastic pipe	310
Sliding aluminum door	456
door	185
glass door	456
glass vinyl-clad door	457

Index

mirror 520
vinyl-clad door 457
window 176, 463
window aluminum 459
wood door 457
Slipform paver barrier 670
paver curb 669
Slop sink 577
Slope grading 621
protection 633
Slot letter 523
Slotted pipe 653
Slow close ball valve 553
Smoke detector 618
Snap-tie formwork 310
Snow guard 439
Soap holder 520
Socket meter 608
Sod 646
Sodding 646
Sodium low pressure fixture 613
Soffit 377
  & canopy framing 377
  aluminum 428
  drywall 489
  metal 431
  molding 402
  plaster 486
  plywood 402
  polyvinyl 402, 431
  PVC 412
  vent 439
  wood 402
Softener water 572
Soil compaction 622
  mix man-made 426
  pipe 569
  tamping 622
  treatment 633
Solid fuel furnace 589
  surface countertops 541
  wood door 450
Sound dampening panel
  gypsum 490
Source heat pump water 593
Spa bath 545
Space heater 667
Spade air 666
  tree 665
Spanish roof tile 425
Special door 456
  stairway 231
SPF roofing 432
Spike grid 362
Spiral pipe supply register 586
  stair 353, 404
Split rib block 328
  ring 362
  system ductless 592
Spray coating 417
  rig foam 666
Sprayed-on insulation 421
Sprayer emulsion 666
  paint 667
Spread footing 314
  footing formwork 308
Spreader aggregate 662, 669
  chemical 669
  concrete 667
Sprinkler ball valve 553
  cabinet 554
  connector flexible 554
  control pressure 555
  dehydrator package 555
  escutcheon 554
  flow control valve 550

head 554
head wrench 555
irrigation system 644
line tester 552
release emergency 554
residential 555
system 554
system accelerator 555
system dry-pipe 555
trim valve 554
underground 644
valve 550, 555
Stabilizer base 670
Staging aids 295
pump 294
Stain cabinet 510
casework 510
door 510
floor 513
lumber 388
shingle 506
siding 429, 506
truss 514
Staining concrete floor 496
Stainless flashing 434
gutter 437
rivet 338
steel countertop 540
steel gravel stop 435
steel hinge 475
Stair basement 404
basement bulkhead 318
brick 330
carpet 501
ceiling 530
climber 548
concrete 315
disappearing 530
electric 530
finish concrete 317
fire escape 351
formwork 310
parts wood 404
precast concrete 318
prefabricated wood 404
railroad tie 636
removal 359
residential 404
retractable 530
slate 331
spiral 353, 404
stringer 373
stringer wood 373
tread 231
tread tile 493
wood 404
Stairway 230
scaffold 294
Stairwork handrails 406
Stake formwork 310
nail 310
Stall shower 574
Stamping slab 317
texture 317
Standard extinguisher 523
Standing seam 433
Standpipe alarm 554
connection 552
Steam bath 545
bath residential 545
boiler 588
boiler electric 588
cleaner 668
Steel anchor 324
beam 340
bin wall 643

bolt 362
bolt & hex nut 337
bridging 371
chain link 640
channel 486
column 340
conduit in slab 601
conduit in trench 602
conduit rigid 600
diffuser 586
door 447, 726
door residential 447
downspout 436
drill 666
drilling 337
edging 636, 648
expansion tank 582
fiber reinforcing 313
fireplace 240
fitting 563
flashing 434
frame 446
furring 483
gravel stop 435
gutter 437
HEX bolt 337
hinge 475
lintel 339
member structural 340
nailer 373
outlet box 600
pipe 655
pipe corrugated 654
plate 340
prestressing 313
rivet 338
screw 361
siding 428
stud NLB 484
underground storage tank . . . 581
window 459
Steeple tip pin 475
Step bluestone 636
brick 636
masonry 636
railroad tie 636
stone 331
Stiffener joist web 347
plate 340
Stiffleg derrick crane 671
Stockade fence 642
Stockpiling of soil 620
Stone aggregate 646
anchor 324
ashlar 331
base 331
crushed 620
cultured 333
curb cut 639
fill 620
floor 331
ground cover 646
mulch 644
paver 330, 638
removal 301
retaining walls 643
simulated 333
step 331
trim cut 331
wall 160, 330, 643
Stool cap 402
window 331, 332
Stop door 402
gravel 435
Storage shelving 523
tank 581

tank concrete aboveground . . . 582
tank steel aboveground 581
tank steel underground 581
tank underground 581
Storm door 190, 446
door residential 446
drainage manhole frames . . . 654
window 190, 467
window aluminum residential . 467
Stove 523
woodburning 523
Strainer downspout 436
gutter 438
roof 436
wire 436
Strap tie 362
Straw 644
bale construction 546
Stringer 231
stair 373
Strip chamfer 310
floor 498
footing 314
lighting 611
shingle 424
soil 620
Striper paint 669, 670
Stripping topsoil 620
Structural columns 339
compaction 632
excavation 622
insulated panel 381
panel 381
steel members 340
Structure log 382
Stucco 220, 487
interior 218
painting 506
wall 158
Stud demolition 359
NLB metal 484
NLB steel 484
NLB wall 484
partition 373
partition framing load-bearing . 342
wall 136, 373
wall blocking load-bearing . . . 341
wall box beam load-bearing . . 342
wall bracing load-bearing . . . 341
wall framing load-bearing . . . 342
wall header load-bearing . . . 342
Stump chipper 664
Subdrainage system 654
Subfloor 384
adhesive 385
OSB 385
plywood 229, 385
wood 385
Submersible pump 572, 651, 667
sump pump 572
Suction hose 667
Sump pump 529
pump submersible 572
Supply ductile iron pipe water . . . 650
Support ceiling 339
drinking fountain 577
lavatory 577
light 339
monitor 339
motor 339
partition 339
X-ray 339
Suppressor lightning 602
Surface landscape 501
preparation exterior 503
preparation interior 504

Index

raceway 609
Surfacing 636
Surround shower 521
   tub 521
Survey crew 297
   monument 300
   property line 300
   topographic 300
Suspended acoustic ceiling tiles . 494
   ceiling 485, 495
Suspension system ceiling 495
Sweeper road 669
Swimming pool 243, 544
   pool enclosure 546
   pool equipment 544
   pool filter 577
   pool filtration 577
   pool heater 588
   pool ladder 544
   pools 731
Swing check valve 558
   check valve bronze 558
Swing-up door 186
   overhead door 457
Switch box 601
   box plastic 601
   decorator 603
   dimmer 602, 609
   electric 609, 610
   general duty 610
   residential 602
   safety 610
   tamper 555
   time 610
   toggle 609
   wiring 280
Switching low-voltage 609
Switchplate low-voltage 609
Synthetic erosion control 633
   fiber 313
   fiber reinforcing 313
   marble 497
System chamber septic 653
   fire extinguishing 550, 553
   septic 652
   sprinkler 550, 554, 555
   subdrainage 654
   T.V. 616
   VHF 616

T

T&G siding 162
Tamper 666
   switch 555
Tamping soil 622
Tandem roller 664
Tank aboveground storage
concrete 582
   expansion 582
   fiberglass 581
   horizontal aboveground 581
   septic 652
   steel aboveground storage . . . 581
   steel underground storage . . . 581
   storage 581
   water 669
Tankless gas water heaters 572
Tap and sleeve 651
Taping and finishing gypsum
board 488
Tapping crosses and sleeves 651
   main 651
Tar kettle 668, 670
   paper 645

pitch emulsion 636
Tarpaulin 296
   duck 296
   polyethylene 296
Tax 292
   sales 292
   social security 292
   unemployment 292
T bar mount diffuser 586
Teak floor 229, 497
   molding 401
   paneling 403
Tee cleanout 560
Telephone receptacle 605
Television receptacle 605
   system 616
Temperature relief valve 558
Tempered glass 478
   hardboard 402
Tempering valve 558
   valve water 558
Temporary construction 296
   heat 308
   protection 297
   shelter 308
   toilet 668
Tennis court fence 114, 640
Terminal A/C packaged 592
   air conditioner packaged 592
Termite control chemical 633
   protection 633
Terne coated flashing 434
Terra cotta coping 333
   cotta demolition 303
Terrazzo base 500
   cove base 500
   curb 500
   floor tile 500
   precast 500
   receptor 520
   wainscot 500
Tester sprinkler line 552
Textile wall covering 502
Texture stamping 317
Textured slab 315
Theater home 616
Thermal & moisture protection
demo 416
Thermostat 581
   integral 596
   wire 607
Thickened edge slab 314
Thin brick veneer 325
Thincoat 214, 216
   plaster 488
Thinset ceramic tile 492
   mortar 493
Threshold 474
   aluminum 474
   cherry 474
   door 408
   oak 474
   ramp aluminum 475
   stone 331
   walnut 474
   wood 402
Thru-wall air conditioner 592
Tie rafter 376
   railroad 636, 648
   strap 362
   wall 324
Tile 492, 494, 499
   aluminum 494
   carpet 501
   ceiling 495
   ceramic 492

clay 425
concrete 426
cork 499
cork wall 502
demolition 482
flue 332
glass 479
grout 492
marble 497
metal 494
porcelain 492
quarry 493
rubber 500
slate 497
stainless steel 494
stair tread 493
vinyl composition 499
wall 492
waterproofing membrane
ceramic 494
   window sill 493
Tiling ceramic 492, 494
Timber connector 361
   fastener 361
   framing 383
   heavy 383
   laminated 388
   retaining wall 643
Time switch 610
Timer clock 606
   interval 603
   switch ventilator 585
Tinted glass 478
Toggle switch 609
Toggle bolt anchor 336
Toilet accessories 520
   accessories commercial 520
   bowl 573
   chemical 668
   temporary 668
Tongue and groove siding 162
Tool powder actuated 338
   rivet 338
Toothed ring 362
Toothing masonry 303
Top demolition counter 360
   vanity 541
Topographical surveys 300
Topsoil 620
   placement and grading 645
   stripping 620
Torch cutting 668
Torrified door 453
Towel bar 520
Tower crane 670, 720
   hoist 671
   light 667
Track drawer 539
   drill 666
   light 613
   lighting 613
   traverse 534
Tractor loader 664
   rake 664
   truck 668
Traffic line remover 670
Trailer floodlight 666
   lowbed 670
   mud 669
   platform 668
   truck 668
   water 668
Tram car 668
Transfer switch automatic 611
Transformer low-voltage 609
Transition molding wood 498

strip floor 500
Transom lite frame 447
   windows 465
Transplanting plant and bulb . . . 648
Trap cast iron 570
   drainage 570
   plastic 571
   PVC 571
Trapezoid windows 466
Trash pump 668
Traverse 534
   track 534
Travertine 331
Tread abrasive 493
   beech 231
   oak 231
   stone 331
   wood stair 405
Treated lumber framing 377
Treatment flooring 496
Tree deciduous 647
   evergreen 646
   guying systems 648
   moving 648
   removal 620
   spade 665
Trench backfill 622
   box 668
   disposal field 653
   excavating 621
   utility 622
Trencher chain 622, 665
   rock 665
   wheel 665
Trenching 106
Trim exterior 395, 396
   exterior paint 509
   molding 401
   painting 514
   PVC 411
   tile 492
   wood 398
Tripod floodlight 666
Trowel coating 417
   concrete 662
   power 662
Truck boom 670
   concrete 662
   crane flatbed 670
   dump 665
   flatbed 665, 668
   holding concrete 316
   loading 622
   mounted crane 670
   pickup 668
   tractor 668
   trailer 668
   vacuum 669
   winch 669
Truss bowstring 388
   demolition 360
   flat wood 387
   framing metal 350
   painting 514
   plate 362
   roof 387, 388
   roof framing 140
   stain 514
   varnish 514
Tub hot 545
   redwood 545
   rough-in 574
   surround 521
Tube and wire kit 593
   fittings copper 562
Tubing connection split system . . 593

Index

copper 561
electric metallic 600
fitting PEX 595
PEX 565, 594
Tub & shower module 574
Tubular scaffolding 294
Tugboat diesel 672
Tumbler holder 520
Tunnel locomotive 670
 muck car 670
 ventilator 670
Turbine wind 658
Turndown slab 314
Turned column 407
T.V. antenna 616
 system 616

U

Undereave vent 480
Underground sprinkler 644
 storage tank 581
 storage tank steel 581
Underlayment hardboard 385
 latex 498
Unemployment tax 292
Unit air-handling 593
 commercial kitchen 530
 line set evap/condensing ... 593
Utility 106
 boxes 650
 excavation 106
 line removal 300
 trench 622
 trench excavating 622

V

Vacuum breaker 568
 catch basin 669
 central 531
 cleaning 531
 pick-up 662
 truck 669
 wet/dry 669
Valance board 537
Valley rafter 376
Valve 571
 assembly dry pipe sprinkler . 555
 bronze 558
 butterfly 650
 cap fire 554
 check 550
 fire hose 553
 mixing 575
 relief pressure 558
 shower by-pass 575
 slow close ball 553
 spout set shower-tub 575
 sprinkler 550, 555
 sprinkler alarm 550
 sprinkler ball 553
 sprinkler flow control 550
 sprinkler trim 554
 swing check 558
 tempering 558
 washing machine 576
 water pressure 558
Vanity base 539
 top 541
 top lavatory 573
Vapor barrier 423
 barrier sheathing 386
 retarder 423

Varnish cabinet 510
 casework 510
 door 510
 floor 513, 516
 truss 514
VCT removal 482
Vehicle charging electric 611
Veneer brick 160, 325
 core paneling 403
 member laminated 388
 removal 323
Venetian plaster 487
Vent cap 571
 cap cast iron 571
 cap plastic 571
 chimney 587
 dryer 529, 576
 eave 439
 flashing 571
 gas 587
 one-way 439
 ridge 439
 ridge strip 480
 soffit 439
Ventilating air conditioning . 585, 592
Ventilation fan 606
 louver 480
Ventilator roof 439
 timer switch 585
 tunnel 670
Verge board 400
Vermiculite insulation 420
Vertical aluminum siding 427
VHF system 616
Vibrating screed, gas engine,
8HP 662
Vibrator concrete 662
 earth 632
 plow 665
Vibratory hammer 664
 roller 664
Vinyl blind 191, 409, 534
 casement window 469
 chain link 640
 clad half round window 465
 clad premium window 466
 composition floor 499
 composition tile 499
 downspout 436
 faced wallboard 489
 fence 640
 floor 499
 flooring 229
 gutter 437
 picture window 470
 replacement window 444
 sheet floor 499
 shutter 191
 siding 166, 429
 wallpaper 502
 window 468
 window double hung 468
 window half round 470
 window single hung 468
 window solid 468
Volume control damper 584

W

Wainscot ceramic tile 492
 molding 402
 quarry tile 493
 terrazzo 500
Walk 636
Walkway luminaire 613

Wall accessories formwork ... 310
 and partition demolition ... 483
 base 498
 blocking load-bearing stud . 341
 box beam load-bearing stud 342
 boxout formwork 309
 bracing load-bearing stud ... 341
 brick 330
 brick shelf formwork 309
 bulkhead formwork 309
 bumper 473
 cabinet 537
 cap residential 584
 cast concrete retaining 642
 ceramic tile 492
 coating 517
 concrete finishing 317
 concrete placement 316
 concrete segmental retaining 642
 covering 730
 covering textile 502
 cutout 302
 drywall 488
 finishes 502
 formwork 309
 foundation 328
 framing 136, 378
 framing load-bearing stud .. 342
 framing removal 303
 furring 380, 484
 grout 324
 header load-bearing stud .. 342
 heater 530
 hung lavatory 573
 insulation 418
 interior 214, 218
 lath 485
 log 382
 louver 480
 masonry 158, 160, 327, 643
 mirror 479
 painting 507, 515, 516
 paneling 403
 plaster 486
 plywood formwork 309
 prefab plywood formwork . 309
 reinforcing 312
 retaining 643
 shear 386
 sheathing 386
 steel bin 643
 stone 330, 643
 stucco 487
 stud 373
 stud NLB 484
 switch plate 609
 tie 324
 tie masonry 324
 tile 492
 tile cork 502
Wallboard repair gypsum 482
Wallcovering 502
 acrylic 502
 gypsum fabric 502
Wall covering natural fiber ... 502
Wallpaper 502, 730
 grass cloth 502
 vinyl 502
Walnut floor 498
 threshold 474
Wardrobe 524
 wood 538
Wash bowl residential 573
 brick 322
 pressure 504
Washable air filter 587

air filter permanent 587
Washer 362
 dryer accessories 576
 pressure 668
Washing machine automatic . 529
 machine valve 576
Water closet 573, 576
 closet removal 558
 closet support 577
 curing concrete 318
 distribution pipe 650, 651
 flow indicator 550
 hammer arrester 569
 heater 529, 607
 heater gas residential 573
 heater insulation 560
 heater residential 529, 573, 607
 heater residential electric .. 572
 heater wrap kit 560
 heating hot 588, 593
 heating wiring 280
 hose 667
 motor alarm 554
 pipe ground clamp 600
 pressure reducing valve ... 558
 pressure relief valve 558
 pressure valve 558
 pump 582, 651, 667
 reducer concrete 316
 repellent 515
 repellent coating 418
 repellent silicone 418
 softener 572
 source heat pump 593
 supply domestic meter 568
 supply ductile iron pipe ... 650
 supply meter 568
 supply PVC pipe 651
 tank 669
 tempering valve 558
 trailer 668
 well 651
Waterproofing coating 417
 demolition 417
Waterstop PVC 310
 ribbed 310
Weather barrier 423
 barrier or wrap 423
 cap entrance 601
Weatherproof receptacle ... 280, 604
Weatherstrip door 476
 window 477
 zinc 476
Weatherstripping 476
Weathervane residential 525
Web stiffener framing 347
Weights and measures 715
Welded frame 447
 wire epoxy 312
 wire fabric 312
 wire galvanized 312
Welding machine 669
Well & accessories 651
 area window 525
 water 651
Wellpoint discharge pipe 671
 equipment rental 671
 header pipe 671
 pump 671
 riser pipe 672
Wet/dry vacuum 669
Wheel trencher 665
Wheelbarrow 669
Wheeled loader 664
Whirlpool bath 545
Widener pavement 669

755

Index

Winch truck 669
Wind turbines 658
Window 170, 172, 174, 176,
                178, 180, 458, 472
   & door demolition 444
   air conditioner 592
   aluminum 458
   aluminum projected 458
   aluminum residential storm . . . 467
   aluminum sliding 459
   and door molding 402
   awning 460
   basement/utility 459
   bay 460
   blind 534
   bow & bay metal-clad 464
   casement 461, 464
   casement wood 461
   demolition 445
   double hung 465
   double hung bay wood 464
   double hung vinyl 468
   double hung wood 460, 462
   double hung 170
   estimates 726
   fiberglass 470
   fiberglass bay 471
   fiberglass single hung 470
   fiberglass slider 471
   frame 463
   glass 479
   grille 466
   half round vinyl 470
   half round vinyl clad 465
   hardware 472
   metal 458
   metal-clad awning 463
   metal-clad casement 463
   metal-clad double hung 463
   muntin 466
   painting 511
   Palladian 465
   picture 462
   picture & sliding metal-clad . . . 463
   plastic 468
   precast concrete sill 318
   removal 445
   replacement 444
   restoration 306
   roof 471
   sash wood 463
   screen 459, 467
   sill 332
   sill marble 331
   sill tile 493
   single hung aluminum 458
   single hung vinyl 468
   sliding 463
   sliding wood 463
   solid vinyl 468
   steel 459
   stool 331, 332
   storm 190, 467
   transom 465
   trapezoid 466
   trim set 402
   vinyl 468
   vinyl casement 469
   vinyl clad awning 464
   vinyl clad half round 465
   vinyl clad premium 466
   vinyl picture 470
   vinyl replacement 444
   weatherstrip 477
   well area 525
   wood 459
   wood awning 460
   wood bow 464
   wood double hung 462
   wood residential 459
Windrow loader 670
Wine cellar 530
Winter concrete 316
   protection 308
Wire copper 599
   electrical 599
   fence 641
   ground 600
   mesh 487
   strainer 436
   thermostat 607
   THW 599
Wiremold raceway 609
Wiring air conditioner 607
   device 609
   fan 606
   power 610
   residential 602, 606
Wood & composite I-joists 387
   awning window 460
   base 391
   beam 371, 383
   blind 191, 535
   block floor demolition 482
   blocking 211, 374
   bow window 464
   bracing 370
   canopy 377
   casework frame 537
   chips 645
   column 372, 383, 407
   composition floor 497
   cupola 525
   deck 244, 384
   demolition 482
   door 224, 226, 448, 449
   door decorator 450
   door frame exterior 408
   door frame interior 408
   door residential 451
   double hung window 462
   drawer 539
   drawer front 539
   fascia 398
   fastener 361
   fence 641
   fiber ceiling 495
   fiber sheathing 386
   fiber underlayment 385
   floor . . 130, 132, 134, 229, 497, 498
   floor demolition 482
   folding panel 535
   foundation 124
   frame 537
   framing 370, 371, 383
   framing demolition 356
   framing miscellaneous 373
   framing open web joist 372
   furnace 589
   furring 380
   gas furnace 589
   girder 371
   gutter 437
   handrail 401, 406
   interior shutter 535
   joist 372, 387
   laminated 388
   louver 406
   nailer 373
   oil furnace 589
   overhead door 457
   panel door 450
   paneling 402
   parquet 497
   partition 373
   plank scaffold 294
   planter 644
   post 372
   rafter 376
   rail fence 642
   railing 401, 404, 406
   residential window 459
   roof deck 384
   sash 462
   screen 467
   screw 361
   shake 425
   shelving 524
   shingle 424
   shutter 191, 409
   sidewalk 636
   siding 162, 428
   siding demolition 417
   sill 377
   soffit 402
   stair 404
   stair baluster 404
   stair newel 405
   stair part 404
   stair railing 405
   stair riser 405
   stair stringer 373
   stair tread 405
   storm door 190, 455
   storm window 190
   strip flooring 498
   subfloor 385
   threshold 402
   trim 398
   truss 387
   veneer wallpaper 502
   wall framing 136
   wardrobe 538
   window 459
   window casement 461
   window demolition 445
   window double hung 460
   window double hung bay 464
   window sash 463
   window sliding 463
Woodburning stove 523
Wool carpet 501
   fiberglass 420
Work boat 672
Workers' compensation . . . 293, 718
Wrecking ball 669
Wrench impact 666
   sprinkler head 555

X

X-ray support 339

Y

Yard fountain 544
Yellow pine floor 498

Z

Z bar suspension 495
Zee bar 491
Zinc plating 361
   roof 433
   weatherstrip 476

Notes

Notes

Notes

Notes

Notes

Other RSMeans Products & Services

RSMeans— a tradition of excellence in construction cost information and services since 1942

Table of Contents
Annual Cost Guides
RSMeans Online and *CostWorks* CD Comparison Mat
Seminars
Reference Books
Order Form

For more information visit the RSMeans website at **www.rsmeans.com**

Unit prices according to the latest MasterForm

Book Selection Guide

The following table provides definitive information on the content of each cost data publication. The number of lines of data provided in each unit price or assemblies division, as well as the number of crews, is listed for each book. The presence of other elements such as reference tables, square foot models, equipment rental costs, historical cost indexes, and city cost indexes, is also indicated. You can use the table to help select the RSMeans book that has the quantity and type of information you most need in your work.

Unit Cost Divisions	Building Construction	Mechanical	Electrical	Commercial Renovation	Square Foot	Site Work Landsc.	Green Building	Interior	Concrete Masonry	Open Shop	Heavy Construction	Light Commercial	Facilities Construction	Plumbing	Residential
1	573	393	410	506		509	191	312	456	572	501	246	1042	403	178
2	777	279	85	733		993	178	399	213	776	734	481	1221	286	257
3	1690	340	230	1085		1471	987	355	2032	1689	1685	481	1788	316	388
4	957	21	0	922		720	179	312	1155	925	612	529	1172	0	441
5	1893	158	155	1090		841	1799	1095	720	1893	1037	979	1909	204	721
6	2452	18	18	2110		110	589	1528	281	2448	123	2140	2124	22	2660
7	1584	215	128	1623		581	757	532	524	1581	26	1317	1685	227	1037
8	2082	81	45	2609		265	1123	1753	105	2077	0	2193	2879	0	1526
9	1971	72	26	1767		313	449	2056	390	1911	15	1659	2220	54	1435
10	1026	17	10	642		215	27	842	158	1026	29	511	1118	235	224
11	1087	208	165	546		124	54	932	28	1071	0	230	1107	169	110
12	544	0	2	319		217	137	1654	14	532	0	352	1704	23	310
13	719	140	137	242		343	111	248	66	695	244	83	746	94	79
14	273	36	0	221		0	0	257	0	273	0	12	293	16	6
21	90	0	16	37		0	0	250	0	87	0	68	426	431	220
22	1157	7544	150	1190		1568	1067	838	20	1121	1677	857	7440	9347	712
23	1177	6971	580	928		158	902	776	38	1101	111	878	5187	1918	469
26	1354	454	10081	1021		793	611	1136	55	1349	562	1298	10030	399	617
27	72	0	297	34		13	0	71	0	72	39	52	279	0	22
28	96	58	136	71		0	21	78	0	99	0	40	151	44	25
31	1498	735	610	803		3217	289	7	1208	1443	3256	600	1560	660	611
32	822	53	8	886		4426	356	405	291	793	1841	420	1701	165	468
33	530	1079	538	252		2178	41	0	237	522	2159	128	1699	1286	154
34	107	0	47	4		190	0	0	31	62	193	0	128	0	0
35	18	0	0	0		327	0	0	0	18	442	0	84	0	0
41	60	0	0	32		7	0	22	0	60	30	0	67	14	0
44	75	79	0	0		0	0	0	0	0	0	0	75	75	0
46	23	16	0	0		274	261	0	0	23	264	0	33	33	0
48	12	0	25	0		0	25	0	0	12	17	12	12	0	12
Totals	24719	18967	13899	19673		19853	10154	16158	8022	24231	15597	15566	49880	16421	12682

Assem Div	Building Construction	Mechanical	Electrical	Commercial Renovation	Square Foot	Site Work Landscape	Assemblies	Green Building	Interior	Concrete Masonry	Heavy Construction	Light Commercial	Facilities Construction	Plumbing	Asm Div	Residential
A		15	0	188	150	577	598	0	0	536	571	154	24	0	1	376
B		0	0	848	2498	0	5658	56	329	1975	368	2089	174	0	2	211
C		0	0	647	926	0	1304	0	1629	146	0	816	250	0	3	588
D		1067	941	712	1859	72	2538	330	825	0	0	1345	1105	1088	4	851
E		0	0	86	260	0	300	0	5	0	0	257	5	0	5	392
F		0	0	0	114	0	114	0	0	0	0	114	3	0	6	357
G		527	447	318	249	3365	729	0	0	535	1350	205	293	677	7	307
															8	760
															9	80
															10	0
															11	0
															12	0
Totals		1609	1388	2799	6056	4014	11241	386	2788	3192	2289	4980	1854	1765		3922

Reference Section	Building Construction Costs	Mechanical	Electrical	Commercial Renovation	Square Foot	Site Work Landscape	Assem.	Green Building	Interior	Concrete Masonry	Open Shop	Heavy Construction	Light Commercial	Facilities Construction	Plumbing	Resi.
Reference Tables	yes	yes	yes	yes	no	yes	yes	yes	yes	yes	yes	yes	yes	yes	yes	yes
Models					111			25					50			28
Crews	571	571	571	550		571		571	571	571	548	571	548	550	571	548
Equipment Rental Costs	yes	yes	yes	yes		yes		yes	yes	yes	yes	yes	yes	yes	yes	yes
Historical Cost Indexes	yes	yes	yes	yes	yes	yes	yes	yes	yes	yes	yes	yes	yes	yes	yes	no
City Cost Indexes	yes	yes	yes	yes	yes	yes	yes	yes	yes	yes	yes	yes	yes	yes	yes	yes

RSMeans data titles are also available in online format! Go to RSMeansOnline.com for more details.

For more information visit the RSMeans website at **www.rsmeans.com**

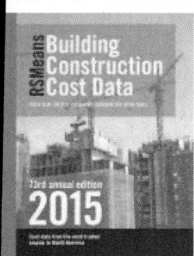

RSMeans Building Construction Cost Data 2015

Offers you unchallenged unit price reliability in an easy-to-use format. Whether used for verifying complete, finished estimates or for periodic checks, it supplies more cost facts better and faster than any comparable source. More than 24,700 unit prices have been updated for 2015. The City Cost Indexes and Location Factors cover more than 930 areas, for indexing to any project location in North America. Order and get *RSMeans Quarterly Update Service* FREE.

$206.95 | Available Sept. 2014 | Catalog no. 60015

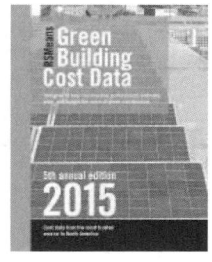

RSMeans Green Building Cost Data 2015

Estimate, plan, and budget the costs of green building for both new commercial construction and renovation work with this fifth edition of *RSMeans Green Building Cost Data*. More than 10,000 unit costs for a wide array of green building products plus assemblies costs. Easily identified cross references to LEED and Green Globes building rating systems criteria.

$170.95 | Available Nov. 2014 | Catalog no. 60555

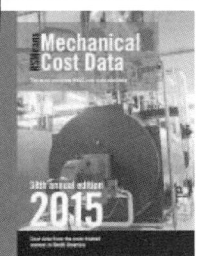

RSMeans Mechanical Cost Data 2015

Total unit and systems price guidance for mechanical construction . . . materials, parts, fittings, and complete labor cost information. Includes prices for piping, heating, air conditioning, ventilation, and all related construction.

Plus new 2015 unit costs for:

- Thousands of installed HVAC/controls, sub-assemblies and assemblies
- "On-site" Location Factors for more than 930 cities and towns in the U.S. and Canada
- Crews, labor, and equipment

$203.95 | Available Oct. 2014 | Catalog no. 60025

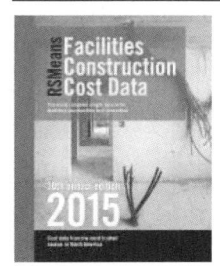

RSMeans Facilities Construction Cost Data 2015

For the maintenance and construction of commercial, industrial, municipal, and institutional properties. Costs are shown for new and remodeling construction and are broken down into materials, labor, equipment, and overhead and profit. Special emphasis is given to sections on mechanical, electrical, furnishings, site work, building maintenance, finish work, and demolition.

More than 49,800 unit costs, plus assemblies costs and a comprehensive Reference Section are included.

$524.95 | Available Nov. 2014 | Catalog no. 60205

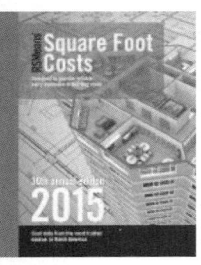

RSMeans Square Foot Costs 2015

Accurate and Easy-to-Use

- **Updated price information** based on nationwide figures from suppliers, estimators, labor experts, and contractors.
- Green building models
- Realistic graphics, offering true-to-life illustrations of building projects
- Extensive information on using square foot cost data, including sample estimates and alternate pricing methods

$218.95 | Available Oct. 2014 | Catalog no. 60055

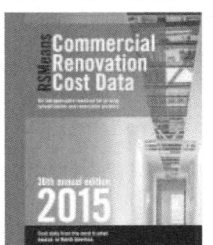

RSMeans Commercial Renovation Cost Data 2015
Commercial/Multi-family Residential

Use this valuable tool to estimate commercial and multi-family residential renovation and remodeling.

Includes: Updated costs for hundreds of unique methods, materials, and conditions that only come up in repair and remodeling, PLUS:

- Unit costs for more than 19,600 construction components
- Installed costs for more than 2,700 assemblies
- More than 930 "on-site" localization factors for the U.S. and Canada

$166.95 | Available Oct. 2014 | Catalog no. 60045

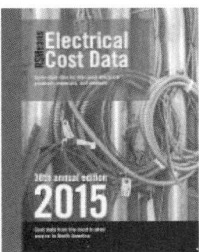

RSMeans Electrical Cost Data 2015

Pricing information for every part of electrical cost planning. More than 13,800 unit and systems costs with design tables; clear specifications and drawings; engineering guides; illustrated estimating procedures; complete labor-hour and materials costs for better scheduling and procurement; and the latest electrical products and construction methods.

- A variety of special electrical systems, including cathodic protection
- Costs for maintenance, demolition, HVAC/mechanical, specialties, equipment, and more

$208.95 | Available Oct. 2014 | Catalog no. 60035

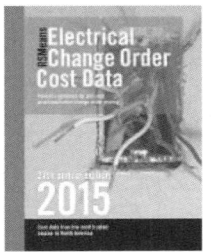

RSMeans Electrical Change Order Cost Data 2015

RSMeans Electrical Change Order Cost Data provides you with electrical unit prices exclusively for pricing change orders—based on the recent, direct experience of contractors and suppliers. Analyze and check your own change order estimates against the experience others have had doing the same work. It also covers productivity analysis, and change order cost justifications. With useful information for calculating the effects of change orders and dealing with their administration.

$199.95 | Available Dec. 2014 | Catalog no. 60235

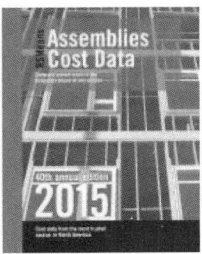

RSMeans Assemblies Cost Data 2015

RSMeans Assemblies Cost Data takes the guesswork out of preliminary or conceptual estimates. Now you don't have to try to calculate the assembled cost by working up individual component costs. We've done all the work for you.

Presents detailed illustrations, descriptions, specifications, and costs for every conceivable building assembly—over 350 types in all—arranged in the easy-to-use UNIFORMAT II system. Each illustrated "assembled" cost includes a complete grouping of materials and associated installation costs, including the installing contractor's overhead and profit.

$334.95 | Available Sept. 2014 | Catalog no. 60065

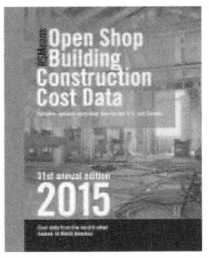

RSMeans Open Shop Building Construction Cost Data 2015

The latest costs for accurate budgeting and estimating of new commercial and residential construction . . . renovation work . . . change orders . . . cost engineering.

RSMeans Open Shop "BCCD" will assist you to:

- Develop benchmark prices for change orders.
- Plug gaps in preliminary estimates and budgets.
- Estimate complex projects.
- Substantiate invoices on contracts.
- Price ADA-related renovations.

$175.95 | Available Dec. 2014 | Catalog no. 60155

RSMeans data titles are also available in online format! Go to RSMeansOnline.com for more details

For more information visit the RSMeans website at www.rsmeans.com

Annual Cost Guides
Unit prices according to the latest MasterFormat!

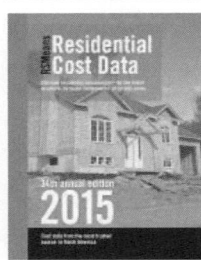

RSMeans Residential Cost Data 2015

Contains square foot costs for 28 basic home models with the look of today, plus hundreds of custom additions and modifications you can quote right off the page. Includes more than 3,900 costs for 89 residential systems. Complete with blank estimating forms, sample estimates, and step-by-step instructions.

Contains line items for cultured stone and brick, PVC trim, lumber, and TPO roofing.

$148.95 | Available Oct. 2014 | Catalog no. 60175

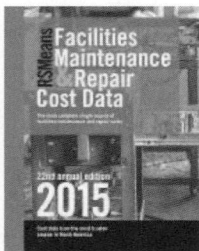

RSMeans Facilities Maintenance & Repair Cost Data 2015

RSMeans Facilities Maintenance & Repair Cost Data gives you a complete system to manage and plan your facility repair and maintenance costs and budget efficiently. Guidelines for auditing a facility and developing an annual maintenance plan. Budgeting is included, along with reference tables on cost and management, and information on frequency and productivity of maintenance operations.

The only nationally recognized source of maintenance and repair costs. Developed in cooperation with the Civil Engineering Research Laboratory (CERL) of the Army Corps of Engineers.

$461.95 | Available Nov. 2014 | Catalog no. 60305

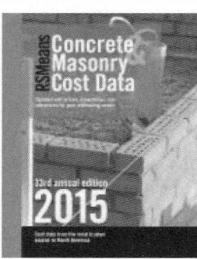

RSMeans Concrete & Masonry Cost Data 2015

Provides you with cost facts for virtually all concrete/masonry estimating needs, from complicated formwork to various sizes and face finishes of brick and block—all in great detail. The comprehensive Unit Price Section contains more than 8,000 selected entries. Also contains an Assemblies [Cost] Section, and a detailed Reference Section that supplements the cost data.

$188.95 | Available Dec. 2014 | Catalog no. 60115

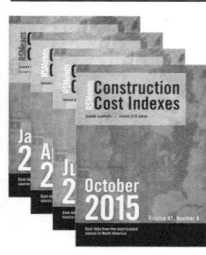

RSMeans Construction Cost Indexes 2015

What materials and labor costs will change unexpectedly this year? By how much?

- Breakdowns for 318 major cities
- National averages for 30 key cities
- Expanded five major city indexes
- Historical construction cost indexes

$359.95 per year (subscription) | Catalog no. 50145
$94.95 individual quarters | Catalog no. 60145 A,B,C,D

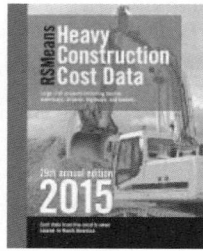

RSMeans Heavy Construction Cost Data 2015

A comprehensive guide to heavy construction costs. Includes costs for highly specialized projects such as tunnels, dams, highways, airports, and waterways. Information on labor rates, equipment, and materials costs is included. Features unit price costs, systems costs, and numerous reference tables for costs and design.

$204.95 | Available Dec. 2014 | Catalog no. 60165

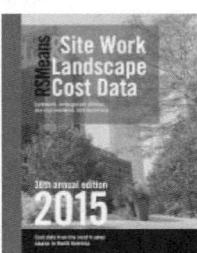

RSMeans Site Work & Landscape Cost Data 2015

Includes unit and assemblies costs for earthwork, sewerage, piped utilities, site improvements, drainage, paving, trees and shrubs, street openings/repairs, underground tanks, and more. Contains more than 60 types of assemblies costs for accurate conceptual estimates.

Includes:

- Estimating for infrastructure improvements
- Environmentally-oriented construction
- ADA-mandated handicapped access
- Hazardous waste line items

$199.95 | Available Dec. 2014 | Catalog no. 60285

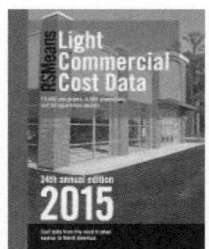

RSMeans Light Commercial Cost Data 2015

Specifically addresses the light commercial market, which is a specialized niche in the construction industry. Aids you, the owner/designer/contractor, in preparing all types of estimates—from budgets to detailed bids. Includes new advances in methods and materials.

Assemblies Section allows you to evaluate alternatives in the early stages of design/planning.

More than 15,500 unit costs ensure that you have the prices you need, when you need them.

$153.95 | Available Nov. 2014 | Catalog no. 60185

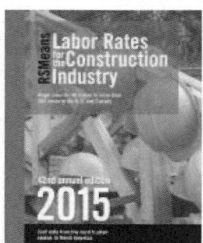

RSMeans Labor Rates for the Construction Industry 2015

Complete information for estimating labor costs, making comparisons, and negotiating wage rates by trade for more than 300 U.S. and Canadian cities. With 46 construction trades in each city, and historical wage rates included for comparison. Each city chart lists the county and is alphabetically arranged with handy visual flip tabs for quick reference.

$450.95 | Available Dec. 2014 | Catalog no. 60125

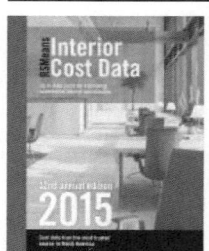

RSMeans Interior Cost Data 2015

Provides you with prices and guidance needed to make accurate interior work estimates. Contains costs on materials, equipment, hardware, custom installations, furnishings, and labor for new and remodel commercial and industrial interior construction, including updated information on office furnishings, and reference information.

$208.95 | Available Nov. 2014 | Catalog no. 60095

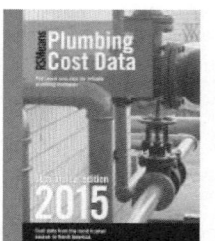

RSMeans Plumbing Cost Data 2015

Comprehensive unit prices and assemblies for plumbing, irrigation systems, commercial and residential fire protection, point-of-use water heaters, and the latest approved materials. This publication and its companion, *RSMeans Mechanical Cost Data*, provide full-range cost estimating coverage for all the mechanical trades.

Contains updated costs for potable water, radiant heat systems, and high efficiency fixtures.

$206.95 | Available Oct. 2014 | Catalog no. 60215

RSMeans Online and RSMeans CostWorks CD Comparison Matrix

Both the RSMeans CostWorks CD package and the RSMeans Online estimating tool provide the same comprehensive, up-to-date, and localized cost information to improve planning, estimating, and budgeting with RSMeans—the industry's most-trusted source of construction cost data.

In addition to key differences noted in the comparison matrix below, both products let you do the following:

- Quickly locate costs in the searchable RSMeans database.
- Instantly adjust the data to your specific location for accurate costs anywhere in the U.S. and Canada.
- Create reliable, updated cost estimates in minutes.
- Manage, share, and export your estimates to MS Excel® with ease.
- Access a wide range of reference materials, including crew rates and national average labor rates.

Both the CD and Online options are available in a convenient one-year subscription!

Use the handy comparison matrix below to help you decide the best product for your cost estimating needs!

Why us?
For more than 70 years, RSMeans has provided the quality cost data construction professionals everywhere depend on to estimate with confidence, save time, improve decision-making, and increase profits.

Improve planning and decision-making
Back your estimates with complete, accurate, and up-to-date cost data for informed business decisions.
- Verify construction costs from third parties.
- Check validity of subcontractor proposals.
- Evaluate material and assembly alternatives.

Increase profits
Use RSMeans to estimate projects quickly and accurately, so you can gain an edge over your competition.
- Create accurate and competitive bids.
- Minimize the risk of cost overruns.
- Reduce variability.
- Gain control over costs.

The following matrix summarizes the key comparison points between RSMeans' two electronic cost data products.

	Question	CostWorks CD	RSMeans Online.com
1	How is the data presented?	• The data is organized by book title and sold at comparable book cost. • Several packaged bundles of five titles are available at a cost savings. • Available in a convenient one-year subscription	• The data is organized by book title and sold at comparable book cost. • Several packaged bundles of five titles are available at a cost savings. • RSMeans Online also offers a Complete Library title with all online data at a cost savings. • Available in a convenient one-year subscription
2	Where is the data stored?	With a typical install, the data is read off of your hard drive or off of the CD.	All data is stored on our secure server.
3	How are updates delivered?	Updates are available on a quarterly basis for download from the Costworks HomePort website.	Data is updated automatically on a quarterly basis, and users are notified when updates occur.
4	Where are my projects and estimates stored?	Upon installation, a "My CostWorks Projects" folder is created within the "My Documents" directory. All projects are defaulted to save here unless otherwise specified.	Estimates are stored on our secure server.
5	Does my IT team need to assist me in installing and running the program?	Yes, depending on the level of your permission. See #11 for system requirements.	While the product is entirely Web-based, you may need IT support with permissions for certain tasks. See #11 for system requirements.
6	Can users collaborate on projects/estimates?	Yes, you can copy project folders between computers.	You can create shared folders that can be viewed, copied, and edited by anyone within your account workgroup.

Comparison matrix continued

#	Question		
7	Can I create and use my own data with the RSMeans database?	Custom lines cannot be added to the CD database. With the Estimator tool, however, you can create and add custom cost lines to any estimate. You can also save custom lines as an Estimator template for use in creating new estimates.	You can create • a custom line within an estimate • a custom cost book consisting of custom unit (not assembly) cost lines, no more than once per quarter **Note** that custom data is searchable using the Include My Custom Data feature.
8	Can I enter my own trade/labor rates and markups?	Although you cannot adjust the trade labor rates, you can add custom markups for each report template using the Estimator tool.	For each quarterly update, you can create one custom unit cost book. You can edit the labor rates by individual trade and the crew equipment costs, and also edit the associated material and equipment markups.
9	What are my reporting options?	The Estimator tool lets you create editable custom reports and report templates. Reports can also be exported to MS Excel®.	You can generate pre-formatted estimate summaries and cost lists in both PDF and Excel formats.
10	How do I find specific cost data?	You can search by CSI line number and by keyword within a division.	You can search by CSI line number and by keyword within a division. There are several additional search options, including keyword and advanced search using special characters.
11	What are the system requirements?	• Operating system – Windows XP or later • PC requirements - 256 MB RAM (recommended), 1024 x 768 High Color Monitor, 800MHz Computer (recommended), 24X CD-ROM, 250 MB Hard Disk Space	• Resolution – Recommended resolution is 1024 x 768 or higher • Compatibility - Internet Explorer version 8.x, 9.x and Mozilla Firefox version 10 and higher • iPad - iOS 5 and higher

2015 RSMeans Seminar Schedule
Note: call for exact dates and details.

Location	Dates	Location	Dates
Seattle, WA	January and August	Jacksonville FL	September
Dallas/Ft. Worth, TX	January	Dallas, TX	September
Austin, TX	February	Charleston, SC	October
Anchorage, AK	March and September	Houston, TX	October
Las Vegas, NV	March and October	Atlanta, GA	November
Washington, DC	April and September	Baltimore, MD	November
Phoenix, AZ	April	Orlando, FL	November
Kansas City, MO	April	San Diego, CA	December
Toronto	May	San Antonio, TX	December
Denver, CO	May	Raleigh, NC	December
San Francisco, CA	June		
Bethesda, MD	June		
Columbus, GA	June		
El Segundo, CA	August		

1-800-334-3509, Press 1

Professional Development

For more information visit the RSMeans website at www.rsmeans.com

eLearning Training Sessions

Learn how to use *RSMeans Online*® or *RSMeans CostWorks*® CD from the convenience of your home or office. Our eLearning training sessions let you join a training conference call and share the instructors' desktops, so you can view the presentation and step-by-step instructions on your own computer screen. The live webinars are held from 9 a.m. to 4 p.m. eastern standard time, with a one-hour break for lunch.

For these sessions, you must have a computer with high speed Internet access and a compatible Web browser. Learn more at www.rsmeansonline.com or call for a schedule: 1-800-334-3509 and press 1. Webinars are generally held on selected Wednesdays each month.

RSMeans Online®
$299 per person

RSMeans CostWorks® **CD**
$299 per person

RSMeans data titles are also available in online format! Go to RSMeansOnline.com for more details.

For more information visit the RSMeans website at www.rsmeans.com

RSMeans Online™ Training

Construction estimating is vital to the decision-making process at each state of every project. RSMeansOnline works the way you do. It's systematic, flexible and intuitive. In this one day class you will see how you can estimate any phase of any project faster and better.

Some of what you'll learn:
- Customizing RSMeansOnline
- Making the most of RSMeans "Circle Reference" numbers
- How to integrate your cost data
- Generate reports, exporting estimates to MS Excel, sharing, collaborating and more

Also available: RSMeans Online® training webinar

Facilities Construction Estimating

In this *two-day* course, professionals working in facilities management can get help with their daily challenges to establish budgets for all phases of a project.

Some of what you'll learn:
- Determining the full scope of a project
- Identifying the scope of risks and opportunities
- Creative solutions to estimating issues
- Organizing estimates for presentation and discussion
- Special techniques for repair/remodel and maintenance projects
- Negotiating project change orders

Who should attend: facility managers, engineers, contractors, facility tradespeople, planners, and project managers.

Construction Cost Estimating: Concepts and Practice

This *one-day* introductory course to improve estimating skills and effectiveness starts with the details of interpreting bid documents, ending with the summary of the estimate and bid submission.

Some of what you'll learn:
- Using the plans and specifications for creating estimates
- The takeoff process—deriving all tasks with correct quantities
- Developing pricing, using various sources; how subcontractor pricing fits in
- Summarizing the estimate to arrive at the final number
- Formulas for area and cubic measure, adding waste and adjusting productivity to specific projects
- Evaluating subcontractors' proposals and prices
- Adding Insurance and bonds
- Understanding how labor costs are calculated
- Submitting bids and proposals

Who should attend: project managers, architects, engineers, owner's representatives, contractors, and anyone who's responsible for budgeting or estimating construction projects.

Maintenance & Repair Estimating for Facilities

This *two-day* course teaches attendees how to plan, budget, and estimate the cost of ongoing and preventive maintenance and repair for existing buildings and grounds.

Some of what you'll learn:
- The most financially favorable maintenance, repair, and replacement scheduling and estimating
- Auditing and value engineering facilities
- Preventive planning and facilities upgrading
- Determining both in-house and contract-out service costs
- Annual, asset-protecting M&R plan

Who should attend: facility managers, maintenance supervisors, buildings and grounds superintendents, plant managers, planners, estimators, and others involved in facilities planning and budgeting.

Practical Project Management for Construction Professionals

In this *two-day* course, acquire the essential knowledge and develop the skills to effectively and efficiently execute the day-to-day responsibilities of the construction project manager.

Covers:
- General conditions of the construction contract
- Contract modifications: change orders and construction change directives
- Negotiations with subcontractors and vendors
- Effective writing: notification and communications
- Dispute resolution: claims and liens

Who should attend: architects, engineers, owner's representatives, project managers.

Mechanical & Electrical Estimating

This *two-day* course teaches attendees how to prepare more accurate and complete mechanical/electrical estimates, avoiding the pitfalls of omission and double-counting, while understanding the composition and rationale within the RSMeans mechanical/electrical database.

Some of what you'll learn:
- The unique way mechanical and electrical systems are interrelated
- M&E estimates—conceptual, planning, budgeting, and bidding stages
- Order of magnitude, square foot, assemblies, and unit price estimating
- Comparative cost analysis of equipment and design alternatives

Who should attend: architects, engineers, facilities managers, mechanical and electrical contractors, and others who need a highly reliable method for developing, understanding, and evaluating mechanical and electrical contracts.

Professional Development

Professional Development

Unit Price Estimating

This interactive *two-day* seminar teaches attendees how to interpret project information and process it into final, detailed estimates with the greatest accuracy level.

The most important credential an estimator can take to the job is the ability to visualize construction and estimate accurately.

Some of what you'll learn:
- Interpreting the design in terms of cost
- The most detailed, time-tested methodology for accurate pricing
- Key cost drivers—material, labor, equipment, staging, and subcontracts
- Understanding direct and indirect costs for accurate job cost accounting and change order management

Who should attend: corporate and government estimators and purchasers, architects, engineers, and others who need to produce accurate project estimates.

RSMeans CostWorks® CD Training

This one-day course helps users become more familiar with the functionality of *RSMeans CostWorks* program. Each menu, icon, screen, and function found in the program is explained in depth. Time is devoted to hands-on estimating exercises.

Some of what you'll learn:
- Searching the database using all navigation methods
- Exporting RSMeans data to your preferred spreadsheet format
- Viewing crews, assembly components, and much more
- Automatically regionalizing the database

This training session requires you to bring a laptop computer to class.

When you register for this course you will receive an outline for your laptop requirements.

Also offering web training for RSMeans CostWorks CD!

Facilities Estimating Using RSMeans CostWorks® CD

Combines hands-on skill building with best estimating practices and real-life problems. Brings you up-to-date with key concepts, and provides tips, pointers, and guidelines to save time and avoid cost oversights and errors.

Some of what you'll learn:
- Estimating process concepts
- Customizing and adapting RSMeans cost data
- Establishing scope of work to account for all known variables
- Budget estimating: when, why, and how
- Site visits: what to look for—what you can't afford to overlook
- How to estimate repair and remodeling variables

This training session requires you to bring a laptop computer to class.

Who should attend: facility managers, architects, engineers, contractors, facility tradespeople, planners, project managers and anyone involved with JOC, SABRE, or IDIQ.

Conceptual Estimating Using RSMeans CostWorks® CD

This *two-day* class uses the leading industry data and a powerful software package to develop highly accurate conceptual estimates for your construction projects. All attendees must bring a laptop computer loaded with the current year *Square Foot Costs* and the *Assemblies Cost Data* CostWorks titles.

Some of what you'll learn:
- Introduction to conceptual estimating
- Types of conceptual estimates
- Helpful hints
- Order of magnitude estimating
- Square foot estimating
- Assemblies estimating

Who should attend: architects, engineers, contractors, construction estimators, owner's representatives, and anyone looking for an electronic method for performing square foot estimating.

Assessing Scope of Work for Facility Construction Estimating

This *two-day* practical training program addresses the vital importance of understanding the SCOPE of projects in order to produce accurate cost estimates for facility repair and remodeling.

Some of what you'll learn:
- Discussions of site visits, plans/specs, record drawings of facilities, and site-specific lists
- Review of CSI divisions, including means, methods, materials, and the challenges of scoping each topic
- Exercises in SCOPE identification and SCOPE writing for accurate estimating of projects
- Hands-on exercises that require SCOPE, take-off, and pricing

Who should attend: corporate and government estimators, planners, facility managers, and others who need to produce accurate project estimates.

Unit Price Estimating Using RSMeans CostWorks® CD

Step-by-step instruction and practice problems to identify and track key cost drivers—material, labor, equipment, staging, and subcontractors—for each specific task. Learn the most detailed, time-tested methodology for accurately "pricing" these variables, their impact on each other and on total cost.

Some of what you'll learn:
- Unit price cost estimating
- Order of magnitude, square foot, and assemblies estimating
- Quantity takeoff
- Direct and indirect construction costs
- Development of contractor's bill rates
- How to use *RSMeans Building Construction Cost Data*

This training session requires you to bring a laptop computer to class.

Who should attend: architects, engineers, corporate and government estimators, facility managers, and government procurement staff.

RSMeans data titles are also available in online format! Go to RSMeansOnline.com for more details.

For more information visit the RSMeans website at **www.rsmeans.com**

Registration Information

Register early... Save up to $100! Register 30 days before the start date of a seminar and save $100 off your total fee. *Note: This discount can be applied only once per order. It cannot be applied to team discount registrations or any other special offer.*

How to register Register by phone today! The RSMeans toll-free number for making reservations is **1-800-334-3509, Press 1.**

Two-day seminar registration fee - $935. One-day *RSMeans CostWorks*® training registration fee - $375. To register by mail, complete the registration form and return, with your full fee, to: RSMeans Seminars, 700 Longwater Drive, Norwell, MA 02061.

Government pricing All federal government employees save off the regular seminar price. Other promotional discounts cannot be combined with the government discount.

Team discount program for two to four seminar registrations. Call for pricing: 1-800-334-3509, Press 1

Multiple course discounts When signing up for two or more courses, call for pricing.

Refund policy Cancellations will be accepted up to ten business days prior to the seminar start. There are no refunds for cancellations received later than ten working days prior to the first day of the seminar. A $150 processing fee will be applied for all cancellations. Written notice of cancellation is required. Substitutions can be made at any time before the session starts. **No-shows are subject to the full seminar fee.**

AACE approved courses Many seminars described and offered here have been approved for 14 hours (1.4 recertification credits) of credit by the AACE International Certification Board toward meeting the continuing education requirements for recertification as a Certified Cost Engineer/Certified Cost Consultant.

AIA Continuing Education We are registered with the AIA Continuing Education System (AIA/CES) and are committed to developing quality learning activities in accordance with the CES criteria. Many seminars meet the AIA/CES criteria for Quality Level 2. AIA members may receive (14) learning units (LUs) for each two-day RSMeans course.

Daily course schedule The first day of each seminar session begins at 8:30 a.m. and ends at 4:30 p.m. The second day begins at 8:00 a.m. and ends at 4:00 p.m. Participants are urged to bring a hand-held calculator, since many actual problems will be worked out in each session.

Continental breakfast Your registration includes the cost of a continental breakfast, and a morning and afternoon refreshment break. These informal segments allow you to discuss topics of mutual interest with other seminar attendees. (You are free to make your own lunch and dinner arrangements.)

Hotel/transportation arrangements RSMeans arranges to hold a block of rooms at most host hotels. To take advantage of special group rates when making your reservation, be sure to mention that you are attending the RSMeans seminar. You are, of course, free to stay at the lodging place of your choice. (**Hotel reservations and transportation arrangements should be made directly by seminar attendees.**)

Important Class sizes are limited, so please register as soon as possible.

Note: Pricing subject to change.

Registration Form

ADDS-1000

Call 1-800-334-3509, Press 1 to register or FAX this form 1-800-632-6732. Visit our website: www.rsmeans.com

Please register the following people for the RSMeans construction seminars as shown here. We understand that we must make our own hotel reservations if overnight stays are necessary.

☐ Full payment of $_____ enclosed.

☐ Bill me.

Please print name of registrant(s).
(To appear on certificate of completion)

_____
_____
_____

P.O. #: _____
GOVERNMENT AGENCIES MUST SUPPLY PURCHASE ORDER NUMBER OR TRAINING FORM.

Firm name_____

Address_____

City/State/Zip_____

Telephone no._____ Fax no._____

E-mail address_____

Charge registration(s) to: ☐ MasterCard ☐ VISA ☐ American Express

Account no._____ Exp. date_____

Cardholder's signature_____

Seminar name_____

Seminar City_____

Please mail check to: RSMeans Seminars, 700 Longwater Drive, Norwell, MA 02061 USA

Professional Development

RSMeans data titles are also available in online format! Go to RSMeansOnline.com for more details

For more information visit the RSMeans website at www.rsmeans.com

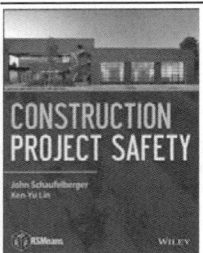

Managing the Profitable Construction Business: The Contractor's Guide to Success and Survival Strategies, 2nd Ed.

by Thomas C. Scheifler, Ph.D.

Learn from a team of construction business veterans led by Thomas C. Schleifer, who is commonly referred to as a construction business "turnaround" expert due to the number of construction companies he has rescued from financial distress. His financial acumen, combined with his practical, hands-on experience, has made him a sought-after private consultant.

$60.00 | 288 pages, hardover | Catalog no. 67370

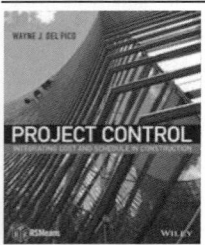

Construction Project Safety

by John Schaufelberger, Ken-Yu Lin

This essential introduction to construction safety for construction management personnel takes a project-based approach to present potential hazards in construction and their mitigation or prevention.

Beginning with an introduction to Accident Prevention Programs and OSHA compliance requirements, the book integrates safety instruction into the building process by following a building project from site construction through interior finish. Reinforcing this applied approach are photographs, drawings, contract documentation, and an online 3D BIM model to help visualize the on-site scenarios.

$85.00 | 320 pages, hardover | Catalog no. 67368

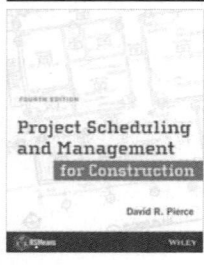

Project Control: Integrating Cost & Schedule in Construction

by Wayne J. Del Pico

Written by a seasoned professional in the field, *Project Control: Integrating Cost and Schedule in Construction* fills a void in the area of project control as applied in the construction industry today. It demonstrates how productivity models for an individual project are created, monitored, and controlled, and how corrective actions are implemented as deviations from the baseline occur.

$65.00 | 240 pages, softcover | Catalog no. 67366

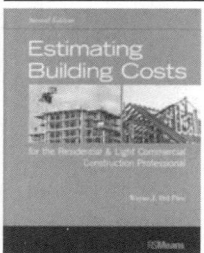

Project Scheduling and Management for Construction, 4th Edition

by David R. Pierce

First published in 1988 by RSMeans, the new edition of *Project Scheduling and Management for Construction* has been substantially revised for both professionals and students enrolled in construction management and civil engineering programs. While retaining its emphasis on developing practical, professional-level scheduling skills, the new edition is a relatable, real world case study.

$95.00 | 272 pages, softcover | Catalog no. 67367

Estimating Building Costs, 2nd Edition
For the Residential & Light Commercial Construction Professional

by Wayne J. Del Pico

Explains, in detail, how to put together a reliable estimate that can be used not only for budgeting, but also for developing a schedule, managing a project, dealing with contingencies, and, ultimately, making a profit.

Completely revised and updated to reflect the CSI MasterFormat 2010 system.

$75.00 | 528 pages, softcover | Catalog no. 67343A

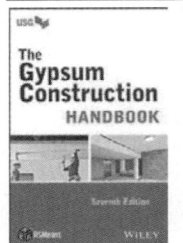

The Gypsum Construction Handbook, 7th Edition

by USG

The tried-and-true Gypsum Construction Handbook is a systematic guide to selecting and using gypsum drywall, veneer plaster, tile backers, ceilings, and conventional plaster building materials. A widely respected training text for aspiring architects and engineers, the book provides detailed product information and efficient installation methodology.

The 7th edition features updates in gypsum products, including ultralight panels, glass-mat panels, paper-faced plastic bead, and ultralightweight joint compound, and modern specialty acoustical and ceiling product guidelines. This comprehensive reference also incorporates the latest in sustainable products.

$40.00 | 576 pages, softcover | Catalog no. 67357B

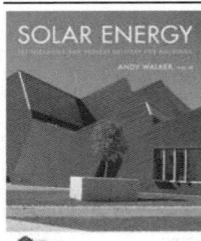

Solar Energy: Technologies & Project Delivery for Buildings

by Andy Walker

An authoritative reference on the design of solar energy systems in building projects, with applications, operating principles, and simple tools for the construction, engineering, and design professionals, the book simplifies the solar design and engineering process and provides sample documentation for the complete design of a solar energy system for buildings.

$85.00 | 320 pages, hardover | Catalog no. 67365

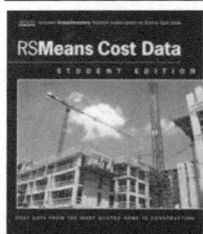

RSMeans Cost Data, Student Edition

Provides a thorough introduction to cost estimating in a print and online package. Clear explanations and example-driven. The ideal reference for students and new professionals. Features include:

- Commercial and residential construction cost data in print and online formats
- Complete how-to guidance on the essentials of cost estimating
- A supplemental website with plans, problem sets, and a full sample estimate

$110.00 | 512 pages, softcover | Catalog no. 67363

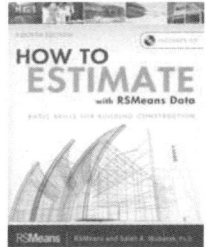

How to Estimate with Means Data & CostWorks, 4th Edition

by RSMeans and Saleh A. Mubarak, Ph.D.

This step-by-step guide takes you through all the major construction items with extensive coverage of site work, concrete and masonry, wood and metal framing, doors and windows, and other divisions. The only construction cost estimating handbook that uses the most popular source of construction data, RSMeans, this indispensible guide features access to the instructional version of *CostWorks* in electronic form, enabling you to practice techniques to solve real-world estimating problems.

$75.00 | 320 pages, softcover | Includes CostWorks CD | Catalog no. 67324C

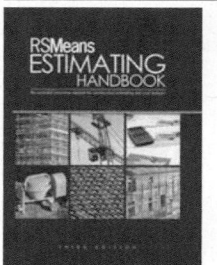

RSMeans Estimating Handbook, 3rd Edition

Widely used in the industry for tasks ranging from routine estimates to special cost analysis projects. This handbook will help construction professionals:

- Evaluate architectural plans and specifications
- Prepare accurate quantity takeoffs
- Compare design alternatives and costs
- Perform value engineering
- Double-check estimates and quotes
- Estimate change orders

$110.00 | Over 976 pages, hardcover | Catalog No. 67276B

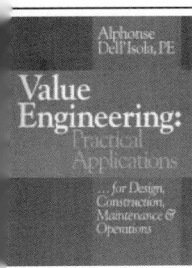

Value Engineering: Practical Applications

For Design, Construction, Maintenance & Operations

by Alphonse Dell'Isola, PE

A tool for immediate application—for engineers, architects, facility managers, owners, and contractors. Includes making the case for VE—the management briefing; integrating VE into planning, budgeting, and design; conducting life cycle costing; using VE methodology in design review and consultant selection; and case studies.

$90.00 | Over 427 pages, illustrated, softcover | Catalog no. 67319A

Risk Management for Design and Construction

Introduces risk as a central pillar of project management and shows how a project manager can prepare to deal with uncertainty. Includes:

- Integrated cost and schedule risk analysis
- An introduction to a ready-to-use system of analyzing a project's risks and tools to proactively manage risks
- A methodology that was developed and used by the Washington State Department of Transportation
- Case studies and examples on the proper application of principles
- Combining value analysis with risk analysis

$125.00 | Over 288 pages, softcover | Catalog no. 67359

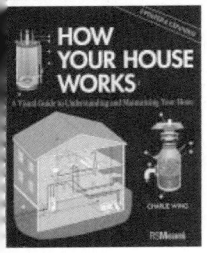

How Your House Works, 2nd Edition

by Charlie Wing

Knowledge of your home's systems helps you control repair and construction costs and makes sure the correct elements are being installed or replaced. This book uncovers the mysteries behind just about every major appliance and building element in your house. See-through, cross-section drawings in full color show you exactly how these things should be put together and how they function, including what to check if they don't work. It just might save you having to call in a professional.

$22.95 | 208 pages, softcover | Catalog no. 67351A

Construction Business Management

by Nick Ganaway

Only 43% of construction firms stay in business after four years. Make sure your company thrives with valuable guidance from a pro with 25 years of success as a commercial contractor. Find out what it takes to build all aspects of a business that is profitable, enjoyable, and enduring. With a bonus chapter on retail construction.

$60.00 | 201 pages, softcover | Catalog no. 67352

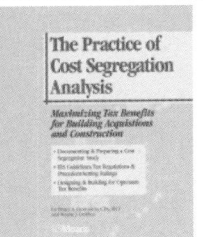

The Practice of Cost Segregation Analysis

by Bruce A. Desrosiers and Wayne J. Del Pico

This expert guide walks you through the practice of cost segregation analysis, which enables property owners to defer taxes and benefit from "accelerated cost recovery" through depreciation deductions on assets that are properly identified and classified.

With a glossary of terms, sample cost segregation estimates for various building types, key information resources, and updates via a dedicated website, this book is a critical resource for anyone involved in cost segregation analysis.

$105.00 | Over 240 pages | Catalog no. 67345

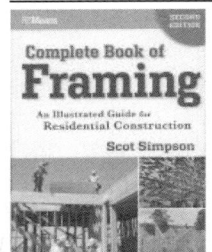

Complete Book of Framing, 2nd Edition

by Scot Simpson

This updated, easy-to-learn guide to rough carpentry and framing is written by an expert with more than thirty years of framing experience. Starting with the basics, this book begins with types of lumber, nails, and what tools are needed, followed by detailed, fully illustrated steps for framing each building element. Framer-Friendly Tips throughout the book show how to get a task done right—and more easily.

$29.95 | 368 pages, softcover | Catalog no. 67353A

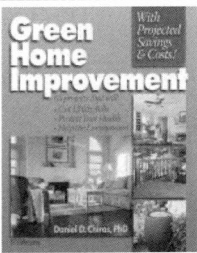

Green Home Improvement

by Daniel D. Chiras, PhD

With energy costs rising and environmental awareness increasing, people are looking to make their homes greener. This book, with 65 projects and actual costs and projected savings, helps homeowners prioritize their green improvements.

Projects range from simple water savers that cost only a few dollars, to bigger-ticket items such as HVAC systems. With color photos and cost estimates, each project compares options and describes the work involved, the benefits, and the savings.

$34.95 | 320 pages, illustrated, softcover | Catalog no. 67355

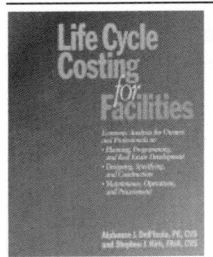

Life Cycle Costing for Facilities

by Alphonse Dell'Isola and Dr. Stephen Kirk

Guidance for achieving higher quality design and construction projects at lower costs! Cost-cutting efforts often sacrifice quality to yield the cheapest product. Life cycle costing enables building designers and owners to achieve both. The authors of this book show how LCC can work for a variety of projects—from roads to HVAC upgrades to different types of buildings.

$110.00 | 396 pages, hardcover | Catalog no. 67341

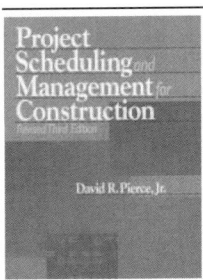

Project Scheduling and Management for Construction, 3rd Edition

by David R. Pierce, Jr.

A comprehensive, yet easy-to-follow guide to construction project scheduling and control—from vital project management principles through the latest scheduling, tracking, and controlling techniques. The author is a leading authority on scheduling, with years of field and teaching experience at leading academic institutions. Spend a few hours with this book and come away with a solid understanding of this essential management topic.

$64.95 | Over 286 pages, illustrated, hardcover | Catalog no. 67247B

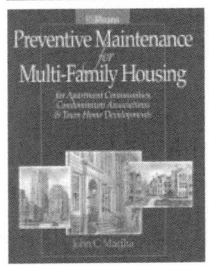

Preventive Maintenance for Multi-Family Housing

by John C. Maciha

Prepared by one of the nation's leading experts on multi-family housing.

This complete PM system for apartment and condominium communities features expert guidance, checklists for buildings and grounds maintenance tasks and their frequencies, and a dedicated website featuring customizable electronic forms. A must-have for anyone involved with multi-family housing maintenance and upkeep.

$95.00 | 290 pages | Catalog no. 67346

RSMeans data titles are also available in online format! Go to RSMeansOnline.com for more details

For more information visit the RSMeans website at **www.rsmeans.com**

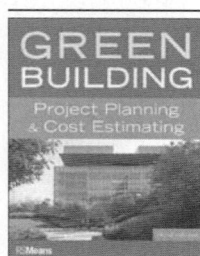

Green Building: Project Planning & Cost Estimating, 3rd Edition

Since the widely read first edition of this book, green building has gone from a growing trend to a major force in design and construction.

This new edition has been updated with the latest in green building technologies, design concepts, standards, and costs. Full-color with all new case studies—plus a new chapter on commercial real estate.

$110.00 | 480 pages, softcover | Catalog no. 67338B

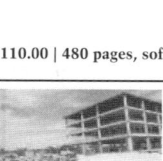

Construction Supervision

Inspires excellence with proven tactics and techniques applied by thousands of construction supervisors. Leadership guidelines carve out a practical blueprint for motivating work performance and increasing productivity through effective communication. Features:

- A unique focus on field supervision and crew management
- Coverage of supervision from the foreman to the superintendent level
- An overview of technical skills whose mastery will build confidence and success for the supervisor
- A detailed view of "soft" management and communication skills

$95.00 | 464 pages, softcover | Catalog no. 67358

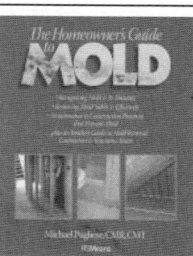

The Homeowner's Guide to Mold

By Michael Pugliese

Mold, whether caused by leaks, humidity or flooding, is a real health and financial issue—for homeowners and contractors. This full-color book explains:

- Construction and maintenance practices to prevent mold
- How to inspect for and remove mold
- Mold remediation procedures and costs
- What to do after a flood
- How to deal with insurance companies

$21.95 | 144 pages, softcover | Catalog no. 67344

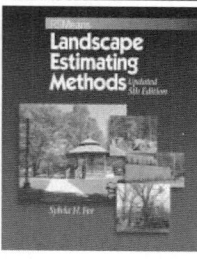

Landscape Estimating Methods, 5th Edition

Answers questions about preparing competitive landscape construction estimates, with up-to-date cost estimates and the new MasterFormat classification system. Expanded and revised to address the latest materials and methods, including new coverage on approaches to green building. Includes:

- Step-by-step explanation of the estimating process
- Sample forms and worksheets that save time and prevent errors

$75.00 | 336 pages, softcover | Catalog no. 67295C

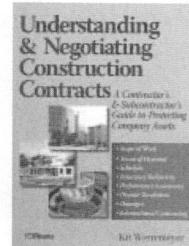

Understanding & Negotiating Construction Contracts

by Kit Werremeyer

Take advantage of the author's 30 years' experience in small-to-large (including international) construction projects. Learn how to identify, understand, and evaluate high risk terms and conditions typically found in all construction contracts—then, negotiate to lower or eliminate the risk, improve terms of payment, and reduce exposure to claims and disputes. The author avoids "legalese" and gives real-life examples from actual projects.

$80.00 | 320 pages, softcover | Catalog no. 67350

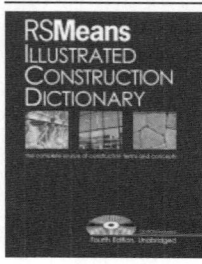

RSMeans Illustrated Construction Dictionary Unabridged 4th Edition, with CD-ROM

Long regarded as the industry's finest, *Means Illustrated Construction Dictionary* is now even better. With nearly 20,000 terms and more than 1,400 illustrations and photos, it is the clear choice for the most comprehensive and current information. The companion CD-ROM that comes with this new edition adds extra features, such as larger graphics and expanded definitions.

$110.00 | Over 880 pages, illust., hardcover | Catalog no. 67292B

Building & Renovating Schools

This all-inclusive guide covers every step of the school construction process—from initial planning, needs assessment, and design, right through moving into the new facility. A must-have resource for anyone concerned with new school construction or renovation. With square foot cost models for elementary, middle, and high school facilities, and real-life case studies of recently completed school projects.

The contributors to this book—architects, construction project managers, contractors, and estimators who specialize in school construction—provide start-to-finish, expert guidance on the process.

$110.00 | Over 412 pages, hardcover | Catalog no. 67342

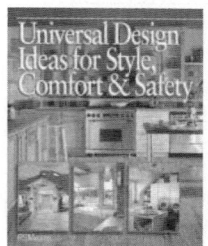

Universal Design Ideas for Style, Comfort & Safety

by RSMeans and Lexicon Consulting, Inc.

Incorporating universal design when building or remodeling helps people of any age and physical ability more fully and safely enjoy their living spaces. This book shows how universal design can be artfully blended into the most attractive homes. It discusses specialized products like adjustable countertops and chair lifts, as well as simple ways to enhance a home's safety and comfort. With color photos and expert guidance, every area of the home is covered. Includes budget estimates that give an idea how much projects will cost.

$21.95 | 160 pages, illustrated, softcover | Catalog no. 67354

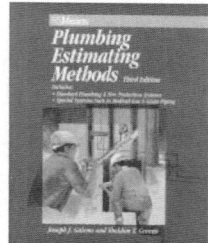

Plumbing Estimating Methods, 3rd Edition

by Joseph J. Galeno and Sheldon T. Greene

Updated and revised! This practical guide walks you through a plumbing estimate, from basic materials and installation methods through change order analysis. *Plumbing Estimating Methods* covers residential, commercial, industrial, and medical systems, and features sample takeoff and estimate forms and detailed illustrations of systems and components.

$65.00 | 381+ pages, softcover | Catalog no. 67283B

2015 Order Form

ORDER TOLL FREE 1-800-334-3509
OR FAX 1-800-632-6732

For more information visit the RSMeans website at www.rsmeans.com

Qty.	Book no.	COST ESTIMATING BOOKS	Unit Price	Total
	60065	Assemblies Cost Data 2015	$334.95	
	60015	Building Construction Cost Data 2015	206.95	
	60045	Commercial Renovaion Cost Data 2015	166.95	
	60115	Concrete & Masonry Cost Data 2015	188.95	
	50145	Construction Cost Indexes 2015 (subscription)	359.95	
	60145A	Construction Cost Index–January 2015	94.95	
	60145B	Construction Cost Index–April 2015	94.95	
	60145C	Construction Cost Index–July 2015	94.95	
	60145D	Construction Cost Index–October 2015	94.95	
	60345	Contr. Pricing Guide: Resid. R & R Costs 2015	39.95	
	60235	Electrical Change Order Cost Data 2015	199.95	
	60035	Electrical Cost Data 2015	208.95	
	60205	Facilities Construction Cost Data 2015	524.95	
	60305	Facilities Maintenance & Repair Cost Data 2015	461.95	
	60555	Green Building Cost Data 2015	170.95	
	60165	Heavy Construction Cost Data 2015	204.95	
	60095	Interior Cost Data 2015	208.95	
	60125	Labor Rates for the Const. Industry 2015	450.95	
	60185	Light Commercial Cost Data 2015	153.95	
	60025	Mechanical Cost Data 2015	203.95	
	60155	Open Shop Building Const. Cost Data 2015	175.95	
	60215	Plumbing Cost Data 2015	206.95	
	60175	Residential Cost Data 2015	148.95	
	60285	Site Work & Landscape Cost Data 2015	199.95	
	60055	Square Foot Costs 2015	218.95	
	62014	Yardsticks for Costing (2014)	189.95	
	62015	Yardsticks for Costing (2015)	201.95	
		REFERENCE BOOKS		
	67329	Bldrs Essentials: Best Bus. Practices for Bldrs	45.00	
	67307	Bldrs Essentials: Plan Reading & Takeoff	50.00	
	67342	Building & Renovating Schools	110.00	
	67261A	The Building Prof. Guide to Contract Documents	75.00	
	67353A	Complete Book of Framing, 2nd Ed.	29.95	
	67146	Concrete Repair & Maintenance Illustrated	75.00	
	67352	Construction Business Management	60.00	
	67364	Construction Law	115.00	
	67358	Construction Supervision	95.00	
	67363	Cost Data Student Edition	110.00	
	67314	Cost Planning & Est. for Facil. Maint.	95.00	
	67230B	Electrical Estimating Methods, 3rd Ed.	75.00	
	67343A	Est. Bldg. Costs for Resi. & Lt. Comm., 2nd Ed.	75.00	
	67276B	Estimating Handbook, 3rd Ed.	110.00	
	67318	Facilities Operations & Engineering Reference	115.00	
	67338B	Green Building: Proj. Planning & Cost Est., 3rd Ed.	110.00	
	67355	Green Home Improvement	34.95	
	67357B	The Gypsum Construction Handbook	40.00	
	67148	Heavy Construction Handbook	110.00	
	67308E	Home Improvement Costs–Int. Projects, 9th Ed.	24.95	
	67309E	Home Improvement Costs–Ext. Projects, 9th Ed.	24.95	
	67344	Homeowner's Guide to Mold	21.95	
	67324C	How to Est.w/Means Data & CostWorks, 4th Ed.	75.00	

Qty.	Book no.	REFERENCE BOOKS (Cont.)	Unit Price	Total
	67351A	How Your House Works, 2nd Ed.	22.95	
	67282A	Illustrated Const. Dictionary, Condensed, 2nd Ed.	65.00	
	67292B	Illustrated Const. Dictionary, w/CD-ROM, 4th Ed.	110.00	
	67362	Illustrated Const. Dictionary, Student Ed.	60.00	
	67348	Job Order Contracting	130.00	
	67295C	Landscape Estimating Methods, 5th Ed.	75.00	
	67341	Life Cycle Costing for Facilities	110.00	
	67370	Managing the Profitable Construction Business	60.00	
	67294B	Mechanical Estimating Methods, 4th Ed.	75.00	
	67345	Practice of Cost Segregation Analysis	105.00	
	67366	Project Control: Integrating Cost & Sched.	95.00	
	67346	Preventive Maint. for Multi-Family Housing	95.00	
	67367	Project Sched. and Manag. for Constr., 4th Ed.	95.00	
	67322B	Resi. & Light Commercial Const. Stds., 3rd Ed.	65.00	
	67359	Risk Management for Design and Construction	125.00	
	67365	Solar Energy: Technology & Project Delivery	85.00	
	67145B	Sq. Ft. & Assem. Estimating Methods, 3rd Ed.	75.00	
	67321	Total Productive Facilities Management	85.00	
	67350	Understanding and Negotiating Const. Contracts	80.00	
	67303B	Unit Price Estimating Methods, 4th Ed.	75.00	
	67354	Universal Design	21.95	
	67319A	Value Engineering: Practical Applications	90.00	

MA residents add 6.25% state sales tax _____
Shipping & Handling** _____
Total (U.S. Funds)* _____

Prices are subject to change and are for U.S. delivery only. *Canadian customers may call for current prices. **Shipping & handling charges: Add 8% of total order for check and credit card payments. Add 8% of total order for invoiced orders.

Send order to:

ADDV-1000

Name (please print) _____

Company _____

☐ Company
☐ Home Address _____
_____

City/State/Zip _____

Phone # _____ P.O. # _____

(Must accompany all orders being billed)

Mail to: RSMeans, 700 Longwater Drive, Norwell, MA 02061

RSMeans Project Cost Report

By filling out this report, your project data will contribute to the database that supports the RSMeans® Project Cost Square Foot Data. When you fill out this form, RSMeans will provide a $50 discount off one of the RSMeans products advertised in the preceding pages. Please complete the form including all items where you have cost data, and all the items marked (☑).

$50.00 Discount per product for each report you submit

Project Description (NEW construction only, please)

☑ Building Use (Office School . . .) _____

☑ Address (City, State) _____

_____

☑ Total Building Area (SF) _____

☑ Ground Floor (SF) _____

☑ Frame (Wood, Steel . . .) _____

☑ Exterior Wall (Brick, Tilt-up . . .) _____

☑ Basement: (check one)   ☐ Full   ☐ Partial   ☐ None

☑ Number of Stories _____

☑ Floor-to-Floor Height _____

☑ Volume (C.F.) _____

% Air Conditioned _____ Tons _____

Total Project Cost $ _____

Owner _____

Architect _____

General Contractor _____

☑ Bid Date _____

☑ Typical Bay Size _____

☑ Occupant Capacity _____

☑ Labor Force: _____ % Union   _____ % Non-Union

☑ Project Description (Circle one number in each line.)

  1. Economy   2. Average   3. Custom   4. Luxury
  1. Square   2. Rectangular   3. Irregular   4. Very Irregular

Comments _____

_____

A	☑	General Conditions	$
B	☑	Site Work	$
C	☑	Concrete	$
D	☑	Masonry	$
E	☑	Metals	$
F	☑	Wood & Plastics	$
G	☑	Thermal & Moisture Protection	$
GR		Roofing & Flashing	$
H	☑	Doors and Windows	$
J	☑	Finishes	$
JP		Painting & Wall Covering	$

K	☑	Specialties	$
L	☑	Equipment	$
M	☑	Furnishings	$
N	☑	Special Construction	$
P	☑	Conveying Systems	$
Q	☑	Mechanical	$
QP		Plumbing	$
QB		HVAC	$
R	☑	Electrical	$
S	☑	Mech./Elec. Combined	$

Please specify the RSMeans product you wish to receive.
Complete the address information.

Product Name _____

Product Number _____

Your Name _____

Title _____

Company _____
  ☐ Company
  ☐ Home   Street Address _____

City, State, Zip _____

Email Address _____

Return by mail or fax 888-492-6770.

Method of Payment:

Credit Card # _____

Expiration Date _____

Check _____

Purchase Order _____

Phone Number _____
(if you would prefer a sales call)

RSMeans
Square Foot Costs Department
700 Longwater Drive
Norwell, MA 02061
www.RSMeans.com

DISCARD